Advances in Lasers and Electro Optics

Edited by

Nelson Costa and Adolfo Cartaxo

Advances in Lasers and Electro Optics
Edited by Nelson Costa and Adolfo Cartaxo

This is a reprint of the book published in 2010 by Intech

Republished by InTech
Janeza Trdine 9, 51000 Rijeka, Croatia

Technical Editor Teodora Smiljanic

Cover Designer Dino Smrekar

Additional hard copies can be obtained from orders@intechopen.com

Advances in Lasers and Electro Optics, Edited by Nelson Costa and Adolfo Cartaxo
p. cm.
ISBN 978-953-307-088-9

For EU product safety concerns:
IN TECH d.o.o., Prolaz Marije Krucifikse Kozulić 3, 51000 Rijeka, Croatia,
info@intechopen.com or visit our website at intechopen.com.

Preface

Lasers and electro-optics is a field of research leading to constant breakthroughs. Indeed, tremendous advances have occurred in optical components and systems since the invention of laser in the late 50s, with applications in almost every imaginable field of science including control, astronomy, medicine, communications, measurements, etc. If we focus on lasers, for example, we find applications in quite different areas. We find lasers, for instance, in industry, emitting power level of several tens of kilowatts for welding and cutting; in medical applications, emitting power levels from few milliwatt to tens of Watt for various types of surgeries; and in optical fibre telecommunication systems, emitting power levels of the order of one milliwatt.

In this book, some advances of lasers and electro-optics in several fields of science, covering quite different subjects, are presented. In order to do so, each chapter is self-contained. Indeed, each chapter is written by different authors who present their research in a given field. Some elementary knowledge is usually assumed.

This book is divided in four sections. The book presents several physical effects and properties of materials used in lasers and electro-optics in the first chapter and, in the three remaining chapters, applications of lasers and electro-optics in three different areas are presented.

The first section of the book is dedicated to the analysis of physical effects and properties of materials used in lasers and electro-optics. The characterization of several materials is performed in the first chapters of the section. Then, some physical effects in optical components are described and demonstrated. An accurate characterization of materials and the knowledge of the physical effects occurring in each material are of extreme importance when the choice of materials for a given application has to be performed. Some innovative laser implementations are described at the two final chapters of this section.

The second section of the book is dedicated to communications. Topics related to optical sampling, all-optical signal processing, optical fibre transmission and data protection are covered in this section. The electronic acquisition of data is quite difficult when data has a broad spectrum, requiring improved sampling schemes. Some of those sampling schemes are discussed. The retiming and reshaping of optical data in the optical domain avoids opto-electrical conversions. Thus, all-optical signal processing is discussed. New modulation formats and enhanced transmission systems, required for increasing the data rate available to each user and reduce the cost of the data transmission, are also presented. A method for data protection is described and analyzed in the final chapter of the section.

Section three focuses on applications of electro-optics in imaging and processing of light. Some background on the physical effects leading to the creation of images is firstly given. Then, materials and strategies for the generation and manipulation of light are presented. Some applications of the analysis performed in each chapter are referred to along the chapters. Examples of such applications are the enhancement of the sensitivity of atomic force microscopes and 3-D displays.

Some applications of electro-optics to biology and medicine are analyzed in section four. This section focuses mainly on technologies for biological tissues imaging. It is shown that electro-optics has the potential to acquire quite good representations of microscopic samples. A chapter related to laser optoperforation for delivering foreign DNA into plants is presented at the end of this section.

We would like to thank to all authors for their contributions. On the behalf of the authors, we would like to acknowledge to Vedran Kordic, who coordinated this project, and to all who made this publication possible. We hope readers enjoy reading this book and that it benefits both novice and experts, providing a thorough understanding of several fields related to lasers and electro-optics.

Editors

Nelson Costa and Adolfo Cartaxo

Instituto de Telecomunicações
Department of Electrical and Computer Engineering
Instituto Superior Técnico
Lisboa, Portugal
nelson.costa@lx.it.pt

Contents

I Physical Effects and Properties of Materials

1

Optical, Photoluminescent, and Photoconductive Properties of Novel High-Performance Organic Semiconductors

Oksana Ostroverkhova, Andrew D. Platt and Whitney E. B. Shepherd
Oregon State University
USA

1. Introduction

Organic semiconductors have been investigated as an alternative to inorganic semiconductors due to their low cost, ease of fabrication, and tunable properties (Forrest, 2004). Applications envisioned for organic semiconductors include xerography, thin-film transistors, light-emitting diodes, solar cells, organic lasers, and many others (Peumans et al., 2003; Murphy & Frechet, 2007; Samuel & Turnbull, 2007). Since most of these applications rely on the conductive and photoconductive properties of the materials, it is important to understand physical mechanisms of charge photogeneration, transport, trapping, and recombination. However, despite numerous theoretical and experimental studies of the optical and electronic properties of organic solids, these mechanisms are not well understood and are still the subject of debate in the literature (Sariciftci, 1997; Podzorov et al., 2004; Moses et al., 2006; Troisi & Orlandi, 2006; Coropceanu et al., 2007; Cheng & Silbey, 2008; Laarhoven et al., 2008). Indeed, it is a complicated task to reveal and utilize the *intrinsic* properties of organic materials, since they are often masked by the influence of impurities, the presence of which is sensitive to the methods of material purification and device fabrication. As a result, measurements performed in the same material using different experimental techniques often provide conflicting results (Nelson et al., 1998; Hegmann et al., 2002; Podzorov et al., 2003; Jurchescu et al., 2004; Lang et al., 2004; Thorsmølle et al., 2004; Ostroverkhova et al., 2005a; Ostroverkhova et al., 2006a; Koeberg et al., 2007; Laarhoven et al., 2008; Najafov et al., 2008; Marciniak et al., 2009). Experimental methods that probe charge carrier dynamics on picosecond (ps) time-scales after a 100-femtosecond (fs) pulsed photoexcitation have had most success in revealing intrinsic properties of organic semiconductors (Hegmann et al., 2002; Thorsmølle et al., 2004; Hegmann et al., 2005; Ostroverkhova et al., 2005a; Ostroverkhova et al., 2005b). In contrast, techniques that probe equilibrium charge transport are much more sensitive to extrinsic effects (Nelson et al., 1998; Knipp et al., 2003; de Boer et al., 2004; Jurchescu et al., 2004). However, since properties under equilibrium conditions are relevant for most devices, it is necessary to understand how they are related to intrinsic properties and find ways to improve materials and device fabrication techniques in order to minimize extrinsic effects. Therefore, one of the emphases of this chapter is on photoexcited charge carrier dynamics from sub-ps (non-equilibrium) to many seconds (equilibrium) after photoexcitation in a variety of organic crystals and thin films (Sections 3.3-3.4).

Of particular technological interest are small-molecular-weight solution-processable materials that can be cast into high-performance (photo)conductive thin films. Functionalized anthradithiophene (ADT) and pentacene derivatives have attracted considerable attention due to their high charge carrier mobility, photoconductivity, and luminescence. In particular, charge carrier (hole) mobilities of over 1.2 cm^2/Vs have been observed in thin-film transistors (TFTs) based on solution-deposited films of these materials (Park et al., 2007; Park et al., 2008). Slight chemical modifications of the side groups of both ADT and pentacene derivatives lead to considerable changes in molecular packing, which affect electronic and optical properties of thin films (Platt et al., 2009a; Platt et al., 2009b; Platt et al., 2009c). Additional changes in these properties may be produced by functionalization of the core of the molecule. In Sections 3.1-3.5, we will summarize optical, photoluminescent (PL), and (photo)conductive properties of several functionalized ADT and pentacene derivatives. In addition to a promising potential of ADT pristine compounds for (opto)electronic applications, the availability of these high-performance, solution-processable, structurally similar, derivatives with different highest occupied molecular orbital (HOMO) and lowest unoccupied molecular orbital (LUMO) energies can be utilized to create composites with optical and electronic properties tailored for specific applications. In Sections 3.1.3 and 3.3.2, we will describe how charge carrier dynamics in ADT films can be manipulated using competition between charge and energy transfer achieved by introducing various guest molecules into an ADT host (Day et al., 2009a). An ultimate understanding of the charge carrier and exciton dynamics in organic semiconductors can be reached, however, only if relationships between charge transfer and energy transfer rates on the molecular level and macroscopic (photo) conductivity in the bulk material are established. Our preliminary studies of properties of individual molecules are presented in Section 3.2.

Finally, most devices that utilize (opto) electronic properties of materials require application of static electric fields with deposited electrodes. The processes occurring at the metal-organic interfaces at the electrodes can significantly affect the performance of a device (Brutting, 2005; Day et al., 2009b), introducing another difficulty in the materials characterization and another variable into the device performance. Effects of the electrode material on the (photo)conductive performance of organic thin-film devices, at various time-scales after photoexcitation, will be briefly reviewed in Section 3.6.

2. Experimental

2.1 Materials and sample preparation

In our studies of non-equilibrium charge carrier dynamics at sub-ps time scales after 100-fs photoexcitation (Section 3.3.1), we used single crystals of ultra-high-purity pentacene (Pc) (Jurchescu et al., 2004), pentacene derivative functionalized with 6,13-bis (triisopropylsilylethynyl) (TIPS) side groups (TIPS-pentacene) (Hegmann et al., 2002; Ostroverkhova et al., 2005a; Ostroverkhova et al., 2005b; Anthony, 2006; Ostroverkhova et al., 2006a), rubrene (Rub) (Podzorov et al., 2003; Podzorov et al., 2004), and tetracene (Tc) (Moses et al., 2006) (Fig.1). The Pc powder was purified using vacuum sublimation under a temperature gradient (Jurchescu et al., 2004). Rub, Tc and Pc single crystals were obtained using physical vapor transport techniques as described in Jurchescu et al., 2004 and Podzorov et al., 2003. The TIPS-pentacene single crystals were grown in a saturated tetrahydrofuran (THF) solution at 4°C (Anthony et al., 2001). Eight TIPS-pentacene crystals and four Pc, Rub and Tc crystals (each) were used in these experiments and yielded similar results (Ostroverkhova et al., 2006a). For the same experiments, we also prepared polycrystalline Pc and TIPS-pentacene thin films. All of the Pc and some TIPS-pentacene films were prepared on mica, glass, or KCl substrates by thermal evaporation of the corresponding powder heated to 200–250 °C in high vacuum (10^{-6}–10^{-7} Torr) at a deposition rate of 0.3 Å/ s. The thickness of the films was measured with a profilometer (Tencor Instruments) and ranged between 150 nm and 1.3 μm (Ostroverkhova et al., 2005a; Ostroverkhova et al., 2005b). TIPS-pentacene films drop-cast from THF solution at room temperature were also prepared.

In our studies of anisotropy of the transient photoconductivity (Section 3.3.1), we used single crystals of pentacene functionalized with: (i) TIPS and (ii) 6,13-bis (triethylsilylethynyl) (TES) side groups. The single crystals had dimensions typically around (1.5-2) mm x (2-4) mm (as illustrated in Fig.9(c)) with a thickness of 300-500 μm. Although both TIPS-pentacene and TES-pentacene crystals are triclinic (with the unit cell parameters listed in Table 1), the molecular packing and resulting π-overlap in these crystals is different, as the TIPS-pentacene crystals assume a more two-dimensional (2D) "brick-wall"-type structure, while the TES-pentacene crystals exhibit a one-dimensional (1D) "slipped-stack"-type structure (Figs.9(a) and (b), respectively)(Anthony, 2006). Our crystallographic analysis showed that the largest area crystal surface corresponds to the *a-b* plane of the crystals, with the *a*-axis parallel (| |) to the long axis in both TIPS-pentacene and TES-pentacene crystals. TIPS-pentacene and TES-pentacene crystals have almost identical absorption spectra in solution, with the absorption edge at around 700 nm, which shifts to ~850 nm in a crystal. Eight TIPS and four TES crystals were used in these experiments (Ostroverkhova et al., 2006b).

Compound	a (Å)	b (Å)	c (Å)	α (deg)	β (deg)	γ (deg)
TIPS-pent.	7.565	7.75	16.835	89.15	78.42	83.63
TES-pent.	7.204	9.994	11.326	80.81	89.13	82.21
ADT-TES-F	7.71	7.32	16.35	87.72	89.99	71.94
ADT-TIPS-F	7.58	8.18	16.15	100.85	92.62	98.79

Table 1. Unit cell parameters for functionalized pentacene and ADT derivatives used in our studies.

In our studies of photophysical and photoconductive properties on time-scales from sub-nanoseconds (ns) after 100 fs excitation to an equilibrium (Sections 3.1, 3.3.2, 3.4, and 3.5), we explored thin films of TIPS-pentacene, of fluorinated ADT derivatives functionalized with TES and TIPS side groups (ADT-TES-F and ADT-TIPS-F, respectively), and of another ADT derivative functionalized with TIPS side-groups, ADT-TIPS-CN (Figure 1) (Day et al., 2008; Day et al., 2009a; Day et al., 2009b; Platt et al., 2009b). In addition to pristine materials, we explored films of ADT-TES-F doped with various concentrations of C_{60}, TIPS-pentacene, or ADT-TIPS-CN. Stock solutions of functionalized ADT derivatives were prepared at ~ 1% by weight in toluene. For solution measurements, solutions were prepared by dilution of stock solutions to ~10^{-4} M. Films with thickness of 1 – 2 μm were prepared by drop-casting stock solutions onto glass substrates at ~60 °C. Composite films were similarly prepared from stock solutions of known mixtures of ADT-TES-F and C_{60}, TIPS-pentacene, or ADT-TIPS-CN. When dealing with pristine compounds, this preparation method yielded polycrystalline ADT-TES-F, ADT-TIPS-F, and TIPS-pentacene and amorphous ADT-TIPS-CN films (as confirmed by X-ray diffraction and transmission electron microscopy (TEM)). The unit cell parameters for the fluorinated ADT derivatives studied are listed in Table 1. For measurements of dark current and photoresponse, glass substrates were prepared by photolithographic deposition of either 5 nm/50 nm-thick Cr/Au or 100 nm-thick aluminum (Al) electrode pairs. Each pair consisted of 10 interdigitated finger pairs, with 1 mm finger length, 25 μm finger width and 25 μm gaps between the fingers of opposite electrodes. Films were drop-cast onto the interdigitated regions. Films on coplanar electrodes with 25 or 50 μm gap were also prepared.

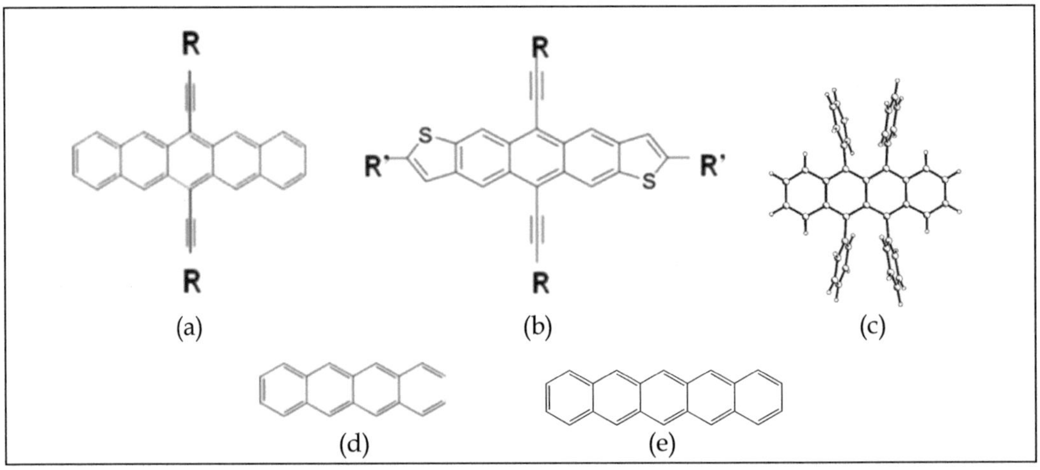

Fig. 1. Molecular structures of (a) functionalized pentacene, (b) functionalized ADT, (c) rubrene (Rub), (d) tetracene (Tc), and (e) pentacene (Pc). R = TIPS or TES, R' = F or CN.

2.2 Measurements of optical and photophysical properties

Optical absorption spectra were measured using a halogen lamp and a fiber-coupled Ocean Optics USB2000 spectrometer. Absorbance A was calculated from the incident (I_0) and transmitted (I) beam intensities as $A = -\mathrm{Log}(I/I_0)$. Reflection losses were taken into account by referencing with respect to cuvettes with pure solvent or clean glass substrates for solution and film measurements, respectively. Emission spectra were acquired in a custom fluorescence measurement setup with laser excitation at wavelengths of either 400 nm (frequency-doubled mode-locked Ti:Sapphire laser from KM Labs) or 532 nm (Nd:YVO_4 laser from Coherent, Inc.). Emitted photons were collected using a parabolic mirror and detected with a fiber coupled spectrometer (Ocean Optics USB2000 calibrated against a 3100 K black-body emitter). Absorption of solutions was measured using a standard 1 cm path length quartz cuvette with a halogen light source fiber-optically delivered to the sample holder and spectrometer. Photoluminescence (PL) quantum yields (QYs) in solution were referenced against standards with known quantum yields and corrected for differences in optical density and solvent refractive index. The ADT derivatives were measured against rhodamine 6G in ethanol ($\Phi = 0.95$) and DCDHF-N-6 in toluene ($\Phi = 0.85$) (Lord et al., 2007). The QY of TIPS pentacene solution was measured against rhodamine 6G in ethanol and

Alexa Fluor 647 in a phosphate buffer solution (pH 7.2, Φ = 0.33). The QYs in films were estimated using DCDHF-N-6 in polymethylmethacrylate (PMMA) (Φ = 0.98) as a reference and assuming a value of 1.7 for the index of refraction (Lord et al., 2007). The detection limit of the setup was estimated to be at $\Phi \approx 0.5\%$ at 650 nm.

PL lifetime measurements were performed using a mode-locked Ti:Sapphire laser frequency-doubled with a beta-barium borate (BBO) crystal with a repetition rate of 93 MHz picked at 9.3 MHz using a home-built pulse picker (based on a TeO_2 acousto-optic modulator from NEOS) and 80-fs pulses as the excitation source. A single-photon avalanche photodiode (SPAD - Molecular Photonic Devices) was used in conjunction with a time-correlated single-photon counter (TCSPC) data analysis board (PicoQuant TimeHarp 200) for detection. The instrument response function (IRF) (~200 ps) was recorded using scattered light from an etched microscope slide. For measurements of the temperature dependence of film spectra, samples on pre-cut microscope slides were mounted on a custom built electrically heated and water cooled stage (range: 278 – 360 K) for temperature control. PL measurements were taken in situ over the entire temperature range, in ambient air. Similar experiments were previously performed under N_2 atmosphere and showed no discernable difference.

For measurements of electric field-induced PL quenching, either 532 nm cw light or pulsed 400 nm light was focused on the samples similar to those used in our (photo)conductivity studies. At 532 nm cw photoexcitation, PL was collected using a Thorlabs amplified photodetector and a SRS830 lock-in amplifier. Keithley 237 source-measure unit was used to apply voltage in the range of 0-500 V and measure photocurrent, simultaneously with measurements of the PL. At 400 nm pulsed excitation, PL transient decay was detected using SPAD and TCSPC, as described above. The experiment was repeated at voltages up to 500 V and at different temperatures.

2.3 Single-molecule-level PL imaging

Samples for PL imaging at the single-molecule level (Section 3.2) were prepared from stock solutions of 1% by weight PMMA (75,000 m.w.) in toluene. ADT-TES-F was doped into the solution at the level of 10^{-10} per PMMA molecule. This solution was then spun coat onto clean glass coverslips at 2000 rpm for 55 s. The samples were imaged with an Olympus IX71 inverted microscope with a 100X UPlanApo objective under wide-field 532 nm cw illumination. The PL was detected by an Andor iXon DU-897 EMCCD camera.

2.4 Transient photocurrent measurements

2.4.1 Optical-pump-terahertz (THz) probe spectroscopy

A detailed description of the optical-pump - THz-probe experimental setup can be found in Lui & Hegmann, 2001. Briefly, an amplified Ti:Sapphire laser source (800 nm, 100 fs, 1 kHz) was used to produce optical pump pulses at wavelengths of 400-800 nm through various wave-mixing schemes and THz probe pulses generated via optical rectification in a 0.5 mm-thick ZnTe crystal. The samples were mounted on 1-2 mm apertures, and both the THz probe and optical pump pulses were at normal incidence to the surface of the films (*a-b* plane) of the single crystal samples. The room-temperature data were taken in air. For temperature dependence measurements, the samples were mounted in an optical cryostat (sample in vapor). The electric field of the THz pulse transmitted through the samples, $T(\mathrm{t})$, was detected by free-space electro-optic sampling in a 2-mm-thick ZnTe crystal and monitored at various delay times (Δt) with respect to the optical pump pulse. The range of optical pump fluences was 0.9-1.5 mJ/cm^2. No transient photoconductivity was observed upon optical excitation of the substrates alone. Optical excitation of all thin film and single crystal samples resulted in a change in the transmitted electric field ($-\Delta T(t)$) due to the transient photoconductivity (i.e. *mobile* photogenerated carriers) (Hegmann et al., 2002; Hegmann et al., 2005; Ostroverkhova et al., 2005a; Ostroverkhova et al., 2005b; Ostroverkhova et al., 2006a; Ostroverkhova et al., 2006b). In the absence of the phase shift between THz waveforms obtained in the unexcited and optically excited sample, as was the case for all our samples, the optically induced relative change in the THz peak amplitude ($-(T-T_0)/T_0 \equiv -\Delta T/T_0$, where T_0 is the amplitude of the THz pulse transmitted through unexcited sample) provides a direct measure of the transient photoconductivity (Hegmann et al., 2002; Thorsmølle et al., 2004; Ostroverkhova et al., 2005a). The time resolution of this experimental setup was ~400 fs. In the approximation of a thin conducting film on an insulating substrate, the differential transmission ($-\Delta T/T_0$) due to optical excitation of mobile carriers at small $|\Delta T/T_0|$ is related to the transient photoconductivity as follows (Lui and Hegmann, 2001; Hegmann et al., 2002; Thorsmølle et al., 2004):

$$\sigma = -(\Delta T/T_0)(1+N)/(Z_0 L), \tag{1}$$

where $Z_0 = 377\ \Omega$ is the impedance of free space, N is the refractive index of the substrate at THz frequencies, and L is the film thickness. Using this expression, and setting the maximum value for the transient response at $\Delta t = 0$ so that $|\Delta T/T_0|_{MAX} = |\Delta T(0)/T_0|$, the product of the charge carrier mobility (μ) and photogeneration efficiency (η) can be calculated as follows (Hegmann et al., 2002; Ostroverkhova et al., 2005a; Ostroverkhova et al., 2005b):

$$\mu\eta = \left|\frac{\Delta T(0)}{T_0}\right| \frac{h\nu(1+N)}{eF(1-R)(1-\exp[-\alpha L])Z_0}, \tag{2}$$

where e is the electric charge, h is Planck's constant, ν is the light frequency, α is the absorption coefficient, F is the incident fluence, and R is the reflection coefficient.

2.4.2 Direct measurements of fast photocurrent using digital sampling oscilloscope

For transient photoconductivity measurements on sub-100 ps to hundreds of microsecond (μs) time-scales after pulsed photoexcitation, an amplified Ti:Sapphire laser (800 nm, 100 fs, 1 kHz) was used in conjunction with a frequency-doubling beta-barium borate (BBO) crystal to excite the samples. Voltage was supplied by a Keithley 237 source-measure unit, and light pulse-induced transient photocurrent was measured with a 50 Ω load by a 50 GHz CSA8200 digital sampling oscilloscope (DSO). The time resolution of this setup was 30-40 ps.

From the peak of the transient photocurrent ($I_{ph,max}$), a product of charge carrier mobility μ and photogeneration efficiency η was calculated using

$$\mu\eta = I_{ph,max}/(eN_{ph}Ed), \tag{3}$$

where N_{ph} is the number density of absorbed photons per pulse, E is the static electric field ($E = V/L$, where V is the applied voltage and L is the gap between the electrodes), e is the charge of the electron, and d is the channel width.

2.5 Dark current and cw photocurrent measurements

For dark current and cw photocurrent measurements, the samples were embedded in a fixture incorporating a thermoelectric unit for temperature control (range: 285−350 K). The Keithley 237 source-measure unit was used to measure current through the sample in the absence and in the presence of cw photoexcitation with a Nd:YVO_4 laser at 532 nm. The photocurrent was calculated as the difference between the two.

2.6 Scanning photocurrent microscopy

Scanning photocurrent microscopy has been previously utilized in probing internal electric field distributions, mapping electronic band structure, measuring mobility-lifetime products, etc. in inorganic and organic films, nanowires, carbon nanotubes, and graphene sheets (Gu et al., 2006; Agostinelli et al., 2007; Ahn et al., 2007; Park et al., 2009). In our experiments, ADT films on coplanar electrodes, either Al or Au, separated by a 50-μm gap, were excited with a focused beam from the glass substrate side, and photoresponse was monitored as the excitation spot was moved across the gap from one electrode to another (Day et al., 2009), under applied voltage. This experiment was performed with either pulsed 100-fs 400-nm excitation or cw 532-nm excitation, using laser sources described above. In the case of pulsed excitation, a lens with 2.5-cm focal distance was used to focus the beam with pulse energy of ~1 nJ to an approximately 4-μm spot at the sample. The lens was translated along the gap using a micrometer-controlled translation stage. At each position of the lens, transient photocurrent resulting from excitation of a localized region of the film was recorded with the DSO. In the case of cw excitation, the sample was placed on an Olympus IX-71 inverted microscope, and a 10x objective with a numerical aperture of 0.6 was used to focus the beam at a power of ~0.4 μW to an approximately 400-nm spot. Position of the localized excitation with respect to the electrodes was monitored by a CCD camera that detected fluorescence (emitted by the photoexcited region of the sample) collected through the same objective. The sample was translated using a closed-loop piezoelectrically controlled x-y stage with sub-nm resolution, with a speed of 1 μm/s. Cw light was chopped at 565 Hz, and the amplitude of the modulated photocurrent signal was measured by a Stanford Research Systems 830 lock-in amplifier. The experiment was repeated as a function of applied voltage and light power, as well as at different voltage polarities, directions of the scan, and lengths of the waiting period between successive scans.

3. Results

3.1 Optical properties

3.1.1 Dilute solutions

Figure 2 shows optical and PL properties of functionalized ADT and pentacene derivatives in toluene solution. In all spectra, vibronic progression due to coupling of the electronic state to ring-breathing vibrational modes (~1400 cm^{-1}), characteristic of oligoacenes, is observed (Pope & Swenberg, 1999). In solutions, optical properties were determined primarily by the core of the molecule and were not affected by TIPS or TES side-groups, which resulted in identical spectra of, for example, ADT-TES-F and ADT-TIPS-F in Fig. 2 or TIPS- and TES-pentacene (only TIPS-pentacene data are shown). Spectra of ADT-TIPS-CN and TIPS-pentacene in solution were both red-shifted with respect to those of ADT-TIPS(TES)-F. Small Stokes shifts of <10 nm, observed in all solutions, are due to rigidity of the molecular core of oligoacenes. PL lifetime decay of solutions was well described by a single-exponential function (~exp[−t/τ], where τ is the PL lifetime). ADT-TIPS-F and ADT-TES-F derivatives exhibited similar lifetimes (τ) of ~9 ns and high PL quantum yields (QYs) (Φ) of ~70% in toluene. Solutions of ADT-TIPS-CN and TIPS-pentacene showed longer lifetimes (~12-12.5 ns) and QYs ~75% (Table 2) (Platt et al., 2009a; Platt et al., 2009b; Platt et al., 2009c).

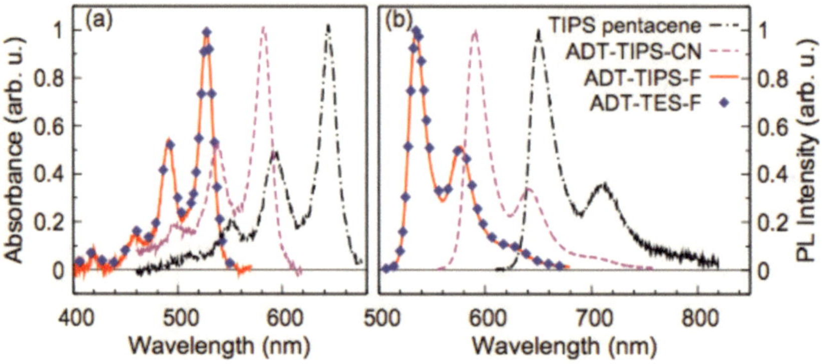

Fig. 2. Normalized optical absorption (a) and PL (b) spectra of functionalized pentacene and ADT derivatives in toluene solution. Spectra of ADT-TES-F and ADT-TIPS-F in solutions are identical. Reprinted from Platt et al., 2009b, with permission. Copyright American Chemical Society (2009).

3.1.2 Thin films

Optical absorption spectra of films (Fig. 3(a)) exhibited a redshift, or displacement Δ, with respect to those in solutions, due to enhanced Coulomb interaction of the molecule with its surrounding and exchange interaction between translationally equivalent molecules (Ostroverkhova et al., 2005b). In general, Δ depends on the molecular-orbital overlap and on the structure and morphology of the film. In contrast to identical absorption and PL spectra of ADT-TIPS-F and ADT-TES-F in solution, those of corresponding films were considerably different (Fig. 3), which we attribute to differences in packing of these molecules in the solid and in film crystallinity. In particular, ADT-TIPS-F and ADT-TES-F films exhibited displacements Δ of ~320 cm^{-1} and ~760 cm^{-1}, respectively. Although Δ exhibited sample-to-sample variation, it was always larger in ADT-TES-F films compared to films of ADT-TIPS-F, indicative of a higher degree of exciton delocalization in ADT-TES-F films. Also redshifted were PL spectra of ADT-TES-F with respect to those of ADT-TIPS-F films (Fig. 3(b)). Vibronic bands in both absorption and PL spectra were broader in films, as compared to solutions, with relative intensities of the bands varied depending on the film thickness and morphology (Platt et al., 2009a; Platt et al., 2009b). Regardless of the film thickness, ADT-TIPS-F films showed a more pronounced vibronic structure of the PL spectra than ADT-TES-F or ADT-TIPS-CN. The differences observed in PL spectra of films compared to those in solutions are due to intermolecular interactions leading to a formation of crystallites and molecular aggregates (whose properties depend on the degree of molecular order, size, and intermolecular coupling). This is further supported by our observations that the PL spectra of molecules under study embedded at low concentrations in the PMMA matrix yielded spectra identical to those of solutions in Fig. 1(b), as expected from non-interacting molecules (Platt et al., 2009b). Although all materials studied could be prone to aggregate formation due to their π-stacking properties, PL properties of aggregates significantly

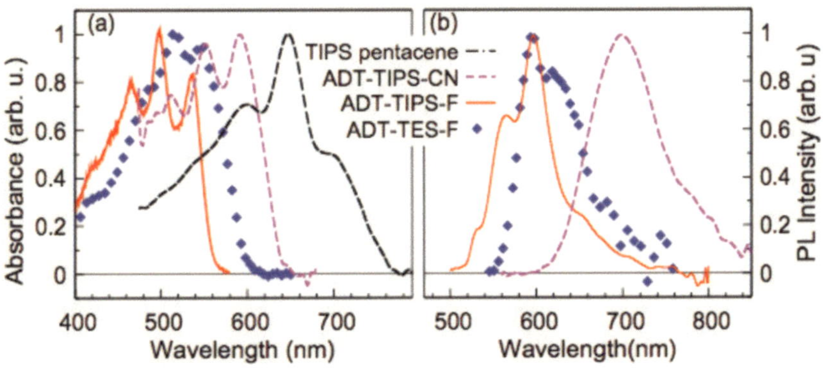

Fig. 3. Normalized optical absorption (a) and PL (b) spectra in TIPS-pentacene and ADT films. No PL response was observed in TIPS-pentacene films. Reprinted from Platt et al., 2009b, with permission. Copyright American Chemical Society (2009).

depended on the material. For example, at room temperature, thin films of ADT-TIPS-F and ADT-TES-F were highly luminescent, with PL QYs reaching 40 – 50% depending on the film thickness and morphology. (These values represent a lower limit, since effects of self-absorption were significant in even the thinnest of our films). In contrast, PL in TIPS-pentacene films was so low that it could not be detected (QYs of < 0.5%).

In all materials studied, the PL decay dynamics in films were faster than those in solutions and could be described by a bi-exponential function ($\sim a_1 \exp[-t/\tau_1] + a_2 \exp[-t/\tau_2]$, where $\tau_{1(2)}$ and $a_{1(2)}$ are shorter (longer) lifetimes and their relative amplitudes, respectively, and $a_1 + a_2 = 1$), characteristic of molecular aggregates. Both τ_1 and τ_2 were shorter than lifetimes τ measured in solutions of the same molecules, and the weighted average lifetimes in films, $\tau_{av} = a_1\tau_1 + a_2\tau_2$, were typically on the order of 0.4 – 4 ns at room temperature, depending on the material, and varied with film quality (e.g. from 1.1 to 2.5 ns in ADT-TIPS-F films).

Temperature dependence. In all ADT films, the PL response was strongly temperature dependent, and PL quantum yields decreased by a factor of 3–6 as the temperature increased from 5 °C to 80 °C, depending on the sample. Figure 4 (a) shows (i) PL QY calculated from the integrated PL spectrum measured at 400 nm excitation, (ii) the peak PL intensity measured under pulsed 80 fs 400 nm excitation with the time resolution of ~ 200 ps, and (iii) integrated time-resolved fluorescence decay measured under pulsed 80 fs 400 nm excitation, all normalized by their values at room temperature of 20 °C, as a function of temperature in an ADT-TIPS-F film. Considerable temperature dependence of the peak PL intensity suggests significant contribution of processes occurring on sub-200 ps time scales, not resolved in our experiments, to the overall temperature dependence of the QYs. The remaining contribution is due to processes occurring on time scales of <2-4 ns. Since no temperature dependence of PL emission of our molecules in solution was observed, the strong temperature dependence observed in films is due to temperature-dependent intermolecular interactions in films (Platt et al., 2009b). In order to quantify the observed temperature dependence, we consider PL quantum yield to be inversely proportional to a sum of temperature independent radiative rate and thermally activated non-radiative rate, so that

$$1/\Phi \sim 1 + a \exp[-\Delta_{PL}/k_BT] , \quad (4)$$

where k_B is the Boltzmann constant, T is temperature, and *a* is a fitting parameter related to the ratio between radiative and temperature-independent non-radiative rate prefactor. PL quenching activation energies Δ_{PL}, obtained from fits of data to Eq.(4) yielded values between 0.11 and 0.27 eV, depending on the sample and on the material (e.g. 0.15 eV in an ADT-TIPS-CN film in Fig.4(b)) (Platt et al., 2009b).

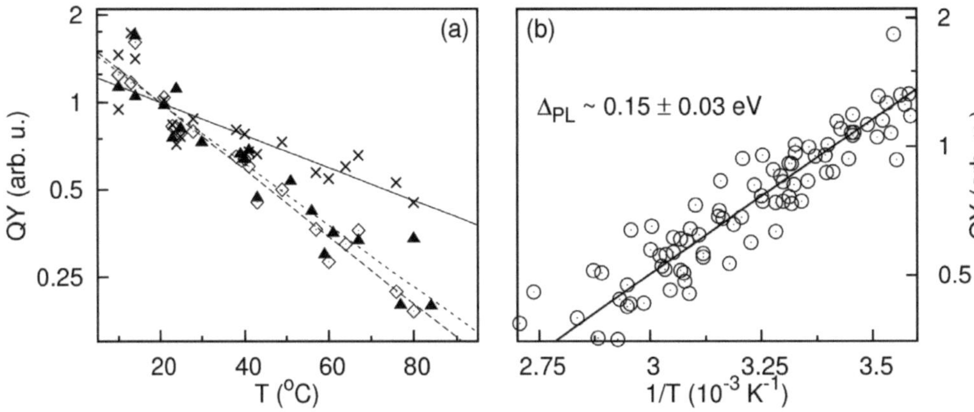

Fig. 4. (a) Temperature dependence of (i) the QYs calculated from the integrated PL spectrum measured at 400 nm excitation (triangles), (ii) peak amplitude of the PL transient measured under pulsed 80 fs 400 nm excitation with 200 ps resolution (×), and (iii) integrated PL decays measured under pulsed 80 fs 400 nm excitation (diamonds) in ADT-TES-F film. All are normalized by their values at room temperature of 20 °C.
(b) Temperature dependence of the QY, normalized at its value at room temperature, obtaine in an ADT-TIPS-CN film. The fit of the QY by Eq. (4) is also shown. Reprinted from Platt et al., 2009b, with permission. Copyright American Chemical Society (2009).

Compound	HOMO[a] (eV)	LUMO[a] (eV)	λ_{abs}[b] (nm)	λ_{PL}[b] (nm)	Φ[c]	τ[c] (ns)	$\lambda_{abs,film}$[d] (nm)	$\lambda_{PL,film}$[d] (nm)
ADT-TES-F	-5.35	-3.05	528	536	0.70	9.4	550	595
ADT-TIPS-F	-5.35	-3.05	528	536	0.72	9.1	537	600
ADT-TIPS-CN	-5.55	-3.49	582	590	0.76	12.7	592	~688
TIPS-pent.	-5.16	-3.35	643	650	0.75	11.8	~705	--

Table 2. Electrochemical and photophysical properties of TIPS-pentacene and ADT derivatives.
[a] Measured using differential pulse voltammetry (Platt et al., 2009b)
[b] Wavelengths of maximal absorption or PL corresponding to 0→0 transitions in toluene solutions
[c] PL QY and PL lifetime in toluene solutions
[d] Wavelengths of maximal absorption or PL in thin films. PL from TIPS-pentacene films could not be detected (QY<0.5%)

Electric field dependence. In order to address a possibility of electric field-induced dissociation of the radiative state, we measured PL spectra and lifetimes upon either cw 532 nm or pulsed 80 fs 400 nm photoexcitation of ADT films, under applied voltage. No changes in the PL QYs, peak PL intensity, or PL lifetimes have been detected at the applied voltages of 0-500 V, i.e., up to average electric fields of 2×10^5 V/cm, at any temperature in the range of 5-80 °C.

3.1.3 Composite thin films

Figure 5 shows absorption and PL spectra of the composite films containing ADT-TES-F host molecules and 2 wt% C_{60}, 10 wt% of TIPS-pentacene or 10 wt% of ADT-TIPS-CN guest molecules. Based on the absorption spectra in Fig. 5(a)), no ground state charge transfer occurs upon addition of guest molecules at these concentrations. In contrast to absorption spectra, addition of any guest molecules to ADT-TES-F in film produced a dramatic effect on the PL spectra (Fig. 5(b)), due to photoinduced charge or energy transfer, depending on the guest (Day et al., 2009a). In particular, very strong photoinduced energy transfer was observed in ADT-TES-F/ADT-TIPS-CN composites: even at a concentration of ADT-TIPS-CN in ADT-TES-F as low as 0.1%, almost no PL emission was observed from the ADT-TES-F host, and the PL response of the film was dominated by that of the ADT-TIPS-CN guest molecules (Platt et al., 2009c). In contrast, photoinduced charge transfer, leading to the PL quenching, is more favored in ADT-TES-F/C_{60} and ADT-TES-F/TIPS-pentacene composites. As we discuss in Section 3.3.2, PL properties of the composites are correlated with photoconductive properties of the same samples on short time-scales after a pulsed 80 fs photoexcitation.

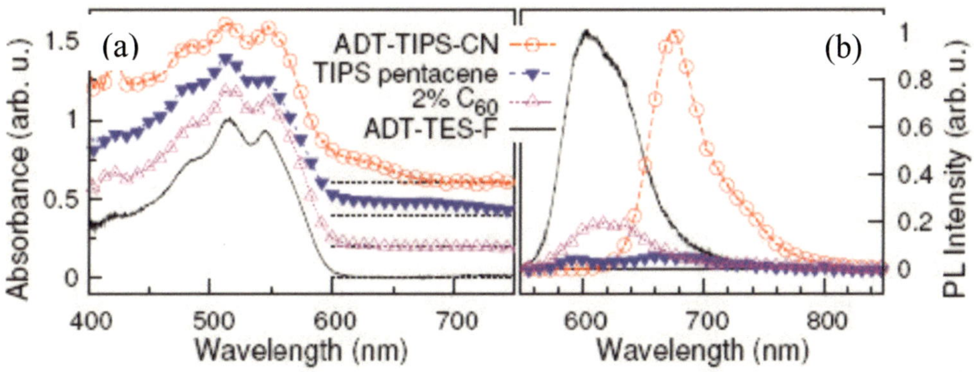

Fig. 5. Optical absorption (a) and PL (b) spectra of a pristine ADT-TES-F film and composites ADT-TES-F/C_{60} (2%), ADT-TES-F/TIPS-pentacene (10%), and ADT-TES-F/ADT-TIPS-CN (10%). Absorption spectra are shifted along y-axis for clarity. Reprinted from Day et al., 2009a, with permission. Copyright American Institute of Physics (2009).

3.2 Properties of individual molecules

Figure 6 shows a typical time trajectory of the PL photon count obtained from an individual ADT-TES-F molecule in a PMMA matrix, as confirmed by a single-step photobleaching at time t = 32 s. This demonstrates that PL quantum yields and photostablity of ADT-TES-F molecules are good enough to enable imaging at the single molecule level at room temperature. Signal–to–noise ratios of over 20:1 were obtained at moderate excitation levels at 532 nm in the epi-illumination geometry. Moreover, ADT-TES-F exhibited remarkably good stability as a single molecule fluorophore, comparable to DCDHF-N-6, which is currently one of the best fluorophores utilized in single-molecule fluorescence spectroscopy. This opens up new possibilities to study charge- and energy-transfer processes in these materials at nanoscales, which are currently underway in our laboratory.

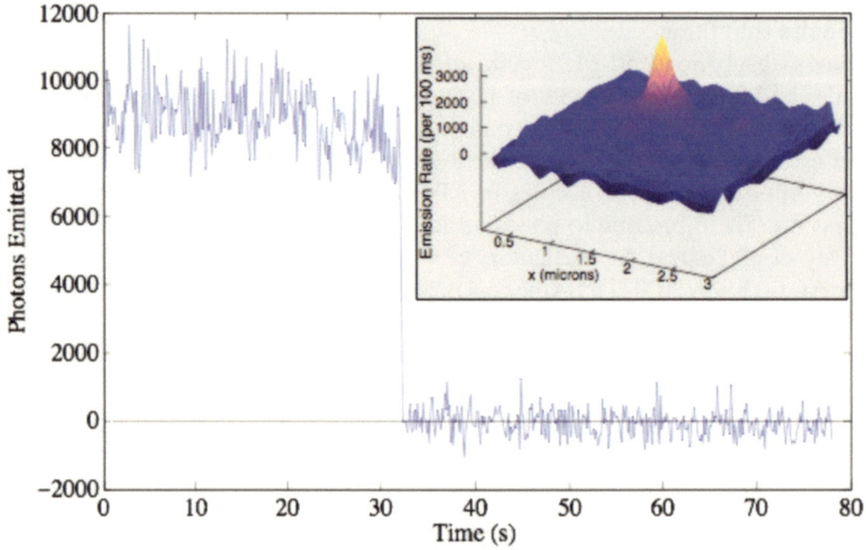

Fig. 6. Time trace of the fluorescence of a single molecule of ADT-TES-F under 700 μW wide-field 532 nm cw illumination. Single-step photobleaching at ~32 s confirms that the trace belongs to a single-molecule. Inset shows a surface plot of the emission of the same molecule with cubic interpolation. Integration time was 100 ms.

3.3 Transient photoconductivity

3.3.1 Non-contract measurements on 400 fs-1 ns time-scales using optical pump-THz probe spectroscopy

Figure 7(a) illustrates the differential THz transmission ($-\Delta T/T_0$) under optical excitation at 580 nm as a function of delay time (Δt) between the optical pump and THz probe pulses, obtained in Pc, TIPS-pentacene, Rub and Tc crystals and Pc and TIPS-pentacene thin films (Ostroverkhova et al., 2005a; Ostroverkhova et al., 2005b; Ostroverkhova et al., 2006a). All transients, except that for the Tc, which was taken at 10 K, were measured at room temperature in air. The onset of the photoresponse reveals a fast photogeneration process for mobile carriers with characteristic times below ~400 fs limited by the time resolution of our setup in all samples. This suggests that free carriers, and not only excitons, can be created in these materials fast, and in the absence of applied electric field. At room temperature (RT), measurements of the peak value of $-\Delta T/T_0$ yielded $\mu\eta$ values (calculated from Eq.(2)) of ~ 0.3-0.35, 0.15-0.2, 0.05-0.06, and 0.03 $cm^2/(Vs)$ for Pc, TIPS-pentacene, Rub and Tc crystals, respectively, and 0.02-0.03 and 0.01-0.06 in Pc and TIPS-pentacene thin films, depending on their structure. These values (in particular the photogeneration efficiency η) include initial carrier trapping and recombination occurring within 400 fs after photoexcitation, not resolved in our experiment. If we assume $\eta = 1$, then these $\mu\eta$ values provide a *lower estimate* for the carrier mobility μ. Since $\eta < 1$, the mobility value is higher.

Likewise, if we assume that RT intrinsic mobilities in these materials are on the order of <10 cm^2/Vs in TIPS-pentacene and Tc and ~20-30 $cm^2/(Vs)$ in Rub and Pc (Jurchescu et al., 2004; Podzorov et al., 2004), then the lower limit for the photogeneration efficiency in single crystals is ~1-2% in Pc and TIPS-pentacene and ~0.3% in Rub and Tc. The observed difference in $\mu\eta$ could be due to the differences in: (i) *intrinsic* carrier mobility, (ii) *initial* photogenerated free carrier yield, and (iii) carrier loss due to charge trapping and recombination. Factor (iii) is likely not to be a significant contributor at such short time-scales after photoexcitation. This is supported by our observation of the $\mu\eta$ values measured in good-quality TIPS-pentacene thin films reaching 30 - 40% of those obtained in TIPS-pentacene single crystals, in spite of the abundance of deep traps at the grain boundaries in thin films (Ostroverkhova et al., 2005b). Therefore, the differences in the trapping properties of TIPS-pentacene, Pc, Rub and Tc crystals most likely account only for a factor up to ~2 in the differences in $\mu\eta$ values obtained in these crystals at RT, while further differences are due to factors (i) and (ii). For example, the lower transient photoconductivity observed in Rub compared to Pc single crystals at RT at 400 nm could be mostly due to the lower photogeneration efficiency η (factor (ii)) in Rub, in qualitative agreement with Podzorov et al., 2004 and Lang et al., 2004.

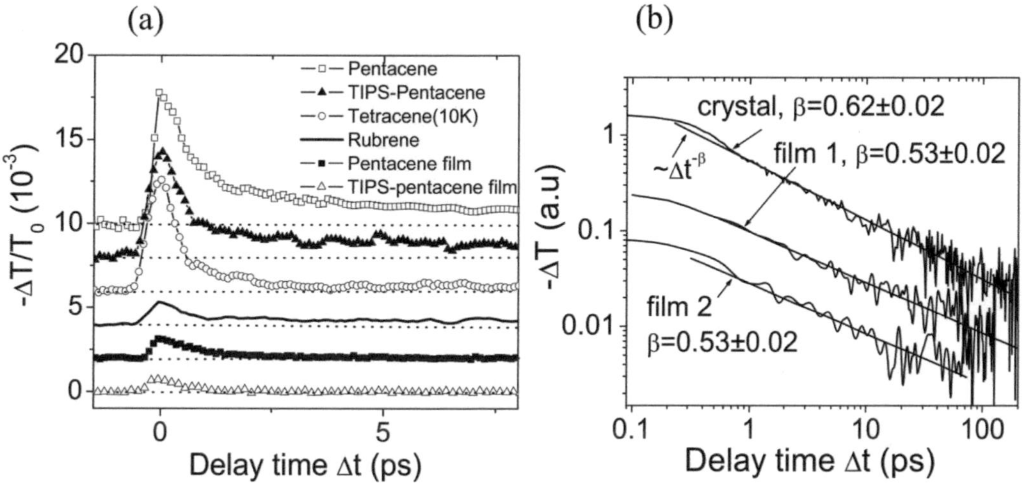

Fig. 7. (a) Differential THz transmission due to transient photoconductivity in various organic crystals and thin films as a function of pump-probe delay time. (b) Decay dynamics of the transient photoconductivity in TIPS-pentacene crystal and two films. Fits with a power-law function are also shown. Adapted from Ostroverkhova et al., 2005b and Ostroverkhova et al., 2006a, with permission. Copyright American Institute of Physics (2005, 2006).

The decay dynamics of the transient photoconductivity yield information about the nature of charge transport, trapping and recombination (Etemad et al., 1981; Moses et al., 1987; Yu et al., 1990). In TIPS-pentacene crystals, the transient photoconductivity exhibited a fast initial decay during the first few picoseconds, followed by a slow decay best described by a power law function ($\sim\Delta t^{-\beta}$) with β= 0.5-0.7 over many orders of magnitude in time, which has been attributed to dispersive transport (Fig.7(b)) (Etemad et al., 1981; Hegmann et al., 2002). Interestingly, these dynamics in TIPS-pentacene crystals did not change appreciably over a wide temperature range of 5-300 K, which suggests tunneling, rather than thermally-activated hopping, mechanism of charge transport. Similar decay dynamics was observed in Pc crystals at RT and in best TIPS-pentacene films (Ostroverkhova et al., 2005b; Ostroverkhova et al., 2006a). In contrast, photoconductivity in lower-quality TIPS-pentacene and Pc films exhibited fast bi- or single-exponential decay dynamics (~$\exp[-\Delta t/\tau_c]$) with time constants τ_c~1 ps for both materials, independent of the temperature, most likely due to deep-level traps at the interfaces between crystallites.

Wavelength dependence. In order to gain insight into the photogeneration process, we repeated the optical pump-THz probe experiment at various pump wavelengths in TIPS-pentacene crystals and Pc and TIPS-pentacene thin films. Due to the short temporal width of the optical pulses (~100 fs), the illumination is not monochromatic, as illustrated in Fig. 8(a), which shows the spectra of some of the optical pump pulses utilized in this experiment. Typical optical absorption spectra of the thin film and single crystal samples are shown in Fig. 8(b). No transient photoresponse was obtained from TIPS-pentacene and Pc thin films upon excitation at 800 nm - the spectral region in which there is too little absorption (Fig. 8(b)). At all wavelengths of optical excitation *within* the absorption spectra, we observed a photoinduced change in THz transmission with a fast onset and decay dynamics similar to that shown in Fig. 7 in all our samples.

Furthermore, $\mu\eta$ calculated from our data was wavelength-independent within our experimental error, as shown in Fig. 8(c). Wavelength-independent photocarrier generation has also been observed in other organic semiconductors using ultrafast techniques (Moses et al., 2000). However, conventional steady-state photoconductivity measurements typically exhibit wavelength-dependent mobile carrier photogeneration efficiencies (Silinsh & Capec, 1994; Pope & Swenberg, 1999), as reported for pentacene single crystals (Lang et al., 2004).

Temperature dependence. The temperature dependence of the photoresponse provides valuable information about the mechanism of photoconductivity (Moses, 1989; Karl, 2001; Karl, 2003). In both single crystals and thin films, the transient photoconductivity *increased* as the temperature *decreased*, as demonstrated in Fig. 8(d). As the temperature *decreased* from 297 K to about 20 K, $\mu\eta$ *increased* in all crystals by a factor of 3-10, depending on the crystal. Assuming η does not increase as the temperature decreases, we can attribute the temperature dependence of $\mu\eta$ shown in Fig. 8(d) to band-like charge carrier transport (Hegmann et al., 2002; Thorsmølle et al., 2004; Ostroverkhova et al., 2005a; Ostroverkhova et al., 2005b), which has not been previously observed over such a *wide* temperature range in Tc and Rub single crystals (de Boer et al., 2004; Newman et al., 2004; Podzorov et al., 2004). The initial increase of $\mu\eta$ by a factor of 1.5 - 3 (depending on the crystal) as the temperature is lowered from 297 K to about 150 K (Fig.8(d)) is consistent with the increase in charge carrier mobility observed from field-effect measurements in Rub (Podzorov et al., 2004) and from space-charge-limited-current measurements in Pc (Jurchescu et al., 2004) and Tc (de Boer et al., 2004) single crystals over the same temperature range. In films, the $\mu\eta$ increased from ~0.02 $cm^2/(Vs)$ to ~0.07 $cm^2/(Vs)$ as the temperature decreased from 297 K to 5 K, the trend which has not been previously observed in polyacene thin films. In addition, even though such band-like mobility has been obtained in several organic crystals (Moses, 1989; Karl, 2001; Hegmann et al., 2002; Karl, 2003; de Boer et al., 2004; Jurchescu et al., 2004; Podzorov et al., 2004; Thorsmølle et al., 2004), only a few studies reported the band-like behavior over a wide temperature range (Karl, 2001; Hegmann et al., 2002; Karl, 2003; Thorsmølle et al., 2004), as observed in our experiments (Fig.8). The thin film data in Fig.

8(d) can be fitted with the power law function $\mu\eta \sim T^{-n}$, with the exponent n~0.32, similar to n~0.27 reported from transient photoconductivity measurements of Pc single crystals. (Thorsmølle et al., 2004).

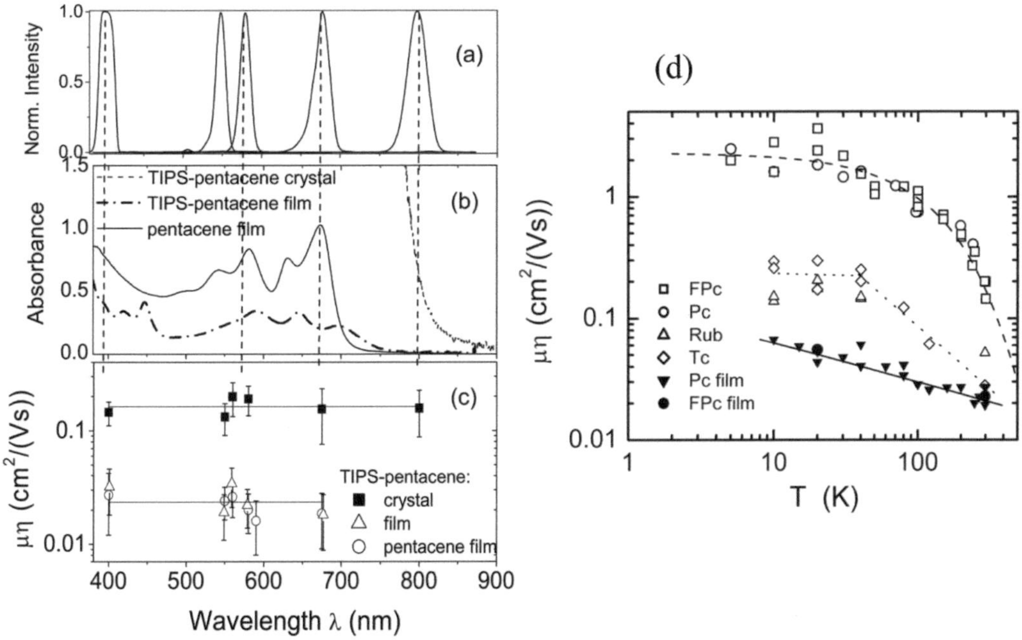

Fig. 8. (a) Normalized intensity spectra of selected optical pump pulses. Spectra of pulses centered at 550 and 590 nm are not included for clarity. (b) Absorption spectra of a TIPS-pentacene crystal and TIPS-pentacene and Pc thin-film samples. (c) Product of mobility and photogeneration efficiency (μη) measured in a TIPS-pentacene crystal and TIPS-pentacene and Pc thin-film samples at room temperature. (d) Temperature dependence of the μη obtained in Pc, Rub, Tc, and TIPS-pentacene (the latter labeled as FPc on the figure) single crystals and Pc and TIPS-pentacene (labeled as FPc) films. Adapted from Ostroverkhova et al., 2005a and Ostroverkhova et al., 2006a, with permission. Copyright American Physical Society (2005) and American Institute of Physics (2006).

Photoconductivity anisotropy. In order to probe the effect of molecular packing on the transient photoconductivity, we performed optical pump-THz probe experiments in TIPS-pentacene and TES-pentacene single crystals (Fig.9(a)-(c)) (Ostroverkhova et al., 2006b). The $-\Delta T/T_0$ transients observed in both TIPS-pentacene and TES-pentacene crystals had very similar shapes and exhibited sub-picosecond charge photogeneration and power-law decay dynamics discussed above (Fig.7). The μη product, calculated using Eq.(2) from the peak of the $-\Delta T/T_0$ transient obtained with both the electric field of the THz probe pulse (E_{THz}) and that of the optical pump pulse (E_{pump}) parallel to the *a*-axis, yielded ~0.15-0.2 $cm^2/(Vs)$ and ~0.05-0.06 $cm^2/(Vs)$ for the charge carrier mobility in TIPS-pentacene and TES-pentacene crystals, respectively, depending on the sample. As discussed above, these numbers represent lower limits for the charge carrier mobility along the *a*-axis. The triclinic symmetry group of TIPS-pentacene and TES-pentacene crystals (Table 1) leads to a complex picture of charge transport described by six components of the mobility tensor μ_{ij}, where i,j = *x*, *y*, *z*, are components in an orthogonal coordinate system (choice of which is somewhat arbitrary), and $\mu_{ij}=\mu_{ji}$. In general, the principal axes of the mobility tensor do not coincide with the crystallographic axes, are different for hole and electron transport, and are temperature-dependent (Karl, 2001). In order to measure charge carrier mobility anisotropy, separately from that of the photogeneration efficiency, we performed the following experiments (Ostroverkhova et al., 2006b). The crystal was rotated in the azimuthal (*a-b*) plane, so that the angle (φ) between the direction of E_{THz} and the *a*-axis of the crystal changed (at $\varphi = 0°$ E_{THz} was parallel to the *a*-axis). The polarization of the pump (E_{pump}) was rotated with the crystal using a half-wave plate and a polarizer combination in order to maintain E_{pump} parallel to the *a*-axis and ensure the same fluence at all angles. At every angle, the peak value of $-\Delta T/T_0$ was measured and the $\mu\eta$ product was calculated using Eq.(2). In this case, the observed angular dependence of the photoconductivity was purely due to that of the charge carrier mobility, and not due to the combined effects of the mobility and photogeneration efficiency anisotropy. Figures 9 (d) and (e) show the angular dependence of the charge carrier mobility (averaged over all samples measured at 800-nm, 400-nm and 580-nm-excitation) normalized to the value at $\varphi = 0°$ in TIPS-pentacene and TES-pentacene crystals, respectively. The charge carrier mobility along a certain direction $\boldsymbol{l}=(l_i, l_j, l_k)$ is given by $\mu_{ll} = \mu_{ij} l_i l_j$. For convenience, we chose the orthogonal coordinate system (x,y,z) so that $x \,||\, a$-axis, $y \perp x$ and is in the *a-b* plane, and $z \perp a\text{-}b$ plane. Then, in the *a-b* plane, $\boldsymbol{l}$ = (cosφ, sinφ, 0), and the mobility along $\boldsymbol{l}$ is:

$$\mu_{ll} = \mu_{xx} (\cos\varphi)^2 + \mu_{yy} (\sin\varphi)^2 + \mu_{xy} \sin(2\varphi) \quad (5)$$

For example, the mobility along the *a*-axis is $\mu_{aa}= \mu_{xx}$. The equation above normalized by μ_{xx} was used to fit the data in Fig. 9 (d) and (e) and yielded $\mu_{yy}/\mu_{xx} = 0.34 \pm 0.05$ (0.09 ± 0.05) and $\mu_{xy}/\mu_{xx} = -0.17 \pm 0.04$ (0.06 ± 0.04) in TIPS-pentacene (TES-pentacene). Using these components and substituting $\varphi = \gamma$ from Table 1 (for μ_{bb}) in Eq.(5), we obtained the ratio of the mobilities along *a*- and *b*-axis (μ_{aa}/μ_{bb} = 3.2 and 8 in TIPS-pentacene and TES-pentacene, respectively). Assuming that μ_{yz} and μ_{xz} components of the mobility tensor are small, we diagonalized the *x-y* part of the tensor and determine the directions of the principal axes 1 and 2, as well as the corresponding mobilities μ_{11} ($1.04\mu_{aa}$ and $1.004\mu_{aa}$ in TIPS-pentacene and TES-pentacene, respectively) and μ_{22} ($0.30\mu_{aa}$ and $0.086\mu_{aa}$ in TIPS-pentacene and TES-pentacene, respectively), whose ratio μ_{22}/μ_{11} yielded 3.5 ± 0.6 and 12 ± 6 for TIPS-pentacene and TES-pentacene, respectively. The mobility anisotropy of 3.5 in the TIPS-pentacene crystal is very similar to that obtained in Rub and Pc single crystals using a field-effect transistor geometry (Podzorov et al., 2004). Considerable difference in the in-plane mobility anisotropy between the TIPS-pentacene and TES-pentacene crystals supports a theoretical prediction of much stronger mobility anisotropy in the case of the TES-pentacene-type crystals that favor 1D charge transport based on the crystal structure and molecular packing (Figs.9(a) and (b)). In TIPS-pentacene, the principal axes 1 and 2 constitute angles $\varphi_1 = -14°$ and $\varphi_2 = 76°$ with respect to the *a*-axis, respectively (Fig.9(d)). Interestingly, the highest mobility axis does not exactly coincide with the direction of maximum π-overlap along the *a*-axis, which highlights the contribution of factors unrelated to band structure, such as fluctuations of the intermolecular coupling, to charge carrier mobility in organic molecular crystals. In TES-pentacene, the principal (highest mobility) axis 1 is coincident with the direction of maximum π-overlap along the *a*-axis (Figs. 9(b) and (e)). However, it is possible that due to the above mentioned fluctuations in intermolecular electronic coupling, the observed in-plane anisotropy of ~12 is smaller than that expected based solely on the band structure. The observed mobility anisotropy in these samples, which is one of the signatures of band-like charge transport in organic crystals, further supports the occurrence of band-like transport in functionalized pentacene single crystals.

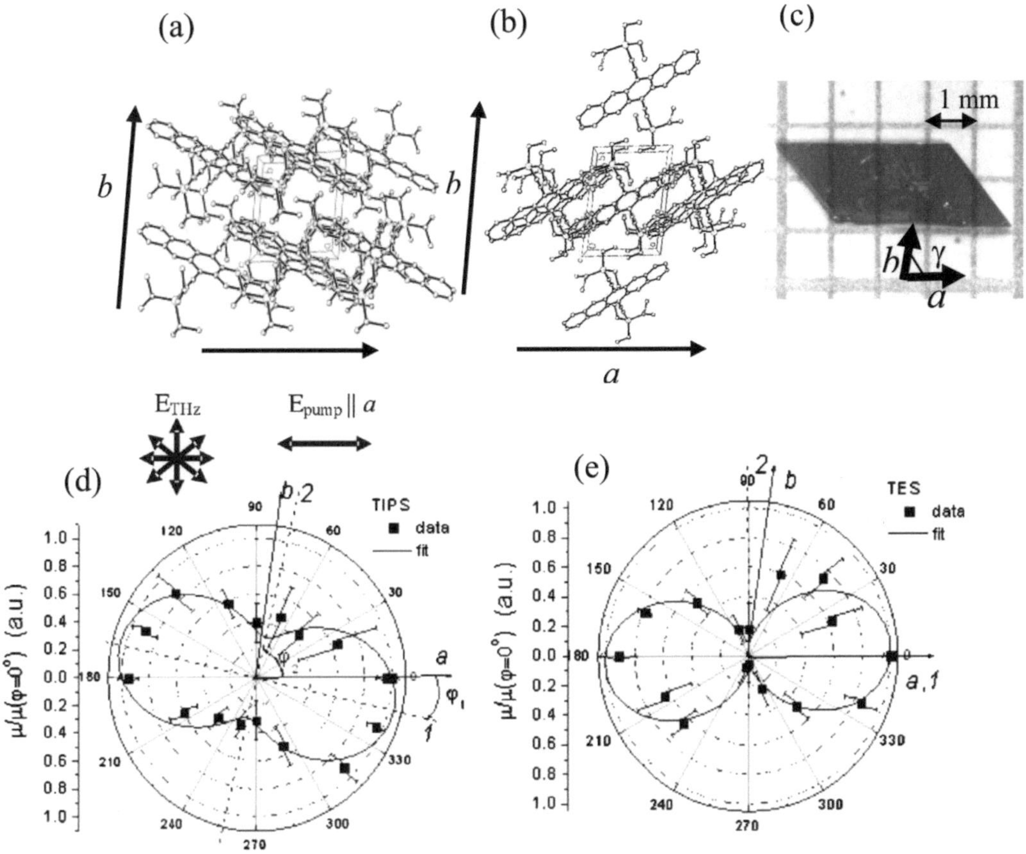

Fig. 9. Molecular packing in (a) TIPS-pentacene and (b) TES-pentacene crystals. (c) Typical TIPS-pentacene crystal. Dependence of the charge carrier mobility on the azimuthal angle φ obtained in (d) TIPS-pentacene and (e) TES-pentacene crystals. In both crystals, φ=0° corresponds to the *a* axis. Lines correspond to the fit with a function described in the text. Crystallographic *a* and *b* axes as well as the principal axes 1 and 2 are also shown. Adapted from Ostroverkhova et al., 2006b, with permission. Copyright American Institute of Physics (2006).

3.3.2 Transient photocurrent on 30 ps to 1 ms time-scales in thin-film device structures

Pristine Materials. Upon excitation with 400 nm 100 fs pulses under applied voltage, thin-film devices based on TIPS-pentacene and ADT derivatives (ADT-TES-F and ADT-TIPS-F, Section 3.1) showed fast photoresponse, with the rise time of 30-40 ps, limited by the time resolution of the sample fixture and DSO, which supports our earlier observations (using noncontact optical pump-THz probe techniques, Section 3.3.1) of fast charge carrier photogeneration in acenes. Transient photocurrents obtained in ADT-TIPS-F and TIPS-pentacene films upon excitation with 400 nm pulses at the electric field *E* of 1.2 × 10^4 V/cm are shown in Fig. 10. In all films, transient photocurrents exhibited fast initial decay, most likely due to initial carrier trapping and recombination, followed by a slow component that can be fitted with a power-law function ($I_{ph} \sim t^{-\beta}$) with β ~0.2 – 0.6, depending on the sample and on the material, over at least three orders of magnitude (inset of Fig. 10) (Day et al., 2008; Day et al., 2009a; Platt et al., 2009b). Among ADT-TIPS-F, ADT-TES-F, and TIPS-pentacene films, the decay of the transient photocurrent was, on average, slowest in ADT-TES-F, followed by ADT-TIPS-F and TIPS-pentacene, when measured under the same conditions. Power-law exponents β describing the transient photocurrent decay dynamics on time scales from sub-ns to at least tens of μs after photoexcitation were also slightly different, with β of ~ 0.2 – 0.3 in ADT-TES-F and of ~0.4 – 0.6 in ADT-TIPS-F and TIPS-pentacene films. Interestingly, the power-law exponents observed in these experiments in TIPS-pentacene films were similar to those obtained at shorter time-scales in the optical pump-THz probe experiments in similar films and TIPS-pentacene single crystals (Fig.7(b)) (Ostroverkhova et al., 2005a; Ostroverkhova et al., 2005b), which suggests that the same physical mechanism is responsible for charge carrier dynamics observed on time-scales from ps to at least tens of μs after photoexcitation.

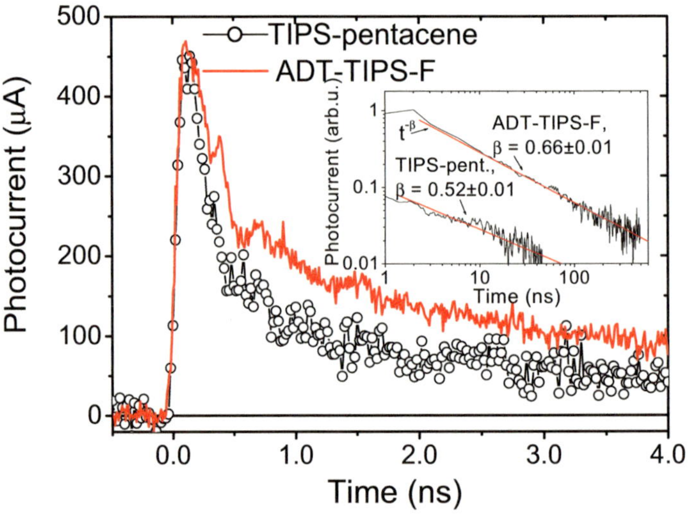

Fig. 10. Transient photocurrent obtained in ADT-TIPS-F and TIPS-pentacene films on Al interdigitated electrodes at 1.2x10^4 V/cm and 30 μJ /cm^2 at 400 nm. Inset: normalized transient photocurrents (I_{ph}) in ADT-TIPS-F and TIPS-pentacene films, at longer time-scales. Power-law fits ($I_{ph} \sim t^{-\beta}$) are also shown. Offset along the *y* axis is for clarity. Adapted from Day et al., 2008, with permission. Copyright American Institute of Physics (2008).

The μη products calculated from Eq.(3) yielded values between 0.01 and 0.025 cm^2/Vs at 1.2×10^4 V/cm in ADT-TES-F, ADT-TIPS-F and TIPS-pentacene films (Table 3). Here η includes trapping and recombination that occurred during the first tens of picoseconds after photoexcitation, not resolved in these experiments, and therefore, $\eta < \eta_0$, where η_0 is the initial photogeneration efficiency. If intrinsic mobilities in ADT materials are on the order of 1 cm^2/Vs, then the lower limit for the initial photogeneration efficiency η_0 is ~1-2%, which is comparable to that of 5% estimated from the transient photocurrents measured in tetracene single crystals under similar conditions (Moses et al., 2006). Similarly to the results of the THz spectroscopy (Section 3.3.1, Fig.8(b)), the $\mu\eta$ obtained from the transient photocurrent amplitudes in these experiments decreased as the temperature increased, at least in the temperature range between 285 K and 350 K (Day et al., 2008).

Composites. Both amplitude and decay dynamics of the transient photocurrent can be manipulated by addition of guest molecules in the host material (Section 3.1). Inset of Figure 11 shows peak amplitude of the transient photocurrent as a function of concentration of C_{60} in the ADT-TES-F/C_{60} composite thin film, at 40, 70, and 100 V (Day et al., 2009a). Addition of C_{60} at a concentration of 2 and 5% increased the photocurrent amplitude by a factor of ~3 and 10, respectively, at 40 V. This increase is most likely due to fast photoinduced electron transfer (a mechanism similar to that leading to sensitization of C_{60}-containing polymer composites and organic glasses (Ostroverkhova and Moerner, 2004)), based on our observation of significant PL quenching in ADT-TES-F/C_{60} composite films compared to pristine ADT-TES-F films (Fig.5(b)) (Day et al., 2009a). Further addition of C_{60} lowered the measured photocurrent amplitude (e.g. by a factor of ~2 at 10% of C_{60}, at 100 V), most likely due to enhanced initial recombination, occurring at times below ~100 ps, not resolved in our experiments. Addition of either TIPS-pentacene or ADT-TIPS-CN to ADT-TES-F lowered the amplitude of the transient photocurrent by a factor of ~3-4 at all applied voltages in the range studied. The most interesting effect of adding these molecules to the ADT-TES-F host was, however, the change in the initial photocurrent decay dynamics (Fig. 11). In particular,

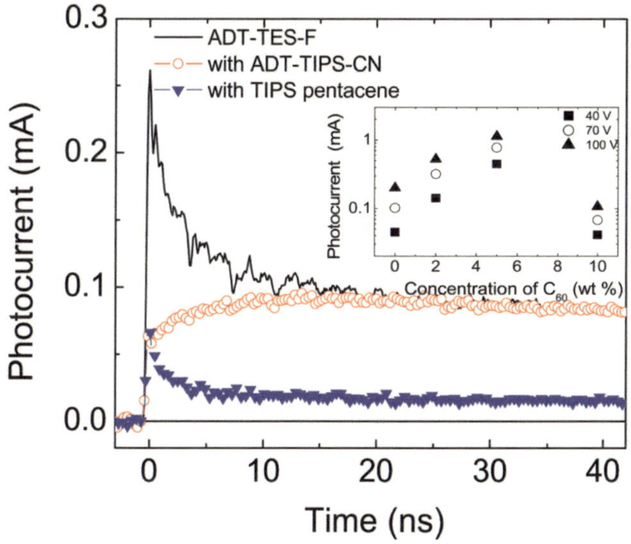

Fig. 11. Transient photocurrent obtained in pristine ADT-TES-F film and in ADT-TES-F/TIPS-pentacene and ADT-TES-F/ADT-TIPS-CN composites at a fluence of 5 μJ /cm^2 at 100 V applied across 25 μm gap. Inset shows amplitudes of the transient photocurrent as a function of concentration of C_{60} in ADT-TES-F/C_{60} composites, measured at 40, 70, and 100 V. Adapted from Day et al., 2009a, with permission. Copyright American Institute of Physics (2009).

upon addition of TIPS-pentacene, the initial decay became faster, the effect similar to that achieved by the addition of C_{60}. For example, only ~32% of "initially" photogenerated carriers remained mobile at 5 ns after photoexcitation in the ADT-TES-F/TIPS-pentacene composite. We attribute this to fast hole transfer from ADT-TES-F to TIPS-pentacene followed by trapping at the TIPS-pentacene molecules (Day et al., 2009a). In contrast, addition of ADT-TIPS-CN to ADT-TES-F introduced a slow component into the photocurrent *rise* dynamics and completely removed the fast initial decay (Fig. 11). In particular, the fast rise of the photocurrent, limited by the time resolution of our setup, accounted only for ~70% of all photogenerated carriers, whereas the other 30% were generated over ~0.1-20 ns. As a result, the peak of the photocurrent in the ADT-TES-F/ADT-TIPS-CN composite was achieved at about 20 ns after photoexcitation, after which a slow decay, characterized by a power-law function with $\beta < 0.1$, persisted to at least 1 ms after photoexcitation. Our observations of complete quenching of the PL of ADT-TES-F, while magnifying that of the ADT-TIPS-CN in the ADT-TES-F/ADT-TIPS-CN composite (Fig.5(b)) (Day et al., 2009a) suggest efficient energy transfer from ADT-TES-F to ADT-TIPS-CN. Therefore, it is possible that the slow component in the rise dynamics of the transient photocurrent in Fig. 11 is due to a multistep process that involves excitation of ADT-TES-F, followed by energy transfer to ADT-TIPS-CN, which then leads to energetically favorable hole transfer back to ADT-TES-F, while the electron remains trapped on the ADT-TIPS-CN molecules.

3.4 Dark current

Space-charge-limited currents (SCLCs) were observed in the ADT-TIPS-F, ADT-TES-F and TIPS-pentacene films on Au electrodes. Effective charge carrier mobilities (μ_{eff}) were calculated from the slope of the fits of the dark current as a function of applied voltage squared. In the case of planar electrode geometry used here, the current flows along a layer of unknown thickness, and there is no analytical solution for the relationship between SCLC (linear) density ($j = I_d/d$, where I_d is the dark current and d is the channel width) and voltage (V) in a film of finite thickness. In the approximation of the infinitely thin film, valid when the film thickness is much lower than the gap width (L) between the electrodes (which is the case here),

$$j = 2\mu_{eff}\varepsilon\varepsilon_0 V^2/(\pi L^2), \tag{6}$$

where ε is a relative dielectric constant of the film and ε_0 is the dielectric permittivity of vacuum. SCLC effective mobilities (μ_{eff}), which represent a lower bound of charge carrier mobilities in these films, showed sample-to-sample variation, especially significant in the case of ADT-TIPS-F and TIPS-pentacene (Table 3). On average, however, μ_{eff} in ADT-TES-F was at least a factor of ~3 higher than that in ADT-TIPS-F, and a factor of ~7 higher than in TIPS-pentacene films.

3.5 Cw photocurrent

Figure 12 shows cw photocurrent obtained at 532 nm excitation of ADT-TIPS-F, ADT-TES-F, and TIPS-pentacene films on interdigitated Au electrodes. In all samples, cw photocurrent measured at light intensity of 0.58 mW/cm^2 was higher than the dark current. Especially strong cw photoresponse was observed in best ADT-TES-F films, with photocurrents of over 200 μA at the average electric field of 4×10^4 V/cm at low light intensities (Fig. 12). This

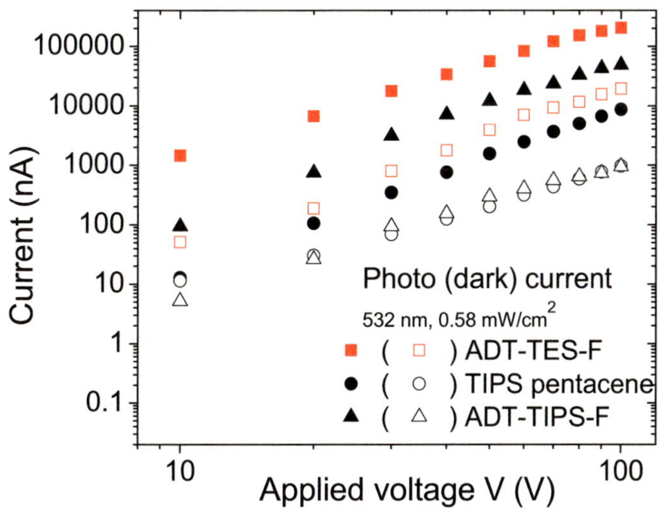

Fig. 12. Dark current and cw photocurrent at 532 nm excitation in ADT-TES-F, ADT-TIPS-F, and TIPS-pentacene films on interdigitated electrodes.

corresponds to linear photocurrent densities ($j_{cw} = I_{cw}/d$) of over 0.1 mA/cm, which, under assumption that photons absorbed throughout the entire thickness of a 1 μm thick film may contribute to the photocurrent, yields (area) current densities (J_{cw}) of over 1 A/cm^2. If photoconductivity (σ_{ph}) is calculated using $\sigma_{ph} = J_{cw}/E$, then values of $\sigma_{ph} \sim 2.5 \times 10^{-5}$ S/cm are obtained at 4×10^4 V/cm at 0.58 mW/cm^2 in ADT-TES-F, which are considerably higher than those reported in conductive polymers such as PPV and in unsubstituted pentacene films or tetracene crystals under cw excitation (Platt et al., 2009b). As the temperature increased, the cw dc photocurrent increased, with the increase described by the function $I_{cw} \sim \exp[-\Delta_{cw}/k_BT]$, with the activation energies Δ_{cw} of 0.05-0.17 eV, depending on the sample and on the material (Day et al., 2008l; Platt et al., 2009b). This is in contrast to the behavior observed on ps time-scales after photoexcitation in the same materials (Sections 3.3.1-3.3.2), which suggests a change in charge photogeneration and/or transport mechanisms as the time progresses.

Photoconductive gain G was calculated from the cw photocurrents, absorption coefficients, and light intensity as the ratio between the number of carriers flowing in the film and the number of absorbed photons. In the case of hole-transporting materials and hole-injecting electrodes (such as Au in the case of ADT-TIPS(TES)-F and TIPS-pentacene), bulk photoconductive gain

$$G \approx \eta_0 \tau_c / t_{tr} \quad (7)$$

(where τ_c is the carrier lifetime, and t_{tr} is the time for the hole to transit through the film), and G can be much larger than 1. Gain values G that are much larger than unity were indeed observed in fluorinated ADT and TIPS-pentacene films, as summarized in Table 3. The highest photoconductive gains, up to 130 at 4×10^4 V/cm at 0.58 mW/cm^2, were achieved in ADT-TES-F films, consistent with highest effective mobility μ_{eff} (shorter transit time t_{tr}) and longest carrier lifetimes (as observed in SCLC and in the transient photocurrent measurements, respectively) in ADT-TES-F films, as compared to other materials under study. The values of G measured in ADT-TES-F films were similar to those in GaN photodetectors (Munoz et al.,1997) at similar light intensity levels and at least an order of magnitude higher than those in unsubstituted pentacene and in functionalized pentacene films (Gao & Hegmann, 2008; Lehnherr et al., 2008).

3.6 Influence of organic-metal interfaces on the current

3.6.1 Transient photocurrent

Figure 13 (a) shows transient photocurrents, normalized at their peak values, measured in ADT–TIPS–F films on Au and Al electrodes. Interestingly, the faster decay component was more pronounced in films on Al electrodes, whereas the slower one was independent of the electrode material and could be described by a power-law function ($I_{ph} \sim t^{-\beta}$), as discussed above (Section 3.3.2). In samples with a 25-μm gap at voltages above ~30 V, the difference in the peak transient amplitudes for films on Au and films on Al electrodes was comparable to that due to morphology-related sample-to-sample variation of approximately a factor of 2 (Fig. 13 (b)). The relation between the peak amplitude and applied voltage ($I_{ph,max} \sim V^{\alpha}$) was sample-dependent, with α = 1.3–1.8. Analysis of the behavior of many samples, however, yielded that on average, α in samples with Al electrodes was higher than in those with Au

Compound	$\mu\eta$,THz[a] (cm²/(Vs)	$\mu\eta$[b] (cm²/(Vs)	μ_{eff}[c] (cm²/Vs)	μ_{FET} (cm²/(Vs)	$\eta_{0, min}$[d]	G[e]
ADT-TES-F	--	0.02-0.025	0.033-0.092	1.7 (film)[f] 6 (crystal)[g]	0.013-0.024	70-130
ADT-TIPS-F	--	0.018-0.025	0.002-0.029	0.1 (crystal)[h]	0.018-0.033	16-30
TIPS-pent.	0.02-0.06 (film) 0.15-0.2 (crystal)	0.01-0.022	0.002-0.007	1.5 (film)[i]	0.0003-0.001	9-28

Table 3. (Photo)conductive properties of TIPS-pentacene and ADT thin films.

[a] Calculated from the peak THz transmission using Eq.(2) (Section 3.3.1 and Ostroverkhova et al., 2005b)

[b] Calculated from the peak of the transient photocurrent measured at 400 nm excitation in thin-film device structures using DSO (Section 3.3.2 and Day et al., 2008)

[c] Calculated from SCLC currents in thin-film devices (Section 3.4 and Platt et al., 2009b)

[d] Calculated from cw photocurrents in thin-film devices on Al electrodes (Section 3.5 and Platt et al., 2009b)

[e] Calculated from cw photocurrents in thin-film devices on Au electrodes (Section 3.4 and Platt et al., 2009b)

[f] Calculated from TFT characteristics of ADT-TES-F spin-coated onto Au electrodes treated with PFBT (Park et al., 2008)

[g] Calculated from FET characteristics of ADT-TES-F single crystal (Jurchescu et al., 2008)

[h] Calculated from FET characteristics of ADT-TIPS-F single crystal (Subramanian et al., 2008)

[i] Calculated from TFT characteristics of spin-coated TIPS-pentacene film (Park et al., 2007)

electrodes (e.g. α = 1.76±0.06 and 1.33±0.01 in ADT–TIPS–F films on Al and Au electrodes, respectively, in Fig. 13 (b)). Similar dependencies were observed for all electrode geometries, regardless of film-side or glass substrate-side illumination. Since measurements of the photocurrent were performed in the presence of applied electric field, the photoresponse was necessarily measured after dark current flow had been established in the film, or, in other words, after the sample had been pre-conditioned by the dark current. In samples with Au and Al electrodes, very different charge injection conditions obtain and, therefore, different pre-conditioning of the samples is realized. In particular, samples with Au electrodes, in which SCLC regime is established (Section 3.4), are expected to have a high density of filled hole traps. In contrast, samples with Al electrodes are expected to be in a close-to-pristine condition (empty traps) at low voltages, and have partially filled traps at higher voltages, after hole injection via, for example, Fowler-Nordheim tunneling is enabled (Day et al., 2009b). If fast initial decay is related to bulk material properties and is due to fast trapping of photogenerated holes, then in films on Au electrodes, the density of empty traps is reduced, which in turn reduces initial trapping of the photogenerated holes. In contrast, most traps in films on Al electrodes are empty and are readily filled by photogenerated holes, thus leading to a fast initial decay of the photocurrent. This is also consistent with previously observed slowing down of the initial decay of the transient photocurrent upon increasing applied electric field in ADT–TIPS–F films on Al electrodes (Day et al., 2008).

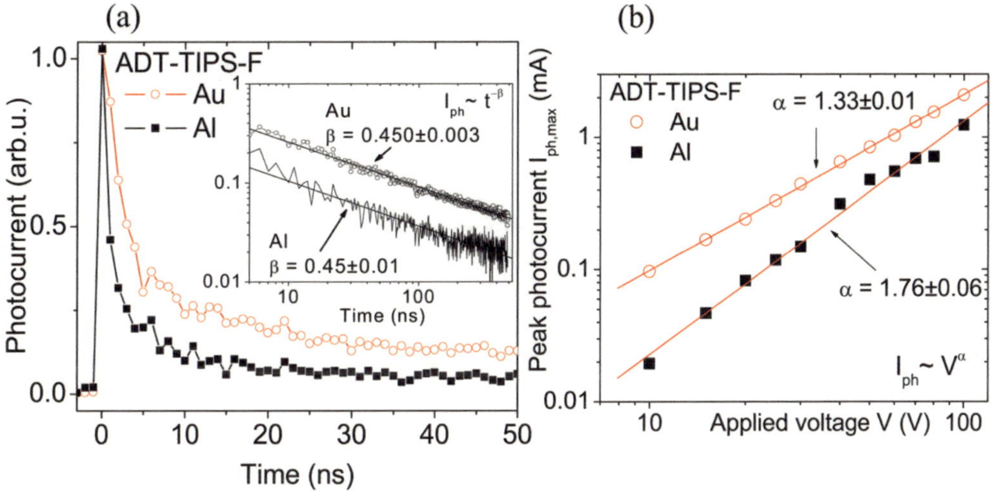

Fig. 13. (a) Transient photocurrents, normalized at their peak values, obtained in ADT-TIPS-F films on Au and Al electrodes under the same experimental conditions. Inset: long time-scale dynamics of the same transient photocurrents. Power-law fits ($I_{ph} \sim t^{-\beta}$) are also shown. (b) Transient photoresponse peak amplitude ($I_{ph;max}$) as a function of applied voltage obtained in ADT-TIPS-F samples with Al or Au interdigitated electrodes. Reprinted from Day et al., 2009b, with permission. Copyright American Institute of Physics (2009).

Scanning microscopy. The peak amplitudes of the transient photocurrent measured at different positions of the localized pulsed photoexcitation of ADT-TES-F and ADT-TIPS-F films on Au and Al electrodes are shown in Fig. 14 (a). No signal was detected at any position in the absence of applied voltage. Change in the transient photocurrent amplitude upon scanning of the pulsed focused beam across a planar device has been previously related to internal electric field distribution in InP planar devices (Day et al., 2009b). In ADT films, although the exact distribution of transient photocurrent amplitudes in the gap somewhat depended on the measurement protocol (such as direction of the scan and waiting time between the data points), these distributions in samples with Au and Al electrodes were similar within the experimental error and, with our spatial resolution, relatively uniform across the gap. In particular, the amplitudes of transient photocurrents obtained upon excitation of near-electrode regions and of a mid-gap region were within a factor of 2 of each other. Transient photocurrent decay dynamics under localized excitation in samples with Au and Al electrodes were similar to those in Fig. 13 (a) obtained under uniform illumination, with the initial decay faster in samples with Al electrodes. No dependence of the decay dynamics of the transient photocurrent on the beam position in the gap was observed.

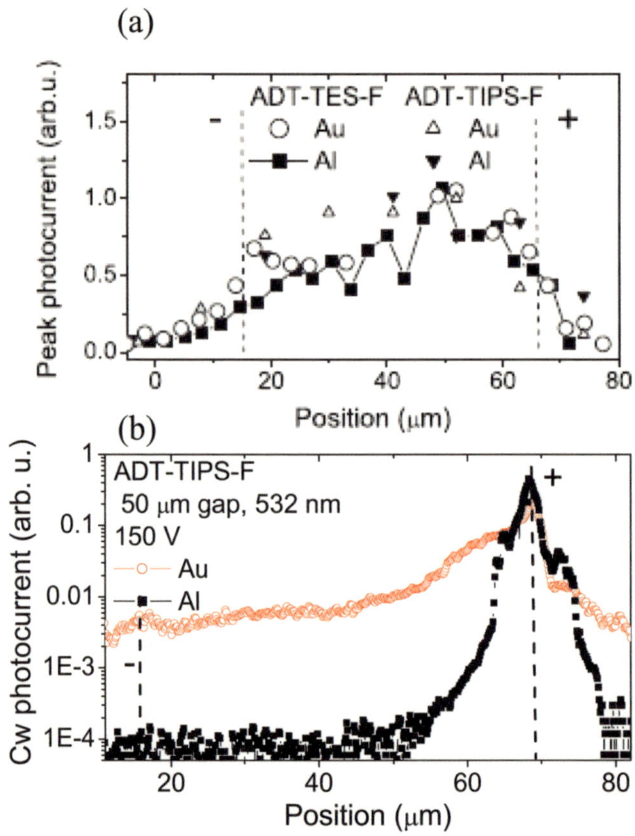

Fig. 14. (a) Peak amplitudes of the transient photocurrent, normalized at their maximal values in the gap, at different positions of the localized beam spot obtained in ADT-TES-F and ADT-TIPS-F films on Au and Al electrodes, in coplanar electrode geometry with 50-μm gap at 100 V and 150 V, respectively. (b) Cw photoresponse as a function of the beam position obtained in an ADT-TIPS-F film in coplanar electrode geometry with 50-μm gap with Al and Au electrodes at 150 V. Dashed lines correspond to the geometrical edges of the electrodes. Adapted from Day et al., 2009b, with permission. Copyright American Institute of Physics (2009).

3.6.2 Dark current and cw photocurrent

Unlike the amplitudes of transient photocurrent, the cw photocurrents (I_{cw}) and dark currents (I_d) for films on Au and on Al electrodes differed by more than an order of magnitude over a wide range of applied voltages, with much higher currents measured in samples with Au electrodes (Fig. 15). Voltage dependencies of the dark current and of the cw photocurrent observed in samples with Au and Al electrodes were drastically different, indicative of different mechanisms responsible for observed currents, depending on the electrode material. In samples with Au electrodes, the relationship between dark current and voltage ($I_d \sim V^{\alpha_d}$) was close to linear at very low voltages and close to quadratic at higher voltages (e.g. α_d = 1.09±0.03 at voltages below 2 V and 1.830±0.006 from 2 to 300 V in an ADT-TES-F film in Fig. 15(a)). In contrast, in samples with Al electrodes, dark current was weakly voltage-dependent at lower voltages, followed by a steep increase at higher voltages. Despite this sharp increase, even at the highest applied voltage of 300 V used in our experiments, dark current in samples with Al electrodes was much lower than that in samples with Au electrodes at the same voltage. In films on either Au or Al electrodes, the cw photocurrent was higher than the dark current in the entire range of light intensities used. Regardless of the material, in samples with Au electrodes, α_{cw} obtained from the fit of voltage dependence of the photocurrent ($I_{cw} \sim V^{\alpha_{cw}}$) ranged between 1.5 and 2.2, depending on the sample. In any given sample, however, a single value of α_{cw} was sufficient to describe voltage dependence of the cw photocurrent over a large voltage range (e.g. α_{cw} = 1.62±0.02 from 5 to 300 V in an ADT-TES-F film in Fig. 15(b)). In samples with Al electrodes, however, α_{cw} was ~1.3–3 at lower voltages (e.g. α_{cw} = 1.32±0.05 in Fig. 15(b)), depending on the sample, followed by steep transition described by α_{cw} ~3.3–5 (e.g. α_{cw} = 4.6±0.2 in Fig. 15(b)) at higher voltages. Similar trends were observed in ADT-TIPS-F films. (Day et al., 2009b).

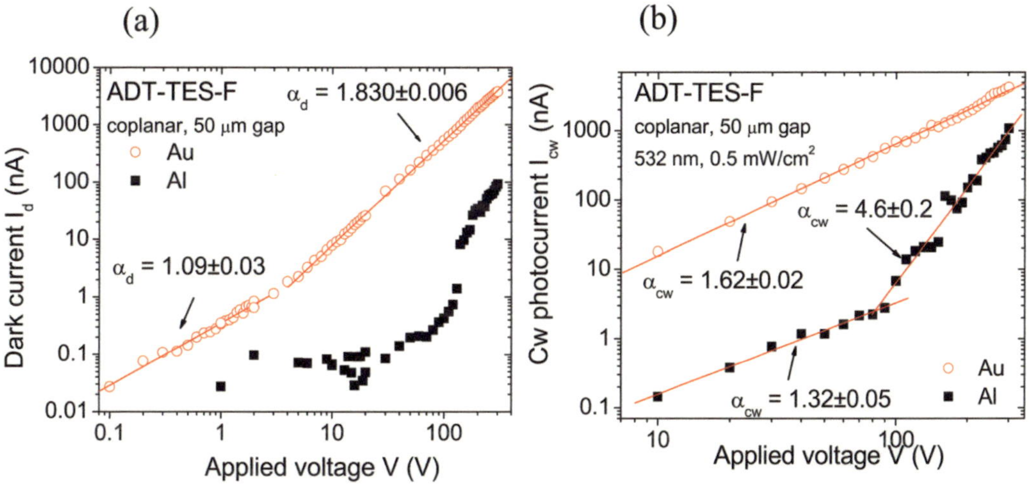

Fig. 15. (a) Dark current (I_d) and (b) cw photocurrent (I_{cw}) obtained in ADT-TES-F films with Al and Au contacts in coplanar geometry with 50-μm gap. Power-law fits $I_d \sim V^{\alpha_d}$ and $I_{cw} \sim V^{\alpha_{cw}}$ in (a) and (b), respectively, are also shown.

The photoresponse under cw illumination is strongly affected by the conditions imposed by the electrode material (Fig. 15(b)) because the steady-state photocurrent (I_{cw}) is proportional to the photoconductive gain (G), which is much larger than 1 in our samples on Au electrodes (Table 3) due to long carrier lifetimes (Day et al., 2009b). In samples with non-injecting electrodes, such as ADT films on Al electrodes at low voltages, the maximal achievable gain cannot exceed the initial photogeneration efficiency η_0 ($\eta_0 < 1$) and therefore, the cw photocurrent I_{cw} is much lower. This is in contrast to the transient photocurrent, the amplitude of which does not significantly depend on the carrier lifetime, thus leading to comparable transient photoresponse of films on Au and Al electrodes.

In samples with blocking electrodes (such as Al in the case of ADT-TIPS(TES)-F and TIPS pentacene), *G* cannot exceed η_0. Thus, the photoconductive gain calculated from cw photocurrents measured in samples with Al electrodes at low voltages represents a lower limit of the photogeneration efficiency $\eta_{0,min.}$ In ADT-TES-F and ADT-TIPS-F films, comparable values of $\eta_{0,min} \approx 0.01$-$0.03$ were obtained at 1.2×10^4 V/cm (Table 3). Interestingly, these values are similar to those estimated from transient photocurrents measured in the same films. In constrast, in TIPS-pentacene films, the $\eta_{0,min}$ values were considerably lower than those calculated from the amplitudes of the transient photocurrents (Table 3). This is most likely due to extensive deep trapping in TIPS pentacene, which reduces carrier lifetime, leading to an underestimation of the initial photogeneration efficiency η_0 from *G*.

Scanning microscopy. Figure 14 (b) shows the cw photoresponse measured under localized excitation of ADT-TIPS-F films on Al and Au electrodes, respectively, as a function of the beam position in the gap. Samples with either Al or Au electrodes showed a marked peak in photoresponse at the excitation near the positively biased electrode, which became more pronounced as the voltage increases. This effect was independent of the direction of the scan, was observed at all light powers ranging between 0.03 and 4 μW used in our experiments, and was attributed to a dominant photogenerated hole transport, accompanied by strong electron trapping (Day et al., 2009b). As seen from Fig. 14(b), at 150 V, in the case of Al electrodes, photoresponse was mostly limited to the several micron-wide region. Under the same conditions, photoresponse in the sample with Au electrodes was more uniform within the gap (Fig. 14(b)), yielding measurable photocurrents upon excitation of the mid-gap region, as well as of the region close to the negatively biased electrode. Interestingly, if the waiting period between the successive scans was short compared to the time needed to thermally empty the filled charge traps, the photocurrent distribution in the gap became more and more uniform with each scan, which could be due to increased carrier lifetime under trap-filled conditions. No photoresponse was observed in the absence of an applied voltage (Day et al., 2009b). Such pronounced differences in cw photoresponse behavior of samples on Au and Al electrodes are in agreement with results obtained under uniform illumination in the same devices (Fig. 15).

4. Discussion

One of the challenging aspects of dealing with organic semiconductors, especially thin films, is that intrinsic properties are often masked by impurities, grain boundaries, and defects. As a result, it is often difficult to reconcile results of different experiments performed on the same material. One of our motivations for the studies described above was to develop a picture of charge photogeneration and transport in acenes that would be consistent with results of a variety of experiments that probe carrier and exciton dynamics from sub-ps to many seconds after photoexcitation.

First of all, it seems clear that charge carriers in acenes can be created via photoexcitation on time-scales of ps and faster, without either thermal or electric-field assistance. However, in keeping with $\mu\eta$ values extracted from the transient photoconductivity experiments (Sections 3.3.1 and 3.3.2) and assuming intrinsic mobilities at room temperature in most of our materials to be below 10 $cm^2/(Vs)$ (Park et al., 2007; Jurchescu et al., 2008; Park et al., 2008, Subramanian et al., 2008), the efficiency of this process is rather low, with initial photogeneration efficiencies η_0 of <5%. This is also consistent with the estimates of the η_0 from cw photocurrent measurements (Table 3). The other >95% of the primary photoexcitation results in exciton formation. The origin of the photogenerated carriers observed in transient photocurrent experiments on time scales from sub-ps to ns after photoexcitation does not appear to be excitonic (i.e. due to exciton dissociation) due to the following observations: (i) decrease of $\mu\eta$ with the temperature, as opposed to an increase expected from the Onsager model of exciton dissociation; (ii) weak, if any, wavelength dependence of the photogeneration efficiency above the absorption edge; (iii) similar $\mu\eta$ values for TIPS-pentacene and functionalized ADTs, despite very different exciton behavior in films (high PL QYs in ADTs and ultra-low ones in TIPS-pentacene); (iv) different charge carrier and exciton dynamics (long power-law and fast bi-exponential, respectively), assessed using time-resolved photocurrent and PL, respectively; and (v) no apparent electric field-induced exciton dissociation up to electric fields that are more than an order of magnitude higher than those used in photocurrent measurements. Photogeneration efficiency can be improved by introducing a strong acceptor such as C_{60} as a guest molecule in the acene host due to fast photoinduced electron transfer from the host molecule to C_{60}, which releases a free hole into the host. Other dopants can affect the photoexcited charge carrier dynamics as well by initiating competition between energy and charge transfer.

It is also interesting to note that $\mu\eta$ values obtained from direct measurements of the transient photocurrent (Section 3.3.2) in TIPS-pentacene solution-deposited films at electric fields of ~10^4 V/cm are only about a factor of 2-3 lower than those measured at earlier times in the optical pump-THz probe experiments (Section 3.3.1) in similar films at zero static electric field (Table 3). Since at low electric fields neither mobility nor photogeneration efficiency are expected to strongly depend on the electric field, this difference would be mostly due to a carrier loss that occurs between ~400 fs and ~30 ps after photoexcitation, which correspond to the time resolution of these two experiments.

Regarding charge transport mechanism, the following picture seems to apply. At ps time-scales after photoexcitation, charge transport is band-like, which is characterized by mobility increasing as the temperature decreases and by mobility anisotropy in crystals. As the time progresses, dispersive transport takes over and persists over many orders of magnitude in time. Interestingly, in functionalized acenes, it appears that non-activated carrier tunneling, rather than hopping, occurs at least up to ns after photoexcitation, since no significant temperature dependence of the transient photoconductivity power-law decay dynamics was observed in the wide temperature range. The fact that this behavior is seen in both single crystals and thin films suggests that it is an intrinsic property of these materials, possibly related to molecular π-stacking. In the equilibrium, the transport is thermally activated, consistent with hopping mechanism. An interesting feature of these materials is persistent photoconductivity due to long carrier lifetimes, which lead to high photoconductive gains.

Equilibrium transport is characterized by the effective mobility $\mu_{eff} = \mu\theta$, where θ is the ratio between free and trapped charge carriers ($\theta \leq 1$). In our films, μ_{eff} extracted from SCLC measurements (Section 3.4) is lower than the intrinsic mobility μ, since most likely the trap-free limit has not been achieved in our samples (Table 3). Indeed, although μ_{eff} values of 0.033−0.092 cm²/Vs obtained in ADT-TES-F films were similar to those obtained in ADT-TES-F TFTs with channel lengths similar to ours (i.e. 25 μm), mobilities of over 1.5 cm²/Vs and 6 cm²/Vs have been achieved in short-channel ADT-TES-F films spin-coated on the Au electrodes treated with pentafluorobenzenethiol (PFTB) (Park et al., 2008) and in ADT-TES-F single crystals (Jurchescu et al., 2008), respectively. Therefore, θ is below ~0.015 in our ADT-TES-F films. It is lower yet in TIPS-pentacene films, since μ_{eff} values of 0.002−0.007 cm2/Vs obtained in our SCLC measurements are much lower than 1.2 cm²/Vs observed in TIPS-pentacene TFTs, indicative of high trap density in our TIPS-pentacene films.

Also, under equilibrium, there is a possible interesting connection between the PL and cw photocurrent. In particular, temperature dependencies of the PL QYs and cw photoconductivity, characterized by Δ_{PL} and Δ_{cw}, respectively (Sections 3.1 and 3.4), seem to be close to each other in the same samples (Platt et al., 2009b). It has been shown theoretically that in materials with π-stacking, such as functionalized pentacene and ADT derivatives, thermal fluctuations lead to dramatic changes in intermolecular coupling (Troisi and Orlandi, 2006), which could be important for both PL and photoconductivity in films. In particular, it appears that thermally activated nuclear motions can both enhance nonradiative recombination, leading to a decrease in the PL, and facilitate charge carrier detrapping from shallow traps and transport of *localized* carriers, which would improve cw photoconductivity. This does not seem to affect transient photocurrents measured at ps time-scales after photoexcitation.

Finally, the electrode material plays a significant role in the dark current and cw photoresponse, since both rely on efficient carrier injection from the electrode. In contrast, transient photocurrents are only weakly affected by the choice of the electrodes.

5. Conclusions and outlook

Over the past decade, tremendous progress has been made towards understanding of physical mechanisms that determine optical and electronic properties of organic semiconductors, in development of novel experimental methods that probe intrinsic properties of materials, in organic synthesis and purification of materials, and in fabrication and characterization of various devices. Novel solution-processable high-purity materials with tunable properties are now available, and a variety of methods of high-quality crystal growth and thin-film deposition have been developed. A number of devices that utilize organic thin-film transistors and light-emitting diodes are already on the market, and performance of many more devices, such as photovoltaic cells and various photorefractive devices, keeps steadily improving and has already almost reached the point of commercial feasibility. Possibility to fine-tune optical and electronic properties by creating solution-deposited organic semiconductor composite films offers an extraordinary flexibility to tailor the materials for the specific application demands. Another exciting possibility is to utilize properties of individual highly photoluminescent molecules (such as ADT-TES-F) as nanoprobes of conduction, nanoswitches, and nano optoelectronic devices.

Despite considerable progress, many issues require further attention. In particular, the exact nature of charge photogeneration and transport in small-molecular-weight organic semiconductors is still not well understood, and new experimental methods reveal additional layers of complexity of their behavior. Also, more solution-processable materials with improved charge-transfer properties are in demand. Moreover, optical and electronic properties strongly depend on the structure and morphology of organic films and, therefore, fabrication methods that yield consistent films, *and over a large substrate area*, are needed. Finally, influence of ambient conditions, including temperature, humidity, the presence of oxygen and other gases, etc. on device performance is strong and, ideally, has to be minimized by creative design of device structures. Nevertheless, the field of organic photonics and electronics is developing fast and will undoubtedly produce new and exciting results in the near future.

6. Acknowledgements

We are indebted to Profs. F. A. Hegmann and D. G. Cooke for the facilities guidance, and contribution to THz spectroscopy work. We thank Profs. J. E. Anthony, V. Podzorov, M. E. Gershenson, O. D. Jurchescu, and T. T. M. Palstra for organic materials used in our work. Technical assistance of Prof. Y.S. Lee, Dr. J. Danielsen, J. Tomaino, and R. Presley is acknowledged. Also thanks to Dr. J. Day for carrying out initial measurements of photoconductive properties of materials in our group. This work was supported by the Killam Trust, Office of Naval Research (grant #N00014-07-1-0457 via ONAMI Nanometrology and Nanoelectronics Initiative) and National Science Foundation via CAREER program (DMR-0748671).

7. References

Agostinelli, T., Caironi, M., Natali, D., Sampietro, M., Biagioni, P., Finazzi, M. and Duo, L. (2007). "Space charge effects on the active region of a planar organic photodetector." *Journal of Applied Physics* 101, 114504.

Ahn, Y. H., Tsen, A. W., Kim, B., Park, Y. W. and Park, J. (2007). "Photocurrent imaging of p-n junctions in ambipolar carbon nanotube transistors." *Nano Letters* 7(11), 3320-3323.

Anthony, J. E., Brooks, J. S., Eaton, D. L. and Parkin, S. R. (2001). "Functionalized pentacene: Improved electronic properties from control of solid-state order." *Journal of the American Chemical Society* 123(38), 9482-9483.

Anthony, J. E. (2006). "Functionalized acenes and heteroacenes for organic electronics." *Chemical Reviews* 106, 5028-5048.

Brutting, W., Ed. (2005). *Physics of Organic Semiconductors*. Weinheim, Wiley-VCH.

Cheng, Y. C. and Silbey, R. J. (2008). "A unified theory for charge-carrier transport in organic crystals." *Journal of Chemical Physics* 128, 114713.

Coropceanu, V., Cornil, J., da Silva, D., Olivier, Y., Silbey, R. J. and Bredas, J. L. (2007). "Charge transport in organic semiconductors." *Chemical Reviews* 107, 926-952.

Day, J., Subramanian, S., Anthony, J. E., Lu, Z., Twieg, R. J. and Ostroverkhova, O. (2008). "Photoconductivity in organic thin films: from picoseconds to seconds after excitation." *Journal of Applied Physics* 103, 123715.

[a]Day, J., Platt, A. D., Ostroverkhova, O., Subramanian, S. and Anthony, J. E. (2009). "Organic semiconductor composites: influence of additives on the transient photocurrent." *Applied Physics Letters* 94(1), 013306.

[b]Day, J., Platt, A. D., Subramanian, S., Anthony, J. E. and Ostroverkhova, O. (2009). "Influence of organic semiconductor-metal interfaces on the photoresponse of functionalized anthradithiophene thin films." *Journal of Applied Physics* 105, 103703.

de Boer, R. W. I., Jochemsen, M., Klapwijk, T. M., Morpurgo, A. F., Niemax, J., Tripathi, A. K. and Pflaum, J. (2004). "Space charge limited transport and time of flight measurements in tetracene single crystals:a comparative study." *Journal of Applied Physics* 95(3), 1196-1202.

Etemad, S., Mitani, T., Ozaki, M., Chung, T. C., Heeger, A. J. and MacDiarmid, A. G. (1981). "Photoconductivity in polyacetylene." *Solid State Communications* 40, 75-79.

Forrest, S. R. (2004). "The path to ubiquitous and low-cost organic electronic appliances on plastic." *Nature* 428, 911-918.

Gao, J. and Hegmann, F. A. (2008). "Bulk photoconductive gain in pentacene thin films." *Applied Physics Letters* 93(22), 223306.

Gu, Y., Romankiewicz, J. P., David, J. K., Lensch, J. L. and Lauhon, L. J. (2006). "Quantitative measurement of the electron and hole mobility-lifetime products in semiconductor nanowires." *Nano Letters* 6(5), 948-952.

Hegmann, F. A., Ostroverkhova, O. and Cooke, D. G. (2005). Probing Organic Semiconductors with Terahertz Pulses. *Photophysics of molecular materials*. G. Lanzani. Weinheim, Wiley-VCH.

Hegmann, F. A., Tykwinski, R. R., Lui, K. P. H., Bullock, J. E. and Anthony, J. E. (2002). "Picosecond transient photoconductivity in functionalized pentacene molecular crystals probed by terahertz pulse spectroscopy." *Physical Review Letters* 89(22), 227403.

Jurchescu, O. D., Baas, J. and Palstra, T. T. M. (2004). "Effect of impurities on the mobility of single crystal pentacene." *Applied Physics Letters* 84, 3061-3063.

Jurchescu, O. D., Subramanian, S., Kline, R. J., Hudson, S. D., Anthony, J. E., Jackson, T. N. and Gundlach, D. J. (2008). "Organic single-crystal field-effect transistors of a soluble anthradithiophene." *Chemistry of Materials* 20(21), 6733-6737.

Karl, N. (2001). Charge-carrier mobility in organic crystals. *Organic Electronic Materials*. R. Farchioni and Grosso, G. New York, Springer, 449.

Karl, N. (2003). "Charge carrier transport in organic semiconductors." *Synthetic Metals* 133, 649-657.

Knipp, D., Street, R. A., Volkel, A. and Ho, J. (2003). "Pentacene thin film transistors on inorganic dielectrics: Morphology, structural properties, and electronic transport." *Journal of Applied Physics* 93(1), 347-355.

Koeberg, M., Hendry, E., Schins, J. M., Laarhoven, H. A. v., Flipse, C. F. J., Reimann, K., Woerner, M., Elsaesser, T. and Bonn, M. (2007). "Simultaneous ultrafast probing of intramolecular vibrations and photoinduced charge carriers in rubrene using broadband time-domain THz spectroscopy." *Physical Review B* 75, 195216.

Laarhoven, H. A. v., Flipse, C. F. J., Koeberg, M., Bonn, M., Hendry, E., Orlandi, G., Jurchescu, O. D., Palstra, T. T. M. and Troisi, A. (2008). "On the mechanism of charge transport in pentacene." *Journal of Chemical Physics* 129, 044704.

Lang, D. V., Chi, X., Siegrist, T., Sergent, A. M. and Ramirez, A. P. (2004). "Amorphouslike density of gap states in single-crystal pentacene." *Physical Review Letters* 93(8), 086802.

Lehnherr, D., Gao, J., Hegmann, F. A. and Tykwinski, R. R. (2008). "Synthesis and electronic properties of conjugated pentacene dimers." *Organic Letters* 10(21), 4779-4782.

Lord, S. J., Lu, Z., Wang, H., Willets, K. A., Schuck, P. J., Lee, H. D., Nishimura, S. Y., Twieg, R. J. and Moerner, W. E. (2007). "Photophysical properties of acene DCDHF fluorophores: long-wavelength single-molecule emitters designed for cellular imaging." *Journal of Physical Chemistry A* 111, 8934-8941.

Lui, K. P. H. and Hegmann, F. A. (2001). "Ultrafast carrier relaxation in radiation-damaged silicon-on-sapphire studied by optical pump-terahertz-probe experiments." *Applied Physics Letters* 78, 3478-3480.

Marciniak, H., Pugliesi, I., Nickel, B. and Lochbrunner, S. (2009). "Ultrafast singlet and triplet dynamics in microcrystalline pentacene films." *Physical Review B* 79, 235318.

Moses, D. (1989). "Charge quantum yield and transient transport properties of anthracene and amorphous selenium." *Solid State Communications* 69(7), 721-725.

Moses, D., Dogariu, A. and Heeger, A. J. (2000). "Ultrafast detection of charged photocarriers in conjugated polymers." *Physical Review B* 61(14), 9373-9379.

Moses, D., Sinclair, M. and Heeger, A. J. (1987). "Carrier Photogeneration and Mobility in Polydiacetylene - Fast Transient Photoconductivity." *Physical Review Letters* 58(25), 2710-2713.

Moses, D., Soci, C., Chi, X. and Ramirez, A. P. (2006). "Mechanism of carrier photogeneration and carrier transport in molecular crystal tetracene." *Physical Review Letters* 97, 067401.

Munoz, E., Monroy, E., Garrido, J. A., Izpura, I., Sanchez, F. J., Sanchez-Garcia, M. A., Calleja, E., Beaumont, B. and Gibart, P. (1997). "Photoconductor gain mechanisms in GaN ultraviolet detectors." *Applied Physics Letters* 71(7), 870-872.

Murphy, A. R. and Frechet, J. M. J. (2007). "Organic semiconducting oligomers for use in thin film transistors." *Chemical Reviews* 107, 1066-1096.

Najafov, H., Lyu, B. and Biaggio, I. (2008). "Investigating the origin of the high photoconductity of rubrene single crystals." *Physical Review B* 77, 125202.

Nelson, S. F., Lin, Y. Y., Gundlach, D. J. and Jackson, T. N. (1998). "Temperature-independent transport in high-mobility pentacene transistors." *Applied Physics Letters* 72(15), 1854-1856.

Newman, C. R., Frisbie, C. D., da Silva Filho, D., Bredas, J. L., Ewbank, P. C. and Mann, K. R. (2004). "Introduction to organic thin film transistors and design of n-channel organic semiconductors." *Chemistry of Materials* 16, 4436-4451.

Ostroverkhova, O. and Moerner, W. E. (2004). "Organic photorefractives: mechanisms, materials and applications." *Chemical Reviews* 104(7), 3267-3314.

[a]Ostroverkhova, O., Cooke, D. G., Shcherbyna, S., Egerton, R. F., Hegmann, F. A., Tykwinski, R. R. and Anthony, J. E. (2005). "Band-like transport in pentacene and functionalized pentacene thin films revealed by transient photoconductivity." *Physical Review B* 71, 035204.

[b]Ostroverkhova, O., Shcherbyna, S., Cooke, D. G., Egerton, R. F., Hegmann, F. A., Tykwinski, R. R., Parkin, S. R. and Anthony, J. E. (2005). "Optical and transient photoconductive properties of pentacene and functionalized pentacene thin films: Dependence on film morphology." *Journal of Applied Physics* 98, 033701.

[a]Ostroverkhova, O., Cooke, D. G., Hegmann, F. A., Anthony, J. E., Podzorov, V., Gershenson, M. E., Jurchescu, O. D. and Palstra, T. T. M. (2006). "Ultrafast carrier dynamics in pentacene, functionalized pentacene, tetracene and rubrene single crystals." *Applied Physics Letters* 88, 162101.

[b]Ostroverkhova, O., Cooke, D. G., Hegmann, F. A., Tykwinski, R. R., Parkin, S. R. and Anthony, J. E. (2006). "Anisotropy of transient photoconductivity in functionalized pentacene single crystals." *Applied Physics Letters* 89, 192113.

Park, J., Ahn, Y. H. and Ruiz-Vargas, C. (2009). "Imaging of photocurrent generation and collection in single-layer graphene." *Nano Letters* 9(5), 1742-1746.

Park, S. K., Jackson, T. N., Anthony, J. E. and Mourey, D. A. (2007). "High mobility solution processed 6,13-bis(triisopropyl-silylethynyl)pentacene organic thin film transistors." *Applied Physics Letters* 91, 063514.

Park, S. K., Mourey, D. A., Subramanian, S., Anthony, J. E. and Jackson, T. N. (2008). "High-mobility spin-cast organic thin film transistors." *Applied Physics Letters* 93, 043301.

Peumans, P., Yakimov, A. and Forrest, S. R. (2003). "Small molecular weight organic thin-film photodetectors and solar cells." *Journal of Applied Physics* 93(7), 3693-3723.

[a]Platt, A. D., Day, J., Shepherd, W. E. B. and Ostroverkhova, O. (2009). Photophysical and photoconductive properties of novel organic semiconductors, in *Organic Thin Films for Photonic Applications* (Ed.S. H. Foulger), Washington, D.C., ACS.

[b]Platt, A. D., Day, J., Subramanian, S., Anthony, J. E. and Ostroverkhova, O. (2009). "Optical, fluorescent, and photoconductive properties of high-performance functionalized pentacene and anthradithiophene derivatives." *Journal of Physical Chemistry C* in press.

[c]Platt, A. D., Shepherd, W. E. B., Anthony, J. E. and Ostroverkhova, O. (2009). "Photophysical and photoconductive properties of organic semiconductor composites." Proc. of SPIE 7413, 7413-OS.

Podzorov, V., Pudalov, V. M. and Gershenson, M. E. (2003). "Field-effect transistors on rubrene single crystals with parylene gate insulator." *Applied Physics Letters* 82(11), 1739-1741.

Podzorov, V., Menard, E., Borissov, A., Kiryukhin, V., Rogers, J. A. and Gershenson, M. E. (2004). "Intrinsic charge transport on the surface of organic semiconductors." *Physical Review Letters* 93(8), 086602.

Pope, M. and Swenberg, C. E. (1999). Electronic processes in organic crystals and polymers. New York, Oxford University Press.

Samuel, I. D. W. and Turnbull, G. A. (2007). "Organic semiconductor lasers." *Chemical Reviews* 107, 1272-1295.

Sariciftci, N. S. (1997). Primary Photoexcitations in Conjugated Polymers: Molecular Exciton versus Semiconductor Band Model. Singapore, World Scientific.

Silinsh, E. A. and Capec, V. (1994). Organic molecular crystals: interaction, localization and transport phenomena. New York, American Institute of Physics.

Subramanian, S., Park, S. K., Parkin, S. R., Podzorov, V., Jackson, T. N. and Anthony, J. E. (2008). " Chromophore fluorination enhances crystallization and stability of soluble anthradithiophene semiconductors." *Journal of the American Chemical Society* 130, 2706-2707.

Thorsmølle, V. K., Averitt, R. D., Chi, X., Hilton, D. J., Smith, D. L., Ramirez, A. P. and Taylor, A. J. (2004). "Ultrafast conductivity dynamics in pentacene probed using terahertz spectroscopy." *Applied Physics Letters* 84(6), 891-893.

Thorsmølle, V. K., Averitt, R. D., Demsar, J., Smith, D. L., Tretiak, S., Martin, R. L., Chi, X., Crone, B. K., Ramirez, A. P. and Taylor, A. J. (2009). "Morphology effectively controls singlet-triplet exciton relaxation and charge transport in organic semiconductors." *Physical Review Letters* 102(1), 017401.

Troisi, A. and Orlandi, G. (2006). "Charge-transport regime of crystalline organic semiconductors: diffusion limited by thermal off-diagonal electronic disorder." *Physical Review Letters* 96, 086601.

Yu, G., Phillips, S. D., Tomozawa, H. and Heeger, A. J. (1990). "Subnanosecond Transient Photoconductivity in Poly(3-Hexylthiophene)." *Physical Review B* 42(5), 3004-3010.

2

Nonlinear Optical Absorption of Organic Molecules for Applications in Optical Devices

Leonardo De Boni, Daniel S. Corrêa and Cleber R. Mendonça
Instituto de Física de São Carlos, Universidade de São Paulo
Brazil

1. Introduction

Nonlinear optical materials play a major role in the technology of photonics. To further advance the performance of optical devices, researchers have sought for materials with enhanced nonlinear optical properties, including inorganic semiconductors, polymeric systems and other organic molecules. The latter, specifically, are of great interest because they present high nonlinear optical properties combined with versatility of available routes of synthesis, used to alter and optimize molecular structure to maximize nonlinear responses and other properties. To select a material for a specific application, it is important to characterize its nonlinear optical absorption in a broad spectral range, in order to choose the optimized excitation wavelength. This book chapter describes resonant nonlinear optical absorption spectrum of the organic molecules Chlorophyll A, Indocyanine Green, Ytterbium Bisphthalocyanine and Cytochrome C, which are materials with potential applications in nonlinear optical devices. These organic molecules present distinct nonlinear optical absorption, such as saturable absorption (SA) and reverse saturable absorption (RSA), depending on the laser excitation wavelength and pulse duration. In order to investigate the excited state absorption and the population dynamics of the molecules, two distinct Z-scan methodologies were used, both based on measurements of the sample's transmittance under laser excitation. The main difference between the methods employed is the possibility to monitor nonlinear effects in distinct temporal regimes, and then observe effects related with singlet and triplet electronic transitions, which, in most cases, can only be discriminated by the temporal dynamic of the system. While one technique uses a sequence of picoseconds pulses to induce and probe nonlinear absorption dynamics between singlet and triplet states, the other one employs femtosecond white-light supercontinuum pulses to map the spectrum of the excited state absorption in the visible and near infrared. By using both methods, one can determine and quantify nonlinear effects from femtoseconds up to hundreds of nanoseconds, including singlet excited state absorption cross-section spectrum and other spectroscopic features, such as decaying time (ranging from femtoseconds to a few nanoseconds) and triplet excited state cross-section. Once one is able to temporally and spectrally map the nonlinearity of materials, it can be predicted how a given nonlinear optical device will act in distinct regimes (temporal, spectral and of intensity). Specifically, Chlorophyll A and Ytterbium Bisphthalocyanine present large RSA or SA depending on the excitation wavelength. Indocyanine Green presents high singlet and triplet excited state absorptions. Cytochrome C also presents saturable absorption, with short singlet-state lifetime and a relatively fast intersystem-crossing time. Therefore, information on the temporal dynamics combined with the excited state absorption cross-section and intensity regime can guide the development of applications based on the organic molecules presented here. Some notable applications of these materials in optical devices include all-optical switches, nonlinear optical limiters, and also bio-related applications, such as Photodynamic Therapy (PDT).

2. Nonlinear optics

Nonlinear optics processes are responsible for the nonlinearity of materials, phenomena which include nonlinear absorption, nonlinear refraction, and induced scattering, among others. The origin of such nonlinearities varies considerably. For instance, nonlinear absorption may be related to two-photon absorption, excited state absorption and free-carrier absorption. There are good reviews in the literature about two-photon absorption (He et al., 2008, Prasad, 1991) and free-carrier absorption (Banfi et al., 1998, He et al., 2008). This chapter will specifically focus on excited state absorption.

Classification of optical nonlinearities usually falls in two main categories: instantaneous and accumulative (slow) nonlinearities. In the instantaneous case, the instantaneous polarization induced by the applied electromagnetic field in the medium can be expressed by an expansion in a Taylor series of the electric field amplitude, according to (Shen, 1984):

$$\vec{P} = \chi^{(1)} \cdot \vec{E} + \chi^{(2)} \cdot \vec{E}\vec{E} + \chi^{(3)} \cdot \vec{E}\vec{E}\vec{E} \tag{1}$$

In Eq. (1), the quadratic term with the electric field describes the first nonlinear effect. The coefficient $\chi^{(2)}$ is named complex susceptibility tensor of order 2. Its magnitude describes the intensity of the 2nd order nonlinear effect, which, in turn, is related to second harmonic generation, optical rectification and sum and difference frequency mixing. Such effects are only observed in noncentrosymmetric materials.

The coefficient $\chi^{(3)}$ is named complex susceptibility tensor of order 3 and describes the intensity of the 3rd order nonlinear effect, such as self-focusing action, third harmonic generation and two-photon absorption. Materials presenting non-zero $\chi^{(3)}$ (*e.g.*, materials exhibiting two-photon absorption) are candidates as active medium for several technological applications, for which there are good reviews available in literature (Bhawalkar et al., 1996, He et al., 2008, Prasad & Willians,1991). However, such materials are out of the scope of this book chapter, which will describe organic molecules' nonlinear effects arising only from resonant absorption. In the case of resonant absorption, the accumulative nonlinearities may occur in a time scale longer than the pulse duration. For instance, the polarization density generated by an applied field may develop or decay in a time scale longer than the excitation duration. Such interactions are normally dissipative, causing energy transfer from the absorbed field to the medium, which triggers the nonlinearity. Therefore, on the contrary to instantaneous nonlinearity, accumulative nonlinearities are strictly dependent on the amount of energy stored into the medium. Examples of such accumulative nonlinearities include resonant excited state absorption, such as saturation of absorption (SA) and reverse saturable absorption (RSA), which find applications in the construction of nonlinear optical devices.

2.1 Nonlinear effects arising from excited states

Nonlinear optical effects can also be associated to one-photon absorption (resonant) processes when they cause significant population change of the material's electronic states. When a medium is excited to a higher energy level, the linear polarization of material, $\vec{P} = \chi^{(1)} \cdot \vec{E}$, cannot be held constant. In this case, let us consider a simplified picture of a medium possessing two energy levels (ground and excited states), so that $\chi = n_g \chi_g + n_e \chi_e$, in which χ_g and χ_e represent the electrical susceptibilities of ground and excited state respectively. The linear electrical susceptibility of excited molecules is other than that in the ground state and can be expressed as $\chi = \chi_g + n_e(\chi_e - \chi_g)$, where $n_g + n_e = 1$. The term n_e describes the population fraction lying in the excited state (N_e / N_0), in which $N_0 = N_g + N_e$ is the total number of molecules of the system, and N_e and N_g are the excited and ground state population, respectively. This description of polarization is not linear with the electric field anymore, because the excited state population fraction depends on the light intensity, which is equivalent to the square of the electric field, according to (Pang et al.,1991):

$$\vec{P} = \chi_g^{(1)} \cdot \vec{E} + \frac{N_e}{N_0}(\chi_e^{(1)} - \chi_g^{(1)}) \cdot \vec{E} \tag{2}$$

The second term of Eq. (2) presents a nonlinear relation to the electric field and depends on the linear susceptibility of the ground and excited states. Therefore, the nonlinear susceptibility in Eq. (1) presents resonant terms that are dependent on the population fraction of molecules. Considering a case of a centro-symmetric material (all even susceptibilities are null), the nonlinear polarization can be rewritten as:

$$\vec{P} = \chi_g^{(1)} \cdot \vec{E} + \chi_g^{(3)} : \vec{E}\vec{E}\vec{E} + \frac{N_e}{N_0}(\chi_e^{(1)} - \chi_g^{(1)}) \cdot \vec{E} + \ldots \tag{3}$$

This new polarization contains resonant and non-resonant effects induced by the electric field of a laser beam. The evolution of the nonlinear effect related to the population of excited states can be monitored by rate equations that describe every electronic transition involved in the absorption. The temporal population evolution of a given excited state can be described by (Pang et al.,1991):

$$\frac{dn_e(t)}{dt} = -\frac{n_e(t)}{\tau} + (1 - n_e(t))\frac{\sigma(\nu)I}{h\nu} \tag{4}$$

in which $\sigma(\nu)$ is the one-photon absorption cross-section, $h\nu$ is the photon energy , and τ is the excited state relaxation time. $\sigma(\nu)I/h\nu$ describes the molecules transition from the ground to the excited state. The solution to this equation is given by:

$$n_e(t) = \frac{\sigma(\nu)}{h\nu} e^{-t/\tau} \int_{-\infty}^{t} I(t')e^{t/\tau'} dt' \tag{5}$$

This integral contains the light intensity I and τ', defined as $\tau' = \tau/(1 + I/I_{sat})$, where $I_{sat} = h\nu/\sigma\tau$ is the saturation intensity of the medium, which occurs as the lowest energy state is depopulated. The temporal evolution of the population from each state provides a dynamic absorption coefficient to the material. For instance, for a material that possesses two excited states, the absorption coefficient can be defined as $\alpha(t) = N_0(n_0(t)\sigma_{01} + n_1(t)\sigma_{12})$, in which σ_{01} and σ_{12} are the one-photon absorption cross-sections related to absorption from the ground to the first excited state, and from the first excited state to the second one, respectively.

3. Applications in nonlinear optical devices

The integration of all-optical and electro-optical devices into the current technology has led to the development of a large number of schemes for controlling and manipulating the phase, direction, polarization and amplitude of optical beams. The ability to control light intensity is of prime relevance in optical-related applications, such as optical communications, optical computing and light-driven chemical reactions. The methods to control the amplitude of light can be generally divided into two major groups: dynamic and passive methods. The first method is accomplished by a device that receives an external feedback, such as an iris or a power optical filter to control the light intensity incident on an optical system. Such method requires other devices to work along in the acquisition (*e.g.*, a sensor, a processor, etc), leaving the process slower and more complex. On the other hand, passive control is obtained by using a nonlinear optical material, in which sensing, processing and actuation are inherent, making the devices simpler and faster. Within the classification of passive devices to control the amplitude of an optical signal, two of great importance are all-optical switches and optical limiters. An ideal optical switch is a nonlinear optical device that is activated at a specific intensity or fluence threshold, becoming totally opaque. In contrast, an ideal optical limiter is a device that exhibits a linear transmittance below a specific intensity or fluence level but, above this level, its output intensity becomes constant. Optical limiters have been used in several applications, such as passive mode locking, pulse compression, and the most popular application: eye and sensor protection in optical system (*e.g.*, telescopes and night vision systems).

For instance, one can construct optical limiters with responses that are insensitive to the incident pulse duration. The resonant nature of the accumulative nonlinearities (such as SA and RSA), however, frequently results in a narrow bandwidth of operation for devices utilizing these mechanisms. By contrast, optical limiters that rely on instantaneous (nonresonant) nonlinearities can be broad band. These nonlinearities, however, require high intensities and typically operate effectively only for very short optical pulses. Therefore, by using an appropriate optical limiter in terms of nature and threshold value, one can extend the lifetime of a device and allow it to continue to operate under severe conditions.

4. Techniques employed to determine the materials' nonlinear optical properties

4.1 Z-scan technique in the nano- and picosecond regime

In order to characterize the nonlinear optical absorption of organic molecules in the nano- and picosecond regime, the open aperture Z-scan technique, developed originally by Sheik-Bahae et. al. (Sheik-Bahae et al.,1989), can be used. Basically, this technique monitors the change in the nonlinear transmittance as the sample is scanned through the Z-axis, which contains a focused Gaussian laser beam. An extension of this technique, called Z-scan technique with pulse train (Misoguti et al.,1999), can also be employed, allowing the investigation of the time evolution of nonlinear processes. The excitation source is a frequency-doubled, Q-switched and mode-locked Nd:YAG laser, delivering pulses at 532 nm and 100 ps. Each pulse train contains about 20 pulses separated by 13 ns at a 10 Hz repetition rate. This low repetition rate is generally used to avoid cumulative thermal nonlinearities. The beam is focused onto a quartz cell, yielding diameters of tens of μm at the focal plane. A photodetector placed in the far field coupled with a digital oscilloscope and a computer are used to acquire the pulse train signal. Each peak height is proportional to the corresponding pulse fluence, once the detection system has a rise time slower than the 100 ps pulse duration. By measuring the beam waist and the pulse train average power, one can find out the pulse fluency. The intensity can be determined by carrying out Z-scan measurements with CS_2. When the sample is located at the focus, the pulse train signal is acquired. Then, this signal is normalized to the one obtained when the sample is far from the focus, yielding the normalized transmittance as a function of pulse number. All optical measurements were carried out with the sample placed in a quartz cuvette. Figure 1 schematically shows the experimental setup.

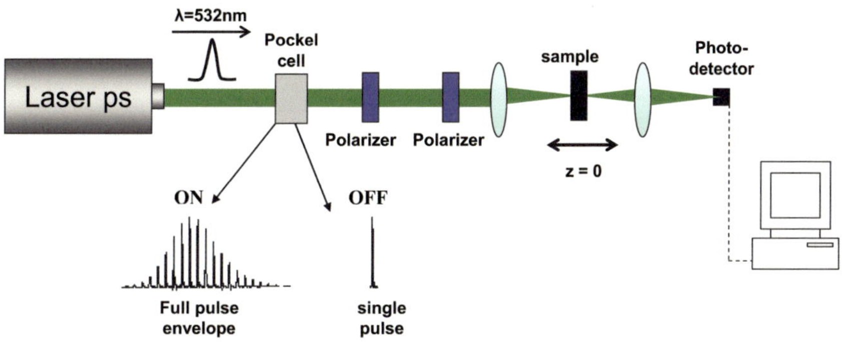

Fig. 1. Experimental setup of the Z-scan technique with pulse trains, used to characterize the material's nonlinear response in the pico- and nanosecond regime.

4.2 Z-scan technique in the femtosecond regime

The nonlinear optical absorption of organic molecules in the femtoseconds regime in a large spectral range may be carried out by means of two methodologies: (a) Single wavelength Z-scan technique and (b) White-Light Continuum Z-scan technique, described in more details as follows.

(a) Single wavelength Z-scan technique

This methodology uses a Ti:sapphire chirped pulse laser amplified system that produces pulses of 150 fs centered in 775 nm, with a repetition rate of 1 kHz, to pump an optical parametric amplifier (OPA), which, in turn, generates wavelengths in the spectral region from 460 nm to 2200nm of nearly 100 fs. Figure 2 (a) schematically displays the details of the Single wavelength Z-scan technique experimental setup.

(b) White-Light Continuum (WLC) Z-scan technique

In this methodology, whose full details can be found elsewhere,(Balu et al., 2004, De Boni et al., 2004), the White-light Continuum (WLC) is produced by focusing a femtoseconds laser beam (Ti:sapphire chirped pulse laser amplified system that produces 150 fs centred in 775 nm, with a repetition rate of 1 kHz,) with a lens onto a quartz cell containing distilled water. A low-pass filter is used to remove the strong pump pulse and the infrared part of the WLC spectrum. The use of typically 0.3 mJ laser pulses generates about 10 μJ of WLC, spanning from 420 up to 750 nm. After re-collimation, the WLC beam is focused onto the sample, which is scanned along the beam propagation z-direction, as usually done in the traditional Z-scan method. The WLC transmitted through the sample is completely focused onto a portable spectrometer with a resolution of ~1 nm. The spectra are acquired for each z position as the sample is scanned along the z-direction and then normalized to the one obtained far from the focal plane. By selecting a particular wavelength from the complete set of measured spectra, a Z-scan signature is obtained according to the nonlinear response at that wavelength. Figure 2 (b) schematically shows the experimental apparatus of the WLC Z-scan technique. When using this technique under resonant conditions, the white-light continuum pulse chirp must be considered, since distinct spectral components will reach the sample at distinct times. Consequently, cumulative effects can occur as result of absorption by excited molecules, which are then promoted to a higher excited state.

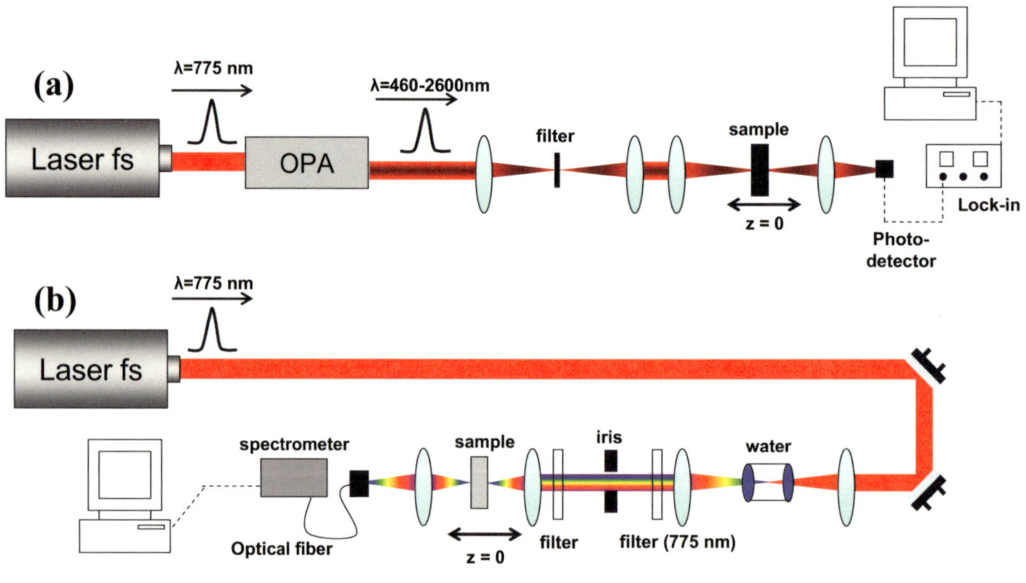

Fig. 2. Experimental setup of the (a) Single wavelength and (b) WLC Z-scan techniques, used to characterize the material's nonlinear response in the femtosecond regime.

5. Nonlinear optical absorption (NLOA) of organic molecules

In this section, the results of the nonlinear optical absorption (NLOA) of the molecules Chlorophyll A, Indocyanine Green, Ytterbium Bisphthalocyanine and Cytochrome C are presented. The molecules are characterized in the nano, pico and femtoseconds regimes and present Reverse Saturable Absorption (RSA) and Saturable Absorption (SA), with potential applications in nonlinear optical devices.

5.1 Chlorophyll A

5.1 (a) NLOA in the nano and picosecond regimes

Chlorophyll A, belonging to the class of porphyrins, is a biomolecule of prime importance in the photophysical processes of plants, acting in the conversion of light into chemical energy in several biological systems (Michel-Beyerle, 1985, Scheidt & Reed, 1981) by taking part in the light absorption and electron transfer in the photosynthetic reaction center (Baker & Rosenqvist, 2004, Carter & Spiering, 2002, Michel-Beyerle, 1985). Due to its relevance in biological processes, Chlorophyll A has been the subject of extensive theoretical and experimental studies (Gouterman, 1961, Hasegawa et al., 1998, Parusel & Grimme, 2000, Sundholm, 1999). Furthermore, porphyrins have been proposed for medical and photonics applications such as optical limiters (Calvete et al., 2004, Neto et al., 2003, Neto et al., 2006, O'Flaherty et al., 2003), optical switches (Loppacher et al.,2003), and sensitizers for photodynamic therapy (Fisher et al.,1995). Hence, studying Chlorophyll A excited states properties is essential to understand biological processes aiming at possible applications in photonics and medicine.

The electronic transitions of Chlorophyll A are usually characterized by two regions: the Q-band, which is relatively weak and occurs in the visible region; and the intense Soret or B-band, which appears in the near UV region and is often accompanied by an N-band of lower intensity (see Figure 3). The linear absorption spectrum of Chlorophyll A has been understood in terms of the four-orbital model applied by Gouterman (Gouterman, 1961), which although very simple reproduces all the major features of this system. There are several theoretical studies carried out using distinct methods to further understand the electronic excited states of Chlorophyll A (Hasegawa et al., 1998, Parusel & Grimme, 2000, Sundholm, 1999). In general, these works assign more than one electronic excited state to describe the experimentally observed features of Chlorophyll A spectrum (Q and B-band). In this book chapter, the choice was based on the electronic states reported by Parusel *et al.* (Parusel & Grimme,2000) obtained through the DFT/MRCI method (density functional theory and multireference configuration interaction), which gives the best interpretation for the linear absorption spectrum of Chlorophyll A, as the basis for the energy diagram employed here to understand the results. The Q-band at 670 nm is the main transition

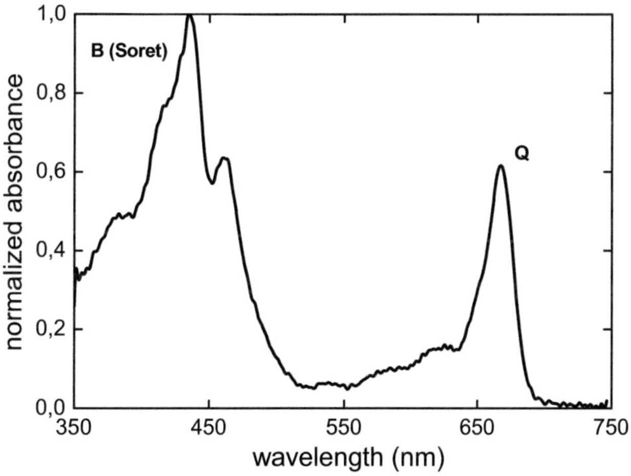

Fig. 3. Absorption spectrum of Chlorophyll A/chloroform solution.

excited by the 532 nm light used in this investigation. This molecule has considerable absorption in the 600-700 nm region, in which human tissues are more transparent. In terms of medical therapy, for instance, light can reach the dye molecule adsorbed in the cells and undergo a photoreaction, i.e. Chlorophyll A satisfies an important requirement for possible use as a sensitizer in PDT.

The emission spectrum at room temperature for excitation at the Q-band presents a strong fluorescence peak at 669 nm, which means that the Q-band is the predominant excitation path. The fluorescence lifetime (τ_f) reported in the literature is 4 ns (Vernon & Seely, 1996).

Based on the absorption and emission spectra and on models traditionally used for other porphyrins, a simplified five-level energy diagram can be sufficient to describe the dynamics of the nonlinear absorption in the picosecond regime, as illustrated in Figure 4.

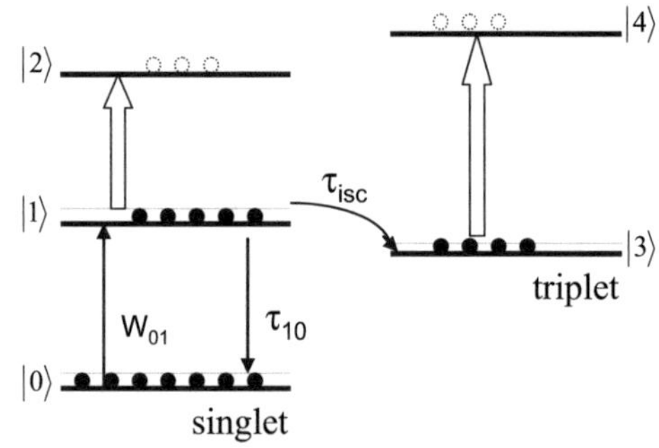

Fig. 4. Five-level energy diagram used to simulate the experimental results.

Figure 5 shows experimental results (open circles) for the nonlinear absorption obtained with the Z-scan technique with pulse train at 532 nm (Correa et al., 2002) and theoretical fitting (solid line) using the five-level energy diagram depicted in Figure 4. The strongest peak in the pulse train was arbitrarily labeled "0." The irradiance is $I_{(0)}$ = 0.35 GW/cm^2.

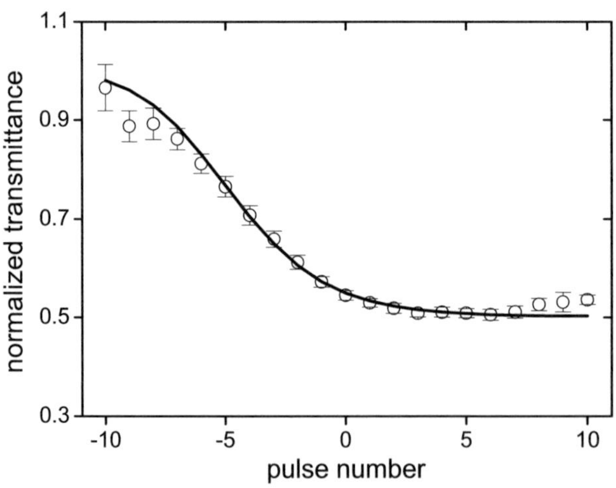

Fig. 5. Normalized transmittance of Chlorophyll A/chloroform solution along the pulse train for a $I_{(0)}$ =0.35 GW/cm^2. Solid line is the theoretical curve obtained by using the five-level energy diagram.

To understand the changes in the nonlinear effect during the train of pulses, basically, one needs to comprehend how the population dynamic is produced by the pulse train. When the first pulse of the pulse train is absorbed by the sample, it will promote molecules from the ground state $|0\rangle$ to the excited singlet state $|1\rangle$. The fraction of population on the singlet excited state $|1\rangle$ may decay radiatively to level $|0\rangle$, with the characteristic lifetime of the state (τ_{10}), or relax to an excited triplet state $|3\rangle$, with the lifetime τ_{isc}, known as intersystem-crossing time. Also, because the lifetimes involved in this nonlinear process have the same order of the time between two consecutives pulses (13 ns), molecules already in $|1\rangle$ and $|3\rangle$ do not have enough time to completely relax back to the ground singlet state. Based on this fact, the next pulse of the pulse train will probe a different population in the electronic states than the first pulse did. If the absorption cross-sections are different, the transmittance of such pulse will be proportional to the new absorption coefficient. This mechanism will be present to the other pulses, as an accumulative effect. In addition, because the higher excited states, $|2\rangle$ and $|4\rangle$, are short-lived, their populations can be neglected. On the basis of this energy diagram, the set of rate equations that describe the fraction of molecules (n_i) at each level is:

$$\frac{dn_0}{dt} = -n_0 W_{01} + n_1\left(\frac{1}{\tau_f} - \frac{1}{\tau_{isc}}\right) \tag{6}$$

$$\frac{dn_1}{dt} = n_0 W_{01} - \frac{n_1}{\tau_f} \tag{7}$$

$$\frac{dn_3}{dt} = \frac{n_1}{\tau_{isc}} \tag{8}$$

where $W_{01} = \sigma_{01} I/h\nu$ is the transition rate. This set of equations was numerically solved using the actual temporal intensity pattern of the Q-switched/mode-locked pulse train of our experiment, yielding the population dynamics, $n_i(t)$. The time evolution of the nonlinear absorption can be calculated according to:

$$\alpha(t) = N\{n_0\sigma_{01} + n_1\sigma_{12} + n_3\sigma_{34}\} \tag{9}$$

where N is the sample concentration, and σ_{12} and σ_{34} are the excited state cross-sections. The ground state cross-section, σ_{01}, was determined by measuring the linear absorption at 532 nm ($\alpha = N\sigma_{01}$). This procedure resulted in σ_{01} = 3.1 x 10^{-18} cm^2. The numerical calculation was carried out with τ_f = 4 ns. In Figure 5, the solid line represents the theoretical fittings obtained with σ_{12} = 4 x 10^{-18} cm^2, σ_{34} = 8 x 10^{-18} cm^2, and τ_{isc} = 1.5 ns. The absorption cross-section of the triplet state is higher than that of the singlet, although with a low ratio (only 2 times). On the other hand, the intersystem-crossing lifetime (1.5 ns) is shorter than the typical values reported for porphyrins and phthalocyanines.(Frackowiak et al., 2001, Shirk et al., 1992). This short intersystem-crossing lifetime indicates an efficient singlet-triplet conversion, which makes Chlorophyll A suitable for applications as a PDT sensitizer. This efficient intersystem-crossing (singlet-triplet) conversion is consistent with those found for Mg phthalocyanine, which has a yield of triplet formation higher than for most phthalocyanines.(Frackowiak et al., 2001)

5.1 (b) NLOA in the femtosecond regime

This section presents the study of the excited state absorption of Chlorophyll A in the femtosecond regime by measuring its nonlinear absorption spectrum from 460 nm to 700 nm using the WLC Z-scan technique. Its resonant nonlinear absorption spectrum presents saturable absorption (SA) and reverse saturable absorption (RSA) depending on the excitation wavelength (De Boni et al., 2007). Figure 6 displays Z-scan curves of Chlorophyll A for some pump wavelengths of the WLC spectrum. An inversion of the normalized transmittance is observed as the nonlinear process changes from RSA (shorter wavelengths) to SA (longer wavelengths).

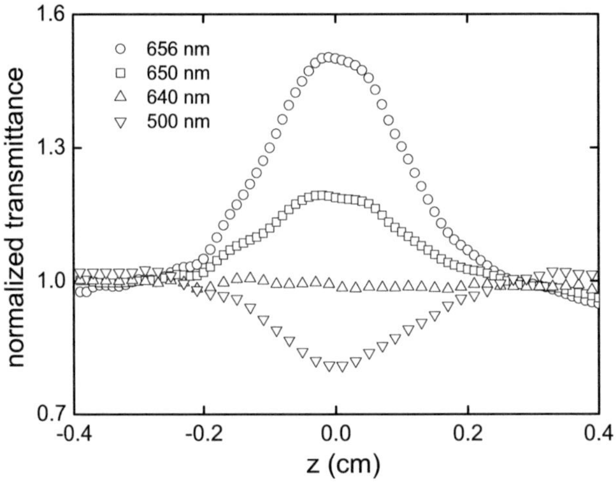

Fig. 6. Experimental Z-scan curves for Chlorophyll A obtained with the WLC Z-scan technique. An inversion of the normalized transmittance is observed according to the dominant nonlinear process (SA or RSA).

Because the white-light continuum pulse temporal width is around 5 ps, only the singlet levels of Figure 4 were used to establish the population dynamics of Chlorophyll A. In this case, molecules in the ground state $|0\rangle$ can be promoted to the first excited state $|1\rangle$ (Q-band) by one-photon absorption, being subsequently excited to a higher excited level. This level does not correspond to the B-band (Linnanto & Korppi-Tommola, 2000, Rivadossi et al., 2004, Wehling & Walla, 2005, Zigmantas et al., 2002) but to a distinct electronic state in the UV region, since the photons used to transition an electron from $|1\rangle$ to a higher excited state belong to the blue spectral region of the WLC pulse and, therefore, are more energetic than those required to promote a transition from $|1\rangle$ to the B-band. The relaxation from level $|1\rangle$ to the ground state can be neglected because of the short pulse temporal width of the WLC pulse. The upper energy levels (located above $|1\rangle$) are assumed to be too short-lived and, therefore, present no appreciable population (Shank et al.,1977). As a consequence, molecules are accumulated only in the first excited state and the absorption cross-section between states $|1\rangle$ and upper energy levels (located in the UV region) can be determined. In this case, no triplet state was considered, since the intersystem-crossing time of Chlorophyll A is in the order of nanoseconds (Correa et al.,2002), which is much longer than the duration of the WLC pulse used. Based on these considerations, the rate equation used to describe the dynamic change of absorption, in accordance with the energy-level diagram, is:

$$\frac{dn_0(t)}{dt} = -n_0(t)W_{01}(\lambda) + \frac{1-n_0(t)}{\tau_{10}} \qquad (10),$$

in which $n_1(t) = 1 - n_0(t)$ and $W_{01}(\lambda) = \sigma_{01}(\lambda)I/h\nu$ is the transition rate, where $\sigma_{01}(\lambda)$ is the ground state cross-section. I is the excitation intensity, $n_i(t)$ is the population fraction in each state, h is the Planck constant, and ν is the photon frequency. Due to the WLC pulse chirp, its red portion (resonant with the Q-band) promotes part of the population to the first excited state $|1\rangle$ and consequently the other spectral components of the WLC pulse probe the excited state absorption (ESA), once the first excite state has a lifetime longer than the pulse duration. The time evolution of the nonlinear absorption, $\alpha(\lambda,t)$, was calculated according to:

$$\alpha(\lambda,t) = N[n_0(t)\sigma_{01}(\lambda) + n_1(t)\sigma_e(\lambda)] \qquad (11),$$

where N is the number of molecules/cm³ and $\sigma_e(\lambda)$ is the excited state cross-section correspondent to the transition $|1\rangle$ to a higher excited state. The first and the second terms in Eq. (11) provide the absorption coefficient of the ground and excited states respectively. Since the ground state absorption cross-section for every spectral component is determined through the linear absorption spectrum, the only adjustable parameters are the excited state cross-sections. By fitting the normalized transmittance spectrum, it is possible to determine the excited state cross-sections of Chlorophyll A for each wavelength within the WLC spectrum. These values are displayed in Figure 7 (closed triangles). The region below 450 nm was omitted because the white-light spectrum generated in the experiment starts around this wavelength. The difference between the values of ground and excited state cross-sections ($\sigma_{01} - \sigma_e$) is also displayed in Figure 7 (open triangles). From these data, one can observe the singlet excited state processes of Chlorophyll A. When $\sigma_{01} - \sigma_e > 0$, there is a decrease in the total absorption coefficient, α, characterizing SA. For Chlorophyll A, this process was observed from 700 nm up to 640 nm. Around 635 nm, the values of σ_{01} and σ_e are the same, giving rise to no appreciable change in the normalized transmittance at this wavelength.

It can be observed that σ_e values (closed triangles) are zero from 700 nm up to 665 nm, indicating that, for this range, there is no transition to a higher excited state. The red portion of the WLC, which is resonant with the Q-band, causes ground state depletion, responsible for the SA. Therefore, up to 665 nm, the WLC is populating state $|1\rangle$, which is then probed by the remaining components of WLC pulse. Consequently, for wavelengths shorter than 665 nm, the values of σ_e are not zero, due to the transition from $|1\rangle$ to the higher excited state, which is allowed according to DFT/MRCI calculations presented in the literature (Parusel & Grimme, 2000). If $\sigma_{01} - \sigma_e < 0$, the material has its absorption coefficient increased with the intensity (RSA process), as shown by open triangles in Figure 7 for wavelengths below 640 nm. The excited state population build-up generated with the WLC Z-scan technique can be advantageously used to shape the pulse intensity spectrum in order to match the most intense linear absorption band of the material. As a consequence, it is possible to obtain an enhancement of the nonlinear absorption in a transparent region through excited state absorption. In practical terms, WLC pulses could be used in applications where a high RSA process is needed in the blue region of the spectrum.

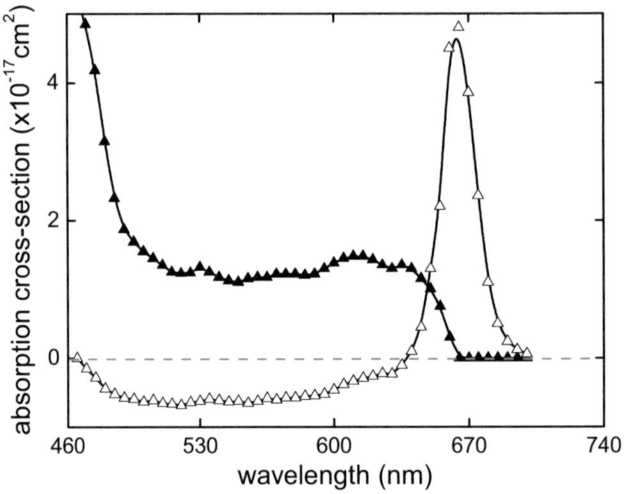

Fig. 7. Excited state (σ_e: closed triangles) cross-sections as a function of the excitation wavelength for Chlorophyll A obtained with the WLC Z-scan technique. The difference between the excited and ground state cross-section ($\sigma_{01} - \sigma_e$: open triangles) is also displayed.

5.2 Ytterbium Bisphytallocyanine

5.2 (a) NLOA in the nano and picosecond regimes

Phthalocyanines are planar organic molecules that can exhibit large third-order susceptibilities due to their high π-conjugation. To further increase the conjugation, and consequently enhance the nonlinear optical properties, one can augment the molecular size by adding peripheral rings or constructing sandwich compounds, known as Bisphthalocyanines ($YbPc_2$), where two phthalocyanine rings are coordinated to a central metal ion. Owing to their excellent environmental stability and optical properties, that can be tuned by varying the central metal ion, or a peripheral side-group, phthalocyanines and bisphthalocyanines are promising for manufacturing optical devices, such as optical-limiting devices. The basic principle of optical-limiting devices is the reverse saturable absorption (RSA), which is normally caused by an efficient intersystem-crossing process from a higher excited singlet state to an excited triplet state, competing with direct radiative decay to the singlet ground-state. This section reports on the dynamic optical nonlinearities of Ytterbium Bisphthalocyanine ($YbPc_2$)/chloroform solution obtained with the Z-scan technique with pulse trains. The dependence of the nonlinear absorption on the pulse fluence presents first SA, and subsequently RSA behavior. A six-energy-level diagram is used to establish the population dynamics and the mechanisms that contribute to the nonlinear refraction and absorption. (Mendonça et al., 2000)

Figure 8 shows that the absorption spectrum of $YbPc_2$ in chloroform solution is similar to those reported in the literature for other phytallocyanines containing metal-ions, and agrees with the energy-level diagram, shown in the inset, obtained from the valence-effective Hamiltonian (VEH) calculation.

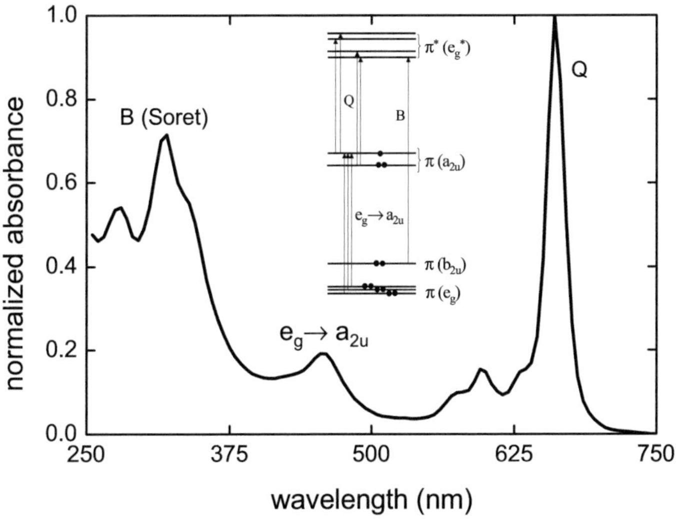

Fig. 8. Absorbance spectrum of $YbPc_2$ /chloroform solution. The inset depicts VEH one-electron energies of the molecular p-orbitals.

The structure around 650 nm, known as Q-band, is attributed to transitions from the split $\pi(a_{2u})$ orbital to the upper $\pi^*(e_g^*)$ orbital. The band around 460 nm corresponds to transitions from the deeper $\pi(e_g)$ level to the half occupied $\pi(a_{2u})$ orbital, while the B (Soret) band, which appears in the ultraviolet region (320 nm), is attributed to the transitions between $\pi(b_{2u})$ and $\pi^*(e_g^*)$ levels. According to the absorption spectrum, both the Q-band and the $e_g \rightarrow a_{2u}$ transition can, at first, be excited when 532 nm is employed. However, time-resolved fluorescence measurements for a pump at this wavelength resulted mostly in an emission centered on 550 nm, with a 4 ns lifetime, indicating that the $e_g \rightarrow a_{2u}$ transition is the main excitation path. A weaker 5 ns lifetime fluorescence (about 15% of the total) centered around 692 nm (Q-band) was also observed, indicating a secondary path for the excitation mechanism.

Figure 9 shows experimental results for the nonlinear absorption obtained with pulse trains Z-scan technique. To explain these results, the six-energy-level diagram depicted in Figure 10 is considered, which is a simplification of the one shown in the inset of Figure 8. Two possible ground state levels can be considered, $|0\rangle$ and $|1\rangle$, because two distinct bands ($a_{2u} \rightarrow e_g^*$ and $e_g \rightarrow a_{2u}$) can absorb photons of the excitation employed. According to the present model, molecules in state $|0\rangle$ can be promoted to level $|1\rangle$, when pumped by excitation at 532 nm, while molecules at level $|1\rangle$ can be excited to level $|2\rangle$. A two-photon absorption process ($e_g \rightarrow e_g^*$) could also be considered, but it was found to have little influence on the theoretical fitting. On the other hand, molecules excited to level $|1\rangle$ can decay radiatively to level $|0\rangle$, and those excited to level $|2\rangle$ can either decay radiatively to level $|1\rangle$ or undergo an intersystem-crossing to the triplet state $|4\rangle$. The upper excited singlet and triplet levels, $|3\rangle$ and $|5\rangle$ respectively, are assumed to be too short-lived to present any significant population build up.

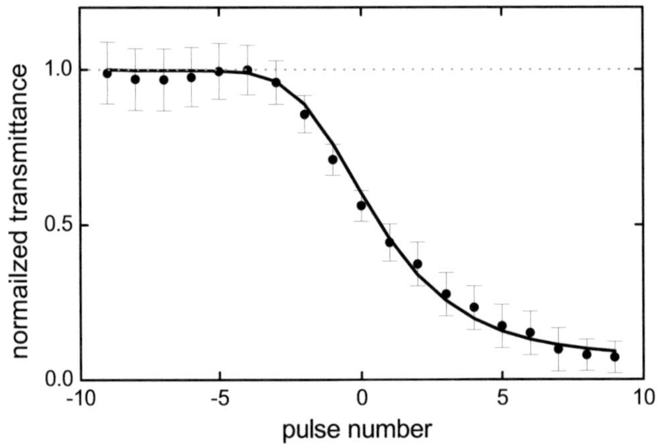

Fig. 9. Nonlinear absorption of $YbPc_2$ along the pulse train. Solid line is the theoretical curve obtained by using the six-energy-level diagram.

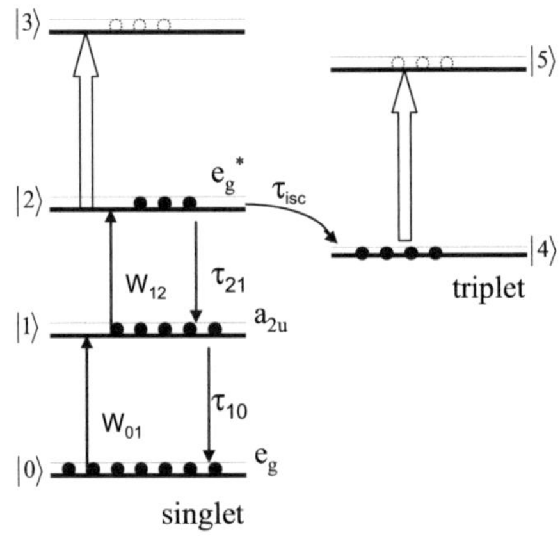

Fig. 10. Six-energy-level diagram used to simulate the experimental result of Ytterbium Bisphytallocyanine.

The rate equations used to describe the fractions of molecules, n_i, at each energy level are:

$$\frac{dn_0}{dt} = -n_0 W_{01} + \frac{n_1}{\tau_{10}} \tag{12}$$

$$\frac{dn_1}{dt} = n_0 W_{01} - n_1 W_{12} + \frac{n_2}{\tau_{21}} - \frac{n_1}{\tau_{10}} \tag{13}$$

$$\frac{dn_2}{dt} = n_1 W_{12} - \frac{n_2}{\tau_{21}} - \frac{n_2}{\tau_{isc}} \tag{14}$$

$$\frac{dn_4}{dt} = \frac{n_2}{\tau_{isc}} \tag{15}$$

where $W_{01} = \sigma_{01}I/h\upsilon$ and $W_{12} = \sigma_{12}I/h\upsilon$ are the transition rates, with σ_{01} and σ_{12} being the ground and excited state cross-sections, respectively. τ_{10} and τ_{21} are the lifetimes of levels $|1\rangle$ and $|2\rangle$, and τ_{isc} is the intersystem-crossing time. This set of equations was numerically solved using the actual temporal intensity pattern of the Q-switched and mode-lock pulse train of the experiment, yielding the population dynamics, $n_i(t)$. The time evolution of the nonlinear absorption can be calculated according to:

$$\alpha(t) = N\{n_0\sigma_{01} + n_1\sigma_{12} + n_2\sigma_{23} + n_4\sigma_{45}\} \tag{16}$$

where N is the concentration, and σ_{23} and σ_{45} are the excited state cross-sections. The excited state cross-sections, σ_{01}, determined by measuring the linear absorption at 532 nm, resulted in $\sigma_{01} = 2.4\times10^{-18} cm^2$. The numerical calculation was carried out with $\tau_{10} = 4$ ns and $\tau_{21} = 5$ ns, values obtained through time-resolved fluorescence measurements. The solid line in Figure 9 represents the theoretical fitting obtained with $\sigma_{23} = 1.0\times10^{-17} cm^2$, $\sigma_{45} = 4\times10^{-17} cm^2$ and $\tau_{isc} = 25\,ns$. A very small saturation for the first few pulses can be observed, which is related to the population buildup in level $|1\rangle$. After this initial step, level $|2\rangle$ starts to be populated, allowing a population transfer to the triplet state. Since this state has an absorption cross-section higher than that of level $|2\rangle$, a reverse saturation occurs. If the transition $e_g \rightarrow a_{2u}$ is not taken into account in the model, the plateau observed for the first few pulses does not appear.

5.2 (b) NLOA in the femtosecond regime

This section reports the resonant nonlinear absorption spectrum of Ytterbium Bisphthalocyanine (YbPc2) from 500 up to 675 nm in the femtoseconds regime determined through the WLC Z-scan. The results indicate the presence of SA, at the Q-band region, and a RSA, around 530 nm (De Boni et al., 2006). The line with circles in Figure 11 shows the nonlinear spectrum (transmittance change (ΔT) spectrum) of $YbPc_2$ obtained through the WLC Z-scan technique. Three distinct behaviors can clearly be observed: (i) a strong SA process that follows the Q-band, indicated by the positive ΔT values, (ii) an excited state absorption which gives an effective SA process below the Q-band and (iii) the negative ΔT values due to a RSA mechanism.

Due to the WLC-pulse chirp, the red portion of the pulse, which is resonant with the Q-band, promotes the population to the first excited state. In this case, a simplification of the diagram showed in Figure 10 can be used, which consists in considering only the first three levels (0, 1 and 2). From this assumption, the population dynamics is established to understand the experimental results. According to this consideration, molecules at the ground state $|0\rangle$ (a_{2u}) can be promoted to the first excite state $|1\rangle$ (e_g^*) by one-photon absorption (Q-Band; $a_{2u} \rightarrow e_g^*$), being subsequently excited to level $|2\rangle$. Molecules at level $|1\rangle$ decay radiatively to the ground state with a relaxation time τ_{10}= 4 ns, which is much longer than the WLC-pulse duration. The upper excited singlet level, $|2\rangle$, is assumed to be too short-lived to present any significant population buildup. In this case, molecules are accumulated in the first excited state and the absorption cross-section between the states $|1\rangle$

and $|2\rangle$ can be determined. As the intersystem-crossing time for this molecule is around 25 ns, no triplet states need to be considered for the temporal regime of the pulses employed. The rate equations used to describe the fraction of molecules, n_i, at each level are obtained from Eq. (12-15) but considering only terms related to levels 0, 1 and 2. The time evolution of the nonlinear absorption can be calculated according to:

$$\alpha(\lambda,t) = N\left[n_0(t)\sigma_{01}(\lambda)+n_1(t)\sigma_{12}(\lambda)\right] \tag{17}$$

where N is the sample concentration. When $\sigma_{01} > \sigma_{12}$, the sample presents a decrease in the effective absorption as the excited state is populated (SA). On the other hand, if $\sigma_{01} < \sigma_{12}$, the sample becomes more opaque, characterizing a RSA process. The occurrence of SA or RSA depends on the contribution of different electronic states, excitation wavelength and pulse width. For the Q-band region (660 nm for instance), the model gives a σ_{12} of approximately zero, which leads to a SA that follows the absorption band, due to the population accumulated in the first excited state. Right below the Q-band (600 nm), SA does not follow the linear absorption. At 600 nm, for example, the theoretical fitting was obtained with $\sigma_{12} = 0.5\text{x}10^{-18}$ cm^2, which is smaller than σ_{01} (SA). Around 530 nm, RSA was observed with $\sigma_{12} = 10\text{x}10^{-18}$ cm^2, which is about four times higher than σ_{01}. A similar behavior in six distinct wavelengths was observed by Unnikrishnan *et al.* (Unnikrishnan et al., 2002), even though they used much longer pulses (nanoseconds). Furthermore, due to the ultrashort pulses regime employed here, no triplet state is being populated and only the singlet state contributes to the observed RSA. The excited state absorption cross-section at 530 nm determined here ($\sigma_{12}/\sigma_{01} \approx 4$) is in agreement with a previous one obtained at 532 nm using the Z-scan technique with picosecond pulses (Mendonça et al., 2000,Mendonça et al., 2001, Misoguti et al., 1999). In that work, RSA was found to be related to singlet and triplet

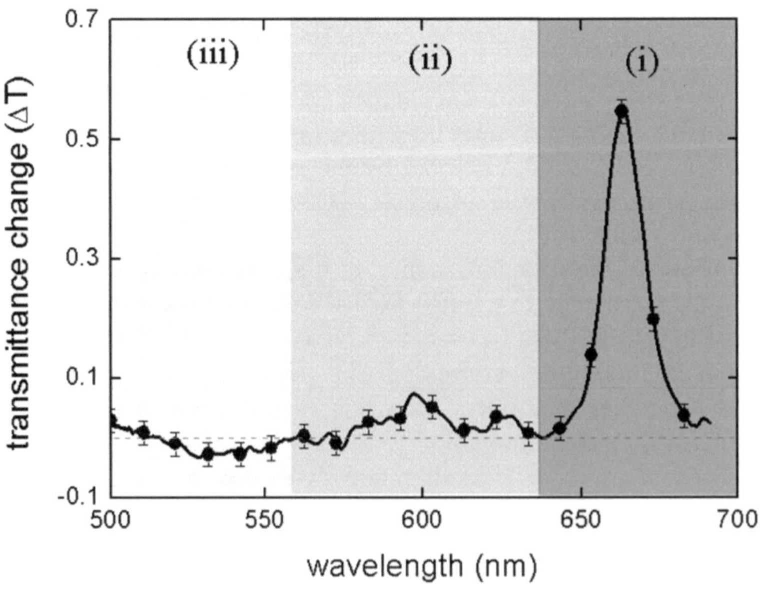

Fig. 11. Normalized transmittance change of $YbPc_2$ solution obtained with WLC Z-scan.

states, being mainly due to the last one, whose cross-section was found to be sixteen times higher than that of the ground state. The smaller singlet state contribution to RSA was comparable to the one presented here.

5.3 Indocyanine Green

5.3.1 NLOA in the nano and picosecond regime

The organic dye Indocyanine Green (ICG) presents high nonlinear optical properties, such as an efficient RSA (O'Flaherty et al., 2003), which makes it an interesting candidate for optics-related applications, such as optical limiting devices. Indocyanine Green can also be used as laser dye and saturable absorber. In medicine, ICG has been used for diagnosis and photo-dynamic therapy (PDT) of cancer. The intersystem-crossing time and quantum yield of triplet formation of ICG in different solvents have already been investigated (Reindl et al., 1997). These results revealed that the conversion efficiency to the triplet state is diminished by increasing the solvent polarity. The same behavior was observed for τ_{01}. For instance, in DMSO (apolar solvent), τ_{01} is 30 times greater than that observed in polar solvents. This section presents the nonlinear absorption of ICG obtained using single pulse and pulse train Z-scan techniques, both at 532 nm. Using the single pulse Z-scan and a theoretical analysis employing a three-energy level diagram, the excited singlet absorption cross-section was determined. Additionally, with the PTZ-scan technique and a five-energy level diagram, the intersystem-crossing time and the triplet absorption cross-section were obtained (De Boni et al.,2007).

Figure 12 shows the linear absorption spectrum of ICG diluted in DMSO. It has a strong band around 800 nm, related to the $\pi \rightarrow \pi^*$ transition. At 532 nm, wavelength employed in the nonlinear optical measurements, only a small absorption was measured.

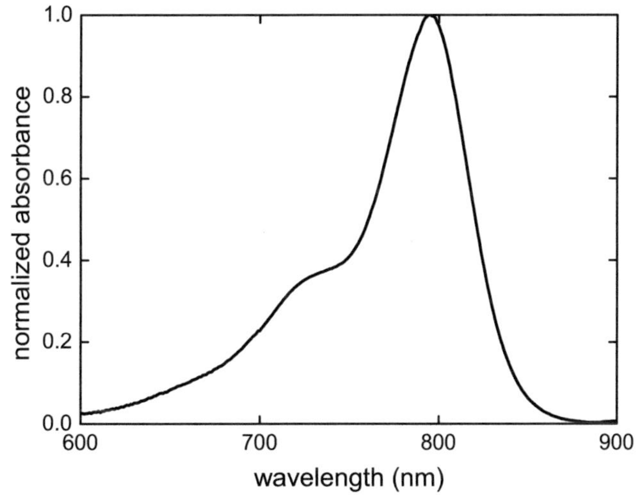

Fig. 12. Absorption spectrum of Indocyanine Green in DMSO.

Figure 13 (a) shows the decrease of the normalized transmittance (NT) for ICG as a function of the pulse irradiance, characterizing a RSA process. From this figure, it is possible to see saturation of the NT due to the accumulation of molecules in the first singlet excited state ($|1\rangle$) and to the depletion of the ground state ($|0\rangle$).

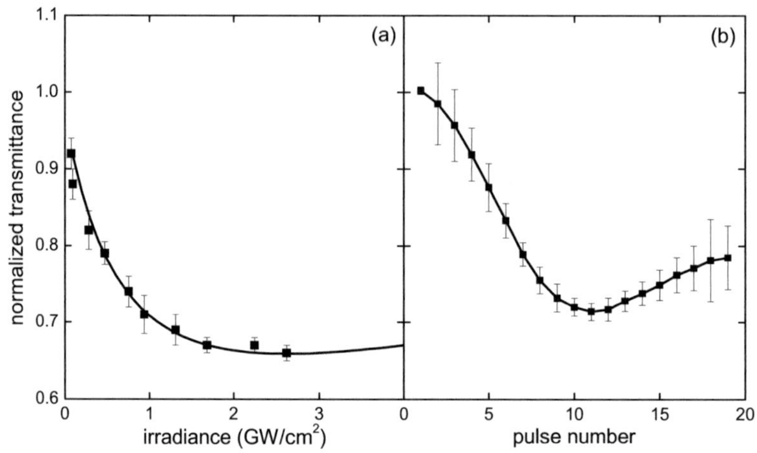

Fig. 13. (a) Normalized transmittance as a function of pulse irradiance for ICG in DMSO. The solid line represents the fitting obtained with three-energy-level diagram.
(b) Normalized transmittance along of the Q-switch envelope (pulse number) for the same sample. The solid line represents the theoretical curve obtained with parameters given in the text, using the five-energy-level diagram.

As seen in Figure 13 (a), the saturation for ICG in DMSO occurs at ~ 2 GW/cm², a relatively low intensity for this type of nonlinear process. This low saturation intensity for ICG is related to its $|1\rangle \rightarrow |0\rangle$ transition lifetime (τ_{10} ~ 700 ps)(Reindl et al., 1997), which allows a considerable accumulation of ICG molecules in the singlet excited state $|1\rangle$. With more molecules in the first excited state, more transitions occur to the second excited state $|2\rangle$, which presents an absorption cross-section approximately null. This process can be visualized by the increase in the NT curve fitting that occurs after 3 GW/cm². In order to fit the experimental data obtained with the single pulse Z-scan technique (Figure 13 (a)), the three-energy-level diagram shown in Figure 14 (a), representing only the singlet states of the molecule, was employed. As the band gap of ICG is around 1.5 eV, the internal conversion (IC) process must be taken into account in the rate equations used to describe the population dynamics. The triplet states were neglected because the duration of each single pulse is shorter than the intersystem-crossing time. In this case, only the singlet states contribute to the nonlinear absorption process. The transition lifetime (τ_{10}) can be described by $1/\tau_{10} =/1/\tau_r + 1/\tau_{IC}$, where $\tau_r \approx 5\,\text{ns}$ and $\tau_{ic} \approx 840$ ps (Reindl et al., 1997) are singlet radiative lifetime and internal conversion time respectively.
It was also assumed that the lifetime of the second excited singlet state, τ_{21}, is in the order of a few femtoseconds; therefore, the population of this state is small at low irradiances. Hence, to describe the fraction of molecules in each state, the rate equations used are given by:

$$\frac{dn_0}{dt} = -w_{01}n_0 + \frac{n_1}{\tau_{10}} \tag{18}$$

$$\frac{dn_1}{dt} = +w_{01}n_0 - w_{12}n_1 - \frac{n_1}{\tau_{10}} + \frac{n_2}{\tau_{21}} \tag{19}$$

$$\frac{dn_2}{dt} = +w_{12}n_1 - \frac{n_2}{\tau_{21}} \tag{20}$$

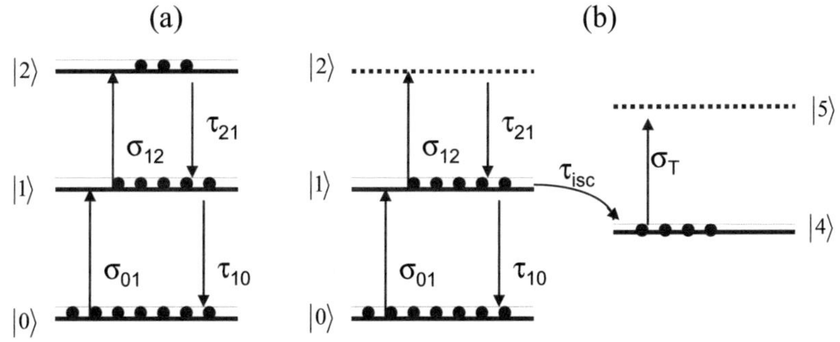

Fig. 14. Three- (a) and five- (b) energy-level diagrams used to model the single pulse and pulse train Z-scan results for ICG.

where n_i's are the population fractions of the singlet states with $n_0 + n_1 + n_2 = 1$. The terms in these equations have already been described in the previous sections. The time dependence of absorption coefficient during the excitation is given by:

$$\alpha(t) = N\{n_0(t)\sigma_{01} + n_1(t)\sigma_{02}\} \tag{21}$$

As mentioned in the previous sections, α_{01} is obtained from the linear absorption spectrum ($\sigma_{01} = \alpha_{01}N$) and, therefore, the only adjustable parameter in this fitting procedure is σ_{12}. The value determined from the fitting was $\sigma_{12} = (12 \pm 1)\times 10^{-17} cm^2$, which is 75 times higher than the ground state cross-section ($\sigma_{01} = 0.16\times 10^{-17} cm^2$).

Figure 13 (b) displays the accumulative nonlinearity for ICG obtained with pulse trains Z-scan technique. As seen, NT decreases with the pulse number up to about pulse 10, after which a small increase can be observed. This behavior could be understood by using a five-energy-level diagram, shown in Figure 14 (b). When excited by a pulse of the train to level $|1\rangle$, the molecule can undergo an intersystem-crossing to the triplet state $|4\rangle$, return to the ground state $|0\rangle$, or be promoted to a second excited state $|2\rangle$. With the arrival of the next pulse of the envelope, accumulative contributions to the optical nonlinearity, due to population built up in the long-lived (~ μs) $|4\rangle$ state, start to appear. The molecules in this state can be promoted to a second triplet state, $|5\rangle$, resulting in a change in the molecule absorption. Given the low irradiance of each individual pulse of the train and the short lifetime of levels $|2\rangle$ and $|5\rangle$, their population can be neglected. Considering this model, the fractions of molecules in each state are given by:

$$\frac{dn_0}{dt} = -w_{01}n_0 + \frac{n_1}{\tau_{10}} \tag{22}$$

$$\frac{dn_1}{dt} = +w_{01}n_{01} - \frac{n_1}{\tau_f} \tag{23}$$

$$\frac{dn_4}{dt} = \frac{n_1}{\tau_{isc}} \tag{24}$$

in which n_4 is the population fraction of the first triplet state. The $|1\rangle \rightarrow |0\rangle$ transition lifetime is given by $1/\tau_{01} = 1/\tau_f - 1/\tau_{isc}$, where τ_f and τ_{isc} are the fluorescence lifetime and the intersystem-crossing time respectively. This set of equations was numerically solved, yielding the time evolution of the absorption as:

$$\alpha(t) = N\{n_0(t)\sigma_{01} + n_1(t)\sigma_{12} + n_4(t)\sigma_T\} \tag{25}$$

where σ_T is the triplet state transition absorption cross-section. The only adjustable parameters are σ_T and τ_{isc}, once σ_{01} and σ_{12} are already known from the single pulse Z-scan analysis. The solid line in Figure 13(b) represents the best fitting obtained.
The intersystem-crossing time obtained through the fitting was $\tau_{isc} \approx (4 \pm 1)ns$, which is in good agreement with the one reported in the literature (Reindl et al., 1997). The quantum yield of triplet formation, ϕ_T, was calculated using $\phi_T = \tau_f / \tau_{isc}$ and τ_{isc} values, providing $\phi_T \approx 15\%$. The absorption cross-section of the triplet state found through the fitting procedure was $\sigma_T = (5 \pm 1) \times 10^{-17} cm^2$. This value is 31 times higher than that of the ground state cross-section ($\sigma_T = 0.16 \times 10^{-17} cm^2$). It was observed that σ_{12} is higher than σ_T ($\sigma_{12} / \sigma_T \approx 2.4$), indicating that the excited singlet state gives a higher contribution to the RSA process for ICG. In table 1 are the spectroscopic parameters obtained by fitting single and pulse train Z-scan data. This table also shows other ICG spectroscopic parameters obtained from the literature.

σ_{01}	σ_{12}	σ_T	τ_f	ϕ_{fl}	ϕ_T	ϕ_{ic}	k_{isc}	k_r	k_{ic}
0.16	12 ± 1	5 ± 0.5	580 ± 40	0.106	0.15 ± 0.04	0.74 ± 0.05	2.5 ± 0.6	1.82 ± 0.07	11 ± 1

Table 1. Cross-section values (x10^{-17} cm^2) for ground (σ_{01}), excited singlet (σ_{12}) and excited triplet (σ_T) states at 532 nm. Fluorescence lifetime (τ_f) (ps) (Reindl et al., 1997), fluorescence (ϕ_{fl}) (Reindl et al., 1997), triplet (ϕ_T) and internal conversion (ϕ_{ic}) quantum yields and rates constants (x10^8 s^{-1}) of intersystem-crossing (k_{isc}), radiative (k_r) (Reindl et al., 1997) and internal conversion (k_{ic}) of ICG/DMSO solution.

5.4 Cytochrome C

5.4.1 NLOA in the nano and picosecond regime

Cytochrome C (cyt c) is one of the most intensively investigated redox proteins, which act as electron carriers in the respiratory chain. It contains a covalent heme group linked to polypeptide chains, which prevent aggregation, feature desirable, for instance, in Photodynamic therapy (PDT). The heme group is an iron porphyrin, the same that is found in hematoporphyrins, with peripheral groups bonded to pyrrole rings, while the polypeptide chains are polymers made by amino acid residues linked by peptide bonds. This section presents some results of Z-scan technique employed to characterize the spectroscopic parameters and the dynamics of excited states of Fe^{3+} cyt c molecules, combined to pump-probe (Shapiro, 1977) measurements at 532 nm. The results clearly show that the nonlinearity origin can be ascribed to population effects of the Q-band followed by a fast relaxation back to the singlet ground state. The saturable absorption process observed has an intensity dependence and time evolution that can be described with a three-energy-level diagram, yielding the excited state parameters of cyt c (Neto et al., 2004).

Figure 15 shows the UV-Vis absorption spectrum for oxidized cyt c water solution. The strong band at 400 nm corresponds to the B (Soret) band, while the transition around 530 nm is attributed to the Q-band of the metalloporphyrin complex. The origin of these bands is related to π-π* and charge transfer transitions. According to the absorption spectrum, only the Q-band is excited when light at 532 nm is used.

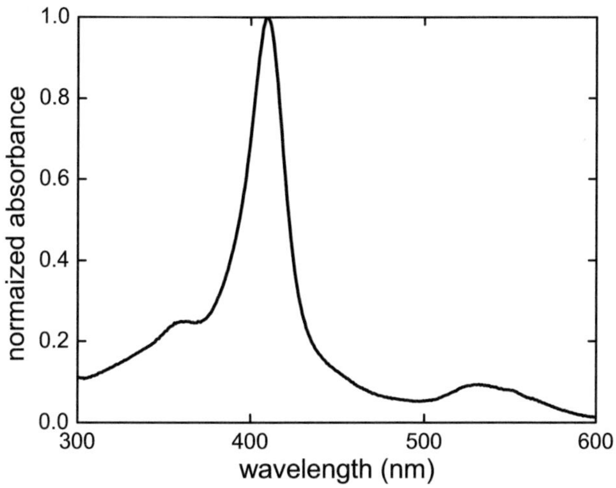

Fig. 15. Normalized absorbance spectrum of oxidized cyt c in water solution.

The results of the Z-scan measurements as a function of the pulse irradiance, in distinct temporal regimes (ps and fs), are depicted as solid circles in Figure 16 (a) and (b). To explain the behavior observed, the three-energy-level diagram presented in the inset of Figure 16 was considered, assuming that only the singlet states contribute to the nonlinear absorption process. This assumption is based on the fact that the pulse duration is faster than the intersystem-crossing time, which avoids any appreciable triplet state population buildup during the light-matter interaction time. In addition, the excited singlet state $|2\rangle$ was assumed to be too short-lived to present an appreciable population buildup.

According to the three-energy-level diagram proposed, molecules at the ground state $|0\rangle$ can be promoted to level $|1\rangle$ when excited by laser pulses of 70 ps at 532 nm, then decaying back to $|0\rangle$ with a relaxation time $\tau_{10.}$ Two-photon absorption processes were neglected because, under resonant conditions, excited state processes (saturable absorption) prevail. (Andrade et al., 2004). The rate equations used to describe the fraction of molecules remaining at ground state are:

$$\frac{dn_0}{dt} = -W_{01}n_0 + \frac{1-n_0}{\tau_{10}} \tag{25}$$

where n_0 and n_1 are the population fractions of the ground and first excited singlet state respectively and $W_{01} = \sigma_{01}I/h\nu$ is the one-photon transition rate. All the terms in Eq. 25

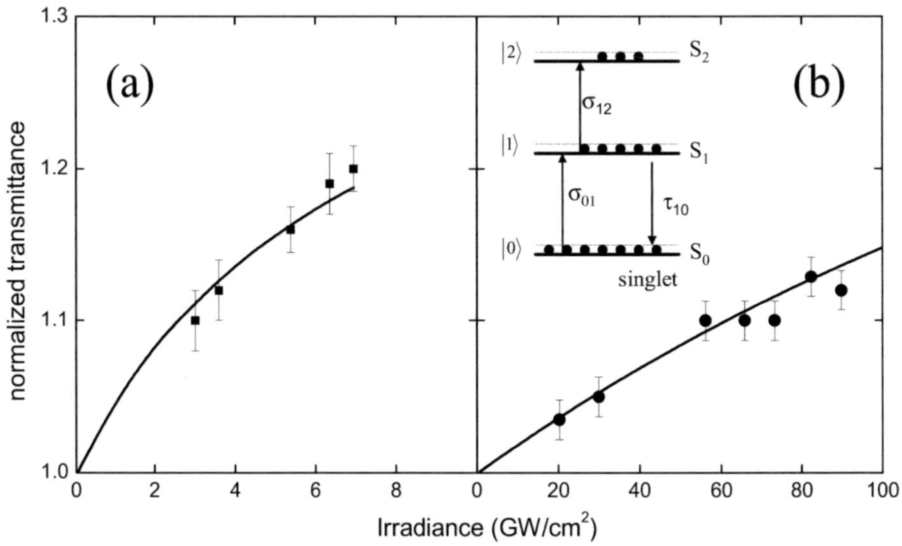

Fig. 16. (a) Normalized transmittance as a function of the 70 ps pulse irradiance at 532 nm. (b) Normalized transmittance as a function of the 120 fs pulse irradiance at 530 nm. The solid line in (a) and (b) are the fitting obtained with the three-energy-level model (inset) with the parameters given in the text.

have already been defined. $n_0 + n_1 = 1$, because the population of the $|2\rangle$ state is neglected. σ_{01} was determined as 4.1 x 10^{-17} cm^2.

In addition, an independent measurement was performed to determine the decay time of level $|1\rangle$, τ_{10}, with the degenerate pump-probe technique at 532 nm, yielding a characteristic time of 2.7 ps (Neto et al., 2004). Therefore, since all parameters of Eq. (25) are determined, it can be numerically solved using a Gaussian temporal intensity pattern for the laser pulse, yielding the population dynamics within the laser pulse. The time-dependent absorption coefficient in this case is:

$$\alpha(t) = N\{n_0(t)\sigma_{01} + n_1(t)\sigma_{12}\} \qquad (26)$$

where N is the concentration, σ_{12} is the excited state cross-section, and n_0 and n_1 are the population fraction in each state. By fitting the experimental data, the best value found for σ_{12} was 3.7 x 10^{-17} cm^2, which is on the same order of magnitude of those reported in the literature for metallporphyrins (Blau et al., 1985, Neto et al., 2003). Since the intersystem-crossing time is unknown, the possibility of the triplet state acquiring some population cannot be disregarded *a priori*. In order to confirm the excited singlet state cross-section value and the three-energy-level model assumed, Z-scan measurements using 120 fs pulses at 532 nm were carried out. In this case, one can safely state that the pulse duration is faster than the intersystem-crossing time and that there is no triplet state population during the pulse interaction, which certainly allows the use of the three-energy-level diagram. Again, an increase due to a saturable absorption mechanism is observed in the normalized transmittance as a function of irradiance, displayed in Fig 16(b), indicating that laser pulses are populating the excited state. The solid line represents the theoretical fitting obtained with the model described previously, resulting in σ_{12} = 3.7 x 10^{-17} cm^2, which is the same value found in the picosecond Z-scan experiment. This result indicates that, even when picosecond pulses are used, the triplet state is not populated, supporting the assumption made on the three-level energy model used to explain the experimental results. It also implies that the intersystem-crossing time of cyt c should be in the order of a few hundred picoseconds (Sazanovich et al., 2003).

The short singlet state lifetime is a clear indication of the fast intersystem-crossing time, which is a characteristic of porphyrins with open shell ions (Kalyanasundaram, 1992). This short intersystem-crossing time, compared with those of closed shell porphyrins (Kalyanasundaram, 1984), indicates an efficient singlet-triplet conversion, making hematoporphyrins suitable for applications as a PDT sensitizer. Besides, cyt c is a biocompatible molecule, which is a requirement for medical applications.

6. Conclusion

This chapter aimed to describe the resonant nonlinear optical properties of four important organic molecules: Chlorophyll A, Indocyanine Green, Ytterbium Bisphthalocyanine and Cytochrome C, which are materials that present interesting optical nonlinearities for applications in optical devices. It was shown that Chlorophyll A solution exhibits a RSA process for Q-switched and mode-locked laser pulses, with an intersystem-crossing time relatively fast and a triplet state cross section value twice higher than that of the singlet. Such features are desired for applications in PDT. However, due to the low triplet–singlet cross-section ratio, Chlorophyll A is not expected to be efficient as an optical limiter. In addition, the excited state population buildup generated with the WLC Z-scan technique can be advantageously used to shape the pulse intensity spectrum in order to match the most intense linear absorption band of the material. As a consequence, one can obtain an enhancement of the nonlinear absorption in a transparent region through excited state absorption. In practical terms, WLC pulses could be used in applications where a high RSA process is needed in the blue region of the spectrum. RSA at 532 nm for ICG solution was also described. For single pulse experiments, it was determined that the excited singlet state cross-section is 75 times higher than that of the ground state. However, when pulse trains are employed, triplet population is identified, with an intersystem-crossing time in the nanosecond time scale. In this case, the triplet absorption cross-section found is 31 times higher than the ground state one. These results indicate ICG as a candidate for applications requiring high RSA, such as optical limiters and all-optical switches. Regarding Ytterbium Bis-phtalocyanine, it was shown that this molecule presents two possible ground state levels and both can absorb the excitation light for some wavelength range. When using femtosecond laser pulses, it was also possible to observe distinct resonant nonlinear absorption behaviours (SA and RSA) depending on the wavelength. Basically, the excited state absorption cross-section is approximately zero in the Q-band region, giving origin to a strong SA process. Oxidized Cytochrome C in water solution exhibits a saturable absorption process when resonant excitation at 532 nm (Q-band) is employed. Its short singlet state lifetime indicates a relatively fast intersystem-crossing time that can lead to an efficient formation of the triplet state. Such feature prompts this molecule as an efficient sensitizer for PDT applications. Therefore, organic molecules presenting high nonlinear optical absorption processes are potential candidates as active media for applications in optical devices.

7. Acknowledgment

We acknowledge financial support from FAPESP and CNPq (Brazil) and AFOSR (FA9550-07-1-0374)

8. References

Andrade, A. A.; Neto, N. M. B.; Misoguti, L.; De Boni, L.; Zilio, S. C. & Mendonca, C. R. (2004). Two-photon absorption investigation in reduced and oxidized cytochrome c solutions. *Chemical Physics Letters*, 390, 4-6(Jun 1):506-510.

Baker, N. R. & Rosenqvist, E. (2004). Applications of chlorophyll fluorescence can improve crop production strategies: an examination of future possibilities. *Journal of Experimental Botany*, 55, 403(Aug):1607-1621.

Balu, M.; Hales, J.; Hagan, D. J. & Van Stryland, E. W. (2004). White-light continuum Z-scan technique for nonlinear materials characterization. *Optics Express*, 12, 16(Aug 9):3820-3826.

Banfi, G.; Degiorgio, V. & Ricard, D. (1998). Nonlinear optical properties of semiconductor nanocrystals. *Advances in Physics*, 47, 3(May-Jun):447-510.

Bhawalkar, J. D.; He, G. S. & Prasad, P. N. (1996). Nonlinear multiphoton processes in organic and polymeric materials. *Reports on Progress in Physics*, 59, 9(Sep):1041-1070.

Blau, W.; Byrne, H.; Dennis, W. M. & Kelly, J. M. (1985). Reverse Saturable Absorption in TetraPhenylPorphyrins. *Optics Communications*, 56, 1: 25-29.

Calvete, M.; Yang, G. Y. & Hanack, M. (2004). Porphyrins and phthalocyanines as materials for optical limiting. *Synthetic Metals*, 141, 3(Mar 25):231-243.

Carter, G. A. & Spiering, B. A. (2002). Optical properties of intact leaves for estimating chlorophyll concentration. *Journal of Environmental Quality*, 31, 5(Sep-Oct):1424-1432.

Correa, D. S.; De Boni, L.; dos Santos, D. S.; Neto, N. M. B.; Oliveira, O. N.; Misoguti, L.; Zilio, S. C. & Mendonca, C. R. (2002). Reverse saturable absorption in chlorophyll A solutions. *Applied Physics B-Lasers and Optics*, 74, 6(Apr):559-561.

De Boni, L.; Andrade, A. A.; Misoguti, L.; Mendonca, C. R. & Zilio, S. C. (2004). Z-scan measurements using femtosecond continuum generation. *Optics Express*, 12, 17(Aug 23):3921-3927.

De Boni, L.; Correa, D. S.; Pavinatto, F. J.; dos Santos, D. S. & Mendonca, C. R. (2007). Excited state absorption spectrum of chlorophyll a obtained with white-light continuum. *Journal of Chemical Physics*, 126, 16(Apr 28):165102.

De Boni, L.; Gaffo, L.; Misoguti, L. & Mendonca, C. R. (2006). Nonlinear absorption spectrum of ytterbium bis-phthalocyanine solution measured by white-light continuum Z-scan technique. *Chemical Physics Letters*, 419, 4-6(Feb 26):417-420.

De Boni, L.; Rezende, D. C. J. & Mendonca, C. R. (2007). Reverse saturable absorption dynamics in indocyanine green. *Journal of Photochemistry and Photobiology a-Chemistry*, 190, 1(Jul 5):41-44.

Fisher, A. M. R.; Murphree, A. L. & Gomer, C. J. (1995). Clinical and Preclinical Photodynamic Therapy. *Lasers in Surgery and Medicine*, 17, 1:2-31.

Frackowiak, D.; Planner, A.; Waszkowiak, A.; Boguta, A.; Ion, R. M. & Wiktorowicz, K. (2001). Yield of intersystem (singlet-triplet) crossing in phthalocyanines evaluated on the basis of a time in resolved photothermal method. *Journal of Photochemistry and Photobiology a-Chemistry*, 141, 2-3(Jul 2):101-108.

Gouterman, M. (1961). Spectra of Porphyrins. *Journal of Molecular Spectroscopy*, 6, 1:138-163.

Hasegawa, J.; Ozeki, Y.; Ohkawa, K.; Hada, M. & Nakatsuji, H. (1998). Theoretical study of the excited states of chlorin, bacteriochlorin, pheophytin a and chlorophyll a by the SAC/SAC-CI method. *Journal of Physical Chemistry B*, 102, 7(Feb 12):1320-1326.

He, G. S.; Tan, L. S.; Zheng, Q. & Prasad, P. N. (2008). Multiphoton absorbing materials: Molecular designs, characterizations, and applications. *Chemical Reviews*, 108, 4(Apr):1245-1330.

Kalyanasundaram, K. (1984). Photochemistry of water-soluble porphyrins: Comparative study of isomeric tetrapyridyl- and tetrakis (N - methylpyridiniumyl) porphyrins. *Inorganic Chemistry*, 23, 16, 2453-2459.

Kalyanasundaram, K. (1992). *Photochemistry of polypyridine and porphyrin complexes*, Academic PressSan Diego.

Linnanto, J. & Korppi-Tommola, J. (2000). Spectroscopic properties of Mg-chlorin, Mg-porphin and chlorophylls a, b, c(1), c(2), c(3) and d studied by semi-empirical and ab initio MO/CI methods. *Physical Chemistry Chemical Physics*, 2, 21:4962-4970.

Loppacher, C.; Guggisberg, M.; Pfeiffer, O.; Meyer, E.; Bammerlin, M.; Luthi, R.; Schlittler, R.; Gimzewski, J. K.; Tang, H. & Joachim, C. (2003). Direct determination of the energy required to operate a single molecule switch. *Physical Review Letters*, 90, 6(Feb 14):066107.

Mendonça, C. R.; Gaffo, L.; Misoguti, L.; Moreira, W. C.; Oliveira, O. N. & Zilio, S. C. (2000). Characterization of dynamic optical nonlinearities in ytterbium bis-phthalocyanine solution. *Chemical Physics Letters*, 323, 3-4(Jun 16):300-304.

Mendonça, C. R.; Gaffo, L.; Moreira, W. C.; Oliveira, O. N. & Zilio, S. C. (2001). Optical properties of ytterbium bis-phthalocyanine solution. *Synthetic Metals*, 121, 1-3(Mar 15):1477-1478.

Michel-Beyerle, M. E. (1985). *Antennas and Reaction Centers of Photosynthetic Bacteria*, Springer-VerlagBerlin.

Misoguti, L.; Mendonca, C. R. & Zilio, S. C. (1999). Characterization of dynamic optical nonlinearities with pulse trains. *Applied Physics Letters*, 74, 11(Mar 15):1531-1533.

Misoguti, L.; Mendonça, C. R. & Zilio, S. C. (1999). Characterization of dynamic optical nonlinearities with pulse trains. *Applied Physics Letters*, 74, 11: 1531-1533.

Neto, N. M. B.; Andrade, A. A.; De Boni, L.; Misoguti, L.; Zilio, S. C. & Mendonca, C. R. (2004). Excited-state absorption in oxidized cytochrome c solution. *Applied Physics B-Lasers and Optics*, 79, 6(Oct):751-754.

Neto, N. M. B.; Correa, D. S.; Dos Santos, D. S.; Misoguti, L.; Oliveira, O. N.; Zilio, S. C. & Mendonca, C. R. (2003). Influence of photodegradation on the optical limiting process of chlorophyll A. *Modern Physics Letters B*, 17, 2(Jan 30):83-87.

Neto, N. M. B.; De Boni, L.; Rodrigues, J. J.; Misoguti, L.; Mendonca, C. R.; Dinelli, L. R.; Batista, A. A. & Zilio, S. C. (2003). Dynamic saturable optical nonlinearities in free base tetrapyridylporphyrin. *Journal of Porphyrins and Phthalocyanines*, 7, 6:452-456.

Neto, N. M. B.; Oliveira, S. L.; Misoguti, L.; Mendonca, C. R.; Goncalves, P. J.; Borissevitch, I. E.; Dinelli, L. R.; Romualdo, L. L.; Batista, A. A. & Zilio, S. C. (2006). Singlet excited state absorption of porphyrin molecules for pico- and femtosecond optical limiting application. *Journal of Applied Physics*, 99, 12(Jun 15):123103.

O'Flaherty, S. M.; Hold, S. V.; Cook, M. J.; Torres, T.; Chen, Y.; Hanack, M. & Blau, W. J. (2003). Molecular engineering of peripherally and axially modified phthalocyanines for optical limiting and nonlinear optics. *Advanced Materials*, 15, 1(Jan 3):19-+.

Pang, Y.; Samoc, M. & Prasad, P. N. (1991). 3rd-Order Nonlinearity and 2-Photon-Induced Molecular-Dynamics - Femtosecond Time-Resolved Transient Absorption, Kerr Gate, and Degenerate 4-Wave-Mixing Studies in Poly(Para-Phenylene Vinylene) Sol-Gel Silica Film. *Journal of Chemical Physics*, 94, 8(Apr 15):5282-5290.

Parusel, A. B. J. & Grimme, S. (2000). A theoretical study of the excited states of chlorophyll a and pheophytin a. *Journal of Physical Chemistry B*, 104, 22(Jun 8):5395-5398.

Prasad, P. N. (1991). Polymeric Materials for Nonlinear Optics and Photonics. *Polymer*, 32, 10(1746-1751.

Prasad, P. N. & Willians, D. J. (1991). *Introduction to Nonlinear Optical Effects in Molecules and Polymers*, Wiley-InterscienceNew York.

Reindl, S.; Penzkofer, A.; Gong, S. H.; Landthaler, M.; Szeimies, R. M.; Abels, C. & Baumler, W. (1997). Quantum yield of triplet formation for indocyanine green. *Journal of Photochemistry and Photobiology a-Chemistry*, 105, 1(May 15):65-68.

Rivadossi, A.; Zucchelli, G.; Garlaschi, F. M. & Jennings, R. C. (2004). Light absorption by the chlorophyll a-b complexes of photosystem II in a leaf with special reference to LHCII. *Photochemistry and Photobiology*, 80, 3(Nov-Dec):492-498.

Sazanovich, I. V.; Ganzha, V. A.; Dzhagarov, B. M. & Chirvony, V. S. (2003). Dichroism of the triplet-triplet transient absorption of copper(II) porphyrins in liquid solution. New interpretation of the subnanosecond relaxation component'. *Chemical Physics Letters*, 382, 1-2(Nov 28):57-64.

Scheidt, W. R. & Reed, C. A. (1981). Spin-State Stereochemical Relationships in Iron Porphyrins - Implications for the Hemoproteins. *Chemical Reviews*, 81, 6:543-555.

Shank, C. V.; Ippen, E. P. & Teschke, O. (1977). Sub-Picosecond Relaxation of Large Organic-Molecules in Solution. *Chemical Physics Letters*, 45, 2:291-294.

Shapiro, S. L.,(1977), *Topics in Applied Physics*, New York.

Sheik-Bahae, M.; Said, A. A. & Van Stryland, E. W. (1989). High-Sensitivity, Single-Beam N2 Measurements. *Optics Letters*, 14, 17(Sep 1):955-957.

Shen, Y. R. (1984). *The Principles of Nonlinear Optics*, John Wiley,0471889989 New York.

Shirk, J. S.; Lindle, J. R.; Bartoli, F. J. & Boyle, M. E. (1992). 3rd-Order Optical Nonlinearities of Bis(Phthalocyanines). *Journal of Physical Chemistry*, 96, 14(Jul 9):5847-5852.

Sundholm, D. (1999). Density functional theory calculations of the visible spectrum of chlorophyll a. *Chemical Physics Letters*, 302, 5-6(Mar 26):480-484.

Unnikrishnan, K. P.; Thomas, J.; Nampoori, V. P. N. & Vallabhan, C. P. G. (2002). Wavelength dependence of nonlinear absorption in a bis-phthalocyanine studied using the Z-scan technique. *Applied Physics B-Lasers and Optics*, 75, 8(Dec):871-874.

Vernon, L. P. & Seely, G. R. (1996). *The Chlorophylls*, AcademicNew York.

Wehling, A. & Walla, P. J. (2005). Time-resolved two-photon spectroscopy of photosystem I determines hidden carotenoid dark-state dynamics. *Journal of Physical Chemistry B*, 109, 51(Dec 29):24510-24516.

Zigmantas, D.; Hiller, R. G.; Sundstrom, V. & Polivka, T. (2002). Carotenoid to chlorophyll energy transfer in the peridinin-chlorophyll-a-protein complex involves an intramolecular charge transfer state. *Proceedings of the National Academy of Sciences of the United States of America*, 99, 26(Dec 24):16760-16765.

3

Optical and Spectroscopic Properties of Polymer Layers Doped with Rare Earth Ions

Vaclav Prajzler[1], Oleksiy Lyutakov[2], Ivan Huttel[2],
Jiri Oswald[3] and Vitezslav Jerabek[1]
[1]*Czech Technical University in Prague, Department of Microelectronics*
Technicka 2, 166 27 Prague 6,
[2]*Institute of Chemical Technology in Prague*
Technicka 2, 166 28 Prague 6,
[3]*Institute of Physics of the Academy of Sciences Czech Republic v.v.i.*
Cukrovarnicka 10, 162 00 Prague,
Czech Republic

1. Introduction

This chapter deals with description properties of a number of Rare-Earth (RE) ions in polymer materials. The list of the RE elements with some of its basic properties are shown in Table 1. The electronic structure of each trivalent RE element consists of partially filled 4f subshell, and outer $5s^2$ and $5p^6$ subshell. With increasing nuclear charge electrons enter into the underlying 4f subshell rather than the external 5d subshell. Since the filled $5s^2$ and $5p^6$ subshells screen the 4f electrons, the RE elements have very similar chemical properties. The screening of the partially filled 4f subshells, by the outer closed $5s^2$ and $5p^6$ subshell, also gives rise to sharp emission spectra independent of the host materials. The intra-subshell transitions of 4f electrons lead to narrow absorption peaks in the ultra-violet, visible, and near-infrared regions.

Atomic number	n*	Element	Electron configuration RE^{3+}	Ground term RE^{3+}
58	1	Cerium - Ce	$4f^1 5s^2 5p^6$	$^2F_{5/2}$
59	2	Praseodymium - Pr	$4f^2 5s^2 5p^6$	$^3H_4$
60	3	Neodymium - Nd	$4f^3 5s^2 5p^6$	$^4I_{9/2}$
61	4	Promethium - Pm	$4f^4 5s^2 5p^6$	$^5I_4$
62	5	Samarium - Sm	$4f^5 5s^2 5p^6$	$^6H_{5/2}$
63	6	Europium - Eu	$4f^6 5s^2 5p^6$	$^7F_0$
64	7	Gadolinium - Gd	$4f^7 5s^2 5p^6$	$^8S_{7/2}$
65	8	Terbium - Tb	$4f^8 5s^2 5p^6$	$^7F_6$
66	9	Dysprosium - Dy	$4f^9 5s^2 5p^6$	$^6H_{15/2}$
67	10	Holmium - Ho	$4f^{10} 5s^2 5p^6$	$^5I_8$
68	11	Erbium - Er	$4f^{11} 5s^2 5p^6$	$^4I_{15/2}$
69	12	Thulium - Tm	$4f^{12} 5s^2 5p^6$	$^3H_6$
70	13	Ytterbium - Yb	$4f^{13} 5s^2 5p^6$	$^2F_{7/2}$

*Number of electros (n) in the 4f shell of three-valence Rare Earth ions.

Table 1. The Rare Earth elements and some of its properties

Trivalent RE ions can be used for many photonics applications. Erbium (Er), neodymium (Nd), praseodymium (Pr) ions are well-known, because these elements have transitions used in telecommunications systems. Except these RE ions other elements are intensively studied. Europium (Eu), terbium (Tb) and cerium (Ce) produce red, green and blue light which is used for full colour displays. Thulium (Tm) and holmium (Ho) lasers has received a large amount of interest during last ten years because these elements are the best candidates for a wide range of applications including medicine and eye-safe remote sensing systems such as laser ranging, coherent Doppler lidar for wind sensing, wind-shear detection and etc (Kenyon A.J.; 2002; Bourdet et al. 2000).
The choice of host materials for RE ions hardly influences the energy levels of the RE ions. Therefore in principle any materials that have little absorption at the pump and emission wavelength can be used. Up to now there have been published many papers describing properties of RE-doped with many different materials. Photonics materials such as glasses, optical crystals ($LiNbO_3$, Al_2O_3, Y_2O_3) and semiconductors (Si, SiGe, SiC, GaN, etc.) are conventional materials for accomplishing lasing action (Polman A; 1997; Wong, et al. 2002; Zavada, et al. 1995). Recently there has been considerable interest in the development of new photonics materials such as polymers which have better properties and a lower price. It is due to fact that a number of properties make polymers attractive hosts for RE ions including high transparency in the visible and near-infrared spectra, well controlled refractive indices, good thermal stabilities, offering simple fabrication process and low cost (Liang et al., 2003; Slooff et al., 2002; Sosa et al., 2003; Tung et al., 2005) .
Polymers doped with RE-ions are still a new area and there is only a small number of groups active in this field as for example: L.H. Slooff from the FOM Institute for Atomic and Molecular Physics, Amsterdam, The Netherlands, W.H. Wong from the Department of Electronic Engineering and Department of Physics and Material Sciences, City University of Hong Kong, H. Liang from Structure Research Laboratory and Department of Polymer Science and Engineering, University of Science and Technology of China and X. Xu from Optical Physics Laboratory, Institute of Physics, Chinese Academy of Sciences, Beijing, China.
For our research we chose two types of polymers. As first polymer we chose Polymethylmethacrylate (PMMA) polymer because it is the most common used polymer and we also used new of type polymer Epoxy Novolak Resin (ENR) due to its low optical losses 0.2 dB/cm at 1090 nm, 0.77 dB/cm at 1310 nm, 1.71 dB/cm at 1550 nm and due to easy fabrication process (Beche et al., 2005). We doped these two polymers with erbium (Er), ytterbium (Yb), europium (Eu), neodymium (Nd), thulium (Tm), holmium (Ho), praseodymium (Pr) and dysprosium (Dy) ions (Prajzler et al., 2007; Prajzler et al., 2008). We chose these RE ions because Er^{3+} doped materials can emit at 1530 nm and Tm^{3+} doped photonics materials can have emission bands around 1470 nm and from 1600 to 2100 nm. Yb^{3+} and Ho^{3+} ions were used as co-dopants. Tm^{3+} doped polymers were co-doped with Ho^{3+} ions and Er^{3+} doped polymers were co-doped with Yb^{3+} ions. Trivalent Dy^{3+} ions are studied for emission at 1300 nm. Other RE ions were chosen for photoluminescence study in visible region.

2. Experimental part

2.1 PMMA layers

Fabrication process of PMMA layers doped with RE ions is following: Small pieces of PMMA (Goodfellow) were left to dissolve in chloroform for a few days before being used in the fabrication of PMMA layers. The layers were formed by the solution either being spin-coated onto silicon and glass substrates or by being poured into bottomless molds placed on a glass substrate and left to dry. For RE doping, solutions whose content ranged from 1.0 at. % to 20.0 at. % RE-ions were added to the PMMA. For RE co-doping rare earth chloride or Rare Earth fluoride were together dissolved in C_5H_9NO or C_2H_6OS. Samples containing 1.0 at. % erbium were co-doped with ytterbium in amounts also ranging from 1.0 at. % to 20.0 at. %.

2.2 Epoxy Novolak Resin

Commercially available polymer Epoxy Novolak Resin (NANO™ Su-8 10) supported by Micro Resist Technology GmbH was used for fabrication of the RE doped samples. Chemical structure of Epoxy Novolak Resin (ENR) polymer is shown in Fig.1.

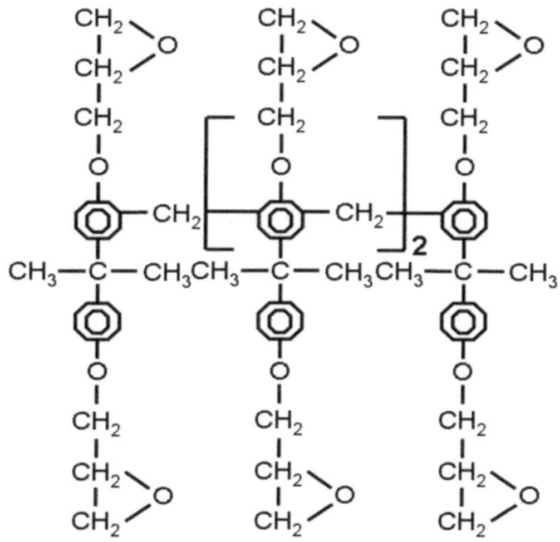

Fig. 1. Structure of the Epoxy Novolak Resin polymer

Polymer layers were formed by the solution either being spin-coated onto silicon or by being poured into bottomless molds placed on a quartz substrate and let to dry. After the deposition the samples were baked at 90°C for 45 min and then UV light was used for hardening. Finally hard baking at 90°C for 60 min was applied. The doping occurred using anhydrous RE chloride or RE fluoride dissolved in C_2H_6OS (Sigma-Aldrich). For the doping, solutions where the RE content ranged from 1.0 at. % to 20.0 at. % were added to the ENR polymer.

3. Results and discussion

3.1 Infrared spectra

The fabricated samples were investigated by infrared spectroscopy (FT-IR). Infrared reflectance and ATR spectra were obtained using a Bruker IFS 66/v FTIR spectrometer equipped with a broadband MCT detector, to which 128 interferograms were added with a resolution of 4 cm^{-1} (Happ-Genzel apodization). Fig. 2a displays the FT-IR spectra of PMMA layers doped with Er^{3+} ions. Fig. 2b shows the FT-IR spectra of ENR layers doped with Nd^{3+} ions in the wavelength range from 3900 to 2600 cm^{-1}.

Fig. 2a shows the three strong broad bands occurring at 2843 cm^{-1}, 2953 cm^{-1}, 2994 cm^{-1} correspond to the aliphatic C-H bands. These bands are assigned to the stretching vibrations of CH_3 and CH_2, and indicate a high content of hydrogen-rich CH_x. The absorption band at 3349 cm^{-1} corresponds to the O-H stretching vibrations of the PMMA layers. Fig.2a also shows that increasing the Er^{3+} content also increases the intensity of the O-H vibrations. This can be explained by the fact that $ErCl_3$, as very hygroscopic substance, not only dopes the polymer samples but also brings a certain amount of water. It is a well-known fact that the presence of O-H groups in a matrix containing rare earth ions unfortunately causes problems by hindering emission in the infrared region.

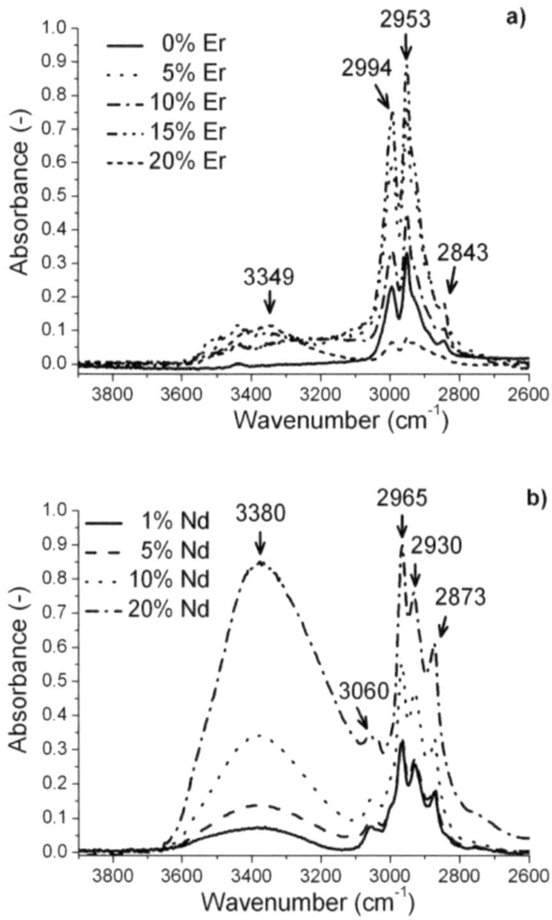

Fig. 2. Infrared spectra a) PMMA samples doped with Er^{3+} using $ErCl_3$ and b) ENR samples doped with Nd^{3+} ions using $NdCl_3$.

Fig. 2b shows also three strong broad bands occurring at 2873 cm^{-1}, 2930 cm^{-1} and 2965 cm^{-1} correspond to aliphatic C-H bands while that one corresponding to aromatic C-H band is at 3060 cm^{-1}. These bands are assigned to the stretching vibrations of CH_3 and CH_2, and indicate a high content of hydrogen-rich CH_x. The absorption bands at 3380 cm^{-1} correspond to the O-H stretching vibrations of the ENR layers. Similar results were obtained for all the samples with the only difference that the O-H stretching bands were for particular rare earths slightly shifted (while the C-H stretching bands remained un-affected). Fig. 2b also shows that increasing RE content caused the O-H vibrations intensity increased as well. This can be explained by the fact that RE chloride and fluoride are very hygroscopic substances.

3.2 Transmission spectra

The transmission measurements were performed using a UV-VIS-NIR Spectrometer (UV-3600 Shimadzu) in the spectral range from 350 to 1800 nm. The transmission spectra of the Er^{3+} doped PMMA layers in the spectral range from 350 nm to 700 nm are shown in Fig.3a and the transmission spectra of the Er^{3+} (1.0 at.%) doped polymers co-doped with Yb^{3+} ions in the spectral range from 900 nm to 1040 nm are shown in Fig. 3b.

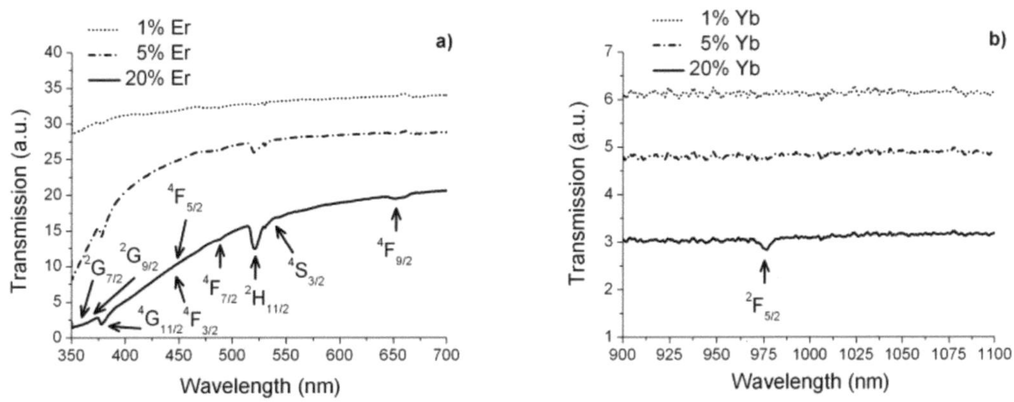

Fig. 3. Transmission spectra a) Er^{3+} doped PMMA layers and b) Yb^{3+} doped PMMA layers

Samples contain 5.0 at. % of Er^{3+}, two bands appeared that were attributed to the following transitions: $^4G_{11/2}$ (377 nm) and $^2H_{11/2}$ (519 nm). Samples contain 20.0 at.% of Er^{3+} another band appeared at $^4F_{9/2}$ (650 nm). We did not observe bands $^2G_{7/2}$ (355 nm), $^2G_{9/2}$ (363 nm), $^2H_{9/2}$ (405 nm), $^4F_{3/2}$ (441 nm), $^4F_{5/2}$ (448 nm), $^4F_{7/2}$ (485 nm) and $^4S_{3/2}$ (539 nm). Sample contains 20.0 at. % of Yb^{3+} ions has a typical Yb^{3+} $^2F_{5/2}$ transition with maximum at 977 nm. Samples with a lower concentration have weaker $^2F_{5/2}$ transition, and the samples with concentration 1 at.% Yb^{3+} ions have no visible Yb^{3+} ($^2F_{5/2}$) transition.

Transmission spectra of the Ho^{3+} doped ENR layers are shown in Fig. 4. Fig. 4a shows transmission spectra in spectral range from 400 nm to 1000 nm and samples contain 10.0 at.% of Ho^{3+} appeared following three strong bands $^5G_6$ (448 nm), $^5F_4$ (535 nm) and $^5F_5$ (639 nm). Fig. 4a also shows three week bands $^5G_5$ (415 nm), $^5F_2$ (466 nm), and $^5F_3$ (482 nm), which can be assigned to the holmium ions. Fig. 4b shows transmission spectra in spectral range from 900 nm to 1300 nm and we found out band at 1152 nm corresponds to $^5I_6$ transitions.

Transmission spectra of ENR polymer doped with Ho^{3+} ions co-doped with Tm^{3+} ions are shown in Fig. 5. Fig. 5a shows four bands $^1G_4$ (463 nm), $^3F_2$ (658 nm), $^3F_3$ (683 nm), and $^3H_4$ (790 nm), and Fig. 5b shows one band $^3H_5$ (1210 nm) corresponds to Tm^{3+} ions. Fig. 5a also shows tree bands $^5G_6$ (448 nm), $^5F_4$ (535 nm), and $^5F_5$ (639 nm) correspond to Ho^{3+} ions.

ENR polymer doped with Nd^{3+} and Dy^{3+} ions is shown in Fig. 6.

Fig. 6a shows four strong bands corresponding to the Nd^{3+} transitions: $^5G_{5/2}$ (574 nm), $^4F_{7/2}$ (740 nm), $^4F_{5/2}$ (794 nm), $^4F_{7/3}$ (865 nm) and Fig. 6b shows six bands corresponds to the Dy^{3+} ions. We observed transitions $^6F_{3/2}$ (758 nm), $^6F_{5/2}$ (807 nm), $^6F_{7/2}$ (906 nm), $^6F_{9/2}$ (1100 nm), $^6F_{11/2}$ (1280 nm) and $^6H_{11/2}$ (1685 nm).

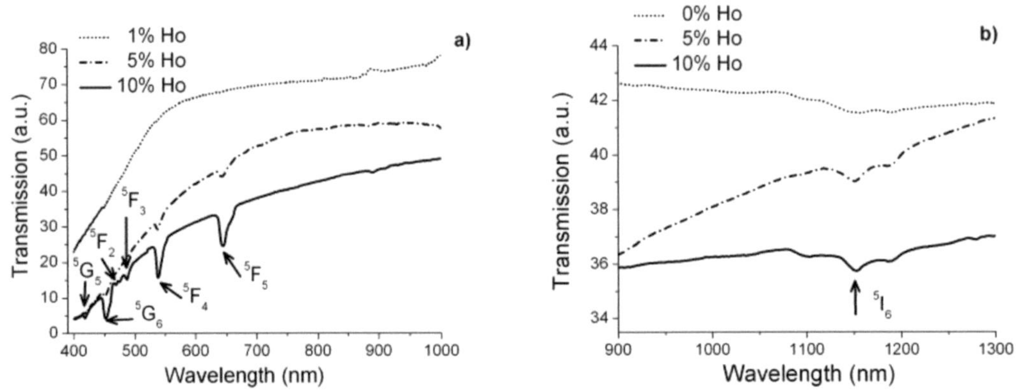

Fig. 4. Transmission spectra of ENR polymer doped with Ho^{3+} ions (1.0 at.%) and co-doped with Tm^{3+} ions a) 10.0 at.% Tm^{3+}, b) 20.0 at.% Tm^{3+}

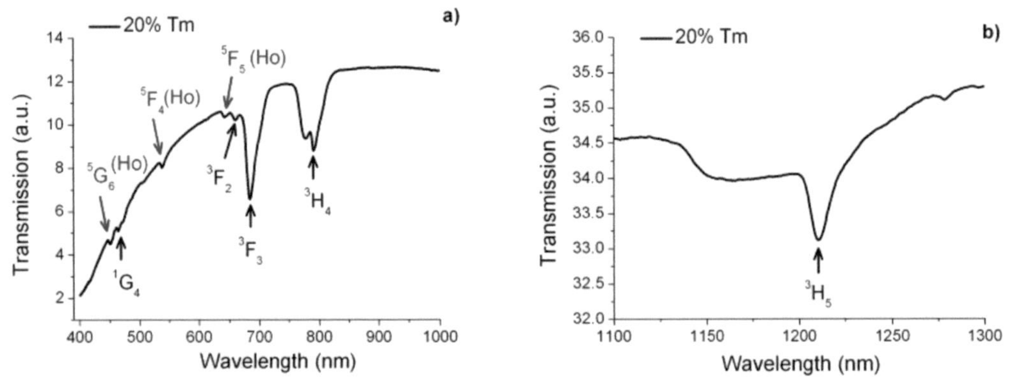

Fig. 5. Transmission spectra of ENR polymer doped with Ho^{3+} ions (1.0 at.%) and co-doped with Tm^{3+} (20.0 at.%) ions

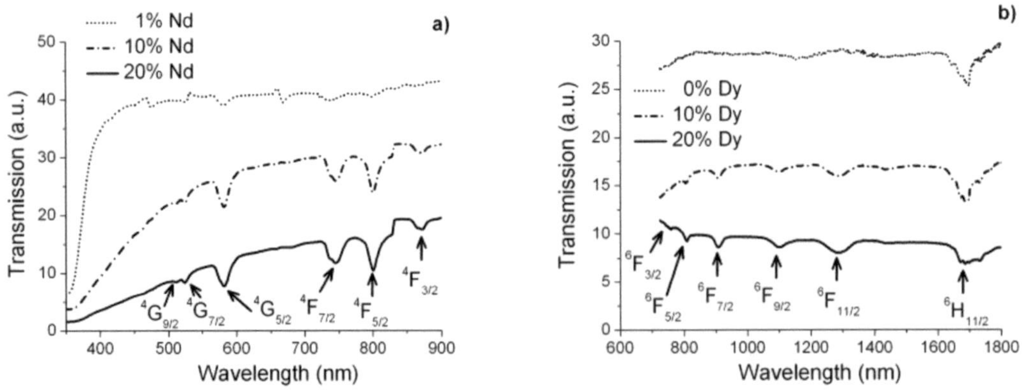

Fig. 6. Transmission spectra of ENR polymer doped with a) Nd^{3+} ions and b) Dy^{3+} ions

The obtained transmission spectra of the ENR layers doped with Eu^{3+} ions are shown in Fig. 7a and ENR polymer doped with Pr^{3+} is shown in Fig. 7b.

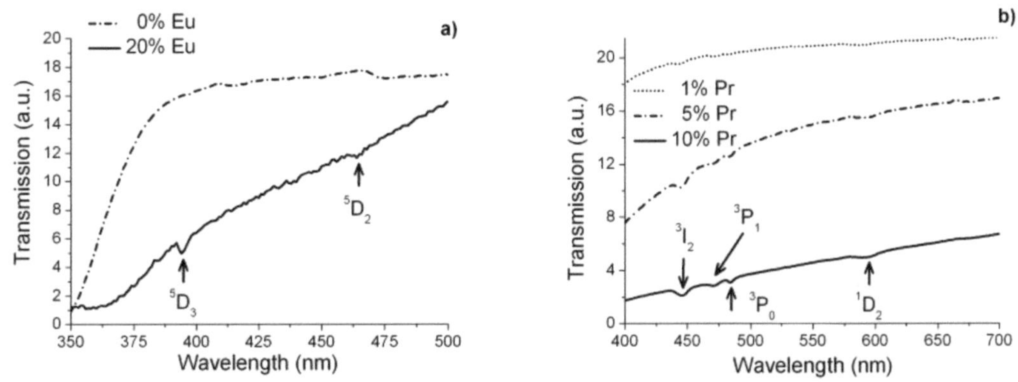

Fig. 7. Transmission spectra of ENR polymer doped with a) Eu^{3+} ions and b) Pr^{3+} ions

Fig. 7a shows only two weak bands correspond to Eu^{3+} ions - $^5D_3$ (393 nm) and $^5D_2$ (464 nm). Fig. 7b shows four bands correspond to Pr^{3+} $^3I_2$ (443 nm), $^3P_1$ (466 nm), $^2P_0$ (479 nm) and $^1D_2$ (587 nm) in the sample containing 10.0 at.% Pr^{3+}. We found out very strong bands at 1440 nm and at 1540 nm, which correspond to $^3F_4$ and $^4F_4$ bands, respectively. The bands at 1950 nm and at 2340 nm (that would be attributed to $^3F_2$ and $^3H_6$ transitions) did not appear not even in the sample with the highest Pr^{3+} concentrations (20.0 at.%).

3.3 Photoluminescence

Semiconductor lasers (operating at λ_{ex}= 980 nm, λ_{ex}= 827 nm) and He-Ne laser excitation (λ_{ex}= 632.8 nm) were used to detect sample luminescence in the range from 1450 nm to 1650 nm.

Fig. 8 shows Photoluminescence spectra (PL) of ENR layers doped with Dy^{3+} ions. Fig. 8a shows PL spectra pumped at 632.8 nm (6 mW) under temperature of 4 K and this Fig. clearly shows the band at 1340 nm caused by the Dy^{3+} transition $^6H_{9/2}$ - $^6F_{11/2} \rightarrow {}^6H_{15/2}$. This transition appeared only in the samples containing 10 at.% of dysprosium. Photoluminescence of the samples with higher content (15 at.%) of dysprosium is much weaker. Fig. 8a also shows that maximum is around 1340 nm and Fig. 8b shows PL spectra

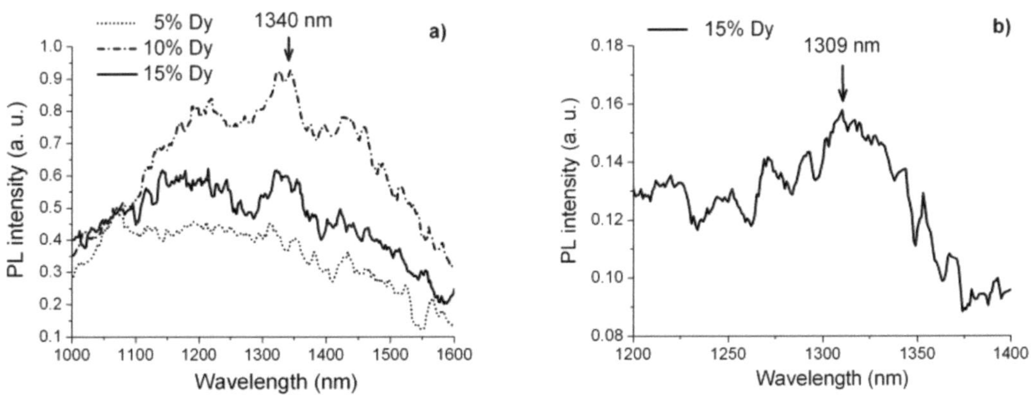

Fig. 8. Photoluminescence spectra of ENR layers doped with Dy^{3+} a) excitation 632 nm and b) excitation 827 nm

obtained by using optical pumping at 827 nm (temperature 4 K). PL spectra caused by Dy^{3+} ions was observed only at samples containing 15 at.% and PL maximum was found out around 1310 nm. Measurements at two additional excitations - 827 nm and 980 nm - were also applied but the photoluminescence occurred only on the background level.

Fig. 9a shows photoluminescence spectra of ENR layers doped with Tm^{3+} ions is spectral range from 1000 nm to 1200 nm. Fig.9b shows PL spectra of ENR doped with Tm^{3+} ions in spectral range from 1300 nm to 1600 nm under excitation 632.8 nm (He-Ne laser) under temperature 4 K. Fig. 9a shows maximum PL intensity around 1047 nm and Fig. 9b shows maximum PL intensity around 1420 nm but this photoluminescence is very weak.

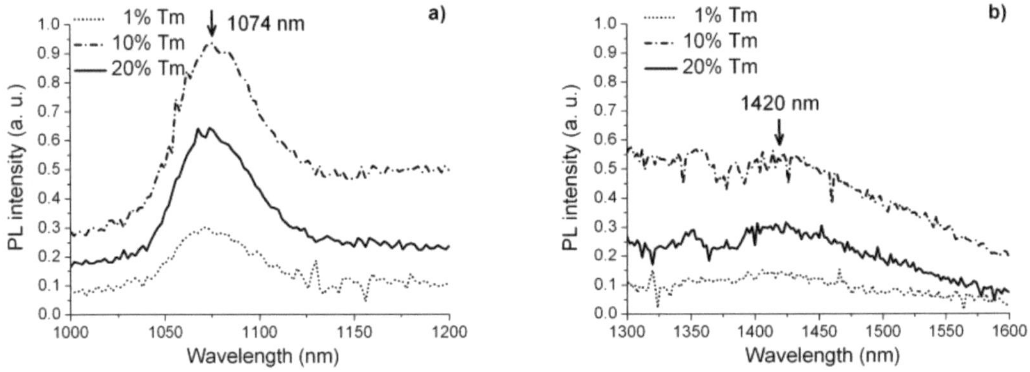

Fig. 9. Photoluminescence spectra of Tm^{3+} doped ENR

Fig. 10 shows photoluminescence spectra of ENR layers doped with Er^{3+} (Fig.10a) and Er^{3+} doped ions co-doped with Yb^{3+} ions (Fig.10b) in spectral range from 1450 nm to 1650 nm. Fig. 10a shows that only the samples with higher Er^{3+} concentrations showed strong photoluminescence bands at 1530 nm attributed to the erbium transition $^4I_{13/2} \rightarrow {}^4I_{15/2}$. The highest emission intensity was found out in the sample containing 10.0 at.% erbium. Fig. 10b shows typical 1530 nm Er^{3+} emission but co-doping with ytterbium ions had no effect on the photoluminescence spectra at all.

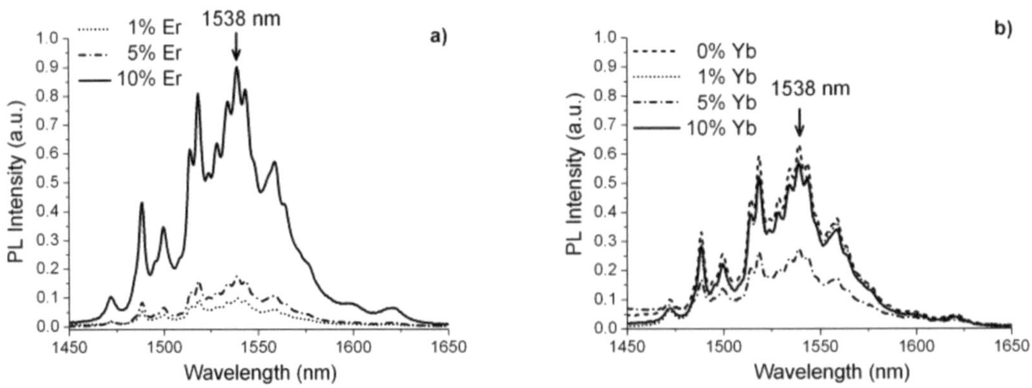

Fig. 10. Photoluminescence spectra of ENR layers doped with a) Er^{3+} ions and b) Er^{3+} ions co-doped with Yb^{3+} ions

Fig. 11 shows photoluminescence spectra of PMMA layers doped with Er^{3+} (Fig.11a) and Er^{3+} doped ions co-doped with Yb^{3+} ions (Fig.11b) under excitation 980 nm using semiconductor laser under room temperature. Fig. 11a shows the typical photoluminescence bands around 1540 nm attributed to the erbium transition $^4I_{13/2} \rightarrow {}^4I_{15/2}$ and Fig. 11a also shows that the increasing content of Er^{3+} ions increases also the photoluminescence intensity. The highest PL intensity observed so far had the sample containing 10.0 at.% of erbium. Fig. 11b shows that Yb^{3+} co-doping have positive influence.

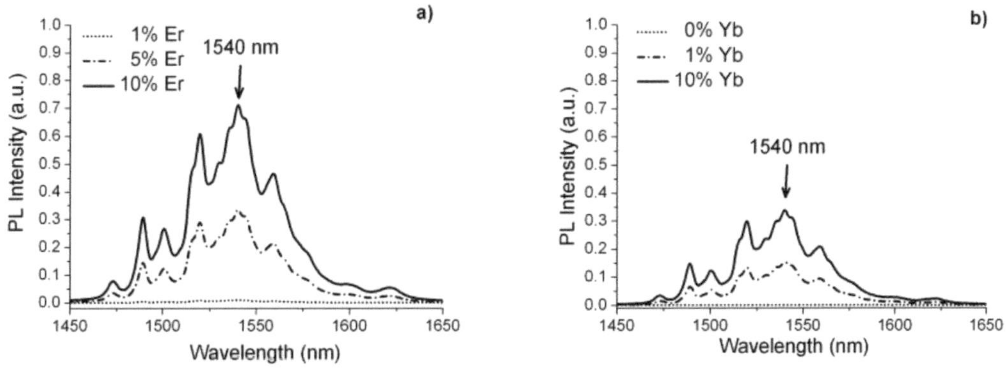

Fig. 11. Photoluminescence spectra of PMMA layers doped with a) Er^{3+} ions and b) Er^{3+} (1.0 at.%) ions co-doped with Yb^{3+} ions

4. Conclusion

We report on spectroscopic properties of the Polymethylmethacrylate and Epoxy Novolak Resin polymer doped with Rare Earth ions. Polymer layers were fabricated by a spin coating or by pouring the polymer into a bottomless mould placed on a quartz substrate.

The fabricated polymer layers doped with RE ions were examined by infrared spectroscopy and IR spectroscopy of the samples revealed absorption bands corresponding to the O-H vibrations in the region from 3340 cm^{-1} to 3380 cm^{-1}. Transmission measurements were performed in the spectral range from 350 nm to 1800 nm. The content of RE ions had a significant effect on the occurrence of the bands attributed to the RE transitions: while they were rather strong in the samples with higher RE concentration they almost vanished in the background in the case of the samples with low RE concentration.

We observed photoluminescence bands around 1300 nm of the ENR samples doped with Dy^{3+} ions caused by the dysprosium transition. ENR and PMMA layers doped with Er^{3+} ions had typical 1 530 nm emission due to the $^4I_{13/2} \rightarrow {}^4I_{15/2}$ transition. It was also found out that the addition of ytterbium did not substantially affect at PL spectra at ENR layers. One possibility how to explain this result is the presence of the O-H groups. The O-H groups are well-known as strong quencher of the Er^{3+} photoluminescence and fabrication samples with lower content of O-H groups will be the aim of the future research.

5. References

Beche, B.; Pelletier, N.; Gaviot, E.; Zyss, J.; (2005). Single-mode TE_{00}-TM_{00} optical waveguides on ENR polymer, *Optics Commmunications,* Vol. 230, (Jan 2005) page numbers 91-94, ISSN 0030-4018.

Bourdet, G.L.; Muller, R.A. (2000). Tm,Ho:YLF microchip laser under Ti:sapphire and diode pumping, *Applied Physics B,* Vol. 70, (Mar 2000) page numbers 345-349, ISSN 0946-2171.

Kenyon, A.J. (2002). Recent developments in rare-earth doped materials for optoelectronics, *Progress in Quantum Electronics,* Vol. 26, No. 4-5, (2002) page numbers 225-284, ISSN 0079-6727.

Liang, H.; Zheng, Z.Q.; Zhang, Q.J.; Ming, H.; Chen, B.; Xu, J.; Zhao, H.; (2003). Radiative properties of Eu(DBM)(3) Phen-doped Poly(methyl methacrylate), *Journal of Materials Research,* Vol. 18, No. 8, (Aug 2003) page numbers 1895-1899, ISSN 0884-2914.

Polman, A.; (1997). Erbium implanted thin film photonic materials, *Journal of Applied Physics,* Vol. 82, No. 1 (Jul 1997) page numbers 1-39, ISSN 0021-8979.

Prajzler, V.; Jerabek, V.; Lyutakov, O.; Huttel, I.; Spirkova, J.; Machovic, V.; Oswald, J.; Chvostova, D.; Zavadil, J.; (2008). Optical properties of erbium and erbium/ytterbium doped polymethylmethacrylate, *Acta Polytechnica Journal of Advanced Engineering,* Vol. 48, (2008) page numbers 14-21, ISSN 1210-2709.

Prajzler, V.; Huttel, I.; Lyutakov, O.; Spirkova, J.; Oswald, J.; Jerabek, V.; (2007). Infrared photoluminescence of Er^{3+} and Er^{3+}/Yb^{3+} doped Epoxy Novolak Resin, *International Journal of Microwave and Optical Technology,* Vol. 2, (2007) page numbers 236-241, ISSN 1553-0396.

Slooff, L.H.; van Blaaderen, A.; Polman, A.; Hebbink, G.A.; Klink, S.I.; (2002). Rare-earth doped polymers for planar optical amplifiers, *Applied Physics Reviews,* Vol. 91, No. 7 (Apr 2002) page numbers 3955-3980, ISSN 0021-8979.

Sosa, R.; Flores, M.; Rodriguez, R.; Munoz, A.; (2003). Optical properties and Judd-Ofelt intensity parameters of Eu^{3+} in PMMA : PAAc copolymer samples, *Revista Mexicana de Fisica,* Vol. 46, No. 6 (Dec 2003) page numbers 519-524, ISSN 0035-001X.

Tung, K.K.; Wong, W.H.; Pun, E.Y.B.; (2005). Polymer optical waveguides using direct ultraviolet photolithography process, *Applied Physics A-Materials Science&Processing,* Vol. 80, (Feb 2005) page numbers 621-626, ISSN 0947-8396.

Wong, S.F.; Pun, E.Y.B. (2002). Er^{3+}-Yb^{3+} codoped phosphate glass waveguide amplifier using Ag^{+}-Li^{+} ion exchange, *IEEE Photonics Technology Letters,* Vol. 14, (Jan 2002), page numbers 80-82, ISSN 1041-1135.

Zavada, J.M.; Zhang, D.; (1995). Luminescence properties of erbium in III-V compound semiconductors, *Solid-State Electronics,* Vol. 38, No. 7, (Jul 1995) page numbers 1285-1293, ISSN 0038-1101.

4

Pure $\chi^{(3)}$ Third-Harmonic Generation in Noncentrosymmetric Media

Kentaro Miyata
Chitose Institute of Science and Technology
Japan

1. Introduction

Recently, higher-order nonlinear effects have attracted great attention with the development of ultrafast laser technology. Using high peak-power lasers, phase-matched third-harmonic generation (THG) ($\omega + \omega + \omega \rightarrow 3\omega$) has been demonstrated in solids to directly obtain frequency-tripled output of the fundamental light in a simple way. The beginning of THG experiments using crystals was as early as 1960s. In this early stage of THG experiments, the centrosymmetric $CaCO_3$, calcite, was investigated by using a Q-switched ruby laser (Terhune et al., 1962, 1963).

Thanks to the invention of the mode-lock technique, ultrashort-pulse lasers became available, which made it possible to obtain stronger nonlinear interaction in crystals without laser-induced damage, since the damage threshold of materials increases as the laser pulse duration becomes short. With a mode-locked Nd-doped laser, Akhmanov et al. achieved phase-matched interactions for forth-harmonic generation (FHG) in $LiCOOH{\cdot}H_2O$ (Akhmanov et al., 1974) and fifth-harmonic generation in calcite (Akhmanov et al., 1975). Comparing the determined nonlinear susceptibilities relative to those for the lower-order ones, they have shown that electric susceptibilities for the higher-order processes decrease more rapidly with increasing the nonlinear order. In these works, the cascade processes of the lower-order nonlinearities in higher-harmonics generation were discussed in detail.

In addition, materials having large nonlinear electric susceptibilities have been developed for noncentrosymmetric crystals. Okada (1971) observed phase-matched THG of a Q-switched Nd:YAG laser in $LiIO_3$, reporting the effective third-order susceptibilities that are two orders of magnitude larger than those for KDP and ADP. The first observations of phase-matched THG (Chemla et al., 1974) and FHG (Kildal et al., 1979) in the transparent mid-IR region were made in $CdGeAs_2$ by using a CO_2 laser at 10.6 μm, where the effective third-order and fourth-order nonlinearities determined from the harmonic signals were compared with the theoretical estimates of the pure $\chi^{(3)}$ and $\chi^{(4)}$ components, respectively. Efficient THG was expected to be obtained from the determined $\chi^{(3)}$ while the large discrepancy encountered for the magnitudes of $\chi^{(4)}$ indicated the significant contribution from the cascade processes.

Finally, conversion efficiencies for THG in solids reached ~1% with the most widely used nonlinear crystals, β-BaB_2O_4 (BBO) and $KTiOPO_4$ (KTP). Qiu and Penzkofer (1988) attained a conversion efficiency of 0.8% in BBO for THG of a Nd:glass laser with a pulse duration of 5 ps, for which the crystal was irradiated by the intensity of 50 GW/cm^2. It was found that

compared with calcite (Penzkofer et al., 1988), this material possesses ~40 times larger effective nonlinearity with smaller walk-off and larger angular acceptance, indicating the superior properties for this application. With a chirped pulse amplification of the Nd:glass laser, Banks et al. (1999) have attempted to use higher pulse intensity with shorter pulse duration in BBO, taking into account the increase of the damage threshold. Focusing the 350-fs-pulse beam into the nonlinear crystal resulted in a highest conversion efficiency of ~6% for third-order frequency conversions in crystals, where the intensity of the fundamental beam was as high as ~200 GW/cm^2. In addition, they have succeeded to distinguish the cascade process from the direct process through the azimuth angle dependence of the output signals, and concluded that the contributions from the cascade processes to the phase-matched THG are significant in the overall conversion efficiencies (Banks et al., 2002).
The large third-harmonic (TH) conversion efficiencies in KTP have been achieved by two different groups, using tunable picosecond optical parametric systems. Boulanger and co-workers (Feve et al., 2002) produced the angularly noncritical phase-matched THG of 1.618 μm radiation with 2.4% efficiency while under the phase-matching condition for sum-frequency generation (SFG) ($\omega + 2\omega \rightarrow 3\omega$) in the similar spectral range, Takagi and Muraki (2000) have achieved a single-crystal TH efficiency of 5% that was five times larger than the phase-matched THG observed with the same fundamental source. According to the work of Boulanger et al (1999), the contribution from the cascade processes to phase-matched THG in KTP is much smaller compared to direct process. Recently, to completely eliminate the involved cascade process for the future quantum correlation experiment based on the three-photon downconversion, the same group used the centrosymmetric TiO_2 rutile for the single-crystal THG, demonstrating a significant enhancement of the cubic TH efficiency at the week input-power level (Gravier & Boulanger, 2006, 2007).
In this study, phase-matched THG in noncentrosymmetric media have been further investigated by using BiB_3O_6 (BIBO) (Miyata et al., 2008, 2009). The symmetry and birefringence analysis has revealed the existence of the phase-matching condition for the direct cubic process, where the cascading quadratic processes are precluded by zero effective nonlinearity. To understand the proposed pure cubic process, Section 2 discusses the cascade process that is generally involved in THG of noncentrosymmetric crystals. The next section presents the third-order frequency conversions in BIBO, including the first realized pure $\chi^{(3)}$ THG in such media. The discussion is separately given in Section 4 to generalize the present result to other materials. Finally, Section 5 concludes this chapter.

2. Cascade third-harmonic generation

In general, frequency tripling of a laser is carried out with two nonlinear crystals both under the phase-matching conditions. First crystal is used for generating the second harmonic (SH) of the fundamental source, and the second crystal is used for mixing the SH with the residual fundamental to create its TH. Since these crystals can be independently adjusted in the system, phase-matching conditions for the second-harmonic generation (SHG) and SFG processes are easily achieved by the angle or temperature tuning at the given fundamental wavelength. This two-step process, $\chi^{(2)}(3\omega, \omega, 2\omega):\chi^{(2)}(2\omega, \omega, \omega)$, exists also for THG observed in a single nonlinear crystal and occurs simultaneously with the direct third-order process mediated by the pure cubic nonlinearity $\chi^{(3)}(3\omega, \omega, \omega, \omega)$. Its phase-matching condition is identical with that for the direct THG as a whole and given by

$$\Delta k_{\mathrm{THG}} = \Delta k_{\mathrm{SHG}} + \Delta k_{\mathrm{SFG}} \tag{1}$$

where $\Delta k_{\mathrm{THG}} = k_{1a} + k_{1b} + k_{1c} - k_3$, $\Delta k_{\mathrm{SHG}} = k_{1a} + k_{1b} - k_2$, and $\Delta k_{\mathrm{SFG}} = k_{1c} + k_2 - k_3$. Thus, when the phase-matching condition for the overall process is fulfilled (i.e. $\Delta k_{\mathrm{THG}} = 0$), the individual SHG and SFG processes are generally not phase-matched inside the nonlinear crystal. Under this condition, the direct third-order and cascading second-order processes have the same order of magnitude and mutually interferes in a constructive or destructive way, provided the effective second-order nonlinearity is nonzero. Also under the phase-matching conditions for either SHG or SFG process, the same order of magnitude of THG becomes feasible. However, in this case, the contribution of the pure cubic nonlinearity to THG is negligibly small because the phase-matching conditions for THG are not fulfilled inside the nonlinear crystal.
Since our interest in the present study is the nonlinear interactions via cubic nonlinearity, consider the specific case, $\Delta k_{\mathrm{THG}} \to 0$, and equivalently, $\Delta k_{\mathrm{SHG}} \sim -\Delta k_{\mathrm{SFG}}$, for the THG process. Solving the nonlinear wave equations that are coupled among the fundamental, SH, and TH waves, under the slowly varying amplitude and fixed-field approximations, the overall TH conversion efficiency is found to be proportional to the square of phase-matching factor

$$\mathrm{sinc}\left(\frac{\Delta k_{\mathrm{THG}}\ell}{2}\right) \tag{2}$$

and effective nonlinear constant

$$c_{\mathrm{eff}} = c_{\mathrm{eff}}^{(3)} + \sum_{\mathrm{i}} c_{\mathrm{i,eff}}^{(2)} \tag{3}$$

with

$$c_{i,\mathrm{eff}}^{(2)} = \frac{D_{\mathrm{i}}^{(2)}}{D^{(3)}}\left(\frac{2\omega\, d_{\mathrm{i,eff}}^{\mathrm{SHG}}\, d_{\mathrm{i,eff}}^{\mathrm{SFG}}}{c n_{\mathrm{i},2\omega}\Delta k_{\mathrm{i,SHG}}}\right) \tag{4}$$

where ℓ is the crystal length and ω is the fundamental wavelength. The first term of Eq. (3) is the effective nonlinear constant calculated by third-order susceptibilities, and the second term is the sum of the cascading contributions calculated by Eq. (4), where $D^{(2)}$ and $D^{(3)}$ represent the degeneracy factors for SHG and THG, respectively, i.e. $D^{(2)} = 1$ or 2, and $D^{(3)} = 1$ or 3 for parallel or orthogonal polarization configuration of the fundamental, respectively. The d_{eff} is the effective second-order nonlinear constants at the given fundamental propagation direction. The subscript i indicates the cascade processes listed in Table 1, where the two eigenmodes for refractive indices n_s and n_f are defined as $n_s > n_f$ for the slow and fast waves, respectively. The typical value of $c_{\mathrm{eff}}^{(2)}$ with $d_{\mathrm{eff}} \neq 0$ and $\Delta k \neq 0$ are calculated to be in the range of 10^{-22}–10^{-24} $\mathrm{m^2/V^2}$ for oxide materials, which is almost same order of magnitude as that of $c_{\mathrm{eff}}^{(3)}$.

3. Third-order frequency conversions in BIBO

3.1 Effective nonlinear constants

Because of the lack of inversion symmetry, third-order frequency conversions in BIBO are also accompanied by cascading quadratic processes. To estimate the magnitude of those

Direct process		Cascade process		
Type	THG	i	SHG	SFG
1	$s_1 + s_1 + s_1 \rightarrow f_3$	1	$s_1 + s_1 \rightarrow s_2$	$s_1 + s_2 \rightarrow f_3$
		2	$s_1 + s_1 \rightarrow f_2$	$s_1 + f_2 \rightarrow f_3$
2	$f_1 + s_1 + s_1 \rightarrow f_3$	1	$s_1 + s_1 \rightarrow s_2$	$f_1 + s_2 \rightarrow f_3$
		2	$s_1 + s_1 \rightarrow f_2$	$f_1 + f_2 \rightarrow f_3$
		3	$f_1 + s_1 \rightarrow s_2$	$s_1 + s_2 \rightarrow f_3$
		4	$f_1 + s_1 \rightarrow f_2$	$s_1 + f_2 \rightarrow f_3$
3	$f_1 + f_1 + s_1 \rightarrow f_3$	1	$f_1 + s_1 \rightarrow s_2$	$f_1 + s_2 \rightarrow f_3$
		2	$f_1 + s_1 \rightarrow f_2$	$f_1 + f_2 \rightarrow f_3$
		3	$f_1 + f_1 \rightarrow s_2$	$s_1 + s_2 \rightarrow f_3$
		4	$f_1 + f_1 \rightarrow f_2$	$s_1 + f_2 \rightarrow f_3$

Table 1. Cascade process coupled with the direct type-1, type-2, and type-3 THG processes. The subscripts 1, 2, and 3 denote the fundamental, SH, and TH waves, respectively. The order of i is arbitrary.

processes in this noncentrosymmetric crystal, we first derive the effective second-order nonlinear constants.

The BIBO belongs to the monoclinic system with point symmetry 2, and the principal optical axis x ($n_x < n_y < n_z$) coincides with the crystallographic two-fold rotation axis b (Hellwig et al., 2000). Here, the tensor elements for the second-order susceptibilities $\chi_{ijk}(\omega_1 + \omega_2, \omega_1, \omega_2)$ are expressed with the optical coordinate system xyz for convenient use in the frequency conversion experiments, i.e. each subscript i, j, and k is defined to take the value, 1 = x, 2 = y, or 3 = z. Applying the symmetry operation of the two-fold rotation axis and using a contracted notation for the last two indices, i.e. l = (1, 2, 3, 4, 5, 6) = (xx, yy, zz, yz, zx, xy), the zero and nonzero tensor elements for second-order nonlinear constants ($d_{ijk} = 1/2\chi_{ijk}$) of BIBO are determined as follows:

$$d_{il} = \begin{pmatrix} d_{11} & d_{12} & d_{13} & d_{14} & 0 & 0 \\ 0 & 0 & 0 & 0 & d_{25} & d_{26} \\ 0 & 0 & 0 & 0 & d_{35} & d_{36} \end{pmatrix} \tag{5}$$

When the Kleinman symmetry is applied, $d_{12} = d_{26}$, $d_{13} = d_{35}$, and $d_{14} = d_{25} = d_{36}$ hold in the above expressions. The number of independent coefficients is, therefore, reduced from 8 to 4. Note that although Kleinman symmetry is sometimes reported to be violated in crystals, this symmetry becomes an excellent approximation in the transparent range.

The absolute values and relative signs of the nonzero second-order nonlinear constants for BIBO have been investigated by Hellwig et al. (1999, 2000) with a Maker fringe method using a quasi-cw Nd:YAP laser at 1.0795 μm. Since each tensor element presented in their work is represented by the crystallophysical system XYZ, the transformation of these tensors to the optical coordinate system xyz is required with the consideration of orientation of the principal optical axes (Hellwing et al., 2000). For instance, the corrected results for SHG of the 1.0795 μm radiation at 20°C in the xyz frame are listed in Table 2.

d_{il} (pm/V)		d_{il} (pm/V)	
d_{11}	± 2.54	d_{25}	± 1.70
d_{12}	± 2.95	d_{26}	± 3.49
d_{13}	∓ 1.94	d_{35}	∓ 1.58
d_{14}	± 1.64	d_{36}	± 1.73

Table 2. Second-order nonlinear constants of BIBO for SHG of the 1.0795 μm radiation in the optical coordinate system *xyz*.

To obtain the effective nonlinear constant d_{eff} for the given propagation direction, the projection of light polarization vectors to the independent nonlinear constants d_{il} has to be first performed. Then, multiplying d_{il} with the determined electric-field components, the effective nonlinear constants are calculated for each interaction type. For biaxial crystals, their analytical forms for arbitrary propagation directions are always complicated without approximations, but simplified by restricting the propagation directions in the principal planes. When Klienman symmetry and no spatial walk-off approximations are applied to this case, $d_{\mathrm{eff}}(\omega_3{}^f = \omega_1{}^s + \omega_2{}^s) = d_{\mathrm{eff}}(\omega_3{}^s = \omega_1{}^s + \omega_2{}^f) = d_{\mathrm{eff}}(\omega_3{}^s = \omega_1{}^f + \omega_2{}^s)$ and $d_{\mathrm{eff}}(\omega_3{}^f = \omega_1{}^s + \omega_2{}^f) = d_{\mathrm{eff}}(\omega_3{}^f = \omega_1{}^f + \omega_2{}^s) = d_{\mathrm{eff}}(\omega_3{}^s = \omega_1{}^f + \omega_2{}^s)$ hold at the given propagation direction, i.e. the permutation of polarizations becomes possible. Finally, introducing the notations *o* and *e* for the interacting waves with polarizations normal and parallel to the principal plane, respectively, the expressions for effective nonlinear constants of BIBO are reduced as

$$\begin{aligned} d_{\mathrm{eff}}^{sss}(=d_{\mathrm{eff}}^{ooo}) &= 0 \\ d_{\mathrm{eff}}^{ssf}(=d_{\mathrm{eff}}^{ooe}) &= d_{13}\sin\phi \\ d_{\mathrm{eff}}^{sff}(=d_{\mathrm{eff}}^{oee}) &= d_{14}\sin 2\phi \\ d_{\mathrm{eff}}^{fff}(=d_{\mathrm{eff}}^{eee}) &= (d_{11}\sin^2\phi + 3d_{12}\cos^2\phi)\sin\phi \end{aligned} \tag{6-1}$$

in the *xy* plane,

$$\begin{aligned} d_{\mathrm{eff}}^{sss}(=d_{\mathrm{eff}}^{eee}) &= 0 \\ d_{\mathrm{eff}}^{ssf}(=d_{\mathrm{eff}}^{eeo}) &= \pm d_{13}\sin^2\theta \pm d_{12}\cos^2\theta - d_{14}\sin 2\theta \\ d_{\mathrm{eff}}^{sff}(=d_{\mathrm{eff}}^{eoo}) &= 0 \\ d_{\mathrm{eff}}^{fff}(=d_{\mathrm{eff}}^{ooo}) &= \pm d_{11} \end{aligned} \tag{6-2}$$

in the *yz* plane ($\phi = \pm 90°$),

$$\begin{aligned} d_{\mathrm{eff}}^{sss}(=d_{\mathrm{eff}}^{ooo}) &= 0 \\ d_{\mathrm{eff}}^{ssf}(=d_{\mathrm{eff}}^{ooe}) &= d_{12}\cos\theta \\ d_{\mathrm{eff}}^{sff}(=d_{\mathrm{eff}}^{oee}) &= -/+ d_{14}\sin 2\theta \\ d_{\mathrm{eff}}^{fff}(=d_{\mathrm{eff}}^{eee}) &= (d_{11}\cos^2\theta + 3d_{13}\sin^2\theta)\cos\theta \end{aligned} \tag{6-2}$$

in the *zx* plane for ($\phi = 0°/180°$, $0° \le \theta < \Omega_z$ or $180° - \Omega_z < \theta \le 180°$), and

$$\begin{aligned} d_{\mathrm{eff}}^{sss}(=d_{\mathrm{eff}}^{eee}) &= +/-(d_{11}\cos^2\theta + 3d_{13}\sin^2\theta)\cos\theta \\ d_{\mathrm{eff}}^{ssf}(=d_{\mathrm{eff}}^{eeo}) &= d_{14}\sin 2\theta \\ d_{\mathrm{eff}}^{sff}(=d_{\mathrm{eff}}^{eoo}) &= +/- d_{12}\cos\theta \\ d_{\mathrm{eff}}^{fff}(=d_{\mathrm{eff}}^{ooo}) &= 0 \end{aligned} \tag{6-4}$$

in the zx plane for ($\phi = 0°/180°$, $\Omega_z < \theta < 180° - \Omega_z$), where θ is the polar angle referring to z in the range of $0° \le \theta \le 180°$ and ϕ is the azimuth angle referring to x in the range of $-180° \le \phi \le 180°$. The general forms for the type-1 and type-2 interactions outside the principal planes have been obtained by Tzankov and Petrov (2005) by neglecting the dispersion of optic axis angles. Note that the signs of the above equations are in an opposite relation with respect to those given by them, owing to the different definition of the electric polarization state.

Similarly, apart from the cascading effects of the second-order processes given in the preceding, the effective nonlinearity for the third-order frequency conversion is directly determined from the third-order nonlinear susceptibilities χ_{ijkl} ($\omega_1 + \omega_2 + \omega_3$; ω_1, ω_2, ω_3), where each subscripts i, j, and k is defined to take the value, $1 = x$, $2 = y$, or $3 = z$. The zero and nonzero tensor elements for the third-order nonlinear constants ($c_{ijkl} = 1/4\chi_{ijkl}$) of BIBO, with a contracted notation, m = (1, 2, 3, 4, 5, 6, 7, 8, 9, 0) = (xxx, yyy, zzz, yzz, yyz, xzz, xxz, xyy, xxy, xyz), is given by

$$c_{\text{im}} = \begin{pmatrix} c_{11} & 0 & 0 & 0 & 0 & c_{16} & 0 & c_{18} & 0 & c_{10} \\ 0 & c_{22} & c_{23} & c_{24} & c_{25} & 0 & c_{27} & 0 & c_{29} & 0 \\ 0 & c_{32} & c_{33} & c_{34} & c_{35} & 0 & c_{37} & 0 & c_{39} & 0 \end{pmatrix}, \tag{7}$$

indicating the nonzero 9 and 16 elements with and without Klienman symmetry condition, respectively (i.e. $c_{16} = c_{37}$, $c_{18} = c_{29}$, $c_{10} = c_{27} = c_{39}$, $c_{23} = c_{34}$, $c_{24} = c_{35}$, $c_{25} = c_{32}$ under the Klienman symmetry condition). From the reported nonlinear refractive indices (Miller et al., 2008), these values are expected to be larger than those of BBO and is comparable to those of KTP and $LiNbO_3$.

Again, with Klienman symmetry and no spatial walk-off approximations, the corresponding effective nonlinear constants in the principal plane of BIBO are given by

$$\begin{aligned} c_{\text{eff}}^{ssss} (= c_{\text{eff}}^{oooo}) &= c_{33} \\ c_{\text{eff}}^{sssf} (= c_{\text{eff}}^{oooe}) &= c_{23} \cos\phi \\ c_{\text{eff}}^{ssff} (= c_{\text{eff}}^{ooee}) &= c_{16} \sin^2\phi + c_{24} \cos^2\phi \\ c_{\text{eff}}^{sfff} (= c_{\text{eff}}^{oeee}) &= (c_{25} \cos^2\phi + 3c_{10} \sin^2\phi) \cos\phi \\ c_{\text{eff}}^{ffff} (= c_{\text{eff}}^{eeee}) &= c_{11} \sin^4\phi + c_{22} \cos^4\phi + \frac{3}{2} c_{18} \sin^2 2\phi \end{aligned} \tag{8-1}$$

in the xy plane,

$$\begin{aligned} c_{\text{eff}}^{ssss} (= c_{\text{eff}}^{eeee}) &= c_{22} \cos^4\theta + c_{33} \sin^4\theta + \frac{3}{2} c_{24} \sin^2 2\theta \\ &\quad \mp 2(c_{23} \sin^2\theta + c_{25} \cos^2\theta) \sin 2\theta \\ c_{\text{eff}}^{sssf} (= c_{\text{eff}}^{eeeo}) &= 0 \\ c_{\text{eff}}^{ssff} (= c_{\text{eff}}^{eeoo}) &= c_{16} \sin^2\theta + c_{18} \cos^2\theta \mp c_{10} \sin 2\theta \\ c_{\text{eff}}^{sfff} (= c_{\text{eff}}^{eooo}) &= 0 \\ c_{\text{eff}}^{ffff} (= c_{\text{eff}}^{oooo}) &= c_{11} \end{aligned} \tag{8-2}$$

in the yz plane ($\phi = \pm 90°$),

$$\begin{aligned}
&c_{\text{eff}}^{ssss}(=c_{\text{eff}}^{oooo})=c_{22}\\
&c_{\text{eff}}^{sssf}(=c_{\text{eff}}^{oooe})=-/+c_{25}\sin\theta\\
&c_{\text{eff}}^{ssff}(=c_{\text{eff}}^{ooee})=c_{24}\sin^2\theta+c_{18}\cos^2\theta\\
&c_{\text{eff}}^{sfff}(=c_{\text{eff}}^{oeee})=-/+(c_{23}\sin^2\theta+3c_{10}\cos^2\theta)\sin\theta\\
&c_{\text{eff}}^{ffff}(=c_{\text{eff}}^{eeee})=c_{11}\cos^4\theta+c_{33}\sin^4\theta+\frac{3}{2}c_{16}\sin^2 2\theta
\end{aligned} \tag{8-3}$$

in the zx plane ($\phi = 0°/180°$, $0° \leq \theta < \Omega_z$ or $180° - \Omega_z < \theta \leq 180°$), and

$$\begin{aligned}
&c_{\text{eff}}^{ssss}(=c_{\text{eff}}^{eeee})=c_{11}\cos^4\theta+c_{33}\sin^4\theta+\frac{3}{2}c_{16}\sin^2 2\theta\\
&c_{\text{eff}}^{sssf}(=c_{\text{eff}}^{eeeo})=+/-(c_{23}\sin^2\theta+3c_{10}\cos^2\theta)\sin\theta\\
&c_{\text{eff}}^{ssff}(=c_{\text{eff}}^{eeoo})=c_{24}\sin^2\theta+c_{18}\cos^2\theta\\
&c_{\text{eff}}^{sfff}(=c_{\text{eff}}^{eooo})=+/-c_{25}\sin\theta\\
&c_{\text{eff}}^{ffff}(=c_{\text{eff}}^{oooo})=c_{22}
\end{aligned} \tag{8-4}$$

in the zx plane for ($\phi = 0°/180°$, $\Omega_z < \theta < 180° - \Omega_z$), where $c_{\text{eff}}^{sssf}=c_{\text{eff}}^{ssfs}=c_{\text{eff}}^{sfss}=c_{\text{eff}}^{fsss}$, and $c_{\text{eff}}^{ssff}=c_{\text{eff}}^{sfsf}=c_{\text{eff}}^{fssf}=c_{\text{eff}}^{fsfs}=c_{\text{eff}}^{ffss}$, $c_{\text{eff}}^{sfff}=c_{\text{eff}}^{fsff}=c_{\text{eff}}^{ffsf}=c_{\text{eff}}^{fffs}$. The signs of the above equations are compatible with those of Eqs. (6). Note that while the absolute signs are not important for the nonlinear applications, the relative signs among interaction types play a crucial role in a situation where more than one nonlinear processes are optically coupled such as the cascading second-order and direct third-order processes for THG described in this chapter.

3.1 Third-harmonic generation at 0.3547 μm

The THG experiment was carried out in the three principal planes of BIBO by using a Q-switched Nd:YAG laser operating at 1.0642 μm at 10 Hz (Miyata et al., 2008). The 5 × 5 × 5 mm³ BIBO samples (Fujian Castech Crystal Inc.) used in the experiments were cut at $\theta = 90°$, $\phi = 35.0°$ in the xy plane, $\theta = 59.8°$, $\phi = +90°$ ($\theta = 30.2°$, $\phi = -90°$) in the yz plane, and $\theta = 55.3°$, $\phi = 0°$ in the zx planes, respectively, where θ and ϕ are polar and azimuth angles referring to z and x, respectively. The cut angle for the yz plane was obtained at 1.0642 μm at room temperature. A fused silica prism was used to separate the generated TH beam from the other beams.

Type	Plane	Process	PM loci (deg) (θ, ϕ)		$\Delta\theta_{ext}\cdot\ell$	$\Delta T\cdot\ell$ (°C·cm)	
			Cal.	Exp.	(mrad·cm)	Cal.	Exp.
1	yz	$e_1 + e_1 + e_1 \rightarrow o_3$	(45.6, +/−90)	-	0.44	1.1/0.8	- / -
	zx	$e_1 + e_1 + e_1 \rightarrow o_3$	(67.5, 0)	(67.9, 0)	0.48	2.3	2.1
2	xy	$e_1 + o_1 + o_1 \rightarrow e_3$	(90, 46.7)	(90, 46.7)	1.75[a]	1.3	1.3
	yz	$o_1 + e_1 + e_1 \rightarrow o_3$	(66.4, +/−90)	(66.7, +/−90)	0.83	1.1/0.9	1.2/ -

Table 3. Phase-matching loci and the corresponding angular and temperature acceptance bandwidths (FWHM) for THG at 0.3547 μm in BIBO at 20°C. [a] $\Delta\phi_{ext}\cdot\ell$ (mrad·cm).

Increasing the input power of the Nd:YAG laser and adjusting the polarization with a retardation plate, the UV signal was easily observed below the damage threshold of this material. The results for the phase-matching loci and the corresponding temperature bandwidths (FWHM) are tabulated in Table 3 together with the theoretical values calculated by the revised Sellmeier and thermo-optic dispersion formulas presented by Miyata et al. (2009). For the present processes, type-3 interactions are not phase-matchable. Some experimental data in the *yz* plane were not taken because of no effective nonlinearity or no availability of suitable BIBO sample. This table also lists the calculated angular bandwidths, which are used for deriving the temperature bandwidths from the variations of the measured phase-matching angles between 20 and 110 °C. As can be seen, the agreement between theory and experiment is good.

It should be pointed out that in this experiment, unphase-matched SHG at 0.5321 μm was observed with the TH signal, indicating the cascade process involved in the present THG processes. Table 4 lists the magnitude of each cascade process. It is found that there is no direct third-order and cascading second-order contribution to type-1 THG in the *yz* plane. For the other processes, cascading contributions exist and therefore, the observed THG outputs at 0.3547 μm are considered as sum of the direct and cascade processes, as is evident from the observed SHG. From this table, total effective nonlinear constants for the cascade processes are determined to be $\Sigma c_{eff}^{(2)}$ = 51.1 pm²/V² for type-1 THG in the *zx* plane, and $\Sigma c_{eff}^{(2)}$ = 12.8 and −30.9/1.4 pm²/V² for type-2 THG in the *xy* and *yz* (ϕ = +90°/−90°) planes, respectively.

Direct cubic process				Cascade process parameters			
Type	Plane	Process	PM loci (deg) (θ, ϕ)	i	Δk_{SHG} (cm⁻¹)	$d_{eff}^{SHG}d_{eff}^{SFG}$ (pm²/V²)	$c_{eff}^{(2)}$ (pm²/V²)
1	*yz*	$e_1 + e_1 + e_1 \rightarrow o_3$	(45.6, +/−90)	1	–4840	0	0
				2	7065	0	0
	zx	$e_1 + e_1 + e_1 \rightarrow o_3$	(67.9*, 0)	1	–5171	–2.76	32.5
				2	8354	2.39	18.6
2	*xy*	$e_1 + o_1 + o_1 \rightarrow e_3$	(90, 46.7*)	1	–5501	0	0
				2	13450	–7.27	–11.8
				3	–14196	2.45	–6.9
				4	4755	3.43	31.5
	zx	$o_1 + e_1 + e_1 \rightarrow o_3$	(66.7*, +/−90)	1	–5271	0	0
				2	12427	–6.84/1.10	–12.1/2.0
				3	–13314	6.16/0.18	–18.8/–0.6
				4	4384	0	0

Table 4. Cascade process parameters for THG at 0.3547 μm with Δk_{THG}=0 in the principal planes of BIBO at 20°C. * Experimental value.

3.2 Pure $\chi^{(3)}$ third-harmonic generation

Next, 90° phase-matched THG was investigated along the principal axes (Miyata et al., 2009). It is found from Eqs. (5) to (8) that at the propagation direction along *x*, all the second-order susceptibilities vanish whereas all the involved third-order ones remain to be non-zero, indicating the existence of pure cubic processes along this direction, i.e. any cascading quadratic processes are effectively precluded from the nonlinear processes as in the case of centrosymmetric crystals. In addition, since the large birefringence between n_y and n_z has been found to exist along x, it also indicates the possibility of phase-matching for the pure cubic processes in BIBO. The corresponding effective third-order nonlinear constants for the type-1, type-2, and type-3 processes are expressed as $c_{eff}^{(3)} = c_{23}$, c_{24}, and c_{25} respectively.

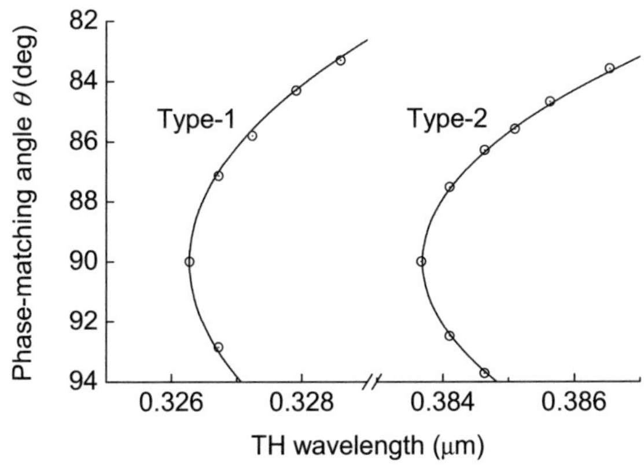

Fig. 1. Phase-matching curves for direct type-1 and type-2 THG in the zx plane of BIBO at 20°C. The circles are our experimental points.

Using the outputs of a KTP OPO pumped by the SH of the same Nd:YAG laser, the phase-matching conditions for the direct type-1 and type-2 THG were checked in the zx plane. Fig. 1 shows the experimental results obtained with a 1.5-cm-long, x-cut BIBO crystal (Crystech Inc.) at 20°C. The solid curve is calculated with the Sellmeier equations and correctly reproduces the experimental points. As can be seen, along the proposed direction x ($\theta = 90°$), phase-matched THG was realized at 0.3263 and 0.3837 μm for the type-1 and type-2 interactions, respectively, and no signal of unphase-matched SH was detected, which indicates the absence of possible energy transfer through the SFG ($\omega + 2\omega \rightarrow 3\omega$) process assisted by the non-phase-matched SH, and confirms the generation of the TH purely mediated by cubic nonlinearity.

To investigate further the phase-matching properties for the pure cubic process, the temperature-tuning of phase-matched THG wavelengths was next performed by heating the crystal up to ~120°C with a temperature-controlled copper oven. The experimental points for the type-1 and type-2 processes measured along x of the same sample are plotted in Fig. 2 together with the theoretical curves calculated with the Sellmeier and thermo-optic dispersion formulas. The temperature-tuning rates derived from these experimental points are $d\lambda_{THG}/dT$ = 0.019 and 0.021 nm/°C for the type-1 and type-2 processes, respectively, which are in excellent agreement with the calculated values of $d\lambda_{THG}/dT$ = 0.019 and 0.023 nm/°C. For the type-3 process, these formulas predict retracing behaviour and two phase-matching wavelengths of λ_{THG} = 0.5994 and 0.8186 μm with $d\lambda_{THG}/dT$ = 0.070 and –0.073 nm/°C, respectively. Unfortunately, the experimental verification was precluded because of the limited tunability of the light source. Note that unphase-matched SHG was not observed during the measurement. The phase-matching properties for these pure cubic processes are summarized in Table 5.

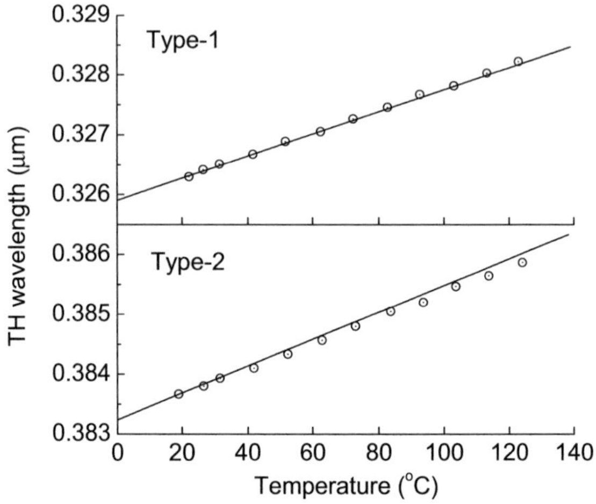

Fig. 2. Temperature-tuned phase-matching curves for direct type-1 and type-2 THG along x of BIBO. The circles are our experimental points.

Type	Process	λ_{THG} (μm)	$\Delta\theta_{ext}\cdot\ell^{1/2}$ (mrad·cm)	$\Delta\phi_{ext}\cdot\ell^{1/2}$ (mrad·cm)	$\Delta\lambda_1\cdot\ell$ (nm·cm)	$\Delta T\cdot\ell$ (°C·cm)
1	$z_1 + z_1 + z_1 \rightarrow y_3$	0.3263*	1.96	3.77	0.08	1.4*
2	$y_1 + z_1 + z_1 \rightarrow y_3$	0.3837*	2.55	4.98	0.20	3.1*
3	$y_1 + y_1 + z_1 \rightarrow y_3$	0.5994	4.45	8.70	3.11	14.9
		0.8186	5.28	10.13	6.01	27.6

Table 5. Phase-matched pure $\chi^{(3)}$ THG wavelengths and the corresponding angular, spectral, and temperature acceptance bandwidths (FWHM) along x of BIBO at 20°C. * Experimental value.

3.3 Three-photon downconversion

From the results given in the preceding, phase-matching interactions for the other types of pure cubic processes are expected to be obtained along the x axis. Here, we consider the three-photon downconversion processes ($\omega_a + \omega_b + \omega_c = \omega_p$). Fig. 3 shows the corresponding phase-matching curves for the type-1 interaction at 20°C. As can be seen, the spectral range of the pump beam is somewhat limited, which is attributed to the UV transmission cutoff wavelength of ~ 0.270 μm (Teng et al., 2001) and the symmetrical relations of the curves among three down-converted waves. The longest pump wavelength is fixed at 0.9789 μm, corresponding to the phase-matched fundamental wavelength for type-1 THG.

In contrast, the tuning curves for the type-2 and type-3 processes give asymmetry relations, resulting in the wide spectral range of the pump beam, as shown in Fig. 4. Thus, for example, using the Nd:YAG laser at 1.0642 μm as the fundamental source in the system, it is possible to pump with the SH (λ_p = 0.5321 μm) or TH (λ_p = 0.3547 μm) beams for the type-2 process, and with the fundamental, SH, or TH beams for the type-3 process. Similarly, the Ti:sapphire laser can be also used as the fundamental source for these processes.

It should be pointed out that at the given pump wavelength, the tuning curves cover the very broad spectral ranges of down-converted waves, except the vicinity of the three-wavelength

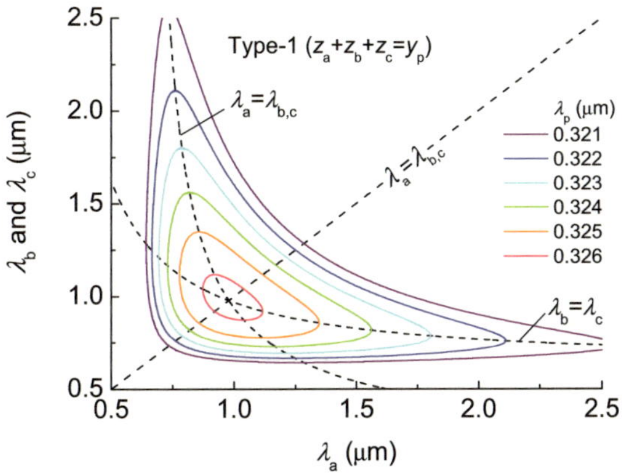

Fig. 3. 90° phase-matching curves for type-1 downconversion processes ($\omega_a + \omega_b + \omega_c = \omega_p$) along x of BIBO at 20°C. The crossing point of the dashed curves corresponds to the degeneracy point.

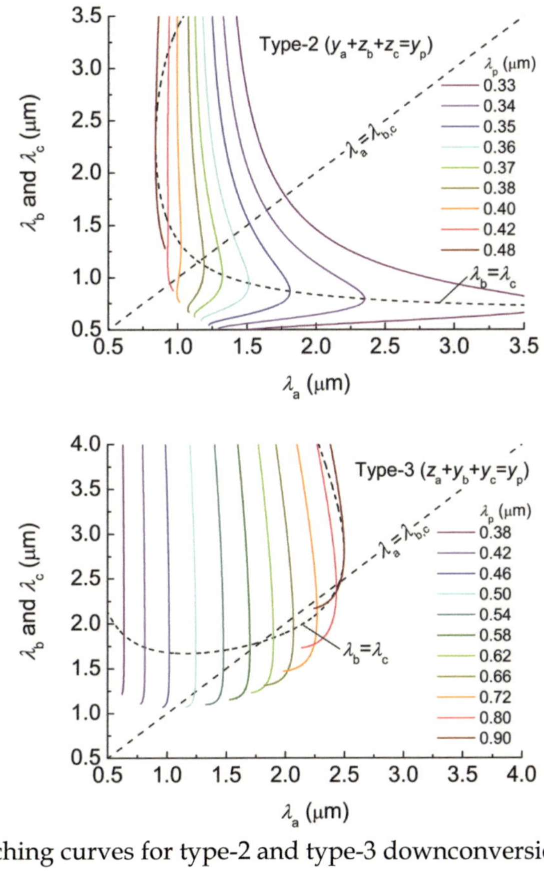

Fig. 4. 90° phase-matching curves for type-2 and type-3 downconversion processes ($\omega_a + \omega_b + \omega_c = \omega_p$) along x of BIBO at 20°C. The crossing point of the dashed curves corresponds to the degeneracy point.

degeneracy point ($\lambda_a = \lambda_b = \lambda_c = 3\lambda_p$) for the type-1 process. This indicates the possibility of unltrabroadband pulse generation with the proper selection of the wavelength of the seed beam. Especially, the interesting points are located at the two-wavelength degeneracy points that are shown with the dashed curves in the figures. It is found that in contrast to the second-order process, two-wavelength degeneracy points are generally obtained at the given crystal orientation and pump wavelength for the third-order processes.

4. Discussion

Recently, some attempts for reducing the cascading quadratic contribution in noncentrosymmetric crystals have been made by Boulanger and co-workers. Their motivation is the quantum properties of three photons that are created by direct cubic downconversion process (Felbinger et al., 1998). They have stated that the cascading quadratic processes are detrimental factor in the quantum correlation experiments based on the cubic nonlinearity, because of the different quantum properties of cascade and direct processes. With the low cascading contribution relative to the direct cubic process, i.e. $[\chi^{(2)}:\chi^{(2)}/\chi^{(3)}]^2$ = 10% (Boulanger et al., 1999), they have obtained the efficient THG along the propagation direction x of KTP (Feve et al., 2000). Their further research has led to the nearly pure cubic difference-frequency generation with $[\chi^{(2)}:\chi^{(2)}/\chi^{(3)}]^2$ = 0.5% in the same material (Douady & Boulanger, 2004, 2005), which is attributed to the relative sign and amount of the phase-mismatching factor for each non-zero second-order process (see Eqs. (3) and (4)). However, this approach gives the pure $\chi^{(3)}$ condition only for a specific frequency- conversion process either with a specific phase-matching configuration (Douady & Boulanger, 2005) or with the help of periodical poling to the material (Feve & Boulanger, 2002).

In this study, pure $\chi^{(3)}$ THG was realized along x of BIBO. The proposed condition, which simultaneously satisfies the two requirements, (1) nonlinear optical coupling of zero second-order and non-zero third-order processes and (2) birefringence property, was given as a result of the symmetry operation of two-fold rotation axis. Provided that the corresponding cubic nonlinearity doesn't vanish owing to the symmetry operations of the other symmetry elements, any frequency-conversion schemes are considered to be purely cubic for propagation along the two-fold rotation axis. It is clear that this is equivalent to using a non-zero cubic interaction in a centrosymmetric crystal under the 90° phase-matching condition, as was demonstrated in TiO_2 rutile by Gravier and Boulanger (2006, 2007) for the direct type-2 THG.

Finally, note that third-order frequency downconversions have been also achieved in centrosymmetric CaF_2 and BaF_2 with non-collinear geometry for widely tunable, ultrashort-pulse IR generation. (Okamoto & Tasumi, 1995; Nienhuys et al., 2001). Their phase-matching conditions in the transparent range can be fulfilled only by four-wave mixing, $\omega_1 + \omega_2 = \omega_3 + \omega_4$, owing to the absence of birefringence property. Regardless of the presence of inversion centre, the birefringence requirement of phase-matching for the present third-order processes, $\omega_1 + \omega_2 + \omega_3 = \omega_4$, does not allow the use of isotropic crystals and an optic axis direction of anisotropic crystals without manipulation of the material structure. Consequently, it is found from the symmetry requirements that with the suitable birefringence, the pure cubic process under the phase-matching condition, $k_1 + k_2 + k_3 - k_4 = 0$ ($k > 0$), can be attained in nine and eight point symmetry classes of centrosymmetric and noncentrosymmetric crystals, respectively (see Table 6), while for the latter, the approach proposed by Douady and Boulanger (2005) may give additional material choices without restriction of symmetry.

stem	Point symmetry	
	Centrosymmetric	Noncentrosymmetric
Triclinic	$\overline{1}$	-
Monoclinic	$2/m$	2
Orthorhombic	mmm	$222, mm2$
Tetragonal	$4/m, 4/mmm$	$422, \overline{4}2m$
Trigonal	$\overline{3}, \overline{3}m$	32
Hexagonal	$6/m, 6/mmm$	$622, \overline{6}m2$

Table 6. Point symmetry classes of centrosymmetric and non-centrosymmetric, non-cubic crystals with zero second-order and nonzero third-order nonlinearities.

5. Conclusion

Because of the multi-photon interactions, numerous interesting applications can be found for higher-order processes. While the cascade process of the lower-order nonlinearity interferes with the direct process in noncentrosymmetric media, its magnitudes can be controlled by the phase-mismatching factor and effective nonlinear constants, and it can be a beneficial or harmful effect, depending on the purpose. Enhancement of the cascade process provides high conversion efficiency in the overall process, whereas its suppression or its separation from the direct process becomes important whenever the quantum properties for higher-order processes are investigated.

This study revealed that even under no inversion symmetry, phase-matched pure $\chi^{(3)}$ THG can be obtained along the two-fold rotation axis, and other pure cubic interactions expressed by $\omega_1 + \omega_2 + \omega_3 = \omega_4$ are also possible at the same propagation direction, with the suitable birefirengence. This new result in nonlinear optics is important not only from the fundamental point of view, but also for its practical applications to the measurements on the nonlinear susceptibilities and quantum properties for higher-order processes in noncentrosymmetric media.

6. References

Akhmanov S. A.; Duvobic A. N.; Saltiel S. M.; Tomov I. V. & Tunkin V. G. (1974). Nonlinear optical effects of fourth order in the field in a lithium formate crystal, *JETP Lett.*, Vol. 20, pp. 117-118.

Akhmanov S. A.; Martynov X. A.; Saltiel S. M.; & V. G. Tunkin (1975). Observation of nonresonant six-photon processes in a calcite crystal, *JETP Lett.*, Vol. 22, pp. 65-67.

Banks P. S.; Feit M. D.; & Perry M. D. (1999). High-intensity third-harmonic generation in beta barium borate through second-order and third-order susceptibilities, *Opt. Lett.*, Vol. 24, pp. 4-6.

Banks P. S.; Feit M. D. & Perry M. D. (2002). High-intensity third-harmonic generation, *J. Opt. Soc. Am. B*, Vol. 19, pp. 102-118.

Boulanger B.; Feve J. P.; Delarue P.; Rousseau I. & Marnier G. (1999). Cubic optical nonlinearity of $KTiOPO_4$, *J. Phys. B: Atm. Mol. Opt. Phys.*, Vol. 32, pp. 475-488.

Chemla D. S.; Begley R. F.; & R. L. Byer (1974). Experimental and theoretical studies of third-harmonic generation in the chalcopyrite $CdGeAs_2$, *IEEE J. Quantum Electron.* QE-10, pp. 71-81.

Douady J. & Boulanger B. (2004). Experimental demonstration of a pure third-order optical parametric downconversion process, *Opt. Lett.* Vol. 29, pp. 2794-2796.

Douady J. & Boulanger B. (2005). Calculation of quadratic cascading contributions associated with a phase-matched cubic frequency difference generation in a $KTiOPO_4$ crystal, *J. Opt. A: Pure and Applied Optics,* Vol. 7, pp. 467-471.

Feve J. P.; Boulanger B. & Guillien Y. (2000). Efficient energy conversion for cubic third-harmonic generation that is phase matched in $KTiOPO_4$, *Opt. Lett.* Vol. 25, pp. 1373-1375.

Feve J. P. & Boulanger B. (2002). Suppression of quadratic cascading in four-photon interactions using periodically poled media, *Phys. Rev. A*, Vol. 65, pp. 063814-063814-6.

Felbinger T.; Schiller S. & Mlynek J. (1998). Oscillation and generation of nonclassical states in three-photon down-conversion, *Phy. Rev. Lett.*, Vol. 80, pp. 492-495.

Gravier F. & Boulanger B. (2006). Cubic parametric frequency generation in rutile single crystal, *Opt. Express*, Vol. 14, pp. 11715-11720.

Gravier F. & Boulanger B. (2007). Third order frequency generation in TiO_2 rutile and $KTiOPO_4$, *Opt. Mater.*, Vol. 30, pp. 33-36.

Hellwig H.; Liebertz J. & Bohaty L. (1999). Exceptional large nonlinear optical coefficients in the monoclinic bismuth borate BiB_3O_6 (BIBO), *Solid State Commun.*, Vol. 109, pp. 249-251.

Hellwig H.; Liebertz J.; & Bohaty L. (2000). Linear optical properties of the monoclinic bismuth borate BiB_3O_6, *J. Appl. Phys.*, Vol. 88, pp. 240-244.

Kildal H. & Iseler G. W. (1979). Higher-order nonlinear processes in $CdGeAs_2$, *Phys. Rev. B* Vol. 19, pp. 5218-5222.

Miller S.; Rotermund F.; Xu G.; Noack F.; Panyutin V. & Petrov V. (2008). Polarization-dependent nonlinear refractive index of BiB_3O_6, *Opt. Mater.*, Vol. 30, pp. 1469-1472.

Miyata K.; Mikami T.; Umemura N. & Kato K. (2008). Direct third-harmonic generation in BiB_3O_6, *Proceedings of SPIE*, Vol. 6875, 687518-687518-5.

Miyata K.; Umemura N. & K. Kato (2009). Phase-matched pure $\chi^{(3)}$ third-harmonic generation in noncentrosymmetric BiB_3O_6, *Opt. Lett.*, Vol. 34, 500-502.

Nienhuys H. K.; Planken P. C. M.; Santen R. A. V. & Bakker H. J. (2001). Generation of mid-infrared generation in CaF_2 and BaF_2, *Opt. Lett.*, Vol. 26, pp. 1350-1352.

Okada M. (1971). Third-order nonlinear optical coefficients of $LiIO_3$, *Appl. Phys. Lett.*, Vol. 18, pp. 451-452.

Okamoto H. & Tasumi M. (1995). Generation of ultrashort light pulses in the mid-infrared (3000–800 cm^{-1}) by four-wave mixing, *Opt. Commun.*, Vol. 121, pp. 63-68.

Penzkofer A.; Ossig F.; & P. Qiu (1988). Picosecond third-harmonic light generation in calcite, *Appl. Phys. B*, Vol. 47, pp. 71-81.

Qiu P. & Penzkofer A. (1988). Picosecond third-harmonic light generation in β-BaB_2O_4, *Appl. Phys. B,* Vol. 45, pp. 225-236.

Takagi Y. & Muraki S. (2000). Third-harmonic generation in a noncentrosymmetrical crystal: direct third-order or cascaded second-order process?, *J. Luminesc.*, Vol. 87-89, pp. 865-867.

Teng B.; Wang J.; Wang Z.; Jiang H.; Hu X.; Song R.; Liu H.; Liu Y.; Wei J. & Shao Z. (2001). Growth and investigation of a new nonlinear optical crystal: bismuth borate BiB_3O_6, J. Cryst. Growth, Vol. 224, pp. 280-283.

Terhune R. W.; Maker P. D.; & Savage C. M. (1962). Optical harmonic generation in calcite, *Phys. Rev. Lett.*, Vol. 62, pp. 404-406.

Terhune R. W.; Maker P. D. & Savage C. M. (1963). Observation of saturation effects in optical harmonic generation, *Appl. Phys. Lett.*, Vol. 2, pp. 54-55.

Tzankov P. & Petrov V. (2005). Effective second-order nonlinearity in acentric optical crystals with low symmetry, *Appl. Opt.*, Vol. 44, 6971-6985.

5

Semiconductor Ridge Microcavities Generating Counterpropagating Entangled Photons

Xavier Caillet, Adeline Orieux, Ivan Favero, Giuseppe Leo and Sara Ducci
Laboratoire Matériaux et Phénomènes Quantiques UMR 7162
Université Paris Diderot Paris 7- CNRS
France

1. Introduction

In the last 25 years entangled photon pairs have been used first to test the foundations of quantum mechanics and then as building blocks in quantum information protocols : from the demonstration of the violation of Bell inequalities (Aspect et al., 1982; Tittel et al., 1998; Weihs et al., 1998) to the recent experiments in the domain of quantum-key distribution (Gisin et al., 2002), quantum computing (Deutch & Ekert, 1998; Wahther et al., 2005), teleportation (Bouwmeester et al., 1997) and absolute metrology (Migdall, 1999; Sergienko & Jaeger, 2003).

The first process used to produce entangled two-photon states have been atomic radiative cascades (Aspect et al., 1982) and parametric fluorescence in birefringent dielectric materials (Kwiat et al., 1985).

In order to achieve a good source, high collection efficiency is a key element because it affects the number of available photon pairs per unit time, and also because the presence of single photons having lost their twin is a source of noise in the detection. In this context we can mention several demonstrations of twin-photon generation based on either parametric down-conversion in periodically poled dielectric waveguides (Tanzilli et al., 2001; Banaszek et al., 2001) or four-wave mixing in optical fibers (Wang et al., 2001; Rarity et al. 2005; Fan & Migdall, 2007).

An attractive alternative is provided by semiconductor materials, which exhibit a huge potential in terms of integration of novel optoelectronic devices. The first semiconductor source of entangled photons was based on the bi-exciton cascade of a quantum dot (Stevenson et al., 2006). With respect to this technique, parametric generation in semiconductor waveguides allows room-temperature operation and a high directionality of the emission, which dramatically enhances the collection efficiency. Several phase-matching schemes have been demonstrated in these systems (Ducci et al., 2005): form birefringence, modal phase matching, counterpropagating phase matching. In particular, the last one has attracted a deal of attention because of its unusual flexibility in the control of the quantum properties of the emitted photons (Walton et al., 2003; Walton et al., 2004; Perina, 2008).

Here we present a semiconductor ridge microcavity emitting counterpropagating entangled photons; in Section 2 we explain the working principle of this device giving the details of the phase matching scheme and the effects of the microcavity. Section 3 is devoted to the description of the quantum properties of the generated two-photon state, which illustrate the versatility of the source. The experimental results of Surface Emitted Second Harmonic Generation and Parametric Fluorescence are presented in Section 4. Finally, we draw our conclusions in Section 5.

2. Working principle and microcavity effect

The twin-photon source presented here is a multilayer AlGaAs waveguide designed to allow a counterpropagating phase matching scheme. The lateral confinement is provided by a wet-etched ridge. In this geometry (Figure 1), a pump field (775 nm) impinges on top of the waveguide generating two counterpropagating, orthogonally polarized waveguided twin photons (~1550 nm) through parametric down-conversion. The frequencies of the emitted fields are fixed by the energy ($\omega_p = \omega_s + \omega_i$) and momentum ($k_p \sin\theta = n_s k_s - n_i k_i$) conservation, where ω_p, ω_s and ω_i (k_p, k_s and k_i) are the frequencies (wavevectors) of pump, idler and signal; θ is the angle of incidence of the pump beam, and n_s and n_i are the effective indices of the signal and idler modes. Momentum conservation in the epitaxial direction is satisfied by alternating AlGaAs layers with different Al concentration (having nonlinear coefficients d_{14} as different as possible) to obtain a Quasi Phase Matching (QPM).

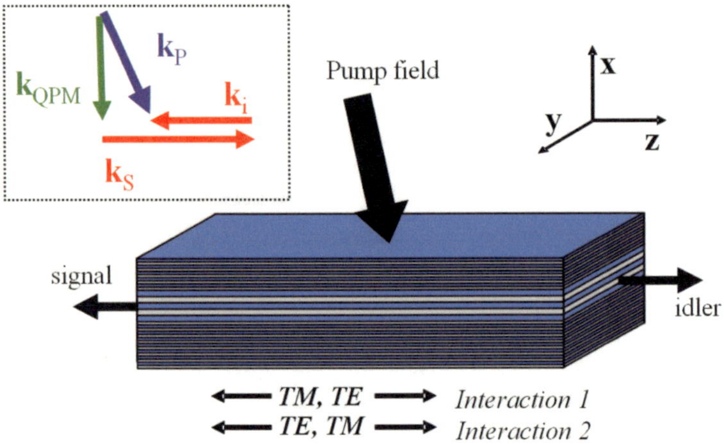

Fig. 1. Counterpropagating phase-matching scheme. Phase matching is automatically obtained in the z direction; Quasi Phase Matching (QPM) is provided by a periodic modulation of the nonlinear susceptibility tensor in the epitaxial direction in the core of the waveguide.

As a consequence of the different propagation directions for the signal and the idler, there are two simultaneously phase-matched processes: one where the signal is TE polarized and the idler is TM polarized, and the other where the signal is TM polarized and the idler is TE polarized. These two processes are expected to exhibit different characteristics because of the z→-z symmetry breaking exerted by the pump beam for $\theta \neq 0$. In the following, subscripts 1 and 2 will refer to these two processes. The central frequencies for the signal and the idler (corresponding to phase matching) are determined through the conservation of energy and momentum in the z direction, giving:

$$\begin{cases} \omega_p = \omega_s + \omega_i \\ \omega_p \sin\theta = \omega_s n_{\mathrm{TE}}(\omega_s) - \omega_i n_{\mathrm{TM}}(\omega_i) \end{cases} \quad \text{(interaction 1)}$$
$$\begin{cases} \omega_p = \omega_s + \omega_i \\ \omega_p \sin\theta = \omega_s n_{\mathrm{TM}}(\omega_s) - \omega_i n_{\mathrm{TE}}(\omega_i) \end{cases} \quad \text{(interaction 2)} \tag{1}$$

where n_{TE} and n_{TM} are the effective indices of the guided modes.
Several advantages derive from such a geometry, in the aim of fabricating a source for quantum information: absence of the pump beam in the guided direction, automatic separation of the down-converted photons, and possibility of direct coupling into two optical fibers through standard pig-tailing process.
A preliminary demonstration of a twin-photon source based on this principle has been reported by (Lanco et al., 2006); however the performances of that device were affected by low conversion efficiency and low signal/noise ratio due to photo-luminescence from the substrate.
An interesting possibility to improve the source consists in the addition of Distributed Bragg Reflectors (DBRs) on top of and under the waveguide, in order to create a microcavity for the pump beam (Andronico et al., 2008). In this case, at resonance wavelength, a nearly standing wave is obtained inside the cavity, whose amplitude can be largely greater than outside (see Figure 2).
In the case of the waveguide without microcavity presented above, the guided field amplitude of the pump beam can be roughly estimated using the Fresnel equations. Assuming, for the sake of simplicity, a homogeneous waveguide with refractive index n_{guide} and air as incident medium, the relation between the internal and external pump field is $E_p^{int} = tE_p^{ext}$, with $|t| = 2/(1 + n_{guide})$.

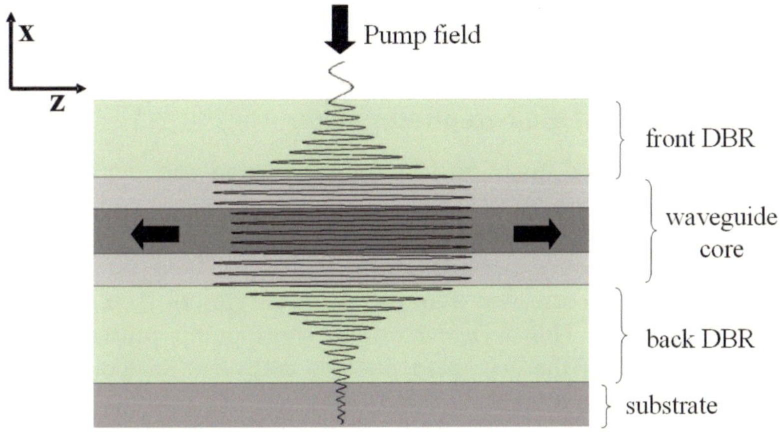

Fig. 2. View in the (xz) plane of the waveguide with the addition of the DBRs. In black is plotted the pump beam profile within the structure at cavity resonance. The field enhancement induces a strong improvement of the conversion efficiency.

In the case of the ridge microcavity showed in Figure 2, we can use the transfer matrix method to compute the amplitude of the standing wave inside the cavity. This is found to be:

$$\frac{E_p^{\text{int}}}{E_p^{ext}} = 2\sqrt{\frac{2F}{\pi\left(1+\left|t_2/t_1\right|^2\right)}} \tag{2}$$

Here, F is the finesse of the cavity, and t_i are the transmission coefficients associated to the front (i = 1) and back (i = 2) mirrors. It can be shown that the conversion efficiency enhancement factor due to the addition of the microcavity is:

$$\frac{\eta_{cavity}}{\eta_0} = \frac{2\left(1+n_{guide}\right)^2}{\pi n_{guide}} \frac{F}{\left(1+\left|t_2/t_1\right|^2\right)} \tag{3}$$

This expression indicates how to optimize the conversion efficiency.
Firstly, the ratio $|t_2/t_1|$ has to be minimized, reflecting the obvious fact that the front mirror should be less reflective than the back one, in order to let the light enter the microcavity.
Secondly, a maximum finesse is desired. It is interesting, for the rest of the discussion, to introduce the quality factor of the cavity: $Q = m_{eff}F$. Here, m_{eff} is the effective order of the cavity, given by $m_{eff} = m + m_0$, where m is the cavity order and m_0 accounts for the penetration of the field in the DBR. For a $\lambda/4$ DBR stack, $m_0 = n_H/(n_H\text{-}n_L)$, where n_H and n_L are the refractive indices of the high- and low-index materials. Equation (3) shows that the higher the quality factor, the higher the efficiency.
However, a practical limitation on Q stems from technical reasons. Firstly, the large amount of energy stored in the cavity can possibly induce thermorefractive effects, leading to undesirably bistable devices. Moreover, the inhomogeneity of the sample inherent to the growth process induces a variation of the cavity resonance wavelength λ along the direction of the waveguide. To obtain an efficient interaction throughout the structure, this variation has to be smaller than the mean cavity linewidth $\delta\lambda$. This, in return, sets an upper limit to the desired quality factor, expressed as $Q = \lambda/\delta\lambda$.

3. Quantum properties of the two-photon state

In this Section we derive the quantum state of the photon pair emitted by our source, showing how the pump beam parameters, in particular its shape and bandwidth, can be used to control the degree of frequency correlation.
In a typical down-conversion experiment using a narrow-band pump beam, the twin photons have perfectly anti-correlated frequencies (as the sum of their energies is fixed by the pump photon energy). However the use of short pump pulses relaxes the strict frequency anti-correlation of the generated photons and open the possibility to generate frequency correlated or frequency uncorrelated states.
The Hamiltonian describing the parametric down conversion process is:

$$H(t) = \int dV\left(\chi_{eff}^{(2)}(r)E_p^{(+)}(r,t)E_s^{(-)}(r,t)E_i^{(-)}(r,t) + h.c.\right) \tag{4}$$

where $\chi_{eff}^{(2)}$ is the effective electric susceptibility tensor (periodically modulated along the epitaxial axis (x) to provide quasi phase matching as described in the previous section), $E_p^{(+)}$ the positive-frequency part of the pump beam electric-field operator, $E_s^{(-)}$ and $E_i^{(-)}$ the negative-frequency part of the signal and idler electric-field operators. The symbol $h.c.$

stands for hermitian conjugate term. The integration is performed over the interaction volume which, in our case, is the illuminated portion of the waveguide.

The suitability of a photon pair source for a given quantum optics application largely depends on the joint spectral intensity (JSI) of the generated photons, $S(\omega_s, \omega_i)$, defined as the probability that the signal (idler) photon frequency is ω_s, (ω_i). In the following we make explicit the link between the JSI and the characteristics of the pump beam in order to clarify the physical parameters that can be used to tailor the two-photon state.

A classical pump field on the air side of the air/semiconductor interface is given by:

$$E_p(z,t) = \int d\omega \tilde{E}(\omega) A(z) e^{i[k(\omega)\sin\theta(\omega) z - \omega t]} \tag{5}$$

where we have neglected the pump variation along the y axis.

In this expression: the spatial profile $A(z)$ can be engineered through linear optics; $\tilde{E}(\omega)$ is the spectral distribution, characterized by a center frequency ω_p and a bandwidth σ_p; $\theta(\omega)$ is the angle of incidence of the ω component of the field, which can be engineered through a dispersive element, such as a quartz wedge or a diffraction grating.

Following the derivation of (Grice et al., 2001) and assuming incident signal and idler fields in the vacuum states, the first-order perturbation solution of the Schrödinger equation using Equation 4 gives the generated two photon state:

$$|\psi\rangle = \propto \iint d\omega_s d\omega_i f(\omega_s, \omega_i) \hat{a}_s^+(\omega_s) \hat{a}_i^+(\omega_i) |vac\rangle \tag{6}$$

Here $\hat{a}_s^+$ and $\hat{a}_i^+$ are the photon creation operators for the signal and idler beams, and the probability amplitude $f(\omega_s, \omega_i)$ is given by:

$$f(\omega_s, \omega_i) \propto \tilde{E}(\omega_s + \omega_i) \phi(\omega_s, \omega_i) \tag{7}$$

with the phase-matching function:

$$\phi(\omega_s, \omega_i) = \int dz \chi_{eff}^{(2)}(z) A(z) e^{i\Delta k(\omega_s, \omega_i) z} \tag{8}$$

$\Delta k(\omega_s, \omega_i) = k(\omega_s + \omega_i)\sin\theta(\omega_s + \omega_i) - \beta_s(\omega_s) + \beta_i(\omega_i)$ being the z component of the wave vector mismatch, and $\beta_{s,i}(\omega)$ the effective propagation constant of the signal and idler photon respectively.

As the JSI is related to the two-photon probability amplitude by the simple relation $S(\omega_s, \omega_i) = |f(\omega_s, \omega_i)|^2$, it is clear that the pump beam characteristics are a useful means to control the JSI.

The possibilities to tailor the JSI offered by the counterpropagating geometry are numerous and a complete review is beyond the scope of this chapter; let us focus here on some simple examples of pump-beam shaping that are sufficient to provide a large class of JSI. In particular we will show that the pump spectral profile can be used as the only parameter to generate generalized frequency-correlated states. Different techniques and geometries have been reported in the purpose of generating frequency-uncorrelated photon pairs in guided-wave co-propagating schemes. All these techniques require to fulfil particular conditions on group-velocity matching and hence on material dispersion. Here we show that there is no need for such requirements for our source: frequency uncorrelated, as well as frequency

correlated and anti-correlated states, are easily obtained by properly choosing the pump pulse duration.
In order to investigate a feasible experimental situation, let us consider a Gaussian spectral pump distribution:

$$\tilde{E}(\omega) \propto \exp\left[-\left(\frac{\omega-\omega_p}{\sigma_p}\right)^2\right] \tag{9}$$

Moreover, let us assume a uniform profile of the pump field along the waveguide, and a fixed angle of incidence θ (independent of the pump frequency components). Under these conditions, we obtain:

$$f(\omega_s,\omega_i) \propto \exp\left[-\left(\frac{\omega_s+\omega_i-\omega_p}{\sigma_p}\right)^2\right]\phi(\omega_s,\omega_i) \tag{10}$$

with

$$\phi(\omega_s,\omega_i) = \text{sinc}\left(\frac{L}{2}(\Delta k(\omega_s,\omega_i))\right)\exp\left(-i\frac{L}{2}(\Delta k(\omega_s,\omega_i))\right) \tag{11}$$

The probability amplitude of the two-photon state is thus the product of two functions: one depending on the pump spectral properties, and the other, the phase matching function ϕ, on the spatial pump properties. In the following we will consider a perfect phase matching situation with ω_s^0 and ω_i^0 be the signal and idler frequencies. The elimination of frequency correlations can easily be shown if we approximate the $\text{sinc}(x/2)$ function by the Gaussian $\exp(-\gamma x^2)$, with $\gamma = 0.0482$ (this value of γ is chosen to obtain the same width at half maximum for the two functions). By introducing the reciprocal group velocities:

$$\beta'_{s,i} = \frac{d\beta_{s,i}(\omega)}{d\omega}\bigg|_{\omega=\omega_{s,i}^0} \quad \text{and} \quad k' = \frac{k(\omega)}{d\omega}\bigg|_{\omega=\omega_p} = \frac{1}{c} \tag{12}$$

and performing a first-order development of the phase-matching function we obtain for the joint spectral density:

$$\begin{aligned} S(\omega_s,\omega_i) \propto & \exp\left[-2(\omega_s-\omega_s^0)^2\left(\frac{1}{\sigma_p^{\,2}}+\gamma L^2(k'\sin\theta-\beta'_s)^2\right)\right] \\ & \times\exp\left[-2(\omega_i-\omega_i^0)^2\left(\frac{1}{\sigma_p^{\,2}}+\gamma L^2(k'\sin\theta+\beta'_i)^2\right)\right] \\ & \times\exp\left[-4(\omega_s-\omega_s^0)(\omega_i-\omega_i^0)\left(\frac{1}{\sigma_p^{\,2}}+\gamma L^2(k'\sin\theta-\beta'_s)(k'\sin\theta+\beta'_i)\right)\right] \end{aligned} \tag{13}$$

In this expression the condition:

$$\frac{1}{\sigma^2}+\gamma L^2(k'\sin\theta-\beta'_s)(k'\sin\theta+\beta'_i)=0 \tag{14}$$

allows the factorization of the JSI and thus the generation of a frequency-uncorrelated state. We notice that this condition can always be fulfilled in our geometry by a proper choice of the pump pulse duration, Figure 3 reports numerical simulations of the JSI obtained for three different sets of waveguide lengths and pump pulse durations. In a 2D representation with signal and idler frequency as coordinates, the JSI level curves of ideal frequency correlated and anti-correlated states are segments parallel to the $\omega_s+\omega_i$ and $\omega_s-\omega_i$ directions respectively. When the correlation is not perfect, these segments become ellipses, the higher the degree of correlation the higher the eccentricity. The JSI level curves of uncorrelated frequency states are ellipses with axes in the ω_s and ω_i directions; in this case the JSI can be factorized as a product of independent probabilities for the signal and idler photons, $S(\omega_s,\omega_i) = p(\omega_s)p(\omega_i)$: the measurement of the frequency of one of the emitted photons does not yield any information about the frequency of the other one.

The ellipse in Figure 3 (left) represents a state with frequency anti-correlation: the length of its major axis is set by the phase matching function, which depends on the waveguide length L. The anticorrelation can readily be maximized as the ellipse minor axis is made arbitrarily small by increasing the pump pulse duration ($\sigma \rightarrow 0$).

A frequency correlated state is depicted in Figure 3 (centre): here the length of the major axis is set by the pump pulse duration, whereas that of the minor axis depends on the phase-matching function and can be reduced by increasing the waveguide length. It is worth noting that the counterpropagating geometry allows an easy generation of such a state thanks to the sharply peaked phase-matching versus $\omega_s-\omega_i$. We stress that alternative geometries require more stringent conditions, on either group velocities or other pump properties (extended phase matching (Giovannetti et al., 2002), achromatic phase matching (Torres et al., 2005)).

Finally a frequency uncorrelated case is shown in Figure 3 (right). Here σ and L have been chosen to equal the major and minor axes length of the ellipse. The JSI appears as a circle and is a special case of frequency uncorrelated state with equal individual bandwidth of the generated photons.

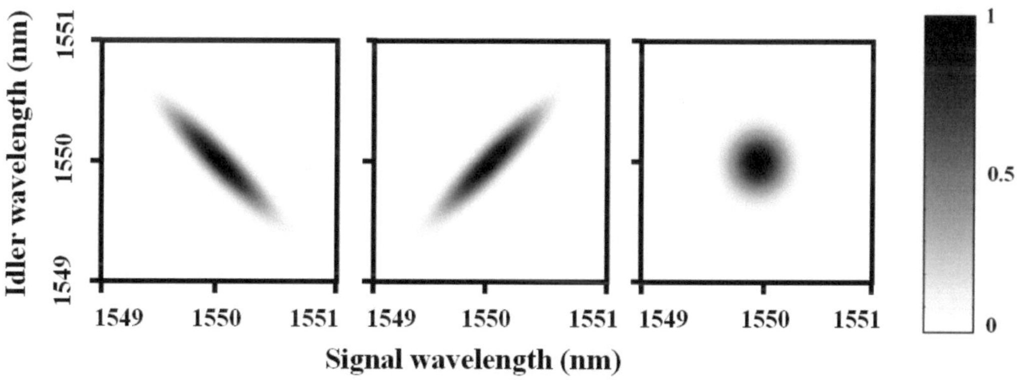

Fig. 3. Two-dimensional representation of the two-photon joint spectral intensity for three pumping configurations. Here L is the waveguide length (completely illuminated) and τ the pulse duration of the pump beam: (a) L=0,22 mm and τ=11 ps (b) L=1,1 mm and τ= 2.2 ps, (c) L=0.5 mm and τ= 5 ps. These configurations correspond to an anti-correlated, correlated and uncorrelated state, respectively.

In order to quantify the separability of the generated state one has to perform a Schmidt decomposition (Law et al., 2000) of the two-photon probability amplitude $f(\omega_s, \omega_i)$, i.e. a basis transformation into a set of orthonormal Schmidt modes $\{\psi_n, \varphi_n\}$.
The function f is then expressed as:

$$f(\omega_s,\omega_i)=\sum_n \sqrt{\lambda_n}\psi_n(\omega_s)\varphi_n(\omega_i) \quad (15)$$

where λ_n, φ_n, ψ_n are solutions of the eigenvalue problem:

$$\iint f(\omega_s,\omega'')f^*(\omega',\omega'')d\omega''\psi_n(\omega')d\omega' = \lambda_n\psi_n(\omega_s) \quad (16)$$

$$\iint f(\omega'',\omega_i)f^*(\omega'',\omega')d\omega''\varphi_n(\omega')d\omega' = \lambda_n\varphi_n(\omega_i) \quad (17)$$

with

$$\sum \lambda_n = 1 \quad (18)$$

If the sum in expression (15) has a unique term, f is factorizable and the photons of the pair are not entangled. If the sum in expression (15) contains a large number of terms the state of the pair is strongly entangled. The measure of the correlation degree is provided by the entropy S:

$$S = -\sum_{n=1}^{N} \lambda_n \log_2(\lambda_n) \quad (19)$$

where N is the number of Schmidt modes used in the decomposition. S is equal to zero for a non entangled state and is the more important the more the state is entangled. To give an example, Figure 4 reports the Schmidt decomposition of a frequency correlated and a frequency anti-correlated state.

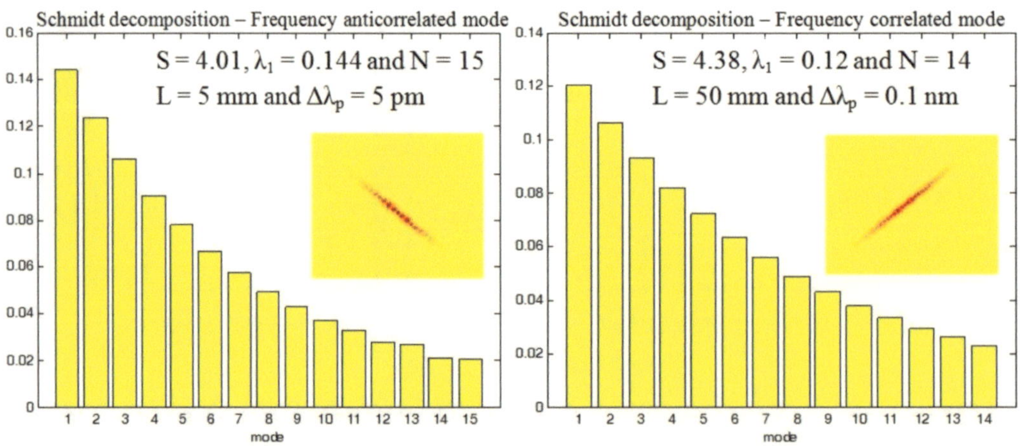

Fig. 4. Schmidt decomposition for a frequency anti-correlated and a frequency correlated mode. L is the length of the guide (completely illuminated) and $\Delta\lambda_p$ the spectral width of the pump beam.

The relative simplicity of frequency entanglement control by a suitable choice of the pump bandwidth is peculiar of the counterpropagating geometry. Further theoretical developments have been made towards the generation of photon pairs with arbitrary joint spectrum (Walton et al., 2004; Perina, 2008). In particular, a more refined shaping of the pump beam using achromatic phase matching (i.e. allowing the angle θ to vary with ω) has been shown to generate uncorrelated twin photons with independent spectral bandwidth (Walton et al., 2004).

4. Experimental results

4.1 Surface emitted second harmonic generation

A practical mean to characterize the efficiency enhancement due to the integration of a vertical microcavity for the pump beam is Surface Emitted Second Harmonic Generation (SESHG) (Caillet et al., 2009). SESHG, which was first demonstrated in 1979 (Normandin & Stegeman, 1979), is the reverse of SPDC at degeneracy ($\omega_s=\omega_i=\omega_p/2$): the non-linear overlap of two counterpropagating modes yields a second harmonic field radiating from the upper surface of the waveguide. Because of its intrinsic higher produced signal, this interaction is well suited to provide a characterization of the effect of the added vertical microcavity and a quantitative estimation of the parametric gain.

The sample resulted by our numerical simulations was grown by metal organic chemical vapor deposition on (100) GaAs substrate. The planar structure was then chemically etched to create 2.5-3.5 μm-thick ridges with 6-9 μm widths. The epitaxial structure consists in 4.5 period $Al_{0.25}Ga_{0.75}As$/ $Al_{0.80}Ga_{0.20}As$ QPM waveguide core, 41-period asymmetrical $Al_{0.25}Ga_{0.75}As$/ $Al_{0.80}Ga_{0.20}As$ DBR (serving as lower cladding + back mirror) and 18-period asymmetrical $Al_{0.25}Ga_{0.75}As$/ $Al_{0.80}Ga_{0.20}As$ DBR (serving as upper cladding + top mirror). Figure 5 reports the tuning curves calculated on the nominal structure.

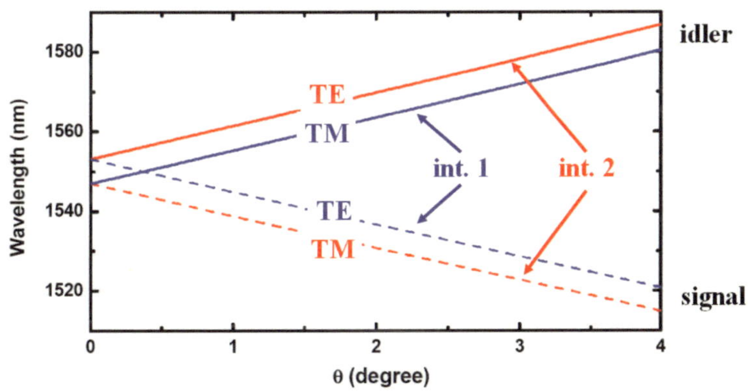

Fig. 5. Calculated tuning curves as a function of the pump incident angle for a pump wavelength of 775 nm.

The SESHG measurements were performed employing one end-fire coupled fundamental frequency (FF) beam relying on Fresnel reflection at the opposite facet to obtain back propagating modes (see Fig. 6).

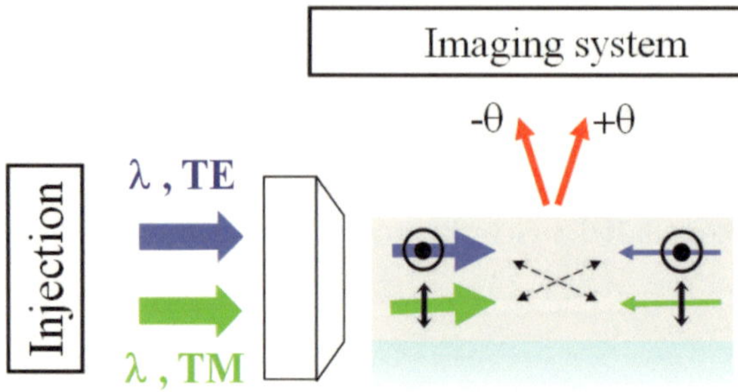

Fig. 6. Scheme of SESHG set up. TE and TM modes are end-fire excited through the same input facet, relying on their Fresnel reflection at the opposite facet. By doing so, two nonlinear processes occur: the interaction of the TE mode with the reflected part of the TM one, and the interaction of the TM mode with the reflected part of the TE one. This symmetric configuration results in the generation of two second harmonic waves of comparable amplitudes, which radiate with angles θ and -θ, respectively: an interference pattern can thus be observed at the waveguide surface.

The fundamental frequency was linearly polarized at 45° relative to the substrate so that the input power was equally divided between TE and TM eigenfields. The input beam was supplied by an external-cavity cw tunable laser beam, amplified with an Er^{3+}-doped fiber. The SH field was acquired with an optical system mounted over the waveguide, perpendicularly to its plane. This consists of a CCD camera Bosch LTC 0335/50 1/3 inch format 512×582 pixels and an Edmund video lens trinocular VZM microscope for near field acquisition. In order to study the dependence of SESHG power vs the FF power, the generated signal was measured with a large area Si photodiode plus a lock-in synchronous detection.

The near-field image of the SH field obtained for a FF injected at 1.565 μm is shown in Figure 7 ; we note that the difference between the resonance wavelength calculated for the nominal structure and the measured one is only 7,5 nm (which is totally compatible with the technological tolerances). The period Λ of the pattern provides a straightforward measure of θ through the relation $|\sin\theta| = \lambda_p/2\Lambda$. In this case θ = 0.4° which is in excellent agreement with the value obtained by numerical predictions.

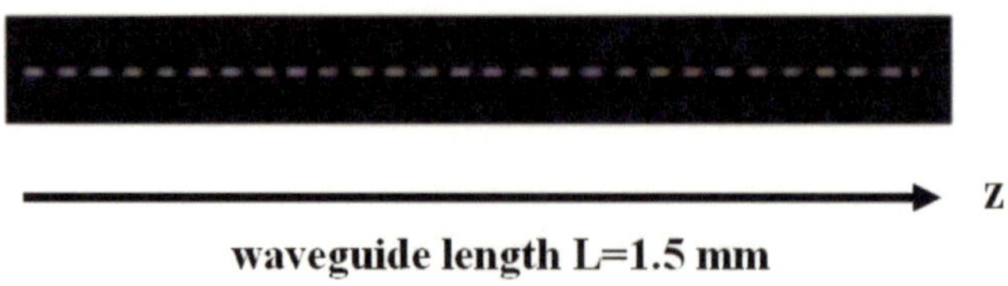

Fig. 7. Complete view of SESHG near field for a FF injected at 1.56 μm.

In Figure 8 the detected SESHG power is plotted versus the guided FF power: SESHG power data are in good agreement with the parabolic fit curve, as expected for a quadratic nonlinear process.

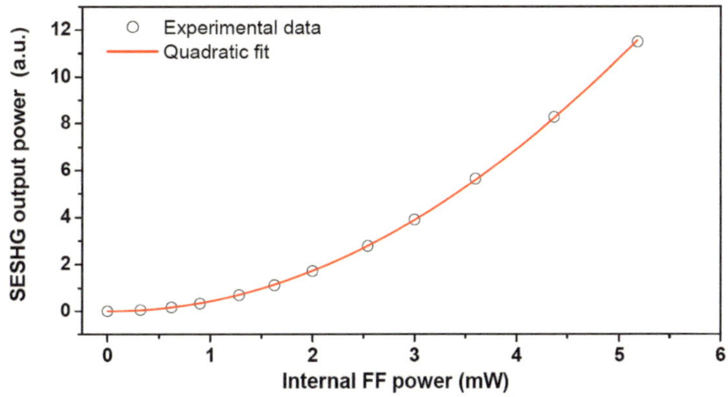

Fig. 8. SESHG power vs. FF power at 1.56 μm (photodiode acquisition).

As we have pointed out, the net frequency dependence of the enhancement factor is dominated by the cavity; the spectrum of the second harmonic field allows thus the characterization of the effect of the integrated vertical cavity.
Figure 9 reports the experimental spectrum obtained with our sample; the experimental points are fitted with a Lorentzian, which is the expected theoretical shape close to the resonance wavelength. The calculated full width at half maximum of the Lorentzian, in the limit of perfect reflectivity of the bottom mirror, is:

$$\gamma = \frac{T\lambda^2}{2\pi n d} \tag{16}$$

with T the transmission coefficient of the upper mirror. The calculated value for our structure is γ= 0.55 nm, which is in excellent agreement with the experimental data (γ= 0.54 nm).

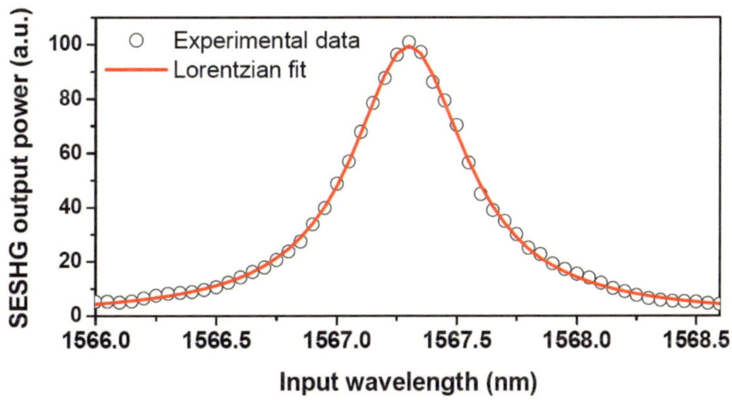

Fig. 9. Experimental spectrum of the SESHG signal (dots) and fit with a Lorentzian (solid line).

4.2 Parametric fluorescence and coincidence histogram

Since the sample described in the previous section presented elevated optical losses that have been imputed to the growth technique, a second sample was grown using the technique of molecular beam epitaxy. The epitaxial structure consists in 4.5 period $Al_{0.25}Ga_{0.75}As$/ $Al_{0.80}Ga_{0.20}As$ QPM waveguide core, 41-period asymmetrical $Al_{0.25}Ga_{0.75}As$/ $Al_{0.80}Ga_{0.20}As$ DBR (serving as lower cladding + back mirror) and 18-period asymmetrical $Al_{0.25}Ga_{0.75}As$/ $Al_{0.80}Ga_{0.20}As$ DBR (serving as upper cladding + top mirror).

Figure 10 shows our first parametric fluorescence spectrum; the pump beam is provided by a TE polarized, pulsed Ti: Sa laser with λ_p=759,5 nm and a 3 kHz repetition rate. The pulse peak power is P_p=10 W and its duration is 150 ns. The pump beam is focused on top of the waveguide ridge using a cylindrical lens with an angle θ in the xz plane. The generated photons are collected from one of the facets of the sample with a microscope objective, spectrally analyzed with a monochromator, and then coupled into a fibered InGaAs single-photon avalanche photodiode (IdQuantique).

The spectra obtained for θ=3°, show the existence of the two expected processes (see Figure 5) and demonstrate the possibility of direct generation of polarization-entangled states. The amplitude difference in the observed signal is due to the fact that the long wavelength photons are collected after their reflection on the facet opposite to the detection. An anti-reflection coating to both facets of the sample would allow an automatic separation of the photons of each pair and their direct coupling into two optical fibers, through standard pigtailing process. The amount of detected photons allows deducing the brightness of our twin photon source, which we estimate around 10^{-11} W^{-1}, which represents an improvement of at least two orders of magnitude with respect to (Lanco et al., 2006).

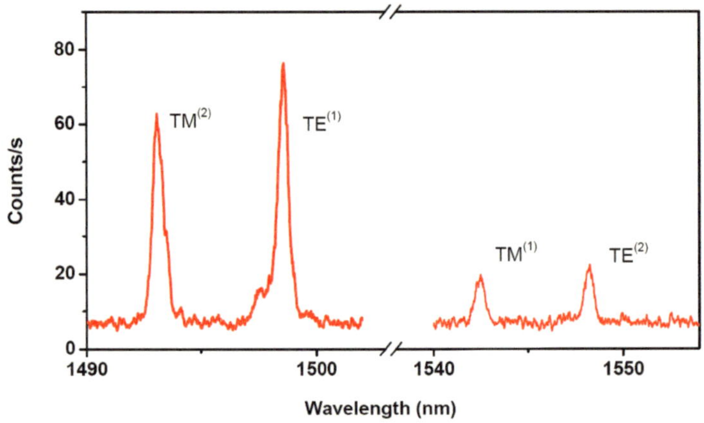

Fig. 10. Experimental spectrum of the parametric fluorescence for the two parametric interactions. The background noise here is due to the dark counts of the detectors.

To further assess the twin character of the emitted photons, the time correlations between the detected counts have been analyzed. The scheme of the setup is shown in Figure 11 and the histogram of the time delays is shown in Fig. 12, for the case of interaction 1 with $\theta = 3°$. With a sampling interval of 40 ps, the histogram results from an acquisition time of 25 minutes. The peak observed for $t_s = t_i$ demonstrates unambiguously the twin character of the generated photons; the 500 ps full width at half maximum of the histogram corresponds to the timing jitter of both detectors. The flat background is produced by the accidental coincidences essentially dues to dark counts: indeed, switching the pump polarization from TE to TM leads to the suppression of the $t_s = t_i$ peak, without modifying this background. Finally, no time-correlation is found between photons that are generated with different interactions: this agrees with the expectations, since these photons are not generated within the same nonlinear process.

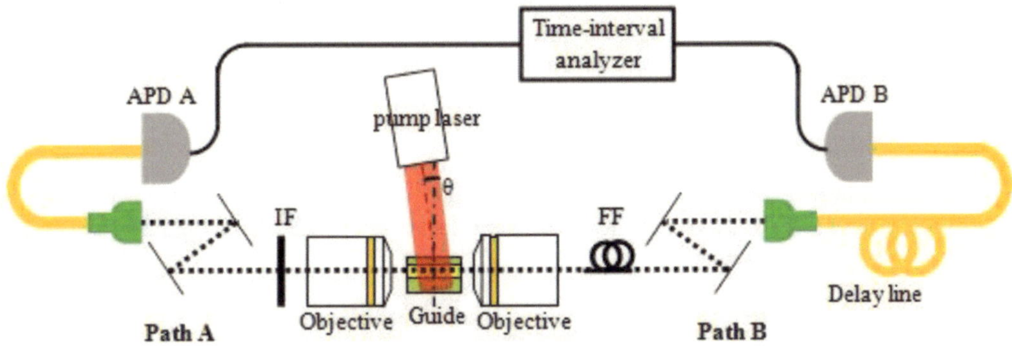

Fig. 11. Experimental set-up for the coincidence measurement. IF: interferential filter. FF: fibered filter. APD: single photon avalanche photodiode. The signal collected by the detectors is sent to a time-interval analyzer to built the time-correlation histogram.

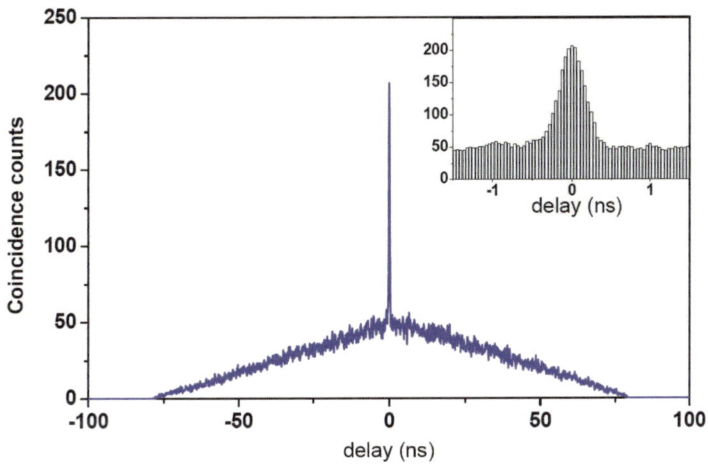

Fig. 12. Time-correlation histogram between counterpropagating photons.

5. Conclusion and perspectives

These results open the way to the demonstration of several interesting features associated to the counterpropagating geometry, as the direct generation of polarization entangled Bell state or the control of the generated two-photon state via an appropriate choice of the spatial and spectral pump beam profile. Indeed recent developments in quantum information theory have arisen a growing interest on 'generalized' states of frequency correlation (like frequency-correlated or frequency-uncorrelated photons). For example: i) quantum teleportation and entanglement swapping require the synchronized creation of multiple photon pairs, which is achieved by using a short pump pulse (thus relaxing the strict frequency anti-correlation of the generated photons mentioned above); ii) linear optical quantum computation requires uncorrelated photons in order to guarantee their

indistinguishability; iii) improvements on clock synchronization need frequency correlation to overcome media dispersion.
The efficiency of this room temperature working device, along with the high-quality quantum properties of the generated photons and their telecom wavelength, makes this source a serious candidate for integrated quantum photonics.

6. Acknowledgments

The authors thank Isabelle Sagnes and Aristide Lemaître (Laboratoire de Photonique et Nanostructures CNRS UPR20, France) for sample growth and Pascal Filloux for ridge processing.

7. References

Andronico, A.; Caillet, X.; Favero, I.; Ducci, S. & Berger, V. (2008). Semiconductor microcavities for enhanced nonlinear optics interactions. *J. of European Optical Society – Rapid Publications*, 3, (September 2008) (08030-1-08030-7), 1990-2573.

Aspect, A.; Grangier, P. & Roger, G. (1982). Experimental Realization of Einstein-Podosky-Rosen-Bohm Gedankenexperiment: A New Violation of Bell's Inequalities. *Phys. Rev. Lett.*, 49, 2, (July 1982) (91-94), 0031-9007.

Banaszek, K.; U'Ren, A. B. & Walmsley, I. A. (2001). Generation of correlated photons in controlled spatial modes by downconversion in nonlinear waveguides. *Optics Letters*, 26, 17, (September 2001) (1367-1369), 0146-9592.

Bouwmeester, D.; Pan, J.W.; Mattle, K.; Eibl, M.; Weinfurter, H. & Zeilinger, A. (1997). Experimental quantum teleportation. *Nature*, 390, (December 1997) (575-579), 0028-0836.

Caillet, X.; Berger, V.; Leo, G.; Sagnes, I. & Ducci, S. (2009). A semiconductor ridge micro cavity to generate counterpropagating twin photons, *Proceedings of SPIE 7222 72221J*, 0277-786X, pp. 72221J1-8, San Jose (CA, USA), January 2009, SPIE.

Deutsch, D. & Ekert, A. (1998). Quantum Computation. *Phys. World*, 11, 3, (March 1998) (47-52), 0953-8585.

Ducci, S.; Lanco, L.; Seurin, Y.; Leo, G.; Berger, V.; De Rossi, A. & Marcadet, X. (2005). Semiconductor sources of twin photons for quantum information. *J. of Optics B: Quantum and Semiclassical Optics*, 7, (June 2005) (S158-S165), 1464-4266.

Fan, J. & Migdall, A. (2007). A broadband high spectral brightness fiber-based two-photon source. *Opt. Express*, 15, 6, (March 2007) (2915-2920), 1094-4087.

Giovannetti, V.; Maccone, L.; Shapiro, J. H. & Wong, F. N. C. (2002). Extended phase-matching conditions for improved entanglement generation. *Phys. Rev. A*, 66, 4, (October 2002) (043813-1-043813-9), 1050-2947.

Gisin, N.; Ribordy, G.; Tittel, W. & Zbinden, H. (2002). *Rev. Mod. Phys.* 74, 1, (March 2002) (145-195), 0034-6861.

Grice, W. P.; U'Ren, A. B. & Walmsley, I. A. (2001). Eliminating frequency and space-time correlations in multiphoton states. *Phys. Rev. A*, 64, 6, (November 2001) (063815-1-063815-7), 1050-2947.

Kwiat, P. G.; Mattle, K.; Weinfurter, H.; Zeilinger, A.; Sergienko, A.V. & Shih, Y. H. (1995). New High-Intensity Source of Polarization-Entangled Photon Pairs. *Phys. Rev. Lett.*, 75, 24, (December 1995) (4337-4341), 0031-9007.

Lanco, L.; Ducci, S.; Likforman, J.-P.; Marcadet, X.; van Houwelingen, J. A. W.; Zbinden, H.; Leo, G. and Berger, V. (2006). Semiconductor Waveguide Source of Counterpropagating Twin Photons. *Phys. Rev. Lett.*, 97, 17, (October 2006) (173901-1-173901-4), 0031-9007.

Law, C. K.;Walmsley, I. A. & Eberly, J. H. (2000) Continuous Frequency Entanglement: effective Finite Hilbert Space and Entropy Control. *Phys. Rev. Lett.* 84, 23, (June 2000) (5304-5307), 0031-9007.

Migdall, A. (1999). Correlated-Photon Metrology Withiout Absolute Standards. *Phys. Today*, 52, 1 (January 1999) (41-46), 0031-9228.

Normandin, R. & Stegeman, G. I. (1979). Nondegenerate four-wave mixing in integrated optics. *Opt. Lett.*, 4, 58 (February 1979) (58-59), 0146-9592.

Perina, J. (2008). Quantum properties of counterpropagating two-photon states generated in a planar waveguide. *Phys. Rev. A*, 77, 1, (January 2008) (013803-1-013803-14), 1050-2947.

Rarity, J. G.; Fulconis, J.; Duligall, J.; Wadsworth, W. J. & Russel, P. S. J. (2005). Photonic crystal fiber source of correlated photon pairs. *Opt. Express*, 13, 2, (January 2005) (534-544), 1094-4087.

Sergienko, A.V. & Jaeger, G.S. (2003). Quantum information processing and precise optical measurement with entangled-photon pairs. *Contemporary Phys.*, 44, 4, (July 2003) (341-356), 0010-7514.

Stevenson, R. M.; Young, R. J.; Atkinson, P.; Cooper, K.; Ritchie, D. A. & Shields, A. J. (2006). A semiconductor source of triggered entangled photon pairs. *Nature*, 439, (January 2006) (179-182), 0028-0836.

Tanzilli, S.; De Riedmatten, H.; Tittel, W.; Zbinden, H.; Baldi, P.; De Micheli, M.; Ostrowsky, D. B. & Gisin, N. (2001). Highly efficient photon-pair source using a Periodically Poled Lithium Niobate waveguide. *Electronics Letters*, 37, 1, (January 2001) (26-28), 0013-5194.

Tittel, W.; Brendel, J.; Zbinden, H. & Gisin, N. (1998). Violation of Bell Inequalities by Photons More Than 10 km Apart. *Phys. Rev. Lett.*, 81, 17, (October 1998) (3563-3566), 0031-9007.

Torres, J. P.; Macia, F.; Carrasco, S. & Torner, L. (2005). Engineering the frequency correlations of entangled two-photon states by achromatic phase matching. *Optics Letters*, 30, 3, (February 2005) (314-316), 0146-9592.

Walther, P.; Resch, K. J.; Rudolf, T.; Schenck, E.; Weinfurter, H.; Vedral, V.; Aspelmeyer, M. & Zeilinger, A. (2005). Experimental one-way quantum computing. *Nature*, 434, (March 2005) (169-176), 0028-0836.

Walton, Z. D.; Booth, M. C.; Sergienko, A. V.; Saleh B. E. A. & Teich, M. C. (2003). Controllable frequency entanglement via auto-phase-matched spontaneous parametric down-conversion. *Phys. Rev. A*, 67, 5 (May 2003) (053810-1-053810-4), 1050-2947.

Walton, Z. D.; Sergienko, A. V.; Saleh, B. E. A. & Teich, M. C. (2004). Generation of polarization-entangled photon pairs with arbitrary joint spectrum. *Phys. Rev. A*, 70, 5, (November 2004) (052317-1-052317-5), 1050-2947.

Wang, L. J.; Hong, C. K. & Friberg, S. R. (2001). Generation of correlated photons via four-wave mixing in optical fibres. *J. Opt. B: Quantum and Semiclass. Opt.*, 3, (October 2001) (346-352), 1464-4266.

Weihs, G.; Jennewein, T.; Simon, C.; Weinfurter, H. & Zeilinger, A. (1998). Violation of Bell's Inequality under Strict Einstein Locality Conditions. *Phys. Rev. Lett.*, 81, 23, (December 1998) (5039-5043), 0031-9007.

6

Two-Wave Mixing in Broad-Area Semiconductor Amplifier

Mingjun Chi[1], Jean-Pierre Huignard[2] and Paul Michael Petersen[1]
[1]*Department of Photonics Engineering, Technical University of Denmark*
[2]*Thales Research & Technology*
[1]*Denmark*
[2]*France*

1. Introduction

Two-wave mixing (TWM) is an interesting area in nonlinear optics and has been intensively investigated in the past three decades. TWM can take place in many different nonlinear media, such as second-order nonlinear media like photorefractive materials (Marrakchi et al., 1981; Huignard & Marrakchi, 1981; Yeh, 1983; 1989; Garrett et al., 1992), third-order nonlinear materials like Kerr media (Silberberg &Bar-Joseph, 1982; 1984; Yeh, 1986; 1989; Grandclément et al., 1987; McGraw, 1992), and in gain media like YAG (Brignon & Huignard, 1993). The microcosmic physical process for TWM in different nonlinear media is different. But in general the TWM process can be explained as: two coherent beams are incident on a nonlinear medium and a interference pattern is formed in the medium, such a pattern is characterized by a periodic spatial variation of the intensity; thus a refractive index and/or a gain (absorption) periodic variations will be induced because of the nonlinear response of the medium, and these refractive index and gain variations are usually called volume refractive index (or phase) grating and gain (or absorption) grating; the two beams propagate through the volume gratings formed by them and they undergo Bragg scattering (the Bragg condition is satisfied automatically); one beam scatters into the other and vice versa, so the energy and phase exchanges may occur between these two beams, i.e., the TWM takes place.

Nonlinear four-wave mixing in narrow-stripe and broad-area semiconductor lasers and amplifiers is of interest as a method to obtain high phase conjugate reflectivity (Nakajima & Frey, 1985; 1986; Frey, 1986; Agrawal, 1987; Kürz et al., 1996). The nonlinear four-wave mixing can also be used to measure carrier dynamics and gain behaviour directly in the devices, as well as for understanding device physics and application (Lucente et al., 1988a; 1988b; Zhu, 1997a; 1997b; 1997c). The gain and refractive index gratings created in broad-area semiconductor lasers by coherent four-wave mixing are very interesting nonlinear interactions which may be applied to realize high brightness semiconductor lasers as well as to study the carrier dynamics and the physics of the devices (Petersen et al., 2005). But no work on TWM was done in broad-area semiconductor amplifiers previously.

In this chapter, we present both the theoretical and experimental results of TWM in broad-area semiconductor amplifier. For the generality, we assume that the frequencies of the

pump beam and the signal beam are different, i.e., a moving gain grating and a moving refractive index grating are induced in the broad-area semiconductor amplifier. The coupled-wave equations of TWM are derived based on Maxwell's wave equation and rate equation of the carrier density. The analytical solutions of the coupled-wave equations are obtained in the condition of small signal when the total light intensity is far below the saturation intensity of the amplifier. The results show that the optical gain of the amplifier is affected by both the moving phase grating and the moving gain grating. The different contributions from both the refractive index grating and the gain grating to the TWM gain are analyzed. Depending on the moving direction of the gratings and the anti-guiding parameter, the optical gain of the amplifier may increase or decrease due to the TWM.
As a special case, the degenerate TWM (the frequencies of the pump beam and the signal beam are the same, i.e., a static gain grating and a static refractive index grating are induced in the semiconductor amplifier) in an 810 nm, 2 mm long and 200 μm wide GaAlAs broad-area amplifier is investigated experimentally. In this case, the theoretical results show that when the amplifier is operated below transparency the optical gain of both beams is increased due to the induced gain grating, and when the amplifier is operated above transparency the optical gain of both beams is decreased due to the gain grating. The refractive index grating does not affect the optical gain of both beams; and there is no energy exchange between the pump and the signal beams. The dependence of the TWM gain on the output power of the pump and angle between the two beams is measured. The experimental results show good agreement with the theory. A diffusion length of 2.0 μm for the carrier is determined from the experiment.

2. Theory of TWM in broad-area semiconductor amplifier

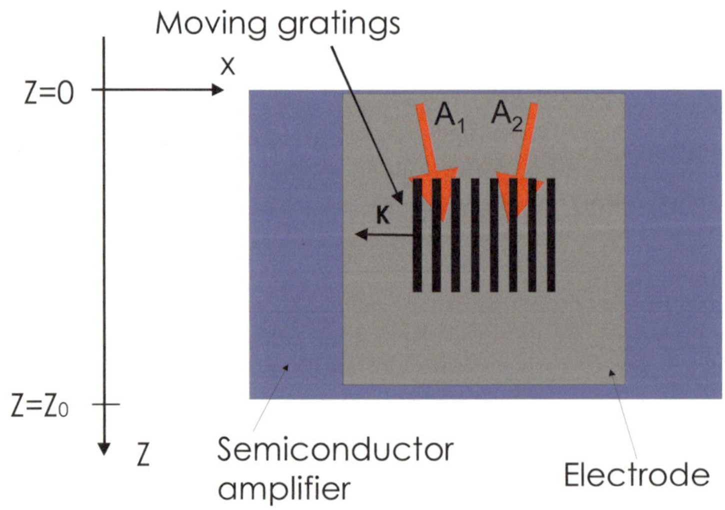

Fig. 1. Configuration of the TWM in a broad-area semiconductor amplifier with moving gratings, **K** shows the direction of the grating vector.

The TWM geometry is shown in Fig. 1, the pump beam of amplitude A_1 and the signal beam of amplitude A_2 are coupled into the broad-area amplifier. Both beams are linearly polarized along the Y direction, and the frequencies are ω_1 and ω_2 respectively. The two beams interfere in the medium to form a moving interference pattern, and a moving modulation of the carrier density in the active medium is caused, thus both a moving gain and a moving phase gratings are created. The nonlinear interaction in the gain medium is governed by Maxwell's wave equation:

$$\nabla^2 E - \frac{n^2}{c^2}\frac{\partial^2 E}{\partial t^2} = \frac{1}{\varepsilon_0 c^2}\frac{\partial^2 P}{\partial t^2}, \tag{1}$$

where n is the refractive index of the semiconductor material at transparency, c is the velocity of light in vacuum, and ε_0 is the vacuum permittivity. The total electric field is given by (Agrawal, 1987; Chi et al., 2006; 2008):

$$E = A_1 e^{i(\mathbf{K}_1\cdot\mathbf{r}-\omega_1 t)} + A_2 e^{i(\mathbf{K}_2\cdot\mathbf{r}-\omega_2 t)}, \tag{2}$$

where $\mathbf{K}_1$ and $\mathbf{K}_2$ are the wave vectors of the pump and the signal in the amplifier. P is the induced polarization in the semiconductor amplifier. It is given by (Agrawal, 1987; Chi et al., 2006; 2008):

$$P = \varepsilon_0 \chi(N) E, \tag{3}$$

where the susceptibility χ is given by (Agrawal, 1987; Chi et al., 2006; 2008):

$$\chi(N) = -\frac{nc}{\omega}(\beta + i)g(N), \tag{4}$$

the quantity β is the anti-guiding parameter accounting for the carrier-induced refractive index change in semiconductor amplifier, and $g(N)$ is the gain for the light intensity that is assumed to vary linearly with the carrier density N, i.e., $g(N) = \Gamma a(N - N_0)$ where a is the gain cross-section, Γ is the confinement factor, and N_0 is the carrier density at transparency. The carrier density N is governed by the following rate Eq. (Petersen et al., 2005; Chi et al., 2006; 2008):

$$\frac{dN}{dt} = \frac{I}{qV} - \frac{N}{\tau} + D\nabla^2 N - g(N)\frac{|E|^2}{\hbar\omega}, \tag{5}$$

where I is the injected current, q is the electron charge, V is the active volume, τ is the spontaneous recombination lifetime and D is the ambipolar diffusion constant. In the TWM process the origin of the gain and the refractive index gratings is the modulation of the carrier density due to the interference between A_1 and A_2. Thus the carrier density that leads to the formation of the moving gratings may be written as:

$$N = N_B + \Delta N \exp[i(-Kx + \delta t)] + \Delta N^* \exp[i(Kx - \delta t)], \tag{6}$$

where N_B is the average carrier density, ΔN is the induced carrier modulation. $K = |\mathbf{K}_2 - \mathbf{K}_1| = 4\pi \sin[(\theta_1 - \theta_2)/2]/\lambda_1$ is the grating vector; θ_1 is the angle between the pump beam and the Z axis, and θ_2 is the angle between the signal beam and the Z axis; we assume $\theta_1 = -\theta_2$, thus the direction of the grating vector is in the X direction; λ_1 is the wavelength of the pump beam in the amplifier (where we assume that $\lambda_1 \cong \lambda_2$, since usually the frequency difference of the signal and the pump beams is less than a few gigahertz). $\delta = \omega_2 - \omega_1$ is the frequency difference between the signal and pump beams. In the following perturbation analysis it is assumed that $\Delta N << N_B$. Inserting Eqs. (2) and (6) into Eq. (5), we find after some simple calculations that the average carrier density N_B and the carrier modulation ΔN are given by:

$$N_B = \frac{I\tau/qV + N_0 |E_0|^2/P_s}{1 + |E_0|^2/P_s}, \tag{7}$$

$$\Delta N = -\frac{(N_B - N_0) A_1 A_2^* / P_s}{1 + D\tau K^2 + |E_0|^2/P_s + i\delta\tau}, \tag{8}$$

where $|E_0|^2 = |A_1|^2 + |A_2|^2$ is the average intensity, and $P_s = (\hbar\omega)/(\Gamma a \tau)$ is the saturation intensity of the amplifier.

Inserting Eqs. (2) and (3) into Eq. (1), and using the obtained results of the average carrier density N_B and the carrier modulation ΔN, after some calculations, the coupled-wave equations for TWM with moving gratings in a broad-area semiconductor amplifier are obtained:

$$\cos\theta_1 \frac{\partial A_1}{\partial z} - i\left[-\frac{\alpha(\beta + i)}{1 + |E_0|^2/P_s}\right]\left(1 - \frac{|A_2|^2/P_s}{1 + D\tau K^2 + |E_0|^2/P_s + i\delta\tau}\right) A_1 = 0, \tag{9}$$

$$\cos\theta_2 \frac{\partial A_2}{\partial z} - i\left[-\frac{\alpha(\beta + i)}{1 + |E_0|^2/P_s}\right]\left(1 - \frac{|A_1|^2/P_s}{1 + D\tau K^2 + |E_0|^2/P_s - i\delta\tau}\right) A_2 = 0, \tag{10}$$

where $\alpha = \Gamma a (I\tau/qV - N_0)/2$ is the small-signal gain coefficient of the amplifier. Since the refractive index of the semiconductor material is high, normally the angles θ_1 and θ_2 are less than 2° in experiment (Chi et al., 2006); so the cosine factor in Eqs. (9) and (10) is neglected below.

In the small signal approximation, and if we assume that the total intensity of the two beams is much less than the saturation intensity, i.e., $|A_2|^2 << |A_1|^2 << P_s$, the terms accounting for saturation in the denominator and the term accounting for the coupling in Eq. (9) may be neglected. Thus the coupled-wave equations can be solved analytically. The solutions are:

$$A_1 = A_{10} \exp[(1 - i\beta)\alpha z], \tag{11}$$

$$A_2 = A_{20} \exp\{(1 - i\beta)[\alpha z - \gamma_1 (e^{2\alpha z} - 1)/2]\}, \tag{12}$$

where A_{10} and A_{20} are the amplitudes of the pump and the signal beams at the front facet of the amplifier. γ_1 is a parameter defined as:

$$\gamma_1 = \frac{|A_{10}|^2}{P_s}\left[1 + \frac{1}{(1 + D\tau K^2 - i\delta\tau)}\right]. \tag{13}$$

The first term in Eq. (13) is for the saturation effect, the second term is for the beam coupling.
Define the TWM gain of the signal beam g_{TWM} as the natural logarithm of the ratio of the output intensity of signal with the coherent pump to that with the non-coherent pump:

$$g_{\mathrm{TWM}} = \ln\left(\frac{\left|A_2(z_0)_{\text{coherent pump}}\right|^2}{\left|A_2(z_0)_{\text{non-coherent pump}}\right|^2} \right) = -\frac{\left|A_1(z_0)\right|^2 - \left|A_{10}\right|^2}{P_s} \frac{1 + D\tau K^2 + \beta\delta\tau}{(1 + D\tau K^2)^2 + (\delta\tau)^2}, \tag{14}$$

where z_0 is the length of the semiconductor amplifier. The non-coherent pump means the pump beam is not coherent with the signal beam, but the intensity is the same as the coherent pump, thus the term accounting for saturation in Eq. (14) vanishes. In experiment, the coherent pump and the non-coherent pump can be achieved by changing the polarization direction of the pump beam (Chi et al., 2006). Eq. (14) shows that g_{TWM} changes linearly with the output intensity (power) of the pump, and it decreases quickly when the angle between the two beams increases because the diffusion of carriers washes out the gratings as the angle between the two beams increases. Eq. (14) also shows that depending on the detuning frequency δ, the TWM gain can be positive or negative no matter the amplifier is operated above or below the transparency (i.e., $\left|A_1(z_0)\right|^2 = \left|A_{10}\right|^2$). These phenomena will be discussed below.

3. Experiment of the degenerate TWM in a broad-area amplifier

In order to verify the theory described in Section 2, a special case of TWM, i.e., degenerate TWM, in a broad-area semiconductor amplifier is investigated experimentally. For this case, the frequencies of the pump and the signal are the same, i.e., δ = 0; thus a static refractive index grating and a static gain grating are induced in the amplifier. The coupled-wave equations (9) and (10) in this case are changed to:

$$\frac{\partial A_1}{\partial z} - i\left[-\frac{\alpha(\beta + i)}{1 + \left|E_0\right|^2 / P_s} \right]\left(1 - \frac{\left|A_2\right|^2 / P_s}{1 + D\tau K^2 + \left|E_0\right|^2 / P_s} \right) A_1 = 0, \tag{15}$$

$$\frac{\partial A_2}{\partial z} - i\left[-\frac{\alpha(\beta + i)}{1 + \left|E_0\right|^2 / P_s} \right]\left(1 - \frac{\left|A_1\right|^2 / P_s}{1 + D\tau K^2 + \left|E_0\right|^2 / P_s} \right) A_2 = 0. \tag{16}$$

The equations show that the coupling term between the two beams decreases the optical gain (above transparency) or absorption (below transparency) for both beams simultaneously. This is different to the situation in photorefractive materials, where one beam is amplified and the other is decreased at the same time (Marrakchi et al., 1981; Huignard & Marrakchi, 1981; Yeh, 1983; 1989). This is also different to the situation in Kerr media, where the intensity of one beam is not affected by the other beam in the degenerate TWM case (Yeh, 1986; 1989; Chi et al., 2009).
To clarify this phenomenon, the relative position of the intensity pattern, the carrier density grating, the refractive index grating and the gain grating is shown in Fig. 2 when the amplifier is operated above the transparency. Because of the spatial hole-burning effect, the

carrier density grating is π out of phase with the intensity pattern. Since the gain varies linearly with the carrier density, the gain grating is also π out of phase with the intensity pattern. The refractive index grating is in phase with the interference intensity pattern due to the anti-guiding effect. The refractive index grating has no contribution to the energy coupling between the two beams, when it is in phase or π out of phase with the interference pattern (Yeh, 1989). The gain grating will decrease the optical gain of both beams simultaneously because it is π out of phase with the interference pattern. Reversely, the gain grating will decrease the absorption of both beams when the amplifier is operating below the transparency since the gain grating is in phase with the interference pattern in that case (Chi et al., 2009).

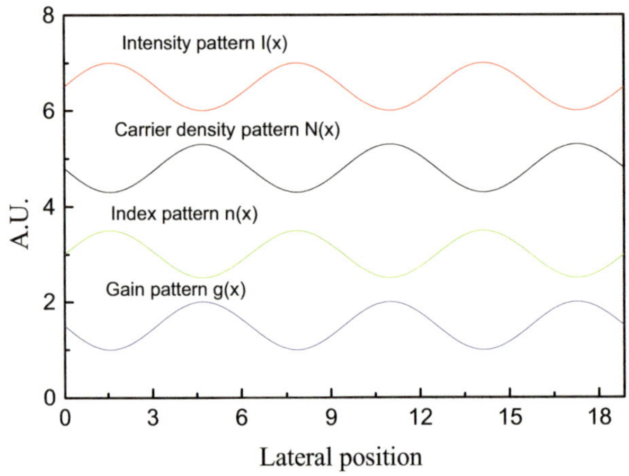

Fig. 2. The relative position of the interference pattern, the carrier density grating, the refractive index and the gain gratings formed in the broad-area amplifier.

Since $\delta = 0$, the TWM gain for the degenerate TWM is changed to:

$$g_{\text{TWM}} = \ln\left(\frac{\left|A_2(z_0)_{\text{coherent pump}}\right|^2}{\left|A_2(z_0)_{\text{non-coherent pump}}\right|^2}\right) = -\frac{\left|A_1(z_0)\right|^2 - \left|A_{10}\right|^2}{\left(1 + D\tau K^2\right)P_s}. \tag{17}$$

Eq. (17) shows that the g_{TWM} is negative when the amplifier is operated above the transparency, is positive when it is operated below the transparency, and is zero when it is operated at transparency. It agrees with the analyse above. Eq. (17) also shows that the g_{TWM} decreases linearly with the output intensity (power) of the pump, and it decreases quickly when the angle between the two beams increases because the diffusion of carriers washes out the gratings as the angle between the two beams increases. These analyses will be verified by experiments of TWM in a semiconductor amplifier below.

The experimental set-up is shown in Fig. 3. The set-up is arranged like a Mach-Zehnder interferometer. The pump beam A_1 and the signal beam A_2 are derived from a tunable diode laser system based on a tapered amplifier (Chi et al., 2005). The wavelength used in the experiment is 813.5 nm. We use the same method as Goldberg (Goldberg et al., 1993) to couple the two beams into the broad-area amplifier. In each arm, a combination of a cylindrical lens of 150 mm focal length and an aspherical lens of 8.0 mm focal length with a N.A. of 0.5 (this lens is shared by the two arms) is used as an afocal telescope to inject the two beams into the amplifier. The input coupling efficiency of this setup is around 50%. The two cylindrical lenses can be translated in the arrow direction to vary the injection angle in the junction plane. After the amplifier, a cylindrical lens of 5.0 mm focal length is used to collimate the output beam in the fast axis. A half-wave plate is inserted in the pump arm to change the polarization direction of the pump. All the components are antireflection coated for the near infrared wavelength.

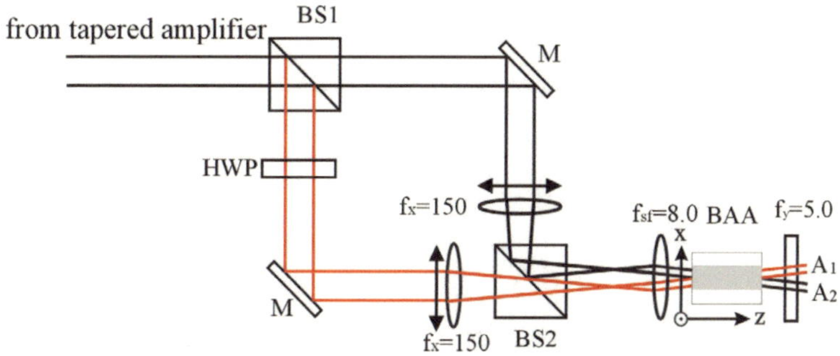

Fig. 3. Experimental set-up for the TWM in a broad-area amplifier. M: mirror, BSs: beam splitter, HWP: half-wave plate, BAA: broad-area amplifier (the units are in mm).

The broad-area amplifier is an 810 nm, 2 mm long and 200 μm wide GaAlAs amplifier. It was grown by the Metallorganic Chemical Vapor Phase Deposition (MOCVD) technique on a GaAs substrate by Alcatel Thales III-V Lab. The structure contains a Large Optical Cavity (LOC), which has a thickness of approximately 1 μm, and which consists of a tensile-strained GaInP quantum well, two GaInP barriers and two AlGaInP claddings. Both facets of the amplifier are antireflection coated; the reflectivity is less than 0.1%.

First, the dependence of the g_{TWM} on the output power of the pump is measured. The input powers of the pump and the signal measured before the aspherical lens are 21.0 and 4.1 mW. The angle between the two beams is around 4°. The output power of the signal was measured at different injected current of the amplifier with a co-polarized pump (the polarization direction of both beams is perpendicular to the chip of the amplifier) and an orthogonally-polarized pump. The output power of the pump was measured when it is coherent with the signal. The experimental results are shown in Fig. 4. It is clearly seen that the g_{TWM} decreases linearly with the output pump power. Fitting the experimental data with Eq. (17), the two parameters: the input power of the pump $|A_{10}|^2$ and the $1/(1+D\tau K^2)P_s$ are obtained. The $|A_{10}|^2$ is round 9.1 mW, corresponding to a coupling efficiency of 43%; and using the result of $D\tau$ obtained later, the saturation power P_s is found to be around 220 mW, which is much larger than the output power of the pump in this experiment. Using the value of $|A_{10}|^2$ and Eq. (11), the optical gain of 1.7 is obtained for the highest output power of the pump. The g_{TWM} is about 5% of the optical gain.

The dependence of the g_{TWM} on the grating vector is also measured by changing the angle between the two beams. The direction of the pump beam is fixed during the experiment; the angle is changed by changing the direction of the signal beam. The injected powers of the pump and the signal measured before the aspherical lens are 21.0 and 4.1 mW; the output

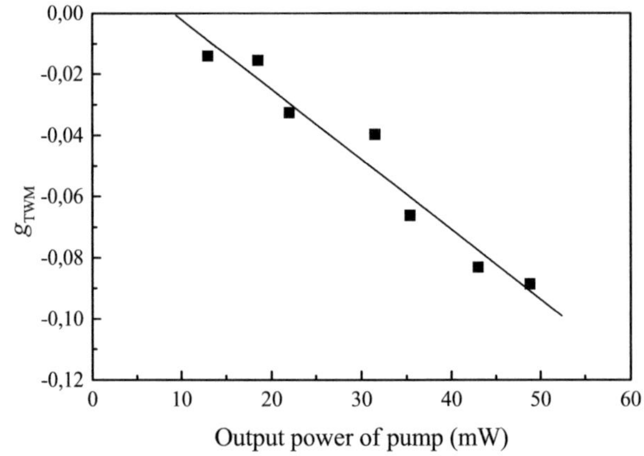

Fig. 4. The g_{TWM} versus the output power of the pump. The squares are the measured data; the line is the fitted result with Eq. (17).

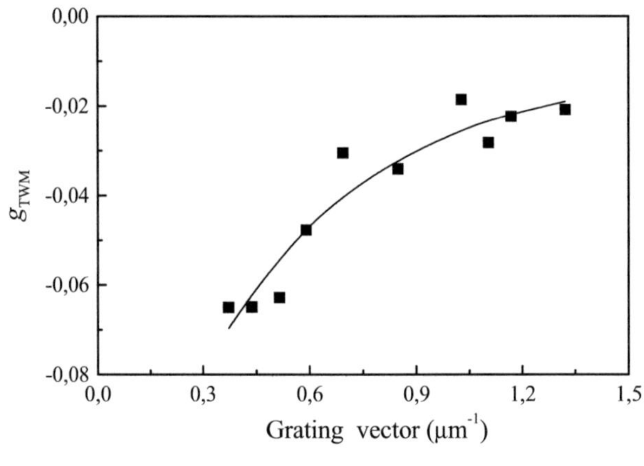

Fig. 5. The g_{TWM} versus the grating vector in the BAA. The squares are the measured data: the curve is the fitted result with Eq. (17).

power of the pump is around 35 mW. The experimental results are shown in Fig. 5. Fitting the experimental data with Eq. (17), $D\tau$ is obtained to be 4.1 μm², leading to a diffusion length $L = \sqrt{D\tau}$ of 2.0 μm. Assuming that τ is 5 ns (Marciante & Agrawal, 1996), D is calculated to be 8.2 cm²/s. This is in good agreement with the direct measured value of 9.5 cm²/s (Lucente et al., 1988b). We should mention that the output power of the pump beam is decreased a little when the polarization direction of it is changed from perpendicular to the chip to parallel to the chip. We do not know the reason of this decrease but the effect of this decrease on the measured g_{TWM} is small.

To obtain the coupled-wave equations of TWM, three assumptions are made. Here we should discuss the validity of these assumptions in our experiment. The first is the plane-wave assumption. In the experiment, since the two beams are coupled into the amplifier from an external laser, the mode of the two beams in the slow axis is determined by the external laser and the focusing optics. The two beams are nearly Gaussian beams in the slow axis, they are collimated by the aspherical lens and the width of the beams is around 140 μm. We believe the plane-wave assumption is a good approximation for these two beams in this direction. The wave guiding mode of the field distribution in the fast axis does not affect the derivation of the equations (Marciante & Agrawal, 1996). The second is the linear variation of the material gain $g(N)$ on the carrier density. The transparent current of the amplifier used here is around 1.1 A, and the highest current used in our experiment is 1.8 A, according to Eq. (7), the carrier density N_B is calculated to be around 1.5 N_0, not much higher than the transparent carrier density. The third assumption is the small population modulation in Eq. (6). With the injected current of 1.8 A, according to Eq. (8), $|\Delta N|$ is calculated to be around 2% of N_0 (≈1.3% of N_B), it is much less than the average carrier density N_B. Therefore, we believe the assumptions made in the theory are valid in our experiment.

In conclusion, the degenerate TWM in broad-area semiconductor amplifier is investigated experimentally. The experimental results show good agreement with the theory. The validity of the theory is discussed.

4. Calculations and discussion

Unlike the condition of degenerate TWM, where only static gratings are generated; the coupling term between the two beams has different contribution to the optical gain of these two beams for the nondegenerate TWM (Chi et al., 2008). The nondegenerate TWM may increase the power of one beam and decrease the power of another beam in this case, i.e., energy exchange occurs.

According to Eq. (14), the dependence of g_{TWM} on the frequency difference δ with different anti-guiding parameter β is calculated; here we assume that the amplifier is operated above the transparent current. The calculated results are shown in Fig. 6. In the calculation, we use the same parameters used in and obtained from the TWM experiment in a GaAlAs broad-area semiconductor amplifier with static gratings described in Section 3 (Chi et al., 2006); i.e., $|A_1(z_0)|^2$ = 48.8 mW, $|A_{10}|^2$ = 9.1 mW, P_s = 220 mW, $D\tau$ = 4.1 μm², K = 0.51 μm⁻¹ (the K value corresponds to a 4.0° angle between the two beams). Assuming that τ is 5 ns (Marciante & Agrawal, 1996). From Fig. 6 we can find that when $\delta = 0$, the g_{TWM} is negative and independent of β; if $\beta = 0$, the g_{TWM} is always negative and the curve of the TWM gain versus δ is symmetric around the axis of $\delta = 0$. If $\beta \neq 0$, however, the g_{TWM} is negative when $\delta > 0$, and the g_{TWM} can be negative or positive when $\delta < 0$. These properties can be explained by analyzing the different contributions from the refractive index grating and the gain grating formed in the broad-area semiconductor amplifier to the TWM gain.

Since the frequencies of the pump and the signal are different, a moving interference pattern is generated in the amplifier: $|E|^2 = |E_0|^2 + \left[A_1 A_2^* e^{i(-Kx+\delta t)} + c.c.\right]$. Inserting Eq. (8) into Eq. (6), the carrier density is obtained:

$$N = N_B + \left\{ -\frac{(N_B - N_0) A_1 A_2^* / P_s}{1 + D\tau K^2 + |E_0|^2 / P_s + i\delta\tau} \exp[i(-Kx + \delta t)] + c.c. \right\}. \quad (18)$$

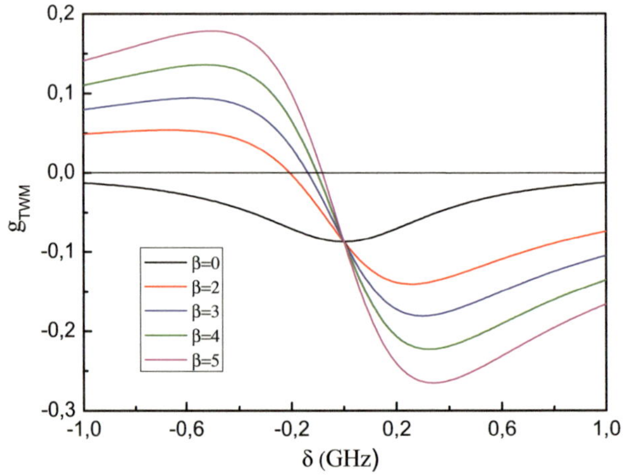

Fig. 6. The calculated TWM gain g_{TWM} versus δ with different anti-guiding parameter β according to Eq. (14).

The modulation part N_{m} of the carrier density for the generating of the gain and the phase gratings is:

$$N_m = \frac{A_1 A_2^*(N_B - N_0)/P_s}{\sqrt{\left(1 + D\tau K^2 + |E_0|^2/P_s\right)^2 + (\delta\tau)^2}} \exp[i(-Kx + \delta t + \pi - \theta)] + c.c., \tag{19}$$

where

$$\theta = arctg\frac{\delta\tau}{1 + D\tau K^2 + |E_0|^2/P_s} \quad (-\pi/2 < \theta < \pi/2). \tag{20}$$

Eq. (19) shows, because of the hole-burning effect and the finite response time of the broad-area amplifier, there is a phase difference π-θ between the interference pattern and the carrier density grating. Since the gain varies linearly with the carrier density, the gain grating Δg is also π-θ out of phase with the intensity pattern, i.e.,

$$\Delta g = \frac{\Gamma a(N_B - N_0)}{2P_s} \frac{A_1 A_2^*}{\sqrt{\left(1 + D\tau K^2 + |E_0|^2/P_s\right)^2 + (\delta\tau)^2}} \exp[i(-Kx + \delta t + \pi - \theta)] + c.c.. \tag{21}$$

The refractive index grating is π out of phase with the gain grating because of the anti-guiding effect, so the refractive index grating Δn is $-\theta$ out of phase with the intensity pattern and proportional to the anti-guiding parameter β, i.e.,

$$\Delta n = \frac{\Gamma a \lambda_1 (N_B - N_0)}{4\pi\, P_s} \frac{\beta A_1 A_2^*}{\sqrt{\left(1 + D\tau K^2 + |E_0|^2/P_s\right)^2 + (\delta\tau)^2}} \exp[i(-Kx + \delta t - \theta)] + c.c.. \tag{22}$$

The relative position of the interference pattern, the carrier density grating, the refractive index grating and the gain grating formed in the broad-area amplifier is shown in Fig. 7.

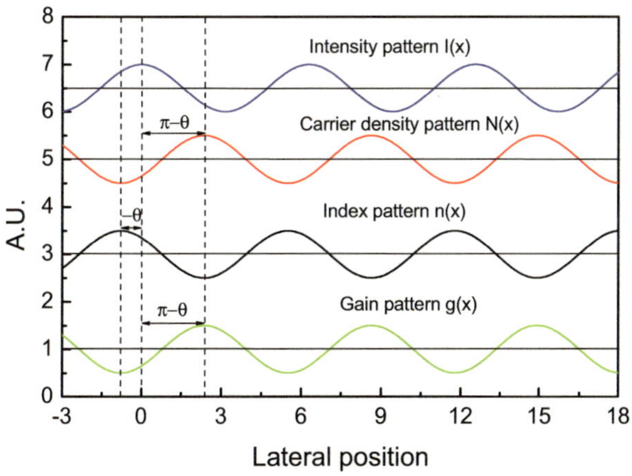

Fig. 7. The relative position of the interference pattern, the carrier density grating, the refractive index grating and the gain grating formed in the BAA, assuming $1+D\tau K^2+|E_0|^2/P_s=\delta\tau$, i.e., $\theta=\pi/4$.

The TWM gain caused by the gain grating g_{gain} is (Chi et al., 2009):

$$g_{gain} \propto \frac{\cos(\pi-\theta)}{\sqrt{\left(1+D\tau K^2+|E_0|^2/P_s\right)^2+(\delta\tau)^2}}. \tag{23}$$

Here we should notice that the effect of the gain grating is the same for both beams, i.e., to increase (below transparent current) or decrease (above transparent current) the intensity of the pump and the signal beams simultaneously, thus it will not cause the energy exchange between the two beams. The TWM gain caused by the phase grating g_{phase} is (Yeh, 1989; Chi et al., 2009):

$$g_{phase} \propto \frac{\beta\sin(-\theta)}{\sqrt{\left(1+D\tau K^2+|E_0|^2/P_s\right)^2+(\delta\tau)^2}}. \tag{24}$$

When $\delta \neq 0$, the refractive index grating will cause energy exchange between the two beams, since there is a phase difference $-\theta$ ($\theta \neq 0$) between the intensity pattern and the refractive index grating (Yeh, 1989). The TWM gain g_{TWM} is the sum of g_{gain} and g_{phase}.
When $\delta = 0$, (i.e., static gratings are induced in the amplifier), θ is equal to zero; thus the gain grating is π out of phase with the interference pattern, and the phase grating is in phase with the interference pattern. According to Eqs (23) and (24), the gain of the phase grating g_{phase} is zero; and the g_{TWM} equal to g_{gain}, is negative and independent of β (Chi et al., 2006). If $\beta = 0$, only the gain grating is generated; according to Eqs. (20) and (23), the TWM gain g_{TWM} is always negative and is symmetric around the axis of $\delta = 0$. If $\beta \neq 0$, both a gain grating and a

phase grating are generated. When $\delta > 0$ ($\theta > 0$), according to Eqs. (23) and (24), both the g_{gain} and the g_{phase} are negative, so the g_{TWM} is negative; when $\delta < 0$ ($\theta < 0$), the g_{gain} is negative and the g_{phase} is positive, so the g_{TWM} can be positive or negative.

The parameters β and τ can be obtained by fitting the measured results of g_{TWM} versus δ. The optimal δ to achieve the maximum TWM gain depends on the device parameters τ, D, β and the grating vector K. From Eq. (14), the optimal δ is $\delta_{opt} = -(1 + D\tau K^2)(1 \pm \sqrt{1+\beta^2})/\beta\tau$.

5. Conclusion

In conclusion, the TWM in broad-area semiconductor amplifier in nondegenerate condition is investigated theoretically. The coupled-wave equations are derived and analytical solutions are obtained when the intensity of the pump is much larger than that of the signal, but much less than the saturation intensity of the amplifier. A special case of TWM, degenerate TWM, is investigated experimentally in a GaAlAs broad-area semiconductor amplifier. The experimental results show good agreement with the theory, and the validity of the theory for this experiment is discussed. A diffusion length of 2.0 µm is determined from the experiment. The TWM gain in broad-area semiconductor amplifier is calculated as a function of the frequency difference between the pump and the signal based on the data obtained from the degenerate TWM experiment; and the calculated results are discussed based on the different contributions from the refractive index grating and the gain grating to the TWM gain. Depending on δ and β, the TWM gain in semiconductor broad-area amplifier can be positive or negative. The energy exchange between the pump and signal beams occurs when $\delta \neq 0$.

6. Acknowledgements

The authors acknowledge the financial support of the European community through the project WWW.BRIGHTER.EU (Grant No. FP6-IST-035266). Mingjun Chi wishes to acknowledge the Danish Research Agency under grant no. 26-04-0229.

7. References

Agrawal, G.P. (1987). Four-wave mixing and phase conjugation in semiconductor laser media. *Opt. Lett.*, Vol. 12, No. 4, 260-262

Brignon, A. & Huignard, J.-P. (1993). Two-wave mixing in Nd:YAG by gain saturation. *Opt. Lett.*, Vol. 18, No. 19, 1639-1641

Chi, M.; Jensen, O.B.; Holm, J.; Pedersen, C.; Andersen, P.E.; Erbert, G.; Sumpf, B. & Petersen P.M. (2005). Tunable high-power narrow-linewidth semiconductor laser based on an external-cavity tapered amplifier. *Opt. Express*, Vol. 13, No. 26, 10589-10596

Chi, M.; Jensen, S.B.; Huignard, J.-P. & Petersen, P.M. (2006). Two-wave mixing in a broad-area semiconductor amplifier. *Opt. Express*, Vol. 14, No. 25, 12373-12379

Chi, M.; Huignard, J.-P. & Petersen, P.M. (2008). Nonlinear gain amplification due to two-wave mixing in a broad-area semiconductor amplifier with moving gratings. *Opt. Express*, Vol. 16, No. 8, 5565-5571

Chi, M.; Huignard, J.-P. & Petersen, P.M. (2009). A general theory of two-wave mixing in nonlinear media. *J. Opt. Soc. Am. B*, Vol. 26, No. 8, 1578-1584

Frey, R. (1986). On-axis intracavity nearly degenerate four-wave mixing in lasers. *Opt. Lett.*, Vol. 11, No. 2, 91-93

Garrett, M.H.; Chang, J.Y.; Jenssen, H.P. & Warde, C. (1992). High beam-coupling gain and deep- and shallow-trap effects in cobalt-doped barium titanate, $BaTiO_3$:Co. *J. Opt. Soc. Am. B*, Vol. 9, No. 8, 1407-1415

Goldberg, L.; Mehuys, D.; Surette, M.R. & Hall, D.C. (1993). High-power, near-diffraction-limited large-area travelling-wave semiconductor amplifiers. *IEEE J. Quantum Electron.*, Vol. 29, No. 6, 2028-2042

Grandclément, D.; Grynberg, G. & Pinard, M. (1987). Observation of continuous-wave self-oscillation due to pressure-induced two-wave mixing in Sodium. *Phys. Rev. Lett.*, Vol. 59, No. 1, 40-43

Huignard, J.-P. & Marrakchi, A. (1981). Coherent signal beam amplification in two-wave mixing experiments with photorefractive $Bi_{12}SiO_{20}$ crystals. *Opt. Commun.*, Vol. 38, No. 4, 249-254

Kürz, P.; Nagar, R. & Mukai, T. (1996). Highly efficient phase conjugation using spatially nondegenerate four-wave mixing in a broad-area laser diode. *Appl. Phys. Lett.*, Vol. 68, No. 9, 1180-1182

Lucente, M.; Carter, G.M. & Fujimoto J.G. (1988a). Nonlinear mixing and phase conjugation in broad-area diode lasers. *Appl. Phys. Lett.*, Vol. 53, No. 6, 467-469

Lucente, M.; Fujimoto J.G. & Carter, G.M. (1988b) Spatial and frequency dependence of four-wave mixing in broad-area diode lasers. *Appl. Phys. Lett.*, Vol. 53, No. 20, 1897-1899

Marciante, J.R. & Agrawal, G.P. (1996). Nonlinear mechanisms of filamentation in broad-area semiconductor lasers. *IEEE J. Quantum Electron.*, Vol. 32, No. 4, 590-596

Marrakchi, A.; Huignard, J.-P. & Günter, P. (1981). Diffraction efficiency and energy transfer in two-wave mixing experiments with $Bi_{12}SiO_{20}$ crystals. *Appl. Phys.*, Vol. 24, No. 2, 131-138

McGraw, R. (1992). Light-scattering-noise limits to two-wave-mixing gain in Kerr media. *J. Opt. Soc. Am. B*, Vol. 9, No. 1, 98-103

Nakajima, H. & Frey, R. (1985). Intracavity nearly degenerate four-wave mixing in a (GaAl)As semiconductor laser. *Appl. Phys. Lett.*, Vol. 47, No. 8, 769-771

Nakajima, H. & Frey, R. (1986). Collinear nearly degenerate four-wave mixing in intracavity amplifying media. *IEEE J. Quantum Electron.*, Vol. 22, No. 8, 1349-1354

Petersen, P.M.; Samsøe, E.; Jensen, S.B. & Andersen, P.E. (2005). Guiding of laser modes based on self-pumped four-wave mixing in a semiconductor amplifier. *Opt. Express*, Vol. 13, No. 9, 3340-3347

Silberberg, Y. & Bar-Joseph, I. (1982). Instabilities, self-oscillation, and chaos in a simple nonlinear optical interaction. *Phys. Rev. Lett.*, Vol. 48, No. 22, 1541-1543

Silberberg, Y. & Bar-Joseph, I. (1984). Optical instabilities in a nonlinear Kerr medium. *J. Opt. Soc. Am. B*, Vol. 1, No. 4, 662-670

Yeh, P. (1983). Contra-directional two-wave mixing in photorefractive media. *Opt. Commun.*, Vol. 45, No. 5, 323-326

Yeh, P. (1986). Exact solution of a nonlinear model of two-wave mixing in Kerr media. *J. Opt. Soc. Am. B*, Vol. 3, No. 5, 747-750

Yeh, P. (1989). Two-wave mixing in nonlinear media. *IEEE J. Quantum Electron.*, Vol. 25, No. 3, 484-519

Zhu, D.X.; Dubovitsky, S.; Steier, W.H.; Uppal, K.; Tishinin, D.; Burger, J. & Dapkus, P.D. (1997a). Noncollinear four-wave mixing in a broad area semiconductor optical amplifier. *Appl. Phys. Lett.*, Vol. 70, No. 16, 2082-2084

Zhu, D.X.; Dubovitsky, S.; Steier, W.H.; Burger, J.; Tishinin, D.; Uppal, K. & Dapkus, P.D. (1997b). Ambipolar diffusion coefficient and carrier lifetime in a compressively strained InGaAsP multiple quantum well device. *Appl. Phys. Lett.*, Vol. 71, No. 5, 647-649

Zhu, D.X.; Dubovitsky, S.; Steier, W.H.; Uppal, K.; Tishinin, D.; Burger, J. & Dapkus, P.D. (1997c). A novel all-optical switch: the wavelength recognizing switch. *IEEE Photonics Tech. Lett.*, Vol. 9, No. 8, 1110-1112

7

Frequency Conversion based on Three-Wave Parametric Solitons

Fabio Baronio[1], Matteo Conforti[1], Costantino De Angelis[1], Antonio Degasperis[2], Sara Lombardo[3,4] and Stefan Wabnitz[1]
[1]CNISM, Universitá degli Studi di Brescia,
[2]Universitá di Roma La Sapienza,
[3]Vrije Universiteit,
[4]Manchester University,
[1,2]Italy
[3]The Netherland
[4]England

1. Introduction

Parametric three-wave mixing in quadratic nonlinear crystals provides a versatile means of achieving widely tunable frequency conversion of laser light, thus permitting a substantial extension of the wavelength coverage of laser sources across broad wavelength regions. For mode-locked lasers, the effective parametric interaction length among short pulses at different wavelengths is limited by crystal dispersion, which leads to group velocity walk-off. Nevertheless, as well known since the 1970s, the frequency conversion of short pulses may be significantly enhanced by means of the optical soliton concept (Zakharov & Manakov, 1973). Indeed, the collision of two soliton input pulses at different frequencies, with proper durations and peak powers, leads to a time compressed pulse at the sum-frequency (Ibragimov & Struthers, 1996). However such pulse is unstable, since it rapidly decays into two, time-shifted replicas of the input pulses. On the other hand, two resonant bright pulses and a kink (or phase jump across a CW background) pulse may propagate locked together as an optical simulton, in spite of their different linear group velocities, owing to their nonlinear mutual trapping (Nozaki & Taniuti, 1973; Degasperis et al., 2006).

In this chapter we present at first a self-contained derivation of the three-wave equations in dispersive and quadratic nonlinear crystals. We outline the generality as well as the limitations to the application of the basic three-wave parametric interaction equations, in which intra-pulse chromatic dispersion is neglected. Next we review a recently discovered class of exact simulton solutions of the parametric three-wave interaction (Degasperis et al., 2006; Conforti et al., 2006; Degasperis et al., 2007), which have as potential applications the stable sum and frequency difference generation of short optical pulses and pulse trains. These nonlinear waves are characterised by the property that the group velocity of the two bright pulses and the kink may be controlled by varying the input energy of the wave-packets, or the relative energy distribution among the three waves (Degasperis et al., 2006). It is particularly significant that the common velocity and the relative energy distribution of these mutually trapped pulses may gradually evolve upon propagation through the crystal

(Conforti et al., 2007). Such unique feature may enable one to achieve a stable frequency conversion of short optical pulses, as well as a continuous control of their frequency conversion efficiency by adjusting the input intensity of a continuous wave background wave (Conforti et al., 2007). In addition, the control of the CW power level also permits the highly efficient generation of a frequency converted train of short optical pulses, which represent a time-periodic copy at a different optical frequency of the input short signal pulse (Baronio et al., 2008). We provide simple analytical relations between the time duration, amplitude and repetition rate of the generated pulse sequence, and the parameters of the input short pulse and CW. Finally, we describe the practical conditions for the experimental observation of the above described pulse shaping, generation and speed control effects via parametric three wave mixing in commonly available quadratic nonlinear crystals. In particular, a simple and easily achievable condition must be respected for the linear group velocities of the three interacting waves, namely that the group velocity of the sum-frequency wave must be intermediate between the group velocities of the two fundamental waves.

2. Equations

The resonant interaction of three waves (TWI) is an important process which appears in various contexts of physics, such as fluid dynamics and optics. It describes the mixing of three quasi–monochromatic waves whose wave numbers $\vec{k}_1, \vec{k}_2, \vec{k}_3$ and frequencies ω_1, ω_2, ω_3 satisfy the resonant conditions

$$\vec{k}_1 + \vec{k}_2 = \vec{k}_3 \ , \ \omega_1 + \omega_2 = \omega_3 \tag{1}$$

in weakly nonlinear and dispersive media. In optics this resonant interaction may occur in any nonlinear medium where nonlinear effects are small and can be considered as a perturbation of the linear wave propagation, the lowest-order nonlinearity is quadratic in the field amplitudes and the dispersion relation $\omega(\vec{k})$ allows for the special choice of the wave numbers $\vec{k}_\alpha$, $\alpha = 1, 2, 3$, such that the phase matching (or resonance) condition (1) is satisfied, where $\omega_\alpha = \omega(\vec{k}_\alpha)$. The proper description of the resonant interaction is provided by a relatively simple model which can be derived from Maxwell equations in a dielectric with quadratic nonlinear susceptibility $\chi^{(2)}$. Here below we first sketch the way to obtain the system of three coupled equations which model the resonant interaction, and then we discuss some of its properties together with special solutions of applicative interest.

If $\vec{P}(\vec{r},t)$ is the dielectric polarization vector field induced by the electromagnetic wave, the charge and current densities in the medium are (hereafter a subscripted variable stands for partial differentiation with respect to that variable, and a dot between two vectors indicates their scalar product)

$$\rho = -\vec{\nabla}_r \cdot \vec{P} \ , \ \vec{J} = \vec{P}_t \ , \tag{2}$$

so that Maxwell's equations reduce in a standard way to the single vector equation for the electric vector field $\vec{E}(\vec{r},t)$

$$\vec{E}_{tt} - c^2[\Delta_r \vec{E} - \vec{\nabla}_r(\vec{\nabla}_r \cdot \vec{E})] = -\frac{1}{\epsilon_0}\vec{P}_{tt} . \tag{3}$$

Here we assume that the medium is homogeneous but not isotropic, and that only its linear response, namely $\chi_{jn}^{(1)}$, is frequency dependent; moreover losses are assumed to be negligibly small. Thus the dielectric polarization field components P_j are related to the electric field by the standard formula (summation over repeated indexes is understood)

$$P_j(\vec{r},t) = \epsilon_0 \int_{-\infty}^{t} dt_1 \chi_{jn}^{(1)}(t-t_1) E_n(\vec{r},t_1) + \epsilon_0 \chi_{jnm}^{(2)} E_n(\vec{r},t) E_m(\vec{r},t) \quad . \tag{4}$$

The way to proceed further is via the widely used multiscale (or slowly varying amplitude approximation) method (see for instance (Degasperis, 2009) and references quoted there). The starting point is the following approximate expression of the solution (here *c. c.* means complex conjugate)

$$\begin{aligned} E_j(\vec{r},t) &= \eta[A^{(1)}(\vec{\xi},\tau)B_j^{(1)}e^{i(\vec{k}_1\cdot\vec{r}-\omega_1 t)} + A^{(2)}(\vec{\xi},\tau)B_j^{(2)}e^{i(\vec{k}_2\cdot\vec{r}-\omega_2 t)} \\ &+ A^{(3)}(\vec{\xi},\tau)B_j^{(3)}e^{i(\vec{k}_3\cdot\vec{r}-\omega_3 t)}] + c.c. + O(\eta^2) \ , \end{aligned} \tag{5}$$

where η is a small parameter which enters also in the scaled spatial co-ordinates $\vec{\xi} = \eta\vec{r}$ and in the slow time variable $\tau = \eta t$. This expression shows that the electric field is the superposition of three quasi-monochromatic waves, namely of three carrier waves which are modulated by the slowly varying amplitudes $A^{(\alpha)}$, α =1, 2, 3. Moreover, we have additionally assumed that only the wave amplitudes are affected by the nonlinear dynamics but not their state of polarization, so that the three vectors $\vec{B}^{(\alpha)}$, α = 1, 2, 3 are taken co-ordinate and time independent while the three dynamical variables $A^{(\alpha)}(\vec{\xi},\tau)$, α = 1, 2, 3 are just scalars. The wave-number vectors k_{1j}, k_{2j}, k_{3j}, their corresponding frequencies ω_1, ω_2, ω_3, together with the polarization state vectors $B_j^{(1)}, B_j^{(2)}, B_j^{(3)}$ satisfy the dispersion relation

$$(\omega^2 - c^2\vec{k}\cdot\vec{k})B_j + c^2\vec{k}\cdot\vec{B}k_j + \omega^2\hat{\chi}_{jn}^{(1)}(\omega)B_n = 0 \, , \tag{6}$$

where the function $\hat{\chi}_{jn}^{(1)}(\omega)$ is the Fourier transform

$$\hat{\chi}_{jn}^{(1)}(\omega) = \int_0^{+\infty} dt e^{i\omega t} \chi_{jn}^{(1)}(t) \quad . \tag{7}$$

It now remains to insert the expression (5) into the Maxwell's equation (3) and to expand the resulting formula in powers of the small perturbative parameter η. We omit detailing this computation and we limit ourselves to the following remarks. At first order, $O(\eta)$, one merely finds that wave numbers, frequencies and polarization state vectors of the three fundamental harmonics (see Eq. (5)) satisfy the dispersion relation (6). The resonance conditions (1) play a crucial role at the next order, $O(\eta^2)$, as they imply the coupling between the three wave amplitudes. Moreover, since the quadratic nonlinearity implies that all harmonics of the form $\exp\{i[\alpha_1(\vec{k}_1\cdot\vec{r}-\omega_1 t)+\alpha_2(\vec{k}_2\cdot\vec{r}-\omega_2 t)\}$ come into play, with any integer α_1 and α_2, one has to assume also that the dispersion relations of the medium are satisfied only for the three cases ($\alpha_1 = \pm1$, $\alpha_2 = 0$), ($\alpha_1 = 0$, $\alpha_2 = \pm1$) and ($\alpha_1 = \alpha_2 = \pm1$). This condition guarantees that all higher harmonics are in the $O(\eta^2)$ term in the expression (5) and can therefore be omitted. The resulting $O(\eta^2)$ equations which are derived this way are then

$$\frac{\partial A_1}{\partial \tau} + \vec{v}_1 \cdot \vec{\nabla}_\xi A_1 = i\gamma_1 A_2^* A_3$$

$$\frac{\partial A_2}{\partial \tau} + \vec{v}_2 \cdot \vec{\nabla}_\xi A_2 = i\gamma_2 A_1^* A_3 \qquad (8)$$

$$\frac{\partial A_3}{\partial \tau} + \vec{v}_3 \cdot \vec{\nabla}_\xi A_3 = i\gamma_3 A_1 A_2 .$$

Here the vectors $\vec{v}_1, \vec{v}_2, \vec{v}_3$ are the group velocities corresponding to the three carrier waves while the coefficients γ_1, γ_2, γ_3 are expressions in terms of the three fundamental harmonics wave numbers and polarization state vectors, and, of course, also of the linear and quadratic susceptibility coefficients $\chi_{jn}^{(1)}$ and $\chi_{jnm}^{(2)}$. This fully tridimensional three wave resonant interaction model has been investigated by Kaup (Kaup, 1981). Here we consider the reduced case in which the three amplitudes A_1, A_2, A_3 depend upon only two independent variables. Since the partial differential equations of the system (8) are of the first order, the four variables ξ_1, ξ_2, ξ_3, τ play a similar role. Hence the choice of two independent variables out of the four may be dictated by the specific physical context under consideration. Thus one may describe temporal dynamics by choosing for instance the space and time coordinates $\xi_1 = \xi$ and τ, or spatial dynamics by choosing instead the variables ξ_1 and ξ_2. In this section we adopt the first notation by giving the variable τ the meaning of time. Finally, we prefer to transform amplitudes and co-ordinates so as to display the characteristic coefficients δ_1, δ_2, δ_3 which are proportional to the inverse of the group velocities, and to set the coupling constants equal to unit, $\gamma_\alpha = 1$, in (8) so as to write the TWI model equations in the dimensionless form

$$\frac{\partial A_1}{\partial \xi} + \delta_1 \frac{\partial A_1}{\partial \tau} = iA_2^* A_3$$

$$\frac{\partial A_2}{\partial \xi} + \delta_2 \frac{\partial A_2}{\partial \tau} = iA_1^* A_3 \qquad (9)$$

$$\frac{\partial A_3}{\partial \xi} + \delta_3 \frac{\partial A_3}{\partial \tau} = iA_1 A_2 .$$

With no loss of generality, we shall write Eqs. (9) in a coordinate system such that $\delta_1 = 0$. Moreover, we consider the case $0 < \delta_3 < \delta_2$. This system of equations is not only of great interest because of its broad applicability, but it exhibits quite special mathematical properties. In fact it has been discovered (Zakharov & Manakov, 1973) that it is an *integrable* Hamiltonian system with the implication that it can be investigated by the spectral methods of soliton theory both in the class of solutions which are well localised (as for bright solitons) and in the class of solutions with nonvanishing values at infinity (as for dark solitons). In particular, this model has explicitly known solutions and infinitely many conservation laws (after early studies (Armstrong et al., 1970; Bers, 1975), for a review paper see (Kaup, 1979) and, more recently (Calogero & Degasperis, 2004; Degasperis & Lombardo, 2006; 2009)). Conservation laws are related to symmetries, namely to transformations of the two independent variables ξ and τ, and the dependent variables A_1, A_2, A_3, which leave the TWI equations (9) invariant. As these transformations depend upon one or more arbitrary parameters, one may insert additional parameters into the expression of an explicit solution by just transforming this solution. Because of this practical use of symmetries, we report here a few of them. Moreover, we also report below, for some of these symmetries, their associated conservation laws which are derived by standard technique from the Lagrangian density

$$\mathcal{L} = \frac{1}{4i}\sum_{j=1}^{3} A_j^*(A_{j\xi} + \delta_j A_{j\tau}) - \frac{1}{2} A_1 A_2 A_3^* + c.c.. \tag{10}$$

To each of these symmetries it corresponds the local conservation law

$$\rho_\tau + \beta_\xi = 0 \tag{11}$$

which implies that, if the evolution variable is τ, the integral

$$R = \int_{-\infty}^{+\infty} d\xi \rho(\xi,\tau) \tag{12}$$

is the conserved quantity since it is τ – independent. On the other hand, if instead the variable ξ is the evolution variable (as in the case this is the spatial transverse co-ordinate), then the integral

$$B = \int_{-\infty}^{+\infty} d\tau \beta(\xi,\tau) \tag{13}$$

is the conserved quantity since it is ξ – independent. The conserved quantity among R and B which should be used depends of course on the physical process under investigation. The simplest symmetries are:

1. translation of the variable ξ, $A_j(\xi,\tau) \to A_j'(\xi,\tau) = A_j(\xi + a,\tau)$ whose associated conserved quantity is

$$P = \frac{1}{4i}\int_{-\infty}^{\infty} \sum_{j=1}^{3} \delta_j A_j^* A_{j\xi} d\xi + c.c., \tag{14}$$

if the evolution variable is τ, or, if the evolution variable is instead ξ, the appropriate conserved quantity is

$$J = \int_{-\infty}^{\infty} \left(\frac{1}{4i}\sum_{j=1}^{3} \delta_j A_j^* A_{j\tau} - \frac{1}{2} A_1 A_2 A_3^*\right) d\tau + c.c.. \tag{15}$$

2. translation of the variable τ, $A_j(\xi,\tau) \to A_j'(\xi,\tau) = A_j(\xi,\tau + b)$ whose associated conserved quantity is

$$E = \int_{-\infty}^{\infty} \left(\frac{1}{4i}\sum_{j=1}^{3} A_j^* A_{j\xi} - \frac{1}{2} A_1 A_2 A_3^*\right) d\xi + c.c. \ , \tag{16}$$

if the evolution variable is τ, or, if the evolution variable is instead ξ, the appropriate conserved quantity is

$$H = \frac{1}{4i}\int_{-\infty}^{\infty} \sum_{j=1}^{3} \delta_j A_j^* A_{j\tau} d\tau + c.c. \ . \tag{17}$$

3. phase translations, $A_1(\xi,\tau) \rightarrow A_1'(\xi,\tau) = \exp(i\theta_1)A_1(\xi,\tau)$, $A_2(\xi,\tau) \rightarrow A_2'(\xi,\tau) = \exp(i\theta_2)A_2(\xi,\tau)$, $A_3(\xi,\tau) \rightarrow A_3'(\xi,\tau) = \exp(i\theta_1 + i\theta_2)A_3(\xi,\tau)$ whose two associated conserved quantities are the well known Manley-Rowe invariants

$$M_{13} = M_1 + M_3 = \frac{1}{2}\int_{-\infty}^{\infty}\left(\delta_1 |A_1|^2 + \delta_3 |A_3|^2\right)d\xi,$$

$$M_{23} = M_2 + M_3 = \frac{1}{2}\int_{-\infty}^{\infty}\left(\delta_2 |A_2|^2 + \delta_3 |A_3|^2\right)d\xi \ , \tag{18}$$

if the evolution variable is τ, or, if the evolution variable is instead ξ, the appropriate Manley-Rowe invariants read

$$I_{13} = I_1 + I_3 = \frac{1}{2}\int_{-\infty}^{\infty}\left(|A_1|^2 + |A_3|^2\right)d\tau,$$

$$I_{23} = I_2 + I_3 = \frac{1}{2}\int_{-\infty}^{\infty}\left(|A_2|^2 + |A_3|^2\right)d\tau \ . \tag{19}$$

Two more symmetries which may turn out to be useful are the scale transformation $A_j(\xi,\tau) \rightarrow A_j'(\xi,\tau) = pA_j(p\xi, p\tau)$, where p is a real arbitrary parameter, and the characteristic phase transformation $A_1(\xi,\tau) \rightarrow A_1'(\xi,\tau) = \exp[iq(\delta_2 - \delta_3)\xi_1]A_1(\xi,\tau)$, $A_2(\xi,\tau) \rightarrow A_2'(\xi,\tau) = \exp[iq(\delta_3 - \delta_1)\xi_2]A_2(\xi,\tau)$, $A_3(\xi,\tau) \rightarrow A_3'(\xi,\tau) = \exp[-iq(\delta_1 - \delta_2)\xi_3]A_3(\xi,\tau)$, where q is an arbitrary real parameter and $\xi_j = \tau - \delta_j\xi$ are the characteristic co-ordinates.

Several explicit soliton solutions, of both bright and dark type, of the TWI equations (9) have been computed by algebraic construction methods, see (Degasperis & Lombardo, 2009) and references quoted there, and some of them are detailed in the following sections with the purpose of displaying their relevance to special optical processes.

On the applicative side, the three-wave resonant interaction has been extensively studied in the context of nonlinear optics, since it applies to parametric amplification, second harmonic generation and frequency conversion, stimulated Raman and Brillouin scattering and light speed control (Taranenko & Kazovsky, 1992; Ibragimov & Struthers, 1996; Ibragimov et al., 1998; Picozzi & Haelterman, 1998).

3. Frequency conversion

Optical parametric amplification in quadratic nonlinear crystals has been studied since the invention of the laser, as it provides a versatile means of achieving widely tunable frequency conversion (Cerullo & De Silvestri, 2003). In parametric processes, the effective interaction length of short optical pulses is limited by temporal walk-off owing to chromatic dispersion, or group velocity mismatch (GVM) (Armstrong et al., 1970; Akhmanov et al., 1992).

Compression and amplification of ultra-short laser pulses in second harmonic and sum-frequency (SF) generation in the presence of GVM was theoretically predicted (Wang & Dragila, 1990; Stabinis et al., 1991) and observed in several experiments (Wang & Luther-Davies, 1992; Chien et al., 1995). The conversion efficiency of generated SF pulses may be optimised (Ibragimov & Struthers, 1996; 1997; Ibragimov et al., 1998; 1999; Fournier et al., 1998) by operating in the soliton regime (Zakharov & Manakov, 1973; Kaup, 1976). In fact, the temporal collision of two short soliton pulses in a quadratic nonlinear crystal may efficiently generate a short, time-compressed SF pulse (Ibragimov & Struthers, 1996). However this SF pulse is unstable: its energy decays back into the two incident pulses after a relatively short distance.

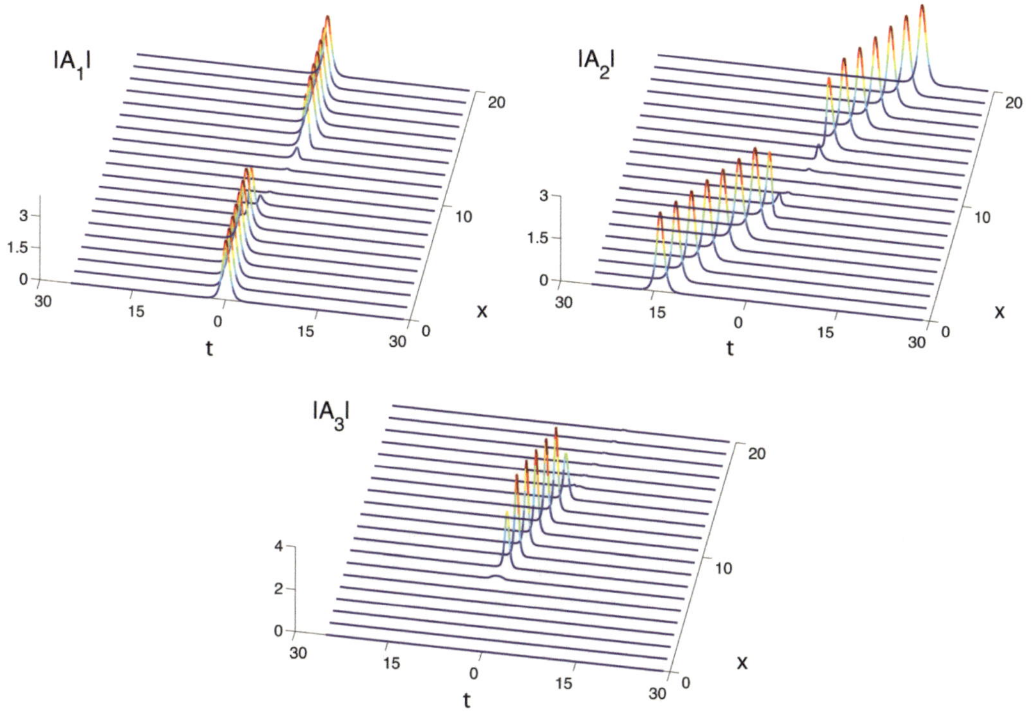

Fig. 1. Sum-frequency generation at ω_3 from the parametric interaction of two short optical signals at ω_1 and ω_2. The characteristic delays are $\delta_1 = 0$, $\delta_2 = 2$, $\delta_3 = 1$.

Figure 1 illustrates a typical example of the efficient SF parametric interaction of two short optical pulses in the soliton regime (Ibragimov & Struthers, 1996). At the crystal input, two isolated soliton pulses A_1 and A_2 with frequencies ω_1 and ω_2 propagate with speeds v_1 and v_2. Whenever the faster pulse overtakes the slower one, an idler pulse A_3 at the SF $\omega_1 + \omega_2$ is generated and propagates with the linear speed v_3. Depending on the time widths and intensities of the input pulses, the temporal duration of the SF pulse is reduced with respect to the input pulse widths. Correspondingly, the SF pulse peak intensity grows larger than the input pulse intensities. Figure 1 shows that, eventually, the SF idler pulse decays back into the two original isolated pulses at frequencies ω_1 and ω_2. Note that the shapes, intensities and widths of the input pulses are left unchanged in spite of their interaction. As shown in Ref. (Ibragimov & Struthers, 1996), the above discussed SF pulse generation process may be analytically described in terms of soliton solutions of Eqs. (9) (Zakharov & Manakov, 1973; Kaup, 1976). In the notation of Eqs. (9), the complete three-wave dynamics reads as:

$$A_1 = -i\frac{\Delta}{4p}\frac{\delta_2 - \delta_3}{\delta_2\delta_3}A_2^* A_3,$$

$$A_2 = i\frac{4p\delta_3}{\Delta}\sqrt{\frac{\delta_2}{\delta_2-\delta_3}}e^{iq_2\tau_2}exp[-2i\frac{\delta_3}{\delta_2-\delta_3}(k-ip)\tau_2],$$

$$A_3^* = i\frac{4p\delta_2}{\Delta}\sqrt{\frac{\delta_3}{\delta_2-\delta_3}}e^{iq_3\tau_3}exp[2i\frac{\delta_2}{\delta_2-\delta_3}(k+ip)\tau_3], \quad (20)$$

where

$$\Delta = 1 + exp(-\frac{4p\delta_3}{\delta_2-\delta_3}\tau_2) + exp(-\frac{4p\delta_2}{\delta_2-\delta_3}\tau_3) \quad (21)$$

$$q_n = q(\delta_{n+1} - \delta_{n+2}), n = 1,2,3 mod(3),$$

$$\tau_n = -\tau + \delta_n\xi. \quad (22)$$

For a given choice of the characteristic linear group velocities, we are left with the three parameters p, k, q. The parameter p is associated with the re-scaling of the wave amplitudes, and of the coordinates τ and ξ. Here the value of k and q adds a phase shift which is linear in both τ and ξ.

The decay of the SF pulse which is shown in Fig. 1 may be a significant drawback in practical applications, since it implies that a given nonlinear crystal length yields efficient conversion for a limited range of input pulse intensities and time widths only. The parametric sum frequency conversion of an ultra-short signal and a quasi-CW background pump-control may be exploited as a means to reduce or even eliminate the decay of the generated idler wave (Conforti et al., 2007). In the presence of GVM, the parametric SF conversion of an ultra-short optical signal and a quasi-CW pump generally leads to the generation of a low-intensity and relatively long idler pulse, whose duration is associated with the interaction distance in the crystal.

This scenario changes dramatically in the soliton regime. Figure 2 illustrates the efficient generation of a stable, ultra-short SF idler pulse from the parametric SF conversion of a properly prepared ultra-short signal and a CW background control with an arbitrary intensity level. In Fig. 2 we injected in the quadratic nonlinear crystal the short signal at frequency ω_2, along with a delayed and relatively long pump-control pulse at frequency ω_1. Initially, the two pulses propagate uncoupled; as soon as the faster pulse starts to overlap in time with the slower quasi-CW control, their nonlinear mixing generates a short SF idler pulse. The sum frequency process displayed in Fig. 2 can be analytically explained and explored in terms of stable TWIS (Three Wave Interaction Soliton) solutions (Degasperis et al., 2006).

In the notation of Eqs. (9), the properly prepared ultra-short pulse is a ZM single-wave soliton pulse (Zakharov & Manakov, 1973) at frequency ω_2; it reads as

$$A_1 = 0, \quad A_2 = 2P\sqrt{\rho\delta_2\delta_3}\frac{e^{i\rho\phi\tau_2}}{\cosh(2P\rho\tau_2)}, \quad A_3 = 0, \quad (23)$$

where

$$\rho = \delta_3/(\delta_2-\delta_3), \quad \tau_2 = -\tau + \delta_2\xi. \quad (24)$$

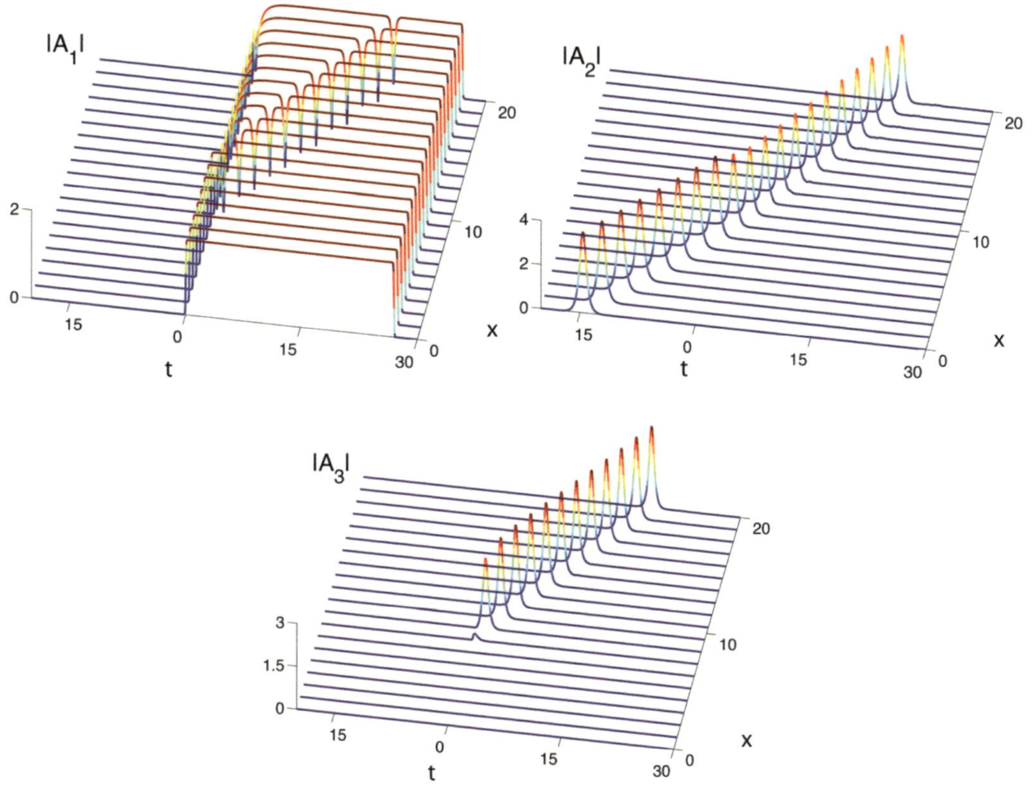

Fig. 2. Sum-frequency generation at ω_3 from the parametric interaction of a short pulse at ω_2 and a quasi-CW control at ω_1. The characteristic delays are $\delta_1 = 0$, $\delta_2 = 2$, $\delta_3 = 1$.

For a given choice of the three linear group velocities, or characteristic delays δ_j, the above ZM soliton is determined in terms of the two real parameters $P > 0$, ϕ. The parameter P fixes both the soliton peak amplitude and its temporal width. Whereas the parameter ϕ corresponds to a phase shift which is linear in both τ and ξ. On the other hand, the generic input quasi CW signal at frequency ω_1 may be described as

$$A_1 = \frac{Ce^{-i\gamma\tau}}{2}[\tanh(\frac{\tau-\tau_i}{\tau_r}) - \tanh(\frac{\tau-\tau_f}{\tau_r})], \quad A_2 = 0, \quad A_3 = 0, \tag{25}$$

where C is the complex amplitude and γ is the frequency shift with respect to ω_1 of the quasi-CW signal; τ_i (τ_f) and τ_r are the switch-on (switch-off) time and the rise/fall time of the quasi-CW signal, respectively. We consider the case in which $|C|^2 < P^2\delta_2\delta_3$.

When the faster short pulse, pre-delayed with respect to the slower quasi-CW pump at frequency ω_1, overtakes the background (at $\tau = 0$, in Fig. 2), their collision leads to the generation of a short idler pulse at the SF ω_3. Additionally, a dip appears in the quasi CW control; whereas the intensity, duration and propagation speed of the input wave at frequency ω_2 are modified. Indeed, the signal-pump interaction generates a stable TWIS simulton (Degasperis et al., 2006). In the notation of Eqs. (9), the TWIS simulton solution reads as

$$A_1 = \{1 + \frac{2pb^*}{|b|^2 + a^2}[1 - \tanh[B(-\tau + \delta\xi)]]\}\frac{iag_1 \exp(iq_1\tau_1)}{g(\delta_2 - \delta_3)}$$

$$A_2 = \frac{2pa}{\sqrt{|b|^2 + a^2}} \frac{g_2}{g(\delta_2 - \delta_3)} \frac{\exp[i(q_2\tau_2 + \chi\tau + \omega\xi)]}{\cosh[B(-\tau + \delta\xi)]},$$

$$A_3 = \frac{-2pb^*}{\sqrt{|b|^2 + a^2}} \frac{g_3}{g(\delta_2 - \delta_3)} \frac{\exp[-i(q_3\tau_3 - \chi\tau - \omega\xi)]}{\cosh[B(-\tau + \delta\xi)]}, \tag{26}$$

where

$$b = (Q-1)(p + ik/Q), \quad r = p^2 - k^2 - |a|^2,$$

$$Q = \frac{1}{p}\sqrt{\frac{1}{2}[r + \sqrt{r^2 + 4k^2p^2}\,]},$$

$$B = p[\delta_2 + \delta_3 - Q(\delta_2 - \delta_3)]/(\delta_2 - \delta_3),$$

$$\delta = 2\delta_2\delta_3/[\delta_2 + \delta_3 - Q(\delta_2 - \delta_3)],$$

$$\chi = k[\delta_2 + \delta_3 - (\delta_2 - \delta_3)/Q]/(\delta_2 - \delta_3),$$

$$\omega = -2k\delta_2\delta_3/(\delta_2 - \delta_3)\ , \ \tau_n = -\tau + \delta_n\xi$$

$$q_n = q(\delta_{n+1} - \delta_{n+2}),\ ,\ g_n = |(\delta_n - \delta_{n+1})(\delta_n - \delta_{n+2})|^{-1/2}$$

$$g = g_1 g_2 g_3\ , n = 1,2,3\ mod(3). \tag{27}$$

For a given choice of the characteristic linear group velocities, we are left with the four parameters p, a, k, q. The parameter p is associated with the re-scaling of the wave amplitudes, and of the coordinates τ and ξ. Whereas a measures the amplitude of the CW background in wave A_1 (namely $a\sqrt{\delta_2\delta_3}$). The value of k is related to the soliton wavenumber. The parameter q simply adds a phase shift which is linear in both τ and ξ (see (Degasperis et al., 2006) for parameter details).
It is remarkable that we may analytically predict the parameters p, k, q, a of the generated TWIS from the corresponding parameters of the input single wave TWIS and the complex amplitude of the pump control. This can be achieved by supposing that the input TWIS adiabatically (i.e., without emission of radiation) reshapes into a new TWI simulton after its collision with the quasi-CW pump at a given point in time (say, at $\tau = 0$). Under this basic hypothesis, the conservative nature of the three-wave interaction permits us to suppose that: i) the energy I_{23} (19) of the input TWI soliton is conserved in the generated TWI simulton; ii) the phase of the ω_2 frequency components of the input TWI soliton and of the generated TWI simulton is continuous across their time interface (i.e., at $\tau = 0$); iii) the amplitude and phase of the control pump C coincide with the corresponding values of the asymptotic plateau of the generated TWI simulton component at frequency ω_1. By imposing the above three conditions, after some straightforward calculations we obtain the following relations that relate the parameters of the incident and of the transmitted TWIS

$$p = P,\ a\sqrt{\delta_2\delta_3} = |C|,\ q(\delta_2 - \delta_3) = \gamma,\ q(\delta_2 - \delta_3) - 2k = \phi, \tag{28}$$

with the restriction $|C| < P\sqrt{\delta_2\delta_3}$.

As an example, in Fig. 2 the input TWI soliton at frequency ω_2 is described by Eqs. (23) with $P = 1.3$, $\phi = 0$, and the background control amplitude is $C = 1.7$. After the collision with the CW background, the above equations predict that the generated TWIS is described by Eqs. (26), with $p = 1.3$, $q = 0$, $k = 0$, and $a = 1.2$. The accuracy of this prediction is well confirmed by its comparison with the numerical solutions of the TWI Eqs. (9). Indeed, Fig. 3 compares the numerical and the analytical evolutions (along the crystal length ξ) of the energy, the pulse duration and the velocity of the idler and signal pulses which correspond to the case shown in Fig. 2. We performed further extensive numerical simulations, which confirmed the general validity of the above described adiabatic transition model for TWIS generation upon collision with a CW background.

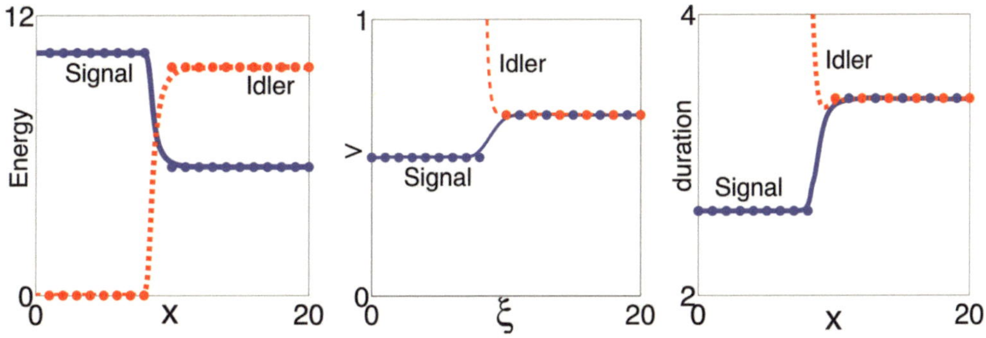

Fig. 3. Numerical evolution (lines) and theoretical predictions (circles) of energy, pulse duration and velocity of idler and signal waves reported in Fig.2.

Indeed, by increasing or decreasing the CW background amplitude $|C|$ in the range $[0, p\sqrt{\delta_2\delta_3}]$, we observed that stable TWISs with different velocity, duration and energy distributions may be adiabatically shaped. The important consequence of this result is that, by means of Eqs. (26)-(28), we may analytically predict and control the characteristics of the generated idler pulse (namely, its velocity, time duration and energy) simply as a function of the intensity level of the CW pump. Moreover, we would like to emphasize that the stability of the whole SF idler conversion process is ensured by the underlying stability of the generated TWIS (Conforti et al., 2006).

4. Ultrafast pulse train generation

Efficient mechanisms for the generation of high-contrast, ultra-short pulse trains are of interest in a broad range of domains, such as time-resolved spectroscopy and microscopy, selective femtosecond chemistry, quantum coherent control in high-field physics, and optical communications. Relevant application examples include quantum-path control of high-harmonic generation (Zhang et al., 2007); multi-pulse excitation of atoms, molecules, and solids (Weiner et al., 1990); multiple-laser-pulse excitation of high-gradient plasma accelerators (Umstadter et al., 1994); high-fluence THz wave-train generation for radar, microwave (Liu et al., 1996) and optical communication systems. Several techniques for generating trains of ultra-short pulses at repetition rates beyond those achievable by active laser mode locking or by means of electrically-controlled modulation have been explored in recent years. For example, linear pulse train generation techniques include the repetition-rate multiplication of a lower rate source by applying amplitude (Petropoulos et al., 2000; Yiannopoulos et al., 2003) or phase-spectral filtering (Longhi et al., 2000; Azana & Muriel, 2001; Caraquitena et al., 2007). Moreover, different all-optical techniques that may result in up to THz pulse train rates have also been proposed by using both quadratic and cubic nonlinear media. Consider for example induced modulation instability (Hasegawa, 1984; Coen & Haelterman, 2001), multiple four-wave-mixing (Pitois et al., 2002; Inoue et al., 2007), and backward quasi-phase-matched second harmonic generation (Conti et al., 1999; Conforti et al., 2005).

Here, we present a TWI nonlinear technique for the flexible generation of a train of ultra-short optical pulses with light-controlled time width, amplitude and repetition rate. This technique originates from the interaction of three waves at resonance (Baronio et al., 2008).

Let us first present numerical evidence of the train generation process. Figure 4 illustrates the possibility of obtaining an efficient modulation of the input CW signal into a train of ultra-short pulses, as a result of its parametric mixing with an ultra-short ZM soliton pulse at a different carrier wavelength. In particular, in Fig. 4 we numerically solved Eqs. (9), where the initial excitation conditions at $\xi = 0$ were provided by a quasi-CW signal at frequency ω_1 (a standing signal in a coordinate system where $\delta_1 = 0$) and a short ZM soliton pulse at frequency ω_2 (with delay parameter $\delta_2 > 0$). The two input pulses were pre-delayed, so that they could collide at some distance inside the crystal. Figure 4 shows that, as soon as the faster ZM soliton pulse starts to overlap in time with the slower quasi-CW signal, their parametric mixing leads to the generation of an idler short pulse at frequency ω_3. Additionally, Fig. 4 also shows that the TWI leads to a periodic oscillation with distance ξ of the intensity of the two short pulses at carrier frequencies ω_2 and ω_3. As a result, it turns out that the centre of mass of these two pulses periodically oscillates in time around an average value, which grows larger with distance according to a common average group velocity; this velocity is clearly (see Fig. 4) different from the group velocity of the incoming short pulse. As for the incoming CW signal, Fig. 4 also shows that a train of short pulses is carved on the CW background at frequency ω_1. Finally, Fig. 4 also illustrates that at the end of the three-wave interaction process the quasi-CW background at ω_1 is modulated into a sequence of ultra-short pulses, the soliton pulse at ω_2 returns back to its original shape, and the generated sum-frequency pulse at ω_3 vanishes.

Quite remarkably, we shall demonstrate that the entire three-wave interaction process which is displayed in Fig. 4 may be analytically represented in terms of particular analytical TWI solutions. As in the previous section, in the notation of Eqs. (9), the arbitrary input quasi CW signal at frequency ω_1 is given by expression (25) and the input ZM single-wave soliton pulse at frequency ω_2 by expression (23). We consider the case in which $|C|^2 > P^2\delta_2\delta_3$ and $\gamma = \phi$.

After the collision of the ZM pulse with the CW background, we surmise that their parametric interaction generates a three-wave trapponic soliton, which consists of a group-velocity locked bright-bright-dark triplet. Similar triplet solutions have been already considered in nonlinear optics in the form of simultons (Degasperis et al., 2006; Conforti et al., 2007) or boomerons (Conforti et al., 2006). Trappon soliton solutions of the TWI Eqs. (9) may be found as discussed in the recent analysis of Ref. (Calogero & Degasperis, 2005;

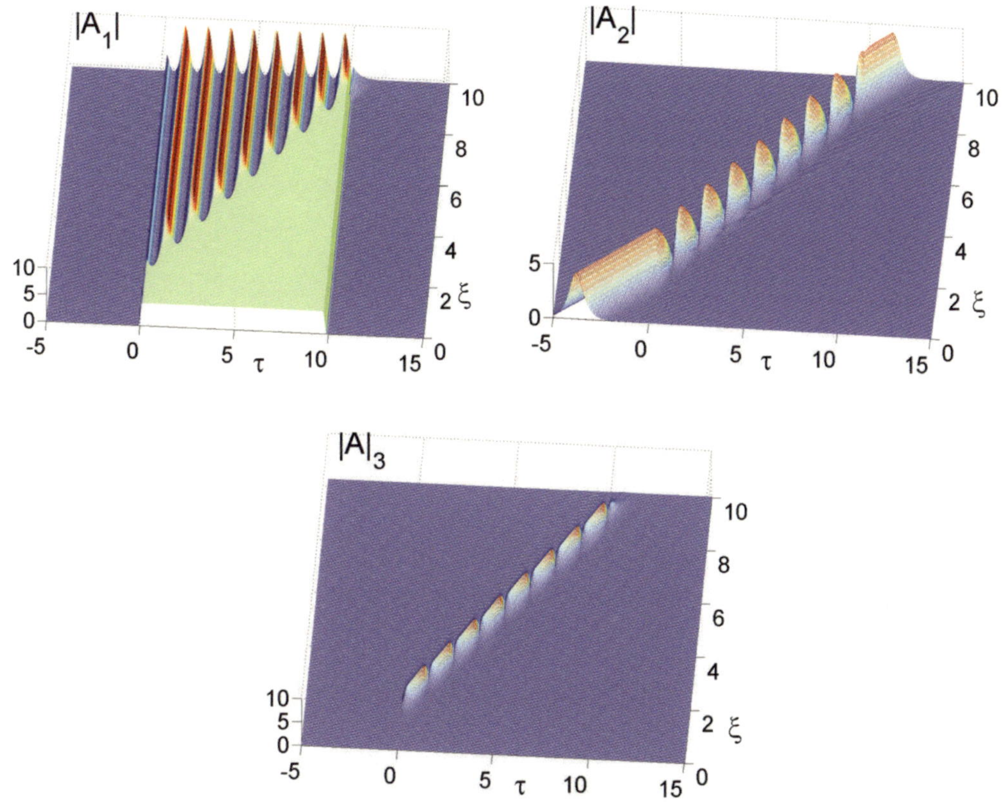

Fig. 4. Sum-frequency generation at ω_3 from the parametric interaction of a quasi-CW control at ω_1 and a short pulse at ω_2. The characteristic delays are $\delta_1 = 0$, $\delta_2 = 2$, $\delta_3 = 1$.

Degasperis et al., 2006). TWI trappons are characterized by a periodic (or trapped) spatio-temporal evolution of the energy distribution of the three waves, as well as of their locked propagation speed. As a result, the center of mass of a trappon also periodically oscillates within a limited time interval. As far as the potential applications are concerned, the most significant property of the trappon dynamics consists in the carving of a train of short pulses into the CW-background at frequency ω_1. Indeed the entire three-wave interaction dynamics after the collision of the short pulse with the CW background as shown by Fig. 4 may be analytically described in terms of the trappon solution

$$A_1 = ia\sqrt{\delta_2\delta_3}e^{iq_1\tau_1} - i\frac{\Delta}{4p}(\frac{\delta_2-\delta_3}{\delta_2\delta_3})A_2^*A_3,$$

$$A_2 = -\frac{2p\delta_3}{\Delta}\sqrt{\frac{2\delta_2}{\delta_2-\delta_3}}e^{iq_2\tau_2}(H_+^* - H_-^*),$$

$$A_3^* = \frac{2p\delta_2}{\Delta}\sqrt{\frac{2\delta_3}{\delta_2-\delta_3}}e^{iq_3\tau_3}(H_+e^{i\beta} - H_-e^{-i\beta}), \tag{29}$$

where

$$\Delta = 1 + |H_+|^2 + |H_-|^2 - 2\cos(\beta)\mathcal{R}e(H_+ H_-^* e^{i\beta}),$$

$$H_\pm = e^{-(B\pm i\chi)\tau} e^{\frac{2\delta_2\delta_3}{\delta_2-\delta_3}p\xi},$$

$$\chi = \sqrt{a^2 - p^2},$$

$$B = p(\frac{\delta_2+\delta_3}{\delta_2-\delta_3}), \ \tan(\beta) = \sqrt{a^2-p^2}\,/\,p,$$

$$\delta = \frac{2\delta_2\delta_3}{\delta_2+\delta_3},$$

$$q_n = q(\delta_{n+1} - \delta_{n+2}) \ , \ n = 1,2,3 mod(3),$$

$$\tau_n = -\tau + \delta_n \xi. \tag{30}$$

For a given choice of the characteristic linear velocities in Eqs. (9), the trappon solution of Eqs. (29) is completely determined by just three independent real parameters, namely p, q, a (with the restrictions that $p > 0$, $a > 0$ and $a^2 > p^2$). The parameter p is associated with the rescaling of the wave amplitudes, and of coordinates τ and ξ. Whereas a measures the amplitude of the CW background in wave A_1. The parameter q adds a phase shift which is linear in both τ and ξ.
The trappon solution (29) extends in time from $\tau = -\infty$ to $\tau = +\infty$ and at large distances (i.e., for $\xi \to +\infty$) it exhibits a time-periodic behavior. Nevertheless, as we shall see, Eqs. (29) exactly describe the mixing of the ZM soliton (23) with the quasi-CW (25) background signal over the finite interval $[\tau_i, \tau_f]$ (which correspond to the times of turning on and off the CW wave at frequency ω_1, respectively).
Indeed, one may analytically predict the characteristic parameters p, q, a of the generated TWI trappon from the corresponding P, ϕ parameters of the isolated input ZM soliton, and the complex amplitude C and the frequency shift γ of the CW signal. As we discussed in the previous section, it is sufficient to suppose that the input ZM soliton adiabatically (i.e., without emission of radiation) reshapes into a trappon soliton after its collision with the quasi-CW background at a given point in time (say, at $\tau = \tau_i$). In the spectral domain (Zakharov & Manakov, 1973) this hypothesis is equivalent to imposing that the input ZM soliton, after the collision with the CW background, transfers its discrete eigenvalue to the trappon soliton. In the frame of this basic hypothesis, the conservative nature of the three-wave interaction permits us to suppose that: i) the energy I_{23} (19) of the input TWI soliton is conserved in the generated trappon soliton; ii) the phase of the ω_2 frequency components of the input TWI soliton and of the generated TWI soliton varies in a continuous manner across their time interface (i.e., at $\tau = \tau_i$); iii) the amplitude and the phase of the CW background coincide with the corresponding values of the asymptotic plateau of the component at frequency ω_1 of the generated TWI trappon. The above three conditions, after some straightforward calculations, permit us to obtain the following equations that relate the parameters of the incident and the transmitted solitons (i.e., before and after the collision, respectively)

$$p = P,\ a\sqrt{\delta_2\delta_3} = |C|,\ q(\delta_2 - \delta_3) = \phi, \tag{31}$$

with the restrictions that $|C|^2 > P^2\delta_2\delta_3$ and $\gamma = \phi$. The above conditions define the matching relations among the amplitudes and the phases of both the input ZM soliton and of the CW signal which should be satisfied in order to excite the trapponic soliton. As an example, consider the case of Fig. 4, where the CW background control is described by Eqs. (25) with $C = 3$, $\gamma = 1$, and the input ZM soliton is obtained from Eqs. (23) with $P = 1.5$, $\phi = 1$. After the collision with the CW background, Eqs. (31) predict that the generated trappon is obtained from Eqs. (29) with $p = 1.5$, $q = 1$, and $a = 3/\sqrt{2}$. We confirmed the accuracy of this prediction by means of extensive comparisons of the analytical expression (29) with numerical solutions of the TWI Eqs. (9). For example, Fig. 5 displays both numerically computed and analytical evolutions for the energy I_2 (along the crystal length ξ) and for the output transverse profile of wave A_1 (along τ) (these correspond to the same interaction process previously shown in Fig. 4).

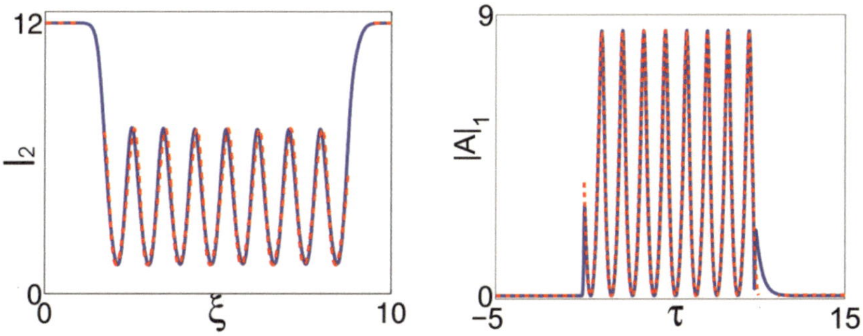

Fig. 5. Numerical evolution (continuous lines) and theoretical prediction from Eqs. (29) (dotted lines) of the energy at ω_2 and of the output pulse profile at ω_1 for the case of Fig. 4.

The amplitude profile of the generated train of ultra-short pulses at frequency ω_1 may be exactly described by performing the limit for $\xi \to +\infty$ of the first of Eqs. (29), which yields

$$|A_{1\infty}(\tau)| = |C|\,|1 - 4\cos(\beta)\frac{\sin(\pi\tau/T)\sin(\pi\tau/T - \beta)}{1 - \cos(\beta)\cos(2\pi\tau/T - \beta)}| \tag{32}$$

where $|C| = a\sqrt{\delta_2\delta_3}$ (see (28)), $\tan(\beta) = \sqrt{a^2 - p^2}/p$ (see (30)) and period $T = \pi/\sqrt{a^2 - p^2}$. Two different examples of the generated pulse trains as described by Eqs. (32) are shown in Fig. 6 a)-b).

Equation (32) provides relatively simple expressions that relate the parameters of the injected soliton pulse and CW background to the main features of the generated output pulse train. Indeed, $|A_{1\infty}|$ is a time periodic signal (with period T) with average value approximatively equal to $|C|$. Moreover, from Eqs. (32) one obtains that the peak amplitude of the generated pulses is $M = |C| + 2\sqrt{\delta_2\delta_3}\,P$; whereas the minimum amplitude is $m = |C| - 2\sqrt{\delta_2\delta_3}\,P$. Hence the maximum intensity contrast for the generated train is obtained for $|C| = 2P\sqrt{\delta_2\delta_3}$: in this case, the field amplitude at the minima vanishes and the peak amplitude reads as $M = 4\sqrt{\delta_2\delta_3}\,P$.

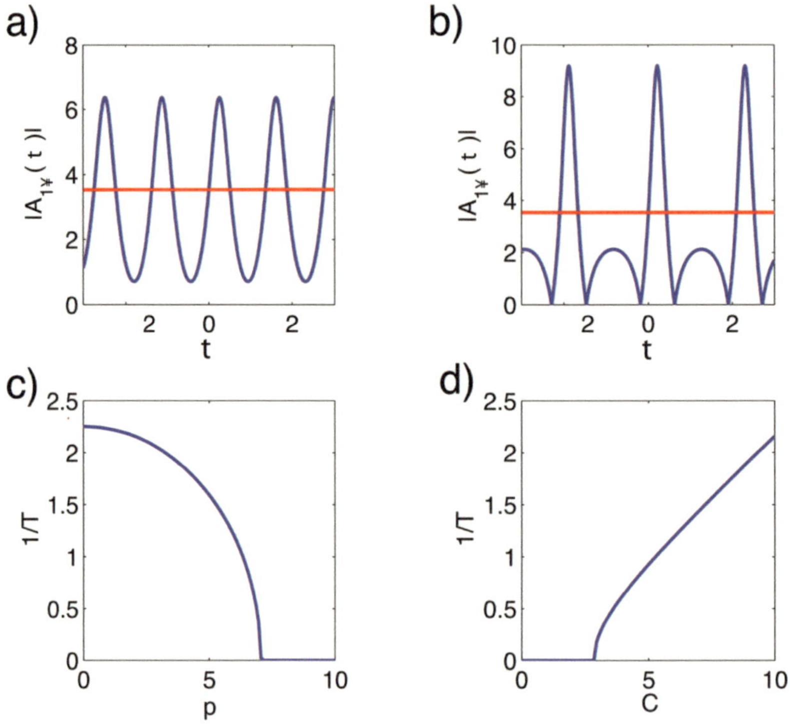

Fig. 6. Output pulse train (32) amplitude (blue line) and its time–average value (red line) for: a) $C = 2.5\sqrt{2}$,$P = 1$ and b) $C = 2.5\sqrt{2}$,$P = 2$; Pulse train frequency $f = 1/T$ c) vs. P ($C = 10$); and d) vs. C ($p = 2$).

In Figure 6 c)–d) we illustrate the analytically predicted dependence of the repetition rate of the ultra-short pulse train $R = 1/T = \sqrt{|C|^2/(\delta_2\delta_3) - p^2}/\pi$, as a function of the amplitudes of the input short pulse and of the CW signal, respectively. As it can be seen in Fig. 6 c), the repetition rate is the largest whenever the amplitude of the input soliton vanishes. Whereas there is a critical input soliton amplitude, above which the period of the generated pulse train diverges to infinity. On the other hand, Fig. 6 d) shows that when varying the CW background amplitude, a finite repetition rate is only obtained for amplitudes $|C|$ above a certain threshold level, and this rate grows larger with $|C|$ in a linear fashion. In fact, these thresholds define the borderlines of existence of TWI simulton triplets or TWI trappons excitations.

The above results show that the peak amplitude, the time width and the repetition rate of the pulse train which is generated at ω_1 may be controlled in a stable manner by simply adjusting the intensity level of the input CW signal $|C|$ and/or the amplitude level (hence the time width) P of the input short soliton pulse.

It is worth noting that the proposed method for generating pulse trains has a very high efficiency: in principle, the entire energy of the pump pulse A_1 may be converted into the pulse train. Indeed from (i) the energy conservation relation (19) and (ii) the fact that at the end of interaction $A_3 = 0$, it follows that the energy of the quasi-CW pulse is conserved before and after the collision with the control pulse A_2.

It should be pointed out that the generation process that we have described so far may be thought of as a way of "writing" a pulse train on a quasi CW signal. The opposite process of "erasing" such wave train to go back to a flat, quasi CW, signal is also possible. Indeed, it is easy to realize that this reverse process is well described in a similar way by the formation of the trappon soliton solution of the TWI Eqs. (9) which is obtained from the expression (29) via the invariance transformation $\xi \rightarrow -\xi$, $\tau \rightarrow -\tau$, $A_1 \rightarrow A_1$, $A_2 \rightarrow A_2$, $A_3 \rightarrow -A_3$.

5. Experimental observability

In this section, we discuss the experimental conditions for the practical observation of three wave solitons, with particular reference to their applications to frequency conversion and pulse train generation as discussed in the previous sections. It is interesting to notice that the three-wave soliton concept may be applied to describe the interaction between either three optical pulses in the time domain, or three beams in the spatial domain.

5.1 Temporal solitons

To be specific, let us consider the nonlinear *eee* interaction of two pulses, at $\lambda_1 = 1550$ nm and at $\lambda_2 = 3400$ nm, which mix in a 2 *cm* long periodically poled bulk Lithium Niobate crystal with about 28μm periodicity, to generate the sum-frequency pulse at $\lambda_3 = 1064$ nm. For perfect phase-matching with collinear geometry, one obtains the refraction indexes: $n_1 = 2.117$, $n_2 = 2.05$ and $n_3 = 2.134$, respectively. The corresponding group velocities satisfy the inequality $v_1 > v_3 > v_2$, namely $v_1 = 1.386 \times 10^8$ m/s, $v_3 = 1.374 \times 10^8$ m/s and $v_2 = 1.3615 \times 10^8$ m/s. In other words, the sum-frequency pulse group velocity is intermediate between the group velocities of the two input signals. This condition is crucial for the existence and stability of TWI soliton, simulton and trapponinc solutions. The effective nonlinear coefficient is $d_{33} = 25$ pm/V.

Whenever the input pulses are ZM single wave solitons, for instance pulses of about 200 *fs* temporal duration with peak intensity of a few hundreds of MW/cm^2, 100% energy conversion into a time compressed sum-frequency pulse is achievable at a certain point of the nonlinear crystal, for a proper ratio among the intensities of the two input fundamental pulses (Ibragimov & Struthers, 1996; 1997). The drawback is that, after a relatively short propagation distance, the generated pulse decays again in the two original fundamental solitons.

Whenever an input pulse is a ZM single wave soliton, i.e. a pulse of about 200 *fs* temporal duration with peak intensity of a few hundreds of MW/cm^2, and the other input signal is a quasi-continuous wave (say, with a 3*ps* time duration), their interaction leads to the generation of an ultra-short stable sum-frequency pulse of approximately the same time width (200 *fs*) of the incident short pulse. It is particularly significant that the peak intensity (and the group velocity) of the generated pulse may continuously evolve upon propagation through the crystal, from zero to a few hundreds of MW/cm^2 (v from v_3 to v_2), simply by varying the intensity of the quasi-CW background (Conforti et al., 2007). Such unique feature may enable to achieve the stable frequency conversion of short signal optical pulses, as well as the continuous control of the frequency conversion efficiency by adjusting the input intensity of the quasi-CW background pump wave.

In addition, the control of the CW pump power level also permits the highly efficient generation of a sum-frequency converted train of idler pulses which represent a time-periodic copy of the input short signal pulse (Baronio et al., 2008). Whenever the peak intensity of the quasi–CW signal is higher than the peak intensity of the short ZM soliton pulse, the parametric mixing leads to the generation of a train of 250 *fs* pulses with a repetition rate of about 2*THz* and peak power levels of the order of a few hundreds of MW/cm^2. Again the peak amplitude, the time width and the repetition rate of the pulse train which is generated may be controlled in a stable manner by simply adjusting the intensity level of the input qusi-CW signal.

In practical terms, the upper limit to the frequency conversion and train generation processes offered by the present technique will be set by the limit of validity of Eqs. (9), which hold true in all three–wave physical systems whenever the presence of group velocity dispersion within each wave can be neglected.

5.2 Spatial solitons

As discussed in section 2, one could write the same TWI model Eqs. (9) where the independent coordinates are, instead of $\xi_1 = \xi$ and τ, the two spatial variables ξ_1 and ξ_2. In order to clearly distinguish this situation, involving the interaction between three spatial beams, from the previous cases involving short temporal pulses, and to maintain the same notation used in (Baronio et al., 2009), we name the longitudinal spatial coordinate $\xi_1 = z$ and the transverse spatial coordinate $\xi_2 = x$. Moreover, we also set in this case $A_{1,2,3} = \phi_{1,2,3}$. Note that the spatial beams under consideration are supposed to be wide enough so that diffraction (in analogy with group-velocity dispersion in the temporal case) may be neglected in the propagation of each individual beam, at least over their interaction distance. The interest in studying the spatial case is twofold: from one side it appears to be the simplest situation for providing experimental evidence of three wave soliton dynamics; from the practical viewpoint, spatial solitons in this framework might reveal interesting properties for beam shaping and switching.

Here we describe the set-up which led to the first experimental demonstration of laser beam reshaping by means of the three-wave ZM soliton effect in optics (Baronio et al., 2009).

To be precise, Fig. 7 illustrates the numerical optical scenario exploited to demonstrate the existence of ZM solitons. Here we show numerical simulations, with $\delta_1 = -\delta_2$ and $\delta_3 = 0$, illustrating the typical wave dynamics in the spatial $x - z$ plane in the presence of two electromagnetic beams ϕ_1 at λ_1, ϕ_2 at λ_2 and no beam ϕ_3 at λ_3 in $z = 0$ interacting in a quadratic nonlinear medium; that is, $\phi_1(x,0) \neq 0$, $\phi_2(x,0) \neq 0$, $\phi_3(x,0) = 0$.

Fig. 7.a shows the dynamics of the low–intensity linear regime: the beams at the two fundamental wavelengths λ_1 and λ_2 do not interact and propagate with their own characteristic directions associated with the respective spatial velocities v_1 and v_2.

Next, Fig. 7.b shows that, at intermediate input intensities, the beams at wavelengths λ_1 and λ_2 interact and generate a beam at λ_3. As the two fundamental waves overlap in space, the sum frequency wave ϕ_3 is generated. After the collision, the energy remaining in each input beam keeps propagating along the original direction corresponding to their linear wavevectors. Note that the collision between the input waves does not lead to any spatial shift in their propagation directions. In fact, in this situation it can be shown that no three-wave soliton is generated by the interaction process. According to the definition of a soliton as a discrete component of the spectrum that is obtained by means of the spectral transform method (Zakharov et al., 1980; Zakharov, 1991), in Fig. 7.b the initial data $\phi_1(x, z = 0)$, $\phi_2(x, z = 0)$, $\phi_3(x, z = 0)$ is only composed of a continuum spectrum component (radiation) without any discrete spectrum component (solitons).

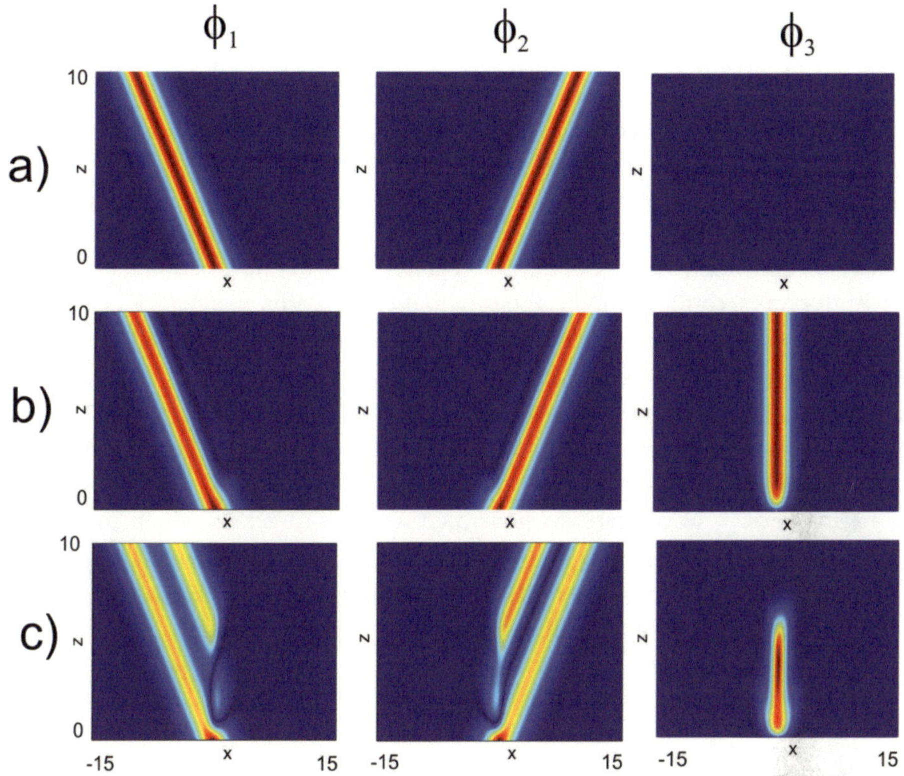

Fig. 7. Numerical $x - z$ TWI dynamics of waves at wavelengths λ_1 and λ_2 which mix to generate a field at the wavelength λ_3. (a) Linear, (b) frequency conversion, and (c) solitonic regime.

Finally, Fig. 7.c shows that, at relatively high input intensities, the interaction between the two input beams and the generated beam is qualitatively different from the previous cases. In fact, in this case the input fields contain a discrete soliton component along with a residual radiation spectrum. The solitonic behavior of the three wave interaction is not visible in the first part (i.e., the signal generation) of the interaction picture of Fig. 7.c, which is similar to what happens in Fig. 7.b: as in this case the residual input beams which keep propagating after generating the sum-frequency wave are composed by pure radiation. In fact, the presence of a soliton in the generated beam is revealed in its subsequent complete decay into the two waves, which is the spatial analog of the temporal soliton decay. From the point of view of the spectral transform of the optical field, we may say that in this case the initial data $\phi_1(x, z = 0)$, $\phi_2(x, z = 0)$, $\phi_3(x, z = 0)$ is composed of both a continuum spectrum component, or radiation, and a discrete spectrum component or soliton.

Another signature of the soliton presence in Fig. 7.c is the characteristic spatial shift that is observed between the two input soliton fundamental beams, and the output soliton fundamental beams that are generated from the decay of the soliton pump beam, with no change of the propagation direction before and after the three-wave interaction process. From another perspective, we may point out that if the generated beam was not a three-wave soliton, then its evolution could be predicted from a simple CW analysis. In this case one would not observe a sharp wave decay at a well-defined point along the crystal, but rather its continuous back-conversion into the two fundamental waves which would emerge from the process substantially wider with respect to the original input beams.

As a matter of fact, as in the case of temporal solitons, in the spatial case we may identify three separate interaction regimes. Namely, the linear, the frequency conversion, and the solitonic regime. In order to provide the experimental demonstration of the above discussed nonlinear dynamics, Ref. (Baronio et al., 2009) considered the optical spatial non-collinear scheme with type II second harmonic generation (SHG) in a birefringent KTP crystal. Two orthogonally polarized beams at frequency $\lambda_1 = \lambda_2 = \lambda$=1064nm, E_λ^e and E_λ^o , were injected into the nonlinear birefringent medium to cross and overlap at the input face of the crystal. Each field was linearly polarized and aligned with a polarization eigenstate of the crystal. In the present scheme the two input waves required for the parametric processes to occur, and for the subsequent observation of solitonic decay, had tilted wave-fronts. In the overlapping area the harmonic $E_{\lambda/2}^e$ at 532nm was generated along a direction which maximized the conversion efficiency as it was fixed by the noncollinear phase-matching conditions. The sum-frequency harmonic propagation direction was found to be in between the directions of the two input waves. As the intensities of the input fields were varied in a suitable range ($1MW/cm^2 - 2.5GW/cm^2$), the TWI linear, frequency conversion, and solitonic regimes in the ordinary KTP plane were experimentally observed as reported in ref.(Baronio et al., 2009).

6. Conclusions

Introduced by a self-contained derivation in section 2 of the 1+1 dimensional equations ruling three-wave parametric interactions in dispersive and quadratic nonlinear optical crystals, in this chapter we presented an extensive review of their exact simulton solutions. Exploiting the properties of such solutions may have important practical applications to the parametric frequency conversion (e.g., sum and frequency difference generation) of short optical pulses. As discussed in section 3, a key property of simulton solutions consisting of two bright and one dark wave packet is that their common group velocity may be controlled in all-optical manner by adjusting the input energy distribution of its three wave components. We pointed out in section 3 that the simulton speed as well as the energy of its component waves may evolve in an adiabatic manner whenever one of their parameters (such as the input power of one of the waves) is varied at the crystal input. The above property permits to obtain a stable (i.e., independent of the precise crystal length or optical peak power) frequency conversion of relatively short isolated optical pulses when mixing them with quasi-CW pulses. In section 4, we have discussed how, by properly adjusting the power of the quasi-CW pulses, one may achieve a full modulation of this CW in a time-periodic replica of the input short pulse at a different frequency. We could provide simple analytical expressions that permit to relate the time duration, the amplitude and the repetition rate of the generated pulse sequence with the input short pulse and CW. Finally, in section 5 we have outlined under which practical conditions it should be possible to achieve an experimental confirmation in commonly available quadratic nonlinear crystals of the theoretically predicted frequency conversion and pulse train generation effects. A first demonstration of nonlinear beam reshaping associated with three–wave interaction soliton properties was recently carried out in the spatial domain, and its peculiar properties have been discussed at the end of section 5.

7. References

Azana, J.; & Muriel, M. (2001) Temporal self-imaging effects: theory and application for multiplying pulse repetition rates, *IEEE J. Sel. Top. Quantum Electron.*, Vol. 7, pp. 728-744

Akjmanov, S.A.; Vysloukh, V.A.; & Chirkin, A.S. (1992) *Optics of Femtosecond Pulses* AIP, New York

Armstrong, J.A.; Jha, S.S.; & Shiren, N.S. (1970) Some effects of group-velocity dispersion on parametric interactions, *IEEE J. Quantum Electronics*, Vol. QE-6, pp. 123-129

Baronio, F.; Conforti, M.; Degasperis, A. & Wabnitz, S. (2008) Three-wave trapponic solitons for tunable high–repetition rate pulse train generation, *IEEE J. Quantum Electronics*, Vol. 44, pp. 542-546

Baronio, F.; Conforti, M.; Andreana, M; Couderc, V.; De Angelsi, C.; Wabnitz, S.; Barthelemy, A; & Degasperis, A. (2009) Frequency generation and solitonic decay in three-wave interactions, *Optics Express*, Vol. 17, pp 13889-13894

Bers, A. (1975) Linear waves and instabilities, *Plasma Physics-Les Houches 1972*. Edited by De Witt C. and Peyraud J., Gordon and Breach, New York, pp 113-215

Calogero, F. & Degasperis, A. (2004). New integrable equations of nonlinear Schrödinger type, *Studies Appl. Math.* Vol. 113, pp 91-137

Calogero, F. & Degasperis, A. (2005) Novel solution of the system describing the resonant interaction of three waves, *Physica D* , Vol. 200, pp. 242–256

Caraquitena, J.; Jiang, Z.; Leiard, D.E.; & Weiner, A.M. (1995). Tunable pulse repetition-rate multiplication using phase-only line-by-line pulse shaping, *Opt. Lett.*, Vol. 32, pp. 716-718

Cerullo, G. & De Silvestri, S. (2003) Ultrafast optical parametric amplifiers, *Rev. Sci. Instrum.*, Vol. 74, pp. 1–18

Chien, C.Y.; Korn, G.; Coe, J.S.; Mourou, G. & Craxton, R.S. (1995). Highly efficient secondharmonic generation of ultraintense Nd:glass laser pulses, *Phys. Rev. E*, Vol. 20, pp. 353-355

Coen, S. & Haelterman, M. (2001) Continuous-wave ultrahigh repetition rate pulse train generation through modulation instability in a passive fiber cavity, *Opt. Lett.*, Vol. 26, pp. 39–41

Conforti, M.; Locatelli, A.; De Angelis, C.; Parini, A.; & Trillo, S. (2005). Self-pulsing instabilities in backward parametric wave-mixing, *J. Opt. Soc. Am. B*, Vol. 22, pp. 2178-2184

Conforti, M.; Baronio, F.; Degasperis, A. & Wabnitz, S. (2006) Inelastic scattering and interactions of parametric three-Wave solitons, *Phys. Rev. E*, Vol. 74, pp. 0656021-4(R)

Conforti, M.; Baronio, F.; Degasperis, A. & Wabnitz, S. (2007). Parametric frequency conversion of short optical pulses controlled by a CW background, *Optics Express*, Vol. 15, Issue 19, pp. 12246-12251

Conti, C.; Assanto,G.; & Trillo, S. (1999). Cavityless oscillation through backward quasi-phasematched second-harmonic generation, *Opt. Lett.*, Vol. 24, pp. 1139-1141

Degasperis, A. & Lombardo, S. (2006). Exact solutions of the 3-wave resonant interaction equation, *Physica D*, Vol. 214, 157-168

Degasperis, A.; Conforti, M.; Baronio, F. &Wabnitz, S. (2006), Stable Control of Pulse Speed in Parametric Three-Wave Solitons, *Phys. Rev. Lett.* Vol. 97, No. 9, 093901-04

Degasperis, A.; Conforti, M.; Baronio, F. & Wabnitz, S. (2007). Effects of nonlinear wave coupling: Accelerated solitons. Boomerons: From curiosity to nonlinear optics, *Eur. Phys. J. Special Topics*, Vol. 147, pp. 233-252

Degasperis, A. (2009). Multiscale expansion and integrability of dispersive wave equations. In: Mikhailov, A.. Integrability. *Lecture Notes in Physics*, Vol. 767, Springer, pp. 215-244

Degasperis, A. & Lombardo, S. (2009). Multicomponent integrable wave equations II. Soliton solutions, submitted to Journal of Physics A: Mathematical and Theoretical

Fournier, S.; Lopez-Martens, R.L.; Le Blanc, C.; Baubeau, E. & Salin, F. (1998). Solitonlike pulse shortening in a femtosecond parametric amplifier, *Opt. Lett.*, Vol. 23, pp. 627-629

Gale, G.M.; Cavallari, M.; Driscoll, T.J.; & Hache, F. (1995). Sub-20-fs tunable pulses in the visible from an 82-MHz optical parametric oscillator, *Opt. Lett.*, Vol. 20, pp. 1562-1564

Hasegawa, A. (1984). Generation of a train of soliton pulses by induced modulation instability in optical fibers *Opt. Lett.*, Vol. 9, pp. 288-290

Ibragimov, E.; and A. Struthers, A. (1996). Second-harmonic pulse compression in the soliton regime *Opt. Lett.*, Vol. 21, pp. 1582-1584

Ibragimov, E.; & Struthers, A. (1997). Three-wave soliton interaction of ultrashort pulses in quadratic media *J. Opt. Soc. Am B*, Vol. 14, pp.1472-1479

Ibragimov, E.; Struthers, A & Kaup, D.J. (1998). Soliton pulse compression in the theory of optical parametric amplification *Optics Comm.*, Vol. 32, pp. 101-107

Ibragimov, E.; Struthers, A; Kaup, D.J.; Khaydarov, J.D. & Singer K.D. (1999). Three-wave interaction solitons in optical parametric amplification *Phys. Rev. E*, Vol. 59, pp. 6122- 6137

Inoue, T.; Hiroishi, J.; Yagi, T.; & Mimura, Y. (2007). Generation of in–phase pulse train from optical beat signal *Opt. Lett*, Vol. 32, pp. 1596-1598

Kaup, D.J. (1976). The three-wave interactiona nondispersive phenomenon *Stud. Appl. Math.*, Vol. 55, pp. 9-44

Kaup, D.J.; Reiman, A.& Bers, A. (1979). Space-time evolution of nonlinear three-wave interactions. I. Interaction in a homofeneous medium *Rev. Mod. Phys.*, Vol. 51, pp. 275-309

Kaup, D.J. (1981). The solution of a general initial value problem for the full three dimensional three-wave resonant interaction *Physica D*, Vol. D3, pp. 374–395

Nozaki, K.; & Taniuti, T. (1973). Propagation of solitary pulses in interactions of plasma waves. II, *J. Phys. Soc. Japan*, Vol. 36, pp.591-595

Ohsawa, Y.; & Nozaki, K.(1974). Propagation of solitary pulses in interactions of plasma waves, *J. Phys. Soc. Japan*, Vol. 34, pp.796-800

Liu, Y.; Park, S.G.; & Weiner,A.M. (1996). Enhancement of narrow-band terahertz radiation from photoconducting antennas by optical pulse shaping, *Opt. Lett.*, Vol. 21, pp.1762- 1764

Longhi, S.; Marano, M.; Laporta, P.; Svelto, O.; Belmonte, M.; Agogliati, B.; Arcangeli, L.; Pruneri, V.; Zervas, M.N.; & Ibsen, M. (2000). 40-Ghz pulse-train generation at 1.5 um with a chirped fiber gratings as frequency multiplier, *Opt. Lett.*, Vol. 24, pp.1481- 1483

Petropoulos, P.; Ibsen, M.; Zervas, M.N.; & Richardson, D.J. (2000). Repetition frequency quadruplication through Fabry-Perot filtering, *Opt. Lett.*, Vol. 25, pp.521-523

Picozzi, A.; & Haelterman, M.(1998). Spontaneous formation of symbiotic solitary waves in a backward quasi-phase-matched parametric oscillator, *Opt. Lett.*, Vol. 23, pp.1808-1810

Pitois, S.; Fatome, J.; & Millot, G. (2002). Generation of a 160-Ghz transform limited pedestal free pulse train thorough multiwave mixing compression of a dual-frequency beat signal, *Opt. Lett.*, Vol. 27, pp.1729-1731

Stabinis, A.; Valiulis, G. & Ibragimov, E.A (1991). Effective sum frequency pulse compression in nonlinear crystals, *Optics Comm.*, Vol. 86, pp. 301-306

Taranenko, Y. N., & Kazovsky, L. G. (1992). Three-Wave Envelope Solitons: Can the Speed of Light in the Fiber be Controlled, *IEEE Phot. Tech. Lett.*, Vol. 4, pp. 494-497

Umstadter, D.; Esarey, E.; & Kim, J. (1994). Nonlinear plasma waves resonantly driven by optimized laser pulse trains, *Phys. Rev. Lett.*, Vol. 72, pp. 1224-1227

Wang, Y.; & Dragila, R. (1990). Efficient conversion of picosecond laser pulses into secondharmonic frequency using group-velocity dispersion, *Phys. Rev. A*, Vol. 41, pp.5645- 5649

Wang, Y.; & Luther-Davies, B. (1992). Frequency-doubling pulse compressor for picosecond high-power neodymium laser pulses, *Opt. Lett.*, Vol. 17, pp.1459-1461

Weiner, A.M.; Leaird, D.E., Wiederrecht, G.P.; & Nelson K.A. (2007). Femtosecond pulse sequences used for optical manipulation of molecular motion, *Science*, Vol. 247, pp.1317-1319

Yiannopoulos, K.; Vyrsokinos, K.; Kehayas, E.; Pleros, N.; Vlachos, K.; Avramopoulos, H.; & Guekos, G. (2003). Repetition frequency quadruplication through Fabry-Perot filtering, *IEEE Photon. Technol. Lett.*, Vol. 15, pp.1294-1296

Zakharov, V.E. & Manakov, S.V. (1973). Resonant interaction of wave packets in nonlinear media , *Sov. Phys. JETP Lett.*, Vol. 18, pp.243-245

Zakharov, V.E.; Manakov,S.V.; Novikov,S.P.; & Pitajevski (1980) *The Theory of Solitons: The Inverse Problem Method* Nauka.

Zakharov, V.E. (1991) *What is integrability?*, Springer Verlag.

Zhang, X.; Lytle, A.L.; Popmintchev, T.; Zhou, X.; Kapten, H.C.; Murnane, M.M.; & Cohen, O. (2007). Quasi-phase-matching and quantum-path control of high harmonic generation using counterpropagating light, *Nat. Phys.*, Vol. 3, pp.270-275

8

Analogue of the Event Horizon in Fibers

Friedrich König, Thomas G. Philbin, Chris Kuklewicz, Scott Robertson, Stephen Hill, and Ulf Leonhardt
School of Physics and Astronomy, University of St Andrews, North Haugh, St Andrews, Fife, KY16 9SS
United Kingdom

1. Introduction

In 1974 Stephen Hawking predicted that gravitational black holes would emit thermal radiation and decay (Hawking, 1974). This radiation, emitted from an area called the event horizon, is since known as Hawking radiation. To date it is still one of the most intriguing physical effects and bears great importance for the development of a quantum theory of gravity, cosmology and high energy physics.

The Hawking effect is one of a rich class of quantum properties of the vacuum (Birrell & Davies, 1984; Brout et. al., a; Milonni, 1994). For example, in the Unruh effect (Moore, 1970; Fulling, 1973; Davies, 1975; DeWitt, 1975; Unruh, 1976), an accelerated observer perceives the Minkowski vacuum as a thermal field. The physics of Hawking radiation leaves us with fascinating questions about the laws of nature at transplanckian scales, the conservation of information and physics beyond the standard model. Because of the thermal nature of the radiation, it is characterized by a temperature, the Hawking temperature. For stable astronomical black holes this lies far below the temperature of the cosmic microwave background, such that an observation of Hawking radiation in astrophysics seems unlikely.

Laboratory analogues of black holes have the potential to make the effect observable (Unruh, 1981; Schleich & Scully, 1984). The space-time geometry of the gravitational field can be represented in coordinates that act as an effective flow (Novello et al., 2002; Volovik, 2003; Unruh, 1981; Jacobson, 1991; Rousseaux et al., 2008). The event horizon lies where the flow velocity appears to exceed the speed of light in vacuum. Analogue systems are thus inspired by the following intuitive idea (Unruh, 1981): the black hole resembles a river (Jacobson, 1991; Rousseaux et al., 2008), a moving medium flowing towards a waterfall, the singularity. Imagine that the river carries waves propagating against the current with speed c'. The waves play the role of light where c' represents c, the speed of light in vacuum. Suppose that the closer the river gets to the waterfall the faster it flows and that at some point the speed of the river exceeds c'. Clearly, beyond this point no wave can propagate upstream anymore. The point of no return is the horizon.

In this chapter we are explaining a recent approach to the realization of an event horizon in optics (Philbin et. al, 2008). We start by describing the propagation of light in optical fibers and show the analogy to a curved space-time geometry. In Sec. 4 we quantize the field equation and give a Hamiltonian. Then we can use the geometrical optics approximation in Sec. 5 to find the behavior of light at a horizon, before we describe the scattering process that is the analogue to Hawking radiation (Sec. 6). In Sec. 7 we describe the experimental findings of frequency shifts at the optical horizon and compare them to our predictions before we conclude.

2. Background

Nothing, not even light, can escape from a gravitational black hole. Yet according to quantum physics, the black hole is not entirely black, but emits waves in thermal equilibrium (Hawking, 1974; 1975; Birrell & Davies, 1984; Brout et. al. , a). The waves consist of correlated pairs of quanta, one originates from the inside and the other from the outside of the horizon. Seen from one side of the horizon, the gravitational black hole acts as a thermal black-body radiator sending out Hawking radiation (Hawking, 1974; 1975; Birrell & Davies, 1984; Brout et. al. , a). The effective temperature depends on the surface gravity (Hawking, 1974; 1975; Birrell & Davies, 1984; Brout et. al. , a) that, in our analogue model, corresponds to the flow-velocity gradient at the horizon (Novello et al., 2002; Volovik, 2003; Unruh, 1981; Jacobson, 1991).
Many systems have been proposed for laboratory demonstrations of analogues of Hawking radiation. One type of recent proposal (Garay et al., 2000; Giovanazzi et al., 2004; Giovanazzi, 2005) suggests the use of ultracold quantum gases such as alkali Bose-Einstein condensates or ultracold alkali Fermions (Giovanazzi, 2005). When a condensate in a waveguide is pushed over a potential barrier it may exceed the speed of sound (typically a few mm/s) and is calculated to generate a Hawking temperature of about 10nK (Giovanazzi et al., 2004). Helium-3 offers a multitude of analogues between quantum fluids and the standard model, including Einsteinian gravity (Volovik, 2003). For example, the analogy between gravity and surface waves in fluids (Schützhold & Unruh, 2002) has inspired ideas for artificial event horizons at the interface between two sliding superfluid phases (Volovik, 2002), but, so far, none of the quantum features of horizons has been measured in Helium-3. Proposals for optical black holes (Leonhardt & Piwnicki, 2000; Leonhardt, 2002) have relied on slowing down light (Milonni, 2004) such that it matches the speed of the medium (Leonhardt & Piwnicki, 2000) or on bringing light to a complete standstill (Leonhardt, 2002), but in these cases absorption may pose a severe problem near the horizon where the spectral transparency window (Milonni, 2004) vanishes.
But do we have to physically move the medium for establishing a horizon? Waves in the river may also see a horizon if the river depth changes due to some barrier, as the flow speed is increased above the barrier. There is again a black hole horizon just before the barrier. The situation is indistinguishable if the water of the river is at rest and the barrier is dragged along the river bed. Thus the medium can be locally disturbed and the wave speed can be reduced locally, leading to a situation of moving horizons in a medium at rest. Any Hawking radiation emitted this way will be immensely Doppler shifted to higher frequencies. Such ideas were discussed for moving solitons and domain walls (Jacobson & Volovik, 1998) in superfluid Helium-3 (Volovik, 2003) and more recently for microwave transmission lines with variable capacity (Schützhold & Unruh, 2005), but they have remained impractical so far.
Ultrashort optical pulses seem suited for this scenario as optical frequencies and velocities are very high. Moving a medium at a fraction of the speed of light seems illusive. The novel idea described in this chapter (Philbin et. al, 2008), illustrated in Fig. 1, is based on the nonlinear optics of ultrashort light pulses in optical fibers (Agrawal, 2001) where we exploit the remarkable control of the nonlinearity, birefringence and dispersion in microstructured fibers (Russell, 2003; Reeves et al., 2003). More recently, ultrashort laser pulse filamentation has been shown to exhibit asymptotic horizons based on similar principles (Faccio et al., 2009).

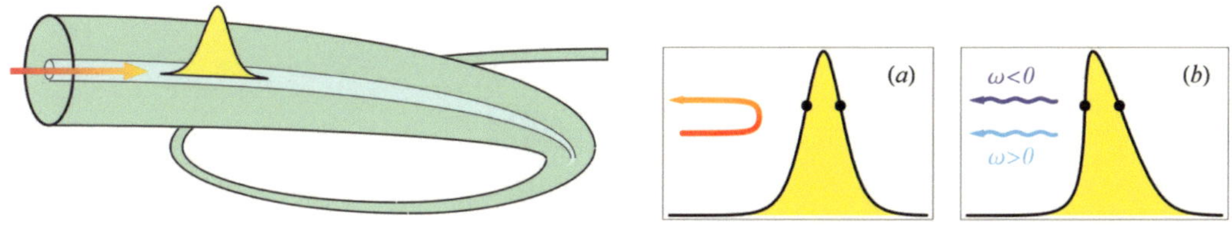

Fig. 1. Fiber-optical horizons. Left: a light pulse in a fiber slows down infrared probe light attempting to overtake it. Right: the diagrams are in the co-moving frame of the pulse. (a) Classical horizons. The probe is slowed down by the pulse until its group velocity matches the pulse speed at the points indicated in the figure, establishing a white hole at the back and a black hole at the front of the pulse. The probe light is blue-shifted at the white hole until the optical dispersion releases it from the horizon. (b) Quantum pairs. Even if no probe light is incident, the horizon emits photon pairs corresponding to waves of positive frequencies from the outside of the horizon paired with waves at negative frequencies from beyond the horizon. An optical shock has steepened the pulse edge, increasing the luminosity of the white hole (Philbin et. al, 2008).

3. Effective moving medium and metric

The fundamental idea behind the fiber-optical event horizon is the nonlinear and local modification of the refractive index of the fiber by a propagating pulse. As we will see later, this refractive index modification has to be ultrafast, i.e. the contributing nonlinearity is the optical Kerr effect (Agrawal, 2001): the (linear) effective refractive index of the fiber, n_0, gains an additional contribution δn that is proportional to the instantaneous pulse intensity I at position z and time t,

$$n = n_0 + \delta n, \quad \delta n \propto I(z,t)\,. \tag{1}$$

This contribution to the effective refractive index n moves with the pulse. It acts as a local modification of the wave speed and thus as an effective moving medium, although nothing material is moving.

In what follows we will review how this nonlinearity arises in a fiber-waveguide, how it forms an effective moving medium, and that the fields follow a metric in analogy to a space-time manifold in the dispersionless case.

3.1 Waveguides

The waveguide confines light in the x and y direction and light propagates along the z direction. We assume a fiber homogeneous in z and with the Fourier-transformed susceptibility

$$\widetilde{\chi}_g = \widetilde{\chi}_g(\omega,x,y)\,. \tag{2}$$

We represent the Fourier-transformed electric field strengths as

$$\widetilde{\mathbf{E}}(\omega,\mathbf{r}) = \widetilde{E}(\omega,z)\,\mathbf{U}(\omega,x,y), \tag{3}$$

where we assume linearly polarized light. Also we require that the fiber modes $\mathbf{U}$ are eigenfunctions of the transversal part of the wave equation for monochromatic light with eigenvalues $\beta^2(\omega)$,

$$\left(-\nabla \times (\nabla \times \mathbf{U}) + (1+\tilde{\chi}_g)\frac{\omega^2}{c^2}\mathbf{U}\right) = \beta^2(\omega)\mathbf{U}\,. \tag{4}$$

For single-mode fibers, only one eigenvalue $\beta^2(\omega)$ exists.
The eigenvalues $\beta^2(\omega)$ of the transversal modes set the effective refractive indices $n(\omega)$ of the fiber for light pulses $E(t, z)$ defined by the relation

$$\beta = \frac{n(\omega)}{c}\,\omega\,. \tag{5}$$

In the absence of losses within the frequency range we are considering, the Fourier-transformed $\tilde{\chi}_g(\omega)$ in the longitudinal mode equation (4) is real for real ω and the longitudinal mode equation (4) is Hermitian and positive. Since the linear susceptibility $\chi_g(t)$ is real, $\tilde{\chi}_g(\omega)$ is an even function, which implies that $n^2(\omega)$ and $\beta^2(\omega)$ are even.

3.2 Effective moving medium

In our case, an intense ultrashort optical pulse interacts with a weak probe field, an incident wave of light or the vacuum fluctuations of the electromagnetic field itself (Milonni, 1994). The vacuum fluctuations are carried by modes that behave as weak classical light fields as well. The pulse is polarized along one of the eigen-polarizations of the fiber; the probe field may be co- or cross polarized. We assume that the intensity profile $I(z, t)$ of the pulse uniformly moves with constant velocity u during the interaction with the probe, neglecting the small deceleration due to the Raman effect and pulse distortions. Since the probe field is weak we can safely neglect its nonlinear interaction with the pulse or itself. As the intensity profile of the pulse is assumed to be fixed, we focus attention on the probe field. We describe the probe by the corresponding component A of the vector potential that generates the electric field E and the magnetic field B, with

$$E = -\partial_t A\,, \quad B = \partial_z A\,. \tag{6}$$

The probe field obeys the wave equation

$$\left(c^2\partial_z^2 + c^2\beta^2(i\partial_t) - \partial_t\chi\partial_t\right)A = 0\,, \quad \chi \propto I(z,t) \tag{7}$$

where χ denotes the susceptibility due to the Kerr effect of the pulse on the probe. β is given by Eq. (5) and we denote the effective refractive index by n_0. Equation (7) shows that the pulse indeed establishes an effective moving medium (Leonhardt, 2003). It is advantageous to use the retarded time τ and the propagation time ζ as coordinates, defined as

$$\tau = t - \frac{z}{u}\,, \quad \zeta = \frac{z}{u}\,, \tag{8}$$

because in this case the properties of the effective medium depend only on τ. τ and ζ play the roles of space and of time, respectively. The z and t derivatives are replaced by

$$\partial_t = \partial_\tau\,, \quad \partial_z = \frac{1}{u}(\partial_\zeta - \partial_\tau)\,, \tag{9}$$

and the wave equation (7) becomes

$$(\partial_\zeta - \partial_\tau)^2 A = \partial_\tau \frac{u^2}{c^2} n^2 \partial_\tau A, \tag{10}$$

where the total refractive index n consists of the effective linear index n_0 and the Kerr contribution of the pulse,

$$n^2 = n_0^2 + \chi. \tag{11}$$

Since $\chi \ll n_0$ we approximate

$$n \approx n_0 + \delta n, \quad \delta n = \frac{\chi}{2n_0}, \tag{12}$$

where we can ignore the frequency dependance of n_0 in $\chi/(2n_0)$. Note that Eq. (8) does not describe a Lorentz transformation to an inertial system, but the τ and ζ are still valid coordinates.

3.3 Dispersionless case and metric

For simplicity, we consider the dispersionless case where the refractive index n_0 of the probe does not depend on the frequency. Note that a horizon inevitably violates this condition, because here light comes to a standstill, oscillating at increasingly shorter wavelengths, leaving any dispersionless frequency window. However, many of the essentials of horizons are still captured within the dispersionless model.

First, we can cast the wave equation (10) in a relativistic form, introducing a relativistic notation (Landau & Lifshitz, 1975) for the coordinates and their derivatives

$$x^\mu = (\zeta, \tau), \quad \partial_\mu = (\partial_\zeta, \partial_\tau) \tag{13}$$

and the matrix

$$g^{\mu\nu} = \begin{pmatrix} 1 & -1 \\ -1 & 1 - u^2 n^2/c^2 \end{pmatrix} \tag{14}$$

that resembles the inverse metric tensor of waves in moving fluids (Unruh, 1981; Visser, 1998). Adopting these definitions and Einstein's summation convention over repeated indices the wave equation (10) appears as

$$\partial_\mu g^{\mu\nu} \partial_\nu A = 0, \tag{15}$$

which is almost the free wave equation in a curved space-time geometry (Landau & Lifshitz, 1975) (In the case of a constant refractive index the analogy between the moving medium and a space-time manifold is perfect[1].). The effective metric tensor $g_{\mu\nu}$ is the inverse of $g^{\mu\nu}$ (Landau & Lifshitz, 1975). We obtain

---

[1] The exact wave equation in a curved space time geometry is $\partial_\mu \sqrt{-g} g^{\mu\nu} \partial_\nu A = 0$. where g is the determinant of the metric tensor (Landau & Lifshitz, 1975). In the case (14) g depends only on the refractive index n and hence g is constant for constant n.

$$g_{\mu\nu} = \frac{c^2}{n^2u^2}\begin{pmatrix} u^2n^2/c^2 - 1 & -1 \\ -1 & -1 \end{pmatrix}. \tag{16}$$

In subluminal regions where the velocity c/n of the probe light exceeds the speed of the effective medium, *i.e.* the velocity u of the pulse, the measure of time $u^2n^2/c^2 - 1$ in the metric (16) is negative. Here both $\partial_\tau$and $\partial_\zeta$are timelike vectors (Landau & Lifshitz, 1975). In superluminal regions, however, c/n is reduced such that $u^2n^2/c^2 - 1$ is positive. A horizon, where time stands still, is established where the velocity of light matches the speed of the pulse.

4. Lagrangian formulation and Hamiltonian

We have now seen that the probe is interacting within an effective moving medium in a way similar to waves in moving fluids, mimicking space-time in general relativity. To find the classical as well as quantum mechanical evolution of the field, we will next find a suitable Lagrangian density and the canonical Hamiltonian. Then we expand the quantized vector potential in terms of creation and annihilation operators.

4.1 Action

The theory of quantum fields at horizons (Hawking, 1974; 1975; Birrell & Davies, 1984; Brout et. al. , a) predicts the spontaneous generation of particles. The quantum theory of light in dielectric media at rest has reached a significant level of sophistication (See e.g. Knöll et. al., 2001), because it forms the foundation of quantum optics (Leonhardt, 2003; See e.g. U. Leonhardt, 1993), but quantum light in moving media is much less studied (Leonhardt, 2003). In optical fibers, light is subject to dispersion, which represents experimental opportunities and theoretical challenges: we should quantize a field described by a classical wave equation of high order in the retarded time. Moreover, strictly speaking, dispersion is always accompanied by dissipation, which results in additional quantum fluctuations (See e.g. Knöll et. al., 2001). Here, however, we assume to operate in frequency windows where the absorption is negligible. To deduce the starting point of the theory, we begin with the dispersionless case in classical optics and then proceed to consider optical dispersion for light quanta.

The classical wave equation of one-dimensional light propagation in dispersionless media follows from the Principle of Least Action (Landau & Lifshitz, 1975) with the action of the electromagnetic field in SI units

$$\begin{aligned} S &= \int\int \frac{\varepsilon_0}{2}(n^2E^2 - c^2B^2)\,dz\,dt \\ &= \int\int \frac{\varepsilon_0}{2}\left[-A\partial_\tau n^2\partial_\tau A - c^2(\partial_z A)^2\right]u\,d\tau\,d\zeta \end{aligned} \tag{17}$$

and hence the Lagrangian density

$$\mathscr{L} = -\frac{\varepsilon_0}{2}\left[A\partial_\tau n^2\partial_\tau A + c^2(\partial_z A)^2\right]. \tag{18}$$

In order to include the optical dispersion in the fiber and the effect of the moving pulse, we express the refractive index in terms of $\beta(\omega)$ and the effective susceptibility $\chi(\tau)$ caused by the pulse, using Eqs. (5) and (11) with $\omega = i\partial_\tau$. We thus propose the Lagrangian density

$$\mathscr{L} = \frac{\varepsilon_0}{2}\left[A\left(c^2\beta^2(i\partial_\tau) - \partial_\tau\chi\partial_\tau\right)A - c^2(\partial_z A)^2\right]. \tag{19}$$

In the absence of losses, $\beta^2(\omega)$ is an even function (Sec. 3.1). We write down the Euler-Lagrange equation (Landau & Lifshitz, 1975) for this case

$$\partial_\zeta \frac{\partial \mathscr{L}}{\partial(\partial_\zeta A)} - \sum_{\nu=0}^{\infty}(-1)^\nu \partial_\tau^\nu \frac{\partial \mathscr{L}}{\partial(\partial_\tau^\nu A)} = 0 \tag{20}$$

and obtain the wave equation (10). Thus the Lagrangian density (19) is correct.

4.2 Quantum field theory

According to the quantum theory of fields (Weinberg, 1999) the component A of the vector potential is described by an operator $\hat{A}$. Since the classical field A is real, the operator $\hat{A}$ must be Hermitian. For finding the dynamics of the quantum field we quantize the classical relationship between the field, the canonical momentum density and the Hamiltonian: we replace the Poisson bracket between the field A and the momentum density $\partial\mathscr{L}/\partial(\partial_\zeta A)$ by the fundamental commutator between the quantum field $\hat{A}$ and the quantized momentum density (Weinberg, 1999). We obtain from the Lagrangian (19) the canonical momentum density

$$\hat{\pi} = -\varepsilon_0 \frac{c^2}{u}\partial_z \hat{A} \tag{21}$$

and postulate the equivalent of the standard equal-time commutation relation (Weinberg, 1999; Mandel &Wolf, 1995)

$$\left[\hat{A}(\zeta,\tau_1), \hat{\pi}(\zeta,\tau_2)\right] = \frac{i\hbar}{u}\delta(\tau_1 - \tau_2). \tag{22}$$

We obtain the Hamiltonian

$$\begin{aligned} \hat{H} &= \int (\hat{\pi}\partial_\zeta \hat{A} - \hat{\mathscr{L}})\, u\, d\tau \\ &= \frac{\varepsilon_0}{2}\int \left(\frac{c^2}{u^2}\left((\partial_\tau \hat{A})^2 - (\partial_\zeta \hat{A})^2\right) - \hat{A}\left(c^2\beta^2 - \partial_\tau\chi\partial_\tau\right)\hat{A}\right) u\, d\tau. \end{aligned} \tag{23}$$

One verifies that the Heisenberg equation of the quantum field $\hat{A}$ is the classical wave equation (10), as we would expect for fields that obey linear field equations.

4.3 Mode expansion

Since the field equation is linear and classical, we represent $\hat{A}$ as a superposition of a complete set of classical modes multiplied by quantum amplitudes $\hat{a}_k$. The mode expansion is Hermitian for a real field such as the electromagnetic field,

$$\hat{A} = \sum_k \left(A_k \hat{a}_k + A_k^* \hat{a}_k^\dagger\right). \tag{24}$$

The modes A_k obey the classical wave equation (15) and are subject to the orthonormality relations (Birrell & Davies, 1984; Brout et. al., a; Leonhardt, 2003)

$$(A_k, A_{k'}) = \delta_{kk'} \quad (A_k^*, A_{k'}) = 0 \tag{25}$$

with respect to the scalar product

$$(A_1, A_2) = \frac{\varepsilon_0 c^2}{i\hbar} \int (A_1^* \partial_z A_2 - A_2 \partial_z A_1^*) \, d\tau \, . \tag{26}$$

The scalar product is chosen such that it is a conserved quantity for any two solutions A_1 and A_2 of the classical wave equation (10),

$$\partial_\zeta (A_1, A_2) = 0, \tag{27}$$

with a prefactor to make the commutation relations between the mode operators particularly simple and transparent. The scalar product serves to identify the quantum amplitudes $\hat{a}_k$ and $\hat{a}_k^\dagger$: the amplitude $\hat{a}_k$ belongs to modes A_k with positive norm, whereas the Hermitian conjugate $\hat{a}_k^\dagger$ is the quantum amplitude to modes A_k^* with negative norm, because

$$(A_1^*, A_2^*) = -(A_1, A_2) \, . \tag{28}$$

Using the orthonormality relations (25) we can express the mode operators $\hat{a}_k$ and $\hat{a}_k^\dagger$ as projections of the quantum field $\hat{A}$ onto the modes A_k and A_k^* with respect to the scalar product (26),

$$\hat{a}_k = (A_k, \hat{A}) \, , \quad \hat{a}_k^\dagger = -(A_k^*, \hat{A}) \, . \tag{29}$$

We obtain from the fundamental commutator (22) and the orthonormality relations (25) of the modes the Bose commutation relations

$$[\hat{a}_k, \hat{a}_{k'}^\dagger] = \delta_{kk'} \, , \quad [\hat{a}_k, \hat{a}_{k'}] = 0. \tag{30}$$

Therefore light consists of bosons and the quantum amplitudes $\hat{a}_k$ and $\hat{a}_k^\dagger$ serve as annihilation and creation operators.

The expansion (24) is valid for any orthonormal and complete set of modes. Consider stationary modes with frequencies ω_k' such that

$$\partial_\zeta A_k = -i\omega_k' A_k \, . \tag{31}$$

We substitute the mode expansion (24) in the Hamiltonian (23) and use the wave equation (10) and the orthonormality relations (25) to obtain

$$\hat{H} = \sum_k \hbar\omega_k' \left(\hat{a}_k^\dagger \hat{a}_k + \frac{1}{2} \right) \, . \tag{32}$$

Each stationary mode contributes $\hbar\omega_k'$ to the total energy that also includes the vacuum energy. The modes with positive norm select the annihilation operators of a quantum field, whereas the negative norm modes pick out the creation operators. In other words, the norm of the modes determines the particle aspects of the quantum field. In the Unruh effect (Moore, 1970; Fulling, 1973; Unruh, 1976; Davies, 1975; DeWitt, 1975), modes with positive norm consist of superpositions of positive and negative norm modes in the frame of an accelerated observer (Birrell & Davies, 1984; Brout et. al. , a). Consequently, this observer perceives the Minkowski vacuum as thermal radiation (Moore, 1970; Fulling, 1973; Unruh, 1976; Davies, 1975; DeWitt, 1975). In the Hawking effect (Hawking, 1974; 1975), the scattering of light at the event horizon turns out to mix positive and negative norm modes, giving rise to Hawking radiation.

5. Field evolution in the geometrical optics approximation

Here we will derive Hamilton's equations in the geometrical optics approximation to understand the frequency shifts of light near a horizon.
To quantitatively describe this effect, we will derive the frequency ω' in a co-moving frame that is connected to the laboratory-frame frequency ω by the Doppler formula

$$\omega' = \left(1 - \frac{nu}{c}\right)\omega. \tag{33}$$

For a stable pulse, ω' is a conserved quantity, whereas ω follows the contours of fixed ω' when δ_n varies with the intensity profile of the pulse, see Fig. 4. If δ_n becomes sufficiently large, the frequency ω completes an arch from the initial ω_1 to the final ω_2; it is blue-shifted by the white-hole horizon. At a black-hole horizon, the arch is traced the other way round from ω_2 to ω_1. For the frequency at the center of the arches an infinitesimal δ_n is sufficient to cause a frequency shift; at this frequency the group velocity of the probe matches the group-velocity of the pulse.

5.1 Geometrical optics

A moving dielectric medium with constant refractive index but nonuniform velocity appears to light exactly as an effective space-time geometry (Leonhardt, 2003)[2]. Since a stationary 1 + 1 dimensional geometry is conformally flat (Nakahara, 2003) a coordinate transformation can reduce the wave equation to describing wave propagation in a uniform medium, leading to plane-wave solutions (Leonhardt & Philbin, 2006). The plane waves appear as phase-modulated waves in the original frame. Consequently, in this case, geometrical optics is exact. In our case, geometrical optics provides an excellent approximation, because the variations of the refractive index are very small.
Consider a stationary mode A. We assume that the mode carries a slowly varying amplitude $\mathcal{A}$ and oscillates with a rapidly changing phase φ,

$$A = \mathcal{A}\exp(i\varphi). \tag{34}$$

We represent the phase as

$$\varphi = -\int \omega(\tau)\, d\tau - \omega'\zeta \tag{35}$$

[2] see footnote 1 in Sec. 3.3

and obtain from the wave equation (10) the dispersion relation

$$(\omega - \omega')^2 = \frac{u^2}{c^2} n^2 \omega^2 \tag{36}$$

by neglecting all derivatives of the amplitude $\mathcal{A}$. Here n includes the additional susceptibility χ due to the Kerr effect of the pulse according to Eq. (11). The dispersion relation has two sets of solutions describing waves that are co- or counter-propagating with the pulse in the laboratory frame. Counter-propagating waves will experience the pulse as a tiny transient change of the refractive index, whereas co-propagating modes may be profoundly affected.

Consider the solution given by Eq. (33). In this case, we obtain outside of the pulse in the laboratory frame $\varphi = n(\omega/c)z - \omega t$, which describes light propagating in the positive z direction. Consequently, the branch (33) of the dispersion relation corresponds to co-propagating light waves. We also see that ω is the frequency of light in the laboratory frame, whereas ω' is the frequency in the frame co-moving with the pulse. Equation (33) describes how the laboratory-frame and the co-moving frequencies are connected due to the Doppler effect.

In order to find the evolution of the amplitude $\mathcal{A}$, we substitute in the exact scalar product (26) the approximation (34) with the phase (35) and the dispersion relation (33). In the limit $\omega_1' \to \omega_2'$ we obtain

$$(A_1, A_2) = \frac{2\varepsilon_0 c}{\hbar} \int \mathcal{A}^2 n\omega \exp\left(i\varphi_2 - i\varphi_1\right) d\tau, \tag{37}$$

which should give $\delta(\omega_1' - \omega_2')$ according to the normalization (25). The dominant, diverging contribution to this integral, generating the peak of the delta function, stems from $\tau \to \pm\infty$ (Landau & Lifshitz, 1977). Hence, for $\omega_1' \to \omega_2'$, we replace φ in the integral by φ at $\tau \to \pm\infty$ where ω does not depend on τ anymore,

$$(A_1, A_2) = \frac{2\varepsilon_0 c}{\hbar} \int \mathcal{A}^2 n\omega \exp\left[i(\omega_2 - \omega_1)\tau\right] d\tau, \quad \omega_2 - \omega_1 = \frac{\partial \omega}{\partial \omega'}(\omega_2' - \omega_1'), \tag{38}$$

which gives $\delta(\omega_1' - \omega_2')$ for

$$|\mathcal{A}|^2 = \frac{\hbar}{4\pi\varepsilon_0 c n\omega} \left|\frac{\partial \omega}{\partial \omega'}\right| \tag{39}$$

and positive frequencies ω in the laboratory frame. Note that positive frequencies ω' in the co-moving frame correspond to negative ω in superluminal regions where the pulse moves faster than the phase-velocity of the probe light.

Hamilton's equations (Landau & Lifshitz, 1976) determine the trajectories of light rays in the co-moving frame, parameterized by the pulse-propagation time ζ. Here τ plays the role of the ray's position. Comparing the phase (35) with the standard structure of the eikonal in geometrical optics (Born & Wolf, 1999) or the semiclassical wave function in quantum mechanics (Landau & Lifshitz, 1977) we notice that $-\omega$ plays the role of the conjugate momentum here. Therefore, we obtain Hamilton's equations with a different sign than usual (Landau & Lifshitz, 1976),

$$\dot{\tau} = -\frac{\partial \omega'}{\partial \omega}, \quad \dot{\omega} = \frac{\partial \omega'}{\partial \tau}. \tag{40}$$

Now we express $\dot{\tau}$ in terms of the group index in the laboratory frame. The group velocity v_g is the derivative of the frequency ω with respect to the wave number $n\omega/c$ (Agrawal, 2001)

$$n_g = n + \omega \frac{\partial n}{\partial \omega}. \tag{41}$$

We obtain from the first of Hamilton's equations (40) and the Doppler formula (33)

$$\dot{\tau} = \frac{u}{c} n_g - 1 = -\frac{n_g}{c} v_g', \quad v_g' = \frac{c}{n_g} - u \tag{42}$$

where v_g' denotes the difference between the group velocity of the probe v_g and the pulse speed u. We see that the velocity $\dot{\tau}$ in the co-moving frame (8) vanishes when the Kerr susceptibility χ reduces the group velocity c/n_g such that it matches the speed of the pulse u. Since $\dot{\omega}$ does not vanish here in general, the ray does not remain there, but changes direction in the co-moving frame.

At such a turning point we expect a violation of the validity of geometrical optics (Landau & Lifshitz, 1977). For example, the amplitude (39) would diverge here. Geometrical optics is an exponentially accurate approximation when

$$\left| \frac{\partial T}{\partial \tau} \right| \ll 1 \quad \text{for} \quad T = \frac{2\pi}{\omega}, \tag{43}$$

as we see from the analogy to the semiclassical approximation in quantum mechanics (Landau & Lifshitz, 1977). Here the cycle T plays the role of the wavelength. We get

$$\frac{\partial T}{\partial \tau} = \frac{\dot{\omega} T}{\omega \dot{\tau}}. \tag{44}$$

Consequently, geometrical optics indeed is no longer valid near a turning point where

$$n_g = \frac{c}{u} \quad v_g' = v_g \frac{\partial \omega'}{\partial \omega} = 0. \tag{45}$$

This turning point defines a *group velocity horizon* where the pulse has slowed down the probe such that it matches the speed of the pulse. At this horizon the incident mode is converted into a mode that represents another solution of the dispersion relation; a red-or blue-shifted wave, depending on the dispersion and the sign of the first derivative of χ with respect to τ at the group velocity horizon. White holes correspond to increasing χ and black holes to decreasing χ. White holes blue-shift, because incident waves freeze in front of the horizon, oscillating with increasing frequency. Black holes red-shift, because they stretch any emerging waves (also because black holes are time-reversed white holes). Due to the dispersion of the fiber, the refractive index changes with frequency. In turn, the dispersion limits the frequency shifting by tuning the light out of the grip of the horizon. In particular, the dispersion limits the blue-shifting at white-hole horizons to respectable but finite frequencies, considering the tiny magnitude of χ, as we discuss in Sec. 7.1.

At the event horizons of astrophysical black holes, similar effects are expected (t'Hooft, 1985; Jacobson, 1991) when, due to the wave-number divergence, the wavelength of light is reduced below the Planck length scale $\sqrt{hG/c^3}$ where G is the gravitational constant. The physics beyond the Planck scale is unknown. This trans-Planckian physics should regularize the logarithmic phase singularities (Birrell & Davies, 1984; Brout et. al. , a) of modes at the event horizon. Studies of trans-Planckian physics indicate, that the Hawking effect of the black hole is not affected (Unruh, 1995; Brout et al. , b). On the other hand, the quantum radiation of white holes is dominated by trans-Planckian physics, because of the extreme blue shift at white-hole horizons. It has been predicted (Corley & Jacobson, 1999; Leonhardt & Philbin, 2007) that black-hole white-hole pairs could act as black hole lasers in a regime of anomalous group velocity dispersion. From a theoretical point of view, trans-Planckian physics regularizes some of the arcane features of quantum black holes and gives a more natural picture of the physics behind the Hawking effect (Brout et al. , b). In our case, the optical analogue of trans-Planckian physics, optical dispersion, is known in principle and turns out to be to the advantage of the experiment.

6. Hawking effect

Imagine instead of a single probe mode a set of probe modes. The modes should be sufficiently weakly excited such that they do not interact with each other, but they experience the cross Kerr effect of the pulse, the presence of the medium (1) moving with the velocity u. Themodes constitute a quantum field of light in a moving medium (Leonhardt, 2003).

A *phase-velocity horizon* is formed if the pulse has slowed down the probe such that its phase velocity is lower than the speed of the pulse. The condition for this horizon is:

$$\frac{\omega'}{\omega} = 0. \tag{46}$$

Here an additional effect occurs: the spontaneous creation of photon pairs, Hawking radiation. In the near ultraviolet around $\lambda = 300$nm, the dispersion of fibers is dominated by the bare dispersion of glass where $n_0(\omega)$ rapidly grows with frequency (Agrawal, 2001), exceeding the group index c/u of the pulse. For such ultraviolet modes, the pulse moves at superluminal speed. According to the Doppler formula (33) these superluminal modes oscillate with negative frequencies ω' in the co-moving frame for positive frequencies ω in the laboratory frame, and vice versa. Moreover, each subluminal mode with positive ω has a superluminal partner oscillating at the same co-moving frequency ω', but with negative laboratory frequency, see Fig. 2. The Kerr susceptibility of the pulse may slow down the subluminal modes such that the pulse moves at superluminal speed. As we will show in this section, in this case suband superluminal modes are partially converted into each other and photon pairs are created, even if the modes were initially in their vacuum states (Birrell & Davies, 1984; Brout et. al., a). This process is the optical analogue of Hawking radiation (Hawking, 1974; 1975). Photons with positive ω' correspond to the particles created at the outside of the black hole (Birrell & Davies, 1984; Brout et. al., a), while the negative-frequency photons represent their partners beyond the horizon.

In the fiber-optical case, the photon pairs are distinguishable from the intense pulse, because their frequencies differ by an octave. Furthermore, one can discriminate the Hawking effect

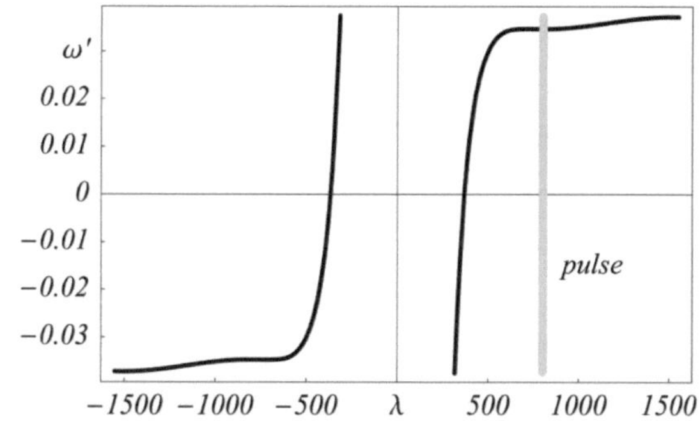

Fig. 2. Doppler-shifted frequency versus wavelength (Philbin et. al, 2008). The figure shows the co-moving frequency (33) (in 10^{15}Hz) for the micro-structured fiber versus the wavelength λ (in nm). We used the fiber dispersion data in Eq. (72), apart from two constants that we fitted to the Sellmeier formula for fused silica at short wavelengths (Agrawal, 2001).

from other nonlinear optical processes, such as Four-Wave Mixing, because it is not subject to their phase-matching conditions (Agrawal, 2001). Moreover, in addition to observing Hawking radiation *per se*, one could detect the correlations of the Hawking partners — a feat that is utterly impossible in astrophysics, because there the partner particles are lost beyond the horizon of the black hole.

Thus in this section we investigate the scattering mechanism which leads to Hawking radiation for our analogue system in a classical as well as quantized treatment.

6.1 Classical Hawking effect

For simplicity, we consider a single white-hole horizon, not the combination of black- and white-hole horizons generated by a moving pulse. We will argue later that in practice the white-hole will dominate the Hawking effect, which *a-posteriori* justifies this simplification. Suppose, without loss of generality, that at τ=0 the Kerr-reduced phase-velocity of the probe, c/n, matches the group velocity of the pulse u. We assume that the mode conversion occurs near this point and expand the Kerr susceptibility χ as a linear function in τ,

$$\chi(\tau) = \chi_h + \alpha''\tau, \quad \alpha'' = \left.\frac{\partial\chi}{\partial\tau}\right|_0 . \tag{47}$$

The group velocity of the incident probe is much lower than the pulse speed u and so both the sub- and the superluminal probe travels from the front of the pulse to the back, from negative to positive retarded time τ. For a white-hole horizon χ increases for decreasing retarded time, and so $\dot{\chi}(0) < 0$.

We proceed similar to (Brout et al. , b) and focus on the conversion region where we Fourier-transform with respect to τ the wave equation (10) with the refractive index (11) for stationary waves in the co-moving frame and using the linear expansion (47). The frequency conjugate to τ is the laboratory-frame frequency ω. We replace τ by $-i\partial_\omega$, ∂_ζ by $-i\omega'$ and ∂_τ by $-i\omega$, denote the Fourier-transformed vector potential by $\tilde{A}$, and obtain

$$\left(n_0^2 + \chi_h - i\alpha''\partial_\omega\right)\omega\tilde{A} = \left(1 - \frac{\omega'}{\omega}\right)^2 \frac{c^2}{u^2}\omega\tilde{A}. \tag{48}$$

This first-order equation has the exact solution

$$\widetilde{A}=\frac{\widetilde{A}_0}{\omega}\,e^{-i\phi}\,,\quad \phi=-\frac{1}{\alpha''}\int\left[\left(1-\frac{\omega'}{\omega}\right)^2\frac{c^2}{u^2}-n_0^2-\chi_h\right]d\omega \tag{49}$$

with constant $\tilde{A}_0$. We introduce

$$\alpha'\equiv-\frac{u^2}{c^2}\frac{\alpha''}{2}\approx-\frac{u}{c}\frac{\dot{\chi}(0)}{2n_0}=-\frac{u}{c}\frac{\partial n}{\partial\tau}\bigg|_0 . \tag{50}$$

Note that the phase ϕ contains a logarithmic contribution,

$$\phi=-\frac{\omega'}{\alpha'}\ln\omega+\phi_0(\omega)\,. \tag{51}$$

This logarithmic asymptotics of the phase will lead to the characteristic mode conversion at the phase velocity horizon. In order to see this, we Fourier-transform $\tilde{A}$ back to the domain of the retarded time,

$$A=\int_{-\infty}^{+\infty}\widetilde{A}\,e^{-i\omega\tau}d\omega=\int_{-\infty}^{+\infty}\frac{\widetilde{A}_0}{\omega}\,e^{-i\phi-i\omega\tau}d\omega \tag{52}$$

and use the saddle-point approximation, *i.e.* we quadratically expand the phase $\phi+\omega\tau$ around the stationary points where $\partial_\omega(\phi+\omega\tau)$ vanishes and perform the integration as Gaussian integrals along the direction of steepest descent. One easily verifies that the stationary points are the solutions of the dispersion relation (36). We denote the two solutions by $\omega_\pm$ indicating their sign. We obtain for the second derivative in the quadratic expansion

$$\partial_\omega^2(\phi+\omega\tau)=-\frac{2}{\dot{\chi}(0)}\frac{n}{\omega}\left(\frac{c}{u}-n_g\right). \tag{53}$$

The Gaussian integrals at $\omega_\pm$ are proportional to the inverse square root of $\partial_\omega^2(\phi+\omega\tau)$. We see from Eqs. (40) and (42) that they are consistent with the amplitudes (39) of geometrical optics. Consequently, we obtain a superposition of the two waves (34) that correspond to the two physically-relevant branches of the dispersion relation (36). We denote the positive-frequency wave by A_+ and the negative-frequency component by A_-^*. The star indicates that this component resembles the complex conjugate of a mode, because a mode predominantly contains positive laboratory-frame frequencies, according to the normalization (39). The coefficient of A_-^* is given by the exponential of the phase integral from the positive branch ω_+ to the negative frequency ω_- on the complex plane. The amplitude of the coefficient is the exponent of the imaginary part of the phase integral, while the phase of the coefficient is given by the real part. We can incorporate the phase of the superposition coefficient in the prefactor (39), but not the amplitude. The imaginary part of the phase integral comes from the logarithmic term (51), giving $\pi\omega'/\alpha'$. Therefore, the relative weight of the negative-frequency component in the converted mode is $\exp(-\pi\omega'/\alpha')$. We thus obtain for $\tau<0$

$$A \sim Z^{1/2}\left(A_+ + A_-^* e^{-\pi\omega'/\alpha'}\right) \tag{54}$$

where Z denotes a constant for given ω'. We determine the physical meaning of Z in Sec. 6.2, but here we can already work out its value by the following procedure: consider a wavepacket with co-moving frequencies around ω' that crosses the horizon. Suppose that this wavepacket is normalized to unity. After having crossed the horizon, the norm of the positive-frequency component is Z, while the negative-frequency component has the negative norm $-Z\exp(-2\pi\omega'/\alpha')$. The sum of the two components must give unity, and so

$$Z = \left(1 - e^{-2\pi\omega'/\alpha'}\right)^{-1}. \tag{55}$$

We represent $Z^{1/2}$ as $\cosh\xi$ and obtain from Eq. (55)

$$Z^{1/2} = \cosh\xi, \quad Z^{1/2}e^{-\pi\omega'/\alpha'} = \sinh\xi. \tag{56}$$

Consequently, the incident wave $A_\pm$ is converted into the superposition $A_\pm\cosh\xi + A_\mp^* \sinh\xi$ when it crosses the horizon from positive to negative τ. Hence we obtain for this process the mode

$$A_{\pm\text{in}} \sim \begin{cases} A_\pm & : \quad \tau > 0 \\ A_\pm\cosh\xi + A_\mp^*\sinh\xi & : \quad \tau < 0 \end{cases}. \tag{57}$$

Equation (57) describes the fate of a classical wave that crosses the horizon. A negative-frequency component is generated with weight $\sinh^2\xi$ relative to the initial wave, but, since $\cosh\xi > 1$, the positive-frequency wave has been amplified. The mode conversion at the horizon is thus an unusual scattering process where the concerted modes are amplified, at the expense of the energy of the driving mechanism, the pulse in our case. (It is also mathematically unusual — the Hawking effect corresponds to scattering without turning points in the complex plane.) Wherever there is amplification of classical waves, *i.e.* stimulated emission of waves, there also is spontaneous emission of quanta (Caves, 1982) — in the case of horizons, Hawking radiation.

6.2 Hawking radiation

Suppose that no classical probe light is incident; the modes $A_{\pm\text{in}}$ are in the vacuum state. The incident modes are characterized by the asymptotics $A_\pm$ for $\tau > 0$ while outgoing modes are required to approach $A_\pm$ for $\tau < 0$. We perform the superposition

$$A_{\pm\text{out}} = A_{\pm\text{in}}\cosh\xi - A_{\mp\text{in}}^*\sinh\xi \tag{58}$$

and see that $A_{\pm\text{out}}$ obeys the asymptotics

$$A_{\pm\text{out}} \sim \begin{cases} A_\pm\cosh\xi - A_\mp^*\sinh\xi & : \quad \tau > 0 \\ A_\pm & : \quad \tau < 0 \end{cases}, \tag{59}$$

as required for outgoing modes. The modes (57) and (58) describe two sets of mode expansions (24) of one and the same quantum field; for a given ω' the sum of $A_{\pm\text{in}}\,\hat{a}_{\pm\text{in}}$ and

$A^*_{\pm in}\hat{a}^\dagger_{\pm in}$ over the two signs ± of ω must give the corresponding sum of $A_{\pm out}\hat{a}_{\pm out}$ and $A^*_{\pm out}\hat{a}^\dagger_{\pm out}$. Consequently,

$$\hat{a}_{\pm in} = \hat{a}_{\pm out}\cosh\xi - \hat{a}^\dagger_{\mp out}\sinh\xi \tag{60}$$

and by inversion

$$\hat{a}_{\pm out} = \hat{a}_{\pm in}\cosh\xi + \hat{a}^\dagger_{\mp in}\sinh\xi\,. \tag{61}$$

The vacuum state $|\text{vac}\rangle$ of the incident field is the eigenstate of the annihilation operators $\hat{a}_{\pm in}$ with zero eigenvalue (the state that the $\hat{a}_{\pm in}$ annihilate),

$$\hat{a}_{\pm in}|\text{vac}\rangle = 0\,. \tag{62}$$

To find out whether and how many quanta are spontaneously emitted by the horizon, we express the in-coming vacuum in terms of the out-going modes. We denote the out-going photon-number eigenstates, the out-going Fock states (Leonhardt, 2003), by $|n_+,n_-\rangle$ with the integers $n_\pm$. Using the standard relations for the annihilation and creation operators

$$\hat{a}|n\rangle = \sqrt{n}\,|n-1\rangle\,,\quad \hat{a}^\dagger|n\rangle = \sqrt{n+1}\,|n+1\rangle\,, \tag{63}$$

one verifies that $\hat{a}_{\pm in}$ vanishes for the state

$$|\text{vac}\rangle = Z^{-1/2}\sum_{n=0}^{\infty} e^{-n\pi\omega'/\alpha'}\,|n,n\rangle\,. \tag{64}$$

This is the remarkable result obtained by Hawking (Hawking, 1974; 1975) for the horizon of the black hole. First, it shows that the event horizon spontaneously generates radiation from the incident quantum vacuum. Second, the emitted radiation consists of correlated photon pairs, each photon on one side is correlated to a partner photon on the other side, because they are always produced in pairs. The total quantum state turns out to be an Einstein-Podolski-Rosen state (Leonhardt, 2003), the strongest entangled state for a given energy (Barnett & Phoenix, 1989; 1991). Third, light on either side of the horizon consists of an ensemble of photon-number eigenstates with probability $Z^{-1}e^{-2n\pi\omega'/\alpha'}$. This is a Boltzmann distribution of n photons with energies $n\hbar\omega'$. Consequently, the horizon emits a Planck spectrum of blackbody radiation with the Hawking temperature of (Philbin et. al, 2008):

$$T' = \frac{\hbar\alpha'}{2\pi k_B}\,, \tag{65}$$

where k_B denotes Boltzmann's constant. Fourth, this Planck spectrum is consistent with Bekenstein's black-hole thermodynamics (Bekenstein, 1973): black holes seem to have an entropy and a temperature.

In our case, the spectrum of the emitted quanta is a Planck spectrum for the frequencies ω' in the co-moving frame, as long as a phase-velocity horizon exists. We performed our analysis for the white-hole horizon, but, since black holes are time-reversed white holes, we arrive at the same result for the black hole, except that the roles of the incident and outgoing modes are reversed. In the laboratory frame, the spectrum is given by the dependance of ω' on the laboratory frequency ω outside of the pulse, *i.e.* by the dispersion relation (33) for $\chi = 0$. In our case, $\omega(\omega')$ is single-valued for the spectral region where phase-velocity horizons are established, see Fig. 2, and so the spectra of black- and white-hole horizons are identical for identical α'. For evaluating α' we consider δn at $\tau = 0$, where

$$\alpha' = -\frac{u}{c}\frac{\partial n}{\partial \tau}\bigg|_0 = -\frac{u}{c}\frac{\partial \delta n}{\partial \tau}\bigg|_0 . \tag{66}$$

Note that T' denotes the Hawking temperature in the co-moving frame, defined with respect to the Doppler-shifted frequencies ω', a temperature that is characterized by the Doppler-shifted Hawking frequency α' in regions away from the pulse. We use the Doppler formula (33) with the refractive index (1) and the linearization (47) taken at $\tau = 0$, and obtain

$$\alpha' = \left(1 - n_0\frac{u}{c}\right)\alpha = \frac{u}{c}\,\delta n\Big|_0\,\alpha . \tag{67}$$

Consequently, the Hawking temperature T in the laboratory frame is

$$T = \frac{\hbar\alpha}{2\pi k_B}, \quad \alpha = -\frac{1}{\delta n}\frac{\partial \delta n}{\partial \tau}\bigg|_0 . \tag{68}$$

T does not depend on the magnitude of δn, as long as a phase velocity horizon is established. Even the small refractive index variations of nonlinear-fiber optics (Agrawal, 2001) may lead to a substantial Hawking temperature when δn varies on the scale of an optical wavelength. Thus the particle-production rate depends primarily on the sharpness of the pulse. This important feature makes the experimental observation of Hawking radiation in optical fibers feasible, as modern few-cycle pulses are only a couple of wavelengths long (F. X. Kärtner, 2004; Brabez & Krausz, 2000). In addition, an optical shock can form and lead to significant further steepening of the trailing (white hole) edge of the pulse at the expense of the luminosity of the black hole (Agrawal, 2001; Philbin et. al, 2008). Assuming that the steepness of this shock front is comparable to twice the frequency of the pulse carrier, 8×10^{14}Hz, the Hawking temperature (68) reaches 10^3K, many orders of magnitude higher than condensed-matter analogues of the event horizon (Garay et al., 2000; Giovanazzi et al., 2004; Giovanazzi, 2005; Jacobson & Volovik, 1998).

7. Experiment

We will now describe the experimental observation of frequency shifting of light at the group velocity horizon. Based on the theory of Secs. 3 - 6, we also derive mathematical expressions for the amount of blue shifting, for the spectral shape, and for estimating the efficiency of this process. We discuss the experimental proceedings and findings and compare them with the theory.

Soliton interactions in fibers are well studied. The connection to event horizons, however, had not been made and thus experiments merely focussed on related nonlinear effects such as optical pulse trapping in fibers (Efimov et al., 2005; Nishizawa & Goto, 2002; Gorbach & Skryabin, 2007; Hill et al., 2009) and pulse compression in fiber gratings (optical push

broom) (de Sterke, 1992; Steel et al., 1994; Broderick et al., 1997). The measurements described here for the first time focussed on the frequency shifts connected to analogues of the event horizon.
In order to observe the frequency shifts at the horizon we launch a pulse into the fiber to establish a group velocity horizon as explained in Sec. 5. We also launch a continuous wave of light, a probe, that follows the pulse with slightly higher group velocity, attempting to overtake it. The probe wavelength λ is chosen much longer than the pulse wavelength λ_0 in order to separate the two. While approaching the pulse, the Kerr contribution δ_n slows down the probe until the probe's group velocity reaches the speed of the pulse. The trailing end of the pulse establishes a white-hole horizon, an object that light cannot enter, unless it tunnels through the pulse. Conversely, the front end creates a black-hole horizon for probe light that is slower than the pulse. Since δ_n is small, the initial group velocity of the probe should be sufficiently close to the speed of the pulse.
In microstructured fibers (Russell, 2003) the group velocity dispersion can be engineered such that the group velocity of pulses of 800nm carrier wavelength matches the group velocities of probe light in the infrared around 1500nm. At the horizon of an astrophysical black hole light freezes, reaching wavelengths shorter than the Planck scale where the physics is unknown (The Planck length is given by $\sqrt{2\pi\hbar G/c^3}$ where G is the gravitational constant.) (t'Hooft, 1985; Jacobson, 1991). In our case, the fiber-optical analogue of Trans-Planckian physics is known and simple — it is contained in the frequency-dependance of the refractive index n, the dispersion of the fiber. At the trailing end of the pulse the incoming probe modes are compressed, oscillating with increasing frequency; they are blue-shifted. In turn, the dispersion limits the frequency shifting by tuning the probe out of the horizon. In the case of normal group-velocity dispersion the blue-shifted light falls behind.

7.1 Dispersion

The creation of artificial event horizons in optical fibers critically depends on the properties of these fibers. Microstructured fibers (Knight et. al., 1996; Russell, 2003) have an arrangement of holes close to the fiber core along the fiber. In the simplest picture, the holes lower the local refractive index in the transverse plane of the fiber, leading to substantially larger index variations compared to conventional fibers. In particular, the anomalous group velocity dispersion required for solitons can be generated at wavelengths reaching the visible. To create an artificial event horizon, an intense optical pulse has to be formed inside the fiber. Optical solitons (Hasegawa & Tappert, 1973; Mollenauer et al, 1980; Agrawal, 2001) offer a unique possibility for nondispersive stable pulses in fibers. These can be ultrashort, allowing for very high peak powers to drive the nonlinearity of the fiber.
The dispersion parameter D of optical fibers is defined as the change of group delay per wavelength change and fiber length. Its units are usually ps/(nmkm). Since the group delay per length is given by n_g/c and $n_g/c=\partial\beta/\partial\omega$, we have (Agrawal, 2001)

$$D = \frac{\partial}{\partial\lambda}\frac{n_g}{c} = \frac{\partial^2\beta}{\partial\lambda\partial\omega}, \quad \lambda = \frac{2\pi c}{\omega}. \tag{69}$$

The group velocity dispersion is often also characterized by β_2,

$$\beta_2 = \frac{\partial^2\beta}{\partial\omega^2} = -\frac{\lambda}{\omega}D. \tag{70}$$

The group velocity dispersion is normal for positive β_2 and negative D, and anomalous for negative β_2 and positive D.

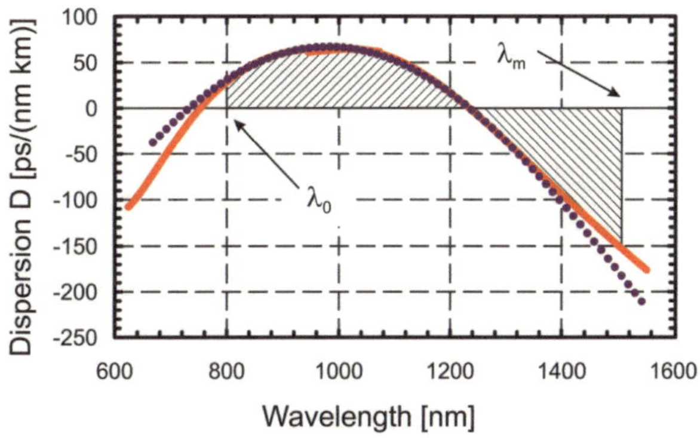

Fig. 3. Two measurements of the dispersion parameter D for the fiber used in the experiments (Philbin et. al, 2008). Red: technical specification; dotted: result of Alexander Podlipensky & Philip Russell, Max Planck Research Group in Optics, Information and Photonics, Erlangen, Germany. As the total shaded area vanishes, the two wavelengths at either end are group velocity matched.

For the creation of a horizon we chose a commercial microstructured fiber, model NL-PM-750B by Crystal Fiber A/S. Figure 3 shows the dispersion of the single mode fiber. The red curve is the manufacturer technical specification; the dotted line was measured for our particular fiber sample by Alexander Podlipensky and Philip Russell at the Max Planck Research Group in Optics, Information and Photonics in Erlangen, Germany. The fiber dispersion is anomalous between ≈ 740nm and ≈ 1235nm wavelength and normal otherwise. Further fiber properties are summarized in Table 1. Solitons can be created in this fiber using ultrashort pulses from Ti:Sapphire lasers. Light that would probe the horizon and experience blue shifting as a result, will have to be slowed down by the Kerr effect of the pulse such that its group velocity matches the speed u of the pulse. The Kerr nonlinearity is small, and so the initial group velocity of the probe should be only slightly higher than u. Integrating Eq. (69) we obtain

$$\int_{\lambda_0}^{\lambda} D\, d\lambda = \beta_1(\lambda) - \beta_1(\lambda_0) = \frac{1}{v_g(\lambda)} - \frac{1}{v_g(\lambda_0)}, \quad v_g(\lambda_0) = u. \tag{71}$$

Here λ_0 and λ denote the center wavelengths of the pulse and the probe light, respectively. Therefore, the probe light travels at the speed of the pulse if the integral of D vanishes, as illustrated by the shaded areas in Fig. 3. This probe wavelength is called the group velocity-matched wavelength λ_m (and ω_m the group velocity-matched frequency). For a pulse carrier-wavelength of 800nm and this fiber we obtain $\lambda_m \approx 1500$nm. This value of λ_m is useful and justifies the choice of fiber, because it is a standard wavelength for optical communication equipment and it is clearly separated from our spectrally broad pulses. Integrating Eq. (69) twice we obtain the propagation constant

$$\beta \;=\; \beta(\omega_0) + \frac{\omega - \omega_0}{u} + \int_{\omega_0}^{\omega}\int_{\lambda_0}^{\lambda} D\, d\lambda\, d\omega = n\frac{\omega}{c}, \tag{72}$$

where n is the linear effective refractive index of the fiber and ω_0 and ω denote the carrier frequency of pulse and probe, respectively.
In general, the two eigenmodes of polarization of the fiber have slightly different propagation constants β. This birefringence creates a refractive index difference Δn between the polarization modes. Our fiber exhibits strong birefringence Δn of a few times 10^{-4}. This leads to non-negligible changes in the group velocity as we will see later.

Property	Crystal Fiber \ Erlangen
Dispersion D_0	$28\backslash 36\,\mathrm{ps/(nm\,km)}$
Dispersion D_m	$-150\backslash -180\,\mathrm{ps/(nm\,km)}$
Third order dispersion $dD_m/d\lambda$	$-0.6\backslash -0.75\,\mathrm{ps/(nm^2km)}$
Dispersion $\beta_2(\lambda_0)$	$-9.5\backslash -12\,\mathrm{ps^2/km}$
Dispersion $\beta_2(\lambda_\mathrm{m})$	$180\backslash 210\,\mathrm{ps^2/km}$
Group velocity-matched wavelength λ_m	$1508\backslash 1494\,\mathrm{nm}$
Birefringence $\Delta n_0\backslash\Delta n_\mathrm{m}$	$7.5\backslash 5.7\times 10^{-4}$
Nonlinearity γ (780 nm)	$0.1\,\mathrm{W^{-1}m^{-1}}$
Fiber length L	$1.5\,\mathrm{m}$

Table 1. Properties of fiber NL-PM-750B (Philbin et. al, 2008). Dispersion data according to Crystal Fiber \ Alexander Podlipensky & Philip Russell, Max Planck Research Group in Optics, Information and Photonics, Erlangen. Nonlinearity according to Crystal Fiber. The fiber birefringence Δn and length L were measured by the authors. The symbols are defined in the text.

7.2 Frequency shifts

Let us now consider the frequency shifts at a group velocity horizon. During the pulse-probe interaction, the co-moving frequency ω' is a conserved quantity and so the probe frequency ω follows a contour line of ω' as a function of the nonlinear susceptibility χ induced by the pulse, see Fig. 4. The maximal χ experienced by the probe is proportional to the peak susceptibility χ_0 experienced by the pulse: assuming perfect mode overlap of pulse and probe, $\chi_{\max}$ reaches $2\chi_0$ when the probe and the pulse are co-polarized and $2\chi_0/3$ when they are cross-polarized (Agrawal, 2001). If the pulse is a soliton, the peak susceptibility and dispersion are connected as (Agrawal, 2001)

$$\chi_0 = \frac{2n_0 c\lambda_0 D_0}{(\omega_0 T_0)^2} = \frac{2n_0 c|\beta_2(\lambda_0)|}{\omega_0 T_0^2} \tag{73}$$

where D_0 denotes the dispersion parameter at the carrier wavelength λ_0. For example, for a soliton at λ_0=800nm whose full width at half maximum (FWHM) is 70fs (corresponding to T_0=40fs), for n_0=1.5, D_m=30ps/(nmkm) the peak susceptibility χ_0 is 2×10^{-6}. Nevertheless, we show that this small variation in the optical properties is sufficient to generate a significant wavelength shift at the horizon.
We obtain the contours of ω' from the Doppler formula (33). We use relations (12) and (72), but integrate from the group velocity-matching point,

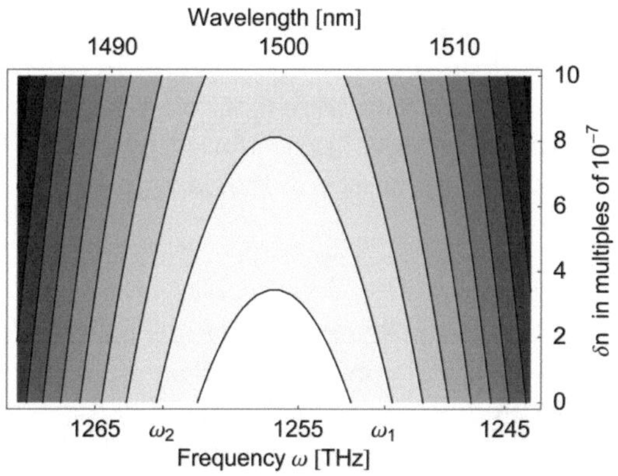

Fig. 4. Doppler contours. The pulse shifts the laboratory frequency ω (or the wavelength λ) of a probe along the contour line of constant ω' as a function (74) of the refractive-index change $\delta n = \chi/(2n_0)$. For a sufficiently intense pulse δn reaches the top of a contour. In this case the probe light completes an arch on the diagram while leaving the pulse; it is red- or blue-shifted, depending on its initial frequency (Philbin et. al, 2008).

$$\begin{aligned} \omega' &= \omega'_m - u\int_{\omega_m}^{\omega}\int_{\omega_m}^{\omega}\beta_2\, d\omega\, d\omega - \frac{\chi}{2n_0}\frac{u}{c}\omega \\ &\approx \omega'_m - \frac{u}{2}\beta_2(\omega_m)(\omega-\omega_m)^2 - \frac{\chi}{2n_0}\frac{u}{c}\omega_m \end{aligned} \tag{74}$$

$$\approx \omega'_m + \frac{\pi u}{\lambda_m}\left(\frac{D_m c(\lambda-\lambda_m)^2}{\lambda_m} - \frac{\chi}{n_0}\right). \tag{75}$$

The contours of ω' do not depend on ω'_m nor on the scaling factor $\pi u/\lambda_m$. Because $D_m = D(\lambda_m) < 0$, they form inverted parabolas with a maximum at λ_m for the corresponding $\chi_{\max}$. They intersect the axis of zero χ at the incident and the emerging wavelengths. Here $|D_m|c(\lambda - \lambda_m)^2/\lambda_m$ equals $\chi_{\max}/n_0$, and so we get

$$\lambda = \lambda_m \pm \delta\lambda, \quad \delta\lambda = \sqrt{\frac{\lambda_m \chi_{\max}}{|D_m| n_0 c}}. \tag{76}$$

Using again that the pulse is a soliton, we obtain

$$\delta\lambda = \frac{\sqrt{2r}\,\lambda_0}{T_0\sqrt{\omega_0\omega_m}}\sqrt{\left|\frac{D_0}{D_m}\right|} = \frac{\sqrt{2r\lambda_m\lambda_0}}{T_0\,\omega_m}\sqrt{\left|\frac{\beta_2(\lambda_0)}{\beta_2(\lambda_m)}\right|} \tag{77}$$

with r=2 for co-polarized and r=2/3 for cross-polarized pulse and probe light. According to Fig. 4 the probe light can maximally be wavelength-shifted from $+\delta\lambda$ to $-\delta\lambda$ over the range $2\delta\lambda$. For the soliton mentioned above the group velocity dispersion D_0 is about 30ps/(nmkm). Using $\lambda_m \approx$ 1500nm and $D_m \approx$ -160ps/(nmkm), the wavelength shift $2\delta\lambda$ is 20nm in the co-polarized case and $2\delta\lambda$=12nm in the cross-polarized case.

We also derive a simple estimate of the efficiency of the frequency shifting. The probe undergoes frequency conversion at the horizon. However, because the group velocities of the probe v_g and of the pulse u are similar, only a small fraction of the total probe light can be converted within the finite length of the fiber. The pulse and the slightly faster probe light travel through the fiber in $t=L/u$ and $t_p=L/v_g$ with $t > t_p$. The time difference multiplied with the probe power P_{probe} is the energy E_{coll} converted by pulse collision: $E_{\text{coll}}=P_{\text{probe}}\, L(1/u - 1/v_g)$. Therefore, the fraction η of probe power that is frequency converted is

$$\eta = \nu_{\text{rep}}\, L\,(1/u - 1/v_g) \approx \nu_{\text{rep}}\, L\,|D_m|\,\delta\lambda, \tag{78}$$

where ν_{rep} is the repetition rate of the pulses and $1/v_g=\partial\beta/\partial\omega\approx 1/u + D_m\delta\lambda$ was used. For L=1.5m and ν_{rep}=80MHz the maximal conversion efficiency η is on the order of 10^{-4}. Note that this model assumes that all of the probe light that interacts with the pulse is frequency shifted. However, in reality this rate can be lower due to the effect of tunnelling of probe light through the pulse.

7.3 Experimental results

The experiment is arranged as displayed in Fig. 5. A modelocked Ti:Sapphire laser (Mai Tai, Spectra Physics) delivers 70-fs pulses (FWHM) in the near infrared (NIR) at 80MHz repetition rate. These linearly polarized pulses are coupled to either one of the principal axes of the microstructured fiber of length L=1.5m. The polarization is rotated by a half-wave plate. Note that the polarizing beam splitter (PBS) only acts on the probe light. At the fiber output temporal autocorrelation traces and spectra are taken to determine the pulse energy necessary to create a fundamental soliton. For the center wavelength of 803nm, a dispersion D_0=30ps/(nmkm) and a nonlinearity γ of $0.1\text{W}^{-1}\text{m}^{-1}$, 70-fs solitons are generated at 5pJ pulse energy corresponding to 400μW average power for the repetition rate ν_{rep} = 80MHz.

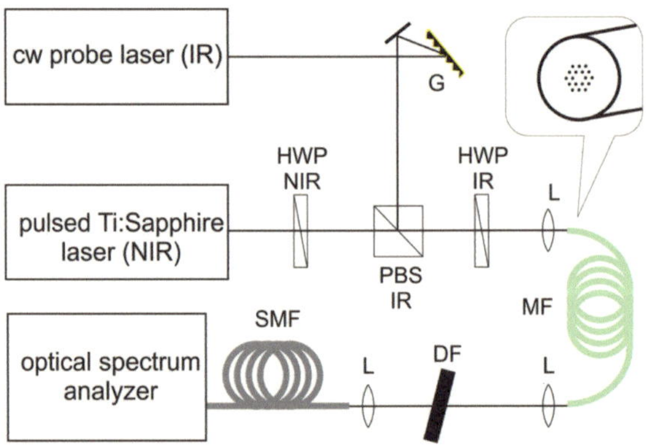

Fig. 5. Light from a continuous infrared (IR) laser is filtered by a diffraction grating (G) and is steered into a microstructured fiber (MF) by a polarizing beam splitter (PBS), a half wave plate (HWP), and a coupling lens (L). Near-infrared (NIR) pulses are launched as well. After the fiber the pulses are removed by a dichroic filter (DF) and the probe spectrum is taken through a standard single mode fiber (SMF) (Philbin et. al, 2008).

The output pulse length equalled the 70-fs input pulse length at an input power of approximately 320μW. This indicates that a soliton has formed. The observed power in comparison with the predicted power of 400μW illustrates the uncertainty in the actual fiber dispersion and nonlinearity. The observed Raman-induced soliton self-frequency shift

(Agrawal, 2001; Mitschke & Mollenauer, 1986; Gordon, 1986) was $\lesssim$ 4nm. Note that this shift decelerates the pulse and hence is changing the group velocity-matched wavelength λ_m in the infrared (IR). To calculate how much λ_m is shifted, we use Eq. (71), replacing λ_0 and λ with $\lambda_0 + \delta\lambda_0$ and $\lambda_m + \delta\lambda_r$ and linearize. In this way we get

$$\delta\lambda_r = \frac{D_0}{D_m}\,\delta\lambda_0 \,. \tag{79}$$

For the dispersion data shown in Fig. 3, a wavelength change of 4nm of the pulse changes λ_m by $\delta\lambda_r$ = -0.75nm. Since the probe light is wavelength-shifted symmetrically around λ_m, there is a change of the wavelength shift of up to -1.5nm.

The probe light is derived from a tunable external grating diode laser (Lynx Series, Sacher Lasertechnik). It delivers up to 20mW of continuous-wave light, tunable from 1460 to 1540nm. The probe light is reflected off a diffraction grating to reduce fluorescence emitted near lasing bandwidth. With another half-wave plate the probe light is coupled into the fiber onto one of the principal axes. Depending on wavelength, 100 to 600μW of probe power were coupled through the fiber. After the fiber we use a dichroic optic to filter out all of the pulse light and couple the IR light into a single-mode fiber connected to an optical spectrum analyzer.

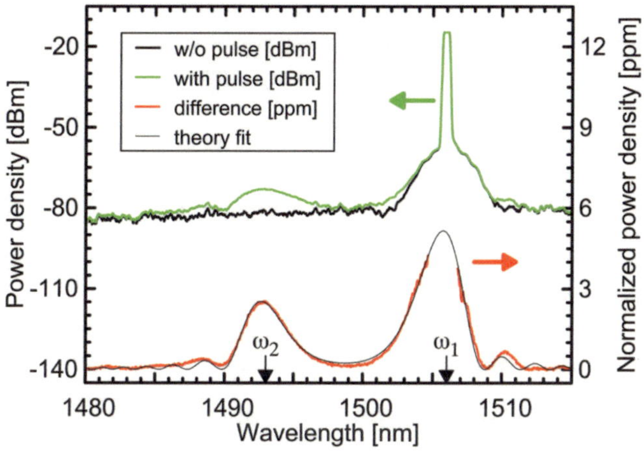

Fig. 6. Measurement of blue-shifting at a white-hole horizon (Philbin et. al, 2008). Spectrum of the blue-shifted light for a copolarized input probe of λ = 1506nm. Traces with (green) and without (black) pulses are shown on the logarithmic scale. Their difference is the normalized signal on the linear scale (red). The peak at ω_2 is the blue-shifted wave.

Figure 6 shows a typical output spectrum. This spectrum was taken with pulse and probe aligned to the slow axis of the fiber. At λ=1506nm the diode-laser input line is visible as a strong signal. From λ=1502nm to λ=1510nm we detect residual weak spontaneous emission from the laser that was not completely eliminated by the diffraction grating. Traces with and without pulses present in the fiber are taken and subtracted, leading to the signal displayed on a linear scale (red color). The signal is normalized by the amount of probe power and by the resolution bandwidth of 0.5nm. With the pulses present, a clear peak appears on the blue side of the input probe light near 1493nm. Since the blue-shifted light is generated from the part of the probe light that overlapped with the pulse during fiber propagation, it constitutes itself a pulse of finite length. Hence, this length is determined by the relative group velocity of probe light and the pulse, see for example Eq. (78). In turn, the unshifted probe light is partially depleted, forming a gap in intensity. These features lead to a spectral broadening of both the shifted and unshifted probe light by a few nanometers. The spectra display clearly the blueshifting of waves as predicted by the theory of optical event horizons. In what follows we look at the influence of probe detuning as well as polarization changes, before we perform a direct fit of the spectrum in Fig.6 to the theory.

From the measurements shown in Fig. 6, the efficiency of the blue-shifting is 1.1×10^{-5}, less than the estimated 10^{-4}. This indicates that a significant part of the probe light tunnels through the pulse; the pulse is too short /weak to establish a nearly perfect barrier. In the tunneling region of the pulse the laboratory frequency $\omega - \omega_m$ is imaginary. In order to estimate the maximal imaginary part of ω we consider the extreme case where the initial frequency of the probe reaches the group velocity-matched frequency ω_m characterized by $\omega' = \omega'_m$. We solve Eq. (74) for ω and obtain

$$\mathrm{Im}\omega = \sqrt{\frac{\chi\omega_m}{n_0\beta_2(\omega_m)c}} = \frac{\sqrt{\chi}\,\omega_m}{\sqrt{n_0 c\lambda_m|D_m|}}. \tag{80}$$

Assuming $\chi \approx 2 \times 10^{-6}$ at the soliton peak, $n_0 \approx 1.5$, $\lambda_m = 1500$nm and $D_m = -160$ps/(nmkm) the imaginary part of ω reaches about 5THz. This is insufficient to significantly suppress tunnelling through a 70-fs pulse, because the product of Imω and T_0 is much smaller than unity. For longer or more intense pulses we expect perfectly efficient frequency conversion at the horizon.

Increasing the probe wavelength further away from λ_m is shifting light further to the blue side of the spectrum, because the wavelength shifts symmetrically around the group velocity-matched wavelength, according to Eq. (76) and Fig. 4. Figure 7 displays the spectra of shifted light for three detunings of the probe light from the group velocity-matched wavelength λ_m. As expected, the spectra move towards shorter wavelengths by the same amount as the probe laser was tuned towards longer wavelengths.

We also measured how the signal strength, the spectrum integrated over the signal peak, evolves with increasing probe power (Philbin et. al, 2008). A clear linear dependence was found in agreement with our theoretical model. This measurement proves that the probe indeed is a probe, not influencing the pulses via nonlinear effects.

Changing the input polarizations changes the group velocities of pulse and probe and therefore the group velocity-matched wavelength λ_m shifts by an amount $\delta\lambda_m$. If we change for example the pump polarization from the fast to the slow axis, the inverse group velocity $\beta_1(\lambda_0)$ increases as n_{g0}/c is replaced by $(n_{g0} + \Delta n_0)/c$. To maintain group velocity matching, $\beta_1(\lambda_m)$ has to change accordingly by $\Delta n_0/c$. We use Eq. (71), linearizing around λ_m, and get

$$\Delta n_0/c \approx D_m \delta\lambda_m. \tag{81}$$

For D_m=-160ps/(nmkm) and $\Delta n_0 = 7.5 \times 10^{-4}$ we obtain $\delta\lambda_m \approx -16$nm $\approx 2\delta\lambda$. This means that the polarization change induces a velocity change that is too large to maintain the optical horizon. Therefore, when changing polarizations, the probe laser has to be retuned to a wavelength were frequency shifting can be observed.

Figure 8 shows spectra for all four different polarization combinations. As expected, the group velocity-matched wavelength changes. Note that there also is a difference in λ_m for the two co-polarized cases, indicating small changes in the dispersion profile for the two polarization axes, a dispersion of the birefringence.

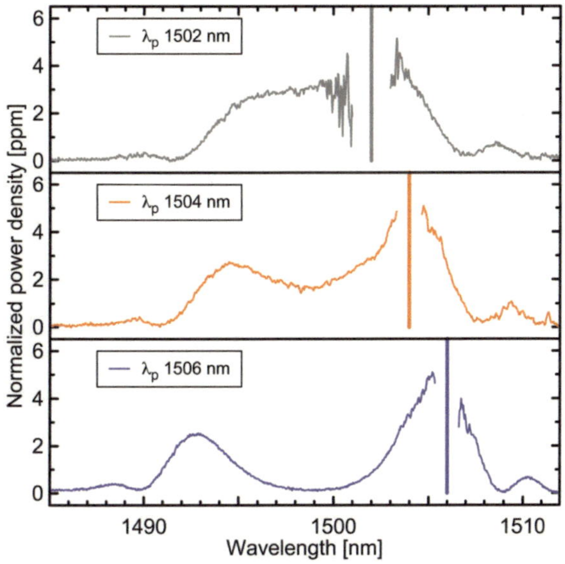

Fig. 7. Spectra for different input probe wavelengths. Since the probe mode is mirrored around the group velocity-matched wavelength λ_m, increasing probe wavelengths experience increasing blue shifting, as is also illustrated by the contours of Fig. 4 (Philbin et. al, 2008).

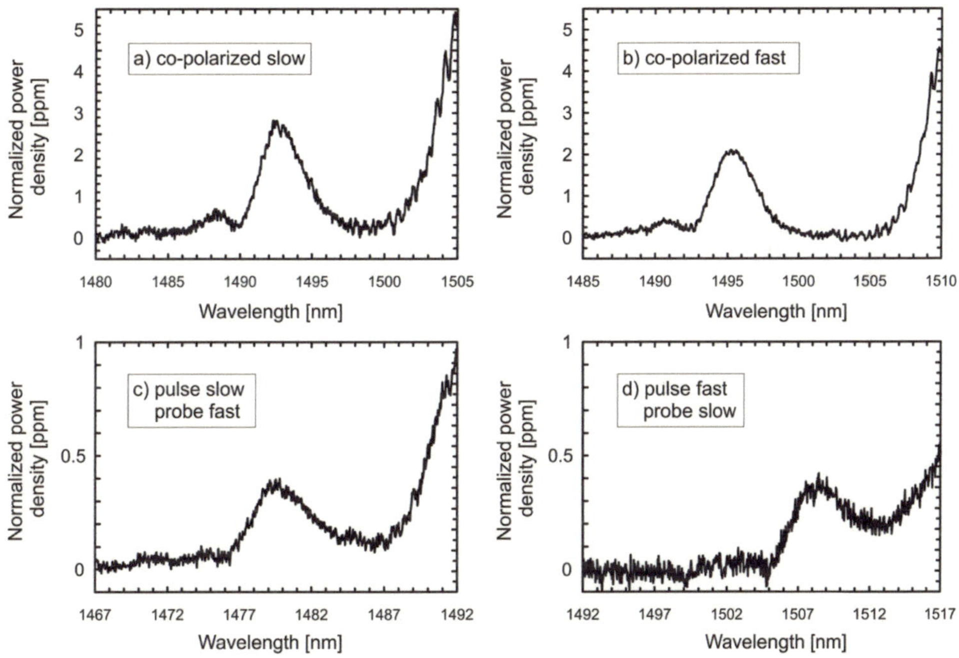

Fig. 8. Blue-shifted spectra for all four polarization combinations. Co-polarized spectra on the slow and fast axis in (a) and (b) and cross-polarized spectra with pulses slow (c) and fast (d). Group velocity-matched wavelengths are (a) 1499.5nm (b) 1503.2nm (c) 1486.4nm and (d) 1513.3nm (Philbin et. al, 2008).

7.4 Theoretical curve fitting of spectra

With the theory at hand, a functional expression for the spectra of the frequency-shifted probe light can be found. For brevity we are giving the results, the full calculation can be found in (Philbin et. al, 2008). The Fourier component of the shifted probe field is given by:

$$\begin{aligned} \widetilde{A}_s &= \frac{\eta_r A_0}{\pi(\omega+\omega_s-2\omega_m)}\left[\exp\left(iq(\omega-\omega_m)^2\right)-\exp\left(iq(\omega_s-\omega_m)^2\right)\right] \\ &- \frac{A_0}{\pi(\omega-\omega_s)}\left[G\left(\sqrt{-iq}\,(\omega-\omega_m)\right)-G\left(\sqrt{-iq}\,(\omega_s-\omega_m)\right)\right]., \end{aligned} \tag{82}$$

Here A_0 is the overall amplitude, ω_s is the shifted probe frequency, η_r is a phenomenological inclusion of the soliton self-frequency shift (Mitschke & Mollenauer, 1986; Gordon, 1986) in terms of a reduced blue-shifted part of the spectrum. The parameter $q = \pi c|D_m|L/\omega_m^2$ combines the experimental parameters fiber length L, dispersion D_m, and group velocity-matched frequency ω_m. Finally, $G(x) = \int_{-\infty}^{+\infty} e^{-\xi^2}/\pi(x - \xi)d\xi$. The power spectrum measured corresponds to $|\widetilde{A}|^2$.

Figure 6 also shows the fit of the observed spectrum with the theoretical curve (82). As fitting parameters we used the overall amplitude A_0, the shifted probe frequency ω_s and group velocity-matched frequency ω_m, the dispersion D_m and η_r. We obtain a very good fit for $A_0 = 4.1 \times 10^{12}\sqrt{Ws}$, $\lambda_s = 1505.31$nm, $\lambda_m = 1499.38$nm, $D_m = -187$ps/(nmkm) and $\eta_r = 0.80$. The shift in λ_p to λ_s is consistent with the effect (79) of the soliton self-frequency shift. The fitted values for D_m and λ_m agree with the independently measured dispersion and the group velocity-matched frequency calculated from the dispersion curve of Fig. 3.

8. Conclusion

In conclusion, we have developed a theory for artificial event horizons in fibers. Experimentally, light was blue-shifted by a near group velocity-matched pulse. The measured data was explained by the presence of an optical group velocity horizon inside the fiber. A very good agreement between theory and experiment was observed. The blue shifting corresponds to the optical analogue of trans-Planckian frequency shifts in astrophysics (t'Hooft, 1985; Jacobson, 1991). The temperature of analogue Hawking radiation was calculated to show that the system gives a realistic chance for experimental observation. In this way, we have demonstrated classical optical effects of the event horizon in our analogue system, a first step towards tabletop astrophysics (Ball, 2001).

Aside from the prospect of analogue Hawking radiation, the group velocity horizons could be used for ultrafast delay lines, dispersion management, frequency conjugation, or quantum frequency conversion.

We are indebted to Govind Agrawal, Malcolm Dunn, Theodor Hänsch, Alan Miller, Renaud Parentani and Wilson Sibbett for discussions and technical support. We thank Alexander Podlipensky and Philip Russell for measuring the dispersion of our fiber. Our paper is supported by the Leverhulme Trust, EPSRC, COVAQIAL, the Ultrafast Photonics Facility at St Andrews, and Leonhardt Group Aue.

9. References

M. J. Ablowitz and A. S. Fokas (1997). *Complex Variables* (Cambridge University Press, Cambridge, 1997).

G. Agrawal (2001). *Nonlinear Fiber Optics* (Academic Press, San Diego, 2001).
P. Ball (2001). Nature 411, 628.
S. M. Barnett and S. J. D. Phoenix (1989). Phys. Rev. A 40, 2404.
S. M. Barnett and S. J. D. Phoenix (1991). Phys. Rev. A 44, 535.
G. Barton and C. Eberlein (1993). Ann. Phys. (New York) 227, 222.
J. D. Bekenstein (1973). Phys. Rev. D 7, 2333; see also arXiv:gr-qc/0009019 for a review.
N. D. Birrell and P. C. W. Davies (1984). *Quantum fields in curved space* (Cambridge University Press, Cambridge, 1984).
M. Born and E. Wolf (1999). *Principles of Optics* (Cambridge University Press, Cambridge, 1999).
T. Brabez and F. Krausz (2000). Rev. Mod. Phys. 72, 545.
N. G. R. Broderick, D. Taverner, D. J. Richardson, M. Ibsen, and R. I. Laming (1997). Phys. Rev. Lett. 79, 4566.
R. Brout, S. Massar, R. Parentani, and Ph. Spindel (1995) (a)., Phys. Rep. 260, 329.
R. Brout, S. Massar, R. Parentani, and Ph. Spindel (1995) (b). Phys. Rev. D 52, 4559.
C. M. Caves (1982). Phys. Rev. D 26, 1817.
S. Corley and T. Jacobson (1999). Phys. Rev. D 59, 124011.
P. C.W. Davies (1975). J. Phys. A 8, 609.
F. DeMartini, C. H. Townes, T. K. Gustafson, and P. L. Kelley (1967). Phys. Rev. 167, 312.
C. N. de Sterke (1992). Opt. Lett. 17, 914.
B. S. DeWitt (1975). Phys. Rep. 19, 295.
A. Efimov, A.V. Yulin, D.V. Skryabin, J. C. Knight, N. Joly, F. G. Omenetto, A. J. Taylor, and P. Russe (2005). Phys. Rev. Lett. 95, 213902.
D. Facciol, S. Cacciatori, V. Gorini, V.G. Sala, A. Averchi, A. Lotti, M. Kolesik, J.V. Moloney (2009). arXiv:0905.4426v1 [gr-qc] 27 May 2009.
S. A. Fulling (1973). Phys. Rev. D 7, 2850.
L. J. Garay, J. R. Anglin, J. I. Cirac, and P. Zoller (2000). Phys. Rev. Lett. 85, 4643.
S. Giovanazzi, C. Farrell, T. Kiss, and U. Leonhardt (2004). Phys. Rev. A 70, 063602.
S. Giovanazzi (2005). Phys. Rev. Lett. 94, 061302.
A. V. Gorbach and D. V. Skryabin (2007). Nature Photonics 1, 653.
J. P. Gordon (1986). *ibid.* 11, 662.
M. B. Green, J. H. Schwarz, and E. Witten (1987). *Superstring Theory* (Cambridge University Press, Cambridge, 1987).
A. Hasegawa and F. Tappert (1973). Appl. Phys. Lett. 23, 142.
S. M. Hawking (1974). Nature 248, 30.
S. M. Hawking (1975). Commun. Math. Phys. 43, 199.
S. Hill, C. E. Kuklewicz, U. Leonhardt, and F. Knig (2009). Opt. Express, 17 13588.
T. Jacobson (1991). Phys. Rev. D 44, 1731.
T. A. Jacobson and G. E. Volovik (1998). Phys. Rev. D 58, 064021.
Few-Cycle Laser Pulse Generation and Its Applications, edited by F. X. Kärtner (Springer, Berlin, 2004).
J. C. Knight, T. A. Birks, P. S. Russell, and D. M. Atkin (1996). Opt. Lett. 21, 1547; P. Russell (2003). Science 299, 358.
See e.g. L. Knöll, S. Scheel, and D.-G.Welsch (2001). *QED in dispersing and absorbing media*, in *Coherence and Statistics of Photons and Atoms* ed. by J. Perina (Wiley, New York, 2001), pp.1-63.
L. D. Landau and E. M. Lifshitz (1975). The Classical Theory of Fields (Pergamon, Oxford, 1975).
L. D. Landau and E. M. Lifshitz (1976). *Mechanics* (Pergamon, Oxford, 1976).
L. D. Landau and E. M. Lifshitz (1977). *Quantum Mechanics* (Pergamon, Oxford, 1977).
L. D. Landau and E. M. Lifshitz (1984). *Electrodynamics of Continuous Media* (Pergamon, Oxford, 1984).

See e.g. U. Leonhardt (1993). *Quantum Theory of Simple Optical Instruments*, PhD thesis, Humboldt University Berlin.
U. Leonhardt, M. Munroe, T. Kiss, Th. Richter, and M. G. Raymer (1996), Opt. Commun. 127, 144.
U. Leonhardt and P. Piwnicki (2000). Phys. Rev. Lett. 84, 822.
U. Leonhardt (2002). Nature 415, 406.
U. Leonhardt (2003). Rep. Prog. Phys. 66, 1207.
U. Leonhardt and T. G. Philbin (2006). New J. Phys. 8, 247.
U. Leonhardt and T. G. Philbin (2007). *Black-hole lasers revisited*, in *Quantum Analogues: From Phase Transitions to Black Holes and Cosmology* edited by W. G. Unruh and R. Schützhold (Springer, Berlin, 2007).
L. Mandel and E. Wolf (1995). *Optical Coherence and Quantum Optics*, (Cambridge University Press, Cambridge, 1995).
R. Meinel, G. Neugebauer, and H. Steudel (1991). *Solitonen* (Akademie Verlag, Berlin, 1991).
P. W. Milonni (1994). *The Quantum Vacuum: An Introduction to Quantum Electrodynamics* (Academic Press, San Diego, 1994).
P. W. Milonni (2004). *Fast Light, Slow Light and Left Handed Light* (Institute of Physics, Bristol, 2004).
F. M. Mitschke and L. F. Mollenauer (1986). Opt. Lett. 11, 659.
L. F. Mollenauer, R. H. Stolen, and J. P. Gordon (1980), Phys. Rev. Lett. 45, 1095.
G. Moore (1970). J. Math. Phys. 11, 2679.
Any two-dimensional Riemannian manifold is conformally flat, see M. Nakahara (2003). *Geometry, Topology and Physics* (Institute of Physics, Bristol, 2003), Sec. 7.6.
N. Nishizawa and T. Goto (2002). Opt. Lett. 27, 152.
(2002). *Artificial black holes*, edited by M. Novello, M. Visser, and G. E. Volovik (World Scientific, Singapore, 2002).
T. G. Philbin, C. Kuklewicz, S. Robertson, S. Hill, F. K¨onig, and U. Leonhardt (2008). Science 319, 1367.
W. H. Reeves, D. V. Skryabin, F. Biancalana, J. C. Knight, P. S. Russell, F. G. Omenetto, A. Efimov, and A. J. Taylor (2003). Nature 424, 511.
G. Rousseaux, C. Mathis, P. Maïssa, T. G. Philbin, and U. Leonhardt (2008). New J. Phys. 10, 053015.
C. Rovelli (1998). Living Rev. Rel. 1, 1.
P. Russell (2003). Science 299, 358.
W. Schleich and M. O. Scully (1984). in *New trends in atomic physics*, edited by G. Grynberg and A. Stora, (Les Houches, Session XXXVIII, 1982, Elsevier Science Publishers B. V.)
R. Schützhold and W. G. Unruh (2002). Phys. Rev. D 66, 044019.
R. Schützhold and W. G. Unruh (2005). Phys. Rev. Lett. 95, 031301.
J. Schwinger (1951). Phys. Rev. 82, 664.
M. J. Steel, D. G. A. Jackson, and S. M. de Sterke (1994), Phys. Rev. A 50, 3447.
G. t'Hooft (1985). Nucl. Phys. B 256, 727.
J. S. Toll (1956). Phys. Rev. 104, 1760.
W. G. Unruh (1976). Phys. Rev. D 14, 870.
W. G. Unruh (1981). Phys. Rev. Lett. 46, 1351.
W. G. Unruh (1995). Phys. Rev. D 51, 2827.
M. Visser (1998). Class. Quantum Grav. 15, 1767.
G. E. Volovik (2002). JETP Lett. 76, 240.
G. E. Volovik (2003). *The Universe in a Helium Droplet* (Clarendon Press, Oxford, 2003).
S. Weinberg (1999). *The Quantum Theory of Fields* (Cambridge University Press, Cambridge, 1999), Volume I.

9

Ultrafast Semiconductor Quantum Optics

Rudolf Bratschitsch and Alfred Leitenstorfer
Department of Physics and Center for Applied Photonics, University of Konstanz
Germany

1. Introduction

Since the "photon" was introduced by Albert Einstein in 1905, a large number of quantum optical experiments has been successfully performed with these quanta of light (Zeilinger et al., 2005). Also the interaction of single photons with atomic or molecular systems has been studied in great detail (Weiner & Ho, 2003). However, from an applications point of view, devices operating under high vacuum conditions are difficult to handle. Therefore, highly integrated quantum optical systems based on the solid-state are worthwhile studying. In this review we are focusing on semiconductor quantum dots.

2. Semiconductor quantum dots

Semiconductor quantum dots are sometimes also called "artificial atoms" due to their discrete energy states (Weisbuch & Vinter, 1991; Bányai & Koch, 1993; Bimberg et al., 1999; Bratschitsch & Leitenstorfer, 2006). If nanocrystals made of semiconductors become so small that the de Broglie wavelength of an electron is comparable to the crystal dimensions, a "quantum dot" is created. The electron is confined in the narrow potential well and resembles a particle in a box with discrete energy levels. The electronic energy states of the quantum dot may be populated with electrons and holes. If the energetically lowest lying electron and hole form a bound complex - similar to an electron and proton in a hydrogen atom - a so called "exciton" is formed. If electron and hole recombine, energy in form of a photon is released, i.e. a semiconductor quantum dot constitutes a nanoscopic light emitter. At the beginning of the 21st century it has been shown that a quantum dot is a single photon source (Michler et al., 2000; Santori et al., 2002). Recently, it has been demonstrated that photons emitted by a single quantum dot may be entangled (Akopian et al., 2006; Stevenson et al., 2006). These two experimental findings are basic building blocks for quantum optical experiments with semiconductor quantum dots. Up to now, most quantum optical experiments have been performed in the steady state with narrowband lasers (Atature et al., 2006; Xu et al., 2007; Kroner et al., 2008).

2.1 Ultrafast quantum optics with a single semiconductor quantum dot

Semiconductor quantum dots are also promising systems for robust and scalable quantum information processing (Li et al., 2003; Petta et al., 2005). Ultrafast sequences of coherent quantum operations may be envisioned with femtosecond light pulses, if the involved quantum states are separated by at least tens of meV. Therefore, small quantum dots with high confinement potentials are favorable. Due to their large Coulomb correlation energies, CdSe quantum dots are ideal candidates.

We have performed two-color femtosecond pump-probe spectroscopy on a single self-assembled CdSe/ZnSe quantum dot (Sotier et al., 2009). The transient quantum dynamics was probed with resonant excitation and detection (Fig. 1).

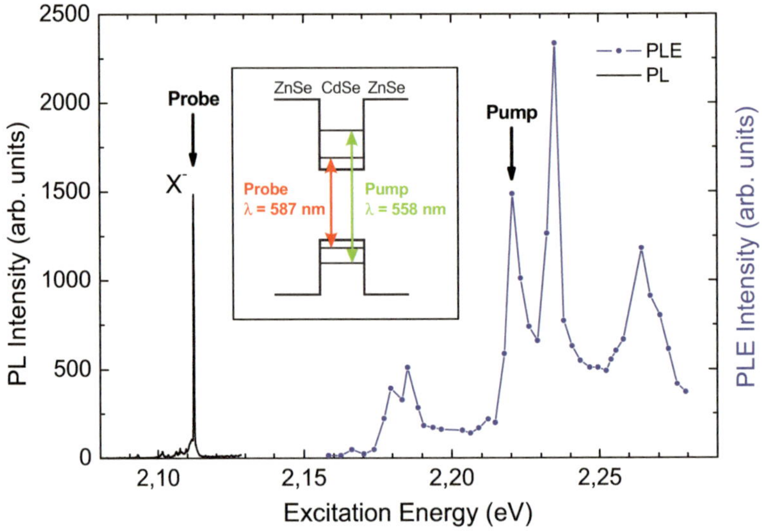

Fig. 1. Photoluminescence (PL) and photoluminescence excitation spectrum (PLE) of a single CdSe/ZnSe quantum dot, Inset: Schematic energy level structure of the quantum dot with indicated pump and probe center wavelength.

We were able to resonantly excite, manipulate and read out a single semiconductor quantum dot with femtosecond time resolution. A two-color Er:fiber laser with excellent noise performance (Tauser et al., 2003; Moutzouris et al., 2006; Moutzouris et al., 2007) is key to these first resonant pump-probe measurements on a single-electron system. The quantum dot was resonantly pumped into the p-shell with laser pulses of a duration of 750 fs. The transient transmission at the charged exciton ("trion") transition was probed by a time-delayed 190 fs pulse (Fig. 2). At a negative delay of -2 ps small oscillatory signatures around the fundamental exciton resonance are discernible, which are related to the perturbed free induction decay (Fig. 2). At +2 ps a bleaching of the exciton transition is observed due to Coulomb renormalization (Fig. 3). In the following 20 ps, the positive transmission signal increases by a factor of 1.5 and reaches a constant plateau that slowly decays with the radiative lifetime of the trion of 480 ps. As the hot exciton relaxes to its ground state (Fig. 3), the original transition is restored, but now under inversion conditions. A time delayed probe pulse causes stimulated emission, corresponding to an increase in differential transmission. The thermalization rate is governed by the electron spin (Fig. 3): the 20-ps onset of stimulated emission corresponds to the relaxation of a singlet electron which is generated by 50% of the pump pulses.

Operating the single semiconductor quantum dot in the nonlinear regime, the ability to change the number of quanta in a femtosecond light pulse by exactly ±1 is demonstrated (Sotier et al., 2009).

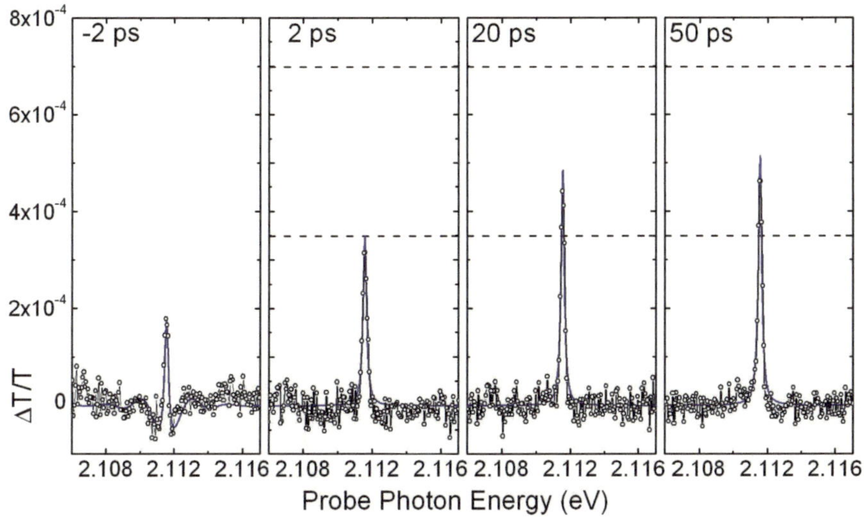

Fig. 2. Spectrally resolved transmission change through a single CdSe/ZnSe quantum dot for different pump-probe time delays t_D. The experimental data are represented by the hollow circles, while fits based on the optical Bloch equations for a two-level system are given by the blue lines. At negative time delays the perturbed free induction decay is visible. At positive time delay ultrafast Coulomb renormalization and spin-dependent electron relaxation is observed.

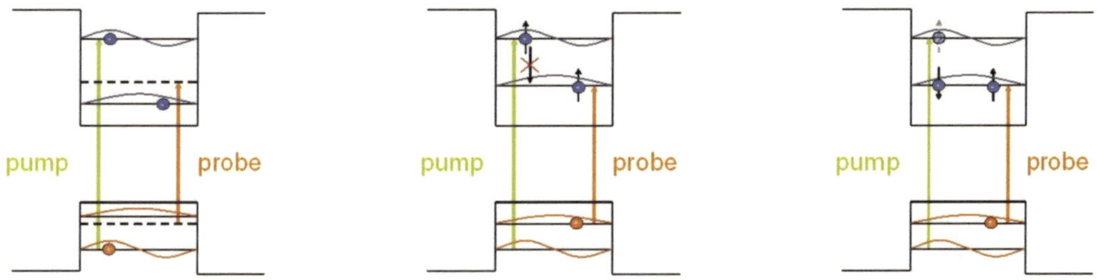

Fig. 3. Schematic drawing of the electronic configurations relevant for the differential transmission signals at positive time delays.

3. Nanophotonic elements for increased light-matter coupling

The experiments presented so far have been performed with at least thousands of photons. To reach the limit of only a single photon manipulating a single electron, the coupling of light with a wavelength of a few hundred nanometers into the object of nanometer dimensions, such as a semiconductor quantum dot, a color center in diamond, or a molecule has to be optimized. We follow two main routes to reach this goal: optical microcavities and metallic nanoantennas.

3.1 Dielectric optical pillar microcavities

We succeeded in embedding colloidal CdSe/ZnSe quantum dots in a planar and pillar microresonator operating in the visible regime (Kahl et al., 2007). A planar dielectric cavity is formed by two Bragg mirrors, each consisting of sputtered pairs of alternating TiO_2 and SiO_2 layers. The semiconductor nanocrystals are embedded in the central cavity layer in SiO_2 surroundings via a spin-on glass technique (Fig. 4).

Fig. 4. Schematic drawing of a planar cavity with embedded colloidal semiconductor quantum dots.

Subsequently, micropillars with diameters in the range of 5 μm down to 600 nm are milled out of the planar resonator via a focused ion beam (Fig. 5).

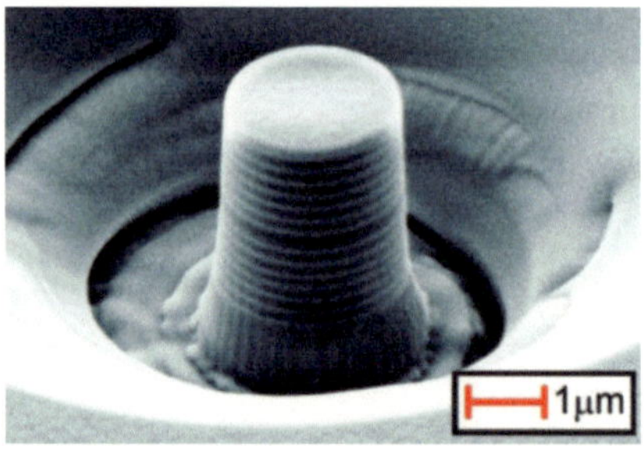

Fig. 5. Scanning electron microscopy (SEM) image of a dielectric micropillar with embedded colloidal CdSe/ZnS quantum dots.

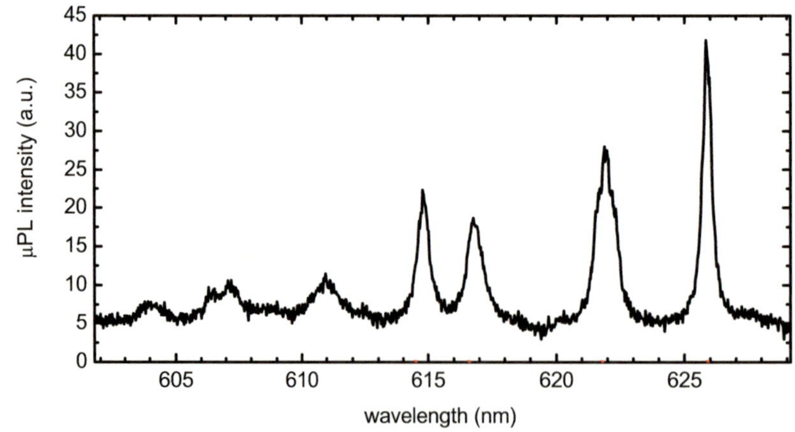

Fig. 6. Photoluminescence emission from a pillar resonator of 3 μm diameter with embedded colloidal CdSe/ZnS quantum dots, showing distinct microcavity modes.

A broadband light transmission measurement through a single micropost cavity shows the lowest cavity mode blueshifting with decreased pillar diameter. This effect directly demonstrates three-dimensional light confinement. Higher cavity modes are identified in the photoluminescence emission from an ensemble of quantum dots (Fig. 6).
The spectral position of these resonances may be calculated by modelling the pillar cavity as a waveguide with an effective refractive index. The theoretical results are in excellent agreement with the experimentally observed pillar mode positions and patterns.
We have also extended our studies to the ultraviolet regime of the electromagnetic spectrum. To minimize absorption in the mirrors of the resonator, HfO_2 and SiO_2 were used as high and low refractive index materials. We manufactured the first high-quality dielectric pillar microcavities with embedded colloidal ZnO quantum dots in the ultraviolet (Thomay et al., 2008). The semiconductor ZnO possesses excellent optical properties (Klingshirn, 2007). Both the large bandgap of ZnO of 3.4 eV and its exciton binding energy of 60 meV will be strongly beneficial to room-temperature operation of future single-photon devices. ZnO has a low refractive index as compared to other semiconductors, which facilitates light extraction. This compound is also a promising candidate for spintronics applications (Janssen et al., 2008). ZnO quantum dots have shown long electron spin dephasing times even at room temperature (Liu et al., 2007) and have been successfully doped with magnetic ions (Kittilstved & Gamelin, 2006). In the future, the quantum dot ensemble used in this study may be replaced by sputtered ZnO quantum dots (Mayer et al., 2009), magnetically doped ZnO nanocrystals, or other UV emitters.

3.2 Metallic optical nanoantennas

An alternative way of coupling single nanoobjects to the light field is metallic nanostructures. These so-called "optical antennas" are one million times smaller than the radio antenna analogues. Successful fabrication and operation of these devices has only been demonstrated recently (Muhlschlegel et al., 2005; Schuck et al., 2005).
We succeeded in tuning the length and feedgap of a single gold bowtie antenna by precise nanomanipulation ("tuneable nanoantenna") with the tip of an atomic force microscope (AFM) while at the same time monitoring the optical response of the nanostructure via dark-field scattering spectroscopy (Merlein et al., 2008). The gold bowtie optical nanoantenna was fabricated via a colloidal nanomask. The bowtie nanoantenna consists of two gold nanotriangles separated by a tip-to-tip distance of 85 nm (Fig. 7a). One arm of the nanoantenna was moved with the tip of an AFM, so that the total length of the nanoantenna and hence its feedgap were gradually reduced (Fig. 7a and c).
In contrast to all previous experiments on nanoantenna structures, we were able to study the evolution of the optical properties of the same single nanoobject as a function of the geometry. In our experiment, the feedgap was precisely controlled on a nanometer scale, while the other experimental parameters were kept constant. Most importantly, the shape of the nanoantenna arms was not altered. Reducing the antenna gap dramatically changed the scattering spectrum. The original single plasmon peak split into two distinct resonances (Fig. 7b). The optical properties of the metal nanoantenna with tuneable gap may be understood in the following way. At a large feedgap of 85 nm, the dark field scattering spectrum is the superposition of the spectra of the two individual nanotriangles. When the gap decreases, the metal antenna arms begin to couple electrodynamically. The charge density distribution in one triangle at any given time acts on the other arm and vice versa.

As a result, collective plasmon modes in direction of the long antenna axis are formed. In the experiment, two resonances were clearly observed. The exact three-dimensional shape of the nanoantenna is the reason for this effect. Breaking the symmetry of the antenna arms in vertical direction due to tilted sidewalls causes a splitting of the dipole mode into two distinct resonances, which are both of dipole type. In contrast to earlier work on nanoantennas, the full three-dimensional shape has to be taken into account to understand their optical properties. This finding will be particularly important if these antenna structures are resonantly coupled to single nanoemitters, such as semiconductor quantum dots, diamond nanocrystals, or single molecules. These results have also opened the door to new nano-opto-mechanical devices, where mechanical changes on the nanometer scale control the optical properties of photonic structures.

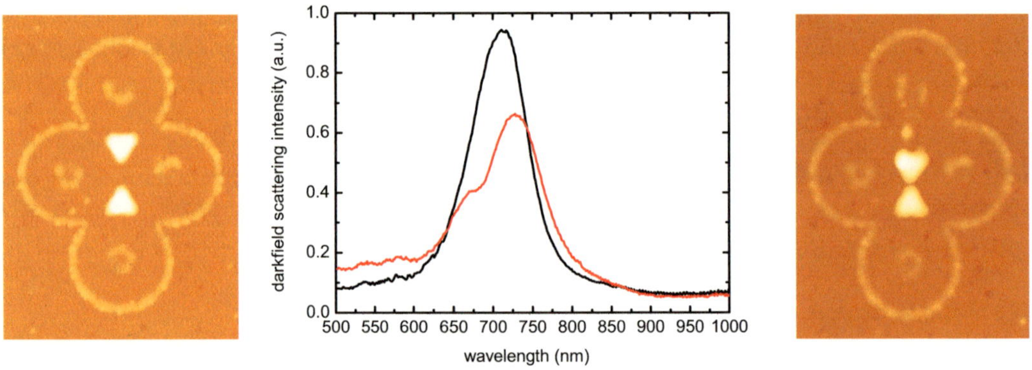

(a) (b) (c)

Fig. 7. Atomic force microscope images (a and c) and dark field scattering spectra (b) of a single gold bowtie optical nanoantenna with feedgap of 85 nm (black line) and 5 nm (red line).

4. Conclusion

In conclusion, we have presented a semiconductor quantum dot as a solid-state system to perform femtosecond quantum optics experiments. The ultrafast dynamics in a single semiconductor quantum dot reveals instantaneous Coulomb renormalization, perturbed free induction decay, and unexpected single-photon gain. It is now possible to prepare, manipulate, and readout this single-electron system on a femtosecond timescale. When driving the dot in the nonlinear regime, it is possible to add single photons to a femtosecond laser pulse. Finally, we have demonstrated means of increasing the coupling of light with objects of nanometer dimensions. We presented dielectric microcavities operating both in the visible and ultraviolet. Alternatively, metal nanoantennas were fabricated. They work as optical counterparts of the well-known radio antennas and concentrate light in the antenna feedgap. We have demonstrated a tunable bowtie nanoantenna and investigated its linear optical properties. We believe that progress in nanotechnology in the next years will further enable us to tailor the interaction of light with single nanoemitters to enter the regime of few-photon femtosecond physics.

5. References

Akopian, N.; Lindner, N. H.; Poem, E.; Berlatzky, Y.; Avron, J.; Gershoni, D.; Gerardot, B. D. & Petroff, P. M. (2006). Entangled photon pairs from semiconductor quantum dots. *Physical Review Letters*, 96, 13, 130501

Atature, M.; Dreiser, J.; Badolato, A.; Hogele, A.; Karrai, K. & Imamoglu, A. (2006). Quantum-dot spin-state preparation with near-unity fidelity. *Science*, 312, 5773, 551-553

Bányai, L. & Koch, S. W. (1993). *Semiconductor quantum dots*, World Scientific, 9810213905, Singapore ; River Edge, NJ

Bimberg, D.; Grundmann, M. & Ledentsov, N. N. (1999). *Quantum dot heterostructures*, John Wiley, 0471973882, Chichester, [Eng.]; New York

Bratschitsch, R. & Leitenstorfer, A. (2006). Quantum dots: Artificial atoms for quantum optics. *Nature Materials*, 5, 11, 855-856

Janssen, N.; Whitaker, K. M.; Gamelin, D. R. & Bratschitsch, R. (2008). Ultrafast spin dynamics in colloidal ZnO quantum dots. *Nano Letters*, 8, 7, 1991-1994

Kahl, M.; Thomay, T.; Kohnle, V.; Beha, K.; Merlein, J.; Hagner, M.; Halm, A.; Ziegler, J.; Nann, T.; Fedutik, Y.; Woggon, U.; Artemyev, M.; Perez-Willard, F.; Leitenstorfer, A. & Bratschitsch, R. (2007). Colloidal quantum dots in all-dielectric high-Q pillar microcavities. *Nano Letters*, 7, 9, 2897-2900

Kittilstved, K. R. & Gamelin, D. R. (2006). Manipulating polar ferromagnetism in transition-metal-doped ZnO: Why manganese is different from cobalt (invited). *Journal of Applied Physics*, 99, 8, 08M112

Klingshirn, C. (2007). ZnO: Material, physics and applications. *Chemphyschem*, 8, 6, 782-803

Kroner, M.; Govorov, A. O.; Remi, S.; Biedermann, B.; Seidl, S.; Badolato, A.; Petroff, P. M.; Zhang, W.; Barbour, R.; Gerardot, B. D.; Warburton, R. J. & Karrai, K. (2008). The nonlinear Fano effect. *Nature*, 451, 7176, 311-314

Li, X. Q.; Wu, Y. W.; Steel, D.; Gammon, D.; Stievater, T. H.; Katzer, D. S.; Park, D.; Piermarocchi, C. & Sham, L. J. (2003). An all-optical quantum gate in a semiconductor quantum dot. *Science*, 301, 5634, 809-811

Liu, W. K.; Whitaker, K. M.; Smith, A. L.; Kittilstved, K. R.; Robinson, B. H. & Gamelin, D. R. (2007). Room-temperature electron spin dynamics in free-standing ZnO quantum dots. *Physical Review Letters*, 98, 18, 186804

Mayer, G.; Fonin, M.; Rudiger, U.; Schneider, R.; Gerthsen, D.; Janssen, N. & Bratschitsch, R. (2009). The structure and optical properties of ZnO nanocrystals embedded in SiO2 fabricated by radio-frequency sputtering. *Nanotechnology*, 20, 7, 075601

Merlein, J.; Kahl, M.; Zuschlag, A.; Sell, A.; Halm, A.; Boneberg, J.; Leiderer, P.; Leitenstorfer, A. & Bratschitsch, R. (2008). Nanomechanical control of an optical antenna. *Nature Photonics*, 2, 4, 230-233

Michler, P.; Imamoglu, A.; Mason, M. D.; Carson, P. J.; Strouse, G. F. & Buratto, S. K. (2000). Quantum correlation among photons from a single quantum dot at room temperature. *Nature*, 406, 6799, 968-970

Moutzouris, K.; Adler, F.; Sotier, F.; Trautlein, D. & Leitenstorfer, A. (2006). Multimilliwatt ultrashort pulses continuously tunable in the visible from a compact fiber source. *Optics Letters*, 31, 8, 1148-1150

Moutzouris, K.; Sotier, F.; Adler, F. & Leitenstorfer, A. (2007). Sum frequency generation of continuously tunable blue pulses from a two-branch femtosecond fiber source. *Optics Communications*, 274, 2, 417-421

Muhlschlegel, P.; Eisler, H. J.; Martin, O. J. F.; Hecht, B. & Pohl, D. W. (2005). Resonant optical antennas. *Science*, 308, 5728, 1607-1609

Petta, J. R.; Johnson, A. C.; Taylor, J. M.; Laird, E. A.; Yacoby, A.; Lukin, M. D.; Marcus, C. M.; Hanson, M. P. & Gossard, A. C. (2005). Coherent manipulation of coupled electron spins in semiconductor quantum dots. *Science*, 309, 5744, 2180-2184

Santori, C.; Fattal, D.; Vuckovic, J.; Solomon, G. S. & Yamamoto, Y. (2002). Indistinguishable photons from a single-photon device. *Nature*, 419, 6907, 594-597

Schuck, P. J.; Fromm, D. P.; Sundaramurthy, A.; Kino, G. S. & Moerner, W. E. (2005). Improving the mismatch between light and nanoscale objects with gold bowtie nanoantennas. *Physical Review Letters*, 94, 1, 017402

Sotier, F.; Thomay, T.; Hanke, T.; Korger, J.; Mahapatra, S.; Frey, A.; Brunner, K.; Bratschitsch, R. & Leitenstorfer, A. (2009). Femtosecond few-fermion dynamics and deterministic single-photon gain in a quantum dot. *Nature Physics*, 5, 5, 352-356

Stevenson, R. M.; Young, R. J.; Atkinson, P.; Cooper, K.; Ritchie, D. A. & Shields, A. J. (2006). A semiconductor source of triggered entangled photon pairs. *Nature*, 439, 7073, 179-182

Tauser, F.; Leitenstorfer, A. & Zinth, W. (2003). Amplified femtosecond pulses from an Er : fiber system: Nonlinear pulse shortening and self-referencing detection of the carrier-envelope phase evolution. *Optics Express*, 11, 6, 594-600

Thomay, T.; Hanke, T.; Tomas, M.; Sotier, F.; Beha, K.; Knittel, V.; Kahl, M.; Whitaker, K. M.; Gamelin, D. R.; Leitenstorfer, A. & Bratschitsch, R. (2008). Colloidal ZnO quantum dots in ultraviolet pillar microcavities. *Optics Express*, 16, 13, 9791-9794

Weiner, J. & Ho, P.-T. (2003). *Light-matter interaction*, Wiley, 0471253774, Hoboken, N.J.

Weisbuch, C. & Vinter, B. (1991). *Quantum semiconductor structures : fundamentals and applications*, Academic Press, 0127426809, Boston

Xu, X. D.; Sun, B.; Berman, P. R.; Steel, D. G.; Bracker, A. S.; Gammon, D. & Sham, L. J. (2007). Coherent optical spectroscopy of a strongly driven quantum dot. *Science*, 317, 5840, 929-932

Zeilinger, A.; Weihs, G.; Jennewein, T. & Aspelmeyer, M. (2005). Happy centenary, photon. *Nature*, 433, 7023, 230-238

10

Artificial Intelligence Tool and Electronic Systems Used to Develop Optical Applications

Margarita Tecpoyotl-Torres[1], Alberto Ochoa[2],
Jesús Escobedo-Alatorre[1], Miguel Basurto-Pensado[1],
Arturo García-Arias[1] and Jessica Morales-Valladares[1],
[1]*Research Center for Engineering and Applied Sciences (CIICAp),*
Autonomous University of Morelos,
[2]*University of Juarez City*
México

1. Introduction

As we know, the computer science and electronics have a very big incidence in several research areas; optics and photonics are not the exception. For our group, the utilization of computers; and electronic systems and devices has permitted to develop several projects to control processes. Some of them will be described in this chapter.
In section 1, the development of irradiance meters based on electronic automation are shown. The design details and the results produced by the analysis of illumination sources in visible range are discussed. In the second section of this chapter, a description of the computer tool called Micro Engraved with Laser (MGL) to characterize materials is realized. The Reasoning Based on Cases (RBC) and its implementation in the software using Java are presented. In order to guaranty the lithography precision, a control system based on a microcontroller was developed and coupled to the mechanical system. On the other hand, in section 3, an alternative of engraving, considering the use of a Personal Digital Assistant (PDA), instead of a Personal Computer (PC) is described. In this case, C language is used for programming. With RBC, the application has the capacity to keep information in the cases library. Its use optimizes the materials characterization, recovering information of materials previously characterized. The communication between the PDA and the displacement table is achieved by means of a system based on a micro-controller DSPIC. The developed computers tool permits us to obtain lithography with channels narrower than an optical fiber, of approximately 145 µm, with minimum equipment. In section 4, perspectives of micro drilling are also analyzed. Micro drilling with smaller channel diameters is a challenge, with a complete analysis of mechanical material properties. The benefits of Laser-induced plasma machining technique are mentioned. Finally, in section 5 our conclusions are provided.

2. Applications of electronics systems to irradiance meters

The light detection constitutes a wide field of analysis in optics. As we know, the light can be detected by the eye, but it has several disadvantages, compared with electronic devices designed to this purpose, such as a very slow response, a not adequate sensitivity to low-level signals, and it is not easy connected to electronic receiver for amplification, or other signal processing (Palais, 1984). The selection of the appropriate detector is very important in the irradiance meter design. It must be considered the costs, spectral responsivity, noise levels, and the type of meter where it will be inserted.

On the other hand, the type of meter can be determined in accordance to the objective that could be the determination of the total irradiance or the distribution produced by an illumination source. For the first case it is well known the Ulbrich Sphere, is generally used to determine the lamps performance and the energy emitted in visible, infrared, and ultraviolet radiation. For the second case, different types of meters have been developed in order to provide an empirical base of information for specific characteristics, such as: photometer and portable intensity light meters. The use of electronics and programming tools has been determinant in the development of these devices. For example, some of the portable meters use microprocessors to increase the accuracy and to give them special characteristics. Other ones are equipped with memory or datalogger for the data recording. In both cases, the addition of these characteristics represent a considerable increase in the cost.

Basic concepts

A high interest in illumination intensity measurement has been shown in several fields, for example in architecture design and public illumination (Westinhouse, 2000), where the selection of the appropriate sources is fundamental and constitutes one of the biggest reasons for the realized studies. Another area of interest is generated by the necessity of manufactures of illumination sources to provide complete information for users, making necessary the characterization of the irradiance profiles. In research activities, the interest in the energy propagation produced by an illumination source has lead to widest studies in the total irradiance and the corresponding distribution.

The power of optic beam is proportional to the light intensity (defined as the square of the electric field). Intensity is proportional to *irradiance*, the power density; its units are watts per square meter.

2.1 Semi-spherical irradiance meter

Among the current projects on this area, at first, we developed a semi-spherical irradiance meter prototype (Roman G., et al, 2006), figure 1, based on Light Emitting Diodes (LEDs) detectors array, distributed in order to produce an uniform covered area (figure 2).

The selection of the optical detector was realized in this case considering basically the costs, because its number in the static array is considerably big.

The optic power generated by a LED is linearly proportional to the forward driving current (Palais, 1984). The linear relationship can be understood by the following argument: The current I is the injected charge per second is then N=i/e, where e is the magnitude of the charge on each electron. If η is the fraction of these charges that will recombine and produce photons, the optic power output will be:

$$P = \eta N W_g = \frac{\eta W_g}{e} i \tag{1}$$

For emission, the diode is forward biased, and charges injected into the junction recombine to produce photons. For detection, the process is reversed: the inversed biased and incoming photons generate electron-hole pairs, producing electrical current.

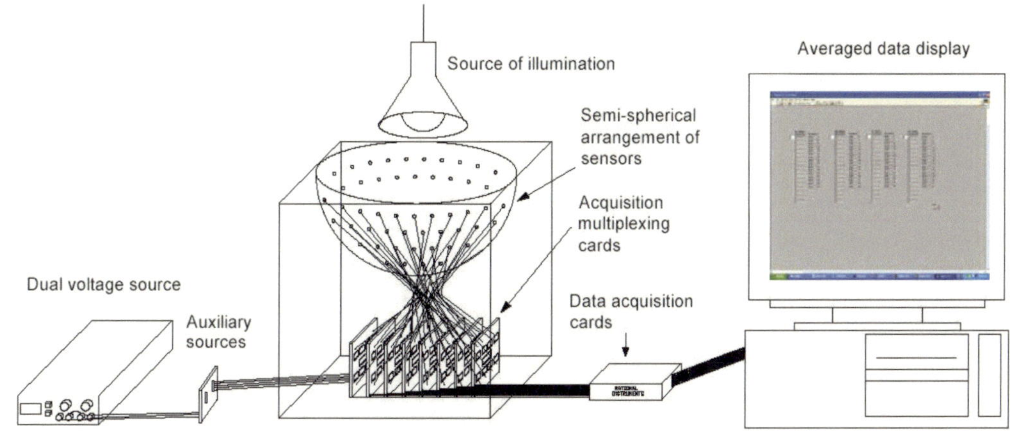

(a)

(b)

Fig. 1. (a) Complete irradiance meter, based on a detectors array. (b) Photograph.

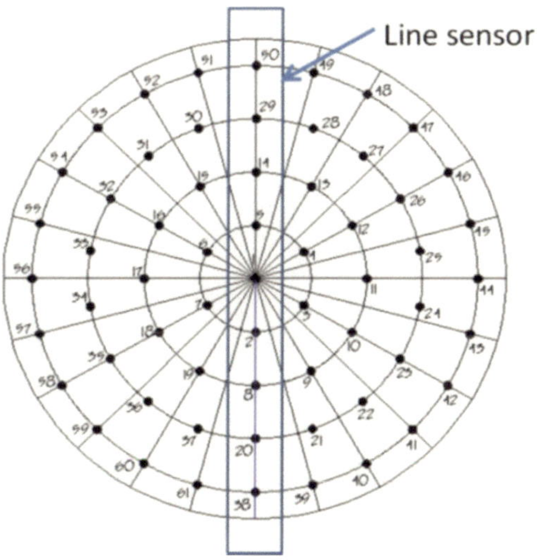

Fig. 2. Detectors distribution.

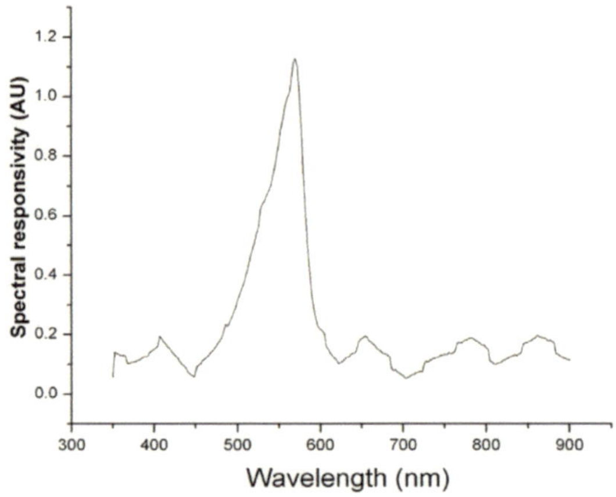

Fig. 3. Spectral responsivity of LED E5/AMB-C.

The largest response proving LEDs of different colors, as detectors was given by an amber LED used as detector, so, it was chosen to form part of the semi-spherical arrangement, in order to analyze white illumination sources profiles.
As the LEDs are commonly acquired without a data sheet, it is necessary to characterize them, this action can be realized by means of a monochromator, in this case, the ACTON 300. The spectral responsivity of this detector (E5/AMB-C) is shown in figure 3, from 350 up to 900 nm. The setup designed to characterize the detectors (LEDs) is shown in figure 4(a) and (b). As the response to illumination on a semiconductor junction, is a very small current, we use the amplification circuit shown in figure 4(b), with an amplification factor determined by the back-feed resistance (Horowitz, 2001). The capacitor was implemented to establish the analyzed signals. The linearity of the circuit can be appreciated in figure 4c, which was kept until the saturation region that means, approximately until 4.15 μA. The maximal response was obtained at 570 nm (figure 3). The linear relation between the irradiance and the voltage is considered due to the linearity of the detection circuit.

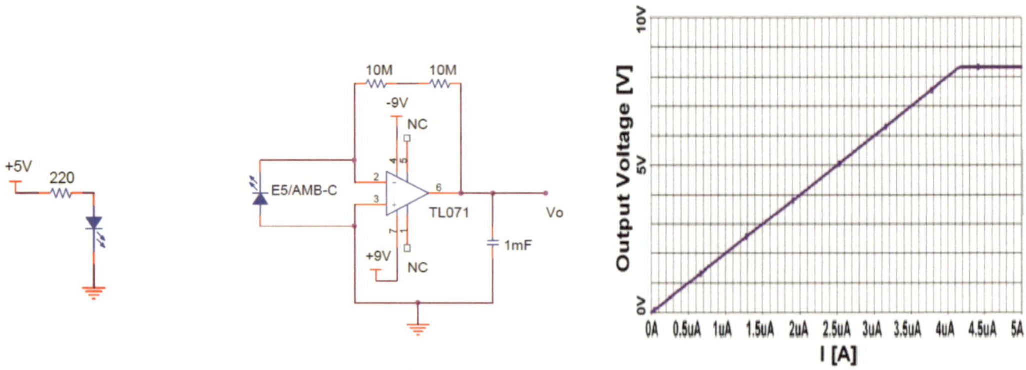

(a) (b) (c)

Fig. 4. Linear response of I-V converter circuit.

A multiplexing stage was necessary due to the number of detectors (61), which is bigger than the number of the analogical inputs (16) of the used acquisition card (PCI-MIO-16E-1 de National Instruments). The data collection rate was of 1.25/16 MS/s, using a dual

(74HC4052) multiplexer. Each of the 8 conditioner cards is formed by the amplification and the multiplexing circuits (figure 5). The 8 conditioner cards were placed on the meter base (figure 1).

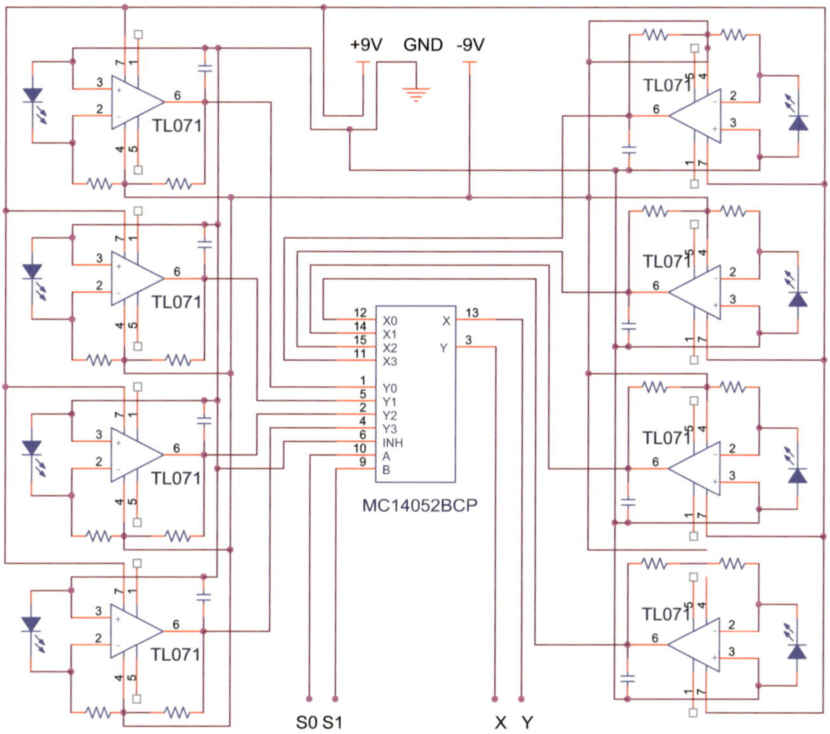

Fig. 5. Signal conditioner circuit.

Data visualization

The flowchart shown in figure 6 was used in the development of the collection and average data program. The data collection is realized by the acquisition data card and displayed using a program realized in LABVIEW, the data are storage, and this process is repeated 4 times, completed this cycle all is repeated again. When 100 data are collected in each detector, the corresponding average is calculated, which forms part of the useful data. After, the process is repeated again, and the new averaged data replace to the previous ones, this process continues until the user gives a stop.

The irradiance profiles, generated with this semi-spherical prototype, were presented using bubbles schemes. An X-lamp LED (figure 7) of warm white light, with an emission diameter of 1.9 cm, was tested. Its irradiance profile and irradiance pattern are shown figures 8 and 9, respectively. This meter can be used as a quality control device for illumination sources such as lamps and bulbs. In the obtained discrete profiles, the following fact was considered: according to the Pointing vector physics, the irradiance is proportional to the voltage intensity detected in each sensor of the meter. These voltage values permit to obtain the corresponding profile of each source and give the capability to choice of more adequate sources for specifics tasks.

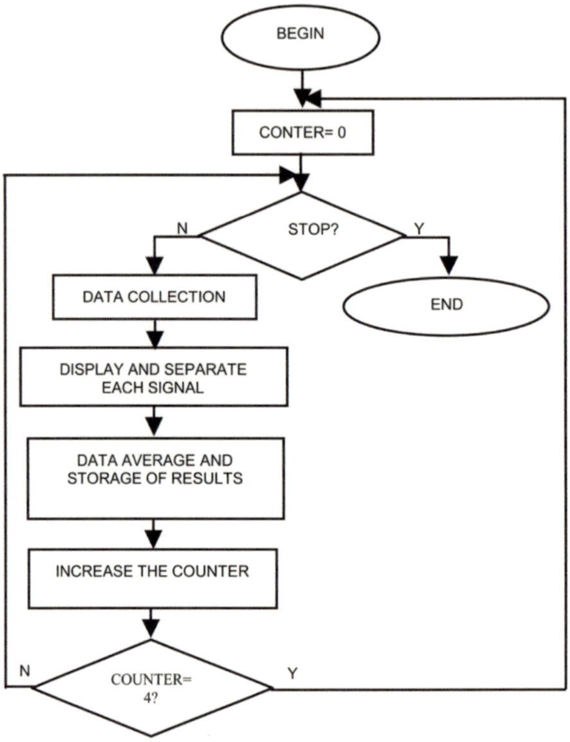

Fig. 6. flowchart of the developed program.

Fig. 7. Photograph of X-lamp LED.

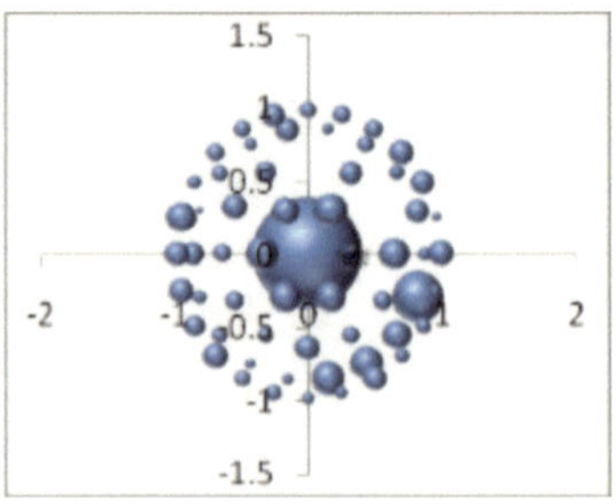

Fig. 8. Irradiance profile of the X-lamp LED.

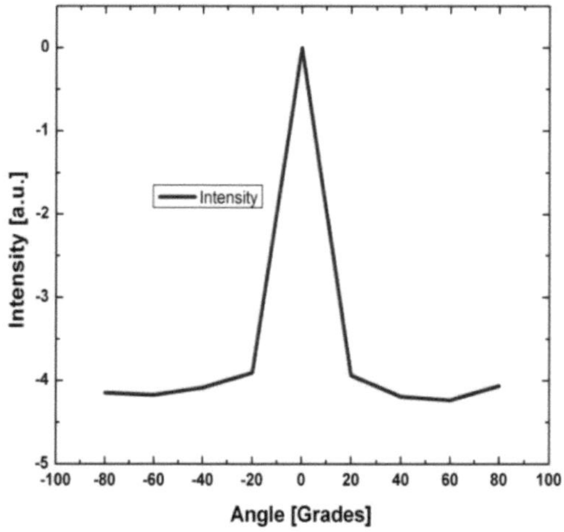

Fig. 9. Irradiance pattern of the X-lamp LED.

Some disadvantages of this static meter are given by the use of the detectors array produce, which produces a not uniform response also at the same co-latitude angle, as can be appreciated in figure 6, especially in the last two external trajectories of detectors. These differences in the spectral response were produced basically for the fabrication processes. As it is known, even in LEDs from the same manufacturing batch, a difference in response exists, although it could be minimal. This fact produces errors in the determination of the irradiance profile from 12% up to 90% in critical cases. The error could be reduced being more careful with the sensor selection, and replacing them when it is necessary, but always it would be present due to their inherent differences. This fact shows the importance of the individual characterization before the selection final elements of the array. The alignment of the illumination source is another important source of error, that can be appreciated in the corresponding data fitting using Zernike polynomials (Sanchez-M. et al., 2003).
The irradiance pattern (figure 7) generated considering the line sensors shown in figure 2 has another inconvenient, the appearance of a piece linear graph produced by the small number of sensors (only 9), and the variations in symmetry produced by their differences in the spectral responsivity. In order to reduce these inconvenient, the use of a single detector is suggested, coupled to a mobile mechanical structure. Although the XY table is very expensive, its availability in several optical labs permits to think in them as a very practical solution, after a certain adaptation, and considering its precision. The automation of the movements could be realized, using step motors or servomotors, without an excessive inversion. The use of recyclable convenient surfaces are recommended. In this case, the use of a CD ROM platform is analyzed.

2.2 Rectangular irradiance meters for convergent sources. Another application of the XY table automation

Manufacturers employ photometric curves for electric lamps, referring them at 1000 lumens, when their emitted fluxes are different of this quantity a normalization is used. These curves are obtained as a section of 3D intensity distribution, for all the solid angles, generally are indicated in polar coordinates, but for specific applications are also provided in Cartesian coordinates, such as in the case of projectors. The attention in this section is focused on the

last representation, convenient to convergent sources, for whom the utilization of rectangular irradiance meters is suitable.
The use of XY tables for different applications has been widely reported (Gwirc, S., et al, 2007, Jaimes-V., et al, 2005) and the case of three grades of liberty has also analyzed (Yañez V., et al, 2008). In this section, a rectangular prototype, based on an XY table, used as irradiance meter is proposed. The use of only one detector makes necessary the implementation of a mechanical mobile support. The implemented detector is a photodiode OPT301. The semiconductor photodiodes have some advantages, above the LEDs used as detectors, that must be mentioned: they are small, light, sensitive, fasts, and can operate with just few bias volts. In order to increase the response, a preamplifier must be integrated onto the same chip as the photodiode, producing an Integrated Detector Preamplifier (IDT), which is the case of the OPT301. The spectral responsivity and its linear response to irradiance can be consulted in www. Alldatasheet.com. Clearly this response is bigger than the LED one, which generally have widths of 20-50 nm (Palais, 1984).
The rectangular prototype was designed on a CD ROM unit, in order to build a very low cost mechanical structure, with the following characteristics:

- The mobile part of the XY table is used to describe the linear trajectories on X and Y axes. It has a total area of 19x14 cm^2. The scanning area is of 4.5x3.5 cm^2. The total height of the base is of 7 cm.
- The square grid was established for simplicity of 14x15 detection points.
- The displacement was controlled by means of servomotors.
- The detector is assembled at one corner of the mobile base (figure 10).
- The illumination source is located over the central part of the mobile area.

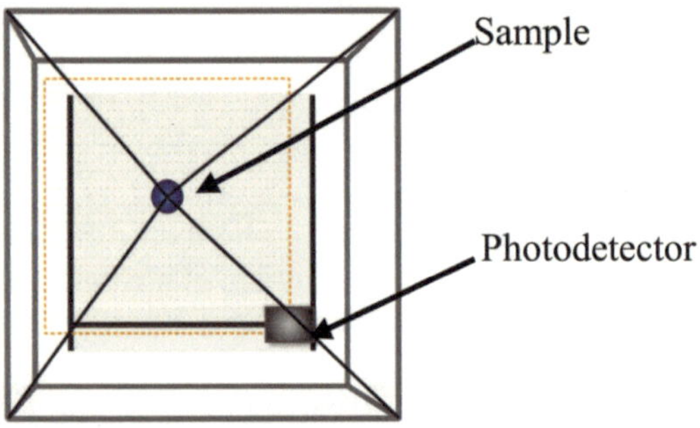

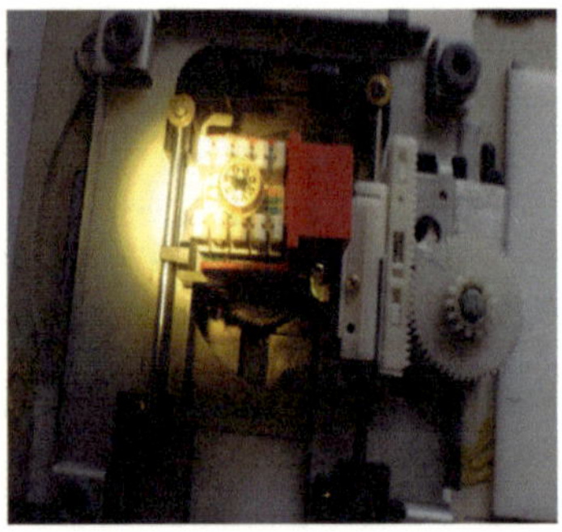

Fig. 10. Meter (a) diagram and (b) photograph.

The XY movements were programmed on the base of the diagram flux shown in figure 11. A photograph of the complete detection circuit and the diagram, using an OPT301M as detector, a dsPIC30F4011 for the movements programming and a display as a graphical interface to the user are shown in figure 12 and 13, respectively.

Servomotor movement

The servomotors normally have a capability to move from 0 to 180 grades in accordance to a signal control (figure 14). The control program is based on the generation of a very precise rectangular signal, with a period of 200 ms, the width of the pulse is modified depending of the desired position of the rotor, if a position of 90 degrees is required, the sent pulse must be of 1.5 ms. In this application, for the control pulse generation, two timers of 16 bits were implemented, one for the period and the other one for the pulse duration (figure 15).

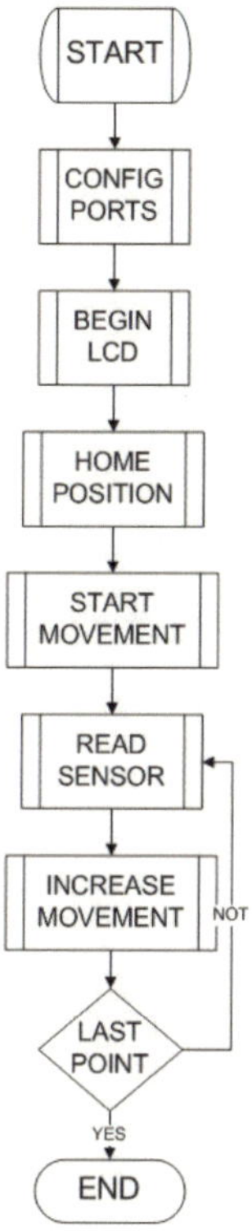

Fig. 11. Flow diagram

Fig. 12. Photograph of the rectangular meter composed by a XY table and the detection circuit.

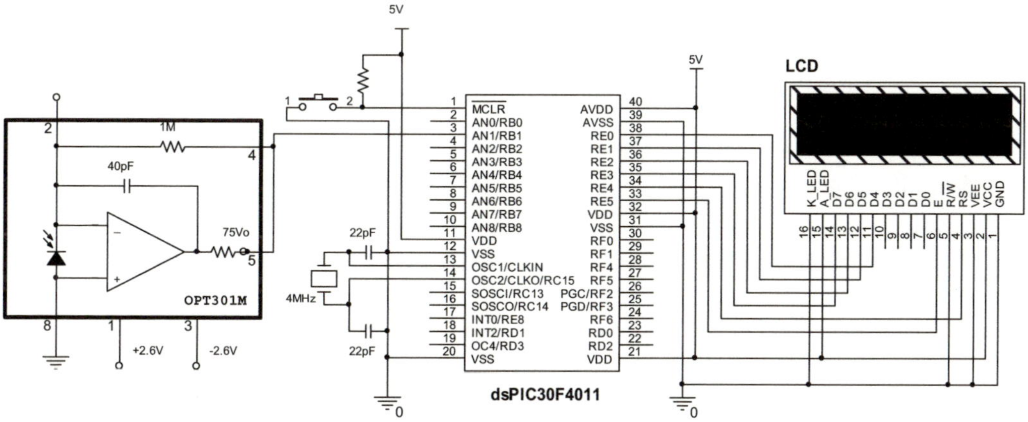

Fig. 13. Schematic of the detection circuit.

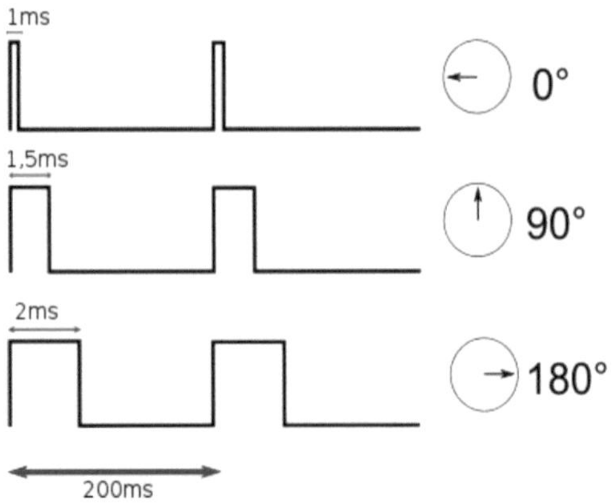

Fig. 14. Relationship among the control signal and the movement of the servomotor.

```
void Scanning_table_xy(void)
{
      int i,j,data_RD,;
      int long retardo;
      delay=10000;            //useg
      for(i=0;j<30;j++)
              {
                         motor_X=0;                     //move Y
                         time_pulse_Y=time_pulse_Y+50;
                         _delay_l(retardo);
                         motor_X=1;                     //move X
                         for(i=0;i<26;i++)
                                 {
                                       time_pulse_X=time_pulse_X+50;
                                       _delay_l(retardo);
                                       data_RD =Read_ ADC(); //Read data of optical sensor
                                       WriteUART2(i,j,data_RD);
while(BusyUART2()); //// Wait for transmission to complete
                                       _delay_l(retardo);
                         }
                         time_pulso_X=900;
                         _delay_l(80000);               //Delay function
              }
}
//Delay function
Void  _delay_l(int long time)
{
      while(time!=0) time--;
}
```

Fig. 15. Main program of scanning, reading and sending data to PC.

They were also implemented interruptions (figure 16), which permit to each timer "to notify" to the CPU when each temporization finishes. The control signals for the two motors of the system XY are implemented at the pines 3 and 1 of the D port, respectively.

Communication system between the mecatronic system and the computer

The serial port US-232 is used to transfer all data read by the optical sensor in each swept point of the table (x, y coordinate) to the computer. A program developed in C, and compiled in the compiler for C of Microchip for family DSPIC30F was implemented.

The obtained data are storage in a txt file in the PC, and after, the corresponding graphs are generated using suitable software, as ORIGIN.

```
void __attribute__((__interrupt__,no_auto_psv)) _T1Interrupt(void)
{
     IFS0bits.T1IF = 0; /* clear interrupt flag */
     if ( motor_X == 1)
            {
                   PORTDbits.RD3=1;    //motor X
                   ini_tmr2(time_pulso_X);
            }
     else
            {
                   PORTDbits.RD1=1;    //motor Y
                   ini_tmr2(time_pulso_Y);
            }
     return;
}
//interruption to define the period time
void __attribute__((__interrupt__,no_auto_psv)) _T2Interrupt(void)
{
     IFS0bits.T2IF = 0; // clear interrupt flag
     if ( motor_X == 1)  PORTDbits.RD3=0;    //motor X
     else PORTDbits.RD1=0;     //motor Y
     return;
}
```

Fig. 16. Interruptions.

```
// Send data to PC
WriteUART2(Txdata_x,y);
while(BusyUART2()); //// Wait for complete transmission
Config_uart2 (void)
{
     unsigned int baudvalue;   // Rate value
     unsigned int config1 ;
     unsigned int config2;                    CloseUART2();
     ConfigIntUART2(UART_RX_INT_EN & UART_RX_INT_PR6 &
     UART_TX_INT_DIS & UART_TX_INT_PR5);  //Interrup config
          // config frecuency of transmition
     // Baud = Fosc/(16(UBRG-1)); UBRG=(Fosc/(16*Frec Baud))+1
     baudvalue = 25;          / /for 9600bauds

     // Config1 UART2 for 8 bit transmission with one bit stop
     // Also Enable loopback mode
     config1 =    UART_EN & UART_IDLE_CON &
                        UART_DIS_WAKE & UART_DIS_LOOPBACK &
                        UART_EN_ABAUD & UART_NO_PAR_8BIT &
                        UART_1STOPBIT;
     config2 =    UART_INT_TX_BUF_EMPTY &
                        UART_TX_PIN_NORMAL &
                        UART_TX_ENABLE & UART_INT_RX_CHAR &
     //UART_TX_ENABLE & UART_INT_RX_3_4_FUL
                        UART_ADR_DETECT_DIS &
                        UART_RX_OVERRUN_CLEAR;
     OpenUART2(config1, config2, baudvalue);
}
```

Fig. 17. Data sending and configuration function for UART (Universal Asynchronous Receiver-Transmitter).

Results

The photodetector was biased with 13V, and without a source of illumination, a lecture of 0.606 V was registered. Three luminaries were analyzed with this meter (figure 18), a dirigible X-lamp LED IL51, a lamp conformed by an array of three LEDs IL1 LED3; and a single ultra bright white LED. The distance between the X-lamp LED and the photodetector was fixed at 13 cm, and due to the different sizes of the lamp with three LEDs, the distance was reduced at 10 cm. In the case of the single white LED, the distance was of 17 cm.

Fig. 18. (a) Dirigible X-lamp LED IL51, as a sample and the XY table. (b)Lamp IL/1 L3 (with a 3 LED array; and (c) single white LED

As can be appreciated in figure 19b, the illuminated area produced by a dirigible X-lamp is almost regular, showing a deviation of only 0.25 units considering a radius of 4.35 distance units. These deviations can be attributed basically to the differences in the gear teeth driven by each servomotor. The high directivity of this illumination source permits to suggest its use in architecture and landscape lighting, for visual inspection, among other possible applications. The level of analysis achieved with this prototype permits us to observe very little details, which are very difficult to observe with other devices, as the case shown in figure 20, where we present the irradiance pattern generated by a spherical mobile prototype developed also by our research group, using the same detector and the same detection circuit. In the last case, the beamwidth is easily determined, and complements the information about the source under test.

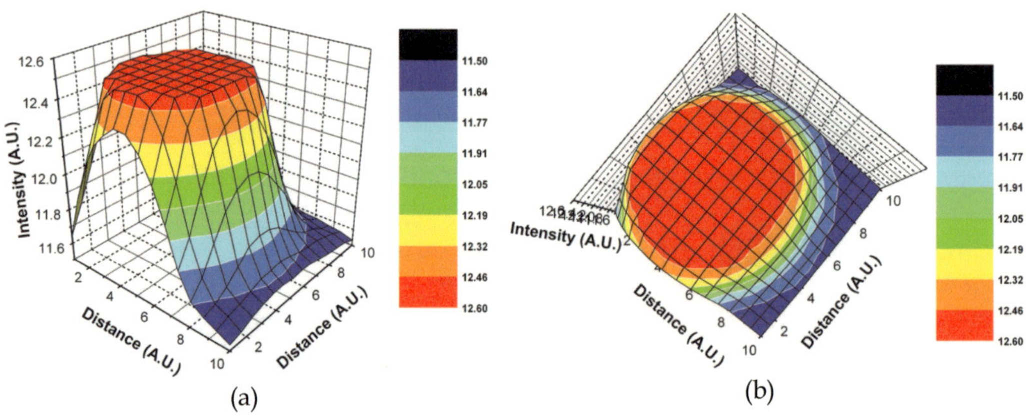

Fig. 19. Irradiance profile of dirigible X-lamp LED (a) complete and (b) top view.

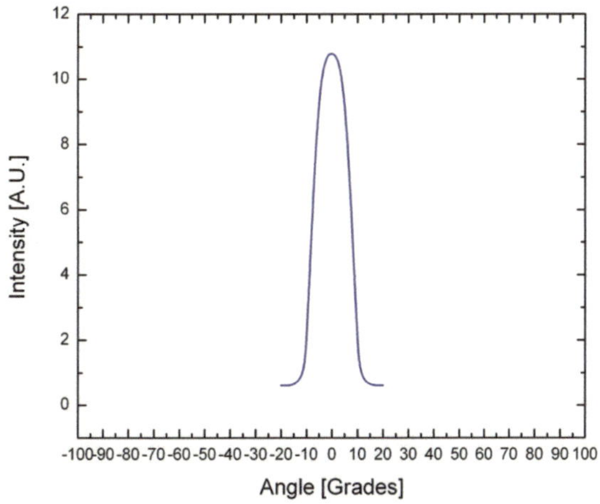

Fig. 20. Irradiance pattern of the dirigible X-lamp LED.

The use of high brightness LEDs in lamps improves several fields of interest in illumination, such as energy-saving, uniform covering and longer lifetime. The level of directivity of lamps based on LED technology has been modified with the use of special metallic covers, as in the case of the dirigible X-lamp LED IL51. For other applications, such as for decorative exterior in gardens, or for paths illumination, lamps like the IL/1 L3 can be used. Its corresponding radiance profile is given in figure 21. As can be observed, it provides a wider illumination angle than the dirigible X-lamp, covering almost uniformly to the total scanning area. In figure 22, the irradiance pattern obtained with the same spherical prototype of the figure 20 is presented. In both cases, the flat surface on the illuminated area is evident.

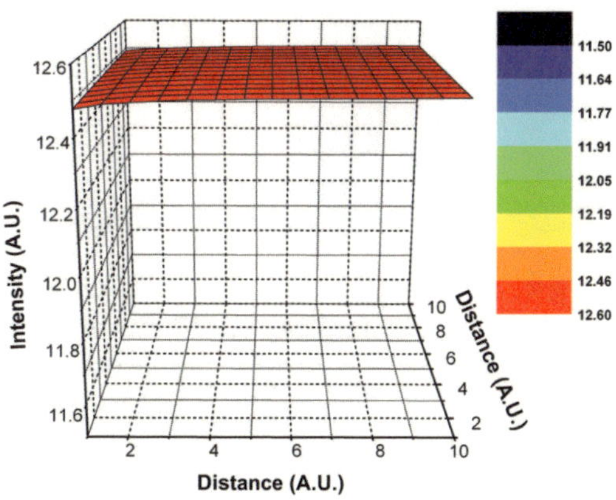

(a)

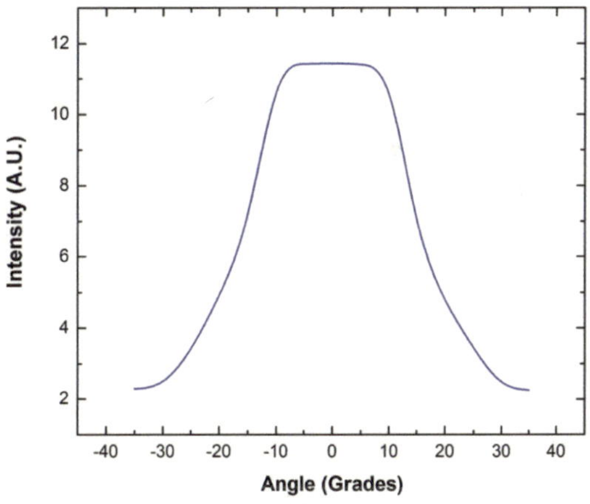

(b)

Fig. 21. Irradiance profile of lamp with three LEDs (a) complete and (b) a zoom in on the left side.

Fig. 22. Irradiance pattern of the single ultra bright white LED generated by the spherical prototype.

Finally, a single ultra bright white LED was tested with the prototype. The corresponding irradiance profile is shown in figure 23. This LED showed problems of stability that means variability in the voltage measurements, and a little asymmetrical response. The same behavior was observed using the spherical irradiance meter and a lux-meter. Its maximum illuminance was of 84 lux at a distance equal to 27 cm, very lower compared with the illuminance produced by the lamps based on a power star LEDs of high brightness, such as the dirigible X-lamp LED, where 1700 lux was measured on the top.

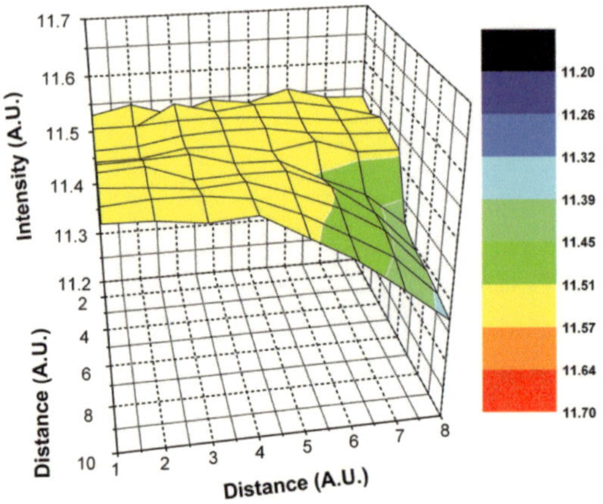

Fig. 23. Irradiance profile of the single ultra bright white LED.

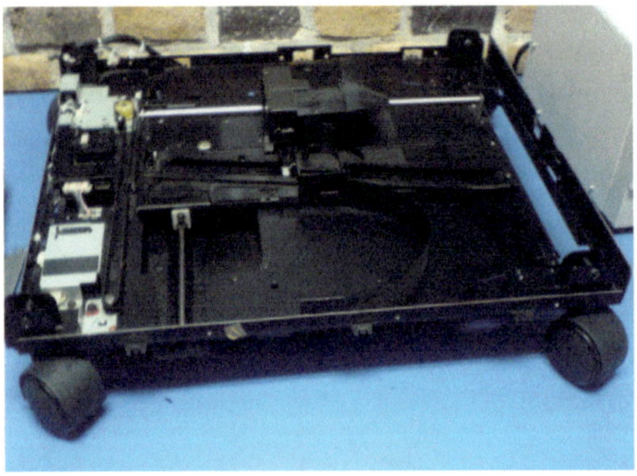

Fig. 24. XY table used as irradiance meter.

In order to increase the scanning area other suitable recyclable surface could be used, such as the conformed by the photocopiers rails (figure 24). Its optical sensor, photodiodes for visible light like BS520 (www.datasheetcatalog.com), designed to applications in precise optical instruments, is useful to use for our purposes, due to its higher stability compared to the semiconductor photodiode used in the CD ROM platform. The BS520 photodiode is most expensive, but the possibility of recycling permits to consider it as a good option. Generally, the absence of the serial numbers of the elements in the photocopier diagrams makes necessary to characterize the detector spectral responsivity before to use it. The data fitting for rectangular geometries can be realized by means of Chevyshev Polynomials (Tecpoyotl, 2006). As we known, the disperse data may lead to large errors unless we use a discrete reduction, as it happens commonly in image analysis.

It can be concluded that the implementation of a single detector in irradiance meters reduces the measurements error due to variations in the spectral responsivity. The precision in this case is determined by the control of the XY displacements. In absence of the servo motors implementation, manual control can be also carried out, keeping the reliability of the measurements, due to the worm drives coupled to the XY table. Similar structures can be realized on the base of the meter presented here, for example, using recycled printer rails or making mechanical structures of low cost. The bigger inversion would be realized on the detection circuit, which involves the DSPIC programming. A total scanning area, at least at 20x20 cm^2 is suitable in order to realize measurements of the luminaries of larger sizes. Also it is suggested to be very careful in the selection of all parts of the prototype, because as we mentioned, the little variations for example in the gear teeth can produce deviations in the generated profiles.

The implementation of the rectangular meter permits to observe with a great detail the profile produced as a result of the beamwidth of the illumination source under test. The generated information permits us to realize a comparison with other spherical prototype developed by our research group, and to have complete information about the total irradiance profile. The rectangular prototype provides empirical information necessary not only for manufacturers, but also for research activities. We found this prototype very useful due to the controllability of position, which increases the feasibility of the measurements, and provides a very complete irradiance pattern for the case of directive illumination sources. The application of the XY table automation shows the high potential of this type of devices.

3. An artificial intelligence development tool to micro engraved with laser (MGL) to control and optimization of the laser engrave process

The purpose of this project is to make devices that can be used in the optical fiber sensors, which requires micrometric dimension engrave. The process consists of two AutoCAD design that acts as reproduction by controlling 2 steps motors adapted to move X, Y axis mechanism. Case Based Reasoning (CBR) methodology is used to optimize the process. In a simple way, CBR resolves a new problems (new case) by a comparison with the other resolved problems (case library), it takes one or more solutions from the most similar cases, the proposed solution is evaluated and if it is necessary, this is adapted. Finally, if the proposed solution solves the problem, the new case is saved in the case library, in contrary case, it is not saved and the comparison continues. In this way, the systems infers knowledge or experience, given better results in accordance with its case library extension. The laser power approach is obtained under this procedure, as a function of the new material properties (per example, the hardness). The comparison of the specific properties with other cases or materials already characterized, makes possible to optimize the process by reducing engraving probes in new materials.

Nowadays, any line of development or research depends on the existing materials and equipment in other areas or development lines very near to it. This it is the case of the area of sensors and optical fibers, which depends mainly on the development of equipment in the communication area, where they are mutilated or modified with other pieces to be able to be used. This kind of problem can be reduced, if each device, considering its material, is designed and made in accordance to the proposed procedure.

The main problem is to make devices in order to apply them in the area of sensors from common materials. This entails two new problems:

- To characterize the materials
- Method of engraving

The devices fabrication can be realized by several methods, such as (Trimmer, 2005):

- Micro mechanized by ablation laser
- Micro mechanized by diamond
- Micro perforated
- Stereolithography or micro molded photo
- LEAGUE (X-ray lithography with metalized)
- Using Excimer laser

From the previous methods, if the purpose is to reduce the costs of the project to the minimum, the most viable for us is the one of engraving laser. For its utilization, it is necessary to account with a high-power laser and the possibility of using different materials from waste (like wood, plastic, paper, among others) to realize the engraving tests. The laboratory of Optics of CIICAp has all these conditions, making possible to realize here all test for the engrave laser process.

Based on the previous research, two lines of work are considered:

1. To characterize the materials to use (like the dimension and depth of penetration of the channel) by means of the CBR Technique (Software), and
2. To design a mechanical device to control the displacements in X and Y axes

In CBR systems, in order to adapt and evaluate a possible solution, frequently it is necessary to consider new recovered cases (representing the problem as a case). There are many cycles in the process (figure 25). Each case typically contains a description of the problem (attributes or characteristics of the problem), a solution and its result (García et al., 2005).

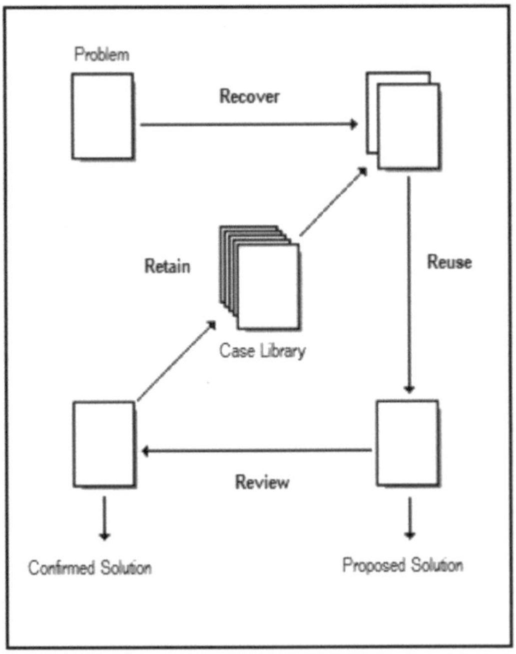

Fig. 25. Basic Cycle of CBR.

In order to find the laser power to use in a new material: the new case (material) is compared against the cases in the case library by means of the near neighbor technique based on the equation 2.

$$\text{Similarity} \qquad (T,S) = \sum_{i=1}^{n} f(Ti,Si) \times Wi \qquad (2)$$

where:
T is the new case
S is the case source
n is the number of attributes in each case
i is an individual attribute from 1 until n
f is the function of similarity for attributes i in cases T and S
W is the importance (the weight) of attribute i
The weights of each attribute are assigned by the expert (a person that assigns the weight to the attributes based on the quality of the engraving), which are designated by an annotation generally going from 0 to 1. For example: the hardness of the material has a weight of 0.60 whereas the translucence has a weight of 0.05. But, not all the attributes (characteristic of the material) are not taken into account, such as the case of the material color, since it is not useful for the analysis.
The values of similarity between the materials change whenever a new material is added to the case library. While greater it is the number of attributes (n) and it is counted on an extensive Case Library, the time in calculating this similarity will be greater for example, if we have 5 attributes in each case and a case library with 100 cases, 500 calculations cases will be realized (5*100). Some authors recommend having a base of cases smaller than 100 cases (Lake, 1996).
The recovered more similar cases are used to suggest a solution that is reused and tried on successfully. In case of being necessary, the solution will be reviewed and adapted by the expert. In addition the expert can make a suggestion like adding wet paper or other techniques that help to obtain an engraving with greater quality. Finally, the present problem and the final solution are conserved as a new case (material characterized). Any solution and/or characteristic of some material can be modified later by the user; while more cases have the system will be able to approach an ideal solution for the engraving of a new more case.
The calculation tool was developed with the programming language Java SDK standard edition 1.42 with more than 3.000 classes (Chan, 2002), along with JBuilder X, they are used to create applications in graphics mode multiplatform (Easttom, 2003). It was necessary to use usesPort and parport-win32 libraries for the shipment and reception of data by the parallel port. The user interface is based on the principle of easy and friendly software (Schildt, 2001).
The system is based on the following process for the accomplishment of the engraving (see Figure 26):

1. The user uses software to interchange the design created in AutoCAD to another one with DXF extension, in ASCII code (Tajadura, 1999). At the moment, designs with lines are only processed (command line).
2. The software only has the data necessary to realize traces (lines made in Autocad), reducing the size of the DXF archive. It transforms the simpler archive MGL (with the same name but with the extension mgl).
3. The user selects the material of the materials base. In case of being a new material, it is added to the system providing its characteristics. The system realizes the CBR process

to suggest the power for the new material, on the base of the resolute cases (material characterized) and the tests of the new material. The fundamental parameters considered for the engraving quality are: the power for engraving and the focal distance, while for the material are: basically the hardness and the roughness, for future analysis the new cases also would consider: information about translucent, heat resistance, and metallic or not metallic characteristics.

4. Continuing with the process, the user selects the lens for the engraving considering the focal length and the diameter of the focal point.
5. The software based on the Autocad file of the design, kept with the extension mgl, realizes the outlines through a communication stage (parallel port), to control the displacement of a milli-machined table. The objective is to control the rotation of the two motors. An improve in the mechanical system was realized by adding the a variable height, and a third motor, which functions as an shutter in order to avoid not desired engravings.
6. At the end the tool, the description of the engraving will be required to the user, using the RBC. The information is stored in the bookstore of cases to make future comparisons between engravings, in order to find an optimized design giving a solution improved for the task that the user wishes to realize. The RBC contribution in the control software is to suggest the power required for engraving in order to reduce the range of the necessary tests for the characterization of the new material. The characterized cases are considered to realize the similarity process.

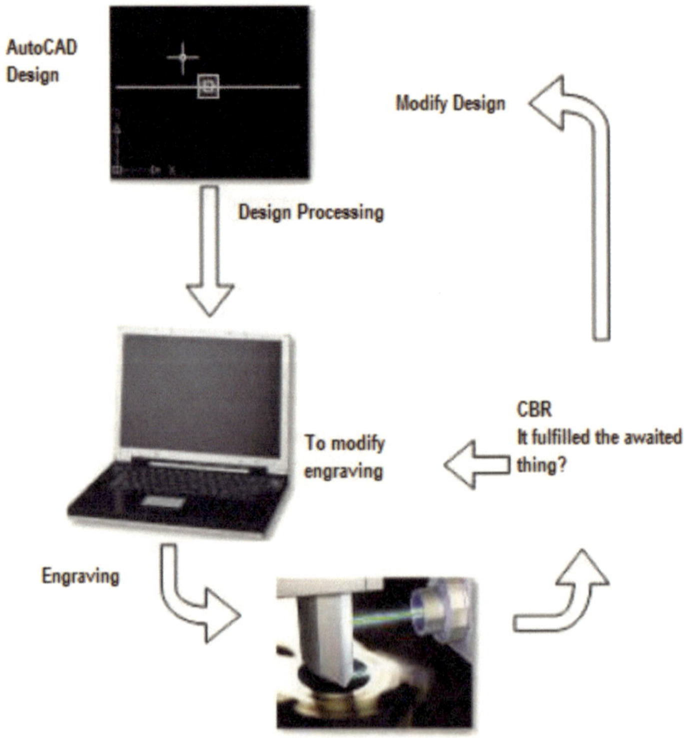

Fig. 26. Schematic diagram of the operation of the calculation tool for the engraving with laser.

The material characterization has been realized considering 4 different tests, based on:

- The focal distance
- The laser power
- Variations on the exposition time
- Traces and sizes (using an Autocad template)

The focal distance tests permits to determine the smallest possible diameter of the focal point. The variable power test produced, depending of the material characteristics, a proportional dependence between the damage and the applied power; the utilization of RBC, by the accumulated experience permits to reduce the range of realized tests. The variation in the exposition time produced the best defined channels, but it takes a lot of time. The variations in trace and sizes tests provide satisfactory results, in spite of some mechanical problems.

The application of CBR to the system produced satisfactory results, better than of those than it was expected, such as the realization of engravings with channels less wide than the same optical fiber (approx. 145 μm) and engravings become attached to the dimensions in the design obtaining in this way a significant advance in spite of the used equipment. The quality of the engraving also depends of the used material. The system continues under development. The present work outlines a prototype with currently obtained results.

It can be concluded that the system has wide possibilities to be more than a tool used in the devices design for sensing area, due to its capacity for both, to engrave and to realize cuts in different materials. The following stage of this project will be the accomplishment of a communication interface between a Palm and a mechanical system by means of electronics to control the precise movements of servo motors, as well as a more reliable and precise mechanical system. Another stage would be to control a new parameter, the displacement in Z axis, which would permit to control the depth of the engraving as well as arcs, ellipses and other more complex geometric figures.

4. Computer tool for engraving by means of PDA (Personal Digital Assistant) based on RBC

This section describes a computing tool on the Interface Development Environment (IDE), that was developed in an environment of development Code Warrior V9, in "C" language. This IDE accounts with an emulator, which allows to make tests before installing the program in the PDA. The goal of this tool is to characterize materials through the Artificial Intelligence technique named Case-Based Reasoning (CBR), with the help of this technique and a few of instructions, the characterization of the materials can be optimized. The development consist in choosing a few of parameters that allow us to do the search in the case library through the CBR and then process the data in the PDA. The information is send to the serial port, which after is sent to the receptor that transmits it to the motors.

The engraving techniques in their origins were realized by equipment of great sizes. Nowadays the computers are more and more small and economic, and so the use of the CNC (Computerized Numerical Control) has been extended to all type of machinery: winches, rectifiers, machines to sew, among others.

Development of a measure tool, that through CBR, handles the information of different materials

The objective is to characterize materials besides executing the program in a portable device that allows the adaptation of the user to the work area. The device that will serve like control is a PDA that has capacity of processing and data storage. These devices count with series and infrared ports. The difference between them is that first one depends on a physical connection, while the infrared does not. The series port requires of wires, but they can be larger than the corresponding to parallel port, and the number of wires is smaller. The pines Tx, Rx and GND of the connector DB9 are used to connect the PDA to the microcontroller. The data conversion is realized by means of the UART module.

The final mission is to characterize material through a complete system that includes our proposed calculation tool, a control module and actuators, as well as a laser.

The CBR is a technique of Artificial intelligence, its methodology is used in our computer tool. The CBR allows having an optimization in the characterization of the materials; some of the data that are being handled to be keeping in the case library are: material name, thickness, translucence, and the figure to be engraved. The CBR accounts with a basic cycle (See Figure 27) that includes the four r's:

- To recover the case(s) more similar;
- To reuse the case(s) to try to solve the problem;
- To review the propose solution in case of necessity, and
- To retain the new solution as it leaves from a new case.

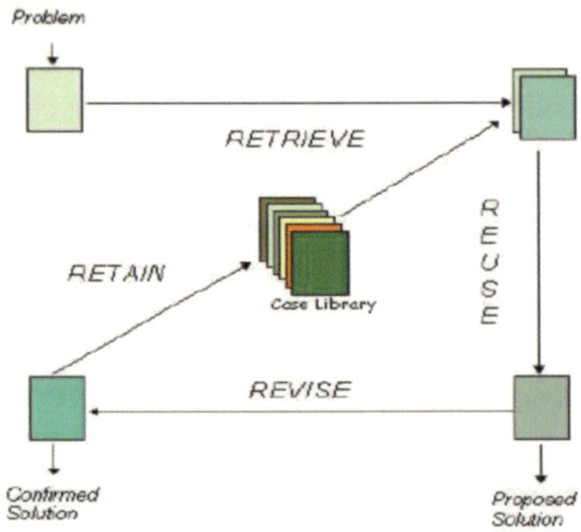

Fig. 27. The Basic Cycle of RBC [adapted of the proposed by (Aamodt & Plaza, 1994)].

A new problem is compared against cases in the cases library and the more similar cases are recovered. A solution is suggested as a result of the similarity analysis, which later is reused and tested in order to achieve the success of the solution. Unless the recovered case is a very similar case, the solution will probably have to be reviewed producing a new case that can be conserved. This cycle happens, currently, rarely without the human intervention (see figure 28). For example many tools of CBR (Ochoa et al., 2004) act mainly as recovery of the case and reutilizing systems. The revision of the case (also call adaptation) is realized often by the ones in charge of the Case Library. Nevertheless, this does not have to be seen like debility of the CBR, since it permits to work as a tool of making decision, with the human collaboration in aid of the best decision (Wainer et al., 2005).

The Code Warrior V9, under the "C" programming language, provides an atmosphere, which allows to build an interface of easy handling for the user. The diagram of figure 28 shows how the system works.

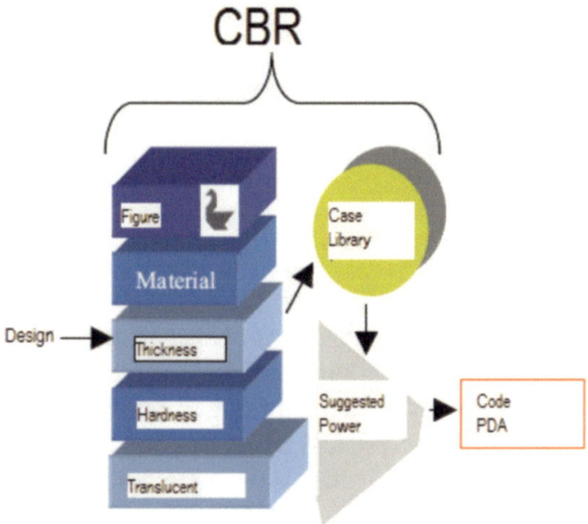

Fig. 28. Blocks under which the developed tool works.

This process begins when choosing the kind of the design to realize from:
1. Predetermined design or
2. Design at free hands,

If the first option is chosen, a window is opened, where the user has a series of designs previously made, after the selection of one of them is realized, the next step is to introduce data of the material to use. A new window with a list of materials is displayed; if the desired material is not previously registered previously in the list, the program provide a series of fields for fulfill its corresponding information. With this information the system realizes tests if it is desired, to acquire a suggested power for engraving. This last data is obtained with all the previously acquired data and processed by the methodology of CBR (Morales et al., 2005).

In this way, the program accounts with all necessary information to complete the process and send the information through the series port to the control module of the actuators. The procedure concludes with the laser engraving (figure 29).

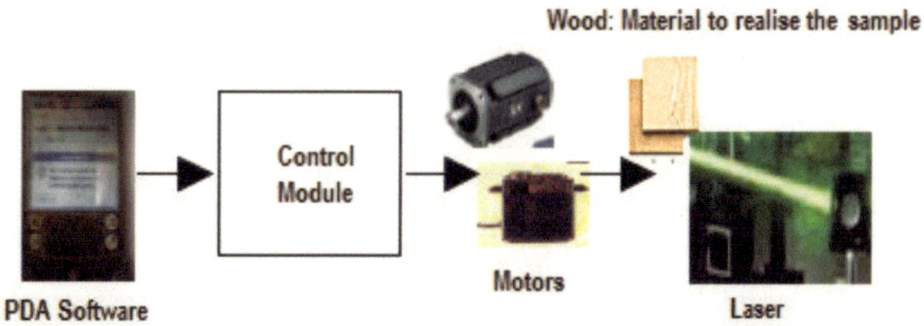

Fig. 29. Conceptual diagram, showing the Process of control by means of PDA.

The computer tool was development to realize different tests of engraving, as in the case of the system shown in section 2. About the hardware, the motors are moved bi-directionally, they are controlled by a PALM and have the capability to realize different types of trajectories to engrave surfaces. Several tests were realized with texts engraving. The use of recyclable material permits the development of the system with minimum costs.

The system is in the stage of finals tests and runs in a PALM 100m (Wilding-McBride, 2004), the electronic module already is in operation; and the bookstore of cases accounts with information of different materials. It is tried that this system can be used from any place where the user is located by means of Internet (WWW). In addition, the possibility to storage more parameters of the materials provides more precision in their characterization.

5. Future trends

The technique called Electrochemical Machining (ECM) is an anodic dissolution process. It utilizes an electrolytic cell formed by a cathode tool and an anode workpiece with a suitable electrolyte flowing between them. The anode workpiece is dissolved according to Faraday's law when a sufficient voltage is applied across the gap between the anode and the cathode in which electrolyte is filled. Electrochemical processes for drilling small and fine holes by controlled anodic dissolution invariably use a weak acidic solution as electrolyte (Shan, 2004). These include electrochemical drilling (ECD) and acid based ECM drilling processes: shaped tube electrolytic machining (STEM), capillary drilling (CD), electro-stream drilling (ESD), and jet electrolytic drilling (JED). The advantages of acid based electrochemical hole drilling processes are:

- Good surface finish;
- Absence of residual stress;
- No tool wear;
- No burr and no distortion of the holes;
- Simultaneous drilling of large number of holes.

The use of acid electrolytes in ECM hole drilling processes facilitate dissolution of metals and the removed material is carried away as metal ions thus making it possible to achieve smooth finish with closer tolerances and deep holes of high aspect ratio (Bellows and Kohls, 1982).
Another Laser (continuous or pulsed) technique named Laser Beam Machining (LBM) is one of the most used techniques, based on thermal energy, of type non-contact. This process can be applied for almost whole range of materials. The lasers used for machining in the industries are CO_2 and Nd:YAG. This two lasers are the most stable. In this kind of work, the use of power and focusing are everything needed to monitor.
The mechanism of material removal during LBM includes different stages such as

a. Melting,
b. Vaporization, and
c. Chemical degradation (chemical bonds are broken which causes the materials to degrade).

If a high energy density laser bream is focused on the work surface the thermal energy is absorbed, which heats and transforms the work volume into a molten, vaporized or chemically changed state that can easily be removed by flow of high pressure assist gas jet (which accelerates the transformed material and ejects it from machining zone) (Hirao et al., 2001). The schematic of LBM is shown in figure 30.
The interesting part is that each material possess a different response to the light. Some interesting materials have been studies as drilling of gamma-titanium aluminide (Biswas et al, 2009). Among the most important materials in optoelectronic devices, the Glass is found, with a high transmission from the UV to IR wavelength region, excellent thermal and electrical properties, and high chemical resistivity (Weber, 2003; Hirao et al., 2001). Furthermore, the glass properties are controllable by adjusting the composition during

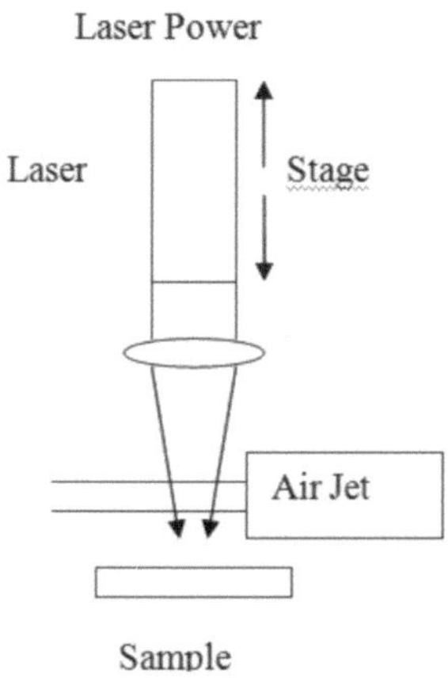

Fig. 30. Basic setup of LBM

development and fabrication. However, these properties make the glass a challenging material to machine (Herman, 2000). Due to poor thermal properties, fabrication of finely machined features using laser-based processes e.g. grooves, channels, microholes, stand-alone levers, etc., in glass materials has been quite a difficult task.

Laser technologies based on nanosecond (ns), femtosecond (fs) and laser-induced plasma processing were investigated in order to obtain high quality laser micro-fabricated features on glass materials. Direct write laser using short pulses from ns and fs lasers influences the quality of processed glass significantly and produces crack-free, clean machining with careful control of the associated thermal processes. Laser-induced plasma machining technique allows fabrication of small-size, shallow features along with superfine surface finishes within the channels. The potential of these technologies have benefits in the fabrication of complex features for biomedical, microfluidic, MEMS and optoelectronic devices.

6. Conclusions

This work confirms the relevance of the automation in optics applications. For the realization of all the prototype, it was required of a multidisciplinary team, involving basically programming, electronics and optics knowledge.

7. References

Aamodt, A. & Plaza, E. "Case-Based Reasoning: Foundational Issues, Methodological Variations, and System Approaches. AI Communications, 7(i): pp 39-59.

Bellows G., Kohls J.B., Drilling without drills, American Machinist, Special Report 743 (1982) 173–188.

Biswas R., Kuar A.S., Sarkar S., Mitra S. "A parametric study of pulsed Nd:YAG lases micro-drilling of gamma-titanium aluminide". Optics & Laser Technology, In Press, Corrected Proof, Available online 20 May 2009.

Chan, Patrick "The Java Developers ALMABAC 1.4, volume 1", Addison Wesley 2002.

Easttom, Check "JBuilder 8.0 JFC and Swing", Wordware Publishing; 2003.

García, Luis A.; Basurto-Pensado, Miguel & Ochoa, Alberto. "Herramienta de Cómputo para Grabado con Láser", Universidad Autónoma de Zacatecas, ENINVIE 2005.
Gonzalez-Roman A., Tecpoyotl-Torres M., Escobedo-Alatorre J., Pal-Verma S. and Sánchez-Mondragón J. "A semi-spherical Irradiance meter used as a quality control device". Proceedings of the First Multiconference on Electronics and Photonics. MEP 2006. Pp. 253- 256 (2006).
Gwirc, S.; Rigotti, J.; Federico, A.;Acquaticci, F. 6o. Jornada de desarrollo e innovación tecnológica. Imágenes Ultrasónicas con Transductor Piezoeléctrico de Película. Instituto Nacional de Tecnología Industrial, (2007).
Herman P.R., Marjoribanks R.S., Oettl A., Chen K., Konovalov I., Ness S., Appl. Surf. Sci. 154/155 (2000) 577.
Hirao K., Mitsuyu T., Si J., Qiu J. (Eds.), Active Glass for Photonic Devices: Photoinduced Structures and Their Application, Springer-Verlag, 2001.
James C. Maida. "An illumination modeling system for human factor analyses". Space human factors laboratory/Flight crew support division/NASA Johnson Space Center.
Jaimes-Vera Edith Alíne, Basurto– Pensado M.A., Escobedo-Alatorre J. Jesus. Diseño y programación de una mesa para mili-maquinado., ENINVIE 2005. Encuentro de Investigación en Ingeniería Eléctrica, 2005.
Lake, David B., "Case-Based Reasoning- Experiences, Lessons, & Future Directions", Edit. The MIT Press 1996.
Morales, J.; Basurto, M. & Ochoa, A. "Herramienta de Cómputo para Grabado mediante PDA", Zacatecas, Zacatecas 2005.
Ochoa A. et al. "Proceedings of Doctoral Forum, PRICAI 2004, Auckland University of Technology, August 2004.
Palais, Joseph C., Fiber Optic Communications, Fourth Edition, Prentice Hall, (1984).
Paul Horowitz, Winfield Hill. The Art of Electronics. 2nd Edition. Cambridge University Press. Pp. 996. 2001.
Sánchez-Mondragón, J., Tecpoyotl-Torres M., Andrade-Lucio J. A., Torres-Cisneros M., Dávila-Alvarez A. and Carpio-Valadez M.. "Data fitting on a spherical shell". Proceedings of SPIE proceeding Vol. 5181. Pp.51-55 (2003).
Schildt, H. "Java 2: The Complete Reference, Fourth Edition", Edit. McGraw Hill 2001.
Shan H.S., Advanced Manufacturing Methods, New Delhi, 2004.
Tajadura, J.A. et Al. "Autocad 2000 Avanzado", Edit. Mc Graw Hill, 1999.
Tecpoyotl-Torres M, Partida-Rivera E., Gonzalez-Roman I. A., Ibarra-Manzano O. and Sánchez- Mondragón J.. "Reconstruction of atmospheric vertical reflectivity profile images". Proceedings of the First Multiconference on Electronics and Photonics. MEP 2006. Pp. 262-265. IEEE CN 06 Ex1524. (2006).
Trimmer, William "Micromechanics and MEMS" Wiley & IEEE book, ISBN 0-7803-1085-3, 2005.
Yañez Valdez, R. *, M. Ruiz Torres, E. Morales Sánchez, E. Castillo Castañeda. Diseño y Construcción de una Mesa de Trabajo XYθ basada en un Mecanismo Paralelo Planar 3RRR. Tecnólog@ Vol. 1, No. 2, may- ago/2008 © 2007 CICATA-Querétaro-IPN. ISSN en trámite. México.
Wainer, J.; Borgonovi, Luana & Ochoa, A. "Ornithological Classification using case-based reasoning for discovered new species", UNICAMP Postdoctoral program; Radamaelli, Brazil. January 2005.
Weber M.J. (Ed.), Handbook of Optical Materials, The CRC Press, Boca Raton, Fl. 2003.
Westinhouse. Manual del alumbrado. 4ª edición. Limusa. Noriega Editores. Madrid (2000).
Wilding-McBride, Darly. Java Development on PDAs, Addison-Wesley 2001.

11

Theory of Unitary Spin Rotation and Spin State Tomography for a Single Electron and Two Electrons

T. Takagahara
Department of Electronics and Information Science,
Kyoto Institute of Technology, Matsugasaki, Kyoto 606-8585
CREST, Japan Science and Technology Agency,
4-1-8 Honcho, Kawaguchi, Saitama 332-0012,
Japan

1. Introduction

Coherent control of quantum states is a critical step toward many novel technological applications ranging from manipulation of quantum bits (qubits) in quantum logic gates to controlling the spin degrees of freedom of electrons [1–13]. A qubit with a longer coherence time is desirable for the application to the quantum information processing. Electron spins in semiconductor nanostructures are considered as one of the most promising candidates of the building blocks for quantum information processing due to their robustness against decoherence effects [14–18]. A quantum media converter between a photon qubit and an electron spin qubit was proposed for the use in quantum repeaters [19–22]. Quantum information can take several different forms and it is preferable to be able to convert among different forms. One form is the photon polarization and another is the electron spin polarization. Photons are the most convenient medium for sharing quantum information between distant locations. Electrons are the most efficient medium for manipulating the quantum information by electrical and optical means. The fundamental operations are the initialization, unitary rotation and measurement of a qubit. The initialization of a single electron spin was demonstrated by the efficient optical method [23–25]. Also, the coherent rotation of a single electron spin has been realized by the microwave ESR (Electron Spin Resonance) method [26] and by the optical STIRAP (STImulated Raman Adiabatic Passage) method with coherence times up to several μs [27–33] in III-V semiconductor nanostructures and up to several tens of ms in the localized electron systems in IV elemental semiconductors [34–36]. The optical STIRAP method is advantageous because of its ultrafast operation. However, the precise control of the spin rotation without leaving behind any population in the intermediate excited states has not yet been realized. It is also important to achieve the unitary spin rotation of two electrons, because the singlet-triplet subspace of two electrons was utilized as a qubit space and the electrical manipulation of the qubit was realized [16]. At the same time, it is absolutely necessary to confirm the quantum state of the electron after the spin state manipulation or the quantum state transfer from a photon, namely, to examine whether the electron spin is prepared in the desired state or not. This requires the full state tomography, namely the measurement of the density matrix of the electron. This state tomography is also important to estimate the fidelity of relevant quantum operations.

Thus it is a challenging task to establish the precise spin rotation and the spin state tomography for both cases of a single electron and two electrons. We review the general aspects of the unitary spin rotation of a single electron by the STIRAP method and develop the scheme to rotate the pseudo-spin formed by the singlet state and the triplet states of two electrons based on the optical STIRAP process, discussing the optimal conditions for the precise control. Also we propose and analyze optical methods to achieve the electron spin state tomography based on the Faraday/Kerr rotation, referring to the recent experiments [37, 38].

2. Optical STIRAP method for spin rotation of a single electron

As mentioned in the Introduction, the spin rotation of a single electron is a crutial ingredient in the quantum information processing. It is desirable to accomplish the spin rotation along an arbitrary direction for an arbitrary rotation angle in a single-shot process. So far, the spin rotation of a single electron was demonstrated by optical and electrical means in the proofof- principle experiments. With respect to the required time for the spin rotation, the optical method based on the STIRAP (stimulated Raman adiabatic passage) process is preferable because of its ultrafast response. But the precise control of the spin rotation is yet to be pursued. Here several characteristics of this STIRAP process will be investigated.

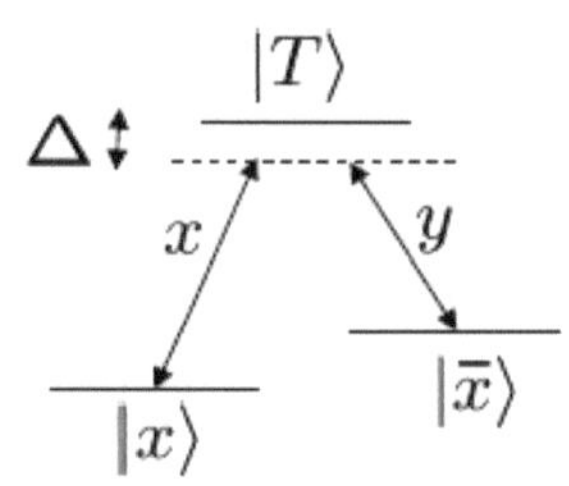

Fig. 1. Schematic energy level structure for the STIRAP process. Allowed optical transitions are depicted by x and y, which represent the mutually orthogonal polarizations. Δ denotes the off-resonance energy of the excitation lights relative to the transition energies. The Raman condition for the excitation lights is imposed.

In order to carry out the STIRAP process, a Λ-type transtion is necessary, as depicted in Fig. 1. The lowest two levels denoted by $|x\rangle$ and $|\bar{x}\rangle$ are the ground doublet states with close energies, e.g., the spin up and spin down states of a single electron or the ground and excited vibrational states of a single molecule. A pseudospin is composed of these doublet states and can be rotated by optical transitions via the intermediate excited state denoted by $|T\rangle$. The important point is that the selection rules of the left and right optical transitions are orthogonal to each other, which are depicted typically as x and y in Fig. 1. The doublet states are not directly connected optically. Then the relevant Hamiltonian is written as

$$H = H_0 + V \, , \tag{1}$$

$$H_0 = E_x|x\rangle\langle x| + E_{\bar{x}}|\bar{x}\rangle\langle\bar{x}| + E_T|T\rangle\langle T| \,, \tag{2}$$

$$\begin{aligned} V = &-\hbar\Omega_x(t)[e^{-i\omega_x t}|T\rangle\langle x| + e^{i\omega_x t}|x\rangle\langle T|] \\ &-\hbar\Omega_y(t)[e^{-i\omega_y t - i\delta}|T\rangle\langle\bar{x}| + e^{i\omega_y t + i\delta}|\bar{x}\rangle\langle T|] \,, \end{aligned} \tag{3}$$

where H_0 represents the unperturbed part, V the optical transitions, Ω_x and Ω_y the Rabi frequencies, δ the relative phase shift of the y-polarized light and the energy E_x is put as $E_x = 0$ for the origin of energy. Then the time evolution proceeds as follows:

$$i\hbar\dot{\psi} = H\psi \,, \ \psi = c_x|x\rangle + c_{\bar{x}}|\bar{x}\rangle + c_T|T\rangle \,, \tag{4}$$

$$\frac{d}{dt}\begin{pmatrix} c_x \\ c_{\bar{x}} \\ c_T \end{pmatrix} = \begin{pmatrix} i\Omega_x(t)e^{i\omega_x t}c_T \\ -i\frac{E_{\bar{x}}}{\hbar}c_{\bar{x}} + i\Omega_y(t)e^{i\omega_y t + i\delta}c_T \\ -i\frac{E_T}{\hbar}c_T + i\Omega_x(t)e^{-i\omega_x t}c_x + i\Omega_y(t)e^{-i\omega_y t - i\delta}c_{\bar{x}} \end{pmatrix} . \tag{5}$$

In order to single out the rapidly oscillating part, we put as

$$c_{\bar{x}}(t) = e^{-i\frac{E_{\bar{x}}}{\hbar}t}\tilde{c}_{\bar{x}}(t) \,, \ c_T(t) = e^{-i\frac{E_T}{\hbar}t}\tilde{c}_T(t) \,, \tag{6}$$

obtaining

$$\frac{d}{dt}\begin{pmatrix} c_x \\ \tilde{c}_{\bar{x}} \\ \tilde{c}_T \end{pmatrix} = \begin{pmatrix} i\Omega_x(t)e^{i(\omega_x - \frac{E_T}{\hbar})t}\tilde{c}_T \\ i\Omega_y(t)e^{i(\omega_y - \frac{E_T}{\hbar} + \frac{E_{\bar{x}}}{\hbar})t + i\delta}\tilde{c}_T \\ i\Omega_x(t)e^{-i(\omega_x - \frac{E_T}{\hbar})t}c_x + i\Omega_y(t)e^{-i(\omega_y - \frac{E_T}{\hbar} + \frac{E_{\bar{x}}}{\hbar})t - i\delta}\tilde{c}_{\bar{x}} \end{pmatrix} . \tag{7}$$

Now we postulate the Raman condition for the x and y polarized lights:

$$\frac{E_T}{\hbar} - \omega_x = \frac{E_T}{\hbar} - \frac{E_{\bar{x}}}{\hbar} - \omega_y = \Delta \tag{8}$$

and also assume the same pulse shape for the x- and y-polatized lights with arbitrary relative intensity ratio determined by θ:

$$\Omega_x(t) = \Omega_0(t)\cos\theta \,, \ \Omega_y(t) = \Omega_0(t)\sin\theta \,. \tag{9}$$

Introducing the bright and dark state amplitudes defined by

$$\begin{pmatrix} c_B \\ c_D \end{pmatrix} = \begin{pmatrix} \cos\theta & e^{-i\delta}\sin\theta \\ -e^{i\delta}\sin\theta & \cos\theta \end{pmatrix}\begin{pmatrix} c_x \\ \tilde{c}_{\bar{x}} \end{pmatrix} = U\begin{pmatrix} c_x \\ \tilde{c}_{\bar{x}} \end{pmatrix} , \tag{10}$$

we can simplify the equations of motion as

$$\frac{d}{dt}\begin{pmatrix} c_B \\ \tilde{c}_T \end{pmatrix} = \begin{pmatrix} i\Omega_0(t)e^{-i\Delta t}\,\tilde{c}_T \\ i\Omega_0(t)e^{i\Delta t}\,c_B \end{pmatrix} , \ \frac{d}{dt}c_D = 0 \,. \tag{11}$$

Thus the dark state does not change. The amplitudes of the bright state and the $|T\rangle$ state satisfy the following equations:

$$\frac{d^2}{dt^2}c_B = \left(\frac{\dot{\Omega}_0(t)}{\Omega_0(t)} - i\Delta\right)\frac{d}{dt}c_B - \Omega_0^2(t)c_B\ , \tag{12}$$

$$\frac{d^2}{dt^2}\tilde{c}_T = \left(\frac{\dot{\Omega}_0(t)}{\Omega_0(t)} + i\Delta\right)\frac{d}{dt}\tilde{c}_T - \Omega_0^2(t)\tilde{c}_T\ . \tag{13}$$

To develop the analytical solutions of these equations [39, 40], we assume a sech pulse envelope:

$$\Omega_0(t) = \Omega\ \mathrm{sech}(\sigma t)\ . \tag{14}$$

Introducing a dimensionless time variable by

$$\zeta = \frac{1}{2}(1 + \tanh(\sigma t))\ , \tag{15}$$

we have

$$\zeta(1-\zeta)\frac{d^2}{d\zeta^2}c_B + \left(\frac{1}{2}(1 + i\frac{\Delta}{\sigma}) - \zeta\right)\frac{d}{d\zeta}c_B + \frac{\Omega^2}{\sigma^2}c_B = 0\ , \tag{16}$$

$$\zeta(1-\zeta)\frac{d^2}{d\zeta^2}\tilde{c}_T + \left(\frac{1}{2}(1 - i\frac{\Delta}{\sigma}) - \zeta\right)\frac{d}{d\zeta}\tilde{c}_T + \frac{\Omega^2}{\sigma^2}\tilde{c}_T = 0\ . \tag{17}$$

This is a hypergeometric differential equation. General solutions are given by

$$\begin{pmatrix} c_B(t) \\ \tilde{c}_T(t) \end{pmatrix} = \begin{pmatrix} F(\alpha, -\alpha, \gamma|\zeta) & e^{i\Delta t}\ \frac{i\alpha}{\gamma^*}\ \zeta^{\gamma^*} F(\alpha+\gamma^*, -\alpha+\gamma^*, 1+\gamma^*|\zeta) \\ e^{-i\Delta t}\ \frac{i\alpha}{\gamma}\ \zeta^{\gamma} F(\alpha+\gamma, -\alpha+\gamma, 1+\gamma|\zeta) & F(\alpha, -\alpha, \gamma^*|\zeta) \end{pmatrix} \times \begin{pmatrix} c_B(-\infty) \\ \tilde{c}_T(-\infty) \end{pmatrix} \tag{18}$$

$$\text{with}\quad \alpha = \frac{\Omega}{\sigma}\ ,\ \beta = -\frac{\Omega}{\sigma}\ ,\ \gamma = \frac{1}{2}(1 + i\frac{\Delta}{\sigma})\ , \tag{19}$$

where $F(\alpha, \beta, \gamma|\zeta)$ is the hypergeometric function. In the rotation of the pseudospin, we start with the initial state in which

$$c_B(-\infty) = \text{finite}\ ,\ c_D(-\infty) = \text{finite}\ ,\ \ \tilde{c}_T(-\infty) = 0 \tag{20}$$

and after the pulse we prefer to have

$$\tilde{c}_T(\infty) = 0 \tag{21}$$

in order to leave no excitation in the intermediate excited state $|T\rangle$. To satisfy this condition, we should have

$$\frac{\alpha}{\gamma} F(\alpha+\gamma, -\alpha+\gamma, 1+\gamma|1) = 0 \tag{22}$$

because the asymptotic behavior ($t\rightarrow\infty$) is given by putting as $\zeta = 1$. Using the formula [41]

$$F(a,b,c|1) = \frac{\Gamma(c)\Gamma(c-a-b)}{\Gamma(c-a)\Gamma(c-b)} \tag{23}$$

which is valid under the condition that $\mathrm{Re}(c - a - b) > 0$ and $c \neq 0, -1, -2, \ldots$, we have

$$\frac{\alpha}{\gamma} F(\alpha+\gamma, -\alpha+\gamma, 1+\gamma|1) = \frac{\alpha}{\gamma}\frac{\Gamma(1+\gamma)\Gamma(1-\gamma)}{\Gamma(1-\alpha)\Gamma(1+\alpha)} \, . \tag{24}$$

Further, using the formulas

$$x\Gamma(x) = \Gamma(x+1) \, , \quad \Gamma(x)\Gamma(1-x) = \frac{\pi}{\sin \pi x} \, , \tag{25}$$

we obtain

$$\frac{\alpha}{\gamma}\frac{\Gamma(1+\gamma)\Gamma(1-\gamma)}{\Gamma(1-\alpha)\Gamma(1+\alpha)} = \frac{\alpha}{\gamma}\frac{\gamma\Gamma(\gamma)\Gamma(1-\gamma)}{\Gamma(1-\alpha)\alpha\Gamma(\alpha)} = \frac{\sin \pi\alpha}{\sin \pi\gamma} \, . \tag{26}$$

Putting in the expression of γ, we finally have

$$\frac{\alpha}{\gamma} F(\alpha+\gamma, -\alpha+\gamma, 1+\gamma|1) = \mathrm{sech}\left(\frac{\pi\Delta}{2\sigma}\right) \sin \pi\alpha \, . \tag{27}$$

This quantity vanishes only when α = 1, 2, … . This condition is nothing but the condition that the pulse area is 2π, 4π, … . This is quite reasonable because the Bloch vector rotates and returns to the initial state for the pulse area of integer times 2π.

Furthermore, under this condition, the amplitude of the bright state receives after a pulse an additional factor given by

$$F(\alpha, -\alpha, \gamma|1) = -\frac{\gamma^*}{\gamma} \, , \; \frac{\gamma^*(\gamma^*+1)}{\gamma(\gamma+1)} \, , \; -\frac{\gamma^*(\gamma^*+1)(\gamma^*+2)}{\gamma(\gamma+1)(\gamma+2)} \cdots \, , \tag{28}$$

where the expressions on the right hand side correspond to α = 1, 2, 3, … , respectively and their absolute magnitude is obviously unity and thus they can be put as

$$F(\alpha, -\alpha, \gamma|1) = e^{-i\phi} \, . \tag{29}$$

This phase ϕ determines the rotation angle of the pseudospin, as will be shown shortly. Now the effect of the pulse can be summarized as

$$\begin{pmatrix} c_B(\infty) \\ c_D(\infty) \end{pmatrix} = \begin{pmatrix} e^{-i\phi} & 0 \\ 0 & 1 \end{pmatrix} \begin{pmatrix} c_B(-\infty) \\ c_D(-\infty) \end{pmatrix} . \tag{30}$$

This relation can be rewritten in terms of $(c_x, \tilde{c}_{\bar{x}})$ amplitudes:

$$\begin{pmatrix} c_x(\infty) \\ \tilde{c}_{\bar{x}}(\infty) \end{pmatrix} = U^\dagger \begin{pmatrix} e^{-i\phi} & 0 \\ 0 & 1 \end{pmatrix} U \begin{pmatrix} c_x(-\infty) \\ \tilde{c}_{\bar{x}}(-\infty) \end{pmatrix} , \tag{31}$$

where the operation of the pulse is calculated as

$$\begin{aligned} U^\dagger \begin{pmatrix} e^{-i\phi} & 0 \\ 0 & 1 \end{pmatrix} U &= e^{-i\phi/2} \left[\cos\frac{\phi}{2} \overset{\leftrightarrow}{1} - i \sin\frac{\phi}{2} \left(\vec{n}\cdot\vec{\sigma}\right)\right] = e^{-i\phi/2} \exp\left[-i\frac{\phi}{2}\,\vec{n}\cdot\vec{\sigma}\right] \\ &= e^{-i\phi/2} \exp\left[-i\phi\,\vec{n}\cdot\vec{S}\right] \end{aligned} \tag{32}$$

$$\text{with} \quad \vec{n} = (\sin 2\theta \cos\delta, \sin 2\theta \sin\delta, \cos 2\theta) , \quad \tan\theta = \frac{\Omega_y}{\Omega_x} . \tag{33}$$

This relation indicates that the pseudospin vector composed of $|x\rangle$ and $|\bar{x}\rangle$ states is rotated by an angle ϕ around the direction vector $\vec{n}$. The rotation angle ϕ can be tuned by the off-resonance energy Δ in Eq. (8), whereas the direction vector $\vec{n}$ can be adjusted by the intensity ratio and the relative phase shift between the orthogonally polarized lights with the same temporal envelope.

In order to estimate the fidelity of this spin rotation, we prepare an arbitrary initial state, follow the time evolution to obtain the asymptotic state, calculate the overlap with the ideal state and average over the initial states. In order to take into account relaxation processes, we consider the density matrix for the system composed of three states. We prepare an initial state:

$$\rho(-\infty) = |\psi(-\infty)\rangle\langle\psi(-\infty)| , \tag{34}$$

$$\psi(-\infty) = c_x(-\infty)|x\rangle + c_{\bar{x}}(-\infty)|\bar{x}\rangle \tag{35}$$

$$\text{with} \quad \begin{pmatrix} c_x(-\infty) \\ c_{\bar{x}}(-\infty) \end{pmatrix} = \begin{pmatrix} \cos\theta_i/2\; e^{-i\varphi_i/2} \\ \sin\theta_i/2\; e^{i\varphi_i/2} \end{pmatrix} , \tag{36}$$

where θ_i and φ_i indicate the initial direction of the pseudospin. The time evolution of the density matrix is given by

$$\dot{\rho}(t) = -\frac{i}{\hbar}[H_0 + V, \rho] + \Gamma\rho , \tag{37}$$

where ¡ includes the population relaxation and decoherence terms. After the time evolution we have the asymptotic state $\pi(\infty)$, which is actually $\pi(T_f)$ for a large enough time T_f, and calculate the fidelity defined by the overlap of the actual density matrix with the ideal density matrix which is obtained without any relaxation terms:

$$F = \langle \mathrm{Tr}_{|x\rangle,|\bar{x}\rangle}\, \rho_{\mathrm{ideal}}(\infty)\rho(\infty)\rangle_{c_x,c_{\bar{x}}} , \tag{38}$$

where the angular bracket means the average over the initial spin direction, namely:

$$\langle \cdots \rangle_{c_x, c_{\bar{x}}} = \frac{1}{4\pi} \int_0^{\pi} d\theta_i \sin\theta_i \int_0^{2\pi} d\varphi_i \ \cdots \ . \tag{39}$$

Some numerical results will be presented for the fidelity and the residual population in the excited state $|T\rangle$. Because of the energy-time duality the following results can be applied for an arbitrary pulse width by scaling appropriately the off-resonance energy. But, for the definiteness, the optical pulse is assumed as sech (t/t_p) with t_p = 5ps and the time evolution is integrated over the time range of $-6t_p \le t \le 6t_p$. The relaxation parameters are chosen as

$$\hbar\Gamma_{T\to x} = \hbar\Gamma_{T\to\bar{x}} = 0.01\text{meV} \ , \ \hbar\gamma_{T\,x} = \hbar\gamma_{T\,\bar{x}} = 0.05\text{meV} \ , \tag{40}$$

where $\Gamma(\gamma)$ indicates the population relaxation (decoherence) rate. The equations of motion for the density matrix elements are similar to those given from Eq. (58) to Eq. (64) in the later section. First of all, the rotation angle ϕ is plotted in Fig. 2 as a function of the off-resonance energy. The normalized off-resonance energy is defined by Δt_p, where Δ is given in Eq. (8), and is dimensionless. For the 2π pulse the rotation angle is monotonically increasing with increasing off-resonance. The fidelity of the spin rotation is exhibited in Fig. 3. The fidelity is improved with increasing off-resonance in general for the 2π and 4π pulses. For the 6π pulse, a strange behavior is seen. But it can be understood that a fidelity peak appears around the off-resonance energy where the rotation angle is almost 360 degrees, namely, the spin returns to the initial state and the deviation from the ideal time evolution is suppressed. Another important quantity is the residual population in the excited state $|T\rangle$ and is exhibited in Fig. 4. This is monotonically decreasing with increasing off-resonance.

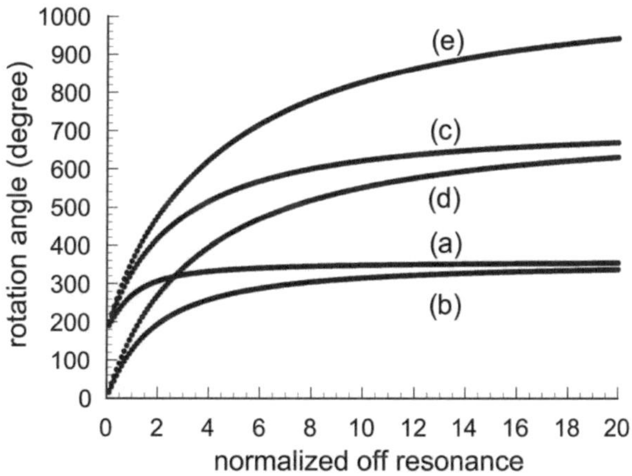

Fig. 2. Angles of the spin rotation are plotted as a function of the normalized off-resonance energy Δt_p for the pulse areas of (a) 2π, (b) 4π, (c) 6π, (d) 8π, and (e) 10π.

Analytically exact solutions are possible only for the sech pulses. In order to see the effect of the pulse shape, a Gaussian pulse is examined for the case of 2π pulse area. Results are exhibited in Figs. 5 and 6 and show that the sech pulse is better for the higher fidelity and the smaller population left in the excited state after the pulse.

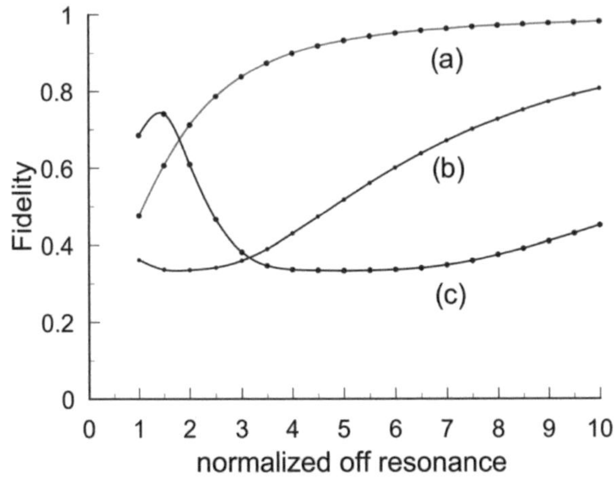

Fig. 3. Fidelity of the spin rotation of a single electron is plotted as a function of the normalized off-resonance. Curves (a), (b) and (c) correspond to the pulse area 2π, 4π and 6π, respectively.

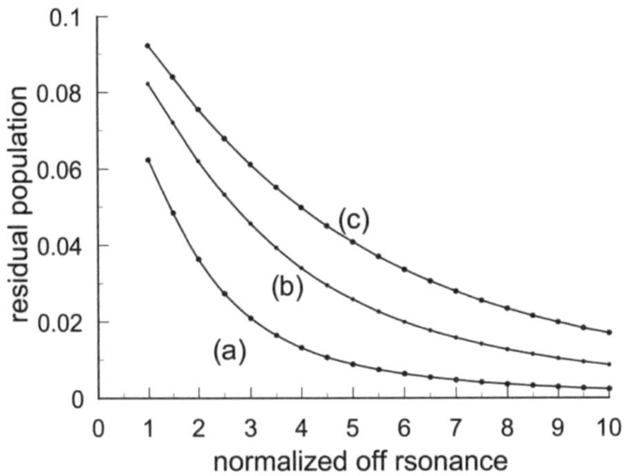

Fig. 4. Residual population in the excited state $|T\rangle$ after the spin rotation of a single electron is plotted as a function of the normalized off-resonance. Curves (a), (b) and (c) correspond to the pulse area 2π, 4π and 6π, respectively.

So far we have considered a typical Λ-type system composed of three energy levels. However, in the case of a singly charged semiconductor quantum dot, there are at least two excited states, namely the trion states, associated with two spin directions of the hole state. Thus, the four level system, as depicted in Fig. 7, is more appropriate. The fidelity of the spin rotation for the four level system is examined using the parameters:

$$\hbar\omega_T - \hbar\omega_{\bar{T}} = 0.05\mathrm{meV}\ , \hbar\omega_{\bar{x}} - \hbar\omega_x = 0.05\mathrm{meV}\ , \hbar\omega_x - \hbar\omega_y = 0.05\mathrm{meV}\ , \tag{41}$$

$$\hbar\Gamma_{T\to x} = \hbar\Gamma_{T\to\bar{x}} = \hbar\Gamma_{\bar{T}\to x} = \hbar\Gamma_{\bar{T}\to\bar{x}} = 0.01\mathrm{meV}\ , \tag{42}$$

$$\hbar\gamma_{T\,x} = \hbar\gamma_{T\,\bar{x}} = \hbar\gamma_{\bar{T}\,x} = \hbar\gamma_{\bar{T}\,\bar{x}} = 0.05\mathrm{meV}\ , \tag{43}$$

where the Raman condition is applied to the left Λ-type transition. Results are given in Fig. 8 and show that the fidelity is not degraded by an additional Λ-type transition, especially for the 2π pulse area. Thus the spin rotation is expected to be robust against the overlap of several Λ-type transitions.

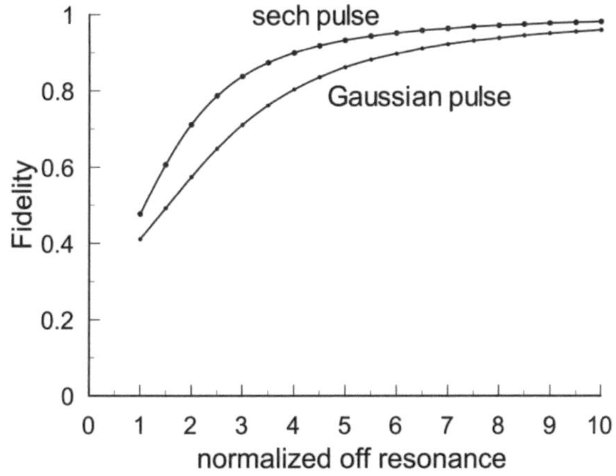

Fig. 5. Fidelity of the spin rotation of a single electron is compared between the cases of a Gaussian pulse and a sech pulse for the 2π pulse area.

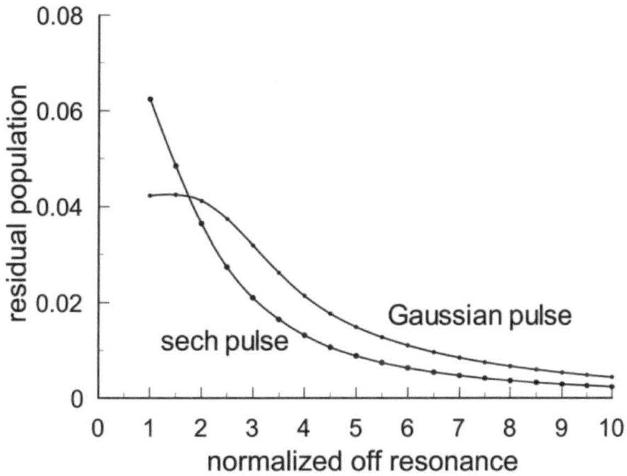

Fig. 6. Residual population in the excited state $|T\rangle$ after the spin rotation of a single electron is compared between the cases of a Gaussian pulse and a sech pulse for the 2π pulse area.

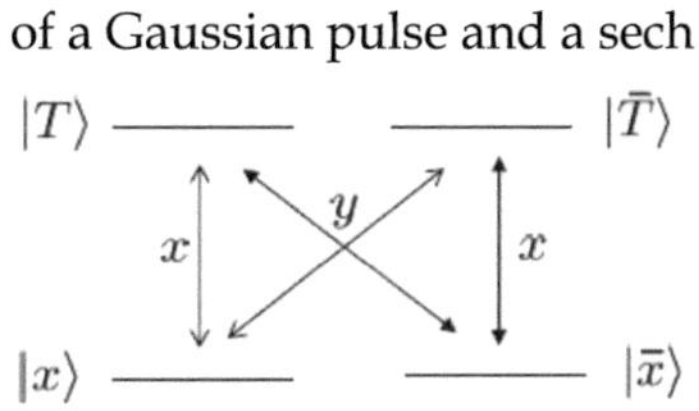

Fig. 7. Four level system composed of two electron spin states (lower levels) and two trion states with different hole spin states (upper levels). Allowed optical transitions are indicated by the x and y polarizations.

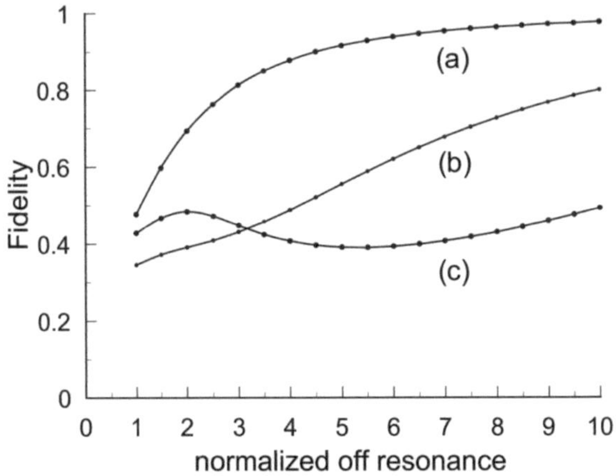

Fig. 8. Fidelity of the spin rotation of a single electron is plotted as a function of the normalized off-resonance in the four-level model. Curves (a), (b) and (c) correspond to the pulse area 2π, 4π and 6π, respectively.

3. Optical STIRAP method for spin rotation of two electrons

Now we extend the above arguments to the spin rotation of two electrons. This spin rotation is important because a qubit composed of the singlet state and one of the triplet states of two electrons confined in a semiconductor quantum dot was established and its electrical manipulation was demonstrated16. Here we examine the possibility of ultrafast spin otation of two electrons by an optical means. As discussed above, the essential ingredient is the Λ-type transition with mutually orthogonal optical selection rules which enables the spin rotation of an arbitrary angle along an arbitrary direction. In the Faraday configuration the allowed optical transitions are exhibited in Fig. 9. The charged exciton state is depicted by X^{2-}. An additional superscript indicates the spin direction of the electron in the excited orbital state and an additional subscript represents the spin direction of the heavy hole in the lowest energy orbital state, namely,

$$|hh+\rangle = |\frac{3}{2}\frac{3}{2}\rangle \ , \ |hh-\rangle = |\frac{3}{2} \ -\frac{3}{2}\rangle \ , \tag{44}$$

where the left hand side indicates the missing state of the valence band electron in the state on the right hand side. There is a Λ-type transition but with the same optical selection rules. Thus the arbitrary spin rotation is not possible.

On the other hand, for the Voigt configuration in which a magnetic field is applied along the quantum dot plane (taken as the x axis), the optical selection rules are exhibited in Fig. 10 for the case associated with the light hole state. Here, an additional superscript attached to X^{2-} indicates the spin direction of the electron in the excited orbital state, namely, +(-) for the $x(-x)$ direction and an additional subscript represents the spin direction of the light hole in the lowest energy orbital state, namely, $\ell h+$ or $\ell h-$ corresponding to

$$|lh+\rangle = \frac{1}{\sqrt{2}}(|\frac{3}{2}\frac{1}{2}\rangle + |\frac{3}{2} - \frac{1}{2}\rangle) \text{ or } |lh-\rangle = \frac{1}{\sqrt{2}}(-|\frac{3}{2}\frac{1}{2}\rangle + |\frac{3}{2} - \frac{1}{2}\rangle) \ , \tag{45}$$

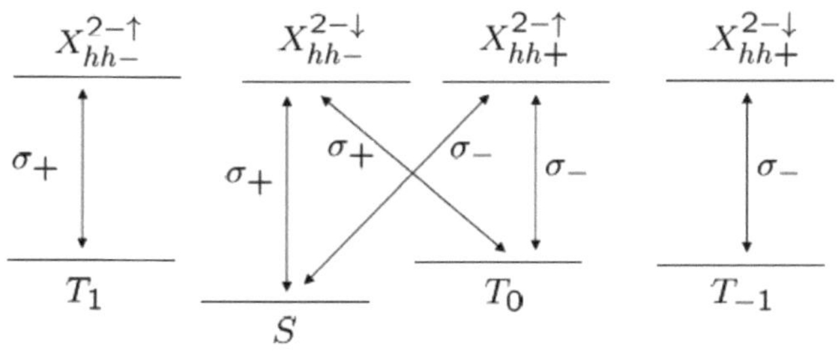

Fig. 9. Allowed optical transitions in the Faraday configuration for two electrons. The lower levels represent the four spin states of two electrons: the singlet (S) and three triplet (T_1, T_0, T_{-1}) states, whereas the upper levels exhibit the negatively doubly charged exciton states (X^{2-}) with indexes indicating the spin state of the electron in the excited orbital and the spin state of the heavy hole.

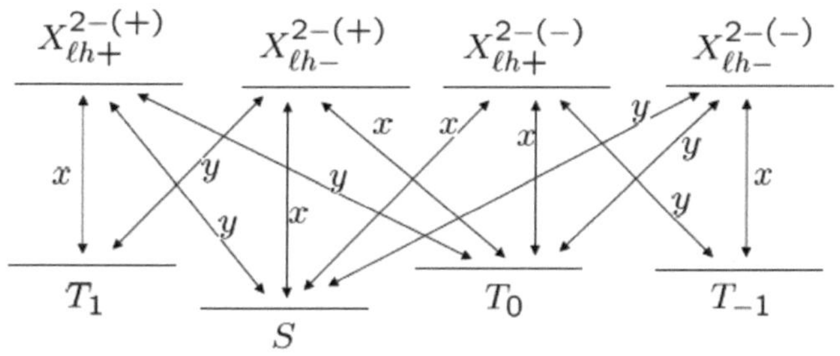

Fig. 10. Allowed optical transitions in the Voigt configuration for two electrons. The lower levels represent the four spin states of two electrons: the singlet (S) and three triplet (T_1, T_0, T_{-1}) states, whereas the upper levels exhibit the negatively doubly charged exciton states (X^{2-}) with indexes indicating the spin state of the electron in the excited orbital and the spin state of the light hole.

where the left hand side indicates the missing state of the valence band electron in the state on the right hand side. Then we find that the spin rotation by STIRAP is possible except for cases of the pseudospin composed of (S, T_0) and (T_1, T_{-1}). The same situation holds also for transitions associated with the heavy hole. As seen in Fig. 10, the four levels in both the ground and excited states are energetically close to each other. In the excited states, they are lying within the range determined by the Zeeman energy difference, which is about several tens of μeV for 1 Tesla. In the ground states, the singlet state lies below the triplet states by the orbital excitation energy and the triplet states are close to each other within the Zeeman energy difference.

It is important to examine the fidelity of the spin rotation under the situation that several Λ-type transitions are overlapping within a similar energy range. As a model system we consider a five-level system as depicted in Fig. 11. Relative energy differences, population relaxation and decoherence rates employed are

$$\hbar\omega_1 - \hbar\omega_3 = 0.05\text{meV}\ , \hbar\omega_2 - \hbar\omega_0 = 0.05\text{meV}\ , \hbar\omega_4 - \hbar\omega_0 = 0.07\text{meV}\ , \tag{46}$$

$$\hbar\Gamma_{1\to 0} = \hbar\Gamma_{1\to 2} = \hbar\Gamma_{1\to 4} = \hbar\Gamma_{3\to 0} = \hbar\Gamma_{3\to 2} = \hbar\Gamma_{3\to 4} = 0.01\text{meV}\ , \tag{47}$$

$$\hbar\gamma_{10} = \hbar\gamma_{12} = \hbar\gamma_{14} = \hbar\gamma_{30} = \hbar\gamma_{32} = \hbar\gamma_{34} = 0.05\text{meV}\ . \tag{48}$$

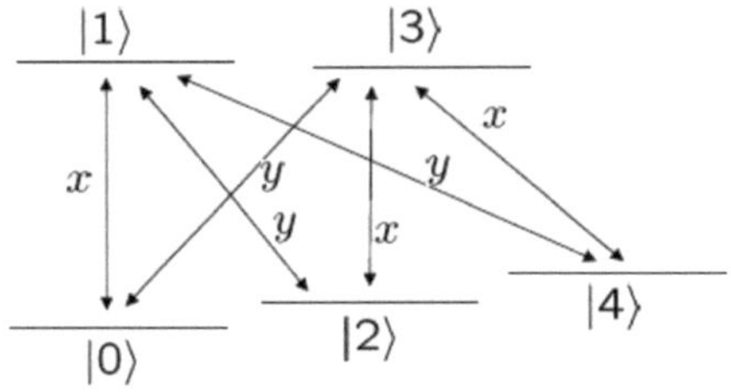

Fig. 11. Five level system composed of three lower levels and two upper levels. This is a simplest idealized model for studying the effect of overlapping Λ-type transitions.

Concerning the four levels composed of |0⟩, |1⟩, |2⟩ and |3⟩, the relevant parameters are the same as for the four-level system in Fig. 7. Thus, the effect of an additional level |4⟩ can be examined. Results are exhibited in Fig. 12. An additional level degrades the coherence of the STIRAP process and reduces the fidelity of the spin rotation. However, in the case of 2π pulse area, the fidelity keeps a good value for large off-resonance energies.

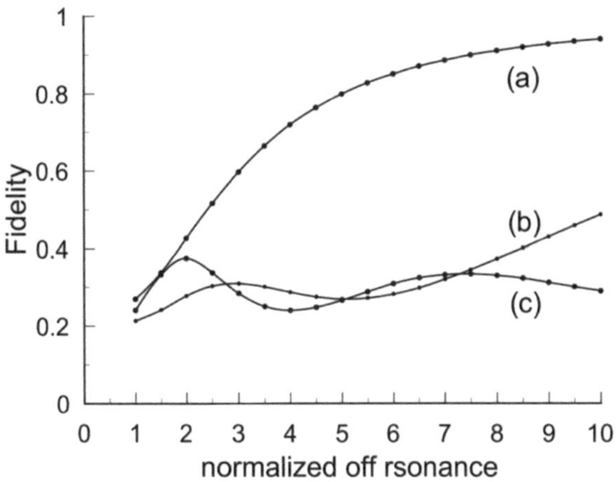

Fig. 12. Fidelity of the spin rotation of two electrons is plotted as a function of the normalized off-resonance in the five-level model. Curves (a), (b) and (c) correspond to the pulse area 2π, 4π and 6π, respectively.

Another important feature is the state initialization within the pseudospin subspace. When we want to rotate the pseudospin composed of |0⟩ and |2⟩ states in Fig. 11, the state should be initialized within this subspace. We examined the effect on the fidelity of the spin rotation of the incomplete state initialization. The fidelity is calculated for the case in which the state is prepared in the subspace spanned by the |0⟩ and |2⟩ states with the weight of 0.9 and in the |4⟩ state with the weight of 0.1. Results are given in Fig. 13 with those for the complete initialization in which the state is prepared in the subspace spanned only by the |0⟩ and |2⟩ states. The fidelity loss proportional to the deviation from the perfect initialization is seen. Thus the state initialization should be carried out as perfect as possible. One possible way of the state initialization is the use of the singlet-triplet level crossing by the magnetic field tuning. At first we prepare the two electrons in the singlet state and then bring the system adiabatically to the crossing point. During the residence period at the crossing point, the state mixing is induced by the spin-orbit interaction and the hyperfine interaction with nuclei, leading to an incoherent mixed state. This incoherent mixed state is sufficient to carry out the spin rotation. When the electron Zeeman energy is sufficiently large and three triplet states are well separated, the state initialization within the subspace composed of two crossing states such as (S, T_1), (S, T_0) and (S, T_{-1}) will be established.

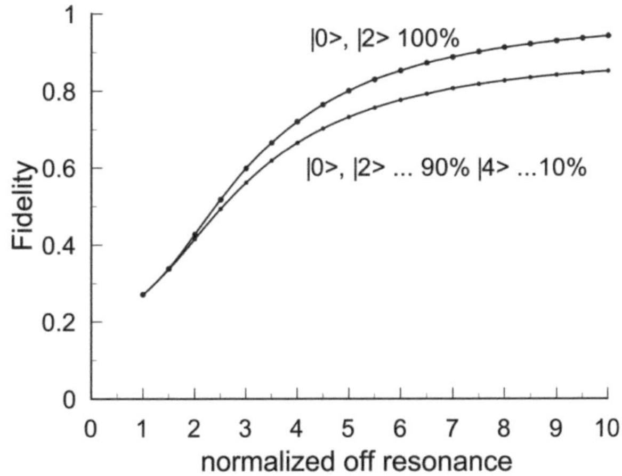

Fig. 13. Fidelity of the spin rotation of two electrons is plotted as a function of the normalized off-resonance in the five-level model for two cases, namely, one case where initially the population is prepared within the states $|0\rangle$ and $|2\rangle$ with 90% weight and in the state $|4\rangle$ with 10% weight and the other case where the population is prepared within the subspace spanned only by $|0\rangle$ and $|2\rangle$. The pulse area is 2π.

4. Spin state tomography of a single electron

The projective measurement of the spin state of a single electron is possible based on the Faraday/Kerr rotation of a linearly polarized light and this has been demonstrated experimentally very recently [42,43]. However, in the spin state tomography, all the components of the spin (s_x, s_y, s_z), namely, the off-diagonal (coherence) components as well as the diagonal components of the density matrix should be measured. The density matrix of a single electron spin in the spin up and down bases is given by

$$\rho = \begin{pmatrix} \rho^0_{\uparrow\uparrow} & \rho^0_{\uparrow\downarrow} \\ \rho^0_{\downarrow\uparrow} & \rho^0_{\downarrow\downarrow} \end{pmatrix} = \frac{1}{2}(1 + \vec{s}\cdot\vec{\sigma}) \quad \text{with} \quad \vec{s} = (s_x, s_y, s_z)\ , \tag{49}$$

$$s_x = \mathrm{Tr}\ \rho\sigma_x = \rho^0_{\downarrow\uparrow} + \rho^0_{\uparrow\downarrow}\ ,\ s_y = \mathrm{Tr}\ \rho\sigma_y = -i(\rho^0_{\downarrow\uparrow} - \rho^0_{\uparrow\downarrow})\ ,\ s_z = \mathrm{Tr}\ \rho\sigma_z = \rho^0_{\uparrow\uparrow} - \rho^0_{\downarrow\downarrow}\ , \tag{50}$$

where σ_i $(i = x, y, z)$ is the Pauli spin matrix. The purity of this state is given by

$$\mathcal{P} = \mathrm{Tr}\rho^2 = \frac{1}{2}(1 + (\vec{s})^2)\ . \tag{51}$$

Thus, by measuring all the components (s_x, s_y, s_z) we can determine whether the state is a pure state or not.

In order to measure all the components (s_x, s_y, s_z) by an optical means, there should be at least one excited state which is connected to both the spin up and spin down states of the electron, in other words, there should be a Λ -type transition. This transition creates the coherence between the spin up and spin down states, rotates the spin and enables the spin state tomography. It is easily shown that such a Λ -type transition is not possible in the Faraday configuration. On the other hand, in the Voigt configuration in which an in-plane magnetic field is applied along, e.g., the x direction, the Λ-type transition is possible as depicted in Fig. 14 for the optical transitions associated with both the heavy hole and light hole states. In Fig.14, the optical polarization selection rules are given in the x and y bases. The excited state is a trion state composed of a spin-singlet electron pair and a hole. The electron and hole states under an in-plane magnetic field are described by

$$|x+\rangle = \frac{1}{\sqrt{2}}(|\uparrow\rangle + |\downarrow\rangle)\,,\ |x-\rangle = \frac{1}{\sqrt{2}}(-|\uparrow\rangle + |\downarrow\rangle)\,, \tag{52}$$

$$|hh+\rangle = \frac{1}{\sqrt{2}}(|\frac{3}{2}\frac{3}{2}\rangle + |\frac{3}{2}-\frac{3}{2}\rangle)\,,\ |hh-\rangle = \frac{1}{\sqrt{2}}(-|\frac{3}{2}\frac{3}{2}\rangle + |\frac{3}{2}-\frac{3}{2}\rangle)\,, \tag{53}$$

$$|lh+\rangle = \frac{1}{\sqrt{2}}(|\frac{3}{2}\frac{1}{2}\rangle + |\frac{3}{2}-\frac{1}{2}\rangle)\,,\ |lh-\rangle = \frac{1}{\sqrt{2}}(-|\frac{3}{2}\frac{1}{2}\rangle + |\frac{3}{2}-\frac{1}{2}\rangle)\,, \tag{54}$$

where for the hole states the left hand side represents the missing state of the valence band electron in the state on the right hand side.

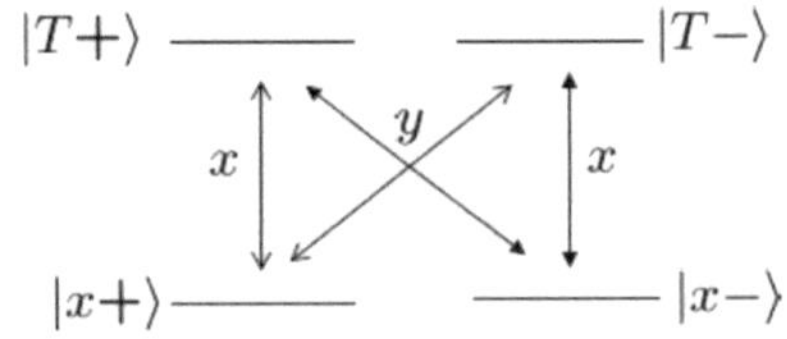

Fig. 14. Λ-type transitions for a single electron in the Voigt configuration. The lower levels indicate the two spin states of the electron, whereas the upper levels represent the trion states associated with the light hole or the heavy hole states. The polarization selection rules are given in terms of the x and y bases, where the in-plane magnetic field is applied in the x direction.

Now we discuss the scheme to measure the spin component of the electron. A probe light propagates along the z axis and its polarization rotation is measured in the transmission or reflection geometry. Thus the dielectric tensor represented in the bases of the electric field components in the x and y directions is relevant. In the theoretical analysis a single Λ-type transition will be considered with the level indexes as depicted in Fig. 15. An external test field is applied to estimate the dielectric tensor and is assumed as

$$\vec{\mathcal{E}}_{\text{test}}(t) = (\mathcal{E}_x\,\vec{e}_x + \mathcal{E}_y\,\vec{e}_y)e^{i\omega t} + \text{c.c.}\,, \tag{55}$$

where $\vec{e}_x(\vec{e}_y)$ is the unit vector in the $x(y)$ direction. The initial density matrix, which is to be fixed from the measurements, is given by

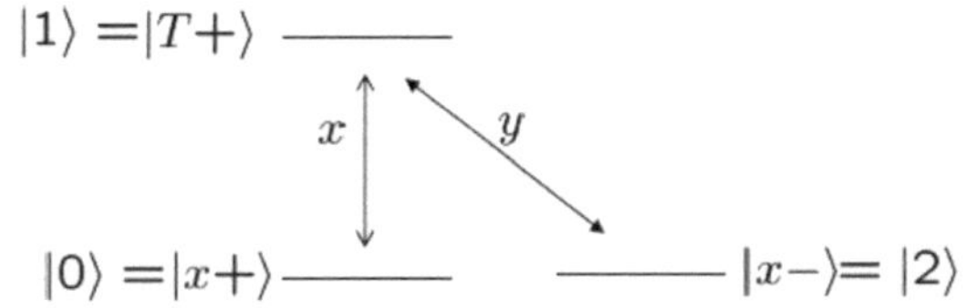

Fig. 15. A Λ-type transition is chosen from the left hand side of Fig. 14 and the levels are numbered to simplify theoretical expressions.

$$\rho(t=0) = \begin{pmatrix} \rho^0_{00} & \rho^0_{02} & 0 \\ \rho^0_{20} & \rho^0_{22} & 0 \\ 0 & 0 & 0 \end{pmatrix} , \tag{56}$$

where the bases are chosen as $|0\rangle$, $|2\rangle$ and $|1\rangle$. The relevant equations of motion for the density matrix take the form:

$$\dot{\rho} = -\frac{i}{\hbar}[H_0 + V, \rho] + \Gamma\rho , \tag{57}$$

where H_0 and V are similar to those in Eqs. (2) and (3) and Γ includes the population relaxation and decoherence terms. Expressions for each matrix element are given below:

$$\dot{\rho}_{00} = i\Omega_x e^{i\omega t}\rho_{10} - i\Omega^*_x e^{-i\omega t}\rho_{01} + \Gamma_{1\to 0}\rho_{11} , \tag{58}$$

$$\dot{\rho}_{11} = -i\Omega_x e^{i\omega t}\rho_{10} - i\Omega_y e^{i\omega t}\rho_{12} + \text{c.c.} - (\Gamma_{1\to 0} + \Gamma_{1\to 2})\rho_{11} , \tag{59}$$

$$\rho_{22} = 1 - \rho_{00} - \rho_{11} , \tag{60}$$

$$\dot{\rho}_{01} = i\Omega_x e^{i\omega t}(\rho_{11} - \rho_{00}) - i\Omega_y e^{i\omega t}\rho_{02} + (i\omega_{10} - \gamma_{01})\rho_{01} , \tag{61}$$

$$\dot{\rho}_{02} = i\Omega_x e^{i\omega t}\rho_{12} - i\Omega^*_y e^{-i\omega t}\rho_{01} + (i\omega_{20} - \gamma_{02})\rho_{02} , \tag{62}$$

$$\dot{\rho}_{12} = i\Omega^*_x e^{-i\omega t}\rho_{02} + i\Omega^*_y e^{-i\omega t}(\rho_{22} - \rho_{11}) + (i\omega_{21} - \gamma_{12})\rho_{12} \tag{63}$$

$$\text{with} \qquad \Omega_x = \frac{\mu^x_{01}\mathcal{E}_x}{\hbar} , \ \Omega_y = \frac{\mu^y_{21}\mathcal{E}_y}{\hbar} , \ \omega_{ij} = (E_i - E_j)/\hbar , \tag{64}$$

where μ^k_{ij} is the optical matrix element between the states $|i\rangle$ and $|j\rangle$ for the light polarization in the k direction, E_i the energy of the state $|i\rangle$, $\Gamma_{i\to j}$ the population decay rate from the state $|i\rangle$ to the state $|j\rangle$ and γ_{ij} is the decay rate of the coherence between the states $|i\rangle$ and $|j\rangle$. In order to facilitate the analysis, the rapidly oscillating parts will be separated out as

$$\rho_{01}(t) = \bar{\rho}_{01}(t)e^{i\omega t} , \ \rho_{12}(t) = \bar{\rho}_{12}(t)e^{-i\omega t} , \tag{65}$$

where $\bar{\rho}_{01}$ and $\bar{\rho}_{12}$ are slowly varying amplitudes. ρ_{02} is also slowly varying because ω_{02} is very small compared with the optical transition energies. Then the equations of motion for these amplitudes become

$$\dot{\bar{\rho}}_{01} = i\Omega_x(\rho_{11} - \rho_{00}) - i\Omega_y \rho_{02} + (i(\omega_{10} - \omega) - \gamma_{01})\bar{\rho}_{01} \ , \tag{66}$$

$$\dot{\rho}_{02} = i\Omega_x \bar{\rho}_{12} - i\Omega_y^* \bar{\rho}_{01} + (i\omega_{20} - \gamma_{02})\rho_{02} \ , \tag{67}$$

$$\dot{\bar{\rho}}_{12} = i\Omega_x^* \rho_{02} + i\Omega_y^*(\rho_{22} - \rho_{11}) + (i(\omega - \omega_{12}) - \gamma_{12})\bar{\rho}_{12} \ . \tag{68}$$

The stationary solutions within the linear response to the test field are given by

$$\rho_{01}^{st} = \frac{i\Omega_x \rho_{00}^0 + i\Omega_y \rho_{02}^0}{i\Delta - \gamma_{01}} \ , \ \rho_{21}^{st} = \frac{i\Omega_x \rho_{20}^0 + i\Omega_y \rho_{22}^0}{i(\Delta - \omega_{20}) - \gamma_{21}} \quad \text{with } \Delta = \omega_{10} - \omega \ . \tag{69}$$

Now the induced polarization and the corresponding susceptibility tensor χ are derived as

$$\vec{P} = \text{Tr}\ \vec{\mu}\rho^{st} = (\mu_{10}^x \rho_{01}^{st} + \mu_{12}^y \rho_{21}^{st})e^{i\omega t} + \text{c.c.} \tag{70}$$

$$= \overleftrightarrow{\chi} \cdot \vec{\mathcal{E}}\, e^{i\omega t} + \text{c.c.} = \begin{pmatrix} \chi_{xx} & \chi_{xy} \\ \chi_{yx} & \chi_{yy} \end{pmatrix} \begin{pmatrix} \mathcal{E}_x \\ \mathcal{E}_y \end{pmatrix} e^{i\omega t} + \text{c.c.} \tag{71}$$

Assuming the large off-resonance: $\Delta \gg \omega_{20}, \gamma_{ij}$, we have

$$\overleftrightarrow{\chi}_A = A \begin{pmatrix} \rho_{00}^0 & i\rho_{02}^0 \\ -i\rho_{20}^0 & \rho_{22}^0 \end{pmatrix} \quad \text{with } A = \frac{|\mu_{01}^x|^2}{v_0 \hbar \Delta} \ , \tag{72}$$

where v_0 is the normalization volume for the polarization density and the subscript A is attached for the later use. The dielectric tensor is given by

$$\overleftrightarrow{\varepsilon} = \varepsilon_0 + 4\pi \overleftrightarrow{\chi}_A \ , \tag{73}$$

where ε_0 is the background dielectric constant. In the case of large off-resonance, another Λ-type transition depicted in Fig. 16 should also be taken into account. After a similar calculation, the corresponding susceptibility $\overleftrightarrow{\chi}_B$ is estimated as

$$\overleftrightarrow{\chi}_B = B \begin{pmatrix} \rho_{22}^0 & i\rho_{20}^0 \\ -i\rho_{02}^0 & \rho_{00}^0 \end{pmatrix} \quad \text{with } B = \frac{|\mu_{2\tilde{1}}^x|^2}{v_0 \hbar \tilde{\Delta}} \text{ and } \tilde{\Delta} = \omega_{\tilde{1}2} - \omega \ . \tag{74}$$

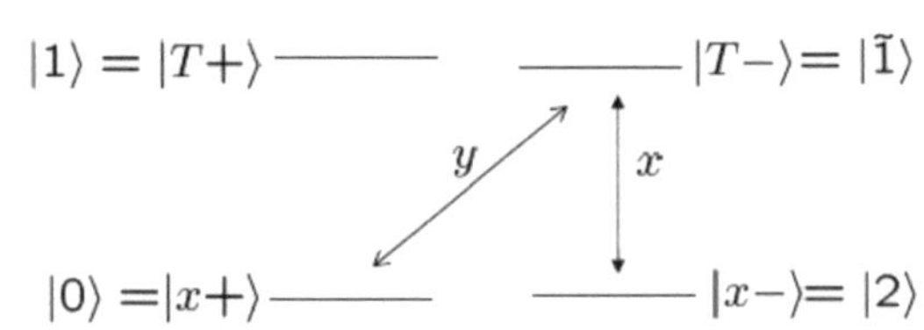

Fig. 16. A Λ-type transition is chosen from the right hand side of Fig. 14 and the levels are numbered to simplify theoretical expressions.

When the trion state associated with the heavy hole is considered, $\Delta \cong \tilde{\Delta}$ and hence $A \cong B$, because the g-factor of the heavy hole is very small. In the absence of the coupling between the heavy hole (hh) and the light hole (lh), the in-plane g-factor of the heavy hole is zero.

Actually, that g-factor is finite due to the hh-lh coupling. However, the g-factor of the heavy hole is one order of magnitude smaller than that of the light hole [25,44]. Then the total susceptibility becomes

$$\overleftrightarrow{\chi}_{\text{tot.}}=\overleftrightarrow{\chi}_A+\overleftrightarrow{\chi}_B\cong A\begin{pmatrix} 1 & i(\rho_{20}^0+\rho_{02}^0) \\ -i(\rho_{20}^0+\rho_{02}^0) & 1 \end{pmatrix}. \tag{75}$$

This indicates that only one component of the spin vector can be monitored. On the other hand, in the case of the light hole, the energy difference between $|1\rangle$ and $|\tilde{1}\rangle$ is rather large and the contribution from either $\overleftrightarrow{\chi}_A$ or $\overleftrightarrow{\chi}_B$ is dominant. Then all the spin components can be measured as discussed below.

Now we discuss the measurement schemes to probe the spin components. For the moment, we consider the transmission geometry, assuming that the $\overleftrightarrow{\chi}_A$ is dominantly contributing to the dielectric tensor. Then the susceptibility tensor can be written in terms of s_x, s_y and s_z in Eq. (50) as

$$\overleftrightarrow{\chi}_A=\frac{A}{2}\begin{pmatrix} 1+s_z & is_x+s_y \\ -is_x+s_y & 1-s_z \end{pmatrix} \tag{76}$$

and the dielectric tensor is written as

$$\overleftrightarrow{\varepsilon}=\varepsilon_0+4\pi\,\overleftrightarrow{\chi}_A=(\varepsilon_0+2\pi A)\,\overleftrightarrow{1}+4\pi\,\overleftrightarrow{\chi}'_A \tag{77}$$

$$\text{with}\quad \overleftrightarrow{\chi}'_A=\frac{A}{2}\begin{pmatrix} s_z & is_x+s_y \\ -is_x+s_y & -s_z \end{pmatrix}=\frac{A}{2}\left[{}^t\vec{s}\cdot\vec{n}\,{}^t\vec{n}\cdot\vec{\sigma}+(\vec{\sigma}\times\vec{s})\cdot\vec{n}\right], \tag{78}$$

where $\vec{n}\,{}^t\vec{n}$ is a dyadic form and the last expression is general for the axially symmetric case with the propagation vector $\vec{n}$ of the probe light. Here $\overleftrightarrow{\chi}'_A$ is a Hermitian matrix and has eigenvectors associated with real eigenvalues χ_1 and χ_2, namely

$$\overleftrightarrow{\chi}'_A\;\vec{u}_1=\chi_1\,\vec{u}_1\,,\;\overleftrightarrow{\chi}'_A\;\vec{u}_2=\chi_2\,\vec{u}_2\,. \tag{79}$$

Then a probe light with an amplitude:

$$\vec{\mathcal{E}}_{\text{probe}}(t,z=0)=(a\,\vec{u}_1+b\,\vec{u}_2)\,e^{i\omega t}+\text{c.c.}=\vec{\mathcal{E}}^{(+)}_{\text{probe}}\,e^{i\omega t}+\text{c.c.} \tag{80}$$

propagates as

$$\vec{\mathcal{E}}_{\text{probe}}(t,z)=\left[a\,\vec{u}_1e^{-ik_0\sqrt{\varepsilon_1}z}+b\,\vec{u}_2\,e^{-ik_0\sqrt{\varepsilon_2}z}\right]e^{i\omega t}+\text{c.c.} \tag{81}$$

$$\text{with}\quad k_0=\frac{\omega}{c}\;\text{and}\;\varepsilon_{1(2)}=\varepsilon_0+2\pi A+4\pi\chi_{1(2)}=\tilde{\varepsilon}_0+4\pi\chi_{1(2)}\,. \tag{82}$$

For a thin sample, e.g., a single quantum dot layer, $k_0z\ll 1$ and the phase factor can be expanded with respect to this smallness parameter. Then the transmitted field is obtained as

$$\vec{\mathcal{E}}_{\rm probe}(t,z) = e^{i\omega t - ik_0\sqrt{\bar{\varepsilon}_0}z}[a\,\vec{u}_1 + b\,\vec{u}_2 - i\frac{2\pi k_0 z}{\sqrt{\bar{\varepsilon}_0}}(a\,\chi_1\vec{u}_1 + b\,\chi_2\vec{u}_2) + \cdots] + {\rm c.c.} \tag{83}$$

$$= e^{i\omega t - ik_0\sqrt{\bar{\varepsilon}_0}z}[\overleftrightarrow{1} - i\eta\ \overleftrightarrow{\chi}'_A + \cdots]\vec{\mathcal{E}}^{(+)}_{\rm probe} + {\rm c.c.\ \ with}\ \eta = \frac{2\pi k_0 z}{\sqrt{\bar{\varepsilon}_0}}\,. \tag{84}$$

Now we can discuss the schemes to measure the spin components. Here the polarization of the probe light is specified in a definite direction and the transmitted probe light is passed through a polarizing beam splitter (PBS). Then the intensity difference between two orthogonally polarized components is measured. In the first measurement, the probe light is polarized along the x direction and the transmitted light is given by

$$\vec{\mathcal{E}}_{\rm trans.} = [\overleftrightarrow{1} - i\eta\ \overleftrightarrow{\chi}'_A]\,\vec{\mathcal{E}}_{\rm probe} = [\overleftrightarrow{1} - i\eta\ \overleftrightarrow{\chi}'_A]\begin{pmatrix}1\\0\end{pmatrix} = \begin{pmatrix}1 - i\eta A s_z/2\\ -\eta A(s_x + i s_y)/2\end{pmatrix}. \tag{85}$$

When this transmitted light is analyzed in the diagonal(D) and cross-diagonal($\bar{D}$) polarization bases, namely,

$$|D\rangle = \frac{1}{\sqrt{2}}\begin{pmatrix}1\\1\end{pmatrix},\ |\bar{D}\rangle = \frac{1}{\sqrt{2}}\begin{pmatrix}1\\-1\end{pmatrix}, \tag{86}$$

the intensity difference is calculated as

$$|\langle D|\vec{\mathcal{E}}_{\rm trans.}\rangle|^2 - |\langle \bar{D}|\vec{\mathcal{E}}_{\rm trans.}\rangle|^2 \cong -\eta A\ s_x + \mathcal{O}((\eta A)^2)\,. \tag{87}$$

Thus the s_x component can be measured. In the second measurement, the probe light is polarized along the x direction and the transmitted light is analyzed in the right-circular(R) and left-circular(L) polarization bases, namely,

$$|R\rangle = \frac{1}{\sqrt{2}}\begin{pmatrix}1\\i\end{pmatrix},\ |L\rangle = \frac{1}{\sqrt{2}}\begin{pmatrix}1\\-i\end{pmatrix}. \tag{88}$$

Then the intensity difference is given by

$$|\langle R|\vec{\mathcal{E}}_{\rm trans.}\rangle|^2 - |\langle L|\vec{\mathcal{E}}_{\rm trans.}\rangle|^2 \cong -\eta A\ s_y + \mathcal{O}((\eta A)^2) \tag{89}$$

and the s_y component can be measured. In the third measurement, the probe light is polarized along the diagonal(D) direction and the transmitted light is given by

$$\vec{\mathcal{E}}_{\rm trans.} = [\overleftrightarrow{1} - i\eta\ \overleftrightarrow{\chi}'_A]\,\vec{\mathcal{E}}_{\rm probe}\ {\rm with}\ \ \vec{\mathcal{E}}_{\rm probe} = \frac{1}{\sqrt{2}}\begin{pmatrix}1\\1\end{pmatrix} \tag{90}$$

$$= \frac{1}{\sqrt{2}}\begin{pmatrix}1 + \eta A s_x/2 - i\eta A(s_y + s_z)/2\\ 1 - \eta A s_x/2 - i\eta A(s_y - s_z)/2\end{pmatrix}. \tag{91}$$

This transmitted light is analyzed in the right-circular(R) and left-circular(L) polarization bases. Then the intensity difference is calculated as

$$|\langle R|\vec{\mathcal{E}}_{\text{trans.}}\rangle|^2 - |\langle L|\vec{\mathcal{E}}_{\text{trans.}}\rangle|^2 \cong \eta A\ s_z + \mathcal{O}((\eta A)^2) \tag{92}$$

and the s_z component can be measured. Thus, all the spin components (s_x, s_y, s_z) are measured and the spin state tomography is completed.

These measurement schemes can be extended to the reflection geometry. Assuming the normal incidence of the probe light, the amplitude of the reflected light is calculated as

$$\vec{\mathcal{E}}_{\text{ref.}} = \frac{1-\sqrt{\varepsilon}}{1+\sqrt{\varepsilon}}\,\vec{\mathcal{E}}_{\text{probe}} = \frac{1-\sqrt{\tilde{\varepsilon}_0}(\overset{\leftrightarrow}{1} + \frac{2\pi \overset{\leftrightarrow}{\chi}{}'_A}{\tilde{\varepsilon}_0} + \cdots)}{1+\sqrt{\tilde{\varepsilon}_0}(\overset{\leftrightarrow}{1} + \frac{2\pi \overset{\leftrightarrow}{\chi}{}'_A}{\tilde{\varepsilon}_0} + \cdots)}\,\vec{\mathcal{E}}_{\text{probe}} \tag{93}$$

$$= \frac{1-\sqrt{\tilde{\varepsilon}_0}}{1+\sqrt{\tilde{\varepsilon}_0}}\,[\overset{\leftrightarrow}{1} + \frac{4\pi\ \overset{\leftrightarrow}{\chi}{}'_A}{(\tilde{\varepsilon}_0 - 1)\sqrt{\tilde{\varepsilon}_0}} + \cdots]\,\vec{\mathcal{E}}_{\text{probe}} \tag{94}$$

$$= \frac{1-\sqrt{\tilde{\varepsilon}_0}}{1+\sqrt{\tilde{\varepsilon}_0}}\,[\overset{\leftrightarrow}{1} + \eta\ \overset{\leftrightarrow}{\chi}{}'_A + \cdots]\,\vec{\mathcal{E}}_{\text{probe}} \quad \text{with} \quad \eta = \frac{4\pi}{(\tilde{\varepsilon}_0 - 1)\sqrt{\tilde{\varepsilon}_0}}\,. \tag{95}$$

Here the first factor is arising from the background dielectric constant and is not relevant in discussing the dependence of the reflected light amplitude on the polarization configuration. The reflected light amplitude denoted by $\vec{\mathcal{E}}_{\text{Kerr}}$ is introduced by

$$\vec{\mathcal{E}}_{\text{Kerr}} = [\overset{\leftrightarrow}{1} + \eta\ \overset{\leftrightarrow}{\chi}{}'_A + \cdots]\,\vec{\mathcal{E}}_{\text{probe}}\,. \tag{96}$$

Then the measurement of the spin components can be achieved as follows. In the first scheme, the polarization of the probe light is chosen along the x direction and the reflected light is given by

$$\vec{\mathcal{E}}_{\text{Kerr}} = [\overset{\leftrightarrow}{1} + \eta\ \overset{\leftrightarrow}{\chi}{}'_A + \cdots]\,\vec{\mathcal{E}}_{\text{probe}} \quad \text{with} \quad \vec{\mathcal{E}}_{\text{probe}} = \begin{pmatrix} 1 \\ 0 \end{pmatrix} \tag{97}$$

$$= \begin{pmatrix} 1 + \eta A s_z/2 \\ \eta A(-i s_x + s_y)/2 \end{pmatrix}\,. \tag{98}$$

This reflected light is analyzed in terms of the right-circular (R) and left-circular (L) polarization bases. Then the intensity difference is given by

$$|\langle R|\vec{\mathcal{E}}_{\text{Kerr}}\rangle|^2 - |\langle L|\vec{\mathcal{E}}_{\text{Kerr}}\rangle|^2 \cong -\eta A\ s_x + \mathcal{O}((\eta A)^2)\,. \tag{99}$$

Thus the s_x component can be measured. In the second scheme, the polarization of the probe light is chosen along the x direction and the reflected light is analyzed in terms of the diagonal (D) and cross-diagonal ($\bar{D}$) polarization bases. Then the intensity difference is given by

$$|\langle D|\vec{\mathcal{E}}_{\text{Kerr}}\rangle|^2 - |\langle \bar{D}|\vec{\mathcal{E}}_{\text{Kerr}}\rangle|^2 \cong \eta A\ s_y + \mathcal{O}((\eta A)^2) \tag{100}$$

and the s_y component can be measured. In the third scheme, the polarization of the probe light is chosen along the diagonal (D) direction, namely

$$\vec{\mathcal{E}}_{\rm probe} = |D\rangle = \frac{1}{\sqrt{2}}\begin{pmatrix}1\\1\end{pmatrix} \tag{101}$$

and the reflected light is analyzed in terms of the horizontal (x) and vertical (y) polarization bases. Then the intensity difference is given by

$$|\langle H|\vec{\mathcal{E}}_{\rm Kerr}\rangle|^2 - |\langle V|\vec{\mathcal{E}}_{\rm Kerr}\rangle|^2 \cong \eta A\ s_z + \mathcal{O}((\eta A)^2) \tag{102}$$

and the s_z component can be measured. In this way, all the spin components (s_x, s_y, s_z) can be measured and the spin state tomography of a single electron is completed.

5. Quantum correlation (Bell) measurement between two electrons

In the scheme of quantum repeater, the primary elements are the quantum state transfer between a photon and an electron and the entanglement swapping through the Bell (correlation) measurement between two electrons which are created through the quantum state transfer from two photons. It is preferable to do the Bell measurement between electrons instead of photons because the mismatch between the photon arrival times can be compensated by the rather long coherence time of electrons, whereas the storage of photons is rather difficult although the techniques for the photon storage are progressing steadily. Thus we start the discussion assuming that two electrons are prepared in a semiconductor nanostructure, e.g., a quantum dot. We propose an optical method to measure the spin state of two electrons based on the Faraday or Kerr rotation. Here we employ a linearly polarized off-resonant probe light and measure the orientation of the transmitted (reflected) light. Thus the method can be nondestructive in the same sense as demonstrated for the case of a single electron [42,43].

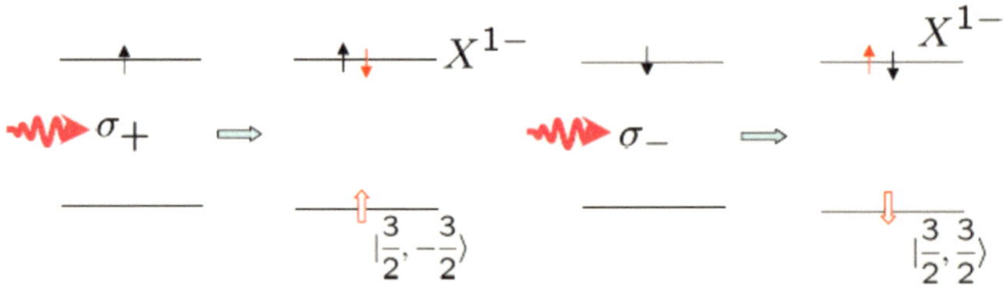

Fig. 17. Elementary processes of the Faraday rotation for the case of a single resident electron. $\sigma_{+(-)}$ denotes the right (left) circularly polarized light. The upper (lower) horizontal line indicates the electron (hole) level. A thin (thick empty) arrow represents an electron (a hole) with the spin direction along the arrow.

Before going into details, let us review briefly the elementary processes of the Faraday rotation for the case of a single electron. We consider a III-V semiconductor quantum dot in which the hole ground state is the heavy hole state and a magnetic field is applied along the crystal growth direction (namely, perpendicular to the quantum well plane). As is well known, the right-circularly polarized light denoted by σ_+ excites a down spin electron from the valence band state $|3/2, -3/2\rangle$ creating a charged exciton or trion, while the left-circularly polarized light denoted by σ_-excites an up spin electron from the valence band state $|3/2, 3/2\rangle$, as exhibited in Fig. 17. When we probe the system with a linearly polarized light along the x direction, i.e.,

$$|x\rangle = \frac{1}{\sqrt{2}}(|\sigma_+\rangle + |\sigma_-\rangle) , \tag{103}$$

where $\sigma_+(\sigma_-)$ may be alternatively denoted by $R(L)$ for the right (left) circular polarization, one of the circular components receives a phase shift and the Faraday rotation occurs. Thus we can distinguish the two spin states of an electron by the sign of the Faraday rotation angle.

Now we extend this argument to the case of two electrons and consider relevant elementary processes for four states of two electrons, namely, the singlet state(S) and the triplet states with the magnetic quantum number 1, 0 and -1 (T_1, T_0, T_{-1}). For the T_1 state, spins of the two resident electrons are aligned in the same direction and a σ_+ polarized light excites a down spin electron from the valence band creating a doubly negatively charged exciton X^{2-}, as shown in Fig. 18, in which the lowest electron orbital state is occupied by a spin-singlet electron pair and the spin direction of the electron in the second lowest orbital state is indicated in the superscript and the spin direction of the heavy hole is depicted in the subscript, namely:

$$|\Uparrow\rangle = |\frac{3}{2}\ -\frac{3}{2}\rangle\ ,\ |\Downarrow\rangle = |\frac{3}{2}\ \frac{3}{2}\rangle\ , \tag{104}$$

where the left hand side represents the missing state of the valence band electron in the state on the right hand side. The T_1 state is optically inactive for the σ_- polarized light. For the T_{-1} state, a σ_- polarized light excites an up spin electron from the valence band creating another doubly negatively charged exciton. This T_{-1} state is optically inactive for the σ_+ polarized light. Thus these two states can be distinguished by the sign of the Faraday rotation angle. On the other hand, the S and T_0 states are optically active for both circular polarizations as exhibited in Figs. 19 and 20, and the sign of the Faraday rotation angle is determined by the competition between the phase shifts for each circular component.

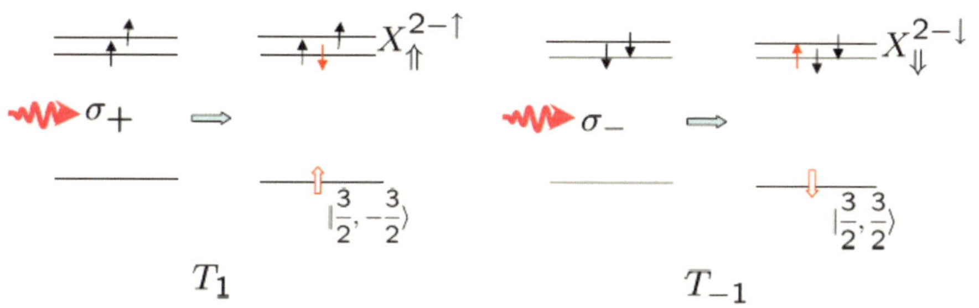

Fig. 18. Elementary processes of the Faraday rotation for the triplet T_1 and T_{-1} states of two resident electrons.

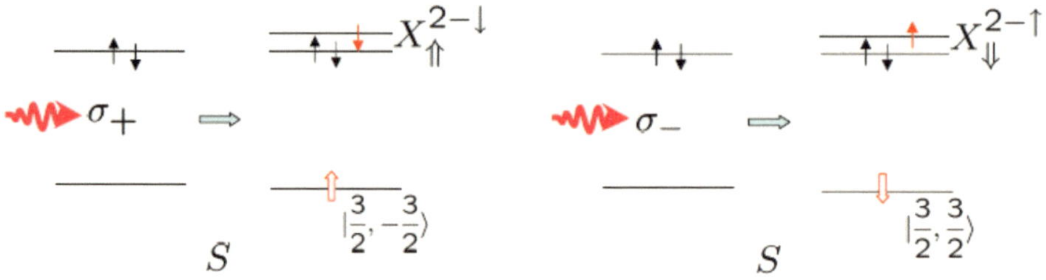

Fig. 19. Elementary processes of the Faraday rotation for the singlet S state of two resident electrons.

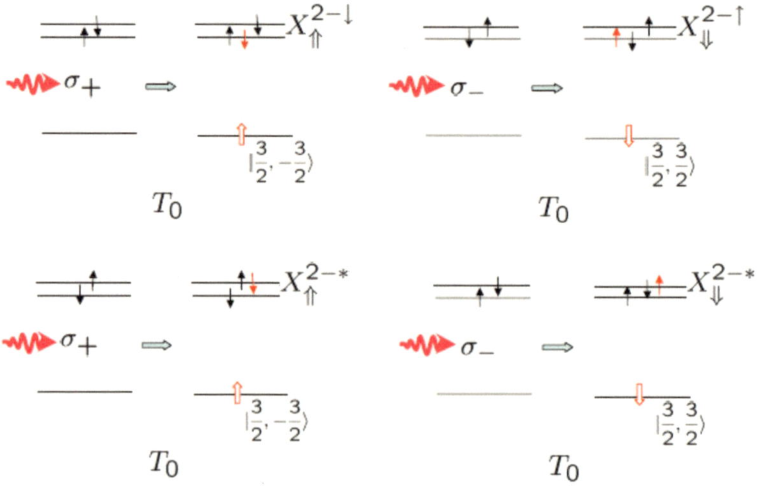

Fig. 20. Elementary processes of the Faraday rotation for the triplet T_0 state of two resident electrons. $X_{\Uparrow}^{2-*}$ and $X_{\Downarrow}^{2-*}$ denote excited states of the doubly negatively charged exciton.

The expression of the Faraday rotation angle is obtained in the perturbation theory and is composed of two terms:

$$\varphi \propto \sum_{j(\sigma+)} \frac{|\langle j|P_{\sigma+}|i\rangle|^2 (E_{j,i} - \hbar\omega)}{(\hbar^2\gamma_{j,i}^2 + (E_{j,i} - \hbar\omega)^2)} - \sum_{k(\sigma-)} \frac{|\langle k|P_{\sigma-}|i\rangle|^2 (E_{k,i} - \hbar\omega)}{(\hbar^2\gamma_{k,i}^2 + (E_{k,i} - \hbar\omega)^2)}, \tag{105}$$

where i indicates the initial state of two electrons, j (k) the final state of the optical transition for the σ_+ (σ_-) component, $E_{a,b} = E_a - E_b$ with E_a being the energy of the a state, $\gamma_{a,b}$ the dephasing rate corresponding to the $a \leftrightarrow b$ transition and $\hbar\omega$ denotes the photon energy of the linearly polarized probe light. As mentioned before, for the T_1 state only the σ_+ transitions contribute, whereas for the T_{-1} state only the σ_- transitions contribute. Thus the two states can be distinguished by the sign of the Faraday rotation angle. For the S and T_0 states, both σ_+ and σ_- transitions contribute and thus more detailed arguments are necessary to determine the sign of the Faraday rotation angle. Now we examine the resonance position of the Faraday rotation angle with respect to the probe photon energy $\hbar\omega$. From the elementary processes exhibited in Figs. 18-20, it is seen that for the triplet states the resonance occurs at around the energy of the doubly charged exciton states ($E(X^{2-})$). On the other hand, for the singlet state the resonance occurs at a higher energy than $E(X^{2-})$ because the lowest orbital state is already occupied by a spin-singlet electron pair and the optical transition should occur to the higher orbital state.

Now we discuss more details of the Faraday rotation angle for the case of T_0 state. As mentioned before, both σ_+ and σ_- circular components contribute to the Faraday rotation. The lowest-energy final state of the optical transition for each circular component is given by

$$j(\sigma+) = X_{\Uparrow}^{2-\downarrow}\,, \qquad k(\sigma-) = X_{\Downarrow}^{2-\uparrow}\,. \tag{106}$$

The energies of these states are different in a magnetic field because the spin configuration is different for these states. In terms of the electron g-factor $g_{c(v)}$ for the conduction (valence) band, these energies are given as

$$E(X_{\Uparrow}^{2-\downarrow}) \cong -\frac{1}{2}(g_c\mu_B B - g_v\mu_B B) + E_0\,, \tag{107}$$

$$E(X_{\Downarrow}^{2-\uparrow}) \cong \frac{1}{2}(g_c\mu_B B - g_v\mu_B B) + E_0\,, \tag{108}$$

where E_0 is the lowest energy of the interband transition. Then the energy difference $|E(X_{\Uparrow}^{2-\downarrow}) - E(X_{\Downarrow}^{2-\uparrow})|$ is typically about one tenth of meV for a magnetic field about 1 Tesla and is comparable to the dephasing rate of the optical transitions. From the formula in Eq. (105) we see that the dependence of the Faraday rotation angle on the probe photon energy is determined by the difference between two dispersive curves with nearly equal resonance energies. Thus the profile is given by the derivative of the dispersive curve as shown in Fig. 21, depending on the sign of the energy difference. The same situation holds for the singlet state S.

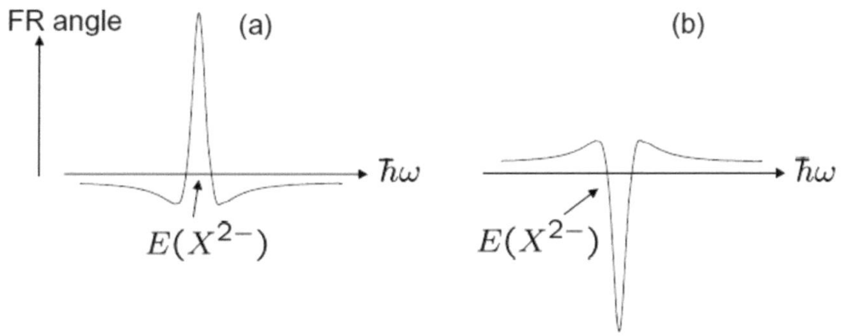

Fig. 21. Dependence on the probe photon energy ($\hbar\omega$) of the Faraday rotation angle for the triplet T_0 state and the singlet S state of two resident electrons. It depends on the sign of the energy difference; namely, (a) $|E(X_{\Uparrow}^{2-\downarrow}) - E(X_{\Downarrow}^{2-\uparrow})| > 0$, (b) $|E(X_{\Uparrow}^{2-\downarrow}) - E(X_{\Downarrow}^{2-\uparrow})| < 0$

Summarizing these considerations, we can show the schematic dependence of the Faraday rotation angle on the probe photon energy in Fig. 22. The triplet states T_1 and T_{-1} exhibit a typical dispersive lineshape. On the other hand, the profile for the triplet T_0 and the singlet S states is given by the derivative of the dispersive curve, where the case of $|E(X_{\Uparrow}^{2-\downarrow})-E(X_{\Downarrow}^{2-\uparrow})| > 0$ is assumed. The resonance occurs at around the energy of the doubly charged exciton state denoted by $E(X^{2-})$ for the triplet states, whereas for the singlet state it occurs at a higher energy than $E(X^{2-})$ by the orbital excitation energy Δ_e. Thus when we choose the probe photon energy at the downward arrow as shown in Fig. 22, the Faraday rotation angle is positive for the T_1 state and is negative for the T_{-1} state. For the T_0 state, the Faraday rotation angle is negative but the magnitude is small. For the singlet S state, the Faraday rotation angle would be vanishingly small because of the large off-resonance. Consequently, we can distinguish between the four states of two electrons by the magnitude and the sign of the Faraday rotation angle.

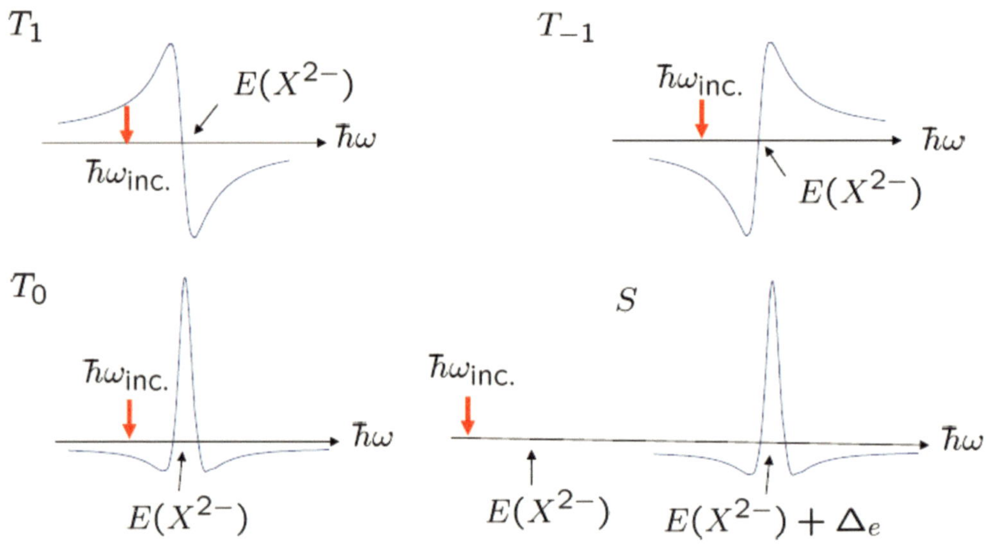

Fig. 22. Dependence on the probe photon energy ($\hbar\omega$) of the Faraday rotation angle for the three triplet states T_1, T_0, T_{-1} and the singlet state S of two resident electrons. Those for T_0 and S are exhibited for the case of $|E(X_{\Uparrow}^{2-\downarrow}) - E(X_{\Downarrow}^{2-\uparrow})| > 0$.

Now we discuss relevant parameters to optimize the Faraday rotation measurement. The essential requirement is the preparation of the lowest two orbital states which are energetically well-separated from higher excited states. We consider a circularly symmetric GaAs quantum dot with parabolic lateral confinement under a magnetic field along the growth direction. Then the orbital eigenstates are represented by the Fock-Darwin states [45,46] whose eigenenergies are given by

$$E_{\nu,n} = (|n| + 1 + 2\nu)\hbar\Omega + \frac{n}{2}\hbar\omega_c \text{ with } \Omega = \sqrt{\omega_0^2 + \frac{\omega_c^2}{4}}, \ \omega_c = \frac{eB}{m^*c}, \tag{109}$$

where ω_0 is the frequency of the harmonic confinement in the lateral direction and m^* is the electron effective mass. When we employ the parameter values: $\hbar\omega_0$ = 5meV, B = 5T and m^* = 0.067m_0 with m_0 being the free electron mass, we have $\hbar\omega_c$ = 8.7meV and $\hbar\Omega$ = 6.63meV. The lowest two orbital levels have the spacing of 2.3 meV and are well-separated from the higher orbital level by 8.7 meV. These parameter values would enable the Faraday rotation measurement to be carried out reliably.

6. Spin state tomography of two electrons

In the last Section, we discussed the projective measurement of the spin state of two electrons in a semiconductor quantum dot based on the Faraday or Kerr rotation. The method can project an arbitrary spin state onto one of the singlet and three triplet states by a single-shot measurement. However, in order to know the spin state precisely, namely, the density matrix in the spin space, we have to achieve the spin state tomography. This tomography cannot be carried out by a single-shot measurement but repeated measurements are necessary under the assumption that exactly the same spin state can be prepared repeatedly. This tomography will be necessary to estimate, for example, the degree of the entanglement between two electrons created at distant locations after photo-absorption of a pair of polarization entangled photons. The tomographic methods for a single photon and two photons were already established [47] and it is instructive to review the tomography of the photon states in considering the spin state tomography.

Photons have two orthogonal polarization states, namely, horizontal (H) and vertical (V) polarizations denoted by

$$|H\rangle = \begin{pmatrix} 1 \\ 0 \end{pmatrix} \;,\; |V\rangle = \begin{pmatrix} 0 \\ 1 \end{pmatrix} \;. \tag{110}$$

The photon density matrix, which is a 2×2 matrix, can be decomposed as

$$\rho = \frac{1}{2}(\overleftrightarrow{1} + \vec{s} \cdot \vec{\sigma}) \quad \text{with} \quad s_i = \mathrm{Tr}\; \rho\sigma_i \quad (i = x,\; y,\; z)\;, \tag{111}$$

where s_i's are the Stokes parameters and σ_i's are the Pauli spin operators. These parameters can be fixed by the apparatus exhibited typically in Fig. 23. The polarizer is assumed to transmit only the vertically polarized photons. The fast axes of the quarter wave plate (QWP) and the half wave plate (HWP) are adjusted to select a particular polarization component. In the $|H\rangle$ and $|V\rangle$ bases, an arbitrary photon state is given by

$$\begin{pmatrix} \bar{a} \\ \bar{b} \end{pmatrix} = a\,|H\rangle + b\,|V\rangle \;, \tag{112}$$

where a and b are arbitrary complex constants satisfying the normalization condition: $|a\rangle^2 + |b\rangle^2 = 1$. Then the role of the HWP is described by

$$\begin{pmatrix} \bar{a} \\ \bar{b} \end{pmatrix} = U_{HWP}(h) \begin{pmatrix} a \\ b \end{pmatrix} \quad \text{with} \quad U_{HWP}(h) = \begin{pmatrix} \cos 2h & -\sin 2h \\ -\sin 2h & -\cos 2h \end{pmatrix} \;, \tag{113}$$

where the left hand side is the photon state after transmission through the HWP and h denotes the angle between the fast axis of the HWP and the vertical direction. In the same way, the effect of the QWP is described by

$$\begin{pmatrix} \bar{a} \\ \bar{b} \end{pmatrix} = U_{QWP}(q) \begin{pmatrix} a \\ b \end{pmatrix} \quad \text{with} \quad U_{QWP}(q) = -e^{i\pi/4} \frac{1}{\sqrt{2}} \begin{pmatrix} i - \cos 2q & \sin 2q \\ \sin 2q & i + \cos 2q \end{pmatrix}, \tag{114}$$

where q denotes the angle between the fast axis of the QWP and the vertical direction. Thus the polarization state of photons which can be transmitted to the detector is given by

$$\begin{pmatrix} a \\ b \end{pmatrix} = U^{\dagger}_{QWP}(q) U^{\dagger}_{HWP}(h) \begin{pmatrix} 0 \\ 1 \end{pmatrix} = \frac{1}{\sqrt{2}} \begin{pmatrix} i \sin 2h + \sin 2(h-q) \\ i \cos 2h - \cos 2(h-q) \end{pmatrix}. \tag{115}$$

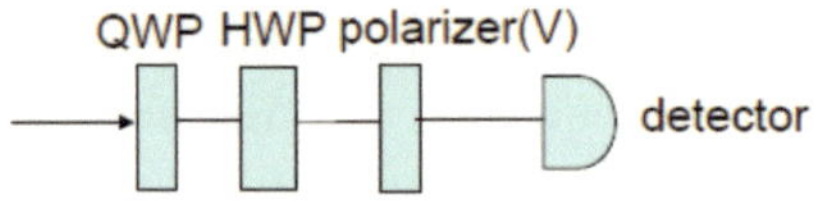

Fig. 23. Apparatus to select a particular polarization component of light. It is composed of a half wave plate (HWP), a quarter wave plate (QWP), a polarizer which is assumed to transmit only a vertically polarized photon, and a photon detector.

polarization	h	q
$\vert H\rangle$	$\pi/4$	0
$\vert V\rangle$	0	0
$\vert D\rangle$	$\pi/8$	$\pi/4$
$\vert R\rangle$	$\pi/8$	0

Table I. Combinations of the angles h and q of the fast axes of the HWP and QWP to select a particular polarization component of light

The four bases of measurement are exhibited as

$$|H\rangle = \begin{pmatrix} 1 \\ 0 \end{pmatrix}, \ |V\rangle = \begin{pmatrix} 0 \\ 1 \end{pmatrix}, |D\rangle = \frac{1}{\sqrt{2}} \begin{pmatrix} 1 \\ 1 \end{pmatrix}, |R\rangle = \frac{1}{\sqrt{2}} \begin{pmatrix} 1 \\ i \end{pmatrix}, \tag{116}$$

where $D(R)$ denotes the diagonally (right-circularly) polarized photon. These photon states can be selected by choosing the angles of the fast axes of the HWP and QWP as given in Table I. Then the photon count rates for these polarizations are calculated as

$$N_H \propto \mathrm{Tr}\rho|H\rangle\langle H| = \frac{1}{2}(1+s_z), \ N_V \propto \mathrm{Tr}\rho|V\rangle\langle V| = \frac{1}{2}(1-s_z), \tag{117}$$

$$N_D \propto \mathrm{Tr}\rho|D\rangle\langle D| = \frac{1}{2}(1+s_x), \ N_R \propto \mathrm{Tr}\rho|R\rangle\langle R| = \frac{1}{2}(1-s_y) \tag{118}$$

and the Stokes parameters can be estimated by

$$s_x = 2\frac{N_D}{N_0} - 1, \ s_y = 1 - 2\frac{N_R}{N_0}, \ s_z = \frac{N_H - N_V}{N_0} \quad \text{with} \quad N_0 = N_H + N_V. \tag{119}$$

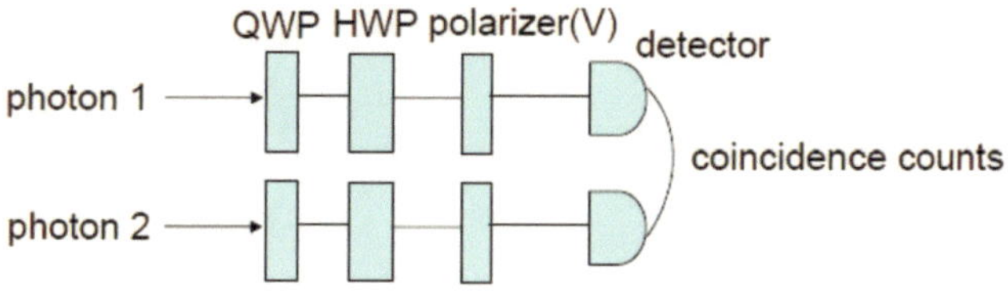

Fig. 24. Parallel combination of two apparatus to select a particular polarization component of photons.

The extension of this scheme to the state tomography of two photons is straightforward [47]. We prepare parallel lines of apparatus as depicted in Fig. 24. By measuring the coincidence counts for sixteen combinations of the angles of the fast axes of two HWPs and two QWPs, we can fix the sixteen parameters $\{r_{ij}\}(I,\ j = 0,\ 1,\ 2,\ 3)$ which specify the density matrix for two photons:

$$\rho = \sum_{i,j=0}^{3} r_{ij}\ \sigma_i^1 \otimes \sigma_j^2\ , \tag{120}$$

where the superscript 1(2) indicates the first (second) photon.
Now we proceed to the spin state tomography of electrons. Extension of the above arguments to the electron spin state tomography is straightforward using the correspondence:

$$|H\rangle \leftrightarrow |\uparrow\rangle\ ,\ |V\rangle \leftrightarrow |\downarrow\rangle\ , \tag{121}$$

where $|\uparrow\rangle(|\downarrow\rangle)$ indicates the spin up (down) state of the electron. But the devices corresponding to the HWP and QWP should be prepared for the electron. This can be realized by making use of the difference in the spin precession angle due to the Zeeman energy splitting. In a magnetic field along the z axis, the spin state evolves in time as

$$\begin{pmatrix} a \\ b \end{pmatrix} \to \begin{pmatrix} e^{-i\omega_z t}a \\ e^{i\omega_z t}b \end{pmatrix} = e^{-i\omega_z t}\begin{pmatrix} a \\ e^{i2\omega_z t}b \end{pmatrix}\ , \tag{122}$$

where $\hbar\omega_z$ is the Zeeman energy for the up-spin electron. The spin HWP (QWP) can be achieved by employing a square-shaped pulsed magnetic field and by adjusting the pulse duration and the field amplitude as $2\omega_z t = \pi(\pi/2)$. The angle of the fast axis corresponds to the tilt angle of the magnetic field. For example, when the tilt angle of the magnetic field is h relative to the z axis, the spin eigenstates are given by

$$|\uparrow\rangle_h = \begin{pmatrix} \cos\frac{h}{2} \\ \sin\frac{h}{2} \end{pmatrix}\ ,\ |\downarrow\rangle_h = \begin{pmatrix} -\sin\frac{h}{2} \\ \cos\frac{h}{2} \end{pmatrix}\ . \tag{123}$$

Then the effect of the "spin" HWP is represented by

$$\begin{pmatrix} \bar{a} \\ \bar{b} \end{pmatrix} = U^s_{HWP}(h)\begin{pmatrix} a \\ b \end{pmatrix} \quad \text{with} \quad U^s_{HWP}(h) = \begin{pmatrix} \cos h & \sin h \\ \sin h & -\cos h \end{pmatrix}\ . \tag{124}$$

In a similar way, the effect of the "spin" QWP is given as

$$\begin{pmatrix} \bar{a} \\ \bar{b} \end{pmatrix} = U^s_{QWP}(q) \begin{pmatrix} a \\ b \end{pmatrix} \quad \text{with} \quad U^s_{QWP}(q) = e^{i\pi/4} \frac{1}{\sqrt{2}} \begin{pmatrix} -i+\cos q & \sin q \\ \sin q & -i-\cos q \end{pmatrix} , \tag{125}$$

where q is the tilt angle of the magnetic field in the "spin" QWP.

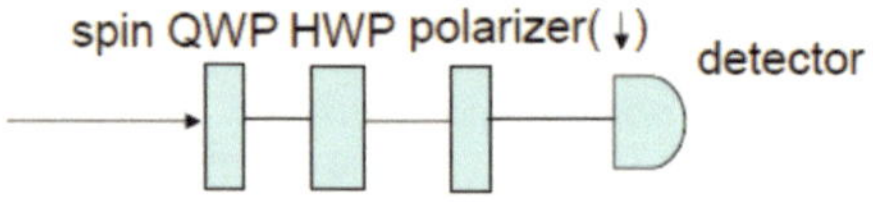

Fig. 25. Apparatus to select a particular spin component of an electron. It is composed of a spin half wave plate (spin HWP), a spin quarter wave plate (spin QWP), a spin polarizer which is assumed to transmit only the down spin electron, and an electron detector.

Now that the basic elements are prepared, we can consider the apparatus for the spin state tomography, as depicted in Fig. 25, which is basically the same as that for the photon case except for the replacement of the HWP and QWP by the spin HWP and spin QWP, respectively. The spin polarizer is assumed to transmit only the down spin electron based on the difference in the Faraday rotation angle. This polarizer should have the ability to measure the single electron spin. Consequently, the apparatus depicted in Fig. 25 filters out the state given by

$$\begin{pmatrix} a \\ b \end{pmatrix} = U^{s\,\dagger}_{QWP}(q) U^{s\,\dagger}_{HWP}(h) \begin{pmatrix} 0 \\ 1 \end{pmatrix} = \frac{1}{\sqrt{2}} \begin{pmatrix} i\sin h + \sin(h-q) \\ -i\cos h + \cos(h-q) \end{pmatrix} . \tag{126}$$

spin direction	h	q
$\mid\uparrow_z\rangle$	$\pi/2$	0
$\mid\downarrow_z\rangle$	0	0
$\mid\downarrow_x\rangle$	$\pi/4$	$\pi/2$
$\mid\downarrow_y\rangle$	$\pi/4$	0

Table II. Combinations of the angles h and q of the fast axes of the spin HWP and spin QWP to select a particular spin component of an electron.

Thus by choosing the angles h and q as exhibited in Table II, the apparatus can filter out those spin states in the Table. The spin state tomography can be carried out in the same manner as that for the photon. The extension of the spin state tomography to the case of two electrons is straightforward. We prepare a parallel set of the apparatus for a single electron as depicted in Fig. 26. Coincidence counts of electrons for sixteen combinations of the angles of the fast axes of two spin HWPs and two spin QWPs can determine, in completely the same way as in the case of two photons, the sixteen parameters $\{r_{ij}\}$ (i, j = 0, 1, 2, 3) of the density matrix of the electron spins:

$$\rho = \sum_{i,j=0}^{3} r_{ij}\, \sigma^1_i \otimes \sigma^2_j , \tag{127}$$

where the superscript 1(2) indicates the first (second) electron.

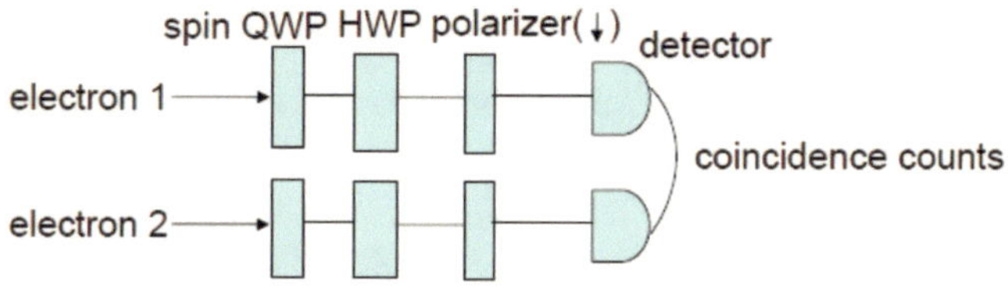

Fig. 26. Parallel combination of two apparatus to select a particular spin component of electrons.

7. Summary

We have investigated theoretically the STIRAP method based on the Λ-type optical transitions for the unitary spin rotation of both a single electron and two electrons and discussed the optimal conditions to minimize the residual population left in the intermediate excited states after the process. The 2π pulse with a large off-resonance energy would be better but the rotation angle is small. Thus, for the rotation of a large angle, multiple pulses would be necessary. For the two-electrons spin qubit, several Λ-type transitions are closely overlapping in the narrow energy range and the Raman condition should be tuned precisely to single out a particular Λ-type transition. It is absolutely necessary to confirm the quantum state of the electron after the spin manipulation, namely, to examine whether the electron spin is prepared in the desired state or not. This requires the full state tomography. We have proposed the methods of the spin state tomography based on the Faraday or Kerr rotation for both cases of a single electron and two electrons. Another fundamental process in the quantum information processing is the Bell (quantum correlation) measurement for the entanglement swapping. This is a single-shot projective measurement in contrast to the full state tomography. We proposed an optical method to distinguish between the four states of two electrons based on the Faraday or Kerr rotation and confirmed the feasibility.

As a prospect in the near future, we can construct hopefully a secure and robust system of the quantum repeater combining the established results, namely, the efficient quantum state transfer between a photon and an electron spin [37, 38, 48], the reliable Bell measurement of two electrons for the entanglement swapping based on the Faraday or Kerr rotation and the longlived quantum memory based on nuclear spins [49, 50].

8. Acknowledgments

We would like to thank Professor H. Kosaka for stimulating discussions and continual encouragements. We appreciate very much the financial supports from the CREST project of the Japan Science and Technology Agency, from the SCOPE program of the Ministry of Internal Affairs and Communications, and from the Ministry of Education, Culture, Sports, Science and Technology, Japan. The numerical calculation in this work has been done using the facilities of the Supercomputer Center, Institute for Solid State Physics, University of Tokyo.

9. References

[1] A. Barenco, D. Deutsch, A. Ekert, and R. Jozsa, "Conditional quantum dynamics and logic gates," *Phys. Rev. Lett.* 74, 4083–4086 (1995).

[2] D. Loss and D. P. DiVincenzo, "Quantum computation with quantum dots," *Phys. Rev. A* 57, 120-126 (1998).

[3] A. Imamoglu, D. D. Awschalom, G. Burkard, D. P. DiVincenzo, D. Loss, M. Sherwin, and A. Small, "Quantum information processing using quantum dot spins and cavity QED," *Phys. Rev. Lett.* 83, 4204-4207 (1999).

[4] C. H. Bennett and D. P. DiVincenzo, "Quantum information and computation," *Nature* 404, 247–255 (2000).

[5] M. A. Nielsen and I. L. Chuang (Eds.), *Quantum Computation and Quantum Information*, Cambridge University Press, Cambridge (2000).

[6] D. Bouwmeester, A. Ekert and A. Zeilinger (Eds.), *The Physics of Quantum Information*, Springer- Verlag, Berlin (2000).

[7] D. D. Awschalom, D. Loss and N. Samarth (Eds.), *Semiconductor Spintronics and Quantum Computation*, Springer-Verlag, Berlin (2002).

[8] T. Takagahara (Ed.), *Quantum Coherence, Correlation and Decoherence in Semiconductor Nanostructures*, Academic Press, Elsevier, New York (2003).

[9] T. Takagahara (Ed.), *Proceedings of the International Symposium on Photonics and Spintronics in Semiconductor Nanostructures* (PSSN 2003, Kyoto) available online at http://www.

[10] X. Li, Y. Wu, D. G. Steel, D. Gammon, T. H. Stievater, D. S. Katzer, D. Park, C. Piermarocchi, and L. J. Sham, "An all-optical quantum gate in a semiconductor quantum dot," *Science* 301, 809-811 (2003).

[11] R. Hanson, L. P. Kouwenhoven, J. R. Petta, S. Tarucha, and L. M. Vandersypen, "Spins in fewelectron quantum dots", *Rev. Mod. Phys.* 79, 1217 (2007).

[12] Dyakonov, M. I. (Ed.), *Spin Physics in Semiconductors*, Springer-Verlag, Berlin (2008).

[13] Henneberger, F. and Benson, O. (Eds.), *Semiconductor Quantum Bits*, Pan Stanford Publishing Pte. Ltd., Singapore (2009).

[14] R. Hanson, B. Witkamp, L. M. K. Vandersypen, L. H. Willems van Beveren, J. M. Elzerman, and L. P. Kouwenhoven, "Zeeman energy and spin relaxation in a one-electron quantum dot," *Phys. Rev. Lett.* 91, 196802 (2003).

[15] M. Kroutvar, Y. Ducommun, D. Heiss, M. Bichler, D. Schuh, G. Abstreiter, and J. J. Finley, "Optically programmable electron spin memory using semiconductor quantum dots," *Nature* 432, 81-84 (2004).

[16] J. R. Petta, A. C. Johnson, J. M. Taylor, E. A. Laird, A. Yacoby, M. D. Lukin, C. M. Marcus, M. P. Hanson, and A. C. Gossard, "Coherent manipulation of coupled electron spins in semiconductor quantum dots," *Science* 309, 2180-2184 (2005).

[17] F. H. L. Koppens, C. Buizert, K. J. Tielrooij, I. T. Vink, K. C. Nowack, T. Meunier, L. P. Kouwenhoven, and L. M. K. Vandersypen, "Driven coherent oscillations of a single electron spin in a quantum dot," *Nature* 442, 766-771 (2006).

[18] T. Meunier, I. T. Vink, L. H. Willems van Beveren, K. J. Tielrooij, R. Hanson, F. H. L. Koppens, H. P. Tranitz, M. Wegscheider, L. P. Kouwenhoven, and L. M. K. Vandersypen, "Experimental signature of phonon-mediated spin relaxation in a two-electron quantum dot," *Phys. Rev. Lett.* 98, 126601 (2007).

[19] H. Kosaka, A. A. Kiselev, F. A. Baron, K. W. Kim, and E. Yablonovitch, "Electron g factor engineering in III-V semiconductors for quantum communications," *Electron. Lett.* 37, 464-465 (2001).

[20] R. Vrijen and E. Yablonovitch, "A spin-coherent semiconductor photo-detector for quantum communication," *Physica E* 10, 569-575 (2001).

[21] E. Yablonovitch, H. W. Jiang, H. Kosaka, H. D. Robinson, D. S. Rao, and T. Szkopek, "Optoelectronic quantum telecommunications based on spins in semiconductors," *Proc. IEEE* 91, 761-780 (2003).

[22] H. Kosaka and E. Yablonovitch, "Quantum media converter from a photon qubit to an electronspin qubit for quantum repeaters," in *Proceedings of the International Symposium on Photonics and Spintronics in Semiconductor Nanostructures* (PSSN 2003, Kyoto), T. Takagahara, Ed., pp. 63-70 (2003).

[23] M. Atatüre, J. Dreiser, A. Badolato, A. Hogele, K. Karrai, and A. Imamoglu, "Quantum-dot spin-state preparation with near-unity fidelity," *Science* 312, 551-553 (2006).

[24] C. Emary, X. Xu, D. G. Steel, S. Saikin, and L. J. Sham, "Fast Initialization of the Spin State of an Electron in a Quantum Dot in the Voigt Configuration", *Phys. Rev. Lett.* 98, 047401 (2007).

[25] X. Xu, Y. Wu, Bo Sun, Q. Huang, J. Cheng, D. G. Steel, A. S. Bracker, D. Gammon, C. Emary, and L. J. Sham, "Fast Spin State Initialization in a Singly Charged InAs-GaAs Quantum Dot by Optical Cooling", *Phys. Rev. Lett.* 99, 097401 (2007).

[26] K. C. Nowack, F. H. L. Koppens, Yu. V. Nazarov, and L. M. K. Vandersypen, "Coherent Control of a Single Electron Spin with Electric Fields", *Science* 318, 1430-1433 (2007).

[27] P. Chen, C. Piermarocchi, L. J. Sham, D. Gammon, and D. G. Steel , "Theory of quantum optical control of a single spin in a quantum dot", *Phys. Rev. B* 69, 075320 (2004).

[28] M. V. G. Dutt, J. Cheng, B. Li, X. Xu, Xiaoqin Li, P. R. Berman, D. G. Steel, A. S. Bracker, D. Gammon, S. E. Economou, R. B. Liu, and L. J. Sham, "Stimulated and spontaneous optical generation of electron spin coherence in charged GaAs quantum dots," *Phys. Rev. Lett.* 94, 227403 (2005).

[29] S. E. Economou and L. J. Sham, Y. Wu and D. G. Steel, "Proposal for optical U(1) rotations of electron spin trapped in a quantum dot", *Phys. Rev. B* 74, 205415 (2006).

[30] Y. Wu, E. D. Kim, X. Xu, J. Cheng, D. G. Steel, A. S. Bracker, D. Gammon, S. E. Economou, and L. J. Sham, "Selective Optical Control of Electron Spin Coherence in Singly Charged GaAs-Al0.3Ga0.7As Quantum Dots", *Phys. Rev. Lett.* 99, 097402 (2007).

[31] S. E. Economou and T. L. Reinecke, "Theory of Fast Optical Spin Rotation in a Quantum Dot Based on Geometric Phases and Trapped States", *Phys. Rev. Lett.* 99, 217401 (2007).

[32] J. Berezovsky, M. H. Mikkelsen, N. G. Stoltz, L. A. Coldren, and D. D. Awschalom, "Picosecond Coherent Optical Manipulation of a Single Electron Spin in a Quantum Dot", *Science* 320 349-352 (2008).

[33] D. Press, T. D. Ladd, B. Zhang, Y. Yamamoto, "Complete quantum control of a single quantum dot spin using ultrafast optical pulses", *Nature* 456, 218-221 (2008).

[34] A. M. Tyryshkin, S. A. Lyon, A. V. Astashkin, and A. M. Raitsimring, "Electron spin relaxation times of phosphorus donors in silicon", *Phys. Rev. B* 68, 193207 (2003).

[35] M. V. Gurudev Dutt, L. Childress, L. Jiang, E. Togan, J. Maze, F. Jelezko, A. S. Zibrov, P. R. Hemmer, and M. D. Lukin, "Quantum Register Based on Individual Electronic and Nuclear Spin Qubits in Diamond", *Science* 316, 1312-1316 (2007).
[36] J. J. L. Morton, A. M. Tyryshkin, R. M. Brown, S. Shankar, B. W. Lovett, A. Ardavan, T. Schenkel, E. E. Haller, J. W. Ager, S. A. Lyon, "Solid-state quantum memory using the 31P nuclear spin", *Nature* 455, 1085-1088 (2008).
[37] H. Kosaka, H. Shigyou, Y. Mitsumori, Y. Rikitake, H. Imamura, T. Kutsuwa, K. Arai, and K. Edamatsu, "Coherent Transfer of Light Polarization to Electron Spins in a semiconductor", *Phys. Rev. Lett.* 100, 096602 (2008).
[38] H. Kosaka, T. Inagaki, Y. Rikitake, H. Imamura, Y. Mitsumori, K. Edamatsu, "Spin state tomography of optically injected electrons in a semiconductor", *Nature* 457, 702-705 (2009).
[39] N. Rosen and C. Zener, "Double Stern-Gerlach Experiment and Related Collision Phenomena", *Phys. Rev.* 40, 502 - 507 (1932).
[40] A. Bambini and P. R. Berman, "Analytic solutions to the two-state problem for a class of coupling potentials", *Phys. Rev. A* 23, 2496 - 2501 (1981).
[41] M. Abramowitz and I. A. Stegun (Ed.), *Handbook of Mathematical Functions,* Dover Publications, Inc., New York (1972).
[42] J. Berezovsky, M. H. Mikkelsen, O. Gywat, N. G. Stoltz, L. A. Coldren, and D. D. Awschalom, "Nondestructive optical measurements of a single electron spin in a quantum dot," *Science* 314, 1916-1920 (2006).
[43] M. Atature, J. Dreiser, A. Badolato, and A. Imamoglu, "Observation of Faraday rotation from a single confined spin," *Nature Phys.* 3, 101-106 (2007).
[44] A. A. Kiselev, K. W. Kim and E. Yablonovitch, "In-plane light-hole g factor in strained cubic heterostructures", *Phys. Rev. B* 64, 125303 (2001).
[45] R. B. Dingle, "Some magnetic properties of metals. I. general introduction, and properties of large systems of electrons," *Proc. R. Soc. London, Ser. A* 211, 500-516 (1952).
[46] R. B. Dingle, "Some magnetic properties of metals. III. diamagnetic resonance," *Proc. R. Soc. London, Ser. A* 212, 38-47 (1952).
[47] D. F. V. James, P. G. Kwiat, W. J. Munro, and A. G. White, "Measurement of qubits", *Phys. Rev. A* 64, 052312 (2001).
[48] T. Takagahara and O. Cakir, "Theoretical aspects of quantum state transfer, correlation measurement and electron-nuclei coupled dynamics in quantum dots", *J. Nanophotonics* 1, 011593 (2007).
[49] J. M. Taylor, C. M. Marcus, and M. D. Lukin, "Long-lived memory for mesoscopic quantum bits," *Phys. Rev. Lett.* 90, 206803 (2003).
[50] O. Cakir and T. Takagahara, "Quantum dynamics in electron-nuclei coupled spin system in quantum dots: Bunching, revival, and quantum correlation in electron-spin measurements", *Phys. Rev. B* 77, 115304 (2008).

12

Stimulated Brillouin Scattering Phase Conjugate Mirror and its Application to Coherent Beam Combined Laser System Producing a High Energy, High Power, High Beam Quality, and High Repetition Rate Output

Hong Jin Kong, Seong Ku Lee[1], Jin Woo Yoon[1],
Jae Sung Shin, and Sangwoo Park
Department of Physics, Korea Advanced Institute of Science and Technology
[1]Advanced Photonics Research Institute, Gwangju Institute of Science and Technology
Republic of Korea

1. Introduction

Stimulated Brillouin scattering (SBS) is a nonlinear optical process that generates backward scattered phase conjugate wave (Zel'dovich et al., 1972; Zel'dovich et al., 1985; Damzen et al., 2003; Brignon & Huignard, 2004). A device that generates the phase conjugate wave by the SBS process is called SBS phase conjugate mirror (PCM). An SBS-PCM can compensate wavefront distortion induced by a phase aberrator, such as a laser gain medium; hence, it is widely used in high-energy laser systems to obtain a high-quality beam. Efficient heat dissipation is a major issue in high-energy laser systems, particularly with regard to the high repetition rate. The combination of beams from small laser systems is a constructive approach to this issue. Of the various beam combined systems using SBS-PCMs, the cross-type beam combined system has many outstanding advantages, such as perfect isolation of leak beam, compensation of thermal birefringence, easy alignment and convenient maintenance (Kong et al., 1997, 2005a). Since the SBS wave generates from a thermal noise, it naturally has a random phase with respect to the incident beam. Therefore the phase controlling of the SBS wave is a key technology in the realization of a coherent beam combined system. For this reason, the self-phase control method was proposed and has been developed by Kong et al. (2005a, 2005b, 2005c), which can control the phase of the SBS wave with the simplest composition as well as ease of alignment, no limitations on the number of combined beams, and excellent phase conjugation. Furthermore, the active phase control with a piezoelectric translator (PZT) enables long-term phase stabilization (Kong et al., 2006, 2008). In addition to a random phase characteristic, the distortion that generally occurs in a pulse waveform of an SBS wave is another negative characteristic in terms of the beam combination. Kong et al. (2005d) has overcome this problem with the SBS waveform preservation technique, which is called the prepulse injection method. These works are expected to boost the development of laser systems in term of a high level of energy and power, a high-quality beam, and a high repetition rate. A laser system with these characteristics has tremendous potential in the fields such as laser machining, particle accelerators, neutron or proton generators, military weapons, and laser fusion drivers.

2. Stimulated Brillouin scattering phase conjugate mirror

2.1 Optical phase conjugation

Optical phase conjugation is a nonlinear optical phenomenon that accurately reverses the propagation direction and phase variation of an incoming light beam. A nonlinear optical device that produces the phase conjugate reflection is called a PCM (Zel'dovich et al., 1972; Eichler et al., 2001). Fig. 1 compares a PCM with a conventional mirror. In case of a conventional mirror, the wavefront is distorted twice when passing through an aberrating medium twice; however, there is no distortion with the PCM. The use of a PCM therefore eliminates phase distortions in optical systems (Andreev et al., 1992; Seidel et al., 1997). For example, in a solid-state laser amplifier, phase distortions arise from thermal refractive index changes in the laser crystal. If a PCM is used to make the incoming beam pass the laser crystal twice, these distortions disappear. Hence, PCMs are widely used in the high-energy laser systems.

Fig. 1. Wavefront reflections at a conventional mirror and a phase conjugate mirror

2.2 Stimulated Brillouin scattering

The most commonly used way of producing optical phase conjugation is SBS (Boyd, 1992; Damzen et al., 2003; Shen, 2003). SBS is normally achieved by focusing a laser beam into a SBS medium. In this medium, spontaneous scattering from a sound wave generates a wave that travels in the opposite direction and that wave interferes with the incoming wave, thereby inducing density modulations. Because the induced density modulations have the same frequency as the initial sound wave, they are amplified and reinforce the backscattering. The phase conjugated backscattered part is dominant because the amplification depends strongly on the extension of the interference area. As a result, there is an exponential rise of the reflected phase conjugated signal. The acoustic grating seems to be a self-adapting mirror because the wavefronts of the acoustic grating match the wavefronts of the incoming beam. If the incident wavefront has any disturbance, it results in a self-adapting mirror curvature with a response time in the nanosecond range. In addition, SBS lowers the frequency of the phase conjugated wave as much as the sound wave frequency in accordance with the energy conservation.

3. Reflectivity of an SBS-PCM

The SBS reflectivity is almost equal to that of an ordinary mirror when the pump bandwidth, $\Delta\nu_p$, is smaller than the Brillouin linewidth, Γ (steady state region) (Boyd, 1992;

Yoshida et al., 1997). However, many SBS-PCM applications necessarily involve a broadband pumped SBS (transient region) because laser systems that use an SBS-PCM usually have a broadband spectrum to obtain high output power and short pulse widths (Dane et al., 1995; Králiková et al., 2000). Several theoretical and experimental investigations have reported on the use of a broadband pump in the SBS reflectivity. For a broadband pump, the SBS reflectivity depends on the relation between four parameters: the coherence length, l_c; the characteristic interaction length, z_0, which is usually equal to the Rayleigh range; the mode spacing, Ω_m; and the Brillouin linewidth, Γ. When the coherence length is longer than the interaction length ($l_c > z_0$), the SBS gain for the broadband pump is as high as the coherence length for the narrowband pump (Narum et al., 1986; D'yakov, 1970; Filippo & Perrone, 1992). Furthermore, if the pump laser mode spacing exceeds the Brillouin linewidth ($\Omega_m > \Gamma$), regardless of the mode structure, the SBS gain is the same as that of a single longitudinal mode pump (Narum et al., 1986). Moreover, even if $\Omega_m < \Gamma$, the off-resonant acoustic waves, which are generated by the beating between the pump laser mode and another Stokes mode, play an important role in enhancing the gain and the reflectivity (Mullen et al., 1987; Bullock et al., 1994). In all the previously mentioned works, however, the influence of the multimode pump was considered only for two or several longitudinal modes and low pump energy near the SBS threshold. For this reason, the characteristics of SBS reflectivity by a multimode pump with numerous modes and high energy have been investigated (Lee et al., 2005; Kong et al., 2005a).

Liquid	Γ (MHz)	g_B (cm/GW)	n_2 (10^{-22} m²/V²)	P_c (MW)	E_b (mJ)
Fluorinert FC-75	350	4.5-5	0.34	7.0	6
Carbon tetrachloride (CCl_4)	528	3.8	5.9	0.4	1.7
Acetone	119	15.8	8.6	0.28	1.5
Carbon disulfide (CS_2)	50	68	122	0.020	0.1

Table 1. Properties of the liquids used for the reflectivity experiments; Γ, Brillouin linewidth; g_B, steady state SBS gain; n_2, nonlinear refractive index; P_c, critical power for self-focusing (calculated); E_b, breakdown threshold energy (measured).

The experimental setup for measuring the reflectivity of the SBS-PCMs is shown in Fig. 2. The pump laser is a Q-switched Nd:YAG laser and, using its single longitudinal mode injection seeder, it can be operated in the single-mode or the multimode. The laser linewidth is approximately 0.09 GHz in the single-mode and approximately 30 GHz in the multimode. Thus, the linewidth of the multimode case is much larger than the Brillouin linewidth of the liquids used in this experiment, listed in Table 1. (Kmetik et al., 1998; Yoshida et al., 1997; Erokhin et al., 1986; Sutherland, 1996). The focal length of the lens used for the SBS-PCM is 15 cm. This length corresponds to a Rayleigh range, z_0, of 0.62 mm. Because the coherence length, l_c, is approximately 1 cm, it satisfies the condition of $l_c >> z_0$. The temporal and spatial pulse width is 6 ns to 8 ns and 4 mm. The pump energy fluctuation is less than 1% for both cases, and the energy is measured for about 30 s at 10 Hz.

The SBS materials used in this experiment are Fluorinert FC-75, carbon tetrachloride (CCl_4), acetone, and carbon disulfide (CS_2). The SBS properties and the nonlinear refractive index, n_2, of each liquid are shown in Table 1. They have a different nonlinear refractive index, n_2, ranging from 0.34×10^{-22} m²/V² to 122×10^{-22}m²/V². Furthermore, each liquid has a different

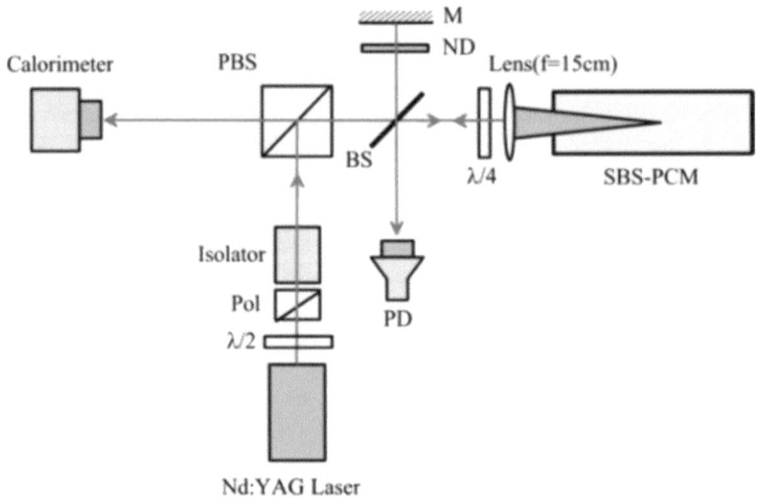

Fig. 2. Experimental setup for the measurement of the SBS reflectivity; λ/2, half wave plate; Pol, polarizer; M, mirror; ND, neutral density filter; λ/4, quarter wave plate; PBS, polarizing beam splitter; BS, beam splitter; PD, photodiode

Brillouin linewidth, ranging from 50 MHz to 528 MHz. The breakdown threshold, E_b, which is listed in Table 1, was measured when a bright spark appeared inside the SBS cell.

Figs. 3(a) and 3(b) show the SBS reflectivity for the single-mode and the multimode cases of CCl4 and Fluorinert FC-75 as a function of the pump energy. Note that CCl_4 and FC-75 have a very similar SBS gain and Brillouin linewidth (see Table 1), which results in a similar reflectivity curve of typical nonlinear variation for the single-mode pump. For the multimode pump, the SBS reflectivity is different for each of the liquids. The fact that the SBS threshold for both pump modes is approximately the same regardless of the liquids indicates that the SBS gain for the multimode pump is as high as that for the single-mode pump in both liquids The peak reflectivity is 30% in CCl4 and more than 65% in FC-75, though the reflectivity decreases as the pump energy increases. Note also that even though the single-mode pump generally has a higher SBS gain (Valley, 1986; Arecchi, 1972), the SBS reflectivity in CCl_4 is slightly higher for the multimode pump than for the single-mode pump near the SBS threshold of the single-mode case. For FC-75, on the other hand, the behavior is exactly the opposite.

Several factors appear to contribute to the reflectivity difference of the multimode pump. We interpret the SBS reflectivity for the multimode pump in terms of the temporal intensity spikes of the multimode pulse, which are absent in the single-mode pulse. A beating between a large numbers of longitudinal mode brings the intensity spikes to rise, and the intensity spikes have enough power to induce nonlinear effects, such as self-focusing and the optical breakdown. The self-focusing that is caused by the intensity spikes is likely to lead to an anomalously high reflectivity of the multimode pump near the SBS threshold in CCl_4. The critical power of the self-focusing is given by

$$P_c = \frac{\pi \varepsilon_o c^3}{n_2 \omega^2}, \tag{1}$$

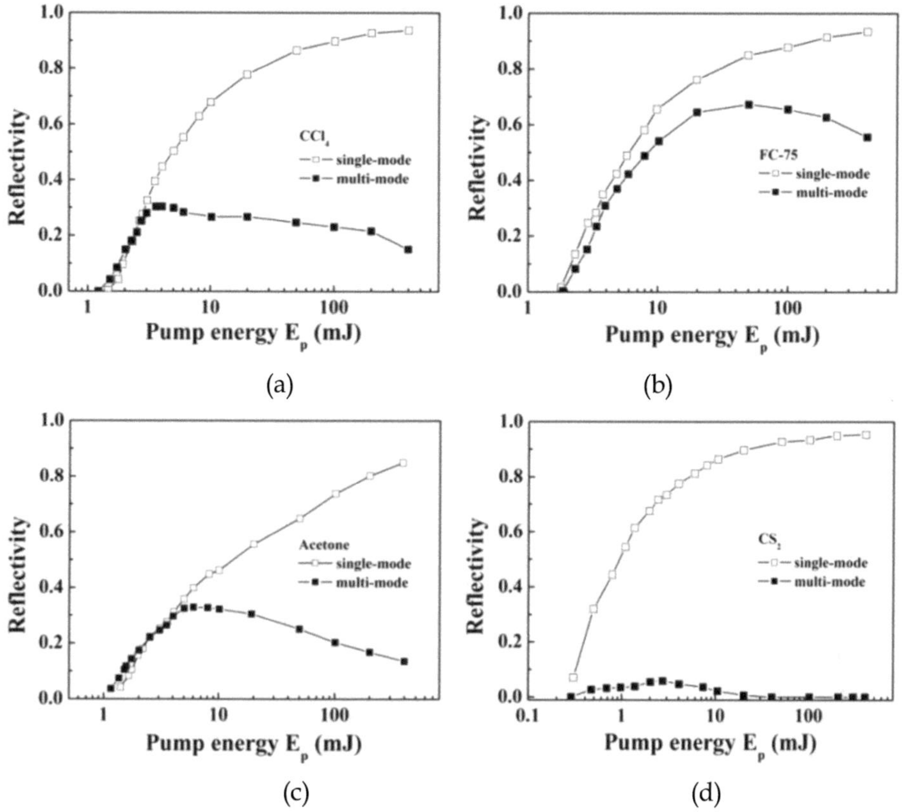

Fig. 3. SBS reflectivity vs. pump energy for various active media in the single-mode and the multimode cases; (a) CCl_4; (b) FC-75; (c) acetone; (d) CS_2

where ε_0 is the permittivity of the vacuum and c is the speed of light (Yariv, 1975). Acording to Eq. (1), the critical power, P_c, is 0.4 MW for CCl_4 and 7.0 MW for FC-75. In the case of the single-mode pump for CCl_4, the critical power of 0.4 MW is slightly larger than the SBS threshold (~5% energy reflection) of approximately 1.8 mJ (0.26 MW). However, the multimode pulse can induce temporal small-scale self-focusing in CCl_4 below the SBS threshold because the high peak power of the intensity spikes can exceed the critical power, P_c. If a good approximation of steady state SBS threshold relation $I_{th}g_B l$=25-30 (where I_{th} is the SBS threshold intensity, g_B is the SBS gain, and l is the interaction length) is maintained, the self-focusing leads to an increase in the intensity of the pump beam in the focal region and, hence, can reduce the SBS threshold energy (Králiková et al., 2000). As shown in Fig. 3(a), the self-focusing consequently results in a lower SBS threshold and a slightly higher reflectivity near the SBS threshold in CCl_4. Fig. 4 represents the temporal pulse shapes of the pump and the Stokes pulse in different energy scales for both types of pump when the pump beam with an E_p value of 1.5 mJ is focused into a CCl_4 cell. As expected, the multimode pulse has large intensity spikes whereas the single-mode pulse has no large intensity spikes. On the other hand, the SBS reflectivity in the FC-75 is not affected by the self-focusing near the SBS threshold because the critical power for FC-75 is approximately 18 times larger than the critical power for CCl_4. Consequently, the SBS reflectivity for the multimode pump is lower than that for the single-mode pump near the SBS threshold.

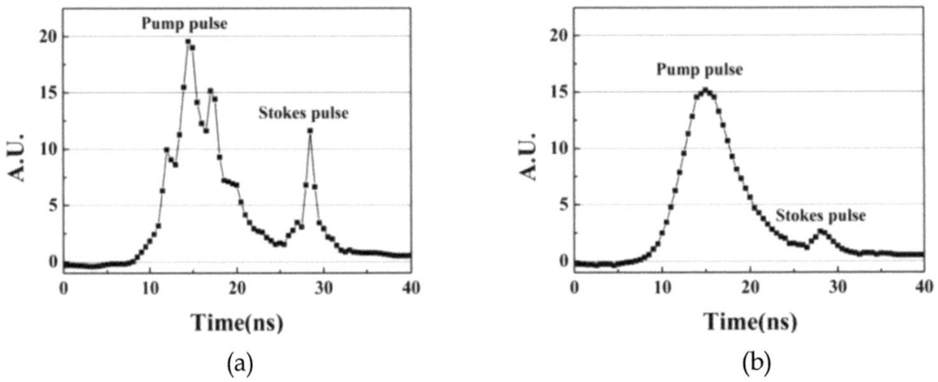

Fig. 4. Pump and reflected pulse shapes in (a) multi-mode and (b) single-mode cases at E_p~1.5 mJ in CCl_4

The self-focusing seems to be deleterious for SBS because it can enhance the optical breakdown. The experimental results in Table 1 confirm that the optical breakdown starts at an E_p value of approximately 1.7 mJ in CCl_4 and at ~6 mJ in FC-75. We observed that the breakdown appears around the focal spot near the breakdown threshold; and when the pump energy increases, the breakdown becomes severe and produces a filament shape consisting of bright sparks. The breakdown disturbs the creation of acoustic phonons.

In addition to the breakdown due to self-focusing, an intensity spike can easily generate an optical breakdown by itself because it has a very steep rising edge. For efficient SBS to occur, temporal fluctuations in the pump must be slow in relation to the acoustic phonon lifetime. If the temporal fluctuations are fast in relation to the acoustic phonon lifetime, the acoustic waves have insufficient time to build up. Hence, the intensity spikes with the steep rising edge and energy levels that exceed the breakdown threshold can reach the focal area without losing their energy since they don't provoke a backward reflection. They can therefore generate an optical breakdown and reduce the SBS reflectivity even at low energy. For the single-mode case, the region of SBS reflection moves fast in the opposite direction to the pump pulse; and the pump pulse is reflected before the focal area from a region in which the optical intensity is too small to induce an optical breakdown (Hon, 1980). Thus, even if the pump energy is large, no optical breakdown is generated for the single-mode pump.

Fig. 3(c) shows the SBS reflectivity for acetone. The multimode pump provides higher reflectivity than the single-mode pump near the SBS threshold, which is very similar to the results of CCl_4. Table 1 shows that acetone has approximately the same nonlinear refractive index as CCl_4. Thus, the SBS reflectivity of the multimode pump with a large number of longitudinal modes is significantly affected by self-focusing induced by the high intensity spikes. The reflectivity for the multimode pump increases as the energy rises to 6 mJ and then decreases strongly because of the severity of the breakdown; in contrast, the reflectivity for the single-mode pump increases monotonically. Fig. 3(d) shows the measured reflectivity of CS_2. Of all the four liquids examined, CS_2 has the lowest SBS threshold energy (approximately 0.3 mJ) and the highest reflectivity (approximately 95%) for the single-mode pump because it has the highest steady state SBS gain (Table 1). On the other hand, the SBS reflectivity for the multimode pump is the lowest and almost zero throughout the entire region. The critical power (20 KW) for the self-focusing is about half the SBS threshold (40 KW). Furthermore, CS_2 has the longest acoustic lifetime (6 ns) of the liquids used (Erokhin et al., 1986), and this life-span is comparable to the pulse-width of the pump beam. As already mentioned, if the temporal fluctuations of the pump pulse are fast in relation to the acoustic phonon lifetime, the acoustic waves lack sufficient time to build up. As a result, CS_2 has a lower breakdown threshold (approximately 0.1 mJ) than the SBS threshold, and this very low threshold can account for the almost zero reflectivity observed. We observed that the optical breakdown produces a filament if the pump energy is almost as weak as the SBS threshold. Note that the stimulated Raman scattering (SRS) may be also responsible for the low reflectivity. CS_2 has a high SRS gain. The very short response time of the SRS process (10^{-11} s) implies that the SRS response to the intensity spikes of the multimode pulse is better than that of the SBS process (Linde et al., 1969).

4. The cross-type double-pass laser amplifier with symmetric SBS-PCMs

Kong et al. proposed a cross-type double-pass laser amplifier with symmetric SBS-PCMs. (Kong et al., 1998; Kong et al., 2001) Fig. 5 shows the conceptual layout. In this section, the noteworthy advantages of the cross-type amplifier with SBS-PCMs will be examined, particularly the use of the threshold of SBS-PCM to attain perfect isolation of the leak beam, the compensation of thermally induced birefringence (TIB), and the alignment-free property.

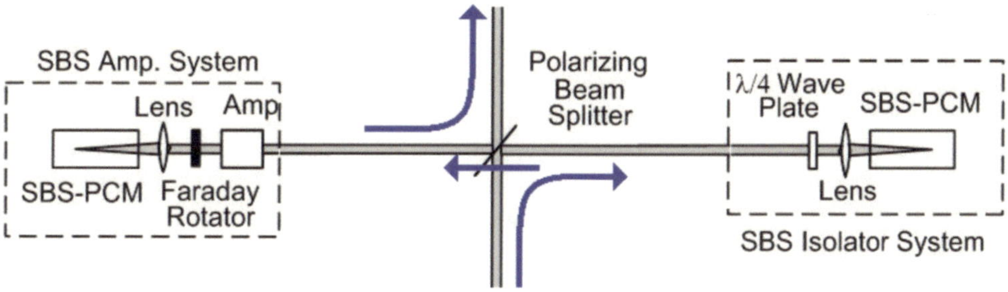

Fig. 5. A cross-type amplifier using SBS-PCM as a basic unit of a beam combined laser

4.1 Optical isolation by the threshold characteristic of SBS

As shown in Fig. 5, the proposed amplifier with an SBS-PCM is cross-shaped and consists of an amplifier and an isolator. The SBS-PCM in the right arm works as an optical isolator. It isolates the master oscillator from the leak beam perfectly if the optical path length, L, between the SBS-PCM and the post-stage is longer than the spatial pulse length. When the energy of the leak beam is lower than the SBS threshold, the leak beam cannot be reflected by the SBS-PCM and consequently the oscillator is isolated from the leak beam (Kong et al., 1998; Kong et al., 2001). Fig. 6(a) shows the experimental setup for testing the isolation when the SBS-PCM is used. An amplifier in the left arm is omitted and a mirror, M1, is in place for the experiment. Fig. 6(b) shows the optical isolator that uses the SBS-PCM can completely cut off the leak beam and the backward propagating beam such as the amplified spontaneous emission from the post-stages.

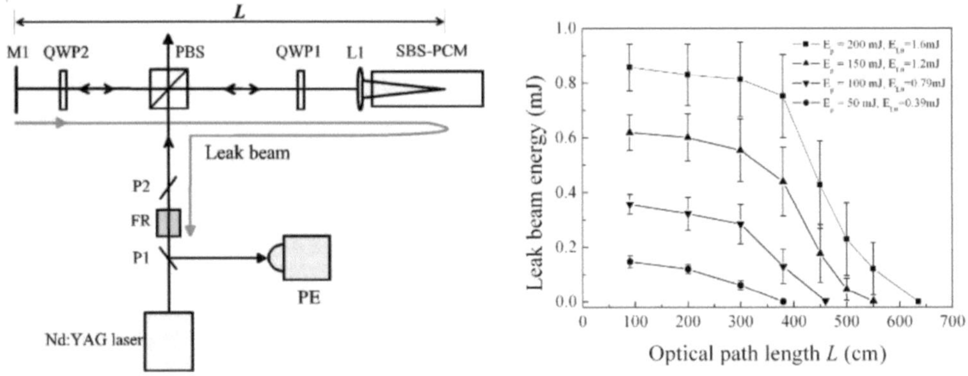

(a) (b)

Fig. 6. (a) Schematic of the experimental set up: P1 & P2, polarizer; FR, Faraday rotator; PE, pyro-electric energy meter; QWP1 & QWP2, quarter-wave-plate; L1, lens; PBS, polarizing beam splitter; M1, conventional mirror. (b) Leak beam energy dependence on the optical path length *L*.

4.2 Alignment-free optical system

The principle of the isolator that uses SBS-PCMs was explained in the previous section. However, the fact that an additional arm uses SBS-PCMs implies that the system has an alignment-free characteristic, which is one of the advantages of the cross-type amplifier system with SBS-PCMs. As shown in Fig. 7, although other systems such as the systems shown in Figs. 7(b) and 7(c) change the beam pointing due to the misalignment of the polarizing beam splitter, the suggested system of Fig. 7(a) has no effect on the beam direction or the beam position. Thus, the beam pointing of the output has the same level as the input (the master oscillator). The insensitivity to the misalignment and tilting of any optical components relies on the fact that the beam reflected by SBS-PCM follows exactly the same path as the incoming beam on account of the phase-conjugate characteristics.

4.3 Compensation for the thermal birefringence in the amplifier medium

Diminishment of the thermal load and the related thermal effects, particularly the compensation for TIB, is a traditional and important topic in the high-energy laser field. As shown in Fig. 8, the amplifying part of a cross-type amplifier with an SBS-PCM has three components: an amplifier rod, a polarization rotator (such as a Faraday rotator or a quarter-wave-plate), and an SBS-PCM (SBS cell and focusing optics). The total system has four available configurations, which are shown in Figs. 8(a) to 8(d). As in similar situations (Han & Kong, 1995), the theoretical prediction of TIB compensation can be given by Jones matrices. When we use the matrix elements of the Faraday rotator, $F=\frac{1}{\sqrt{2}}\begin{bmatrix}1 & 1\\ -1 & 1\end{bmatrix}$, the quarter-wave-plate, $Q=\begin{bmatrix}1 & i\\ i & 1\end{bmatrix}$, the glass rod, $G=\begin{bmatrix}e^{i\phi_r} & 0\\ 0 & e^{i\phi_\theta}\end{bmatrix}$, and the rotation matrix, $R=\begin{bmatrix}\cos\theta & \sin\theta\\ -\sin\theta & \cos\theta\end{bmatrix}$, the output after a double-pass amplification, $\begin{bmatrix}E_x\\ E_y\end{bmatrix}$, can be calculated. In the setup of Fig. 8(a), the output polarization is represented by

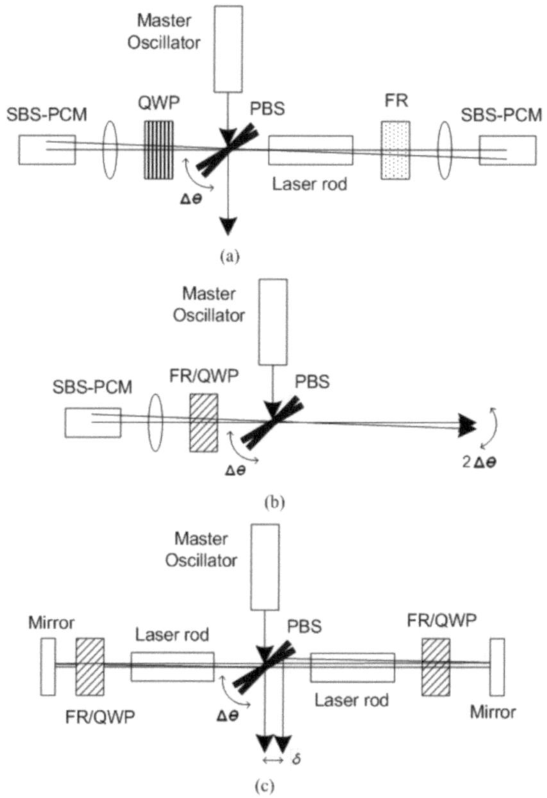

Fig. 7. Change of the beam pointing due to the tilting PBS: (a) gives no change in cross-type amplifier with symmetric SBS-PCMs; (b) gives tilting in the conventional application of SBS-PCM; (c) gives displacement in the combination of conventional mirror and SBS-PCM

$$\begin{bmatrix} E_x \\ E_y \end{bmatrix} = QRGR^{-1}1RGR^{-1}\begin{bmatrix} 1 \\ 0 \end{bmatrix} = -\frac{i}{2}\begin{bmatrix} (\cos\theta - i\sin\theta)^2(e^{2i\phi_r} - e^{2i\phi_\theta}) \\ (e^{2i\phi_r} + e^{2i\phi_\theta}) \end{bmatrix}. \tag{2}$$

In the setup of Fig. 8(b), the output polarization is represented by

$$\begin{bmatrix} E_x \\ E_y \end{bmatrix} = RGR^{-1}Q1QRGR^{-1}\begin{bmatrix} 1 \\ 0 \end{bmatrix} = -\frac{i}{2}\begin{bmatrix} 8\sin\theta\cos\theta(e^{i\varphi_r} - e^{i\phi_\theta})^2(\cos^2\theta e^{i\phi_r} + \sin^2\theta e^{i\phi_\theta}) \\ -(e^{i\phi_r} + e^{i\phi_\theta})^2 + (e^{i\phi_r} - e^{i\phi_\theta})^2\cos 4\theta \end{bmatrix}. \tag{3}$$

In the setup of Fig. 8(c), the output polarization is represented by

$$\begin{bmatrix} E_x \\ E_y \end{bmatrix} = FRGR^{-1}1RGR^{-1}F\begin{bmatrix} 1 \\ 0 \end{bmatrix} = -\frac{i}{2}\begin{bmatrix} \cos 2\theta(e^{2i\phi_r} - e^{2i\phi_\theta}) \\ -(e^{2i\phi_r} + e^{2i\phi_\theta}) + (e^{2i\phi_r} - e^{2i\phi_\theta})\sin 2\theta \end{bmatrix}. \tag{4}$$

In the set up of Fig. 8(d), the output polarization is represented by

$$\begin{bmatrix} E_x \\ E_y \end{bmatrix} = RGR^{-1}F1FRGR^{-1}\begin{bmatrix} 1 \\ 0 \end{bmatrix} = -2e^{i(\phi_r + \phi_\theta)}\begin{bmatrix} 0 \\ 1 \end{bmatrix}. \tag{5}$$

Eq. (5) shows that the setup of Fig. 8(d) gives a perfect 90° rotated output and compensates the TIB.

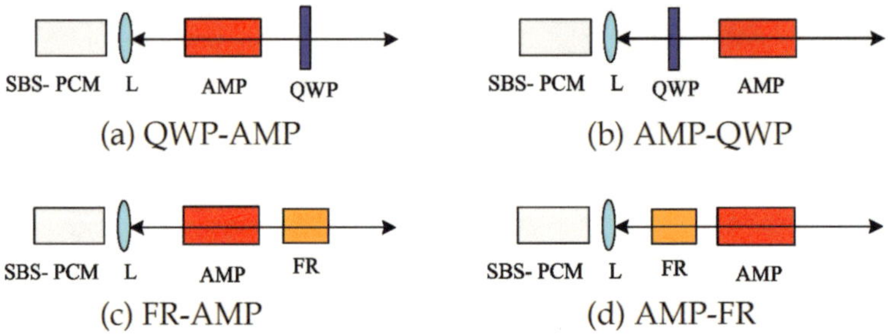

Fig. 8. Four possible optical schemes for rotating the polarization of the backward beam by 90-degree with respect to the input beam (L, lense; QWP, quarter-wave plate; FR, Faraday rotator; AMP, amplifier)

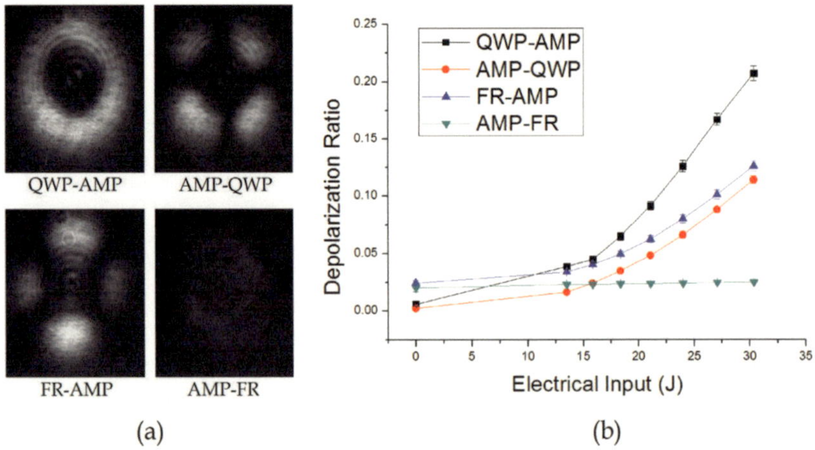

Fig. 9. Experimental results of the depolarization measurement for the four possible optical schemes: (a) leak beam patterns, (b) depolarization ratio versus electrical input energy (Shin et al., 2009)

Fig. 9(a) shows the corresponding leak beam patterns for the four possible optical schemes of Fig. 8. This experimental result shows typical shape for each case. And Fig. 9(b) shows the depolarization ratio versus electrical input energy. The experimental result for the setup of Fig. 8(d) shows that the depolarization ratio is maintained at the low value as the electrical input energy increases, while the results for other setups (Fig. 8(a) - Fig. 8(c)) shows the depolarization ratio rises as the electrical input energy increases (Fig. 9(b)).

5. Waveform preservation of SBS waves via prepulse injection

There are difficulties in a laser system with SBS, particularly when multiple SBS cells are used in series for a high-power laser system. As the pulse is reflected from the SBS cell, the temporal pulse shape is deformed so that the reflected SBS wave has a steep rising edge (Shen, 2003). If SBS cells are used in series, the rising edge of the pulse becomes steeper and can cause an optical breakdown in the optical components. For the SBS-PCM, the steep rising edge leads to low reflectivity and low fidelity of the phase conjugated wave in the SBS medium (Dane et al., 1992). Thus, a suitable technique is needed to preserve the temporal waveform of the reflected SBS wave (Kong et al., 2005d).

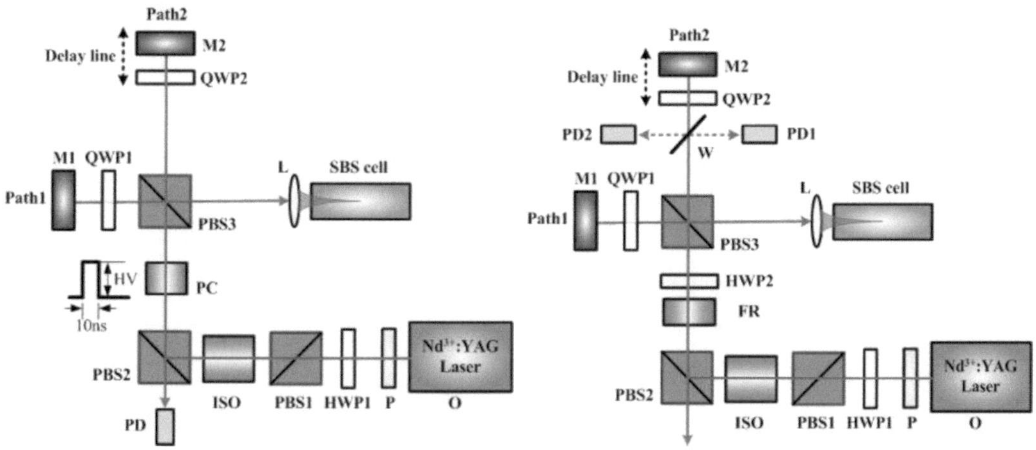

(a) (b)

Fig. 10. (a) Proposed system for preserving a temporal SBS pulse shape; (b) experimental setup for this experiment: O, Nd3+:YAG laser oscillator; P, linear polarizer; HWPs, half-wave plates; PBSs, polarizing beam splitters; ISO, Faraday isolator; FR, Faraday rotator; QWPs, quarter-wave plates; PC, Pockels cell; Ms, full mirrors; W, wedge; L, convex lens (f=15 cm); PDs, photodiodes; SBS cell (FC-75, 30 cm long).

The loss of the front part of the pumping energy to create the acoustic Brillouin grating is one of the main causes of the deformation. As a solution, the prepulse technique can be used to maintain the temporal waveform. In this scheme, the incident wave is divided into two pulses, the prepulse and the main pulse, and the prepulse is sent to the SBS medium before the main pulse with some delay. When the prepulse is injected before the main pulse, the main pulse can be reflected by means of a preexisting acoustic grating and the reflected pulse waveform can be preserved.

The scheme of the proposed setup for the temporal waveform preservation is presented in Fig. 10(a). A single longitudinal mode Nd:YAG laser oscillator is used as a pump source. It has a pulse width of 7~8ns and a repetition rate of 10 Hz. A Pockels cell (PC) is used to adjust the proper ratio of the prepulse energy and the main pulse energy, and the adjustment is made by adapting the high voltage that is applied to a PC for 10 ns, which is the time it takes for an incoming pulse to pass through the PC. The PC is in the off state when the pulse returns. The incident wave is split into two paths after PBS3, namely path 1 (prepulse) and path 2 (main pulse). The prepulse, which is initially s polarized, is reflected when it reaches PBS2 after the SBS process because the PC is in the off state when the pulse returns. The main pulse, which initially has p polarization, follows a process that is very similar to the process of the prepulse and consequently has the p polarization needed to pass through PBS2. There is another variation that uses no active optics. In Fig. 10(b), HWP2 and the Faraday rotator (45° rotator) are used instead of the PC in Fig. 10(a). HWP2 is used to adjust the ratio of the prepulse energy and the main pulse energy. The measurement is taken on path 2. A wedge plate is inserted to monitor the shape of the reflected main pulse and the incoming main pulse waveform and the reflected SBS waveform are obtained. The delay is modulated by the movable mirror, and the FC-75 fluid is used as the SBS medium.

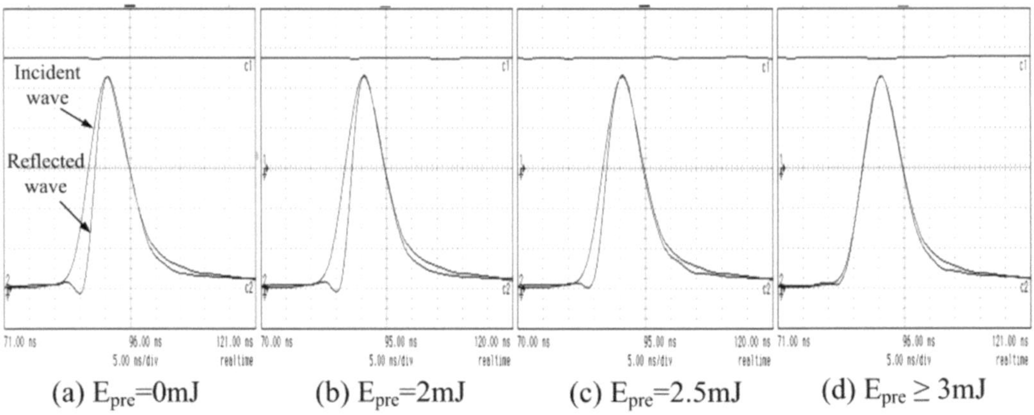

Fig. 11. Incident and reflected waveforms with the prepulse injection; (a) E_{pre} = 0 mJ, (b) E_{pre} = 2 mJ, (c) E_{pre} = 2.5mJ, (d) $E_{pre} \geq$ 3mJ for values of T_{delay} = 8 ns and E_{main} = 10 mJ.

Let us define E_{main} as the energy of the main pulse, E_{pre} as the energy of the prepulse, and T_{delay} as the delay between the prepulse and the main pulse. Fig. 11 shows the waveform measured for values of T_{delay} = 8 ns and E_{main} = 10 mJ. As E_{pre} increases, the temporal waveforms of the reflected wave become similar to that of the incident wave. When E_{pre} exceeds 3 mJ, the experimental data have very similar aspects as the case of E_{pre} = 3 mJ. This similarity implies that if we set the prepulse energy equal to or larger than 3 mJ with a delay of 8 ns, the main pulse need not consume its own energy to build the acoustic grating.

Fig. 12 shows the minimum prepulse energy required to preserve the waveform of reflected pulse for various T_{delay} (Yoon et al., 2009). For small T_{delay}, the main pulse arrives so early that a part of the main pulse energy can play a role in building the acoustic grating, because the integrated energy of the prepulse is insufficient to generate the grating before the main pulse arrives. Therefore the energy required to preserve the waveform of the main pulse is higher than the moderate T_{delay}. For large T_{delay}, most of the acoustic grating disappears before the main pulse arrives at the SBS interaction region so that more energy is required to preserve the waveform.

A theoretical calculation that describes these experimental results was formulated using a simple model. If the pump pulse is focused in the SBS medium, acoustic phonons are generated and then accumulated in the focal area. Considering the phonon decay, the pump pulse energy transferred to acoustic phonons and accumulated by time t, $E_g(t)$, is given by

$$E_g(t) = \int_0^t P(t')e^{-(t-t')/\tau}dt' \tag{6}$$

where $P(t)$ is the temporal pulse shape and τ is the phonon lifetime. If the pulse width is independent of the pulse energy, the temporal pulse shape can be represented as

$$P(t) = E \cdot W(t) \tag{7}$$

where E is the pulse energy, and $W(t)$ is the normalized waveform.

To instigate the stimulated process, an amount of acoustic phonons over the required threshold is required. The accumulated phonon energy needed for SBS ignition, called the critical energy E_c, can be determined by the maximally accumulated energy with a threshold pump energy E_{th}, as follows:

$$E_c = \int_0^{t_m} E_{th} W(t') e^{-(t_m - t')/\tau} dt' \tag{8}$$

where t_m is the time when E_c becomes maximum. If the main pulse arrives at the interaction region when $E_g(t)$ accumulated by the prepulse is larger than Ec, perfect waveform preservation is achievable without energy consumption.

$$E_c \le \int_0^{t_d} E_{pre} W(t') e^{-(t_d - t')/\tau} dt' \tag{9}$$

where t_d is the delay time between the prepulse and the main pulse. For theoretical calculation, 2 mJ threshold energy and 0.9 ns phonon decay time were assumed (Yoshida et al., 1997). Fig. 12 shows experimental results agree with the theoretical predictions qualitatively.

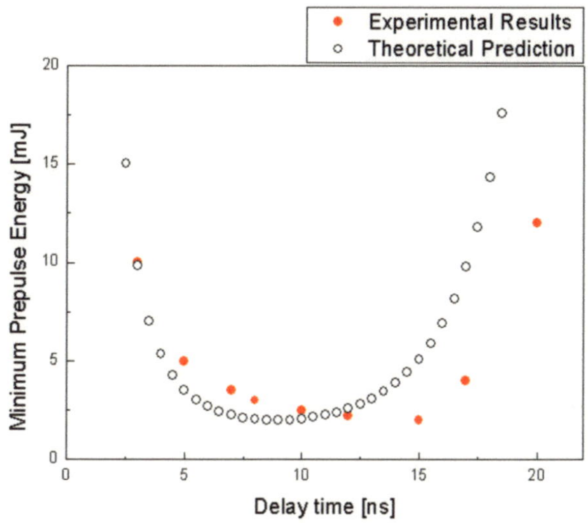

Fig. 12. Minimum prepulse energy required to preserve the waveform of reflected pulse for various T_{delay}; comparison between the experimental results and the theoretical prediction

6. Coherent beam combined laser system with phase stabilized SBS-PCMs

To achieve a high repetition rate in a high-power laser, many researchers have widely investigated several methods, such as a beam combination technique with SBS-PCMs, a diode-pumped laser system with gas cooling, an electron beam-pumped gas laser, and a large ceramic crystal (Lu et al., 2002; Kong et al., 1997, 2005a, 2005b; Rockwell & Giuliano, 1986; Loree et al., 1987; Moyer et al., 1988). The beam combination technique seems to be one of the most practical of these techniques. The laser beam is first divided into several sub-beams and then recombined after separate amplification. With this technique there is no need for a large gain medium; hence, regardless of the output energy, this type of laser can operate at a repetition rate exceeding 10 Hz and can be easily adapted to modern laser technology. However, with conventional SBS-PCMs, the SBS waves have random initial phases because they are generated by noises. For this reason, the phase locking of the SBS wave is strongly required for the output of a coherent beam combination.

6.1 Phase control of the SBS wave by means of the self-generated density modulation

There have been several successful works in the history of the phase locking of SBS waves (Rockwell & Giuliano, 1986; Loree et al., 1987; Moyer et al., 1988). Although these works show good phase locking effects, they have some problems in terms of the practical application of a multiple beam combination. In the overlapping method, all the beams are focused on one common point. The energy scaling is therefore limited to avoid an optical breakdown, and the optical alignment is also difficult. In the back-seeding beam method, the phase conjugation is incomplete if the injected Stokes beam is not completely correlated. Kong *et al.* (2005a, 2005b, 2005c) proposed a new phase control technique involving self-generated density modulation. In this method, which is simply called the self-phase control method, a simple optical composition is used with a single concave mirror behind the SBS cell; furthermore, each beam phase can be independently and easily controlled without destruction of the phase conjugation. Thus, the phase control method obviates the need for any structural limitation on the energy scaling.

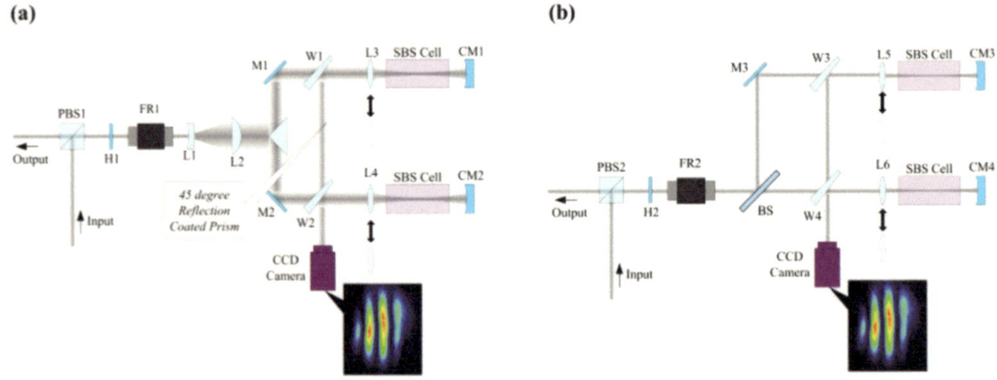

Fig. 13. Experimental setups of (a) ***wavefront division scheme*** and (b) ***amplitude division scheme*** for phase control of the SBS wave by means of the self-generated density modulation: M1,M2&M3, mirrors; W1,W2,W3&W4, wedges; L1&L2, cylindrical lenses: L3,L4,L5&L6, focusing lenses, CM1,CM2,CM3&CM4, concave mirrors; H1&H2, half wave-plates; PBS1&PBS2, polarizing beam splitters.

The wavefront division scheme, which spatially divides the beam, is used to demonstrate the phase control effect with the self-phase control method in the first experiment (Kong et al., 2005a, 2005b, 2005c). The experimental setup is shown schematically in Fig. 13(a). A 1064 nm Nd:YAG laser is used as a pump beam for the SBS generation. The pulse width is 7 ns to 8 ns, and the repetition rate is 10 Hz. The laser beam from the oscillator passes through a 2× cylindrical telescope and is divided into two parts by a prism, which has a high reflection coating for an incident angle of 45°. The two parts of the divided beam pass through separate wedges and are focused into SBS-PCMs. The wedges reflect part of the backward Stokes beams so that they are overlapped onto a CCD camera. Then, the interference pattern of them is generated. The degree of the fluctuation of the relative phase difference between the SBS waves is quantitatively analyzed by measuring the movement of the peaks in the interference pattern.

For the case of the wavefront division, the divided sub-pump beams get fluctuating energies for every shot due to the beam pointing effect of the laser source, which seems to generate the fluctuation of the relative phase difference between the SBS waves, because the phase of the SBS wave depends on the pump energy. This beam pointing problem can be overcome by using an amplitude division method, whereby the sub-pump beams have almost the same level of energy (Lee et al., 2005). The experimental setup of the amplitude division scheme is shown in Fig. 13(b). In the amplitude division scheme, the laser beam from an oscillator is divided into two sub-beams by a beam splitter (BS).

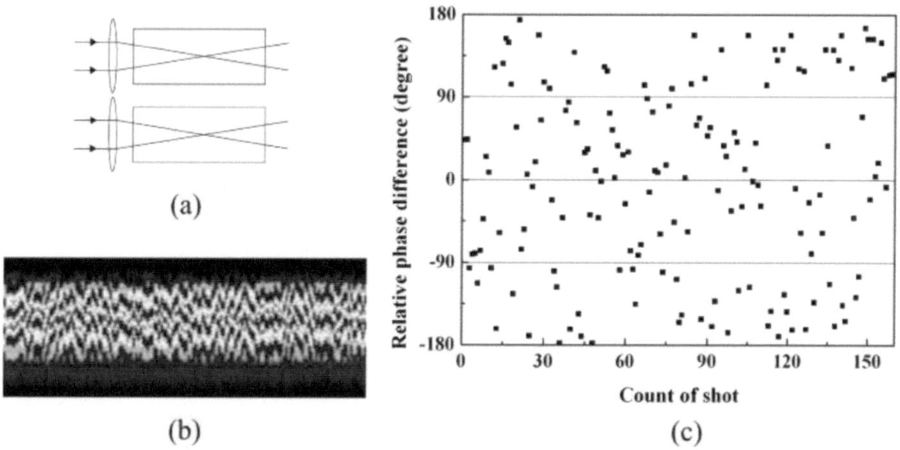

Fig. 14. Experimental result for the ***unlocked case***: (a) schematic; (b) intensity profile of horizontal lines selected from 160 interference patterns; (c) relative phase difference between two beams for 160 laser pulses.

Fig. 14 shows the experimental schematic and experimental results for the unlocked case. Each point in Fig. 14(c) represents one of 160 laser pulses. As expected, δ has random value for every laser pulse. Fig. 14(b) shows the intensity profile of the 160 horizontal lines selected from each interference pattern. The profile also represents the random fluctuation.

Fig 15 shows phase control experimental results in the wavefront division scheme. Fig. 15(a) shows the schematic and the experimental result of the concentric-type self-phase control. A small amount of the pump pulse is reflected by an uncoated concave mirror and then injected into the SBS cell. The standard deviation of the measured relative phase difference is ~ 0.165λ. Moreover, 88% of the data points are contained within a range of $\pm 0.25\lambda$ ($\pm 90°$). This result demonstrates that the self-generated density modulation can fix the phase of the backward SBS wave. Fig. 15(b) shows the schematic and the experimental result of the confocal-type self-phase control, where the pump beams are backward focused by a concave mirror coated with high reflectivity. The standard deviation of the measured relative phase difference is ~ 0.135λ. Furthermore, 96% of the data points are contained in a range of $\pm 0.25\lambda$.

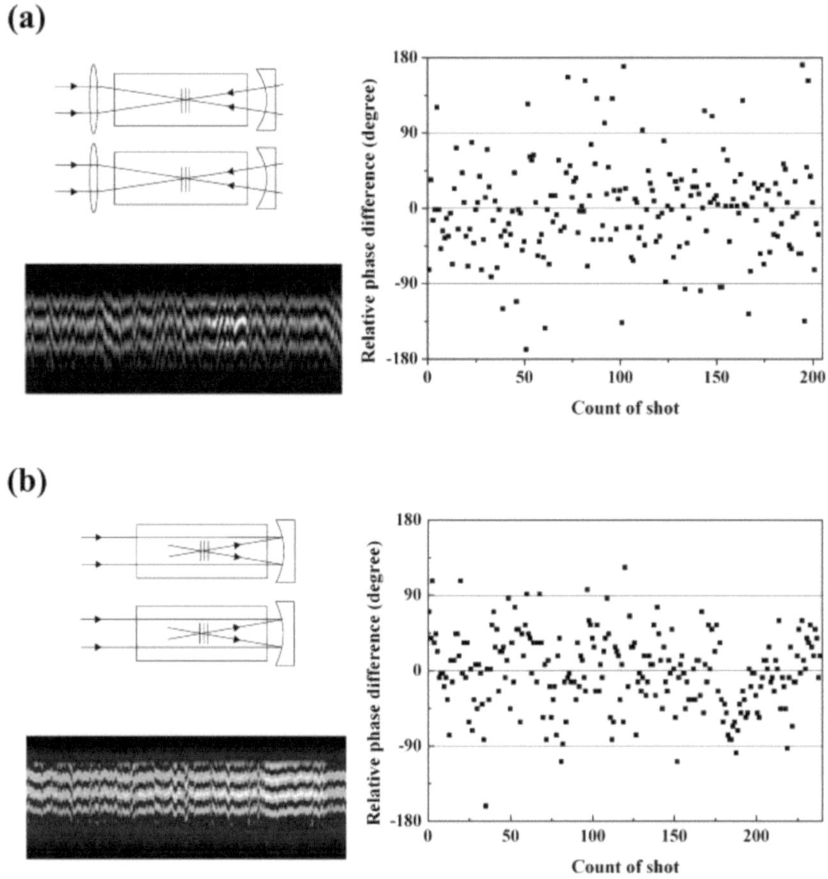

Fig. 15. Phase control experimental results in the ***wavefront division scheme,*** with (a) **concentric-type self-phase control** ((left-up) schematic, (left-down) intensity profile of horizontal lines from interference pattern, (right) relative phase difference between two beams for 203 laser pulses) and (b) **confocal-type self-phase control** ((left-up) schematic, (left-down) intensity profile of horizontal lines from interference pattern, (right) relative phase difference between two beams for 238 laser pulses).

Fig. 16 shows phase control experimental results in the amplitude division scheme. Fig. 16(a) shows the schematic and the experimental result of the concentric-type self-phase control. The standard deviation of the measured relative phase difference is ~ 0.0366λ. And Fig. 16(b) shows the schematic and the experimental result of the confocal-type self-phase control. The standard deviation of the measured relative phase difference is ~ 0.0275λ. By employing the amplitude division scheme, the relative phase difference is remarkably stabilized compared with the wavefront dividing scheme.

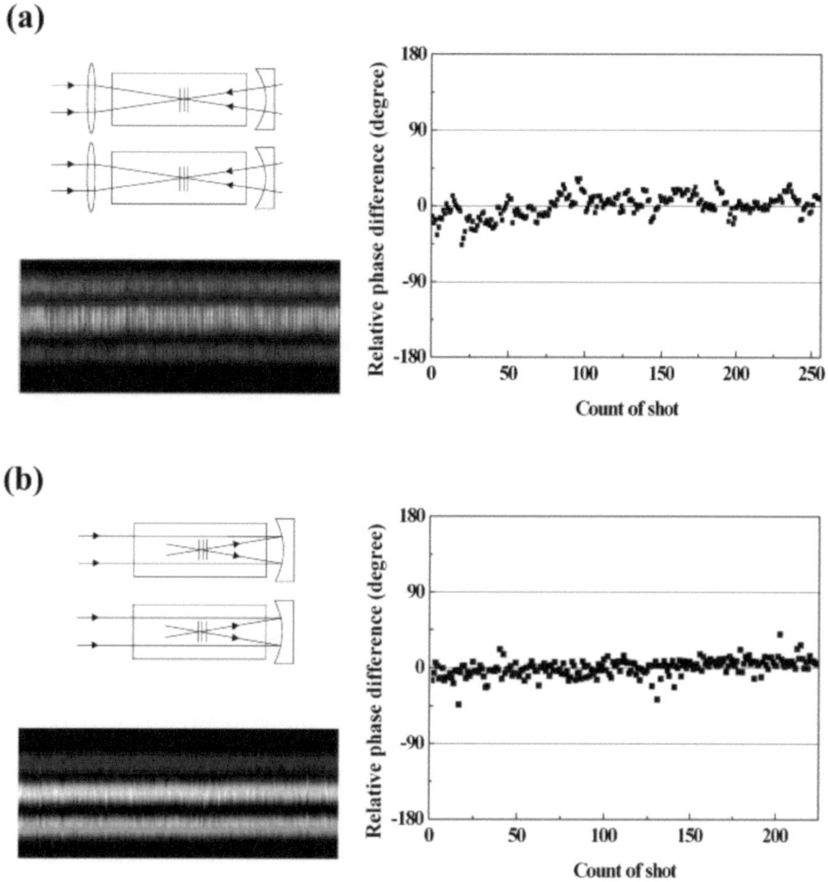

Fig. 16. Phase control experimental result in the ***amplitude division scheme***, with (a) **concentric-type self-phase control** ((left-up) schematic, (left-down) intensity profile of horizontal lines from interference pattern, (right) relative phase difference between two beams for 256 laser pulses) and (b) **confocal-type self-phase control** ((left-up) schematic, (left-down) intensity profile of horizontal lines from interference pattern, (right) relative phase difference between two beams for 220 laser pulses).

6.2 Theoretical modeling on the phase control of SBS waves

In the previous section, the experimental results demonstrate the effect of the self-phase control method. On the basis of the phase control experiments, we present in this section the theoretical model suggested by Kong et al. to explain the principle of the self-phase control (Ostermeyer et al., 2008). Given that the pump wave propagates towards the positive z direction in the SBS medium, the pump wave, E_P, and the Stokes wave, E_S, can be expressed as

$$E_P = A\sin(\omega_P t - k_P z + \phi_P) \quad (10)$$

and

$$E_S = B\sin(\omega_S t + k_S z + \phi_S)\,, \tag{11}$$

where A and B are the amplitudes of E_P and E_S; ω, k and ϕ are the angular frequency, the wave number and the initial phase, respectively; and P and S are the pump wave and the Stokes wave, respectively. The density modulation of the SBS medium is proportional to the total electrical field. The density modulation, ρ, can therefore be represented as

$$\begin{aligned}\rho \propto \left|E_P + E_S\right|^2 = {} & A^2\sin^2(\omega_P t - k_P z + \phi_P) + B^2\sin^2(\omega_S t + k_S z + \phi_S) \\ & -AB\cos[(\omega_P + \omega_S)t - (k_P - k_S)z + (\phi_P + \phi_S)] \\ & +AB\cos[(\omega_P - \omega_S)t - (k_P + k_S)z + (\phi_P - \phi_S)].\end{aligned} \tag{12}$$

Only the final term of Eq. (12) can contribute to the acoustic wave because the first two terms are DC components and the third term denotes the fast oscillating components. The acoustic wave can be also expressed as

$$\rho = \rho_0 \cos(\Omega t + k_a z + \phi_a)\,, \tag{13}$$

where ρ_0 is the mean value of the medium density and Ω, k_a, and ϕ_a are the frequency, the wave number, and the initial phase of the acoustic wave, respectively. From Eqs. (12) and (13), the relations of $\Omega = \omega_P - \omega_S$, $k_a = k_P + k_S$ and $\phi_a = \phi_P - \phi_S$ can be obtained. If ϕ_a and ϕ_P are known values, ϕ_S can be definitely determined in accordance with the phase relation.
If the acoustic wave is assumed to be initially generated at time t_0 and position z_0, the acoustic wave can be rewritten as

$$\rho = \rho_0 \cos[\Omega(t - t_0) - k_a(z - z_0)]$$

$$= \rho_0 \cos[\Omega t - k_a z - \Omega t_0 + k_a z_0]\cdot \tag{14}$$

The phase of the acoustic wave is then given by

$$\phi_a = -\Omega t_0 + k_a z_0\,. \tag{15}$$

In conventional SBS generation, t_0 and z_0 have random values as the SBS wave is generated from a thermal acoustic noise. However, t_0 and z_0 can be locked effectively by the proposed self-phase control method.

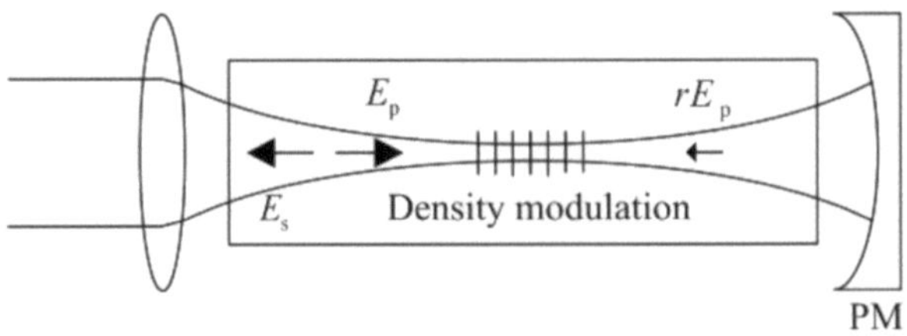

Fig. 17. Concept of phase control of the SBS wave by the self-generated density modulation. PM is a partial reflectance concave mirror whose reflectivity is r. E_P and E_S denote the pump wave and the SBS wave, respectively.

Fig. 17 describes the concept of the self-phase control method. The weak periodic density modulation is generated at the focal point due to the electrostriction by an electromagnetic standing wave that arises from the interference between the main beam, E_P, and the low intensity counter-propagating beam, rE_P. In the suggested theoretical model, the weak density modulation from the standing wave is assumed to act as an imprint for the ignition of the Brillouin grating. Hence, the initial position, z_0, is no longer random but fixed to one of the nodal points of the density modulation. However, there are many candidates of the nodal points in the Rayleigh range because the Rayleigh length, l_R, is much larger than the period of the stationary density modulation, $\lambda_P/2$, where λ_P is the wavelength of the pump wave. The phase differences between the acoustic waves generated at different nodal points have the values of $\Delta\phi_a = k_a(\lambda_P/2)N \cong 2\pi N$ (N: integer) for the relation of $k_a \cong 2k_P = 4\pi/\lambda_P$. Thus, the phase uncertainty of $2\pi N$ does not affect the phase accuracy.
The initial time, t_0, when the acoustic wave is determined should be known. In the research on the preservation of the SBS waveform (Kong et al., 2005d), the front part of the pump energy is consumed to create the acoustic Brillouin grating of the SBS process. This consumed energy is regarded as the SBS threshold energy. The critical time, t_c, when the SBS is initiated can then be determined by the following equation:

$$E_{\mathrm{th}} = \int_0^{t_c} P(t)dt\ , \tag{16}$$

where E_{th} is the SBS threshold energy of the SBS medium and $P(t)$ is the pump power. It is assumed that t_0 is equal to t_c because the SBS waves and the corresponding acoustic wave are generated simultaneously. Eq. (16) suggests suggests that the initial ignition time, t_0, of the acoustic wave changes if the total energy of the pump pulse given by $E_0 = \int_0^{\infty} P(t)dt$ changes under a constant pulse width. In this model, the critical time, t_c, varies with the total energy, E_0. Thus, the change that occurs in the initial phase, $\Delta\phi_0$, as a result of the energy fluctuation, ΔE_0, can be represented as

$$\Delta\phi_0 = \Omega\Delta t_c = \Omega\frac{\Delta t_c}{\Delta E_0}\frac{\Delta E_0}{E_0}E_0 \tag{17}$$

if we assume that z_0 is fixed; $\Delta\phi_0$ can be calculated numerically for FC-75, which has an acoustic wave frequency of 1.34 GHz; and the SBS threshold is about 2 mJ for a 10 ns pulse. Let's assume that the pump pulse, $P(t)$, is given by

$$P(t) = \frac{4E_0}{a^3\sqrt{\pi}}t^2\exp[-(t/a)^2]\ \ (a = 8.66\mathrm{ns})\,. \tag{18}$$

Fig. 18 shows the calculated critical time, t_c, as a function of the pump energy, which ranges from the SBS threshold of FC-75 (2 mJ) to 100 mJ.
When two beams are combined by the SBS-PCM, energy fluctuations of the each input beam give the shot-to-shot change on the critical time difference. Fig. 19 shows the calculated results and the experimental results of the phase fluctuation. Using the measured energy fluction of the each input beam, the phase fluctuation of Fig. 19(a) is simulated. The experimental investigation is conducted for the cases of E_1= 10mJ, 30mJ, 50mJ, and 70mJ

with several E_2 values. In both graphs, the standard deviation of the relative phase fluctuation is shown. The shapes of the graphs are similar but the vertical scales are different.

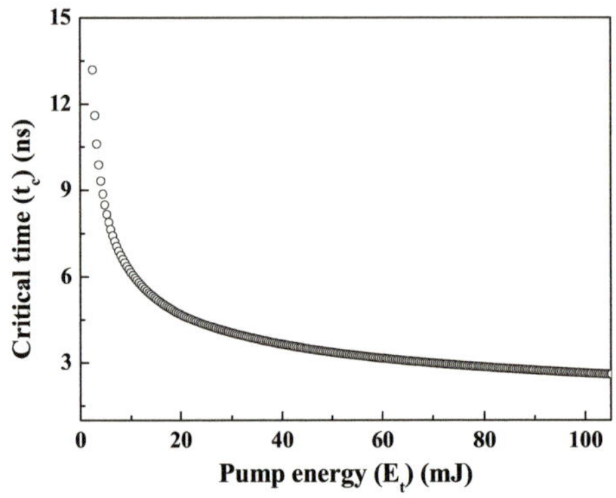

Fig. 18. Critical time t_c as a function of the pump energy (E_{th}=2mJ).

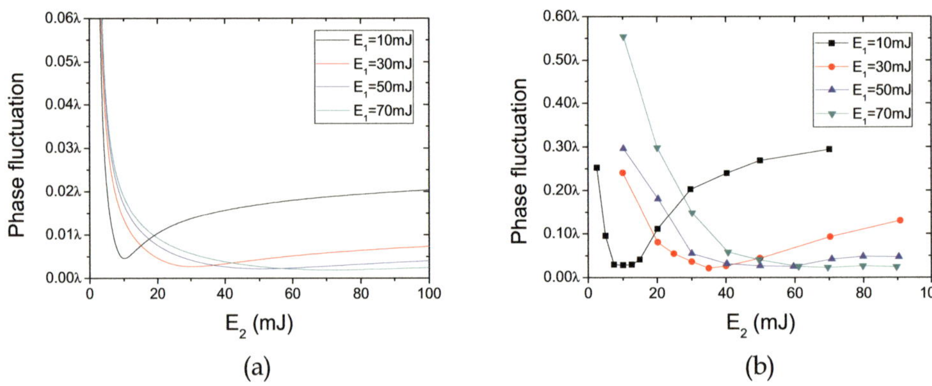

Fig. 19. (a) Calculated results of the relative phase difference for the cases of E_1 = 10 mJ, 30 mJ, 50 mJ, and 70 mJ with E_2 = 2 mJ to 100mJ; the critical time is calculated directly from the energy measurements. (b) Experimental results of the relative phase difference for the cases of E_1 = 10 mJ, 30 mJ, 50 mJ, and 70 mJ with several E_2 values.

6.3 Long-term phase stabilization of SBS wave

The self-phase control method ensures the SBS wave is well stabilized for several hundred shots. However, a thermally induced long-term phase fluctuation occurs when the number of laser shots increases (Kong et al., 2006, 2008). This slowly varying phase fluctuation can be easily compensated through the active control of PZTs attached to one concave mirror of the SBS-PCM. Figs. 20 and 21 show the phase control experimental results for the cases with PZT control and without PZT control, respectively, in a two-beam combination system. The phase difference and the output energy are measured during 2500 laser shots (250 s) for a pump energy level of $E_{p1,2}$≈50 mJ. The case without PZT control showed long-term phase and output energy fluctuations. In the case with the PZT control, the phase difference between the SBS beams is well stabilized with a fluctuation of 0.0214λ(=λ/46.8) by standard deviation; furthermore, the output energy is stabilized with a fluctuation of 4.66%.

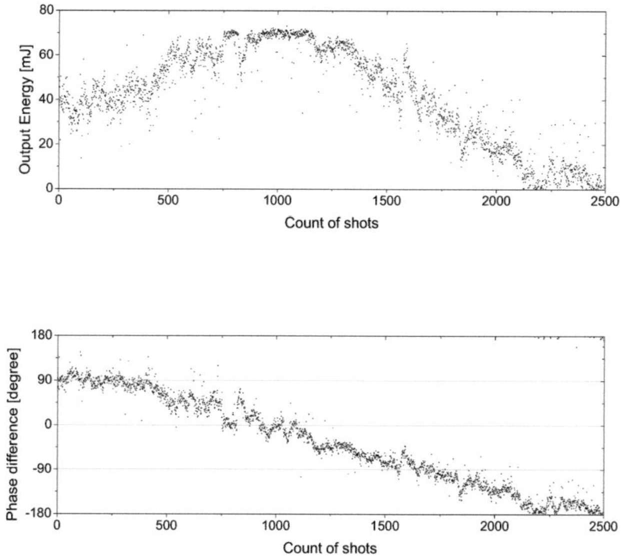

Fig. 20. Experimental results of (a) the output energy and (b) the phase difference between two SBS beams without PZT control during 2500 laser shots (250 s) for the case of $E_{p1,2}≈50$ mJ pump energy.

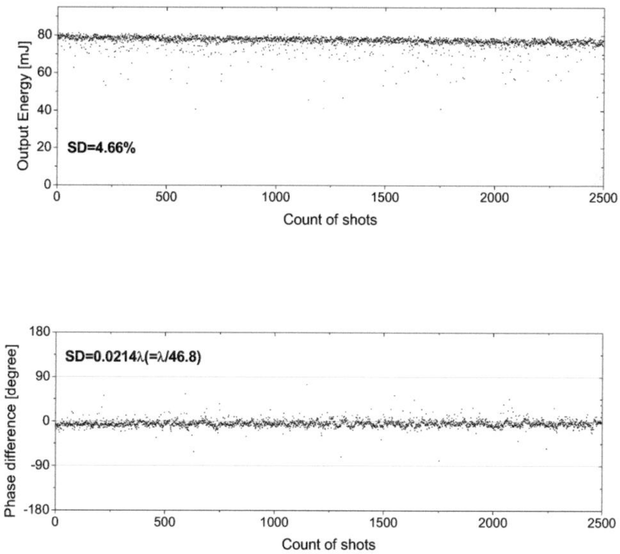

Fig. 21. Experimental results of (a) the output energy and (b) the phase difference between two SBS beams with PZT control during 2500 laser shots (250 s) for the case of $E_{p1,2}≈50$ mJ pump energy.

6.4 Coherent beam combined laser system for high energy, high power, high beam quality, and high repetition rate output

Figs. 22(a) and 22(b) show the conceptual schemes of the coherent beam combination laser system for high energy, high power, high beam quality, and a high repetition rate (Kong et al., 1997, 2005a, 2005b). Fig. 22(a) shows the wavefront division scheme, and Fig. 22(b) shows the amplitude division scheme. In this beam combination laser system, the main beam is divided into many sub-beams for separate amplification; the beam is divided either by prisms in the wavefront division scheme or by polarizing beam splitters in the amplitude scheme. Both schemes include a series of cross-type amplifier stages. Each cross-type amplifier has SBS-PCMs on both sides and is insensitive to the misalignments of the optical components because the reflected phase conjugate waves return to exactly the same path as the incident beam. As a result, the cross-type beam combination system is highly beneficial in terms of alignment, maintenance, and repair. The SBS-PCMs on the right-hand side of each cross-type amplifier stage perform as optical isolators. On the left-hand side of each cross-type amplifier stage, the array amplifier can increase the beam's energy with double pass optical amplification when it is divided by some of the sub-beams. For the reflectors in the array amplifier, we used SBS-PCMs instead of conventional mirrors. The SBS-PCMs can compensate for the thermally induced wavefront distortions, and self-focusing can occur in the active media with the generation of phase conjugate beams. A diffraction-limited high quality beam can therefore be obtained at the output stage. The divided sub-beams are recombined again after the double-pass amplification and become the input beam of the next amplifier stage. By using many amplifier stages of beam combination, we can obtain a high-energy laser output for the fusion. In the array amplifier, Faraday rotators are located on the amplification beam lines to compensate for the thermally induced birefringence, and phase-controlled SBS-PCMs are used with the self phase control method for coherent output.

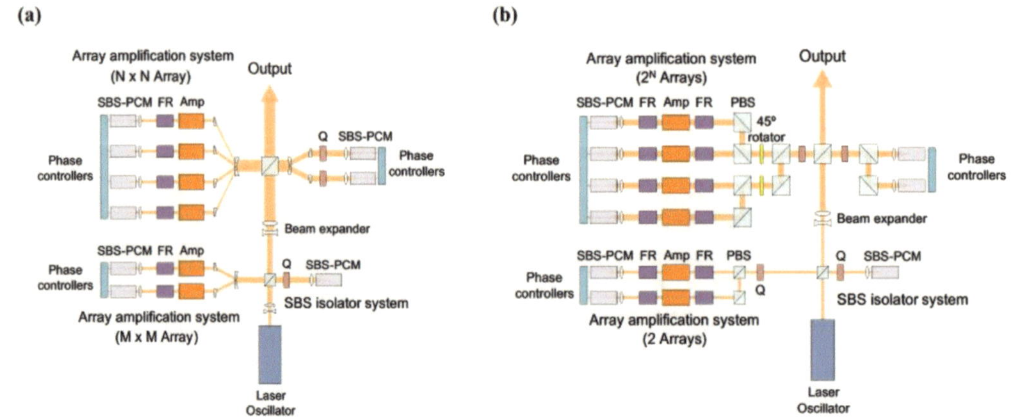

Fig. 22. Conceptual schemes of scalable beam combined laser system for a laser fusion driver: (a) wavefront division scheme (b) amplitude division scheme (QWP, quarter wave plate; SBS-PCM, stimulated Brillouin scattering phase conjugate mirror, FR, Faraday rotator; AMP, optical amplifier)

7. Conclusion

In this chapter, a high-energy, high-power amplifier system using SBS-PCMs is introduced. The system, which is constructed systematically with a cross-type amplifier and SBS-PCMs as a basic unit, has many advantages: for example, it has freely scalable energy and a perfectly isolated leak beam; it also compensates for the thermally induced optical distortion and it has misalignment insensitiveness. For the coherent output of the combined beam, a new phase control method of the SBS wave with self-density modulation has been developed. The principle of this phase control method in the experiments for the wavefront and amplitude division schemes has been also explained and successfully demonstrated, as well as in theoretical modeling, and in the active control of the long-term phase fluctuation. In conclusion, the proposed beam combination laser system with SBS-PCMs, which is based on the cross-type amplifier, contributes to the realization of the a high energy, high power laser that can operate with a repetition rate higher than 10 Hz, even for a huge output energy in excess of several MJ.

8. References

Andreev, F. ; Khazanov, E. & Pasmanik, G. A. (1992). Applications of Brillouin cells to high repetition rate solid-state lasers, *IEEE J. Quantum. Electron.*, Vol. 28, No. 1, 330-341, ISSN 0018-9197.

Arecchi F. T. & Schulz-Dubois E. O. (1972). *Laser Handbook Vol. 2,* Nort-Holland, Amsterdam, ISBN 0-7204-0213-1.

Boyd, R. W. (1992). *Nonlinear Optics,* Academic Press Inc., San Diego., ISBN 0-12-121680-2.

Brignon, A. & Huignard J.-P. (2004). *Phase Conjugate Laser Optics,* John Wiley & Sons, Inc., New Jersey and Canada, ISBN 0-471-43957-6.

Bullock D. L. ; Nguyen-Vo N.-M. & Pfeifer S. J. (1994). Numerical model of stimulated Brillouin scattering excited by a multiline pump, *IEEE J. Quantum Electron.*,Vol. 30, No. 3, 805-811, ISSN 0018-9197.

D'yakov Yu. E. (1970). Excitation of stimulated light scattering by broad-spectrum pumping, *JETP Lett.*, Vol. 11, 243-246, ISSN 0021-3640.

Damzen, M. J. ; Vlad, V. I. ; Babin, V. & Mocofanescu A. (2003). *Stimulated Brillouin Scattering,* Institute of Physics Publishing, Bristol and Philadelphia, ISBN 0-7503-0870-2.

Dane C. B. ; Neuman W. A. & Hackel L. A. (1992). Pulse-shape dependence of stimulated-Brillouin-scattering phase-conjugation fidelity for high input energies, *Opt. Lett.*, Vol. 17, No. 18, 1271-1273, ISSN 0146-9592.

Dane C. B. ; Zapata L. E. ; Neuman W. A. ; Norton M. A. & Hackel L. A. (1995). Design and operation of a 150 W near diffraction-limited laser amplifier with SBS wavefront correction. *IEEE J. Quantum. Electron.*, Vol. 31, No. 1, 148-163, ISSN 0018-9197.

Eichler, H. J. & Mehl, O. (2001). Phase conjugate mirrors, *Journal of Nonlinear Optical Physics & Materials*, Vol. 10, No. 1, 43-52, ISSN 0218-8635.

Erokhin A. I. ; Kovalev V. I. & Faizullov F. S. (1986). Determination of the parameters of a nonlinear response of liquids in an acoustic resonance region by the method of nondegenerate four-wave interaction, *Sov. J. Quantum Electron.*,Vol. 16, No. 7, 872-877, ISSN 0049-1748.

Filippo A. A. & Perrone M. R. (1992). Experimental study of stimulated Brillouin scattering by broad-band pumping, *IEEE J. Quantum Electron.*, Vol. 28, No. 9, 1859-1863, ISSN 0018-9197.

Han, K.G. & Kong, H.J. (1995). Four-pass amplifier system compensation thermally induced birefringence effect using a novel dumping mechanism, *Jpn. J. Appl. Phys*, Vol. 34, Part 2, No. 8A, 994–996, ISSN 0021-4922.

Hon, D. T. (1980). Pulse compression by stimulated Brillouin scattering, *Opt. Lett.*, Vol. 5, No.12 516-518, ISSN 0146-9592.

Kmetik, V. ; Fiedorowicz, H. ; Andreev, A. A. ; Witte, K. J. ; Daido H. ; Fujita H. ; Nakatsuka M. & Yamanaka T. (1998). Reliable stimulated Brillouin scattering compression of Nd:YAG laser pulses with liquid fluorocarbon for long-time operation at 10 Hz, *Appl. Opt.*, Vol. 37, No. 30, 7085-7090, ISSN 0003-6935.

Kong, H. J. ; Beak, D. H. ; Lee, D. W. & Lee, S. K. (2005d). Waveform preservation of the backscattered stimulated Brillouin scattering wave by using a prepulse injection, *Opt. Lett.*, Vol. 30, No. 24, 3401-3403, ISSN 0146-9592.

Kong, H. J. ; Lee, J. Y. ; Shin, Y. S. ; Byun, J. O. ; Park, H. S. & Kim H. (1997). Beam recombination characteristics in array laser amplification using stimulated Brillouin scattering phase conjugation, *Opt. Rev.* Vol. 4, No. 2, 277-283, ISSN 1340-6000.

Kong, H. J. ; Lee, S. K. & Lee, D. W. (2005a). Beam combined laser fusion driver with high power and high repetition rate using stimulated Brillouin scattering phase conjugation mirrors and self-phase locking, *Laser Part. Beams* Vol. 23, 55-59, ISSN 0263-0346.

Kong, H. J. ; Lee, S. K. & Lee, D. W. (2005b). Highly repetitive high energy/power beam combination laser : IFE laser driver using independent phase control of stimulated Brillouin scattering phase conjugate mirrors and pre-pulse technique, *Laser Part. Beams* Vol. 23, 107-111, ISSN 0263-0346.

Kong, H. J. ; Lee, S. K. ; Lee, D. W. & Guo, H. (2005c). Phase control of a stimulated Brillouin scattering phase conjugate mirror by a self-generated density modulation, *Appl. Phys. Lett.* Vol. 86, 051111, ISSN 0003-6951.

Kong, H. J. ; Yoon, J. W. ; Shin, J. S. & Beak, D. H. (2008). Long-term stabilized two-beam combination laser amplifier with stimulated Brillouin scattering mirrors, *Appl. Phys. Lett.* Vol. 92, 021120, ISSN 0003-6951.

Kong, H. J. ; Yoon, J. W. ; Shin, J. S. ; Beak, D. H. & Lee, B. J. (2006). Long term stabilization of the beam combination laser with a phase controlled stimulated Brillouin scattering phase conjugation mirrors for the laser fusion driver, *Laser Part. Beams.* Vol. 24, 519-523, ISSN 0265-0346.

Kong, H.J.; Kang, Y.G.; Ohkubo, A., Yoshida, H. & Nakatsuka, M. (1998). Complete isolation of the back reflection by using stimulated Brillouin scattering phase conjugation mirror. *Review of Laser Engineering*, Vol. 26, 138–140, ISSN 0387-0200.

Kong, H.J. ; Lee, S.K.; Kim, J.J.; Kang, Y.G. & Kim. H. (2001). A cross type double pass laser amplifier with two symmetric phase conjugation mirrors using stimulated Brillouin scattering, *Chinese Journal of Lasers*, Vol. B10, I5-I9, ISSN 1004-2822.

Králiková B. ; Skála J.; Straka P. & Turčičová H. (2000). High-quality phase conjugation even in a highly transient regime of stimulated Brillouin scattering, *Appl. Phys. Lett.*, Vol. 77, 627-629, ISSN 0003-6951.

Lee, S. K. ; Kong, H. J. & Nakatsuka, M. (2005). Great improvement of phase controlling of the entirely independent stimulated Brillouin scattering phase conjugate mirrors by balancing the pump energies, Appl. Phys. Lett., Vol. 87, 161109, ISSN 0003-6951.

Linde D. ; Maier M. & Kaiser, W. (1969). Quantitative Investigations of the Stimulated Raman Effect Using Subnanosecond Light Pulses, *Phys. Rev.*, Vol. 178, No. 1, 11-15, ISSN 0031-899X.

Loree T. R. ; Watkins, D. E. ; Johnson, T. M. ; Kurnit, N. A. & Fisher, R. A. (1987). Phase locking two beams by means of seeded Brillouin scattering, *Opt. Lett.*, Vol. 12, No. 3, 178-180, ISSN 0146-9592.

Lu J. ; Lu J. ; Murai, T. ; Takaichi, K. ; Uematsu, T. ; Xu, J. ; Ueda, K. ; Yagi, H. ; Yanagitani, T. & Kaminskii, A. A. (2002). 36-W diode-pumped continuous-wave 1319-nm Nd:YAG ceramic laser, *Opt. Lett.* Vol. 27, No. 13, 1120-1122, ISSN 0146-9592.

Moyer R. H. ; Valley, M. & Cimolino, M. C. (1988). Beam combination throgh stimulated Brillouin scattering, *J. Opt. Soc. Am. B* Vol. 5, No. 12, 2473-2489, ISSN 0740-3224.

Mullen R. A.; Lind R. C. & Valley G. C. (1987). Observation of stimulated Brillouin scattering gain with a dual spectral-line pump, *Opt. Commun.*, Vol. 63, No. 2, 123-128, ISSN 0030-4018.

Narum P. ; Skeldon M. D. & Boyd R. W. (1986). Effect of laser mode structure on stimulated Brillouin scattering, *IEEE J. Quantum Electron.*, Vol. 22, No. 11, 2161-2167, ISSN 0018-9197.

Ostermeyer, M. ; Kong H. J. ; Kovalev, V. I. ; Harrison R. G. ; Fotiadi A. A. ; P., Mégret, Kalal, M. ; Slezak, O. ; Yoon, J. W. ; Shin, J. S. ; Beak, D. H. ; Lee, S. K. ; Lü, Z. ; Wang, S. ; Lin, D. ; Knight, J. C. ; Kotova, N. E. ; Sträßer, A. ; Scheikh-Obeid, A. ; Riesbeck, T. ; Meister, S. ; Eichler, H. J. ; Wang, Y. ; He, W. ; Yoshida, H. ; Fujita, H. ; Nakatsuka , M. ; Hatae, T. ; Park, H. ; Lim, C. ; Omatsu, T. ; Nawata K. ; Shiba N. ; Antipov O. L. ; Kuznetsov, M. S. & Zakharov, N. G. (2008). Trends in stimulated Brillouin scattering and optical phase conjugation, *Laser Part. Beams* Vol. 26, No. 3, 297-362, ISSN 0265-0346.

Shin, J. S.; Park, S. & Kong, H. J. (2009). Compensation of the thermally induced depolarization in a double-pass Nd:YAG rod amplifier with a stimulated Brillouin scattering phase conjugate mirror, to be published.

Rockwell, D. A. & Giuliano, C. R. (1986). Coherent coupling of laser gain media using phase conjugation, *Opt. Lett.*, Vol. 11, 147-149, ISSN 0146-9592.

Seidel, S & Kugler, N. (1997). Nd:YAG 200-W average-power oscillator-amplifier system with stimulated-Brillouin-scattering phase conjugation and depolarization compensation, *J. Opt. Soc. Am. B*, Vol. 14, No. 7, 1885-1888, ISSN 0740-3224.

Shen Y. R. (2003). *Principles of Nonlinear Optics,* Wiley, New York, ISBN 0471-88998-9.

Sutherland R. L. (1996). *Handbook of Nonlinear Optics,* Marcel Dekker Inc., New York, ISBN 082-474243-5.

Valley G. C. (1986). A review of stimulated Brillouin scattering excited with a broad-band pump laser. *IEEE J. Quantum. Electron.*, Vol. 22, No. 5, 704-712, ISSN 0018-9197.

Yariv, Ammon (1975). *Quantum Electronics,* John Wiley, New York, ISBN 047-160997-8.

Yoon, J. W.; Shin, J. S. ; Kong, H. J. & Lee, J. (2009). Investigation of the relationship between the prepulse energy and the delay time in the waveform preservation of a stimulated Brillouin scattering wave by prepulse injection, to be published.

Yoshida H.; Kmetik V.; Fujita H.; Nakatsuka M.; Yamanaka T. & Yoshida K. (1997). Heavy fluorocarbon liquids for a phase-conjugated stimulated Brillouin scattering mirror, *Appl. Opt.*, Vol. 36, No. 16, 3739-3744, ISSN 0003-6935.

Zel'dovich, B. Y. ; Popovichev, V. I. ; Ragulsky, V. V. & Faizullov, F. S. (1972). Connection between the wave fronts of the reflected and exciting light in stimulated Mandel' shtem-Brillouin scattering, *Sov. Phys. JETP*, Vol. 15, 109, ISSN 0038-5646.

Zel'dovich, B. Y. ; Pilipetsky, N. F. & Shkunov, V. V. (1985). *Principle of Phase Conjugation*, Springer-Verlag, Berlin, Heidelberg, New York and Tokyo, ISBN 3-540-13458-1.

13

The Intersubband Approach to Si-based Lasers

Greg Sun
University of Massachusetts Boston, Massachusetts, U.S.A.

1. Introduction

Silicon has been the miracle material for the electronics industry, and for the past twenty years, technology based on Si microelectronics has been the engine driving the digital revolution. For years, the rapid "Moore's Law" miniaturization of device sizes has yielded an ever-increasing density of fast components integrated on Si chips: but during the time that the feature size was pushed down towards its ultimate physical limits, there has also been a tremendous effort to broaden the reach of Si technology by expanding its functionalities well beyond electronics. Si is now being increasingly investigated as a platform for building photonic devices. The field of Si photonics has seen impressive growth since early visions in the 1980s and 1990s [1,2]. The huge infrastructure of the global Si electronics industry is expected to benefit the fabrication of highly sophisticated Si photonic devices at costs that are lower than those currently required for compound semiconductors. Furthermore, the Si-based photonic devices make possible the monolithic integration of photonic devices with high speed Si electronics, thereby enabling an oncoming Si-based "optoelectronic revolution".

Among the many photonic devices that make up a complete set of necessary components in Si photonics including light emitters, amplifiers, photodetectors, waveguides, modulators, couplers and switches, the most difficult challenge is the lack of an efficient light source. The reason for this striking absence is that bulk Si has an indirect band gap where the minimum of the conduction band and the maximum of the valence band do not occur at the same value of crystal momentum in wave vector space (Fig. 1). Since photons have negligible momentum compared with that of electrons, the recombination of an electron-hole pair will not be able to emit a photon without the simultaneous emission or absorption of a phonon in order to conserve the momentum. Such a radiative recombination is a second-order effect occurring with a small probability, which competes with nonradiative processes that take place at much faster rates. As a result, as marvelous as it has been for electronics, bulk Si has not been the material of choice for making light emitting devices including lasers.

Nevertheless, driven by its enormous payoff in technology advancement and commercialization, many research groups around the world have been seeking novel approaches to overcome the intrinsic problem of Si to develop efficient light sources based on Si. One interesting method is to use small Si nanocrystals dispersed in a dielectric matrix, often times SiO_2. Such nano-scaled Si clusters are naturally formed by the thermal annealing of a Si-rich oxide thin film. Silicon nanocrystals situated in a much wider band gap SiO_2 can effectively localize electrons with quantum confinement, which improves the radiative recombination probability, shifts the emission spectrum toward shorter wavelengths, and

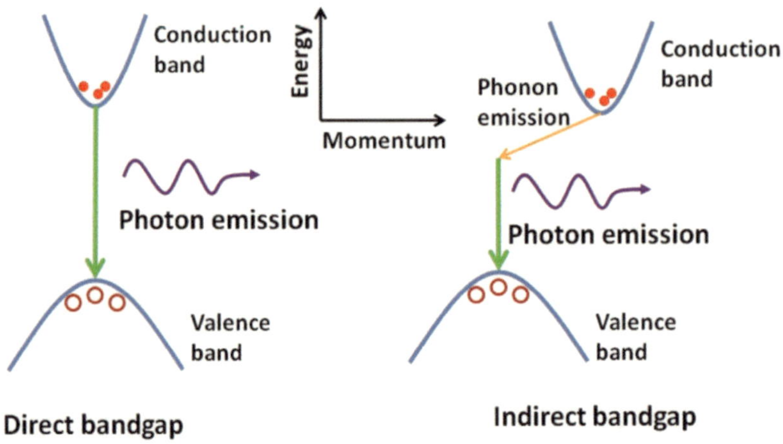

Fig. 1. Illustration of a photon emission process in (a) the direct and (b) the indirect band gap semiconductors.

decreases the free carrier absorption. Optical gain and stimulated emission have been observed from these Si nanocrystals by both optical pumping [3,4] and electrical injection [5], but the origin of the observed optical gain has not been fully understood as the experiments were not always reproducible - results were sensitive to the methods by which the samples were prepared. In addition, before Si-nanocrystal based lasers can be demonstrated, the active medium has to be immersed in a tightly confined optical waveguide or cavity.

Another approach is motivated by the light amplification in Er-doped optical fibers that utilize the radiative transitions in Er ions (Er^{3+}) [6]. By incorporating Er^{3+} in Si, these ions can be excited by energy transfer from electrically injected electron-hole pairs in Si and will subsequently relax by emitting photons at the telecommunication wavelength of 1.55 μm. However, the concentration of Er^{3+} ions that can be doped in Si is relatively low and there is a significant energy back-transfer from the Er^{3+} ions to the Si host due to the resonance with a defect level in Si. As a result, both efficiency and maximum power output have been extremely low [7,8]. To reduce the back transfer of energy, SiO_2 with an enlarged band gap has been proposed as host to remove the resonance between the defect and the Er^{3+} energy levels [9]. Once again, Si-rich oxide is employed to form Si nanocrystals in close proximity to Er^{3+} ions. The idea is to excite Er^{3+} ions with the energy transfer from the nearby Si nanocrystals. Light emitting diodes (LEDs) with efficiencies of about 10% have been demonstrated [10] on par with commercial devices made of GaAs, but with power output only in tens of μW. While there have been proposals to develop lasers using doped Er in Si-based dielectric, the goal remains elusive.

The only approach so far that has led to the demonstration of lasing in Si exploited the effect of stimulated Raman scattering [11-13], analogous to that produced in fiber Raman amplifiers. With both the optical pumping and the Raman scattering below the band gap of Si, the indirectness of the Si band gap becomes irrelevant. Depending on whether it is a Stokes or anti-Stokes process, the Raman scattering either emits or absorbs an optical phonon. Such a nonlinear process requires optical pumping at very high intensities (~100MW/cm^2) and the device lengths (~cm) are too large to be integrated with other photonic and electronic devices in any type of Si VLSI-type circuit [14].

Meanwhile, the search for laser devices that can be integrated on Si chips has gone well beyond the monolithic approach to seek solutions using hybrid integration of III-V compounds with Si. A laser with an AlGaInAs quantum well (QW) active region bonded to a silicon waveguide cavity was demonstrated [15]. This fabrication technique allows for the optical waveguide to be defined by the CMOS compatible Si process while the optical gain is provided by III-V materials. Rare-earth doped visible-wavelength GaN lasers fabricated on Si substrates are also potentially compatible with the Si CMOS process [16]. Another effort produced InGaAs quantum dot lasers deposited directly on Si substrates with a thin GaAs buffer layer [17]. Although these hybrid approaches offer important alternatives, they do not represent the ultimate achievement of Si-based lasers monolithically integrated with Si electronics.

While progress is being made along these lines and debates continue about which method offers the best promise, yet another approach emerged that has received a great deal of attention in the past decade—an approach in which the lasing mechanism is based on intersubband transitions (ISTs) in semiconductor QWs. Such transitions take place between quantum confined states (subbands) of conduction or valence bands and do not cross the semiconductor band gap. Since carriers remain in the same energy band (either conduction or valence), optical transitions are always direct in momentum space rendering the indirectness of the Si band gap irrelevant. Developing lasers using ISTs therefore provides a promising alternative that completely circumvents the issue of indirectness in the Si band gap. In addition, this type of laser can be conveniently designed to employ electrical pumping - the so-called quantum cascade laser (QCL). The pursuit of Si-based QCLs might turn out to be a viable path to achieving electrically pumped Si-based coherent emitters that are suitable for monolithic integration with Si photonic and electronic devices.

In this chapter, lasing processes based on ISTs in QWs are explained by drawing a comparison to conventional band-to-band lasers. Approaches and results towards SiGe QCLs using ISTs in the valence band are overviewed, and the challenges and limitations of the SiGe valence-band QCLs are discussed with respect to materials and structures. In addition, ideas are proposed to develop conduction-band QCLs, among them a novel QCL structure that expands the material combination to SiGeSn. This is described in detail as a way to potentially overcome the difficulties that are encountered in the development of SiGe QCLs.

2. Lasers based on intersubband transitions

Research on quantum confined structures including semiconductor QWs and superlattices (SLs) was pioneered by Esaki and Tsu in 1970 [18]. Since then confined structures have been developed as the building blocks for a majority of modern-day semiconductor optoelectronic devices. QWs are formed by depositing a narrower band gap semiconductor with a layer thickness thinner than the deBroglie wavelength of the electron (~10nm) between two wider band gap semiconductors (Fig. 2(a)). The one-dimensional quantum confinement leads to quantized states (subbands) in the direction of growth z within both conduction and valence bands. The energy position of each subband depends on the band offset ($\Delta E_C, \Delta E_V$) and the effective mass of the carrier. In directions perpendicular to z (in-plane), the carriers are unconfined and can thus propagate with an in-plane wave vector $\boldsymbol{k}$ which gives an energy dispersion for each subband. (Fig. 2(b))

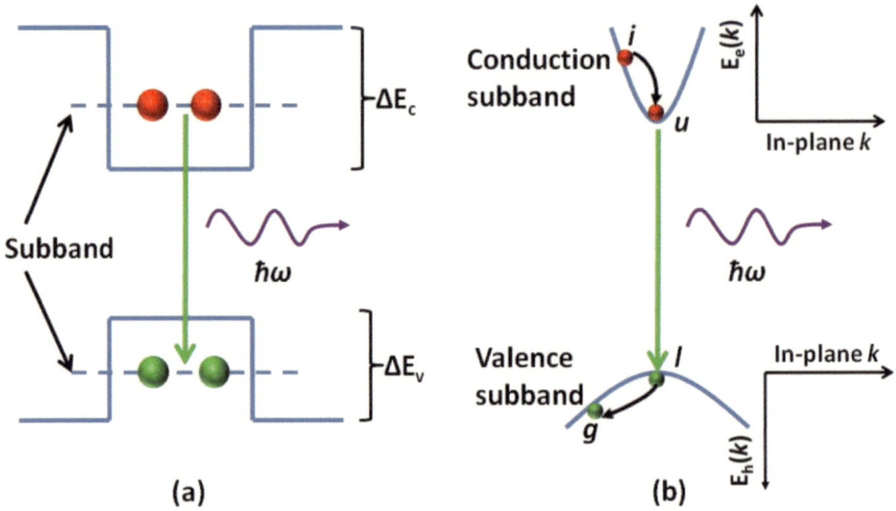

Fig. 2. Illustration of (a) conduction and valence subband formations in a semiconductor QW and (b) in-plane subband dispersions with optical transitions between conduction and valence subbands.

Obviously, if the band offset is large enough, there could be multiple subbands present within either conduction or valence band as shown in Fig. 3 where two subbands are confined within the conduction band. The electron wavefunctions (Fig. 3(a)) and energy dispersions (Fig. 3(b)) are illustrated for the two subbands. The concept of ISTs refers to the physical process of a carrier transition between these subbands within either the conduction or valence band as illustrated in Fig. 3. Carriers originally occupying a higher energy subband can make a radiative transition to a lower subband by emitting a photon. Coherent sources utilizing this type of transition as the origin of light emission are called intersubband lasers.

The original idea of creating light sources based on ISTs was proposed by Kazarinov and Suris [19] in 1971, but the first QCL was not demonstrated until 1994 by a group led by Capasso at Bell Laboratories [20]. In comparison with the conventional band-to-band lasers, lasers based on ISTs require much more complex design of the active region which consists of carefully arranged multiple QWs (MQWs). The reason for added complexity can be appreciated by comparing the very different band dispersions that are involved in these two types of lasers. In a conventional band-to-band laser, it appears that the laser states consist of two broad bands. But a closer look at the conduction and valence band dispersions (Fig. 2(b)) reveals a familiar four-level scheme where in addition to the upper laser states $|u>$, located near the bottom of the conduction band and the lower laser states $|l>$, near the top of the valence band, there are two other participating states - intermediate states $|i>$, and ground states $|g>$. The pumping process (either injection or optical) places electrons into the intermediate states, $|i>$, from which they quickly relax toward the upper laser states $|u>$ by inelastic scattering intraband processes. This process is very fast, occurring on a sub-pico-second scale. But once they reach states $|u>$, they tend to stay there for a much longer time determined by the band-to-band recombination rate which is on the order of nanoseconds. Electrons that went through lasing transitions to the lower laser states $|l>$ will quickly scatter into the lower energy states of the valence band – ground states $|g>$ --

by the same fast inelastic intraband processes. (A more conventional way to look at this is the relaxation of holes toward the top of the valence band.) The population inversion between $|u>$ and $|l>$ is therefore established mostly by the fundamental difference between the processes determining the lifetimes of upper and lower laser states. As a result, the lasing threshold can be reached when the whole population of the upper conduction band is only a tiny fraction of that of the lower valence band.

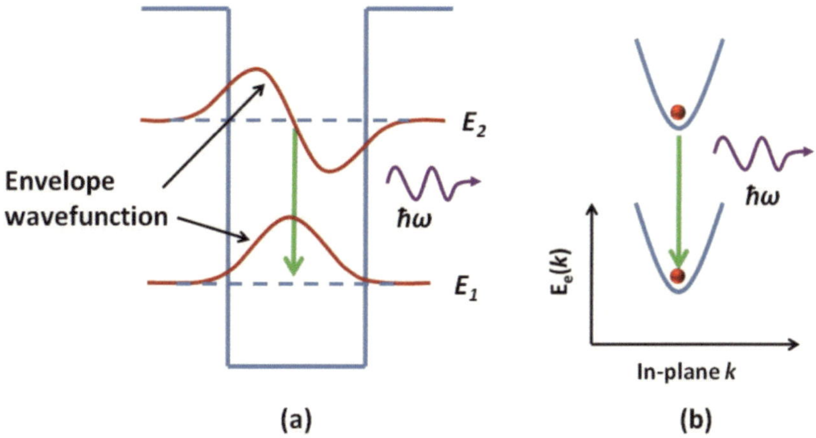

Fig. 3. (a) Two subbands formed within the conduction band confined in a QW and their election envelope functions, (b) in-plane energy dispersions of the two subbands. Radiative intersubband transition between the two subbands is highlighted.

Let us now turn our attention to the intersubband transition shown in Fig. 3(b). The in-plane dispersions of the upper $|u>$ and lower $|l>$ conduction subbands are almost identical when the band nonparabolicity can be neglected. For all practical purposes they can be considered as two discrete levels. Then, in order to achieve population inversion it is necessary to have the whole population of the upper subband exceed that of the lower subband. For this reason, a three- or four-subband scheme becomes necessary to reach the lasing threshold. Even then, since the relaxation rates between different subbands are determined by the same intraband processes, a complex multiple QW structure needs to be designed to engineer the lifetimes of involved subbands.

Still, intersubband lasers offer advantages in areas where the conventional band-to-band lasers simply cannot compete. In band-to-band lasers, lasing wavelengths are mostly determined by the intrinsic band gap of the semiconductors. There is very little room for tuning, accomplished by varying the structural parameters such as strain, alloy composition, and layer thickness. Especially for those applications in the mid-IR to far-IR range, there are no suitable semiconductors with the appropriate band gaps from which such lasers can be made. With the intersubband transitions, we are no longer limited by the availability of semiconductor materials to produce lasers in this long wavelength region. In addition, for ISTs between conduction subbands with parallel band dispersions, the intersubband lasers should therefore have a much narrower gain spectrum in comparison to the band-to-band lasers in which conduction and valence bands have opposite band curvatures.

A practical design that featured a four-level intersubband laser pumped optically was proposed by Sun and Khurgin [21,22] in the early 1990s. This work laid out a comprehensive analysis of various intersubband processes that affect the lasing operation including scattering mechanisms that determine subband lifetimes, conditions for population inversion between two subbands, band engineering to achieve it, and optical gain sufficient to compensate for losses under realistic pumping intensity. The QCLs developed soon thereafter significantly expanded the design in order to accommodate electrical pumping by implementing a rather elaborate scheme of current injection with the use of a chirped SL as the injector region placed in between the active regions (Fig. 4). The QCL has a periodic structure with each period consisting of an active and an injector region. Both active and injector regions are composed of MQWs. By choosing combinations of layer thicknesses and material compositions, three subband levels with proper energy separations and wave function overlaps are obtained in the active region. The injector region, on the other hand, is designed with a sequence of QWs having decreasing well widths (chirped SL) such that they form a miniband under an electric bias that facilitates electron transport. The basic operating principle of a QCL is illustrated in Fig. 4. Electrons are first injected through a barrier into subband 3 (upper laser state) of the active region, they then undergo lasing transitions to subband 2 (lower laser state) by emitting photons, followed by fast depopulation into subband 1 via nonradiative processes. These electrons are subsequently transported through the injector region into the next active region where they repeat the process in a cascading manner, typically 20 to 100 times.

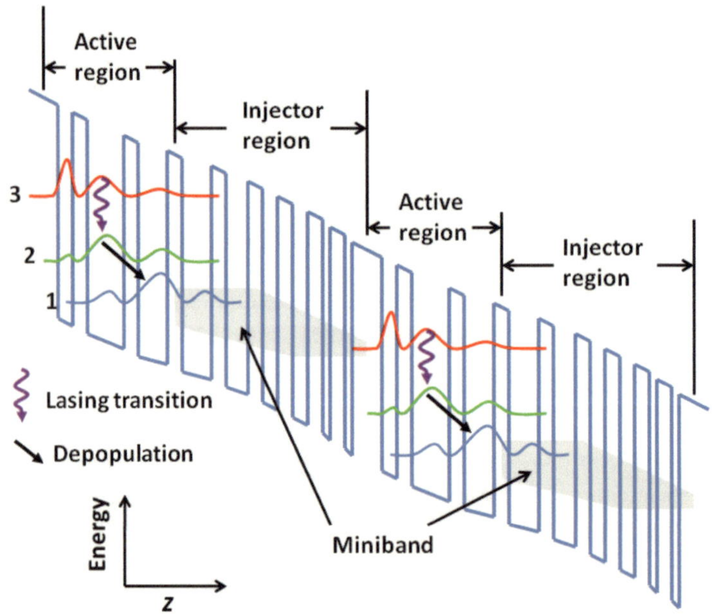

Fig. 4. Schematic band diagram of two periods of a QCL structure with each period consisting of an active and an injector region. Lasing transitions are between the states 3 and 2 in the active regions with rapid depopulation of lower state 2 into state 1 which couples strongly with the minibands formed in injector regions that transport carriers to state 3 in the next period. The magnitude-squared wavefunctions for the three subbands in active regions are illustrated.

Advances of QCLs since the first demonstration have resulted in dramatic performance improvement in spectral range, power and temperature. They have become the dominant mid-IR semiconductor laser sources covering the spectral range of $3 < \lambda < 25$ μm [23-25], many of them operating in the continuous-wave mode at room temperature with peak power reaching a few watts [26,27]. Meanwhile, QCLs have also penetrated deep into the THz regime loosely defined as the spectral region $100\text{GHz} < f < 10$ THz or $30 < \lambda < 3000$ μm, bridging the gap between the far-IR and GHz microwaves. At present, spectral coverage from 0.84-5.0 THz has been demonstrated with operation in either the pulsed or continuous-wave mode at temperatures well above 100K [28].

3. Intersubband theory

In order to better explain the design considerations of intersubband lasers, it is necessary to introduce some basic physics that underlies the formation of subbands in QWs and their associated intersubband processes. The calculation procedures described here follows the envelope function approach based on the effective-mass approximation [29]. The $\boldsymbol{k} \cdot \boldsymbol{p}$ method [30] was outlined to obtain in-plane subband dispersions in the valence band. Optical gain for transitions between subbands in conduction and valence bands is derived. Various scattering mechanisms that determine the subband lifetimes are discussed with an emphasis on the carrier-phonon scattering processes.

3.1 Subbands and dispersions

Let us treat the conduction subbands first. It is well known in bulk material that near the band edge, the band dispersion with an isotropic effective mass follows a parabolic relationship. In a QW structure, along the in-plane direction ($\boldsymbol{k} = k_x\hat{x} + k_y\hat{y}$) where electrons are unconfined, such curvature is preserved for a given subband i, assuming the nonparabolicity that describes the energy-dependent effective mass m_e^* can be neglected,

$$E_{i,\boldsymbol{k}} = E_i + \frac{\hbar^2 k^2}{2m_e^*} \tag{1}$$

where $\hbar$ is the Planck constant and E_i is the minimum energy of subband i in a QW structure. This minimum energy can be calculated as one of the eigen values of the Schrödinger equation along the growth direction z,

$$\left[-\frac{\hbar^2}{2}\frac{d}{dz}\frac{1}{m_e^*(z)}\frac{d}{dz} + V_c(z)\right]\varphi_i(z) = E_i\varphi_i(z) \tag{2}$$

where the z-dependence of m_e^* allows for different effective masses in different layers and $V_c(z)$ represents the conduction band edge along the growth direction z, . The envelope function of subband i, $\varphi_i(z)$, together with the electron Bloch function $u_e(\boldsymbol{R})$ and the plane wave $e^{j\boldsymbol{k}\cdot\boldsymbol{r}}$, gives the electron wavefunction in the QW structure as

$$\Phi_i(\boldsymbol{r}, z) = \varphi_i(z)u_e(\boldsymbol{R})e^{j\boldsymbol{k}\cdot\boldsymbol{r}} \tag{3}$$

where the position vector is decomposed into in-plane and growth directions $\boldsymbol{R} = \boldsymbol{r} + z\hat{\boldsymbol{z}}$. Since we are treating electron subbands, the Bloch function is approximately the same for all subbands and all $\boldsymbol{k}$-vectors. The electron envelope function can be given as a combination of the forward and backward propagations in a given region l of the QW structure (either a QW or a barrier region), $d_l < z < d_{l+1}$

$$\varphi_i(z) = A_l e^{jk_z(z-d_l)} + B_l e^{-jk_z(z-d_l)} \tag{4}$$

where A_l and B_l are constants that need to be fixed with the continuity conditions at each of the interfaces $z = d_l$,

$$\varphi_i(z) \text{ and } \frac{1}{m_e^*(z)} \frac{d\varphi_i(z)}{dz} \text{ continous} \tag{5}$$

in conjunction with the relationship between the subband minimum energy E_i and the quantized wave vector k_z in the z-direction

$$E_i = \frac{\hbar^2 k_z^2}{2m_e^*(z)} + V_c(z) \tag{6}$$

where k_z assumes either real or imaginary value depending on $E_i - V_c(z)$. The continuity conditions in Eq.(5) ensure continuous electron distribution and conservation of electron current across the interface.

In the presence of an electric field $\mathbb{E}$ applied in the z-direction, the potential term $V_c(z)$ in the Schrödinger equation Eq.(2) becomes tilted along the z -direction according to $-e\mathbb{E}z$. If the Coulomb effect due to the distribution of electrons in the subband needs to be taken into consideration, then the potential in region l of the QW structure with the conduction band edge $V_{c,l}$ should be modified as

$$V_c(z) = V_{c,l} - e\mathbb{E}z - e\phi(z) \tag{7}$$

where $e\phi(z)$ takes into account the potential due to electron distributions in all subbands and can be obtained by solving the Poisson equation

$$\frac{\partial^2}{\partial z^2}\phi(z) = -\frac{e}{\varepsilon_0 \varepsilon(z)}\left[\sum_i n_i |\varphi_i(z)|^2 - N_d(z)\right] \tag{8}$$

consistently with Eq.(2), where e is the charge of a free electron, ε_0 is the permittivity of free space, $\varepsilon(z)$ is the z-dependent dielectric constant of the QW structure, n_i is the electron density of subband i, and $N_d(z)$ is the n-type doping profile in the structure.

In comparison with the conduction band, the situation in the valence band is far more complex mostly because of the interactions between subbands of different effective masses that produce strong nonparabolicity. The in-plane dispersion of valence subbands and their associated envelope functions can be obtained in the framework of the effective mass approximation by applying the $\boldsymbol{k} \cdot \boldsymbol{p}$ theory [30] to QWs [31] where, in the most general treatment, an 8×8 Hamiltonian matrix is employed to describe the interactions between the conduction, heavy-hole (HH), light-hole (LH), and spin-orbit split off (SO) bands. Often times, for semiconductors in which the conduction band is separated far in energy from the valence band, the coupling of the conduction band can be ignored. For the group-IV semiconductors Si and Ge with indirect band gaps, this approximation is particularly adequate. In those structures where there is little strain such as GaAs/AlGaAs, the SO band coupling can also be ignored. The 8×8 Hamiltonian matrix can then be reduced to a 4×4 matrix. But for systems with appreciable lattice mismatch, strain induces strong coupling

between LH and SO bands. For the SiGe system with a large lattice mismatch, the SO band should be included and a 6×6 matrix Hamiltonian equation needs to be solved to come up with the dispersion relations and envelope functions. Such a 6×6 matrix Hamiltonian equation can be solved exactly in multiple QW structures under the bias of an electric field. A procedure based on the Luttinger-Kohn Hamiltonian [32,33] is outlined as follows.
The 6×6 Luttinger-Kohn Hamiltonian matrix including the uniaxial stress along (001) is given in the HH ($|\frac{3}{2},\pm\frac{3}{2}>$), LH ($|\frac{3}{2},\pm\frac{1}{2}>$), and SO ($|\frac{1}{2},\pm\frac{1}{2}>$) Bloch function space as

$$
\begin{array}{cccccc}
|\frac{3}{2},\frac{3}{2}> & |\frac{3}{2},\frac{1}{2}> & |\frac{3}{2},-\frac{1}{2}> & |\frac{3}{2},-\frac{3}{2}> & |\frac{1}{2},\frac{1}{2}> & |\frac{1}{2},-\frac{1}{2}>
\end{array}
$$

$$
H=-\begin{bmatrix}
P+Q & -S & R & 0 & -\frac{1}{\sqrt{2}}S & \sqrt{2}R \\
-S^\dagger & P-Q & 0 & R & -\sqrt{2}Q & \sqrt{\frac{3}{2}}S \\
R^\dagger & 0 & P-Q & S & \sqrt{\frac{3}{2}}S^\dagger & \sqrt{2}Q \\
0 & R^\dagger & S^\dagger & P+Q & -\sqrt{2}R^\dagger & -\frac{1}{\sqrt{2}}S^\dagger \\
-\frac{1}{\sqrt{2}}S^\dagger & -\sqrt{2}Q & \sqrt{\frac{3}{2}}S & -\sqrt{2}R & P+\Delta & 0 \\
\sqrt{2}R^\dagger & \sqrt{\frac{3}{2}}S^\dagger & \sqrt{2}Q & -\frac{1}{\sqrt{2}}S & 0 & P+\Delta
\end{bmatrix}+V_v(z) \qquad (9)
$$

where $V_v(z)$ is the valence band edge profile (degenerate for HH and LH bands) of the QW structure,

$$
\begin{aligned}
P &= \frac{\hbar^2}{2m_0}\gamma_1\left(k_x^2+k_y^2+k_z^2\right)-a_v\left(\epsilon_{xx}+\epsilon_{yy}+\epsilon_{zz}\right) \\
Q &= \frac{\hbar^2}{2m_0}\gamma_2\left(k_x^2+k_y^2-2k_z^2\right)-\frac{b}{2}\left(\epsilon_{xx}+\epsilon_{yy}-2\epsilon_{zz}\right) \\
S &= \frac{\hbar^2}{2m_0}2\sqrt{3}\gamma_3\left(k_x-jk_y\right)k_z \\
R &= \frac{\hbar^2}{2m_0}\sqrt{3}\left[-\gamma_2\left(k_x^2-k_y^2\right)+2j\gamma_3k_xk_y\right]
\end{aligned} \qquad (10)
$$

in which m_0 is the mass of a free electron, $\gamma_1, \gamma_2, \gamma_3$ are the Luttinger parameters and a_v, b are the deformation potentials [34] with different values in QWs and barriers, and the lattice mismatch strain

$$
\epsilon_{xx}=\epsilon_{yy}=\frac{a_0-a}{a}, \qquad \epsilon_{yy}=-\frac{2C_{12}}{C_{11}}\epsilon_{xx} \qquad (11)
$$

with a_0, a being the lattice constants of the substrate (or buffer) and the layer material, and C_{11} and C_{12} the stiffness constants.

The Hamiltonian in Eq.(9) operates on wavefunctions that are combinations of six mutually orthogonal HH ($|\frac{3}{2},\pm\frac{3}{2}>$), LH ($|\frac{3}{2},\pm\frac{1}{2}>$), and SO ($|\frac{1}{2},\pm\frac{1}{2}>$) Bloch functions

$$\begin{aligned}\Psi_i(\boldsymbol{r},z) = e^{j\boldsymbol{k}\cdot\boldsymbol{r}}\Big[\chi_1(z)\left|\frac{3}{2},\frac{3}{2}\right> + \chi_2(z)\left|\frac{3}{2},\frac{1}{2}\right> + \chi_3(z)\left|\frac{3}{2},-\frac{1}{2}\right> + \chi_4(z)\left|\frac{3}{2},-\frac{3}{2}\right> \\ + \chi_5(z)\left|\frac{1}{2},-\frac{1}{2}\right> + \chi_6(z)\left|\frac{1}{2},-\frac{1}{2}\right>\Big]\end{aligned} \tag{12}$$

where $\chi_n(z), n = 1,2,\cdots,6$ forms a six-component envelope-function vector $\boldsymbol{\chi}(z)$. Each component in a given region l of the QW structure (either QW or barrier), $d_l < z < d_{l+1}$, is a superposition of the forward and backward propagations identical to Eq.(4) with constants $A_{n,l}$ and $B_{n,l}$, $n = 1,2,\cdots,6$ that can be fixed by the continuity equations that require at each interface $z = d_l$,

$$\boldsymbol{\chi}(z) \text{ and } \begin{bmatrix} p+q & -s & 0 & 0 & -\frac{1}{\sqrt{2}}s & 0 \\ -s^\dagger & p-q & 0 & 0 & -\sqrt{2}q & \sqrt{\frac{3}{2}}s \\ 0 & 0 & p-q & s & \sqrt{\frac{3}{2}}s^\dagger & \sqrt{2}q \\ 0 & 0 & s^\dagger & p+q & 0 & -\frac{1}{\sqrt{2}}s^\dagger \\ -\frac{1}{\sqrt{2}}s^\dagger & -\sqrt{2}q & \sqrt{\frac{3}{2}}s & 0 & p & 0 \\ 0 & \sqrt{\frac{3}{2}}s^\dagger & \sqrt{2}q & -\frac{1}{\sqrt{2}}s & 0 & p \end{bmatrix} \boldsymbol{\chi}(z) \text{ continuous} \tag{13}$$

where

$$\begin{aligned} p &= \gamma_1\frac{\partial}{\partial z} \\ q &= -2\gamma_2\frac{\partial}{\partial z} \\ s &= -\sqrt{3}j\gamma_3\left(k_x - jk_y\right) \end{aligned} \tag{14}$$

to maintain undisruptive carrier distribution and current across the interface.

It is important to point out that when the above described algorithms are used for the situation where an electric field is applied along the growth direction z, it is necessary to digitize the potential term $V_c(z)$ and $V_v(z)$, i.e. the regions that are used in Eq.(4) are no longer defined by the QW and barrier boundaries; instead, there could be many regions within each QW or barrier depending on the number of digitization steps used to satisfy the accuracy requirement.

This procedure applied at each wave vector point ($\boldsymbol{k} = k_x\hat{x} + k_y\hat{y}$) produces the in-plane dispersion for each subband. An example is illustrated in Fig. 5 for a 70Å/50Å

$GaAs/Al_{0.3}Ga_{0.7}As$ SL [35]. In-plane dispersions of three subbands (two for HH and one for LH) are shown where strong nonparabolicity is demonstrated. It can be seen from Fig. 5 that the band nonparabolicity could be so severe that the LH subband maximum is no longer at the Γ-point which leads to useful valence QCL design applications in Section IV.

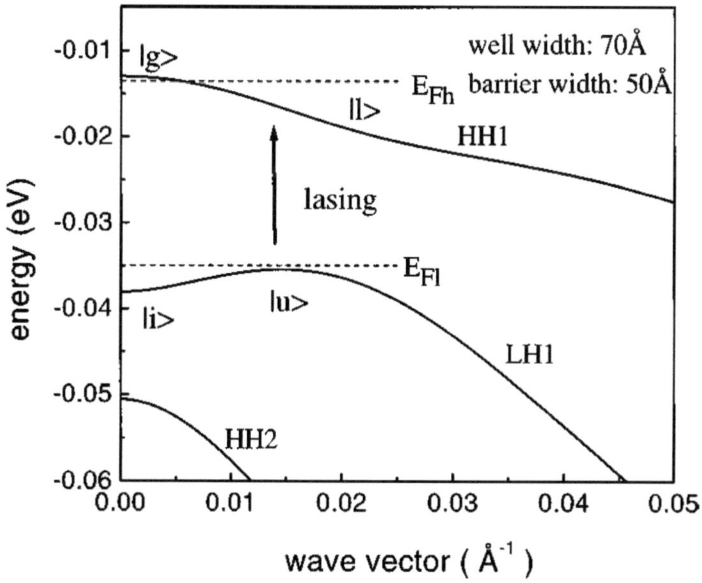

Fig. 5. In-plane dispersions of subbands HH1, LH1, and HH2 for a 70 Å/50 Å GaAs/AlGaAs SL [35].

3.2 Optical gain

For lasing to occur between two subbands, it is necessary to induce stimulated emission between them. To sustain such emission of photons, there must be sufficient optical gain to compensate various losses in the laser structure. The intersubband optical gain can be obtained by analyzing transition rates between two subbands.

According to the Fermi Golden rule, the transition rate between two discrete states 1 and 2 that are coupled by a perturbation electro-magnetic (EM) field with a frequency of ω is

$$g_{12} = \frac{2\pi}{\hbar}|H_m|^2\delta(E_2 - E_1 - \hbar\omega) \tag{15}$$

where $H_m = \langle 1|H_{ex}|2\rangle$ is the transition matrix element under the influence of a perturbation Hamiltonian H_{ex} between the two states with an exact transition energy $E_2 - E_1$ in the absence of any broadening. In reality, the transition line $E_2 - E_1$ is not infinitely sharp and is always broadened. As a result, $E_2 - E_1$ is not known exactly, instead a probability for it to appear in the energy interval $E \to E + dE$ is described. In the case of homogeneous broadening, this probability should be given as $L(E)dE$ with the Lorentzian lineshape centered at some peak transition energy E_0

$$L(E) = \frac{\Gamma/2\pi}{(E - E_0)^2 + \Gamma^2/4} \tag{16}$$

where Γ is the full width at half maximum (FWHM) that characterizes the broadening due to various homogeneous processes that include collisions and transitions. The transition rate in Eq.(15) should thus be modified by an integral that takes into account of this broadening as

$$g_{12} = \frac{2\pi}{\hbar}\int |H_m|^2 \delta(E - \hbar\omega)\, L(E) d = \frac{2\pi}{\hbar}|H_m|^2 \frac{\Gamma/2\pi}{(\hbar\omega - E_0)^2 + \Gamma^2/4} \tag{17}$$

essentially replacing the δ-function in Eq.(15) with the Lorentzian lineshape Eq.(16).
In the presence of an EM field with an optical potential vector $\boldsymbol{A}$ in a medium with isotropic effective mass, the perturbation Hamiltonian H_{ex} that describe the interaction between the field and electron in isotropic subbands is

$$H_{ex} = \frac{e\boldsymbol{A}\cdot\boldsymbol{P}}{m_e^*} \tag{18}$$

where $\boldsymbol{P}$ is the momentum operator.
From Eq.(18), it is not difficult to see that the selection rules for intersubband transitions in the conduction band are such that only those EM fields that are polarized in the growth direction (z) can induce optical transitions. The transition matrix element can then be given as

$$H_m = \frac{eA}{m_e^*}P_{12} \tag{19}$$

where the momentum matrix element

$$P_{12} = \left\langle \varphi_1 \middle| -j\hbar\frac{\partial}{\partial z} \middle| \varphi_2 \right\rangle \tag{20}$$

is evaluated as the envelope function overlap between the two subbands, which is related to the dipole matrix element [36]

$$z_{12} = \langle \varphi_1 | z | \varphi_2 \rangle = \frac{i\hbar}{m_0 E_{12}} P_{12} \tag{21}$$

and to the oscillator strength [37]

$$f_{12} = \frac{2m_0}{m_e^{*2} E_{12}} |P_{12}|^2. \tag{22}$$

It is not difficult to see from Eq.(17) that the transition rate induced by an EM field between two eigen states is the same for upward and downward transitions. Now let us apply Eq.(17) to intersubband transitions between the upper subband 2 and lower subband 1 in the conduction band (Fig. 3). Since momentums associated with photons are negligible, all photon-induced transitions are vertical in $\boldsymbol{k}$-space. It is therefore possible to obtain a net downward transition rate (in the units of number of transitions per unit time per unit sample area) between the two subbands by evaluating the following integral

$$g_{\text{net}} = \int g_{12}\{f_2(E_{2,\boldsymbol{k}})[1 - f_1(E_{1,\boldsymbol{k}})] - f_1(E_{1,\boldsymbol{k}})[1 - f_2(E_{2,\boldsymbol{k}})]\}\rho_r(E_{2,\boldsymbol{k}} - E_{1,\boldsymbol{k}})d(E_{2,\boldsymbol{k}} - E_{1,\boldsymbol{k}}) \tag{23}$$

where $f_1(E_{1,\boldsymbol{k}})$ and $f_2(E_{2,\boldsymbol{k}})$ are the electron occupation probabilities of those states at the same $\boldsymbol{k}$ in subbands 1 and 2, respectively, and $\rho_r(E_{2,\boldsymbol{k}} - E_{1,\boldsymbol{k}})$ is the reduced density of states (DOS) between $E_{1,\boldsymbol{k}}$ and $E_{2,\boldsymbol{k}}$, which is equal to DOS of subbands 1 (ρ_1) and 2 (ρ_2) when they are parallel, $\rho_r = \rho_1 = \rho_2 = m_e^*/\pi\hbar^2$. Since the Lorentzian lineshape in Eq.(23) should be much broader than the spread of the energy transitions between the two parallel subbands which can be approximated as sharply centered at the subband separation at their energy minima $E_{12} = E_2 - E_1$. Thus,

$$\begin{aligned} g_{\text{net}} &= \frac{2\pi}{\hbar}|H_m|^2 \frac{\Gamma/2\pi}{(\hbar\omega - E_{12})^2 + \Gamma^2/4}\left[\int f_2(E_{2,k})\,\rho_2 dE_{2,k} - \int f_1(E_{1,k})\,\rho_1 dE_{1,k}\right] \\ &= \frac{2\pi}{\hbar}|H_m|^2 \frac{\Gamma/2\pi}{(\hbar\omega - E_{12})^2 + \Gamma^2/4}(N_2 - N_1) \end{aligned} \tag{24}$$

where N_1 and N_2 are the total electron densities in subband 1 and 2 per unit area, respectively.

The optical gain coefficient γ that describes the increase of the EM field intensity, I, as $\gamma = I^{-1}dI/dz$ can be defined as power increase per unit volume divided by the intensity, which in turn can be expressed in terms of the net downward transition rate Eq.(24) using the momentum and dipole matrix element relation

$$\gamma(\omega) = \frac{g_{\text{net}}\hbar\omega}{IL_p} \tag{25}$$

where L_p is the length of the QW structure that is equal to the length of one period in case of QCLs. In order to relate the EM field intensity I that propagates in in-plane with the optical potential $\mathbb{A}$ polarized along z, a real expression for the potential $\mathbb{A}$ has to be used

$$\mathbb{A} = A_0\cos(\boldsymbol{\beta}\cdot\boldsymbol{r} - \omega t)\,\hat{\boldsymbol{z}} = \frac{1}{2}A_0\hat{\boldsymbol{z}}\left[e^{j(\boldsymbol{\beta}\cdot\boldsymbol{r}-\omega t)} + e^{-j(\boldsymbol{\beta}\cdot\boldsymbol{r}-\omega t)}\right] \tag{26}$$

where $\boldsymbol{\beta}$ is the in-plane propagation wave vector of the EM field. It is easy to see that only one of the two terms on the right side of Eq.(26) couples with subbands 1 and 2, $E_2 - E_1 = \hbar\omega$. Thus, the optical potential that participates in the transition matrix Eq.(19) is only half of its real amplitude, $A = A_0/2$. Since the EM field intensity I is related to the optical potential amplitude A_0 as $I = \varepsilon_0 c n_{eff} A_0^2\omega^2/2$, Eq.(25) can be written as

$$\begin{aligned} \gamma(\omega) &= \frac{e^2|P_{12}|^2}{2\varepsilon_0 c n_{eff} m_e^{*2}\omega L_p}\frac{\Gamma}{(\hbar\omega - E_{12})^2 + \Gamma^2/4}(N_2 - N_1) \\ &= \frac{e^2 m_0^2\omega z_{12}^2}{2\varepsilon_0 c n_{eff} m_e^{*2} L_p}\frac{\Gamma}{(\hbar\omega - E_{12})^2 + \Gamma^2/4}(N_2 - N_1) \end{aligned} \tag{27}$$

where c is the speed of light in free space and n_{eff} is the effective index of refraction of the QCL dielectric medium. The population inversion $N_2 - N_1 > 0$ is clearly necessary in order to achieve positive gain which peaks at the frequency $\omega_0 = E_{12}/\hbar$ with a value of

$$\gamma(\omega_0) = \frac{2e^2 m_0^2\omega z_{12}^2}{\varepsilon_0 c n_{eff} m_e^{*2}\Gamma L_p}(N_2 - N_1) \tag{28}$$

For transitions between valence subbands with nonparallel dispersions and strong mixing between HH, LH, and SO bands, we have to re-examine the intersubband transition rate.

Consider the intersubband transition in Fig. 5 from the upper state $|u>$ in subband LH1 to the lower state $|l>$ in subband HH1, if the spread of intersubband transitions is wide enough compared to the homogeneous broadening, the Lorentzian lineshape in the net downward transition rate Eq.(23) can be approximated as a δ-function yielding

$$g_{net}^{(v)} = \frac{2\pi}{\hbar}\left|H_m^{(v)}\right|^2 \rho_r(E_l - E_h)[f_{LH}(E_l) - f_{HH}(E_h)]\,|_{E_l-E_h=\hbar\omega} \tag{29}$$

where $\rho_r(E_l - E_h)$ is the reduced DOS for the transition between subbands LH1 and HH1, $f_{LH}(E_l)$ and $f_{HH}(E_h)$ are hole occupation probabilities at states with energy E_l and E_h in subband LH1 and HH1, respectively, at the same in-plane wave vector $\boldsymbol{k}$ separated by a photon energy $\hbar\omega$, and the optical transition matrix between LH1 and HH1 taking into account of the mixing

$$H_m^{(v)} = \sum_{n=1}^{6} \frac{eA}{m_n^*}\left\langle \chi_n^{(l)}\right| -j\hbar\frac{\partial}{\partial z}\left|\chi_n^{(h)}\right\rangle \tag{30}$$

where $\chi_n^{(l)}$ and $\chi_n^{(h)}$ are respectively the n-th component of the envelope function vectors for subband LH1 and HH1 as defined in Eq.(12), and m_n^* are the corresponding hole effective mass in z-direction with $m_{1,4}^* = m_0/(\gamma_1 - 2\gamma_2)$ for HH, $m_{2,3}^* = m_0/(\gamma_1 + 2\gamma_2)$ for LH, and $m_{5,6}^* = m_0/\gamma_1$ for SO. The optical gain can then be expressed

$$\begin{aligned}\gamma^{(v)} &= \frac{\pi e^2}{\varepsilon_0 c n_{eff}\omega L_p}\left|\sum_{n=1}^{6}\frac{P_n^{(lh)}}{m_n^*}\right|^2 \rho_r(E_l - E_h)[f_{LH}(E_l) - f_{HH}(E_h)]\,|_{E_l-E_h=\hbar\omega} \\ &= \frac{\pi e^2 m_0^2\omega}{\varepsilon_0 c n_{eff} L_p}\left|\sum_{n=1}^{6}\frac{z_n^{(lh)}}{m_n^*}\right|^2 \rho_r(E_l - E_h)[f_{LH}(E_l) - f_{HH}(E_h)]\,|_{E_l-E_h=\hbar\omega}\end{aligned} \tag{31}$$

in terms of momentum matrix $P_n^{(lh)} = \left\langle \chi_n^{(l)}\right| -j\hbar\frac{\partial}{\partial z}\left|\chi_n^{(h)}\right\rangle$ as well as dipole matrix elements $z_n^{(lh)} = \left\langle \chi_n^{(l)}\right|z\left|\chi_n^{(h)}\right\rangle$ between the same n-th component of the envelope function vectors of the two valence subbands.

In comparison with the optical gain Eq.(28) for the conduction subbands, we can see that it is not necessary to have total population inversion, $N_l - N_h > 0$, in order to have positive gain between the valence subbands. Instead, all we need is local population inversion $[f_{LH}(E_l) - f_{HH}(E_h)]\,|_{E_l-E_h=\hbar\omega} > 0$ in the region where the intersubband transition takes place (those states near $|u>$ and $|l>$ in Fig. 5).

3.3 Intersubband lifetimes

It has been established in Eqs.(27) and (28) that the population inversion between the upper (2) and lower (1) subbands, $N_2 - N_1 > 0$, is necessary in order to obtain optical gain. But what determines the population inversion? This question is answered with the analysis of lifetimes of these subbands as a result of various intersubband relaxation mechanisms including carrier-phonon, carrier-carrier, impurity, and interface roughness scattering processes. Among them, phonon scattering is the dominant process, especially when the energy separation between the two subbands exceeds that of an optical phonon, in which case the transitions from upper to lower subband are highly efficient with the emission of optical phonons. Different from the optical transitions, these scattering processes do not necessarily occur as vertical transitions in $\boldsymbol{k}$-space. In the case of phonon scattering, the conservation of in-plane momentum can be satisfied by a wide range of momentum of involved phonons as shown in Fig. 6(a) where intersubband as well as intrasubband transitions due to phonon scattering are illustrated.

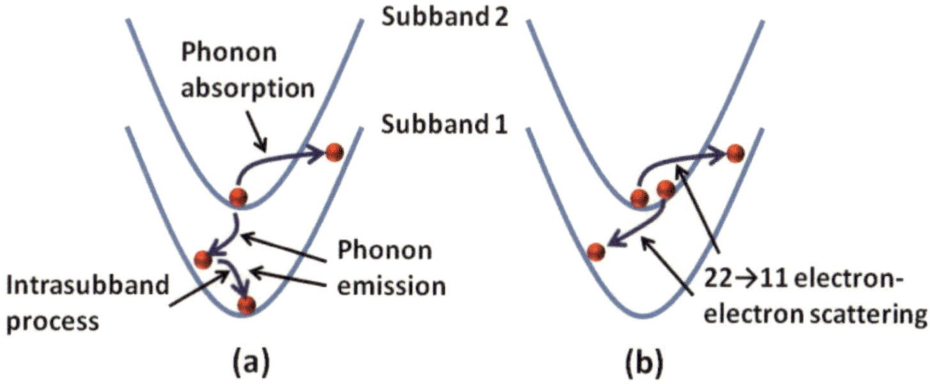

Fig. 6. (a) Intersubband and intrasubband transitions due to electron-phonon scattering (b) the 22→11 transition induced by the electron-electron scattering.

Up to now, practically all approaches in developing Si-based QCLs are based on materials from group-IV, mostly Si, Ge, SiGe alloy, and more recently, SiGeSn alloy. Different from the polar III-V and II-VI semiconductors, group-IV materials are nonpolar. The carrier scatterings by nonpolar optical phonons are much slower than those due to polar optical phonons [38]. Starting from Fermi Gold rule Eq.(15), the scattering rate for a carrier in subband 2 with the in-plane wave vector $\boldsymbol{k}$ to subband 1 with $\boldsymbol{k}'$ by a phonon with an energy $\hbar\omega_{\boldsymbol{Q}}$ and wave vector, $\boldsymbol{Q} = \boldsymbol{q} + q_z\hat{\boldsymbol{z}}$, can be expressed as an integral over all the participating phonon states

$$\frac{1}{\tau_{12}} = \frac{2\pi}{\hbar}\int \left|H_{ep}\right|^2 \delta\left(E_{2,\boldsymbol{k}} - E_{1,\boldsymbol{k}'} \mp \hbar\omega_{\boldsymbol{Q}}\right) dN_f \tag{32}$$

where H_{ep} is the electron-phonon interaction matrix element, the carrier energies $E_{1,\boldsymbol{k}'}$ and $E_{2,\boldsymbol{k}}$ are given by Eq.(1) for conduction subbands, but for valence subbands, they need to be obtained by the $\boldsymbol{k}\cdot\boldsymbol{p}$ method described above. We will proceed with the following approximations: 1) all phonons are treated to be bulk-like by neglecting the phonon confinement effect in QW structures, 2) energies of acoustic phonons are negligible $\hbar\omega_{\boldsymbol{Q}} \approx 0$, and 3) optical phonon energies are taken as a constant $\hbar\omega_{\boldsymbol{Q}} \approx \hbar\omega_0$. The matrix element of carrier-phonon interaction for different type of phonons can be written as [39,40]

$$\left|H_{ep}\right|^2 = \begin{cases} \dfrac{\Xi^2 K_B T}{2c_L\Omega}\delta_{\boldsymbol{q},\pm(\boldsymbol{k}'-\boldsymbol{k})}|G_{12}(q_z)|^2, & \text{acoustic phonon} \\ \dfrac{\hbar D^2}{2\rho\omega_0\Omega}\delta_{\boldsymbol{q},\pm(\boldsymbol{k}'-\boldsymbol{k})}\left[n(\omega_0) + \dfrac{1}{2} \mp \dfrac{1}{2}\right]|G_{12}(q_z)|^2, & \text{nonpolar optical phonon} \end{cases} \tag{33}$$

where the upper sign is for absorption and lower for emission of one phonon, K_B is the Boltzmann constant, Ω is the volume of the lattice mode cavity, c_L is the elastic constant for acoustic mode, Ξ and D are the acoustic and optical deformation potential, respectively, and $n(\omega_0)$ is the number of optical phonons at temperature T,

$$n(\omega_0) = \frac{1}{\exp(\hbar\omega_0/K_B T) - 1}. \tag{34}$$

The wavefunction interference effect between conduction subbands is

$$G_{12}(q_z) = \langle \varphi_1 | e^{jq_z z} | \varphi_2 \rangle \tag{35}$$

and between valence subbands is

$$G_{12}(q_z) = \sum_{n=1}^{6} \left\langle \chi_n^{(1)} \middle| e^{jq_z z} \middle| \chi_n^{(2)} \right\rangle. \tag{36}$$

The Kronecker symbol $\delta_{q,\pm(\boldsymbol{k}'-\boldsymbol{k})}$ in the matrix element Eq.(33) represents the in-plane momentum conservation $\boldsymbol{k}' = \boldsymbol{k} \pm \boldsymbol{q}$.
Since phonon modes have density of states $\Omega/(2\pi)^3$, the participating phonon states in the integral Eq.(32) can be expressed as

$$dN_f = \frac{\Omega}{(2\pi)^3} q dq d\theta dq_z \tag{37}$$

where θ is the angle between $\boldsymbol{k}$ and $\boldsymbol{q}$. For conduction subbands with a parabolic dispersion Eq.(1), the phonon scattering rate Eq.(32) can be evaluated analytically

$$\frac{1}{\tau_{12}} = \begin{cases} \dfrac{\Xi^2 K_B T m_e^*}{4\pi c_L \hbar^3} \displaystyle\int |G_{12}(q_z)|^2 dq_z, & \text{acoustic} \\ \dfrac{D^2 m_e^* \left[n(\omega_0) + \frac{1}{2} \mp \frac{1}{2}\right]}{4\pi\rho\hbar^2\omega_0} \displaystyle\int |G_{12}(q_z)|^2 dq_z, & \text{nonpolar optical.} \end{cases} \tag{38}$$

But for valence subbands where there is a strong nonparabolicity, Eq.(32) can no longer be integrated analytically. However, if we take the wave vector of the initial state in subband 2 to be at the Γ-point $\boldsymbol{k} = 0$, then the phonon wave vector $\boldsymbol{q} = \boldsymbol{k}'$, Eq.(38) can still be used to evaluate the phonon scattering rate between valence subbands by substituting the effective mass with some average effective mass in the final subband 1.
The phonon scattering rate in Eq.(38) has been used to compare the lifetimes of two similar three-level systems, SiGe/Si and GaAs/AlGaAs, as shown in Fig. 7(a) [38]. The lifetime difference between the upper (3) and lower (2) subband is calculated as the function of the transition energy $E_3 - E_2$ which is varied by changing the barrier width between the two QWs that host the two subbands. The main result in Fig. 7(b) is that the lifetimes in the SiGe system can be an order of magnitude longer than in the GaAs/AlGaAs system because of SiGe's lack of polar optical phonons. This property can potentially lead to significantly reduced lasing threshold for the SiGe system. The sudden drops in the lifetimes have to do with the shifting of subband energy separations $E_2 - E_1$ and $E_3 - E_2$, to either below or above the optical phonon energy.

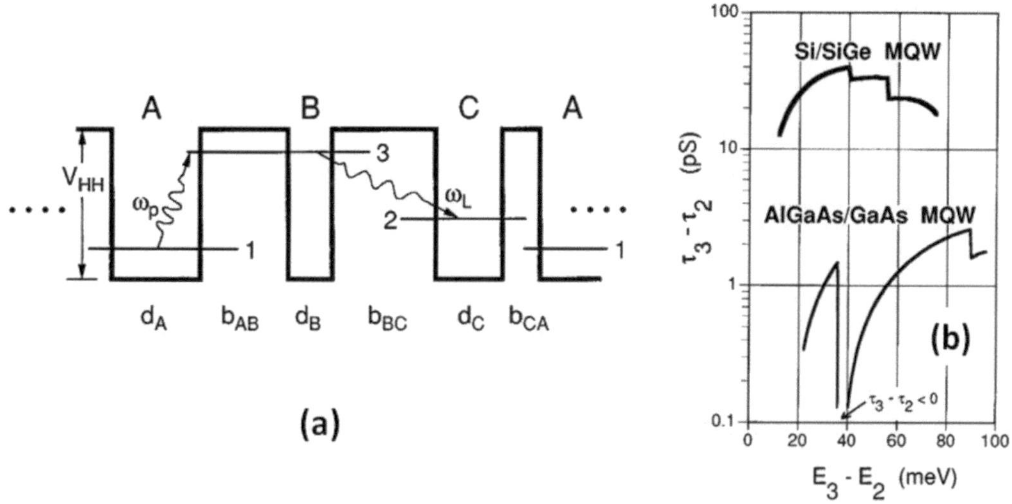

Fig. 7. (a) HH valence band diagram of one period of a SiGe/Si SL with hole energy increase in the upward direction, and (b) comparison of Lifetime difference ($\tau_3 - \tau_2$) between the SiGe/Si and GaAs/AlGaAs SL (in a similar three scheme) as a function of the transition energy ($E_3 - E_2$) [38].

Among the different phonon scattering processes - emission and absorption of acoustic and optical phonons, the emission of an optical phonon is by far the fastest process. But in far-IR QCLs when the subband energy separation is less than the optical phonon energy and the emission of an optical phonon is forbidden, phonon scattering may no longer be the dominant relaxation mechanism. Other scattering mechanisms need to be taken into consideration, such as the carrier-carrier [41], impurity [42], and interface roughness scatterings [43], all of which are elastic processes. The carrier-carrier scattering is a two-carrier process that is particularly important when carrier concentration is high which increases the possibility of two carriers interacting with each other. There are many possible outcomes as a result of this interaction in inducing intersubband as well as intrasubband transitions. Among them, the 22→11 process where both carriers originally in subband 2 end up in subband 1 is the most efficient one in terms of inducing intersubband transitions (Fig. 6(b)). It has been reported in experiment that the intersubband transition times on the order of tens of ps have been observed for carrier densities of $10^9 \sim 10^{11}/cm^2$ in GaAs/AlGaAs QWs [44]. In QCLs, since doping is mostly introduced away from the active region where optical transitions take place, impurity scattering does not seem to play a major role in determining the lifetimes of laser subbands, however, its influence on carrier transport in the injection region can be rather important. Interface roughness depends strongly on the process of structural growth, its impact on scattering should be more significant on narrow QWs, particularly for those transitions between two wavefunctions that are localized in MQWs that span across several interfaces.

4. Valence band SiGe QCLs

Up to now, all of the demonstrated QCLs are based on epitaxially grown III-V semiconductor heterostructures such as GaInAs/AlInAs, GaAs/AlGaAs, and InAs/AlSb, using electron subbands in conduction band. With the promise of circumventing the indirectness of Si band gap, a SiGe/Si laser based on intersubband transitions was first proposed by Sun et al in 1995 [38] where a comparative study was performed between the SiGe/Si and GaAs/AlGaAs systems. Since then there has been a series of theoretical and experimental investigations aimed at producing Si-based QCLs. A natural choice of the material system is SiGe because Si and Ge are both group-IV elements, SiGe alloys have been routinely deposited on Si to produce heterojunction bipolar transistors or as a strain-inducing layer for CMOS transistors [45].

While QCLs based on SiGe alloys can be monolithically integrated on Si if successfully developed, there are significant challenges associated with this material system. First, there is a 4% lattice mismatch between Si and Ge. Layers of $Si_{1-x}Ge_x$ alloys deposited on Si substrates induce strain which can be rather significant in QCLs because a working structure typically consists of at least hundreds of layers with a total thickness that easily exceeds the critical thickness above which the built-in strain simply relaxes to develop defects in the structure. In dealing with the issue of strain in SiGe/Si QC structures, one popular approach is to use strain balanced growth where the compressively strained $Si_{1-x}Ge_x$ and tensile strained Si are alternately stacked on a relaxed $Si_{1-y}Ge_y$ buffer deposited on a Si substrate where the buffer composition ($y < x$) is chosen to produce strains in $Si_{1-x}Ge_x$ and Si that compensate each other, so that the entire structure maintains a neutral strain profile [46,47]. Strain balanced structures have effectively eliminated the limitations of critical thickness and produced high quality SiGe/Si structures consisting of nearly 5000 layers (15 μm) by chemical vapor deposition [48].

Second, the band offsets between compressively strained SiGe and tensile strained Si or between SiGe of different alloy compositions is such that the conduction band QWs are shallow, and nearly all band offsets are in valence band. Practically all of the investigations of SiGe QCLs are focused on intersubband transitions in the valence band. But the valence subband structure is much more complex in comparison with the conduction subbands because of the mixing between the HH, LH and SO bands. Their associated subbands are closely intertwined in energy making the design of valence QCLs extremely challenging. Third, any valence QCL design in general has to inevitably involve HH subbands since they occupy lower energies relative to LH subbands because of their large effective mass. In SiGe, the HH effective mass is high ($\sim 0.2m_e$), which leads to small IST oscillator strength between the laser states and poor carrier transport behavior associated with their low mobilities.

The challenge presented by the valence-band mixing also creates an opportunity to engineer desirable subband dispersions such that total population inversion between the subbands becomes unnecessary in a way analogous to the situation in conventional band-to-band lasers discussed in section II. It was reported in QCLs that the population inversion was only established locally in $\boldsymbol{k}$-space in the large $\boldsymbol{k}$-vector region of the conduction subbands because the interactions between the subbands produced nonparallel in-plane dispersions [49]. In comparison with the conduction-band nonparabolicity, this effect is known to be much stronger in the valence band [31]. As a matter of fact, in the valence band of most diamond and zinc-blende semiconductors, LH and HH subbands usually anti-cross, and near the point of anti-crossing, the LH subband in-plane dispersion becomes electron-like. Thus, an earlier design was accomplished to effectively tailor the dispersions of two valence subbands in a GaAs/AlGaAs QW (Fig. 5) similar to those of the conduction and valence bands, in which one of the subbands is electron-like and the other hole-like, i.e., one of the subbands shall have its effective mass inverted [35]. If we now designate states near the Γ-point of subband LH1 as the intermediate states, $|i>$, states near the valley (inverted-effective-mass region) of subband LH1 as the upper laser states $|u>$, states in subband HH1 vertically below the valley of subband LH1 as the lower laser states $|l>$, and states near the Γ-point of subband HH1 as the ground states $|g>$ (counting the hole energy downward in Fig. 5), we can see that the situation closely resembles the one in the conventional band-to-band semiconductor laser. The upper and lower laser states can now be populated and depopulated through fast intrasubband processes, while the lifetime of the upper laser states is determined by a much slower intersubband process between subbands LH1 and HH1. Such a large lifetime difference between the upper and lower laser states is certainly favorable for achieving population inversion between them.

The inverted mass approach was later on applied to the SiGe system [50,51]. Two slightly different schemes were developed, one utilized the inverted LH effective mass [50], and the other inverted HH mass [51]. In both cases, the effective mass inversion is the result of strong interaction between the valence subbands. The inverted-effective-mass feature requires the coupled subbands to be closely spaced in energy, typically less than all the optical phonon energies in the SiGe material system (37meV for the Ge-Ge mode, 64meV for the Si-Si mode, and 51meV for the Si-Ge mode [52]), which suppresses the nonradiative intersubband transitions due to the optical phonon scattering, but also limits the optical transitions in the THz regime. The structures under investigation were strain balanced with compressively strained $Si_{1-x}Ge_x$ QW layers and the tensile strained Si barrier layers deposited on a relaxed $Si_{1-y}Ge_y$ buffer layer ($0 < y < x$) on Si. The in-plane dispersions of the inverted LH scheme are shown in Fig. 8 for a 90Å/50Å $Si_{0.7}Ge_{0.3}$/Si super lattice (SL). Three lowest subbands are shown. The numbers 1, 2, 3, and 4 indicate how this inverted mass

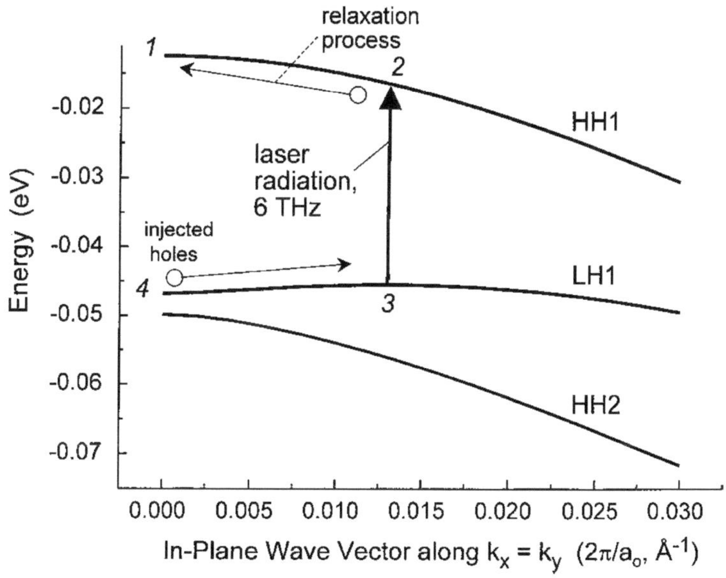

Fig. 8. Dispersions of subbands HH1, LH1 and HH2 in a 90Å/50Å $Si_{0.7}Ge_{0.3}$/Si SL strained balanced on a $Si_{0.81}Ge_{0.19}$ buffer, obtained with a 6 × 6 valence band matrix taking into account HH, LH and SO interactions and strain effect [50].

intersubband laser mimics the operation of a conventional band-to-band laser. The lifetime of the upper laser state 3 is long because the intersubband transition energy at 6THz (~50μm) is below that of optical phonons, allowing only much weaker acoustic phonon scattering between the two subbands. Calculation results have shown that optical gain in excess of 150/cm can be achieved without total population inversion being established between the LH and HH subbands.

The inverted LH effective mass approach utilizes optical transitions between the LH and HH subband. It can be argued from the component overlap of the envelope functions in Eq.(30) that the optical transition matrix between subbands of different types is always smaller than that between subbands of the same type. We therefore tried to engineer the same inverted effective mass feature between two HH subbands. The challenge is to lift the LH subband above the HH2 subband. Once again, a strain balanced SL structure is considered but with different SiGe alloy compositions and layer thicknesses [51]. The band structure for a 90Å/35Å $Si_{0.8}Ge_{0.2}/Si$ SL under an electric bias of 30kV/cm is shown in Fig. 9(a) where each QW has two active doublets formed by bringing HH1 and HH2 subbands in the neighboring QWs into resonance under the bias. There is a 3meV energy split within the doublet. The resulting in-plane dispersions for the two doublets are shown in Fig. 9(b). Simulation results showed that optical gain of 450/cm at 7.3THz can be achieved at a pumping current density of 1.5KA/cm^2 at 77K.

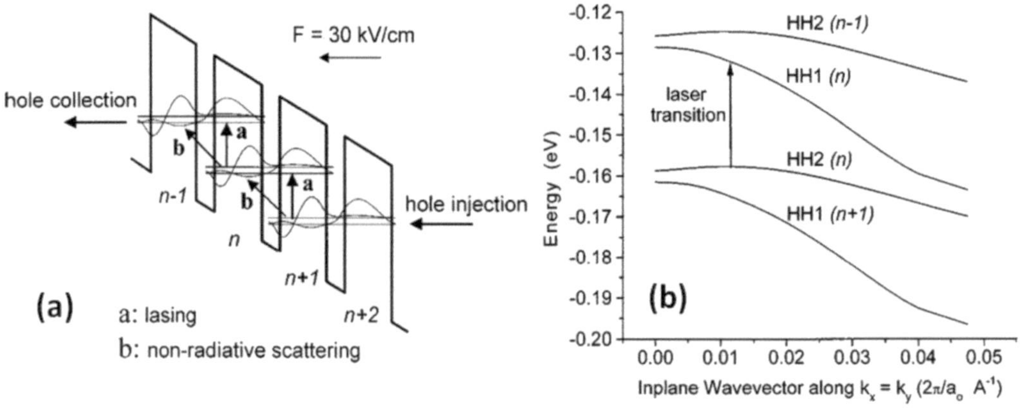

Fig. 9. (a) Band diagram of the $Si_{0.8}Ge_{0.2}/Si$ SL under an electric bias of 30 kV/cm. The labels $(n-1, n, n+1, \ldots)$ represent the QWs in which the wave functions are localized [51], (b) Dispersions of the four levels (two doublets) in a QW.

Electroluminescence (EL) from a SiGe/Si quantum cascade emitter was first demonstrated in a SiGe/Si quantum cascade emitter using HH to HH transitions in the mid-IR range in 2000 [53]. Since then several groups have observed EL from the same material system with different structures. EL emissions have been attributed to various optical transitions including HH-to-HH [54], LH-to-HH [55], and HH-to-LH [56], with emission spectra ranging from mid IR to THz (8 ~ 250μm). But lasing has not been observed. Improvement on the QCL design has been to be made. One of the most successful III-V QCL designs has been the approach of bound-to-continuum where the lower laser state sitting at the top of a miniband is delocalized over several QWs while the upper laser state is a bound state in the minigap as illustrated in Fig. 10 [57,58]. Electrons that are injected into the bound upper state 2 are prevented from escaping the bound state by the minigap, and then undergo lasing transitions to a lower state 1. The depopulation of the lower state 1 is accelerated through the efficient miniband carrier transport. Such a design has led to improved performance in terms of operating temperature as well as output power for III-V QCLs. A similar bound-to-continuum design has been implemented in SiGe with both bound and continuum formed by HH states, once again showing just EL with no lasing [59]. It is believed that in this structure LH states are mixed within the HH states. Although the impact of this intermixing has not been fully understood, these LH states can in principle present additional channels for carriers to relax from the upper laser state reducing its lifetime. An improved version has been sought after by lifting the LH states out of all involved HH states for bound and continuum with the use of strain, as a result, a clear intersubband TM polarized EL is shown suggesting that LH states have been pushed away from the HH radiative transitions [60].

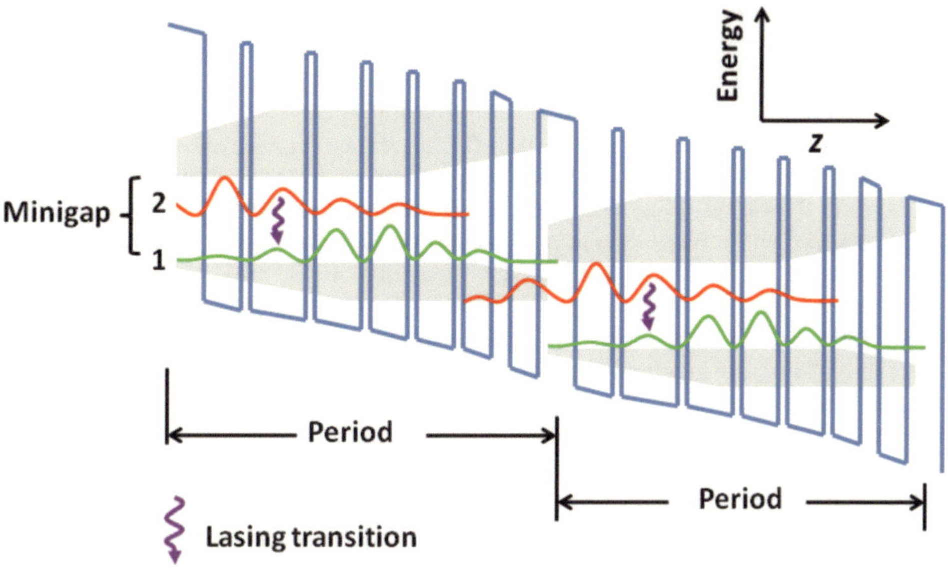

Fig. 10. Illustration of two periods of a bound-to-continuum QCL. Lasing transition occurs between an isolated bound upper state 2 (formed in the minigap) and a delocalized lower state 1 (sitting on top of a miniband).

Nearly a decade has passed since the first experimental demonstration of EL from a SiGe/Si quantum cascade emitter [53]. During this period III-V QCLs have been improved dramatically to allow for commercialization and system integration for various applications, however there are still no SiGe QCLs. The seemingly inherent difficulties with the valence QCL approach have propelled some researchers back into the conduction band to look for solutions.

5. Conduction band Si-based QCLs

Before QCLs can be designed using conduction subbands, there must be sufficient conduction band offset. Contrary to the situation in compressively strained $Si_{1-x}Ge_x$, tensile strained $Si_{1-x}Ge_x$ can have larger conduction band offset, but the conduction band minima occur at the two Δ_2-valleys whose effective mass (longitudinal) along the growth direction is very heavy ($m_l \sim 0.9m_0$) resulting in small oscillator strength and poor transport behavior possibly even worse than the HHs. Any approach of developing Si-based QCLs based on transitions between conduction subbands needs to necessarily go beyond the conventional methods in selecting the material system and growth technique.

Prospects of developing such Si-based QCLs have been investigated theoretically. One approach stayed with the Si-rich SiGe/Si material system but instead of the conventional growth direction in (100), the structural orientation has been rotated to the [111] crystal plane [61]. Conduction band offset was calculated to be 160meV at the conduction band minima consisting of six-degenerate Δ-valleys, sufficient for designing far IR QCLs. The effective mass along (111) direction can be obtained as the geometric average of the longitudinal and transverse effective masses of the Δ -valley, $\sim 0.26m_0$, lower than that of longitudinal $m_l \sim 0.9m_0$ in the (100) structure. Another design relying on the Ge-rich Ge/SiGe material system has been proposed to construct conduction band QCLs using compressively strained Ge QWs and tensile strained $Si_{0.22}Ge_{0.78}$ alloy barriers grown on a relaxed [100] $Si_{1-y}Ge_y$ buffer [62]. The intersubband transitions in this design are within the L-valleys which are the conduction band minima in Ge QWs whose effective mass along (100) direction has been determined to be $\sim 0.12m_0$. Since $Si_{1-x}Ge_x$ alloys with $x < 0.85$ are similar to Si in that the conduction band minima appear in the Δ -valleys, the conduction band lineup in the Ge/$Si_{0.22}Ge_{0.78}$ structure is rather complex with conduction band minima in Ge at L-valleys but in $Si_{0.22}Ge_{0.78}$ at Δ_2-valleys along the (100) growth direction. Although the band offset at the L-valleys is estimated to be as high as 138 meV, the overall band offset between the absolute conduction band minima in Ge and $Si_{0.22}Ge_{0.78}$ is only 41meV. Although the quantum confinement effect helps to lift those electron subbands at Δ_2-valleys, the two Δ_2-valleys are inevitably entangled with the L-valleys in the conduction band, leading to design complexity and potentially creating additional nonradiative decay channels for the upper laser state.

Recently, a new group-IV material system that expands beyond the $Si_{1-x}Ge_x$ alloys has been successfully demonstrated with the incorporation of Sn. These new ternary $Ge_{1-x-y}Si_xSn_y$ alloys have been studied for the possibility of forming direct band gap semiconductors[63-66]. Since the first successful growth of this alloy [67], device- quality epilayers with a wide range of alloy contents have been achieved [68,69]. Incorporation of Sn provides the opportunity to engineer separately the strain and band structure since we can vary the Si (x) and Sn (y) compositions independently. Certain alloy compositions of this material system offer three advantages: (1) the possibility of a "cleaner" conduction band lineup in which the L-valleys in both well and barrier sit below other valleys (Γ, Δ), (2) an electron effective mass along the (001) growth direction that is much lower than the HH mass, and (3) a strain free structure that is lattice matched to Ge. In addition, recent advances in the direct growth of Ge layer on Si provide a relaxed matching buffer layer on a Si substrate upon which the strain-free Ge/$Ge_{1-x-y}Si_xSn_y$ is grown [70]. Based on this material system, a strain free QCL operating in the conduction L-valleys was proposed [71].

Since band offsets between ternary Sn-containing alloys and Si or Ge are not known experimentally, we have calculated the conduction band minima for a lattice- matched heterostructure consisting of Ge and a ternary $Ge_{1-x-y}Si_xSn_y$ based on Jaros' band offset theory [72] which is in good agreement with experiment for many heterojunction systems. For example, this theory predicts an average valence band offset, $\Delta E_{v,av} = 0.48$ eV for a

Ge/Si heterostructure (higher energy on the Ge side), close to the accepted value of $\Delta E_{v,av} = 0.5$ eV. The basic ingredients of our calculation are the average (between HH, LH, and SO bands) valence band offset between the two materials and the compositional dependence of the band structure of the ternary alloy. For the Ge/α-Sn interface, Jaros' theory predicts $\Delta E_{v,av} = 0.69$ eV (higher energy on the Sn side). For the $Ge_{1-x-y}Si_xSn_y$/Ge interface we have used the customary approach for alloy semiconductors, interpolating the average valence band offsets for the elementary heterojunctions Ge/Si and Ge/α-Sn. Thus we used (in eV)

$$\Delta E_{v,av}(x,y) = E_{v,av}(\text{GeSiSn}) - E_{v,av}(\text{Ge}) = 0.48x + 0.69y. \tag{39}$$

Once the average valence band offset is determined, the energies of individual conduction band edges in the $Ge_{1-x-y}Si_xSn_y$ alloy can be calculated relative to those in Ge from the compositional dependence of the spin-orbit splitting of the top valence band states and the compositional dependence of the energy separations between those conduction band edges and the top of the valence band in the alloy [73]. We have assumed that all required alloy energies can be interpolated between the known values for Si, Ge, and α-Sn as

$$\begin{aligned} E_{\text{GeSiSn}}(x,y) = E_{\text{GeSiSn}}(1-x-y) + E_{\text{Si}}x + E_{\text{Sn}}y \\ -b_{\text{GeSi}}(1-x-y)x - b_{\text{GeSn}}(1-x-y)y - b_{\text{SiSn}}xy \end{aligned} \tag{40}$$

The bowing parameters b_{GeSi}, b_{GeSn}, and b_{SiSn} have been discussed in Refs. [74] and [75]. Finally, for the indirect conduction band minimum near the X-point, Weber and Alonso find

$$E_x = 0.931 + 0.018x + 0.206x^2 \tag{41}$$

(in eV) for $Ge_{1-x}Si_x$ alloys [76]. On the other hand, the empirical pseudo-potential calculations of Chelikovsky and Cohen place this minimum at 0.90 eV in α-Sn, virtually the same as its value in pure Ge [77]. We thus assume that the position of this minimum in ternary $Ge_{1-x-y}Si_xSn_y$ alloys is independent of the Sn concentration y. The conduction band minima results are shown in Fig. 11 for Sn concentrations $0 < y < 0.1$. The Si concentration x was calculated using Vegard's law in such a way that the ternary $Ge_{1-x-y}Si_xSn_y$ is exactly lattice-matched with Ge.

It can be seen from Fig. 11 that a conduction-band offset of 150 meV at L-valleys can be obtained between lattice-matched Ge and $Ge_{0.76}Si_{0.19}Sn_{0.05}$ alloy while all other conduction-band valleys (Γ, X, etc) are above the L-valley band edge of the $Ge_{0.76}Si_{0.19}Sn_{0.05}$ barrier. This band alignment presents a desirable alloy composition from which a QCL operating at L-valleys can be designed using Ge as QWs and $Ge_{0.76}Si_{0.19}Sn_{0.05}$ as barriers without the complexity arising from other energy valleys.

Figure 12 shows the QCL structure based upon Ge/ $Ge_{0.76}Si_{0.19}Sn_{0.05}$ QWs. Only L-valley conduction-band lineups are shown in the potential diagram under an applied electric field of 10 kV/cm. In order to solve the Schrödinger equation to yield subbands and their associated envelope functions, it is necessary to determine the effective mass m_z^* along the (001) growth direction (z) within the constant-energy ellipsoids at the L-valleys along the (111) direction which is tilted with respect to (100). Using the L-valley principal transverse effective mass $m_t^* = 0.08m_0$, and the longitudinal effective mass $m_l^* = 1.60m_0$ for Ge, we obtain $m_z^* = (2/3m_t^* + 1/3m_l^*)^{-1} = 0.12m_0$. The squared magnitudes of all envelope functions are plotted at energy positions of their associated subbands. As shown in Fig. 12, each period of the QCL has an active region for lasing emission and an injector region for carrier transport. These two regions are separated by a 30 Å barrier. The active region is constructed with 3 coupled Ge QWs that give rise to three subbands marked 1, 2, and 3. The lasing transition at the wavelength of 49 μm is between the upper laser state 3 and the lower laser state 2. The injector region consists of 4 Ge QWs of decreasing well widths all

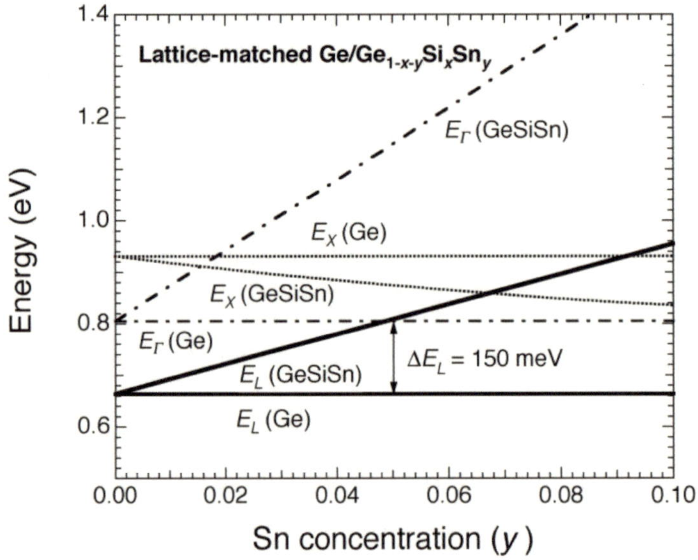

Fig. 11. Conduction band minima at L, Γ, X points of $Ge_{1-x-y}Si_xSn_y$ that is lattice matched to Ge [71].

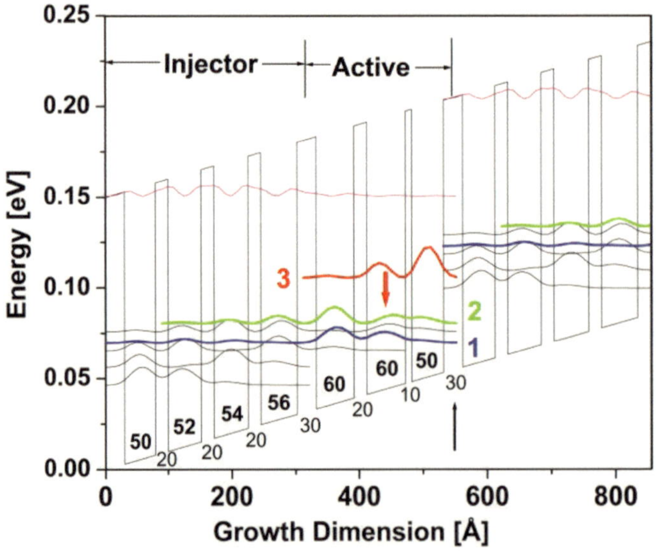

Fig. 12. L-valley conduction band profile and squared envelope functions under an electric field of 10kV/cm. Layer thicknesses in angstrom are marked with bold numbers for Ge QWs and regular for GeSiSn barriers. Array marks the injection barrier [71].

separated by 20Å $Ge_{0.76}Si_{0.19}Sn_{0.05}$ barriers. The depopulation of lower state 2 is through scattering to state 1 and to the miniband downstream formed in the injector region. These scattering processes are rather fast because of the strong overlap between the involved states. Another miniband in the injector region formed of quasi-bound states is situated 45 meV above the upper laser state 3, effectively preventing escape of electrons from upper laser state 3 into the injector region.

The nonradiative transition rates between different subbands in such a low-doped nonpolar material system with low injection current should be dominated by deformation-potential scattering of nonpolar optical and acoustic phonons. For this Ge-rich structure, we have used bulk-Ge phonons for calculation of the scattering rate to yield lifetimes for the upper laser state τ_3 and the lower laser state τ_2, as well as the 3→2 scattering time τ_{32} [78]. The results obtained from Eq.(38) are shown in Fig. 13 as a function of operating temperature. These lifetimes are at least one-order of magnitude longer than those of III-V QCLs owing to the nonpolar nature of GeSiSn alloys. The necessary condition for population inversion $\tau_{32} > \tau_2$ is satisfied throughout the temperature range. Using these predetermined lifetimes in the population rate equation under current injection:

$$\begin{cases} \dfrac{\partial N_3}{\partial t} = \dfrac{\eta J}{e} - \dfrac{N_3 - \bar{N}_3}{\tau_3} \\ \dfrac{\partial N_2}{\partial t} = \dfrac{N_3 - \bar{N}_3}{\tau_{32}} - \dfrac{N_2 - \bar{N}_2}{\tau_2} \end{cases} \tag{42}$$

where $N_i, (i = 2,3)$ is the area carrier density per period in subband i under injected current density J with an injection efficiency η, and $\bar{N}_i$ is the area carrier density per period due to thermal population. Solving the above rate equation at steady state yields population inversion

$$N_3 - N_2 = \tau_3 \left(1 - \frac{\tau_2}{\tau_{32}}\right) \frac{\eta J}{e} - (\bar{N}_2 - \bar{N}_3). \tag{43}$$

which can then be used to evaluate the optical gain of the TM polarized mode following Eq.(28) at the lasing transition energy $\hbar\omega_0 = E_3 - E_2 = 25$ meV as

$$\gamma(\omega_0) = \frac{2e^2 m_0^2 \omega_0 z_{23}^2}{\varepsilon_0 c n_{eff} m_z^{*2} \Gamma L_p} \left[\tau_3 \left(1 - \frac{\tau_2}{\tau_{32}}\right) \frac{\eta J}{e} - (\bar{N}_2 - \bar{N}_3)\right]. \tag{44}$$

For the QCL structure in Fig. 12, the following parameters are used: index of refraction $n_{eff} = 3.97$, lasing transition FWHM $\Gamma = 10$ meV, length of one period of the QCL $L_p = 532$ Å, area doping density per period of 10^{10} cm^2, and unit injection efficiency $\eta = 1$.

Since the relatively small conduction band offset limits the lasing wavelength to the far-IR or THz regime (roughly 30 μm and beyond), the waveguide design can no longer rely on that of conventional dielectric waveguides such as those used in laser diodes and mid IR QCLs. This is mainly because the thickness required for the dielectric waveguide would have exceeded what can be realized with the epitaxial techniques employed to grow the laser structures. One solution is to place the QCL active structure between two metal layers to form the so-called plasmon waveguide [79,80]. While the deposition of top metal is trivial, placing bottom metal requires many processing steps such as substrate removal, metal deposition, and subsequent wafer bonding. The QCL waveguides are typically patterned into ridges as shown in Fig. 14.

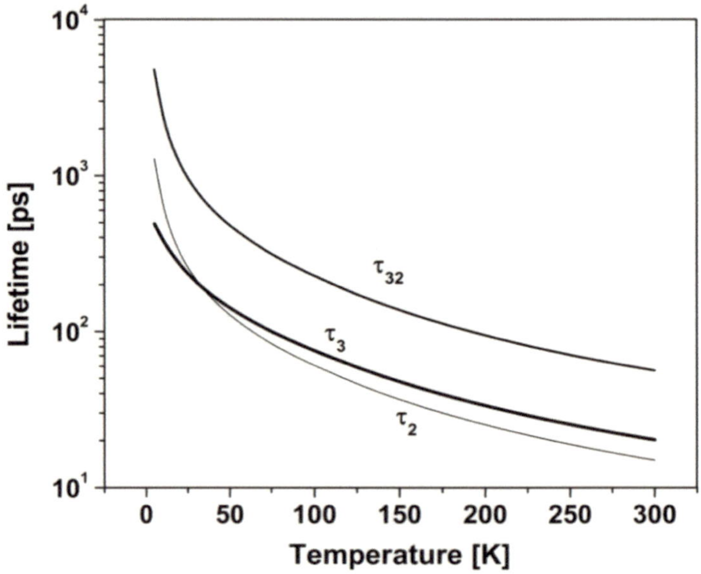

Fig. 13. Upper state lifetime τ_3, lower state lifetime τ_2, and scattering time τ_{32} between them as a function of temperature.

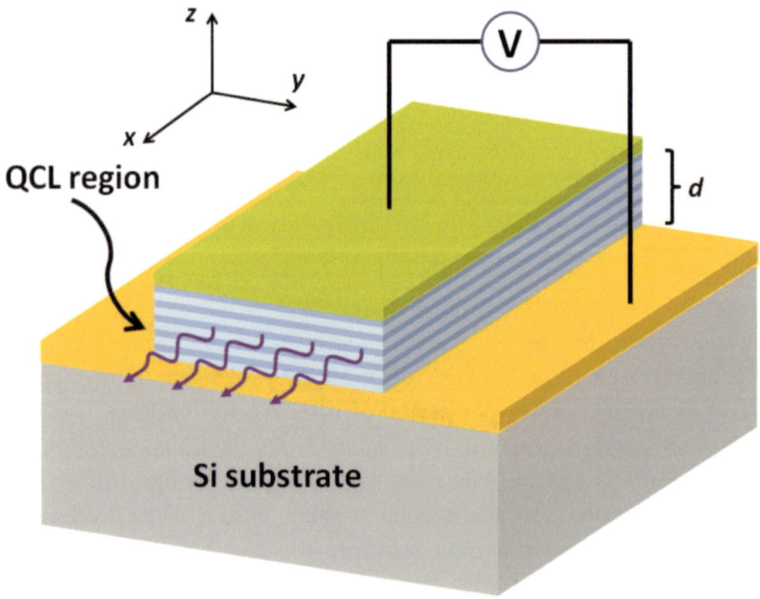

Fig. 14. Schematic of a ridge plasmon waveguide with the Ge/GeSiSn QCL sandwiched between two metal layers.

This plasmon waveguide supports only TM polarized EM mode that is highly confined within the QCL region, $-d/2 < z < d/2$. We can assume Drude model to describe the metal dielectric function

$$\varepsilon_M = 1 - \frac{\omega_p^2}{\omega^2 + j\gamma_m\omega} \tag{45}$$

where ω_p is the metal plasmon frequency, and γ_m is the metal loss ($\hbar\omega_p = 8.11$ eV, $\hbar\gamma_m = 65.8$ meV for Au [81]), and $\varepsilon_D = n_{eff}^2 \approx 16$ for the Ge rich $Ge/Ge_{0.76}Si_{0.19}Sn_{0.05}$ QCL active region. Consider the EM wave propagate along the x-direction as shown in Fig. 14, its electric field can then be obtained as

$$E = \begin{cases} \frac{E_0}{\varepsilon}\cosh(kd/2)\,(j\beta\hat{\mathbf{z}} + q\hat{\mathbf{x}})e^{-q(z-d/2)}e^{j(\beta x-\omega t)}, & z > \frac{d}{2} \\ E_0[j\beta\cosh(kz)\,\hat{\mathbf{z}} - k\sinh(kz)\,\hat{\mathbf{x}}]e^{j(\beta x-\omega t)}, & |z| < \frac{d}{2} \\ \frac{E_0}{\varepsilon}\cosh(kd/2)\,(j\beta\hat{\mathbf{z}} - q\hat{\mathbf{x}})e^{q(z+d/2)}e^{j(\beta x-\omega t)}, & z < -\frac{d}{2} \end{cases} \tag{46}$$

where $\varepsilon = \varepsilon_M/\varepsilon_D$, and E_0 is a constant. The complex propagation constant $\beta = \beta' + j\beta''$ follows these relations, $\beta^2 - k^2 = k_D^2$, $\beta^2 - q^2 = \varepsilon k_D^2$ with $k_D = \sqrt{\varepsilon_D}\omega/c$. It is easy to see that the continuity of the normal component of the electric displacement is satisfied at the boundaries $z = \pm d/2$, the requirement of continuity of tangential electric field leads to

$$k^2\left[\varepsilon^2\tanh^2\left(\frac{kd}{2}\right) - 1\right] = k_D^2(1-\varepsilon) \tag{47}$$

which determines the TM modes that can propagate in this plasmon waveguide. The waveguide loss α_w is dominated by the metal loss, which can be determined from the imaginary part of the propagation constant, as $\alpha_w = 2\beta''$. As a superposition of two surface plasmon modes bound to the two metal-dielectric interfaces at $z = \pm d/2$, this TM mode decays exponentially into the metal, providing an excellent optical confinement factor defined as $\Gamma_w = \int_{-d/2}^{d/2}|E|^2\,dz/\int_{-\infty}^{\infty}|E|^2\,dz$. We have simulated the TM-polarized mode in a QCL structure of 40 periods ($d = 2.13$ μm) that is confined by double-Au-plasmon waveguide and obtained near unity optical confinement $\Gamma_w \approx 1.0$ and waveguide loss $\alpha_w = 110$ /cm. Assuming a mirror loss $\alpha_m = 10$ /cm for a typical cavity length of 1 mm, the threshold current density J_{th} can be calculated from the balancing relationship, $\Gamma_w\gamma_{th} = \alpha_w + \alpha_m$, where γ_{th} is the optical gain Eq.(44) obtained at the threshold J_{th}. The result is shown in Fig. 15 for J_{th} that ranges from 22 A/cm^2 at 5 K to 550 A/cm^2 at 300 K. These threshold values are lower than those of III-V QCLs as a result of the longer scattering times due to nonpolar optical phonons.

While GeSiSn epilayers with alloy compositions suitable for this QCL design have been grown with MOCVD [68,69], implementation of Ge/GeSiSn QCLs is currently challenged by the structural growth of the large number of hetero-layers in the QCL structure with very fine control of layer thicknesses and alloy compositions. Nevertheless, progress is being made towards experimental demonstration.

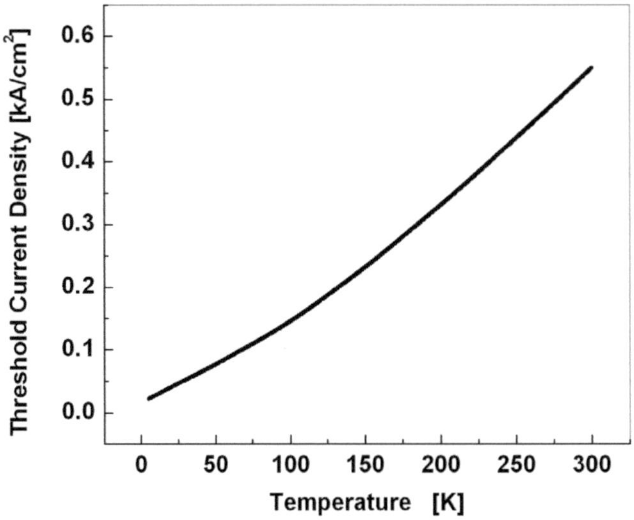

Fig. 15. Simulated threshold current density of the Ge/GeSiSn QCL as a function of temperature.

6. Summary

Silicon-based lasers have been long sought after for the possibility of monolithic integration of photonics with high-speed Si electronics. Many parallel approaches are currently taken to reach this goal. Among them Si nanocrystals and Er-doped Si have been investigated rather extensively. While EL has been demonstrated, lasing has not been observed. The only reported lasing in Si so far has been achieved using stimulated Raman scattering which requires optical pumping at very high intensity on a device of large scale – impractical for integration with Si electronics. The QCLs that have been successfully developed in III-V semiconductors offer an important alternative for the development of Si-based lasers. The salient feature of QCLs is that lasing transitions take place between subbands that are within the conduction band without crossing the band gap. Such a scheme makes the indirect nature of the Si band gap irrelevant. In order to appreciate the QCL designs, some theoretical background underlying the basic operating principles has been introduced here. In particular, subband formation and energy dispersion in semiconductor QWs are described in the framework of envelope functions with the effective-mass approximation for both conduction and valence band taking into account mixing between HH, LH, and SO bands. Optical gain based on ISTs is derived and intersubband lifetimes are discussed with a more detailed treatment of carrier-phonon scattering.

The development of Si-based QCLs has been primarily focused on ISTs between valence subbands in the Si-rich SiGe/Si material system. Such a material system has been routinely used in CMOS-compatible processes. There are two reasons for using holes instead of electrons. One is that the compressively strained $Si_{1-x}Ge_x$ with tensile strained Si grown on a relaxed $Si_{1-y}Ge_y$ has very small conduction band offset - QWs are too shallow to allow for elaborate QCLs. Tensile strained $Si_{1-x}Ge_x$, on the other hand, can have larger conduction band offset, but the conduction band minima occur at the two Δ_2-valleys whose effective mass (longitudinal) along the growth direction is heavy ($m_l \sim 0.9m_0$) resulting in small oscillator strength and poor transport behavior such as reduced tunneling probabilities. It is generally believed that SiGe QCLs have to be pursued within the valence band as a p-type device. But the situation in valence band also presents challenges in several perspectives. First, the strong mixing of HH, LH, and SO bands makes the QCL design exceedingly cumbersome albeit the opportunities presented by the strong nonparabolicity in valence subbands to take advantage of schemes such as the inverted effective mass where the total population inversion between subbands may not be necessary. Second, there is a great deal of uncertainty in various material parameters for the SiGe alloy - often times approximation has to be made to linearly extrapolate parameters from those of Si and Ge , thus, the accuracy of the designs has a great degree of ambiguity. Third, any valence QCLs have no choice but to deal with HH subbands; their large effective mass hinders carrier injection efficiency and leads to small IST oscillator strength between laser states. Fourth, for any significant band offset needed to implement QCLs, lattice-mismatch-induced strain in SiGe QWs and Si barriers even in strain balanced structures is significant, which presents a challenge in structural growth and device processing. While EL was demonstrated from a valence-band SiGe/Si quantum cascade emitter nearly a decade ago, lasing remains elusive.

Recently, several ideas of developing Si-based conduction-band QCLs have emerged to circumvent the hurdles in the SiGe/Si valence-band approach. The proposals offer ways to increase the conduction band offset and to reduce the effective mass along the growth direction. One scheme proposes to orient the structural growth along the (111) direction, and another relies on ISTs in the *L*-valleys of the conduction band in Ge-rich Ge/SiGe material system. The former has accomplished more in increasing the conduction-band offset, and the latter in reducing the effective mass. A third approach that expands the material system beyond SiGe to GeSiSn has been discussed in detail. A $Ge/Ge_{0.76}Si_{0.19}Sn_{0.05}$ QCL that operates at *L*-valleys of the conduction band was designed. According to our estimation of the band lineup, this particular alloy composition gives a "clean" conduction band offset of 150meV at *L*-valleys with all other energy valleys conveniently out of the way. All QCL layers are lattice matched to a Ge buffer layer on a Si substrate and the entire structure is therefore strain free. The electron effective mass along the growth direction is much lighter than that of heavy holes bringing a significant improvement in tunneling rates and oscillator strengths. The lasing wavelength of this device is 49 μm. With different GeSiSn alloy compositions that are lattice matched to Ge, QCLs can be tuned to lase at other desired wavelengths. Lifetimes determined from the deformation potential scattering of nonpolar optical and acoustic phonons are at least an order of magnitude longer than those in III-V QCLs with polar optical phonons, leading to a reduction in threshold current density and the possibility of room temperature operation. While there are considerable challenges in material growth of this QCL design, advances in fine control of structural parameters including layer thicknesses and alloy compositions are moving towards implementation of conduction-band QCLs in the GeSiSn system.

When are we going to realize Si-based lasers that can be integrated with Si electronics? Clearly, breakthroughs in material science and device innovation are necessary before that happens, but with the variety of approaches that are being pursued--driven by the potential pay-off in commercialization--the prospect is promising.

7. References

[1] R. A. Soref and J. P. Lorenzo, "All-silicon active and passive guided-wave components for λ=1.3 and 1.6μm," *IEEE Journal of Quantum Electronics* QE-22, (1986) 873-879.

[2] R. A. Soref, "Silicon-based optoelectronics," *Proceedings of the IEEE* 81, (1993) 1687-1706.

[3] L.Pavesi, L. Dal Negro, C. Mazzoleni, G. Franzo, and F. Priolo, "Optical gain in silicon nanocrystals," *Nature*, 408, (2000) 440-444.

[4] J. Ruan, P. M. Fauchet, L. Dal Negro, C. Mazzoleni, and L.Pavesi, "Stimulated emission in nanocrystalline silicon superlattice," *Applied Physics Letters* 83, (2003) 5479-5481.

[5] M. J. Chen, J. L. Yen, J. Y. Li, J. F. Chang, S. C. Tsai, and C. S. Tsai, "Stimulated emission in a nanostructured silicon pn-junction diode using current injection," *Applied Physics Letters* 84, (2004) 2163-2165.

[6] E. Desurvire, Erbium-doped Fiber Amplifiers: Principles and Applications, Wiley, New York, (1994)

[7] G. Franzò and S. Coffa, "Mechanism and performance of forward and reverse bias electroluminescence at 1.54 μm from Er-doped Si diodes," *Journal of Applied Physics* 81, (1997) 2784-2793.

[8] F. Priolo, G. Franzò, S. Coffa and A. Carnera, "Excitation and nonradiative deexcitation processes of Er^{3+} in crystalline Si," *Physical Review B* 57, (1998) 4443-4455.

[9] A. J. Kenyon, P. F. Trwoga, M. Federighi, and C W Pitt "Optical properties of PECVD erbium-doped silicon-rich silica: evidence for energy transfer between silicon microclusters and erbium ions," *Journal of Physics: Condensed Matter* 6, (1994) L319-324.

[10] M. E. Castagna, S. Coffa, L. carestia, A. Messian, and C. Buongiorno, "Quantum dot materials and devices for light emission in silicon," *Proceedings of 32nd European Solid-State Device Research Conference (ESSDERC 2002)*, Fireze, Italy, (2002) D21.3

[11] R. Calps, D. Dimitropoulos, V. Raghunathan, Y. Han, and B. Jalali, "Observation of stimulated Raman scattering in silicon waveguides," *Optics Express* 11, (2003) 1731-1739

[12] A. Liu, H. Rong, M. Paniccia, O. Cohen, D. Hak, "Net optical gain in a low loss silicon-on-insulator waveguide by stimulated Raman scattering," *Optics Express* 12, (2004) 4261-4268.

[13] O. Boyraz and B. Jalali, "Demonstration of a silicon Raman laser," *Optics Express* 12, (2004) 5269-5273.

[14] B. Jalali, "Silicon photonics," *Journal of Lightwave Technology* 24, (2006) 4600-4615.

[15] H. Park, A. W. Fang, S. Kodama, and J. E. Bowers, "Hybrid silicon evanescent laser fabricated with a silicon waveguide and III-V offset quantum wells," *Optics Express* 13, (2005) 9460-9464.

[16] J. H. Park and A. J. Steckl, "Demonstration of a visible laser on silicon using Eu-doped GaN thin films," *Journal of Applied Physics* 98, (2005) 056108

[17] Z. Mi, P. Bhattacharya, J. Yang, and K. P. Pipe, "Room-temperature self-organized $In_{0.5}Ga_{0.5}As$ quantum dot laser on silicon," *Electronics Letters* 41, (2005) 742-744.

[18] L. Esaki and R. Tsu, "Superlattice and negative differential conductivity in semiconductors", *IBM Journal of Research and Development* 14, (1970), 61-65.

[19] R. F. Kazarinov and R. Suris, "Electric and electromagnetic properties of semiconductors with a superlattice," *Soviet Physics – Semiconductors* 5, (1971) 797.

[20] J. Faist, F. Capasso, D. L. Sivco, A. L. Hutchinson, C. Sirtori, and A. Y. Cho, "Quantum cascade laser," *Science* 264, (1994) 553-556
[21] G. Sun and J. B. Khurgin, "Feasibility of optically pumped four-level infrared lasers," in *Intersubband Transitions in Quantum Wells*, (Edited by E. Rosencher, B. Vinter, and B. Levine, Plenum Press, New York, 1992), pp.219-226
[22] G. Sun and Khurgin, "Optically pumped four-level infrared laser based on intersubband transitions in multiple quantum wells: feasibility study," *IEEE Jounral of Quantum Electronics* QE-29, (1993) 1104-1111.
[23] D. G. Revin, J. W. Cockburn, M. J. Steer, R. J. Airey, M. Hopkinson, A. B. Krysa, L. R. Wilson, and S. Menzel, "InGaAs/AlAsSb/InP quantum cascade lasers operating at wavelengths close to 3 μm," *Applied Physics Letters* 90, (2007) 021108.
[24] M. P. Semtsiv, M. Wienold, S. Dressler, and W. T. Masselink, "Short-wavelength ($\lambda \approx 3.05$ μm) InP-based strain-compensated quantum-cascade laser," *Applied Physics Letters* 90, (2007) 051111.
[25] R. Colombelli, F. Capasso, C. Gmachl, A. L. Hutchinson, D. L. Sivco, A. Tredicucci, M. C. Wanke, A. M. Sergent, and A. Y. Cho, "Far-infrared surface-plasmon quantum-cascade lasers at 21.5 µm and 24 µm wavelengths," *Applied Physics Letters* 78, (2001) 2620-2622.
[26] Y. Bai, S. Slivken, S. R. Darvish, and M. Razeghi, "Room temperature continuous wave operation of quantum cascade lasers with 12.5% wall plug efficiency," *Applied Physics Letters* 93, (2008) 021103.
[27] A. Lyakh, C. Pflügl, L. Diehl, Q. J. Wang, F. Capasso, X. J. Wang, J. Y. Fan, T. Tanbun-Ek, R. Maulini, A. Tsekoun, R. Go, and C. Kumar N. Patel, "1.6 W high wall plug efficiency, continuous-wave room temperature quantum cascade laser emitting at 4.6 μm," *Applied Physics Letters* 92, (2008) 111110.
[28] B. S. Williams, "Tereahertz quantum-cascade lasers," *Nature Phtonics* 1, (2007) 517-525.
[29] S. R. White and L. J. Sham, "Electronic properties of flat-band semiconductor heterostructures," *Physical Review Letters* 47, (1981) 879 - 882.
[30] E. O. Kane, "Band structure of indium antimonide," *Journal Physics and Chemistry of Solids* 1, (1957) 249-261.
[31] G. Bastard, *Wave Mechanics Applied to Semiconductor Heterostructures* (Les Editions de Physique, les Ulis), 1998, Chapter 3.
[32] J. M. Luttinger and W. Kohn, "Motion of Electrons and Holes in Perturbed Periodic Fields," *Physical Rev*iew 97, (1955) 869-883.
[33] J. M. Luttinger, "Quantum Theory of Cyclotron Resonance in Semiconductors: General Theory," *Physical Rev*iew 102, (1956) 1030-1041.
[34] G. L. Bir and G. E. Pikus, *Symmetry and Strain-Induced Effects in Semiconductors* (Wiley, New York, 1974)
[35] G. Sun, Y. Lu, J. B. Khurgin, "Valence intersubband lasers with inverted light-hole effective mass," *Applied Physics Letters* 72, (1998) 1481-1483.
[36] L. C. L. Y. Voon and L. R. Ram-Mohan, "Tight-binding representation of the optical matrix elements: theory and applications," *Physical Rev*iew *B* 47, (1993) 15500-15508.
[37] L. C. West and S. J. Eglash, "First observation of an extremely large-dipole infrared transition within the conduction band of a GaAs quantum well," *Applied Physics Letters* 46, (1985) 1156-1158.

[38] G. Sun, L. Friedman, and R. A. Soref, "Intersubband lasing lifetimes of SiGe/Si and GaAs/AlGaAs multiple quantum well structures," *Applied Physics Letters* 66, (1995) 3425-3427.
[39] B. K. Ridley, "The electron-phonon interaction in quasi-two-dimensional semiconductor quantum-well structures," *Journal of Physics C: Solid state Physics* 15, (1982) 5899-5917.
[40] P. J. Price, "Two-dimensional electron transport in semiconductor layers. I. Phonon scattering," *Annals of Physics* 133, (1981) 217-239.
[41] P. Harrison, *Quantum Wells, Wires, and Dots*, (John Wiley & Sons, 2000) p.272
[42] H. Callebaut, S. Kumar, B.S. Williams, Q. Hu, and J.L. Reno, "Importance of electron impurity scattering for electron transport in terahertz quantum-cascade lasers," *Applied Physics Letters* 84, (2004) 645-647.
[43] R. Ferreira and G. Bastard, "Evaluation of some scattering times for electrons in unbiased and biased single-and multiple-quantum-well structures," *Physical Review B* 40, (1989) 1074-1086.
[44] M. Hartig, J. D. Ganière, P. E. Selbmann, B. Deveaud, and L. Rota, "Density dependence of carrier-carrier induced intersubband scattering in GaAs/AlGaAs quantum wells," *Physical Review B* 60, (1999) 1500-1503.
[45] J. D. Cressler, SiGe and Si Strained-Layer Epitaxy for Silicon Heterostructure Devices, CRC Press, (2008)
[46] L. Friedman, R. A. Soref, G. Sun, and Y. Lu, "Theory of the strain-symmetrized silicon-based Ge-Si superlattice laser," *IEEE Journal of Selected Topics in Quantum Electronics* 4, (1998) 1029-1034.
[47] L. Friedman, R. A. Soref, G. Sun, and Y. Lu, "Asymmetric strain-symmetrized Ge-Si interminiband laser," *IEEE Photonics Technology Letters* 10, (1998) 1715-1717.
[48] D. J. Paul, G. Matman, L. Lever, Z. Ikonić, R. W. Kelsall, D. Chrastina, G. Isella, H. von Känel, E. Müller, and A. Neels, "Si/SiGe bound-to-continuum terahertz quantum cascade emitters," *ECS Transactions*. 16, (2008) 865-874.
[49] J. Faist, F. Capasso, C. Sirtori, D. L. Sivco, A. L. Hutchinson, M. S. Hybertson, and A. Y. Cho, "Quantum cascade lasers without intersubband population inversion," *Physical Review Letters* 76, (1996) 411-414.
[50] L. Friedman, G. Sun, and R. A. Soref, "SiGe/Si THz laser based on transitions between inverted mass light-hole and heavy-hole subbands," *Applied Physics Letters* 78, (2001) 401-403.
[51] R. A. Soref and G. Sun, "Terahertz gain in SiGe/Si quantum staircase utilizing the heavy-hole inverted effective mass," *Applied Physics Letters* 79, (2001) 3639-3641.
[52] A. Fasolino, E. Molinari, and J. C. Maan, "Resonant quasiconfined optical phonons in semiconductor superlattices," *Physical Review B* 39, (1989) 3923-3926.
[53] G. Dehlinger, L. Diehl, U. Gennser, H. Sigg, J. Faist, K. Ensslin, D. Grützmacher, and E. Müller, "Intersubband electroluminescence from silicon-based quantum cascade structures," *Science* 290, (2000) 2277.
[54] I. Bormann, K. Brunner, S. Hackenbuchner, G. Zindler, G. Abstreiter, S. Schmult, and W. Wegscheider, "Midinfrared intersubband electroluminescence of Si/SiGe quantum cascade structures," *Applied Physics Letters* 80, (2002) 2260-2262.
[55] S. A. Lynch, R. Bates, D. J. Paul, D. J. Norris, A. G. Cullis, Z. Ikonić, R. W. Kelsall, P. Harrison, D. D. Arnone, and C. R. Pidgeon, "Intersubband electroluminescence

from Si/SiGe cascade emitters at terahertz frequencies," *Applied Physics Letters* 81, (2002) 1543-1545.
[56] R. Bates, S. A. Lynch, D. J. Paul, Z. Ikonic, R. W. Kelsall, P. Harrison, S. L. Liew, D. J. Norris, A. G. Cullis, W. R. Tribe, and D. D. Arnone, "Interwell intersubband electroluminescence from Si/SiGe quantum cascade emitters," *Applied Physics Letters* 83, (2003) 4092-4094.
[57] J. Faist, M. Beck, T. Aellen, and E. Gini, "Quantum-cascade lasers based on a bound-to-continuum transition," *Applied Physics Letters* 78, (2001) 147-149.
[58] G. Scalar, et al "Far infrared ($\lambda \cong 87\mu m$) bound-to-continuum quantum-cascade lasers operating up to 90K," *Applied Physics Letters* 82, (2003) 3165-3167.
[59] L. Diehl, S. Mentese, E. Müller,, D. Grützmacher, H. Sigg, U. Gennser, I. Sagnes, Y. Campiedelli, O. Kermarrec, D. Bensahel, and J. Faist, "Electroluminescence from strain-compensated $Si_{0.2}Ge_{0.8}$/Si quantum cascade structures based on a bound-to-continuum transitions," *Applied Physics Letters* 81, (2002) 4700-4702.
[60] D. J. Paul, G. Matmon, L. Lever, Z. Ikonić, R. W. Kelsall, D. Chrastina, G. Isella, and H. von Känel, "Si/SiGe bound-to-continuum quantum cascade terahertz emitters," *Proceedings of SPIE* 6996, (2008) 69961C.
[61] L. Lever, A. Valavanis, Z. Ikonić, and R. W. Kelsall, "Simulated [111] Si-SiGe THz quantum cascade laser," *Applied Physics Letters* 92, (2008) 021124.
[62] K. Driscoll and R. Paiella, "Silicon-based injection lasers using electrical intersubband transitions in the L-valleys," *Applied Physics Letters* 89, (2006) 191110.
[63] R. A. Soref and C. H. Perry, "Predicted band gap of the new semiconductor SiGeSn," *Journal of Applied Physics* 69, (1991) 539-541.
[64] D. W. Jenkins and J. D. Dow, "Electronic properties of metastable Ge_xSn_{1-x} alloys," *Physical Review B* 36, (1987) 7994-8000.
[65] G. He and H. A. Atwater, "Interband Transitions in Sn_xGe_{1-x} Alloys," *Physical Review Letters* 79, (1997) 1937-1940.
[66] J. Menéndez and J. Kouvetakis, "Type-I Ge/$Ge_{1-x-y}Si_xSn_y$ strained-layer heterostructures with a direct Ge bandgap," *Applied Physics Letters* 85, (2004) 1175-1177.
[67] M. Bauer, C. Ritter, P. A. Crozier, J. Ren, J. Menéndez, G. Wolf, and J. Kouvetakis, "Synthesis of ternary SiGeSn semiconductors on Si(100) via Sn_xGe_{1-x} buffer layers," *Applied Physics Letters* 83, (2003) 2163-2165.
[68] R. Roucka, J. Tolle, C. Cook, A. V. G. Chizmeshya, J. Kouvetakis, V. D'Costa, J. Menendez, Zhihao D. Chen, and S. Zollner, "Versatile buffer layer architectures based on $Ge_{1-x}Sn_x$ alloys," *Applied Physics Letters* 86, (2005) 191912.
[69] P. Aella, C. Cook, J. Tolle, S. Zollner, A. V. G. Chizmeshya, and J. Kouvetakis, "Optical and structural properties of $Si_xSn_yGe_{1-x-y}$ alloys," *Applied Physics Letters* 84, (2004) 888-890.
[70] M. A. Wistey, Y.-Y. Fang, J. Tolle, A. V. G. Chizmeshya, and J. Kouvetakis, "Chemical routes to Ge/Si(100) structures for low temperature Si-based semiconductor applications," *Applied Physics Letters* 90, (2007) 082108.
[71] G. Sun, H. H. Cheng, J. Menéndez, J. B. Khurgin, and R. A. Soref, "Strain-free Ge/GeSiSn quantum cascade lasers based on *L*-valley intersubband transitions," *Applied Physics Letters* 90, (2007) 251105.
[72] M. Jaros, "Simple analytic model for heterojunction band offsets," *Physical Review B* 37, (1988) 7112-7114.

[73] C. G. Van de Walle, "Band lineups and deformation potentials in the model-solid theory," *Physical Review B* 39, (1989) 1871-1883.

[74] V. R. D'Costa, C. S. Cook, A. G. Birdwell, C. L. Littler, M. Canonico, S. Zollner, J. Kouvetakis, and J. Menéndez, "Optical critical points of thin-film $Ge_{1-y}Sn_y$ alloys: A comparative $Ge_{1-y}Sn_y/Ge_{1-x}Si_x$ study," *Physical Review B* 73, (2006) 125207.

[75] V. R. D'Costa, C. S. Cook, J. Menéndez, J. Tolle, J. Kouvetakis, and S. Zollner, "Transferability of optical bowing parameters between binary and ternary group-IV alloys," *Solid State Communications* 138, (2006) 309-313.

[76] J. Weber and M. I. Alonso, "Near-band-gap photoluminescence of Si-Ge alloys," *Physical Review B* 40, (1989) 5683-5693.

[77] M. L. Cohen and T. K. Bergstresser, "Band Structures and Pseudopotential Form Factors for Fourteen Semiconductors of the Diamond and Zinc-blende Structures," *Physical Review* 141, (1966) 789-796.

[78] B. K. Ridley, *Electrons and Phonons in semiconductor Multiplayers* (Cambridge University Press, Cambridge, 1997), Chapter 1.

[79] B.S. Williams, S. Kumar, H. Callebaut, Q. Hu, and J.L. Reno," Terahertz quantum cascade laser at $\lambda_{\sim}$100μm using metal waveguide for mode confinement," *Applied Physics Letters* 83, (2003) 2124-2126.

[80] K. Unterrainer, R. Colombelli, C. Gmachl, F. Capasso, H.Y. Hwang, A.M. Sergent, D.L. Sivco, and A.Y. Cho, "Quantum cascade lasers with double metal-semiconductor waveguide resonators," *Applied Physics Letters* 80, (2002) 3060-3062.

[81] P. B. Johnson and R. W. Christy, "Optical Constants of the Noble Metals," *Physical Review B* 6, (1972) 4370-4379.

II Applications in Communications

14

Evolution of Optical Sampling

Gianluca Berrettini[1], Antonella Bogoni[2], Francesco Fresi[1], Gianluca Meloni[2] and Luca Potì[2]
[1]*Scuola Superiore Sant'Anna, Pisa,*
[2]*CNIT - Photonic Networks National Laboratory,*
Via G. Moruzzi 1, 56124 Pisa,
Italy

1. Introduction

The need of direct measure and monitoring of ultra-fast signal in the time domain is rapidly increasing, being of interest in a large number of applications such as ultra-fast communication, biophotonics, sensing, large systems synchronization, dynamic characterization and testing of new materials. In particular in the telecommunication field optical sampling can been exploited for high bit rate waveform and eye diagram measurements, time resolved state of polarization monitoring, and investigation of fiber transmission impairments.

Microwave digital sampling techniques are evolved into powerful tools for resolving signals up to 100 GHz (Agoston et al., 2003), but electronic bandwidth limitations still remain.

Nowadays digital sampling operations in the optical domain look like an effective alternative solution for increasing the sampling bandwidth and resolve signals up to 640 Gb/s and beyond. In the optical sampling techniques system, the optical signal is sampled in the optical domain by an optical sampling gate. Only then, the resulting samples are converted to an electrical signal and detected. In this case the need for high bandwidth electronics is circumvented and the bandwidth of the measurement instrument is only limited by the optical sampling gate.

Up to now different kinds of optical sampling techniques have been proposed in order to study the behavior of ultra-fast optical signals. Many solutions implement a synchronous sampling that may enable low jitter, high resolution and high accuracy. This technique needs a clock recovery system to synchronize the optical sampling pulses to the signal under test; however, when the data rate or repetition rate of the analyzed signal is very high, the development of the synchronization circuitry can be very critical and expensive. In particular, in case of repetition rates beyond 100 GHz, some all-optical clock recovery solutions have to be adopted (Yamamoto et al., 2001; Tong et al., 2000; Uhua et al., 2003), but which are still far from being technologically consolidated. Other optical sampling schemes carry out an asynchronous sampling exploiting sophisticated electronics for the generation of a sampling gate (Shake et al., 2003a). Optical asynchronous sampling has been successfully demonstrated for signal up to 160 Gbit/s in (Westlund et al., 2005 a,b), where the capability of the optical sampler to estimate the Q value and the performances of the

exploited nonlinearity are investigated, confirming the high potentialities of an asynchronous sampling technique.
Finally, quasi-synchronous optical sampling has been demonstrated for 640 Gb/s optical time division multiplexing (OTDM) signals without data-post processing (Fresi et al., 2008). Independently on the sampling technique, the optical sampling operation requires a nonlinear interaction between the signal to be resolved and an ultra-short pulse train that act as sampling signal. Such nonlinear process can be generated in different nonlinear elements as highly non linear fiber (HNLF) (Malacarne et al., 2007; Watanabe et al., 2004; Schmidt et al., 2002a), nonlinear crystals and waveguides (Nogiwa et al., 2000; Takara et al., 1996a; Kawanishi et al., 2001) or semiconductor devices (Maguire et al., 2005; Shirane et al., 2000). Fibers, crystals and waveguides may enable ultra-fast dynamics and high efficiency, while semiconductor devices could reduce the power consumption.

2. Digital sampling operation

In contrast to an analog oscilloscope, a sampling oscilloscope acquires the waveform as a series of samples, and stores these samples until it accumulates enough samples to describe a waveform. Conventional electronic digital sampling oscilloscopes use analod to digital convertersi (ADC) to sample the signal at discrete points in time and convert the signal's voltage at these points into digital values called sample points. The sample clock determines how often the ADC takes a sample. This rate is referred to as the sample rate and is expressed in samples per second (S/s). Although there are a number of different implementations of sampling technology, today's digital oscilloscopes utilize two basic sampling methods: real-time sampling and equivalent-time sampling. Equivalent-time sampling can be divided further, into two subcategories: random and sequential. Each method has distinct advantages, depending on the kind of measurements being made.
Real-time sampling is ideal for signals whose frequency range is less than half the oscilloscope's maximum sample rate. Here, the oscilloscope can acquire more than enough points in one "sweep" of the waveform to construct an accurate picture. Real-time sampling is the only way to capture fast, single-shot, transient signals with a digital oscilloscope. High-frequency transient events occur only once, and must be sampled in the same time frame that they occur. If the sample rate isn't fast enough, high-frequency components can "fold down" into a lower frequency, causing aliasing in the display. In addition, real-time sampling is further complicated by the high-speed memory required to store the waveform once it is digitized.
Equivalent-time sampling can be used to accurately acquire signals whose frequency exceeds half the oscilloscope's sample rate. When measuring high-frequency signals, the oscilloscope may not be able to collect enough samples in one sweep. Equivalent time digitizers (samplers) take advantage of the fact that most naturally occurring and man-made events are repetitive. Equivalent-time sampling constructs a picture of a repetitive signal by capturing a little bit of information from each repetition. The waveform slowly builds up like a string of lights, illuminating one-by-one. This allows the oscilloscope to accurately capture signals whose frequency components are much higher than the oscilloscope's sample rate. There are two types of equivalent-time sampling methods: random and sequential. Each one has its advantages. Random equivalent-time digitizers (samplers) utilize an internal clock that runs asynchronously with respect to the input signal and the

signal trigger. Samples are taken continuously, independent of the trigger position, and are displayed based on the time difference between the sample and the trigger. Although samples are taken sequentially in time, they are random with respect to the trigger - hence the name "random" equivalent-time sampling. Sequential equivalent-time sampling provides much greater time resolution and accuracy. It acquires one sample per trigger, independent of the time/div setting, or sweep speed. When a trigger is detected, a sample is taken after a very short, but well-defined, delay. When the next trigger occurs, a small time increment "t" is added to this delay and the digitizer takes another sample. This process is repeated many times, with "t" added to each previous acquisition, until the time window is filled. Sample points appear from left to right in sequence along the waveform when displayed on the oscilloscope screen. Since with sequential sampling the sample is taken after the trigger level is detected, the trigger point cannot be displayed without a variable delay line, which may, in turn, reduce the bandwidth of the instrument. Both require that the input signal be repetitive (Tektronix, 2001).

3. Optical sampling

The optical sampling is a novel method to perform time-resolved measurements of optical data signals at high bit rates with a bandwidth that cannot be reached by conventional photodetectors and oscilloscopes (Schmidt-Langhorst & Weber, 2005). Fig. 1 explains the principle of optical sampling. The upper part of the figure shows an optical data signal with RZ modulation format as example. The data signal is a concatenation of optical data bits, each within a bit slot of 6.25 ps corresponding to a bit rate of 160 Gbit/s. This optical data signal is passed through an optical gate. The gate is closed by default, i.e. it does not transmit the data signal for most of the time except for ultra short periods of time. The period of time for which the gate transmits the data signal is called "gating window". If the width of the gating window is shorter than the bit duration, as shown in Fig. 1, only a fraction of the optical data bit is sliced out. This fraction is called "optical sample". The optical sample is determined by the instantaneous amplitude of the data bit at the sampling time. Ideally, the sampling gate should exhibit a linear transfer function making the amplitude of the optical sample directly proportional to the instantaneous power of the data signal at the sampling time.

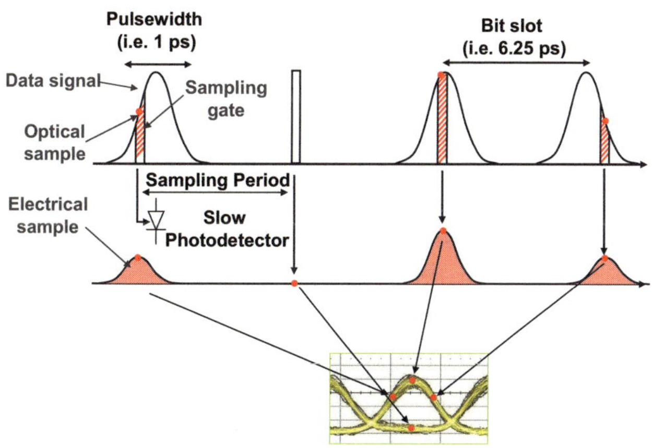

Fig. 1. Operating principle of optical digital sampling

In order to avoid high-speed electronic signal processing for the detection of the optical samples, it is useful to operate the sampling gate with a modest repetition frequency, then only equivalent-time sampling techniques are considered. At the output of the detector, the "electrical sample" appears broadened in time due to the low bandwidth of the detector, but still generates a photocurrent that is proportional to the instantaneous optical power of the optical data bit at the sampling time. After a transimpedance amplifier, the peak voltage V_P of each electrical sample is measured by an ADC. The eye diagram of the optical data signal is obtained from the measured V_P-values if the corresponding sampling times are known.
The optical sampling technique described in Fig. 1 allows to visualize the eye diagram of any data signal that has been encoded by an amplitude modulation format. In the case of a phase modulation format, this sampling technique will not distinguish between logical mark and space levels in the eye diagram. Up to now, only few work has been reported on the monitoring of phase modulated optical data signals. In a simple approach, a phase demodulator was incorporated in an optical sampling system (Schmidt-Langhorst et al., 2005). The demodulator converts the phase modulated data signal into an amplitude modulated data signal before the sampling operation. Another approach is the measurement of constellation diagrams of a phase modulated data signal (Dorrer et al., 2004). Such diagrams represent the amplitude and phase information of the data signal in the complex plane. Moreover FROG technique can be exploited to acquire signal amplitude and phase information.
For sake of shortness we consider here optical sampling techniques for solving only the signal amplitude.

3.1 Sampling gate generation

The generation of the sampling gate is the main subsystem of an optical sampling oscilloscope. It can be performed exploiting nonlinear interactions between the signal to be resolved and an ultra-short pulse train that acts as sampling signal.
The most important parameters of the sampling pulse source are the timing jitter and the pulse width. The timing jitter of the pulse source determines the timing jitter of the whole sampling system, whereas the pulse width limits the temporal resolution of the sampling system. As a rough estimate, a timing jitter of less than 300 fs and a pulse width of about 1 ps are necessary to measure a 160 Gbit/s optical eye diagram. Another important parameter is the repetition frequency of the pulse source. The required sampling rate is typically a few hundred MHz, since the O/E detection frequency in the optical sampling systems is limited to a few hundred MHz.
Different techniques can be considered to generate the sampling signal. Directly modulated, gain-switched laser diodes exhibit high jitter and could require additional compression stage (Ohta et al., 2000). Distributed FeedBack DFB lasers in continuous wave mode cascaded by electroabsorption modulators (EAM) to carve sampling pulses generate broad pulses due to the limited bandwidth of the available EAM's (Otani et al., 1999). Therefore, the pulses had to be compressed for the application as sampling pulses. Hybrid mode-locked semiconductor laser diode offers the potential of monolithic integration but the repetition rate has to be usually reduced using external $LiNbO_3$ amplitude modulator (Schmidt et al., 2002b). Mode-locked Erbium doped fiber lasers are the most widely used sampling pulse sources due to their very low jitter. Also in this case the pulse repetition frequency has to be externally reduced to a few hundred MHz by gating the pulse train with a $LiNbO_3$ modulator (Li et al., 2004; Li et al., 2001; Takara et al., 1996a,b).

The most common techniques used to produce optical nonlinear interaction for the generation of the sampling gate exploit optical fiber, crystals and waveguides or semiconductor devices. A summary of all nonlinear effects used for optical sampling is reported in Tab. 1.

Material	Nonlinear effect
Optical fiber	FWM XPM induced wavelength shifting NOLM Kerr gate
Semiconductor devices	FWM in SOA UNI GT-UNI EAM TPA
Crystals and waveguides	SFG SHF/DFG

Table 1. Summary of nonnlinear effects and nonlinear media exploited for optical sampling.

3.1.1 Optical fiber

The ultra-fast Kerr nonlinearity of the optical fiber provides short gating windows comparable to those of the crystal based gates. The exploited nonlinear processes include four wave mixing (FWM) (Miyazaki & Kubota, 2003), parametric amplification (Li et al., 2001) and cross phase modulation (XPM) induced wavelength shifting (Li et al., 2004). As the efficiency of these processes is rather small, high peak powers were needed for switching. The operational wavelength range of FWM-based all-optical sampling is usually limited by the phase-matching condition. Moreover, in order to cover the whole usable signal wavelength band, the wavelength of the sampling pulses should be set far away from that of the signal, which results in poorer temporal resolution, due to the relatively large walkoff between the signal and sampling pulses. Compared to FWM-based schemes, XPM-based optical sampling can place the sampling pulses just outside the usable signal wavelength region, therefore, the whole usable signal wavelength band can be covered, while the temporal resolution at the order of the sampling pulse width is maintained. However, XPM-based schemes exploiting narrow filtering are more sensitive to the chirp of the data signal, which can introduce some distortions.
Interferometric gates based on highly nonlinear fibers need less peak power of the sampling pulses to achieve gating windows with high on-off contrast. Examples include the nonlinear optical loop mirror (NOLM) (Schmidt et al., 2002a) and the so-called "Kerr gate" (Schmidt-Langhorst et al, 2002). The latter was recently combined with the effect of parametric amplification in (Watanabe et al., 2004).

3.1.2 Crystals and waveguides

For the purpose of obtaining higher sensitivity and higher temporal resolution, some sampling systems employ sum frequency generation (SFG) in the nonlinear crystals (Yamada et al., 2002), periodically poled Lithiumniobate (Yamada et al., 2004; Nogiwa et al., 2000; Ohta et al., 2001) or the organic crystal (Takara et al., 1996b). While these gates provide very short optical gating windows (<1 ps), their main drawback is the required high peak power, which is necessary to generate a sufficient amplitude of the frequency converted signal. In general, the use of periodically poled material reduces the power requirements as long interaction lengths without loss of phase matching can be achieved. Moreover, if the SFG process is performed with sampling pulses at about 780 nm, the generated light has a wavelength of about 520 nm and can be easily separated from the sampling pulse light as was shown in (Jungerman et al., 2002). In addition to the SFG process, the cascaded $\chi(2)$ processes of second harmonic generation and difference frequency generation (SHG/DFG) have been used as well (Kawanishi et al., 2001).

3.1.3 Nonlinear effects in semiconductor devices

The resonant nonlinearity of a semiconductor optical amplifier (SOA) requires less optical power for the gating pulses (Shirane et al., 2000). Devices based on conventional FWM in SOA's suffer from the fact that the conversion efficiency and the signal-to-background ratio drop rapidly for wavelength detunings between control and data wave larger than some nanometers. So, for the sake of high conversion efficiency or high FWM output power (i.e., high optical power of the FWM signal at the SOA output), the signal has to be kept within the gain wavelength region of the SOA. Hence, the data signal will contribute to gain saturation in the SOA. This is unwanted in sampling applications, where a strict linearity of the FWM output power versus the data input power is required for a quantitative assessment of the shape of the data pulses. Additionally, there is a strong contribution of amplified spontaneous emission (ASE) noise to the converted FWM signal, since the data is within the spectral range of the ASE. A higher switching efficiency is obtained by XPM in an SOA in combination with an interferometric configuration as gate. A promising gate of this kind is the so called "ultrafast nonlinear interferometer" (UNI), which was used in (Kang & Dreyer, 2003). However, this gate suffered from the amplified spontaneous emission (ASE) of the SOA. An EAM with sophisticated synchronization and driving electronics was additionally needed to suppress the ASE after the UNI gate. In a more advanced scheme, the superior gating performance of interferometric optical gates was combined with the so called gain-transparent operation of an SOA. In this operation of the SOA, the wavelength of the data signal is outside the gain spectrum of the SOA. Therefore it does not suffer from ASE degradation. Moreover, the gain-transparent SOA gate exhibits a linear transfer function for the data signal. This is in contrast to the conventional SOA gates, which show a strong saturation. Finally, the gain-transparent operation enables a wide wavelength acceptance range of the gate. The gain-transparent configuration was used in the "gain-transparent ultrafast-nonlinear interferometer" (GT-UNI) sampling gate in (Schmidt et al., 2002b). For the sake of completeness it should be mentioned also that even EAM have been used recently as sampling gates in optical sampling experiments. Since the width of the gating window, which can be achieved with an EAM, is rather large (about 5 ps), these gates could be applied only to bit rates up to 40 Gbit/s (Shake et al., 2003b). Recently though, eye diagrams even at a bit rate of 160 Gbit/s were measured using a double-pass EAM configuration (Kang & Dorrer, 2003).

3.1.4 Two photon absorption (TPA) in semiconductor devices:

The phenomenon of TPA is a nonlinear optical-to-electrical conversion process where two photons are absorbed in the generation of a single electron-hole pair (Folliot, 2002). It occurs when photons of energy E_{ph} are incident on the active area of a semiconductor device with a bandgap exceeding E_{ph} but less than $2E_{ph}$. The generated photocurrent is proportional to the square of the intensity, and this nonlinear response enables the use of TPA for optical sampling. As TPA is an instantaneous optical nonlinearity, it may be used for all-optical high-speed sampling in photonic systems. The main difficulty with using TPA for high-speed optical sampling is its inherent inefficiency, which means that such systems either require high optical intensities or very long detectors, making them unsuitable for practical telecommunications applications. One possible way to overcome this efficiency problem is to employ a semiconductor microcavity (Krug et al., 2004), which should significantly enhance the TPA response of the device, and enable the implementation of a practical sampling element for high-speed optical systems.

3.2 Optical sampling parameters

An optical sampling system is characterized by the following main parameters:

Temporal resolution

It is typically specified by the full-width at half maximum (FWHM) of the sampling system gating window The temporal resolution is equivalent to the bandwidth of the sampling system.

Timing jitter

The timing jitter is typically specified as root-mean-square (RMS) jitter.

Wavelength range

It is usually specified as 3 dB bandwidth and defines the range of wavelengths, which a data signal can have to be accepted by the sampling system.

Sensitivity

There is no common definition of this parameter with respect to optical sampling systems. Some authors use the product of the average sampling pulse power and the peak signal power. Another definition of the sensitivity uses the peak power of the signal pulses, which is required to achieve a certain displayed signal-to-noise ratio (Yamada et al., 2002) or in case of SFG the well know conversion efficiency parameter.

Dynamic range

It is the power range in which the signal under investigation must fall.

Polarization sensitivity

It describes the amount by which the amplitude of the generated samples varies, when the polarization of the signal under investigation changes.

Applicable bit rate

It is the range of bit rate, at which a particular sampling system was able to display an open eye diagram.

3.3 Synchronous, quasi-synchronous and asynchronous optical sampling

As reported in Fig. 2, in the synchronous sampling configuration, the sampling gate is opened synchronous to the optical data signal but with an increasing time delay in order to scan over the complete bit slot. Thus a technique is needed for the synchronization of the sampling gate to the optical data signal under investigation. This function is performed by an optical clock recovery (CR).

The quasi-synchronous scheme requires an additional subsystem to impose a frequency mismatch between the repetition rates of the signal under test and the sampling signal. In this way a complete scan of long signal time intervals can be directly achieved. In the asynchronous sampling configuration, the clock processing is omitted and the sampling process is performed at random times. The sampling times are derived from an arrival time measurement of the sampling pulses themselves. As compared to the synchronous sampling configuration, the random sampling configuration requires less components as there is no need for a clock processing circuitry. However, it requires the use of a special O/E-detector, which performs the arrival time measurement of the sampling pulses with high temporal accuracy, as an error in the arrival time measurement directly translates into timing jitter of the sampling system.

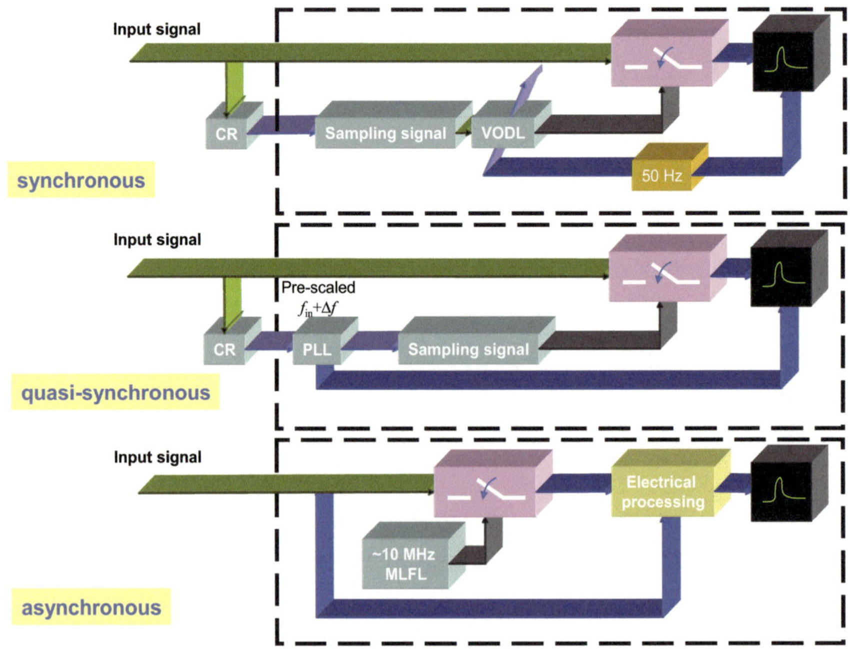

Fig. 2. Scheme of a synchronous (top), quasi-synchronous (center), and asynchronous (bottom) optical digital sampling.

In the following sections the implementation of synchronous, quasi-synchronous and asynchronous optical samplers exploiting optical fiber nonlinearities is presented.

4. Synchronous optical sampler based on Kerr nonlinearities

In the synchronous sampling technique, the sampling signal has to be synchronized with the signal to be sampled, thus requiring some kind of synchronization circuitry. The proposed scheme is based on the nonlinear effects in a highly non linear fibre. The accuracy of this sampling technique strongly depends on the non linear phenomenon efficiency by means of the non linear coefficient, and the fibre length. Nowadays, HNLFs with a flat chromatic dispersion curve and high birefringence are available, thus avoiding efficiency reduction caused by signals walk-off and polarization fluctuations.

The principle of the implemented optical sampler is the same as for commercial electrical sampling oscilloscopes. A schematic about the sampling principle is displayed in Fig. 3. In an ideal sampling, the signal to be measured S(t), is multiplied in the time domain by a delta signal $\delta(t_0)$. The result of the interaction between the two signals, is the power of $S(t)$ measured for $t = t_0$, that is $S(t_0)$. By continuously time shifting the delta signal over $S(t)$, it is possible to measure the signal power at different instants t_i. Such samples represent a replica for the signal under measure. In Fig. 3(a), the principle scheme about the ideal signal sampling is shown. In a real measurement, the sampling signal will have necessarily a non-zero time duration (Fig. 3(b)); thus consisting in a unavoidable sampling error.

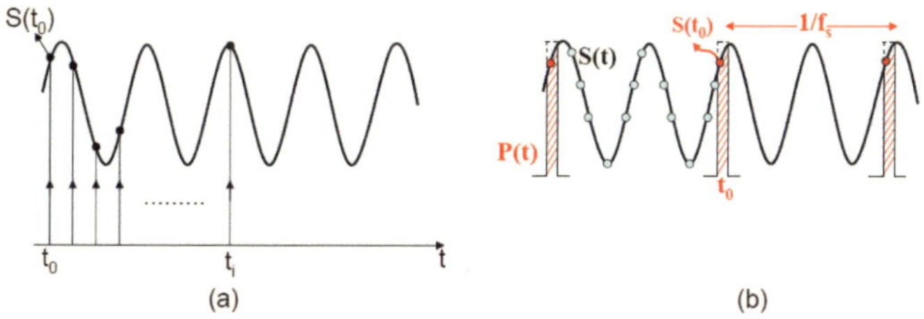

Fig. 3. Ideal operating sampling principle (a), and the real sampling of a periodic signal (b).

If the signal to be sampled is periodic with a period T, it is possible to sample it using another periodic signal whose period T_S is a multiple of T ($T_S = mT$). The accuracy of the sampling process is expressed by:

$$\varepsilon = S(t_0) - f_S \int_0^{1/f_S} S(t)P(t - t_0)dt \quad (4.1)$$

where $P(t)$ is the sampling signal, and f_S is the sampling frequency ($f_S = 1/T_S$). The previous technique, can be implemented exploiting whatever effect between $S(t)$ and $P(t)$, that gives a well known function of the signal power. A non linear effect that can be exploited is the Four Wave Mixing (FWM). The FWM non linear Kerr effect takes place when two or more powerful signals at different wavelength propagate through a non linear medium. When two optical signal $P(t)$ with a wavelength λ_i, and $S(t)$ with a wavelength λ_j, are transmitted through a non linear device, new components (called conjugate components) are generated with wavelengths related to the difference $\Delta\lambda = |\lambda_i - \lambda_j|$. In Fig. 4 it is displayed the FWM effect in the frequency domain. The conjugate component at the wavelength λ_i - $\Delta\lambda$ has a power proportional to $S(t)$, and then it can be filtered out and detected, measuring the original signal. The key parameters in this technique are the non linear coefficient g of the used non linear medium, and the power of the signals.

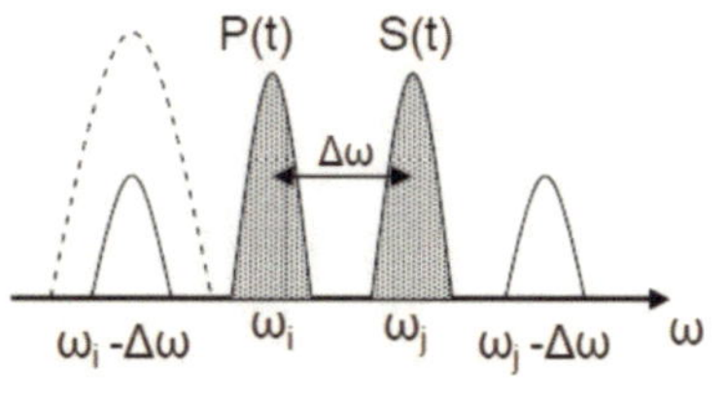

Fig. 4. FWM effect taking place when two powerful signals P(t) and S(t) are launched through a non linear device.

The FWM efficiency is given by:

$$\eta = (\gamma PL)^2 \quad (4.2)$$

where L is the fibre length, and P is the peak power of the sampling signal. The temporal resolution of the fibre based all optical sampler is determined by a combination of the sampling pulse width, and the temporal walk-off due to the chromatic dispersion between the sampling pulses and the signal during the sampling process. Afterwards, in order to obtain a good resolution, it is necessary to sample the signal with short sampling pulses, and to have a low value of dispersion slope. In a first implementation of the sampler, a fibre based active mode locking source has been used in order to generate the sampling pulses, with a repetition frequency of 10 GHz and a pulse width of about 4 ps. The laser source output has been directly used to sample a nonreturn-to-zero (NRZ) modulated signal. As further measure, a return to zero (RZ) signal was sampled, verifying the time resolution of the scheme. For the RZ signal sampling, a higher time resolution was obtained compressing the ML source pulses down to about 1 ps, and using that as sampling signal. For both the measures, the FWM component generated during the transmission was filtered out by means of an optical filter (dotted line in Fig. 4).

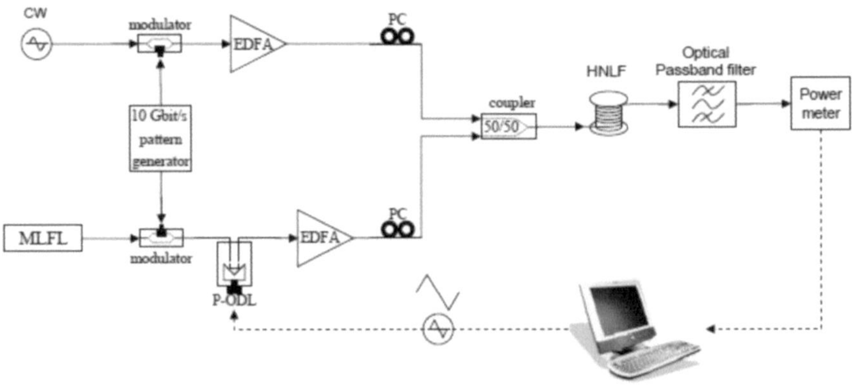

Fig. 5. Setup of the implemented synchronous optical sampler.

In Fig. 5, the setup of the implemented synchronous optical sampler is shown. In order to validate the functioning of the scheme, a NRZ signal, obtained modulating a 1550 nm Continuous Wave (CW), was sampled. For the experiment, a 400 m long HNLF with $\gamma = 10W^{-1}Km^{-1}$ and slope $S = 0.3\ ps/nm^2/Km$, was used as non linear medium. The NRZ signal was coupled together with the ML laser source output, whose wavelength was $\lambda_s = 1557$ nm. The power and the polarization state of both signal and pump (sampling signal) was optimized by means of Erbium Doped Optical Amplifiers and polarization controllers respectively. A Piezo-electric Optical Delay Line (P-ODL) driven by a 100 MHz electric triangular waveform, was used in order to shift the sampling pulses with respect to the signal. In such a way it is possible to visualize the behaviour of the mean power at the output of the optical filter with a refreshing time of the order of 100 ms. A 150 MHz photodiode (PD) and a 500 MHz oscilloscope triggered with the 100 MHz triangular waveform, was used to display the sampling result. An alternate modulation pattern was used, whereas the sampling signal frequency was fixed to 5 GHz in order to have the same periodicity as the signal under test. The mean power measured by changing the sampling pulses time position, directly gives the NRZ signal shape. In Figure 6 the results about the sampling of the NRZ signal are reported. The signal was opportunely pre-distorted in order to evaluate the sampler accuracy. In order to demonstrate the high resolution that can be obtained exploiting the Kerr non linearities (i.e. FWM), the ML laser source pulses was resolved. In this measure, the setup was slightly different from the scheme of Fig. 5. In particular, the ML pulsed signal was split, compressed and used also to generate the sampling signal. The compression was made by means of a higher order soliton compressor, obtaining pulses of about 350 fs. In Fig. 7 the scheme of the compressor and the obtained compressed optical pulse autocorrelation trace are displayed.

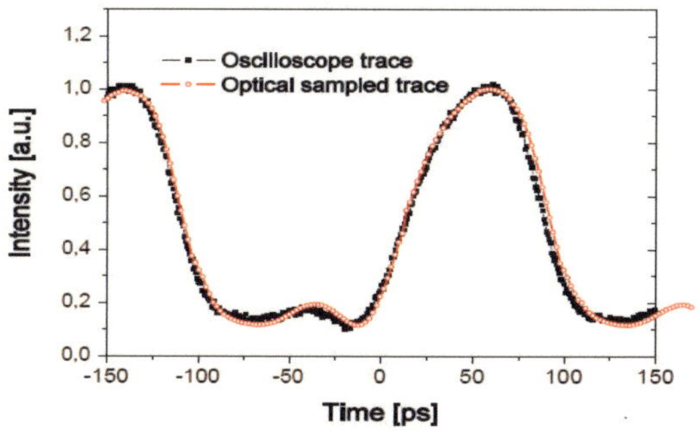

Fig. 6. comparison between the implemented optical sampler and the commercial oscilloscope measures.

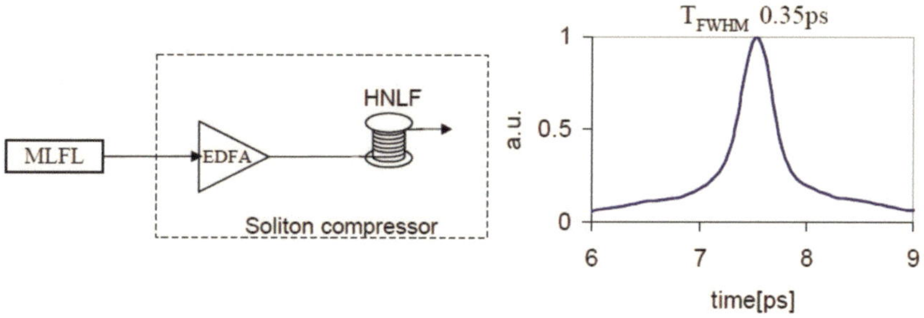

Fig. 7. Scheme of the higher order soliton compressor (left), and autocorrelation trace of the obtained compressed pulse (right).

In Fig. 8, the 10 GHz ML sampled pulses (left) and the comparison between the autocorrelation trace measured by means of a commercial autocorrelator (right) are reported.

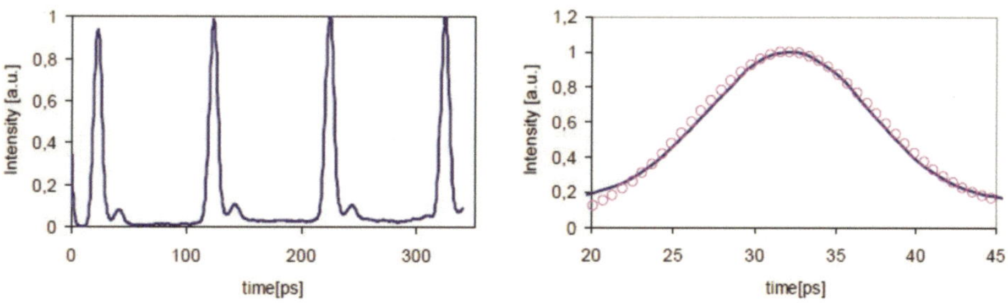

Fig. 8. 10 GHz sampled signal (left), and comparison between the autocorrelation traces (right).

Another experiment was made in order to evaluate the sampler resolution. The signal to be measured was obtained time-multiplying the ML laser source output with an all optical

multiplexer realized with the split-and-delay technique. In Fig. 9 are reported the results of the 80 GHz signal sampling.
Finally, in order to qualitatively estimate the maximum resolution of the implemented optical sampler, the signal to be sampled was distorted changing the power level in the compressor, obtaining an asymmetric and irregular optical pulse. This way, sub-picosecond dynamics was obtained and measured by the sampler, verifying its functioning for the measure of ultra short signals. In Fig. 10 the result about the sampling of the distorted pulse is shown.

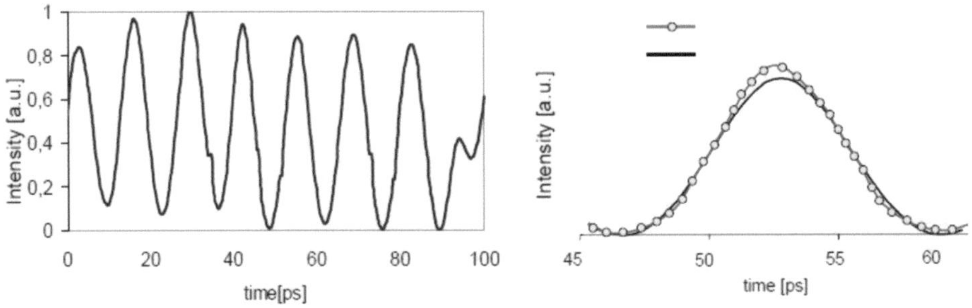

Fig. 9. 80 GHz sampled signal (left), and comparison between the autocorrelation traces (right).

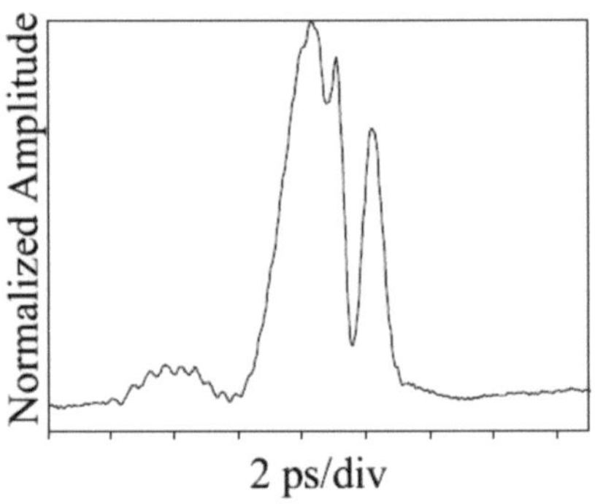

Fig. 10. Intensity trace of the distorted sampled pulse.

A scheme similar to the one described was implemented exploiting the non linear effect of cross phase modulation (XPM). In particular, the exploited effect was the XPM-induced polarization rotation. In Fig. 11 the setup of the optical sampler is reported. As in the previous scheme, an ultra short pump with high peak power interacts with the signal while propagating in the non linear fibre. In this optical sampler, a quasi instantaneous phase rotation (typical response time < 50 fs) is induced by the sampling pulse on the signal portion superimposed to the pump.
By means of a polarization controller and a polarizer at the output of the non linear element, it is possible to extract the portion of the signal time overlapped to the pump. This power portion can be then measured by using a narrow bandwidth photoreceiver. If the pump is short enough, that measured power can be approximated with the instantaneous signal power. As described before, by shifting the pump with respect to the signal its time shape

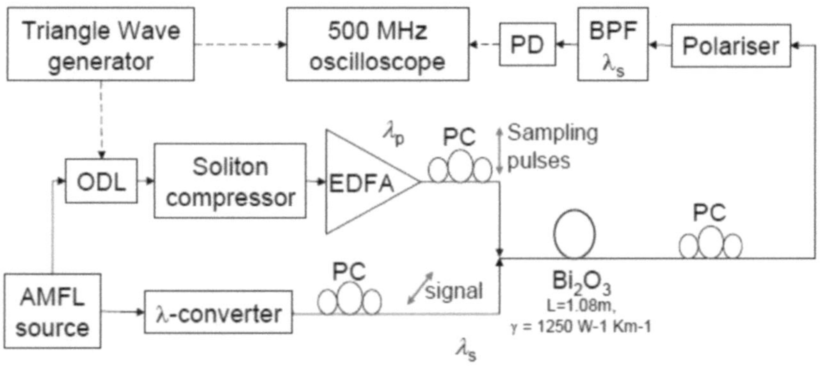

Fig. 11. Experimental setup of the synchronous sampler based on XPM-induced polarization rotation in 1 m of Bi_2O_3 fibre.

can be measured. One difference with respect to the previous scheme, is that in this case, the non linear medium is a Bismuth Oxide fibre (Bi_2O_3) with a very high non linear coefficient $\gamma = 1250\ W^{-1}Km^{-1}$, a total attenuation a = 6 dB and an accumulated chromatic dispersion D_t = - 0.33 ps/nm at a wavelength λ = 1550 nm. Such a high non liner coefficient permits to strongly decrease the fibre length down to L = 1.08 m. The use of short fibre spans reduces at the same time the polarization fluctuation and the scheme footprint, making the system more stable and compact. The sampling pulses was obtained by means of soliton compression on the ML sources and the signal under test was obtained by λ-converting the same ML laser. The wavelength conversion was made exploiting FWM in 250 m of HNLF. Moreover, the signal was coupled with the pump with a 45°-rotated polarization state in order to obtain maximum XPM-induced polarization rotation. Fig. 12 (left) shows the optically sampled pulse shape, whose pulse width of 6 ps was measured. In order to verify the accuracy of the implemented scheme, the obtained sampled trace autocorrelation was compared with the results supplied by a commercial autocorrelator obtaining a very good agreement between the traces, and verifying a picosecond resolution of the sampler (see Fig. 12 (right)).

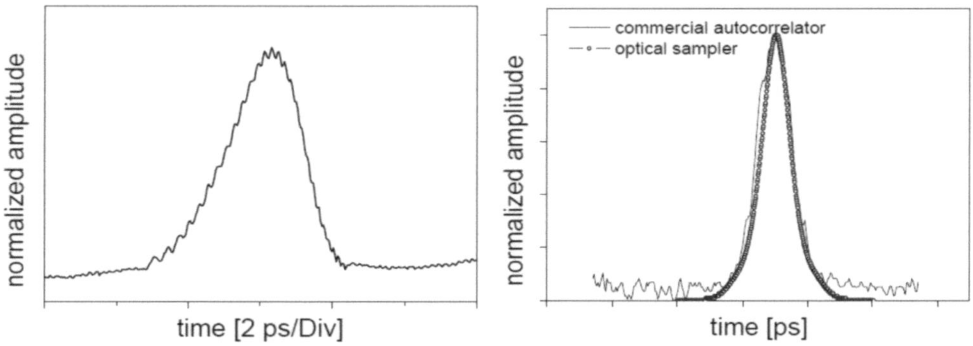

Fig. 12. 6 ps soliton pulse measured by the optical sampler (left) and comparison between the autocorrelation traces supplied by a commercial autocorrelator and the processed data measured by the implemented optical sampler (right).

5. Asynchronous optical sampling

An important issue of synchronous sampling technique is that it requires a synchronization between the sampling signal and the signal under investigation. Unfortunately, the synchronization of fast optical signals it is not easy and it requires complex all optical techniques and fast electronics. For this reason, it is useful an asynchronous sampling operation. An Asynchronous all optical Sampling Oscilloscope based on FWM in HNLF (ASOF) has been also implemented. The asynchronous solution, based on the knowledge of the mean frequency of the signal under test, avoids the use of a clock recovery circuit, with strong advantages in terms of complexity and cost. This kind of sampler has not only the advantage of not requiring synchronization circuits, but also the capability of describing both signal eye diagram and long bit sequences (several ns), thanks to the asynchronous operation. Moreover, a sub-ps resolution can be obtained due to the exploited ultra fast non linear Kerr effects in fiber. A short refresh time was obtained using a commercial oscilloscope able to directly acquire and process a large amount of data. The working principle of the implemented asynchronous sampler is shown in Fig. 13. As in the previous schemes, the asynchronous optical sampling oscilloscope exploits the non linear interaction between the signal under test and the sampling signal. The signal under test was periodic with a frequency $f_S = 1/T_S$, where T_S is its period.

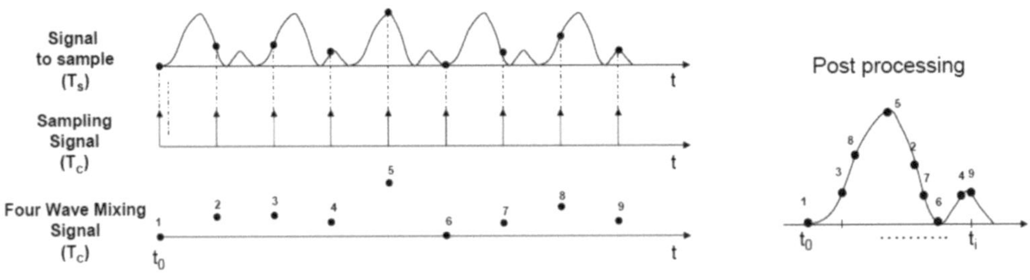

Fig. 13. ASOF working principle.

The sampling frequency f_C is not correlated to f_S, meaning that there is no need for any fixed relation between the two frequencies. Every pulse of the sampling signal generates a pulsed FWM component, whose energy is proportional to the instantaneous power of the signal under test in the corresponding interaction time. Since the frequencies of the involved signals are not correlated, two consecutive samples do not correspond to two consecutive instantaneous power values of the signal; on the opposite the temporal sequence of the samples appears as shuffled sampling of the analyzed signal. Besides, the sample sequence can be stored through an acquisition board, and then the sampled signal can be correctly reconstructed, if the difference of the period T_S - T_C is known. Therefore a post processing can be applied to the sample sequence for re-ordering the collected points and reconstructing the original signal shape. In more details, the position of the *i-th* sample, with respect to the period of the signal under test, is defined by the following equation:

$$t_i = t_0 + Mod\left[i\frac{(T_S - T_C)}{T_S} \right] \tag{5.1}$$

where t_0 is a reference initial time that can be set to zero. The samples can not be equally spaced depending on the relationship between T_S and T_C, and on the acquisition time. Consequently a mean resolution R of the ASOF can be defined as:

$$R = \frac{T}{N} \tag{5.2}$$

where T is the signal time interval to be resolved, and N is the number of acquired samples. One of the most critical issue of an asynchronous sampling technique is the phase noise that can be introduced by the sampling process. For this reason an analytical study of the phase noise generated in the implemented scheme was made. Due to the asynchronous behavior of the sampling operation, the introduced phase noise strongly depends on the phase noise of the signal under test and of the sampling pulse train. For the implemented scheme, it is possible to neglect the phase noise of the sampling pulse since its value was verified to be one order of magnitude lower than the phase noise of the considered signal. The analysis was made considering a sinusoidal waveform at 39.9 GHz as signal to be resolved. In this case, being $S(t)$ the signal under test, it can be modelled in the time domain as the sum of two components: the first one represents the noise-free sinusoidal signal $S_{nf}(t)$, and the second one $H(t)$ includes all amplitude and phase noise contributions:

$$S(t) = S_{nf}(t) + H(t) \tag{5.3}$$

Considering the signal electrical spectrum $P(f)$, the noise term can be modelled as the sum of the sinusoidal functions that represent all the components of the sideband noise of $P(f)$:

$$H(t) = \sum_{i=0}^{M} a_i \sin(2\pi(f_{low} + i\Delta f) + \delta i) \tag{5.4}$$

where the sideband noise of the electrical spectrum has been divided in M slices with a frequency step equal to Δ_f, while f_{low} *and* $f_{high} = (f_{low}+M\Delta f)$ are the upper and lower bounds of the noise bandwidth to be considered; a_i is the spectral power associated to the frequency $f_i = (f_0+f_{low}+i\Delta f)$ being f_0 the carrier frequency of the signal, and δi is a random variable uniformly distributed in the range $0 \div 2\pi$ accounting for the phase of the noise spectral components. The bandwidth $f_{high} - f_{low}$, depends on the time needed for the acquisition of all the optical samples allowing the desired resolution of the signal under test. In the implemented scheme, all the acquisition times are shorter than 10 ms and consequently it is possible to neglect all the noise components at frequencies lower than f_{low} = 10 KHz. The upper limit of the bandwidth to be considered was fixed to f_{high} = 1 MHz, since all higher spectral components present negligible amplitude. $H(t)$ includes both amplitude and phase noise, but it can be demonstrated (Von der Linde, 1986), that for $f_{low} \geq$ 10 KHz the phase noise is dominant. Consequently all the contribution of $H(t)$ can be assumed as phase noise, thus making a worst case approximation. The phase noise can be estimated in terms of timing jitter. The jitter can be extracted from the sideband noise of the signal electrical spectrum according to:

$$J\left[f_{low}, f_{high}\right] = \frac{1}{2\pi f_0}\sqrt{\frac{P_{sb}}{P_C}} \tag{5.5}$$

where P_C is the mean power of the sinusoidal waveform $S_{nf}(t)$, and P_{sb} in the integral of the sideband of the electrical spectrum of $S(t)$:

$$P_{sb} = \int_{f_0+f_{low}}^{f_0+f_{high}} \frac{P(f)}{1.2B} df \tag{5.6}$$

As explained in the previous chapter, the factor 1.2 normalizes the filter response of the electrical spectrum analyzer to a rectangular function, and B is the resolution bandwidth of the electrical spectrum analyzer. In the analytical investigation, $H(t)$ was obtained from the experimental sideband noise of the electrical spectrum reported in Fig. 14 (left), concerning a sinusoidal clock at 39.9 GHz.

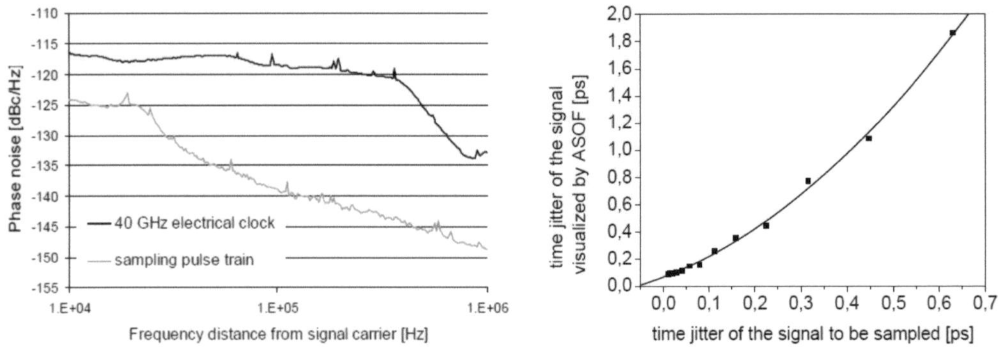

Fig. 14. Frequency distance from signal carrier (left) and analytical time jitter of the sampled signal as a function of the time jitter of the signal to be sampled (right).

The timing jitter of this signal, measured in the range 10 KHz ÷ 1 MHz, is equal to 70 fs. The sampling signal was considered to be a train of ideal delta functions with a repetition rate of 500 MHz, the acquisition time was set to 1 ms, and the time interval to be measured was fixed to 25 ps. The mean resolution can be estimated to be R = 50 fs. An analytical tool extracts the sequence of the instantaneous powers of the signal $S(t)$ corresponding to the interaction times between the signal and the delta function train. The tool then calculates the correct time position for each sample, with respect to the period of the signal to be measured, exploiting eq. 5.1. The timing jitter of the signal measured by the asynchronous sampling operation, could not be evaluated exploiting the previous equations since it was obtained by a data post processing, and its electrical spectrum was not available. Therefore the timing jitter was evaluated directly in the time domain, as it is for any commercial oscilloscope, considering the sampled curve as the sum of an ideal curve and a random variable with a gaussian time distribution. The timing jitter was then measured as the standard deviation of this time distribution, evaluated at half maximum of the curve displayed by the optical sampler. For the implemented scheme, a signal timing jitter of 156 fs was estimated, about two times higher then the timing jitter of the signal under test (76 fs). Consequently, it is possible to assert that the phase noise introduced by the asynchronous sampling operation does not compromise the measure of this 40 GHz signal with a sub-ps resolution. In Fig. 14 (right) the estimated timing jitter of the signal as visualized by the ASOF is plotted as a function of the timing jitter of the signal under test, obtaining a clear parabolic behavior. The increasing timing jitter of the signal to be sampled, was obtained considering in all cases the sideband noise of the electrical spectrum reported in Fig. 14(left), but with a decreasing mean power P_C of $S(t)$. For an input signal with a timing jitter lower than 100 fs in the range 10 KHz÷1 MHz, the implemented asynchronous sampler allows to describe the signal time evolution with a timing jitter lower than 200 fs. For higher values up to 600 fs, the timing jitter of the sampled curve as displayed by the ASOF does not exceed a value of 2 ps. Therefore this analysis has verified that the asynchronous sampling technique, despite avoiding the clock recovery operation, introduces a low amount of additional phase noise, allowing to measure ultra fast signal with high accuracy. The experimental setup for the ASOF implementation is shown in Fig. 15.

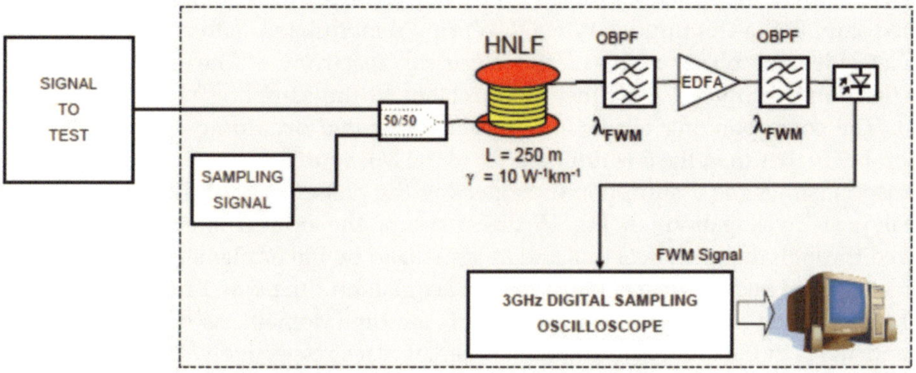

Fig. 15. ASOF experimental setup.

In order to generate the sampling signal, a 10 GHz active ML laser source was exploited, producing sub-ps pulses at a tunable wavelength λ_C. The pulse train repetition rate was then reduced down to 500 MHz by a Mach Zehnder (MZ) intensity modulator. The choice of the sampling rate depends on many factors: the upper value is limited by the bandwidths of the receiver and the electrical oscilloscope used for the acquisition of all the samples. On the other hand, low sampling rates, increase the acquisition time needed to obtain a fixed number of samples, and then require a higher memory capacity in the acquisition board. A sampling rate equal to 500 MHz represented the best trade off, using a 1 GHz receiver and an acquisition board with a memory capacity of 10 MPoints at a rate of 10 GSample/s. The sampling pulses were coupled with the signal under test in a 250 m long span of HNLF (with $\gamma = 10W^{-1}Km^{-1}$) generating the FWM components. An optical tunable filter was then used to extract the FWM components spectrally closer to λ_C thus obtaining a sample power linearly dependent on the instantaneous power of the signal under test. The optical samples were then detected through a 1 GHz bandwidth photo-receiver and measured on a 3 GHz bandwidth oscilloscope working at 10 Gsample/s. The oscilloscope used for the experiment was able to directly acquire and process the data exploiting MatLab scripts. The implemented sampler has the capability to analyze a signal for very long time intervals with high resolution and low refresh time. Considering the measure of a 10 ns long time interval with a desired resolution of 1 ps, the corresponding refresh frequency is about 1 Hz. The accuracy of the sampling technique, depends on the stability of the frequencies f_S and f_C whose mean values have to be measured and used as parameters in the data post processing. As described before, the phase noise was considered in the range

10 KHz÷1 MHz. The lower limit was due to the acquisition time (lower than 10 ms) and assuming that the mean frequency values, which are measured at each acquisition, are not affected by noise components lower than 10 KHz. On the other hand the phase noise components higher than 1 MHz present negligible amplitude and then can be neglected. The resolution of the ASOF is determined by the difference T_S - T_C, by the maximum number of acquired samples, and by the length of the analyzed time interval. According to eq. 5.1, the possibility to finely change the period of the sampling signal allows to significantly modify the sampling resolution. For this purpose, The ML source was provided with a variable optical delay line into the cavity, in order to vary the length of the fibre ring, thus finely changing the repetition rate of the generated sampling pulse train in the range 500 MHz÷1.5 MHz. The previous analytical investigation about the sampler phase noise was validated sampling a Continuous Wave (CW) signal modulated with an electrical 39.9 GHz sine. The sideband phase noise of the electrical spectrum of the clock is reported in Fig. 14 (left) and compared with the curve related to the sampling signal in the range of interest. The corresponding clock timing jitter, measured according to eq. 5.5, was 70 fs, considerably higher than the 5 fs timing jitter of the ML source used as sampling signal. This comparison justifies the assumption of neglecting the phase noise of the sampling signal in the analytical investigation. In Fig. 16 the curve of the optical sinusoidal waveform as measured through the analytical tool and as visualized by the implemented sampler.

In both analytical and experimental cases, the acquisition time was 1 ms and the resolution of 50 fs. The timing jitter, directly measured in the time domain, as described before, was 156 fs and 188 fs in the analytical and experimental cases respectively. In the analytical case, the sampling signal was an ideal delta function train and the amplitude noise, due to the FWM generation, was not considered. Consequently, the comparison in terms of timing jitter of the resolved curve, shows a good agreement between analysis and experiment, confirming the accuracy of the proposed analytical investigation. The potentialities of the ASOF, has been investigated considering a 40 Gbit/s optical NRZ signal. In this case, the optical CW was modulated through the Mach Zehnder modulator driven by a 40 Gbit/s data sequence supplied by a pattern generator.

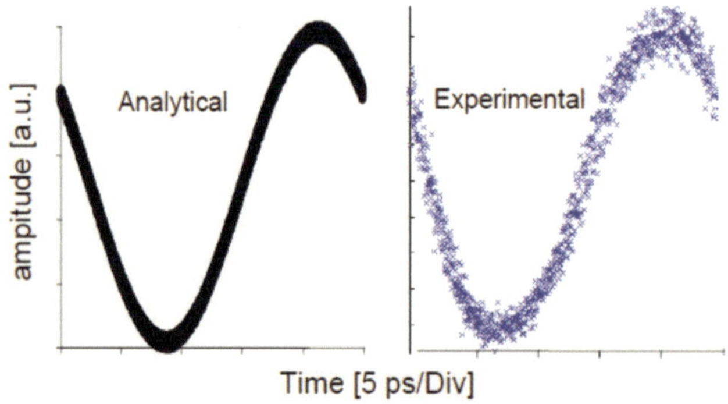

Fig. 16. 40 GHz optical sinusoidal waveform analytically and experimentally sampled.

The electrical clock, as in the previous case, was the 39.9 GHz sinusoidal waveform, whose sideband noise is reported in Fig. 14 (left). In this experiment, the acquisition time was 2 ms. Fig. 17 shows a 32 bit-long sequence (top) and its relative eye-diagram (center), of a signal as visualized by the proposed scheme (right) and by a commercial oscilloscope with a bandwidth of 53 GHz (left). The same comparison is reported also for the eye-diagram modulating the signal with a (2^{31} - 1)-long Pseudo Random Bit Sequence (PRBS) (bottom). In the case of the 32 bit sequence, the time interval to be resolved was 800 ps and consequently the mean resolution of the ASOF, was 0.8 ps. The comparison demonstrates the higher resolution obtained with the ASOF, that can better resolve the isolated ones and zeroes. The slight increase of the amplitude noise on the mark level in the ASOF traces, was mainly due to the FWM term generation and amplification. For both the 32 bit sequence and the PRBS, the higher resolution of the ASOF, allowed to obtain a good eye opening, especially close to bit transitions. The suitability of the ASOF, to measure ultra fast signals in the whole C-band, has been also demonstrated. Depending on the signal wavelength, the sampling pulse train wavelength can be moved in different spectral regions, in order to optimize the FWM generation.

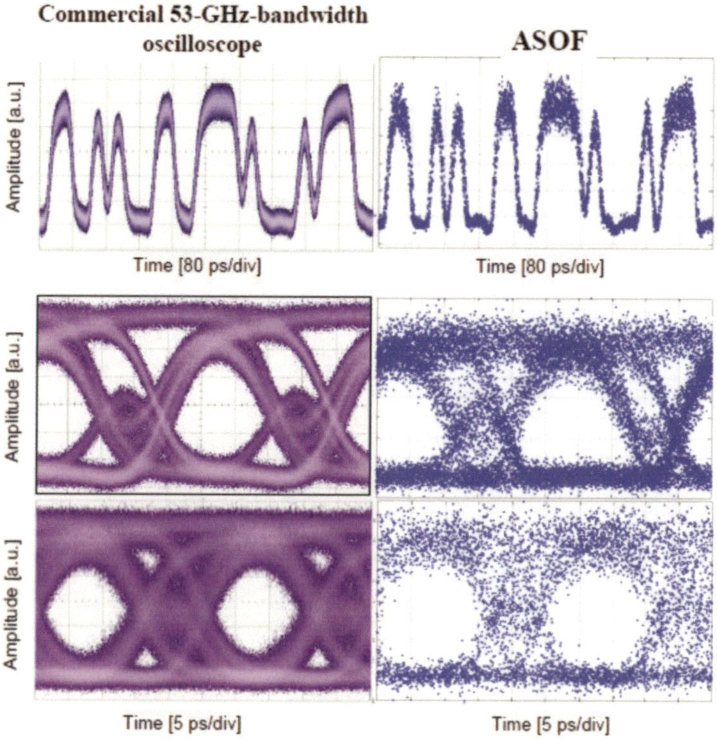

Fig. 17. 40 Gbit/s NRZ signal, visualized by a 53 GHz commercial oscilloscope (left) and by the ASOF (right); Bottom: eye-diagram of a (2^{31} - 1)-long PRBS. The wavelength of the signal under test was 1561 nm, while λ_C = 1552.5 nm.

The tunability of the pulse train source has been obtained inserting into the ML cavity, a tunable optical filter. The main specifications of the laser, such as the pulsewidth and the time jitter, were independent on the selected wavelength in the whole C-band. Fig. 18 (left) shows the optical spectrum at the output of the HNLF for three different wavelengths of the 40 Gbit/s signal under test: 1544 nm, 1556 nm, and 1561 nm. In all cases, the pulse train wavelength was fixed at 1553 nm, corresponding to 9 nm, 3 nm, and 8 nm signal detuning respectively. The same 32 bit-long sequence in the three different cases is reported in

Fig. 18 (right). It can be noticed that the quality of the resolved curves is comparable, making the proposed scheme suitable for applications in the whole C-band. Moreover, the detuning between the signal under test and the sampling pulse train was not a big issue for the FWM generation, for signal spacing up to 9 nm. This value allows to process also ultra fast signals with many frequency components.

As last measure, the high resolution of the ASOF, was demonstrated characterizing the optical pulse generated by a second 10 GHz ML source at 1561 nm. In Fig. 19, the generated optical pulse resolved by the ASOF, is reported. In this case, the timing jitter of the input signal in the range 10 KHz÷1 MHz, was 10 fs. The acquisition time was 800 ns, and the time span was 100 ps; consequently, the mean resolution was equal to about 0.25 ps. The measured 3 dB pulsewidth was 4.2 ps, very close to the value of 4 ps given by a commercial autocorrelator. The accuracy in the sampling of this short pulse, confirmed the high performance of the proposed ASOF measuring signals with bandwidth up to 1 THz.

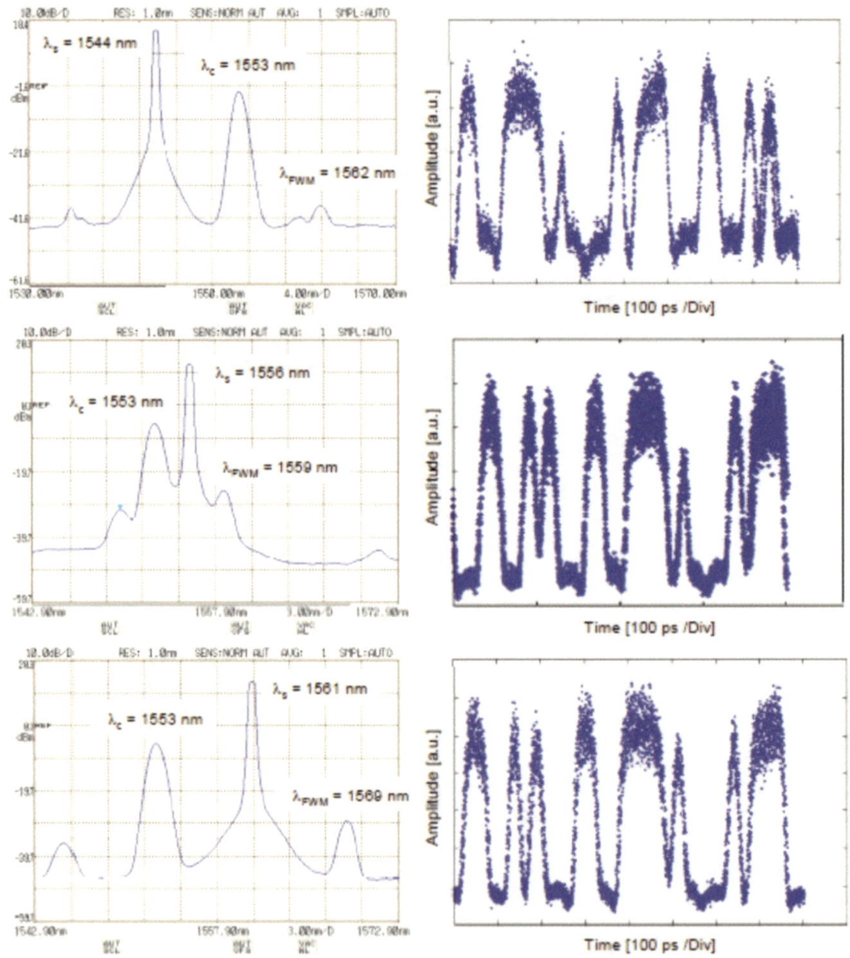

Fig. 18. Optical spectrum at the output of the HNLF (left) and corresponding 32 bit-long sequence resolved by the ASOF (right) in the case of signal wavelength equal to 1544 nm (top), 1556 nm (center), and 1561 nm (bottom).

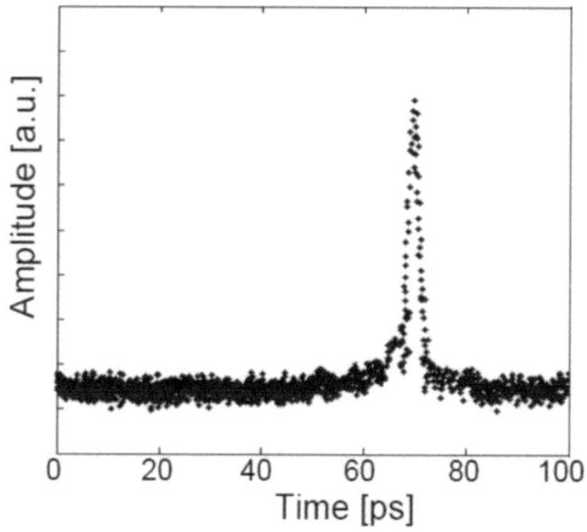

Fig. 19. trace of a 4 ps optical pulse generated by a 10 GHz ML laser source, as resolved by the ASOF.

6. Quasi-Synchronous optical sampler

As shown in the previous sections, synchronous sampling methods can typically reach sub-picosecond resolution, but they are able to analyze limited time intervals allowing to resolve only the eye-diagram of a data signal. Asynchronous optical sampling methods can describe long time intervals (> 1 ms), but with longer refresh time due to data post-processing. Moreover they introduce a higher jitter, intrinsic in the asynchronous operation. A possibility to obtain the advantages of both synchronous and asynchronous solutions is to exploit quasi-synchronous sampling (QS). In particular, the implemented sampler was based on the effect of Four Wave Mixing in high non linear fiber. It exploits the nonlinear interaction between the signal to be resolved and a sampling ultra-short pulse train whose frequencies are locked to a fixed difference. With this solution long bit sequences and eye-diagram for a 640 Gb/s data-frame can be displayed without the need of data post-processing and consequently with a very low refresh time (< 1.6 ms), exploiting low-bandwidth electronics. A double PLL scheme able to maintain a fixed frequency mismatch is described. Finally, sub-ps-resolution has been demonstrated by comparisons with a commercial autocorrelator.

The working principle and the setup of the QS optical sampling oscilloscope are shown in Fig. 20 (a,b). The sampling signal is an ultra-short pulse train at low repetition rate; when a frequency mismatch is imposed between the repetition rates of the signal under test and the sampling signal, a shift of the sampling signal on the signal under test is induced, as shown in Fig. 20(a). If the signal under test is periodic with $T=NT_s$ (being T_s the bit time) and N the sequence bit number, the period of the sampling signal can be chosen as $T_C = NT_s + \Delta t$, where Δt is the desired temporal resolution. This way samples are collected consecutively, and no post processing is required. Due to the low repetition rate, it is possible to reconstruct the sampled signal using a slow PhotoDiode and a low-bandwidth oscilloscope, acquiring just the samples sequence envelope. A piezoelectric delay line inside the optical sampling source cavity makes the source acting as an Optical VCO (OVCO), allowing to finely tune the sampling frequency f_C. Through a double PLL scheme, the difference between the Nth sub-multiple of the signal frequency f_S and the sampling frequency f_C ($f_{LO}=f_S/N-f_C$) is maintained constant. The frequency mismatch is imposed using a local oscillator and is determined by the desired resolution, according to the formula reported in Fig. 20(a). Therefore, it is possible to tune the desired resolution Δt just changing the value f_{LO}. Furthermore, exploiting a wavelength-tunable sampling signal, we obtain a QS sampler working within whole C-band. If N=128, every bit pattern with a period of 2^n bit and $n \leq 7$ can be resolved. For example, when f_S = 10 GHz, choosing a sampling signal repetition rate f_C = 78.124 MHz and a frequency mismatch $f_{LO} \sim$ 600 Hz, a 100 fs temporal resolution can be reached. This solution provides a refresh time of ~ 1.6 ms. At last, tuning the sampling signal repetition rate, it is possible to resolve also bit pattern with a standard length of 2^{n-1} bit.

The optical sampler exploits the FWM nonlinear interaction between the signal under test and the sampling signal (that acts as a pump). The two optical signals are coupled together and launched into a Highly Nonlinear Fiber (HNLF). The FWM-generated signal corresponds to the sequence of samples. The FWM effect is maximized by means of Automatic Polarization Controllers (APC) that polarization align the two signals. An optical BPF (Band-Pass Filter) centred at the FWM wavelength is used to isolate the samples. Each APC includes an electro-optical feedback, making the scheme polarization insensitive.

(a)

(b)

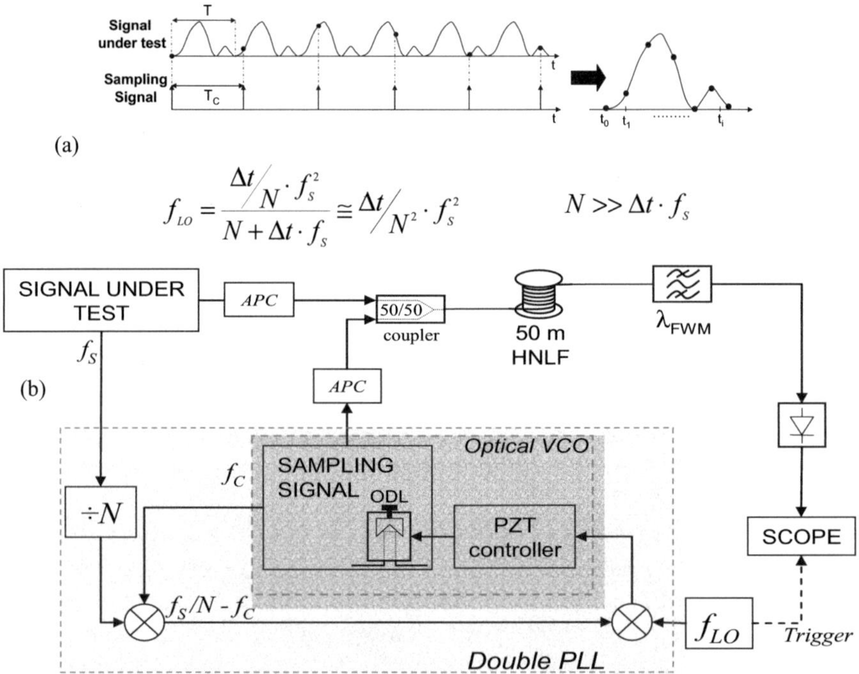

Fig. 20. Working principle of quasi-asynchronous sampling operation and Schematic of the quasi-asynchronous sampling oscilloscope.

Experimental measurements have been carried out using a 640 Gb/s RZ signal as signal under test. Such signal has been obtained time-multiplexing a RZ 10 Gbit/s one (4x10 Gb/s channels spaced 1.56 ps each other), produced by a 10 GHz Actively Mode-Locked Fiber Laser (AMLFL) and modulated by a Mach Zehnder modulator with a N = 8 bit pattern. A pulse compression stage is required to reduce the pulsewidth from the original 4 ps of the AMLFL down to 550 fs. The sampling signal is obtained with a second AMLFL acting as an OVCO, followed by a compression stage to reach a pulsewidth of 500 fs. A Δt of 50 fs is imposed setting a frequency mismatch f_{LO} = 80 KHz. The sampling pulses are then coupled with the signal under test into 50 m of HNLF. The average power of the sampling signal and the signal under test was 11 dBm and 9 dBm respectively. The wavelengths of the two signals have been set at 12 nm distance (λ_S=1560 nm, λ_C=1548 nm) due to their large bandwidth; the FWM signal generated at 1536 nm is filtered by a 1.2 nm BPF, then the sampled signal is photo-detected and viewed on a 600 MHz real-time oscilloscope. When the oscilloscope is directly triggered by f_{LO} the bit pattern is visualized (Fig. 21 (a)), while eye diagram evaluation can be performed using a xN frequency multiplier at the trigger (Fig. 21 (b)). To confirm the accuracy of the acquired trace, a comparison between the autocorrelation supplied by a commercial autocorrelator and the one measured from the optical sampler trace has been carried out (Fig. 21 (c)). The good agreement between the two traces confirms the effectiveness of the sampler in terms of shape and pulsewidth measurement; the measured pulsewidth is 550 fs.

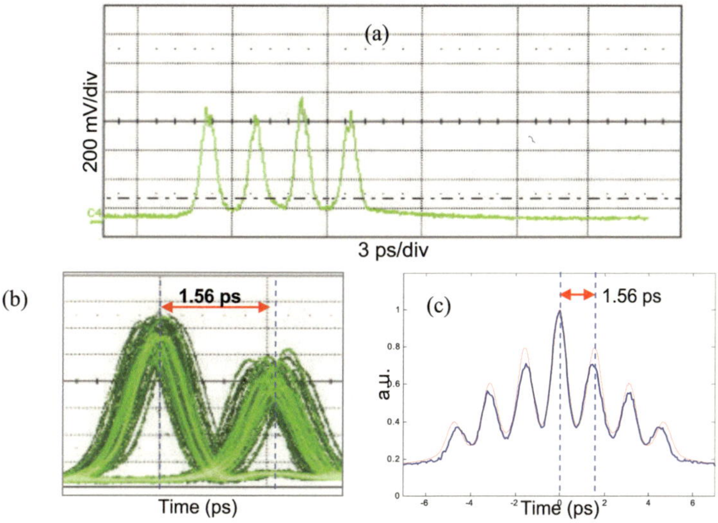

Fig. 21. 640 Gb/s 550 fs RZ pulses sequence evaluated using the optical sampler, visualized on the low-bandwidth oscilloscope after time rescaling (a). Eye-diagram evaluation (b) and comparison between autocorrelation measured through the optical sampler (red thin line) and the one supplied by a commercial autocorrelator (blue thick line) (c).

7. Conclusion

A general view of the optical digital sampling evolution has been reported. Most used optical sampling techniques are described in terms of solutions to obtain timing information and nonlinear interaction to generate sampling gate. In particular the implementation of synchronous, quasi-synchronous and asynchronous schemes exploiting fiber nonlinearities for the sampling gate generation has been described in details.

8. References

Agoston, A.; Pepper, S.; Norton, R.; Ebner, J. & Schoen, K. (2003). 100 GHz through-line sampler system with sampling rates in excess of 10 Gsamples/second, Proceedings of Microwave Symposium Digest, Vol. 3, pp. 1519-1521, ISBN: 0-7803-7695-1, June 2003, IEEE MTT-S International, Philadelphia, PA

Diez, S. Ludwig, R. Schmidt, C. Feiste, U. & Weber, H.G. (1999). 160-Gb/s optical sampling by gain-transparent four-wave mixing in a semiconductor optical amplifier. IEEE Photon. Techonol. Lett., vol. 11, no. 11, Nov. 1999, pp. 1402-1404, ISSN: 1041-1135

Dorrer, C.; Leuthold, J. & Doerr, C.R. (2004). Direct measurement of constellation diagrams of optical sources. Proceedings of OFC 2004, PDP33, ISBN: 1-55752-772-5, Los Angeles, CA, Feb. 2004

Folliot, H.; Lynch, M.; Bradley, A.L.; Krug, T.; Dunbar, L.A.; Hegarty, J.; Donegan, J. F. & Barry, L.P. (2002). Two-photon-induced photoconductivity enhancement in semiconductor microcavities: a theoretical investigation. J. Opt. Soc. Am. B, Vol. 19, No. 10, 2002, pp. 2396-2402.

Fresi, F. Chiuchiarelli, A. Malacarne, A. Potì, L. & Bogoni, A. (2008). 640 Gb/s RZ Eye-Diagram Evaluation by Optical Sampling Oscilloscope w/o Post-Processing and ms Refresh Time. Proceedings of CLEO 2008, CMII5, ISBN: 978-1-55752-859-9, May 2008, San Jose, CA, USA

Jungerman, R.L.; Lee, G.; Buccafusca, O.; Kaneko, Y.; Itagaki, N.; Shioda, R.; Harada, A.; Nihei, Y. & Sucha, G. (2002). 1-THz bandwidth C- and L-band optical sampling with a bit rate agile timebase. IEEE Photon. Technol. Lett., Vol. 14, No. 8, Aug. 2002, pp. 1148-1150, ISSN: 1041-1135.

Kang, I. & Dorrer, C. (2003). Optical sampling source-free simultaneous eye-diagram monitoring and clock recovery up to 160 Gb/s. Proceedings of ECOC 2003, pp. 292–293, Tu4.2.4, Rimini, Italy, Sept. 2003.

Kang, I. & Dreyer, K.F. (2003). Sensitive 320 Gbit/s eye diagram measurements via optical sampling with semiconductor optical amplifier-ultrafast nonlinear interferometer. Electronics Letters, Vol. 39, No. 14, July 2003, pp 1081 – 1083, ISSN: 0013-5194.

Kawanishi, S.; Yamamoto, T.; Nakazawa, M. & Fejer, M.M. 2001. High sensitivity waveform measurement with optical sampling using quasi-phasematched mixing in LiNbO3 waveguide. Electron. Lett., Vol. 37, No. 13, June 2001, pp. 842–844, ISSN: 0013-5194

Krug, T.; Lynch, M.; Bradley, A.L.; Donegan, J.F.; Barry, L.P.; Folliot, H.; Roberts, J.S. & Hill, G. (2004). High-sensitivity two-photon absorption microcavity autocorrelator. IEEE Photonics Technol. Lett., Vol. 16, No. 6, June 2004, pp. 1543–1544, ISSN 1041-1135.

Li, J.; Hansryd, J.; Hedekvist, P.O.; Andrekson, P.A. & Knudsen, S.N. (2001). 300-Gb/s eye-diagram measurement by optical sampling using fiber-based parametric amplification. IEEE Photon. Technol. Lett., Vol. 13, No. 9, Sep 2001, pp. 987-989, ISSN: 1041-1135

Li, J.; Westlund, M.; Sunnerud, H.; Olsson, B.-E.; Karlsson, M. & Andrekson, P.A. (2004).0.5-Tb/s eye-diagram measurement by optical sampling using XPM-induced wavelength shifting in highly nonlinear fiber. IEEE Photon. Technol. Lett., Vol. 16, No. 2, Feb. 2004, pp. 566-568, ISSN: 1041-1135.

Maguire, P.J.; Barry, L.P.; Krug, T.; Lynch, M.; Bradley, A.L.; Donegan, J.F. & Folliot, H. (2005). All-optical sampling utilising two-photon absorption in semiconductor microcavity. Electronics Letters, Vol. 41, No. 8, Apr. 2005, pp.489 – 490, ISSN: 0013-5194.

Malacarne, A. Fresi, F. Meloni, G. Poti, L. & Bogoni, A. (2007). Implementation of a Quasi-Asynchronous sub-ps-resolution all-optical sampler for long bit sequences view without post-processing, Proceedings of LEOS 2007, pp. 920-921, ISBN: 978-1-4244-0925-9, Oct 2007, Lake Buena Vista, Florida

Meloni, G. Bogoni, A. & Potì, L. (2005). Real-time ps-resolution optical sampler based on XPM-induced polarization rotation in 1-meter-long bismuth oxide fibre. Proceedings of ECOC 2005, Vol. 1, pp. 63- 64, ISBN: 0-86341-543-1, Sept 2005, Glasgow

Miyazaki, T. & Kubota, F. (2003). Simultaneous demultiplexing and clock recovery for 160 Gb/s OTDM signal using a symmetric Mach-Zehnder switch in electrooptic feedback loop. IEEE Photon. Technol. Lett., Vol. 15, No. 7, July 2003, pp. 1008–1010, ISSN: 1041-1135

Nogiwa, S.; Kawaguchi, Y.; Ohta, H. & Endo, Y. (2000). Highly sensitive and time-resolving optical sampling system using thin PPLN crystal. Electron. Lett., Vol. 36, No. 20, Sept. 2000, pp. 1727–1728, ISSN: 0013-5194.

Nogiwa, S.; Ohta, H.; Kawaguchi, Y. & Endo, Y. (1999). Improvement of sensitivity in optical sampling system. Electron. Lett., Vol. 35, No. 11, May 1999, pp. 917–918, ISSN: 0013-5194.

Ohta, H. Banjo, N. Yamada, N. Nogiwa, S. & Yanagisawa, Y. (2001). Measuring eye diagram of 320 Gbit/s optical signal by optical sampling using passively modelocked fibre laser. Electron. Lett., Vol. 37, No. 25, Dec. 2001, pp. 1541–1542, ISSN: 0013-5194.

Ohta, H.; Nogiwa, S., Kawaguchi, Y. & Endo, Y. (2000). Measurement of 200 Gbit/s optical eye diagram by optical sampling with gain-switched optical pulse. Electron. Lett., Vol 36, No. 8,Apr. 2000, pp. 737– 739, ISSN: 0013-5194.

Otani, A.; Otsubo, T. & Watanabe, H. (1999). A turn-key-ready optical sampling oscilloscope by using electro-absorption modulators. Proceedings of ECOC 1999, pp. 374–375, Nice, France, September 1999.

Schmidt, C.; Futami, F.; Watanabe, S.; Yamamoto, T.; Schubert, C.; Berger, J.; Kroh, M.; Ehrke, H.-J.; Dietrich, E.; Burner, C.; Ludwig, R. & Weber, H.G. (2002). Complete optical sampling system with broad gap-free spectral range for 160 Gbit/s and 320 Gbit/s and its application in a transmission system. Proceedings of OFC 2002, pp. 528–530, ISBN: 1-55752-701-6, Anaheim, USA, Mar 2002.

Schmidt, C.; Schubert, C.; Watanabe, S.; Futami, F.; Ludwig, R.; & Weber, H.G. (2002). 320 Gb/s All-Optical Eye Diagram Sampling using Gain-Transparent Ultrafast-Nonlinear Interferometer (GT-UNI). Proceedings of ECOC 2002, ISBN: 87-90974-63-8, Copenhagen, Denmark, September 2002.

Schmidt-Langhorst, C & Weber, H. G. (2005). Optical Sampling Techniques, In: J. of Optical and Fiber Communications Research, ed. Springer New York, pp. 86-114, ISSN 1867-3007

Schmidt-Langhorst, C.; Schubert, C.; Boerner, C.; Marembert, V.; Ferber, S.; Ludwig, R. & Weber, H.G. (2005). Optical sampling system including clock recovery for 320 Gbit/s DPSK and OOK data signals. Proceedings of OFC 2005, OWJ6, ISBN 1-55752-783-0, Washington DC, March 2005.

Shake, I Takara, H. & Kawanishi, S. (2003). Simple Q factor monitoring for BER estimation using opened eye diagrams captured by high-speed asynchronous electrooptical sampling. IEEE Photon. Technol. Lett., Vol. 15, No.4, Apr. 2003, pp. 620-622, ISSN: 1041-1135

Shake, I.; Kasahara, R.; Takara, H.; Ishii, M.; Inoue, Y.; Ohara, T.; Hibino, Y. & Kawanishi, S. (2003). WDM signal monitoring utilizing asynchronous sampling and wavelength selection based on thermo-optic switch and AWG. Proceedings of ECOC 2003, pp. 780–781, We4.P112, Rimini, Italy, Sept. 2003.

Shirane, M.; Hashimoto, Y.; Yamada, H. & Yokoyama, H. (2000). A compact optical sampling measurement system using mode-locked laser-diode modules. IEEE Photon. Technol. Lett., Vol. 12, No. 11, Nov. 2000, pp. 1537–1539, ISSN: 1041-1135.

Takara, H.; Kawanishi, S. & Saruwatari, M. (1996). Optical signal eye diagram measurement with subpicosecond resolution using optical sampling," Electron. Lett., Vol. 32, No. 15, Jul 1996, pp. 1399–1400, ISSN: 0013-5194.

Takara, H.; Kawanishi, S.; Yokoo, A.; Tomaru, S.; Kitoh, T. & Saruwatari, M. (1996). 100 Gbit/s optical signal eye-diagram measurement with optical sampling using organic nonlinear optical crystal. Electron. Lett., Vol. 32, No. 24, Nov 1996, pp. 2256–2258, ISSN: 0013-5194.

Tong, D.T.K. Deng, K.-L. Mikkelsen, B. Raybon, G. Dreyer, K.F. & Johnson, J.E. (2000). 160 Gbit/s clock recovery using electroabsorption modulator-basedphase-locked loop. Electron. Lett., vol. 36, no. 23, Nov. 2000, pp. 1951-1952, ISSN: 0013-5194

Uhua, L. Cheolhwan, K. Guifang, L. Kaneko, Y. Jungerman, R.L. & Buccafusca, O. (2003). Wavelength and polarization insensitive all-optical clock recovery from 96-Gb/s data by using a two-section gain-coupled DFB laser. IEEE Photon Technol. Lett., vol. 15, no. 4, Apr. 2003, pp. 590-592, ISSN: 1041-1135

Von der Linde, D. (1986). Characterization of the noise in continuosly operating modelocked lasers, *Journal of Applied Physics B: Lasers and Optics*, Vol. 39, No. 4, April 1986, pp. 201–217, ISSN 0946-2171.

Watanabe, S.; Okabe, R.; Futami, F.; Hainberger, R.; Schmidt-Langhorst, C.; Schubert, C. & Weber, H. G. (2004). Novel fiber kerr-switch with parametric gain: Demonstration of optical demultiplexing and sampling up to 640 Gb/s. Proceedings of ECOC 2004, Th4.1.6, Stockholm, Sweden, Sept. 2004.

Westlund, M. Andrekson, P.A. Sunnerud, H. Hansryd, J. & Jie, L. (2005). High-performance optical-fiber-nonlinearity-based optical waveform monitoring. IEEE J. of Lightwave Technol., Vol. 23, No. 6, June 2005, pp 2012 - 2022, ISSN: 0733-8724

Westlund, M. Sunnerud, H. Karlsson, M. & Andrekson, P.A. (2005). Software-synchronized all-optical sampling for fiber communication systems. IEEE J. of Lightwave Technol., Vol. 23, No. 3, Mar 2003, pp.1088 - 1099, ISBN: 1-55752-746-6

Tektronix Application Notes. (2001). XYZs of Oscilloscope, www.tek.com.

Yamada, N.; Nogiwa, S. & Ohta H. (2004). 640-Gb/s OTDM Signal Measurement With High-Resolution Optical Sampling System Using Wavelength-Tunable Soliton Pulses. IEEE Photon. Technol. Lett., Vol. 16, No. 4, April 2004, pp. 1125-1127, ISSN: 1041-1135.

Yamada, N.; Ohta, H. & Nogiwa, S. (2002). Jitter-free optical sampling system using passively modelocked fibre laser. Electron. Lett., Vol. 38, No. 18, Aug. 2002, pp. 1044–1045, ISSN: 0013-5194.

Yamamoto, T. Oxenlowe, L.K. Schmidt, C. Schubert, C. Hilliger, E. Feiste, U. Berger, J. Ludwig, R. & Weber, H.G. (2001). Clock recovery from 160 Gbit/s data signals using phase-locked loop with interferometric optical switch based on semiconductor optical amplifier", Electron Lett., vol. 37, no. 8, Apr 2001, pp. 509-510, ISSN: 0013-5194

15

NIR Single Photon Detectors with Up-conversion Technology and its Applications in Quantum Communication Systems

Lijun Ma, Oliver Slattery, and Xiao Tang
Information Technology laboratory, National Institute of Standards and Technology
United States of America

1. Introduction

The performance of any quantum communication system is limited by its transmission loss and detection efficiency, both of which must be balanced for optimal overall system performance. For current fiber-optic based systems, the transmission loss is small in the near infrared (NIR) range, and many fiber-based communication systems and devices tend to use this wavelength range. Therefore, the 1310 nm and 1550 nm bands, both of which are in the NIR range, have become mainstream in the telecom industry. However, the most efficient and low cost single photon detectors, such as silicon based avalanche photodiodes (APD), do not work in the NIR wavelength range. Bridging this gap is, of course, essential for an optimal quantum communication system.

In current systems, the preferred types of single photon detectors include photocathodebased detectors, APD-based detectors and superconducting-based detectors. Photocathodebased detectors use an InGaAs/InP photomultiplier tube (PMT) or an InGaAs Microchannel plate (MCP) for single photon detection in the NIR range. APD-based detectors, on the other hand, may only use InGaAs/InP APDs when detecting NIR single photons. Almost all superconducting-based detectors work in the NIR and can be described by two main types, including the Transition Edge Sensor (TES) and Superconducting Single-Photon Detectors (SSPD). In addition to these mainstream detectors, single photon detection at NIR can be achieved using a technique known as frequency up-conversion. We discuss this alternative technique in detail in this chapter.

1.1 Single photon detectors

PMTs, first invented in the 1930's, are used in many scientific applications, especially in those that require very large photosensitive areas. The wavelength sensitivity of PMTs is determined by an electron multiplying coating on the photocathode. While many suitable materials are available for visible light and UV-sensitive photocathodes, NIR sensitivity is not easily attainable. Currently, only InGaAs/InP based PMTs can operate in the NIR range, and its performance is limited by very low quantum efficiency (QE) (1 % at 1600 nm) and large timing jitter (1.5 ns) [Hamamatsu, 2005]. MCPs are micro-capillary electron multipliers coated with an electron-emissive material and multiply photon-excited electrons from a photon cathode [Wiza, 1979]. MCPs usually have faster rise times and lower timing jitter than is achievable with PMTs. However, similar to PMTs, MCPs are most suitable for the visible light range, and only InGaAs MCPs can work in the NIR range. These MCPs, like PMTs, are limited by low QE (~1 %) [Martin, J. & Hink P. 2003].

APDs, initially studied in the 1960s [Goetzberger et al, 1963], are the solid-state counterpart of PMTs. In an APD, a photon is absorbed in a bulk semiconductor, where it generates an electron-hole pair. With a sufficiently high electric field, carriers are accelerated to speeds where they can generate more electron-hole pairs though impact ionization, resulting in an avalanche multiplication. The silicone based APDs (Si-APD) are the most practical and widely used single photon detectors in recent years. Si-APDs have high QE with low noise levels in the visible light range and can work at room temperature. However, its QE decreases rapidly at wavelengths approaching NIR (i.e. longer than 1000 nm) and it does not work at the two telecom bands (1310 nm and 1550 nm). InGaAs/InP based APDs do work in NIR but with significant limitations. The ionization coefficient for electrons and holes in InGaAs are comparable, which leads to higher dark counts, a measure of the noise level in the detector [Lacaita et al., 1996]. To reduce this noise, the APD should be operated at very low temperatures. However, the cut-off wavelength of InGaAs shortens as the temperature decreases and the device loses sensitivity to 1550 nm photons at around -100 °C. Furthermore, trapped carriers in the device cause severe afterpulsing in this type of APD, especially at lower temperatures as the trapping lifetime becomes longer. Therefore, the operating temperature for this type of APD is set between -100 °C and -20 °C where the total dark count rate (combining those due to thermal generation and those due to afterpulsing) remains low and sensitivity to the desired wavelengths still exists. To overcome the severe influence of afterpulsing, commercial single photon counting modules based on InGaAs/InP APDs use active quenching and gated, or Geiger, mode operation to suppress the noise. However, with the gated rate limited to the MHz range, this does not satisfy the requirement of high-speed quantum communications. Recently, a self-difference technique has been developed for InGaAs APDs that suppresses the afterpulsing noise, and it has been successfully applied to a GHz quantum communication system [Yuan et al., 2007]. The InGaAs APD has about 10 % detection efficiency, but also about a 6 % afterpulse probability which contributes extra errors to the quantum communication system.

For some time now, superconducting technology has been used to implement single photon detectors in the NIR [Gol'tsman et al., 2001; Korneev et al., 2004; Hadfield et al., 2007; Takesue et al., 2007; Lita et al., 2008; Ma et al., 2009]. These types of detectors can have extremely low dark count rates and flat wavelength sensitivity extending far into theinfrared (IR) range. A TES, or TES microcalorimeter, consists of a piece of wolfram film, which is cooled below 100 mK. The film is kept at the transition edge of superconducting to normal conduction by Joule heating provided by the current from an associated circuit. In the TES detector, a photon is absorbed in the film producing a photoelectron which heats the electron system, raising its resistance and causing a drop in the current. TES detectors have no intrinsic limitation on QE, and currently achieve almost 100 % QE at the 1550 nm wavelength. However, the timing jitter of a TES detector is quite large (~ 100 ns), and therefore it is not suitable for high speed quantum communication systems. SSPD, or SNSPD (superconducting nanowire single-photon detectors), consists of a thin superconducting nanowire, which is meandered into a certain pattern. The detector is cooled to about 3 °K while the current in the nanowire is biased slightly lower than the critical current. When a photon is absorbed into the detection area, it will generate a hot spot. The current is forced to flow though a smaller cross-section of the nanowire around the hot spot, causing the current density to exceed the critical current density, which results in the loss of superconducting. After a few picoseconds, the hotspot disappears and the detector is restored to the original sate. SSPDs have the advantages of very small timing jitter and a high counting rate in excess of a gigahertz. It also works in free-running mode, which is preferred for the optimal performance of a quantum communication system. However, because it must operate at very low temperatures, the bulky and costly SSPD systems become a considerable impediment for practical applications.

1.2 Up-conversion detectors

A single photon detector using frequency up-conversion technology, also simply called an up-conversion detector, is not a direct detection method. It uses a non-linear optical media to up-convert the frequency of photons in the NIR range to a shorter wavelength by a process known as sum frequency generation (SFG). The emerging photons, at visible wavelengths, are then detected using visible region single photon detectors. Single photon detectors at the visible light region, such as the Si-APD, typically have high efficiency, low noise, can be operated in ambient temperatures, and are compact, inexpensive and practical. While up-conversion technology for strong light is not new [Midwinter & Warner 1967; Gurski, 1973], the up-conversion for single photon levels of light was achieved at the beginning of this century [Kim et al., 2001], and highly efficient up-conversion at single photon levels has only recently been demonstrated by using bulk periodically poled lithium niobate (PPLN) crystals [Vandevender & Kwiat, 2004] or PPLN waveguides [Xu et al., 2007; Diamanti et al., 2005; Langrock et al., 2005; Thew et al., 2006; Tanzilli et al., 2005]. Upconversion detectors work at room temperature, though the non-linear media is usually heated to satisfy a phase matching condition, described later, that is required for optimal conversion within the waveguide. The internal conversion efficiency of the waveguide can reach as high as 100 %, and the total QE for this type of detector is about 20 % ~ 35 %. The frequency up-conversion process does not contribute any significant timing jitter to the system. Instead, the timing jitter for this type of detector is influenced mainly by the Si-APD, and is usually in the region of 40 ps ~ 100 ps. Furthermore, the frequency up-conversion process provides some unique characteristics to the detector, such as narrow-band wavelength acceptance and polarization sensitivity, both of which are very useful for fiberbased quantum systems. To date, several groups have successfully developed highly efficient up-conversion detectors for the NIR range and have employed them in high speed quantum communications systems.

Table 1 summarizes the performance and characteristics of the various single photon detectors in the NIR range and an up-conversion detector, including their working temperature, maximum count rate, QE, dark count rate (DCR) and timing jitter.

In this chapter, we offer a general overview of the theoretical principles and experimental results of single photon detectors using frequency up-conversion technology, and its applications in quantum information systems. We begin with a brief introduction of nonlinear optics and its phenomena, especially the sum frequency generation in a quasi-phase matching (QPM) grating that is the basis of up-conversion technology. We then describe an up-conversion single-photon detector developed at the National Institute of Standards and Technology (NIST), and some key techniques used to implement the detectors with high efficiency, low noise, and low dark count rate. Finally, we introduce an existing quantum information, or quantum key distribution, system using up-conversion detectors.

Single photon detectors	Temperature, °K	Maximum Count rate, Hz	QE, %	DCR, Hz	Timing jitter, ps
PMT (Hamamatsu)	193	10^7	1	160k	1500
MCP (Burle)	210	1×10^6	1	100k	1000
InGaAs APD (id Quantique)	Peltier cooled	4×10^6 (gated)	10	10 k	60
InGaAs APD (Toshiba, UK)	243	4×10^8	10	10 k	60
SSPD (NIST)	<3	10^9	1	10	60
TES (NIST)	<0.1	10^5	95	400	100000
Up-conversion detector (NIST)	Room*	5×10^6	31	2.5 k	100

Table 1. Performance of single photon detectors responsive in the NIR range. The data are from [Hamamatsu, 2005; Martin & Hink 2003, Ma et al., 2009; Takesue et al., 2007; Yuan et al., 2007; Xu et al., 2007; Lita et al. 2008].
* Although the up-conversion system itself is operated at room temperature, the non-linear crystal is heated locally to satisfy the phase matching condition required for optimal conversion.

2. Frequency up-conversion

2.1 Sum frequency generation

Frequency up-conversion technology is based on a second-order non-linear sum frequency generation process, in which two input photons (a signal and a pump photon) at different frequencies annihilate and another photon at their sum frequency is simultaneously generated in a second-order non-linear media as shown in Fig. 1(a). According to non-linear optics theory, this process can happen only if the following two conditions are satisfied:

$$\omega_s + \omega_p = \omega_o \tag{1a}$$

$$\vec{k}_s + \vec{k}_p = \vec{k}_o \tag{1b}$$

where ω_s, ω_p and ω_o are the angular frequencies of the signal, pump and the output light, respectively. $\vec{k}_s$, $\vec{k}_p$ and $\vec{k}_o$ are the wave vectors of the signal, pump and the output light. Eq. 1(a) specifies conservation of energy and Eq. 1(b) specifies conservation of momentum. The frequency sum generation process is illustrated in Fig. 1.

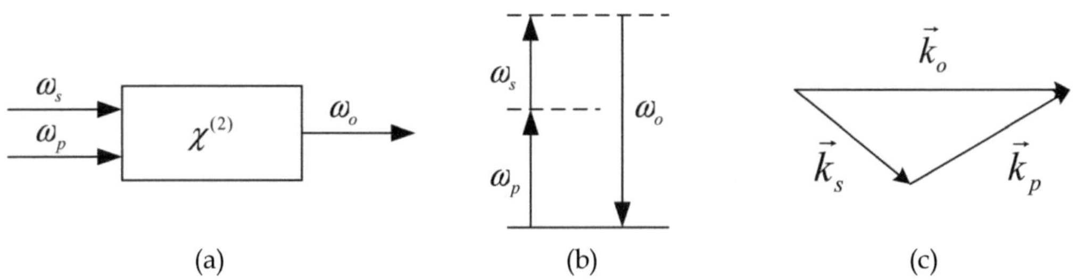

Fig. 1. Frequency up-conversion. (a) Geometry of the interaction. (b) Energy conservation condition. (c) Momentum conservation condition.

The nonlinear field evolution of the SFG process in a non-linear optical media can be described by:

$$\frac{dE_s}{dz} = i\frac{\omega_s d_{eff}}{n_s c} E_o E_p^* \exp(i\Delta k \cdot z) \quad (2a)$$

$$\frac{dE_p}{dz} = i\frac{\omega_p d_{eff}}{n_p c} E_o E_s^* \exp(i\Delta k \cdot z) \quad (2b)$$

$$\frac{dE_o}{dz} = i\frac{\omega_o d_{eff}}{n_o c} E_s E_p \exp(i\Delta k \cdot z) \quad (2c)$$

where E_s, E_p, and E_o are the electric field strengths of the signal, pump and output light, respectively; n_s, n_p and n_o are the indices of refraction at the three wavelengths; d_{eff} is the effective nonlinear coefficient of the crystal; c is the speed of light, and z is the longitudinal position along the propagation direction of the output light within the crystal. Δk represents the phase mismatch, which is defined by:

$$\Delta k = \left|\vec{k}_o - \vec{k}_s - \vec{k}_p\right| \quad (3)$$

At perfect phase-matching condition, Δk equals zero.

Single-photon up-conversion devices use the principle of SFG to convert the single photon signal light to a wavelength that is efficiently detectable by single photon detectors such as a Si-APD. Because the signal light is at single photon levels and the pump power is much stronger than the signal ($E_p >> E_s$), the pump power intensity does not deplete significantly in the up-conversion process, resulting in the approximation $\frac{dE_p}{dz} \approx 0$. Therefore, Eq. 2(a-c) can be reduced to two coupled first-order differential equations.

There is no input at the sum wavelength ($E_{o(z=0)}=0$), which is the initial condition for the equations. By solving the equations with this initial condition, the probability of upconversion (or its transfer function response in general) is given as follows:

$$P_o(z) \approx \sin^2(\alpha\sqrt{I_p}z) \quad (4a)$$

where I_p is the intensity of pump light. α is the conversion coefficient of the non-linear media and can be estimated by the following equation:

$$\alpha \approx \left(\frac{\omega_s \omega_o d_{eff}^2}{n_s n_o c^2}\right)^{1/2} \quad (4b)$$

From Eq. 4 (a, b), one can see that the up-conversion efficiency is a sinusoidal oscillation with respect to pump power. There is an optimal pump power, at which the conversion efficiency reaches its maximum. For perfect phase-matching, the conversion efficiency can be as high as 100%.

According to Eq. 4 (a, b), in order to obtain high conversion efficiency, we need to increase the pump intensity while maintaining or even reducing the optimal pump power since stronger pump power can lead to more noise. Furthermore, it can be seen from Eq. 4 (a) that a longer interaction length, z, will increase the conversion efficiency. Choosing materials with high non-linear coefficients is another option for increasing the conversion efficiency.

2.2 Birefringent phase matching

From Eq. 1(a, b), we can write the phase matching conditions as:

$$\omega_s + \omega_p = \omega_o \tag{5a}$$

$$n_s\omega_s + n_p\omega_p = n_o\omega_o \tag{5b}$$

Because all non-linear crystals have dispersion, (i.e. the refractive index is wavelength dependent), it is impossible to satisfy the Eq. 5(a, b) simultaneously if the three light beams have the same polarization. In practice, we can use the birefringence of the crystal to satisfy the phase matching condition. A beam of light may be decomposed into two rays (the ordinary ray, or o-ray and the extraordinary ray, or e-ray) when it passes through a birefringent material. Many optical materials are birefringent, which means the refractive index depends on the polarization of the light beam. In that case, one can select beams with different polarization direction and align the crystal to satisfy the Eq. 5 (a, b) and thus implement phase matching. The main two types of birefringence phase matching are Type I (e.g. o-ray+o-ray→e-ray) and Type II (o-ray+e-ray→e-ray).

The advantage of birefringence phase matching is that it is perfect phase matching (Δk is zero). However, birefringence phase matching has several limitations, including:

1. Wavelength selection is limited by the materials birefringent refractive index and orientation angles. Therefore, not all wavelengths can find a suitable material and orientation angle to implement phase matching.
2. The most severe problem for birefringent phase-matching is walk-off in which the e-ray and the o-ray travel in different directions. This walk-off limits interaction length and reduces the internal conversion efficiency.
3. The nonlinear coefficients for the birefringent phase-matching conversion process are relatively low. For example, the largest coefficient for the birefringent phasematching in lithium niobate crystal is only -4.64 pm/V (d_{31}) [Dmitriev et al, 1999].

Due to these limitations, birefringent phase matching can not implement high efficiency frequency conversion, and is therefore not suitable for up-conversion detectors.

2.3 Quasi-phase-matching

To overcome the limitations associated with birefringent phase matching, a new approach has been developed. Having the pump beam, signal beam and output beam all be collinear and aligned to the same polarization orientation will overcome the problem of walk-off and will also take advantage of larger non-linear coefficients. Specifically, all beams are in the extraordinarily polarization mode, and therefore the highest nonlinear coefficient, d_{33} = -40 pm/V [Dmitriev et al, 1999], can be used, which is an order of magnitude higher than d_{31}. However, in that case, it is impossible to satisfy the perfect phase-matching condition shown in Eq. 5 (a, b), and therefore we need to use another scheme called quasi-phase-matching.

To explain the quasi-phase-matching technique, we must go back to Eq. 2 (a-c). If all light beams are collinear and have the same polarization orientation, birefringent phase-matching is not achieved and $\Delta k_Q \neq 0$. In this case, the "sign" of dE_o/dz is flipped when z changes by $\pi/\Delta k_Q$, resulting in the periodic cancellation of the electrical filed strength of the output beam E_o. By reversing the domain poles every $\pi/\Delta k_Q$, the quasi-phase-matching technology contributes an alternating "sign" component over the same period, ensuring that there is a positive energy flow from the signal and pump frequencies to the output frequencies, even though all the frequencies involved are not phase locked with each other. The electrical field strength of the output beam versus the longitudinal position along the waveguide, with and without periodical poling, is shown in Fig. 2. Since the typical period for complete conversion to the output electrical field is on the order of thousands of poling periods, the evolution of the output field is well characterized by a sine function of the longitudinal position. Therefore, we can ignore the small fluctuations illustrated in Fig. 2 and approximate the full up-conversion process for the poled system with Eq. 4 (a, b)

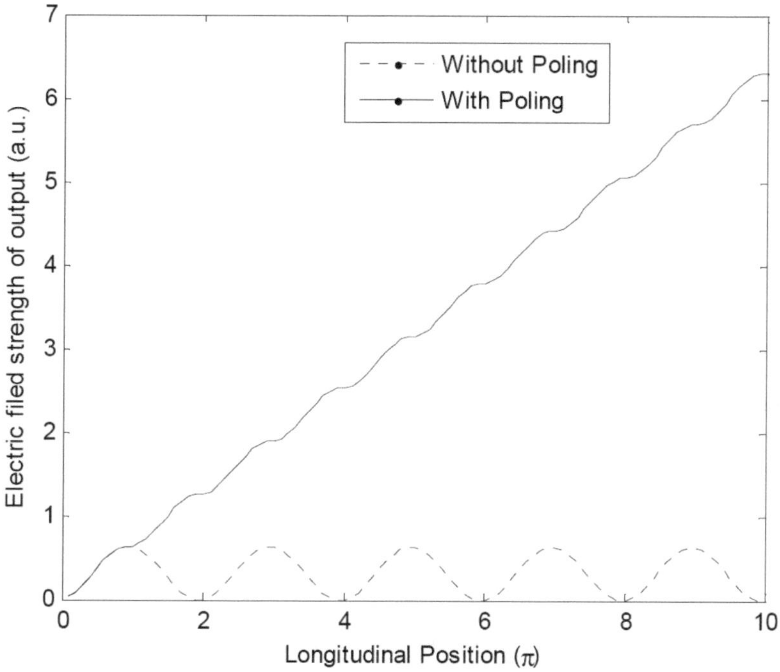

Fig. 2. Electric field strength of the output light versus the longitudinal position with poling and without periodic poling.

In terms of the phase, the periodic "sign" change results in an extra term ($2m\pi/\Lambda$) in Eq. 3. Λ is the poling period for the mth order quasi-phase-matched condition of the nonlinear PPLN waveguide. Because all three beams are collinear, their wavevectors have the same direction and we are able to show the phase relation in one dimension in Fig. 3.

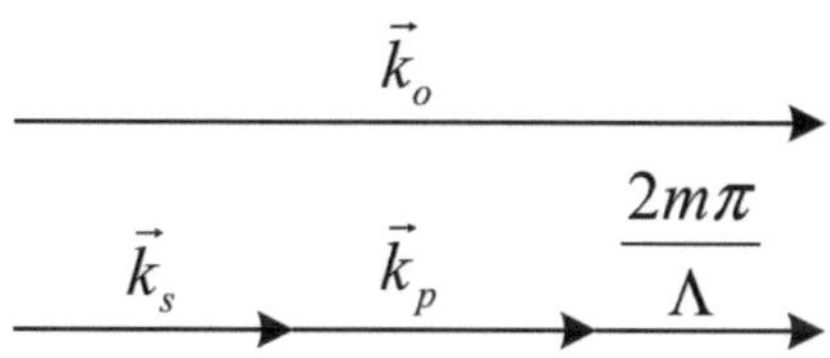

Fig. 3. Phase-match condition in quasi-phase-matching.

The phase mismatching is determined by the Eq. 6:

$$\Delta k \equiv k_o - k_s - k_p - \frac{2m\pi}{\Lambda} \tag{6}$$

where k_s ,k_p ,k_o are the wave numbers of the signal, pump and output beams in the crystal. By properly selecting Λ, the poling period of the crystal, the quasi-phase-matching is satisfied such that Δk =0. In the most desirable first-order (m=1) quasi-phase-matching condition, the ideal period can be calculated by using the following equation:

$$\Lambda = 2\pi / (k_o - k_s - k_p) \tag{7}$$

Once the ideal period, Λ, is calculated, a suitable mask can be designed with the ideal period imprinted and a photo-lithographic technique may then be used to make a periodic pattern on a nonlinear material substrate. When the substrate is heated and an intense electric field applied to the region exposed by the mask, the electrical field flips the direction of the nonlinear susceptibility of the crystal. After annealing, the periodically poled material is ready to use. Fine tuning of the Λ is achieved by adjusting the crystal temperature since the refractive index of the crystal is a function of temperature.

In quasi-phase-matching, the phase is not perfectly matched in each poling period but it ensures that positive energy is converted from the signal and pump frequencies to the output frequencies throughout the whole optical length of the crystal. Due to the imperfect phase matching in each poling period, the effective nonlinearity is reduced and can be estimated by the following equation:

$$d_{eff} = (\frac{2}{\pi})(\frac{1}{m})d_{33} \tag{8}$$

All even number orders (m=2, 4, 6…) of the quasi-phase-matching technique will have no output since periodical cancellation of the electrical filed strength of output light will occur. For all odd number orders (m=1, 3, 5…), according to Eq. 8, the first-order (m=1) quasiphase- matching has the highest effective nonlinear coefficient, though it is relatively hard to fabricate since the periods are shorter. Third-order (m=3) quasi-phase-matching, on the other hand, is often used as the longer periods are easier to fabricate. Because d_{33} is much larger than d_{31} in lithium niobate, the effective nonlinear coefficient in third order quasiphase matching is larger than that in birefringent phase matching.

Quasi-phase-matching can remove constraints on finding wavelengths and beam orientation angles to satisfy phase matching, and allow use of the highest nonlinear coefficient. The greatest advantage of the quasi-phase-matching technique is the elimination of walk-off and subsequently, the longer allowable interaction distance within the crystal. Furthermore, all three beams can be coupled together into a crystal waveguide, in which a higher beam intensity and longer interaction distance leads to significantly higher conversion efficiency.

As mentioned, the quasi-phase-matching in periodically poled lithium niobate allows us to take advantage of the larger d_{33} nonlinear coefficient. With a waveguide implementation of the PPLN, a higher intensity of pump can be provided and a longer interaction distance becomes possible. Currently, PPLN waveguides are the most suitable devices to implement frequency up-conversion with almost 100% internal conversion efficiency achievable with relatively low noise.

3. NIR Up-conversion detector

Recently, several highly efficient up-conversion single photon detectors for the NIR range have been demonstrated using PPLN waveguides [Xu et al., 2007; Diamanti et al., 2005;

Langrock et al., 2005; Thew et al., 2006; Tanzilli et al., 2005] and bulk PPLN crystals [Vandevender & Kwiat, 2004]. In this section, we will describe in detail an up-conversion detector developed at NIST [Xu et al., 2007] and analyze its characteristics.

3.1 NIST Up-conversion detector configuration

The NIST detector uses an up-conversion device to convert single photons at 1310 nm, with a pump at 1550 nm, to photons at 710 nm that are then efficiently detected by a Si-APD.
The configuration of the NIST up-conversion detector is shown in Fig. 4. A 1550 nm continuous wave (CW) laser provides the pump seed. If needed, the seed light can be modulated to an optical pulse train by a synchronized signal. This feature is similar to an optical gate, which is very useful for noise reduction or high speed gating operation in a communications system. The modulated 1550 nm pump seed is then amplified by an erbium-doped fiber amplifier (EDFA) (IPG: EAR-0.5K-C). Two 1310/1550 wavelength division multiplexer (WDM) couplers with a 25 dB extinction ratio are used to clean up the 1550 nm pulsed pump, specifically suppressing any EDFA noise that may extend to 1310 nm. The amplified 1550 nm pump light is then combined with a weak 1310 nm signal by another WDM coupler and the combined pump and signal are then coupled into the PPLN waveguides. The waveguide is a reverse-proton-exchange PPLN waveguide with magnesium oxide doping. The input polarization state of both the signal and the pump are adjusted by the polarization controllers, PC1 and PC2 respectively, to align with the polarization of a waveguide before entering the coupler. From Eq. 2, we know that a longer waveguide will require less pump power to reach the maximum conversion efficiency. The PPLN waveguide in the NIST up-conversion detector is 5 cm, the longest possible with current fabrication capability. The input of the PPLN waveguide is fiber coupled, while the output is free-space with a 710 nm anti-reflection (AR) coating to increase transmission of the converted output signal. The output light of the PPLN waveguide, including the newly generated photons at 710 nm (SFG), the pump at 1550 nm and its second harmonic generation at 775 nm, are separated by two dispersive prisms. The pump light (1550 nm) is clearly separated after the first prism and blocked by a beam block. Because the 775 nm beam is close to the 710 nm being detected, a second dispersive prism is used to further separate them, and an adjustable iris is used to block the 775 nm photons. Because all the light beams are linearly polarized and their polarization is aligned with the p-polarization direction of the prisms, there is no intrinsic loss when the incident angle of the 710 nm light is close to the Brewster's angle. A 20 nm band-pass filter (Omega Optical, Inc.: 3RD700-720) is used to reduce other noise, such as photons leaked into the system from any external sources. The 710 nm photons are then detected by a Si-APD (PerkinElmer: SPCM-AQR-14).

3.2 Detection efficiency

Detection efficiency is the one of the most important metrics in single photon detectors. The overall detection efficiency of an up-conversion detector is determined by the internal conversion efficiency in the PPLN waveguide, the coupling and component insertion losses, as well as the detection efficiency of the Si-APD at the converted wavelength. The overall detection efficiency of an up-conversion detector can be estimated by the following formula:

$$\eta_o = \eta_{loss} \cdot \eta_{con} \cdot \eta_{\det} \approx \eta_{loss} \cdot \eta_{\det} \cdot \sin^2(\alpha \cdot \sqrt{I_p} \cdot L) \tag{9}$$

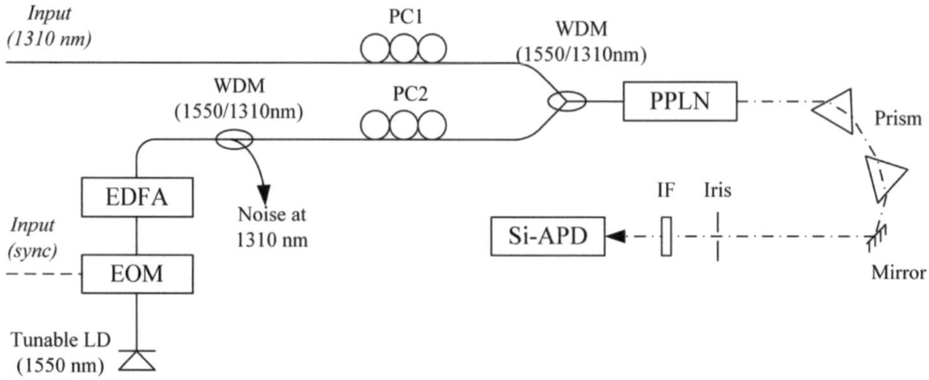

Fig. 4. Schematic diagram of the up-conversion detector. EOM: Electric-optic modulator; EDFA: Erbium-doped fiber amplifier; WDM: Wavelength-division multiplexing coupler; PC: Polarization controller; PPLN: Periodically-poled LiNbO3 waveguides; IF: Interference filter. Solid line: Optical fiber; Dash line: Free space optical transmission.

where η_o is the overall detection efficiency of the up-conversion detector; η_{loss} is the total loss in the detector, including the component insertion loss and waveguide coupling loss; η_{con} is the internal conversion efficiency in the PPLN, and can be estimated by Eq. 4; and η_{det} is the detection efficiency of the Si-APD at the converted wavelength, which at 710 nm is specified to be about 65 %.

When the insertion and coupling losses, the detection efficiency of the Si-APD and the structure of the waveguide are fixed, the overall conversion efficiency of the detector is determined by the internal conversion efficiency of the waveguide, which is dependent on the pump intensity with a $\sin^2(\sqrt{\ })$ relationship according to Eq. 9. The measured conversion efficiency versus pump power in a CW mode and in a pulsed pump mode is shown in Fig. 5. The measured results are in agreement with the estimated value from Eq. 9. The maximum detection efficiency is 32 % for both pump modes, which corresponds to 100 % internal conversion efficiency after excluding the insertion loss and the Si-APD detection efficiency.

In many quantum information systems, the photons arrive along with a classical signal that is synchronized to the photon transmission. That transmission clock can be recovered from the classical signal and used to operate the up-conversion detector in a pulse pump mode. The detection efficiency measured here is from a 625 MHz synchronized classical signal with 600 ps (FWHM) pulses. The quantum optical pulse is pumped with the same synchronized signal but has a shorter 300 ps (FWHM) pulse width. The detector operating in pulse pump mode can reach the maximum conversion efficiency with a lower average pump power, which helps to reduce the noise (discussed in detail in the next section). In cases where there is not a synchronized signal, a CW pump is needed. For pulse pump and CW pump modes, the optimal pump power (average) is about 38 mW and 78 mW, respectively.

3.3 Noise reduction

For a single photon detector, the noise level, or dark count rate, is the most important performance parameter since a higher dark count rate can cause more errors in the quantum information system and degrade the system's fidelity.

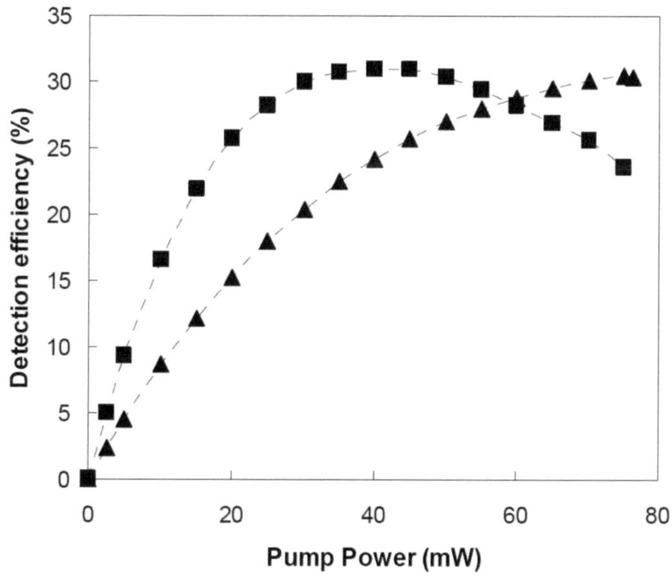

Fig. 5. The detection efficiency as a function of pump power. Two cases are shown: CW pump (triangle) and pulsed pump (square).

The dark count rate has been extensively studied in frequency up-conversion technology [Xu et al., 2007; Diamanti et al., 2005; Langrock et al., 2005; Thew et al., 2006]. The dark counts are contributed mainly by three parts: the intrinsic dark counts of the Si-APD, dark counts caused by the noise in the pump tail at the signal wavelength, and dark counts caused by the Raman scattering. The intrinsic dark count rate is constant, about 100 c/s in the NIST system. The dark counts caused by the noise in the pump tail occur as the pump noise at 1310 nm is up-converted to 710 nm and detected by Si-APD. About a 20 kHz dark count rate caused by the noise from the pump is observed at the maximum conversion efficiency. We use two WDM couplers (a 50 dB extinction ratio in total) to greatly suppress this noise. The dark counts caused by the Raman scattering occur as 1310 nm photons are generated by Raman scattering with the strong pump light in the transmission fiber and waveguide, and then up-converted to 710 nm and detected by Si-APD. In this up-conversion detector unit, we use a 1550 nm laser as a pump, whose wavelength is longer than that of the quantum signal we want to measure. Because the anti-Stokes component of the Raman process is much weaker than the Stokes component, a dark count rate of less than 2400 c/s is achieved when the conversion efficiency is maximized.

As shown in the Fig. 6, the pulse pump generates more dark counts than the CW pump for a given average power, because the peak power of the pulse pump is higher than the average power. However, the pulse pump needs less average power than the CW pump to achieve a given detection efficiency, see Fig. 5, and in the end, the pulse pump can achieve a given detection efficiency with less dark counts than the CW pump. As an example, the maximum detection efficiency is reached at 38 mW pulse pump resulting in a dark count rate of 2400 c/s. For the CW pump, on the other hand, a pump power of 78 mW is required to achieve the maximum detection efficiency, which incurs a dark count rate of 3100 c/s.

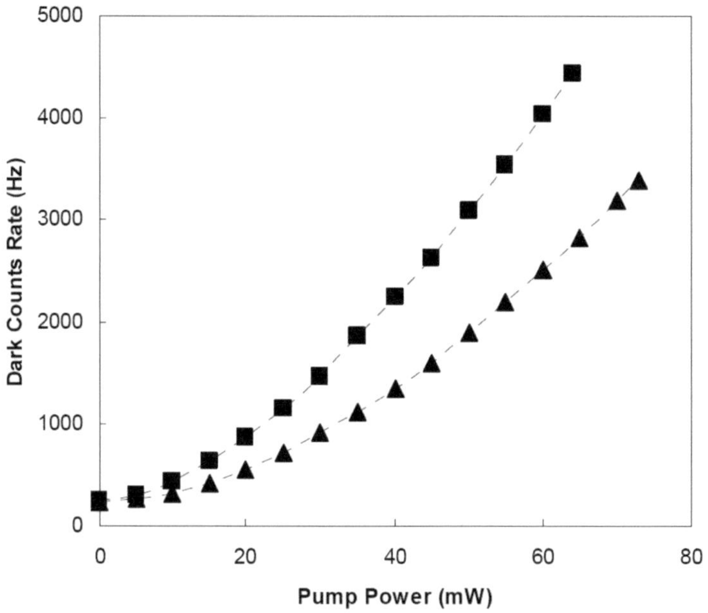

Fig. 6. The dark count rate as a function of pump power at the PPLN input. Two cases are studied: CW pump (triangle) and pulsed pump (square).

3.4 Wavelength and temperature response

When the quasi-phase matching condition in a PPLN waveguide is satisfied at a particular signal wavelength, the maximum up-conversion efficiency is achieved. When the signal is shifted from that wavelength the up-conversion efficiency is reduced. Therefore, the upconversion detectors have a narrow wavelength acceptance width, similar to a narrow band pass filter. While it helps to filter out noise at wavelengths other than the signal wavelength, this may be a drawback when the detector is used to measure photons with wider spectrums. The acceptance spectral width of the up-conversion detector is determined by the transfer function response of the PPLN waveguide. The transfer function response of a finite-length uniform QPM grating in the waveguide is a function of a $\mathrm{sinc}^2(\ )$ as follows [Fejer et al, 1992; Micheli 1997]:

$$P_o(\Delta k_Q) \propto P_p \cdot P_s \cdot \mathrm{sinc}^2(A \cdot \Delta k_Q \cdot L) \tag{10}$$

where P_o, P_p, P_s are the powers of SFG output, pump, and signal beam; A is a constant; L is the waveguide length; and Δk_Q is the phase-mismatch, which can be calculated by the following relation with the system wavelengths:

$$\Delta k_Q = 2\pi \cdot \left(\frac{n_o}{\lambda_o} - \frac{n_p}{\lambda_p} - \frac{n_s}{\lambda_s} - \frac{m}{\Lambda}\right) \tag{11}$$

where λ_o, λ_p and λ_s are the wavelengths for output, pump, and signal, respectively, and n_o n_p and n_s are the reflective index for the three wavelengths. Λ is, as described earlier, the poling period for the mth order quasi-phase-matched condition of the nonlinear PPLN waveguide. From Eq. 2 and 3, the acceptance spectral width is dependent on the length of the waveguide, the longer the waveguide, the narrower the acceptance spectral width will be. Fig. 7 shows the measured detection efficiency as a function of the signal wavelength at a fixed pump wavelength and temperature. From the figure, we can see that the spectrum is similar to the sinc^2( ) function and the acceptance spectral width of the main peak is about 0.25 nm (FWHM). If we use a short waveguide or a pump light with a wider spectrum, the acceptance spectral width can be broadened.

The up-conversion wavelength peak is also temperature sensitive. Therefore, one or both of the pump and signal wavelengths, or the waveguide temperature, must be accurately tuned to achieve the maximum up-conversion efficiency. To investigate the temperature sensitivity of the up-conversion, we sent a 1 mW CW 1310 nm laser beam with a linewidth of less than 10 MHz into the PPLN waveguide. Moreover, we turned off the pump seed laser so that the amplified spontaneous emission (ASE) noise from the EDFA acted as the pump. Using an optical spectrum analyzer, we measured the spectrum at the output of PPLN waveguide at different temperatures from 50 °C to 70 °C. The resulting output spectrum is shown in Fig. 8, normalized to the peak power after we subtracted the ASE spectrum.

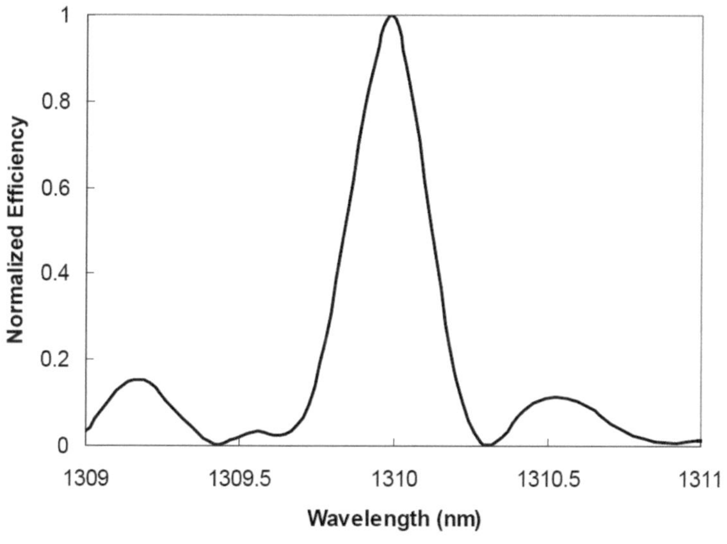

Fig. 7. The normalized detection efficiency as a function of signal wavelength, when the pump wavelength and temperature of the waveguide are fixed.

As shown in Fig. 8(a), the spectral width of the sum frequency at 710 nm is about 0.15 nm. The result is consistent with the spectral width of 0.25 nm for the signal at 1310 nm as shown in Fig. 7. Also from Fig. 8(a), the peak wavelength is shifted as the temperature changes, which means that the quasi-phase matching condition can be achieved by either varying the converted wavelength (via tuning of the pump wavelength and/or signal wavelength), or by varying the waveguide temperature, as shown in Fig. 8 (b). Within a temperature variation range of 20 degrees, the central wavelength for maximum efficiency linearly varies by approximately 1.1 nm. The temperature response of the waveguide can provide a method to fine tune the up-conversion detector, even if the signal and pump wavelengths are fixed.

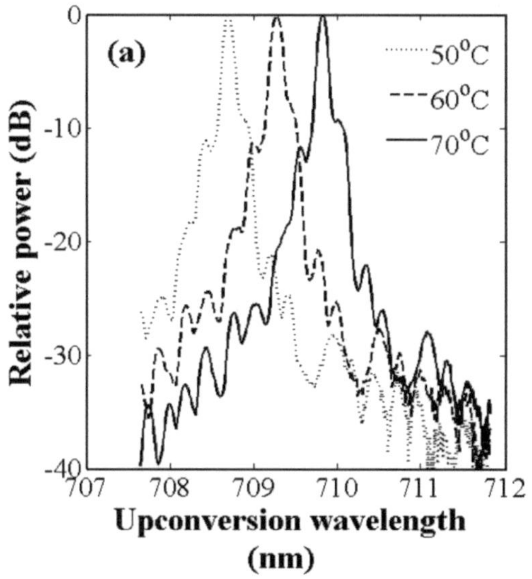

Fig. 8. (a) The normalized output spectrum of the PPLN at different temperatures. (b) The up-conversion wavelength peak as a function of temperature.

3.5 Polarization characteristic

Because the PPLN waveguide is based on proton-exchange, it is effective only for guiding the e-wave, and the o-polarized light is not transmitted. In effect, the device is therefore polarization sensitive. If its polarization extinction ratio is sufficiently high, the device can be used as a polarizer. This feature is very useful in a polarization-encoding quantum communications system. Fig. 9 shows the dependence of the detection efficiency on the polarization direction of an input signal at 1310 nm. The deviation angle is the angle (in Jones space) between the given input polarization state and the one at which the conversion efficiency is maximized. We also compared this measurement result with a $\cos^2(\ )$ curve, the function which represents an ideal polarizer. The curve is in good agreement with the measured data and we believe that the slight difference is caused by the measurement uncertainty of the polarimeter. As shown in Fig. 9, the polarization extinction ratio of the PPLN is over 25 dB. Therefore, an up-conversion detector can be used as a polarizer in a polarization-based quantum information system, eliminating the need for an otherwise required polarizer and avoiding the additional insertion loss that a polarizer would add.

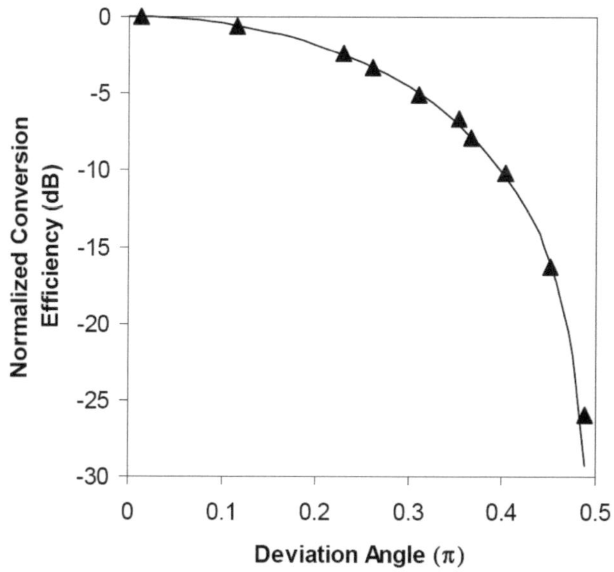

Fig. 9. The normalized conversion efficiency of the PPLN waveguide as a function of deviation angle of the input signal at 1310 nm. The deviation angle is the angle between the given polarization state and the state at which the conversion efficiency is maximized. Triangle: Measurement results; Solid line: $\cos^2()$ curve

4. Application in quantum communication systems

Quantum key distribution (QKD) is an important application of quantum communication, which is a technique for developing shared secret keys over unsecured communication channels that is guaranteed by the fundamental quantum properties of single photons instead of mathematical complexity [Gisin et al., 2002]. It is not possible to make a perfect copy (clone) of an unknown quantum state, thus precise measurement by an eavesdropper is not achievable. The Heisenberg uncertainty principle states that pairs of quantum properties cannot be precisely measured simultaneously; for example, position and momentum. Horizontal-vertical and diagonal polarization of photons is another such pair.

QKD systems use quantum states, such as polarization, to encode information on single photons. An initial random key is established by randomly encoding quantum state information on these photons, transmitting the photons and recovering the encoded state information at the other end of the link. After sifting, error correction and privacy amplification, the three conventional processing procedures required to complete the QKD protocol, the initial (raw) keys become secure keys and are ready for use.

The idea to use quantum states to securely encode information originated with Stephen Wiesner in 1983 [Wiesner, 1983]. This idea was taken forward by Charles Bennett and Gilles Brassard in 1984 [Bennett & Brassard, 1984] who developed the famous QKD protocol called BB84, which uses four quantum states. In 1992, Charles Bennett proposed a simplified version of the protocol, named B92, [Bennett, 1992] that uses only two quantum states. These two protocols are commonly used in most QKD systems today. The first demonstration of a QKD system was completed in 1989, in which the quantum channel was a 30 cm long path of air in a laboratory [Bennett & Brassard, 1989]. Since then, a number of groups have successfully developed many experimental QKD systems, which are described in a comprehensive review article by Nicolas Gisin [Gisin et al., 2002].

Single photon detectors are one of the key elements for a QKD system since information is encoded as the quantum state of single photons. Among the available types of single photon detectors, up-conversion detectors are quite suitable for QKD systems due to their high detection efficiency, low dark count rate and unique characteristics, such as narrow acceptance spectral width and polarization sensitivity. Specifically, the advantages offered by up-conversion detectors include:

1. High detection efficiency: many QKD systems use a narrow linewidth attenuated laser light as the single photon source, which is much narrower than the acceptance bandwidth of up-conversion detection. Therefore, an up-conversion detector can reach its maximum detection efficiency, resulting in a higher secure key rate.
2. Low dark count rate: many QKD systems recover the clock signal from their classical channel, which can also be used as the synchronized trigger for the pulse pump operation in an up-conversion detector. As described earlier, the pulse pump operation of the up-conversion detector can lead to a lower dark count rate and, therefore, a lower system error rate.
3. Narrow acceptance spectral width: each up-conversion detector has a relatively narrow acceptance spectral width that functions as a band-pass filter, rejecting the noise due to crosstalk from strong signals in classical channel that share the same fiber in many QKD systems.
4. Polarization sensitivity: as described earlier, this feature can be used as a polarizer, which avoids the additional insertion loss that an otherwise required polarizer would add.

Because of these outstanding performance characteristics, several research groups have successfully demonstrated fiber-based QKD systems using up-conversion single photon detectors. By way of an example, we introduce a fiber-based QKD system developed at NIST[Xu et al., 2007]. The system uses the B92 protocol [Bennett, 1992] with 1310 nm photons that share a single optical fiber with bi-directional classical signals at the 1550 nm band.

A QKD system using the B92 protocol requires two detectors to detect the photons emerging from the two different measurement bases. Fig. 10 outlines a compact dual up-conversion detector system, and is an all fiber system rather than the free-space output configuration described in Fig. 3. A weak pump seed laser is amplified by an EDFA and split into two parts, each of which will provide the pump for one of the detectors. Each detector in this dual detector system consists of a 5 cm PPLN waveguide, whose input and output are fiber coupled. The detectors use in-line narrow band-pass filters to suppress noise from the pump light and its SHG component. This all fiber detector is easy to use and more compact compared to the free space output detector described earlier, however its overall detection efficiency is reduced to 15~20 %, due to output coupling loss from the waveguide to the fiber and the insertion loss associated with the narrow band-pass filters.

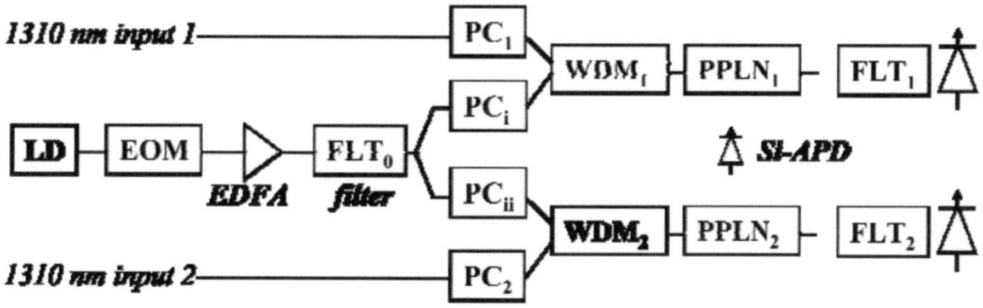

Fig. 10. The configuration of compact dual up-conversion detectors. LD: Laser diode; EOM: Electric-optic modulator; EDFA: Erbium-doped fiber amplifier; FLT: Optical filter; PC: Polarization controller; WDM: Wavelength-division multiplexer for 1310 nm and 1550 nm; PPLN: PPLN waveguide.

The configuration of the QKD system using the up-conversion detector is shown in the Fig. 11. The system is designed to generate a secure key between a sender (Alice) and a receiver (Bob). The QKD system uses a custom printed circuit board with a field-programmable gate array (FPGA) [Mink et al., 2006] to generate a random stream to encode the photon with quantum state information and to transmit and receive classical data. The classical data is carried by a conventional optical signal at 1550 nm.

To polarization-encode the quantum channel with the random data, we first modulate a 1310 nm CW beam into a 625 MHz pulse train which is evenly split into two polarization channels. Each pulse train is further modulated by one of two complementary 625 Mbit/s quantum channel data streams. The two quantum channels are then combined by a 45-degree polarization-maintaining combiner and attenuated to a mean photon number of 0.1 per bit, and then multiplexed with the classical channel before being coupled into a standard single-mode fiber for transmission.

At the receivers end, another WDM is used to demultiplex the quantum and the classical channels. A clock signal is recovered from classical channel and it is sent into the upconversion detectors as the synchronized trigger for the pulse pump operation. Photons in the quantum channel are randomly split into two non-orthogonal measurement bases by a coupler and are then sent to their respective up-conversion detectors. The up-conversion detector's polarizer-like characteristic performs a polarization measurement on the photons which are then converted and detected. For each detection event, the bit position, but not the bit value, of the detected photon is returned to Alice over the classical channel. Alice's FPGA matches each detection event with the corresponding event of the stored bit-stream. The matched positions, again without the bit values, are returned to Bob. Both Bob and Alice send the bit values to their respective CPUs for reconciliation and privacy amplification to generate their shared secret keys, which are then used in a conventional security application.

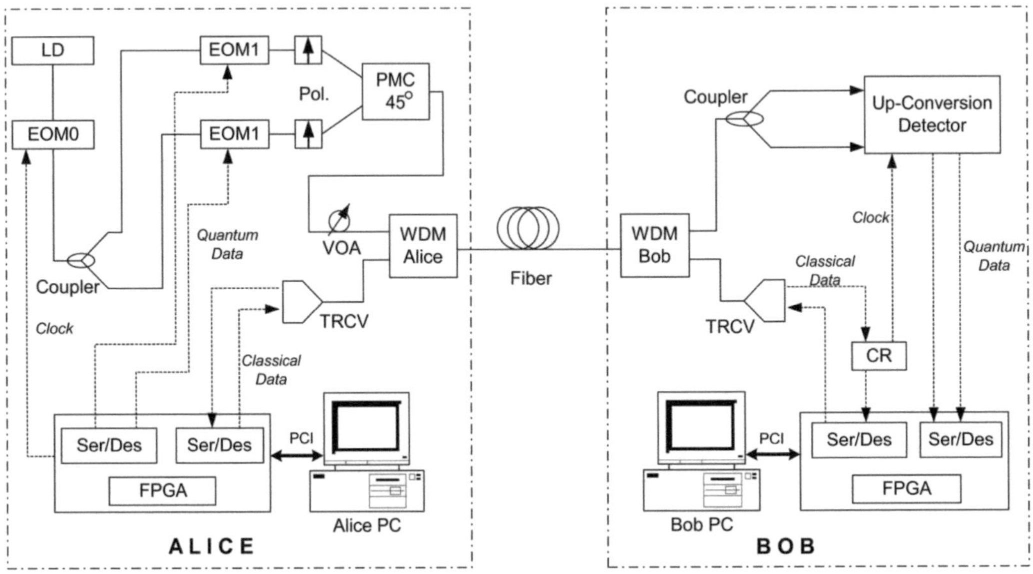

Fig. 11. The B92 polarization coding QKD system. LD: Laser diode; EOM: Electric-optic modulator (LiNbO3); PC: Polarization controller; PMC-45°: Polarization maintaining combiner that combines two light signals that are separated by 45 degrees; VOA: Variable optical attenuator; WDM: Wavelength-division multiplexer; SMF: Standard single-mode fiber; TRCV: Optical transceiver; CR: Clock recovery module; FPGA: Custom printed circuit board controlled by a field-programmable gate array; PCI: PCI connection; Up-conversion detector: See Fig. 8; Dotted line: Electric cable; Solid line: Optical fiber.

In theory, the security of a QKD system is guaranteed by the fundamental principles of physics and cannot be broken. However, in practice its security is limited by its noise level. The noise level, or dark count rate, is therefore the main concern for QKD system development and implementation. Although the up-conversion detector has a low darkcount rate, other sources of noise must be identified when the detector is integrated into a system. In the QKD system, the quantum channel (1310 nm) and the bi-directional classical channel (1510 nm & 1590 nm) share a single standard telecom fiber, so there is a concern that the quantum channel may suffer from noise from the classical channel. In a QKD system that uses up-conversion detectors, the crosstalk from the strong signal in the classical channel (1510 nm & 1590 nm) will be blocked by the narrow acceptance spectral width of the up-conversion detector. In the classical channel, there are two noise sources including a small amount of noise around 1310 nm emitted from the transceivers and noise due to nonlinear effects, which is widely believed to be from Raman scattering. This is similar to the 1550 nm pump noise in the up-conversion detector discussed in section 3.3. The strong signals at 1510 nm and 1590 nm interact with the fiber and, in a nonlinear anti-Stokes process, generate photons around 1310 nm. These 1310 nm fiber generated photons enter the PPLN waveguides and are up-converted to 710 nm. Fig. 12(a) shows the extra dark count rate induced by the classical channel at various fiber link distances. We first measure the dark count rate when one or both of the classical transceivers are on, and then subtract the dark count rate measured when both transceivers are off. The photon leakage can be evaluated by the extra dark counts in a back-to-back (0 km) connection while the nonlinearly induced dark photon effect will vary over the transmission distance. As shown in the Fig. 12(a), the photon leakage noise is small and the dark counts are mainly induced by the Raman anti-Stokes process, particularly from the classical signals at 1510 nm propagating from Alice to Bob (forward anti-Stokes). The backward anti-Stokes noise is generated by the classical signals at 1590 nm propagating from Bob to Alice and is much weaker than the forward anti-Stokes because of the greater separation between the 1590 nm light and the nm signal photons. In general, the classical channel induces a non-negligible

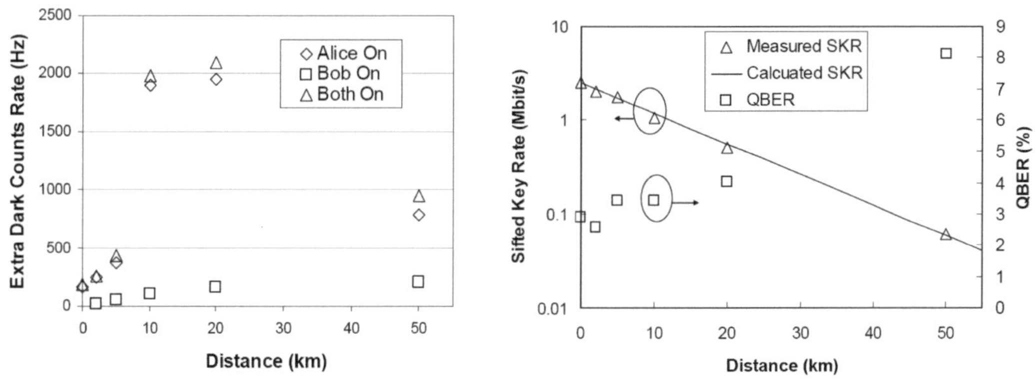

Fig. 12. (a) The extra dark count rate induced by the classical channel in one of the PPLN detector in three cases: Diamond, only the transceiver at Alice is on; square, only the transceiver at Bob is on; triangle, both transceivers are on. The other PPLN detector exhibits similar behaviors. (b) The system performance of the B92 polarization-based QKD system with the 1550 nm pumped up-conversion detector.

dark count rate into the QKD system, particularly from 1510 nm. A longer wavelength transceiver would greatly help to reduce the dark count rate further, but care must be taken to keep it within the standard telecom band.

The system performance is shown in Fig. 12(b). During our measurements, the pump power was fixed at 40 mW. The sifted-key rate is 2.5 Mbit/s for a back-to-back connection, 1 Mbit/s at 10 km, and 60 kbit/s at 50 km. The quantum bit error rate (QBER) is approximately 3 % for the back-to-back configuration, remains below 4 % up to 20 km, and reaches 8 % at 50 km. The finite extinction ratio of the modulator and timing jitter of the system induces a background QBER of approximately 2.5 % and the remaining QBER results from dark counts generated by both the pump light and the classical channel, as we described earlier. We also calculated the theoretical sifted-key rate and QBER and they agree with the measured results. Although we fixed the pump power close to the maximum up-conversion efficiency, the QBER remains small until 20 km due to the low dark count rate of the 1550 nm up-conversion detector. The QKD system can generate sufficient secure keys in real time for one-time-pad encryption of continuous 200 kbit/s encrypted video transmission over 10 km.

5. Conclusion

Frequency up-conversion single photon detectors use the principle of sum frequency generation to up-convert single photons in the near IR range to a shorter wavelength for an efficient detection by low cost detectors, such as Si-APDs. The up-conversion detectors are usually operated at room temperature with high detection efficiency and a low dark count rate. The detectors have a very narrow acceptance spectral width and polarization sensitivity, properties that can be exploited in certain applications requiring narrow linewidth or polarization specific detection. These unique characteristics can be used to enhance system performance in some applications, including fiber-based quantum communications systems.

6. References

Bennett, C. H. & Brassard, G. (1984). Quantum cryptography: Public key distribution and coin tossing. *Proc. IEEE Int. Conf. Comput. Syst. Signal Process.*, pp. 175–179.

Bennett, C. H. & Brassard, G. (1989). The Dawn of a New Era in Quantum Cryptography: the Experimental Prototype is Working .*SIGACT NEWS,* Vol.20, pp. 78-83

Bennett, C. H. (1992). Quantum cryptography using any two nonorthogonal states. *Phys. Rev. Lett.*, Vol. 68, pp 3121-3124

Boyd, R. W. (2008). Nonlinear Optics, Academic Press, ISBN 978-0123694706, New York. Diamanti, E.; Takesue, H.; Honjo, T.; Inoue, K. & Yamamoto, Y. (2005). Performance of various quantum-key-distribution systems using 1.55-μm up-conversion singlephoton detectors. *Phys. Rev. A,* Vol. 72, 052311

Dmitriev, V.G.; Gurzadyan, G. G. & Nikogosyan, D. N. (1999) Handbook of Nonlinear Optical Crystals, Springer, ISBN 978-3540653943

Fejer, M.; Magel, G.; Jundt, D. & Byer, R. (1992). Quasi-phase-matched second harmonic generation: tuning and tolerances. IEEE J. Quantum Electron. Vol.28, pp 2631-2654

Gisin, N.; Ribordy, G.; Tittel, W. & Zbinden, H., (2002). Quantum cryptography. *Rev. Mod. Phys.* Vol. 74, pp 145–195

Goetzberger, A.; McDonald, B.; Haitz, R.H. & Scarlett, R.M. (1963). Avalanche effects in silicon p-n junctions. II. Strucrually perfect junctions. *J. Appl. Phys.* Vol 34: 1591-1600

Gol'tsman, G. N.; Okunev, O.; Chulkova G.; Lipatov, A.; Semenov, A.; Smirnov, K.; Voronov, B. & Dzardanov, A. (2001). Picosecond superconducting single-photon optical detector. *Appl. Phys. Lett.* Vol. 79, pp 705-707

Gurski, T. (1973). High-quantum-efficiency infrared up-conversion. *Appl. Phys. Lett.*, Vol. 23, pp 273-275.

Hadfield, R.; Schlafer, J.; Ma, L.; Mink, A.; Tang, X. & Nam, S. (2007). Quantum key distribution with high-speed superconducting single-photon detectors. *CLEO/QELS Technical Digest,* QML4,

Hamamatsu. (2005). Near infrared photomultiplier tube R5509-73 data sheet.

Kim, Y.; Kulik, S. & Shih, Y. (2001) Quantum teleportation of a polarization state with a complete bell state measurement. *Phys. Rev. Lett.* Vol. 86, pp 1370-1373

Korneev, A.; Kouminov, P.; Matvienko, V.; Chulkova, G.; Smirnov, K.; Voronov, B.; Gol'tsman, G. N.; Currie, M.; Lo, W.; Wilsher, K.; Zhang, J.; Słysz, W.; Pearlman, A.; Verevkin, A. & Sobolewski, R. (2004). Sensitivity and gigahertz counting performance of NbN superconducting single-photon detectors. *Appl. Phys. Lett.* Vol. 84, pp 5338-5340

Langrock, C.; Diamanti, E.; Roussev, R. V.; Yamamoto, Y.; Fejer, M. M. & Takesue, H. (2005). Highly efficient single-photon detection at communication wavelengths by use of upconversion in reverse-proton-exchanged periodically poled LiNbO3 waveguides. *Opt. Lett. Vol.* 30, pp. 1725-1727

Lacaita, A.; Zappa, F.; Cova, S. & Lovati, P. (1996) Single-photon detection beyond 1 µm: performance of commercially available InGaAs/InP detectors. *Appl. Opt.,* Vol. 35, pp 2986-2996

Lita, A. E.; Miller, A. J. & Nam, S. W. (2008). Counting near-infrared single-photons with 95% efficiency," *Opt. Express,* Vol. 16, pp3032-3040

Ma, L.; Nam, S.; Xu, H.; Baek, B.; Chang, T.; Slattery, O.; Mink, A. & Tang, X. (2009). 1310 nm differential phase shift QKD system using superconducting single photon detectors. *New Journal of Physics,* Vol. 11, pp 054020

Martin, J. & Hink P. (2003) Single-Photon Detection with MicroChannel Plate Based Photo Multiplier Tubes. *Workshop on Single-Photon: Detectors, Applications and Measurement Methods, NIST.*

Micheli, M. P. (1997) χ^2 effects in waveguides. *Quantum Semiclassic. Opt,* Vol. 9, pp 155–164.

Midwinter, J. & Warner, J. (1967). Up-conversion of near infrared to visible radiation in lithium-meta-niobate. J. Appl. Phys. Vol 38, pp 519-523

Mink, A.; Tang, X.; Ma, L.; Nakassis, T.; Hershman, B.; Bienfang, J. C.; Su, D.; Boisvert, R.; Clark, C. W. & Williams, C. J. (2006). High speed quantum key distribution system supports one-time pad encryption of real-time video. *Proc. of SPIE,* Vol. 6244, 62440M,

Nakassis, A., Bienfang, J. & Williams, C. (2004). Expeditious reconciliation for practical quantum key distribution. *Proc. of SPIE,* Vol. 5436, pp. 28-35.

Takesue, H.; Nam, S.; Zhang, Q.; Hadfield, R. H.; Honjo, T.; Tamaki, K. & Yamamoto, Y. (2007). Quantum key distribution over a 40-dB channel loss using superconducting single-photon detectors. *Nature Photonics,* Vol. 1, pp 343-348

Tang, X.; Ma, L.; Mink, A.; Nakassis, A.; Xu, H.; Hershman, B.; Bienfang, J.; Su, D.; Boisvert, R.; Clark, C. & C. Williams. (2006). Experimental study of high speed polarizationcoding quantum key distribution with sifted-key rates over Mbit/s. *Optics Express,* Vol. 14, No.6, pp 2062-2070

Tanzilli, S.; Tittel, W.; Halder, M.; Alibart, O.; Baldi, P.; Gisin, N. & Zbinden, H. (2005). A photonic quantum information interface. *Nature,* Vol 437, pp 116-120

Thew, R. T.; Tanzilli, S.;, Krainer, L.; Zeller, S. C.; Rochas, A.; Rech, I.; Cova, S.; Zbinden, H. & Gisin, N. (2006). Low jitter up-conversion detectors for telecom wavelength GHz QKD. *New J. Phys.* Vol. 8, pp 32.

Vandevender, A. P. & Kwiat, P. G. (2004). High efficiency single photon detection via frequency up-conversion. *J. Mod. Opt.,* Vol. 51, 1433-1445

Wiesner, S. (1983). Conjugate coding. *Sigact News,* Vol. 15, pp 78-88

Wiza, J. (1979). Microchannel plate detectors. *Nuclear Instruments and Methods* Vol. 162: pp 587-601

Xu, H.; Ma, L.; Mink, A.; Hershman, B. & Tang, X. (2007). 1310-nm quantum key distribution system with up-conversion pump wavelength at 1550 nm. *Optics Express,* Vol 15, No.12, pp 7247- 7260

Yuan, Z. L.; Dixon, A. R.; Dynes, J. F.; Sharpe, A. W. & Shields, A. J. (2008). Gigahertz quantum key distribution with InGaAs avalanche photodiodes. *Appl. Phys. Lett.* Vol. 92, 201104.

16

All-Optical Signal Processing with Semiconductor Optical Amplifiers and Tunable Filters

Xinliang Zhang, Xi Huang, Jianji Dong, Yu Yu, Jing Xu and Dexiu Huang
Wuhan National Laboratory for Optoelectronics,
Huazhong University of Science and Technology
P.R.China

1. Introduction

All-optical signal processing has been and is receiving more and more attention all over the world because it can increase the capacity of the optical networks greatly in avoiding of the Optical-Electrical-Optical (O/E/O) conversion process, and it can also reduce the system power consumption to a great extend and then increase the system stability. All-optical signal processing can be widely used in optical signal regeneration and switching in next-generation optical networks (Yoo 1996; Danielsen et al. 1998; Saruwatari 2000), such as Optical Time Division Multiplexing (OTDM), Optical Orthogonal Frequency Division Multiplexing (OOFDM), Optical Code Division Multiplexing Accessing (OCDMA), Optical Packet Switching (OPS) and so on. There are many different elemental functions in all-optical signal processing: all-optical wavelength conversion, all-optical logic operation, all-optical 3R regeneration, all-optical format conversion, all-optical sampling, all-optical time demultiplexing, all-optical buffering, etc. It should be mentioned that all-optical wavelength conversion is one of the most important technologies, and it is the basis of other functions.
In past two decades, many schemes have been proposed to demonstrate all-optical signal processing functions, and nonlinearities in passive and active waveguides, such as high nonlinear fiber (Olsson et al., 2001), periodic-poled $LiNbO_3$ (Langrock et al., 2006), silicon-based waveguides (Haché & Bourgeois 2000), chalcogenide-based waveguides (Ta'eed et al., 2006) and semiconductor optical amplifiers (SOAs) (Liu et al., 2006; Stubkjaer 2000) , are elemental mechanisms for these schemes. SOA is one of powerful candidates for all-optical signal processing because of its various nonlinear effects, low power consumption, small footprint and possibility to be integrated, therefore, SOAs have been receiving the most widely attention and have been exploited to realize nearly all functions for all-optical signal processing.
In SOAs, nonlinear effects such as cross-gain modulation (XGM), cross-phase modulation (XPM), four-wave mixing and transient cross-phase modulation can all be exploited to demonstrate all-optical signal processing functions (Durhuus et al., 1996; Stubkjaer 2000). Taking all-optical wavelength conversion as an example, XGM wavelength conversion has some advantages such as simple structure, large dynamic optical power range, high conversion efficiency and large operation wavelength range, but it also has some problems

such as extinction ratio degradation and chirp (Durhuus et al., 1996); XPM wavelength conversion has some characteristics such as good output performance but small dynamic range and difficult to control and fabricate (Durhuus et al., 1996); FWM wavelength conversion (Kelly et al., 1998) is bitrate and format transparent but low conversion efficiency and narrow operation wavelength range; transient XPM conversion is inherent high operation speed but low conversion efficiency.

While used in all-optical signal processing, the input probe signals of SOAs will experience amplitude and phase variations which are induced by carrier density or distribution variations taken by other input pump signals. The optical spectra of the input signals will experience broadening and shifting processes in which the information to be processed is included. Therefore, the SOA can be regarded as spectrum transformer. Combing with appropriate filtering process, all-optical signal processing function can be realized correspondingly. For different filtering processes, we can demonstrate different signal processing functions.

Regarding filtering processes, there are many schemes to realize and demonstrate, such as BPF filters, microring resonators, delay interferometers (fiber-based, silicon waveguide based, $LiNbO_3$ waveguide based, PMF loop mirror, etc.), FP etalons, dispersive fibers, arrayed waveguide grating (AWG) and so on. Usually we should cascade two or more different kinds of filters to get better output results. It is very important to choose and optimize the filtering processes to realize desired functions and improve the output performance.

In this chapter, we theoretical and experimental analyzed all-optical signal processing with SOAs and tunable filters where SOAs were regarded as spectrum transformers and tunable filters were used to realize different filtering processes and then different signal processing functions. In section 2, complicated theoretical model for SOA is presented, and many nonlinear effects are taken into consideration, such as carrier heating, spectral hole burning, etc. On the other hand, a theoretical model for optimizing the filtering process is also presented. These two theoretical models are value for any different signal processing functions. In section 3, experimental research on all-optical wavelength conversion is discussed and analyzed. In section 4, experimental results for all-optical logic operation are presented. Finally, multi-channel all-optical regenerative format conversion is experimental investigated in section 5. Some remarks are also given in final conclusions.

2. Theoretical model

In order to represent the generality for different kinds of signal processing functions, we establish a general theoretical model based on SOA's model and filter's model. As shown in Fig.1, a SOA is cascaded with two basic filters: an optical bandpass filter (OBF) and a delay

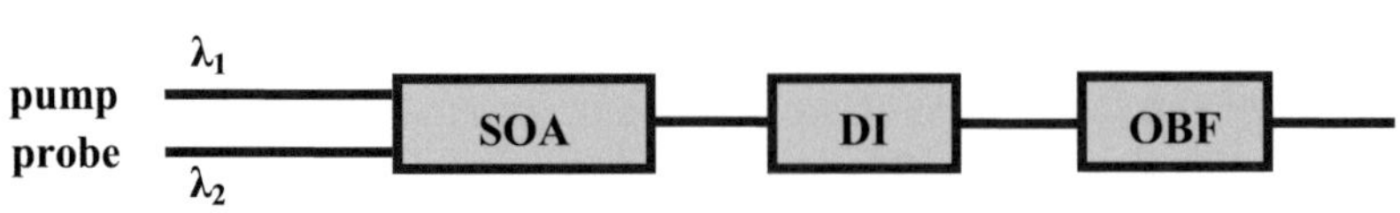

Fig. 1. Schematic diagram for signal processing with SOA and filters

interferometer (DI). These two filters are the most possible to be used to realize signal processing functions. The theoretical model corresponding to Fig. 1 can be exploited to analyze any kinds of signal processing functions. The key point of this model is calculating out the output signal spectrum after the SOA based on a complicated SOA model. Only all kinds of nonlinear effects are taken into account, the accuracy of the output spectrum can be believed. The final output signal spectrum can be analyzed with the help of transmission functions of the cascaded two filters. With iFFT tool, we can get output signal waveform in time domain.

2.1 Theoretical model of SOAs

Based on theoretical models in literatures (Mork & Mark 1995; Mork, et al., 1994; Mork & Mark 1992; Agrawal & Olsson 1989; Mork & Mecozzi 1996), we can derive theoretical model for SOAs in which ultrafast nonlinear effects are taken into account. Firstly, the propagation equation for the input signal in the SOA can be derived as the following equation:

$$\frac{\partial A(z,\tau)}{\partial z}=\left\{\frac{1}{2}\Gamma g(z,\tau)-\frac{1}{2}\Gamma_2\beta_2(1+i\alpha_2)|A(z,\tau)|^2-\frac{1}{2}\Gamma\beta_C n_C(z,\tau)-\frac{1}{2}\Gamma\beta_v n_v(z,\tau)-\frac{1}{2}\alpha_{\text{int}}\right.$$
$$\left.+\frac{i}{2}[\alpha g_N(z,\tau)+\alpha_{CH}\Delta g_{CH}(z,\tau)+\alpha_{SHB}\Delta g_{SHB}(z,\tau)]\right\}A(z,\tau) \quad (1)$$

In Eq.(1), the first to fifth terms on the right hand side represent the linear gain, two-photon absorption (TPA), FCA in conduction band, FCA in valence band and linear absorption loss respectively. The last three terms represent phase modulation process accompanied with linear gain variation, carrier heating and spectral hole burning, which are corresponding to parameters α, α_{CH} and α_{SHB} respectively.

In order to calculate the gain coefficient, the local carrier densities should be calculated out firstly. The local carrier densities satisfy the following two equations (Mork, et al., 1994):

$$\frac{\partial n_c(z,\tau)}{\partial \tau}=-\frac{n_c(z,\tau)-\bar{n}_c(z,\tau)}{\tau_{1c}}-v_g g(z,\tau)S-n_c(z,\tau)\beta_c v_g S \quad (2)$$

$$\frac{\partial n_v(z,\tau)}{\partial \tau}=-\frac{n_v(z,\tau)-\bar{n}_v(z,\tau)}{\tau_{1v}}-v_g g(z,\tau)S-n_v(z,\tau)\beta_v v_g S \quad (3)$$

The first terms on the right hand sides of Eq. (2)and (3) describe the relaxation process of the electrons and holes to their quasi-equilibrium values $\bar{n}_c(z,\tau)$ and $\bar{n}_v(z,\tau)$, respectively. These relaxation processes are driven by the electron-electron and hole-hole interaction with time constant of τ_{1c}, τ_{1v}. The second terms describe carrier consumption due to stimulated emission, and the last terms corresponding to carrier consumption due to two photon absorption.

In this theoretical model, the gain can be expressed as the following equations:

$$\begin{cases} g(z,\tau)=\dfrac{1}{v_g}a(\omega_0)[n_c(z,\tau)+n_v(z,\tau)-N_0] \\ g_N(z,\tau)=\dfrac{1}{v_g}a(\omega_0)[n_c^*(z,\tau)+n_v^*(z,\tau)-N_0] \\ \Delta g_{CH}(z,\tau)=\dfrac{1}{v_g}a(\omega_0)[\overline{n_c}-n_c^*(z,\tau)+\overline{n_v}-n_v^*(z,\tau)] \\ \Delta g_{SHB}(z,\tau)=\dfrac{1}{v_g}a(\omega_0)[n_c-\overline{n_c}+n_v-\overline{n_v}] \end{cases} \tag{4}$$

where $a(\omega_0)$ is the differential gain coefficient, and N_0 is the transition density of states in optically coupled region. g is total gain dynamics, g_N the gain changes accompanied with carrier density variation due to interband recombination, Δg_{CH} the gain changes due to CH, Δg_{SHB} the gain changes due to SHB.

In order to solve Eqs. (2) ~ (4), $\overline{n_R}(z,\tau)$, $n_R^*(z,\tau)$, $R\in[c,v]$ should be got firstly, and they can be defined as:

$$n_R(z,\tau)=N_0F(E_{fR},T_R(z,\tau),E_R) \tag{5}$$

$$n_R^*(z,\tau)=N_0F(E_{fR},T_L,E_R) \tag{6}$$

where E_{fc} and E_{fv} are the quasi-Fermi level in the conduction band and the valence band, respectively. T_C and T_V are the temperature of the carriers in the conduction band and the valence band. T_L is the lattice temperature. $E_C$and E_V are the corresponding transition energies in the conduction band and the valence band. F is the Fermi-Dirac distribution function shown as follows:

$$F(\mu,T,E)=\frac{1}{1+\exp(\dfrac{E-\mu}{k_bT})} \tag{7}$$

To calculate instantaneous carrier temperature (T_R) and quasi-Fermi level (E_{fR}), we need calculate the total electron-hole pair density N and the energy state densities U. The total electron-hole pair density satisfies the following equation:

$$\frac{\partial N(z,\tau)}{\partial\tau}=\frac{I}{eV}-\frac{N}{\tau_s}-v_g g(z,\tau)S+v_g\beta_2S^2 \tag{8}$$

It should be noted that, $N(z,\tau)$ counts all the electron-hole pairs, including those that are not directly available for the stimulated emission.

The energy state densities satisfy the following two questions:

$$\frac{\partial U_c(z,\tau)}{\partial\tau}=\beta_c\hbar\omega_0 n_cS-E_cv_g g(z,\tau)S+E_{2c}v_g\beta_2S^2-\frac{U_c(z,\tau)-\overline{U}_c(z,\tau)}{\tau_{hc}} \tag{9}$$

$$\frac{\partial U_v(z,\tau)}{\partial\tau}=\beta_v\hbar\omega_0 n_vS-E_vv_g g(z,\tau)S+E_{2v}v_g\beta_2S^2-\frac{U_c(z,\tau)-\overline{U}_c(z,\tau)}{\tau_{hv}} \tag{10}$$

In these equations, the first terms describe the change in energy density due to the stimulated emission. The second terms depict the changes due to FCA and the third terms account for the TPA. The last terms represent the relaxation to equilibrium due to carrier-phonon interactions with time constant of τ_{hc} and τ_{hv}. The equilibrium energy densities are defined as:

$$\overline{U}_c = \frac{1}{V}\sum_k \frac{\hbar k^2}{2m_c^*} F(E_{fc}(z,\tau), T_L(z,\tau), \frac{\hbar k^2}{2m_c^*}) \tag{11}$$

$$\overline{U}_v = \frac{1}{V}\sum_k \frac{\hbar k^2}{2m_v^*} F(E_{fv}(z,\tau), T_L(z,\tau), \frac{\hbar k^2}{2m_v^*}) \tag{12}$$

The total carrier density and total energy density need to be self consistently calculated in each time step. We can calculate the quasi-Fermi level and instantaneous temperature of the electrons in conduction band based on self consistently theory.

$$\begin{cases} N(z,\tau) = \frac{1}{V}\sum_k F(E_{fc}(z,\tau), T_c(z,\tau), \frac{\hbar k^2}{2m_c^*}) \\ U_c(z,\tau) = \frac{1}{V}\sum_k \frac{\hbar k^2}{2m_c^*} F(E_{fc}(z,\tau), T_c(z,\tau), \frac{\hbar k^2}{2m_c^*}) \end{cases} \tag{13}$$

Similarly, we can also obtain the instantaneous Fermi levels and temperatures in the valence band.

$$\begin{cases} N(z,\tau) = \frac{2}{V}\sum_k F(E_{fv}(z,\tau), T_v(z,\tau), \frac{\hbar k^2}{2m_v^*}) \\ U_v(z,\tau) = \frac{2}{V}\sum_k \frac{\hbar k^2}{2m_v^*} F(E_{fv}(z,\tau), T_v(z,\tau), \frac{\hbar k^2}{2m_v^*}) \end{cases} \tag{14}$$

It should be noted that, the factor of 2 on the right hand of Eq.(14) is observed, because we consider two sub-bands in valence band including heavy hole band and light hole band. Using Eqs(1-14),we can numerically simulate the dynamics characterization in SOA active region and the signal propagation.

2.2 Theoretical model for filtering

OBFs and DIs are typical filters for all-optical signal processing, especially in ultrahigh speed operation scheme. The transmission function of the BPF and the DI can be described as the following two expressions.

$$\begin{cases} F_1(\omega) = \frac{1}{2}[\exp(i\phi) + \exp(i2\pi\tau\omega)] \\ F_2(\omega) = \exp[-2\ln 2 \cdot (\frac{\omega - \omega_f}{B_0})^2] \end{cases} \tag{15}$$

where F_1 and F_2 are the transmission function of DI and band-pass filters, respectively. ϕ is the phase difference between two arms of the DI, τ is the time delay of two arms of the DI. ω_f is the central angle frequency of the BPF, B_0 is 3 dB bandwidth of the BPF.
The optical field after SOA can be described as:

$$E_{out}(t) = \sqrt{P_{out}} \exp[i(\omega_0 t + \Phi_{NL}(t))] \tag{16}$$

Based on Fast Fourier Transformer (FFT), the optical spectrum of the output signal after the SOA can be obtained as

$$E_{out}(\omega) = FFT[E_{out}(t)] \tag{17}$$

After optical filtering process, the optical spectrum of the output signals after the two cascaded filters can be described as:

$$E_{opt}(\omega) = E_o(\omega) \bullet F_1(\omega) \bullet F_2(\omega) \tag{18}$$

Then, based on inverse Fast Fourier Transformer (iFFT), the output signal waveform in time domain can be calculated out.

$$P_{opt}(t) = \left| F^{-1}[E_{opt}(\omega)] \right|^2 \tag{19}$$

It should be noted that sometimes we should exploit more filters to optimize the output performance, but, the analytical process is identical, adding the transmission function of the new filter in Eq. 17 can get the correct output results.

2.3 Applications in all-optical signal processing

For some applications, the configuration and mechanism are fixed and known to us, we can analyze the output performance based on above theoretical model. The analytical process based on the above SOA model and filter model can be illustrated as the following flow diagram.
As shown in Fig.2, based on above SOA theoretical model, we can get output signal waveforms in time domain from SOA and phase variation information is also included in the output signal filed. Using FFT tool, we can calculate out the signal spectra. Combing with the filter model iFFT tool, we can simulate out the output signal field. We can optimize the SOA parameters or filter parameters to improve the output performance. This process can be used to optimize the SOA structure and filter shape for special applications.
On the other hand, we can also use the above theoretical model to explore some novel schemes for special signal processing functions. The analytical process can be illustrated as following flow diagram. As shown in Fig.3, for special signal processing functions, input signal and output signal are fixed and known to us, their spectra can be calculated out based on FFT tool, so the transmission functions of the potential schemes can be determined by input spectra and output spectra. Usually, the spectrum transformation process of the SOA is fixed and can be determined by the above SOA model. Using some iteration algorithms, the filtering process and related filters can be optimized.

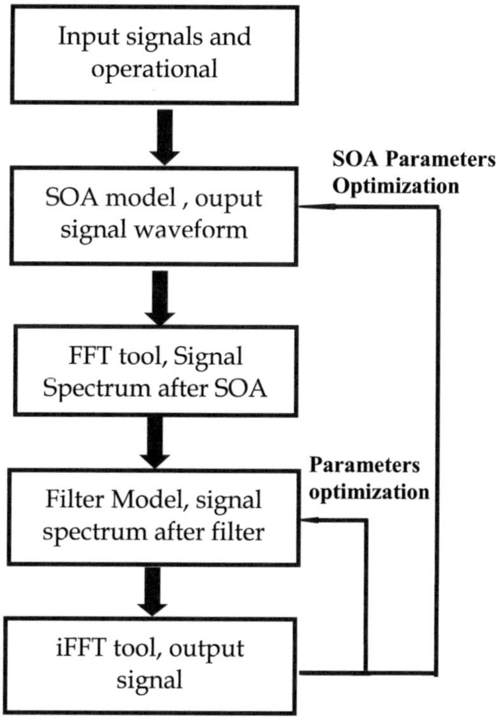

Fig. 2. Analytical process for all-optical signal processing schemes with fixed configurations

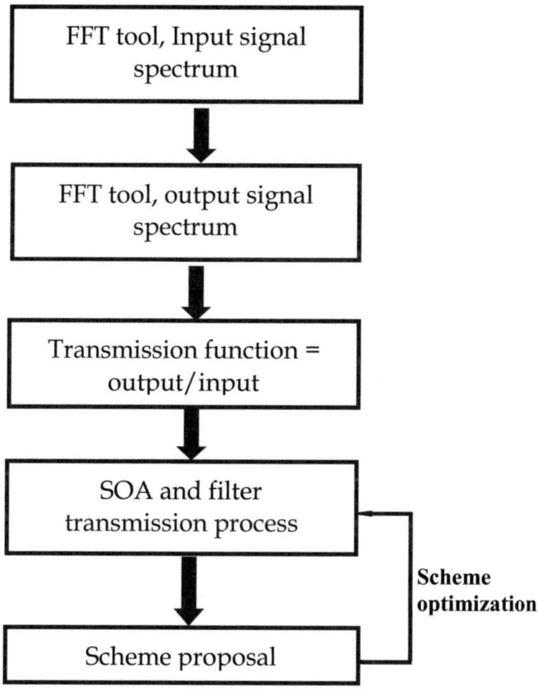

Fig. 3. Analytical process diagram for exploring novel schemes

3. All-optical wavelength conversion with SOAs and filters

All-optical wavelength conversion can be regarded as the most important signal processing function because it is the basis of other signal processing functions. In this section, inverted and non-inverted wavelength conversion at 40Gb/s based on different filter detuning were investigated firstly (Dong et al., 2008), then, experimental results on 80Gb/s wavelength conversion and related filtering optimization process are discussed (Huang et al., 2009).

3.1 Bi-polarity wavelength conversion for RZ format at 40Gb/s

Fig. 4 shows the schematic diagram of both inverted and non-inverted wavelength conversion (Dong et al., 2008). A CW probe signal and a data signal with RZ format are launched into an SOA. The following OBF has some detuning to the probe signal with the central wavelength $\lambda_c + \Delta\lambda_{det}$, where $\Delta\lambda_{det}$ is the detuning value from probe wavelength at λ_c. The input 40Gb/s RZ signal will induce transient nonlinear phase shifts and intensity modulation to the probe signal via cross phase modulation (XPM) and cross gain modulation (XGM) in the SOA. The nonlinear phase shifts will result in a chirped converted signal with the broadened spectrum. The leading edges of the converted probe light are red-shifted, whereas the trailing edges are blue-shifted. Whether the output converted signal is inverted or non-inverted depends on the detuning value.

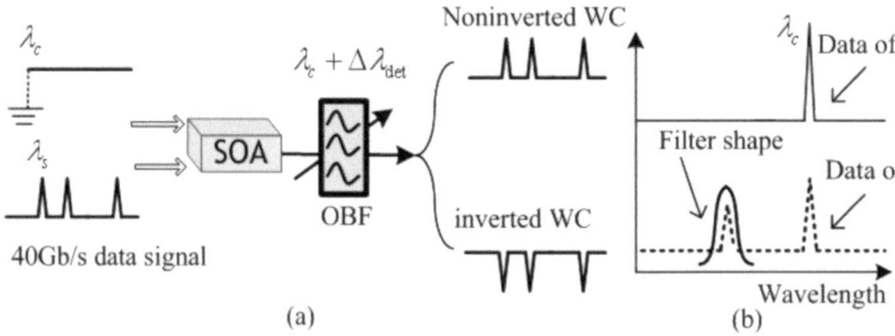

Fig. 4. (a) Operation principle of the bi-polarity wavelength conversion, (b) variation of probe spectrum in the non-inverted wavelength conversion

On the one hand, the wavelength shift of the chirped probe occurs only in the leading/trailing edges of input RZ signals. When the data signal is mark, the probe spectrum will be broadened with sideband energy. If the OBF is detuned far away from the probe wavelength so as to select the sideband energy at $\lambda_c + \Delta\lambda_{det}$, the OBF output will be mark. When the data signal is space, there is no instantaneous frequency shift, and then the OBF output is space, as shown in Fig. 4(b). Therefore, the converted signal will keep in-phase to the input RZ signal. That is non-inverted wavelength conversion.

On the other hand, the XGM will result in the inverted wavelength conversion with relatively slow recovery without the OBF detuning. However, the amplitude recovery can be accelerated and the pattern effects can be eliminated if the OBF is slightly blue shifted. The reason can be explained in Fig. 5. The dotted and dashed lines are the SOA gain and chirp, respectively. When the pulse starts at point A, the SOA carrier depletes and the gain reaches the pit at point B. In time slot from A to B, the probe experiences red chirp and the blue shifted OBF attenuates the probe power. After the pulse duration stops, the gain starts to recover slowly. Assume that the probe signal gets its maximum blue chirp at point C. After point C, the chirp decreases toward zero, then the blue shifted OBF decreases the transmittance. But the gain recovery is going on. Therefore, the blue shifted OBF can balance the power of blue chirped component and the probe power during gain recovery. As a result, the net power at the OBF output is approximately constant (see time slot from C to D). If the SOA and the OBF are treated as a whole system, the amplitude recovery of the system is much faster than the SOA gain. The fast amplitude recovery technique is also suitable for NRZ format. The detail explanation can be found in reference (Liu et al., 2006).

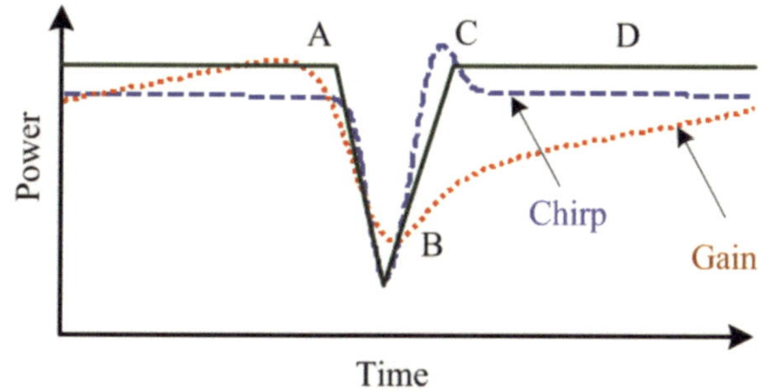

Fig. 5. Principle of accelerating the amplitude recovery.

The experimental setup for bi-polarity wavelength conversion is shown in Fig. 6. Tunable laser diode (LD1) generates a CW probe light at 1557.32nm with the power of 0dBm. Tunable LD2 generates another light source at 1563.5nm, which is modulated by two $LiNbO_3$ Modulators at 40Gb/s to form a 2^{31}-1 RZ pseudo random binary sequence (PRBS) signal, then an erbium-doped fiber amplifier (EDFA) and an attenuator (ATT) are used to fix the RZ output average power at -1.8dBm. The 40Gb/s RZ signal with 8ps-wide pulses is combined with the probe light, and launched into the SOA. The SOA (Kamelian NL-SOA) is biased at 200mA, and its 90%~10% recovery time, defined as the time needed for the gain compression to recover from 90% to 10% of the initial compression, is about 60ps, which is longer than one bit period. The small signal gain@1550nm is 22dB. A tunable OBF1 with bandwidth of 0.32nm follows the SOA. The OBF1 has somewhat detuning to the probe signal to obtain high speed wavelength conversion. Another EDFA and an OBF with 1nm

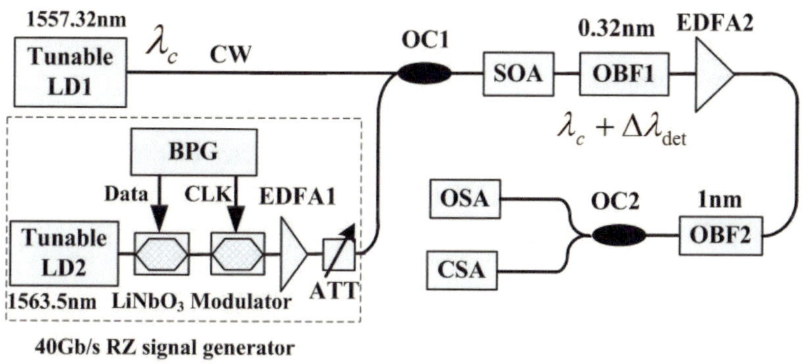

Fig. 6. Experimental setup for bi-polarity wavelength converters at 40Gb/s. BPG: bit pattern generator; ATT: attenuator; OC: optical coupler; OSA: optical spectrum analyzer; CSA: communication signal analyzer.

bandwidth are used to amplify the converted signal power and eliminate the crosstalk. Finally, the optical spectrum analyzer (OSA) and communication signal analyzer (CSA) are used to observe the optical spectrum and waveform of the converted signal.

Fig. 7 shows the experimental results of both inverted and non-inverted wavelength conversion. The left column is the captured waveforms whose time scale is 52ps/div, and the right column is the corresponding eye diagrams whose time scale is 20ps/div. Fig. 7(i) shows the waveform of input 40Gb/s RZ signal. When the OBF1 detuning is -0.3nm (blue shifted) and +0.4nm (red shifted) respectively, the non-inverted wavelength conversion is observed in Fig. 7(ii) and (iv). Good eye diagram is shown in Fig. 7(ii) while some pattern effects occur in Fig. 7(iv). We can see the consecutive marks ①, ②, ③ show a decreasing amplitude. When the OBF1 is slightly blue shifted by 0.1nm, the output waveform becomes inverted and no pattern effects occur, shown in Fig. 7(iii). When the OBF1 has the same central wavelength to the probe carrier, the output waveform has very serious pattern effects, shown in Fig. 7(v). Therefore a slightly blue shifted OBF can accelerate the amplitude recovery in the inverted wavelength conversion.

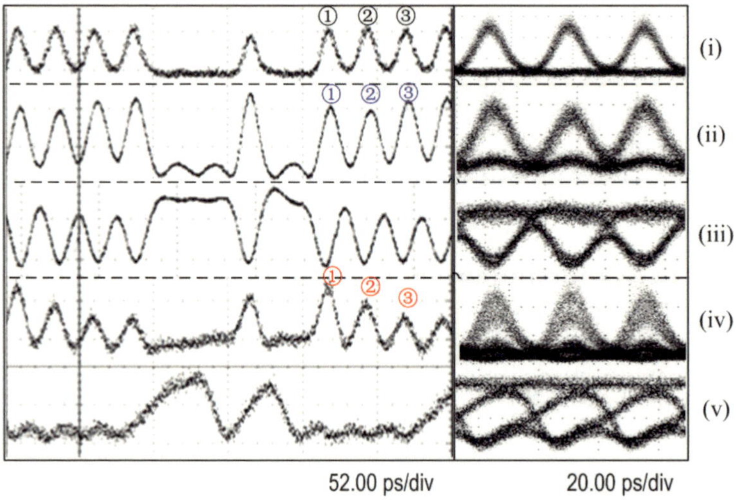

Fig. 7. Waveforms of converted signal with different detuning, (i) the input RZ waveform, (ii)-(v) are the output waveforms of converted signal when the OBF1detuning is -0.3nm, -0.1nm, +0.4nm, and 0nm, respectively. The left column and right column are the captured waveforms and eye diagrams.

The experimental results can be explained from the spectrum. Fig. 8 shows the spectra of converted signal when the OBF1 is detuned. The probe spectra before and after the SOA are shown in Fig. 8(a). At the SOA output, the probe spectrum is broadened asymmetrically due to the XPM. The output spectra of converted signals are shown in Fig. 8(b)-(e) corresponding to the OBF1 detuning -0.3nm, -0.1nm, +0.4nm, and 0nm, respectively. In Fig. 8(b), the blue sideband of converted signal becomes dominant with the assistance of the blue shifted OBF1, therefore good eye diagram could be observed. While in Fig. 8(d), the OBF1 cannot suppress the probe carrier. The crosstalk between red sideband peak and probe carrier will result in the pattern effects in time domain. In Fig. 8(b) and (d), the OBF1 detuning is different for achieving the best non-inverted wavelength conversion because of the asymmetric probe spectrum at the SOA output. Besides, the negative slope of OBF1 is larger than the positive slope, so the blue shifted OBF is easy to suppress the probe carrier, but red shifted OBF is not. In Fig. 8(c), the probe carrier keeps dominant, so the output waveform becomes inverted.

The non-inverted wavelength conversion with blue shifted OBF shows better performance than red shifted OBF. This can be explained with the chirp characteristics. Fig. 9(a) shows the input RZ signal with four consecutive bits "1", and Fig. 9(b) shows the probe phase variation at the SOA output. One can see that the phase increases fast in the leading edge, which corresponds to carrier depletion. However the phase decreases slowly in the trailing edge, which results from the carrier recovery. Fig. 9(c) shows the probe chirp, which is the first order derivative of the phase variation by contrariety. With consecutive "1" pulses injection, the carrier depletion decreases, then the red peak chirp decreases as well. This leads to the decreasing amplitude of the converted pluses by means of the red OBF transfer function (see ①, ②, ③ of the red OBF), and the converted pulses show serious pattern effects. On the other hand, one notices that the blue peak chirp increases very slowly, and remains constant approximately. This results from the similar carrier recovery under consecutive "1" pulses injection. By means of the blue OBF transfer function, the amplitude of converted pulses remains constant (see ①, ②, ③ of the blue OBF). Therefore, the non-inverted wavelength conversion performance is better with blue shifted OBF than with red shifted OBF.

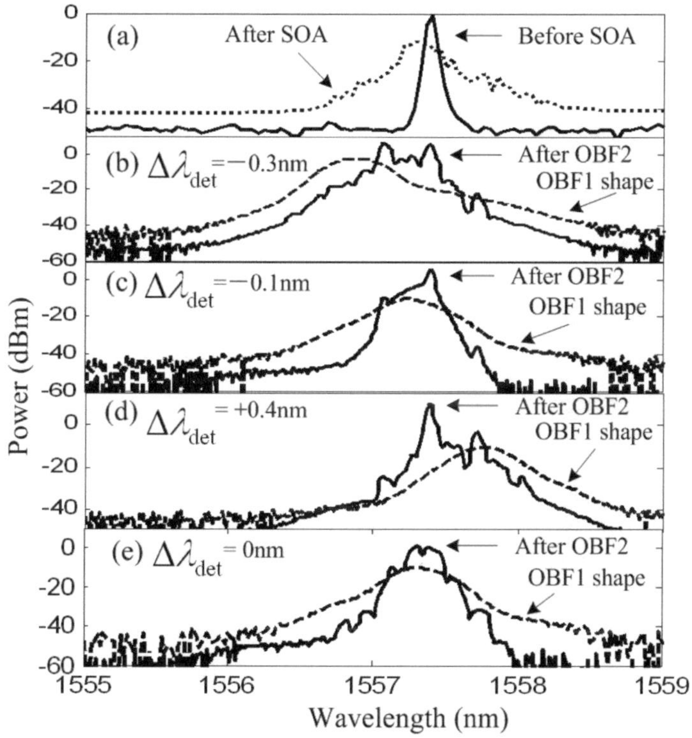

Fig. 8. Spectra of converted signal with different detuning, (a) the probe spectrum before and after SOA, (b)-(e) are the output spectra of converted signal when the OBF1 detuning is -0.3nm, -0.1nm, +0.4nm, and 0nm, respectively.

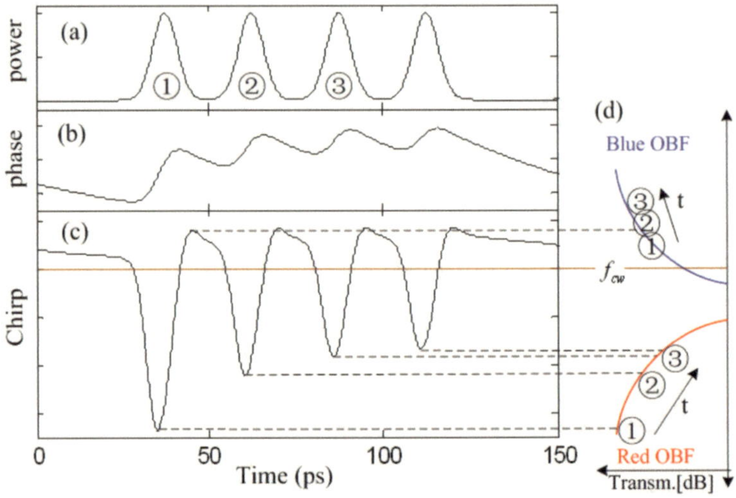

Fig. 9. Comparison of blue shifted OBF and red shifted OBF by frequency-amplitude conversion at the OBF slopes, (a) consecutive "1" pulses, (b) phase evolution, (c) chirp evolution, (d) frequency-amplitude conversion.

The wavelength tunability is further investigated in our experiment. For ease of discussion, we only adjust the wavelength of tunable LD2. We investigate the output extinction ratio (ER) under the optimal OBF1 detuning, as shown in Fig. 10. The output ER fluctuates around 7dB in the whole C-band (1528-1563nm), except the near region of RZ wavelength. The inset of Fig. 10 shows the SOA amplified spontaneous emission (ASE) spectrum, which reveals that the SOA gain is low at the shorter wavelength. Therefore the ER decreases at shorter wavelength. Our experiment scheme cannot complete the wavelength conversion of the same wavelength since the OBF cannot separate the probe and signal channels at the same wavelength.

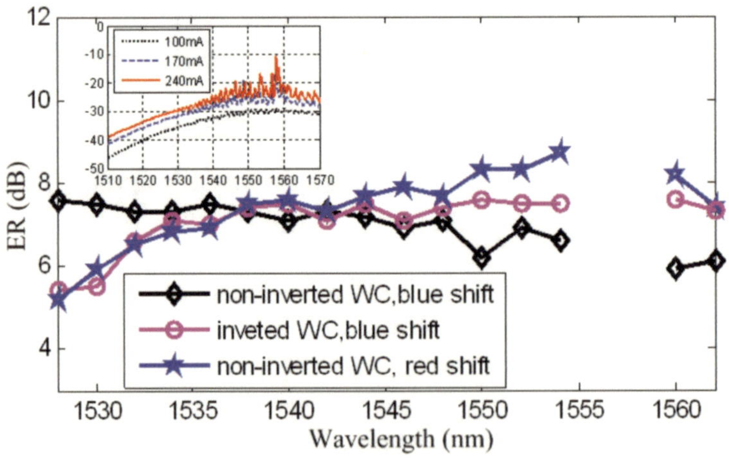

Fig. 10. Output ER as a function of the input signal wavelength when the OBF1 detuning is -0.3nm, -0.1nm, and +0.4nm, respectively. The inset is the SOA ASE spectra at different bias currents.

From Fig. 10, we can see that the output ER is not very high in the three kinds of wavelength converters. The reasons resulting in low ER are quite different between inverted wavelength conversion and non-inverted wavelength conversion. For non-inverted wavelength conversion, the OBF1 does not have a sharp slope, which could not separate the sideband signal from the probe spectrum completely, as shown in Fig. 8(b) and (d). Therefore, the crosstalk between the sideband signal and the probe carrier will degrade the output ER. For inverted wavelength conversion, we need ultrashort pulse injection to enhance the T-XPM effect and to generate large chirp of the probe signal. However, the 8ps-wide input pulses are not narrow enough to obtain inverted wavelength conversion with large ER. We believe the output ER could be improved if the OBF slope is optimized and the input RZ pulses are compressed as narrow as possible.

3.2 80Gb/s wavelength conversion with SOA and cascaded filters

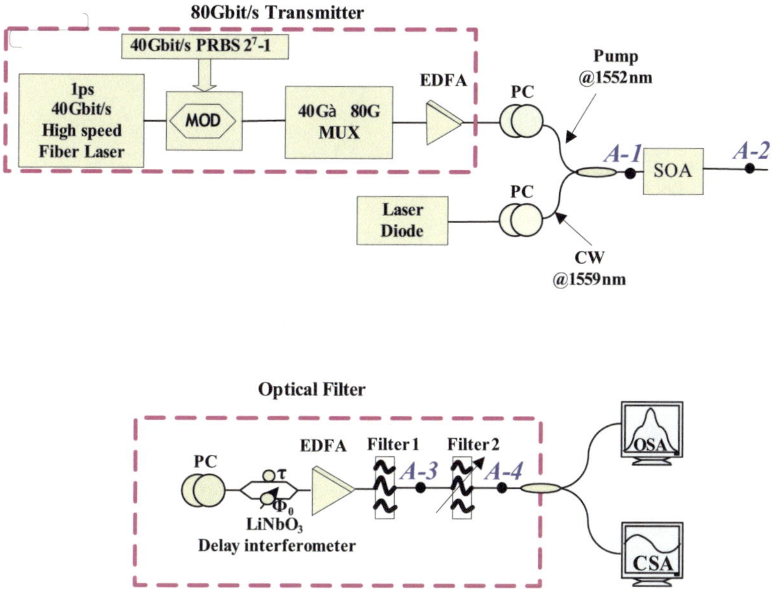

Fig. 11. Experimental setup for 80Gb/s wavelength conversion with SOA and cascaded filters: PC: Polarization controller, EDFA: Erbium-doped fiber amplifier, OSA: optical spectrum analyzer; CSA: communication signal analyzer.

The experimental setup is shown in Fig.11 (Huang et al., 2009). A 40GHZ 1.0-ps wide (FWHM) optical pulse is modulated by an external amplitude modulator (MOD) at 40Gbit/s to generate a 2^7-1 RZ-PRBS signal. This data stream is then optical time multiplexed (MUX) to 80Gbit/s. After amplification, the average optical power of the 80Gbit/s data stream is 4.8mW and the continuous wave (CW) probe signal is 3mW. After the polarization controller, the 80Gbit/s signal is combined with the CW probe and fed into an SOA via 3 dB coupler. As shown in Fig.12, the cascaded filtering model is consisted of a 3.125 ps delay $LiNbO_3$-DI, an optical band-pass filter 1 with bandwidth of 3 nm and the tunable optical band-pass filter 2 with bandwidth of 1 nm which is detuned 1.2 nm to the blue side of the probe carrier wavelength. An inverted 80Gbit/s signal can be obtain at the output of the SOA. The converted signal is subsequently injected into the $LiNbO_3$ DI, where the inverted signal is converted into a non-inverted signal. At the output of the tunable optical band pass filter 2, the non-inverted probe signal is monitored by using an optical sampling scope; the optical spectrum is analyzed by using an optical spectrum analyzer (OSA) with a resolution of 0.050 nm, simultaneously. In our experimental setup, the SOA is biased at 250mA.

It should be noted that, the sampling frequency of the OSA used in our experiment is 40GHz, while the data stream is modulated at 80 Gb/s. Thus the short pulse monitored by the OSA is broadened. However, we are still able to distinguish the eye opening and ER of the output waveform which are shown in Fig.13 (b), (c), (d).

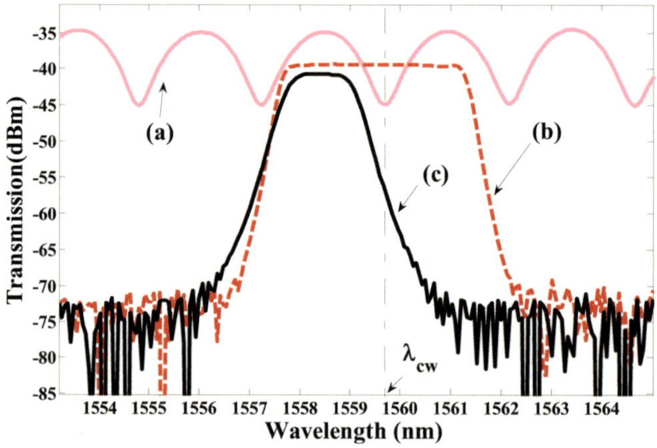

Fig. 12. Transmission spectra the DI (a-1), the band pass filter 1(a-2), the band pass filter2 (a-3).

The operation principle for optical spectrum filtering base on SOA is described as follows. An ultra-fast pulse-stream is combined with a CW probe light and launched into the SOA. The pump signal will induce nonlinear phase shift to the probe signal via T-XPM in the SOA, As a result, the spectrum of the probe signal is broadened. To obtain the high output quality, we optimize the spectrum filter using the $LiNbO_3$ DI and band-pass filters. As seen curve (a) in Fig.12, the "notch" characteristic of $LiNbO_3$ DI is clearly visible. An important feature to be noted is that for the non-inverted output, the wavelength of the notch is set to be the center wavelength of the converted probe signal, ensuring a high attenuation of DC component corresponding to the "1"level in the inverted signal and a larger transmittance of the "0"level.On the other hand, the $LiNbO_3$ DI modifies the spectrum of the output probe of SOA. The central wavelength of the filter 1 is fixed at carrier wavelength of the probe signal .Thus, the pump signal is suppressed and the power ratio of the probe and pump signal is about 30dB (seen in Fig.13 (A-3)).Another low-noise EDFA 2 is applied to amplify the output signal. Then, we use the filter 2 to extract out the component at the central wavelength $\lambda_C+\Delta\lambda$, where λ_C is the central wavelength of the probe signal, $\Delta\lambda$ is the detuning value from λ_C. In this experiment, probe wavelength λ_C is 1559.89nm, and the wavelength detuning $\Delta\lambda$ is -1.2nm.

Fig.13 (a) depicts the optical spectrum measurement at the different position of the experimental setup.Fig.13 (b-d) shows the measured eye diagrams. Fig.13 (b) is the input pump signal at 1541nm, Fig.13(c) shows the eye diagram of the output signal after the OBF1, and Fig.13 (d) depicts output signal after OBF2. They all show good eye-opening performance, the ER of input pump signal is 13.529dB, while the ER of output signal after the BPF 1 is only 3.291dB, and the ER of output signal after the OBF 2 is as high as 20.00dB.

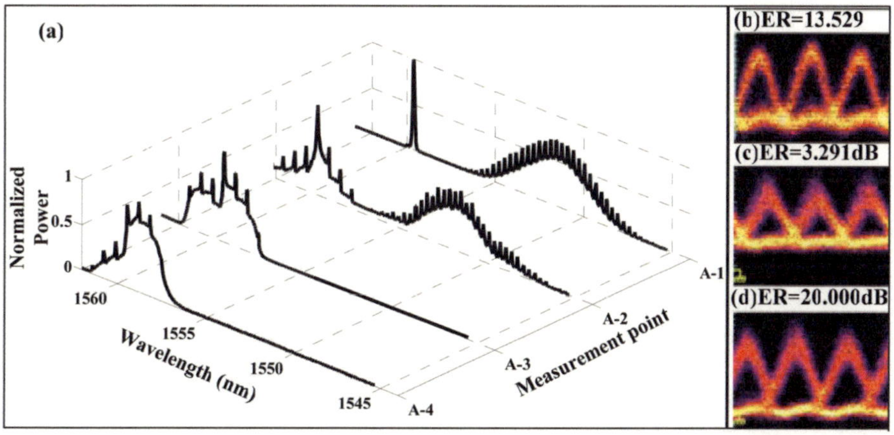

Fig. 13. Experimental results for 80Gb/s wavelength conversion (a) the optical spectrum measurement at different position corresponding to Fig.11; Eye diagram for (b) input pump signal at 1552nm, (c) output probe signal after OBF 1, (d) output probe signal after OBF 2

For wavelength conversion with SOAs, XGM, XPM, FWM and Transient XPM effects can all be exploited. However, for different operation conditions, one main effect dominates over other effects which maybe improve or degrade the output signal performance. Therefore, optimization of SOA parameters, filtering parameters and operational conditions is very important to get better output performance, and this optimization process can be achieved based on theoretical model presented in section 2.

4. All-Optical logic operation with SOAs and filters

In this section, we will focus on experimental study for all-optical logic operation based on SOAs and filters. Three schemes for all-optical logic operation were introduced. Firstly, All-optical logic AND gate at 40Gb/s based on XGM in cascaded SOAs was presented (Xu et al., 2007), and operation condition and output performance were analyzed. Secondly, based on single SOA and different filtering processing, five different logic gates were demonstrated (Dong(b) et al., 2007; Dong et al., 2008; Wang et al., 2007), different nonlinear effects such as XGM, FWM, Transient XPM are exploited in different logic gates respectively. Thirdly, a flexible scheme for all-optical minterms generation was proposed and demonstrated (Xu(a) et al., 2008; Xu(b) et al., 2008). Based on DI and XGM of SOAs, all-optical minterms for two input signals and three input signals were realized respectively.

4.1 All-optical logic AND gate based on cascaded SOAs

It is known that the logic function of inverted wavelength conversion can be written as $\overline{A} \cdot B$ given that data signal A and B are used as pump and probe light respectively. Particularly, it degenerates into a NOT gate when a continuous-wave (CW) serves as the probe light. Therefore, AND gate can be realized by cascading two sets of SOA and filter and configuring the first one as a NOT gate, i.e. $A \cdot B = \overline{\left(\overline{A}\right)} \cdot B$ (Zhang et al., 2004).

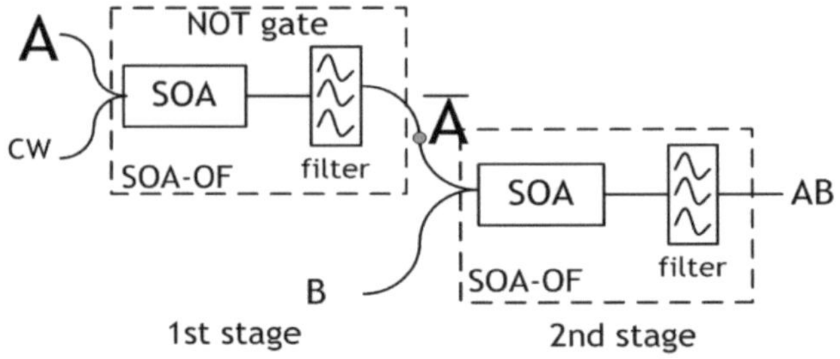

Fig. 14. Schematic diagram for all-optical logic AND gate with cascaded SOAs

As shown in Fig. 14 (Xu et al., 2007), a continuous wave (CW) beam is used as an intermediate wavelength connecting two stages. As probe light at the first stage, it is converted into the negated signal of data A at the output of first stage and serves as pump light at the second stage. Note that the optical filter mentioned above particularly refers to the one who effectively reshapes the spectrum of the modulated probe light. If pump wavelength can not be blocked by such OF, additional optical filter should be used to set the pump and probe wavelength apart.

The experimental setup for the ultrafast AND gate is shown in Fig. 15. In this experiment, three wavelengths generated by LD1, LD2, LD3 are 1560nm(λ_1), 1549.32nm(λ_2) and 1555.75nm(λ_C) respectively. λ_1, λ_2 are modulated by Transmitter simultaneously with 2^7-1 pseudo-random binary sequence (PRBS) RZ data streams at 40Gb/s. The duty cycle of these RZ pulses is 33%. Two wavelengths are separated by a demultiplexer (DMUX) and the optical delay line (ODL) is used to synchronize the input data sequences at the second stage. Thus, two quasi-independent data signals at $\lambda_1$and λ_2 are obtained at the input of SOAs. λ_C is used as intermediate wavelength. The time delay of DI is 25ps which equals to the single bit period of 40Gb/s data rate. The optical BPF following the DI is used to extract the probe light. The filtered probe light is amplified before coupled into the second SOA. The 3dB bandwidth of the Tunable BPF is 0.32nm. The average optical power measured at the input of SOA1 are 7.93dBm(λ_1) and 5.92dBm(λ_2), while 3.10dBm(λ_2) and -17.92dBm (λ_C) at the

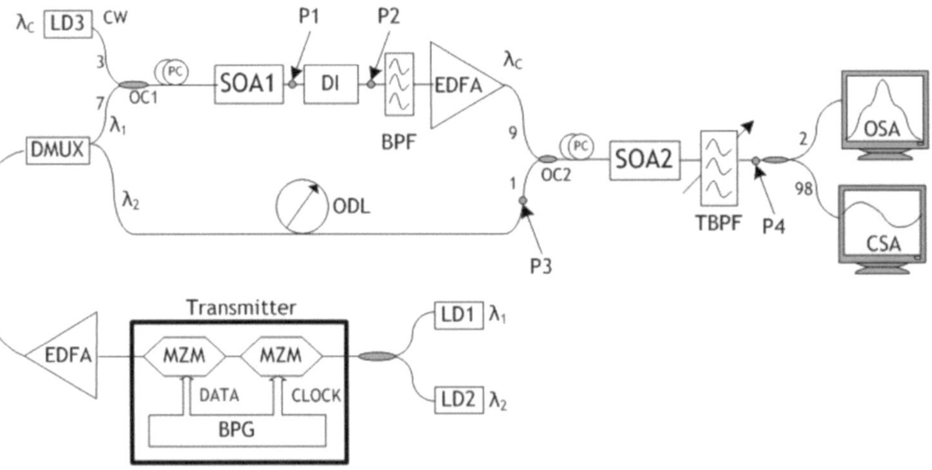

Fig. 15. Experimental setup for all-optical logic AND gate at 40Gb/s with cascaded SOAs

input of SOA2. AND logic results can be achieved through properly tuning the notches of DI and the center wavelength of tunable BPF. In our experiment, the transmission spectrum of the DI can be tuned by adjusting the operational temperature of DI.

Fig. 16 shows the AND logic results (R6) of data signal R7 and R5. R3 is the negated signal of R7, which is NRZ format due to the equivalency between the time delay of DI and the single bit period. The ER of measure AND results is 8.8dB. The bottom trace shows the eye diagram of derived AND results which display open and clear eyes. The QF of the measured eyes is 6.3.

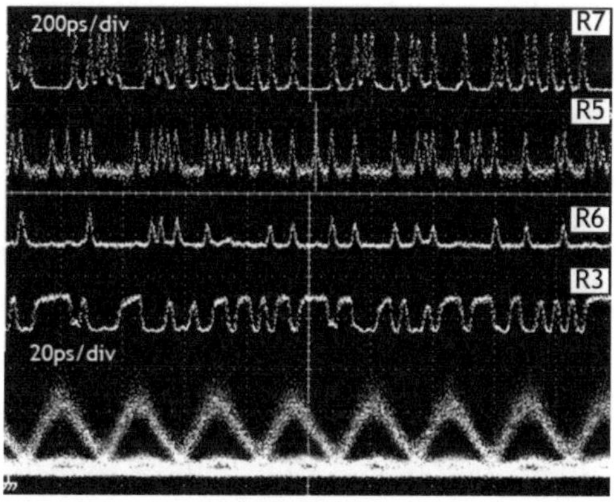

Fig. 16. Output experimental results for all-optical logic AND gate

It should be noted the the SOA1 is a slow recovery bulk material SOA which carrier recovery time is about 500ps. In this SOA, XPM effect is very strong which dominate the output performance. The DI is used to demodulate the phase modulation process, the time delay equals to the bit period, therefore, RZ input signal is wavelength converted to a NRZ signal. The SOA2 is a fast recovery ultrafast SOA which carrier recovery time is about 60ps. The followed the narrow bandpass filter is detuning from the signal B, the detuning process can be optimized to get the best output performance according to the theoretical model in section 2 and analysis of accelarating mechanism in section 3. On the other hand, the most important factor for good AND results is the extinction ratio of the converted signal from stage 1. If we want to improve the output performance or increase the operation speed, the parameters of SOAs and filtering processes should be optimized.

4.2 Configurable all-optical logic gates based on single SOA and tunable filter

In this subsection, we propose and experimentally demonstrate reconfigurable all-optical logic gates based on various nonlinearities in single SOA (Dong(b) et al., 2007; Dong et al., 2008). The operation principle of the configurable logic gates is described in Fig. 17. Data *A* and *B* are the data signals to be processed, whose wavelengths are λ_A and λ_B, respectively. The probe signal is a CW at wavelength λ_C, which will be gain- and phase-modulated by the data signals through the SOA. Thus the output optical spectrum of the probe signal will be broadened. Different logic gates can be realized at different OBF setting.

When both data signals are presented in the SOA, the conjugated light is generated due to FWM effect. The converted signal can be optically filtered out to implement AND logic. When either data *A* or *B*, or both are presented, the probe signal is gain-modulated with polarity-inverted output, which is logic NOR gate. Whereas, the slow gain recovery of SOA

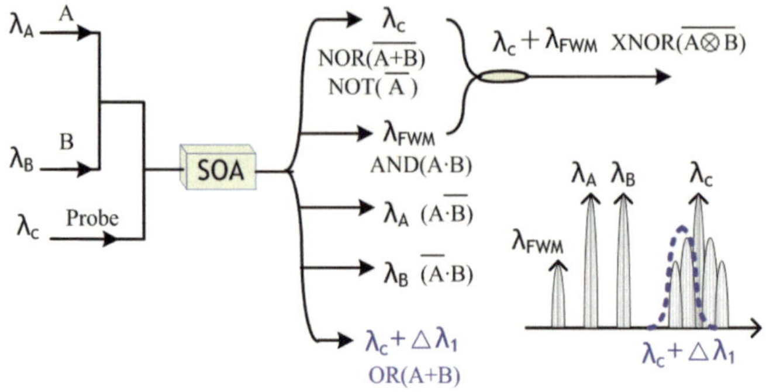

Fig. 17. Illustration of operational principle of the configurable logic gates

degrades the output logic with serious pattern effects. In order to accelerate the SOA gain recovery, the blue shifted OBF with small detuning to the probe carrier is necessary. On the other hand, when the OBF is blue shifted by properly large detuning (i.e., $\lambda_c + \Delta\lambda_1$), the OBF is used to reject the probe carrier and select the blue-shifted spectrum. Either data *A* or *B* or both launched into the SOA will induce blue shifted spectrum, which fits in the OBF passband. If both data signals are absent, the OBF will block the probe carrier. Therefore the output is logic OR gate, which is based on the principle of SOA T-XPM. The XNOR can be obtained by coupling the AND output and NOR output with proper power equalization. The NOR logic gate can be simply changed to NOT logic, merely turning off one data signal.

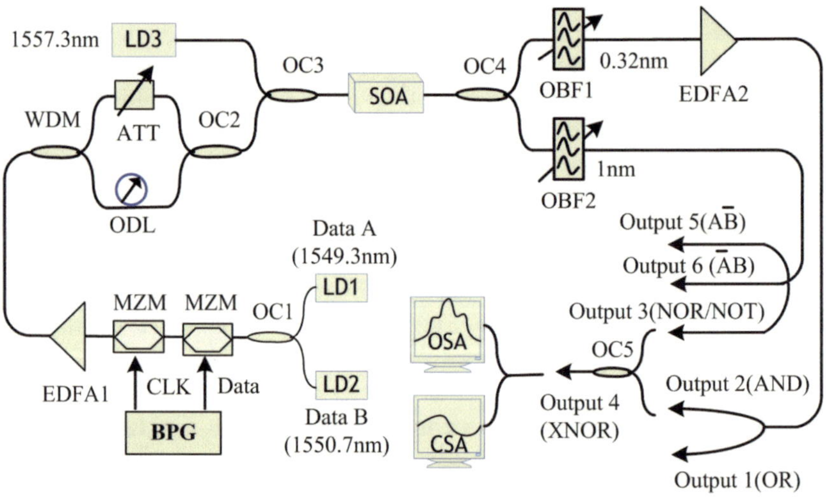

Fig. 18. Experimental setup of the configurable logic gates

The experimental setup for configurable logic gates are described in Fig. 18. The wavelengths of three CW beams generated by LD1, LD2, and LD3 are 1549.3nm (λ_A), 1550.7nm (λ_B), and 1557.3nm (λ_C), respectively. The data signals (λ_A and λ_B) are modulated by two Mach-Zehnder Modulators (MZMs) at 40Gb/s to form 2^{31}-1 return-to-zero (RZ) pseudo random binary sequence (PRBS) signals. The duty cycle of these RZ pulses is 33%. Two data signals will be separated by the wavelength division multiplexer (WDM) and one of them is delayed for several bits by an optical delay line (ODL), therefore, two data signals with different data pattern are obtained. The employed SOA is the same to that of Fig. 6. A tunable narrow OBF1 with 0.32nm bandwidth is used to filter the OR logic and AND logic. Another 1nm-bandwidth tunable OBF2 is used to filter the probe signal with NOR/NOT output, or filter the data *A* with $A\overline{B}$ output, or filter data *B* with $\overline{A}B$ output. Both $A\overline{B}$ and $\overline{A}B$ should be obtained with large power contrast between data *A* and *B*. EDFA2 is used to amplify the AND power, and the coupler (OC5) can combine it with NOR power to realize XNOR logic. Finally, the optical spectrum analyzer (OSA) and communication signal analyzer (CSA) are used to observe the optical spectrum and waveform of the converted signal.

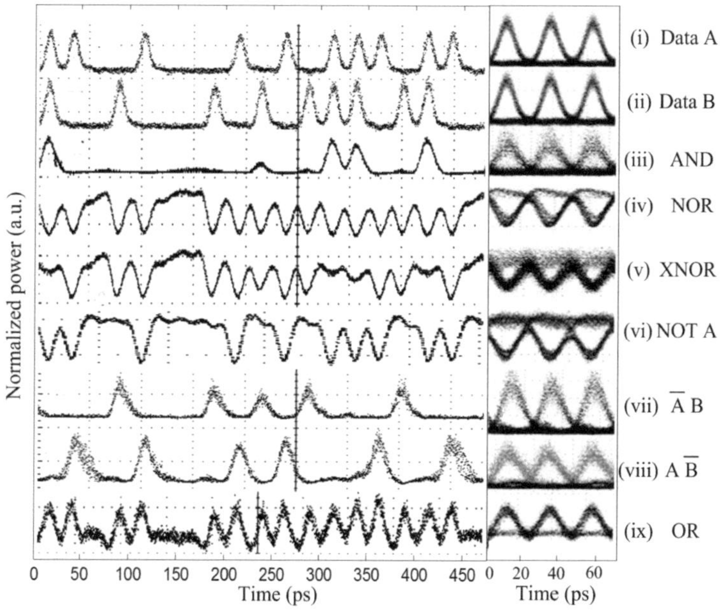

Fig. 19. Output waveforms for different logic gates, (i) and (ii) are input data signals, (iii)-(ix) are logic AND, NOR, XNOR, NOT, $\overline{A}B$, $A\overline{B}$, and OR, respectively.

The input data *A* and *B* before entering the SOA are shown in Fig. 19(i) and (ii), respectively. Both waveforms have a peak power of 2.6mW with extinction ratio (ER) over 13dB. The probe signal has a power of 0.6mW. The conjugated light appears at 1548nm at the SOA output. The conjugated light is filtered out by OBF1 and amplified by EDFA2, then the output signal is the logic AND with good eye pattern, as shown in Fig. 19(iii). The output ER is 8.04. In fact, the input probe signal has additional function to accelerate gain recovery speed of SOA and eliminate pattern dependent distortions. When the central wavelength of OBF2 is blue-shifted by 0.1nm with respect to the probe wavelength, the output signal is NOR logic, as shown in Fig. 19(iv). The ER of NOR logic operation is 10dB. The AND output has a low power level due to low conversion efficiency of FWM, while the NOR output has a high power level. With the assistance of EDFA2, the AND output and NOR output have an equal power level with peak power of 1.7mW, which are combined by optical coupler (OC5), thus the mixed signal is XNOR logic, shown in Fig. 19(v). We can observe much noise appears in level "one", which is caused by different modulation intensity in the NOR and AND outputs. As a result, there is a small eye opening ratio with ER of 6dB. When LD2 is turned off, the NOR gate can be simply changed to the NOT gate, as shown in Fig. 19(vi). Good eye pattern can be observed and the ER reaches 11.5dB.

Based on these five logic gates, all-optical digital encoder and comparator could be demonstrated (Wang et al., 2007). As shown in Fig. 20(a) and Fig. 20(c), digital encoder consists of four logic outputs Y0, Y1, Y2, Y3, which are corresponding to four different input conditions. These four different outputs are achieved by four different logic gates: $\bar{A}\cdot\bar{B}$, $\bar{A}B$, $A\bar{B}$ and AB, respectively. For input signal A and B with bits "00", "01", "10" and "11", output bit "1" appears only at port Y0, Y1, Y2 and Y3, respectively.

For digital comparator, three logic outputs are needed to represent three results after comparing the two digital signals. When A is bit "0" and B is bit "1", only the A<B output port is bit "1", and this operation can be represented by $\bar{A}B$ logic. When A and B are both bit "0" or bit "1", only A=B output port is bit "1", and this operation can be represented by $A\odot B$ or XNOR logic. When A is bit "1" and B is bit "0", only the output A>B port is bit "1", this operation can be represented by $A\bar{B}$ logic. From above discussions, we can find that Y1 output in digital encoder is identical with A<B output in comparator and Y2 output is identical with A>B output. In other words, all-optical digital encoder and comparator can be achieved by five different logic functions: $\bar{A}\cdot\bar{B}$, $\bar{A}B$, $A\bar{B}$, AB and $A\odot B$.

Fig. 20(b) shows the principle diagram of proposed scheme for all-optical digital encoder and comparator. Three SOAs are exploited in this scheme. Signal A and B are input signals with wavelength λ_A and λ_B, respectively. SOA1 is used to achieve $\bar{A}B$ logic function at wavelength λ_B based on XGM effect while the optical power of signal *A* is much larger than signal B. Contrarily, SOA2 is used to achieve $A\bar{B}$ logic function at wavelength λ_A while signal B is much stronger than signal A. Signal A and B are injected into SOA3 together with a continuous wave λ_{cw}. FWM and XGM effects occur simultaneously in SOA3. Based on XGM effect, we can get NOR logic at wavelength λ_{cw}. On the other hand, we can achieve

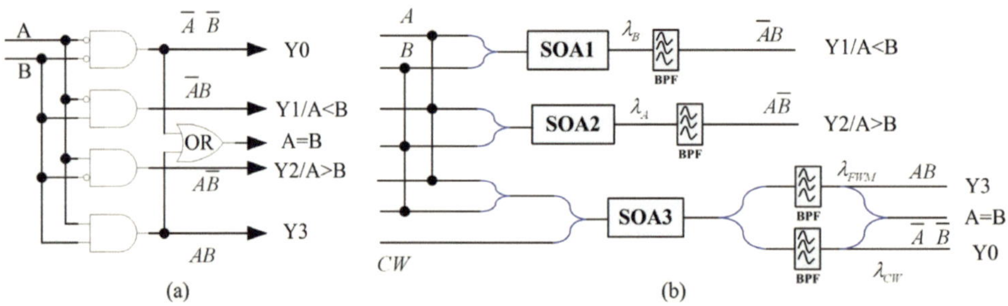

Input data bits		Encoder (four outputs)				Comparator(three outputs)		
A	B	Y0	Y1	Y2	Y3	A<B	A=B	A>B
0	0	1	0	0	0	0	1	0
0	1	0	1	0	0	1	0	0
1	0	0	0	1	0	0	0	1
1	1	0	0	0	1	0	1	0

(c)

Fig. 20. Concept and operation principle of digital encoder and comparator, (a) digital gate-level diagram of encoder/comparator; (b) optical implementation of encoder/comparator; (c) logical truth table for the encoder/comparator logic AND at the new generated channel based on FWM effect while the optical power of two data signals is nearly equal. Based on the output AND and NOR gates, we can get the XNOR gate by coupling the two outputs together with proper power equalization. Therefore, we can obtain five different logic gates based on XGM or FWM effects in three SOAs, which can be exploited to achieve all-optical digital encoder and comparator simultaneously.

4.3 All-optical minterms generation based on delay interferometer and SOAs

In this sub-section, a general scheme for reconfigurable logic gates for multi-input DPSK signals with integration possibility is proposed (Xu(a) et al., 2008; Xu(b) et al., 2008). Benefiting from the optical logic minterms developed by two kinds of optical devices, i.e., optical delay interferometers and SOAs, target logic functions can be realized by combining specific minterms together. The scheme is reconfigured by changing the phase control of the delay interferometers or the input wavelengths.

In our scheme, DIs and SOAs are used to develop NOT gates and NOR gates, respectively. A DI is a Mach-Zehnder interferometer which has a differential delay τ in one arm and a tunable phase controller Φ_0 in the other, as shown in Fig. 21. τ must equal the bit interval of the given bitrate in order to correctly demodulate DPSK signals while Φ_0 must be tunable to ensure accurate demodulation.

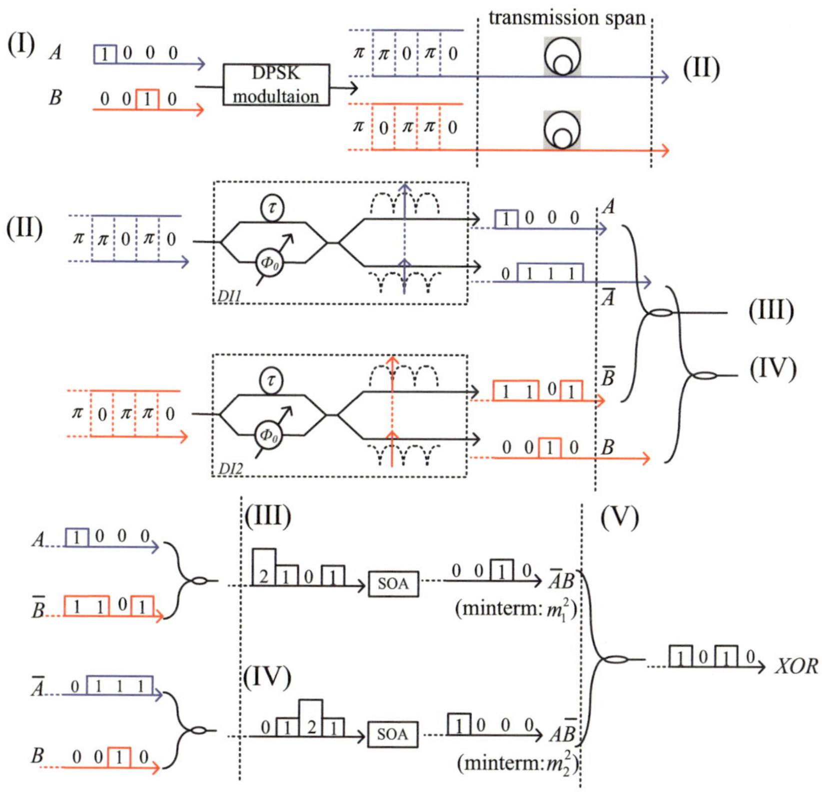

Fig. 21. Logic evolution of two DPSK signals in the generation of optical logic gates from optical minterms based on DIs and SOAs.

In order to explain the operation principle of the scheme, the logic evolutions of DPSK signals through the entire system is briefly described, as shown in Fig. 21. In the first stage (I), two DPSK signals are generated from two absolute binary data A and B respectively. The coding rule is assumed that '1' is encoded as no phase shift between adjacent bits while '0' is encoded as π shift. After transmission, as shown in the second stage (II), DIs are used to demodulate DPSK signals and recover the original binary data (i.e., A or B in this case). Note that either the original data or its inversion can be obtained at a certain output of the DI, depending on whether the interference at that port is constructive or destructive. This can be seen from the frequency domain by checking whether the signal wavelength is located on the transmission peak or notch of the spectrum of the concerned output. If the signal wavelength is located on the transmission notch, the spectrum will features as two main peaks with a noticeable notch at its central wavelength. On the other hand, only one main peaks is observed. Based on the illustrated locations of the signal wavelengths on the transmission spectra of the DI (as shown in Fig. 21), A (orignial data) shows up in the upper output port of DI1 and $\overline{A}$ (inverted data) in the lower output. Oppositely, B is obtained in the lower output of DI2 and $\overline{B}$ in the upper output. In fact, DIs offer a large degree of flexibility of the scheme besides carrying out NOT operation, as will be shown later.

The demodulated signals are combined by optical couplers before launching into the SOAs. It is well known that the cross-gain modulation (XGM) of SOA can be used to carry out NOR operation of nonreturn-to-zero (NRZ) OOK signals. Fig. 22 shows the output probe (λ_2) power of the SOA versus the input pump (λ_1) power. Due to the gain-saturation characteristics of the SOA, the CW probe light will be switched off at the output of the SOA if the input signal power is larger than $P_{in,\,H}$, corresponding to '0' in the output. On the other hand, CW probe light is switched on at the output of the SOA if the input pump power is smaller than $P_{in,\,L}$, corresponding to '1' in the output. For input power between $P_{in,\,L}$ and $P_{in,\,H}$, error logic results will occur. Note that SOA can carry out multi-input NOR operation as well. This is because when one tributary is at ON-state, no matter what states other tributaries are, the total input power during that bit period will exceed $P_{in,\,H}$ and saturate the gain of the SOA to generate '0' at the output. The case that the input DPSK signals are return-to-zero (RZ) format needs to be mentioned. Although the logic integrity is kept, the NOR logic results given by the SOAs will be in dark-RZ pulses due to the characteristics of XGM. To avoid this, other kinds of NOR gates that can process RZ signals can be utilized instead, such as logic gates based on SOAs and optical filtering. In the third stage (III), an SOA can carry out NOR operation of data A and $\overline{B}$, creating logic result $\overline{A}B$. Similarly, the other SOA generates $A\overline{B}$ by executing NOR operation of $\overline{A}$ and B in stage (IV). In stage (V), final logic $\overline{A}B + A\overline{B}$ is derived by combining the output of stage (III) and (IV) through an optical coupler which functions as an OR gate due to the fact that the probability of concurrence of '1' in different minterms is zero. Therefore, an exclusive-OR (XOR) logic result has been derived. If we change the connection of the optical couplers before stage (III) and (IV) so that the low output port of DI1 is connected to the upper output port of DI2, an XNOR logic ($\overline{A}\overline{B} + AB$) can be obtained. However, the same result can be achieved without changing any physical connections. This is because the DIs can provide a way to exchange the output signals between its two output ports. That is, we can adjust the location of the signal wavelength on the transmission spectra of the DIs to exchange the interference conditions of their two output ports. This can be achieved by tuning Φ_0 of the DIs or adjusting the wavelengths of the input signals. Note that unlike doing proof-of-concept experiments as what we have done, it is difficult to change the input signals in practical situations. In that case, tuning Φ_0 is the only choice.

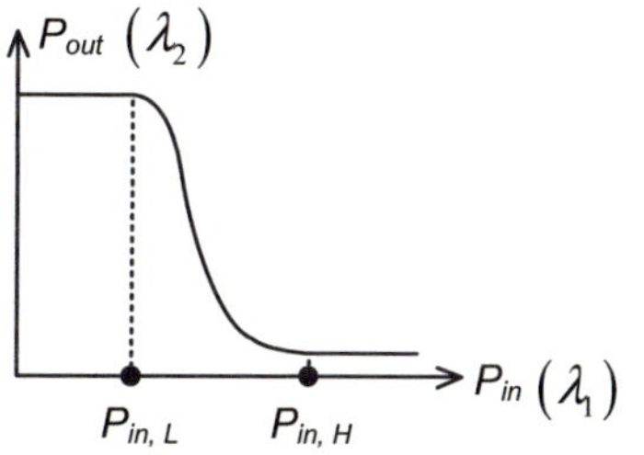

Fig. 22. Output probe power of the SOA versus the input pump power.

Simplified setups are adopted in the experimental trials. That is, a single DI can perform NOT operation for several DPSK signals simultaneously if the input wavelengths can be adjusted. Fig. 23 shows the experimental setup for realizing two-input minterms. Fig. 24 shows the setup for realizing three-input minterms. To facilitate description, important measuring points, i.e., D_{o1}, D_{o2}, S_{i1}, S_{i2}, S_{o1}, S_{o2} and S_{o3}, are marked on Fig. 23 and Fig. 24.

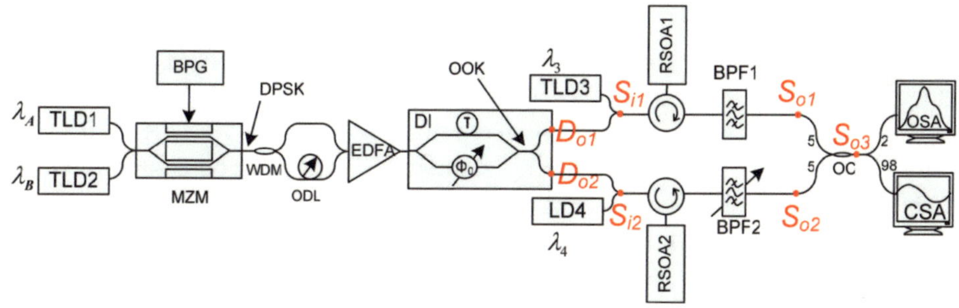

Fig. 23. Experimental setup for two-input NRZ-DPSK logic minterms or logic gates

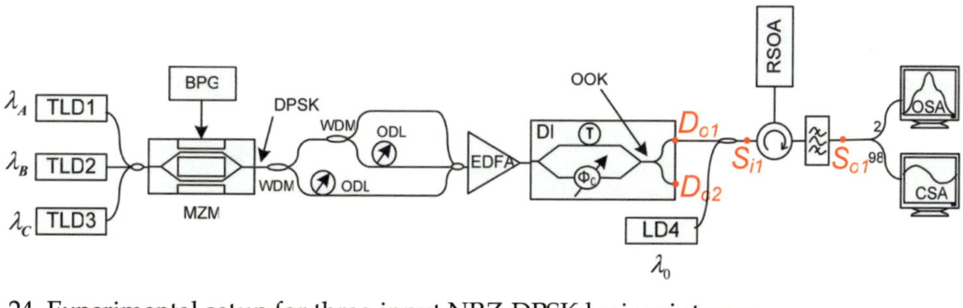

Fig. 24. Experimental setup for three-input NRZ-DPSK logic minterms.

Two-Input Minterms

Due to energy conservation, m_1^2 and m_2^2 appear simultaneously and so does m_0^2 and m_3^2. The signal spectra STO2 shown in Fig. 25(a) are measured at S_{i1} when two-input minterms m_2^2 are derived at S_{o1}. Simultaneously, m_1^2 are obtained at S_{o2} and the spectrum measured at S_{i2} are shown by STO1. In this case, signal at λ_A and λ_B are destructively and constructively demodulated at S_{i1}, respectively. Using the same setup but shifting λ_A downwards by 0.4nm, both λ_A and λ_B are constructively demodulated at S_{i1}. m_0^2 and m_3^2 are derived at the same time and the spectrum measured at S_{i1} and S_{i2} are shown by STO3 and STO0 in Fig. 25(b).

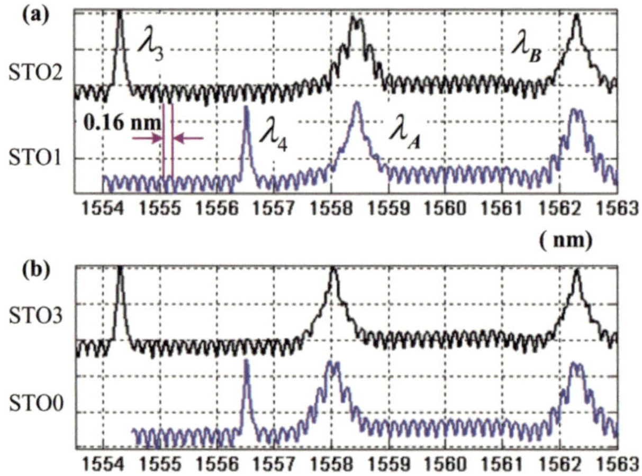

Fig. 25. STO0~STO3: signal spectra measured at S_{i1} or S_{i2} in Fig. 23 when two-input minterms ($m_0^2 \sim m_3^2$) are derived at S_{o1} or S_{o2}.

Fig. 26 shows the measured trace of $m_0^2 \sim m_3^2$ (MTO0 ~MTO3), observed at S_{o1} (MTO1 and MTO2) and S_{o2} (MTO0 and MTO3), respectively. Also shown in Fig. 26 are combined traces CTO0~CTO3 measured at D_{o1} or D_{o2}, corresponding to $\sum(\overline{A},\overline{B})$, $\sum(\overline{A},B)$, $\sum(A,\overline{B})$, and $\sum(A,B)$, respectively. The symbol $\sum$ stands for power superposition. Note that m_0^2 equals NOR logic and m_3^2 equals AND logic. Since m_0^2 and m_3^2 appear at the same time, a simultaneous AND and NOR logic operation can be obtained.

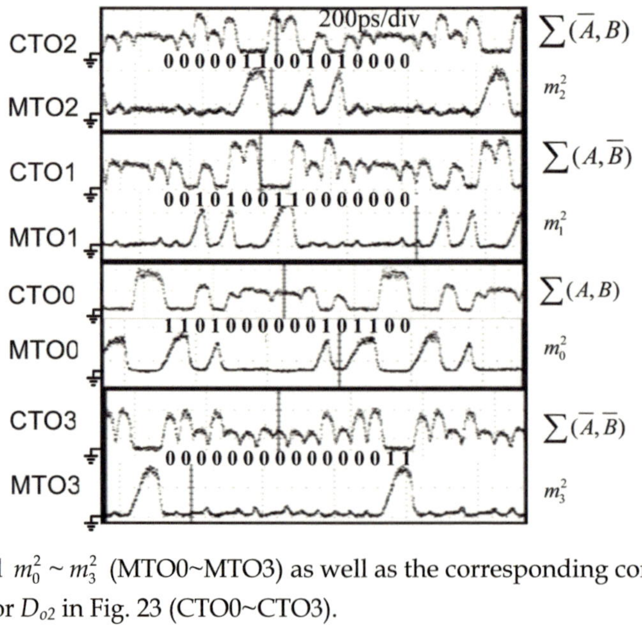

Fig. 26. Measured $m_0^2 \sim m_3^2$ (MTO0~MTO3) as well as the corresponding combined signal measured at D_{o1} or D_{o2} in Fig. 23 (CTO0~CTO3).

Three-Input Minterns

Using the set up shown in Fig.24, eight minterms for three-input DPSK signals are obtained. The temporal waveforms of $m_0^3 \sim m_7^3$ are shown by MTE0~MTE7 in Fig. 26. The temporal waveforms of combined signals measured at D_{o1} when m_7^3 are derived at S_{o1} are shown by CTE7 in Fig. 26 as well. As shown by STE7, $\sum\left(\overline{A},\overline{B},\overline{C}\right)$ are derived at the input of the RSOA, resulting in m_7^3 after the NOR operation which can be verified by comparing CTE7 and MTE7 in Fig. 26. The full logic integrity of the scheme can be verified by calculating the logic results manually.

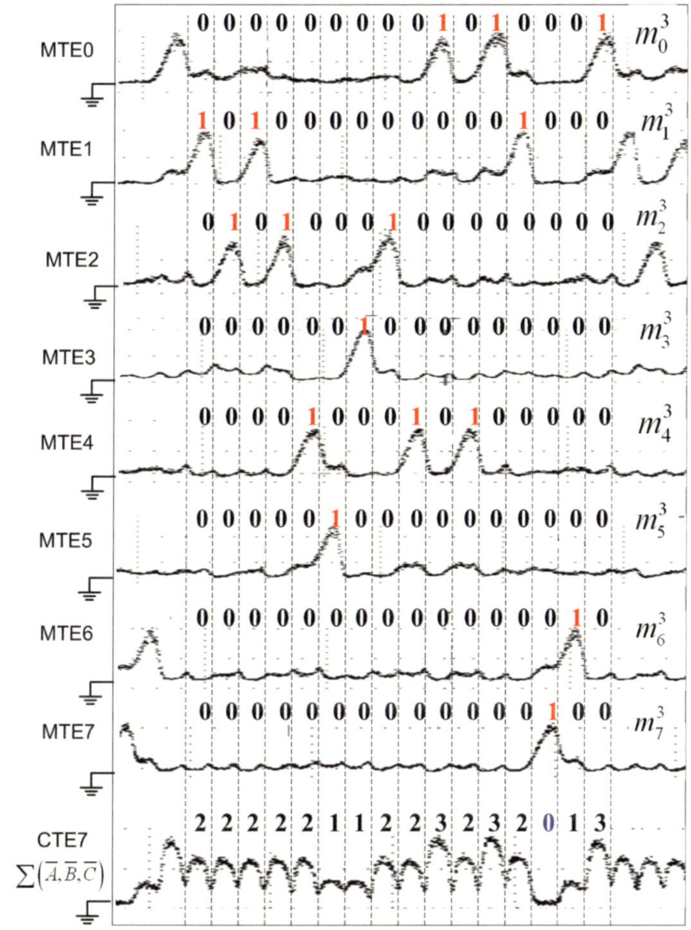

Fig. 26. MTE0~MTE7: measured temporal waveforms of $m_{0\sim7}^3$

In this section, different logic operation functions were demonstrated. Especially, all-optical minterms generation is very important for arbitrary logic functions. On the other hand, these schemes all have potential to be integrated based on SOA and filters. Further investigation should be carried out in order to improve the output performance, increase the operation speed, increase the optical power dynamic range and enhance the stability.

5. Multi-channel regenerative pattern conversion based on SOA and DI

As we know, based on SOA and filter, NRZ-to-RZ conversion for signal channel can be realized while the filter is a little detuning from the central wavelength of the probe NRZ signal (Dong(a) et al., 2007). As shown in Fig.27(a), if multi input channels aim to the transmission spectrum of the DI and are all detuning from the transmission peaks of the comb-like filter (DI) simultaneously, multi-channel NRZ-to-RZ conversion can be realized. As shown in Fig.27(b), the SOA, acting as nonlinear element, causes the broadened spectrum of input NRZ signal due to the XPM effect, and the DI is used to extract the specific spectra from the broadened spectra, for all the 16 channels at the same time. An additional tunable filter with 0.3nm bandwidth is used to filter out one of the converted channels for evaluation (Yu et al., 2008). Simply speaking, the NRZ signals will generate a transient frequency shifts within the control signal duration, and the filter with proper detuning will only transmit frequency shifted components caused by the XPM. For multi-channel operation, in order to minimize the XGM induced inter-channel crosstalk, the SOA should be deeply saturated by the control signal (clock signal).

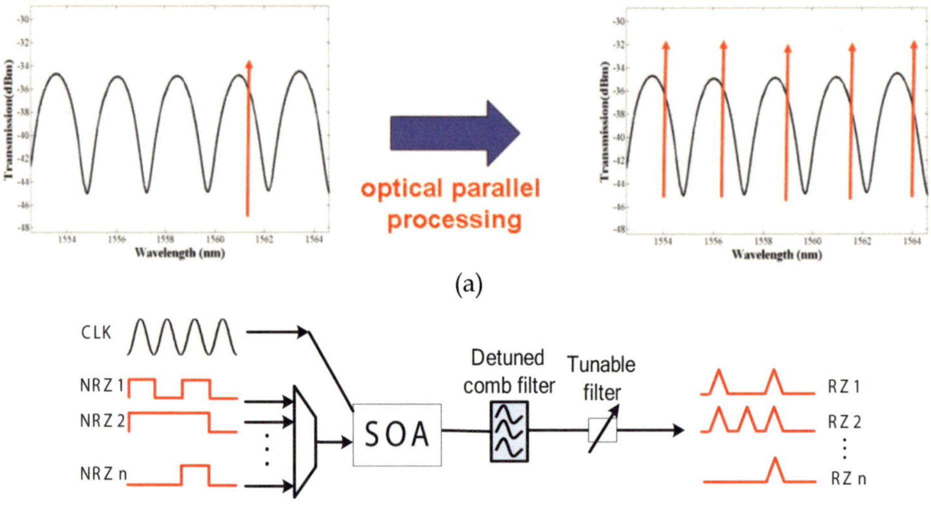

Fig. 27. Operation principle of the multi-channel format conversions: (a) parallel processing principle based on comb filter; (b) schematic diagram for multi-channel pattern conversion scheme.

The experimental setup is shown in Fig. 28 (Yu et al., 2008). 16 channel CW lights from tunable lasers (wavelength from 1547.79 to 1559.79nm with spacing of 0.8nm) are coupled into two MZMs with two AWGs. The odd channels are driven by the RF data (PRBS 2^{31}-1) from Anritsu 10G pattern generator; while the even channels are driven by $\overline{\text{data}}$. An optical delay line is added for odd channels to ensure the decorrelation for two data streams. Another CW light (1546.99 nm) is fed into a third MZM, which is driven by the 10GHz RF sine clock signal, to obtain an optical clock signal as the control signal. The delay of the RF clock can be adjusted to synchronize the NRZ signals. Then, the signals are lunched into SOA via a WDM coupler. The average powers of the NRZ signals and the clock signal are about 3 and 13 dBm before the SOA. The SOA is Kamelian pre-amplifier bias at 160mA. A fiber based DI with free spectral range (FSR) of 0.8nm is used to extract part of each of the broadened DWDM channel spectra. By controlling the operating temperature, its transmission peaks are adjusted to be offset from each carrier wavelength with optimal detuning. Thus, the chirp induced on each channel is passed, while the original spectral components are suppressed to some extent, and hence format conversions from NRZ to RZ can be achieved. A subsequent tunable filter with a 0.3nm 3dB bandwidth is used to filter out one of the converted channels for evaluation.

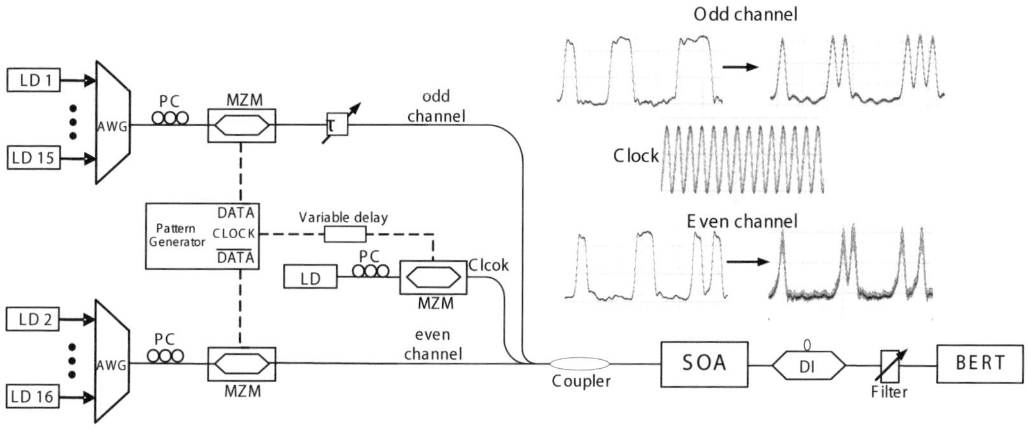

Fig. 28. Experimental setup for 16 channels pattern conversion with SOA and DI

For practical all-optical format conversion, the clock can be recovered from one of the NRZ signals as we have demonstrated in reference (Yu et al., 2006). Since clock extraction is not the purpose of this experiment, for simplicity we obtained the optical clock by modulating a MZM with the RF clock signal directly from the pattern generator.

Since the SOA is deeply saturated by the clock signal, input NRZ signals are only modulated by the XPM, almost without amplitude modulation. Thus, the 16 channels can be with different patterns, which has confirmed by simulation. However, for experimental convenience, only two patterns are used here.

The insets in Fig. 28 show the measured bit stream, both odd and even channels with optimal filter detuning. Results show that the proposed converter works well for different bit pattern simultaneously. The measured spectra before the SOA, after the SOA, after the DI and one of the converted spectra (channel 4) are shown in Fig. 29. It is obvious that all the NRZ spectra are broadened due to the modulation of the clock signal after the SOA, regardless of the spacing between the clock signal and each NRZ signal. The DI filters out the specific part of the spectra periodically.

Taking one of the 16 channels (channel 4) for example; Fig. 30 shows the evolution of the obtained duty cycle as the increase of the DI and the following filter detuning. It can be seen clearly that the pulse width of the converted RZ without detuning is 78 ps, which is wider than the pulse width of the pump clock signal. In other word, the modulation induced by the clock signal is very slight. The pulse width can be significantly compressed to 46 ps with filter detuning of 0.25nm. Although larger detuning can further compress the pulse width, the cross talk from adjacent channel becomes serious.

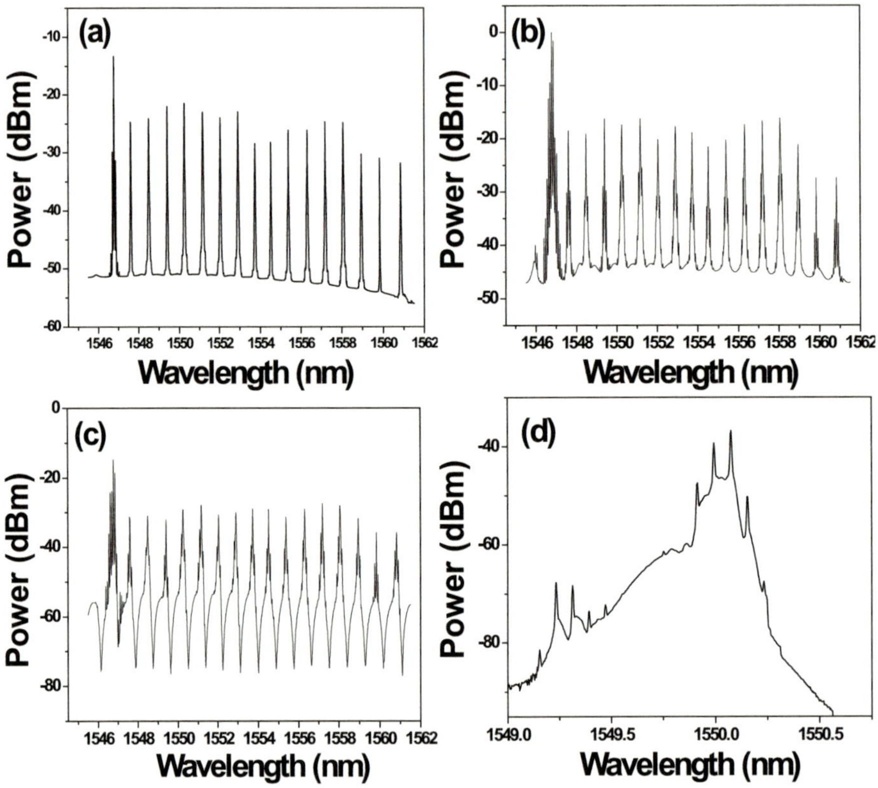

Fig. 29. Spectra of 16 DWDM channels (a) before SOA (b) after SOA (c) after DI and (d) one of the converted channels.

Fig. 31 plots the BER measurements for the format conversions, which are taken for channel 1 (odd channel) and 16 (even channel) and both for three different duty cycles. We can see that negative power penalty can be achieved, for obtained RZ signals with duty cycles of 46, 58 and 69%, respectively. The eye diagram of one channel NRZ signal at back to back is also shown in Fig. 31.

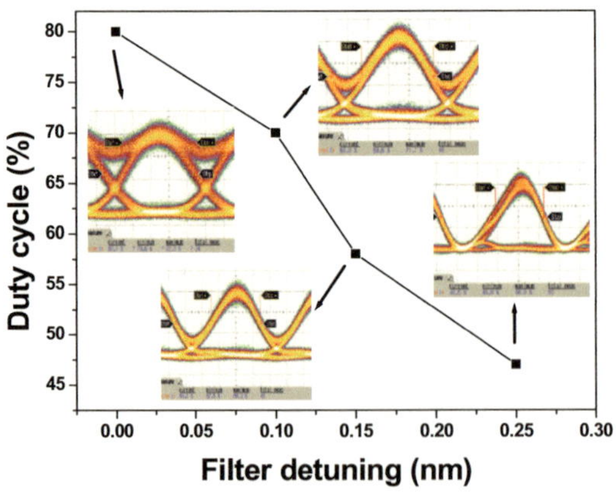

Fig. 30. Duty cycle evolution versus filter detuning

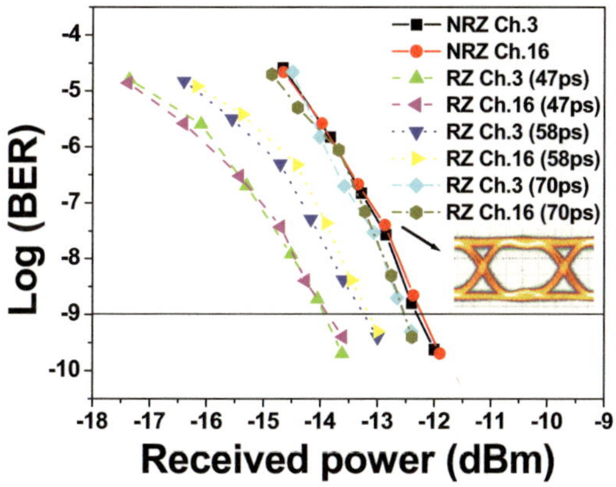

Fig. 31. BER measurements for the conversions

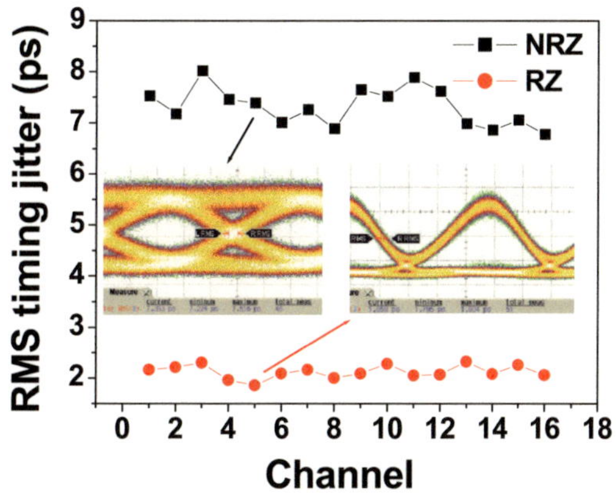

Fig. 32. Measured timing jitter reduction before and after the conversions

Since the clock signal is introduced in the scheme, our proposed converter is regenerative with suppression of timing jitter. To further investigate the retiming performance, the input 16 channel NRZ signals are distorted by attenuating their power and passing them through an EDFA. Fig. 32 presents the RMS timing jitter measurements, more than 3 times jitter reduction can be found for all channels.

6. Conclusions

All-optical signal processing functions are key enabling technologies for signal regeneration and switching in next generation optical networks. SOA has been receiving much attention in all-optical signal processing for its various nonlinear effects and potential to be integrated, and it has been exploited to realize all-kinds of signal processing functions such as all-optical wavelength conversion, all-optical logic operation, all-optical 3R regeneration, all-optical pattern conversion, all-optical sampling, all-optical demultiplexing, etc.

In this chapter, all-optical signal processing based on SOAs and filters was experimental and theoretical investigated. Complicated theoretical model for SOAs is presented, in which besides those conventional effects such as XGM and XPM related to interband recombination process, those ultrafast nonlinear effects such as carrier heating, spectral hole burning, two-photon absorption are also taken into account. Based on FFT and iFFT tool, filtering process based on different filters can be theoretical analyzed. For those applications with fixed schemes and mechanism, the theoretical model can be used to analyze the output signal performance and optimize the related parameters and operational conditions. Simultaneously, this model can also be used to explore some novel schemes for special signal processing functions. All-optical wavelength conversion for 40Gb/s conventional RZ signals and 80Gb/s RZ signal with ultra-short pulse were experimental realized with SOA and filters. For different filtering process, we can get different wavelength conversion output results, and filtering optimization is very important for improving the output performance. Three schemes for all-optical logic operation were analyzed. Based on cascaded SOAs and related filters, 40Gb/s logic AND gate was demonstrated, in which a slow recovery SOA was exploited in the first stage. Based on various nonlinear effects and different filtering process, configurable five logic gates were achieved. This scheme demonstrated powerful function of SOA with the assistance of filters. Based on demodulation process of DIs and XGM in SOAs, all-optical minterms for two input signals and three input signals were generated. It is possible to demonstrate arbitrary logic functions based on these minterms. Finally, with SOA and comb filter DI, 16 channels pattern conversion at 10Gb/s were effectively demonstrated. Multi-channel signal processing functions are very important for signal regeneration in DWDM optical networks. In further and future investigation, there exist some key points to be noted. First is SOA device optimization, we should find effective and feasible way to accelerate the carrier recovery and enhance the ultrafast nonlinear effects which are the basis for high speed operation, decreasing the needed input power and increasing the flexibility of the scheme. Second is device integration, especially for the scheme based on SOA and DI. InP-based DI is possible to be fabricated, so monolithically integrated SOA and DI structure can be fabricated out, which has potential to be used in multi-channel signal regeneration and all-optical minterms generation. On the other hand, we should pay much attention to all-optical signal processing functions for novel modulation formats, such as DPSK or DQPSK signals. Multi-channel signal processing functions should also receive much more attention because multi-channel optical signal are transmitted in DWDM optical networks.

7. Acknowledgements

Related researches on all-optical signal processing based on SOAs and filters were supported by the National High Technology Developing Program of China (Grant No. 2006AA03Z414) and the National Natural Science Foundation of China (Grant No. 60407001 and 6087705), the Science Fund for Distinguished Young Scholars of Hubei Province (Grant No. 2006ABB017) and the Program for New Century Excellent Talents in Ministry of Education of China (Grant No. NCET-04-0715). I would also express great thanks to many graduates for their fruitful works, many professors for their helpful discussions, Vera for her encouragement and my family for their support and love.

8. References

Agrawal, G. & Olsson, N. (1989). Self-Phase Modulation and Spectral Broadening of Optical Pulses in Semiconductor Laser Amplifiers. *IEEE Journal of Quantum Electronics*, vol. 25, No.11, pp 2297-2306, ISSN 0018-9197.

Danielsen, S.; Hansen, P. & Stubkjaer, K. (1998). Wavelength conversion in optical packet switching. *IEEE Journal of Lightwave Technology*, Vol. 16, No.12, pp2095~2108, ISSN 0733-8724.

Dong(a), J.; Zhang, X.; Xu, J.; Huang, D.; Fu, S. & Shum, P. (2007). 40 Gb/s all-optical NRZ to RZ format conversion using single SOA assisted by optical bandpass filter. *Optics Express*, Vol. 15, No.6, pp2907-2914, ISSN 1094-4087.

Dong(b), J.; Zhang, X.; Wang, Y.; Xu, J. & Huang, D. (2007). 40Gb/s reconfigurable photonic logic gates based on various nonlinearities in single SOA. *Electronics Letters*, Vol.43, No.16, pp884-886, ISSN 0013-5194.

Dong, J.; Zhang, X. ; Fu, S.; Xu, J.; Shum, P. & Huang, D. (2008). Ultrafast all-optical signal processing based on single semiconductor optical amplifier and optical filtering. *IEEE Journal of Selected Topics in Quantum Electronics*, Vol.14, No.3, pp770-778, ISSN 1077-260X.

Durhuus, T.; Mikkelsen, B.; Joergensen, C.; Lykke D. & Stubkjaer, K. (1996). All-optical wavelength conversion by semiconductor optical amplifiers. *IEEE Journal of Lightwave Technology*, Vol.14, No.6, pp942-954, ISSN 0733-8724.

Haché, A. & Bourgeois M. (2000). Ultrafast all-optical switching in a silicon-based photonic crystal. *Applied Physics Letters* Vol.77, No.25, pp4089-4090, ISSN 0003-6951.

Huang, X.; Zhang, Y.; Zhang, Y.; Huang, D. & Zhang, X. (2009). Extinction ratio enhanced 80-Gbit/s wavelength conversion based on optimization of spectrum filtering. *ECOC'2009*, Hongkong, Jul. 2009, FE4.

Kelly, A.; Ellis, A. & Nesset, D. (1998). 100Gbit/s wavelength conversion using FWM in an MQW semiconductor optical amplifier. *Electronics Letters*. Vol. 34, No. 20, pp1955-1956, ISSN 0013-5194.

Langrock, C.; Kumar, S.; McGeehan, J.; Willner, A. & Fejer, M. (2006). All-optical signal processing using $\chi^{(2)}$ nonlinearities in guided-wave devices. *IEEE Journal of Lightwave Technology*, Vol.24, No.7, pp 2579~2592, ISSN 0733-8724.

Liu,Y.; Tangdiongga, E.; Li, Z.; Zhang, S.; DeWaardt, H.; Khoe, G. & Dorren, H. (2006), Error-free all-optical wavelength conversion at 160 Gb/s using a semiconductor optical amplifier and an optical bandpass filter. *IEEE Journal of Lightwave Technology*, Vol. 24, No. 1, pp230-236, ISSN 0733-8724.

Mark, J. & Mørk, J. (1992). Subpicosecond gain dynamics in InGaAsP optical amplifiers: Experiment and theory. *Applied Physics Letters*, Vol. 61, No.19, pp 2281-2283, ISSN 0003-6951.

Mork, J. & Mecozzi, A.(1996). Theory of the ultrafast optical response of active semiconductor waveguide. *J. Opt. Soc. Am. B.*, Vol.13, No.8, pp1803-1816, ISSN 0740-3224.

Mork, J.; Mark, J. & Seltzer, C. P. (1994). Carrier heating in InGaAsP laser amplifiers due to two-photon absorption. *Applied Physics Letters*, Vol. 64, No. 17, pp 2206-2208, ISSN 0003-6951.

Mork, J. & Mark, J. (1995). Time-resolved spectroscopy of semiconductor laser devices: experiments and modeling. *Proceeding of SPIE*, Vol. 2399, pp146-159, Physics and Simulation of Optoelectronic Devices III, Marek Osinski; Weng W. Chow; Eds. Monday 06 February 1995, San Jose, CA, USA.

Olsson, B. & Blumenthal, D. (2001). All-optical demultiplexing using fiber cross-phase modulation (XPM) and optical filtering. *IEEE Photonics Technology Letters,* Vol.13 No.8, pp875~878, ISSN 1041-1135.

Saruwatari M. (2000). All-optical signal processing for terabit/second optical transmission. *IEEE Journal of Selected Topics in Quantum Electronics,* Vol.6, No. 6, pp1363~1374, ISSN 1077-260X.

Stubkjaer K E. (2000). Semiconductor optical amplifier-based all-optical gates for high-speed optical processing. *IEEE Journal of Selected Topics in Quantum Electronics,* Vol.6, No. 6, pp1428~1435, ISSN 1077-260X.

Ta'eed, V.; Shokooh-Saremi, M; Fu, L; Littler, I.; Moss D.; Rochette, M.; Eggleton, B.; Yuan, Y. &Luther-Davies B. (2006). Self-phase modulation-based integrated optical regeneration in chalcogenide waveguides. *IEEE Journal of Selected Topics in Quantum Electronics,* Wol.12, No.3, pp360-370, ISSN 1077-260X.

Wang, Y.; Zhang, X.; Dong, J. & Huang, D.(2007). Simultaneous demonstration on all-optical digital encoder and comparator at 40 Gb/s with semiconductor optical amplifiers. *Optics Express,* Vol.15, No.23, pp15080-15085, ISSN 1094-4087.

Xu, J.; Zhang, X.; Dong, J.; Liu, D. & Huang, D. (2007). Ultrafast all-optical AND gate based on cascaded SOAs with assistance of optical filters. *Electronics Letters,* Vol. 43, No. 10, pp 585-587, ISSN 0013-5194.

Xu(a), J.; Zhang, X.; Dong, J.; Liu, D. & Huang, D. (2008). Simultaneous All-Optical and and nor Gates for NRZ Differential Phase-Shift-Keying Signals. *IEEE Photonics Technology Letters,* Vol.20, No.8, pp596-598, ISSN 1041-1135.

Xu(b), J.; Zhang, X.; Dong, J.; Liu, D. & Huang, D. (2008). All-optical minterm generator for three-input NRZ-DPSK signals based on SOAs and delay interferometers. *OFC/NFOEC.* USA, San Diego, 2008. JWA78.

Yoo S. (1996). Wavelength conversion technologies for WDM network applications. *IEEE Journal of Lightwave Technology,,* Vol.14, No.6, pp955~966, ISSN 0733-8724.

Yu, Y.; Zhang, X. & Huang, D. (2006). All-optical clock recovery from NRZ-DPSK signal. *IEEE Photon. Technol. Lett.,* Vol.18, No.11, pp 2356-2358, ISSN 1041-1135.

Yu, Y.; Zhang, X.; Rosas-Fernández, J.; Huang, D; Penty, R. & White, I. (2008) Single SOA based 16 DWDM channels all-optical NRZ-to-RZ format conversions with different duty cycles. *Optics Express,* Vol.16, No.20, pp16166- 16171, ISSN 1094-4087.

Zhang, X.; Wang, Y.; Sun, J.; Liu, D. & Huang, D. (2004). All-optical AND gate at 10 Gbit/s based on cascaded single-port-coupled SOAs. *Optics Express,* 2004, Vol. 12, No.3, pp 361-366, ISSN 1094-4087.

17

Nonlinear Photonic Signal Processing Subsystems and Applications

Chi-Wai Chow[1] and Yang Liu[2]
[1]*Department of Photonics, National Chiao Tung University,*
[2]*Material and Packaging Technologies Group, Hong Kong Applied Science and Technology Research Institute Company limited,*
[1]*Taiwan*
[2]*Hong Kong*

1. Introduction

Over the last decade, the world has seen a great transformation in telecommunication, storage, multimedia, and entertainment. This is mainly due to the advances in computers and Internet, which were made possible by the various advances in lasers and high-capacity optical communication technologies. Without the invention of lasers and fiber optic communication systems, current global telecommunication infrastructure and Internet would have been impossible. The deployment of the optical communication systems started at the 1970s. The bit rate of the early stage fiber optic communication systems in 1980 was ~100 Mb/s, increasing to 1.25 Gb/s, 2.5 Gb/s and 10 Gb/s over the past three decades. By using the wavelength division multiplexing (WDM) technology developed in the 1990s, systems supporting > 100 wavelength channels increase the total capacities up to 1 Tb/s (Kehayas, 2009). Recently, commercial deployments of WDM and dense WDM (DWDM) systems with 40 Gb/s per wavelength channel have started.

This increase in traffic capacity will generate problems. One of them is the huge power consumption required by the network equipments. The DWDM requires optical-toelectrical (O/E) and electrical-to-optical (E/O) signal conversions with the same number as that of the wavelength, resulting in an increase in power consumption. Besides, the high speed electrical signal processing of packet routing and switching at the router consumes huge amounts of power. Apart from the huge power consumption, recent advances in optical signal generations, modulation formats, detections and post-detection signal processing have made the single channel bit rate > 100 Gb/s. This creates the so-called "electronic bottle-neck" between the data transmission rate of optical systems and the limited processing speed of electronic systems. The mismatch between the processing speeds of electronics and photonics stimulates many researches and developments of the nonlinear photonic signal processing technologies in an attempt to remove the electronic bottle-neck. Ultra-fast optical switching using a nonlinear fiber-loop mirror (NOLM) has been demonstrated as early as 1988. And the research activities of the nonlinear photonic signal processing have been continued with the aims to reduce the switching energy and increase the processing speed. Although the processing speed of electronic circuit has increased a lot in the past decades, nonlinear photonic signal processing still plays an important role (Willner, 2008). Nonlinear photonic signal processing techniques, such as alloptical demultiplexing, all-optical sampling, all-optical signal regeneration and all-optical wavelength conversion are more effective (particularly at high speed) and require less power than their electronic counterparts. Owing to the envision of high capacity optical time division multiplexing (OTDM)/WDM networks with transparent connectivity in the future, recently an European Union funded project TRIUMPH (http://www.ihq.unikarlsruhe.de/research/projects/TRIUMPH/) has been established. This project is to develop systems and switch node with optical grooming and multi-wavelength regeneration functionalities for transparent metro and core networks interconnection. Internet has provided a huge impact on our daily life. We have gained the benefit of networks; however, we are also facing the problem of increased power consumption and limited processing speeds of the electronics. We are required to establish new technologies, such as nonlinear photonic signal processing, that enable transmission of huge capacity data with minimum power consumption.

This chapter is organized as following: In Section 2, we will discuss various functions of nonlinear photonic signal processing that may be needed inside the switch node in the future Internet, including wavelength conversion, WDM-to-OTDM and OTDM-to-WDM conversions, all-optical de-multiplexing, and all-optical regeneration. In Section 3, we will discuss the nonlinear photonic signal processing using silicon-on-insulator (SOI). The potential for low cost mass production using complementary metal oxide semiconductor (CMOS) fabrication techniques makes SOI an attractive platform for the monolithic integration of optical waveguides and photonic signal processing functional devices. Finally, a conclusion will be given in Section 4.

2. Functions of photonic signal processing

2.1 Wavelength conversion

Wavelength conversion is one of the most important functions of photonic signal processing in future optical networks. All-optical wavelength routing is a technique which allows optical signals to be routed according to their wavelength, without the need of O/E and E/O conversions. The realization of all-optical networks (AON) can potentially introduce significant cost savings for high speed data networks (Ramaswami & Sivarajan, 1994, Raghavan & Upfal, 1994, Birman & Kershenbaum, 1995). In 2000, field trials of all-optical networks in North America (Froberg et al., 2000) and Europe (Berthelon et al., 2000) have been carried out. All-optical wavelength converters can allow data at one wavelength to be transferred to another wavelength, in order to resolve collisions which would block the routing of a channel which has already been occupied by a wavelength. Besides, wavelength conversion is also important in the trans-multiplexing between the WDM-to-OTDM and OTDM-to-WDM systems (Willner, 2008). In this sub-section, we will discuss several wavelength conversion schemes based on semiconductors and fibers.

Here, we first start with the wavelength conversion scheme using semiconductor laser diode, such as Fabry-Perot laser diode (FP-LD). FP-LD is the simplest and low-cost laser that can be used as the nonlinear medium for all-optical wavelength conversion. Dual-wavelength injection locking (DWIL) technique can be used for this purpose. Injection locking is a way of synchronizing one (or several) free-running oscillator(s) to a stabilized master oscillator. The technique was described in the classical papers of (van der Pol, 1927) and (Adler, 1946) for the case of electrical oscillators. Injection locking of microwave oscillators was reported by (Kurokawa, 1973) and injection locking of gas lasers was demonstrated by He-Ne laser (Stover & Steier, 1966) and CO_2 laser (Buczek & Frejberg, 1972). The injection locking technique has been widely used to improve the static and dynamic performances of semiconductor lasers (Lang, 1982, Mogensen et al., 1985, Murakami et al., 2003). The waveform of the laser output response to a modulated current can be greatly improved with light injection because the transient relaxation oscillation can be severely damped (Lang & Kobayashi, 1977). Single mode operation can be achieved (Kobayashi, 1980), a reduction of the spectral linewidth (Mogensen, 1985) can also be observed. A reduction of frequency chirping of an injection locked laser is theoretically analyzed (Piazzolla et al., 1986) and demonstrated (Lin et al., 1985), and the mode partition noise can be suppressed (Iwashita & Nakagawa, 1982). The dispersion penalty for a signal generated by a current modulated laser can be greatly reduced by injection locking (Cartledge, 1990) and the crosstalk tolerance is also improved (Caspar et al., 1996). Moreover, the modulation bandwidth of the injection locked laser is enhanced (Chen, et al., 2000, Simpson & Liu, 1997, Simpson & Liu, 1996, Simpson & Liu, 1995, Liao et al., 2009). Although the injection locking of semiconductor laser has been studied for two decades, new aspects and applications have only become apparent with recent advances in fiber optic technology. Recently, injection locked lasers have been used for different kinds of nonlinear photonic signal processing, including signal regeneration (Weich et al., 1994, Kakagawa et al., 2001, Inoue & Oda, 1995, Yamashita & Suzuki, 2004, Kuramoto & Yamashita, 2003, Chan et al, 2002), optical repolarization (Chung et al., 2002, Chan et al., 2002), optical packet demultiplexing (Chan et al., 2003), modulation format conversion (Chow et al., 2003, Chow et al., 2003, Chow et al., 2002, Chang et al., 2004, Chow et al., 2004, Yeh et al., 2009). Injection locking also shows promising applications in optical packet switched networks (Chow et al., 2004), Cable-TV networks (Lu et al., 2006), passive optical networks (PONs) (Lu et al., 2006, Wang et al., 2009) and radioover- fiber (ROF) networks (Lu et al., 2008).

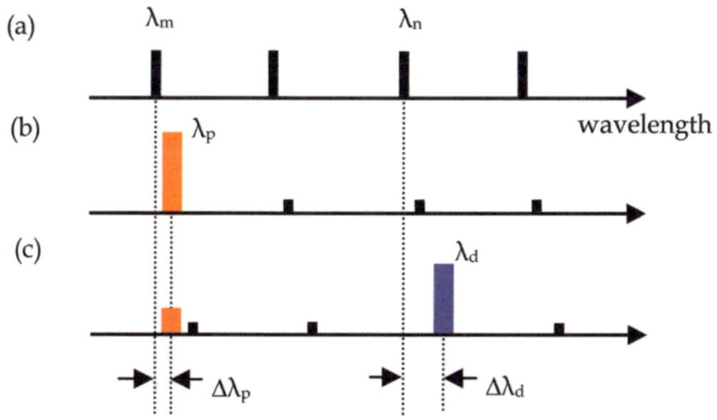

Fig. 1. Schematic of DWIL of FP-LD for all-optical wavelength conversion. (a) Longitudinal mode spectrum of the free-run FP laser, (b) spectrum under injection of only the probe signal, and (c) spectrum under injection of both probe and data signals (Horer & Patzak, 1997).

In the scheme of DWIL (Horer & Patzak, 1997), the slave laser was a FP-LD. The probe wavelength was a continuous-wave (CW) signal injected to an FP-LD longitudinal mode; and the pump wavelength was a modulated signal injected to another longitudinal mode.

The output signal at the wavelength of the CW light was switched by the modulated signal, and thereby the information can be transferred from one wavelength to the other. Fig. 1 shows the schematic of DWIL using a single, direct current (dc) biased FP-LD. First, the CW probe light was injection locked one of the longitudinal modes of the FP-LD (Fig. 1(a-b)). Strong amplification took place if the wavelength of the probe light (λ_p) was tuned to one of the laser modes, e.g., the λ_m mode. Then the mode comb was slightly red-shifted compared to its position when no light was injected and other laser modes were strongly suppressed. The data-bearing pump signal with wavelength λ_d was adjusted within the locking range of another FP-LD longitudinal mode. Fig. 1 (c) shows the situation when the pump (λ_d) was switched "ON". That laser mode was then resonantly amplified. This process further decreased the carrier density by stimulated emission, and further increased the refractive index in the active area of the FP-LD. The increase in refractive index led to the decrease in the FP cavity resonance frequency. The FP mode comb and simultaneously the transmission curve for the probe signal were further red-shifted. Now the probe frequency was outside the high transmission range resulting in low probe output power. Hence, the probe light can be switched "ON" and "OFF" by the data signal, and a logically inverted wavelength conversion can be achieved. The modes λ_m and λ_n used for conversion can be freely chosen within the gain spectrum of the FP-LD. Non-inverting operation can also be achieved. For this purpose, in the absence of data light, the probe light was adjusted with a small red-shift relative to its transmission maximum. Then, an incoming data pump signal shifted the mode toward the maximum. The amplification of the probe light increased simultaneously with the power of the data signal so that the logically non-inverting wavelength conversion was achieved.

Due to the red-shift of the laser longitudinal modes, the injection locking rang was asymmetric (Horer & Patzak, 1997). During light injection, the laser output of a slave laser increased, and the carrier density in the active area of the laser diode decreased correspondingly. Decrease in the carrier density will increase the refractive index of the active area, resulting in the lowering of the cavity resonance frequency or red-shift of the longitudinal modes of the slave laser. When the injected signal wavelength was detuned towards the shorter wavelength side from the cavity resonance wavelength, the laser output decreased. The carrier density increasing and the refractive index decreasing caused the down-shift of the resonance wavelength, resulting in a partial compensation of the wavelength detuning. The FP-LD was still locked when the injected light wavelength was detuned towards the shorter wavelength side of a mode. On the other hand, as the injected wavelength was detuned towards the longer wavelength side, it first enhanced the laser output power due to the further depletion of carrier and further red-shift of the laser mode. Subsequent, when the injected light was outside the locking range, the downshift of the resonance wavelength of the FP-LD happened, resulting in a rapid decrease of locked output and the slave laser became unlocked.

DWIL is potentially capable of high speed all-optical switching. The speed is related to the relaxation oscillation frequency. We can increase the relaxation frequency by increasing the biased current; however for high bias current, the dynamically stable locking range is very small. This drawback can be overcome by using high external injected optical power. Hence, by using appropriate injection of optical pump, probe signal and bias current, the speed of all-optical wavelength conversion can be increased. Since injection locking is polarization sensitive, configuration using TWINS-FP (Two Wavelength Injection locked slave Fabry-

Perot) (Chow et al., 2003) has been proposed to solve the polarization sensitivity issue of the DWIL.

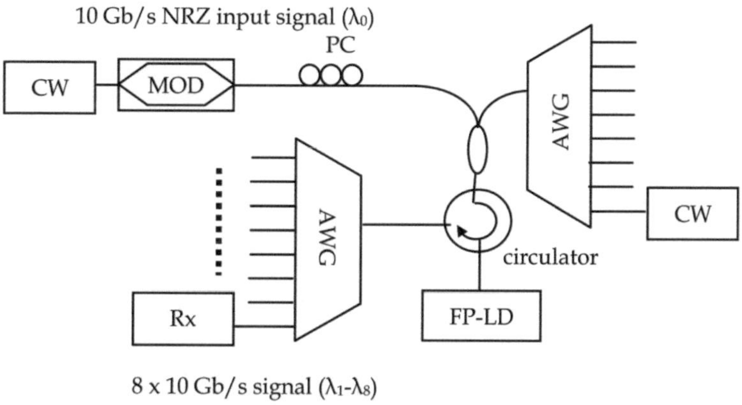

Fig. 2. Experimental setup of all-optical multicasting using a FP-LD. MOD: modulator, PC: polarization controller, AWG: arrayed waveguide grating, Rx: receiver (Chow et al., 2004).

Fig. 2 shows the experimental setup of the 8 x 10 Gb/s all-optical multicasting wavelength conversion using a FP-LD. The pump signal (λ_0: 1547.38 nm) was encoded with 2^{31}-1 pseudo random binary sequence 10 Gb/s non-return-to-zero (NRZ) data. Eight CW probe signals (from λ_1: 1549.03 nm to λ_8: 1560.38 nm, 200 GHz spacing) were wavelength-multiplexed by an arrayed waveguide grating (AWG). They were launched into the FP-LD, which was dc biased at 17 mA ($2I_{th}$) and had a central wavelength of 1545.7 nm with the longitudinal mode spacing of about 1.6 nm. Fig. 3 shows the optical spectrum of the multicasting wavelength conversion obtained at the output port. The side-mode suppression ratio (SMSR) of the converted signals was ~30 dB. Bit-error rate (BER) performance of the wavelength converter was performed for the 8 x 10 Gb/s as shown in Fig. 4. The average power penalty was about 2.5 dB at BER of 10^{-9}, with channel-to-channel variation by about ± 0.4 dB. The eye diagrams of the multicast channel are shown in the insets of Fig. 4. A larger number of output channels may be possible; and the maximum number of wavelength channels is determined by the gain spectrum of the FP-LD.

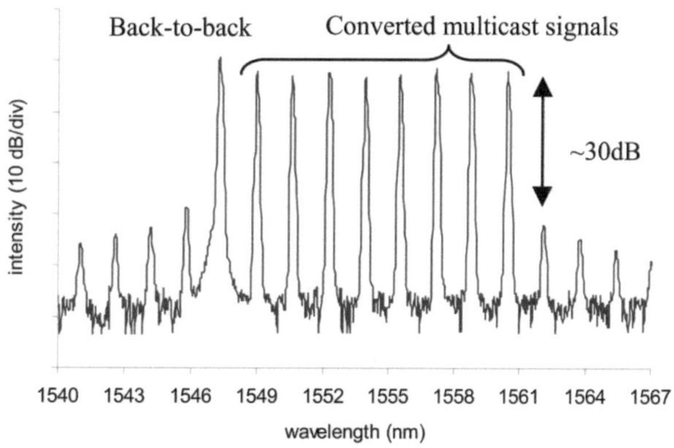

Fig. 3. Optical spectrum of the all-optical multicasting wavelength conversion using a FPLD (Chow et al., 2004).

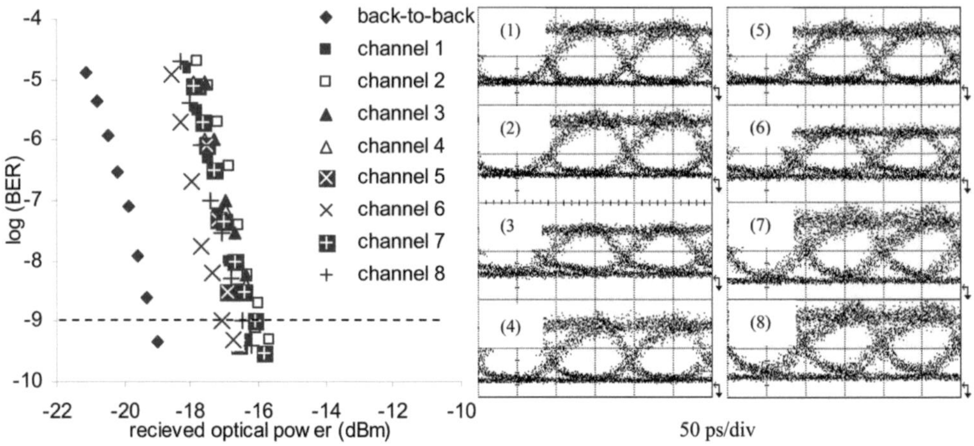

Fig. 4. Plot of 10 Gb/s BER measurements for the back-to-back and each wavelength converted channel (channel 1-channel 8 represents $\lambda_1 - \lambda_8$ respectively) (Chow et al., 2004)

Besides using FP-LD, semiconductor optical amplifier (SOA) is another essential semiconductor device for nonlinear photonic signal processing (Yoo, 1996), due to its high gain, compact size and possibilities for integration. Before studying the wavelength conversion characteristics of a SOA, we first study the SOA carrier dynamics. When an optical pulse is injected into a SOA, the majority of the electrons in the excited state are depleted due to stimulated emission. The reduction in the density of excited electrons has two consequences. First, the SOA gain is reduced, and consequently the refractive index changes as described by Kramers-Kronig relations. This causes a change in wavelength of the output pulse as a function of time (frequency chirp). The gain recovery processes following depletion by an injected optical pulse are described in this sub-section.

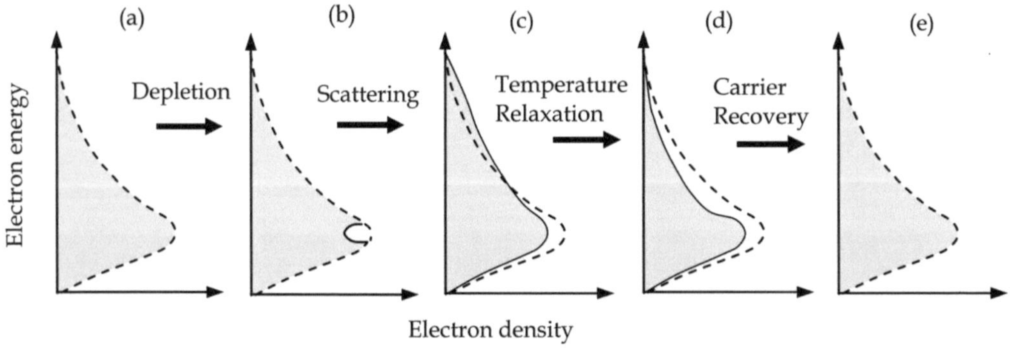

Fig. 5. Gain recovery of an SOA with the electron density versus electron energy.

Fig. 5 shows the gain recovery process of a SOA. When an optical beam is injected into a SOA, electrons in the excited states are depleted, leading to a reduction in gain. Spectral hole burning (SHB) is a localised reduction in gain at the lasing wavelength (Gomatam & DeFonzo, 1990) as shown in Fig. 5(b). The magnitude of gain compression generated by this effect depends on the intraband relaxation times of the carriers. There is a subsequent redistribution of carrier energies due to carrier-carrier scattering as shown in Fig. 5(c), having time scales of < 100 fs. For wavelength conversion using a SOA, the probe wavelength will not be affected by SHB as the pump signal is located at a different wavelength.

When a pump pulse is transmitted through a SOA, it reduces the carrier population but "heats" the plasma (by removing "cool" carriers), thus the gain reduces. This process is called carrier heating (CH), which is a transient heating of electron and hole temperatures (Willatzen, 1991). Thus carriers are excited into the high momentum states. The carrier temperatures will then relax towards the lattice temperature by electron-phonon scattering on the time scales of 1-2 ps as shown in Fig. 5(c-d). The main mechanism of CH within the lattice structure is stimulated emission; however there are other contributions such as freecarrier absorption (FCA) and two-photon absorption (TPA). FCA is an effect where a free carrier (arising from lattice vibrations) can absorb a photon and move to a higher energy level within the band, thus increasing the temperature of the lattice (Stix et al., 1986). TPA is a process where two photons are absorbed in the generation of a single electron-hole carrier pair. It becomes dominant when the input power of the signal is high (Mork & Mark, 1994).

Finally, interband carrier recovery takes place. It is the carriers depleted from the conduction band replenished by the applied bias current, as shown in Fig. 5(d-e). It has a time scales of tens to hundreds of ps. The interband carrier recovery time is dependent on the SOA length, the bias current and the injected optical power (Manning et al., 1994, Girardin et al., 1998).

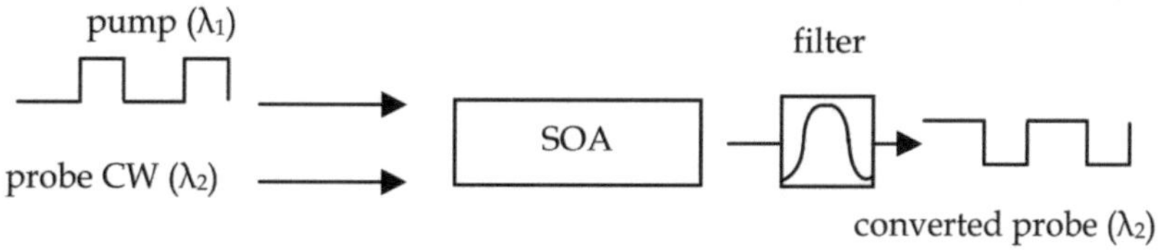

Fig. 6. Cross gain modulation (XGM) in a SOA.

After studying the carrier dynamics of a SOA, we discuss the simplest wavelength conversion using a SOA: the cross gain modulation (XGM). XGM requires a power data signal (pump) and a CW signal (probe), as shown in Fig. 6. The pump modulates the gain of the SOA; and the probe is then modulated by the gain variation of the SOA. The optical filter used will be positioned at the centre of the probe wavelength in the XGM. XGM is simple and polarisation insensitive; but the converted signal is logic inverted, having a relatively high frequency chirp, and limited extinction ratio. The speed of XGM is limited by the interband gain recovery time of SOA. The extinction ratio of the converted signal depends on the variation of the differential gain (Durhuus et al., 1996). When the SOA gain is saturated, the gain peak of the amplifier shifts to longer wavelengths, resulting in a higher slope on the shorter wavelength side of the gain peak. Hence, the probe channels at the shorter wavelengths will experience larger gain variations. As a result, down-conversion (translation from longer wavelength to shorter wavelength) can produce higher output extinction ratio in the converted signal. The extinction ratio can also be improved by increasing the input pump signal power, since this will induce a larger gain modulation of the CW probe signal.

Due to the speed limitation of XGM, modifications of the setup in Fig. 6 can be made to achieve cross phase modulation (XPM) in a SOA. As described in previous paragraph, the refractive index of a SOA is dependent on the gain, and this will cause a phase shift of the probe signal first to the longer wavelength side (red-shift) and then to the shorter wavelength side (blue-shift). Fig. 7 shows the probe signal has a positive phase shift corresponding to the generation of red-chirped components owing to the SOA gain depletion. Then, the SOA gain recovers resulting in a negative phase shift and the generation of blue-chirped components. Using offset filtering of wavelength converted signal in the SOA can greatly enhance the operation speed (Ellis et al., 1998). Recently, improved performance has been demonstrated up to 320 Gb/s (Liu et al., 2006, Liu et al., 2007, Tangdiongga et al., 2006). The disadvantage of this scheme is that large portion of the optical spectrum is lost due to the offset filtering. Similar effect can also be achieved by using an asymmetric Mach–Zehnder interferometer (MZI), which is commonly referred to as a delayed-interference signal converter (DISC) (Ueno et al., 1998, Leuthold et al., 2001). It has been demonstrated that a wavelength converter based on a delayed-interferometric configuration can achieve 168 Gb/s (Nakamura et al., 2001). A theoretical analysis for this concept of the delayed-interferometric configuration has been given in (Leuthold et al., 2004). The delayed-interferometer acts as an optical filter, and (Ueno et al., 2002, Nielsen & Mørk, 2004, Nielsen et al., 2006) show theoretically how the optical filtering can increase the modulation bandwidth of SOA-based switches. The speed of the wavelength conversion can also be enhanced by using two cascaded SOAs, called 'Turbo-Switch' (Manning et al., 2006). Wavelength conversion can also be achieved by arranging two or more SOAs in the MZI configuration (Ratovelomanana et al., 1995, Durhuus et al., 1994, Idler et al., 1995), as shown in Fig. 8. In this scheme, the CW input is split to each arm of the MZI by a coupler. The databearing signal induces different phase shift in the input CW signal via carrier density induced refractive index changes. The output coupler then recombines the split CW signals, where they can interfere constructively or destructively. Commercial wavelength converter based on this structure has been available, such as from Alcatel (Idler et al., 1996).

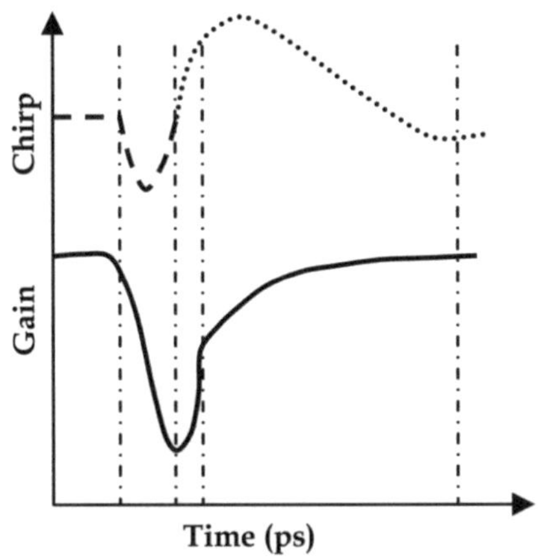

Fig. 7. The change in gain and chirp of the probe signal in a SOA when injected by a short picosecond pump pulse (Liu et al., 2006).

The mechanism of cancelling out the slow relaxation of the carrier density change can be implemented in other configurations. Using one SOA and the interference of orthogonally polarized probe light beams, called an ultrafast nonlinear interferometer (UNI) (Patel et al., 1996). Other nonlinear photonic signal processing applications using an SOA in a NOLM; also known as a semiconductor laser amplifier in a loop mirror (SLALOM) (Eiselt et al.,

1995) and terahertz optical asymmetric demultiplexer (TOAD) (Sokoloff et al., 1993) have also been reported.

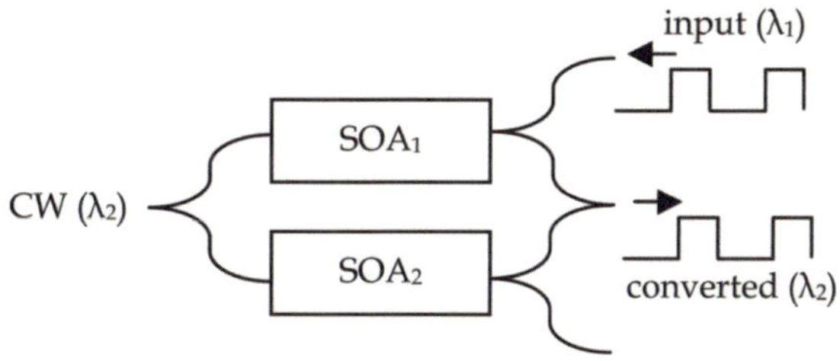

Fig. 8. Cross phase modulation wavelength conversion using SOAs in MZI configuration.

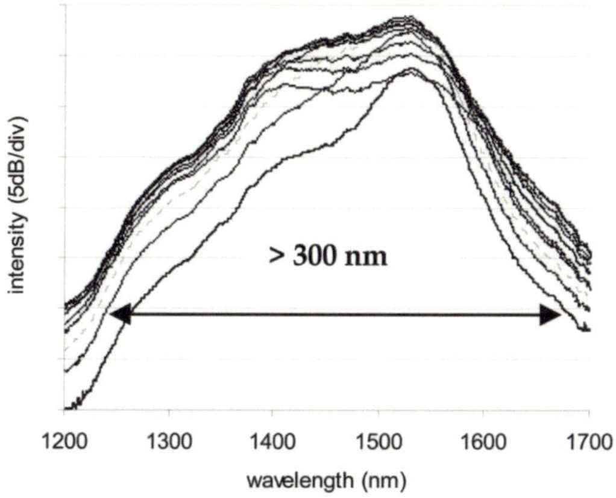

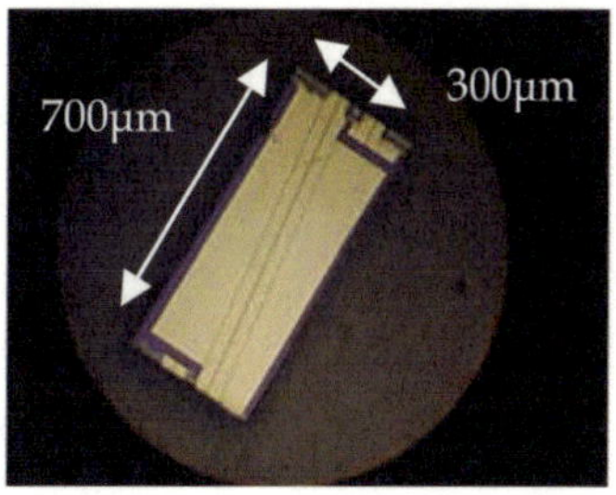

Fig. 9. (a) Spectra of the NWQW-SOA at the bias current from 40 mA to 400 mA, (b) and its dimensions (Wan et al., 2005).

Wavelength conversion based on a SOA can also be implemented by using cross polarization birefringent switching (XPolS). Fig. 9(a) shows the optical spectra of a nonidentical multiple quantum well (NMQW) SOA, and its photo and dimensions is shown in Fig. 9(b). Due to a more uniform carrier distribution in the quantum well sequence, the NMQW-SOA exhibits an extremely broadband spectrum (1300 nm to 1600 nm). Fig. 10(a) shows the experimental setup of XPolS using the NMQW-SOA. The pump signal (λs: 1557.38 nm) and the probe signal (λp: 1439.60 nm) were first launched into the SOA. The probe signal was initially coupled into the SOA at 45° with respect to the transverse-electric (TE) axis of the SOA. The probe was minimized and blocked by the polarizer at the output of the SOA. The pump signal was then injected into the SOA at TE or transverse-magnetic (TM) polarization. The pump depleted the carriers in the SOA and caused refractive index change, which induced a change in birefringence for the probe beam in the SOA and the probe beam was switched "ON" after the polarizer (Wan et al., 2005), as shown in the experimental optical spectra in Fig. 10(b).

Four-wave-mixing (FWM) wavelength conversion can preserve phase information of the input pump signal; and it is an ultrafast process suitable for bit rates beyond 100 Gb/s (Kelly et al., 1998). When optical waves mix within the SOA, carrier density pulsation (CDP) occurs due to the beating between the two optical waves which are detuned by less than the frequency response of the carriers (Agrawal, 1988). The conversion efficiency of conventional FWM in SOA drops rapidly with wavelength detuning and it is polarization dependent. The drop in conversion efficiency with detuning can be solved by using the broadband orthogonal pump (BOP) FWM (Contestabile et al., 1998, Morgan et al., 1998). One pump beam (P_1) is polarized parallel to the signal and detuned by a small amount from the signal wavelength (to generate CDP), and another pump beam (P_2) is polarized orthogonal to P_1 and interacts with the CDP to produce the desired output wavelength. The problem of polarization dependence in BOP FWM was solved by the polarization diversity loop (Mak et al., 2000). It is possible to eliminate both external laser sources in the BOP FWM by placing the SOA in a double-ring fiber cavity (Chow et al., 2002). High speed wavelength conversion at 160-Gb/s using FWM in quantum dot (QD) - SOA has been demonstrated (Otsubo et al., 2005).

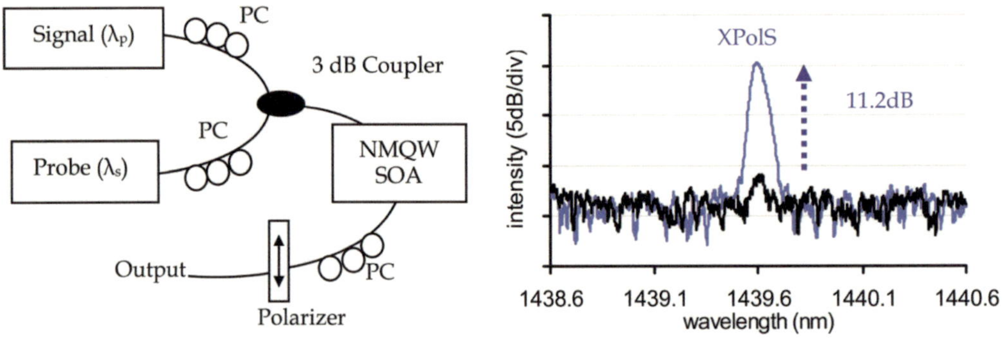

Fig. 10. (a) Experimental setup of XPolS in NMQW-SOA. PC: polarization controller. (b) Optical spectra of the probe signal after the polarizer (1439.60 nm) with and without the pump signal (1557.38 nm) (Wan et al., 2005).

Fiber based devices are also promising candidates for nonlinear photonic signal processing. High-speed switching is possible because of the non-resonant nature of the fiber nonlinearity. Due to its passive in nature, the fiber length needed to produce the nonlinear effect is usually much longer that that of a semiconductor devices. Commonly used specialty fibers for nonlinear photonic signal processing include dispersion-shifted fiber (DSF), highly nonlinear fiber (HNLF), photonic crystal fiber (PCF) and bismuth oxide fiber. The potential of using specialty fibers for all-optical signal processing have been reviewed (Lee, 2008).

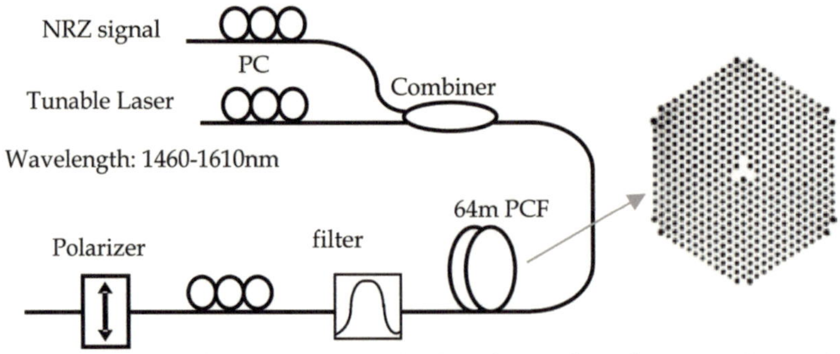

Fig. 11. Experimental setup for the S-, C-, and L-band wavelength conversion us XPolS in a PCF. Inset: cross section of the PCF (Kwok et al., 2006).

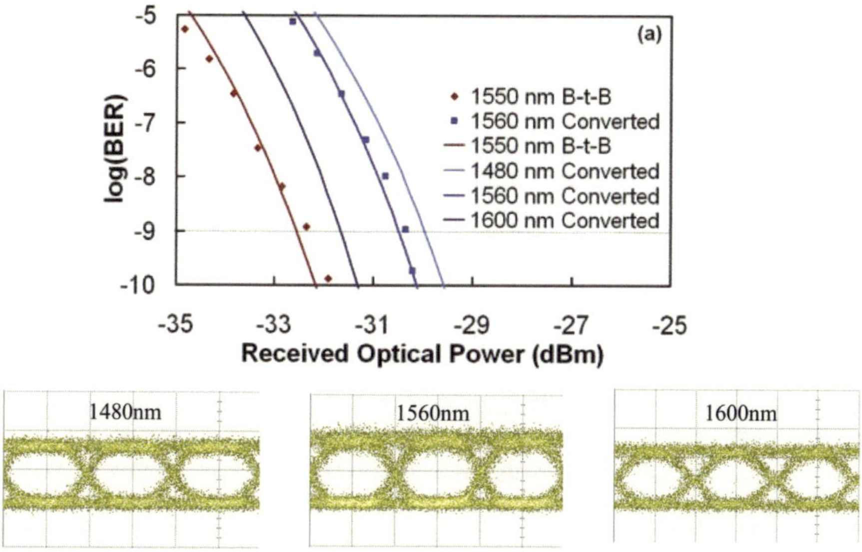

Fig. 12. BER performance of the S-, C-, and L-band wavelength conversion us XPolS in a PCF. Inset: eye diagrams of converted signals at 1480 nm, 1560 nm and 1600 nm (Kwok et al., 2006).

Here, we describe the XPolS in a dispersion-flattened PCF for ultra-broadband (> 100 nm) all-optical wavelength conversion (Kwok et al., 2006). The 64 m dispersion-flattened PCF (inset of Fig. 11) has a mode field diameter of 3.5 mm and a non-linear coefficient of 11.2 $(W\ km)^{-1}$ with dispersion of about -3 ps/(km nm) in the range 1500-1600 nm. The dispersion variation is less than 1 ps/(km nm) in the range of 1465-1655 nm with dispersion slope less than 10^{-3} ps/(km nm^2). A CW probe light was launched at an angle of 45° with respected to the pump signal. At the output of the PCF, a polarizer was used to block the linearly polarized CW probe light in the absence of the pump signal. The linearly polarized pump signal introduced nonlinear phase shifts to the two orthogonal components of the CW probe light with different XPM efficiency. The difference in the overall phase shifts of the two orthogonal components changed the state of polarization of the probe light, and it will turn into a power change at the output of the polarizer. Hence, the intensity modulated pump signal will modulate the intensity of the probe signal after the polarizer. The BER performances of the 1550 nm back-to-back and the converted signals are shown in Fig. 12. Figs. 12 also show the experimental eye diagrams of the converted signal at 1480 nm, 1560 nm and 1600 nm. It was estimated in simulation that the conversion bandwidth for 10 Gb/s NRZ signal can be larger than 200 nm using this approach, while the conversion bandwidth of 90 nm can also be achieved for 40 Gb/s NRZ signals.

2.2 WDM-to-OTDM and OTDM-to-WDM conversions and demultiplexing

As mentioned in the introduction, the scalability of the optical technologies and the elimination of expensive O/E and E/O conversions are expected to significantly reduce the cost and power consumption in future Internet. Initial commercial activities happen, consisting of several regional systems that are interconnected via reconfigurable optical add-drop multiplexers (ROADMs) and optical cross connects (OXCs) to provide transparency extending over a substantial portion of a continental area (Sygletos et al., 2008, Fishman et al., 2006). Recently, an European funded project TRIUMPH (Transparent Ring Interconnection Using Multi-wavelength Photonic switches) has been carried out. It proposes the multi-wavelength photonic switches that perform optical processing including transparent optical grooming/aggregation and multi-wavelength optical regeneration. The proposed network node can support high capacity with transparent connectivity between core-rings (> 100 Gb/s) and metro-rings (up to 40 Gb/s). Inside the photonic node (Ellis et al., 2008), traffic from the metro-rings is switched by an OXC to either metro-ring or via the add-path to the core-ring. In the add-path, asynchronous data packets from the metro-ring are launched into the WDM-to-OTDM converter (Ellis et al., 2008, Norte & Willner, 1996), where each optical packet is re-timed to a local clock and wavelength converted to the desired wavelength. Afterward, the re-timed optical packets are time interleaved (transmultiplexed) to form the OTDM signal for the core-ring. The re-timing can be performed by using an asynchronous digital optical regenerator (ADORE) (Cotter & Ellis, 1998, Chow et al., 2007, Ibrahim et al., 2008, Ibrahim et al., 2008, Chow et al., 2009). For exiting the core-ring, the OTDM signal is simultaneously extracted to WDM signals using multi-wavelength probe signals applied to an OTDM-to-WDM converter. Each demultiplexed WDM signal then passes to the OXC, where it is transmitted to either of the metro-rings or back to the core-ring via the ADORE modules. A field trial of the WDM-to-OTDM trans-multiplexing employing the ADORE has been demonstrated recently (Zarris et al., 2009).

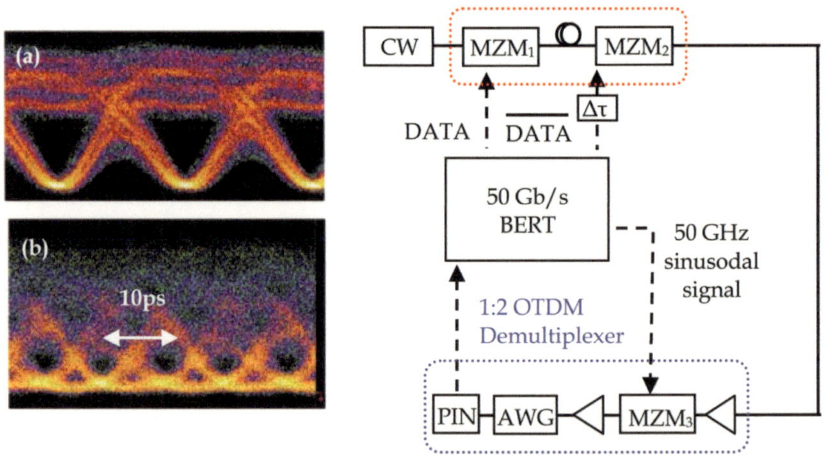

Fig. 13. Experiment setup of 100 Gb/s transmitter and demultiplexer. MZM: Mach-Zehnder modulator, AWG: arrayed waveguide grating. (a) Experimental 50 Gb/s eyes after MZM_1 and (b) 100 Gb/s SDRZ eyes after MZM_2 (Ellis & Chow, 2006).

We then describe the generation of a single channel 100 Gb/s signal. Several groups have proposed and demonstrated 100 Gb/s transmitters, such as using NRZ (Schuh et al., 2007), duo-binary (Winzer et al., 2005), and differential quadrature phase shift keying (DQPSK) (Daikoku et al., 2006) and serial dark-return-to-zero (SDRZ) (Ellis & Chow, 2006). Fig. 13 shows the SDRZ optical transmitter, which was constructed by using commercially available 40 Gb/s components. A CW signal was launched into a single-drive balanced 33 GHz Mach-Zehnder modulator (MZM_1). A 50 Gb/s NRZ was applied to MZM_1. The MZM was biased at the transmission minimum and the NRZ drive voltage switched the modulator towards the two adjacent maxima. At a transition in the NRZ signal, a dark optical pulse with opposite phases between adjacent maxima was produced. The NRZ signal was differentially pre-coded. A second 50 Gb/s electrical signal was modulated by MZM_2 onto the residual CW background between the optical dark pulses, so generating a 100 Gb/s SDRZ signal. At the receiver side, a 1:2 OTDM demultiplexer converted the 100 Gb/s signal down to 50 Gb/s using MZM_3. The demultiplexed signal was filtered by a standard flat-top 100-GHz AWG with an 80 GHz bandwidth, and then coupled into a 45 GHz PIN photodiode for BER measurements.

Fig. 14 shows an OTDM-to-WDM conversion from 129 Gb/s OTDM signal to 3 x 43 Gb/s WDM signals (Vorreau et al., 2008) by using a single NOLM switch (Doran & Wood, 1988). The scheme is using the ultra-fast Kerr based XPM in a NOLM constructed by a specially designed HNLF. The three WDM probe pulses ($\lambda_1, \lambda_2, \lambda_3$) were aligned to different time slots of the OTDM data signal ($\lambda_{control}$) and were launched into the NOLM via FC_2. If the OTDM data pulse ($\lambda_{control}$) was logic 'one', the control pulse caused nonlinear phase shift generated by XPM to the co-propagating probe signals. When the two counter-propagating probe signals interfered at the FC_2 after they had traversed the loop, the probe signal was transmitted at the output of the NOLM and logic 'one' of the probe signal appeared. If the OTDM data pulse ($\lambda_{control}$) was logic 'zero', the two counter-propagating probe signals had identical phase shifts when they traversed the loop and interfered constructively at the coupler with no signal transmitted at the output port of the NOLM ('zero'). By using an optical filter at the output port of the NOLM, each WDM switched signal can then be properly filtered. For multi-wavelength operation of the NOLM, an accurate control of the dispersion and dispersion slope of the fiber is needed to minimize the walk-off between the control and the WDM channels. OTDM-to-WDM conversions based on SPM (Doran & Wood, 1988, Morais et al., 2008) and FWM (Awaji et al, 2004) have also been reported. Besides, we can consider OTDM-to-WDM conversion is an example of all-optical demultiplexing, which allows ultra-high speed OTDM data to be time-division demultiplexed into lower bit-rate channels, and recently, 1.28 Tb/s and 2.56 Tb/s all-optical demultiplexing based on NOLM have been demonstrated (Mulvad et al., 2009, Weber et al., 2006).

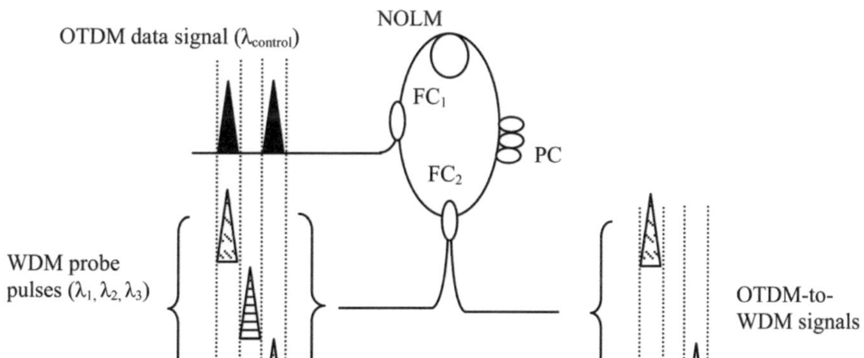

Fig. 14. OTDM-to-WDM conversion using a NOLM. FC: fiber coupler, PC: polarization controller (Vorreau et al., 2008).

Optical Regeneration

All-optical regeneration, such as 2R (re-amplification, re-shaping) and 3R (re-amplification, re-shaping, re-timing), is another key technology for future Internet to improve the signal quality (Provost et al., 2009). When the data rate is becoming higher and higher (towards 100 Gb/s), optoelectronic regeneration schemes will be very hard to implement or even impossible. Besides, optoelectronic regeneration generally can only support a single channel, hence, all-optical regeneration supporting multi-wavelength operation can be considered as an economical upgrade. Simple 2R optical regeneration can be implemented by using an optical nonlinear fiber with an offset optical filter (Mamyshev, 1998), as shown in Fig. 15. Recent studies (Provost et al., 2006, Provost et al., 2007) also allow us to increase the understanding of its operation principle. It was based on self-phase modulation (SPM) that arises when a degraded optical signal is launched into a HLNF, causing the signal spectral broadening. Since the method can produce a steep transfer function, it can reshape the optical pulses by suppressing the amplitude fluctuations in the logic 'one' and 'zero'. An offset optical filter after the HNLF was used to obtain the regenerated output. This scheme can support very high data rates (> 160 Gb/s) and it does not suffer from any carriertransport limitation. This scheme can support single and multi-wavelength channels. However, in the case of multi-wavelength operation, the performance limitation will appear due to the XPM and the FWM produced by the multi-wavelength channels. An improved version of this technique where the XPM can be suppressed is to use an appropriately designed dispersion map by ensuring sufficient walk-off between pulses and minimizing their interaction (Vasilyev & Lakoba, 2005, Cuenot et al., 2006). Other schemes of multiwavelength all-optical regenerations have also been demonstrated recently using special fibers (Parmigiani et al., 2009, Provost et al., 2008, Provost et al., 2008, Kouloumentas et al., 2008) and QD-SOA (Sygletos et al., 2007, Spyropoulou et al., 2007). 10 Gb/s all-optical signal regeneration over 20,000 km has been demonstrated by Kailight Photonics (Cbayet et al., 2004).

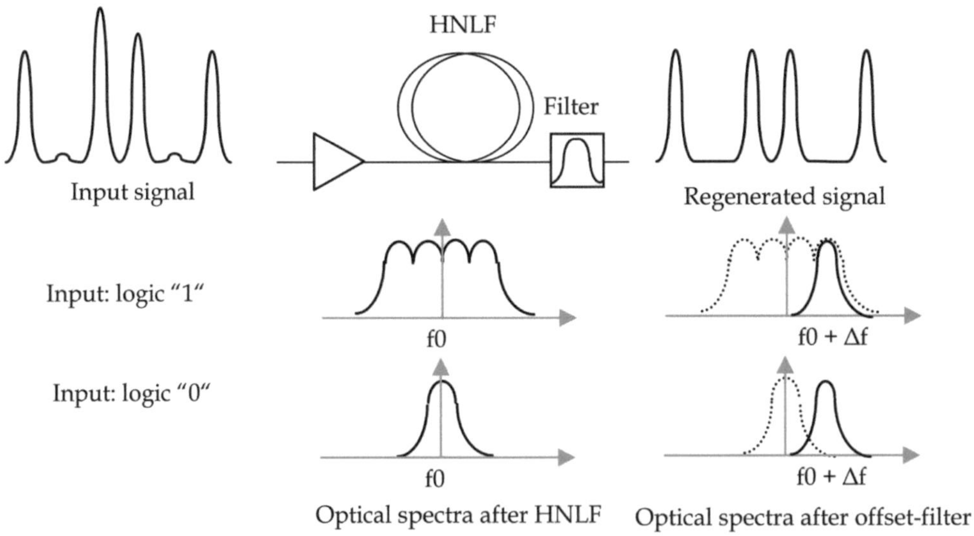

Fig. 15. 2R all-optical regeneration based on SPM, using a HNLF and an offset filter (Mamyshev, 1998).

3. Nonlinear photonic signal processing using SOI

The potential of low-cost mass production using mature CMOS fabrication facilities makes silicon an attractive platform for the monolithic integration of optical waveguides and alloptical signal processing functional devices. The high index contrast of SOI waveguides (Soref & Lorenzo, 1985) allows small nano-wire waveguides (core dimensions ~400 nm) to be fabricated. Hence the input optical powers at mW level can produce peak intensities of $\sim10^{13}$ W m^{-2}. Silicon wire waveguides can have nonlinear parameters (γ) of $\sim 10^{6}$ $(W\ km)^{-1}$ (Koos et al., 2007), which is about 1000 times greater than that of HNLF. Moreover, the Raman gain coefficient in silicon is 1,000 - 10,000 times greater than that in ordinary fiber. All these mean that many of the concepts and applications already developed nowadays that are based on nonlinear effects in semiconductors and fibers can also be adapted for SOI waveguides. Thus, nonlinear optical effects such as SPM (Tsang et al., 2002, Boyraz et al, 2004, Liu et al., 2007), TPA (Liu & Tsang, 2007, Liang & Tsang, 2004), stimulated Raman scattering (SRS) (Claps et al., 2002, Claps et al., 2003), XPM (Liang et al., 2005), slow light for tunable optical delay (Okawachi et al., 2006) and optical bistability (Priem et al., 2005) have all been demonstrated.

Although silicon has a higher Raman gain coefficient when compared with silica fiber, the optical loss in silicon is much higher than that in fiber, particularly at high input power. Nonlinear losses due to TPA-induced free-carrier absorption (FCA) (Claps et al., 2004, Rong et al., 2004) become dominant at high input power. Net gain was first achieved in SOI waveguides using pulsed pumping (Liang & Tsang, 2004). By using a reverse biased p-i-n structure or ion implantation (Liu & Tsang, 2006), the nonlinear absorption can be significantly reduced and CW net Raman gain can be realized (Jones et al., 2005, Sih et al., 2007). After the realization of optical Raman amplification, silicon Raman lasers can be constructed by forming a laser cavity using hybrid fiber loop (Boyraz & Jalali, 2004), chip facet coatings (Rong et al, 2005) or a fully monolithic integrated ring cavity (Rong et al., 2006). Besides, hybrid silicon evanescent laser utilizes a silicon waveguide bonded to AlGaInAs quantum wells is demonstrated (Park et al., 2005). All-optical wavelength conversion can be implemented in SOI using FWM (Raghunathan et al., 2004, Rong et al., 2006), and recently error-free 40 Gb/s wavelength conversion using SOI has been implemented (Kuo et al., 2006). In 2009, 107.8 Gb/s all-optical demultiplexing using FWM in the silicon-organic hybrid (SOH) waveguides has been demonstrated (Koos et al., 2009). This paves the way to the realization of integrated silicon photonic chip for future ultra-fast photonic signal processing.

4. Conclusion

Nonlinear photonic signal processing towards terabit per second is steadily growing in maturity. In the previous sections, we have briefly described various applications of nonlinear photonic signal processing based on semiconductors, fibers and SOI. Various other nonlinear materials can also be employed for these functionalities, including chalcogenide waveguides, periodicallypoled lithium niobate (PPLN) etc. Although many nonlinear photonic devices have been evaluated in transmission experiments and their usefulness has been demonstrated, there are still many challenges for realizing commercial ultrafast communication networks. However, these challenges so far have given researchers and engineers precious knowledge for the present fruitful and diversify applications of ultra-fast photonic signal processing. As the data rate of optical communication will be > 1 Tb/s in the near future, nonlinear photonic signal processing techniques are irreplaceable technologies for the future high speed Internet.

5. Acknowledgement

The authors would like to thank Prof. Hon Tsang, Prof. Chinlon Lin, Dr. Andrew Ellis, Prof. Paul Townsend, Dr. Chien-Hung Yeh and Prof. Sien Chi for supports and discussions.

6. References

Adler, R. (1946). A study of locking phenomena in oscillators. *Proc. IRE*, Vol. 34, (1946) pp. 351-357

Agrawal, G. P. (1988). Population pulsations and nondegenerate four-wave mixing in semiconductor lasers and amplifiers. *J. of Optic. Soc. of Am. B*, Vol. 5, No. 1, (1988) pp.147-158

Awaji, Y.; Miyazaki, T. & Kubota, F. (2004). 160/4 x 40Gb/s OTDM/WDM conversion using FWM fibre. *European Conference on Optical Communication (ECOC)*, Paper We1.5.6, Sept. 2004, Stockholm, Sweden

Berthelon, L.; Audouin, O.; Bonno, P.; Drion, C.; Grard, E.; Rofidal, O.; Martin, C. M.; Ringoot, E.; Van de Voorde, I.; Eilenberger, G. J.; Raeymaekers, F.; Chauvin, J.; Hoa, D. H. B.; Morin, M.; Qiu, X. Z.; Vandewege, J.; Daeleman, T.; Angelopoulos, J. & Taille, C. (2000). Design of a cross-border optical core and access networking field trial: First outcomes of the ACTS-PELICAN project. *J. Lightwave Technol.*, Vol.18, No.12, (2000) pp.1939-1954

Birman, A. and Kershenbaum, A. (1995). Routing and wavelength assignment methods in single-hop all-optical networks with blocking. *Proceedings of the Fourteenth Annual Joint Conference of the IEEE Computer and Communication Societies*, pp. 431, ISBN:0-8186-6990- X, 1995, IEEE Computer Society Washington, DC, USA

Boyraz, O. & Jalali, B. (2004). Demonstration of a silicon Raman laser. *Opt. Express*, Vol. 12, (2004) pp. 5269

Boyraz, O.; Indukuri, T. & Jalali, B. (2004). Self-phase-modulation induced spectral broadening in silicon waveguides. *Opt. Express*, Vol. 12, (2004) pp. 829–834

Buczek, C. J. and Frejberg, R. J. (1972). Hybrid injection locking of higher power CO2 lasers. *IEEE J. Quantum Electron.*, Vol. QE-8, (1972) pp. 641-650

Cartledge, J. C. (1990). Theoretical performance of multigigabit-per-second lightwave systems using injection locked semiconductor lasers. *J. Lightwave Technol.*, Vol. 8, (1990) pp. 1017-1022

Caspar, C.; Foisel, H. M.; Patzak, E.; Strebel, B. & Weich, K. (1996). Improvement of crosstalk tolerance in optical cross-connects by regenerative frequency converters. *Electron. Lett.*, Vol. 32, (1996) pp. 1801-1802

Cbayet, H.; Ben Ema, S.; Sbacbar, N.; Tradok, S.; Tsadka, S. & Leuthold, J. (2004). Regenerative all-optical wavelength converter based on semiconductor optical amplifier and sharp frequency response filter. *Optical Fiber Communication Conference and Exposition (OFC)*, ThS2, Feb. 2004, Anaheim, USA

Chan, L. Y.; Chung, W. H.; Wai, P. K. A.; Moses, B.; Tam, H. Y. & Demokan, M. S. (2002). Simultaneous repolarization of two 10-Gb/s polarization-scrambled wavelength channels using a mutual-injection-locked laser diode. *IEEE Photon. Technol. Lett.*, Vol. 14, (2002) pp. 1740-1742

Chan, L. Y.; Chung, W. H.; Wai, P. K. A.; Tam, H. Y. & Demokan, M. S. (2002). All-optical stabilization of state of polarization of high speed pulse train using injection-locked laser diode. *Electron. Lett.*, Vol. 38, (2002) pp. 1116-1118

Chan, L. Y.; Wai, P. K. A.; Lui, L. F. K.; Moses, B.; Chung, W. H.; Tam, H. Y. & Demokan, M. S. (2003). Demonstration of an all-optical switch by use of multiwavelength mutual injection-locked laser diode. *Opt. Lett.*, Vol. 28, (2003) pp. 837-839

Chang, Y. C.; Lin, Y. H.; Chen, J. & Lin, G. R. (2004). All-optical NRZ-to-PRZ format transformer with an injection-locked Fabry-Perot laser diode at un-lasing condition. *Opt. Express*, Vol. 12, (2004) pp.4449-4456

Chen, H. F.; Liu, J. M. & Simpson, T. B. (2000). Response characteristics of direct current modulation on a bandwidth-enhanced semiconductor laser under strong injection locking. *Opt. Commun.*, Vol. 173, (2000) pp. 349-355

Chow, C. W.; Ellis, A. D. & Cotter, D. (2007). Asynchronous digital optical regenerator for 4 x 40 Gbit/s WDM to 160 Gbit/s OTDM conversion. *Opt. Express*, Vol. 15, (2007) pp. 8507–8512

Chow, C. W.; Ellis, A. D. & Parmigiani, F. (2009). Trans-multiplexer using pulse position locking for 100 Gb/s applications. *Opt. Express*, Vol. 17, (2009) pp. 6562-6567

Chow, C. W.; Wong, C. S. & Tsang, H. K. (2002). All-optical NRZ to RZ format and wavelength converter by dual-wavelength injection locking. *Opt. Comm.*, Vol. 209, (2002) pp. 329-334

Chow, C. W.; Wong, C. S. & Tsang, H. K. (2003). All-optical data-format and wavelength conversion in two-wavelength injection locked slave Fabry-Perot laser diodes. *Electron. Lett.*, Vol. 39, (2003) pp. 997-999

Chow, C. W.; Wong, C. S. & Tsang, H. K. (2003). All-optical RZ to NRZ data format and wavelength conversion using an injection locked laser. *Opt. Comm.*, Vol. 223, (2003) pp. 309-313

Chow, C. W.; Wong, C. S. & Tsang, H. K. (2004). All-optical ASK/DPSK label-swapping and buffering using Fabry-Perot laser diodes. *J. Select. Top. in Quantum Electron.*, Vol.10, (2004) pp. 363-370

Chow, C. W.; Wong, C. S. & Tsang, H. K. (2004). All-optical modulation format conversion and multicasting using injection-locked laser diodes. *J. Lightwave Technol.*, Vol. 22, No. 11, (2004) pp. 2386-2392, Nov., 2004

Chow, K. K.; Shu, C.; Mak, M. W. K. & Tsang, H. K. (2002). Widely tunable wavelength converter using a double-ring fiber laser with a semiconductor optical amplifier" *IEEE Photon. Technol. Lett.*, Vol. 14, No. 10, (2002) pp. 1445-1447

Chung, W. H.; Chan, L. Y.; Tam, H. Y.; Wai, P. K. A. & Demokan, M. S. (2002). Output polarization control of fiber DFB laser using injection locking. *IEEE Photon. Technol. Lett.*, Vol. 14, (2002) pp. 920-922

Claps, R.; Dimitropoulos, D.; Han, Y. & Jalali, B. (2002). Observation of Raman emission in silicon waveguides at 1.54 μm. *Opt. Express*, Vol. 10, (2002) pp. 1305–1313

Claps, R.; Dimitropoulos, D.; Raghunathan, V.; Han, Y. & Jalali, B. (2003). Observation of stimulated Raman amplification in silicon waveguides. *Opt. Express*, Vol. 11, pp. 1731- 1739

Claps, R.; Raghunathan, V.; Dimitropoulos, D. & Jalali, B. (2004). Role of nonlinear absorption on Raman amplification in silicon waveguides. *Opt. Express*, Vol. 12, (2004) pp. 2774–2780

Contestabile, G.; Martelli, F.; Mecozzi, A.; Graziani, L.; D'Ottavi, A.; Spano, P.; Guekos, G.; Dall'Ara, R. & Eckner, J. (1998). Efficiency flattening and equalization of frequency upand down-conversion using four-wave mixing in semiconductor optical amplifiers. *IEEE Photon. Technol. Lett.*, Vol. 10, No. 10, (1998) pp.1398-1400

Cotter, D. & Ellis, A. D. (1998). Asynchronous digital optical regeneration and networks. *J. Lightwave Technol.*, Vol. 16, (1998) pp.2068–2080

Cuenot, B.; Ellis, A. D. & Richardson, D. J. (2006). Optical regeneration using self phase modulation and Quasi-continuous filtering. *IEEE Photon. Technol. Lett.*, Vol. 18, (2006) pp.1350-1352

Daikoku, M.; Morita, I.; Taga, H.; Tanaka, H.; Kawanishi, T.; Sakamoto, T.; Miyazaki, T.; & Fujita, T. (2006). 100Gbit/s DQPSK transmission experiment without OTDM for 100G Ethernet Transport. *Optical Fiber Communication Conference and Exposition (OFC)*, PDP36, March 2006, Anaheim, USA

Doran, N. J. & Wood, D. (1988). Nonlinear-optical loop mirror. *Optics Lett.*, Vol.13, No. 1, (1988) pp. 56-58

Durhuus, T.; Joergensen, C.; Mikkelsen, B.; Pedersen, R. J. S. & Stubkjaer, K. E. (1994). Alloptical wavelength conversion by SOA's in a Mach-Zehnder configuration. *IEEE Photon. Technol. Lett.*, Vol. 6, (1994) pp. 53-55

Durhuus, T.; Mikkelsen, B.; Joergensen, C.; Danielsen, S. L. & Stubkjaer, K. (1996). All-optical wavelength conversion by semiconductor optical amplifiers. *J. of Lightwave Technol.*, Vol. 14, (1996) pp. 942-954

Eiselt, M.; Pieper, W. & Weber, H. G. (1995). SLALOM: semiconductor laser amplifier in a loop mirror. *J. Lightwave Technol. Lett.*, Vol. 13, (1995) pp. 2099-2112

Ellis, A. D. & Chow, C. W. (2006). Serial OTDM for 100 GB-ethernet applications. *Electron. Lett.*, Vol. 42, (2006) pp. 485-486

Ellis, A. D.; Cotter, D.; Ibrahim, S.; Weerasuriya, R.; Chow, C. W.; Leuthold, J.; Freude, W.; Sygletos, S.; Vorreau, P.; Bonk, R.; Hillerkuss, D.; Tomkos, I.; Tzanakaki, A.; Kouloumentas, C.; Richardson, D. J.; Petropoulos, P.; Parmigiani, F.; Zarris, G. & Simeonidou, D. (2008). Optical interconnection of core and metro networks [Invited]. *J. Optical Networking*, Vol. 7, (2008) pp. 928-934

Ellis, A. D.; Kelly, A. E.; Nesset, D.; Pitcher, D.; Moodie, D. G. & Kashyap, R. (1998). Error free 100Gbit/s wavelength conversion using grating assisted cross-gain modulation in 2mm long semiconductor amplifier. *Electron. Lett.*, Vol. 34, (1998) pp. 1958-1559 EU project Triumph, http://www.ihq.uni-karlsruhe.de/research/projects/TRIUMPH/

Fishman, D.; Correa, D. L.; Goode, E. H.; Downs, T. L.; Ho, A. Y.; Hale, A.; Hofmann, P.; Basch, B. & Gringeri, S. (2006). The rollout of optical networking: LambdaXtreme national network deployment. *Bell Lab. Tech. J.*, Vol. 11, (2006) pp.55–63

Froberg, M. N.; Henion, S. R.; Rao, H. G.; Hazzard, B. K.; Parikh, S.; Romkey, B. R. & Kuznetsov, M. (2000). The NGI ONRAMP test bed: reconfigurable WDM technology for next generation regional access networks. *J. Lightwave Technol.*, Vol.18, No.12, (2000) pp.1697-1708

Girardin, F.; Guekos, G. & Houbavlis, H. (1998). Gain recovery in bulk semiconductor optical amplifiers. *IEEE Photonics Technol. Lett.*, Vol. 10, (1998) pp. 784-786

Gomatam, B. N. and DeFonzo, A. P. (1990). Theory of hot carrier effects on nonlinear gain in GaAs-GaAlAs lasers and amplifiers. *IEEE J. Quantum Electron.*, Vol. 26, (1990) pp. 1689- 1703

Horer, J. and Patzak, E. (1997). Large-signal analysis of all-optical wavelength conversion using two-mode injection-locking in semiconductor. *IEEE J. Quantum Electron.*, Vol. 33, (1997) pp. 596-608

Ibrahim, S. K.; Hillerkuss, D.; Weerasuriya, R.; Zarris, G.; Simeonidou, D.; Leuthold, J. & Ellis, A. D. (2008). Novel 42.65 Gbit/s dual gate asynchronous digital optical regenerator using a single MZM. *European Conference on Optical Communication (ECOC)*, paper Tu4.D.3., Brussels, Belgium, Sept. 21-25, 2008

Ibrahim, S. K.; Weerasuriya, R.; Hillerkuss, D.; Zarris, G.; Simeonidou, D.; Leuthold, J.; Cotter, D. & Ellis, A. D. (2008). Experimental demonstration of 42.6 Gbit/s asynchronous digital optical regenerators. *International Conference on Transparent Optical Networks (ICTON)*, invited paper We.C3.3., Athens, Greece, Jun. 22- 26, 2008

Idler, W.; Daub, K.; Deppisch, B.; Dutting, K.; Klenk, M.; Lach, E.; Laube, G.; Luz, G.; Schilling, M.; Ulmer, M. & Wunstel, K. (1996). All optical wavelength converter modules for operation up to 10 Gbit/s. *European Conference on Optical Communication (ECOC)*, ThB. 2.4, Oslo, Norway, September 15-19,1996

Idler, W.; Schilling, M.; Daub, K.; Baums, D.; Korner, U.; Lach, E.; Laube, E. & Wunstel, K. (1995). Signal quality and BER performance improvement by wavelength conversion with an integrated three-port Mach-Zehnder interferometer. *Electron. Lett.*, Vol. 31, (1995) pp. 454-455

Inoue, K. and Oda, K. (1995). Noise suppression in wavelength conversion using a lightinjected laser diode. *IEEE Photon. Technol. Lett.*, Vol. 7, (1995) pp. 500-502

Iwashita, K. and Nakagawa, K. (1982). Suppression of mode partition noise by laser diode light injection. *IEEE J. Quantum Electron.*, Vol. QE-18, (1982) pp. 1669-1674

Jones, R.; Rong, H.; Liu, A.; Fang, A.; Paniccia, M.; Hak, D.; & Cohen, O.; (2005). Net continuous wave optical gain in a low loss silicon-on-insulator waveguide by stimulated Raman scattering. *Opt. Express*, Vol. 13, (2005) pp. 519-525

Kakagawa, J.; Marhic, M. E. & Kazovskty, L. G. (2001). All-optical 3R regeneration technique using injection-locking in gain-switched DFB-LD. *Electron. Lett.*, Vol. 37, (2001) pp. 231- 232

Kehayas, E. (2009). EUROFOS Workshop on Photonic Systems (WORPS): Technologies for developing terabit capacity switching systems, *International Conference on Transparent Optical Networks (ICTON)*, Azores, June 2009

Kelly, A. E.; Ellis, A. D.; Nesset, D.; Kashyap, R. & Moodie, D. G. (1998). 100Gbit/s wavelength conversion using FWM in an MQW semiconductor optical amplifier. *Electron. Lett.*, Vol. 34, No. 20, (1998) pp. 1955-1956

Kobayashi, S.; Yamada, J.; Machida, S. & Kimura, T. (1980). Single mode operation of 500 Mbit/s modulated AlGaAs semiconductor laser. *Electron. Lett.*, Vol. 16, (1980) pp. 746- 747

Koos, C.; Jacome, L.; Poulton, C.; Leuthold, J. & Freude, W. (2007). Nonlinear silicon-oninsulator waveguides for all-optical signal processing. *Opt. Express*, Vol. 15, (2007) pp. 5976–5990

Koos, C.; Vorreau, P.; Vallaitis, T.; Dumon, P.; Bogaerts, W.; Baets, R.; Esembeson, B.; Biaggio, I.; Michinobu, T.; Diederich, F.; Freude, W. & Leuthold, J. (2009). All-optical high-speed signal processing with silicon-organic hybrid slot waveguides. *Nat. Photon.*, Vol. 3, (2009) pp.216 - 219

Kouloumentas, C.; Provost, L.; Parmigiani, F.; Tsolakidis, S.; Petropoulos, P.; Tomkos, I. & Richardson, D. J. (2008). Four-channel all-fiber dispersion-managed 2R regenerator. *IEEE Photon. Technol. Lett.*, Vol. 20, (2008) pp. 1169-1171

Kuo, Y. H.; Rong, H.; Sih, V.; Xu, S.; Paniccia, M. & Cohen, O. (2006). Demonstration of wavelength conversion at 40 Gb/s data rae in silicon waveguides. *Opt. Express*, Vol. 14, (2006) pp. 11721,

Kuramoto, A. and Yamashita, S. (2003). All-optical regeneration using a side-mode injectionlocked semiconductor laser. *J. Select. Topics in Quantum Electron.*, Vol. 9, (2003) pp. 1283-1287

Kurokawa, K. (1973). Injection locking of microwave solid-state oscillators. *Proc. IEEE*, Vol. 61, (1973) pp. 1386-1410

Kwok, C. H.; Chow, C. W.; Tsang, H. K.; Lin, Chinlon & Bjarklev, A. (2006). Nonlinear polarization rotation in a dispersion-flattened photonic-crystal fiber for ultra-wide band (> 100 nm) all-optical wavelength conversion of 10 Gb/s NRZ signals. *Opt. Lett.*, Vol. 31, No. 12, (June, 2006) pp. 1782-1784

Lang, R. (1982). Injection locking properties of a semiconductor laser. *IEEE J. Quantum Electron.*, Vol. QE-18, (1982) pp. 976-983

Lang, R. and Kobayashi, K. (1977). Suppression of relaxation oscillation in the modulated output of semiconductor lasers. *IEEE J. Quantum Electron.*, Vol. QE-12, (1977) pp. 520- 525

Lee, J. H. (2008). All-optical signal processing using specialty fibers. *Optical Fiber Communication Conference and Exposition (OFC)*, OTuD7, San Diego, USA, February 24- 28, 2008

Leuthold, J.; Mikkelsen, B.; Raybon, G.; Joyner, C. H.; Pleumeekers, J. L.; Miller, B. I.; Dreyer, K. & Behringer, R. (2001). All-optical wavelength conversion between 10 and 100 Gb/s with SOA delayed-interference configuration. *Opt. Quantum Electron.*, Vol. 33, (2001) pp.939–952

Leuthold, J.; Moller, L.; Jaques, J.; Cabot, S.; Zhang, L.; Bernasconi, P.; Cappuzzo, M.; Gomez, L.; Laskowski, E.; Chen, E.; Wong-Foy, A. & Griffin, A. (2004). 160 Gb/s SOA alloptical wavelength converter and assessment of its regenerative properties. *Electron. Lett.*, Vol. 40, (2004) pp.554-555

Liang, T. K. & Tsang, H. K. (2004). Efficient Raman amplification in silicon-on-insulator waveguides. *Appl. Phys. Lett.*, Vol. 85, (2004) pp. 3343-3345.

Liang, T. K. and Tsang, H. K. (2004). The role of free-carriers from two-photon absorption in Raman amplification in silicon-on-insulator waveguides. *Appl. Phys.Lett.*, Vol. 84, (2004) pp.2745–2747

Liang, T. K.; Nunes, L.; Sakamoto, T.; Sasagawa, K.; Kawanishi, T.; Tsuchiya, M.; Priem, G.; Van Thourhout, D.; Dumon, P.; Baets, R.; & Tsang, H. K. (2005). Ultrafast all-optical switching by cross-absorption modulation in silicon wire waveguides. *Opt. Express*, Vol. 13, pp. 7298-7303

Liao, Y. S,; Kuo, H. C.; Chen, Y. J. & Lin, G. R. (2009). Side-mode transmission diagnosis of a multichannel selectable injection-locked Fabry-Perot Laser Diode with anti-reflection coated front facet. *Opt. Express* , Vol. 17, (2009) pp.4859-4867

Lin, C.; Andersen, J. K. & Mengel, F. (1985). Frequency chirp reduction in a 2.2 Gbit/s directly modulated InGaAsP semiconductor laser by CW injection. *Electron. Lett.*, Vol. 21, (1985) pp. 80-81

Liu, Y. & Tsang, H. K. (2006). Nonlinear absorption and Raman gain in helium-ionimplanted silicon waveguides. *Opt. Lett.*, Vol. 31, (2006) pp. 1714–1716

Liu, Y. & Tsang, H. K. (2007). Time dependent density of free carriers generated by two photon absorption in silicon waveguides. *Appl. Phys. Lett.*, Vol. 90, (2007) 211105

Liu, Y.; Chow, C. W.; Tsang, H. K. & Wong, S. P. (2007). Enhancement of self phase modulation induced spectral broadening in silicon waveguides by ion implantation. *Conference on Lasers and Electro-Optics (CLEO)*, CThQ2, May 2007, Baltimore, USA

Liu, Y.; Tangdiongga, E.; Li, Z.; de Waardt, H.; Koonen, A. M. J.; Khoe, G. D.; Dorren, H. J. S.; Shu, X. & Bennion, I. (2006). Error-free 320 Gb/s SOA-based wavelength conversion using optical filtering. *Optical Fiber Communication Conference and Exposition (OFC)*, PDP28, March 2006, Anaheim, USA

Liu, Y.; Tangdiongga, E.; Li, Z.; de Waardt, H.; Koonen, A. M. J.; Khoe, G. D.; Shu, X.; Bennion, I. & Dorren, H. J. S. (2007). Error-free 320-Gb/s all-optical wavelength conversion using a single semiconductor optical amplifier. *J. Lightwave Technol.*, Vol. 25, (2007) pp.103–108

Lu, H. H.; Chuang, Y. W.; Chen, G. L.; Liao, C. W. & Chi, Y. C. (2006). Fiber optical CATV system performance improvement employing light injection and optoelectronic feedback techniques. *IEEE Photon. Technol. Lett.*, Vol. 18, (2006) pp. 1789-1791

Lu, H. H.; Ho, W. J.; Lin, W. I.; Peng, H. C.; Lai, P, C. & Yee, H. (2008). Radio-on-DWDM transport systems based on injection-locked Fabry-Perot laser diodes. *IEICE Transactions on Communications*, Vol. E91-B(3), (2008) pp. 848-853

Lu, H. H.; Tzeng, S. J.; Chuang, C. P.; Chi, Y. C.; Tsai, C. C.; Chen, G. L. & Chuang, Y. W. (2006). HDTV/ Gigabit ethernet over bidirectional WDM-PON based on injectionlocked Fabry-Perot laser diodes. *Opt. Commun.*, Vol. 267, (2006) pp.102-107

Mak, M. W. K.; Tsang, H. K. & Chan, K. (2000). Widely tunable polarization-independent all-optical wavelength converter using a semiconductor optical amplifier. *IEEE Photon. Technol. Lett.*, Vol.12, (2000) pp.525-527

Mamyshev, P. V. (1998). All-optical data regeneration based on self-phase modulation effect. *European Conference on Optical Communication (ECOC)*, pp. 475-476, Sept 20-24, 1998, Madrid, Spain

Manning, R. J.; Davies, D. A. O. & Lucek, J. K. (1994). Recovery rates in semiconductor laser amplifiers: optical and electrical bias dependencies. *Electron. Lett.*, Vol. 30, (1994) pp. 1233-1234

Manning, R. J.; Yang, X.; Webb, R. P.; Giller, R.; Gunning, F.C. Garcia & Ellis, A. D. (2006). The 'Turbo-switch' - a novel technique to increase the high-speed response of SOAs for wavelength conversion," *Optical Fiber Communication Conference and Exposition (OFC)*, Paper OWS8, March 5-10, 2006, Anaheim, USA

Mogensen, F.; Olesen, H. & Jacobsen, G. (1985). FM noise suppression and linewidth reduction in an injection-locked semiconductor laser. *Electron. Lett.*, Vol. 21, (1985) pp. 696-697

Mogensen, F.; Olesen, H. & Jacobsen, G. (1985). Locking conditions and stability properties for a semiconductor laser with external light injection. *IEEE J. Quantum Electron.*, Vol. QE-21, (1985) pp. 784-793

Morais, R.; Meleiro, R.; Monteiro, P.; & Marques, P. (2008). OTDM-to-WDM conversion based on wavelength conversion and time gating in a single optical gate. *Optical Fiber Communication Conference and Exposition (OFC)*, OTuD5, February 24-28, 2008, San Diego, USA

Morgan, T. J.; Lacey, J. P. R. & Tucker, R. S. (1998). Widely tunable four-wave mixing in semiconductor optical amplifiers with constant conversion efficiency. *IEEE Photon. Technol. Lett.*, Vol. 10, No. 10, (1998) pp.1401-1403

Mork, J. and Mark, J. (1994). Carrier heating in InGaAsP laser amplifiers due to two photon absorption. *Appl. Phys. Lett.*, Vol. 64, (Apr. 1994) pp. 2206-2208

Mulvad, H. C.; Oxenløwe, L. K.; Galili, M.; Clausen, A. T.; Gru¨ner-Nielsen, L. & Jeppesen, P. (2009). 1.28 Tbit/s single-polarisation serial OOK optical data generation and demultiplexing. *Electron. Lett.*, Vol. 45, (2009) pp. 280-281

Murakami, A.; Kawashima, K. & Atsuki, K. (2003). Cavity resonance shift and bandwidth enhancement in semiconductor lasers with strong light injection. *IEEE J. Quantum Electron.*, Vol. 39, (2003) pp. 1196-1204

Nakamura, S.; Ueno, Y. & Tajima, K. (2001). 168-Gb/s all-optical wavelength conversion with a symmetric-Mach-Zehnder-Type switch. *IEEE Photon. Technol. Lett.*, Vol. 13, (2001) pp.1091-1093

Nielsen, M. L. & Mørk, J. (2004). Increasing the modulation bandwidth of semiconductoroptical- amplifier-based switches by using optical filtering. *J. Opt. Soc. Am. B*, Vol. 21, (2004) pp.1606-1619

Nielsen, M. L.; Mørk, J.; Suzuki, R.; Sakaguchi, J. & Ueno, Y. (2006). Experimental and theoretical investigation of the impact of ultra-fast carrier dynamics on high-speed SOA-based all-optical switches. *Opt. Express*, Vol. 14, (2006) pp.331-347

Norte, D. & Willner, A. E. (1996). All-optical data format conversions and reconversions between the wavelength and time domains for dynamically reconfigurable WDM networks. *J. Lightwave Technol.*, Vol. 14, (1996) pp. 1170–1182

Okawachi, Y.; Foster, M.; Sharping, J.; Gaeta, A.; Xu, Q.; & Lipson, M. (2006). All-optical slow-light on a photonic chip. *Opt. Express*, Vol. 14, pp. 2317-2322

Otsubo, K.; Akiyama, T.; Kuwatsuka, H.; Hatori, N.; Ebe, H. & Sugawara, M. (2005). Automatically controlled C-band wavelength conversion with constant output power based on four-wave mixing in SOA's. *IEICE Trans. Electron.*, Vol. E88-C, (2005) pp. 2358–2365

Park, H.; Fang, A. W.; Kodamaa, S.; & Bowers, J. E. (2005). Hybrid silicon evanescent laser fabricated with a silicon waveguide and III-V offset quantum wells. *Opt. Express*, Vol. 13, No. 23, (2005) pp. 9460-9464

Parmigiani, F.; Vorreau, P.; Provost, L.; Mukasa, K.; Takahashi, M.; Tadakuma, M.; Petropoulos, P.; Richardson, D. J.; Freude, W. & Leuthold, J. (2009). 2R regeneration of two 130 Gbit/s channels within a single fiber. *Optical Fiber Communication Conference and Exposition (OFC)*, JThA56, March 22.–26, 2009, San Diego, USA

Patel, N. S.; Rauschenbach, K. A. & Hall, K. L. (1996). 40-Gb/s demultiplexing using an ultrafast nonlinear interferometer (UNI). *IEEE Photon. Technol. Lett.*, Vol. 8, No. 12, (1996) pp. 1695-1697

Piazzolla, S.; Spano, P. & Tamburrini, M. (1986). Small signal analysis of frequency chirping in injection-locked semiconductor laser. *IEEE J. Quantum Electron.*, Vol. QE-22, (1986) pp. 2219-2223

Priem, G.; Dumon, P.; Gogaerts, W.; Van Thourhout, D.; Morthier, G. & Beats, R. Optical bistability and pulsating behaviour in silicon-on-insulator ring resonator structures. *Opt. Express*, Vol. 13, (2005) pp. 9623

Provost, L.; Finot, C.; Mukasa, K.; Petropoulos, P. & Richardson, D. J. (2007). Generalisation and experimental validation of design rules for self-phase modulation-based 2Rregenerators. *Optical Fiber Communication Conference and Exposition (OFC)*, OThB6, March 25 - 29, 2007, Anaheim, USA

Provost, L.; Finot, C.; Petropoulos, P. & Richardson, D. J. (2006). Design scaling laws for selfphase modulation-based 2R-regenerators. *European Conference on Optical Communication (ECOC)*, We4.3.2, Sept 2006, Cannes, France

Provost, L.; Parmigiani, F.; Petropoulos, P. & Richardson, D. J. (2008). Investigation of simultaneous 2R regeneration of two 40 Gb/s channels in a single optical fiber. *IEEE Photon. Technol. Lett.*, Vol. 20, (2008) pp.270-272

Provost, L.; Parmigiani, F.; Petropoulos, P.; Finot, C.; Mukasa, K.; Petropoulos P. & Richardson, D. J. (2008). Analysis of a two-channel 2R all-optical regenerator based on a counter-propagating configuration. *Opt. Express*, Vol. 16, (2008) pp. 2264-2275

Provost, L.; Petropoulos, P. & Richardson, D. J. (2009). Optical WDM regeneration: status and future prospects. *Optical Fiber Communication Conference and Exposition (OFC)*, OFD7, Mar 22-26, 2009, San Diego, USA

Raghavan, P. and Upfal, E. (1994). Efficient routing in all-optical networks. *Proceedings of the ACM Symposium on Theory of Computing*, pp.134--143, 1994

Raghunathan, V.; Claps, R.; Dimitropoulos, D. & Jalali, B. (2004). Wavelength conversion in silicon using Raman induced four-wave mixing. *Appl. Phys. Lett.*, Vol. 85, (2004) pp. 34- 36

Ramaswami, R. and Sivarajan, K. N. (1994). Routing and wavelength assignment in alloptical networks. *IEEE/ACM Transactions on Networking (TON)*, Vol.3, No. 5, (October 1995) pp.489-500, ISSN: 1063-6692

Ratovelomanana, F.; Vodjdani, N.; Enard, A.; Glastre, G.; Rondi, D.; Blondeau, R.; Joergensen, C.; Durhuus, T.; Mikkelsen, B.; Stubkjaer, K.E.; Jourdan, A. & Soulage, G. (1995). An all-optical wavelength-converter with semiconductor optical amplifiers monolithically integrated in an asymmetric passive Mach-Zehnder interferometer. *IEEE Photon. Technol. Lett.*, Vol. 7, (1995) pp. 992-994

Rong, H.; Jones, R.; Liu, A.; Cohen, O.; Hak, D.; Fang, A.; & Paniccia, M.; (2005). A continuous-wave Raman silicon laser. *Nature*, Vol. 433, (2005) pp. 725-728

Rong, H.; Kuo, Y. H.; Liu, A.; Paniccia, M. & Cohen, O. (2006). High efficiency wavelength conversion of 10 Gb/s data in silicon waveguides. *Opt. Express*, Vol. 14, (2006) pp. 1182

Rong, H.; Kuo, Y. H.; Xu, S.; Liu, A.; Jones, R.; Paniccia, M.; Cohen, O. & Raday, O. (2006). Monolithic integrated Raman silicon laser. *Opt. Express*, Vol. 14, (2006) pp. 6705

Rong, H.; Liu, A.; Nicolaescu, R.; & Paniccia, M. (2004). Raman gain and nonlinear optical absorption measurement in a low loss silicon waveguide. *Appl. Phys. Lett.*, Vol. 85, (2004) pp. 2196-2198

Schuh, K.; Junginger, B.; Lach, E.; Veith, G.; Lutz, J. & Möller, M. (2007). Serial 107Gbit/s ETDM NRZ transmission over 320km SSMF. *Optical Fiber Communication Conference and Exposition (OFC)*, OWE2, Mar 25 - 29, 2007, Anaheim, USA

Sih, V.; Xu, S.; Kuo, Y. H.; Rong, H.; Paniccia, M.; Cohen, O. & Raday, O. (2007). Raman amplification of 40 Gb/s data in low-loss silicon waveguides. *Opt. Express*, Vol. 15, (2007) pp.357

Simpson, T. B. and Liu, J. M. (1995). Bandwidth enhancement and broadband noise reduction in injection-locked semiconductor lasers. *IEEE Photon. Technol. Lett.*, Vol. 7, (1995) pp. 709-711

Simpson, T. B. and Liu, J. M. (1996). Small-signal analysis of modulation characteristics in a semiconductor laser subject to strong optical injection. *IEEE Photon. Technol. Lett.*, Vol. 32, (1996) pp. 1456-1468

Simpson, T. B. and Liu, J. M. (1997). Enhanced modulation bandwidth in injection-locked semiconductor lasers. *IEEE Photon. Technol. Lett.*, Vol. 9, (1997) pp. 1322-1324

Sokoloff, J. P.; Prucnal, P. R.; Glesk, I. & Kane, M. (1993). A terahertz optical asymmetric demultiplexer (TOAD). *IEEE Photon. Technol. Lett.*, Vol. 5, (1993) pp. 787–790

Soref, R. A. & Lorenz, J. P. (1985). Single-crystal silicon—a new material for 1.3 and 1.6 μm integrated-optical components. *Electron. Lett.* Vol. 21, (1985) pp. 953–954

Spyropoulou, M.; Sygletos, S. & Tomkos, I. (2007). Investigation of multi-wavelength regeneration employing quantum dot semiconductor optical amplifiers beyond 40 Gb/s. *Intern. Conf. on Transparent Optical Networks (ICTON)*, Tu.A1.7, July 1–5, 2007, Rome, Italy

Stix, M. S.; Keslar, M. P. & Ippen, E. P. (1986). Observations of subpicosecond dynamics in GaAlAs laser diodes. *Appl. Phys. Lett.*, Vol. 48, (June 1986) pp. 1722-1724

Stover, H. L. and Steier, W. H. (1966). Locking of laser oscillators by light injection. *Appl. Phys Lett.*, Vol. 8, (1966) pp. 91-93

Sygletos, S.; Spyropoulou, M.; Vorreau, P.; Bonk, R.; Tomkos, I.; Freude; W. & Leuthold, J. (2007). Multi-wavelength regenerative amplification based on quantum-dot semiconductor optical amplifiers (invited)," *Intern. Conf. on Transparent Optical Networks (ICTON)*, We.D2.5, July 1–5, 2007, Rome, Italy

Sygletos, S.; Tomkos, I. & Leuthold, J. (2008). Technological challenges on the road toward transparent networking. *J. Optical Networking*, Vol. 7, (2008) pp.321–350

Tangdiongga, E.; Liu, Y.; deWaardt, H.; Khoe, G. D. & Dorren, H. J. S. (2006). 320-to-40-Gb/s demultiplexing using a single SOA assisted by an optical filter. *IEEE Photon. Technol. Lett.*, Vol. 18, (2006) pp.908–910

Tsang, H. K.; Wong, C. S.; Liang, T. K.; Day, I. E.; Roberts, S. W.; Harpin, A.; Drake, J. & Asghari, M. (2002). Optical dispersion, two-photon absorption and self-phase modulation in silicon waveguides at 1.5 μm wavelength. *Appl. Phys. Lett.*, Vol. 80, (2002) pp. 416–419

Ueno, Y.; Nakamura, S. & Tajima, K. (2002). Nonlinear phase shifts induced by semiconductor optical amplifiers with control pulses at repetition frequencies in the 40–160-GHz range for use in ultrahigh-speed all-optical signal processing. *J. Opt. Soc. Am. B*, Vol. 19, (2002) pp. 2573-2589

Ueno, Y.; Nakamura, S.; Tajima, K. & Kitamura, S. (1998). 3.8-THz wavelength conversion of picosecond pulses using a semiconductor delayed-interference signal-wavelength converter (DISC). *IEEE Photon. Technol. Lett.*, Vol. 10, (1998) pp.346–348

Van der Pol, B. (1927). Forced oscillations in a circuit with non-linear resistance. *Philosophical Mag.*, Vol. iii, (1927) pp. 65-80

Vasilyev, M. & Lakoba, T. (2005). Fibre-based all-optical 2R regeneration for multiple WDM channels. *Optical Fiber Communication Conference and Exposition (OFC)*, OME62, March, 2005, Anaheim, USA

Vorreau, P.; Parmigiani, F.; Mukasa, K.; Ibsen, M.; Petropoulos, P.; Richardson, D. J.; Ellis, A. D.; Freude, W. & Leuthold, J. (2008). TDM-to-WDM conversion from 130 Gbit/s to 3 & 43 Gbit/s using XPM in a NOLM switch. *International Conference on Transparent Optical Networks (ICTON)*, Jun 22- 26, 2008, Athens, Greece

Wan, S. M.; Chow, C. W.; Tsang, H. K.; Su, Y. S. & Lin, C. F. (2005). Broadband wavelength conversion in semiconductor optical amplifier with non-identical multiple quantum wells. *Conference on Lasers and Electro-Optics Pacific Rim (CLEO-PR)*, CThC3-P18, 2005, Tokyo, Japan

Wang, C. H.; Shih, F. Y.; Yeh, C. H.; Chow, C. W. & Chi, S. (2009). 10 Gb/s TDM passive optical networks using four wavelengths multiplexed channels. *Opt. Commun.*, Vol. 282, (2009) pp. 2476-2479

Weber, H. G.; Ferber, S.; Kroh, M.; Schmidt-Langhorst, C.; Ludwig, R.; Marembert, V.; Boerner, C.; Futami, F.; Watanabe, S. & Schubert, C. (2006). Single channel 1.28 Tbit/s and 2.56 Tbit/s DQPSK transmission. *Electron. Lett.*, Vol. 42, (2006) pp. 178-179

Weich, K.; Eggemann, R.; Horer, J.; As, D. J.; Mohrle, M. & Patzak, E. (1994). 10 Gbit/s alloptical clocked decision circuit using two-section semiconductor lasers. *Electron. Lett.*, Vol. 30, (1994) pp. 784-785

Willatzen, M.; Uskov, A.; Mork, J.; Olesen, H.; Tromborg, B. & Jauho, A. P. (1991). Nonlinear gain suppression in semiconductor lasers due to carrier heating. *IEEE Photonics Technol. Lett.*, Vol. 3, (July 1991) pp. 606-609

Willner, A. E. (2008). All-optical signal processing in next-generation communication systems, *Optical Fiber Communication Conference and Exposition (OFC)*, February 2008, San Diego, USA

Winzer, P. J.; Raybon, G. & Duelk, M. (2005). 107-Gb/s optical ETDM transmitter for 100G Ethernet transport. *European Conference on Optical Communication (ECOC)*, Paper Th4.1.1, 2005, Glasgow, USA

Yamashita, S. and Suzuki, J. (2004). All-optical 2R regeneration using a two-mode injectionlocked Fabry-Perot laser diode. *IEEE Photon. Technol. Lett.*, Vol. 16, (2004) pp. 1176-1178

Yeh, C. H.; Chow, C. W.; Wang, C. H.; Shih, F. Y.; Wu Y. F. & Chi, S. (2009). Using multimode Fabry-Perot laser without external-injection for wavelength conversion. *Electron. Lett.*, Vol. 45, (2009) pp. 327-329

Yoo, S. J. B. (1996). Wavelength conversion technologies for WDM network applications. *J. Lightwave Technol.*, Vol. 14, (Jun. 1996) pp. 955-966

Zarris, G.; Parmigiani, F.; Hugues-Salas, E.; Weerasuriya, R.; Hillerkuss, D.; Gonzalez, N.A.; Spyropoulou, M.; Vorreau, P.; Morais, R.; Ibrahim, S.K.; Klonidis, D.; Petropoulos, P.; Ellis, A.D.; Monteiro, P.; Tzanakaki, A.; Richardson, D.; Tomkos, I.; Bonk, R.; Freude, W.; Leuthold, J. & Simeonidou, D. (2009). Field trial of WDM-OTDM transmultiplexing employing photonic switch fabric-based buffer-less bit-interleaved data grooming and all-optical regeneration. *Optical Fiber Communication Conference and Exposition (OFC)*, PDPC10, Mar 22-26, 2009, San Diego, USA

18

Wavelength Conversion and 2R-Regeneration in Simple Schemes with Semiconductor Optical Amplifiers

Napoleão S. Ribeiro[1], Cristiano M. Gallep[2], and Evandro Conforti[1]

[1]*Department of Microwave and Optics (DMO) -University of Campinas – Unicamp*

[2]*Division of Telecommunication Technology (DTT) of FT/Unicamp*

Brazil

1. Introduction

Future optical networks may require both wavelength conversion and bit shape regeneration in an all-optical domain. The possibility of pulse reshaping while providing wavelength conversion may support new demands over medium and large distances links (Kelly, 2001). Indeed, during propagation the optical data signal suffers deterioration due to the amplified spontaneous emission (ASE) from optical amplifiers, pulse distortion from intrinsic dispersion, crosstalk, and attenuation. All-optical regenerators may be important components for the restoration of these signals, providing complexity and cost reductions with the avoidance of optoelectronic conversions. The regeneration could be 2R (*re-amplification* and *reshaping*) or 3R, which also provide *retiming* to solve jitter (Simon et al., 2008). Several 3R regenerators using the ***semiconductor optical amplifier*** (SOA) have been proposed, such as cascaded SOAs setups (Funabashi et al., 2006) or SOA based Mach-Zehnder interferometers (MZI) (Fischer et al., 1999).

However for small and medium distances systems, where the signal amplitude noise and distortions form the main problem and where jitter has fewer magnitudes, the simpler 2R processes can be adequate to keep signal quality (Simon et al., 1998). In addition, the SOA is a helpful device for both 2R-regeneration (Ohman et al., 2003) and wavelength conversion (Durhuus et al., 1996). Several techniques for 2R-regeneration based on SOAs have been proposed and tested, for example by using *four-wave mixing* (FWM) (Simos et al., 2004), *cross-gain modulation* (XGM) (Contestabile et al., 2005), integration within MZI (Wang et al., 2007), multimode interferometric SOA (Merlier et al., 2001), *cross-phase modulation* (XPM) with filtering (Chayet et al., 2004), and *feed forward technique* (Conforti et al., 1999). However, these techniques require complex designs and involve critical operation points, even the simplest ones based on XGM features. In addition, most of these techniques are not capable of wavelength conversion and regeneration simultaneously.

Recently a regenerator based on cross-gain modulation was proposed using one SOA for wavelength conversion (in a counter-propagating mode) and another deeply saturated SOA (synchronized by an optical delay line) to achieve cross-gain compression (Contestabile 2005). This efficient approach has similarities with the all-optical feed-forward techniques. In addition, this regenerator could not done wavelength conversion if the wavelength of the input signal is chosen in the output. Although good results can be obtained, this technique demonstrates to be complex since it used optical delay line and two SOAs. In this chapter, we introduce a more simple technique based on XGM (using just one SOA, an optical isolator, an optical circulator, and a CW laser) with easy robust operation at high speed reconfiguration (Ribeiro et al., 2008; Ribeiro et al., 2009a). The regeneration is based on the abrupt profile of the SOA cross-gain modulation efficiency, which is compressed at high input optical powers. The two optical carriers are amplified in the counter-propagating mode allowing conversion to another or to the same wavelength.

In addition, we present 2R-regeneration and conversion results for different kinds of deteriorated input signals. Experimental results such as eye diagrams and measured Q-factors are also shown, for various optical input powers, carriers detuning, bit rates and optical polarizations. Moreover, the estimative of the bit error rates (BER) are presented. Finally, the regenerator extinction ratio (ER) deterioration and its relation with the Q-factor improvements are discussed.

2. Experimental setup

The single-SOA all-optical 2R-regenerator setup is presented in Figure 1. This regenerator will be called 2R-converter. The experimental scheme is divided in blocks. In the first block, the optical carrier at λ_1 is modulated by pseudo random bit sequence (PRBS) data. In most cases a non–return to zero (NRZ) modulation was used, and the polarization of the input signal was controlled to maximize de modulator response.

In the *Deterioration Block,* the signal was degenerated by different types of deterioration processes to analyze the regenerative effects of this device. In Figure 1, between point 1 and 2, the three elements used to deteriorate the input signal are presented: another SOA as a booster, a buried fiber link (*KyaTera-Fapesp* Project) and an erbium doped fiber amplifier (EDFA). Different *deterioration cases* were obtained by combining these elements.

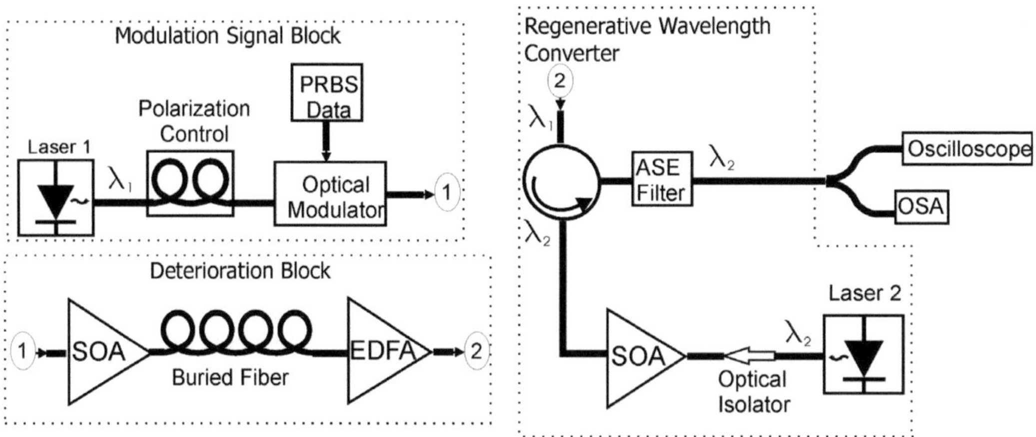

Fig. 1. All-optical 2R-regenerator and wavelength converter experimental setup.

The block for regeneration and wavelength conversion is the last one (Ribeiro et al., 2009a). In this block the modulated signal at λ_1 was converted to the wavelength of the laser 2 (λ_2), occurring regeneration and wavelength conversion simultaneously. This 2R-converter is a very simple device with just a laser CW (Continuous Wave), a non-linear SOA, an optical

circulator and an optical isolator. These last two components are needed for operation in a counter-propagating mode of the wavelength converter based on XGM. The optical filter presented in this block was used to reduce the ASE noise added by the non-linear SOA, and to allow better eye diagrams visualization at the oscilloscope. If an oscilloscope with higher sensitivity was used, this optical filter might not be needed. In this way, this optical filter after the regenerator is not considered here as a regenerator component.
The SOA features are presented in Table 1. The non-linear commercial SOA was biased at 300 mA (near the maximum supported current of 400 mA) to obtain the regenerative effects.

Item	Test condition	Values
Small signal gain	I = 200 mA	25 - 30 dB
Polarization dependent saturated gain (PDG)	I = 300 mA, P_{in}> 0 dBm	0.5 – 1 dB
Saturated output power	I = 200 mA	6 – 8 dBm
Gain peak	I = 200 mA	1550 – 1570 nm
Saturated gain recovery time	I = 300 mA, P_{in} > 0 dBm, 1555 nm	16 - 25 ps
3 dB optical bandwidth	I = 200 mA	45 nm
Active cavity length		2 mm
Bias current		up to 400 mA

Table 1. Parameters of the non-linear type encapsulated SOA.

In some *deterioration cases*, an optical attenuator was used before the oscilloscope to maintain the output signal power at the same level of the input signal, in order to carry out a bit reshaping comparison, excluding the regenerator gain.
Regenerator characterization was made for different parameter variations as for example: the optical power of lasers 1 and 2; bit rates; detuning; polarization angle of the input signal; and the extinction ratio (ER).
The different modulated input signal *deteriorations cases* are presented in the following subsections. Theses deteriorations were quantified by the signal Q-factor. This parameter is calculated by (Agrawal, 2002):

$$Q = \frac{I_1 - I_0}{\sigma_1 + \sigma_0} \qquad (1)$$

In (1), I_1 and I_0 are the current level of the bits levels "1" and "0", respectively; σ_1 and σ_0 are standard deviation of the level '1'and '0', respectively.

2.1 Case "SOA"

In this first *deterioration case*, another SOA was used to deteriorate the modulated input signal at λ_1. This SOA acted as a booster, amplifying and adding ASE noise. Depending on the power level of the input signal of this SOA, an overshoot related to the saturation of this device could happen. The 2R-converter performance is better for higher overshoot levels since this device totally removes the overshoot.
An optical band-pass filter was needed due to the higher level of ASE noise added to the signal. The modulated input signal Q-factor could be changed by varying the laser 1 power level and/or the bias current of the SOA used as a booster.

2.2 Case "LINK+SOA"

In this *deterioration case* the modulated signal is degenerated by dispersion and attenuation of an 18-km standard buried fiber link of the *KyaTera-Fapesp* project (www.kyatera.fapesp.br). The fiber Corning SMF-28 Standard was used. The Table 2 shows some features of this fiber. Due to the attenuation, another SOA was used to amplify the signal in order to achieve the power level (at the entrance of the 2R-converter) enough to reach regenerative effects, besides better visualization on the oscilloscope. The modulated input signal Q-factor could be changed in a similar way of the previous case.

Item	Values
Attenuation	≤ 0.22 dB/km
Dispersion	16 to 19 ps/km-nm
Effective Area	80 μm^2
Numerical aperture	0.14
Zero-dispersion wavelength	1313 nm
Polarization mode dispersion	≤ 0.2 $ps/km^{1/2}$

Table 1. Fiber Corning SMF-28 features.

2.3 Case "EDFA"

In this *deterioration case* an EDFA was used to amplify the modulated input signal adding ASE noise. The higher ASE noise addition cause higher bit level (at both "1" and "0") variance. The modulated input signal Q-factor could be modified by varying the laser 1 power and/or EDFA pump laser power.

2.4 Case "LINK+EDFA"

This *deterioration case* is similar to the *case* "LINK+SOA". The 18-km standard buried fiber link of the KyaTera-Fapesp project was used again. The dispersion effects, attenuation, and ASE noise are the deterioration effects added to the modulated input signal.

2.5 Case "SOA+LINK+EDFA"

This *deterioration case* is the more complex case. It involves the three elements used to degenerate the modulated input signal. The overshoot appearance, the noise adding to bit levels "1" and "0" and high variance in those levels turn the modulated input signal presented in this case as the most deteriorated case. Optical filters were needed to reduce the ASE noise.

3. 2R-regenerator and wavelength converter working principle

The quality of a 2R regenerator depends on its ability to suppress optical noise and to improve the extinction ratio. The ideal regenerative properties are provided by a system with a transfer function as close as possible to an ideal "S" like behavior. This refers to a characteristic function with the following properties: wide and flat dynamic range gain for bit levels "1" and "0" in order to suppress the noise, and a linear increasing curve which determines the discrimination between bit levels "0" and "1" (Simos et al., 2004).

The 2R-regenerator presented in this chapter has a characteristic function similar to the "S" like behavior. The response of the device here is similar, not equal to the "S", since it presents gain compression only for the bit level "1". This is due to the fact that just the power level of this bit (at the high level "1") can saturate the SOA gain. The compression of the level "1" is noticed in cases where an overshoot is presented. After the regenerator, the overshoot is removed because SOA gain is deeply saturated. In this way, the 2R-converter works as a "low-pass filter", removing "higher frequencies" present in the overshoot. Therefore, saturated gain acts as a power equalizer for the fluctuation in the bit level "1" reducing the noise. On the other way, the bit level "0" has much less improvement than bit level "1", since its low signal power level cannot saturate the SOA.

The 2R-converter of this chapter is based on XGM effect in SOAs, and this non-linear effect always degrades the extinction ratio due to the ASE noise added by the SOA. Consequently, there is no extinction ratio improvement. However, it will be showed in this chapter that the improvements caused by the 2R-converter and quantified by Q-factor improvement can be higher than the ER degradation, at least for some cases.

The pattern dependence reduction caused by operating the 2R-converter in its optimum optical input powers could improve both the format of the bit levels "1" and "0". As it will be shown, there is an optimum relationship between the modulated input signal power (at λ_1) and the CW signal power (at λ_2), which should be maintained for different cases in order to kept the pattern dependence effects at the same level.

As mentioned before, the regeneration in the 2R-converter occurs simultaneously with the wavelength conversion via XGM. The regenerative effects are associated with the wavelength conversion efficiency, and XGM is the simplest technique using SOA to implement wavelength conversion. In this process, a strong modulated input signal at λ_1 saturates the amplifier. A continuous wave pump signal at λ_2, injected simultaneously with the modulated signal, is modulated by the gain saturation while being concurrently amplified. The output signal, properly filtered, is a replica of the modulated input signal at a different wavelength with a phase inversion of 180^0. In this way, methods to obtain the data without phase inversion are needed. One way is to use a proper software control. Commands given to the receiver force it to associate a bit "1" (in its input) to a bit "0" (in the original data). Another way is to use others devices like an additional SOA, but it will be enlarge the ER degradation. The exploiting of the phase changes provoked by the XGM effect using an optical filter could invert the output signal without ER degradation (Leuthold et al., 2003). Others forms are for example, the use of: delay interferometer (Liu et al., 2007) and quantum dots SOAs with narrow optical filters (Raz et al., 2008).

The eye diagrams obtained after the 2R-converter presented in this chapter are inverted in phase. In this manner, the commentaries made before about the deterioration of the bit levels "1" and "0" refer to bit levels of modulated input signal. Otherwise, the improvement observed in the output signal is commented about the inverted bit levels, that is: for deterioration in the bit level "1" of the modulated input signal, the improvement is noted in the level "0" after the 2R-converter; and in a similar way for the deterioration in the level "0" of the modulated input signal.

The gain saturation of the SOA is related to the XGM wavelength conversion and to the gain compression. The last one is responsible for the noise reduction in the bit level "1" and for the overshoot elimination. The SOA gain must be deeply saturated to obtain the gain compression effects. This behavior can be noted in Figure 2. The eye diagram of the modulated input signal (λ_1) deteriorated by the *case* "SOA" is presented in Figure 2(a). This eye diagram presents overshoot caused by the gain saturation of the SOA used as a booster that deteriorated the input signal. The eye diagram of the modulated signal (λ_1) after the 2R-converter is presented in Figure 2(b). This much clear output signal was obtained for the regeneration case where it was observed a Q-factor improvement for the converted signal at λ_2 equal to 1.5. Due to the higher level power of the modulated and CW input signals, and the high SOA bias current (300 mA), the SOA in the 2R-converter presents a saturated gain. In this manner, the eye diagram for the output signal after the regenerator presents compressed eye. The overshoots and undershoots observed in the output signal might be associated to a self-phase modulation (SPM) and/or changes in the signal phase around the narrow band optical filter. These results of compressed eye diagrams were observed for other values of the modulated input signal power. Therefore it confirms that the gain of SOA used in the 2R-converter is saturated, inducing the compression needed to the regenerative effects occurrence.

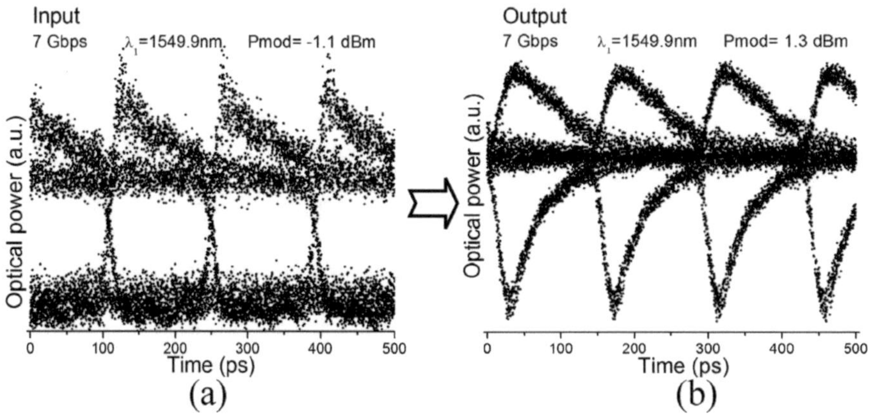

Fig. 2. Eye diagrams: (a) modulated input signal for the *deterioration case* "SOA"; (b) modulated signal after 2R-converter.

4. Optical spectrum and signal to noise ratio

The optical spectrum of the input and converted signals in those particular points of the experimental setup (Figure 1) are presented. The spectrums were obtained for the *deterioration case* "SOA" as an example, since the optical spectrum is similar to other *deterioration cases*. The *optical signal to noise ratio* (OSNR) was calculated as well as the eye diagrams were obtained corresponding to the optical spectrums illustrated in Figure 3.

The case of Figure 3 employed a wavelength conversion from 1550 to 1551 nm with a modulation rate of 7 Gbps NRZ. The Figure 3(a) shows the modulated input signal without deterioration and an eye diagram without distortions. A Q-factor of 9 and OSNR of 55.28 dB are observed in this case. Then this modulation input signal is inserted into the SOA used as a booster. The gain of this SOA is saturated due to the power level of the input signal of -2 dBm and a bias current of 130 mA (higher than current threshold). Therefore the eye diagrams presented in Figure 3(b) are obtained with much noise in the bit levels "1" and "0" as well as overshoots. After the SOA used as a booster, the Q-factor decrease to 4.3 as well as the OSNR to 37.6 dB. In this manner, this SOA presented a noise figure of 18.3 dB caused by deeply saturated gain and by extra noise added to the signal. An optical band-pass filter is needed to filter the ASE noise added. Consequently, the signal in Figure 3(c) appeared filtered with an improvement in the Q-factor to 4.8.

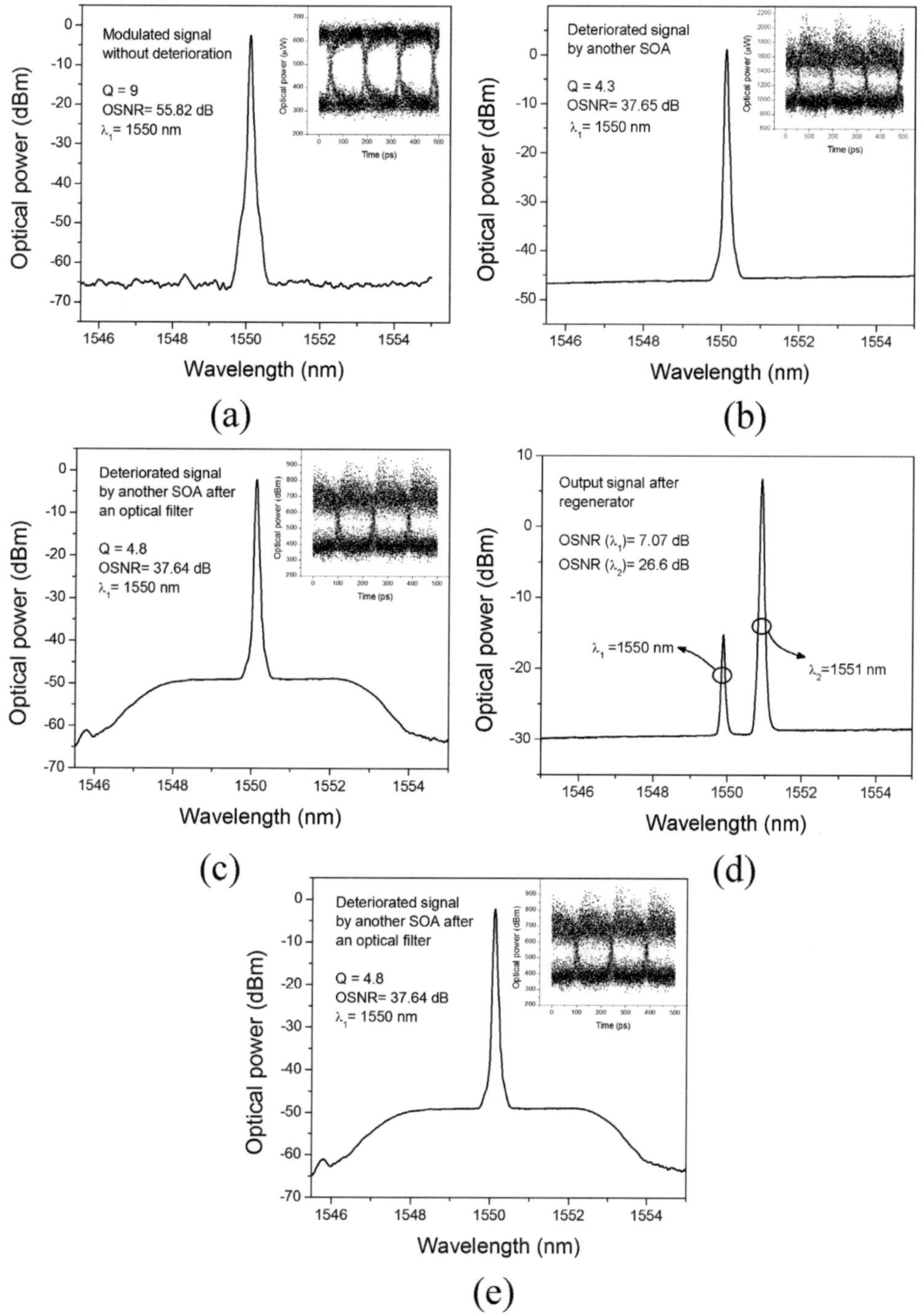

Fig. 3. Optical spectrums and eye diagrams of the *deterioration case* "SOA": (a) modulated signal without deterioration; (b) modulated signal deteriorated by another SOA; (c) modulated signal deteriorated by another SOA and filtered; (d) output signal after the 2R-converter; and (e) regenerated signal and filtered.

The optical spectrum after the 2R-converter is presented in Figure 3(d). The input signal initially at 1550 nm (λ_1) is still present but with an OSNR of 7 dB. According to the experimental setup of Figure 1, the original signal at λ_1 should not be present because the conversion scheme is a counter-propagating mode with an optical isolator that should eliminate this original signal. Nevertheless, the original signal presence after the 2R-converter could be explained by possible internal reflections in the optical isolator and in the SOA used to regenerate the signal. The regenerated and converted signal at 1551 nm (λ_2) of Figure 3(d) presents an OSNR of 26.6 dB, i. e., the SOA used in the regenerator presented a noise figure of 11 dB. As the same case mentioned before, this noise figure higher than commons values (7 a 8 dB) could be justified by the presence of both the modulated input signal power (-2 dBm) and the CW signal power (-6 dBm) as well as higher bias current (300 mA) which saturated the SOA gain, adding a lot of noise to the signal. Eye diagram is not illustrated in Figure 3(d) due to the high power level of the output signal (7 dBm). Indeed, the bit level "1" of the output signal eye diagram was in the upper part of the oscilloscope scale limit, not allowing the acquisition of points by the software Labview (using GPIB port). Despite this limitation, a reduction of the noise in bit levels "1" and "0" and an overshoot elimination were noted in the eye diagrams (quantified by Q-factor improvement from 4.8 to 7.2). These results evince that the original signal still present at λ_1 does not decrease the regenerative effects.

The regenerated and converted signal at 1551 nm (λ_2) after an optical narrow filter is illustrated in Figure 3(e). The optical filter allows an OSNR improvement to 63.2 dB. The eye diagram observed in this figure presents improvements already mentioned in the previous case. These improvements increase the Q-factor to 7.5. Through calculating the Q-factor variation from the modulated input signal, an improvement of 2.7 can be observed.

For the cases presented in this section and in Figure 3, the results after the regenerator were not attenuated to guarantee the same power level of the modulated input signal. This was made to allow the observation of the 2R-converter performance as a whole, analyzing the re-amplification and reshaping.

The calculation of the OSNR was made following application notes published by the manufacturer of the optical spectrum analyzer used here. Therefore the optical spectrums presented in Figure 3 are just illustrations since the accurate OSNR calculations need an operation of the optical spectrum analyzer with higher resolution and smaller span.

The optical spectrums and the OSNR were presented just for the *deterioration case* "SOA". The optical spectrums for the others *deterioration cases* are similar, presenting differences in the optical power values. In relation to the OSNR calculation, the values obtained for the other *deterioration cases* are very close, with variations due to: the optical signal power used; bias current of the SOA; and the pump laser power used in the EDFA. A study of the ER degradation will be presented in following sections in order to analyse the signal degradation after the 2R-converter that was caused by the ASE noise addition of the SOA. This study will help to understand how the noise degenerate the signal for different cases, associating these results with the OSNR deterioration. In this manner, it will be possible to estimate the OSNR behavior for the different deterioration cases not presented in this section.

5. Re-amplification

The 2R-converter presented in this chapter provides re-amplification and bit reshape. Thus, the first improvement caused by this regenerator is the signal re-amplification that will be present in this section. The *deterioration case* "SOA" is used to illustrate the optical gain originated by the 2R-converter. A wavelength conversion from 1550 nm (λ_1) to 1551 nm (λ_2) with a bit rate of 10.3 Gbps was used. The optical gain was calculated as the difference between the output signal power at λ_2 and input modulated signal power at λ_1. The results are presented in Figure 4.

The optical gain versus CW signal power at λ_2 is illustrated in Figure 4(a). A gain increase with the CW signal power can be noted. This happens because the output signal was kept at λ_2. Thus, by increasing the CW signal power, the output signal power at λ_2 increases too. Since the optical gain was calculated as a function of the modulated input signal power, which is fixed for each curve in Figure 4 (a), the optical gain increases linearly with the CW signal power. In Figure 4(b), an optical gain decreasing with the modulated signal power increasing is noted, presenting higher optical gain values for the modulated signal power around -7.5 dBm. This result is associated with the SOA gain saturation. Some Q-factor improvements (figured by ΔQ) are showed in Figure 4(a) and (b) just to illustrate the dependence of this parameter with the power relation, which will be commented in other section. Here, ΔQ is defined as the difference between the Q-factor of converter signal at λ_2 and Q-factor of the modulated input signal at λ_1.

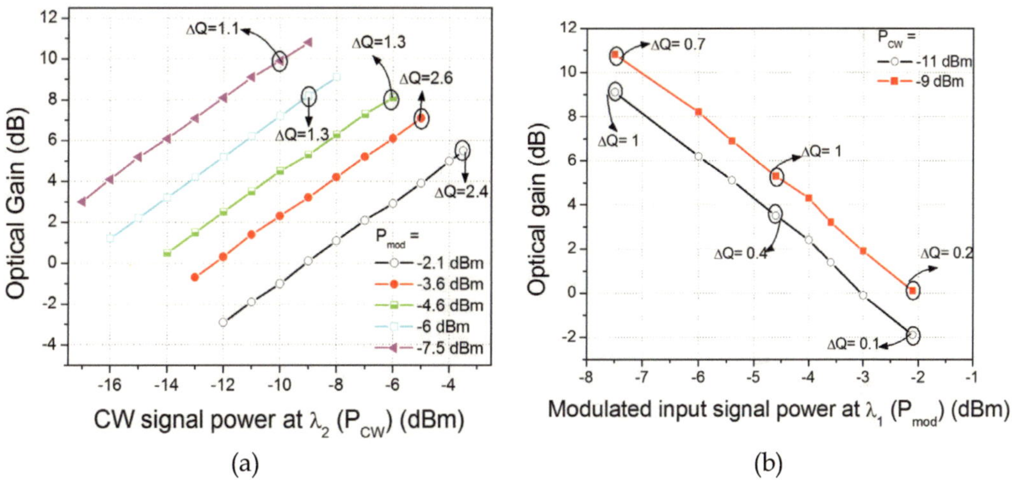

Fig. 4. 2R-converter optical gain of the *deteriorated case* "SOA": (a) optical gain versus CW signal power for different input modulated signal powers; (b) optical gain versus input modulated signal power for different CW signal power.

The 2R-converter presented an optical gain varying from -3 to 12 dB. In most of the cases, the better values of Q-factor improvement occurred for higher optical gain values. In this manner, it is clear that the 2R-converter is capable to re-amplify the signal, presenting optical gain up to 12 dB, together with the bit reshape quantified by the Q-factor improvement.

These results of optical gain presented here are proper of the SOA and can be associated to the input modulated signal power and CW signal power. In this way, if the same values of the input optical powers in Figure 4(a) and (b) is used for the others *deterioration cases*, the results should be similar to the ones presented here.

6. Eye diagrams

The eye diagrams obtained from the oscilloscope clarify the improvements obtained by the 2R-converter. In this section, some eye diagrams of the different *deteriorated cases* are presented to illustrate the improvements of the bit shape. An up-conversion (1550 to 1551 nm) was employed to obtain the eye diagrams. This type of conversion causes higher ER deterioration, but in other way it also provides higher SOA gain saturation, which contributes to a better signal regeneration. Therefore, all the eye diagrams presented in this section as well as most of the results presented in this chapter were obtained from up-conversion. In addition, it is valid to comment that the output eye diagrams are inverted in relation to input signal.

Initially, input and output eye diagrams and the respective Q-factors for the bit rate of 10.3 Gbps NRZ are illustrated in Figure 5, where the output signal was not attenuated to the same level of the input modulated signal power. Therefore the illustrated eye diagrams present two regeneration effects: re-amplification and reshaping. An arbitrary unit for the optical power was considered to allow the comparison between input and output eye diagrams in same proportion.

Two *deterioration cases* are studied. The first case is "LINK+SOA", which presented a medium quality input signal (Figure 5(a)) with Q-factor of 5.7, presenting intense pulse distortion due to the intrinsic dispersion caused by the 18-km standard buried fiber link (Ribeiro et al., 2009a). The dispersion effect could be noted by the triangular form of the pulse. In Figure 5(b), the regenerated output signal presents a higher eye opening, a reduction of the overshoots, and of the bit level (at both "1" and "0") variance, facts quantified by the Q-factor increasing to Q=10.

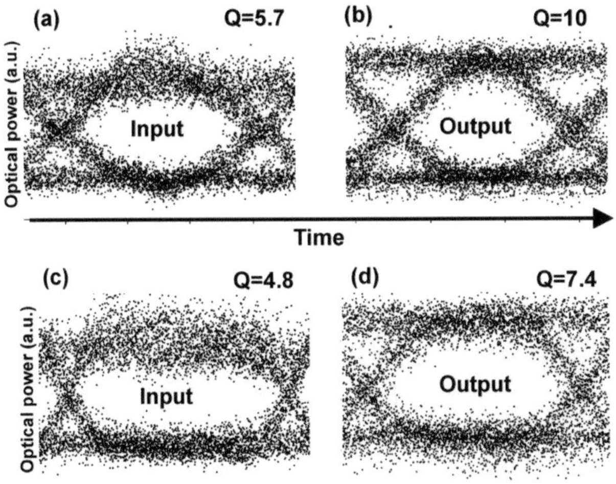

Fig. 5. Eye diagrams (NRZ, 10.3 Gbps): (a) *case* "LINK+SOA" input signal with Q=5.7; (b) output signal with Q=10; (c) *case* "SOA" input signal with Q=4.8; (d) output signal with Q=7.4 (adapted from Ribeiro et al., 2009a).

The second *deterioration case* is "SOA" (Ribeiro et al., 2009a). In Figure 5(c), the eye diagram presents low quality (Q=4.8) due to the pattern dependence effect, overshoots, and the great amount of noise added to both bit levels by the SOA used as a booster. As the case mentioned before, an improvement in the eye opening can be observed as well as an overshoot elimination is noted by the small variance of the bit level "0" of the regenerated signal. In addition, the bit levels "1" and "0" of the regenerated signal present lower width (reduction of the bit level variance) if be compared to the inverted bit level of the input signal. These improvements are quantified by the increasing of the Q-factor to 7.4.

Eye diagrams for bit rate of 7 Gbps for the same cases mentioned before are presented in Figure 6. An important difference is that in Figure 6 the output signal is attenuated to guarantee the same level of modulated input signal power. In this manner, the unit of μW could be used. This study of output signal attenuated analyzes just the improvement provoked by the bit reshaping.

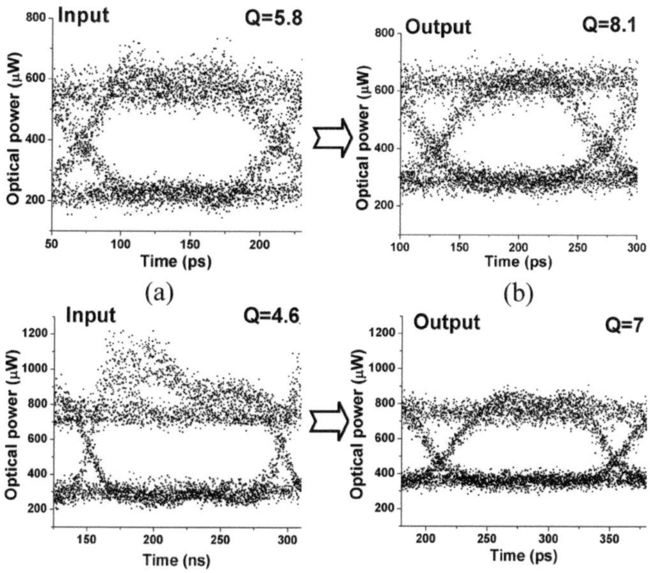

(a) (b) (c) (d)

Fig. 6. Eye diagrams (NRZ, 7 Gbps): (a) *case* "LINK+SOA" input signal with Q=5.8; (b) output signal with Q=8.1; (c) *case* "SOA" input signal with Q=4.6; (d) output signal with Q=7.

Despite the output signal attenuation, the behavior is similar to the cases mentioned in Figure 5. In *case* "LINK+SOA", the deterioration effects presented in the input signal of Figure 6(a) are the same presented in previous figure, as well as the improvements in the output signal (Figure 6(b)). These similarities are quantified by the Q-factors 5.8 and 8.1 for the input and output signal respectively. A decreasing of the Q-factor improvement can be noted by comparing to the previous case. This result is due to the attenuation of the output signal. Besides, this is another situation where the modulated and CW signals powers are different from the cases of Figure 5. The eye diagrams illustrations are used to observe the improvement caused by the 2R-converter, comparing the input and output signal in each case, and not to make comparisons between the different *deterioration cases* where different parameters are used.

The *deterioration case* "SOA" is presented in Figure 6(c). The input signal presented a higher overshoot as well as deterioration caused by the pattern dependence effect and ASE noise added by the SOA used as a booster. The output signal presents overshoot elimination and lower variance of both bit levels. The improvements are quantified by the Q-factor increasing from 4.6 to 7.

The others *deterioration cases* like "LINK+EDFA" and "EDFA" are illustrated in Figure 7. These eye diagrams were obtained for a bit rate of 7 Gbps NRZ. Besides, the output signal was attenuated to guarantee the same level of the modulated input signal power. In the *case* "LINK+EDFA", the input signal illustrated in Figure 7(a) presents a high amount of ASE noise added by the EDFA, and deterioration caused by the dispersion of the buried fiber link. The bit levels of "1" and "0" present a large width, i.e., high variance. In the output signal (Figure 7(b)) the increasing of the eye opening is noticeable. It is caused by the decrease of the noise present in bit levels "1" and "0", visualized by the variance reduction of these levels. A higher noise reduction is noted to the bit level "0" of the regenerated signal. The Q-factor was increased from 4.5 to 6.2.

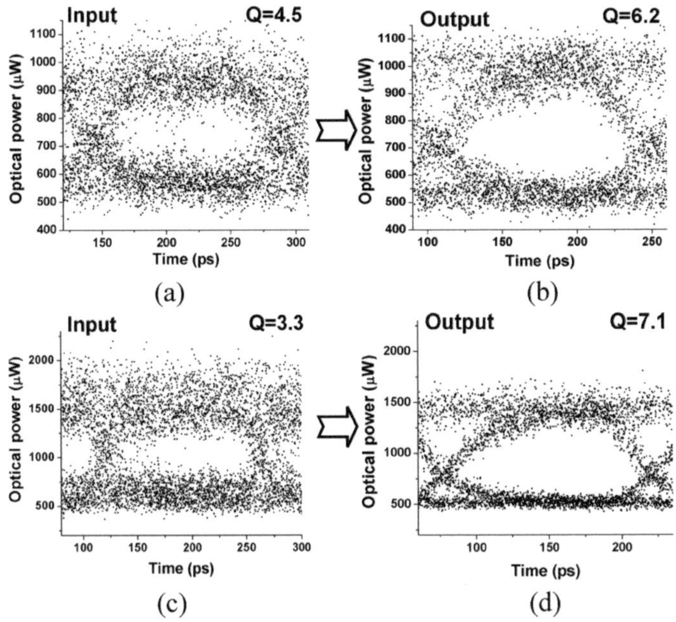

Fig. 7. Eye diagrams (NRZ, 7 Gbps): (a) *case* "LINK+EDFA" input signal with Q=4.5; (b) output signal with Q= 6.2; (c) *case* "EDFA" input signal with Q=3.3; (d) output signal with Q=7.1.

The *deterioration case* "EDFA" is illustrated in Figure 7(c). A great amount of ASE noise deteriorating the input signal with a low eye opening can be noted. This higher deterioration presented in the input signal is quantified by the low Q-factor of 3.3. In Figure 7(d), the improvement caused by the 2R-conveter can be observed. Due to the SOA gain saturation, the noise is reduced in both bit levels "1" and "0". In the last one bit level, a lower variance can be noted. With the ASE noise reduction, the eye opening increase as well as the Q-factor to 7.1.

The last *deterioration case* illustrated involves all the degeneration effects: "SOA+LINK+EDFA". For this last case, a bit rate of 7 Gbps NRZ as well as output signal attenuation are used (Ribeiro et al., 2009a). The Figure 8(a) illustrates the input signal which presents a higher overshoot caused by the SOA used as a booster. Besides, the input signal presents higher variance in both bit levels provoked by the ASE noise addition by the SOA and EDFA. The input signal also presents a bit enlargement caused by the intrinsic dispersion of the buried fiber link. The output signal illustrated in Figure 8(b) presents a noise reduction in both bit levels. The overshoot was reduced as well as the fluctuations presented in the bit level "1". Nevertheless, the output signal presents a lower difference between the bit levels "0" and "1", i. e., lower extinction ratio (ER). The improvement observed in Q-factor was from 5.3 to 8.6.

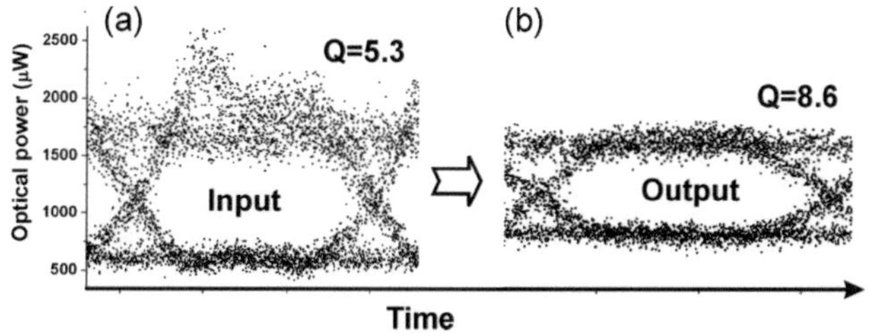

Fig. 8. Eye diagrams (NRZ, 7 Gbps): (a) *case* "SOA+LINK+EDFA" input signal with Q=5.3; (b) output signal with Q= 8.6 (adapted from Ribeiro et al., 2009a).

Due to the availability of equipments, the same bit rate could not be used for all the cases studied. In this manner, the use of modulation rate higher than 7 Gbps NRZ was used just for some measurements, but for most cases the characterization was limited to 7 Gbps.

By comparing the eye diagram of the output signal with the input, a reduction of cross point level between "0" and "1" can usually be noted. This reduction is more pronounced for cases where the output signal is attenuated. The ER degeneration is the main reason for this cross point level reduction. Another reason is the SOA gain recovery time. The rising time of the bit level "1" is slower than the falling time, decreasing the cross-point level.

The modulation RZ (Return to zero) was used in the 2R-converter characterization either. The Figure 9 presents the eye diagrams for R1 modulation that correspond to inverted RZ. A wavelength conversion from 1550 to 1551 nm was used without attenuation in the output signal. Figure 9(a) illustrates the input signal for the *deterioration case* "LINK+SOA". The pulse presents a triangular shape due to the buried fiber link. A decreasing in the variance of the bit level "1" of the input signal can be noted in Figure 9(b). An estimation of the regenerative effects was done using the variance of both bits levels. An improvement of 51% was obtained for this *deterioration case*.

The *deterioration case* "SOA" is illustrated in Figure 9(c), presenting pulse shape more rectangular and more ASE noise in the bit level "1". The output signal (Figure 9(d)) presents narrower pulses due to the gain response time of the SOA. Besides, there is a reduction in the bit level "1" variance. As the previous case, an estimate was calculated to eye opening improvement being obtained 44%.

These results proved that the 2R-converter is capable to regenerate RZ signals. Nevertheless, the results present in this chapter use just NRZ modulation since this modulation type is more complex. Furthermore the Q-factor used to quantify the regenerative effects is just provided by the oscilloscope for the NRZ modulation type.

The overshoot elimination presented for the cases in which another SOA was used to amplify the modulated input signal is a good feature of the 2R-converter. In optical systems, the overshoot could be added to the signal from different forms, one of them is the use of the PISIC technique used to increase the speed of the electro-optical switching using SOAs (Gallep & Conforti, 2002) (Ribeiro et al., 2009b).

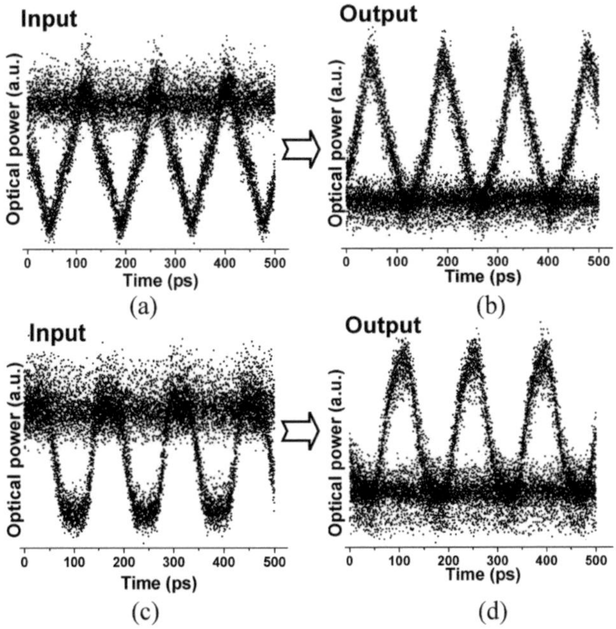

Fig. 9. Eye diagrams (R1, 7 Gbps): (a) *case* "LINK+SOA" input signal; (b) output signal; (c) *case* "SOA" input signal; (d) output signal.

The overshoots presented in the *deterioration case* "SOA", "LINK+SOA", and "SOA+LINK+EDFA"are eliminated by the 2R-converter as could be observed in Figure 5, 6, and 8. This elimination occurs due to the saturation effects of the SOA gain. The SOA does not maintain the gain level for those higher power values present in the overshoots. In this manner, the overshoot is not transferred to the CW signal by the wavelength conversion.
Therefore, the 2R-converter is a possible solution to eliminate the overshoot of an optical signal. Despite the overshoot elimination, the signal after the regenerator will present ER decreasing, as it will be shown in future sections. Thus, the analysis if the overshoot elimination can compensates for the ER degeneration is necessary.

7. Optical polarization

The input light polarization dependence of the wavelength conversion is very important since polarization is an unpredictable factor in real optical systems and an automatic polarization controller can be expensive. The SOA used in the 2R-converter presents a *polarization dependent saturated gain* (PDG) of less than 1 dB. Studies of the input polarization angle influence in the 2R-converter performance were done to confirm this value.
By adjusting de polarization controller, different polarization angles of the modulated input signal were obtained to analyse the Q-factor and gain variation. The Figure 10 presents eye diagrams of different polarization angles for the *deterioration case* "SOA". The modulated input signal changes very little for each polarization angle. In this manner, the eye diagram presented as *input* is representing all the input eye diagrams. The eye diagrams following in the time scale illustrate the variations noted in the output signal for some input polarization angles. The regenerative effects are presented in all the output signal with overshoot elimination and the reduction of the bit levels "1" and "0" variance.

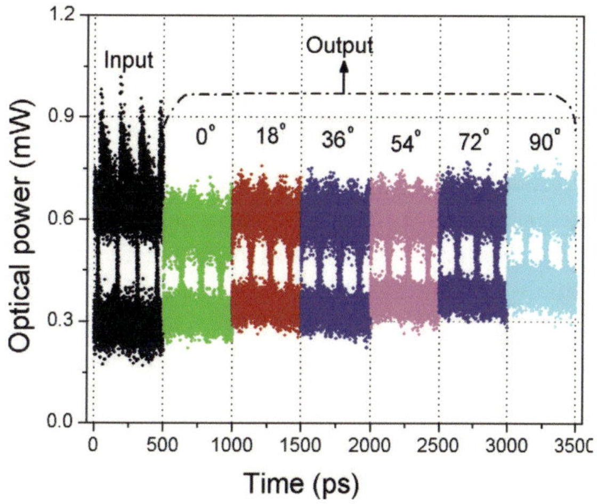

Fig. 10. Eye diagrams for different polarization angles of the input signal for the *deterioration case* "SOA".

Observing the eye diagrams for different polarization angles of the input signal, regenerative and power variations could be noted. These variations can be better observed in Figure 11 (a) where the optical gain varying from 0.6 to 1.5 dB is observed. In addition, it was observed that for this deterioration case, the Q-factor improvement (ΔQ) varied from 0.7 to 1.9.

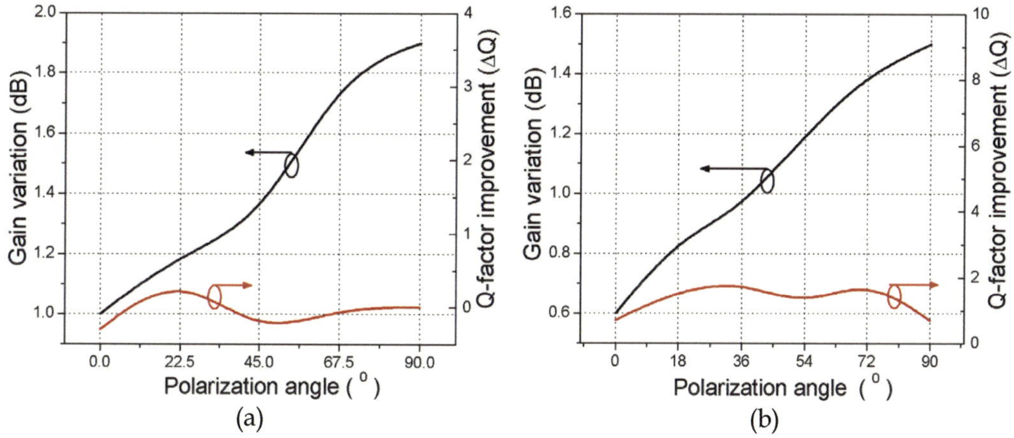

Fig. 11. Gain and Q-factor improvement variation as function of different input polarization angles for the *deterioration case*: (a) "SOA" and (b) "LINK+SOA" (adapted from Ribeiro et al., 2009a).

A study of the *case* "LINK+SOA" was done in a similar manner, obtaining the results presented in Figure 11(b). In this case, a gain variation of 0.9 dB and a Q-factor improvement variation of 0.9 were obtained.

In general, the 2R-converter presented low dependence with the input signal polarization, fact proved by the results in Figure 11, with a gain variation of less than 1 dB. This behavior is explained by the use of a SOA with low PDG. Thus, the 2R-converter has this advantage of low polarization dependence, with a practical effect in real systems where the fiber income signal polarization is not known or cannot be controlled.

8. Input optical power: the modulated and the CW signals

The power levels of the modulated (λ_1) and CW (λ_2) signals are important to achieve the SOA gain saturation used in the 2R-converter, necessary for regeneration and conversion by XGM effect. Moreover, the power levels of these input signals affect the power level of the output signal and the quality of the input signal. A study of the Q-factor improvement for different power level pairs of CW and modulated channels, for each different case of deterioration was done. For the cases presented in this section, up-conversions were performed from 1550 to 1551 nm at a rate of 7 Gbps, with attenuation of the output power to guarantee the same level in relation to the modulated input signal power.

In the Figure 12(a) (*case* "SOA") it is noted that there is a Q-factor improvement increasing while the CW signal power increases, with a maximum ΔQ value for a power value of the CW signal above from 2 to 3 dB, compared to the modulated signal input power. After this maximum, there is a decrease of ΔQ with the CW signal power increasing. In the case of the -9 dBm modulated signal power, there is a different behavior with two ΔQ value peaks, which occur for values of CW signal power equal to or greater than 1 dB in relation to the modulated signal power. However, the decrease after these peaks is maintained.

The relation between power and Q-factor improvement can be understood considering the cases of -6.7 dBm and -7.6 dBm of modulated signal power. In these cases, for levels of CW signal power lower than the modulated input signal, the SOA does not reach the desired saturation to attain greater conversion efficiency and regenerative effects. Therefore, higher ΔQ values do not exist. However, for CW signal power values above from 2 to 3 dB, the ideal saturation is accomplished. Moreover, the power level of the output signal at λ_2 becomes more influenced by this CW signal power. From this value of CW power, the SOA gain saturation becomes very intense due to the higher power injected in this device, without efficiency in the XGM conversion, leading to a ΔQ decreasing.

Another factor that influences the behavior of these curves is the input signal quality, which is dependent of the modulated signal power. For example, the curve of the -9 dBm modulated signal power has a different behavior and lower ΔQ values, because it has the highest Q-factor initial value, which is 6 while the other cases vary from 4.2 to 5. This curve has the highest Q-factor even with the lowest power level, because the signal has low amplification from the SOA used as a boster, adding less noise and overshoot. Thus, the ΔQ improvement is small because it has a large input Q value. A Q-factor improvement up to 3.3 for a total input power (P_{mod} + P_{CW}) of -9.3 dBm was observed. Besides that, it was obtained a Q-factor improvement for input power values from -15.8 dBm to -1.68 dBm.

The *case* "LINK+SOA" is shown in the Figure 12(b), where is possible to observe a behavior like the *case* "SOA", with maximums of ΔQ for the CW signal power above 1 to 2.5 dB compared to the modulated signal input power. The justification for this behavior is also associated with the SOA gain saturation. For this case, lower ΔQ values were obtained. However, it was observed a Q-factor improvement for values of total input power from -19 dBm to -6 dBm. Since there is link attenuation, the power values of the modulated signal could be extended to small power values (-12.2 dBm).

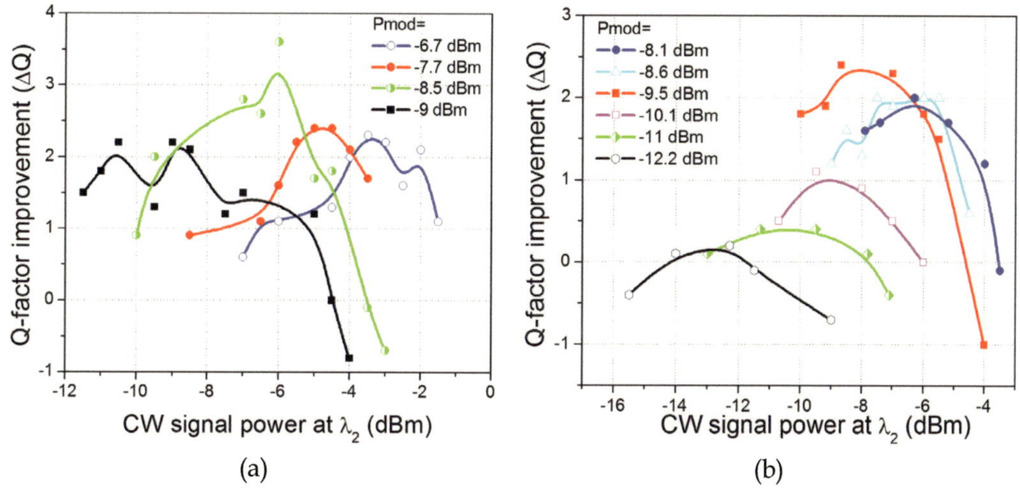

Fig. 12. Q-factor improvement versus input optical power (modulated and CW) for the *deterioration cases*: (a) "SOA" and (b) "LINK+SOA".

The Figure 13(a) (*case* "EDFA") shows the Q-factor improvement as a function of CW signal power for three cases of modulated signal input power. It reveals a similar behavior, but with less ΔQ variation, since the scale is represented from 3.2 to 4.8 units. Moreover, there are more oscillations between ΔQ minimums and maximums. This behavior may be partly explained by the utilized scale, since these variations between minimums and maximums are actually just from 0.2 to 0.4 units. Another reason may be inaccuracies when the Q-factor values were obtained directly from the oscilloscope.

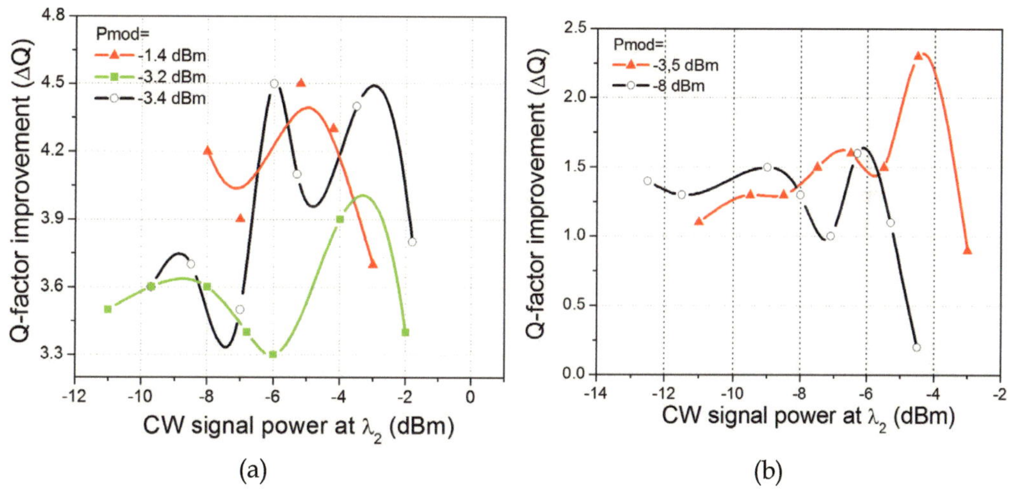

Fig. 13. Q-factor improvement versus input optical power (modulated and CW) for the *deterioration cases*: (a) "EDFA" and (b) "LINK+EDFA".

The ΔQ maximums occurred for values of the CW signal power from 0.8 to 3.6 dB below the values of the modulated signal input power. This occurs since the power levels of the modulated signal already have high values; therefore the CW signal powers cannot assume values higher than the power levels of the modulated signal, otherwise more SOA gain saturation will occur with consequent ΔQ reduction. For this case, it was attained a Q-factor improvements for total input power values from -5.8 dBm to 2 dBm, with a ΔQ maximum of 4.8.

The *deterioration case* "LINK+EDFA" is illustrated in the Figure 13(b), showing power values of the modulated input signal lower than the previous case, owing to attenuation caused by the buried fiber link. These curves also present oscillations as a result of inaccuracies. However, disregarding the minimum ΔQ values, the behavior is similar to that observed previously. The ΔQ maximums have different behavior for the two power values of the modulated input signal. For the case of -3.5 dBm, the ΔQ maximum occurs for the CW power level of -4.5 dBm. Nevertheless, for the case of the modulated signal input power of -8 dBm, the ΔQ maximum occurs for power values of the CW signal of -6 dBm. These results are also related to the SOA gain saturation. It was achieved a Q-factor improvements for values of total input power from -15.6 dBm to -1.1 dBm, with ΔQ maximum of 2.3.

The last *deterioration case* ("SOA+LINK+EDFA") is presented in the Figure 14. It is possible to note a behavior similar to the others, presenting a ΔQ increasing with the CW signal power up to a maximum, and then a decrease for the higher values of CW signal power. The ΔQ maximums occur for power values of the CW signal below from 0.5 to 1.5 dB compared to the modulated input signal. It was obtained Q-factor improvements for values of total input power from -12 dBm to 1.76 dBm, with ΔQ maximum of 4.

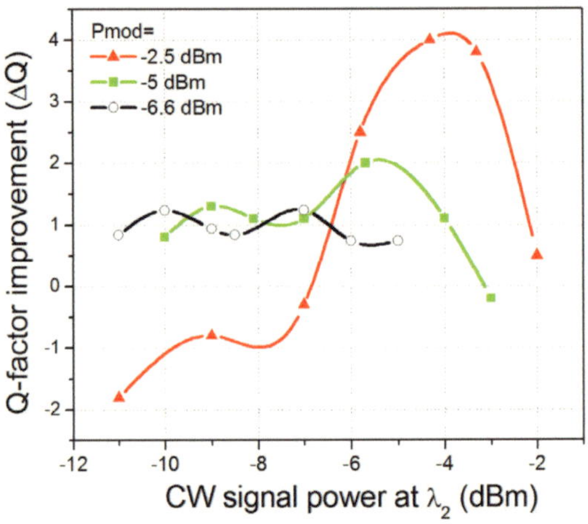

Fig. 14. Q-factor improvement versus input optical power (modulated and CW) for the *deterioration case* "SOA+LINK+EDFA".

From the results presented in the previous figures, it is possible to observe that there is an optimum relation between the powers of the modulated input signal and the CW signal. This relation is associated to the SOA saturation levels, varying for each deterioration case. In addition, it is possible to note that the higher values of Q-factor improvements occur for the *case* "EDFA", where the power levels of the modulated signal are the highest with the lowest Q-factor values for the modulated input signal, because the EDFA provides large amplification. As a result, small values of initial Q-factor allow more possibility for improvement. In the cases where another SOA is used to amplify the signal, causing overshoot, high ΔQ values were also obtained. As seen in the eye diagrams presented in the

Section 6, the 2R-converter eliminates overshoot, consequently providing more improvements in the Q-factor.

Analyzing the use of the 2R-converter in real systems, one can see from the figures above that it can regenerates degraded signals even with low input powers (<-12 dBm). Conclusively it is not necessary to use other amplifiers or SOAs to amplify the signal before the regenerator. It is important to comment that the SOA used as a booster and the EDFA were just used in this setup just to degrade the input signal.

9. Wavelength detuning

The wavelength detuning between the modulated input (λ_1) and CW (λ_2) signals is another factor which influences the XGM conversion. It is calculated as $\Delta\lambda = \lambda_1 - \lambda_2$. The wavelength detuning also influences the regenerative effects, because the regeneration is associated to the conversion efficiency. The results shown so far were obtained for $\Delta\lambda = 1$ nm, from 1550 nm to 1551 nm. In this section, $\Delta\lambda$ is varied and the Q-factor improvement is observed. For the results presented here, the output signal of the regenerator was attenuated to guarantee the same power level of the modulated input signal.

In order to analyze the detuning influence in the Q-factor improvement, the CW signal at λ_2 was kept fixed at 1551 nm due to the narrow-band filter used before the oscilloscope input. If the wavelength of this signal had been modified, it would be necessary adjustments of the filter for each case. Thus, the wavelength of the modulated input signal (λ_1) was varied to obtain different detuning.

The results for the *deterioration cases* "SOA" and "LINK+SOA" are illustrated in the Figure 15 (a) for a 10.3 Gbps NRZ. For these cases, the power values of the CW and modulated input channels were kept fixed. Only the modulated signal wavelength (λ_1) was varied. As seen in the Figure 3(c), a band-pass filter was used after the SOA used as a booster. As a consequence, the wavelength variation of the modulated input signal is limited by the filter bandwidth, which is about 4 nm. Hence the detuning varies just from -3 to 1 nm in the curves of the Figure 15(a).

Analyzing the *case* "SOA", it is seen a ΔQ decreasing while the detuning increases, with high ΔQ values for the up-conversions. As stated before, although this case presents more extinction rate degradation, it causes more SOA gain saturation and can generate more regenerative effects. The ΔQ values ranged from 1.6 to 2.4. Furthermore, there was a Q-factor improvement for conversions to the same wavelength, which is an interesting fact because the device presented here also can be used in a system where is necessary the signal regeneration without wavelength conversion.

The *case* "LINK+SOA" shows a behavior like the previous, but with higher ΔQ values (2.5 to 4.2). The behavior of this case has a minimum for a 0 (zero) detuning (conversion to the same wavelength) with ΔQ increasing again for positive detuning. It is not possible to show higher positive detuning values because of the band limiting of the filter, but due to the curves behavior, it is expected that ΔQ also increases with the positive detuning increasing. By the behavior of these two cases, it is expected that for higher absolute values of detuning, the Q-factor improvements still have good results but there will be limitations for very large detuning values.

The Figure 15(b) presents the variation of the Q-factor improvement as a function of the wavelength detuning for the *deterioration cases* "EDFA", "LINK+EDFA" and "SOA+LINK+EDFA" for a 7 Gbps NRZ . The *case* "EDFA" has a behavior similar to that in

the Figure 15(a) for the same detuning range from -3 to 1 nm, with a ΔQ decreasing while the detuning increases, and with a minimum for the conversion to the same wavelength, besides better results for up-conversion. However, without the band-pass filter cited previously, it was possible to extend the detuning values from -15 to +15 nm, observing that for high absolute values of detuning there is a ΔQ decrease. Considerable values for detuning of -15 nm (2.8) and +15 nm (3.3) were attained even though with the ΔQ reduction. This *deterioration case* shows the highest ΔQ values, varying from 2.8 to 4.8.
The *case* "LINK+EDFA" has a slight different behavior, with higher ΔQ values for down-conversions. There is an almost continuous ΔQ increasing with the wavelength detuning increasing, and the decrease occurs only for very high detuning values. In the curve of the *case* "LINK+EDFA" is also seen lower ΔQ values, varying from 0.3 to 2.9. For the extreme cases, a Q-factor improvement was attained from 0.9 to -5 nm, and from 1.1 to +10 nm.
The *case* "SOA+LINK+EDFA"involves all the deterioration effects discussed and it is also showed in the Figure 15(b). The detuning values were limited by the band-pass filter. The behavior of this case is similar to the *case* "LINK+EDFA", with ΔQ values ranging from 2.1 to 3.5.

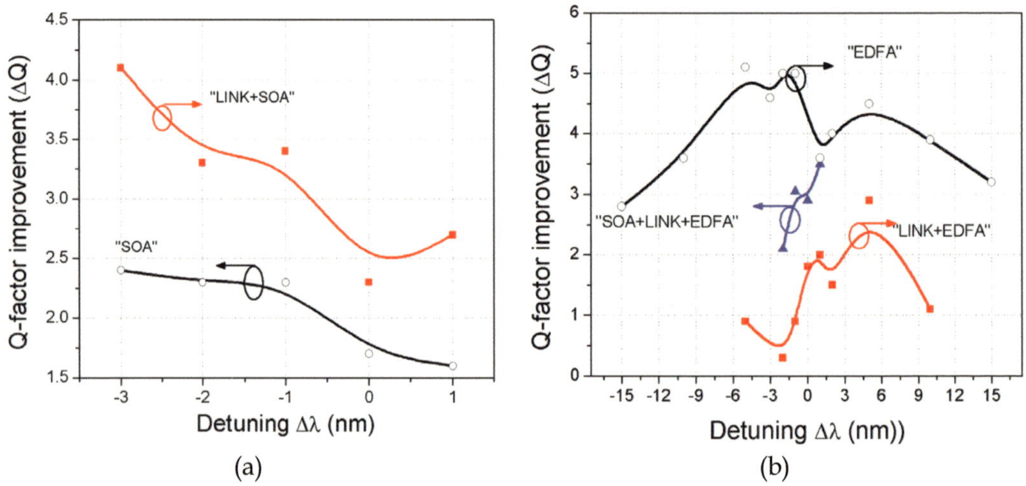

Fig. 15. Q-factor improvement versus detuning for the *deterioration cases*: (a) "SOA" and "LINK+SOA"; (b) "EDFA", "LINK+EDFA" and "SOA+LINK+EDFA".

In a broader context, it is possible to observe that 2R-converter can keep the regenerative effects even for large detuning, assuming different behaviors in accordance with the *deterioration case* and showing better performance for up-conversion in some cases and for down-conversion in others. Good ΔQ values were attained for conversions to the same wavelength, ranging from 1.6 to 4.2. Additionally, in the cases where a band-pass filter does not limit the detuning, there were good regenerative effects for up-conversion or down-conversion for detuning up to 15 nm.

10. Bit rate

High bit rates are interesting for any practical optical communication system. Therefore, it is necessary to study the response of the 2R-converter presented in this chapter at distinct bit rates. For this study, the modulated input signal and the CW signal wavelengths were kept fixed at 1550 and 1551 nm, respectively. Furthermore, the powers of these signals were also kept fixed and the regenerator output signal was not attenuated to the same power level of the modulated input signal. Hence, the signal re-amplification and re-shaping are considered in the results of this section.

Studies were conducted for the *cases* "SOA" and "LINK+SOA", with values ranging from 0.622 to 13.5 Gbps (Ribeiro et al., 2009a). The Figure 16(a) shows the Q-factor improvement as a function of the bit rate for the *case* "LINK+SOA". It is possible to observe that ΔQ increases almost linearly with a bit rate around 5 Gbps. The best ΔQ values (> 4 units) were obtained from 5 to 10.7 Gbps, showing a slight decreasing while the rate increasing. For values upper than 10.7 Gbps occur an abrupt ΔQ decreasing until the minimum value 1 at 13.5 Gbps.

The analysis of the *deterioration case* "SOA" is illustrated in the Figure 16(b) with a similar behavior. The main differences are the extension of linear ΔQ increasing up to 7 Gbps and the lower ΔQ values, with a maximum of 2.5 units. Lower ΔQ values occur due to the low overshoots from the modulated input signal used in this analysis, which reduce the ΔQ improvement provided by the regenerator.

The behavior of those curves can be explained by the operation of the 2R-converter. As seen before, the regenerator acts like a "low-pass filter". Hence, "high-frequency components", like the overshoot that arises in some cases, are filtered after the regenerator. The speed of the variance presented in each bit level varied with the modulation bit rate. Thus, for low frequencies (up to 5 Gbps) the speed of the variance is lower, and by the low-pass behavior of the regenerator variance is not very reduced, showing lower ΔQ values. With the bit rate increasing, the speed of the variance is higher, what increases the variance reducing, thereby increasing ΔQ.

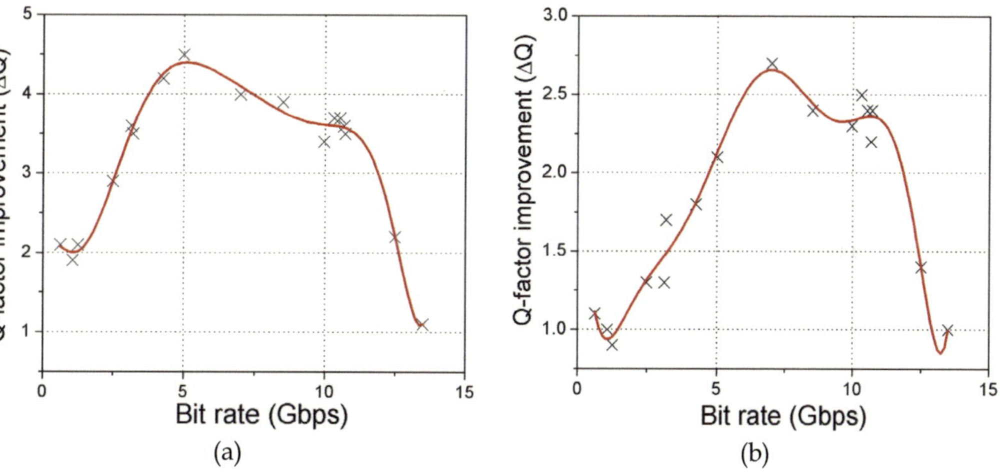

Fig. 16. Q-factor improvement versus bit rates for the *deterioration cases*: (a) "LINK+SOA" (adapted from Ribeiro et al., 2009a) and (b) "SOA".

In the range from 5 to 10.7 Gbps, ΔQ has a little variation because speed of the variance at these rates is similar. But there is a small ΔQ decreases in this range, influenced by another effect, which becomes relevant for rates upper than 10.7 Gbps, the pattern-dependence. At higher rates, the recovery time of the SOA gain is not enough to keep the bit pattern and distortion could occur in cases where there are large sequences of bits "1", for example. This pattern-dependence is stronger in the NRZ modulation case at higher rates. Thus, at rates upper than 10.7 Gbps, the pattern-dependence has more influence in the Q-factor improvement than the low-pass filter behavior, resulting in a reduction of the Q-factor improvement. The SOA gain recovery time is associated with the carrier lifetime in the SOA active region, which allows assign those bit rate limitations to the utilized SOA. To conclude if the limitation is either due to the SOA or it is from the setup of the 2R-converter used, it would be necessary another SOA with faster response. The discussion above is related only to the regenerator performance, which decays at rates upper than 10.7 Gbps. Despite this decrease in the Q-factor improvement, there are positive ΔQ for higher rates, confirming that the 2R-converter regenerates even at rates above 10.7 Gbps.

11. Conversion efficiency

The device presented in this chapter reshapes the optical pulses simultaneously with the wavelength conversion. So, it is necessary an analysis about the conversion efficiency, which was done using the regenerator in a different setup from that one showed in the Figure 1. The results presented until this section were obtained using 13.5 Gbps as the maximum bit rate. Then, in order to analyze the wavelength conversion efficiency, the modulation rate was augmented with a sinusoidal signals generator up to 20 GHz (Ribeiro et al., 2009a), with corresponds to the bandwidth of a 40 Gbps NRZ signal.

Since the purpose of this section is the analysis of the wavelength conversion efficiency based on the XGM effect in a counter-propagating mode, the methods to degrade the input signal were not used in these measures. In order to study the conversion efficiency it is necessary to analyze the modulation index. To analyze the modulation frequency components, both the input and the output regenerator signals were converted to the electric field by a photodetector and observed in an electrical spectrum analyzer. As a result of the sinusoidal modulation, the spectrum shows only one peak in the modulation frequency. Thus, the conversion efficiency is determined as the difference between the values of this peak before and after the regenerator.

The results were obtained for two situations. In the first one, the modulated input signal and the CW signal were kept fixed at 1550 and 1551 nm, respectively. Moreover, the CW carrier power was maintained at -16 dBm and the modulated signal input power was varied to values of -16, -12.3, and -8.5 dBm. The efficiency conversion results for this case are illustrated in the Figure 17(a), where is seen conversion efficiency values up to 25 dB for a -16 dBm modulated signal input power. In addition, there is a conversion efficiency increase from 4 to 7 dB for values of the modulation frequency up to 4.5 GHz, and decreasing beyond this value. Still, there is positive conversion efficiency for power values lower than -12.3 dBm, even at 20 GHz. The best conversion efficiency results were obtained for the lowest modulated signal powers, and they are associated with the SOA gain saturation, which is more intense for higher values of modulated signal power.

In the second situation, the modulated signal input power and the CW signal power were kept fixed in -9.5 dBm and -9 dBm, respectively. However, the modulated signal wavelength was varied while the CW signal was fixed at 1551 nm. The results of the conversion efficiency as a function of the modulation frequency are presented for distinct detuning values (already defined above as $\Delta\lambda = \lambda_1 - \lambda_2$) in Figure 17(b). As the previous case, the efficiency increases up to 4.5 GHz, decreasing after this value. Nevertheless, positive conversion efficiency values occur for the most of detuning cases, except for the -10 nm detuning. According to the results, there is higher conversion efficiency for down-conversion, exhibiting very similar behavior between the +10 and +5 nm detuning cases, with an efficiency peak of 15 dB. The conversion efficiency for the detuning values of -1, 0 and +1 have similar behavior at conversion efficiency ranging from 13 dB (4 GHz) to 5.5 dB (20 GHz) (not shown here).

The good results attained for the conversion efficiency of the 2R-converter show that this device can be used both as a regenerator and a wavelength converter. Furthermore, in some cases, good conversion efficiency up to 20 GHz shows that this device will perform well as a wavelength converter even at high frequencies of modulation.

The behavior observed in Figure 17(a) and (b) may help to elucidate the results of the previous section. Clearly, the results of these figures indicate that the device has good results in wavelength conversion, and they are not related to regenerative features. Besides, the Figure 17(a) and (b) cases are quite different from the Figure 16(a) and (b). In the first case, it is a sinusoidal signal and input signal without deterioration. In the second, the signal is a PRBS pulse stream and the input signal is deteriorated. Despite the differences, there are common points in the results: an increase in the Q-factor improvement up to 5 Gbps and an increase in the conversion efficiency up to 4.5 GHz, followed by a decrease of these parameters. Thus, the behavior of ΔQ with the bit rate can be said as dependent of the conversion efficiency, which complements the previous explanations about the regenerator behavior like a "low-pass filter", and the pattern-dependence effects. Despite this similarity, there is still some doubt whether the regenerator performance is limited by the SOA rate or the setup used, because from this section results it is just possible to know that as a converter, the device can gets good values of conversion efficiency up to 20 GHz sinusoidal modulation, but nothing might be concluded from the regenerative features.

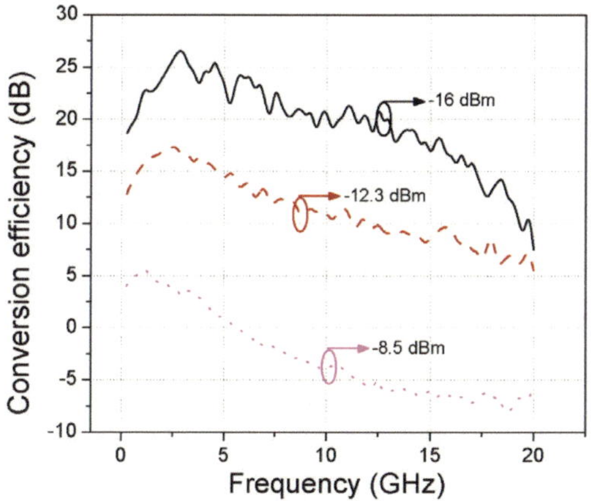

(a)

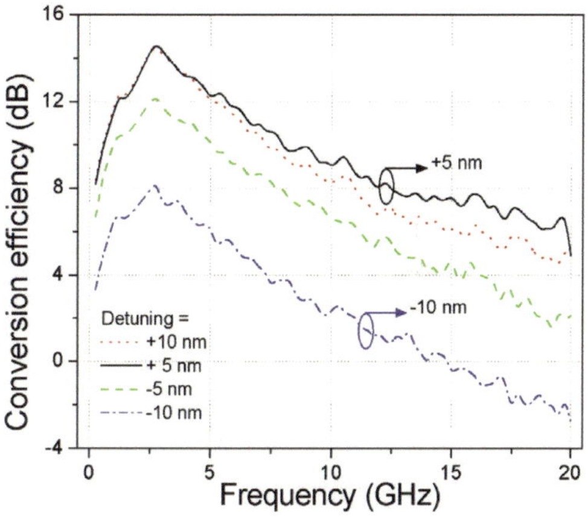

(b)

Fig. 17. Conversion efficiency versus frequency: (a) for fixed detuning and CW carrier power, varying the modulated signal optical power; (b) for fixed total input power, varying detuning (adapted from Ribeiro et al., 2009a).

12. Bit error rate

As observed in eye diagrams presented before, the bit reshape is visible and quantified by the Q-factor improvement. These results were analyzed only on the oscilloscope, without considering the real improvement that these results could cause in communications systems. In this way, it is interesting to study if the 2R-converter can reduce the error of the receptor when it decides what kind of bit is arriving. The probability of incorrect identification of a bit by the decision circuit of the receiver is the BER definition.

The BER results showed in this work are just an estimation, because they were obtained from measured eye diagrams using equations, in which some approximations were assumed. The equations consider just the thermal noise, well described by Gaussian statistics with zero mean and the shot noise, with a ordinary approximation, treating it as a Gaussian random variable for both p-i-n and APD receivers, but with different variances. The sum of two Gaussian random variables (the two noise types considered) is also a Gaussian random variable. As a consequence the BER is given by (Agrawal, 2002):

$$BER = \frac{1}{4}\left[erfc\left(\frac{I_1 - I_D}{\sigma_1\sqrt{2}} \right) + erfc\left(\frac{I_D - I_0}{\sigma_0\sqrt{2}} \right) \right] \quad (2)$$

where I_1 and I_0 are the current level of the level '1' and '0', respectively; σ_1 and σ_0 are standard deviation of the level '1' and '0', respectively; I_D is the decision threshold; and erfc stands for the complementary error function. Other approximation is $\ln(\sigma_1/\sigma_0)=0$.

The BER with optimum setting of the decision threshold is obtained by using (1) and (2) and depends only on the Q-factor as (Agrawal, 2002):

$$BER = \frac{1}{2} erfc\left(\frac{Q}{\sqrt{2}} \right) \approx \frac{\exp(-Q^2/2)}{Q\sqrt{2\pi}} \quad (3)$$

Given that the Q-factor can be obtained directly from eye diagrams visualized at the oscilloscope, it is possible to use (3) and obtain the BER estimation. As the BER values presented are just estimatives, lower values were obtained. Therefore, the graphics we presented illustrates only results that could be obtained in experiments (up to 10^{-13}). Besides, the BER floor similar to one obtained in experiments is difficult to observe due to the lower BER values attained.

The analysis of the receptor bit error rate for different received signal power is necessary to study the BER estimation. In this way, the input BER estimation (before the 2R-converter) was calculated from the Q-factors attanied for different modulated input signal powers (λ_1=1550 nm) for each deterioration cases. After the input BER estimation, the CW signal power (λ_1=1551 nm) was optimized to obtain better values of output signal Q-factor, and consequently, better BER estimations. The output signal was attenuated to guarantee the same power level of the modulated input signal power, in order to compare the input and output BER estimations. The measurements were done at 7 Gbps NRZ.

The obtained results are illustrated in Figure 18, presented as -Log(BER) versus the modulated signal input power that was the same level after the regenerator due to the use of an attenuator, and so was represented in the figures just as power. By the results of Figure 18, the absence of low values and large variation of the modulated signal power can be observed. This fact is due to the insufficient receiver sensitivity of the used oscilloscope. In addition, the curves presented in Figure 18 are fitting by polynomials to better represent the bevahior of the attained points.

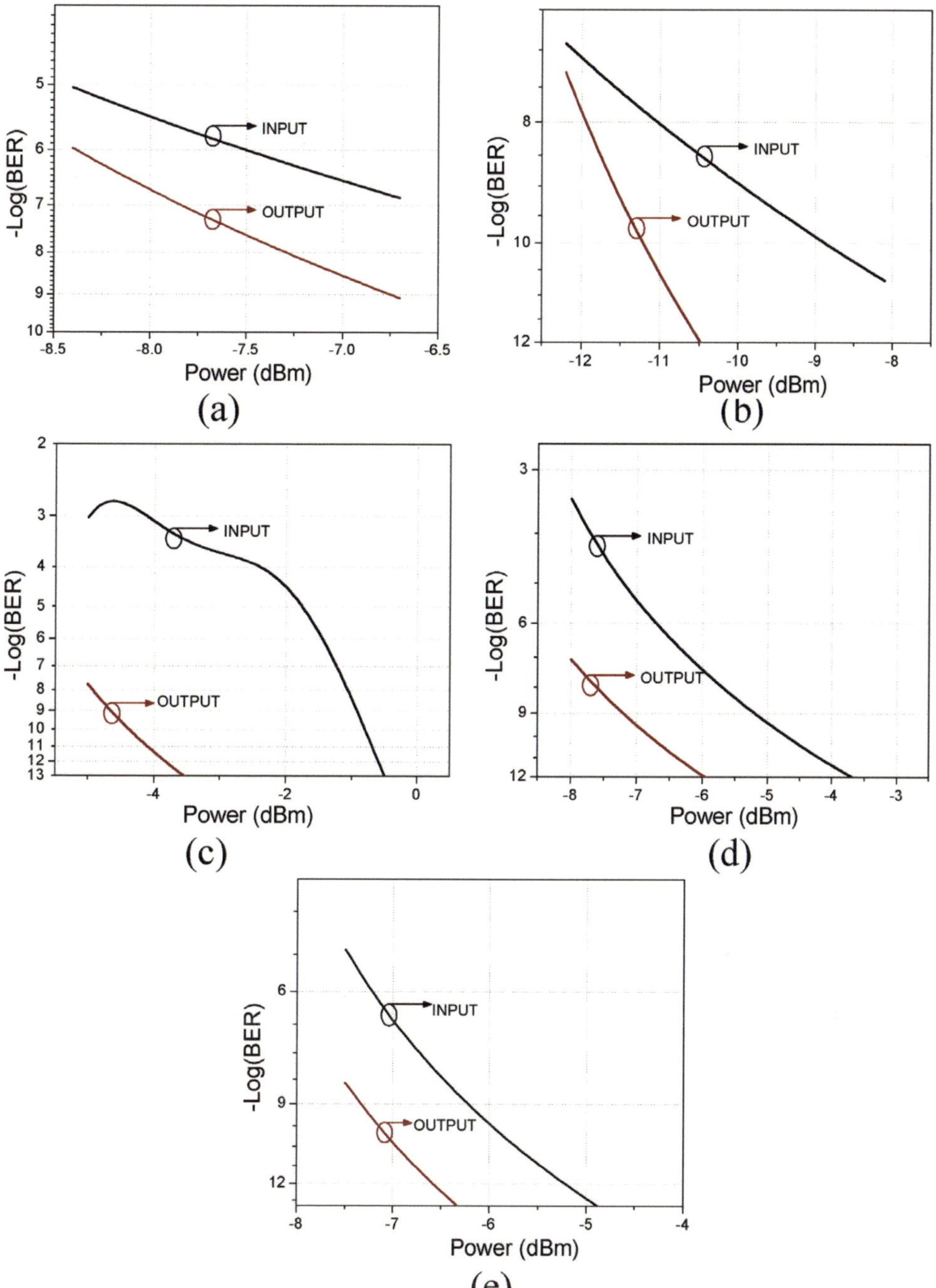

Fig. 18. BER estimation for different *deterioration cases*: (a) "SOA"; (b) "LINK+SOA"; (c) "EDFA"; (d) "LINK+EDFA", and (e) "SOA+LINK+EDFA".

The *case* "SOA" is illustrated in Figure 18(a). In this figure, it is observed an improvement of 1 dB in the power needed at the receiver in order to present the same BER, and BER values of 10^{-10}. The *case* "LINK+SOA" is shown in Figure 18(b). For this case, low BER values were

obtained due to the high Q-factor values and an improvement of 2 dB was obtained in the power level. Therefore, it could be note a trend that for higher power values, this improvement can be higher.
For the Figure 18(c) (*case* "EDFA"), it can be observed higher modulated signal power and a stranger behavior for the input BER estimation. The BER output corresponds to the expected BER estimation behavior, presenting BER values that starts in 10^{-8} due to the high Q-factor values obtained. An improvement of 4 dB in the power needed to have the same BER was obtained.
The performance of the *case* "LINK+EDFA" in Figure 18(d) is similar to the other cases with improvement of 2.5 dB. The last case in Figure 18(e) involves the various deterioration types ("SOA+LINK+EDFA"). It can be observed an improvement of 1.5 dB in the power.
The BER estimation values presented here could be modified, depending on the modulated input signal deterioration. In this manner, the objective of this BER estimation is to compare the input and output BER results for each deterioration case, observing the improvement that the 2R-converter could provoked in optical systems, without comparison between different *deterioration cases*. As mentioned before, the BER values presented here are just an estimation calculated from Q-factor attained from measured eye diagrams, which justified the lower BER values. Excepting these lower BER values, the curves behavior (disregarding the BER input of the *case* "EDFA") and the improvement in power level to have the same BER value are feasible. Nevertheless, the study of extinction ratio (ER) is necessary since the BER analysis were obtained just from Q-factors. By considering a real system and measuring the BER experimentally, the ER migth be responsible for a BER results different from the presented ones in this section.

13. Extinction ratio

In the previous sections, the improvement of the bit format, of the Q-factor, and consequently of the BER estimation caused by the 2R-converter was demonstrated. However, this regenerator is based on the XGM effect and so it presents ER deterioration. This figure of merit is defined as the relation between the power levels of the bits "1" and "0". In this way, lower ER means that power levels of the bit "1" and "0" are near, being more difficult for the receptor make a decision about which bit was received. Besides, the ER is related to optical signal to noise ratio.
In Figure 19, the study of ER deteriorations after the regenerator for the all cases of deterioration present in this chapter are reported. This study was optimized just for the ER, others parameters were fixed, and the input ER was varied through the deviation of the modulator voltage. The ER input and output were obtained directly from the oscilloscope for the *deterioration cases* studied in this chapter. An up-conversion from 1550nm to 1551 nm was done to study the ER deterioration. As mentioned before, despite this type of wavelength conversion presents higher ER degeneration, it also presented higher SOA gain saturation, enabling higher regenerative effects.
The curves in Figure 19 were obtained by polynomial adjustment. It could be noted that there is an increase of the ER degeneration (input ER – output ER) with an input ER increase in all the *deterioration cases*. The *case* "LINK+EDFA" presents the better perfomance with lower ER degeneration values, and in some cases with ER improvement. Nevertheless, these ER improvements should be associated with small inaccuracies of the values obtained from the oscilloscope.

As mentioned before, these ER degeneration results are higher because up-conversion was used. If down-conversion were used, lower ER degeneration values would be attained. The worst ER degeneration value obtanied was 5.35 dB for the deterioration *case* "LINK+SOA" and an input ER of 10.7 dB.

Despite this deterioration, it is necessary to compare the ER degradation with the Q-factor improvement to validate the obtained results for BER estimation. Thus, if the Q-factor improvements (ΔQ), converted to dB, present higher values than the ER degeneration, the obtained values of output BER will be higher than input BER values, proving the regenerative capacity of the 2R-converter.

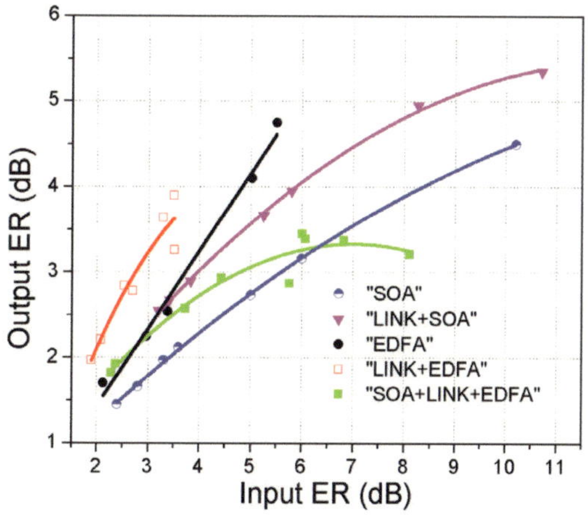

Fig. 19. Output ER as function of the input ER for different cases of input signal deterioration.

The Figure 20 shows the comparison between the Q-factor improvement (ΔQ) and the ER degeneration for different *deterioration cases*, considering different values of the output signal power, which were attenuated to guarantee the same values of the modulated input signal power. For the *case* "SOA" showed in Figure 20(a), it could be noted that the Q-factor improvement (ΔQ) is lower than ER degeneration for all the power values, differently of the *case* "LINK+SOA" (Figure 20(b)) in which, for the power values higher than -10.8 dBm, the Q-factor improvement (ΔQ) is higher than ER degeneration.

In Figure 20(c), the *case* "EDFA" is illustrated. In this figure, the Q-factor improvement surpass the ER degeneration for all power values as well as the *case* "LINK+EDFA"(Figure 20(d)). The difference between these cases is the higher Q-factor improvement for the first case. Moreover, some ER degeneration positive values were observed for the *case* "EDFA". Nevertheless, these results are affected by inaccuracies in the values obtained from the oscilloscope.

In Figure 20(e), the *case* "SOA+LINK+EDFA" is reported. This case presented a similar behavior with Figure 20(b), presenting power values where Q-factor improvement surpass ER deterioration and other values where this result does not occur. For the range of -7 to -4 dBm, the ΔQ values are higher then ER degeneration as well as the range of -2.7 to -1.2 dBm. For power value between these ranges, the ER degeneration surpass ΔQ.

These obtanied results are relative to the used parameters. Note that for different modulated input signal power values, or even for different deterioration levels (input Q-factor) for the deterioration cases presented, the curves behavior could be different. One example is the *case* "SOA", for other deterioration level, the improvement Q-factor can surpass the ER degeneration.

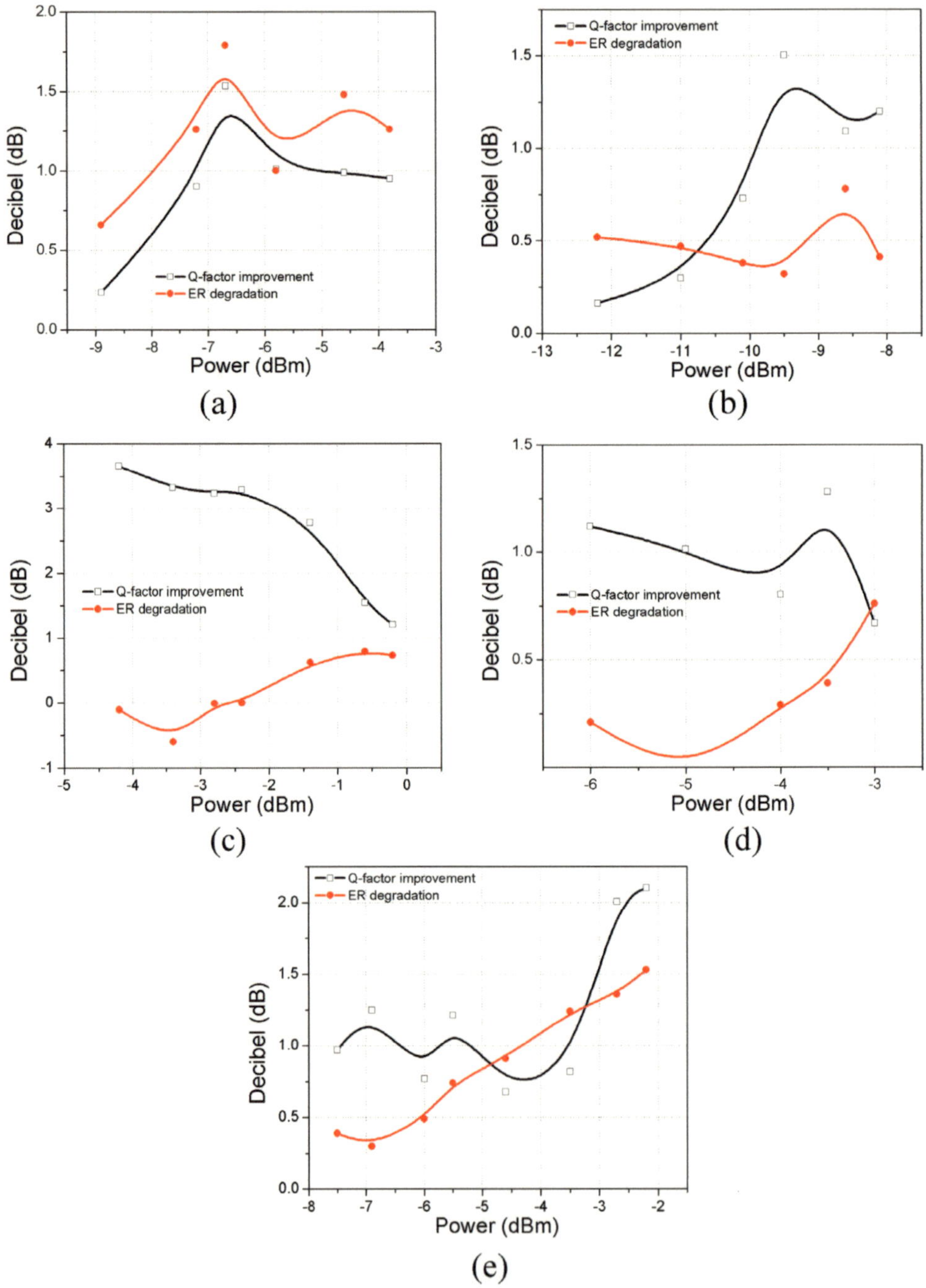

Fig. 20. Q-factor improvement versus ER degradation for different *deterioration cases*: (a) "SOA"; (b) "LINK+SOA"; (c) "EDFA"; (d) "LINK+EDFA", and (e) "SOA+LINK+EDFA".

In this manner, observing these comparison between Q-factor improvement and ER degeneration, it is possible to prove that, in the most of the cases, the output BER (despite the decrease of the BER values due to ER degeneration not considered) will be higher than input BER, confirming the regenerative effects presented by the 2R-converter.

14. Integration possibility

A great advantage to use setups based on SOA is the integration capacity of this amplifier due to its small dimensions. There are many devices that use SOA integrated to other optical components as for example: wavelength conversion based on SOA-MZI (Kehayas et al., 2006), regenerator and wavelength converter using SOA integrated to distributed-feedback laser (Stephens et al., 1999), optical clock recovery with SOA integrated to mode-locking laser (Koch et al., 2007), and others.

The 2R-converter is very simple, using just one non-linear SOA, an optical circulator, an optical isolator and a CW laser. The integration between SOA and laser is already used as the examples mentioned before. Integrated optical circulators that use nonreciprocal phase shifters already exist a long time ago (Okamura et al., 1984). In addition, there is waveguide polarization-independent optical circulator using MZI (Sugimoto et al., 1999) or MMI with Faraday rotator (Zaman et al., 2006). Optical isolator also been obtained using nonreciprocal coupled waveguides (Bahlmann et al., 1999), based on efficient nonreciprocal radiation mode conversion (Shintaku, 1998) or other ways. Besides, if the narrow filter used to reduce the ASE noise before the oscilloscope input is considered essential to the 2R-converter, there are optical filters based on ring resonators (Rabus et al., 2002), on MZI with two ring resonators (Rasras et al., 2007), and on photonic crystals (Zhang et al., 2007). In this section, just some techniques used to obtain the integration of the optical components mentioned are present, existing others technique not reported in this chapter.

The results mentioned about SOA, laser, optical circulator, isolator, and filter do not considered the simultaneously integration of these several components. In this manner, it is necessary the study of techniques to obtain the integration of these different optical component like the experimental setup in Figure 1 to prove the 2R-converter integration possibility. Thus, optical components integration area should be asserted higher integration possibility to the 2R-converter.

15. Conclusion

The 2R-converter presented in this chapter demonstrated good performance for regeneration and wavelength conversion of signals deteriorated by different effects. Considering these different deterioration cases, the 2R-converter presented improvement in the degenerated signal as: signal amplification with an optical gain of 12 dB, reduction of the ASE noise presented in bit levels "1" and "0" with variance decreasing, overshoot elimination and reduction of the intrinsic dispersion degenerative effects. These results were quantified by the Q-factor improvement (ΔQ) presenting values up to 4.5.

Furthemore, an existence of an optimum relation between the power of modulated input signal and the CW signal that maximize the regenerative effects was observed. This relation is associated to the SOA gain saturation level. Regeneration for lower input power (<-12 dBm) was observed too. Additionally, the 2R-converter demonstrated to be practically independent of the polarization angle of modulated input signal with gain variation of just

0.9 dB. Regeneration was obtained for up-conversion, down-conversion and conversion to the same wavelength, with good results even for large detuning (15 nm).
The 2R-converter presented good conversion efficiency values for rate of 20 GHz sinusoidal. In relation to the regenerative effects, the device presented good results up to 13.5 Gbps NRZ with better perfomance from 5 to10.7 Gbps. The use of another SOA with lower gain recovery time is a possiblity to future works to determine if the bit rate limitation is caused by the used SOA or by the experimental setup configuration.
The BER estimations from measured eye diagrams were presented, confirming the good 2R-converter performance. Good results were obtained for different cases of input signal deterioration with improvement in the receiver sensitivity up to 4 dB. Moreover, for most cases, the improvement caused by the 2R-converter surpass the ER degeneration, which is presented in all wavelength conversion based on XGM effect.
In this way, the 2R-converter showed good regenerative capacity for different deterioration types. Due to the use of a SOA, this setup can be integrated, decreasing the cost and volume. Another important features of the 2R-converter are: simplicity; robust operation; high-speed reconfiguration, limited just by the CW laser tuning (since the narrow filter in the oscilloscope input is not considered). In this manner, wavelength conversion and 2R-regeneration in a simple scheme with SOA was presented and characterized.

16. Acknowledgments

This work was supported by the Brazilian agency FAPESP (Fundação de Amparo à Pesquisa do Estado de São Paulo), under CEPOF and KyaTera-FAPESP project, and by CNPq.

17. References

Agrawal, G.P. (2002). *Fiber-optic communication systems,* Willey-Interscience, ISBN 0-471-21571-6, New York, USA.

Bahlmann, N.; Lohmeyer, M.; Zhuromskyy, O.; Dotsch, H. & Hertel, P. (1999). Nonreciprocal coupled waveguides for integrated optical isolators and circulators for TM-modes," Elsevier Optics Communications, Vol. 161, (March 1999) pp. 330-337, DOI: 10.1016/S0030-4018(99)00027-9.

Chayet, H.; Ezra, S.B.; Shachar, N.; Tzadok, S.; Tsadha, S. & Leuthold, J. (2004). Regenerative all-optical wavelength converter based on semiconductor optical amplifier, *Proceedings of the Conference on Lasers and Electro-Optics (CLEO 2004)*, ISBN: 1-55752-772-5, Baltimore-USA, February 2004.

Conforti, E.; Bordonalli, A.C.; Ho, S. & Kang, S. (1999). Optical 2R remodulator using feed-forward control of semiconductor optical amplifier, *Microwave and Optical Technology Letters*, Vol. 21, No. 1, (March 1999) pp. 39-42.

Contestabile, G.; Proietti, R.; Calabretta, N. & Ciaramella, E. (2005). Reshaping capability of cross-gain compression in semiconductor amplifiers, *IEEE Photonics Technology Letters*, Vol. 17, No. 12, (December 2005) pp. 2523-2525, ISSN: 1041-1135.

Durhuus, T.; Mikkelsen, B.; Joergensen, C.; Danielsen, S.L. & Stubkjaer, K.E. (1996). All-optical wavelength conversion by semiconductor optical amplifiers, *Journal of Ligthwave Technology*, Vol. 14, No. 6, (June 1996) pp. 942-954, ISSN: 0733-8724.

Fischer, S.; Dulk, M.; Gamper, E.; Vogt, W.; Gini, E.; Mechior, H.; Hunziker, W.; Nesset , D. & Ellis, A.D. (1999). Optical 3R regenerator for 40 Gbit/s networks, *Electronics Letters*, (November 1999), Vol. 35, No. 23, ISSN: 0013-5194.

Funabashi, M.; Zhu, Z.; Pan, Z., Paraschis, L. & Yoo, S.J.B. (2006). Optical clock recovery and 3R regeneration for 10-Gb/s NRZ signal to achieve 10 000-hop cascadability and 1 000 000-km transmission, *IEEE Photonics Technology Letters*, Vol. 18, No. 20, (December 2006) pp. 2078-2080, ISSN: 1041-1135.

Gallep, C.M. & Conforti, E. (2002). Reduction of semiconductor optical amplifier switching times by preimpulse step-injected current technique, *IEEE Photonics Technology Letters*, Vol. 14, No. 7, (July 2002) pp. 902-904, ISSN: 1041-1135.

Kehayas, E.; Tsiokos, D.; Bakopoulos, P.; Apostolopoulos, D.; Petrantonaski, D.; Stampoulidis, L.; Poustie, A.; McDougall, R.; Maxwell, G.; Liu, Y.; Zhang, S.; Dorren, H.J.S.; Seoane, J.; Nielsen, P.H.; Jeppesen, P. & Avramopoulos, H. (2006). 40 Gb/s all-optical processing systems using hybrid photonic integration technology, *Journal of Ligthwave Technology*, Vol. 24, No. 12, (December 2006) pp. 4903-4911, ISSN: 0733-8724.

Kelly, A.E. (2001). Ultra high-speed wavelength conversion and regeneration using semiconductor optical amplifiers, *Proceedings of the Optical Fiber Communication Conference (OFC 2001)*, ISBN:1-55752-655-9, Anaheim-USA, March 2001.

Koch, B.R.; Barton, J.S.; Masanovic, M.; Hu, Z.; Bowers, J.E. & Blumenthal, D.J. (2007). Monolithic mode-locker laser and optical amplifier for regenerative pulsed optical clock recovery, *IEEE Photonics Technology Letters*, Vol. 19, No. 9, (May 2007) pp. 641-643, ISSN: 1041-1135.

Leuthold, C.M.; Marom, D.; Cabot, S.; Ryf, R.; Bernasconi, P.; Baumann, F.; Jaques, J.; Neilson, D.T. & Giles, C.R. (2003). All-optical wavelength converter based on a pulse reformatting optical filter, *Proceedings of the Optical Fiber Communication Conference (OFC 2003)*, ISBN: 1-55752-746-6, March 2003.

Liu, Y.; Herrera, J.; Raz, O.; Tangdiongga, E.; Ramos, F.; Marti, J.; Waadt, H.; Koonen, A.M.J.; Khoe, G.D. & Dorren, H.J.S. (2007). 160 Gbits/s all optical SOA based wavelength conversion and error-free transmisssion through two 50 km fibre links, *Electronics Letters*, Vol. 43, No. 25, (December 2007).

Merlier, J.; Morthier, G.; Caenegen, T.; Baets, R.; Moerman, I. & Daele, P. (2001). Experimental demonstration of 15 dB extinction ratio improvement in a new 2R optical regenerator based on an MMI-SOA, *Proceedings of the European Conference on Optical Communication (ECOC 2001)*, ISBN: 0-7803-6705-7, Amsterdam, Vol. 4, pp. 574-575. February 2008.

Okamura, Y.; Negami, T. & Yamamoto, S. (1984). Integrated optical isolator and circulator using nonreciprocal phase shifters: a proposal, *Appied Optics*, Vol. 23, No. 11, (June 1984) pp 1186-1189, DOI:10.1364/AO.23.001886.

Ohman, F.; Bischoff, S.; Tromborg, B. & Mork, J. (2003). Semiconductor devices for all-optical regeneration, *Proceedings of the International Conference on Transparent Optical Networks (ICTON' 03)*, Vol. 2, pp-41-46, ISBN: 0-7803-7816-4, June 2003.

Rabus, D.G.; Hamacher, M.; Troppenz, U. & Heidrich, H. (2002). Optical filters based on ring resonators with integrated semiconductor optical amplifiers in GaInAsP-InP, *IEEE Journal of Selected Topics in Quantum Electronics*, Vol. 8, No. 6, (November/December 2002) pp.1405, ISSN: 1077-260X.

Rasras, M.S.; Gill, D.M.; Patel, S.S.; Tu, k.y.; Chen, Y.K.; White, A.E.; Pomerene, A.T.S; Carothers, D.N.; Grove, M.J.; Sparacin, D.K.; Michel, J.; Beals, M.A. & Kimerling, L.C. (2007). Demonstration of a fourth-order pole-zero optical filter integrated using CMOS processes, *Journal of Ligthwave Technology*, Vol. 25, No. 1, (January 2007) pp. 87, ISSN: 0733-8724.

Raz, O.; Herrera, J.; Calabretta, N.; Tangdiongga, E.; Anantathanasarn, S.; Notzel, R. & Dorren, H.J.S. (2008). Non-inverted multiple wavelength converter at 40 Gbits/s using 1550 nm quantum dot SOA, *Electronics Letters*, Vol. 44, No. 16, (July 2007).

Ribeiro, N.S.; Gallep, C.M. & Conforti, E. (2008). Wavelength conversion and 2R-regeneration using one semiconductor optical amplifier with cross-gain modulation compression, *Proceedings of the Conference on Lasers and Electro-Optics (CLEO 2008)*, San jose -USA, ISBN:978-1-55752-859-9, May 2008.

Ribeiro, N.S.; Gallep, C.M. & Conforti, E. (2009a). High eye-quality improvement by a single SOA regenerative wavelength converter, accepted to *Microwave and Optical Technology Letters*.

Ribeiro, N.S.; Toazza, A.L.; Gallep, C.M. & Conforti, E. (2009b). Rise time and gain fluctuations of an electrooptical amplified switch based on multi pulse injection in semiconductor optical amplifiers, *IEEE Photonics Technology Letters*, Vol. 21, No. 10, (2009) pp. 769-771, ISSN: 1041-1135.

Shintaku, T. (1998). Integrated optical isolator based on efficient nonreciprocal radiation mode conversion, *Applied Physics Letters*, Vol. 73, No. 14, (October 1998), DOI: 10.1063/1.122331.

Simon, J.C.; Gay, M.; Bramerie, L.; Roncin, V.; Joindot, M.; Chartier, T.; Lobo, S.; Girault, G.; Le, Q.T.; Nguyen, T.N. & Ngo, M.N. (2008). Long distance using optical regeneration, *Proceedings of the Optical Fiber Communication Conference (OFC 2008)*, ISBN: 978-1-55752-856-8, San Diego-USA, February 2008.

Simos, H.; Bogris, A. & Syvridis, D. (2004). Investigation of a 2R all-optical regenerator based on four-wave mixing in a semiconductor optical amplifier, *Journal of Ligthwave Technology*, Vol. 22, No. 2, (February 2004) pp. 595-604, ISSN: 0733-8724.

Stephens, M.F.C.; Penty, R.V. & White, I.H. (1999). All-optical regeneration and wavelength conversion in an integrated semiconductor optical amplifier/distributed-feedback laser, *IEEE Photonics Technology Letters*, Vol. 11, No. 8, (August 1999) pp. 979-981, ISSN: 1041-1135.

Sugimoto, N.; Shintaku, T.; Tate, A.; Terui, H.; Shimokozono, M.; Kubota, E.; Ishii, M. & Inoue, Y. (1999). Waveguide polarization-independent optical circulator, *IEEE Photonics Technology Letters*, Vol. 11, No. 3, (March 1999) pp. 335-337, ISSN: 1041-1135.

Zaman, T.R.; Guo, X. & Ram, R.J. (2006). Proposal for a polarization-independent integrated optcial circulator, *IEEE Photonics Technology Letters*, Vol. 18, No. 12, (June 2006) pp. 1359-1361, ISSN: 1041-1135.

Zhang, Z.; Dainese, M.; Wosinski, L.; Xiao, S. & Qiu, M. (2007). Optical filter based on two-dimension photonic crystal surface-mode cavity in amorphous silicon-on-silica structure, *Applied Physics Letters*, Vol. 90, No. 4, (January 2007) pp. O41108-1, DOI: 10.1063/1.2432228

Wang, J.P.; Savage, S.J.; Robinson, B.S.; S.A. Hamilton, S.A; Ippen, E.P; Mu, R.; Wang, H.; Spiekman, L. & Stefanov, B.B. (2007). Regeneration using SOA-MZI in a 100 pass 10,000-km recirculating fiber loop, *Proceedings of the Conference on Lasers and Electro-Optics (CLEO 2007)*, ISBN: 978-1-55752-834-6, Baltimore -USA, May 2007.

19

Optical DQPSK Modulation Performance Evaluation

Nelson Costa and Adolfo Cartaxo
Group of Research on Optical Fiber Telecommunication Systems,
Instituto de Telecomunicações,
Dept. of Electrical and Computer Engineering, Instituto Superior Técnico
Portugal

1. Introduction

The always increasing amount of internet traffic in optical networks led to the need of using efficient modulation formats. Conventional on-off-keyed (OOK) signals have been extensively employed in optical communication systems. However, OOK modulation is inadequate for transmission of 40 Gbit/s per channel, or higher, bit rates (Winzer & Essiambre, 2006) mainly because of its reduced robustness to fiber nonlinearity, chromatic dispersion and optical filtering at such bit rates. In order to overcome such impairments, several advanced modulation formats received particular attention in the last few years. Some of these modulation formats still carry the information in the amplitude of the signal. However, they also modulate the phase of the signal to increase its robustness to transmission impairments. Some examples of such modulation formats are duobinary and alternate mark inversion.

The most promising modulation formats for future optical networks make use of the phase of the signals to carry information. Among such formats, differential phase-shift-keying (DPSK) and differential quadrature phase-shift-keying (DQPSK) are the ones more often referred. These modulation formats led already to several landmark experimental results confirming their potential (Ho, 2005), (Winzer et al., 2008). The main advantages of DPSK are approximately 3 dB improvement on optical signal-to-noise ratio (OSNR) when compared with conventional OOK, and improved dispersion and polarization mode dispersion tolerance (Xu et al., 2004). DQPSK shows also improved spectral efficiency (Morita & Yoshikane, 2005).

The evaluation of DQPSK system performance is usually performed using Monte-Carlo (MC) simulation or Karhunen-Loève series expansion (Bosco & Poggiolini, 2006). The complexity of evaluating the performance of DQPSK system where noise added by optical amplifiers is the main noise source is the main reason for using these performance evaluation methods. However, while MC simulation has the time consumption disadvantage, Karhunen-Loève series expansion derivation is usually of complex nature when rigorous DQPSK system performance evaluation is the goal. Furthermore, these methods provide reduced insight on the effects impairing the transmission system performance unless extensive analysis of different sets of parameters is performed.

Transmission system optimization using such methods is also much time consuming. Thus, much faster, but still accurate, methods for performance evaluation are highly desirable. With this objective, the Gaussian approximation (GA) for the probability density function (PDF) of current at the decision circuit input was investigated. This approximation has been extensively used in OOK modulated system performance evaluation, leading to quite good accuracy (Rebola & Cartaxo, 2001). However, its use leads to high discrepancies in DQPSK system performance evaluation (Bosco & Poggiolini, 2006). In order to overcome such limitation, the GA for the phase difference between consecutive symbols was proposed for DPSK system (Xu et al., 2004). This approximation was generalized in (Nelson & Cartaxo, 2007) in order to take into account the influence of the electrical filter of RX on the phase difference between consecutive symbols, leading to the proposal of an "equivalent" differential phase (EDP). Quite good accuracy on the performance evaluation of DPSK system was achieved in (Nelson & Cartaxo, 2007).
In this chapter, the use of the GA for the EDP of DQPSK system is analysed. The chapter is organized as follows. In section 2, the DQPSK transmitter (TX) and receiver (RX) structures are analyzed. In section 3, the PDF of the EDP is analysed. In section 4, the mean and standard deviation (STD) of the EDP are derived as a function of the received signal and power spectral density (PSD) of optical noise at RX input and are compared with the ones obtained from MC simulation. In section 5, a semi-analytical simulation method (SASM) for DQPSK system performance evaluation based on the GA for the EDP is proposed. In section 6, the DQPSK system performance estimates obtained with MC simulation are compared with those obtained by the SASM, with the objective of validating the SASM. In section 7, the main conclusions and future research work are outlined.

2. DQPSK transmitter and receiver structures

The DQPSK TX and RX are presented in this section. Firstly, the TX structure and operation are briefly discussed. Then, the RX structure and operation are described. Finally, the mathematical description of the RX operation is presented.

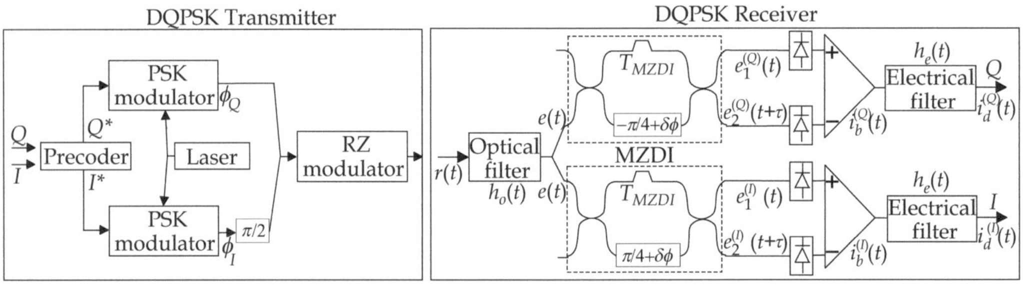

Fig. 1. Scheme of the DQPSK transmitter and receiver.

2.1 DQPSK transmitter

Fig. 1 shows the scheme of the DQPSK TX and RX (Ho, 2005). The input data is firstly precoded at the TX. The logic operations of the DQPSK precoder presented in (Ho, 2005) are considered in this chapter. The precoder eliminates error propagating at RX (Ho, 2005). Furthermore, the precoder allows also direct mapping of the bit sequence of each DQPSK component from the input to the output (Bosco & Poggiolini, 2006). Assuming that the first bit is transmitted in the in-phase (I) component and the second bit is transmitted in the quadrature (Q) component of the DQPSK signal, direct mapping is achieved by coding the input data (I and Q) 11, 01, 00 and 10 with the phase differences of 0, $\pi/2$, π and $-\pi/2$, respectively (Ho, 2005). The coded data (I^* and Q^*) is firstly modulated by phase modulators. Then, a $\pi/2$ phase shift is applied to the I component of the DQPSK signal. The combination of the I and Q components generates the DQPSK signal. Table 1 shows an example of the logical value and phase of the DQPSK signal at several points of the TX.

I	1	1	1	1	0	0	0	1	0	0	1	0	1	1	0	0
Q	1	0	1	0	1	1	1	1	0	0	0	0	1	0	1	0
I^*	0	1	1	1	1	0	0	0	1	0	0	1	1	0	1	0
Q^*	0	0	0	1	0	0	1	1	0	1	0	1	1	1	1	0
ϕ_I	0	π	π	π	π	0	0	0	π	0	0	π	π	0	π	0
ϕ_Q	0	0	0	π	0	0	π	π	0	π	0	π	π	π	π	0
Phase at TX output	$\frac{\pi}{4}$	$-\frac{\pi}{4}$	$-\frac{\pi}{4}$	$-\frac{3\pi}{4}$	$-\frac{\pi}{4}$	$\frac{\pi}{4}$	$\frac{3\pi}{4}$	$\frac{3\pi}{4}$	$-\frac{\pi}{4}$	$\frac{3\pi}{4}$	$\frac{\pi}{4}$	$-\frac{3\pi}{4}$	$-\frac{3\pi}{4}$	$\frac{3\pi}{4}$	$-\frac{3\pi}{4}$	$\frac{\pi}{4}$
Phase difference	0	$-\frac{\pi}{2}$	0	$-\frac{\pi}{2}$	$\frac{\pi}{2}$	$\frac{\pi}{2}$	$\frac{\pi}{2}$	0	π	π	$-\frac{\pi}{2}$	π	0	$-\frac{\pi}{2}$	$\frac{\pi}{2}$	π

Table 1. Logical value and phase of the DQPSK signal at several points of the TX.

The DQPSK signal at the return-to-zero (RZ) modulator input is a non-return-to-zero (NRZ)-DQPSK signal. The RZ modulator is used for generating a RZ-DQPSK signal, if that is the goal. Other TX schemes for generating the DQPSK signals are referred in (Ho, 2005), such as the cascading of two phase modulators or the driving of a single phase modulator by a four level signal. However, the TX scheme shown in Fig. 1 is the one more often referred in the literature mainly because of its simpler implementation. Thus, only the TX scheme shown in Fig. 1 is considered in this chapter. Moreover, as the optical intensity is ideally a constant pulse train without ripple between symbols after the RZ modulator, independently of the TX structure (Ho, 2005), the results shown in this chapter are similar for other TX schemes.

Fig. 2 shows the optical intensity at the output of an ideal TX in case of transmitting a 33% duty-cycle RZ-DQPSK signal or a NRZ-DQPSK signal considering the symbol sequence shown in Table 1. The same average power is imposed for both signals. Fig. 2 shows that the shape of optical intensity is independent of the transmitted symbol sequence.

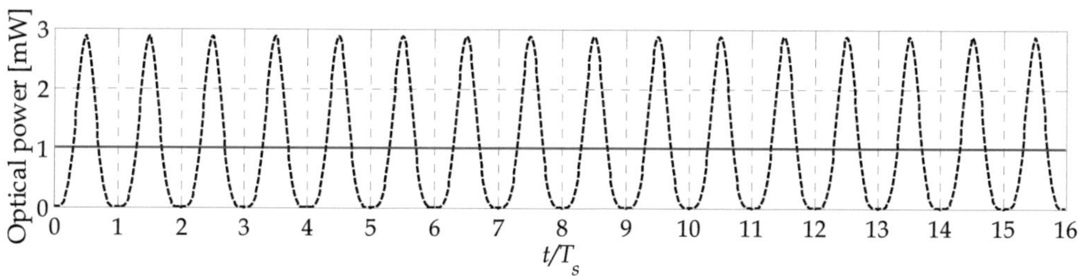

Fig. 2. Optical intensity at the TX output obtained by simulation. Continuous line: NRZ-DQPSK signal; dashed line: 33% duty-cycle RZ-DQPSK signal.

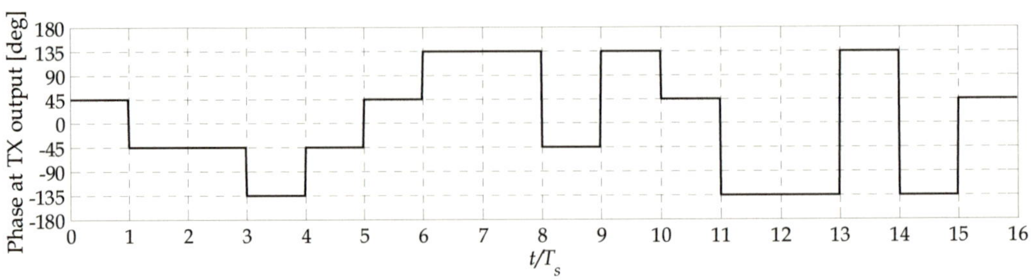

Fig. 3. Phase of the DQPSK signal at the TX output obtained by simulation.

Fig. 3 shows the phase at the output of an ideal TX. Fig. 3 shows that the phase changes accordingly to the symbol sequence, being constant for the duration of each symbol. The shape of the phase is independent of the duty-cycle of the DQPSK signal.

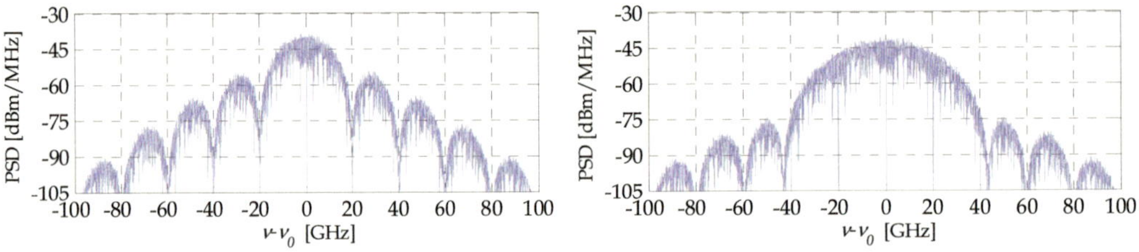

Fig. 4. PSD of the optical field at the output of an ideal TX obtained by simulation. Left: NRZ-DQPSK signal; right: 33% duty-cycle RZ-DQPSK signal.

Fig. 4 shows the PSD of the optical field at the TX output. As a symbol rate of 20 Gbaud/s is imposed, the first null, in case of the NRZ-DQPSK signal, occurs for $\nu - \nu_0 = 20\,\text{GHz}$, where ν is the optical frequency and ν_0 is the optical carrier frequency. In case of the 33% duty-cycle RZ-DQPSK signal, a much wider signal spectrum is observed, as expected. A detailed discussion on the spectral characteristics of RZ signals can be found in (Ip & Kahn, 2006).

2.2 DQPSK receiver

At the RX side, the DQPSK signal is firstly filtered by an optical filter. Then, the resulting signal is split to a pair of Mach-Zehnder delay interferometers (MZDIs) to perform the detection of the I and Q components of the DQPSK signal. Both output ports of each MZDI carry the full information. Indeed, the signal at the output of the two ports of each MZDI is logically conjugated. Thus, the detection of the output of a single port of the MZDI is sufficient to recover the information. However, the use of a balanced detector allows improving the OSNR by about 3 dB and, therefore, the use of the balanced detector is usually considered. The electrical current at the decision circuit input results from filtering the electrical current at the balanced detector output.

Assuming an ideal DQPSK RX, neglecting the filtering impact and defining the received signal as $E_r(t) = Ae^{j\phi(t)}$, where A and $\phi(t)$ are the amplitude and phase of the received signal, respectively, the electrical current at the decision circuit input is $i_d^{(I,Q)}(t) = A^2/2\cos(\phi(t) - \phi(t-T_s) + \theta)$, where T_s is the symbol duration and $\theta = \pi/4$ for the I component of the DQPSK signal, and $\theta = -\pi/4$ for the Q component of the DQPSK signal.

This expression shows that the combined effect of the MZDI and balanced detector translates in creating dependence of the received electrical current on the phase difference between two consecutive symbols. Thus, the transmitted information can be extracted from the EDP, $\Delta\phi_e^{(I,Q)}(t)$, defined as $\Delta\phi_e^{(I,Q)}(t) = \phi(t) - \phi(t - T_s) + \theta$.

Phase difference	0	$-\frac{\pi}{2}$	0	$-\frac{\pi}{2}$	$\frac{\pi}{2}$	$\frac{\pi}{2}$	$\frac{\pi}{2}$	0	π	π	$-\frac{\pi}{2}$	π	0	$-\frac{\pi}{2}$	$\frac{\pi}{2}$	π
EDP (I comp.)	$\frac{\pi}{4}$	$-\frac{\pi}{4}$	$\frac{\pi}{4}$	$-\frac{\pi}{4}$	$\frac{3\pi}{4}$	$\frac{3\pi}{4}$	$\frac{3\pi}{4}$	$\frac{\pi}{4}$	$-\frac{3\pi}{4}$	$-\frac{3\pi}{4}$	$-\frac{\pi}{4}$	$-\frac{3\pi}{4}$	$\frac{\pi}{4}$	$-\frac{\pi}{4}$	$\frac{3\pi}{4}$	$-\frac{3\pi}{4}$
I	1	1	1	1	0	0	0	1	0	0	1	0	1	1	0	0
EDP (Q comp.)	$-\frac{\pi}{4}$	$-\frac{3\pi}{4}$	$-\frac{\pi}{4}$	$-\frac{3\pi}{4}$	$\frac{\pi}{4}$	$\frac{\pi}{4}$	$\frac{\pi}{4}$	$-\frac{\pi}{4}$	$\frac{3\pi}{4}$	$\frac{3\pi}{4}$	$-\frac{3\pi}{4}$	$\frac{3\pi}{4}$	$-\frac{\pi}{4}$	$-\frac{3\pi}{4}$	$\frac{\pi}{4}$	$\frac{3\pi}{4}$
Q	1	0	1	0	1	1	1	1	0	0	0	0	1	0	1	0

Table 2. EDP of the I and Q components and corresponding bit sequence.

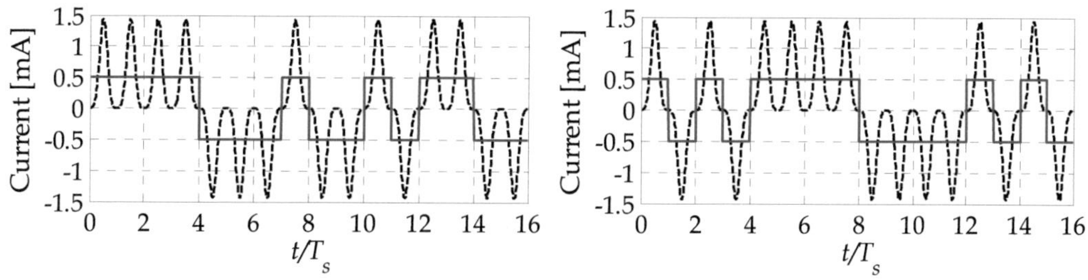

Fig. 5. Electrical current at the decision circuit input in the I component (left) and in the Q component (right) of the DQPSK signal obtained by simulation. Continuous line: NRZ-DQPSK signal; dashed line: 33% duty-cycle RZ-DQPSK signal.

Table 2 shows the EDP of the I and Q components after balanced detection for the symbol sequence presented in Table 1, and the corresponding detected bit sequence. The analysis of Table 2 shows that, for both I and Q components, two EDP give rise to the same bit decision: $\pm\pi/4$ for the "1" bit and $\pm 3\pi/4$ for the "0" bit. This result is a consequence of the electrical current being given as a function of a cosine function. Thus, to decide correctly, two EDP thresholds levels should be set: one threshold level between the symbols transmitted with nominal means of $-3\pi/4$ and $-\pi/4$ and another threshold level between the symbols transmitted with nominal means of $\pi/4$ and $3\pi/4$. However, the RX performs bit decision on the detected electrical current where there is only one electrical current level threshold. Both EDP threshold levels must correspond to that electrical current threshold level. Thus, as the electrical current is given as a function of a cosine function with constant absolute value, the positive and the negative EDP threshold levels must be symmetric. For example, the electrical current level threshold of 0 corresponds to the EDP threshold level of $\pm\pi/2$.

Fig. 5 shows the electrical current at the decision circuit input in the I and Q components when the symbol sequence shown in Table 1 is transmitted. The comparison of Fig. 5 with Fig. 2 reveals that the detected signal has only half the amplitude of the signal at the TX output. This is a consequence of splitting the optical signal into the pair of MZDIs at the RX.

2.3 Mathematical characterization of DQPSK receiver and of noise

The mathematical characterization of the RX operation is important for assessing the impact of the RX on the received signal analytically.

The signal field at the optical filter input is assumed to be completely polarized while optical noise is assumed to be unpolarized and additive with known PSD. When no nonlinear fiber transmission is considered, the amplified spontaneous emission (ASE) noise of optical amplifiers is usually the main noise source and is modelled as additive white Gaussian noise (AWGN) at RX input over each polarization direction.

For additive optical noise, the optical field at the RX input is written as $\mathbf{r}(t)=\mathbf{r}_{in}(t)+\mathbf{r}_n(t)$ where $\mathbf{r}_{in}(t)=r_{in}(t)\mathbf{u}_{\|}$ and $\mathbf{r}_n(t)=r_{n,\|}(t)\mathbf{u}_{\|}+r_{n,\perp}(t)\mathbf{u}_{\perp}$. $\mathbf{r}_{in}(t)$ and $\mathbf{r}_n(t)$ are the Jones vectors of the DQPSK signal and noise fields at the RX input, respectively, and $r_{in}(t)$, $r_{n,\|}(t)$ and $r_{n,\perp}(t)$ are the DQPSK signal and the noise components, at the RX input, over the parallel (to the signal) and perpendicular polarization directions, respectively, defined by the unit vectors $\mathbf{u}_{\|}$ and $\mathbf{u}_{\perp}$. The lowpass equivalent of the DQPSK signal and noise fields, $\mathbf{s}_{in}(t)$ and $\mathbf{n}_n(t)$, respectively, are defined by $\mathrm{r}_{in}(t)=\sqrt{2}\Re\left(\mathrm{s}_{in}(t)e^{j2\pi\nu_0 t}\right)=\sqrt{2}\Re\left(s_{in}(t)e^{j2\pi\nu_0 t}\right)\mathrm{u}_{\|}$ and $\mathrm{r}_n(t)=\sqrt{2}\Re\left(\mathrm{n}_n(t)e^{j2\pi\nu_0 t}\right)=\sqrt{2}\Re\left(\left[n_{n,\|}(t)\mathrm{u}_{\|}+n_{n,\perp}(t)\mathrm{u}_{\perp}\right]e^{j2\pi\nu_0 t}\right)$, where $s_{in}(t)$, $n_{n,\|}(t)$ and $n_{n,\perp}(t)$ are the lowpass equivalents of $r_{in}(t)$, $r_{n,\|}(t)$ and $r_{n,\perp}(t)$, respectively, and $\Re\{x\}$ is the real part of x.

By assuming that the optical filter is polarization-maintaining with impulse response $h_o(t)$, the optical signal at each MZDI input is given by $\mathbf{e}(t)=\mathbf{e}_s(t)+\mathbf{e}_n(t)$, with $\mathbf{e}_s(t)=\mathbf{r}_{in}(t)*h_o(t)/\sqrt{2}$ and $\mathbf{e}_n(t)=\mathbf{r}_n(t)*h_o(t)/\sqrt{2}$. The $\sqrt{2}$ factor quantifies the power splitting into the pair of MZDIs. Ideally, the central frequency of the optical filter, ν_c, is the same of the optical carrier. However, a frequency detuning may occur. Thus, the impact of frequency detuning, defined as $\Delta\nu=\nu_c-\nu_o$, on the DQPSK system performance is assessed. Defining the lowpass equivalent of the optical filter impulse response, $h_{o,l}(t)$, as $h_o(t)=2\Re\left(h_{o,l}(t)e^{j2\pi\nu_0 t}\right)$, the lowpass equivalent of the signal at each MZDI input, $\mathbf{e}_l(t)$, defined as $\mathbf{e}(t)=\sqrt{2}\Re\left(\mathbf{e}_l(t)e^{j2\pi\nu_0 t}\right)$, is $\mathbf{e}_l(t)=\mathbf{s}_f(t)+\mathbf{n}_f(t)$ where

$$
\begin{aligned}
\mathbf{s}_f(t)&=s_{in}(t)*h_o(t)/\sqrt{2}\,\mathbf{u}_{\|}=s(t)\mathbf{u}_{\|}\\
\mathbf{n}_f(t)&=\left(n_{n,\|}(t)*h_o(t)\mathbf{u}_{\|}+n_{n,\perp}(t)*h_o(t)\mathbf{u}_{\perp}\right)/\sqrt{2}=n_{\|}(t)\mathbf{u}_{\|}+n_{\perp}(t)\mathbf{u}_{\perp}.
\end{aligned}
\tag{1}
$$

The total optical signal over each polarization direction is $\mathbf{e}_l(t)=e_{\|}(t)\mathbf{u}_{\|}+e_{\perp}(t)\mathbf{u}_{\perp}$ with

$$
\begin{aligned}
e_{\|}(t)&=s(t)+n_{\|}(t)=s_r(t)+n_{\|,r}(t)+j\left(s_i(t)+n_{\|,i}(t)\right)\\
e_{\perp}(t)&=n_{\perp}(t)=n_{\perp,r}(t)+jn_{\perp,i}(t)
\end{aligned}
\tag{2}
$$

where the indexes *r* and *i* refer to the real and imaginary part of the signals, respectively.

Polarization-maintaining MZDIs are assumed. These devices are mathematically modelled by (Bosco & Poggiolini, 2006):

$$
\begin{aligned}
e_{1,(\|,\perp)}^{(I,Q)}(t)&=1/2\left(\gamma e_{(\|,\perp)}(t-T_{MZDI})+e_{(\|,\perp)}(t)e^{j\theta}\right)\\
e_{2,(\|,\perp)}^{(I,Q)}(t)&=1/2\left(\gamma e_{(\|,\perp)}(t-T_{MZDI})-e_{(\|,\perp)}(t)e^{j\theta}\right)
\end{aligned}
\tag{3}
$$

where θ quantifies the phase rotation performed by the MZDI. Ideally, $\theta=\pi/4$ for the I component of the DQPSK signal, and $\theta=-\pi/4$ for the Q component of the DQPSK signal. However, an imperfect phase rotation with a phase-shifting of $\delta\phi$ from the ideal case may occur. Thus, $\theta=\pm\pi/4+\delta\phi$. A frequency detuning parameter, Δf, can be associated with the imperfect phase rotation by $\Delta f/B_R=\delta\phi/(4\pi)$ (Bosco & Poggiolini, 2006). B_R is the bit rate which is related with the symbol duration by $B_R=2/T_s$. T_{MZDI} is the delay introduced by the MZDI between its two arms. Ideally, T_{MZDI} is equal to the symbol duration. However, a delay error, defined as $\delta T_{MZDI}=T_{MZDI}-T_s$ may occur. The γ parameter, which is in the range [0,1], allows taking into account the extinction ratio of the MZDI. $\gamma=0$ means that the delayed signal is eliminated at the MZDI while $\gamma=1$ means that the delayed signal suffers from the same loss as the original signal at the MZDI. The extinction ratio of the MZDI, ε, is related with γ by $\varepsilon=(1+\gamma)^2/(1-\gamma)^2$. A balanced detector is used at the output of each MZDI. The electrical current at a positive-intrinsic-negative (PIN) photodetector output is related with the optical field at its input, $\mathbf{f}(t)$, by $i(t)=R|\mathbf{f}(t)|^2=R\left(|f_{||}(t)|^2+|f_{\perp}(t)|^2\right)$, where R is the PIN photodetector responsivity. Thus, the electrical current of each component of the DQPSK signal at the balanced detector output is given by

$$\begin{aligned} i_b^{(I,Q)}(t) &= R_1\left|\mathbf{e}_1^{(I,Q)}(t)\right|^2 - R_2\left|\mathbf{e}_2^{(I,Q)}(t+\tau)\right|^2 \\ &= R_1\left(\left|e_{1,||}^{(I,Q)}(t)\right|^2+\left|e_{1,\perp}^{(I,Q)}(t)\right|^2\right) - R_2\left(\left|e_{2,||}^{(I,Q)}(t+\tau)\right|^2+\left|e_{2,\perp}^{(I,Q)}(t+\tau)\right|^2\right) \end{aligned} \tag{4}$$

where τ quantifies the time-misalignment between the signals at the input of the balanced detector. The amplitude-imbalance of the balanced detector is assessed by the k factor, defined as $k=(R_1-R_2)/(R_1+R_2)$. The electrical current resulting from the balanced detector is filtered by a low-pass electrical filter with impulse response $h_e(t)$. Thus, the electrical current at the decision circuit input is $i_d^{(I,Q)}(t)=i_b^{(I,Q)}(t)*h_e(t)$. The electrical current at the decision circuit input may also be expressed in a complex representation as

$$i_d^{(I,Q)}(t)=\Re\left(A_e^{(I,Q)}(t)e^{j\Delta\phi_e^{(I,Q)}(t)}\right) \tag{5}$$

where $A_e^{(I,Q)}$ is the equivalent amplitude of the electrical current. Using eqs. (3), (4) and (5), the EDP can be written as a function of the optical signal at the MZDI input as:

$$\begin{aligned} \Delta\phi_e^{(I,Q)}(t) = \arg\Bigg(& \frac{R_1}{2}\left[\gamma e_{||}(t)e_{||}^*(t-T_{MZDI})e^{j\theta}+\gamma e_{\perp}(t)e_{\perp}^*(t-T_{MZDI})e^{j\theta}\right]*h_e(t) \\ & +\frac{R_1}{4}\left[|\gamma e_{||}(t-T_{MZDI})|^2+|e_{||}(t)|^2+|\gamma e_{\perp}(t-T_{MZDI})|^2+|e_{\perp}(t)|^2\right]*h_e(t) \\ & +\frac{R_2}{2}\left[\gamma e_{||}(t+\tau)e_{||}^*(t-T_{MZDI}+\tau)e^{j\theta}+\gamma e_{\perp}(t+\tau)e_{\perp}^*(t-T_{MZDI}+\tau)e^{j\theta}\right]*h_e(t) \\ & -\frac{R_2}{4}\left[|\gamma e_{||}(t-T_{MZDI}+\tau)|^2+|e_{||}(t+\tau)|^2+|\gamma e_{\perp}(t-T_{MZDI}+\tau)|^2+|e_{\perp}(t+\tau)|^2\right]*h_e(t)\Bigg) \end{aligned} \tag{6}$$

where $e^{*}(t)$ is the complex conjugate of $e(t)$. In case of an ideal RX, where $\Delta\upsilon$=0, $R_1 = R_2$, $T_{MZDI} = T_s$, $\theta = \pm\pi/4$, $\tau = 0$ and $\gamma = 1$, the EDP is given by $\Delta\phi_e^{(I,Q)}(t) = \arg\left(\left[\left(e_{\|}(t)e_{\|}^{*}(t-T_s) + e_{\perp}(t)e_{\perp}^{*}(t-T_s)\right)e^{j\theta}\right] * h_e(t)\right)$ which is the result shown in (Costa & Cartaxo, 2009). Expression (6) shows that the EDP is a nonlinear function of the signal at the MZDI input. Therefore, the rigorous statistical characterization of the EDP may be quite complex. However, approximated solutions may be used in order to get closed-form expressions for the moments of the EDP. The derivation of closed-form expressions for the moments of the EDP is quite important because it allows the assessment of DQPSK system performance in a time-efficient manner.

3. Probability density function of the equivalent differential phase

In this section, the PDF of the EDP is assessed. The PDF of the EDP at the time instant leading to higher eye-opening in the absence of noise is approximately Gaussian-distributed when 33% duty-cycle RZ-DQPSK pulses are considered, even in presence of RX imperfections (Costa & Cartaxo, 2009b). However, the PDF of the EDP when NRZ-DQPSK pulses are considered has not been assessed yet, as far as the authors are aware of. Thus, this analysis is performed in this section. Unless otherwise stated, an OSNR of 15 dB, measured in a 0.1 nm bandwidth, is imposed at the RX input. The bit rate is set to 40 Gbit/s. The frequency response of the modulator of the TX is modelled by a five-pole Bessel electrical filter with a -3 dB bandwidth equal to the symbol rate. A second-order super-Gaussian optical filter with a -3 dB bandwidth of 40 GHz and a five-pole Bessel electrical filter with a -3 dB bandwidth equal to 18 GHz are used at the RX side. These parameters lead to a bit error probability (BEP) of about 10^{-5} in case of ideal RX. Only the PDF of the EDP and eye-diagram of the I component of the DQPSK signal are shown because the Q component of the DQPSK signal showed similar results in all considered cases.

Fig. 6 shows the eye-diagram of electrical current at the electrical filter output when the ideal RX and when several different RX imperfections are considered. The corresponding PDF of the EDP at the time instant leading to higher eye-opening in the absence of noise is also shown. Only the PDF of the EDP of the first symbol transmitted with each nominal mean is shown because the PDFs of the EDP of the remaining symbols transmitted with each nominal mean is similar. Both the actual PDF of the EDP (marks) and the GA for the PDF of the EDP estimated from the mean and STD of the actual PDF (lines) are shown. An ideal RX is considered in Fig. 6 a). In this case, the eye-diagram is quite symmetric at the time instant leading to the higher eye-opening. This symmetry occurs for moderate and wide optical filtering. The PDF of the EDP is approximately Gaussian-distributed with nominal means of $\pm\pi/4$ and $\pm 3\pi/4$. The illustration of imperfect phase rotation is shown in Fig. 6 b). Fig. 6 b) shows that an imperfect phase rotation leads to the closure of the eye-diagram. Its impact on the PDF of the EDP is, however, just a shift on the nominal means equal to the imperfection of the phase rotation. The STD of the EDP remains the same. The illustration of the time-misalignment of signals at the balanced receiver input is shown in Fig. 6 c). This RX imperfection leads to quite asymmetric eye-diagrams. Nevertheless, the EDP is still approximately Gaussian-distributed. However, a slight underestimation of the PDF of the EDP at the area of interest (where the PDF of the EDP of the symbols transmitted with nominal means of $\pm\pi/4$ intersects with the PDF of the EDP of the symbols transmitted

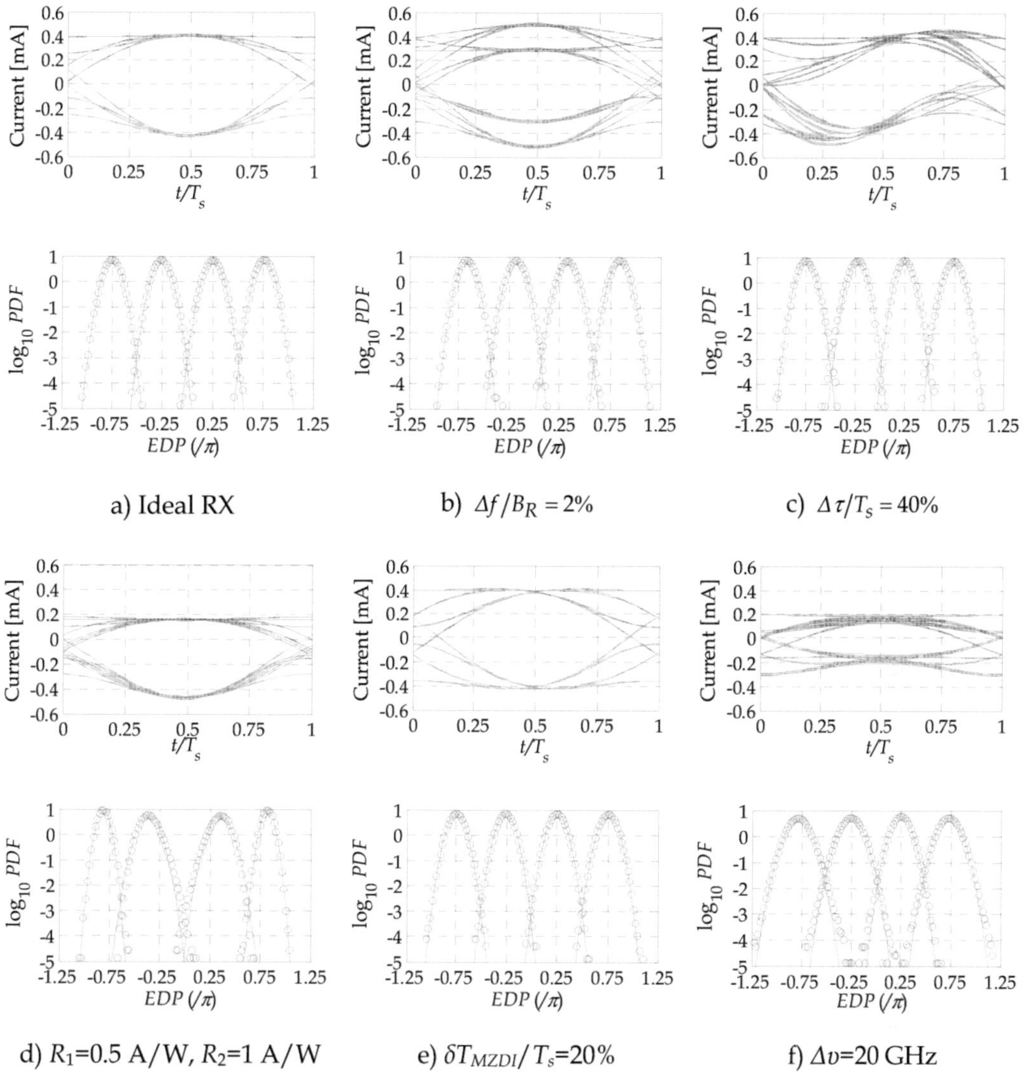

Fig. 6. Eye-diagram of electrical current and corresponding PDF of the EDP in presence of several different RX imperfections. Marks: MC simulation; lines: GA estimated from the results of MC simulation.

with nominal means of $\pm 3\pi/4$) is performed by the GA. Fig. 6 d) illustrates the amplitude-imbalance of detector. This RX imperfection leads to quite asymmetric eye-diagrams and to some inaccuracy in the GA for the PDF of the EDP. However, the EDP at the area of interest is still approximately Gaussian-distributed. The illustration of delay errors of MZDI is shown in Fig. 6 e). This RX imperfection leads to some distortion of the eye-diagram. Nevertheless, the EDP is still approximately Gaussian-distributed. The illustration of the optical filter detuning is shown in Fig. 6 f). The optical filter detuning leads to considerable degradation of the eye-diagram. The EDP at the area of interest is still approximately Gaussian-distributed. However, the GA tends to slight underestimate the PDF of the EDP.

Another RX imperfection is the finite extinction ratio of the MZDIs. This imperfection affects only the DQPSK system performance when combined with amplitude-imbalanced detectors (Bosco & Poggiolini, 2006). In such case, the performance degradation is mainly imposed by the amplitude-imbalance unless much reduced extinction ratios are considered. Thus, both the eye-diagram and PDF of the EDP in presence of finite extinction ratios of the MZDIs are usually similar to those shown in Fig. 6 d).

Fig. 7 shows the eye-diagram of electrical current and the corresponding PDF of the EDP at the decision circuit input when Butterworth electrical filters are considered at the RX side. This analysis allows assessing the impact of the group delay of electrical filters on the eye-diagram and PDF of the EDP because the group delay of Butterworth electrical filters is quite different from the one of Bessel electrical filters. The analysis of Fig. 7 shows that the PDF of the EDP remains approximately Gaussian-distributed even when Butterworth electrical filters are considered.

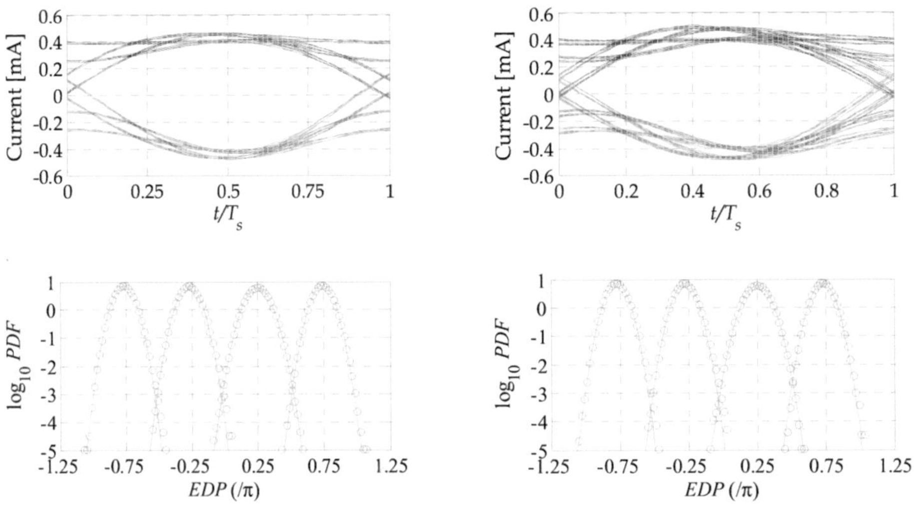

Fig. 7. Eye-diagram of electrical current and corresponding PDF of the EDP when a three (left-hand side) or a five (right-hand side) pole Butterworth electrical filter with $B_e = 18$ GHz is considered at the RX side. An ideal RX is considered. Marks: MC simulation; lines: GA estimated from the results of MC simulation.

The PDF of the EDP has also been assessed for 67% duty-cycle RZ-DQPSK signals for both types of electrical filters, leading to similar conclusions to those presented in this section.

4. Gaussian approximation for equivalent differential phase

The GA consists in approximating a given PDF by a Gaussian PDF. In order to do so, the mean and STD of the Gaussian PDF are set equal to the mean and STD of the PDF it is approximating. The mean and STD of the EDP are derived in this section as a function of the received DQPSK signal and PSD of optical noise at the RX input in order to obtain closed-form expressions for the mean and STD of the EDP. Substituting eq. (2) in eq. (6) and setting $d = t - T_{MZDI}$ to simplify the expressions, we get:

$$\begin{aligned}\Delta\phi_e^{(I,Q)}(t) = \arg\Big\{\Big(& {}^{R_1}\!/_2 \Big[\gamma\Big(s(t)s^*(d) + s(t)n_{\|}^*(d) + n_{\|}(t)s^*(d) + n_{\|}(t)n_{\|}^*(d) + n_{\perp}(t)n_{\perp}^*(d)\Big)e^{j\theta}\Big] \\
& + {}^{R_1}\!/_4 \Big[\gamma^2\Big(|s(d)|^2 + 2\Re\{s(d)n_{\|}^*(d)\} + |n_{\|}(d)|^2 + |n_{\perp}(d)|^2\Big) \\
& \quad + |s(t)|^2 + 2\Re\{s(t)n_{\|}^*(t)\} + |n_{\|}(t)|^2 + |n_{\perp}(t)|^2\Big] \\
& + {}^{R_2}\!/_2 \Big[\gamma\Big(s(t+\tau)s^*(d+\tau) + s(t+\tau)n_{\|}^*(d+\tau) + n_{\|}(t+\tau)s^*(d+\tau) \\
& \quad + n_{\|}(t+\tau)n_{\|}^*(d+\tau) + n_{\perp}(t+\tau)n_{\perp}^*(d+\tau)\Big)e^{j\theta}\Big] \\
& - {}^{R_2}\!/_4 \Big[\gamma^2\Big(|s(d+\tau)|^2 + 2\Re\{s(d+\tau)n_{\|}^*(d+\tau)\} + |n_{\|}(d+\tau)|^2 + |n_{\perp}(d+\tau)|^2\Big) \\
& \quad + |s(t+\tau)|^2 + 2\Re\{s(t+\tau)n_{\|}^*(t+\tau)\} + |n_{\|}(t+\tau)|^2 + |n_{\perp}(t+\tau)|^2\Big]\Big) * h_e(t)\Big\}\end{aligned} \tag{7}$$

In order to obtain closed-form expressions for the mean and STD of the EDP, the dependence of the EDP on noise is linearized. This approximation should lead to only very small discrepancies in the mean and STD of the EDP as the EDP conditioned on the transmitted symbols is approximately Gaussian-distributed. The linearization of the EDP is performed expressing the argument of eq. (7) as an arctangent function. Thus, the several beat terms of eq. (7) are decomposed in their real and imaginary parts. The several beat terms can be written and defined as shown in eq. (14) and eq. (15) (Appendix 9.1). The time dependence of the DQPSK signal and noise is omitted in order to simplify the notation. By substituting the results shown in eqs. (14) and (15) in eq. (7) and by approximating the EDP by a first order Taylor series we get

$$\begin{aligned}\Delta\phi_e^{(I,Q)}(t) = \arctan(k) + \frac{k}{1+k^2}\Big(& \gamma\frac{R_1}{2}\Big(c_2\big[sn_{1,r} + sn_{2,r} + nn_{\|,r} + nn_{\perp,r}\big] \\
& \quad + c_1\big[sn_{1,i} + sn_{2,i} + nn_{\|,i} + nn_{\perp,i}\big]\Big) \\
& + \gamma\frac{R_2}{2}\Big(c_2\big[sn_{\tau,1,r} + sn_{\tau,2,r} + nn_{\tau,\|,r} + nn_{\tau,\perp,r}\big] \\
& \quad + c_1\big[sn_{\tau,1,i} + sn_{\tau,2,i} + nn_{\tau,\|,i} + nn_{\tau,\perp,i}\big]\Big) \\
& - \frac{R_1}{4B}\Big(2\gamma^2 sn_d + \gamma^2 nn_{d,\|} + 2sn_t + nn_{t,\|} + \gamma^2 nn_{d,\perp} + nn_{t,\perp}\Big) \\
& + \frac{R_2}{4B}\Big(2\gamma^2 sn_{\tau,d} + \gamma^2 nn_{\tau,d,\|} + 2sn_{\tau,t} + nn_{\tau,t,\|} + \gamma^2 nn_{\tau,d,\perp} + nn_{\tau,t,\perp}\Big)\Big)\end{aligned} \tag{8}$$

where

$$\begin{aligned}& A = \gamma\Big(\frac{R_1}{2}\big[ss_i\cos(\theta) + ss_r\sin(\theta)\big] + \frac{R_2}{2}\big[ss_{\tau,i}\cos(\theta) + ss_{\tau,r}\sin(\theta)\big]\Big); \\
& B = \gamma\Big(\frac{R_1}{2}\big[ss_r\cos(\theta) - ss_i\sin(\theta)\big] + \frac{R_2}{2}\big[ss_{\tau,r}\cos(\theta) - ss_{\tau,i}\sin(\theta)\big]\Big) \\
& \quad + \frac{R_1}{4}\big[\gamma^2 ss_d + ss_t\big] - \frac{R_2}{4}\big[\gamma^2 ss_{\tau,d} + ss_{\tau,t}\big]; \\
& k = A/B; \quad c_1 = \cos(\theta)/A + \sin(\theta)/B; \quad c_2 = \sin(\theta)/A - \cos(\theta)/B;\end{aligned} \tag{9}$$

From eq. (8), the mean of the EDP is

$$\mu(t) = \arctan(k) + \frac{k}{1+k^2}\left(\gamma\frac{R_1}{2}\left[c_2\left(\mathrm{E}\{nn_{\|,r}\} + \mathrm{E}\{nn_{\perp,r}\}\right) + c_1\left(\mathrm{E}\{nn_{\|,i}\} + \mathrm{E}\{nn_{\perp,i}\}\right)\right] \right.$$
$$+\gamma\frac{R_2}{2}\left[c_2\left(\mathrm{E}\{nn_{\tau,\|,r}\} + \mathrm{E}\{nn_{\tau,\perp,r}\}\right) + c_1\left(\mathrm{E}\{nn_{\tau,\|,i}\} + \mathrm{E}\{nn_{\tau,\perp,i}\}\right)\right]$$
$$-\frac{R_1}{4B}\left(\gamma^2\mathrm{E}\{nn_{d,\|}\} + \mathrm{E}\{nn_{t,\|}\} + \gamma^2\mathrm{E}\{nn_{d,\perp}\} + \mathrm{E}\{nn_{t,\perp}\}\right)$$
$$\left. +\frac{R_2}{4B}\left[\gamma^2\mathrm{E}\{nn_{\tau,d,\|}\} + \mathrm{E}\{nn_{\tau,t,\|}\} + \gamma^2\mathrm{E}\{nn_{\tau,d,\perp}\} + \mathrm{E}\{nn_{\tau,t,\perp}\}\right]\right) \quad (10)$$

Assuming uncorrelated noise over both polarization directions, i.e., $\mathrm{E}\left(nn_{\|,x}nn_{\perp,y}\right) = \mathrm{E}\left(nn_{\|,x}\right)\,\mathrm{E}\left(nn_{\perp,y}\right)$, where x and y represent the real or imaginary part of noise-noise beat terms and, as odd order moments of Gaussian processes with zero mean are null, the variance of the EDP is given by:

$$\sigma^2(t) = \left(\frac{k}{1+k^2}\right)^2\left(\sum_{l=1}^{8}\left[\sigma^2_{N,s-ASE,l} + \sigma^2_{N,ASE-ASE,l}\right]\right) \quad (11)$$

where $\sigma^2_{N,s-ASE,l}$ and $\sigma^2_{N,ASE-ASE,l}$ are the contributions to the signal and noise-noise beat variance, respectively, presented in Appendices 9.2 and 9.3. The variance of the EDP (eq. (11)) is given by a lenghty expression. However, the evaluation of the several terms of eq.

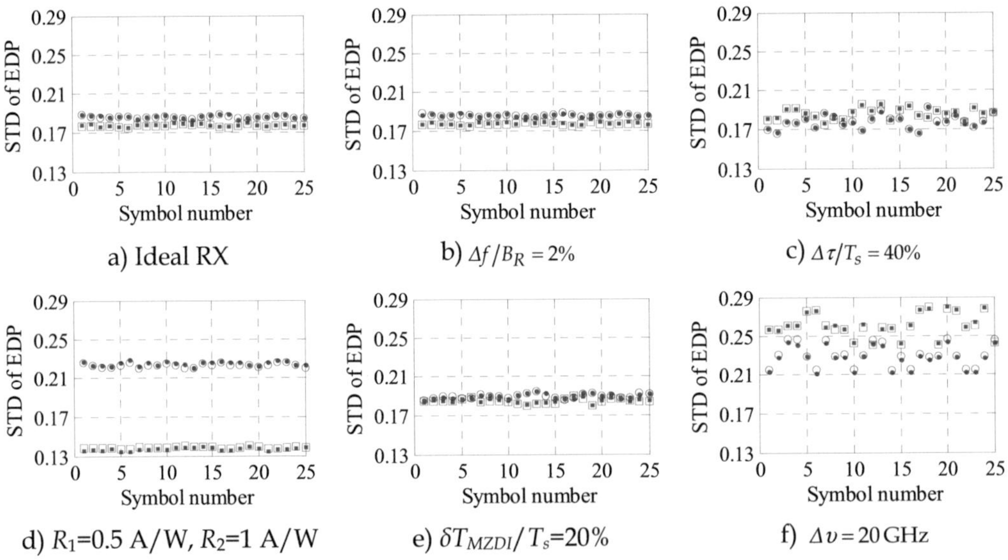

Fig. 8. Standard deviation of the EDP. Only the STD of the EDP of some symbols transmitted with two of the four nominal means (circles: $\pi/4$; squares: $-3\pi/4$) is shown in order to make the figures clearer. Filled symbols: estimates from MC simulation results, obtained considering 15000 noise realizations; empty symbols: estimates from the GA (eq. (11)).

(11) is quite simple which makes the evaluation of the variance of the EDP of quite reduced complexity. Furthermore, if no RX imperfections are considered, eq. (11) is quite simplified, leading to the result shown in (Costa & Cartaxo, 2009). The derivation of the mean and variance of EDP as a function of the received DQPSK signal and PSD of optical noise after optical filtering is shown in (Costa & Cartaxo, 2009b)
Fig. 8 shows the STD of the EDP estimated using the results from MC simulation and the GA (eq. (11)). Analysis of Fig. 8 shows that the estimates of the STD of the EDP obtained using eq. (11) are quite accurate in presence of the majority of RX imperfections. The accuracy of the estimates for the mean of the EDP, estimated using eq. (10), has also been assessed showing that the mean of the EDP is always quite well estimated by eq. (10). The quite good accuracy achieved in the estimation of the mean and STD of the EDP using eqs. (10) and (11) shows that the linearization of the EDP leads only to very small discrepancies on the evaluation of the mean and STD of the EDP and that the impact of noise on the mean and STD of the EDP is correctly estimated.

5. Bit error probability computation by semi-analytical simulation method

A SASM for performance evaluation of DQPSK systems is proposed in this section. The DQPSK signal at the RX input is evaluated by simulation. This permits evaluating the impact of the transmission path, e.g. the nonlinear fiber transmission, the optical add-drop multiplexer concatenation filtering, on the waveform of the DQPSK signal. A quaternary deBruijn sequence with total length N_S is used in the simulation. DeBruijn sequences include all possible symbol sequences with a given length using the lower number of symbols (Jeruchim et al., 2000). This characteristic is important since it assures that all possible cases of inter-symbol interference (ISI) for a given sequence length occur. On the other hand, as the EDP is approximately Gaussian-distributed when the optical noise is modelled as AWGN at the RX input, the impact of noise on the DQPSK system performance is assessed analytically.
As the precoding performed in the TX allows direct mapping of the bit sequence from the TX input to the RX output, the overall BEP is given by $BEP = \left(BEP^{(I)} + BEP^{(Q)}\right)/2$, where $BEP^{(I,Q)}$ is the BEP of each component of the DQPSK signal. In order to take accurately into account the impact of ISI on the DQPSK system performance, separate Gaussian distributions with different means and STDs are associated with each one of the transmitted bits. This approach has already proved to be accurate to estimate the ISI impact on OOK modulation (Rebola & Cartaxo, 2001). The BEP of each component of the DQPSK signal can be seen as the mean of four BEPs associated with the four nominal means for the PDF of the EDP. Thus, defining F as the EDP threshold level, with $F \geq 0$, the BEP of the I and Q components of the DQPSK signal is given by

$$BEP^{(I,Q)} = \frac{1}{2N_s}\left(\sum_{\substack{n=1 \\ a_n=\pm\pi/4}}^{N_s} \operatorname{erfc}\left[\frac{F - \left|\mu_{a_n,n}\right|}{\sqrt{2}\sigma_{a_n,n}}\right] + \sum_{\substack{n=1 \\ a_n=\pm 3\pi/4}}^{N_s} \operatorname{erfc}\left[\frac{-F + \left|\mu_{a_n,n}\right|}{\sqrt{2}\sigma_{a_n,n}}\right] \right) \tag{12}$$

where erfc(x) is the complementary error function and $\mu_{a_n,n}$ and $\sigma_{a_n,n}$ are the mean and STD of the EDP at the sampling time for the n-th received symbol with nominal mean a_n.

$\mu_{a_n,n}$ and $\sigma_{a_n,n}$ are obtained from eq. (10) and eq. (11), respectively, by evaluating these expressions at the sampling time and by associating each sampling time with each transmitted symbol. The optimal threshold level of the EDP, F_{opt}, is assessed by setting to zero the derivative of eq. (12) with respect to F, leading to the transcendental equation

$$\sum_{\substack{n=1 \\ a_n=\pm\pi/4}}^{N_s} \frac{1}{\sigma_{a_n,n}} \exp\left(-\frac{1}{2}\left[\frac{F_{opt}-\left|\mu_{a_n,n}\right|}{\sigma_{a_n,n}}\right]^2\right) = \sum_{\substack{n=1 \\ a_n=\pm 3\pi/4}}^{N_s} \frac{1}{\sigma_{a_n,n}} \exp\left(-\frac{1}{2}\left[\frac{-F_{opt}+\left|\mu_{a_n,n}\right|}{\sigma_{a_n,n}}\right]^2\right) \tag{13}$$

that can be numerically solved using the Newton-Raphson method.

6. Accuracy of the SASM based on the GA for the EDP

In this section, the accuracy of the SASM for DQPSK system performance evaluation based on the GA for the EDP is assessed. This analysis is performed comparing the results obtained using eq. (12) with those obtained using MC simulation. A $BEP = 10^{-4}$ is set as the target BEP mainly because MC simulation is much time consuming for lower BEP and the use of forward error correction (FEC), such as Reed-Solomon codes, allows to achieve much lower BEP at the expense of only a slight increase on the bit rate. The accuracy of the SASM is firstly assessed in presence of RX imperfections. Then, the accuracy of the SASM is assessed considering nonlinear fiber transmission. The bit error ratio estimates obtained using MC simulation are only accepted after at least 100 errors occurring in each component of the DQPSK signal. The threshold level is optimized and the time instant leading to higher eye-opening in the absence of noise is chosen as sampling time. The TX and RX parameters are the same as the ones considered in section 3, unless otherwise stated.

6.1 Accuracy of the SASM in presence of RX imperfections

When the ideal RX is considered, the MC simulation estimates that an OSNR of about 14 dB is required to achieve $BEP = 10^{-4}$. The SASM estimates a required OSNR of only about 13.8 dB. This small difference is attributed mainly to the difference between the GA for the PDF of the EDP and its actual PDF. This conclusion results from having very good agreement between the estimates of the mean and STD of the EDP obtained using eq. (10) and eq. (11) with the corresponding ones obtained using MC simulation. Indeed, the SASM leads to the correct required OSNR (14 dB) by increasing the STD of the EDP, calculated using eq. (11), by only about 2.5%.

Fig. 9 shows the impact of several different RX imperfections on the OSNR penalty at $BEP = 10^{-4}$. The considered RX imperfections cover all expected values for each imperfection. The impact of the RX imperfections on the DQPSK system performance has been assessed by MC simulation and by SASM in order to assess the accuracy of the SASM. The analysis of Fig. 9 shows that the SASM is quite accurate in presence of the majority of the typical RX imperfections leading usually to a discrepancy on the OSNR penalty not exceeding 0.2 dB. Among the cases analysed in Fig. 9, the higher discrepancies occur for high time-misalignment of signals at the balanced detector input $\left(\left|\Delta\tau/T\right| > 30\,\%\right)$ and for high frequency detuning of the optical filters $\left(\left|\Delta\nu\right| > 15\text{ GHz}\right)$. Indeed, the SASM leads to an underestimation of the OSNR penalty in both cases that may attain about 0.5 dB.

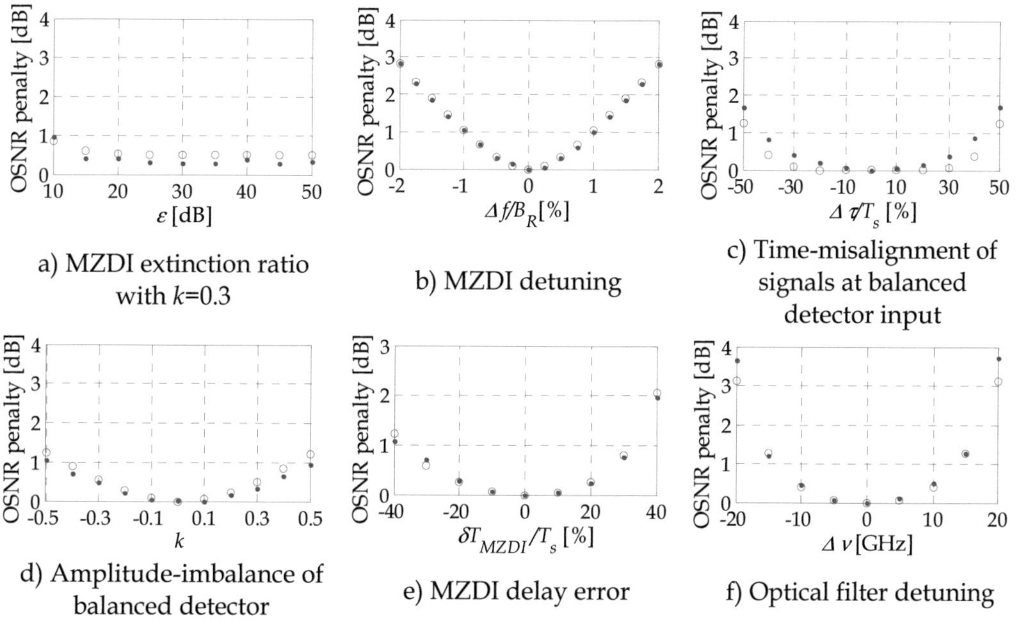

Fig. 9. OSNR penalty at $BEP=10^{-4}$ as a function of several different RX imperfections. Filled circles: MC simulation; empty circles: SASM.

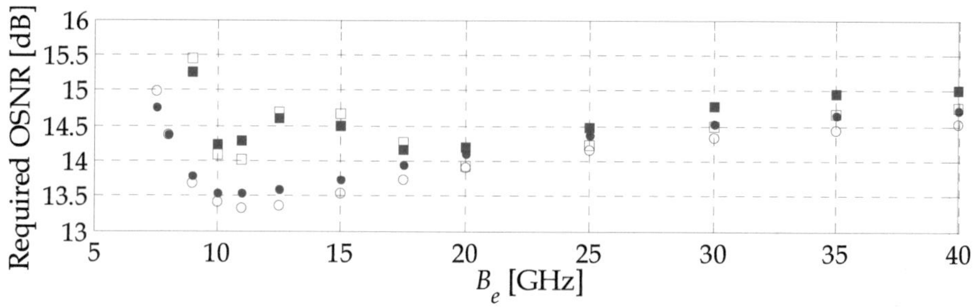

Fig. 10. Required OSNR at $BEP=10^{-4}$ as a function of the electrical filter type and bandwidth, considering an ideal RX. Empty marks: SASM; filled marks: MC simulation. Circles: five-pole Bessel electrical filter; squares: five-pole Butterworth electrical filter.

Fig. 10 illustrates the accuracy of the SASM when different bandwidths and types of electrical filter are considered. Fig. 10 shows that the required OSNR is quite well estimated independently of the type and bandwidth of the electrical filter. Indeed, the discrepancy of the required OSNR does not usually exceed 0.2 dB. This small discrepancy is mainly attributed to the difference between the GA for the PDF of the EDP and its actual PDF. Fig. 10 shows also that the behavior of the required OSNR as a function of the electrical filter bandwidth depends on the electrical filter type. The different behaviors illustrated in Fig. 10 for filter bandwidths around 12 GHz can be explained by observing the eye-opening. Indeed, we find that the eye-opening is more reduced for B_e around 12.5 GHz than for B_e around 11 GHz when the Butterworth electrical filter is used, which does not occur in case of the Bessel electrical filter.

6.2 Accuracy of the SASM in presence of nonlinear fiber transmission

To reach long-haul cost-efficient transmission, as required in core networks, the fiber spans should be quite long to reduce the number of required optical amplifiers. The power level at the input of each span should also be as high as possible to achieve high OSNR. On the other hand, when high power levels are used, the fiber nonlinearity imposes a severe power penalty. Thus, a compromise between the optical power level and the power penalty imposed by the fiber nonlinearity has to be accomplished. Standard single-mode fiber (SSMF) is the transmission fiber type more commonly used in these networks. Despite its many advantages, it introduces high distortion in the transmitted signal due to its high dispersion. Thus, the use of dispersion compensation along the transmission path is required.

In an ideal single-mode optical fiber, the two orthogonal states of polarization are degenerated, i. e. they propagate with identical propagation constants (Iannone et al., 1998). Thus, the input light-polarization would remain constant over the whole propagation length. In reality, optical fibers may have a slightly elliptical core which leads to birefringence, i. e. the propagation constants of the two orthogonal states of polarization differ slightly. External perturbations such as stress, bending and torsion lead also to birefringence (Hanik, 2002). Thus, the impact of fiber birefringence, group velocity dispersion (GVD) and self-phase modulation (SPM) are considered to assess the accuracy of the SASM in presence of nonlinear fiber transmission.

The MC simulation is performed by solving the coupled nonlinear Schrödinger propagation equation, also known as the vector version of the nonlinear Schrödinger propagation equation, instead of the scalar version of the nonlinear Schrödinger propagation equation, in order to take into account the impact of fiber birefringence. However, the solution of the coupled nonlinear Schrödinger propagation equation is much more complex than the one of the scalar version (Iannone et al., 1998). Nevertheless, the split-step Fourier method, which is usually used to solve the scalar version of the nonlinear Schrödinger propagation equation, can be applied to its vector version when the so-called high-birefringence condition (Iannone et al., 1998) is verified. In this case, the exponential term in the vector version of the nonlinear Schrödinger propagation equation that depends on the birefringence fluctuates rapidly and its effect tends to average out (Iannone et al., 1998). This approximation is usually verified in single mode optical fibers and has been commonly used in the literature where negligible loss of accuracy is usually achieved (Marcuse et al., 1997). Furthermore, by choosing an adequate integration step, the coupling between the polarization modes can be neglected when solving the propagation within a single step. After each step, the eigenpolarizations are randomly rotated and a random phase shift is added. A more detailed explanation of how the simulation of fiber nonlinear transmission is performed can be found in (Iannone et al., 1998). In our MC simulation, the birefringence is assumed constant over successive integration steps of 100 meters. The eigenpolarizations are uniformly distributed over the birefringence axes and the phase shift, which corresponds to 2π over the beat length, has a Rician distribution with mean value $0.1 \cdot 2\pi\ \mathrm{m}^{-1}$ and variance $0.01 \cdot 2\pi\ \mathrm{m}^{-1}$ (Carena et al., 1998).

The DQPSK system performance evaluation by the SASM requires assessing the DQPSK noiseless waveform and PSD of optical noise at RX input after nonlinear fiber transmission. The noiseless waveform of the DQPSK signal is assessed by performing noiseless transmission of the DQPSK signal using the scalar version of the nonlinear Schrödinger equation, but with the fiber nonlinearity coefficient reduced by a 8/9 factor. Indeed, the scalar version of the nonlinear Schrödinger propagation equation leads to similar results to those of its vector version when it is solved with the fiber nonlinearity coefficient set to 8/9 of its real value (Carena et. al., 1998), (Hanik, 2002). The PSD of optical noise depends on the polarization direction. Indeed, the AWGN approximation for optical noise at the RX input over the same polarization direction as the DQPSK signal may be quite inaccurate when nonlinear fiber transmission is considered. Indeed, when a strong signal (the DQPSK signal) propagates along a transmission fiber, it creates a spectral region around itself where a small signal (the optical amplifier's ASE noise) experiences gain. This phenomenon is known as parametric gain (Carena et al., 1998). Furthermore, the nonlinear phase noise due to the amplitude-to-phase noise conversion effect arising from the interaction of the optical amplifier's ASE noise and the nonlinear Kerr effect must also be taken into account. The evaluation of the parametric gain and nonlinear phase noise can be performed considering the nonlinear fiber transmission of only one polarization direction but with the fiber nonlinearity coefficient reduced by the 8/9 factor. This approximation allows evaluating the PSD of optical noise after nonlinear fiber transmission over the DQPSK signal polarization direction using the method proposed in (Demir, 2007). The method proposed in (Demir, 2007) evaluates the PSD of optical noise in a quite time efficient manner by deriving a linear partial-differential equation for the noise perturbation. In order to do so, the nonlinear Schrödinger equation is linearized around a continuous-wave signal. The AWGN approximation for optical noise at RX input over the perpendicular polarization direction is still quite accurate (Carena et al., 1998). Thus, the PSD of optical noise over the perpendicular polarization direction is obtained by adding the individual ASE noise contributions of each optical amplifier, each affected by the total gain from the optical amplifier till the RX.

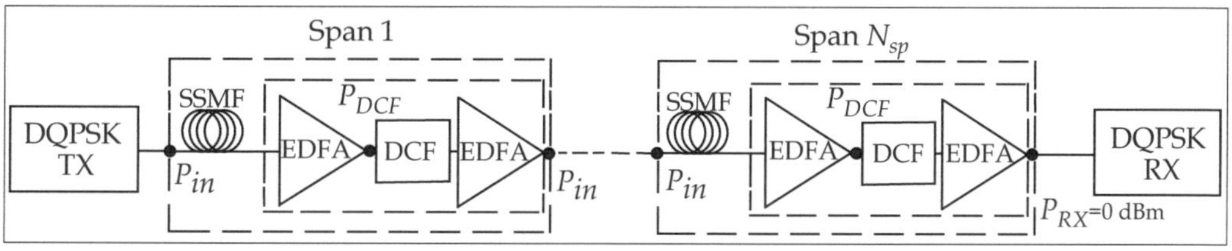

Fig. 11. Scheme of the DQPSK transmission system.

Fig. 11 shows the schematic configuration of the DQPSK transmission system. The total link is composed by N_{sp} spans, with N_{sp} =20. Each span is composed by 100 km of SSMF followed by a double-stage erbium-doped fiber amplifier (EDFA). Dispersion compensating fibers (DCFs) are used for total compensation of the dispersion accumulated in the SSMF of each span. To assure that all DCFs operate nearly in linear regime, the power level denoted by P_{DCF} is imposed at the DCFs input. The average power level at the input of each SSMF is denoted by P_{in}. The total gain of both EDFAs' stages compensates for the power loss in each span, except in the last span. In this case, the second stage EDFA is used to impose a power level of 0 dBm at the RX input. The SSMF has an attenuation parameter of 0.21 dB/km, a dispersion parameter of 17 ps/nm/km, an effective core area of 80 μm^2 and a nonlinear index-coefficient of 0.025 nm^2/W. The EDFA's noise figure is 7 dB. The dispersion parameter of the DCF is -100 ps/nm/km and its attenuation parameter depends on the transmission scenario that is being considered. Indeed, in order to keep the BEP high enough to perform MC simulation in a reasonable amount of time and, as 33% duty-cycle RZ-DQPSK pulses show better performance than NRZ-DQPSK pulses, the attenuation parameter of DCF is 0.5 dB/km when NRZ-DQPSK pulses are considered and 0.6 dB/km when 33% duty-cycle RZ-DQPSK pulses are considered. Furthermore, $P_{DCF} = -8$ dBm when NRZ-DQPSK pulses are considered and $P_{DCF} = -12$ dBm when 33% duty-cycle RZ-DQPSK pulses are considered.

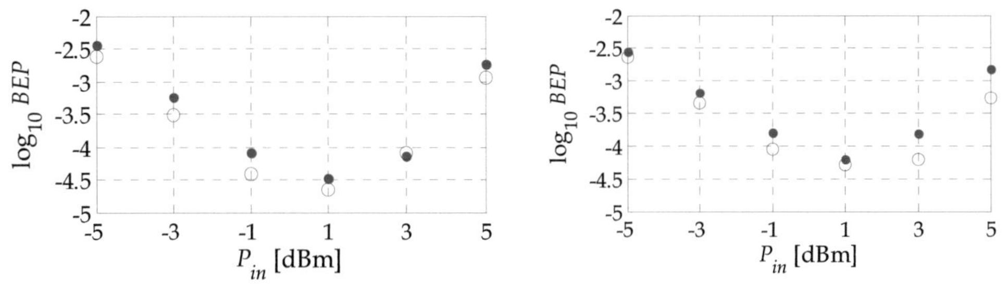

Fig. 12. Performance of NRZ-DQPSK (left) and 33% duty-cycle RZ-DQPSK (right) for the transmission system of Fig. 11. Filled circles: MC simulation; empty circles: SASM.

Fig. 12 shows the performance of DQPSK modulation in presence of nonlinear fiber transmission. When NRZ-DQPSK signals are considered (figure on the left), a good accuracy is achieved both when ASE noise is the main transmission impairment (lower power levels) and when the transmission is mainly limited by fiber nonlinearity (higher power levels). This result leads to the conclusion that methods for DQPSK system performance evaluation based on the GA for the EDP lead to quite good accuracy even in presence of nonlinear fiber transmission when NRZ-DQPSK signals are considered. The analysis of the figure on the right-hand side, where the transmission of a 33% duty-cycle RZ-DQPSK signal is considered, shows that the SASM estimates the performance of the DQPSK system quite accurately when ASE noise is the main impairment. However, the accuracy of the SASM tends to decrease with the increase of the impact of the fiber nonlinearity. This loss of accuracy is not a consequence of the loss of accuracy of the GA for the EDP. Indeed, quite good accuracy is achieved when the mean and STD of the EDP are estimated from the results of MC simulation. The decrease of accuracy of the SASM is a consequence of the inaccuracy in the evaluation of the PSD of optical noise. Indeed, the linearization of the nonlinear Schrödinger equation around a continuous-wave signal does not provide an acceptable description of noise statistics when 33% duty-cycle RZ-DQPSK signals are transmitted and the impact of fiber nonlinearity is important.

Computation time gains of about 15000 times have been achieved by the SASM when compared with MC simulation for $BEP = 10^{-4}$.

7. Conclusion and work in progress

The performance evaluation of simulated optical DQPSK modulation has been analysed. The EDP of DQPSK signals is approximately Gaussian-distributed. Thus, a SASM for DQPSK systems performance evaluation based on the GA has been proposed. The SASM relies on the use of the closed-form expressions derived for the mean and STD of the EDP for assessing the performance of the DQPSK system in a time-efficient manner. Quite good agreement between MC simulation and the results of the SASM is usually achieved, even in presence of RX imperfections and nonlinear fiber transmission. Indeed, although the SASM leads usually to an underestimation of the required OSNR of about 0.2 dB, the discrepancy of the OSNR penalty at $BEP = 10^{-4}$ is usually below 0.2 dB for the majority of the typical RX imperfections.

Several subjects on the performance evaluation of DQPSK signals using the GA for the EDP are still to be addressed. Indeed, the evaluation of the PSD of optical noise at RX input, after nonlinear transmission, admitting a modulated signal and the accuracy improvement of the SASM achieved by using the more accurate description for the PSD of optical noise is still to be performed. The validation of the SASM for BEP around 10^{-12} and the proposal of a scheme for evaluating the EDP experimentally are also still to be addressed.

8. Acknowledgments

This work was supported in part by Fundação para a Ciência e a Tecnologia from Portugal under Ph.D. contract SFRH/BD/42287/2007.

9. Appendix

9.1 List of beat terms at decision circuit Input

The current and the EDP at the decision circuit input are given as a function of the following beat terms:

$$
\begin{aligned}
&\left(s(t)s^*(d)\right)*h_e(t) = \left(s_r(t)s_r(d) + s_i(t)s_i(d) + j\left[s_i(t)s_r(d) - s_r(t)s_i(d)\right]\right)*h_e(t) \equiv ss_r + jss_i \\
&\left(s(t)n_{\|}^*(d)\right)*h_e(t) = \left(s_r(t)n_{\|,r}(d) + s_i(t)n_{\|,i}(d) + j\left[s_i(t)n_{\|,r}(d) - s_r(t)n_{\|,i}(d)\right]\right)*h_e(t) \equiv sn_{1,r} + jsn_{1,i} \\
&\left(n_{\|}(t)s^*(d)\right)*h_e(t) = \left(n_{\|,r}(t)s_r(d) + n_{\|,i}(t)s_i(d) + j\left[n_{\|,i}(t)s_r(d) - n_{\|,r}(t)s_i(d)\right]\right)*h_e(t) \equiv sn_{2,r} + jsn_{2,i} \\
&\left(n_{\|}(t)n_{\|}^*(d)\right)*h_e(t) = \left(n_{\|,r}(t)n_{\|,r}(d) + n_{\|,i}(t)n_{\|,i}(d) + j\left[n_{\|,i}(t)n_{\|,r}(d) - n_{\|,r}(t)n_{\|,i}(d)\right]\right)*h_e(t) \\
&\qquad \equiv nn_{\|,r} + jnn_{\|,i} \\
&\left(n_{\perp}(t)n_{\perp}^*(d)\right)*h_e(t) = \left(n_{\perp,r}(t)n_{\perp,r}(d) + n_{\perp,i}(t)n_{\perp,i}(d) + j\left[n_{\perp,i}(t)n_{\perp,r}(d) - n_{\perp,r}(t)n_{\perp,i}(d)\right]\right)*h_e(t) \\
&\qquad \equiv nn_{\perp,r} + jnn_{\perp,i} \\
&|s(d)|^2 * h_e(t) = \left(s_r^2(d) + s_i^2(d)\right)*h_e(t) \equiv ss_d \\
&\Re\left\{s(d)n_{\|}^*(d)\right\}*h_e(t) = \left(s_r(d)n_{\|,r}(d) + s_i(d)n_{\|,i}(d)\right)*h_e(t) \equiv sn_d \\
&|s(t)|^2 * h_e(t) = \left(s_r^2(t) + s_i^2(t)\right)*h_e(t) \equiv ss_t \\
&|n_{\|}(d)|^2 * h_e(t) = \left(n_{\|,r}^2(d) + n_{\|,i}^2(d)\right)*h_e(t) \equiv nn_{d,\|} \\
&\Re\left\{s(t)n_{\|}^*(t)\right\}*h_e(t) = \left(s_r(t)n_{\|,r}(t) + s_i(t)n_{\|,i}(t)\right)*h_e(t) \equiv sn_t \\
&|n_{\|}(t)|^2 * h_e(t) = \left(n_{\|,r}^2(t) + n_{\|,i}^2(t)\right)*h_e(t) \equiv nn_{t,\|} \\
&|n_{\perp}(d)|^2 * h_e(t) = \left(n_{\perp,r}^2(d) + n_{\perp,i}^2(d)\right)*h_e(t) \equiv nn_{d,\perp} \\
&|n_{\perp}(t)|^2 * h_e(t) = \left(n_{\perp,r}^2(t) + n_{\perp,i}^2(t)\right)*h_e(t) \equiv nn_{t,\perp}
\end{aligned}
\tag{14}
$$

and, similarly:

$$
\begin{aligned}
&\left(s(t+\tau)s^*(d+\tau)\right)*h_e(t) \equiv ss_{\tau,r} + jss_{\tau,i}; \quad \left(s(t+\tau)n_{||}^*(d+\tau)\right)*h_e(t) \equiv sn_{\tau,1,r} + jsn_{\tau,1,i} \\
&\left(n_{||}(t+\tau)s^*(d+\tau)\right)*h_e(t) \equiv sn_{\tau,2,r} + jsn_{\tau,2,i}; \quad \left(n_{||}(t+\tau)n_{||}^*(d+\tau)\right)*h_e(t) \equiv nn_{\tau,||,r} + jnn_{\tau,||,i} \\
&\left(n_{\perp}(t+\tau)n_{\perp}^*(d+\tau)\right)*h_e(t) \equiv nn_{\tau,\perp,r} + jnn_{\tau,\perp,i}; \quad |s(d+\tau)|^2 *h_e(t) \equiv ss_{\tau,d} \\
&\Re\left\{s(d+\tau)n_{||}^*(d+\tau)\right\}*h_e(t) \equiv sn_{\tau,d}; \quad |s(t+\tau)|^2 *h_e(t) \equiv ss_{\tau,t} \\
&|n_{||}(d+\tau)|^2 *h_e(t) \equiv nn_{\tau,d,||}; \quad \Re\left\{s(t+\tau)n_{||}^*(t+\tau)\right\}*h_e(t) \equiv sn_{\tau,t} \\
&|n_{||}(t+\tau)|^2 *h_e(t) \equiv nn_{\tau,t,||}; \quad |n_{\perp}(d+\tau)|^2 *h_e(t) \equiv nn_{\tau,d,\perp} \\
&|n_{\perp}(t+\tau)|^2 *h_e(t) \equiv nn_{\tau,t,\perp}
\end{aligned} \tag{15}
$$

9.2 Contributions to the signal-noise beat variance

The variance of the signal-noise beat may be separated in several contributions. To illustrate the impact of RX imperfections, the contributions to the signal-noise beat variance of ideal RX, shown in (Costa & Cartaxo, 2009b), are used as reference. Thus, we find that the $c_2^2\left(\mathrm{E}\left\{(sn_{1,r})^2\right\} + 2\mathrm{E}\left\{sn_{1,r}sn_{2,r}\right\} + \mathrm{E}\left\{(sn_{2,r})^2\right\}\right)$ contribution to the signal-noise beat variance results from

$$
\begin{aligned}
\sigma_{N,s-ASE,1}^2 = {} & \gamma^2 \frac{R_1^2}{4} c_2^2\left(\mathrm{E}\left\{(sn_{1,r})^2\right\} + 2\mathrm{E}\left\{sn_{1,r}sn_{2,r}\right\} + \mathrm{E}\left\{(sn_{2,r})^2\right\}\right) \\
& + \gamma^2 \frac{R_2^2}{4} c_2^2\left(\mathrm{E}\left\{(sn_{\tau,1,r})^2\right\} + 2\mathrm{E}\left\{sn_{\tau,1,r}sn_{\tau,2,r}\right\} + \mathrm{E}\left\{(sn_{\tau,2,r})^2\right\}\right) \\
& + \gamma^2 \frac{R_1R_2}{2} c_2^2\left(\mathrm{E}\left\{sn_{1,r}sn_{\tau,1,r}\right\} + \mathrm{E}\left\{sn_{1,r}sn_{\tau,2,r}\right\} + \mathrm{E}\left\{sn_{2,r}sn_{\tau,1,r}\right\} + \mathrm{E}\left\{sn_{2,r}sn_{\tau,2,r}\right\}\right)
\end{aligned} \tag{16}
$$

by considering an ideal RX. Similarly, the $c_1^2\left(\mathrm{E}\left\{(sn_{1,i})^2\right\} + 2\mathrm{E}\left\{sn_{1,i}sn_{2,i}\right\} + \mathrm{E}\left\{(sn_{2,i})^2\right\}\right)$ and $2c_1c_2\left(\mathrm{E}\left\{sn_{1,r}sn_{1,i}\right\} + \mathrm{E}\left\{sn_{2,r}sn_{1,i}\right\} + \mathrm{E}\left\{sn_{1,r}sn_{2,i}\right\} + \mathrm{E}\left\{sn_{2,r}sn_{2,i}\right\}\right)$ contributions to the signal-noise beat variance result from

$$
\begin{aligned}
\sigma_{N,s-ASE,2}^2 = {} & \gamma^2 \frac{R_1^2}{4} c_1^2\left(\mathrm{E}\left\{(sn_{1,i})^2\right\} + 2\mathrm{E}\left\{sn_{1,i}sn_{2,i}\right\} + \mathrm{E}\left\{(sn_{2,i})^2\right\}\right) \\
& + \gamma^2 \frac{R_2^2}{4} c_1^2\left(\mathrm{E}\left\{(sn_{\tau,1,i})^2\right\} + 2\mathrm{E}\left\{sn_{\tau,1,i}sn_{\tau,2,i}\right\} + \mathrm{E}\left\{(sn_{\tau,2,i})^2\right\}\right) \\
& + \gamma^2 \frac{R_1R_2}{2} c_1^2\left(\mathrm{E}\left\{sn_{1,i}sn_{\tau,1,i}\right\} + \mathrm{E}\left\{sn_{1,i}sn_{\tau,2,i}\right\} + \mathrm{E}\left\{sn_{2,i}sn_{\tau,1,i}\right\} + \mathrm{E}\left\{sn_{2,i}sn_{\tau,2,i}\right\}\right)
\end{aligned} \tag{17}
$$

and

$$\begin{aligned}\sigma^2_{N,s-ASE,3} &= \gamma^2\frac{R_1^2}{2}c_1c_2\left(\mathrm{E}\{sn_{1,r}sn_{1,i}\}+\mathrm{E}\{sn_{2,r}sn_{1,i}\}+\mathrm{E}\{sn_{1,r}sn_{2,i}\}+\mathrm{E}\{sn_{2,r}sn_{2,i}\}\right)\\ &+\gamma^2\frac{R_1R_2}{2}c_1c_2\left(\mathrm{E}\{sn_{\tau,1,r}sn_{1,i}\}+\mathrm{E}\{sn_{\tau,2,r}sn_{1,i}\}+\mathrm{E}\{sn_{\tau,1,r}sn_{2,i}\}+\mathrm{E}\{sn_{\tau,2,r}sn_{2,i}\}\right)\\ &+\gamma^2\frac{R_1R_2}{2}c_1c_2\left(\mathrm{E}\{sn_{1,r}sn_{\tau,1,i}\}+\mathrm{E}\{sn_{1,r}sn_{\tau,2,i}\}+\mathrm{E}\{sn_{2,r}sn_{\tau,1,i}\}+\mathrm{E}\{sn_{2,r}sn_{\tau,2,i}\}\right)\\ &+\gamma^2\frac{R_2^2}{2}c_1c_2\left(\mathrm{E}\{sn_{\tau,1,r}sn_{\tau,1,i}\}+\mathrm{E}\{sn_{\tau,2,r}sn_{\tau,1,i}\}+\mathrm{E}\{sn_{\tau,1,r}sn_{\tau,2,i}\}+\mathrm{E}\{sn_{\tau,2,r}sn_{\tau,2,i}\}\right)\end{aligned} \tag{18}$$

respectively, when the ideal RX is considered. Other contributions to the signal-noise beat variance arise from the imperfections of the RX and are cancelled when the ideal RX is considered. These contributions are:

$$\begin{aligned}\sigma^2_{N,s-ASE,4} &= \gamma\frac{R_1R_2}{2B}c_2\left(\gamma^2\mathrm{E}\{sn_{1,r}sn_{\tau,d}\}+\mathrm{E}\{sn_{1,r}sn_{\tau,t}\}+\gamma^2\mathrm{E}\{sn_{2,r}sn_{\tau,d}\}+\mathrm{E}\{sn_{2,r}sn_{\tau,t}\}\right)\\ &-\gamma\frac{R_1^2}{2B}c_2\left(\gamma^2\mathrm{E}\{sn_{1,r}sn_{d}\}+\mathrm{E}\{sn_{1,r}sn_{t}\}+\gamma^2\mathrm{E}\{sn_{2,r}sn_{d}\}+\mathrm{E}\{sn_{2,r}sn_{t}\}\right)\end{aligned} \tag{19}$$

$$\begin{aligned}\sigma^2_{N,s-ASE,5} &= \gamma\frac{R_1R_2}{2B}c_1\left(\gamma^2\mathrm{E}\{sn_{1,i}sn_{\tau,d}\}+\mathrm{E}\{sn_{1,i}sn_{\tau,t}\}+\gamma^2\mathrm{E}\{sn_{2,i}sn_{\tau,d}\}+\mathrm{E}\{sn_{2,i}sn_{\tau,t}\}\right)\\ &-\gamma\frac{R_1^2}{2B}c_1\left(\gamma^2\mathrm{E}\{sn_{1,i}sn_{d}\}+\mathrm{E}\{sn_{1,i}sn_{t}\}+\gamma^2\mathrm{E}\{sn_{2,i}sn_{d}\}+\mathrm{E}\{sn_{2,i}sn_{t}\}\right)\end{aligned} \tag{20}$$

$$\begin{aligned}\sigma^2_{N,s-ASE,6} &= \gamma\frac{R_2^2}{2B}c_2\left(\gamma^2\mathrm{E}\{sn_{\tau,1,r}sn_{\tau,d}\}+\mathrm{E}\{sn_{\tau,1,r}sn_{\tau,t}\}+\gamma^2\mathrm{E}\{sn_{\tau,2,r}sn_{\tau,d}\}+\mathrm{E}\{sn_{\tau,2,r}sn_{\tau,t}\}\right)\\ &-\gamma\frac{R_1R_2}{2B}c_2\left(\gamma^2\mathrm{E}\{sn_{\tau,1,r}sn_{d}\}+\mathrm{E}\{sn_{\tau,1,r}sn_{t}\}+\gamma^2\mathrm{E}\{sn_{\tau,2,r}sn_{d}\}+\mathrm{E}\{sn_{\tau,2,r}sn_{t}\}\right)\end{aligned} \tag{21}$$

$$\begin{aligned}\sigma^2_{N,s-ASE,7} &= \gamma\frac{R_2^2}{2B}c_1\left(\gamma^2\mathrm{E}\{sn_{\tau,1,i}sn_{\tau,d}\}+\mathrm{E}\{sn_{\tau,1,i}sn_{\tau,t}\}+\gamma^2\mathrm{E}\{sn_{\tau,2,i}sn_{\tau,d}\}+\mathrm{E}\{sn_{\tau,2,i}sn_{\tau,t}\}\right)\\ &-\gamma\frac{R_1R_2}{2B}c_1\left(\gamma^2\mathrm{E}\{sn_{\tau,1,i}sn_{d}\}+\mathrm{E}\{sn_{\tau,1,i}sn_{t}\}+\gamma^2\mathrm{E}\{sn_{\tau,2,i}sn_{d}\}+\mathrm{E}\{sn_{\tau,2,i}sn_{t}\}\right)\end{aligned} \tag{22}$$

$$\begin{aligned}\sigma^2_{N,s-ASE,8} &= \frac{R_1^2}{4B^2}\left(\gamma^4\mathrm{E}\{(sn_d)^2\}+2\gamma^2\mathrm{E}\{sn_d sn_t\}+\mathrm{E}\{(sn_t)^2\}\right)\\ &+\frac{R_2^2}{4B^2}\left(\gamma^4\mathrm{E}\{(sn_{\tau,d})^2\}+2\gamma^2\mathrm{E}\{sn_{\tau,d}sn_{\tau,t}\}+\mathrm{E}\{(sn_{\tau,t})^2\}\right)\\ &-\frac{R_1R_2}{2B^2}\left(\gamma^4\mathrm{E}\{sn_d sn_{\tau,d}\}+\gamma^2\mathrm{E}\{sn_d sn_{\tau,t}\}+\gamma^2\mathrm{E}\{sn_t sn_{\tau,d}\}+\mathrm{E}\{sn_t sn_{\tau,t}\}\right).\end{aligned} \tag{23}$$

9.3 Contributions to the noise-noise beat variance

The variance of the noise-noise beat may be separated in several contributions. To illustrate the impact of RX imperfections, the contributions to the noise-noise beat variance of ideal RX, shown in (Costa & Cartaxo, 2009b), are used as reference. Thus, we find that the $c_2^2\left(\mathrm{E}\left\{(nn_{\|,r})^2\right\}-\mathrm{E}^2\left\{nn_{\|,r}\right\}+\mathrm{E}\left\{(nn_{\perp,r})^2\right\}-\mathrm{E}^2\left\{nn_{\perp,r}\right\}\right)$ contribution to the noise-noise beat variance results from

$$\begin{aligned}\sigma^2_{N,ASE-ASE,1} &= \gamma^2\frac{R_1^2}{4}c_2^2\left(\mathrm{E}\left\{(nn_{\|,r})^2\right\}-\mathrm{E}^2\left\{nn_{\|,r}\right\}+\mathrm{E}\left\{(nn_{\perp,r})^2\right\}-\mathrm{E}^2\left\{nn_{\perp,r}\right\}\right)\\ &+\gamma^2\frac{R_2^2}{4}c_2^2\left(\mathrm{E}\left\{(nn_{\tau,\|,r})^2\right\}-\mathrm{E}^2\left\{nn_{\tau,\|,r}\right\}+\mathrm{E}\left\{(nn_{\tau,\perp,r})^2\right\}-\mathrm{E}^2\left\{nn_{\tau,\perp,r}\right\}\right)\\ &+\gamma^2\frac{R_1R_2}{2}c_2^2\left(\mathrm{E}\left\{nn_{\|,r}nn_{\tau,\|,r}\right\}-\mathrm{E}\left\{nn_{\|,r}\right\}\mathrm{E}\left\{nn_{\tau,\|,r}\right\}+\mathrm{E}\left\{nn_{\perp,r}nn_{\tau,\perp,r}\right\}-\mathrm{E}\left\{nn_{\perp,r}\right\}\mathrm{E}\left\{nn_{\tau,\perp,r}\right\}\right)\end{aligned} \tag{24}$$

by considering an ideal RX. Similarly, the $c_1^2\left(\mathrm{E}\left\{(nn_{\|,i})^2\right\}-\mathrm{E}^2\left\{nn_{\|,i}\right\}+\mathrm{E}\left\{(nn_{\perp,i})^2\right\}-\mathrm{E}^2\left\{nn_{\perp,i}\right\}\right)$ and $2c_1c_2\left(\mathrm{E}\left\{nn_{\|,r}nn_{\|,i}\right\}-\mathrm{E}\left\{nn_{\|,r}\right\}\mathrm{E}\left\{nn_{\|,i}\right\}+\mathrm{E}\left\{nn_{\perp,r}nn_{\perp,i}\right\}-\mathrm{E}\left\{nn_{\perp,r}\right\}\mathrm{E}\left\{nn_{\perp,i}\right\}\right)$ contributions to the noise-noise beat variance result from

$$\begin{aligned}\sigma^2_{N,ASE-ASE,2} &= \gamma^2\frac{R_1^2}{4}c_1^2\left(\mathrm{E}\left\{(nn_{\|,i})^2\right\}-\mathrm{E}^2\left\{nn_{\|,i}\right\}+\mathrm{E}\left\{(nn_{\perp,i})^2\right\}-\mathrm{E}^2\left\{nn_{\perp,i}\right\}\right)\\ &+\gamma^2\frac{R_2^2}{4}c_1^2\left(\mathrm{E}\left\{(nn_{\tau,\|,i})^2\right\}-\mathrm{E}^2\left\{nn_{\tau,\|,i}\right\}+\mathrm{E}\left\{(nn_{\tau,\perp,i})^2\right\}-\mathrm{E}^2\left\{nn_{\tau,\perp,i}\right\}\right)\\ &+\gamma^2\frac{R_1R_2}{2}c_1^2\left(\mathrm{E}\left\{nn_{\|,i}nn_{\tau,\|,i}\right\}-\mathrm{E}\left\{nn_{\|,i}\right\}\mathrm{E}\left\{nn_{\tau,\|,i}\right\}+\mathrm{E}\left\{nn_{\perp,i}nn_{\tau,\perp,i}\right\}-\mathrm{E}\left\{nn_{\perp,i}\right\}\mathrm{E}\left\{nn_{\tau,\perp,i}\right\}\right)\end{aligned} \tag{25}$$

and

$$\begin{aligned}\sigma^2_{N,ASE-ASE,3} &= \gamma^2\frac{R_1^2}{2}c_1c_2\left(\mathrm{E}\left\{nn_{\|,r}nn_{\|,i}\right\}-\mathrm{E}\left\{nn_{\|,r}\right\}\mathrm{E}\left\{nn_{\|,i}\right\}+\mathrm{E}\left\{nn_{\perp,r}nn_{\perp,i}\right\}-\mathrm{E}\left\{nn_{\perp,r}\right\}\mathrm{E}\left\{nn_{\perp,i}\right\}\right)\\ &+\gamma^2\frac{R_1R_2}{2}c_1c_2\left(\mathrm{E}\left\{nn_{\tau,\|,r}nn_{\|,i}\right\}-\mathrm{E}\left\{nn_{\tau,\|,r}\right\}\mathrm{E}\left\{nn_{\|,i}\right\}+\mathrm{E}\left\{nn_{\tau,\perp,r}nn_{\perp,i}\right\}-\mathrm{E}\left\{nn_{\tau,\perp,r}\right\}\mathrm{E}\left\{nn_{\perp,i}\right\}\right)\\ &+\gamma^2\frac{R_1R_2}{2}c_1c_2\left(\mathrm{E}\left\{nn_{\|,r}nn_{\tau,\|,i}\right\}-\mathrm{E}\left\{nn_{\|,r}\right\}\mathrm{E}\left\{nn_{\tau,\|,i}\right\}+\mathrm{E}\left\{nn_{\perp,r}nn_{\tau,\perp,i}\right\}-\mathrm{E}\left\{nn_{\perp,r}\right\}\mathrm{E}\left\{nn_{\tau,\perp,i}\right\}\right)\\ &+\gamma^2\frac{R_2^2}{2}c_1c_2\left(\mathrm{E}\left\{nn_{\tau,\|,r}nn_{\tau,\|,i}\right\}-\mathrm{E}\left\{nn_{\tau,\|,r}\right\}\mathrm{E}\left\{nn_{\tau,\|,i}\right\}+\mathrm{E}\left\{nn_{\tau,\perp,r}nn_{\tau,\perp,i}\right\}-\mathrm{E}\left\{nn_{\tau,\perp,r}\right\}\mathrm{E}\left\{nn_{\tau,\perp,i}\right\}\right)\end{aligned} \tag{26}$$

respectively, when the ideal RX is considered. Other contributions to the noise-noise beat variance arise from the imperfections of the RX and are cancelled when the ideal RX is considered. These contributions are:

$$
\begin{aligned}
\sigma^2_{N,ASE-ASE,4} = {} & \gamma \frac{R_1 R_2}{4B} c_2 \Big(\gamma^2 \big[E\{nn_{\parallel,r} nn_{\tau,d,\parallel}\} - E\{nn_{\parallel,r}\}E\{nn_{\tau,d,\parallel}\} + E\{nn_{\perp,r} nn_{\tau,d,\perp}\} - E\{nn_{\perp,r}\}E\{nn_{\tau,d,\perp}\}\big] \\
& + E\{nn_{\parallel,r} nn_{\tau,t,\parallel}\} - E\{nn_{\parallel,r}\}E\{nn_{\tau,t,\parallel}\} + E\{nn_{\perp,r} nn_{\tau,t,\perp}\} - E\{nn_{\perp,r}\}E\{nn_{\tau,t,\perp}\}\Big) \\
& - \gamma \frac{R_1^2}{4B} c_2 \Big(\gamma^2 \big[E\{nn_{\parallel,r} nn_{d,\parallel}\} - E\{nn_{\parallel,r}\}E\{nn_{d,\parallel}\} + E\{nn_{\perp,r} nn_{d,\perp}\} - E\{nn_{\perp,r}\}E\{nn_{d,\perp}\}\big] \\
& + E\{nn_{\parallel,r} nn_{t,\parallel}\} - E\{nn_{\parallel,r}\}E\{nn_{t,\parallel}\} + E\{nn_{\perp,r} nn_{t,\perp}\} - E\{nn_{\perp,r}\}E\{nn_{t,\perp}\}\Big)
\end{aligned}
\quad (27)
$$

$$
\begin{aligned}
\sigma^2_{N,ASE-ASE,5} = {} & \gamma \frac{R_1 R_2}{4B} c_1 \Big(\gamma^2 \big[E\{nn_{\parallel,i} nn_{\tau,d,\parallel}\} - E\{nn_{\parallel,i}\}E\{nn_{\tau,d,\parallel}\} + E\{nn_{\perp,i} nn_{\tau,d,\perp}\} - E\{nn_{\perp,i}\}E\{nn_{\tau,d,\perp}\}\big] \\
& + E\{nn_{\parallel,i} nn_{\tau,t,\parallel}\} - E\{nn_{\parallel,i}\}E\{nn_{\tau,t,\parallel}\} + E\{nn_{\perp,i} nn_{\tau,t,\perp}\} - E\{nn_{\perp,i}\}E\{nn_{\tau,t,\perp}\}\Big) \\
& - \gamma \frac{R_1^2}{4B} c_1 \Big(\gamma^2 \big[E\{nn_{\parallel,i} nn_{d,\parallel}\} - E\{nn_{\parallel,i}\}E\{nn_{d,\parallel}\} + E\{nn_{\perp,i} nn_{d,\perp}\} - E\{nn_{\perp,i}\}E\{nn_{d,\perp}\}\big] \\
& + E\{nn_{\parallel,i} nn_{t,\parallel}\} - E\{nn_{\parallel,i}\}E\{nn_{t,\parallel}\} + E\{nn_{\perp,i} nn_{t,\perp}\} - E\{nn_{\perp,i}\}E\{nn_{t,\perp}\}\Big)
\end{aligned}
\quad (28)
$$

$$
\begin{aligned}
\sigma^2_{N,ASE-ASE,6} = {} & \gamma \frac{R_2^2}{4B} c_2 \Big(\gamma^2 \big[E\{nn_{\tau,\parallel,r} nn_{\tau,d,\parallel}\} - E\{nn_{\tau,\parallel,r}\}E\{nn_{\tau,d,\parallel}\} + E\{nn_{\tau,\perp,r} nn_{\tau,d,\perp}\} - E\{nn_{\tau,\perp,r}\}E\{nn_{\tau,d,\perp}\}\big] \\
& + E\{nn_{\tau,\parallel,r} nn_{\tau,t,\parallel}\} - E\{nn_{\tau,\parallel,r}\}E\{nn_{\tau,t,\parallel}\} + E\{nn_{\tau,\perp,r} nn_{\tau,t,\perp}\} - E\{nn_{\tau,\perp,r}\}E\{nn_{\tau,t,\perp}\}\Big) \\
& - \gamma \frac{R_1 R_2}{4B} c_2 \Big(\gamma^2 \big[E\{nn_{\tau,\parallel,r} nn_{d,\parallel}\} - E\{nn_{\tau,\parallel,r}\}E\{nn_{d,\parallel}\} + E\{nn_{\tau,\perp,r} nn_{d,\perp}\} - E\{nn_{\tau,\perp,r}\}E\{nn_{d,\perp}\}\big] \\
& + E\{nn_{\tau,\parallel,r} nn_{t,\parallel}\} - E\{nn_{\tau,\parallel,r}\}E\{nn_{t,\parallel}\} + E\{nn_{\tau,\perp,r} nn_{t,\perp}\} - E\{nn_{\tau,\perp,r}\}E\{nn_{t,\perp}\}\Big)
\end{aligned}
\quad (29)
$$

$$
\begin{aligned}
\sigma^2_{N,ASE-ASE,7} = {} & \frac{R_1^2}{16B^2} \Big(\gamma^4 \big[E\{(nn_{d,\parallel})^2\} - E^2\{nn_{d,\parallel}\} + E\{(nn_{d,\perp})^2\} - E^2\{nn_{d,\perp}\}\big] \\
& + 2\gamma^2 \big[E\{nn_{d,\parallel} nn_{t,\parallel}\} - E\{nn_{d,\parallel}\}E\{nn_{t,\parallel}\} + E\{nn_{d,\perp} nn_{t,\perp}\} - E\{nn_{d,\perp}\}E\{nn_{t,\perp}\}\big] \\
& + E\{(nn_{t,\parallel})^2\} - E^2\{nn_{t,\parallel}\} + E\{(nn_{t,\perp})^2\} - E^2\{nn_{t,\perp}\}\Big) \\
& + \frac{R_2^2}{16B^2} \Big(\gamma^4 \big[E\{(nn_{\tau,d,\parallel})^2\} - E^2\{nn_{\tau,d,\parallel}\} + E\{(nn_{\tau,d,\perp})^2\} - E^2\{nn_{\tau,d,\perp}\}\big] \\
& + 2\gamma^2 \big[E\{nn_{\tau,d,\parallel} nn_{\tau,t,\parallel}\} - E\{nn_{\tau,d,\parallel}\}E\{nn_{\tau,t,\parallel}\} + E\{nn_{\tau,d,\perp} nn_{\tau,t,\perp}\} - E\{nn_{\tau,d,\perp}\}E\{nn_{\tau,t,\perp}\}\big] \\
& + E\{(nn_{\tau,t,\parallel})^2\} - E^2\{nn_{\tau,t,\parallel}\} + E\{(nn_{\tau,t,\perp})^2\} - E^2\{nn_{\tau,t,\perp}\}\Big) \\
& - \frac{R_1 R_2}{8B^2} \Big(\gamma^4 \big[E\{nn_{d,\parallel} nn_{\tau,d,\parallel}\} - E\{nn_{d,\parallel}\}E\{nn_{\tau,d,\parallel}\} + E\{nn_{d,\perp} nn_{\tau,d,\perp}\} - E\{nn_{d,\perp}\}E\{nn_{\tau,d,\perp}\}\big] \\
& + \gamma^2 \big[E\{nn_{d,\parallel} nn_{\tau,t,\parallel}\} - E\{nn_{d,\parallel}\}E\{nn_{\tau,t,\parallel}\} + E\{nn_{d,\perp} nn_{\tau,t,\perp}\} - E\{nn_{d,\perp}\}E\{nn_{\tau,t,\perp}\}\big] \\
& + \gamma^2 \big[E\{nn_{t,\parallel} nn_{\tau,d,\parallel}\} - E\{nn_{t,\parallel}\}E\{nn_{\tau,d,\parallel}\} + E\{nn_{t,\perp} nn_{\tau,d,\perp}\} - E\{nn_{t,\perp}\}E\{nn_{\tau,d,\perp}\}\big] \\
& + E\{nn_{t,\parallel} nn_{\tau,t,\parallel}\} - E\{nn_{t,\parallel}\}E\{nn_{\tau,t,\parallel}\} + E\{nn_{t,\perp} nn_{\tau,t,\perp}\} - E\{nn_{t,\perp}\}E\{nn_{\tau,t,\perp}\}\Big)
\end{aligned}
\quad (30)
$$

$$\begin{aligned}\sigma^2_{N,ASE-ASE,8} =& \gamma\frac{R_2^2}{4B}c_1\Big(\gamma^2\Big[\mathrm{E}\{nn_{\tau,\parallel,i}nn_{\tau,d,\parallel}\}-\mathrm{E}\{nn_{\tau,\parallel,i}\}\mathrm{E}\{nn_{\tau,d,\parallel}\}+\mathrm{E}\{nn_{\tau,\perp,i}nn_{\tau,d,\perp}\}-\mathrm{E}\{nn_{\tau,\perp,i}\}\mathrm{E}\{nn_{\tau,d,\perp}\}\Big]\\ &+\mathrm{E}\{nn_{\tau,\parallel,i}nn_{\tau,t,\parallel}\}-\mathrm{E}\{nn_{\tau,\parallel,i}\}\mathrm{E}\{nn_{\tau,t,\parallel}\}+\mathrm{E}\{nn_{\tau,\perp,i}nn_{\tau,t,\perp}\}-\mathrm{E}\{nn_{\tau,\perp,i}\}\mathrm{E}\{nn_{\tau,t,\perp}\}\Big)\\ &-\gamma\frac{R_1R_2}{4B}c_1\Big(\gamma^2\Big[\mathrm{E}\{nn_{\tau,\parallel,i}nn_{d,\parallel}\}-\mathrm{E}\{nn_{\tau,\parallel,i}\}\mathrm{E}\{nn_{d,\parallel}\}+\mathrm{E}\{nn_{\tau,\perp,i}nn_{d,\perp}\}-\mathrm{E}\{nn_{\tau,\perp,i}\}\mathrm{E}\{nn_{d,\perp}\}\Big]\\ &+\mathrm{E}\{nn_{\tau,\parallel,i}nn_{t,\parallel}\}-\mathrm{E}\{nn_{\tau,\parallel,i}\}\mathrm{E}\{nn_{t,\parallel}\}+\mathrm{E}\{nn_{\tau,\perp,i}nn_{t,\perp}\}-\mathrm{E}\{nn_{\tau,\perp,i}\}\mathrm{E}\{nn_{t,\perp}\}\Big)\end{aligned} \quad (31)$$

9.4 List of acronyms

AWGN – Additive white Gaussian noise

BEP – Bit error probability

DCF – Dispersion compensating fiber

DPSK – Differential phase-shift-keying

DQPSK – Differential quadrature phase-shift-keying

EDFA – Erbium doped fiber amplifier

EDP- Equivalent differential phase

GA- Gaussian approximation

GVD – Group velocity dispersion

I – In-phase

MC - Monte-Carlo

MZDI – Mach-Zehnder delay interferometer

NRZ – Non-return-to-zero

OOK – On-off keying

OSNR - Optical signal-to-noise ratio

PDF – Probability density function

PIN - Positive-intrinsic-negative photodetector

PSD – Power spectral density

Q – Quadrature

RX – Receiver

RZ – Return-to-zero

SASM - Semi-analytical simulation method

SPM – Self-phase modulation

SSMF - Standard single-mode fiber

STD – Standard deviation

TX - Transmitter

10. References

Bosco, G. & Poggiolini, P. (2006). On the joint effect of receiver impairments on direct-detection DQPSK systems, *IEEE/OSA J. Lightwave Technol.*, vol. 24, no. 3, Mar. 2006, pp. 1323-1333.

Carena, C., Curri, V. et al. (1998). On the joint effects of fiber parametric gain and birefringence and their influence on ASE noise, *IEEE/OSA J. Lightwave Technol.*, vol. 16, no. 7, Jul. 1998, pp. 1149-1157.

Costa, N. & Cartaxo, A. (2007). BER estimation in DPSK systems using the differential phase Q taking into account the electrical filtering influence, *IEEE Proc. Intern. Microwave and Optoelectron. Conf.*, Salvador, Brazil, Oct. 2007, pp. 337-340.

Costa, N. & Cartaxo, A. (2009). Novel semi-analytical method for BER evaluation in simulated optical DQPSK systems, *IEEE Photon. Technol. Lett.*, vol. 21, no. 7, Apr. 2009, pp. 447-449.

Costa, N. & Cartaxo, A. (2009b). Optical DQPSK system performance evaluation using equivalent differential phase in presence of receiver imperfections, *IEEE/OSA J. Lightwave Technol.*, submitted paper.

Demir, A. (2007). Nonlinear phase noise in optical-fiber-communication systems, *J. Lightwave Technol.*, vol. 25, no. 8, Aug. 2007, pp. 2002-2032.

Hanik, N. (2002). Modelling of nonlinear optical wave propagation including linear mode-coupling and birefringence, *Optics Communications*, vol. 214, Dec. 2002, pp. 207-230.

Ho, K.-P. (2005). *Phase-Modulated Optical Communication Systems,* Springer, ISBN-13: 978-0-387-24392-4, United States of America, 2005.

Ip, E. & Kahn, J. (2006). Power spectra of return-to-zero optical signals, *IEEE/OSA J. Lightwave Technol.*, vol. 24, no. 3, Mar. 2006, pp. 1610-1618.

Iannone, E., Matera, F. et al. (1998). *Nonlinear Optical Communication Networks,* John Wiley & Sons, ISBN: 0-471-15270-6, United States of America, 1998.

Jeruchim, M., Balaban, P. & Shanmugan, K. (2000). *Simulation of Communication Systems - Modelling, Methodology and Techniques*, Kluwer Academic/Plenum Publishers, ISBN: 0-306-46267-2, United States of America, 2000.

Marcuse, D., Menyuk, C. R. & Way, P. K. (1997). Application of the Manakov-PMD equation to studies of signal propagation in optical fibers with randomly varying birefringence, *IEEE/OSA J. Lightwave Technol.*, vol. 15, no. 9, Sept. 1997, pp. 1735-1746.

Morita, I. & Yoshikane, N. (2005). Merits of DQPSK for ultrahigh capacity transmission, *Proc. IEEE LEOS Annual Meeting*, Sydney, Australia, Oct. 2005, paper We5.

Rebola, J. L. & Cartaxo, A. (2001). Gaussian approach for performance evaluation of optically preamplified receivers with arbitrary optical and electrical filters, *IEE Proc. Optoelectron.*, vol. 148, no. 3, Jun. 2001, pp. 135-142.

Winzer, P. J. & Essiambre, R.-J. (2006). Advanced modulation formats for high-capacity optical transport networks, *IEEE/OSA J. Lightwave Technol.*, vol. 24, no. 12, Dec. 2006, pp. 4711-4728.

Winzer, P. J., Raybon, G. et al. (2008). 100-Gb/s DQPSK transmission: from laboratory experiments to field trials, *IEEE/OSA J. Lightwave Technol.*, vol. 26, no. 20, Oct. 2008, pp. 3388-3402.

Xu, C., Liu, X. & Wei, X. (2004). Differential phase-shift keying for high spectral efficiency optical transmissions, *IEEE J. Select. Topics in Quantum Electron.*, vol. 10, no. 2, Mar./Apr. 2004, pp. 281-293.

20

Fiber-to-the-Home System with Remote Repeater

An Vu Tran, Nishaanthan Nadarajah and Chang-Joon Chae
Victoria Research Laboratory, NICTA Ltd.
Level 2, Building 193, Electrical & Electronic Engineering
University of Melbourne, VIC. 3010,
Australia

1. Introduction

In the last few years, there has been rapid deployment of fixed wireline access networks around the world based on fiber-to-the-home (FTTH) architecture. Passive optical network (PON) is emerging as the most promising FTTH technology due to the minimal use of optical transceivers and fiber deployment, and the use of passive outside plant (OSP) (Dixit, 2003; Kettler et al., 2000; Kramer et al., 2002). However, large scale PON deployment is to some degree still limited by the high cost of the customer's optical network unit (ONU), which contains a costly laser transmitter. The active optical network (AON) architecture is one potential solution that can reduce the ONU cost by utilizing low-cost vertical cavity surface emitting laser (VCSEL) based transmitters. However, this system requires an Ethernet switch at the remote node, which is expensive in terms of cost and maintenance, and needs additional transceiver per customer. Another drawback of traditional PON systems is that it has a split ratio limitation of 1:32, which makes it harder and much more expensive to upgrade the network once more customers are connected.

In this book chapter, a FTTH system to reduce the ONU transmitter cost based on the use of an upstream repeater at the remote node is reported. The repeater consists of standard PON transmitter and receiver and therefore, does not significantly increase the overall system cost. Moreover, by utilizing bidirectional Ethernet PON (EPON) transceiver modules to regenerate the downstream signals as well as the upstream signals, we are able to extend the feeder fiber reach to 60 km and split ratio of the FTTH system to 1:256. The repeater-based system is demonstrated for both standard EPON-based FTTH and extended FTTH systems and shows insignificant performance penalty.

In order to achieve higher user count and longer range coverage in the access network, the repeater-based FTTH system can be cascaded in series. This will result in a lower network installation cost per customer, especially when FTTH take-up rate is low. In this chapter, we also investigate the jitter performance of cascaded repeater-based FTTH architectures via a recirculating loop. Our demonstration shows that we can achieve up to 4 regeneration loops with insignificant penalty and the total timing jitter is within the IEEE EPON standard requirement.

With the presence of the active repeater at the remote node of a FTTH system, we can provide additional functionalities for the network including video service delivery and local internetworking. These systems will be investigated and presented in this chapter together with an economic study of the repeater-based FTTH system compared with other technologies.

2. FTTH system with remote repeater

A schematic of the proposed FTTH architecture with an upstream repeater is shown in Fig. 1 (Tran et al., 2006b). In this architecture, a conventional 1×N star coupler (SC) is replaced by a 2×N SC. One arm of the SC on the optical line terminal (OLT) side is connected to the remote repeater, which could be at the same location as the SC or at a different location for access to commercial power lines with a battery back-up. The other arm of the SC is to transmit downstream signals through the SC and bypass the remote repeater. An isolator is installed on this downstream path to prevent upstream signals from entering. The downstream and upstream signals are separated/combined using a coarse wavelength division multiplexer (CWDM). The upstream signals can be 2R or 3R regenerated at the remote repeater using a burst-mode receiver (BMR), a burst-mode transmitter (BMT) and/or a clock-data recovery (CDR) module. The BMR and BMT can have the same specification as the OLT-receiver and ONU-transmitter, respectively. The CDR should be able to recover the clock and data at a rate of the PON system.

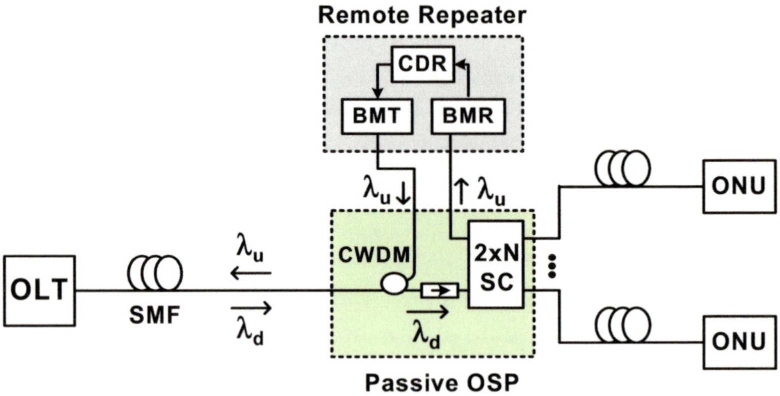

Fig. 1. FTTH system with upstream repeater.

The use of an upstream repeater provides the opportunity for much lower cost implementation of ONU using low power and low cost optical transmitters, such as 0.8/1.3/1.55 μm VCSEL-based transmitters. The ONU transmitters will now need much less output power (up to 10 dB lower) than standard PON system due to feeder fiber and OLT coupling losses. The use of simple and standard PON transceivers at the repeater significantly saves cost, maintenance expenses and power compared to an Ethernet switch as in the case of the AON architecture. Our proposed technique uses the conventional PON fiber plant for both downstream and upstream transmissions. Moreover, it is compatible with any existing media access control (MAC) protocols in the conventional PON systems as the repeater simply regenerates the upstream signals without any modification to the internal frame structure. Another advantage of our proposed FTTH system is that the downstream signals need not to be regenerated at the remote repeater and as a result, downstream channel can be upgraded without any change in the repeater allowing broadcast services to be transmitted transparently through the 1.55 μm wavelength window. PON transmitter and receiver are usually commercially available as a single bidirectional transceiver (TRX) unit. By utilizing the bidirectional property of the transceiver, we can achieve regeneration on the downstream path as well as the upstream path. This downstream regeneration enables the feeder fiber length and the split ratio at the SC to be increased, which in turn extends the coverage area of the PON system. This can offer cost effective broadband service delivery by removing the need for a separate metro network and connecting users directly to core nodes, similar to the long-reach PON structure reported in (Nesset et al., 2005). Our proposed concept is illustrated in Fig. 2. By using standard EPON OLT and ONU transceivers, we can increase the feeder fiber reach from 20 km to approximately 60 km (due to 15 dB loss saving on the SC) and the split ratio from 1:32 to 1:256 (due to 10 dB loss saving on the feeder fiber). This increase in reach and split ratio provides an attractive upgradeability solution for the existing PON deployment using very low cost components. The remote node, which houses the repeater in this FTTH system, can be placed at a location close to the community in the proposed broadband-to-the-community architecture (Jayasinghe et al., 2005a; Jayasinghe et al., 2005b). At this repeater station, digital satellite TV signals and local area network (LAN) interconnection are re-distributed to the PON system. This architecture is useful in situations, where the conventional service provider has restricted right for TV signal broadcasting and the community will have control over the TV signals that they receive. We will be discussing these features with experimental demonstrations in Section 4.

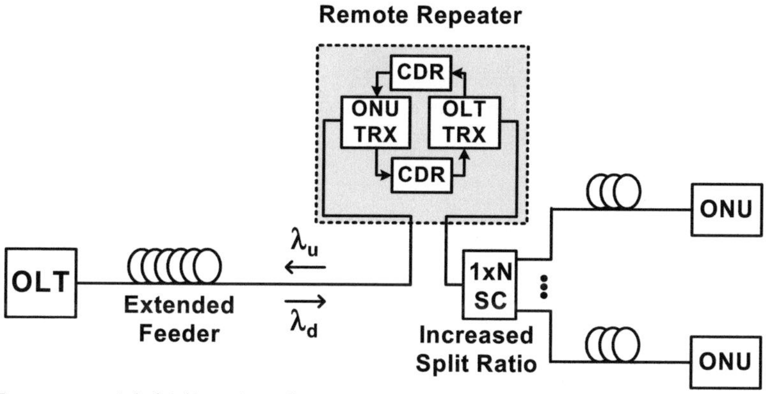

Fig. 2. FTTH system with bidirectional repeater.

2.1 Experiments and results for the upstream repeater

The first experimental setup to demonstrate the proposed FTTH system with upstream repeater is similar to that shown in Fig. 1. In this setup, we used commercially available 1.25 Gb/s EPON OLT and ONU TRXs. The feeder fiber is 20 km long using standard single-mode fiber (SMF) and we only used upstream regeneration. A 1.25 Gb/s BMR at 1310 nm was used at the repeater to receive the bursty signals from two ONUs. The electrical outputs from this BMR were used to drive a BMT at 1490 nm directly without retiming (i.e. no CDR was used in this experiment). The ONU signals were generated using user-defined patterns at 1.25 Gb/s to simulate bursty signals and the OLT signals were generated using continuous pseudo-random binary sequence (PRBS) $2^{23} - 1$.

Fig. 3(a) shows the upstream signals from ONU_1 and ONU_2 received at the OLT when the upstream repeater was used. Fig. 3(b) shows the measured eye diagrams for the upstream signals received at the OLT with and without the upstream repeater. Fig. 3(c) shows the zoomed-in beginning of the upstream burst signals from ONU_1. The waveform clearly shows that the OLT can quickly recover the first few bits from the bursty regenerated upstream signals. As shown in the table, the average total timing jitter from the upstream repeater was measured to be 90 ps and is smaller than 599 ps, which is specified by the IEEE 802.3ah EPON standard (IEEE, 2004). The rise time and fall time of the pulses were measured to be 118 ps and 115 ps, respectively, which are well within the 512 ns rise time and fall time specification of the IEEE 802.3ah standard.

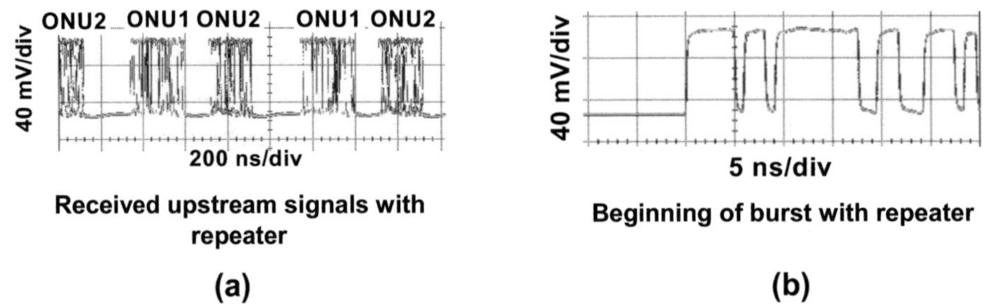

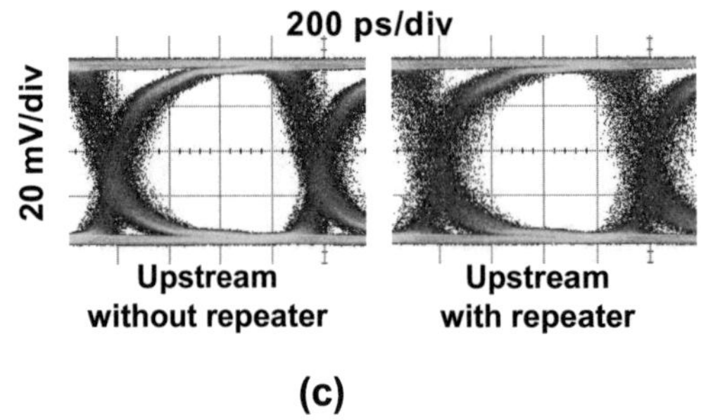

	Measured	IEEE 802.3 EPON
Average total timing jitter	90 ps	599 ps
Rise-time	118 ps	512 ps
Fall-time	115 ps	512 ps

(d)

Fig. 3. Measured eye diagrams and waveforms with and without upstream repeater.

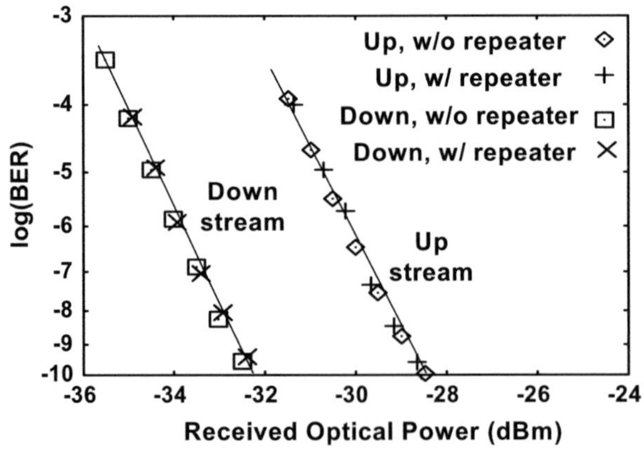

Fig. 4. Measured BERs for upstream and downstream signals with and without upstream repeater.

Fig. 4 shows the measured bit-error-rates (BERs) for downstream and upstream signals with and without the upstream repeater. No power penalty due to the upstream repeater was observed. The measured waveforms and BER results confirm that the upstream repeater can be used to reduce the requirement on the ONU transmit power without introducing penalty to the existing PON system and still conforming to the IEEE EPON standard requirements.

2.2 Experiments for video transmission

We also used this upstream repeater in a commercial EPON evaluation system from Teknovus to test its performance. The Teknovus system implements the IEEE 802.3ah EPON standard for delivery of triple-play services. Fig. 5 shows the experimental setup along with the measured upstream spectrum and captured video when video signals were streamed from the ONU to the OLT through the upstream repeater. No degradation in received upstream video quality was observed in the experiment.

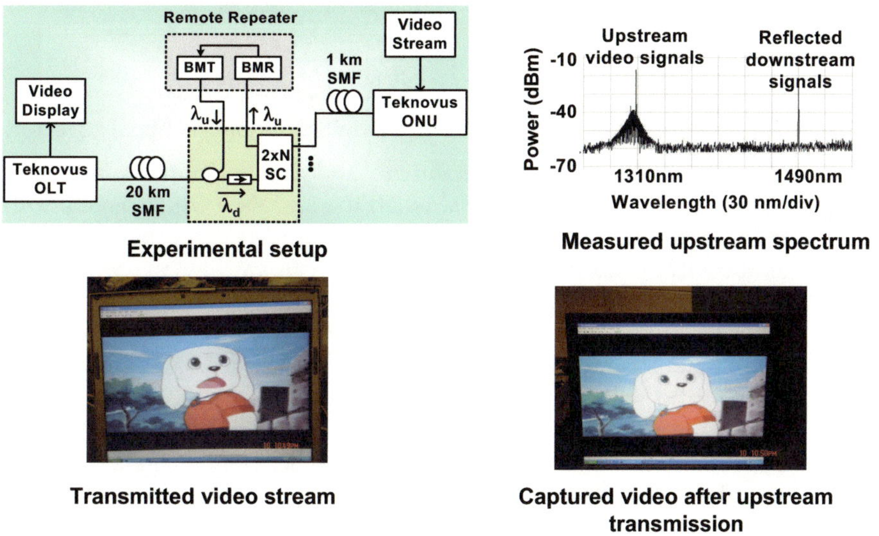

Fig. 5. Experimental setup and observed upstream spectrum and captured video after transmission through Teknovus system with upstream repeater.

2.3 Experiments and results for the bidirectional repeater

An experimental setup to demonstrate the reach extension and split ratio increase of the PON system was constructed and is similar to that shown in Fig. 2. In this case, a pair of bidirectional OLT and ONU TRXs were used at the remote node to provide both upstream and downstream regeneration. The feeder fiber is 50 km long. No CDR modules were used at the repeater. Fig. 6 shows the measured eye diagrams. The total timing jitter due to the repeater for downstream and upstream signals was measured to be 30 ps and 210 ps, respectively, which are still within the jitter specification of 599 ps of the IEEE 802.3ah standard. The rise time and fall time for downstream and upstream signals are 93 ps, 85 ps, 120 ps, and 140 ps, respectively, which are also well within the limit of the IEEE 802.3ah standard. It is expected that by using CDR modules at the remote repeater these jitter values would be further improved.

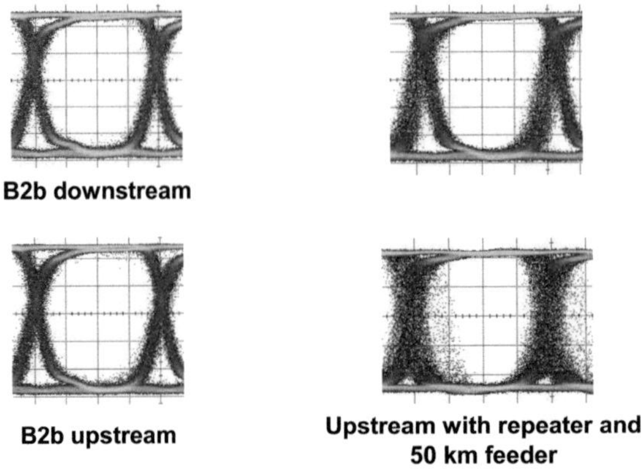

	Downstream	Upstream	IEEE 802.3 EPON
Jitter	30 ps	210 ps	599 ps
Rise-time	93 ps	85 ps	512 ps
Fall-time	120 ps	140 ps	512 ps

Fig. 6. Measured eye diagrams and waveform with and without bidirectional repeater.

Fig. 7 shows the measured BERs for the upstream and downstream signals. No power penalty was observed for downstream signals when the signals were transmitted through the repeater compared to the results when the signals were transmitted without the repeater. A small penalty < 0.2 dB was found for upstream signals, which could be attributed to non-perfect clock synchronization between the BER test-set and the pattern generator as no CDRs were used in the experiments. These results confirm that commercially available EPON transceivers can be used as bidirectional repeater to increase EPON reach and split ratio without introducing significant penalty to the existing system and without violating the IEEE 802.3ah standard. This is a very important feature of the proposed remote repeater-based optical access network scheme as it can certainly support existing interfaces at the OLT and ONU terminals.

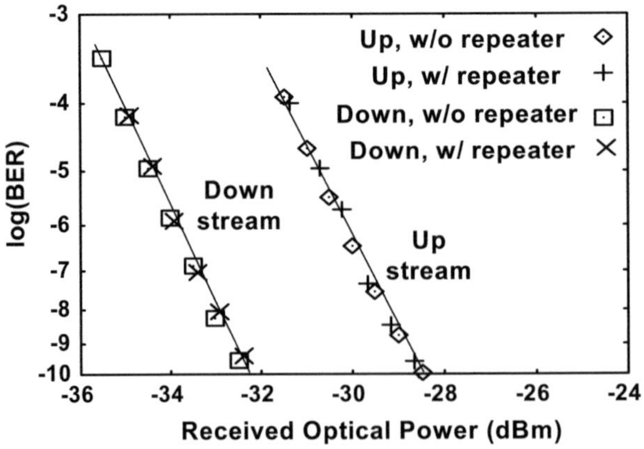

Fig. 7. Measured BERs for upstream and downstream signals with and without bidirectional repeater.

3. Jitter analysis of cascaded repeater-based FTTH system

The use of the remote repeater allows much lower cost implementation of ONU using low power and low cost optical transmitters, such as 0.8/1.3/1.55 μm VCSEL-based transmitters. If standard EPON components are used, the repeater can help increase the feeder fiber reach from 20 km to 50 km (due to 15 dB loss saving on the star coupler (SC)) and the split ratio from 1:32 to 1:256 (due to 10 dB loss saving on the feeder fiber). We can achieve higher user count and longer range coverage in the access network by cascading repeater-based FTTH systems as shown in Fig. 8 (Tran et al., 2006c). This will result in a lower network installation cost per customer and provide more effective broadband service delivery.

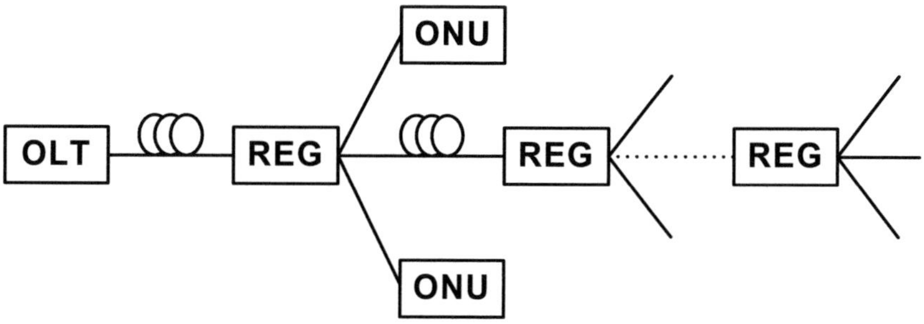

Fig. 8. Cascaded repeater-based FTTH system.

When the FTTH systems are cascaded, there are issues affecting the performance such as media access control (MAC) protocol designs, bandwidth allocation algorithms, jitter and delay parameters, etc. In this section, we investigate the jitter performance of cascaded repeater-based FTTH architectures via a recirculating loop. Other issues are beyond the scope of this work. Fig. 9 shows the experimental setup to investigate the jitter performance of cascaded FTTH systems with remote repeater. We used a 1.25 Gb/s EPON transmitter at 1490 nm with a PRBS of $2^{23} - 1$ to feed signals into the recirculating loop. The two acousto-optic switches and a 2x2 coupler control the signals coming in and out of the loop. Inside the loop, there is 20 km of standard single-mode fiber (SMF) to simulate the feeder fiber in a standard PON. Followed the SMF are the EPON receiver (RX) and transmitter (TX) connected directly to each other without any CDR modules. An attenuator is used inside the loop to simulate the star coupler loss. At the output of the loop, an EPON RX is used to detect the signals after recirculation.

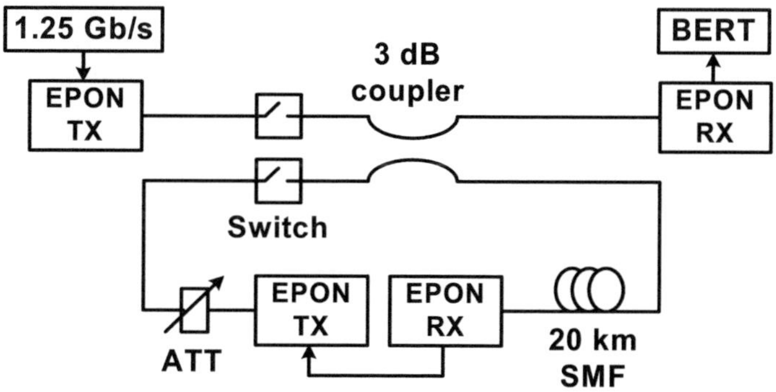

Fig. 9. Experimental setup to demonstrate cascaded remote-repeater-based FTTP systems.

Table 1 shows the measured timing jitter performance of the received signals after each loop. As seen, after 6 regenerations, the total jitter is only 380 ps, still smaller than 599 ps, which is specified by the IEEE 802.3ah EPON standard (IEEE, 2004). Note that, the measured total jitter has contribution from regeneration as well as clock misalignment due to the absence of clock recovery at the receiver.

Loop No.	Mean jitter RMS (ps)	Mean peak-to-peak jitter (ps)
0	18	62
1	18	70
2	24	81
3	49	215
4	70	228
5	76	298
6	90	380

Table 1. Measured timing jitter for different loop numbers.

Fig. 10 shows the measured eye diagrams of the 1.25 Gb/s signals. As seen in Fig. 10(c), the eye is still clear and open after 4 loops, enabling error-free operation. After 6 loops (Fig. 10(d)), the eye becomes distorted and partly closed. This eye closure is due to both regeneration and the absence of clock recovery at the receiver.

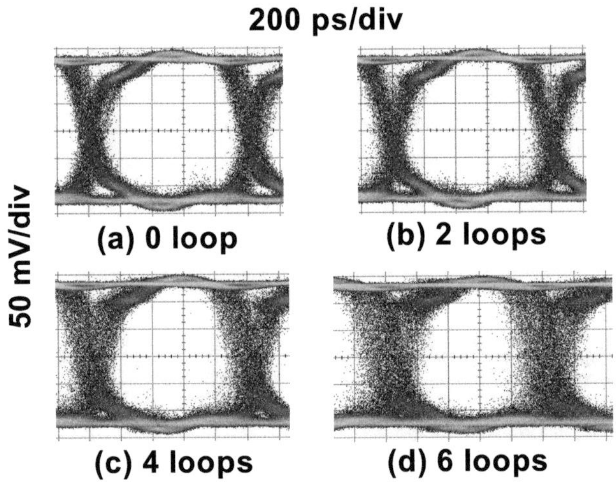

Fig. 10. Measured eyes at 1.25 Gb/s without clock recovery.

We measured the BERs for the 1.25 Gb/s signals after 0 loop as the baseline, and then repeated the measurement after each loop. The results are shown in Fig. 11. Compared with the BER results for the 0 loop, the power penalty is less than 0.5 dB for the cases up to 4 loops. After 5 loops, the power penalty is more than 4 dB at BER = 10^{-8}. After 6 loops, we reached an error floor at 10^{-5}. We can conclude that without clock recovery, our proposed FTTH architecture with remote repeater can be cascaded 4 times without introducing significant penalty to the system performance. This result illustrates that if we take into account a single repeater-based FTTH system with a feeder fiber of 50 km and a customer base of 256, a cascaded system can then theoretically extend the reach to 200 km and 256^4 customers with no jitter performance degradation. However, the actual number of customers will eventually be limited by the MAC protocol and dynamic bandwidth allocation scheme.

In order to investigate the effect of clock recovery on the receiver jitter performance, we used 155 Mb/s PON transceivers at the transmitter, receiver and regenerator. A CDR module was also used at the receiver. Fig. 12 shows the measured eye diagrams after 0 regeneration and 10 regenerations through the recirculating loop with and without clock recovery. Without clock recovery, the timing jitter was measured to be 489 ps and 1.9 ns for

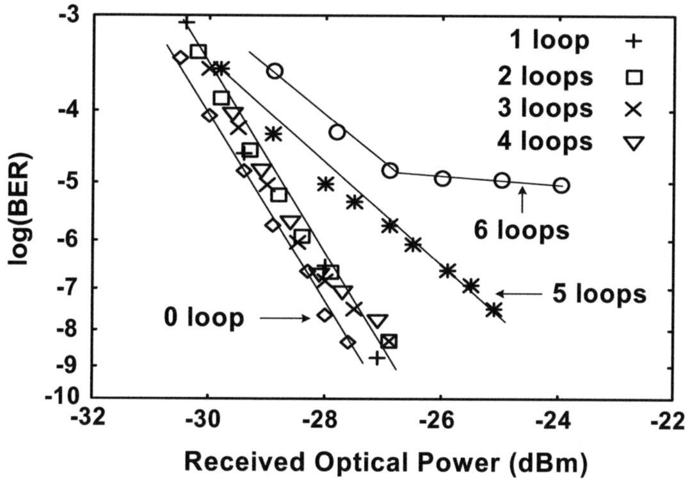

Fig. 11. Measured BER curves at 1.25 Gb/s.

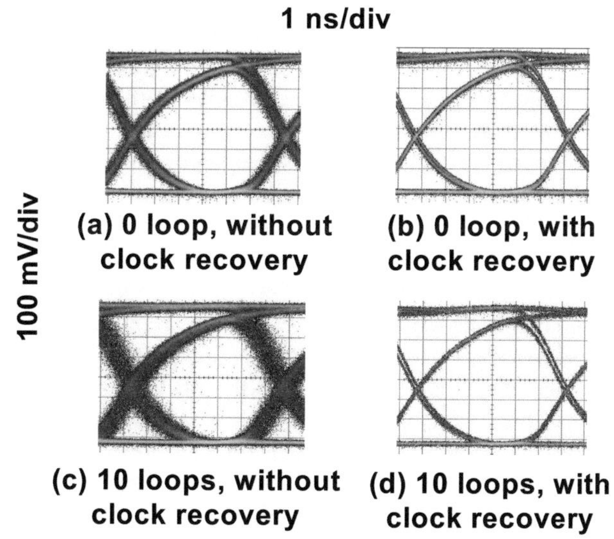

Fig. 12. Eyes at 155 Mb/s with and without clock recovery.

0 and 10 loops, respectively. With clock recovery, the timing jitter was approximately 390 ps for both 0 and 10 loops. This illustrates that with clock recovery at the receiver, we can cascade our FTTH systems for up to 10 times and the total jitter is still within the EPON standard requirement.

4. Video service integration over a repeater-based FTTH system

Integrating video services over a PON has become important since it could reduce the cost of operation. Generally, cable TV and satellite TV services are delivered over coaxial cables within the premises. As the transmission distance and the number of video channels increase, the coaxial cable based infrastructure may not support larger number of video channels. Considering a scenario whereby multiple video channels have to be delivered to densely populated areas such as multi-dwelling units, apartment houses, or hotels, a cost-effective solution is essential. Due to regulatory issues in several countries, video services may not be delivered over a PON from the CO. Moreover, the unbundling of fiber bandwidth in the access networks enables the provisioning of video service delivery by many cable TV providers. However, due to the competition between the cable TV providers and the telecom companies, certain video services may not be delivered through a PON from the CO. Therefore, a cost effective architecture that enables the video service delivery by multiple service providers in an access network is required. As an added feature for the remote repeater, we demonstrate a video service delivery scheme for a large-split repeater-based FTTH network whereby the video channels are carried on a radio frequency (RF) subcarrier multiplexed (SCM) format.
A generic integrated optical access network to support data, voice and video services is shown in Fig. 13(a). The proposed architecture is based on the repeater-based optical access network that could potentially support a larger number of customers (Nadarajah et al., 2007). In this setup, cable TV and satellite TV signal distribution links can be overlaid on the downstream link at the repeater to be delivered to the customers. We proposed the video service delivery using RF SCM transmission, whereby the chosen RF carrier frequency for the delivery of the video channels is placed outside the bandwidth of the downstream data signals as shown in Fig. 13(b). At the repeater, the received video channels are upconverted to the designated RF frequency and then electrically combined with the regenerated downstream data signals before the transmission to the customers. Integration of RF video channels with the downstream signals at the repeater has several advantages. Time multiplexing video channels with the downstream data at the CO increases the total downstream bandwidth. The scalability of this scheme is potentially limited by the power budget and therefore limits the number of video channels. As the number of splits in the star coupler (SC) increases, this issue becomes even more pronounced. In a large-split PON, the delivery of RF video channels from the CO suffers from inadequate signal-to-noise ratio (SNR) to receive the video channels error-free. As the bandwidth requirements and the number of customers increase, it becomes more difficult. However, the SNR requirements can be satisfied in a repeater-based optical access network as the signals are regenerated at the RN enabling the video delivery for a larger number of customers. On the other hand, the video channels that are regenerated at the remote repeater can be optically combined with the downstream wavelength channel. Compared to SCM based scheme, this scheme does not use electrical combination of the RF video channels with the downstream data at the repeater. At the ONUs, the received wavelength channels can be detected using a single

receiver. As the RF video channels are placed outside the bandwidth of the baseband downstream data, one receiver setup at each ONU becomes feasible.

Fig. 14 shows the experimental setup to demonstrate the capabilities of the proposed scheme. A 1.25 Gb/s downstream signal of 2^{31} - 1 pseudo random binary sequence non-

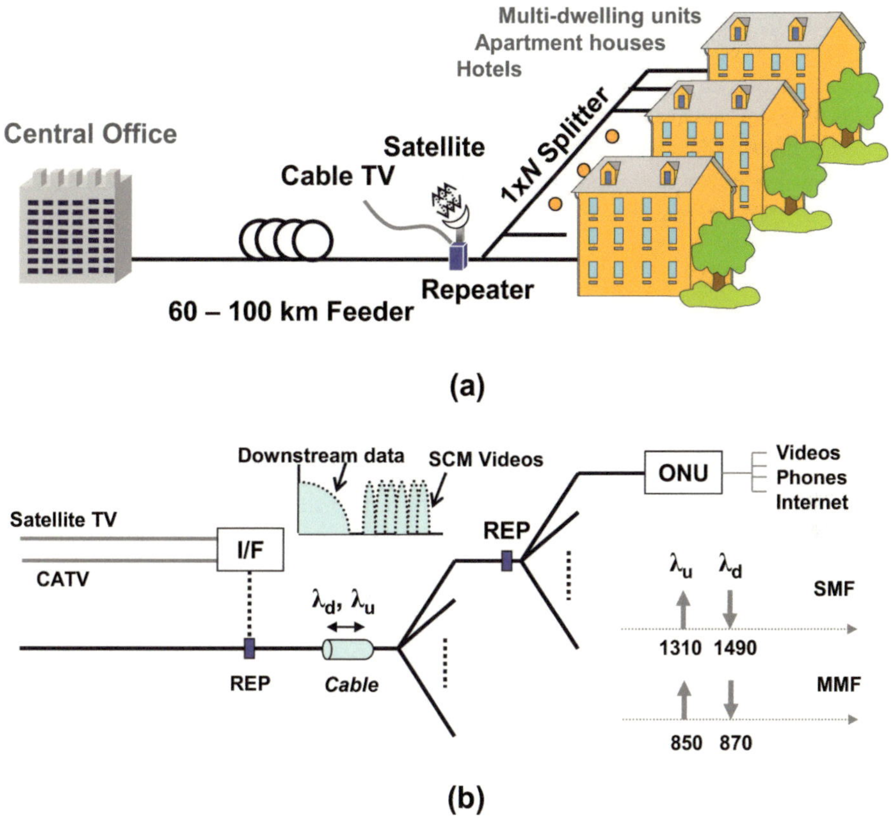

Fig. 13. (a) Generic integrated repeater-based optical access network architecture for video service delivery; (b) Repeater-based distribution network providing simultaneous RF SCM videos and data services.

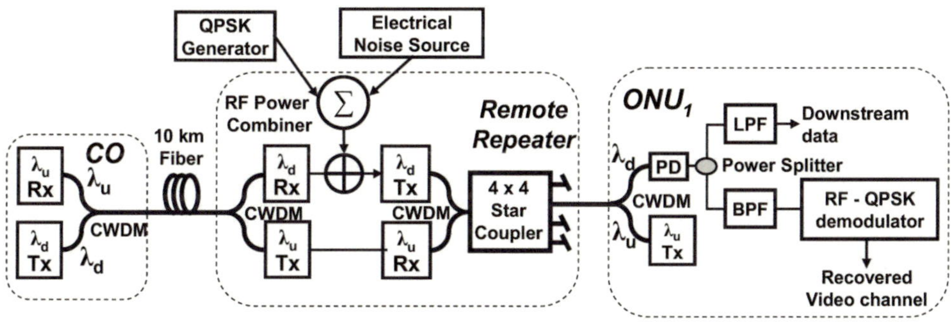

Fig. 14. Experimental setup to demonstrate the video service delivery over a repeater based optical access network.

return-to-zero (PRBS NRZ) data was directly modulated onto downlink carrier, λ_d (1490 nm), and transmitted to the repeater through a 10 km standard single mode fiber. At the repeater, the downstream signal was detected using a commercially available 1.25 Gb/s receiver. A 4.096 Msymbols/s quadrature phase shift keyed (QPSK) data was generated using a vector signal generator and upconverted onto a RF carrier frequency at 1.7 GHz.
As shown in Fig. 15, the generated signal was electrically combined with additive white gaussian noise (AWGN) generated from an electrical noise source that simulated multiple video channels. Before the combination, AWGN was band limited using a band pass filter (BPF) with a center frequency of 2 GHz and bandwidth of 300 MHz to reduce the spectral overlap with the QPSK signal and the downstream data. The detected downstream signal was then sent through a low pass filter (LPF) with a cut-off frequency of 1.25 GHz to avoid crosstalk to the RF signals before electrically combining with the RF signals. The composite signals were directly applied to a distributed feedback laser operating at 1550.92 nm and transmitted to the ONUs. The downstream wavelength channel from the repeater was passed through a 4x4 SC and detected using a 2.5 Gb/s receiver. The received signals were split using a RF splitter and 1.25 Gb/s dowstream data was recovered using a LPF with a cut-off frequency of 1.25 GHz while the RF signals were separated from the downstream data using a BPF and QPSK signal was recovered using a demodulator. In the upstream direction, a 1.25 Gb/s 2^{31} - 1 PRBS NRZ continuous data was directly applied to a Fabry-Perot laser diode operating at 1310 nm wavelength window (λ_u) and transmitted to the repeater, where it was regenerated and transmitted to the CO through the 10 km fiber. It should be noted that longer fiber transmission would not have made significant difference in the results because the combination of video channels with downstream data is performed at the repeater that is closer to the ONUs and the downstream data can be regenerated using cascaded repeaters. Coarse wavelength division multiplexing couplers were used at the repeater to separate λ_u and λ_d before and after the regeneration of the signals.

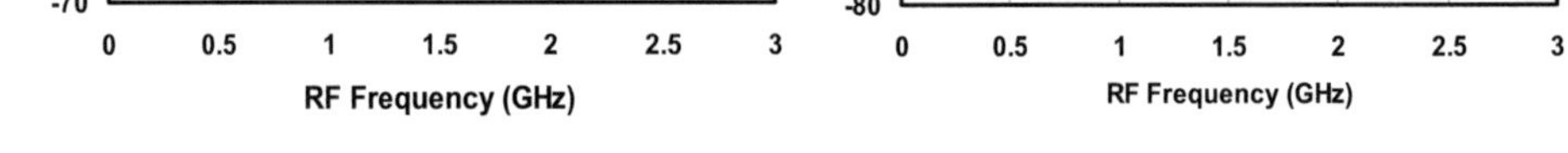

Fig. 15. Observed RF spectra of the downstream signals and RF video channels at the repeater and ONU.

Fig. 16 shows the BER measurements for the upstream and downstream data signals. For the upstream signal, less than 0.4 dB penalty was observed when the signals were passed through the repeater compared to back-to-back (B-B) measurements and this penalty was introduced at the repeater electronics. For the downstream data, no penalty was observed when the RF SCM signals were added with the downstream data signals. The sensitivity of the upstream data signal was approximately 2 dB better than that of downstream data signal. The modulation depth of the downstream data was reduced to avoid intermodulation products as it was modulated with the RF video signals at the repeater.

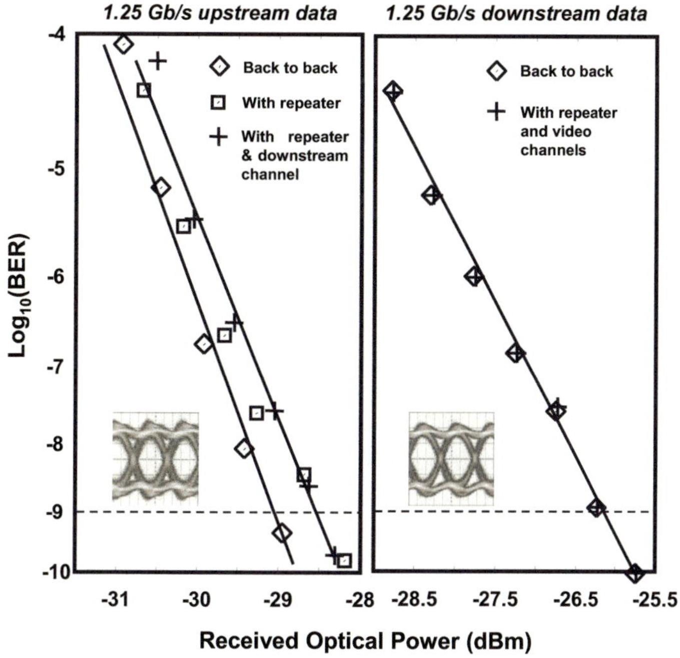

Fig. 16. Measured BER curves for 1.25 Gb/s upstream data and 1.25 Gb/s downstream data signals.

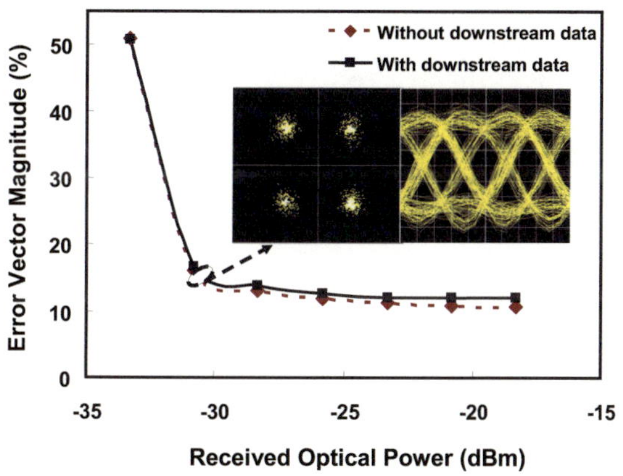

Fig. 17. Measured EVM values for QPSK signals. Insets show the observed constellation & eye diagrams.

Fig. 17 shows the measured error vector magnitude (EVM) for the recovered QPSK signal. An EVM of 16.6% (corresponds to a BER of 10^{-9}) was measured for the recovered QPSK data signal when the received optical power was -31.81 dBm. The graph also shows that QPSK signal did not suffer from crosstalk penalty from to the downstream data signal. The insets of Fig. 17 show the constellation diagram and the eye diagram for the recovered QPSK signal when the received optical power was -31.81 dBm. As the bandwidth requirements for the digital video channels increase, higher order modulation formats can be adopted without changing the allocated frequency spectrum. The experiment was repeated for 4.096 Msymbols/s 16-quadrature amplitude modulation (QAM) video channel transmissions and the resulting constellation diagram is shown in Fig. 18.

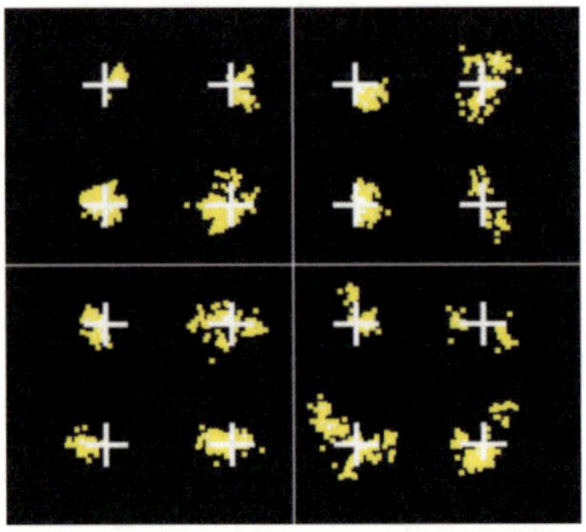

Fig. 18. Observed constellation diagram for the recovered 4.096 Msymbols/s 16-QAM.

A power budget calculation for the video channels was performed based on parameters given below. The transmitted optical power from the repeater, distribution fiber length, attenuation of fiber, WDM coupler loss, sensitivity for 2.5 GHz optical receiver and loss of the 64-split SC are chosen to be 0 dBm, 5 km, 0.2 dB/km, 0.75 dB, -26 dBm and 21 dB respectively. Under these circumstances, a power margin of 2.5 dB can be obtained for the video channels. Assuming the bandwidth of each RF video channel is 6 MHz, more than 100 video channels can be simultaneously delivered. As the bandwidth of the downstream data increases, higher bandwidth receiver is required to receive both signals. Higher bandwidth RF video channel transmissions can be performed using higher order modulation formats that require higher SNR for error-free recovery. As the optical modulation index of the video channels increases, the modulation index of the downstream data decreases. This leads to the reduction in sensitivity of the downstream data and limits the power budget of downstream data signal. Therefore, the modulation index for both signals has to be taken into account considering the power budget requirements.

5. Local area networking over a repeater-based FTTH system

As the access network grows with increasing number of customers and demand for more bandwidth, several value added services also need to be delivered in an efficient way. For example, customers within a PON environment may require private communication links between themselves for various computer applications and telecommunication services, such as distributed data processing, broadcast information systems, teleconferencing, and interactive video games. Moreover, some customers leasing several floors within a building may require their own private network such as LAN apart from the standard communication links with the CO. To serve this purpose, two solutions can be found. The first one is to deploy another optical network interconnecting all customers within the PON to facilitate the local customer networking. Intercommunication between the customers may be realized by overlaying a separate network in which each ONU is connected to all other ONUs via a point-to-point optical link. Deploying a separate network for inter-networking amongst the customers is extremely complex and costly especially in a densely populated area. Moreover it also becomes impractical and inefficient to connect to each customer in the network via this setup (Cohen, 2003; Venkateswaran, 2001). The second solution is to intelligently use the existing PON infrastructure to provide the additional services. Reuse of the PON infrastructure to facilitate intercommunication links between customers of the same PON can greatly reduce the cost and simplify management issues of the network. Overlaying a LAN on the existing PON incurs a minimum additional cost since it utilizes the existing facility. The overlaid network can be used to interconnect several customers in scattered buildings to form a group of community (IEEE, 2004; Park et al., 2004). The resulting PON system enables fiber-to-the-premises so that the tenants in a building can subscribe to telecommunications, internet, and video services individually while keeping their own network. Moreover, the change in bursty data traffic patterns opens up an opportunity for an efficient use of inherent PONs with distributed statistical multiplexing to increase the efficiency of the access networks. Using distributed multiplexing, LAN capabilities amongst the customers can be developed. The overlaid virtual PON technologies bring multiple campuses and multi-tenant business buildings onto a same optical fiber facility and therefore making savings on fiber facilities' capital and ongoing operational costs (Arnaud et al., 2003; Iannone et al., 2000).

A number of higher layer LAN emulation schemes have been proposed to the Ethernet in the First Mile alliance (EFM) IEEE 802.3ah (Dixit, 2003; Hernandez-Valencia, 1997). One of these solutions considered the use of PON tags and a trivial reflector function at the CO to emulate a shared medium and therefore reflecting all upstream traffic back to the ONUs. Even though this scheme is compatible with higher layer, to enable multicasting to several ONUs, the frame transmission should be carried multiple times, which wastes bandwidth. In the advanced upper-layer shared LAN emulation (ULSLE) scheme, the LAN traffic is separated using the bridges and/or routers located at the CO. Using the media access control (MAC) address of the packet, intelligent decisions are made and the LAN traffic is rerouted back to the appropriate ONUs using the downstream wavelength channel. The bridges or routers that are equipped at the CO need to be very complex to obtain relatively higher efficiency of transmission bandwidth of the wavelength channels. These bridges or routers must be capable of supporting higher layer protocols, thereby potentially increasing the cost and complexity of the network. Furthermore, the effective downstream channel bandwidth is reduced as the LAN traffic is routed back to the ONUs on the downstream wavelength channel. Moreover, the redirected LAN traffic needs to be separated from the downstream traffic using complex filtering mechanisms that are employed at the ONUs. Even though using a fiber access facility to communicate with another user in the network is more efficient and practical, using ULSLE schemes for the transmission of LAN traffic amongst the customers increases the complexity at the ONUs. Therefore, a simplified LAN that is overlaid on the existing PON with high flexibility is required for the intercommunications amongst the customers in the PON. By comparison, emulating point-to-point links amongst customers directly on the optical layer in the PON can effectively overcome several drawbacks (Chae et al., 1999; Chae et al., 2001; Nadarajah et al., 2005; Tran et al., 2006a; Wong et al., 2004). Compared to the higher layer LAN emulation schemes, an optical layer LAN emulation can also be provided whereby the LAN traffic is physically redirected back to the ONUs without the use of higher layer bridges or routers located at the

CO. This effectively reduces the required higher layer interfaces at the CO while obtaining higher bandwidth utilization for the downstream wavelength channel. It has been shown that in the optical layer LAN emulation schemes, as the percentage of LAN traffic increases the available upstream and downstream bandwidth per ONU also increases whereas it decreases in the higher layer LAN emulation schemes. Moreover, no further complex filtering mechanisms are required at the ONUs to separate the LAN traffic from the downstream traffic. The use of more complex filtering and packet forwarding schemes that are used in the bridges or routers at the CO are also eliminated and therefore reduces the cost and complexities associated with the operation at the ONUs and CO. We have proposed and experimentally demonstrated a multi-functional high split repeater-based optical access network architecture that enables simpler and bandwidth efficient LAN capabilities amongst the customers.

5.1 System architecture

A generic integrated optical access network to support LAN services is shown in Fig. 19 (Nadarajah et al., 2009). In this network, a 1×N (N < 256) splitter is used at the remote node, whereby the larger split is enabled by a remote repeater that regenerates the upstream and downstream signals. The repeater not only performs regeneration of the signals to support larger number of customers but also provides intelligent LAN capabilities. The LAN traffic that is carried from one ONU to another along with the upstream traffic to the CO is physically separated at the remote repeater. Thereafter, LAN traffic is combined with the downstream traffic and carried to the ONUs. We propose the LAN traffic delivery for the repeater based optical access network using RF subcarrier multiplexed transmission, whereby the chosen RF carrier is placed outside the bandwidth of the baseband upstream traffic and downstream traffic as shown in Fig. 19. The repeater consists of transceiver modules and an electrical LPF, and BPF for the separation and combination of the LAN traffic.

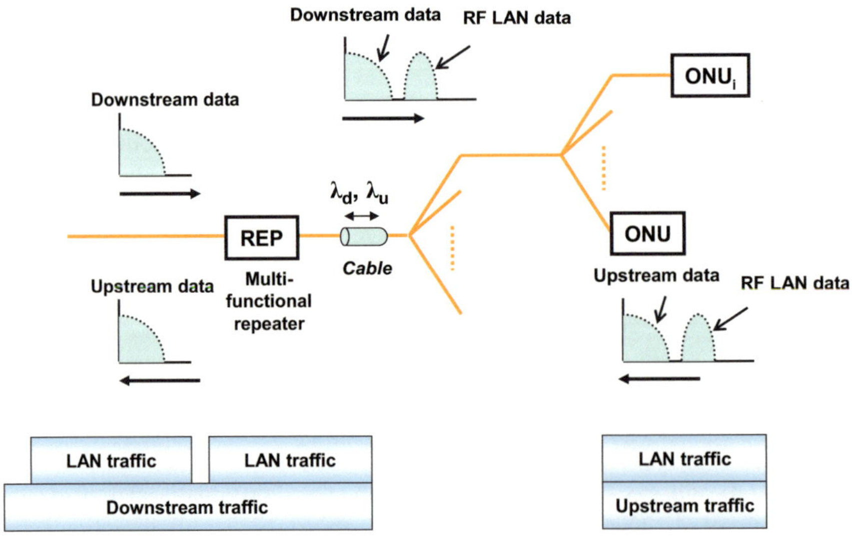

Fig. 19. Integrated repeater based optical access network architecture for local area networking.

5.2 Experiments and results

Fig. 20 shows the experimental setup to demonstrate the capabilities of the proposed scheme. A 1.25 Gb/s downstream signal of 2^{31} - 1 PRBS NRZ data was directly modulated onto downlink carrier, 1490 nm and transmitted to the repeater through a 10 km fiber. At the repeater, the downstream signal was detected using a 1.25 Gb/s receiver. For the upstream transmisisons a 155 Mb/s 2^{23} - 1 PRBS NRZ was upconverted onto the RF frequency at 2.5 GHz using a mixer. The upconverted RF LAN data was then electrically combined with 1.25 Gb/s 2^{23} - 1 PRBS NRZ and directly modulated onto a DFB laser operating on a wavelength at 1550.92 nm and transmitted to the repeater through a 3 km fiber and 4×4 SC.

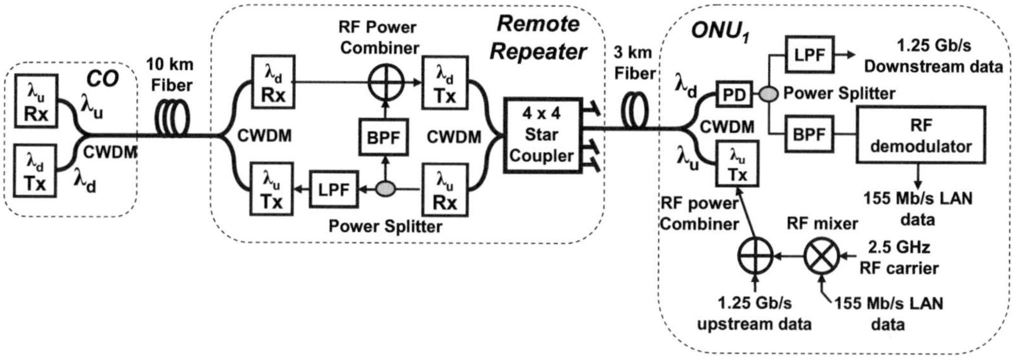

Fig. 20. Experimental setup for the demonstration of the proposed LAN scheme.

Fig. 21 shows the RF spectra of the transmitted composite signals. The received upstream signals at the repeater were detected and RF power split. One portion of the signals was passed through a LPF with a cutoff frequency of 1.25 GHz to separate the 1.25 Gb/s upstream data. This upstream data was then directly modulated onto a FP-LD operating at 1310 nm and transmitted to the CO through a 10 km fiber. The other portion of the signals was then sent through a BPF with a center frequency of 2.5 GHz and bandwidth of 300 MHz to separate the LAN data. The separated LAN data was then combined with the 1.25 Gb/s downstream data and directly modulated onto DFB laser operating at 1554.13 nm and transmitted to the ONUs through a 4×4 SC and 3 km fiber. The received signals at the ONU were split using a RF splitter and 1.25 Gb/s dowstream data was recovered using a 1.25 GHz LPF while the RF signals were separated from downstream data using a BPF and 155 Mb/s LAN data was recovered using a demodulator.

Fig. 22 shows the BER measurements for the upstream data, downstream data, and LAN data signals. For the downstream signal, no significant penalty was observed for the 10 km fiber transmission compared to back-to-back (B-B) measurements. After the RF LAN data was added at the repeater, more than 5 dB penalty was observed for the downstream data. This larger penalty was a result of the reduction of the extinction ratio. This larger penalty can be minimized by optimizing the combination of both downstream data and LAN data signals. For the 1.25 Gb/s upstream data, less than 0.5 dB penalty was observed when the RF LAN data signals were added and transmitted with the upstream signals. For the LAN data signal, less than 0.5 dB penalty was observed in the presence of 1.25 Gb/s upstream data signals. An additional 0.5 dB penalty was observed when the LAN data signals were carried to the ONUs with the downstream data signals. These penalties can be attributed to the crosstalk from the baseband signals.

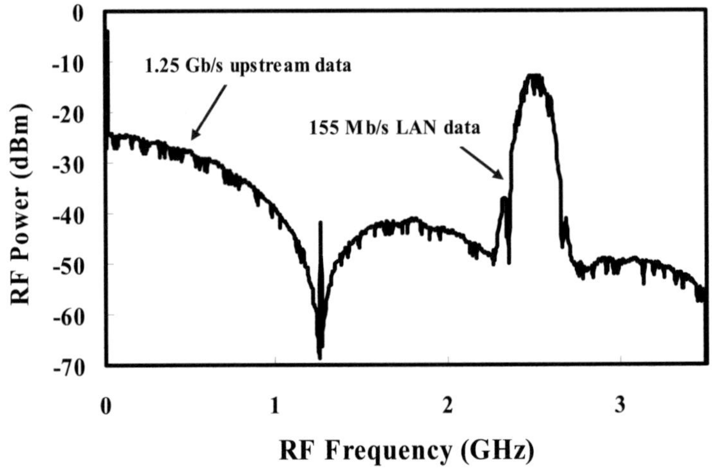

Fig. 21. Observed downstream signals at the repeater.

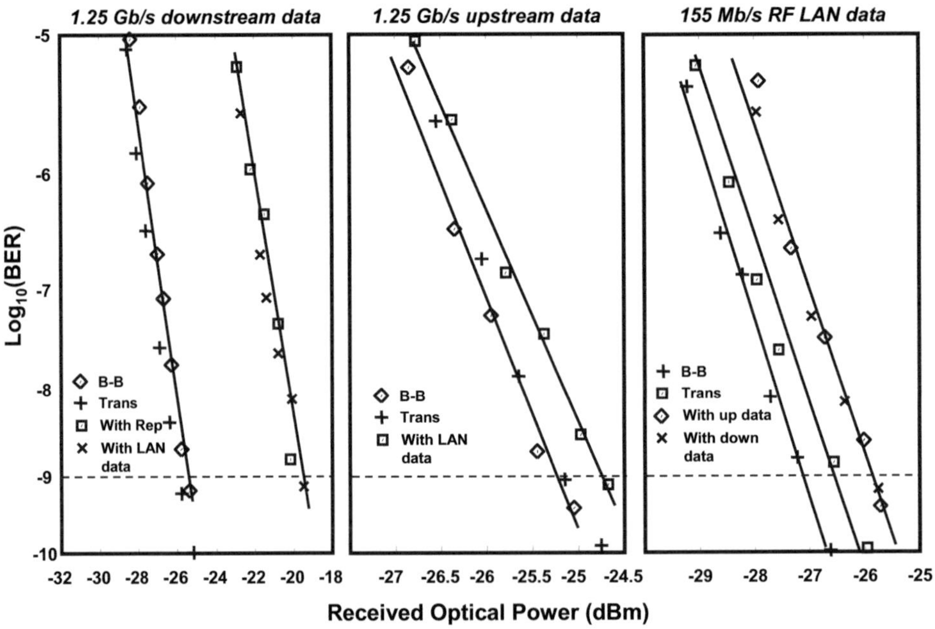

Fig. 22. Measured BER signals for all the signals.

6. Econometric modelling

We have shown the proposed architecture for a repeater-based FTTH system that could potentially increase the number of subscribers served by a single optical access network infrastructure to 256 and extend the feeder fiber transmission distance up to 100 km.

Supporting a larger number of subscribers using a single PON infrastructure is more cost effective and an economic model has been developed to study the cost effectiveness of this particular architecture in comparison with the conventional PON architecture and fiber-to-the-node (FTTN) architecture that incorporates digital subscriber lines (DSL) from the remote node (RN) to the customer units (CUs). At the RN, a digital subscriber line access multiplexer (DSLAM) is placed for the collection and distribution of signal from/to each of the CUs through the already built-in DSL lines. The objective of this section is to identify the essential costs of building passive optical access networks and to perform a comparison of different technologies using varying performance criteria. Simple generic models are used to calculate trenching and cable costs taking into account different deployment cases.

6.1 Network model

A model framework of generic connections, housings, and equipment are considered for this study. In this model, all links between OLT, RN and ONU are via single-mode fiber. The distance between the OLT and the RN is taken as 10 km. Moreover, it is considered that the RN is placed between the subscriber's ONU and the OLT. The RN is the equivalent of the active switch in the AON structure, the SC in the PON structure and the repeater and the SC in the repeater-based optical access network. As these types of FTTH architectures are more suitable for the multi-tenant buildings, it is considered that the RN is placed at the basement of the building while the houses are located on each floor.

Fig. 23 shows the layout of the houses in a floor of the building. There are n (typically n = 8) houses in one row, while the N rows (typically N = 4) in the single floor. The size of each house is 15 x 15 m.

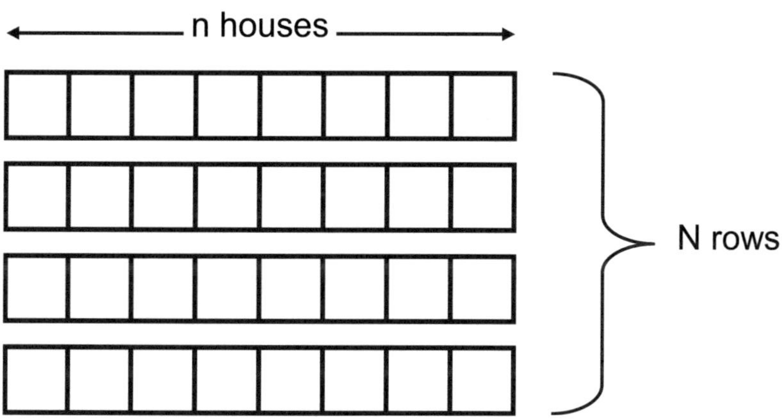

Fig. 23. Layout of the houses in a floor of the building.

If there are m floors in a building, then the average fiber cable length between the RT and the customer unit can be given as

$$\text{Average cable length} = \frac{(n+1)(m+1)}{4} \times 15 \text{ m.} \tag{1}$$

$$\text{Average trench length} = \frac{n(m+1)}{4} \times 15 \text{ m.} \tag{2}$$

6.2 Network economics

Fig. 24 shows the architectures that are considered for the economic model analysis. The conventional PON architecture uses multiple SCs in the RN and they are connected to the OLT using multiple feeder fibers and ONUs using distribution fibers. No active electronics is used in the RN. The repeater-based FTTH network architecture is similar to that of conventional PON architecture, however a simple repeater is used at the RN. Moreover, a higher split (1x256) SC is used instead of multiple (1x32) SCs. The fiber-to-the-node and DSL (FTTH-DSL) architecture uses a point-point link optical fiber link between the OLT and the RN. At the RN, the CUs are connected using DSL lines and a DSLAM is used at the RN for the aggregation of the upstream signal from each subscriber.

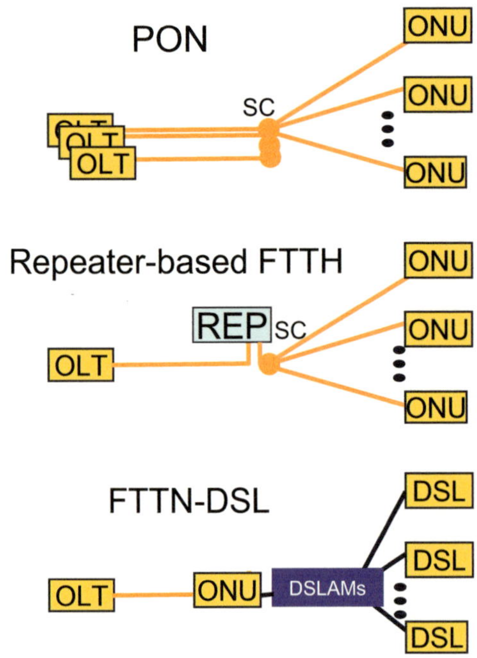

Fig. 24. Conventional PON, repeater-based FTTH, and FTTN-DSL architectures for the economic analysis.

Table 2 shows the parameters that are used for the equipment cost for each network model. The optical transceiver cost for each scheme remains the same even though repeater-based hybrid network supports a larger number of CUs. This is because the downstream and upstream signals are regenerated at the RN and therefore high power lasers and high sensitivity receivers are not required at the OLT. In the conventional PON, 32 CUs are supported through a single OLT interface while repeater-based hybrid optical access network supports 256 CUs through a single interface. Therefore, 8 optical transceivers with interfaces are required for the conventional PON to support 256 CUs using a single infrastructure. The housing cost of the RN is higher for the FTTN-DSL as it contains an active DSLAM requiring larger space with higher installation costs. In repeater-based FTTH, the power requirement and the chassis costs are higher than that of conventional PON; but lower than that of FTTN-DSL. The cost of the optical transceiver used at the RN for repeater-based FTTH architecture is lower than that of FTTN-DSL.

	Repeater-Based FTTH	PON	FTTN-DSL
OLT Parameters			
Housing cost ($)	50000	50000	50000
Transceiver cost ($)	10000	10000	10000
RT Parameters			
Housing cost ($)	1000	500	8000
Chassis cost ($)	500	100	2000
Remote powering cost ($)	250	0	2500
Transceiver cost ($)	1000	0	2000
Output port cost ($)	N/A	160	N/A
Max house per RT	256	32	256
Max RT boxes	1	8	---
SC cost ($)	9000	3000	---
Switch cost per port ($)	---	---	80
Cable Costs ($/km)			
OLT to RT	128	128	128
RT to CU	128	128	---
Cost per splice ($)	24	24	24
Trenching Costs ($/km)			
OLT to RT	5000	5000	5000
RT to CU	1000	1000	0
ONU Costs			
Install cost ($)	100	100	100
CU cost ($)	150	200	100
Interface cost ($)	500	500	500

Table 2. Component and installation costs.

In the repeater-based FTTH system, as the number of splits is higher compared to that of in the conventional PON, the cost of the SC used in the repeater-based FTTH is also higher. In the FTTN-DSL, it is considered that the DSL lines are already in place however requires further rearrangement.

Fig. 25 shows the cost of each CU for the FTTH systems for 256 CUs and 128 CUs for the take rate of 100%. In both scenarios, conventional PON architecture requires higher cost for each CU while the FTTN-DSL requires the lowest cost for each CU of all architectures. All the FTTH architectures with 128 CUs require higher cost for each CU compared to that with 256 CUs as the infrastructure is shared by many CUs. For the repeater-based FTTH architecture with 256 CUs the cost per customer is approximately $2459 while it is $3412 for the FTTN-DSL. Similarly, the cost per customer in the repeater-based FTTH network is approximately $137 higher than that in FTTN-DSL with 128 CUs. We define take rate is the percentage of homes covered by the access network infrastructure that subscribe to the service. As a consequence, all infrastructure costs (e.g. housing, electronics, and trench deployment) are incurred for all homes, even though they can only be recovered from the revenue by those that subscribe.

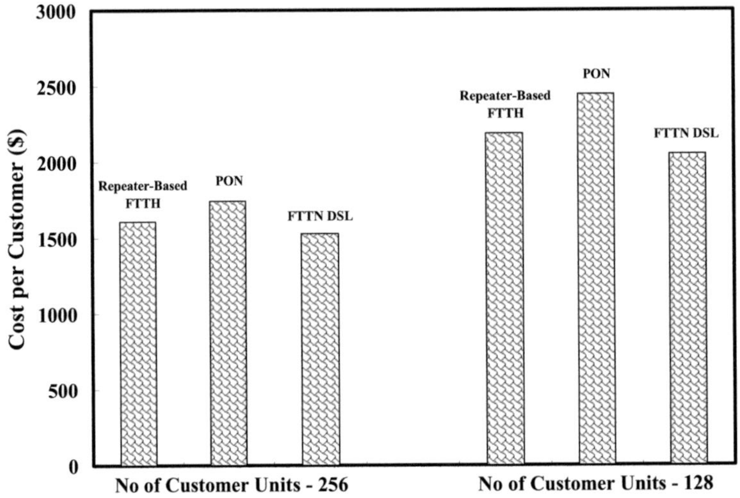

Fig. 25. Cost per customer for the FTTH architectures with 128 and 256 CUs for 100% take rate.

The network cost per subscriber is calculated as:

$$\text{Cost per Subscriber} = \frac{\text{Infrastructure Cost per Home}}{\text{\% Take Rate / 100}} + \text{per-subscriber costs} \quad (3)$$

We now use the model framework with the above parameters, cost elements, and calculation of shared trench and cables to evaluate the deployment costs of various access technologies with/without protection. The protection model for each of the considered FTTH architectures is shown in Fig. 26.

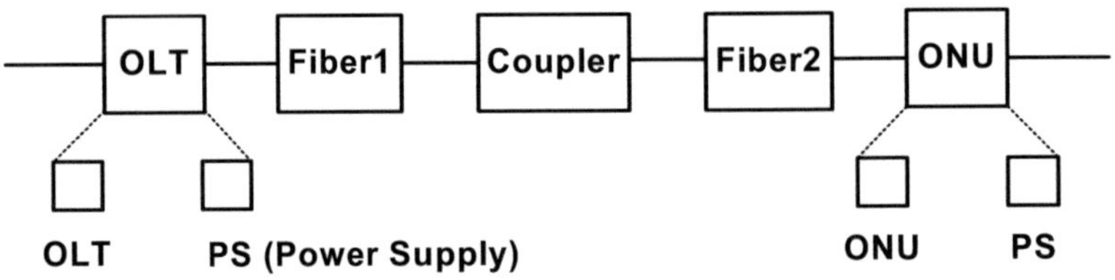

(a) Unprotected architecture

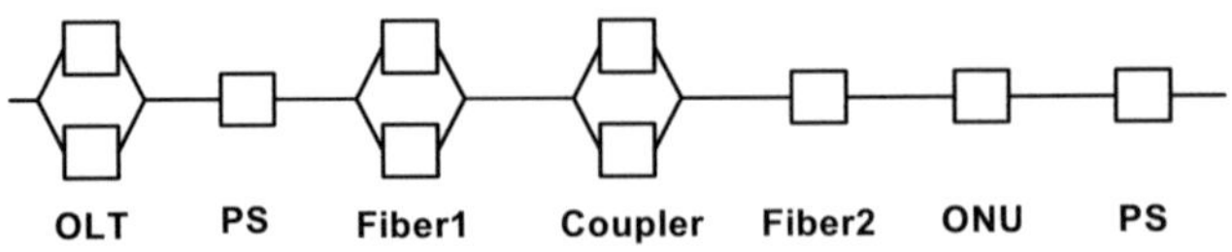

(b) Protected architecture

Fig. 26. Optical access protection models.

In this model, OLT, feeder fiber, and RN costs are duplicated whereas the ONU and the distribution costs are not. Fig. 27 shows the cost per customer for varying take rate for the considered architectures in both protected and unprotected cases. As expected the protected networks cost higher than that of unprotected networks. The protected networks require more than 34% cost increase compared to the unprotected architectures for the take rate of 100%. The conventional PON architectures require more than 40% increase in costs for the protected architecture.

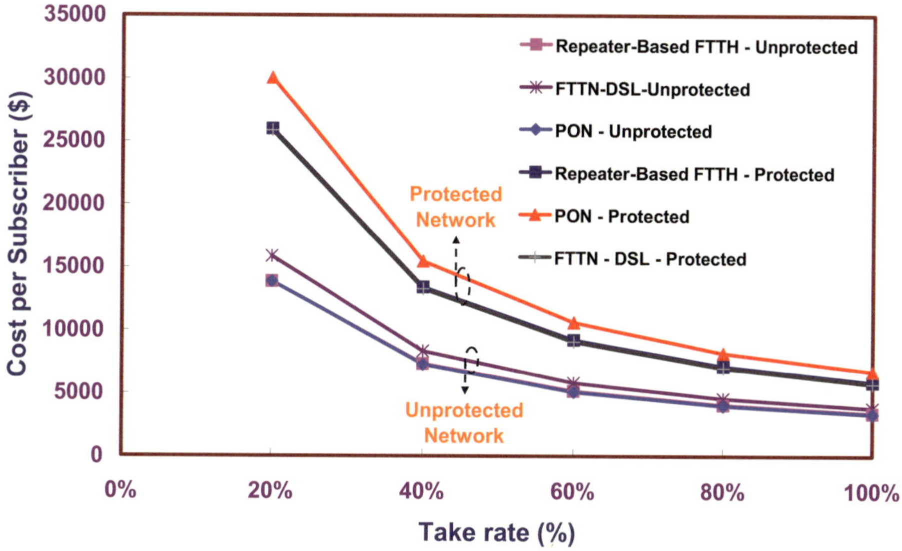

Fig. 27. Cost per subscriber for different access architectures vs. take rates.

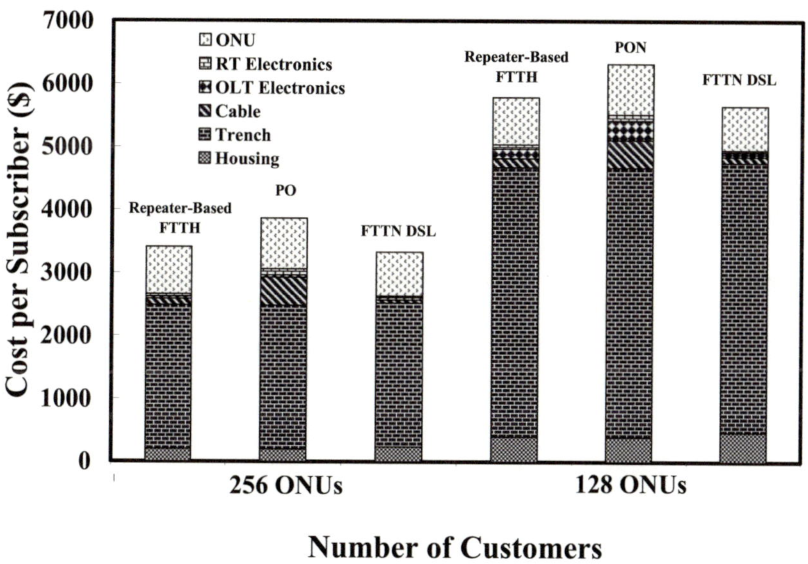

Fig. 28. Cost per subscriber for different access architectures for take rate of 100%.

Fig. 28 shows the cost per subscriber for different architectures for the take rate of 100% showing the cost splits. As can be seen, the ONU costs and the trenching costs dominate the total cost per subscriber for each scheme. As the feeder fiber length is 100 km, the trenching costs dominate the total CU costs and approximately are more than 67% of the total costs, while the ONU cost is more than 21% of the total cost.
For this economic modelling, the cost of the OLT transceiver is considered to be $10,000 while the cost of the ONU transceiver is $500 for all architectures. If the optical transceiver cost used at the CU for the repeater-based FTTH architecture is $453 or less while the optical transceiver cost used at the CU in the FTTN-DSL architecture is $500, then the total cost per subscriber for both repeater-based FTTH and FTTN-DSL architectures is $3459. Therefore, the optical transceiver cost used the CU in the repeater-based FTTH network has to be less than $453 to make this architecture cheaper than other considered solutions.

7. Conclusions

We have reported a new FTTH system to reduce ONU transmitter cost using an upstream repeater at the remote node. The system can be further modified to provide both downstream and upstream regeneration by utilizing bidirectional property of standard transceivers. This is to extend conventional PON system's feeder fiber reach to 60 km and split ratio to 1:256. The use of the repeater introduces insignificant penalty to the existing PON performance and meets the IEEE 802.3ah standard requirements. We show that the system achieves good performance compared to a standard PON using SMFs. The system has a large network coverage and potentially provides a low-cost solution to accelerate broadband access deployment. Furthermore, we investigated the jitter performance of cascaded repeater-based FTTH systems through a recirculating loop. The results show that the system without clock recovery can be cascaded up to 4 times with insignificant performance degradation and the total jitter is within EPON standard specification. When clock recovery at the receiver is used, we can extend the system's coverage area significantly. We have shown that value-added services such as video service integration and LAN emulation can intelligently be added in the repeater-based access networks. We have demonstrated a cost-efficient video service delivery scheme for this type of densely populated repeater-based optical access network, whereby the remote repeater enables the integration of the video services to the downstream traffic transport. The video signals are carried on a RF carrier that is placed outside the bandwidth of the downstream traffic. The BER results and the constellation diagrams measured from the experimental demonstrations show that both signals can be recovered with minimal penalty. For the local area networking amongst the customers in the repeater-based FTTH network, the remote repeater performs intelligent functionalities to provide optical layer LAN capabilities while regenerating the signals. The experimental results show that all signals can be recovered error-free after transmissions. We have also performed an economics study of different FTTH technologies taking into account 1+1 protection. It is shown that the repeater-based FTTH architecture is competitive with the FTTN-DSL architecture in terms of cost per subscriber whereas the conventional PON architecture requires higher costs. The trenching costs and the ONU costs are far more dominant of all costs in all kinds of architectures. It has also been shown that longer feeder fiber incurs more cost per subscriber and therefore to make a feasible and more economical solution to provide broadband services, a larger number of customers have to be supported over a single infrastructure.

8. References

Arnaud, B.; Wu, J. & Kalali, B. (2003). Customer-controlled and -managed optical networks. *Journal of Lightwave Technology,* Vol. 21, No. 11, Nov. 2003, 2804-2810, ISSN 0733-8724.

Chae, C.-J.; Seung-Tak, L.; Geun-Young, K. & Heesang, P. (1999). A PON system suitable for internetworking optical network units using a fiber Bragg grating on the feeder fiber. *IEEE Photonics Technology Letters,* Vol. 11, No. 12, Dec. 1999, 1686-1688, ISSN 1041-1135.

Chae, C.-J; Heesang, P & Jong-Hoon, E. (2001). An ATM PON system overlaid with a 155-Mb/s optical star network for customer networking and fiber to the premises. *IEEE Photonics Technology Letters,* Vol. 13, No. 10, Oct. 2001, 1133-1135, ISSN 1041-1135.

Cohen, R. (2003). On the establishment of an access VPN in broadband access networks. *IEEE Communications Magazine,* Vol. 41, No. 2, Feb. 2003, 156 - 163, ISSN: 0163-6804.

Dixit, S. (2003). *IP over WDM: Building the Next Generation Optical Internet,* John Wiley & Sons, ISBN 0471212482, New Jersey.

Hernandez-Valencia, E.J. (1997). Architectures for broadband residential IP services over CATV networks. *IEEE Network,* Vol. 11, No. 1, Jan.-Feb. 1997, 36-43, ISSN 0890-8044.

Iannone, P.P.; Reichmann, K.C.; Smiljanic, A.; Frigo, N.J.; Gnauck, A.H.; Spiekman, L.H. & Derosier, R.M. (2000). A transparent WDM network featuring shared virtual rings. *Journal of Lightwave Technology,* Vol. 18, No. 12, Dec. 2000, 1955-1963, ISSN 0733-8724.

IEEE (2004). IEEE Standard 802.3ah-2004.

Jayasinghe, T. & Chae, C.-J. (2005a) Implementation of local re-broadcasting and LAN services over an Ethernet Passive Optical Network, *Proceedings of 31st European Conference on Optical Communication, ECOC 2005,* pp. 539-540, ISBN 0-86341-543-1, Sep. 2005.

Jayasinghe, T.; Chae, C.-J. & Tucker, R.S. (2005b). Rebroadcasting of broadband services over passive optical network in residential community. *Electronics Letters,* Vol. 41, No. 24, 24 Nov. 2005, 1343-1344, ISSN 0013-5194.

Kettler, D.; Kafka, H. & Spears, D. (2000). Driving fiber to the home. *IEEE Communications Magazine,* Vol. 38, No. 11, Nov. 2000, 106-110, ISSN 0163-6804.

Kramer, G. & Pesavento, G. (2002). Ethernet passive optical network (EPON): building a next-generation optical access network. *IEEE Communications Magazine,* Vol. 40, No. 2, Feb. 2002, 66-73, ISSN 0163-6804.

Nadarajah, N.; Attygalle, M.; Nirmalathas,A. & Wong, E. (2005). Novel schemes for local area network emulation in passive optical networks with RF subcarrier multiplexed customer traffic. *Journal of Lightwave Technology,* Vol. 23, No. 10, Oct. 2005, 2974-2983, ISSN 0733-8724.

Nadarajah, N.; Chang, C.-J; Tran, A.V. & Nirmalathas, A. (2007). Video service delivery over a repeater-based optical access network. *IEEE Photonics Technology Letters,* Vol. 19, No. 20, 15 Oct. 2007, 1637-1639, ISSN 1041-1135.

Nadarajah, N.; Chang, C.-J; Tran, A.V. & Nirmalathas, A. (2009). Optical Layer Local Area Network Emulation in a Multifunctional Repeater-Based Optical Access Network. *IEEE/OSA Journal of Optical Communications and Networking,* Vol. 1, No. 1, Jun. 2009, 43-49, , ISSN 1943-0602.

Nesset, D.; Davey, R.P.; Shea, D.; Kirkpatrick, P.; Shang, S.Q.; Lobel, M. & Christensen, B. (2005). 10 Gbit/s bidirectional transmission in 1024-way split, 110 km reach, PON system using commercial transceiver modules, super FEC and EDC, *Proceedings of 31st European Conference on Optical Communication, ECOC 2005,* pp. 135-138, ISBN 0-86341-543-1, Sep. 2005.

Park, S.-J.; Lee, C.-H.; Jeong, K.-T.; Park, H.-J.; Ahn, J.-G. & Song, K.-H. (2004). Fiber-to-the-home services based on wavelength-division-multiplexing passive optical network. *Journal of Lightwave Technology,* Vol. 22, No. 11, Nov. 2004, 2582–2591, ISSN 0733-8724.

Tran, A.V.; Chae, C.-J. & Tucker, R.S. (2006a). Bandwidth efficient PON system for broadband access and local customer networking. *IEEE Photonics Technology Letters,* Vol. 18, No. 5, 1 Mar. 2006, 670-672, ISSN 1041-1135.

Tran, A.V.; Chae, C.-J. & Tucker, R.S. (2006b). Low-cost and scalable passive optical network architecture using remote repeater. *Electronics Letters,* Vol. 42, No. 10, 11 May 2006, 589-591, ISSN 0013-5194.

Tran, A.V.; Chae, C.-J. & Tucker, R.S. (2006c). Cascadability performance of remote-repeater-based FTTP systems. *Electronics Letters,* Vol. 42, No. 24, 23 Nov. 2006, 1418-1420, ISSN 0013-5194.

Venkateswaran, R. (2001). Virtual private networks. *IEEE Potentials,* Vol. 20, No. 1, Feb.-Mar. 2001, 11 - 15, ISSN 0278-6648.

Wong, E. & Chae, C.-J. (2004). CSMA/CD-based Ethernet passive optical network with optical internetworking capability among users. *IEEE Photonics Technology Letters,* Vol. 16, No. 9, Sep. 2004, 2195-2197, ISSN 1041-1135.

Photonic Millimeter-wave Generation and Distribution Techniques for Millimeter/sub-millimeter Wave Radio Interferometer Telescope

Hitoshi Kiuchi[1] and Tetsuya Kawanishi[2]
[1]*National Astronomical Observatory of Japan*
[2]*National Institute of Information and Communications Technology*
Japan

1. Introduction

In the field of high-frequency radio interferometer, Atacama Large Millimeter/sub-millimeter Array (ALMA), its highest receiving frequency reaches 950 GHz. To receive such high frequencies, a higher reference frequency for frequency conversion is required (as much as over 100 GHz). To maintain signal coherency, this reference signal has to be highly stable. To address these issues, we have developed a new method to generate and transmit a reference signal in the form of frequency difference between two coherent light waves. One method to generate two optical signals is producing them from a pair of laser sources using optical phase lock loop for feed back control (Cliche & Shillue (2006)), however, optical phase lock loop also requires a stable laser source. A good alternative method to the optical phase lock scheme is the lithium niobate ($LiNbO_3$) Mach-Zehnder optical modulator which is capable of generating two highly stable optical signals (upper sideband and lower sideband components) by applying a sinusoidal microwave signal to an input laser signal. Compared to the current optical phase lock scheme, the Mach-Zehnder modulator has significant advantages in terms of stability (free from the influence of the input laser stability), robustness to mechanical vibration and acoustic noise, and capability of maintaining polarization state of the input laser. During the signal transmission through the fiber cable, the cable length delay fluctuation is caused together with polarization mode dispersion, which will impact the performance of coherent signal distribution. We have developed the phase stabilizer using the dual difference round-trip phase measurement method with Michelson's interferometer. The roundtrip phase measurement is performed on each lightwave signal separately. The Round-trip phase measurement method is helpful for successful delay compensation of the microwave signal which is converted from the two coherent optical signals by a photo mixer. The two transmitted optical signals require phase stability better than 10^{-13} (1sec) in white phase noise in the Allan standard deviation for ALMA project.

2. Astronomical requirements

The modulation signal is transmitted via one optical fiber in the form of frequency difference between two coherent optical signals. These two optical signals are subsequently converted into millimeter-wave signal by the photo mixer at the remote antenna. The region from 27 GHz to 122 GHz (Table 1) is used for operational frequency for ALMA 10-band receivers. The two-tone generator requires polarization maintaining capability as well as mechanical vibration and acoustic noise robustness to avert the impact of the polarization effect on the photo-mixer and that of the polarization mode dispersion on the transmission fiber.

	Reference freq. range (GHz)	Local freq. range (GHz)
Band 1	27.3-33.0	27.3-33.0
Band 2	79.0-94.0	79.0-94.0
Band 3	92.0-108.0	92.0-108.0
Band 4	68.5-75.5	137.0-151.0
Band 5	87.5-99.5	175.0-199.0
Band 6	74.3-87.7	223.0-263.0
Band 7	94.3-121.7	283.0-365.0
Band 8	79.4-97.6	397.0-488.0
Band 9	102.3-118.0	614.0-708.0
Band 10	88.8-104.2	799.0-938.0

Table 1. Required frequency range.

2.1 Phase noise

The behavior of phase noise can be analyzed by the Allan standard deviation (Allan (1966; 1976)). The frequency instability is the frequency change induced by internal or external factors within a given time interval. In other words, the frequency instability is defined as the degree to which the output frequency remains constant over a specified period of time. Characteristics of phase noises (Healey (1972)) are shown in Fig. 1.

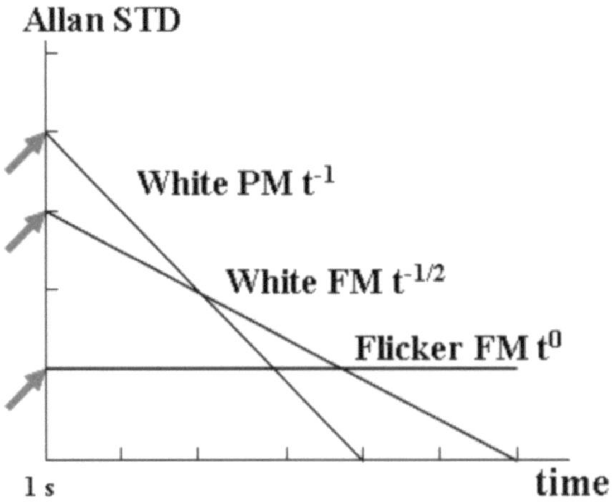

Fig. 1. Phase noises of a highly stable signal.

Noises are classified into the following five types according to the noise generation mechanism: white phase modulation noise (τ^{-1}), flicker phase modulation noise (τ^{-1}), white frequency modulation noise ($\tau^{-1/2}$), flicker frequency modulation noise (τ^{0}) and Random Walk frequency modulation noise ($\tau^{1/2}$) (Healey (1972)). Some of these noises are generated by electronic equipment or by changes in the environment (such as temperature change).

2.2 Estimation of coherence loss and time error by the Allan standard deviation

The coherence loss due to the instability in the frequency standard for *T-sec* integration times is estimated Eq. (1) (Rogers & Moran (1981); Rogers et al. (1984); Kawaguchi (1983)).

$$L_c = \omega_o^2[\frac{\alpha_p}{6} + \frac{\alpha_f}{12}T + \frac{\sigma_y^2}{57}T^2] \tag{1}$$

where
L_c the loss of coherence,
ω_o the angular frequency of local oscillator,
α_p the Allan variance [(*standard deviation*)^2] of white phase noise at 1 sec,
α_f the Allan variance [(*standard deviation*)^2] of white frequency noise at 1 sec,
σ_y^2 the constant Allan variance [(*standard deviation*)^2] of flicker frequency noise,
T the integration time [sec].

Coherence loss and time error are calculated by the Allan standard deviation. One of the stability measurement methods in time domain is the Dual-Mixer Time Difference (DMTD) method (Allan (1976)) which is adopted by NIST (National Institute of Standards and Technology, USA), NICT (National Institute of Communications and Technology, Japan) and other time/frequency standard institutes. Using this method, the phase stability of a device-undertest can be obtained without influence of unstable local frequencies of the measurement system. The Dual-Mixer Time Difference method allows time measurements and frequency and frequency stability measurements for sample times as short as a few milliseconds or longer without dead time. Moreover, the phase noise in the measurement system can be canceled out with this method.

The total system instability is calculated by RSS (root sum square) of the Allan standard deviation of each component. Time error of phase noise is calculated as follows:

$$Time\ error = \frac{T \times \sigma_y(T\ sec)}{\sqrt{3}}$$

$$= \frac{\sigma_y(1\ sec)}{\sqrt{3}} \tag{2}$$

in white Phase Modulation noise

$$= \frac{\sigma_y(1\ sec)}{\sqrt{T}} \tag{3}$$

in white Frequency Modulation noise

$$= \frac{T \times \sigma_y (1\, sec)}{\sqrt{\ln 2}} \tag{4}$$

in flicker Frequency Modulation noise

According to the system-level technical requirements of ALMA, the instrumental delay/phase error of the 1st Local system should be 53 fs in the short time period, and the difference between 10 sec averages at intervals of 300 sec should be 17.7 fs in RMS. When these values are converted to the Allan standard deviation using Equations (2), (3), and (4), it turns out that the noises are white phase modulation noise and flicker frequency modulation noise. The short time stability of white phase modulation noise is obtained Eq. (2): $\sigma_y(\tau = 1) = 9.2 \times 10^{-14}$. Calculating from Eq. (4) and 10 seconds averaging, the required stability is $\sigma_y(\tau = 1) = 1.56 \times 10^{-16}$ (flicker phase modulation noise) in the long-time period.

3. Photonic millimeter-wave generator

In the high-extinction ratio lithium niobate ($LiNbO_3$) Mach-Zehnder intensity modulator (Izutsu et al. (1977), Izutsu et al. (1981), Kawanishi et al. (2006), Kawanishi et al. (2004a), Kawanishi et al. (2004b), Kawanishi et al. (2007), Onillon et al. (2005), Vegas et al. (2003)), the optical frequency difference between two optical signals is exactly twice (or four times) the modulation frequency, and the output signal is equivalent to FSK (frequency shift keying) spectrum. Compared to the optical phase lock scheme, the Mach-Zehnder modulator (shown in Fig. 2) has significant advantages in terms of robustness to mechanical vibration

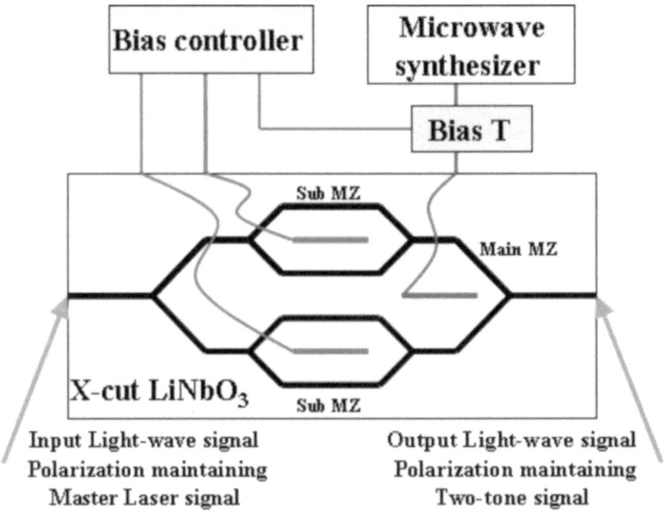

Fig. 2. Simplified structure of an optical modulator with two arms and electrodes. Optical phase of each arm is controlled by applying DC bias to the electrodes. Amplitude imbalance due to fabrication error is compensated with sub-Mach-Zehnder trimmers. When two lightwaves are in phase, the output optical signals are strengthened each other. On the other hand, when the phases of the input lightwaves are shifted, the phase-shifted lightwaves are radiated away as higher-order waves, and do not reach the optical waveguide. This is the main feature of the Mach-Zehnder modulator.

and acoustic noise, stability (free from the influence of the input laser stability), and capability of maintaining polarization state of the input laser. The Mach-Zehnder modulator is so reliable that it has been used for optical submarine cables. The estimated lifetime of the Mach-Zehnder modulator extends to several decades.

The output spectrum depends on the DC bias voltage applied to the electrodes in the Mach-Zehnder structure. The Mach-Zehnder modulator has the following two operation modes (Kawanishi et al. (2005), Kiuchi et al. (2007), Sakamoto et al. (2005)).

3.1 Operation mode

3.1.1 Null-bias point operation mode

When the bias of the Mach-Zehnder modulator is set to a minimum transmission point (nullbias point), the first-order upper side band (USB) and lower side band (LSB) components are strengthened, and the carrier is suppressed (Fig. 3). The frequency difference between the two spectral components is twice the modulation sinusoidal signal frequency. As the spectral components generated by the optical modulation are phase-locked, it is possible to construct a robust system without using any complicated feedback control technique. However, as the modulation frequency is limited by the frequency response of the modulator, the frequency upper limit of the two optical signals can not be higher than 100 GHz in the null-bias point operation mode. For this reason, the null-bias point operation mode is suitable for the low-frequency application.

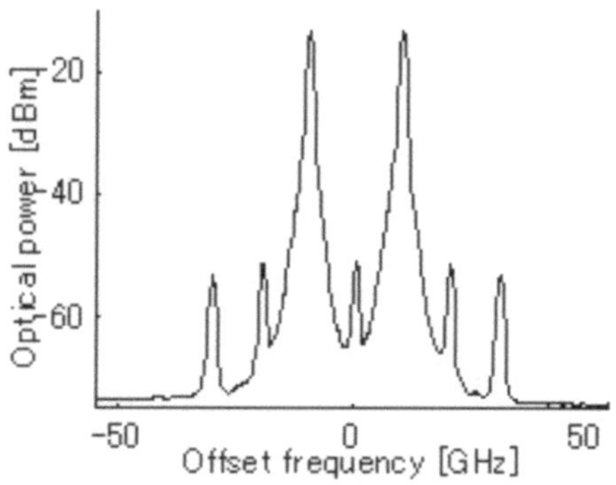

Fig. 3. When the bias of the Mach-Zehnder modulator is set to a minimum transmission point (null-bias point), the first-order USB and LSB components are strengthened, and the carrier is suppressed. The frequency difference between the two spectral components is exactly twice the modulation signal frequency. Each sideband signal spectrum shows a copy of the input laser spectrum. High carrier suppression ratio of 50 dB was demonstrated by the nullbias point operation mode using the integrated Mach-Zehnder modulator with an intensity trimmer in each arm (sub-Mach-Zehnder interferometer).

3.1.2 Full-bias point operation mode

When the bias is set to a maximum transmission point (full-bias point), the second-order USB and LSB are strengthened, and the carrier is not suppressed. If the extinction ratio of the Mach-Zehnder modulator is high, undesired odd-order USB and LSB components can be successfully suppressed with this technique. When the odd-order sideband components are suppressed in this mode, the optical frequency of even-order (zero- and second-order) components is remained (Fig. 4 left side). Eliminating the zero-order component (carrier), the remaining is a two-tone optical spectrum whose frequency is four times the modulation frequency or 4 *fm* (*fm* is the modulation frequency of the RF signal applied to the modulator). The frequency difference between the zero-order and second-order components is 2 *fm*. When 4 *fm* > 50 GHz, the frequency difference is large enough that the zero-order component can be eliminated with a conventional optical filter (Fig. 4 right side). The optical signal filtered by the optical filter is amplified by an optical amplifier. At this point, the first-order components are suppressed by the Mach-Zehnder modulator with high extinction ratio to prevent undesired spurious signals.

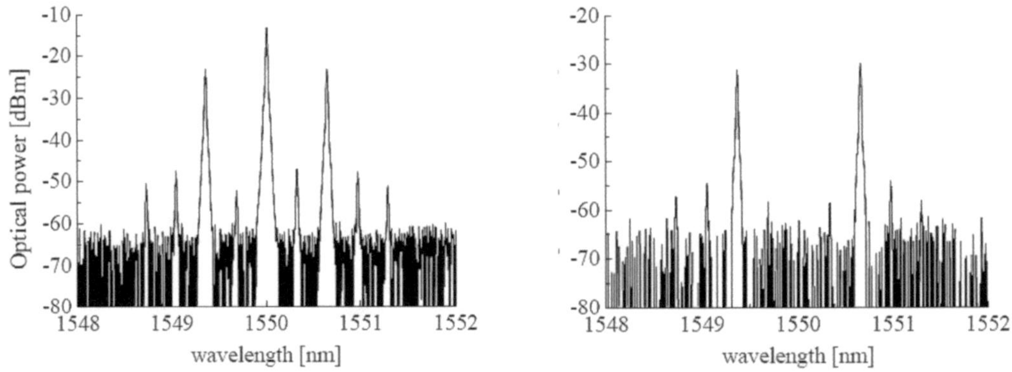

Fig. 4. When the bias of the Mach-Zehnder modulator is at full-bias point, the odd-order sideband components are suppressed. In this case, the optical frequency of even-order (zero-order and second-order) components is remained (left chart). Eliminating the zero-order component (carrier: input lightwave), the remaining is a two-tone optical spectrum whose frequency is four times the modulation frequency (right chart).

3.2 Harmonic generation process

The input light-wave is assumed to be monochromatic, and can be described by $A_{LW}e^{2\pi i f_0 t}$, where A_{LW} is the optical transmittance in the waveguide. Light-waves with RF signal can be obtained by modulation of sinusoidal RF signals into USB and LSB components using an optical intensity modulator. Assuming the RF signal is a sinusoidal signal, it is expressed as $2A_{RF}\sin(2\pi f_m t + \phi_B)$, and the optical output is expressed as shown in Eq. 7(Kawanishi et al. (2007)).

$$R = \frac{1}{2}A_{LW}e^{2\pi i f_0 t}[e^{i(A_{RF}\sin(2\pi f_m t+\phi_B/2))}$$

$$+e^{-i(A_{RF}\sin(2\pi f_m t+\phi_B/2))}] \tag{5}$$

$$= \frac{1}{2}A_{LW}e^{2\pi i f_0 t}\sum_{n=-\infty}^{\infty}J_n(A_{RF})e^{2\pi i[f_0 t+nf_m t]}$$

$$\times[e^{i\phi_B/2}+(-1)^n e^{-i\phi_n/2}] \tag{6}$$

$$= A_{LW} e^{2\pi i f_0 t} [\cos\frac{\phi_B}{2} \sum_{n=-\infty}^{\infty} J_{2n}(A_{RF}) e^{2\pi i [f_0 t + 2n f_m t]}$$

$$+ i \sin\frac{\phi_B}{2} \sum_{n=-\infty}^{\infty} J_{2n+1}(A_{RF}) e^{2\pi i [f_0 t + (2n+1) f_m t]}] \quad (7)$$

The output optical intensity $|R|^2$, which is detected by a high-speed photo-mixer, is expressed by Eq. 8.

$$|R|^2 \simeq |A_{LW}|^2 [J_0^2(A_{RF})\cos^2\frac{\phi_B}{2} + 2J_1^2(A_{RF})\sin^2\frac{\phi_B}{2}$$

$$-4J_0(A_{RF})J_1(A_{RF})\sin\frac{\phi_B}{2}\cos\frac{\phi_B}{2}\sin 2\pi f_m t$$

$$+2(2J_0(A_{RF})J_2(A_{RF})\cos^2\frac{\phi_B}{2}$$

$$-J_1^2(A_{RF})\sin^2\frac{\phi_B}{2})\cos(2\times 2\pi f_m t)] \quad (8)$$

where the high-order components are neglected assuming $A_{RF} \ll 1$, and the high-order components are neglected. By using Taylor's expansion of Bessel function, Eq. 9 is obtained.

$$\frac{|R|^2}{|A_{LW}|^2} = \frac{1}{2} + \frac{1-|A_{RF}|^2}{2}\cos\phi_B$$

$$-A_{RF}\sin\phi_B \sin 2\pi f_m t$$

$$+\frac{1}{2}|A_{RF}|^2 \cos\phi_B \cos(2\times 2\pi f_m t) \quad (9)$$

The intensities of the fundamental component $\sin(2\pi f_m t)$ and the second-order harmonic $\cos(2 \times 2\pi f_m t)$ can be controlled by the DC-bias ϕ_B. The fundamental and second-order components are proportional to $\sin(\phi_B)$ and $\cos(\phi_B)$, respectively. The ratio between the average power and RF signal component depends largely on the conversion efficiency from light-waves to RF signals at the photo-mixer. The ratios for the fundamental and second-order components are expressed in Eqs. 10, 11.

$$D_1 = \left|\frac{2A_{RF}\sin\phi_B}{1+(1-|A_{RF}|^2)\cos\phi_B}\right| \quad (10)$$

$$D_2 = \left|\frac{|A_{RF}|^2\cos\phi_B}{1+(1-|A_{RF}|^2)\cos\phi_B}\right| \quad (11)$$

In the case of $\phi_B = \pi$, the even-order components in the output signal R, including the carrier components $e^{2\pi i f_0 t}$ are suppressed and the average power $|R|^2$ is reduced to minimum of

$|A_{RF}|^2/2$, where the dominant components are first-order USBs and LSBs. In the case of ϕ_B = 0, the odd-order components are suppressed, and the dominant components are zero-order and second-order USBs and LSBs.

3.3 Need for the high-extinction ratio modulator

Three Mach-Zehnder structure LN-modulator can provide high-extinction ratio (more than 55 dB) modulation signals. Simulated signals are shown in Figs. 5 and 6. High-extinction

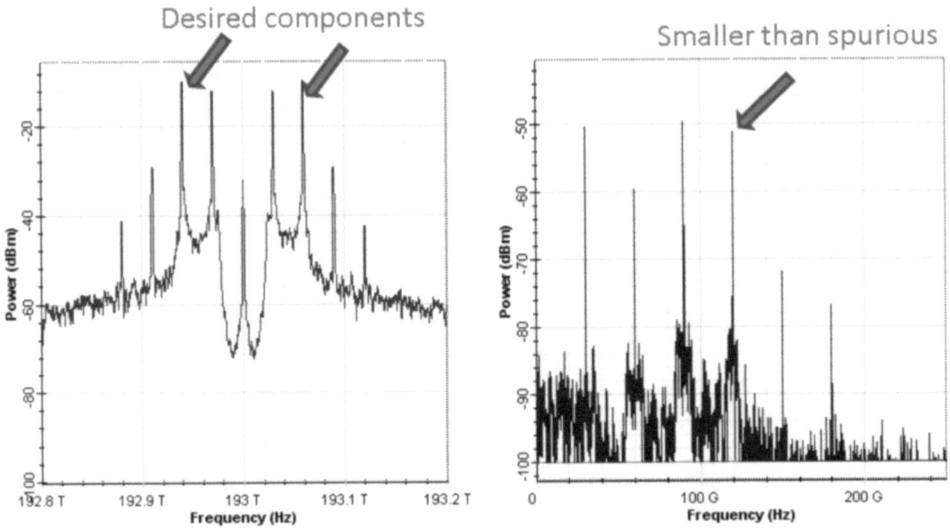

Fig. 5. Simulated low extinction ratio (20 dB) modulation signal. Optical spectrum (left) and micro wave spectrum (right).

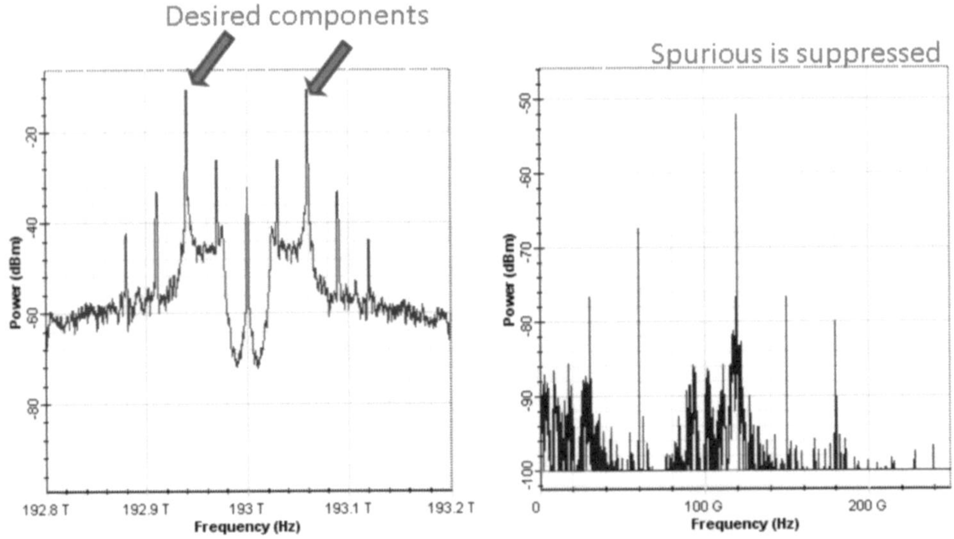

Fig. 6. Simulated high extinction ratio (50 dB) modulation signal. Optical spectrum (left) and micro wave spectrum (right).

ratio performance is effective in suppressing of excessive signals. Suppression of spurious is very important to ensure effective photonic LO signal distribution.

3.4 Stability measurement

In the case of the interferometer, we use the hydrogen maser which has the best short-term stability among existing atomic clocks as the reference signal source if necessary. There is also a method to measure the phase noise of components without using the hydrogen maser. We can estimate the total phase noise of the interferometer system, using the covariance that is obtained by; 1) measuring the phase noise of a single unit independent from the reference signal and the reference signal phase noise that is separately measured and 2) taking the root sum square of these phase noises. We should use time domain Allan standard deviation measurement with DMTD method instead of the frequency domain SSB phase noise measurement method which measures the phase noises of all signals as a whole. The Allan standard deviation in time domain is used to calculate the coherence loss and time error.

3.4.1 Time domain phase measurement method for the null-bias point operation mode

Figure 7 shows a time-domain stability measurement system to measure the differential phase between the second harmonic of the reference synthesizer and the first-order modulated signal (null-bias point operation mode). The figure shows the experimental setup of the Dual-Mixer Time Difference system (mixers, filters, and a Time Interval Analyzer: TSC-5110A) for phase noise measurement using a 22 GHz signal. The origin of the source signal is a 11 GHz synthesizer. The 11 GHz signal is used as a modulation signal, and the 22 GHz signal (spurious signal of 11 GHz, Fig. 7) is used as a reference signal (on the lower arm). These signals are coherent since the 22 GHz signal is a harmonic of the 11 GHz signal. Two coherent optical signals with 22 GHz difference are generated by optical modulation of the optical source signal using the Mach-Zehnder modulator. These two signals are subsequently converted to a 22 GHz microwave signal (on the upper arm) by the photo-

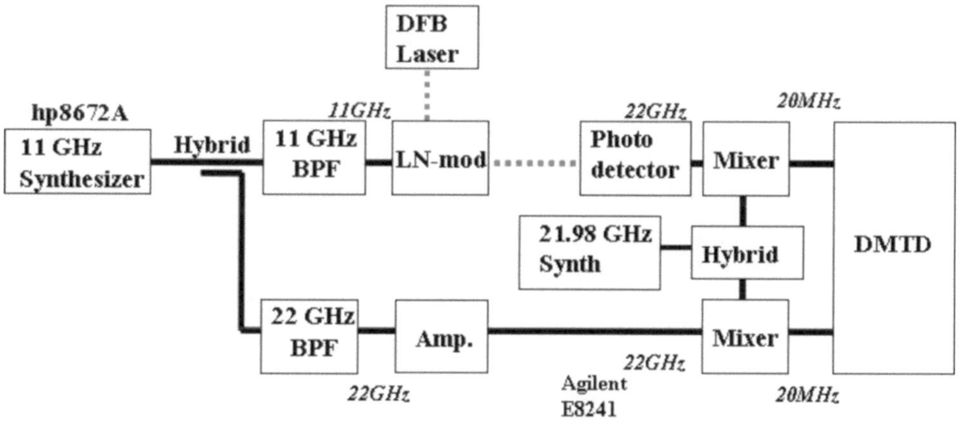

Fig. 7. Block diagram of a time-domain stability measurement system for the null-bias point operation mode (the first-order optical signal). This phase noise measurement system is free from the influence of reference signal phase noise and frequency conversion signal phase noise.

mixer. The frequencies of the two 22 GHz signals (on both arms) are converted to 20 MHz with a common 21.98 GHz signal. After these processes, the phase difference between the two 20 MHz signals is measured by the Dual-Mixer Time Difference system. In this experimental setup, the 21.98 GHz synthesizer, the hybrid, and mixers compose a kind of a Dual-Mixer Time Difference system. During these operations, the 20 MHz signals are free from the instability of the 11 GHz and 21.98 GHz synthesizers.

3.4.2 Time domain phase measurement method for the full-bias point operation mode

Figure 8 shows a time-domain stability measurement system to measure the differential phase between the multiplied (×4) reference signals and the second-order modulated signal (Full-bias point operation mode). In the case of 100 GHz measurement, the source signal is generated from the 25 GHz sinusoidal synthesizer, and the generated 25 GHz signal is used as a modulation signal and a multiplied reference signal. The microwave multiplier generates 100 GHz. Two coherent optical signals with 100 GHz difference are generated by optical modulation of the optical source signal using the Mach-Zehnder modulator. These two signals are subsequently converted to a 100 GHz microwave signal by the photo-mixer. The frequencies of the two 100 GHz signals are converted to 10 MHz by harmonic-mixers (multiplied number is 10) with a common 9.999 GHz synthesizer signal. After these processes, the differential phase between the two 10 MHz signals is measured by the Dual-Mixer Time Difference system. In this experimental setup, the 9.999 GHz synthesizer, the hybrid, and harmonic-mixers in the figure compose a kind of a common noise system. During these operations, the 10 MHz signals are free from the instability of the 25 GHz and 9.999 GHz synthesizers. The measured phase noise is the covariance of the two systems (Mach-Zehnder modulator and multiplier).

We used an NTT photo-mixer, an Uni-traveling-carrier photodiode (UTC-PD)(Hirota et al. (2001), Ito et al. (2000)). Responsibility of the photodiode is approximately 0.4 A/W. The typical output power (100 GHz) is approximately 0.5 mW.

3.5 Measured stability

To make the Dual-Mixer Time Difference method available, it is required that the phase stability of the multiplier be better than that of the Mach-Zehnder modulator, or the stability of the two systems be almost equivalent.

The results of the SSB phase noise measurement method include not only the phase noises of the LN-modulator (or multiplier) but also those of the reference signal generator (Synthesizer). Therefore the measured SSB phase noise heavily depends on the reference signal phase noise. On the other hand, the DMTD method measures differential phase noise between the measurement signal and the reference signal. In our system, the measurement signal and the reference signal are generated from the same source, which means we can offset the phase noise of the signal source, or the common noise, when obtaining the covariance between the modulator and multiplier. If the phase noises of the modulator and multiplier are almost equivalent or that of the modulator is better, we can use the obtained Allan standard deviation as the phase noise after dividing it with the square root of two. If the multiplier has much better phase noise, the obtained covariance should be considered as the phase noise of the modulator.

We made a comparison between single side band (SSB) phase noises of the multiplier and the Mach-Zehnder modulator signals using the SSB phase noise measurement system as shown in Fig. 8.

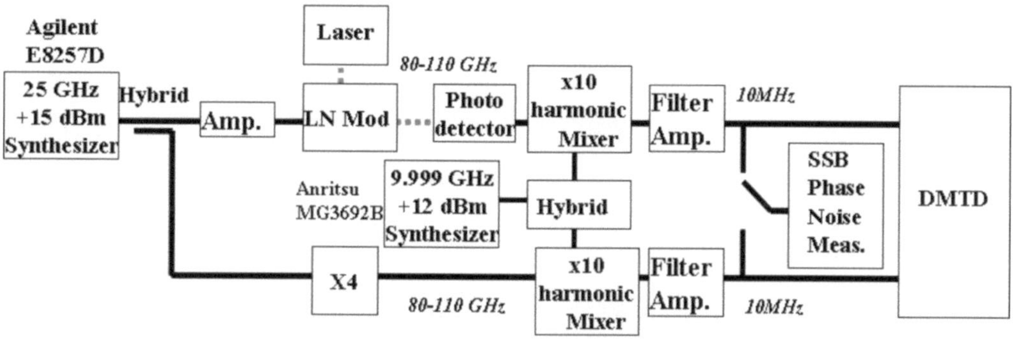

Fig. 8. Block diagram of a time-domain stability measurement system using the multiplier signal for the full-bias operation mode (the second-order optical modulation signal). This phase noise measurement system is free from the influence of reference signal phase noise and frequency conversion signal phase noise. This method is also regarded as a Dual-Mixer Time Difference method. The measured phase stability is the covariance of the Mach-Zehnder modulator and multiplier phase noises.

Since the current system doesn't have two identical LN modulators, we cannot perform the phase noise measurement between two identical LN modulators with the DMTD method. Consequently, it is meaningless to use the DMTD method if the phase noise of the multiplier to be compared is extremely bad.

The obtained results show at least the modulator has phase noise that is equivalent to or better than that of the multiplier in 1 kHz and higher frequency. The lower frequency phase noise is masked by the synthesizer phase noise. The measurement results of SSB phase noise is no more than a criterion for judgment of effectiveness of the measured Allan standard deviation with the DMTD method.

Phase stability of the Mach-Zehnder modulator measured using the Allan standard deviation is shown in Fig. 9. The stability is independent of the input laser line-width for a short fiber cable, the input lasers are a DFB-laser (10 MHz line-width) and a fiber-laser (1 kHz line-width).

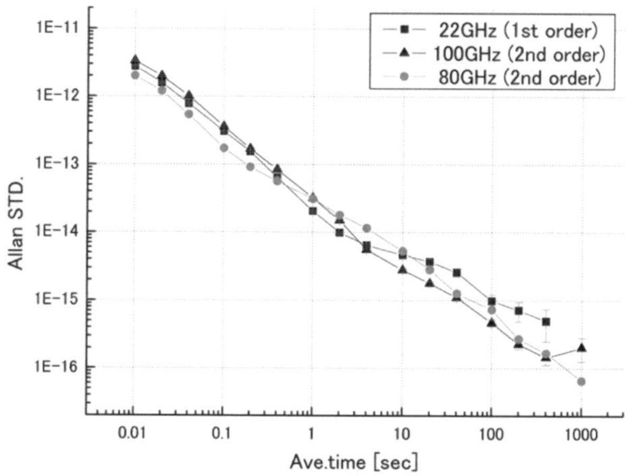

Fig. 9. Measured phase stabilities of the Mach-Zehnder modulator, the first-order 22 GHz signal and the second-order 100 GHz signal.

3.6 Differential polarization angle between two light-waves

The theme of this paper covers optical signal generation, but the ultimate goal of the photonic system is generation of highly-stable optical signal and its transmission with fiber system. The delay compensation must be performed on the delay caused during the optical signal transmission through an optical fiber cable in order to keep the signals coherent. In the photonic LO (Local) system, two optical signals are transmitted and converted by a photo mixer at a remote antenna into a microwave signal. During the signal transmission through the fiber cable, the cable length delay is caused, including Polarization Mode Dispersion (PMD), a bottleneck in performing successful phase compensation (delay change compensation). PMD is the state of polarizations (SOP) dispersing randomly in the cable. PMD is caused when the state of polarization of the two optical signals is absolutely changed by the movement of the cable through which the signals are transmitted. The magnitude of PMD is inversely proportional to the degree of the polarization alignment of the two optical signals. Since the generation of PMD contributes to the emergence of the Differential Group Delay (DGD) (synonymous with LO phase jitter), SOP of the two signals needs to be coincident so as to reduce the second order PMD effect on DGD.

We measured the differential polarization angle between two light-waves generated by the Mach-Zehnder modulator. The measurement block diagram is shown in Fig. 10. In this measurement, the two light-waves are transmitted to the ITU-Grid programmable optical filter (Peleton QTM050C), which selects one of the two light-waves for polarization. The polarization is measured by the polarization meter (Polarimeter). The differential angle is calculated by Eq. (12): spherical trigonometry.

$$\cos d = \sin\delta_1 \times \sin\delta_2 + \cos\delta_1 \times \cos\delta_2 \times \cos(\lambda_1 - \lambda_2) \tag{12}$$

The measured polarization angles in degrees are (δ_1: -29.2 in Azimuth, λ_1: -4.54 in Elevation) and (δ_2: -28.3, λ_2:-4.59). The calculated differential polarization angle: d is 0.90 degrees.

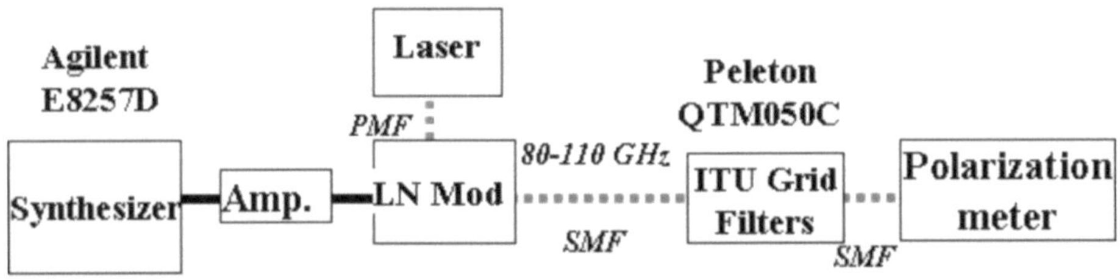

Fig. 10. Block diagram of the Polarization measurement. One of the two optical signals is selected by the ITU grid switch for polarization and transmitted to the Polarization meter.

3.7 Astronomical application

3.7.1 Estimated coherence loss

The measured stability of the null-bias point operation mode is 2.4×10^{-14} (white phase modulation noise) with 1.3×10^{-14} (white frequency modulation noise) at $\tau = 1$ sec, while the stability of the full-bias point operation mode is 3×10^{-14} (white phase modulation noise). With respect to a $\times n$ multiplier, multiplied phase noise (Vanblerkom & Aneman (1966)) should also be considered as shown below:

$$\textit{Multiplied phase noise}$$

$$= \textit{Measured phase noise} \times \sqrt{\textit{Multiplied number}} \qquad (13)$$

The coherence loss calculated from Equation (1) is smaller than 5% at the highest local frequency (938 GHz).

In the Dual-Mixer Time Difference system for the null-bias point operation mode shown in Fig. 7, phase noise of the measurement system (supposedly, white frequency modulation noise) is not canceled out as common noise, because the signal phase becomes unstable and incoherent in the amplification process by the AMP in the figure. The mild peak in 22 GHz around 30 seconds is thought to be due to white frequency modulation noise or instability of the amplifier, as the similar peak is not detected in the full-bias point operation (80 and 100 GHz measurements). Assuming the white frequency modulation noise is caused by any component other than the Mach-Zehnder Modulator, the phase noise of the Mach-Zehnder Modulator will be $\sigma_y(\tau = 1) = 2.4 \times 10^{-14}$. In this case, the coherence loss due to the phase noise will be constant, because the loss due to white phase modulation noise is independent of integration time. However, even if both of these noises are considered, the Mach-Zehnder modulator is still applicable to the most advanced systems such as ALMA and Very Long Baseline Interferometer (VLBI). The photonic millimeter-wave generator has been authorized as the MZM-LS (Mach-Zehnder Modulator scheme Laser Synthesizer) in ALMA project.

4. Round-trip phase stabilizer

Reference microwave signal or reference laser signal transfer via optical fiber have been researching in many fields (Sato et al. (2000), Daussy et al. (2005), Musha et al. (2006), Foreman et al. (2007)).

4.1 Basic concept of the round-trip phase stabilizer

Figure 11 shows the basic concept of the round-trip phase stabilizer(Kiuchi (2008)) for the two coherent-optical-signals. The optical signals are transmitted in one single-mode fiber. Under the effect of polarization mode dispersion (PMD), the transmission line lengths (the length of the signal path in the optical fiber cable) are different between the two coherent-optical-signals which are transmitted as a set.

The phase of these signals (λ_1 and λ_2 in wavelength) at the starting point of the roundtrip transmission is assumed to be zero, and the phase of these signals which have returned to the starting point are obtained from the following equations: $[(2\pi m) + \phi_1]$ for λ_1, and $[(2\pi n)+\phi_1 +2\Phi]$ for λ_2, respectively, where m and n are integers and Φ is the variable which is controlled by a phase shifter. The signal phase at the middle point of the roundtrip transmission (at the other end of the fiber) can be expressed as follows: For λ_1, $(\phi_1/2)$: m is even or $[(\phi_1/2) + \pi]$: m is odd, and: For λ_2, $[(\phi_2/2) + \Phi]$: n is even or $[(\phi_2/2) + \pi + \Phi]$: n is odd. Therefore, the transmitted signal phase is $(\phi_1/2)-[(\phi_2/2)+\Phi]$ or $(\phi_1/2)-[(\phi_2/2)+\Phi]+\pi$.

If we adjust the phase Φ as follows;

$$\phi_1 = \phi_2 + 2\Phi. \qquad (14)$$

the signal phase at the antenna is the same as or just π different from the signal phase at the starting point of the roundtrip transmission.

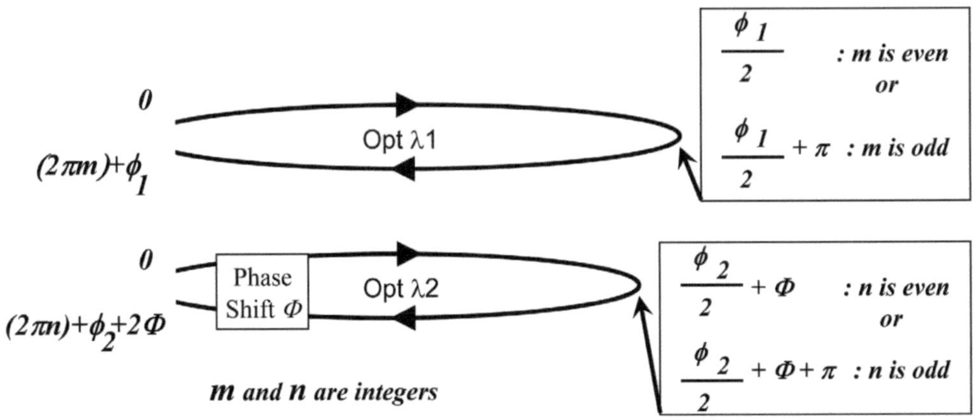

Fig. 11. Basic concept of the round-trip phase stabilizer. The two coherent-optical-signals (λ_1 and λ_2) are transmitted in one single-mode-fiber. Under the effect of PMD, the transmission line lengths (the length of the signal path in the optical fiber cable) are different between the two coherent-optical-signals. The effect of PMD will be expressed in this figure.

4.2 Round-trip optical dual-differential phase measurement scheme

The basic configuration of the system is shown in Figure 12. Signals generated by the two coherent-optical-signals generator in the previous section (Kawanishi et al. (2007),Kiuchi et al. (2007)) are sent to the antennas from the base-station (ground unit), together with PMD

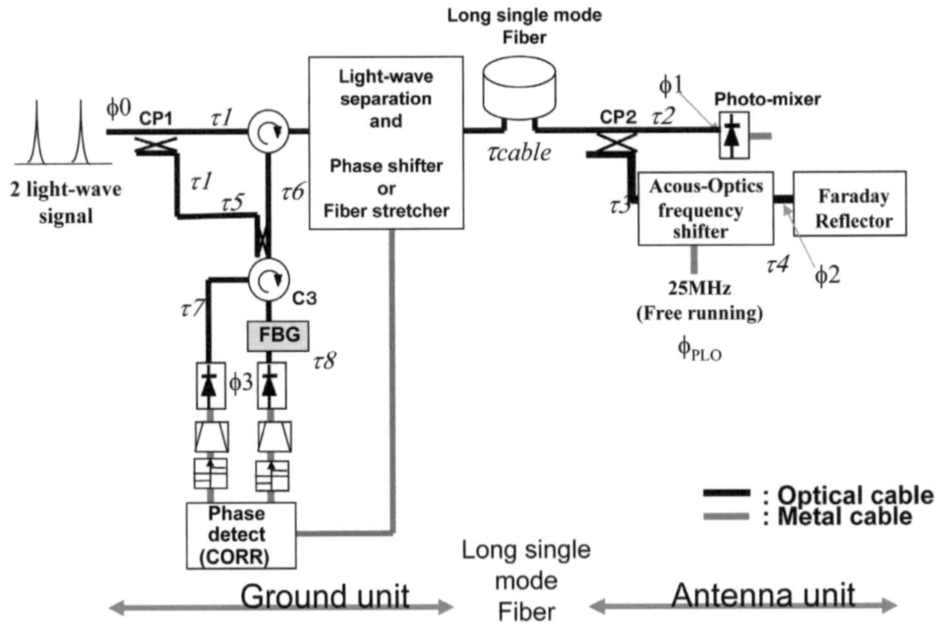

Fig. 12. The round-trip optical phase measurement scheme of the round-trip phase stabilizer.

caused by the rotation and coupling of the fiber cross section signals. At each antenna, frequency-shift modulation (ϕ_{PLO}, angular frequency is ω_t) is performed by the Acoust-Optics frequency shifter for the received optical signals which are then reflected by the optical reflector and returned to the shifter. The signals pass through one path in transmission. The frequency shift modulation is used to distinguish the round-trip signal from back-scattered signals. The phase difference between the signal at the starting point of the roundtrip transmission and the returned signal is detected by Michelson's interferometry to perform correlation of the orthogonal signals which are generated by a 90-degree phase shift of $2\omega_t$ (50 MHz). These orthogonal signals are not required for the phase-lock to the modulation signal at the antenna. Since the modulation frequency ($2\omega_t$) is small, its PMD (the second order PMD) can be ignorable (the estimated deviation value is shown in the next subsection). The round-trip phase measurement method is helpful for successful delay compensation of the microwave signal which is converted from the two coherent-optical-signals by a photo mixer.

In this method, a Faraday-reflector or a mirror both can be used as the reflector at the antenna. In the case of the Faraday reflector, the route of the transmitted and return of light are not completely corresponding. This difference becomes a fixed phase offset. However, the change of the phase offset can be compensated by the phase locked loop. The fixation phase offset does not influence the transmitted phase stability. In the case of using the Faraday rotator and a polarization splitter, it becomes advantageous with respect to the carrier noise ratio. The influence such as back-scattering can be reduced by separating polarization.

4.2.1 Polarization mode dispersion (PMD)

Polarization mode dispersion (PMD) (Agrawal (2002),Derickson (1998)) is the state of polarizations dispersing randomly in the cable. PMD arises from the anisotropic nature of the fiber cross section (θ_x and θ_y). PMD mainly consists of two components 1st and 2nd-order terms. The 1st-order component is differential group delay (DGD), and the 2nd-order components are polarization chromatic dispersion. In contrast to group velocity dispersion, PMD shows temporal change. PMD is caused when the state of polarization of the two coherent-optical-signals is absolutely changed by the movement of the cable through which the signals are transmitted.

We introduce two equations (Eqs. (15) and (17). The variance of differential group delay (Agrawal (2002), Derickson (1998)), can be approximated to be

$$\sigma_\tau = D_p L \tag{15}$$

Where Dp is the fiber PMD parameter of the optical fiber cable [$ps/\sqrt{km}$], and L is the cable length [km]. The variation of the delay will have a standard deviation of 39 fs (15 km fiber) if we choose a fiber with the lowest PMD of 0.01 $ps/\sqrt{km}$.

Second order PMD is the wavelength dependence of the propagation delay in the different polarization modes. The birefringence of the optical fiber cable is wavelength dependent; different wavelengths will cause different types of PMD. The deviation of the propagation delay caused by the second order PMD is as follows (Ciprut et al. (1998));

$$D_2 = \frac{2\pi c D_p^2}{\lambda^2 \sqrt{3}} \Delta_\lambda L \tag{16}$$

Where Δ_λ is the frequency difference between the two coherent-optical-signals. The deviation of the propagation delay caused by the second PMD is calculated as

$$\sigma_{\tau 2} = D_2 \times \Delta_{\lambda max} \times L_{max}. \tag{17}$$

DGD is calculated as the co-variance of the two deviations of the propagation delay. The maximum differential frequency of the two coherent-optical-signals is Δ_{max} = 1.1 *nm*. And when the L_{max} is 15 km, $\sigma_{\tau 2}$ is 0.74 fs.

In the case of the conventional technologies (Cliche & Shillue (2006)), as the round-trip measurement is performed with either one of the two optical signals, the delay on the two signals are compensated commonly by the fiber stretcher using the delay of the measured signal only. On the other hand, in the basic concept of the proposed system (Figures 11 and 12), the delays (σ_τ and $\sigma_{\tau 2}$) of the two signals are considered. The group delay σ_τ acts like a common mode noise to the two coherent-optical-signals. In addition, the round trip delays of the two coherent-optical-signals are measured and compensated independently, taking the differential delay between two coherent-optical-signals into consideration (Figure 11).

4.2.2 Phase relational expression

Firstly, for the phase relationship of the signals in one of the two coherent-optical-signals in Figure 11, the instrumental delay analysis is shown in Figure 12. The suffixes of the equations ($\lambda 1$ and $\lambda 2$) indicate the optical wavelength.

The phase of the optical signal to be transmitted from the two coherent-optical-signals generator is defined as $\phi_0(t)$.

$$\phi_0(t) = \omega_{\lambda 1}(t) + \phi_{\lambda 1}, \tag{18}$$

Where $\omega_{\lambda 1}$ is optical angular frequency, t is time, and $\phi_{\lambda 1}$ is initial/offset phase. If the time delay caused in the roundtrip signal transmission through the optical fiber cable is assumed to be τ_1, τ_{cable} (Figure 12), the received signal phase at the antenna is expressed as $\phi_1(t)$, at the point of the photomixer at antenna.

$$\phi_1(t) = \omega_{\lambda 1}(t - \tau_1 - \tau_{cable} - \tau_2) + \phi_{\lambda 1} \tag{19}$$

At the antenna, the received signals are modulated (frequency-shifted) by a microwave signal ϕ_{PLO} (25 MHz) and sent back to the ground unit through the optical cable.

$$\phi_{PLO}(t) = \omega_c(t) + \phi_c, \tag{20}$$

Where ω_c is a shift angular frequency (25 MHz), and ϕ_c is an initial phase. Frequency-shift of $\phi_{PLO}(t)$ is done by the Acoust-Optics frequency shifter. The signal phase at the reflector on the antenna is as follow;

$$\phi_2(t) = (\omega_{\lambda 1} + \omega_c)t - \omega_{\lambda 1}(\tau_1 + \tau_{cable} + \tau_3 + \tau_4)$$

$$-\omega_c \tau_4 + \phi_{\lambda 1} + \phi_c \tag{21}$$

The signal is reflected by an optical reflector, and returned to the ground-unit via the same cable in reciprocal process.

Differential phase between transmission and reception signals is measured by the Michelson's interferometer. The above equation is established assuming that the signal ($\lambda 1$) is reflected by Fiber Brag Grating (FBG1) and is converted into microwave ϕ_3 by the low-frequency photo mixer to detect $2\omega_c$. The frequency ($2\omega_c$) is selected by a microwave band pass filters.

$$\phi_3(t) = 2\omega_c(t) - \omega_{\lambda 1}(2\tau_{cable} + 2\tau_3 + 2\tau_4 + \tau_6 - \tau_5)$$

$$-2\omega_c(\tau_{cable} + \tau_3 + \tau_4 + \tau_6 + \tau_7) + 2\phi_c \tag{22}$$

This equation means that the roundtrip delay is measured as the optical differential phase of the frequency ($c/\lambda 1$, c: speed of light) after being converted to a microwave angular-frequency ($2\omega_c$).

Secondly, the phase relationship of the other optical signal ($\lambda 2$) can be obtained in conformity with Eqs. (18) to (22). When we use the two coherent-optical-signals, the cable delay is different between $\lambda 1$ and $\lambda 2$ under the effect of PMD. In the following equations, the cable delay in $\lambda 2$ is shown with hat. Initial optical ($\lambda 2$) signal is as follows:

$$\hat{\phi}_0(t) = \omega_{\lambda 2}(t) + \phi_{\lambda 2}, \tag{23}$$

The phase of the optical signal at the antenna is expressed as $\hat{\phi}_1(t)$.

$$\hat{\phi}_1(t) = \omega_{\lambda 2}(t - \hat{\tau}_1 - \hat{\tau}_{cable} - \hat{\tau}_2) + \phi_{\lambda 2} \tag{24}$$

The optical modulation is performed simultaneously for the wavelength of the two signals ($\lambda 1$ and $\lambda 2$) at the antenna, assuming that the optical signal passes through FBG1 and detected as microwave $\hat{\phi}_3(t)$ by the other photo mixer.

$$\hat{\phi}_3(t) = 2\omega_c(t) - \omega_{\lambda 2}(2\hat{\tau}_{cable} + 2\hat{\tau}_3 + 2\hat{\tau}_4 + \hat{\tau}_6 - \hat{\tau}_5)$$

$$-2\omega_c(\hat{\tau}_{cable} + \hat{\tau}_3 + \hat{\tau}_4 + \hat{\tau}_6 + \hat{\tau}_8) + 2\phi_c \tag{25}$$

This equation also means that the roundtrip delay is measured as the optical differential phase of the frequency ($c/\lambda 2$) which is then converted to a microwave angular-frequency $2\omega_c$.

Thirdly, Eq.(26) shows how to obtain the differential phase between $\phi_0(t)$ and $\hat{\phi}_0(t)$ at the starting point of the roundtrip transmission (with the single mode fiber long cable over 10 km).

$$\phi_0(t) - \hat{\phi}_0(t) = \omega_{\lambda 1}(t) - \omega_{\lambda 2}(t) + \phi_{\lambda 1} - \phi_{\lambda 2}, \tag{26}$$

In Eq. (27), the differential phase between $\phi_1(t)$ and $\hat{\phi}_1(t)$ is that of the signal received at the antenna.

$$\phi_1(t) - \hat{\phi}_1(t) = [\omega_{\lambda 1}(t) - \omega_{\lambda 2}(t) + \phi_{\lambda 1} - \phi_{\lambda 2}]$$

$$-\omega_{\lambda 1}(\tau_1 + \tau_{cable} + \tau_2) + \omega_{\lambda 2}(\hat{\tau}_1 + \hat{\tau}_{cable} + \hat{\tau}_2) \tag{27}$$

Comparing Eqs. (26) and (27), it is clear what comprises the instrumental delay. The equation for the phase change (ϕ_d) of the two optical signals caused in the transmission is as follows;

$$\phi_d = -\omega_{\lambda 1}(\tau_1 + \tau_{cable} + \tau_2) + \omega_{\lambda 2}(\hat{\tau}_1 + \hat{\tau}_{cable} + \hat{\tau}_2). \tag{28}$$

On the other hand, half of the differential phase between $\phi_3(t)$ and $\hat{\phi}_3(t)$, or the double-difference between the signals before/after the roundtrip transmission is as follows;

$$\frac{\phi_3(t) - \hat{\phi}_3(t)}{2} = -\omega_{\lambda 1}(\tau_{cable} + \tau_3 + \tau_4 + \frac{\tau_6 - \tau_5}{2})$$

$$+\omega_{\lambda 2}(\hat{\tau}_{cable} + \hat{\tau}_3 + \hat{\tau}_4 + \frac{\hat{\tau}_6 - \hat{\tau}_5}{2})$$

$$-\omega_c(\tau_{cable} + \tau_3 + \tau_4 + \tau_6 + \tau_7$$

$$-\hat{\tau}_{cable} - \hat{\tau}_3 - \hat{\tau}_4 - \hat{\tau}_6 - \hat{\tau}_8). \tag{29}$$

If this differential phase is compensated, the coherent transmission from the ground unit to the antenna can be realized. To compare Eq. (28) and Eq. (29), the term $(-\omega_{\lambda 1}\tau_{cable} + \omega_{\lambda 2}\hat{\tau}_{cable})$ is compensated by Eq. (29) (measured data).
The residual phase in this method is as follows:

$$Residual\ phase =$$

$$-\omega_{\lambda 1}(\tau_3 + \tau_4 + \frac{\tau_6 - \tau_5}{2}) + \omega_{\lambda 2}(\hat{\tau}_3 + \hat{\tau}_4 + \frac{\hat{\tau}_6 - \hat{\tau}_5}{2}) \tag{30}$$

$$-\omega_c(\tau_{cable} - \hat{\tau}_{cable} + \tau_3 - \hat{\tau}_3 + \tau_4 - \hat{\tau}_4 + \tau_6 - \hat{\tau}_6) \tag{31}$$

$$-\omega_c(\tau_7 - \hat{\tau}_8). \tag{32}$$

Lastly, the meanings of these equations are described below.
Eq.(30) shows the second order PMD of the cable whose length is $(\tau_3 + \tau_4 + (\tau_6 - \tau_5)/2)$,
Eq.(31) shows the ω_c(25 MHz) phase drift equivalent to the second order PMD of the cable length obtained by $(\tau_{cable} + \tau_3 + \tau_4 + \tau_6)$,
Eq.(32) shows the ω_c (25 MHz) phase difference equivalent to the phase drift of the cable length obtained by $(\tau_7 - \hat{\tau}_8)$.

Equations (30) and (32) are ignorable: the change of the differential delay is ignorable, because the length of $[\tau_3, \tau_4, \tau_5, \tau_6, \tau_7$ *and* $\tau_8]$ is a few meters and not long enough to cause problems. Equation (31) is almost equal to Eq. (33).

$$\omega_c(\tau_{cable} - \hat{\tau}_{cable}) \tag{33}$$

This value is the inevitable error of this method. According to the Equations (15) and (17), the offset frequency in the round-trip signal ($2 \times \omega_c$=50 MHz) is $\Delta_{\lambda offset}$ = 0.0004 *nm*. The deviation of the propagation delay $\sigma_{\tau 2}$ is less than 0.003 fs, which is very small.

In the process, the effect of the first and second order PMD can be reduced by using double-difference of the independently measured phases of two optical signals of round-trip measurement.

As a result, we can measure the instrumental delay phase (twice of the cable delay phase). Moreover, this method does not require the transmission of the modulation signal (ω_L), which means we do not have to consider any phase delay of the modulation signal (ω_L). The measured phase is used to compensate the instrumental delay change and phase change.

4.3 Two optical signal separation and optical phase control scheme

If we use a fiber stretcher that stretches the two signals together and performs phase shift on both of them, it is hard to get an enough dynamic range of the phase shift. On the other hand, in our basic concept (Fig. 11) using a phase shifter (General Photonics FPS-001) instead of the fiber stretcher, the phase shift is performed on only one of the two optical signals (λ_1 and λ_2). Figure 13 shows the execution example. Transmission delay on the fiber is measured as the differential phase of the optical round-trip delay of each lightwave signal. At first, the two coherent-optical-signals have a vertical and high-extinction ratio polarization. In a series of processing in the ground unit, the polarization is maintaining. The signal flow is shown in Figure 13.

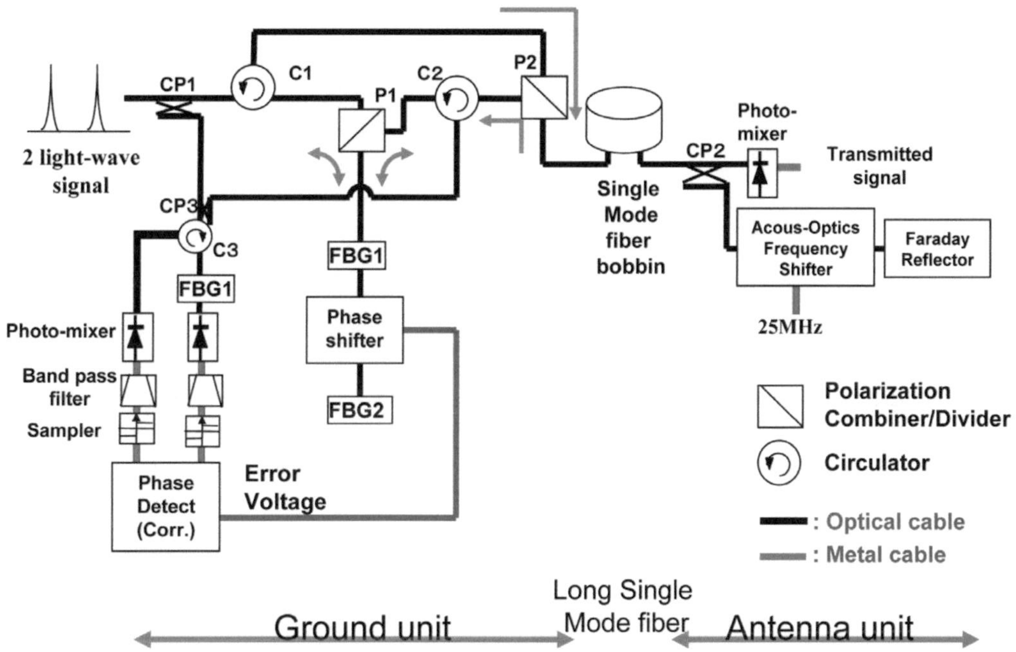

Fig. 13. Two optical signal separation and optical phase control scheme. Where CP1, CP2, CP3: optical coupler, C1, C2, C3: circulator, P1, P2: polarization beam splitter, and FBG1, FBG2: fiber bragg grating

The character in parentheses means an optical device in the figure. The signal, passing through the optical coupler (CP1), circulator (C1), and polarization beam splitter (P1), is divided into two wavelengths (λ_1 and λ_2). Wavelength λ_1 signal is reflected by a fiber bragg

grating (FBG1) and returned to the beam splitter (P1), while wavelength λ_2 signal is reflected by a fiber bragg grating (FBG2) and returned to the beam splitter (P1) via the phase-shifter. The returned light-waves are recombined at the beam splitter (P1) and sent to the circulator (C1), and then, to the polarization beam splitter (P2). The signal is divided into two signals at the optical coupler (CP2) after passing through a long single-mode fiber. One of the divided signals is converted to a millimeter wave by a photo-mixer, and the other signal is reflected by a Faraday reflector after the frequency shift by an optical frequency shifter (Acous-Optics frequency shifter). The reflected signal is converted into a 90-degree different optical polarization signal by the Faraday reflector. The signal, after passing through the frequency shifter again, is returned back to the polarization beam splitter (P2) in the ground unit. As the signal goes through the optical reciprocal process, the received signal has a horizontal (90-degree different polarization angle to the transmission signal) polarization at this point. After passing through the circulator (C2) and the beam splitter (P1), the signal is divided into two wavelengths (λ_1 and λ_2) again. As described above, wavelength λ_1 signal is reflected by the fiber bragg grating (FBG1) and returned to the beam splitter (P1), while wavelength λ_2 signal is reflected by FBG2 and returned to the beam splitter (P1) via the phase-shifter. The returned light-waves are recombined again at the beam splitter (P1) and sent to the circulator (C2) because the optical polarization is horizontal. And, finally the signal is recombined with the divided transmission signal at the optical coupler (CP3).
The differential phases on the angular frequency $2\omega_c$ between transmission and round-trip signals on each light-wave signal are detected by low-frequency photo-mixers after wavelength separation by the FBG1 optical filter (see previous Section). These measured phases are equivalent to the round-trip phases on both lightwave signals. In the proposed method, the transmitted signal will be stabilized by controlling the differential phase on the measurement signals to zero.
According to our experiments, a polarization controller is put into place between P1 and FBG1 produce a good effect.

4.4 Laboratory tests

A block diagram of the performance measurement system is shown in Figure 14. A set of the two coherent-optical-signals generated is divided into two signals: one is transmitted to the phase stabilizer system and the other to the photo mixer (Nippon Telephone and Telegraph (NTT) unitraveling-carrier photo-diode(Hirota et al. (2001),Ito et al. (2000))) as a reference signal. The signal passes through a 10-km Single-Mode Fiber cable with/without the phase stabilizer.

4.4.1 Phase stability measurements (Laboratory test)

We measured the phase stability (Fig. 15) of the transmitted signal (80 GHz) at the antenna through the single mode fiber cable (10 km) in the time-domain Allan STD method(Allan (1966), Allan (1976)) by a time interval analyzer: TSC-5110A. The measurements were conducted with/without the phase stabilizer to check the improvement of the phase stabilizer in the interferometric system.
When the optical signal (80 GHz) is transmitted through the single mode fiber cable (10 km), the phase stability begins to degrade around 10 seconds integration time. In the case of using the phase stabilizer, the degradation of the phase stability is staved off. The measured phase noise is the white phase noise.

Dynamic range of this method was measured by using a manual controlled air-gap stretcher which was inserted between the Ground unit and the single mode fiber spool in Figure 14. The measured dynamic range was larger than 5 cm.

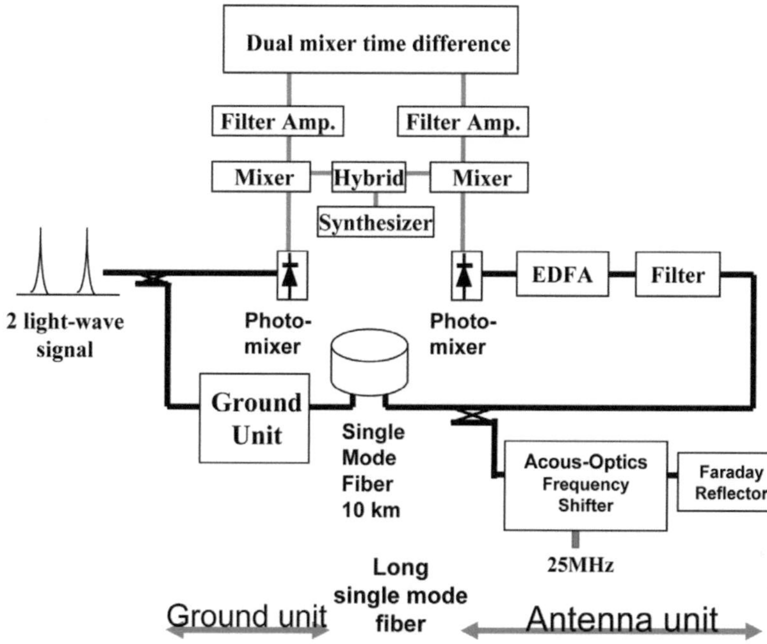

Fig. 14. A block diagram of phase stability measurement system. The signal is provided from the two coherent-optical-signals generator.

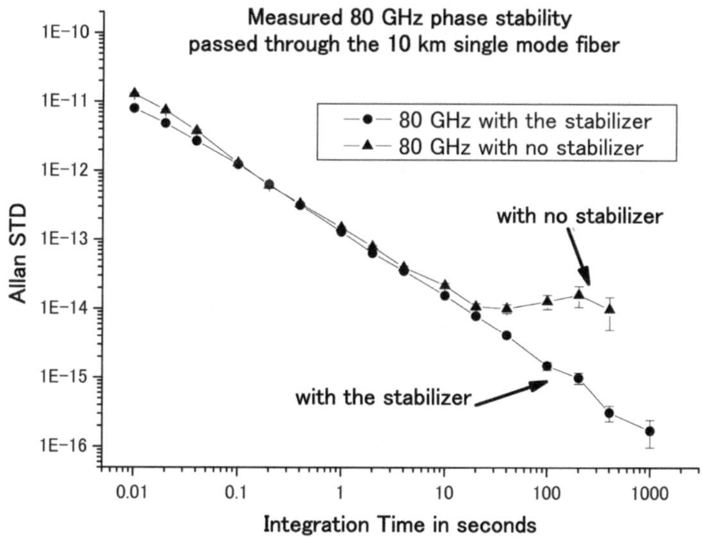

Fig. 15. The 80 GHz phase stability that passed through the 10 km fiber. The phase stability begins to degrade around 10 seconds integration time. In the case of using the phase stabilizer, the degradation of the phase stability is staved off.

4.4.2 Phase stability measurements (Field test)

In ALMA OSF (Operations Support Facility: 2900m sea level), there are built-up antennas and a Holography system which measures the antenna surface accuracy. The Photonic system field test was carried out using the Holography signal and two antennas (Antenna-1 and Antenna-4). The experiment block diagram is shown in Figure 16. The Holography transmitter, Antenna-2, Antenna-1 and Antenna-4 are standing in a low. Therefore the received Holography signals at Anttena-1 and Antenna-4 are blocked by Antenna-2, the received Holography signal levels are very weak. However Anttena-1 and Antenna-4 can receive the Holography signal simultaneously.

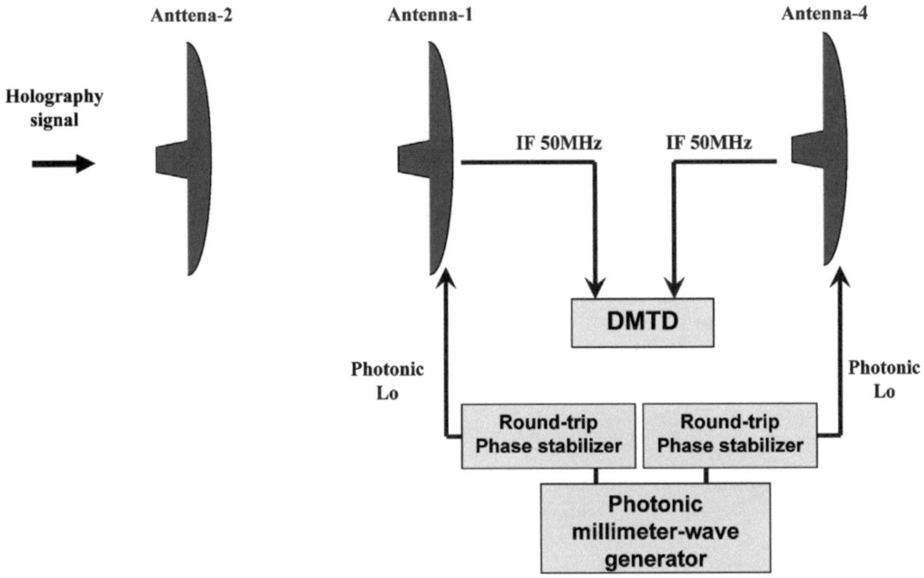

Fig. 16. Block diagram of the phase stability measurement experiment with the Holography transmitter. The Holography transmitter faces Antenna-2, with Antenna-1 and Antenna-4 aligned behind antenna-2. Two antennas can receive the Holography signal simultaneously.

In this experiment, the Holography signal was the common signal. Differential phase of Holography signal between Anttena-1 and Antenna-4 was measured. Received Holography signals (104.02 GHz) were converted down to intermediate frequency (IF: 50 MHz) signals by using the provided photonic signal from the Photonic millimeter-wave generator via the Round-trip phase stabilizer. The differential signal phase of these 50 MHz signals are measured by DMTD method. The measured phase stability is shown in Figure 17. The phase noise of 10^{-13} in White-PM noise was obtained, which is the covariance phase noise of two antenna system.

4.5 Verification results

In the ALMA Specification, instrumental delay/phase error on the 1st Local oscillator should be 53 fs (rms) in the short term, and long term drift should be 17.7 fs between 10 sec averaging at intervals of 300 seconds: $\sigma_y(1\ sec) < 9.2 \times 10^{-14}$. On the other hand, in the very long baseline interferometer (Rogers & Moran (1981), Rogers et al. (1984)) (VLBI), the requirements of 320 GHz are as follows: $\sigma_y(1\ sec) < 2 \times 10^{-13}$, $\sigma_y(100\ sec) < 1.3 \times 10^{-14}$ and $\sigma_y(1000\ sec) < 3 \times 10^{-15}$.

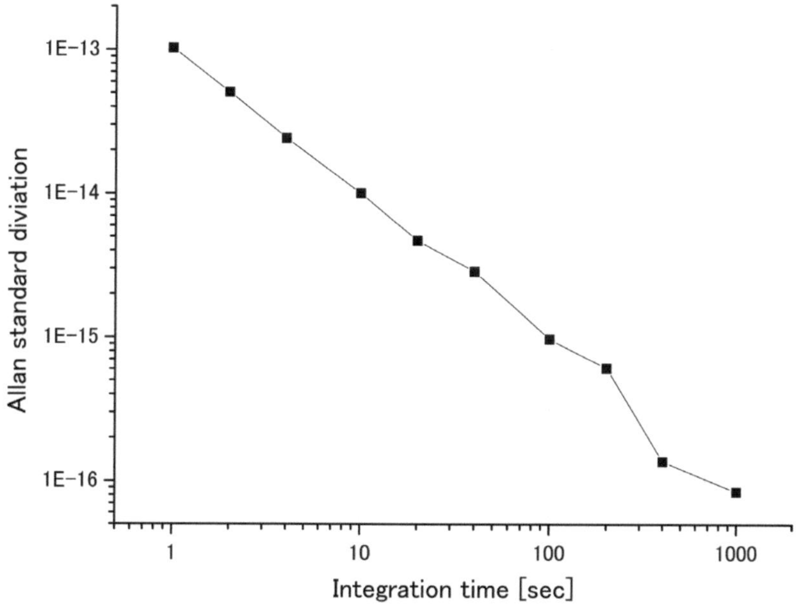

Fig. 17. The measured phase stability is measured by DMTD. The measured stability shows the co-variance of two antenna system stability. Received Holography signal (104.02 GHz) was converted down to 50 MHz by using the provided photonic signal from the Photonic millimeter-wave generator via the Round-trip phase stabilizer.

The verifications matrix is shown in Table 2. The measured values meet the ALMA specifications.

ALMA Specifications	Measured Values
53 fs (rms) in short-term 17.7 fs in long-term	42 fs (short) 13 fs (long)
Required Phase stability for VLBI $\sigma_y(1sec) < 9.2 \times 10^{-14}$ $\sigma_y(100sec) < 1.3 \times 10^{-14}$ $\sigma_y(1000sec) < 3 \times 10^{-15}$	Measured Phase stability $\sigma_y(1sec) < 7.3 \times 10^{-14}$ $\sigma_y(100sec) < 1.79 \times 10^{-15}$ $\sigma_y(1000sec) < 2.29 \times 10^{-16}$

Table 2. Verifications matrix.

5. Conclusion

Based on our experiment results, we propose a new high carrier suppression optical doublesideband intensity modulation technique using the integrated *LiNbO*$_3$ Mach-Zehnder modulator which is capable of compensating the imbalance of the Mach-Zehnder arms with a pair of active trimmers (null-bias operation mode). The full-bias point operation mode introduced in this paper is also a novel modulation technique for the second-order harmonic generation. The Mach-Zehnder modulator can generate two coherent light waves with frequency difference equivalent to four times the modulation frequency. Photonic local signals of 120GHz can also be generated using this technique.

The two spectral components of the two optical signals generated with this technique are phase-locked without using any complicated feedback control. All of the measurements were carried out on a table (without vibration isolation) in a normally air-conditioned room without acoustic noise isolation. In short, all of the measurements were performed under normal environment. Temperature change and mechanical vibrations may have affected the output lightwaves to some degree, however there was no chaotic phenomenon such as mode hopping or mode competition during the experiments. Based on these results, we concluded that the proposed techniques will be useful to construct a robust, low-cost and simple setup for the photonic local signals.

Compensation of the Local signal transmission delay is an indispensable technique for accurate interferometrical observation. PMD delay, which is caused during the signal transmission, needs to be reduced because it deteriorates the accuracy of the delay amount by affecting the signal polarization and wavelength. The two coherent-optical-signals generator (Kawanishi et al. (2007), Kiuchi et al. (2007)) is required to help stabilization of polarization, and to maintain the high extinction ratio, and to keep the signal state of polarizations in stable condition for preventing the delay generation.

We proposed the double difference phase measurement method. The method is also available to use the fiber stretcher instead of the phase shifter. The Double-difference method is more robust to external influences and more accurate than the current scheme which uses one of the two optical signals for measurement. This method can reduce deterioration in the signal phase stability caused by the long fiber signal transmission.

The performance advantages of the system are:

1. The modulation signal (ω_t) transmission and its phase compensation are not required (the modulation signal on the antenna is generated by a free-running oscillator);
2. External noise (acoustic noise, vibration noise) on the long single mode fiber cable is dealt with a common noise; and
3. The PMD problems are reduced, as the round trip delays on the two optical signals are measured and compensated independently.

6. References

Alferness,R.C.(1982), Waveguide electro-optic modulators, *IEEE Trans. Microw. Theory Tech.*, vol.30, no.8, pp.1121-1137.

Agrawal,G.P.(2002), *Fiber optic communication system, third edition*, JohnWiley & Sons Inc.

Allan, D.W. (1966), Statistics of Atomic Frequency Standards, *Proc. IEEE*, vol.54, pp.221.

Allan, D.W.(1976), Report on NBS dual mixer time difference system (DMTD) built for time domain measurements associated with phase 1 of GPS, *NBS IR*, vol.75, pp.827.

Ciprut,P., Gisin,B., Gisin,N., Passy,R., Von der Weid,J.P., Prieto,F., & Zimmer,C.w.(1998), Second-order polarization mode dispersion: Impact on analog and digital transmissions, *IEEE Journal of lightwave technology*, vol.16, no.5, pp.757-771.

Cliche,J., & Shillue,B.(2006), Precision timing control for radioastronomy, maintaining femtosecond synchronization in Atacama Large Millimeter Array, *IEEE control system magazine*, pp.19-26.

Daussy,C., Lopez,O., Amy-Klein,A., Goncharov,A., Guinet,M., Chardonnet,C., Narbonneau, F., Lours,M., Chambon,D., Bize,S., Clairon,A., & Santarelli,G.(2005), Long-Distance Frequency Dissemination with a Resolution of 10^{-17}, *Physical review letters*, vol.94, 203904.

Derickson,D., *Fiber optics test and measurement*, Prentice Hall PTR.

Fonseca,D.D., Monteiro,P., Cartaxo,A.,V.,T., & Fujita,M.(2004), Single sideband demonstration using a four phase-modulators structure, *2004 IEEE/LEOS workshop on advanced modulation formats*, FC2.

Foreman,S.M., Ludlow,A.D., de Miranda,M.H.G., Stalnaker,J.E., Diddams,S.A., & Ye,J.(2007), Coherent optical phase transfer over a 32-km fiber with 1-s instability at 10^{-17}, *Physics review letters*,DOI:10.1103/PhysRevLett.99.153601.

Healey, D.J.III (1972), Flicker of frequency and phase and white frequency and phase; Fluctuations in frequency sources, *Proc. 25th Annu. Sympo. On Frequency Control (ASFC)*, pp.29-42.

Hirota,Y., Ishibashi,T., & Ito,H.(2001), 1.55-um wavelength periodic traveling-wave photodetector fabricated using unitraveling-carrier photodiode structures, *IEEE J. of Lightw. Technol*, vol.19, pp.11.

Ito,H., Furuta,T., Kodama,S., & IshibashiT.(2000), InP/InGaAs uni-travelling-carrier photodiode with 310 GHz bandwidth, *Electron. Lett.*, vol.38, no.21, pp.1809-1810.

Izutsu,M., Yamane,Y., & Sueta,T.(1977), Broad-band travering-wave modulator using *$LiNbO_3$* optical waveguide, *IEEE J. Quantum Electron.*, vol.13, no.4, pp.287-290.

Izutsu,M., Shikamura,S., & Sueta,T.(1981), Integrated optical SSB modulator/frequency shifter, *J. Quantum Electron.*, vol.17, pp.2225-2227.

Jiang,Q., & Kavehrad,M.(1993), A Sub-carrier-Multiplexed Coherent FSK System Using a Mach-Zehnder Modulator with Automatic Bias Control, *IEEE/LEOS Photonics Tech. Let. Journal*, vol.5, no.8, pp.941-943.

Kawaguchi, N. (1983), Coherence loss and delay observation error in Very-Long-Baseline Interferometry, *J. Rad. Res. Labs.*, vol.30, no.129, pp.59-87.

Kawanishi,T., Sakamoto,T., Shinada,S., Izutsu,M., Higuma,K., Fujita,T., & Ichikawa,J.(2004a), *$LiNb_3$* high-speed optical FSK modulator, *Electron. Lett.*, vol.40, pp.691-692.

Kawanishi,T., & Izutsu,M.(2004b), Linear single-sideband modulation for high-SNR wavelength conversion, *Photon. Technol. Lett.*, vol.16, pp.1534-1536.

Kawanishi,T., Kiuchi,H., Yamada,M., Sakamoto,T., Tsuchiya,M., Amagai,J., & Izutsu,M.(2005), Quadruple frequency double sideband carrier suppressed modulation using high extinction ratio optical modulators for photonic local oscillators, *MWP 2005*, PDP03.

Kawanishi,T., Sakamoto,T., Miyazaki,T., Izutsu,M., Fujita,T., Mori,S., Higuma,K., & Ichikawa,J. (2006), High-speed optical DQPSK and FSK modulation using integrated Mach-Zehnder interferometers, *Optics Express*, vol.14, no.10, pp.4496-4478.

Kawanishi,T., Sakamoto,T., & Izutsu,M. (2007), High-speed control of lightwave amplitude phase and frequency by use of electrooptic effect, *IEEE Journal of selected topics in quantum electronics*, vol.13, no.1, pp.79-91.

Kiuchi,H., Kawanishi,T., Yamada,M., Sakamoto,T., Tsuchiya,M., Amagai,J. & Izutsu,M.(2007), High Extinction Ratio Mach-Zehnder Modulator Applied to a Highly Stable Optical Signal Generator, *IEEE Trans. Microwave Theory and Techniques*, vol.55, no.9, pp.1964-1972.

Kiuchi,H.(2008), Highly stable millimeter-wave signal distribution with an optical round-trip phase stabilizer, *IEEE Trans. Microwave Theory and Techniques*, vol.56, no.6, pp.1493-1500.

Lim,C., Nirmalathas,A., Novak,D., & Waterhouse,R.(2000), Optimisation of baseband modulation scheme for millimeter-wave fibre-radio systems, *Electron. Lett.*, vol.36, pp.442-443.

Musha,M., Sato,Y., Nakagawa,K., Ueda,K., Ueda,A., Ishiguro,M., Robust and precise length stabilization of a 25-km long optical fiber using an optical interferometric method with a digital phase-frequency discriminator, *Appl. Phys. B*, vol.82, pp.555-559.

Onillon,B., Constant,S., & Llopis,O., Optical links for ultra low phase noise microwave oscillators measurement, *IEEE frequency control symposium*.

Rogers, A.E.E. & Moran, J.M. (1981), Coherence limits for very-long-baseline interferometry, *IEEE Trans. Instrum. Meas.*, vol.30, no.4, pp.283-286.

Rogers, A.E.E., Moffet, A.T., Backer ,D.C. & Moran, J.M. (1984), Coherence limits in VLBI observation at 3-millimeter wavelength, *Radio Science*, vol.19, no.6, pp.1552-1560.

Sakamoto,T., Kawanishi,T., & Izutsu,M.(2005), Optical minimum-shift-keying with external modulation scheme, *Opt. Exp.*, vol.13, pp.7741-7747.

Sato,K., Hara,T., Kuji,S., Asari,K., Nishio,M., & Kawano,N.(2000), Development of an ultra stable fiber optic frequency distribution system using an optical delay control module, *IEEE Trans. Inst. Meas.*, vol.49, no.1, pp.19-24.

Vanblerkom,R., & Aneman,S.,L.(1966), Considerations for the short term stability of frequency multiplication, *IEEE AES-2*, vol.1, pp.36-47.

Vegas Olmos,J.,J., Tafur Monroy,I., & Koon,A.M.J.(2003), High bit-rate combined FSK/IM modulated optical signal generation by using GCSR tunable laser sources, *Opt. Exp.*, vol.11, pp.3136-3140.

22

Quantum Direct Communication

Gui Lu Long[1,2], Chuan Wang[1,3], Fu-Guo Deng[4], and Wan-Ying Wang[1]
[1]Key Laboratory of Atomic and Molecular Nanosciences and Department of Physics, Tsinghua University, Beijing 100084,
[2]Tsinghua National Laboratory for Information Science and Technology, Beijing 100084,
[3]School of Science and Key Laboratory of Optical Communication and Lightwave Technologies, Beijing University of Posts and Telecommunications, Beijing, 100876,
[4]Department of Physics, Beijing Normal University, Beijing 100875,
People's Republic of China

1. Introduction

Quantum key distribution (QKD) is considered as an ideal method to make secret message unreadable to eaves-dropper but intelligible to the two authorized parties of the communication [1-6]. In the research of experimental quantum key distribution, single photons and entangled photon pairs are used as the carriers. In quantum key distribution, secret keys are generated first between the communication parties. The security of quantum key distribution is guaranteed by the laws of quantum mechanics. After the quantum key distribution is completed, the communication parties should share secret keys, then the sender encrypts the secret message using the secret keys to form the ciphertext and transmits the ciphertext through a classical channel. The receiver receives the ciphertext and then decrypt the ciphertext to get the secret message. Altogether there are four steps in a secret communication process with QKD: key generation, encryption, transmission and decryption.

Here in this review, we review some new development in quantum communication, quantum direct communication (QDC). Quantum direct communication is a form of quantum communication where secret messages can be transmitted through a quantum channel with or without additional classical communications. There are two forms of quantum direct communication, quantum secure direct communication (QSDC) [7{9] and deterministic secure quantum communication (DSQC) [10, 11]. In QSDC, secret messages are transmitted directly between the communication parties, from sender Alice to receiver Bob, without additional classical communication except those for the necessary eavesdropping check. In other words, the quantum key distribution process and the classical communication of ciphertext are condensed into one single quantum communication procedure in QSDC. Deterministic secure quantum communication is another type of quantum direct communication, such as those proposed in Ref. [10, 11], where classical communication is required in order to read out the secret message. As mentioned earlier, to complete a secure communication with the help of QKD, one usually encodes the secret message with an encryption scheme, and the ciphered text is transmitted through a classical channel. With a quantum channel, this procedure can be varied. For instance, Alice can encrypt her secret message with a random key and encodes the ciphertext into the quantum states of the information carriers. The ciphertext is then sent from Alice to Bob deterministically. Alice also sends the random key to Bob through a classical channel. With this knowledge, Bob can decode the message from the ciphertext obtained through the quantum communication. Quantum principle ensures that Eve cannot steal the ciphertext. Because the ciphertext needs be transmitted through a quantum channel deterministically, not all quantum key distribution can be adapted to construct DSQC. Only deterministic QKD schemes can be adapted for DSQC purposes. The fundamental difference between QSDC and DSQC is the need of another round of classical communication. Hence it is always possible to use a QSDC scheme as a DSQC scheme.

The first QSDC protocol is the two-step QSDC protocol where qubits in an EPR pair are sent from one user to another user in two steps [8, 9]. The two-step QSDC protocol was first proposed by Long and Liu in 2001 [8], and standardized and analyzed by Deng, Long and Liu in 2003 [9]. Another QSDC protocol is the ping-pong QSDC protocol where one qubit of an EPR pair is sent from one user to another and then back to the sender again like the ping-pong. While the two-step QSDC protocol uses all four dense coding operations, the ping-pong protocol uses only two of the four dense coding operations. In another development, Shimizu and Imoto proposed the first DSQC protocol using entangled photon pairs [10]. In their scheme, the ciphertext is encoded in the state of entangled pairs, and the photons are transmitted from Alice to Bob. The receiver Bob performs a Bell-basis measurement to read out the partial information. Full information of the ciphertext is read out after Alice notifies him the encoding basis through a classical communication. In 2002, Beige et al. [11] proposed another DSQC scheme based on single photon two-qubit states. The message can be read out only after a transmission of an additional classical information for each qubit. In recent years, quantum direct communication has attracted extensive interests and many interesting and important works have been carried out in QSDC for instance in Refs. [12-30], and in DSQC for instance in Refs. [31-39]. In the following sections, we will focus on the development of these two forms of quantum direct communication. We will also discuss their applications, such as in quantum secret sharing and quantum network.

2. Deterministic secure quantum communication protocols

As mentioned above, there are two kinds of deterministic schemes. One is quantum secure direct communication (QSDC) in which the receiver can read out the secret message directly, and classical information is exchanged between the two parties of quantum communication only for security checking. The other is called deterministic secure quantum communication (DSQC) [31] in which the receiver can read out the secret message by exchanging at least an additional bit for each qubit, i.e. classical communication is needed besides eavesdropping check. To some extent, DSQC process is similar to the QKD protocol which is used to create a random key first and then use it to encrypt the message. In the following, we will describe some DSQC protocols.

A. DSQC with nonmaximally entangled states

We describe here two DSQC protocols without using maximally entangled states which was proposed by Li et al. [31], following some ideas in the delay-measurement quantum communication protocol [40]. It utilizes the pure entangled states as quantum information carriers, called the pure-entanglement-based DSQC, and the other one makes use of the d-dimensional single photons, called the single-photon-based DSQC. Both of them introduce the decoy photons [41, 42] for security checking and only single-photon measurements are required for the two communication parties.

The two parties use pure entangled states as the quantum information carries in the pure-entanglement-based DSQC protocol [31]. Also this protocol assumes that the receiver has the capability of making single-particle measurements. The pure entangled states can be described as

$$|\Psi'\rangle_{AB} = a|0\rangle_A|1\rangle_B + b|1\rangle_A|0\rangle_B \quad (1)$$

where the subscript A and B indicate the two correlated photons in each entangled state. $|0\rangle$ and $|1\rangle$ are the two eigenvectors of the two-level operator σ_z, say the basis Z. a and b satisfy the relation $|a|^2 + |b|^2 = 1$.

Firstly, the sender, say Alice prepares a sequence of ordered N two-photon pairs, and each pair is randomly in one of the two pure entangled states $|\Psi'\rangle_{AB}$, $|\Psi''\rangle_{AB}$, and

$$|\Psi''\rangle_{AB} = a|1\rangle_A|0\rangle_B + b|0\rangle_A|1\rangle_B. \quad (2)$$

Alice picks up A particles to form an ordered sequence S_A and picks up the other partner photons to form the sequence S_B. For security checking, Alice replaces some photons in the sequence S_B with her decoy photons S_{de} which are produced randomly in one of the four states $\{|0\rangle, |1\rangle, |+\rangle, |-\rangle\}$. Here $|\pm\rangle = \frac{1}{\sqrt{2}}(|0\rangle \pm |1\rangle)$ are the two eigenvector of the two-level operator σ_x, say the basis X. The decoy photons is easily prepared from the pure entangled quantum system $|\Psi\rangle_{AB}$ by taking a single-photon measurement on the photon A and manipulating the photon B with some unitary operations. Secret message is encoded on the photons in S_B sequence by performing $I = |0\rangle\langle 0| + |1\rangle\langle 1|$ or $\sigma_x = |1\rangle\langle 0| + |0\rangle\langle 1|$ at Alice's side and the two unitary operations represent classical bits 0 or 1, respectively. Then Alice sends sequence S_B to Bob. After Bob receives S_B sequence, Alice and Bob check the security of communication by measuring the decoy photons and comparing the outcomes. If the error rate is lower than the security bound, Alice and Bob measure their remaining photons with basis Z, and they get the final results R_A and R_B, respectively. Alice announces her results R_A. Then Bob reads out the secret message M_A as $M_A = R_A \oplus R_B \oplus 1$. As this scheme requires only single-photon measurements and pure entangled quantum signals, it is far more convenient than the schemes with entanglement swapping and quantum teleportation, and it is more feasible in practice. In this protocol, the information carriers in two-particle pure entangled states can be prepared in experiment easily with present technology, and a single-photon measurement is simpler than a multi-particle joint measurement at present. This protocol is also generalized to the case with d-dimensional quantum systems [31]. The intrinsic efficiency approaches 100% and the total efficiency exceeds $\frac{1}{3}$ in theory which is larger than congeneric schemes using Einstein-Podolsky-Rosen (EPR) pairs.

B. DSQC with single photons

In the single-photon-based DSQC protocol [31], d- dimensional single-photon quantum systems are utilized as the information carriers. The Z_d basis of a d- dimensional system is

$$|0\rangle\,,\quad |1\rangle\,,\quad |2\rangle\,,\ \cdots,\quad |d-1\rangle\,. \tag{3}$$

The d-dimensional eigenvectors of the measuring basis X_d are

$$\begin{aligned}
|0\rangle_x &= \frac{1}{\sqrt{d}}\left(|0\rangle + |1\rangle \ + \cdots \ + |d-1\rangle\right),\\
|1\rangle_x &= \frac{1}{\sqrt{d}}\left(|0\rangle + e^{\frac{2\pi i}{d}}|1\rangle + \cdots + e^{\frac{(d-1)2\pi i}{d}}|d-1\rangle\right),\\
|2\rangle_x &= \frac{1}{\sqrt{d}}\left(|0\rangle + e^{\frac{4\pi i}{d}}|1\rangle + \cdots + e^{\frac{(d-1)4\pi i}{d}}|d-1\rangle\right),\\
&\cdots\cdots\cdots\cdots\cdots\cdots\\
|d-1\rangle_x &= \frac{1}{\sqrt{d}}(|0\rangle + e^{\frac{2(d-1)\pi i}{d}}|1\rangle + e^{\frac{2\times 2(d-1)\pi i}{d}}|2\rangle\\
&\quad + \cdots + e^{\frac{(d-1)\times 2(d-1)\pi i}{d}}|d-1\rangle).
\end{aligned} \tag{4}$$

At first, sender Alice prepares a sequence of d-dimensional single-photons randomly in the eigen-basis states of Z_d or X_d operators, the sequence is labeled as S. She chooses some photons in the S-sequence as the decoy photons and encrypts her secret message M_A on the other photons with unitary operations U_m, U_m^x, where

$$U_m = \sum_j |j + m \bmod d\rangle\langle j|, \tag{5}$$

$$U_m^x = \sum_j e^{\frac{2\pi i}{d}jm}|j + m \bmod d\rangle\langle j|. \tag{6}$$

In other words, Alice encodes her message with U_m if the photon is prepared with the Z_d basis. Otherwise, she will encode the message with U_m^x. Then Alice sends the S sequence to Bob. After the transmission, they check the eavesdropping by measuring the decoy photons and analyzing the error rate. If the transmission is secure, Alice tells Bob the original states of the photons. Then Bob measures them with the suitable bases and reads out the secret information M_A with his outcomes. This protocol is more convenient in practical applications in virtue of that it only requires the parties to prepare and measure single photons.

C. DSQC with quantum teleportation and entanglement swapping

Quantum teleportation [43] has been studied widely since it was first proposed in 1993, and has been applied in some other quantum communication branches, such as QKD, quantum secret sharing (QSS) and so on. In 2004, Yan et al. put forward a DSQC scheme using EPR pairs and quantum teleportation [32]. In their scheme, the qubits do not carry the secret message when they are transmitted between the two parties, and this makes this communication more secure and convenient for post-processing such as privacy amplification.

At first, the two parties share a set of entangled pairs randomly in one of the four Bell states. Suppose that all the EPR pairs used in the scheme are $|\phi^+\rangle_{AB}$. The sender Alice prepares a sequence of C particles in the X basis $|\psi\rangle_C$ according to her secret message ($|+\rangle$ for "0", $|-\rangle$ for "1"). Then Alice performs Bell-state measurements on her two particles BC. Each outcome will appear with equal probability 0.25 and Bob's particles will be related to the initial states of particles C by a unitary transformation U_{ij} relying on Alice's measurement outcomes. After Alice publicly announces her out-comes, Bob applies the corresponding inverse transformation U_{ij}^{-1} to his particles and measures them with the basis $X \equiv \{|+\rangle, \{|-\rangle\}$. Then Bob can obtain Alice's message. The security of this scheme is ensured because the security of quantum channel is ensured before the trans- mission of secret message, hence it is completely secure.

Subsequently, Gao et al. proposed another direct secure quantum communication scheme using controlled teleportation [20]. Three-particle entangled states are used in this scheme. When the communication starts, the three parties first share a set of entangled states. The sender Alice performs a Bell-state measurement on a information particle and a particle in the entangle state, and the controller Charlie performs a single-particle measurement. According to their measurement outcomes, the receiver Bob chooses a suitable unitary operation and then takes a single-particle measurement on his particle for reading out the secret message.

Entanglement swapping is also exploited to design a deterministic secure quantum communication protocol [35]. The protocol also uses the maximally entangled EPR pairs as the information carriers. The two parties assume that each of the four unitary operation represents a two-bit classical information beforehand. Bob prepares a series of EPR pairs in the state $|\Psi^+\rangle_{A_iB_i} = \frac{1}{\sqrt{2}}(|01\rangle + |10\rangle)_{AB}$ and sends the A sequence to Alice which consists of all the A particles in the EPR pairs. They both store the photons into two groups, i.e. photons A_1 and A_2 as a group and B_1 and B_2 as another group. In the case that the transmission is secure, Alice performs her two-bit encoding via local unitary operation on one photon of each group. Then they perform the Bell-state measurement on each group of their own particles. Alice announces her measurement results to Bob. Bob then concludes Alice's operation according to his measurement outcomes and those published by Alice, and extracts the secret message. This protocol makes use of two EPR pairs for entanglement swapping. For two bits of information, four qubits were prepared and two additional bits are transmitted.

Quantum teleportation or entanglement swapping can be utilized in DSQC schemes because they have the same advantages that the security of communication is based on the security of the process for sharing the entanglements, so that they can ensure the security before the secret message communication. Once entanglement is established, the qubits do not suffer from the noise and the loss aroused by the channel again, the bit rate and the security will very high.

D. DSQC based on the rearrangement of orders of particles

In this part, we describe DSQC protocols based on the rearrangement of orders of particles which uses EPR pairs as the information carriers, following some ideas in the controlled-order-rearrangement-encryption QKD protocol [6].

One DSQC protocol uses EPR pairs [21]. The transmitting order of the particles which ensures the security of communication is secret to anyone except for the sender Bob himself. The two parties agree that the four unitary operations in the dense coding represent two bits of classical information. The receiver Alice prepares a sequence of EPR pairs randomly in one of the four Bell states $\{|\phi^{\pm}\rangle_{AB}, |\psi^{\pm}\rangle_{AB}\}$. Here

$$|\psi^-\rangle_{AB} = \frac{1}{\sqrt{2}}(|0\rangle_A|1\rangle_B - |1\rangle_A|0\rangle_B), \tag{7}$$

$$|\psi^+\rangle_{AB} = \frac{1}{\sqrt{2}}(|0\rangle_A|1\rangle_B + |1\rangle_A|0\rangle_B), \tag{8}$$

$$|\phi^-\rangle_{AB} = \frac{1}{\sqrt{2}}(|0\rangle_A|0\rangle_B - |1\rangle_A|1\rangle_B), \tag{9}$$

$$|\phi^+\rangle_{AB} = \frac{1}{\sqrt{2}}(|0\rangle_A|0\rangle_B + |1\rangle_A|1\rangle_B). \tag{10}$$

Alice divides them into two corresponding sequences, called A sequence and B sequence. A sequence is composed of all the A particles in the EPR pairs. Alice sends the B sequence to Bob. Bob selects a sufficiently large subset of photons as his checking set and performs one of the four unitary operations on them randomly. For the other photons, Bob chooses a suitable unitary operation on each photon, according to his secret message. Before sending back the encoded photon sequence, Bob rearranges the order of the photons in the sequence. After Alice confirms the receipt of the B sequence, Bob tells Alice the positions of the checking photons. Alice performs the Bell-state measurements on the sample pairs and then checks the eavesdropping with the checking set. In the case that the transmission is secure, Bob exposes the secret order and then Alice can obtain the secret message with Bell-state measurements on the other EPR pairs after recovering their original orders.

Subsequently, a DSQC protocol was proposed with single photons based on the secret transmitting order of particles [22], following some ideas in Refs[6, 12]. The receiver Alice prepares a sequence of single photons (i.e., ordered N single photons) which are randomly in one of the four states $\{|0\rangle, |1\rangle, |+\rangle, |-\rangle\}$ and sends the sequence to Bob. Bob selects randomly a sufficiently large subset to perform $U_0 = I = |0\rangle\langle 0|+|1\rangle\langle 1|$ or $U_3 = |0\rangle\langle 1|-|1\rangle\langle 0|$ operation randomly for eavesdropping check later. He performs one of these two operations on the remaining photons according to his secret message, and sends them back to Alice. If the error rate exceeds the threshold they preset, they abort their quantum communication. Otherwise, Bob publishes the secret order of the photons in the sequence. Alice reads out Bob's message with single-photon measurements using the basis she prepared the photons.

These two DSQC protocols [21, 22] using transmitting order rearranging method are simple as they only require one eavesdropping check. However, there is a security loophole because they both are two-way quantum communication protocols. The security of these two quantum communication protocol is based on the secret order of the particles which will be published after the security checking. If Alice and Bob cannot detect the eavesdropper during the checking process, the eavesdropper Eve can get the secret order and the whole message. Recently, Li et al. point out the security leak and present a possible improvement [44]. They indicate that the protocols are insecure with Trojan horse attack strategies. An invisible photon or delayed one are introduced to attack these schemes. The invisible photon proposed by Cai is a photon produced with a wavelength different from the wavelength of the authorized parties. As that the single photon detector is only sensitive to the photons with a special wavelength, the invisible photon will not be detected.

Generally, the invisible photon may obtain nothing if the legitimate users' operation is done by optical device which is wavelength-dependent. However, in the protocols there is no security checking in the line from Alice to Bob. Eve can choose a special wavelength close to the legitimate one to produce the invisible photons without worrying about being detected and the probability that Eve can obtain the correct information is close to 1. The delay-photon Trojan horse attack is inserting a spy photon in a legitimate signal with a delay time, shorter than the time windows of the optical device. The attack strategy is described as follows. (1) Eve prepares a set of spy photons (invisible one or delay one both work) and inserts them into the legitimate signal in the line from Alice to Bob. (2) After Bob performs the unitary operation, Eve sorts her spy photons out in the line from Bob to Alice. As there is no security check at this stag, Eve will not be detected. When Bob performs his unitary operations on the authorized photons, he also performs them on the spy photons. So does the order rearranging manipulation. (3) After Bob publishes the secret order, Eve can perform measurements on the spy photons and get the secret message freely and fully. In order to defeat this kind of attack, another security checking is inserted before Bob's operations. That is, Bob chooses a large subset of photons randomly as sample photons. He splits the sample signals with photon number splitters (PNS) and measures the two signals with bases Z or X randomly, and analyzes the multi-photon rate and the error rate. If both the error rate and the multi-photon rate are very low, they continue to the next step. Otherwise, they terminate the communication. Furthermore, Bob has to insert a filter in front of his devices to filter out the photon signal with an illegitimate wavelength. This improvement will help these DSQC protocols defeating the Trojan horse attack. In a word, the insecurity point of these two DSQC protocols is that there is only one security checking for a two-way quantum communication. The most important point is that for each block of transmission, an eavesdropping check is inevitable for secure communication, no matter what is transmitted with a quantum channel [44].

3. Quantum secure direct communication protocols

QSDC transmits secret messages directly through a quantum channel. QSDC has higher security requirement than both QKD and DSQC. In secret communication with QKD, one can protect secret message by first ensuring the security of the keys in the QKD process. In DSQC one can protect the secret message by checking the security of the DSQC process while holding the classical information. However in QSDC, the secret message is encoded in the information carriers directly, hence it is more stringent to ensure the security. According to Deng-Long criteria [9, 12], a real secure QSDC scheme should satisfy four requirements:

1. The secret message can be read out by the receiver directly and there is no additional classical information exchange between the sender and the receiver in principle except in the process for security checking and error rate estimating.
2. The eavesdropper cannot obtain any useful information about the secret message no matter what kind of attack she will perform.
3. Eve can be detected by the legitimate users before Alice and Bob encode the secret message on the quantum states.
4. The quantum states are transmitted in block by block way.

The last one of the four criteria is a necessary tool in QSDC hence one can use it easily to see if a quantum communication is a QSDC or not. It is a necessary condition hence even though some protocols are not QSDC even though they use block data transmission.

The two-step QSDC scheme [8, 9] is secure as it satisfies all these four requirements. In the following, we will discuss some QSDC protocols in details.

A. Two-step Quantum Secure Direct Communication

The two-step QSDC protocol is the first QSDC protocol using EPR states [8, 9]. The communication utilizes EPR pairs as information carriers in which each is in one of the four Bell states $\{|\phi^{\pm}\rangle, |\psi^{\pm}\rangle\}$.

The two-step QSDC scheme is the first secure model for quantum direct communication. This QSDC principle is shown in Fig.1, and the protocol is described in detail as follows [9].

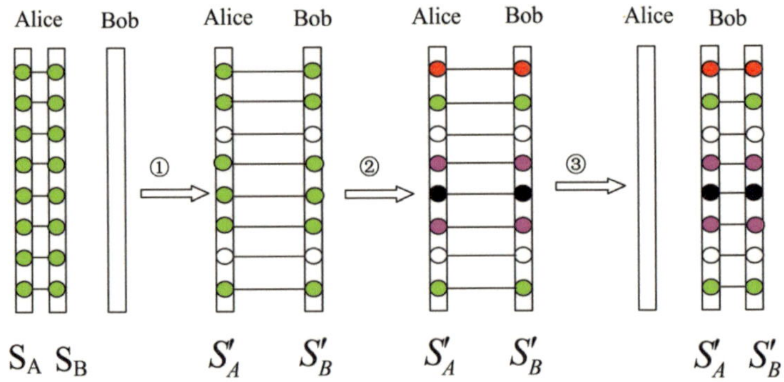

Fig. 1. The principle of the two-step QSDC scheme. Each line connect two photons represents an EPR pair. S_A is the message-coding sequence and S_B is the checking sequence.

In the QSDC process, Alice prepares an ordered N EPR pairs which are in the same state $|\phi^{+}\rangle_{AB} = \frac{1}{\sqrt{2}}(|0\rangle_A|0\rangle_B + |1\rangle_A|1\rangle_B)$. Alice separates the two particles into two parts. Each part is an ordered EPR partner particle sequence. One of the sequence is made up of all the photons marked with A in the ordered N EPR pairs which is called the message-coding (M) sequence S_A. The remaining EPR particles forms another particle sequence which is called the checking (C) sequence S_B. Alice and Bob agree that the four Bell states $|\phi^{+}\rangle$, $|\phi^{-}\rangle$, $|\psi^{+}\rangle$ and $|\psi^{-}\rangle$ correspond to the classical bits 00, 01, 10 and 11, respectively.

The security of two-step QSDC protocol is also considered. The security checking process during QSDC consists of two steps: first, Alice sends the checking sequence S_B to Bob and checks the security of transmission with Bob. Then if the two legitimate users confirm that the transmission of the checking sequence S_B is secure, Alice encodes her secret message on the S_A sequence with four unitary operations U_i (i = 0, 1, 2, 3) and then sends S_A to Bob who reads out the secret message directly by Bell-state measurements. Here the four unitary operations used for coding are described below:

$$U_0 = I = |0\rangle\langle 0| + |1\rangle\langle 1|, \tag{11}$$

$$U_1 = \sigma_z = |0\rangle\langle 0| - |1\rangle\langle 1|, \tag{12}$$

$$U_2 = \sigma_x = |1\rangle\langle 0| + |0\rangle\langle 1|, \tag{13}$$

$$U_3 = i\sigma_y = |0\rangle\langle 1| - |1\rangle\langle 0| \tag{14}$$

In the first step, Alice and Bob perform the security checking procedures. Bob chooses randomly a subset of the photons that he received as the samples for security checking, then he measures them by choosing randomly one of the two basis, Z ≡ {|0i, |1i} and X ≡ {j§i = $\frac{1}{\sqrt{2}}$ (j0i § j1i)}. Then Bob tells Alice the positions of the sample photons he has chosen, also the measuring bases and the outcomes. Alice chooses the same bases to mea- sure the corresponding photons in the M sequence. They compare their results publicly. If there is no eavesdrop- ping attack, their results should be in correspondence with each other, otherwise the eavesdropping behavior will be discovered.

In the second step, Alice selects some photons in the M sequence randomly and performs on them one of the four operations U_i (i = 0, 1, 2, 3). The remaining photons in the M sequence are used for information transmission. After Bob's Bell-state measurements on the EPR pairs, they perform the second step of security checking. Alice first tells Bob the positions of the checking qubits and the type of unitary operations. Bob's measurement will get an estimate of the error rate in the M sequence transmission. If the transmission of the C sequence S_B is secure, Eve can only disturb the transmission of the M sequence S_A and cannot get any information encoded on it as none can read out a useful information from part of maximally entangled quantum system.

If the quantum channel shows low noises or no loss, error correction procedures can be used in the communication. The bits preserving correction code should be used to preserve the integrity of the message.

In 2005, Wang et al. [14] generalized the two-step QSDC scheme based on superdense coding. Using high level particles, each particle could carry more than one bit information than two-step QSDC.

In d-dimension QSDC, the Bell-basis states are described as:

$$|\Psi_{nm}\rangle_{AB} = \sum_j e^{2\pi ijn/d}|j\rangle \otimes |j + m \bmod d\rangle/\sqrt{d} \tag{15}$$

where n,m = 0, 1, …, d –1. The unitary operations in d-dimensional Hilbert space are

$$U_{nm} = \sum_j e^{2\pi ijn/d}|j + m \bmod d\rangle\langle j|. \tag{16}$$

The unitary operations on the particle B can make the following transformations on the d-dimension Bell-basis state

$$|\Psi_{00}\rangle_{AB} = \sum_j |j\rangle \otimes |j\rangle/\sqrt{d} \tag{17}$$

into the Bell-basis state $|\Psi_{nm}\rangle_{AB}$.

When the QSDC starts, Alice and Bob transmit their information directly in two rounds which is shown in Fig.2. The procedures of communication is described in detail as follows:

1. The receiver Bob prepares a sequence of entangled photon pairs which are in the Bell state $|\Psi_{00}\rangle$.

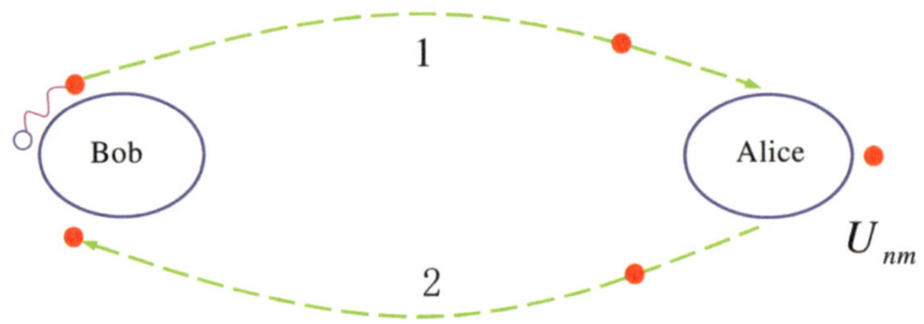

Fig. 2. Schematic description of quantum superdense coding [14].

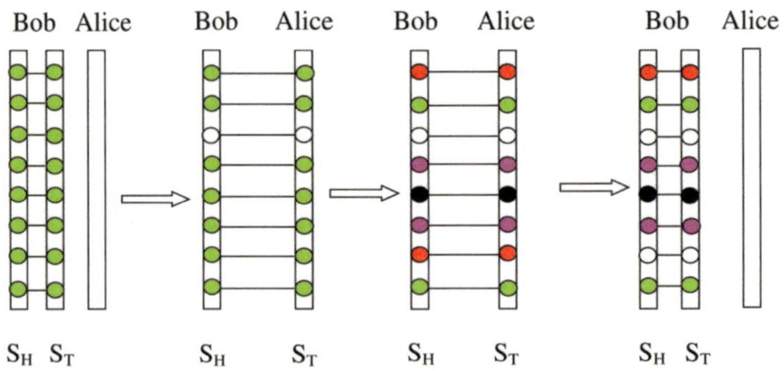

Fig. 3. Illustration of the QSDC protocol with a sequence of d-dimensional EPR pairs [14]. The T sequence is traveling forth and back from Bob to Alice.

2. Bob takes one particle from each entangled particle pair for making up of an ordered partner particle sequence, say [$P_1(H)$, $P_2(H)$, $P_3(H)$, ... , $P_N(H)$]. It is called the home (H) sequence. The remaining partner particles compose another particle sequence [$P_1(T)$, $P_2(T)$, $P_3(T)$, ... , $P_N(T)$], and it is called the traveling (T) sequence, shown in Fig.3. Here the subscript indicates the pair order in the sequence.
3. Bob sends the photon sequence to the sender of the secret message and then they check eavesdropping.

B. Deng-Long quantum one-time pad QSDC scheme

Single photons are easy to be produced than entangled photons for quantum communication. So Deng and Long proposed the QSDC scheme using one-time pad QKD method, called Deng-Long quantum one-time pad quantum secure direct communication. In quantum one-time pad QSDC scheme, Alice and Bob first share a sequence of single-photon quantum states securely, then the sender Alice encodes her secret message and transmits the states to the receiver Bob. The implementation of quantum one-time pad QSDC scheme is described in Fig.4.

Deng-Long quantum one-time pad QSDC scheme are described in details as follows:

(1) The secure doves sending phase.

In this step, Alice and Bob first share a series of single photons. The receiver Bob prepares a sequence of polarized single photons S and sends these photons to Alice.The states of the photons are chosen randomly in one of the four states $\{|0\rangle, |1\rangle, |+\rangle, |-\rangle\}$. Alice and Bob can check the security of the transmission after receiving the photons. Bob chooses randomly the

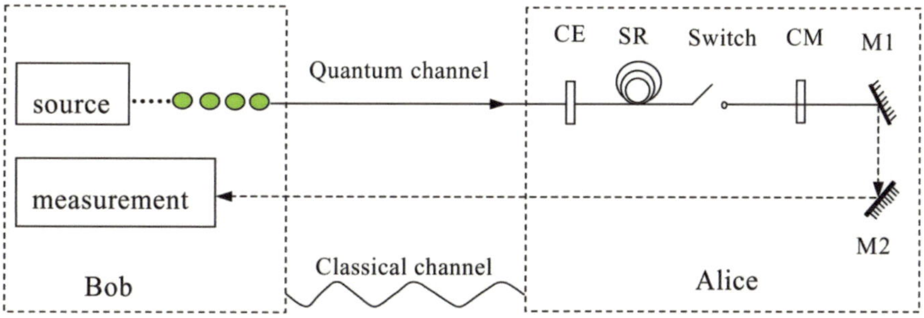

Fig. 4. Implementation of the quantum one-time pad QSDC scheme with optical delays [12]. CE is the eavesdropping check; SR represents an optical delay; Switch is used to control the quantum communication process, if the batch of photons are safe, the switch is on and the message coding is performed; CM encodes the secret message, M1 and M2 are two mirrors for in this simple illustrative set-up.

security checking qubits from the sequence S. Alice checks the security of this transmission by measuring the samples with the randomly chosen MBs and compares the states information with Bob. If they confirm that the transmission is secure, Alice and Bob continue their communication to the second phases; otherwise, they abandon their transmission and begin the communication from the beginning.

(2) The message coding and doves returning phase.

Alice encodes her secret message on each photon in the sequence S' with the unitary operation $U_0 = I = |0\rangle\langle 0| + |1\rangle\langle 1|$ or the operation $U_3 = i\sigma_y = |0\rangle\langle 1| - |1\rangle\langle 0|$ according to the message bit is 0 or 1, respectively. The S' sequence are the remaining photons in the sequence S after the first eavesdropping check. The U_3 operation only flips the two eigenvectors in both MBs Z and X, i.e.,

$$U_3|0\rangle = -|1\rangle, \quad U_3|1\rangle = |0\rangle, \tag{18}$$

$$U_3|+\rangle = |-\rangle, \quad U_3|-\rangle = -|+\rangle. \tag{19}$$

the two unitary operations U_0 and U_3 do not change the MBs of the photons. Alice sends the encoded photon sequence S' back to Bob. Since Bob knows the initial state information completely, he can choose the original MBs to measure each photon for reading out the secret message. Alice picks some photons in the S' sequence to check the security of transmission. She chooses randomly the U_0 and U_3 operations to encode some checking information in the message coding phase. After Bob measures the photons in the sequence S', Alice tells Bob the positions and the coded bit values of these checking photons. These checking photons gives Alice and Bob opportunity to estimate whether there is an Eve in the line to intercept their communication. Eve's eavesdropping in this phase will not get any useful information about the secret message as she does not know the original states of the photons in the sequence S.

Single photons and quantum memories are needed in this practical QSDC scheme. The sender Alice must have the capability of storing quantum states. By now, this technique is not fully developed. However, this technique is a vital ingredient for quantum computation and quantum communication, and there has been great interest in developing techniques for quantum state storage [40]. In this book, we would like to introduce the method of the photon storage realized by optical delays in a fibre, as shown in Fig.4. In practice, there are also losses in the transmission lines, error correcting techniques are necessary.

C. multi-step QSDC with Greenberger-Horne-Zeilinger states

The two-step QSDC can be generalized to multi-step QSDC scheme with multi-particle entangled state, such as Greenberger-Horne-Zeilinger (GHZ) states [15]. The three-particle GHZ state can be described as:

$$|\varphi\rangle_{ABC} = \frac{1}{\sqrt{2}}(|000\rangle_{ABC} + |111\rangle_{ABC}). \tag{20}$$

There are eight independent GHZ states, namely

$$|\varphi\rangle_0 = \frac{1}{\sqrt{2}}(|000\rangle + |111\rangle), \tag{21}$$

$$|\varphi\rangle_1 = \frac{1}{\sqrt{2}}(|000\rangle - |111\rangle), \tag{22}$$

$$|\varphi\rangle_2 = \frac{1}{\sqrt{2}}(|100\rangle + |011\rangle), \tag{23}$$

$$|\varphi\rangle_3 = \frac{1}{\sqrt{2}}(|100\rangle - |011\rangle), \tag{24}$$

$$|\varphi\rangle_4 = \frac{1}{\sqrt{2}}(|010\rangle + |101\rangle), \tag{25}$$

$$|\varphi\rangle_5 = \frac{1}{\sqrt{2}}(|010\rangle - |101\rangle), \tag{26}$$

$$|\varphi\rangle_6 = \frac{1}{\sqrt{2}}(|110\rangle + |110\rangle), \tag{27}$$

$$|\varphi\rangle_7 = \frac{1}{\sqrt{2}}(|110\rangle - |001\rangle). \tag{28}$$

By performing single-particle unitary operations $\{U_i\}$ (i = 0, 1, 2, 3) on two of the three particles, the initial GHZ state can be changed to any of the state in the set.

In the beginning, Alice and Bob make an agreement that each of the states $|\varphi\rangle_k$ (k = 0, 1,..., 7) represents a three bits binary number, namely $|\varphi\rangle_0$, $|\varphi\rangle_1$, ..., and $|\varphi\rangle_7$ corresponds to the classical coded as 000, 001, ..., and 111, respectively.

The sender Alice prepares a sequence of ordered N three-particle GHZ-state quantum systems, labeled as $[P_1(A)P_1(B)P_1(C)$, $P_2(A)P_2(B)P_2(C)$, ..., $P_N(A)P_N(B)P_N(C)]$ which are in the state: $|\varphi\rangle_0 = \frac{1}{\sqrt{2}}(|000\rangle_{ABC} + |111\rangle_{ABC})$. Then Alice divides the sequence into three partner particle sequences, $S_A = [P_1(A), P_2(A), \ldots, P_N(A)]$, $S_B = [P_1(B), P_2(B), \ldots, P_N(B)]$ and $S_C = [P_1(C), P_2(C), \ldots, P_N(C)]$.

Then Alice and Bob complete the multi-step QSDC as follows:

(1) First step entanglement sharing process.

Alice sends the sequence labeled with C to the receiver Bob. After that she checks the security of the transmission with Bob to discover the eavesdropping attack. Bob randomly chooses some particles from S_C sequence as the sample qubits and measures them by choosing one of the two MBs Z and X randomly. Then Bob notice Alice the positions and the MBs of the sample particles. After that Alice chooses a MB in the state $\{|00\rangle, |01\rangle, |10\rangle, |11\rangle\}$ to measure her corresponding partner particles in the sequences S_A and S_B when Bob chooses the MB Z to measure his sample particles; otherwise, Alice performs a Bell-basis measurement on her particles. Alice checks the correspondence of the qubits with Bob and analyzes the error rate η_e of the samples. If they find that η_e is reasonably low then they continues the quantum communication; otherwise, Alice and Bob abandon the communication and repeat their quantum communication from the beginning.

(2) Information coding process.

Alice encodes the secret message on the GHZ states using the U_0 and U_2 operations which can be used on the S_B sequence and the four unitary operations $\{U_0, U_1, U_2, U_3\}$ which are performed on the S_C particles. Alice also chooses some sample particles in the sequences S_B and S_A for the second step of security check which are operated by one of the four operations U_i ($i = 0, 1, 2, 3$) randomly.

(3) Second step entanglement sharing process.

Alice measures the samples chosen from the remaining particles in sequence S_A using the MB Z or X with 50% probabilities. Then she sends the sequence S_B to Bob. When Bob receives the S_B sequence, Alice tells Bob the positions of the samples. Bob measures the corresponding particles in the sequence S_B and S_C based on Alice's MBs' information. If Alice measures her particles with the basis Z, then Bob chooses to measure with the measuring basis $\{|00\rangle, |01\rangle, |10\rangle, |11\rangle\}$; otherwise, he chooses the Bell-basis state measurement on his sampling particles. After that they compare the outcomes of the measurement to analyze the error rate of these particles. If the error rate is lower than the security bound, Alice and Bob continue their quantum communication; otherwise they abort the communication and repeat it again.

(4) Information decoding process.

Alice sends the S_A sequence to Bob in the last step. Bob reads out the secret message with joint three-particle measurements on the particles in the three sequences S_A, S_B and S_C. Then they turn to the third step security checking, Alice tells Bob the positions of the sampling particles and they analyze the error rate of the samples. If the error rate is lower than the security bound, they can accomplish the transmission of the secret message.

During the communication process, Alice and Bob transmit one of the three sequences S_C, S_B and S_A each time. So the eavesdropper Eve can only capture one sequence during eavesdropping. She cannot obtain any information about a GHZ-state quantum system if she only captures one particle. Thus this QSDC scheme is secure as that of the two-step QSDC scheme [8, 9].

Besides Bell states and GHZ states, various entangled sources are used in QSDC and DSQC protocols. In 2008, Lin et al. constructs a QSDC scheme [45] using χ-type entangled states which is in the form

$$|\chi^{00}\rangle_{1234} = \frac{1}{2\sqrt{2}}(|0000\rangle - |0011\rangle - |0101\rangle + |0110\rangle + |1001\rangle + |1010\rangle + |1100\rangle + |1111\rangle)_{1234}. \quad (29)$$

An efficient QSDC process based on the χ-type entangled states is discussed both in ideal and noisy communication channels. Later Dong et al. proposed a QSDC protocol using three-particle W states [46]. Also another DSQC protocol using four-particle entangled states is proposed by Xiu et al. in the same group [47].

D. quantum-encryption-based QSDC scheme

The principle of the quantum-encryption-based QSDC scheme [16] is shown in Fig.5. In this protocol, a controlled-not (CNot) gate is used to encode and decode the secret message. The two parties share privately a sequence of two-photon pure entangled states before-hand, and then use the states as their private quantum key which is reusable with a eavesdropping check before each round. The message can be readout directly by the receivers and each photon transmitted between the two parties can carry one bit of message securely in principle.

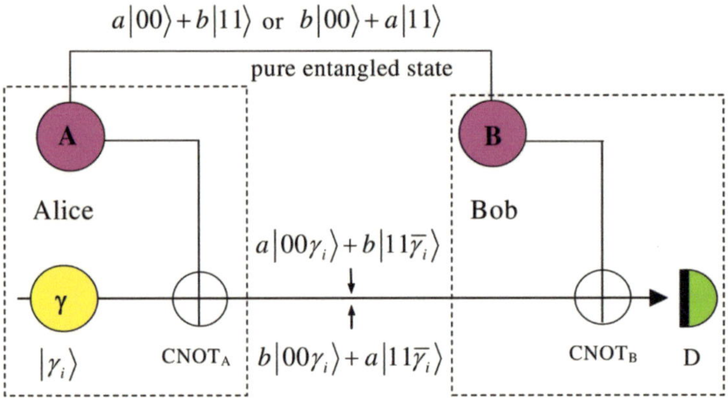

Fig. 5. Illustration of the quantum-encryption-based QSDC scheme [44]. The pure entangled states are used as quantum key which are repeatedly used. D represents the measurement with the MB Z.

In the quantum-encryption-based QSDC protocol, Alice and Bob first share a sequence of two-particle entangled states privately and then use them as their private quantum key which are used to encrypt secret message.

In detail, Alice first prepares n two-photon pairs randomly in one of the two pure entangled states $|\Psi\rangle_{AB} = a|0\rangle_A|0\rangle_B+b|1\rangle_A|1\rangle_B$ and $|\Phi\rangle_{AB} = b|0\rangle_A|0\rangle_B+a|1\rangle_A|1\rangle_B$. For the purpose of secure communication, decoy photons are used in secure sharing the pure entangled states [41, 42]. Alice picks up the photon marked with B in each pair to make up the sequence S_B : [B_1, B_2, ... ,B_n]. The other sequence S_A is made up of particles A_i (i = 1, 2...n). The sequence S_B is sent to Bob and the sequence S_A is kept by Alice. Alice inserts some decoy photons S_{de} into the sequence S_B for checking the security of the transmission. The decoy photons are randomly in one of the four states $\{|0\rangle, |1\rangle, |+\rangle = \frac{1}{\sqrt{2}}(|0\rangle + |1\rangle), |-\rangle = \frac{1}{\sqrt{2}}(|0\rangle - |1\rangle)\}$. Alice can get a decoy photon by measuring one photon in a two-photon pair $|\Psi\rangle_{AB}$ with the basis Z and operating the other photon with σ_x or a Hadamard (H) operation. Bob measures the decoy photons with the suitable bases that Alice told him and analyzes the error rate of those outcomes with Alice. If the error rate is reasonably low, they can obtain a sequence of quantum key privately; otherwise, they discard the qubits and repeat quantum communication from the beginning.

Alice prepares a sequence of traveling particles γ_i in one of the two states $\{|0\rangle, |1\rangle\}$ according to the bit value of her secret message is 0 or 1 (called the traveling particle sequence S_T). As discussed in Refs. [8, 9, 12], Alice randomly inserts some decoy photons in the sequence S_T for security checking. The sequence, say S_D, are randomly in the four states $\{|0\rangle, |1\rangle, |+\rangle, |-\rangle\}$. The quantum key, the pure entangled pairs shared $\{|\Psi\rangle_{AB}, |\Phi\rangle_{AB}\}$ are used by Alice to encrypt the traveling particles in the sequence S_T except for the decoy photons, shown in Fig.5. Alice performs a controlled-not (CNOT) operation on the particles A_i and γ_i ($i = 1, 2, \ldots, n$) by using the particle A_i as the control qubit. Then all the traveling particles are sent to Bob. Alice announces to Bob the positions and the states of the decoy photons and then Bob measures them with the same bases. Bob then takes a CNOT operation on the particles B_i and γ_i with the particle B_i as the control qubit and then he measures the particles γ_i with the basis Z. The outcomes of the measurements are recorded. If the transmission channel is secure, Bob reads out the message directly. The quantum keys are used to transmit the secret message in the next round by repeating the communications. Otherwise, they have to abandon their results and repeat their quantum communication form the beginning.

The quantum key is randomly in one of the two states $|\Psi\rangle_{AB} = a|0\rangle_A|0\rangle_B + b|1\rangle_A|1\rangle_B$ and $|\Phi\rangle_{AB} = b|0\rangle_A|0\rangle_B + a|1\rangle_A|1\rangle_B$ for the eavesdropper Eve, the state of the composite quantum system composed of the two particles A_iB_i in a quantum key and the traveling particle γ_i is randomly in one of the two states $\{a|00\gamma_i\rangle + b|11\overline{\gamma_i}\rangle, b|00\gamma_i\rangle + a|11\overline{\gamma_i}\rangle\}$. So the density matrix of the traveling particle γ_i for Eve is $\rho_i = \frac{1}{2}\begin{pmatrix} 1 & 0 \\ 0 & 1 \end{pmatrix}$. The eavesdropping behavior on the traveling particle reveals no useful information about the secret message. Moreover, Eve's action will leave a trace in the results of the decoy photons and be discovered by the communication parties. So this quantum-encryption-based QSDC scheme is secure in principle.

Noises are inevitably exist in the practical quantum channel, so the users must exploit entanglement purification [49] to keep the entanglement in the quantum key, and quantum privacy amplification [50] on them as well. However, Bell states are not needed in this protocol, just the pure entangled states $|\Psi\rangle_{AB} = a|0\rangle_A|0\rangle_B + b|1\rangle_A|1\rangle_B$ ($|\Phi\rangle_{AB} = b|0\rangle_A|0\rangle_B + a|1\rangle_A|1\rangle_B$) or $|\Psi\rangle'_{AB} = a'|0\rangle_A|0\rangle_B + b'|1\rangle_A|1\rangle_B$ ($|\Phi\rangle'_{AB} = b'|0\rangle_A|0\rangle_B + a'|1\rangle_A|1\rangle_B$) are used in this scheme. Here $|a'|^2 + |b'|^2 = 1$. As the quantum key is just used to encrypt and decrypt the secret message, it is unnecessary for the users to keep the same states as those they used in last time, just the correlation of each pair, which will increase the efficiency of the entanglement purification process largely. On the one hand, the users should do error correction on their results in practical applications. On the other hand, this QSDC scheme can only used to distribute a private key if the loss of the quantum line is unreasonably large. The obvious advantage of the quantum-encryption-based QSDC scheme is that the quantum key is a sequence of pure entangled states, not maximally entangled states, which will make this scheme more convenient than others as an entanglement source usually produces pure entangled signals because of asymmetric features of the quantum source.

E. QSDC using one party quantum error correction code

Quantum error correcting codes(QECC) is a key technique towards protecting quantum system in quantum communication and quantum computation from errors mainly brought by decoherence. Here we introduce the QSDC protocol using the one-party quantum error correcting codes(one-party-QECC) [51]. The use of one- party QECC proves that QSDC is able to tolerate higher error rates in transmission process. As described in the proof of unconditional secure BB84 protocols, the success of error correction in QSDC may lead a path to prove its unconditional security.

Protocol: one-party-QECC QSDC protocol

1. Bob owns the entanglement source and prepares a sequence contains $3n$ EPR pairs in the initial states |©⁺⟩.
2. Bob chooses $3n$ bit binary string b, then he chooses to apply the Hadamard transformation H to the second halves of the pairs in which the corresponding bits of string b are 1. After that he sends the second halves to Alice.
3. Alice announces receiving the qubits to Bob publicly. Bob tells Alice which qubits are operated by the Hadamard operations. Then Alice applies Hadamard transformation H on the corresponding qubits in her part.
4. Alice and Bob choose an n subset of the EPR pairs randomly for security checking. They measure the checking qubits respectively in the Z-basis and compare the results publicly. The results on both Alice's and Bob's sides in each pair will be the same if there are no errors. However, if there are too many inconsistencies, they notice that the transmitting qubits are been eavesdropped and the protocol is aborted.
5. Alice randomly selects m subset of the rest $2m$ logical EPR pairs as the second-round check pairs and the rest are used for coding. A $2m$ bit binary string b' is also chosen randomly. Bob applies the Hadamard transformation $\bar{H}$ to the second halves of the pairs when the corresponding bits of b are 1. Then the second halves are sent to Bob. During the communication process, if Alice wants to send a k bit binary sequence of message M. She first picks a $[[2m, 2k, t]]$ one-party-QECC that are used to correct the errors in the second transmission. There are k logical EPR pairs in the code pairs and she encodes M to her halves of the second-level logical qubits in the code pairs by applying

$$\bar{U}_{2i} = \bar{Z}_{2i}\bar{X}_{2i} \tag{30}$$

on the $2i$-th logical qubit where the i-th bit of M is 1. Actually Alice is able to apply this local operation. Then she returns all her qubits to Bob.

6. Bob receives the qubits and announces his receipt publicly. Then Alice tells Bob the binary string b', and Bob applies the first-level logical $\bar{H}$ to the received qubits where the corresponding bits of b' are 1.
7. Alice announces to Bob the places of second-round check pairs and the one-party-QECC that she chooses. If Bob measures both the qubits in each checking pairs in Z-basis respectively, he will get the same results with Alice without error. Thus if Bob's error rate is high than the security bound, the protocol is aborted.
8. Bob then uses the [[$2m$, $2k$, t]] one-party-QECC to correct the bit errors on the rest m first round transmission logical EPR pairs and obtains k second round transmission logical code pairs.
9. Bob measures both the qubits of the rest k second round transmission logical code pairs in Z-basis. By comparing the measurements on corresponding pairs, Bob can retrieve the full information of M.

F. Quantum secret sharing based on quantum secure direct communication

Classical secret sharing aims to distribute secret keys between the boss and his agents. When the boss expects to generate secret keys with the two agents separately and the two agents cannot reveal the boss's information until they combine their results. Quantum secret sharing(QSS) is a special utilization of quantum mechanics in classical secret sharing. The basic model of QSS permits the boss and two(or more) remote parties to share the secret keys and any eavesdropping behavior will be discovered by the communication parties. QSS was first proposed by M. Hillery, V. Buµzek and A. Berthiaume. In the protocol, three-particle maximally entangled state (Green-Horne-Zeilinger state) is used to realize the secret sharing process.

In 2005, Zhang et al. proposed a (n, n)-threshold multiparty quantum secret sharing protocol [48] of secure direct communication based on the two-step QSDC protocol. In the QSS process, the sender's secure message can be extracted only if all the agents collaboration. Different from the protocol of multi-particle GHZ states QSS, if the agents number is larger than 3, the use and identification of Bell states are enough in their two protocols disregarding completely the sharer number. Later, Li et al. generalized Zhang's protocol and propose a (t, n)- threshold QSS protocol using secure direct communications [27]. In Li's QSS protocol, the boss distributes the classical secret shares to his agents and each agent owns a secret share in advance. His secure direct communication message can be extracted by the collaboration of at least t or more agents can obtain the secret message with the mutual assistances. Any t-1 or fewer agents cannot reveal any information. Compared with the previous multiparty quantum secret sharing protocols in which the sender's secret message can be recovered only if all the agents collaborate and the protocol is more practical and more flexible. Wang et al. generalized Zhang's QSS protocol to the high-dimensional case via quantum superdense coding [28]. The channel capacity and security are improved by the high-dimensional quantum coding, so the protocol shows a better efficiency and security.

4. Quantum secure direct communication network

Quantum communication network is also an important branch of quantum information and there are many related works. But a QSDC network protocol requires high security and almost all the existing point-to-point QSDC schemes cannot be used for QSDC network directly. In a quantum network, servers who prepare and measure the quantum signals are needed, which simplifies the users' devices. On the other hand, it increases the di±culty for the two legitimate users to prevent the server from eaves-dropping. Here in this section, we will introduce three different QDC network protocols.

A. Quantum secure direct communication based on entangled pairs

As we have discussed in Ref.[9], maximally entangled photon states are used as the information carriers in two- step QSDC process. Recently, Li et al. [17] proposed the first QSDC network based on the two-step QSDC protocol with the EPR states. The communication network consists of three parties, the server Alice, the sender Bob and the receiver Carol. Each member in the network is required to exchange message with others securely. Here the three parties are going to finish the network communication by a circular transmission process.

When the communication starts, the sever Alice prepares the initial states which are described as: $|\psi^{+}\rangle_{CM} = \frac{1}{\sqrt{2}}(|01\rangle+|10\rangle)_{CM}$. Then she divides them into two corresponding sequence S_C (checking sequence) and S_M (message sequence). The two sequence are sent to Bob by two steps transmission process. After receiving the S_C sequence, Bob begins his QSDC with Carol by replacing some of the checking qubits with his decoy photons. The decoy photons are prepared randomly in one of the four states $\{|0\rangle, |1\rangle, |+\rangle, |-\rangle\}$ which are used for security checking. Then he sends this sequence to Carol. After Carol confirms the receipt, they check the security of the channel by comparing the states of the decoy photons after measuring the photons in randomly chosen basis.

If the error rate of the publicly comparing results, they confirm that the channel is secure. Bob then performs the unitary operations on his coding qubits in the S_M sequence. These operations are represented by Pauli operators which corresponds to the classical two bits information. Bob also picks out a subset of S_M and performs random operations on them for security checking. Then Bob sends the coding sequence to Carol. After that, they perform the second step of security checking. If the error rate is lower than the security bound, Carol performs one of the four operations randomly on one photon of each EPR pair and sends all the pairs to Alice. Alice performs Bell-state measurements on the EPR pairs and announces the outcomes. Bob and Carol use the remaining photons to estimate the error rate and Carol can read out Bob's message independently.

Another protocol of QSDC network based on entangled photon is proposed by Deng et al. using bidirectional QSDC scheme [13]. The structure of this network protocol is shown in Fig.6.

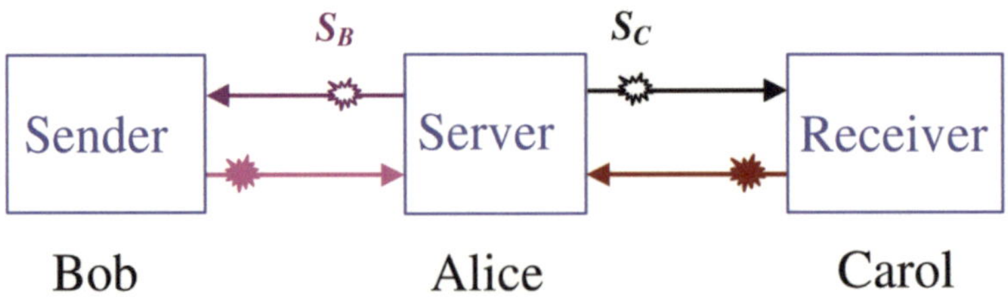

Fig. 6. The subsystem of QSDC network [13].

The sever Alice prepares a set of EPR pairs in the state $|\psi^-\rangle_{BC} = \frac{1}{\sqrt{2}}(|01\rangle - |10\rangle)_{BC}$. The two photons in each state are divided into two sequence S_B and S_C. The S_B and S_C sequences are composed of all the particles marked with B and C in the EPR pairs $|\psi^-\rangle_{BC}$ respectively.
The S_B sequence is sent to Bob and the S_C sequence is sent to Carol. After receiving the two sequences, Bob and Carol perform the security checking. They select a sufficiently large subset of these EPR pairs as samples to check eavesdropping. They choose to measure the sample photons randomly with σ_x and σ_z operators to check the security of transmission. If the transmission process is secure, Bob encodes his message on the S_B sequence by choosing one of the four Pauli operations and Carol performs on the photons in S_C sequence either. Also Bob selects a subset of photons as checking samples in this process, then Bob and Carol both send the sequence back to Alice. After receiving the sequences, Alice performs Bell-state measurements on the EPR pairs and announces the outcomes. Carol can deduce Bob's message with his random operations that he has chosen after the security checking process. Then the network communication is finished.

B. Quantum secure direct communication network based on single photons

Single photons are easy to realized by attenuated laser pulses which exhibit ideal properties for quantum communication. Deng et al. proposed a more practical QSDC network based on single photons. The initial states are prepared in the same state $|0\rangle$ by the sever Alice. Then Alice sends the single-photon sequence S_0 which is formed by the single photons to the receiver Carol. Carol measures photons selected randomly from the S_0 sequence with the basis Z to check the transmitting security and uses beam splitters to check the multi- photon rate.
If she confirms that there is no eavesdroppers, she encodes the information by performing the I or σ_x operations on the single photons randomly. She also inserts some decoy photons which are produced by Hadamard operation on the particles in S, and then she sends them to the sender Bob. Bob checks the states with Carol of all the decoy photons to confirm the security of the communication. If the error rate is lower than the security bound, Bob encodes his message on the photons by choosing the Pauli operators I or σ_x. Bob selects a subset of photons as samples for checking the security and then he sends the photons to Alice. Alice measures the photons with Z basis and announces the outcomes to all the parties. So Carol can read out Bob's message directly. In this protocol, three checking processes are needed for three transmission processes to ensure the security of quantum communication. This QSDC network scheme is easy to be realized as only the single-photon measurement and local unitary operation are needed.

5. Singe qubit quantum privacy amplification for QSDC with single photons

In a practical quantum communications, noises inevitably exit, so the keys obtained from the QKD process are not complete secure. Quantum privacy amplification is often used to generate a key sequence with arbitrarily high security. Privacy amplification with single photons have been used in the BB84 QKD protocol. With entangled photon pairs, the privacy amplification procedure will be different, for instance quantum privacy amplification (QPA) [49, 50] has been used for QKD using entangled quantum systems in the Ekert91 QKD scheme [2].

A quantum privacy amplification for QSDC has recently been designed by Deng et al. for privacy amplification of QSDC with single photons [52]. The circuit of the privacy amplification is shown in Fig.7 which includes two controlled-not (CNOT) gates and one Hadamard (H) gate. Thee neighboring two qubits are sent to the circuit each time, then CNot-Hadamard-CNot (CHC) operation is performed together with a follow-on single qubit measurement on one qubit (the target qubit) by choosing the basis Z. The target photon collapses to the eigen-state of the Z basis. The controlled photon is preserved and it carries the state information of both the two initial photons. After the CHC operation, the state information of the two photons is condensed into a single photon. Hence the privacy of the state of the left-over photon is amplified. After repeating the process many times, the leakage of the state information will be reduced to an arbitrarily low level.

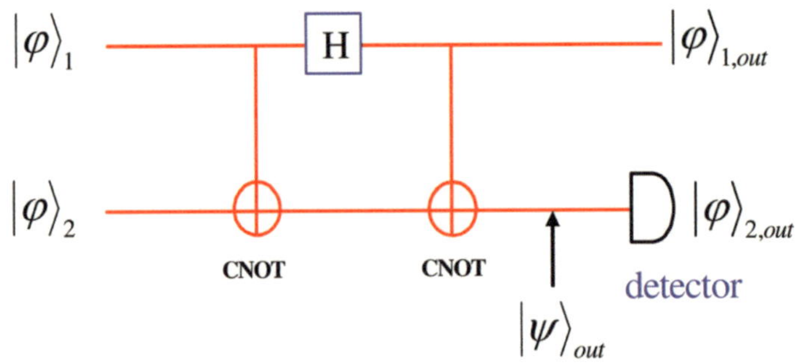

Fig. 7. Quantum privacy amplification operation for two qubits [52]. It consists of two controlled-not (CNOT) gates and a Hadamard (H) gate. $|\varphi\rangle_1$ and $|\varphi\rangle_2$ are the states of the two qubits, respectively. After the three unitary operations, the qubit 2 is measured and the information of the original state of photon 2 is incorporated into photon 1.

In the following, we will discuss the SQ-QPA process in detail. Suppose that Bob prepares a series of single photons which are randomly prepared in one of the four quantum states $|0\rangle$, $|1\rangle$, $(|0\rangle+|1\rangle)/2$ and $(|0\rangle-|1\rangle)/2$ and sends to Alice. An error bit ratio r is known for the photon batch. The SQ-QPA task is to process a portion of photons from the photon batch so that Eve's information about the processed photons is below a desired level.

The basic CHC operations of SQ-QPA is shown in Fig.7 for two qubits. We assume that the quantum states of single photon 1 and 2 are in the general forms:

$$|\varphi\rangle_1 = a_1 |0\rangle + b_1 |1\rangle , \tag{31}$$

$$|\varphi\rangle_2 = a_2 |0\rangle + b_2 |1\rangle , \tag{32}$$

where

$$|a_1|^2 + |b_1|^2 = |a_2|^2 + |b_2|^2 = 1. \tag{33}$$

After the CHC operations, the state of the joint system is changed to

$$|\psi\rangle_{out} = \frac{1}{\sqrt{2}}\{(a_1a_2+b_1b_2)\left|0\right\rangle_1+(a_1b_2-b_1a_2)\left|1\right\rangle_1\}\left|0\right\rangle_2 + \frac{1}{\sqrt{2}}\{(a_1a_2-b_1b_2)\left|1\right\rangle_1+(a_1b_2+b_1a_2)\left|0\right\rangle_1\}\left|1\right\rangle_2. \quad (34)$$

After measuring the second qubit with the Z basis, the state of the control qubit $|\varphi\rangle_{1out}$ will contain the information of the state of the original target qubit. Tables I and II give the output state of control qubit after the measurement on the target qubit with result 0 and 1, respectively. It depends not only on the result of the measurement on the target qubit, but also on the original states of the two input single photons.

	φ_1			
φ_2	$\|+z\rangle$	$\|-z\rangle$	$\|+x\rangle$	$\|-x\rangle$
$\|+z\rangle$	$\|0\rangle$	$\|1\rangle$	$\|-x\rangle$	$\|+x\rangle$
$\|-z\rangle$	$\|1\rangle$	$\|0\rangle$	$\|+x\rangle$	$\|-x\rangle$
$\|+x\rangle$	$\|+x\rangle$	$\|-x\rangle$	$\|0\rangle$	$\|1\rangle$
$\|-x\rangle$	$\|-x\rangle$	$\|+x\rangle$	$\|1\rangle$	$\|0\rangle$

Table I. The state of the output qubit when the result of the second qubit measurement is $|0\rangle$. φ_1 and φ_2 are the states of the original control and target qubit, respectively.

	φ_1			
φ_2	$\|+z\rangle$	$\|-z\rangle$	$\|+x\rangle$	$\|-x\rangle$
$\|+z\rangle$	$\|1\rangle$	$\|0\rangle$	$\|x\rangle$	$\|-x\rangle$
$\|-z\rangle$	$\|0\rangle$	$\|1\rangle$	$\|-x\rangle$	$\|+x\rangle$
$\|+x\rangle$	$\|+x\rangle$	$\|-x\rangle$	$\|0\rangle$	$\|1\rangle$
$\|-x\rangle$	$\|-x\rangle$	$\|+x\rangle$	$\|1\rangle$	$\|0\rangle$

Table II. The state of the output qubit when the result of the second qubit measurement is $|1\rangle$. φ_1 and φ_2 are the states of the original control qubit and target qubit, respectively.

QPA reduces the information leakage to the eavesdropper in quantum communications. For example, if the eavesdropper (say Eve) knows completely the state information of the first qubit, but the second photon is unknown to her, then Eve's knowledge about the output state of the control qubit after the quantum privacy amplification operation becomes

$$\rho = \frac{1}{4}\left(|+z\rangle\langle+z|+|-z\rangle\langle-z|+|+x\rangle\langle+x|+|-x\rangle\langle-x|\right) = \frac{1}{2}\begin{pmatrix} 1 & 0 \\ 0 & 1 \end{pmatrix}. \quad (35)$$

So Eve has no knowledge at the density matrix, all the out put state will appear with the same probabilities. But for Bob who has prepared the original states of the two qubits, he will know completely the output state when Alice tells him the $\sigma_{2,z}$ measurement result. However, if it happens that Eve has complete information about both qubits, the probability is

$$P_2 = r^2, \tag{36}$$

where r is four times of the error bit rate ε detected by Alice and Bob using random sampling. For advanced privacy amplification, we can use the output qubit again as a control qubit and choose a third qubit from the batch as the target qubit and perform SQ-QPA operation on them again. Since more qubits are used in the SQ-QPA process, Eve's information is reduced exponentially to

$$P_m = r^m, \tag{37}$$

where m is the number of qubits that have been used in the SQ-QPA. In this way, Alice can condense a portion of single photons from a batch of N photons with negligibly small information leakage. This condensed single photon sequence can be used to encode secret message and complete the quantum secure direction communication.

This SQ-QPA scheme is a practical method for quantum communication. Single qubit unitary and a two-qubit unitary operations are used for the QPA process. The measurement process consists of only single qubit measurement which is easy to realized using current technology.

6. Conclusion

In this study, we present several deterministic secure quantum communication and quantum secure direct communication protocols. The QSDC protocol permits exchange secret information directly through quantum channel. In QSDC, when the secure channel is established, all the eavesdropping behaviors will be discovered before the information transmission. Besides these protocols, high capacity QSDC protocols, QSDC with quantum error correction codes and QSDC networks are also discussed. DSQC is another form quantum direct communication. In DSQC secret messages can be transmitted between the legitimate users securely with the help of some classical communication.

In these protocols, both single photons and entangled photon states are used as the information processing carriers for secure direct communication. Signals are transmitted both in optical fibers and in free space. Single photon detectors are usually required, and in some protocol, Bell-basis measurements are also required.

At present, technical efforts are concentrated in quantum key distribution. We can see that in the future quantum technology will become more popular and demanding, the need and feasibility of other forms of quantum information processing such as QSDC and DSQC will increase. As we may see from this review, the technical requirements for QSDC and DSQC are almost the same as those for QKD. We expect that in the future intensive research on QSDC and DSQC, especially experimental studies of these subject, will remain and become an active and fruitful area of research.

7. Acknowledgments

Supported by the National Fundamental Research Program Grant Nos. 2006CB921106, 2009CB929402. China National Natural Science Foundation Grant Nos. 10874098, 10775076.

8. References

[1] Bennett C.H. and Brassard G., in *Proceedings of IEEE International Conference on Computers, Systems and Signal Processing, Bangalore, India* (IEEE, New Yoir, 1984), pp.175-179.

[2] Ekert A. K., Quantum cryptography based on Bells theorem. Phys. Rev. Lett., 1991, 67: 661-663.

[3] Bennett C. H., Quantum cryptography using any two nonorthogonal states, Phys. Rev. Lett, 1992, 68: 3121-3124.

[4] Bennett C. H., Brassard G., and Mermin N. D., Quantum cryptography without Bell's theorem, Phys. Rev. Lett., 1992, 68: 557-559.

[5] Bennett C.H. and Wiesner S.J., Communication via one- and two-particle operators on Einstein-Podolsky-Rosen states. Phys. Rev. Lett., 1992, 69: 2881.

[6] Deng F. G. and Long G. L., Controlled order rearrangement encryption for quantum key distribution, Phys. Rev. A, 2003, 68: 042315.

[7] Boström K. and Felbinger T., Deterministic secure direct communication using enganglement, Phys. Rev. Lett., 2002, 89: 187902

[8] Long G.L. and Liu X. S., Theoretically efficient high-capacity quantum-key-distribution scheme, Phys. Rev. A, 2002, 65: 032302

[9] Deng F. G., Long G. L. and Liu X. S., Two-step quantum direct communication protocol using the Einstein-Podolsky-Rosen pair block, Phys. Rev. A, 2003, 68: 042317

[10] Shimizu K. and Imoto N., Communication channels secured from eavesdropping via transmission of photonic Bell states, Phys. Rev. A, 1999, 60: 157-166

[11] Beige A., Englert B. G., Kurtsiefer C. and Weinfurter H., Secure communication with a publicly known key, Acta Phys. Pol. A, 2002, 101 (3): 357.

[12] Deng F. G. and Long G. L., Secure direct communication with a quantum one-time pad, Phys. Rev. A, 2004, 69: 052319.

[13] Deng F. G., Li X. H., Li C. Y., Zhou P. and Zhou H. Y., Quantum secure direct communication network with Einstein-Podolsky-Rosen pairs, Phys. Lett. A, 2006, 359: 359.

[14] Wang C., Deng F. G., Li Y. S., Liu X. S. and Long G. L., Quantum secure direct communication with high-dimension quantum superdense coding, Phys. Rev. A, 2005, 71: 044305.

[15] Wang C., Deng F. G. and Long G. L., Multi-step quantum secure direct communication using mult-particle Green-Horne-Zeilinger state, Opt. Commun., 2005, 253: 15.

[16] Li X. H., Li C. Y., Deng F. G., Zhou P., Liang Y. J. and Zhou H. Y., Quantum secure direct communication with quantum encryption based on pure entangled states, Chin. Phys., 2007, 16 (8): 2149-2153.

[17] Li X. H., Zhou P., Liang Y. J., Li C. Y., Zhou H. Y. and Deng F. G., Quantum secure direct communication network with two-step protocol, Chin. Phys. Lett., 2006, 23: 1080.

[18] Cai Q. Y. and Li B. W. Improving the capacity of the Bostrom-Felbinger protocol Phys. Rev. A, 2004, 69: 054301.

[19] Cai Q. Y. and Li B. W., Deterministic secure communication without using entanglement Chin. Phys. Lett., 2004, 21: 601-603.

[20] Gao T., Controlled and secure direct communication using GHZ state and teleportation, Z. Naturforsch, A, 2004, 59: 597

[21] Zhu A. D., Xia Y., Fan Q. B., and Zhang S., Secure direct communication based on secret transmitting order of particles Phys. Rev. A, 2006, 73: 022338.

[22] Wang J., Zhang Q., and Tang C. J., Quantum secure direct communication based on order rearrangement of single photons, Phys. Lett. A, 2006, 358: 256-258.

[23] Cao H. J. and Song H. S., Quantum secure direct communication with W state, Chin. Phys. Lett., 2006, 23: 290-292.

[24] Long G.L, Deng F.G., Wang C., Li X.H., Wen K. and Wang W.Y., Quantum Secure Direct Communication and Deterministic Secure Quantum Communication, Frontiers of Physics in China, 2007, 2(3) 251.

[25] Wang T. Y., Qin S. J. , Wen Q. Y., Zhun F. C., Analysis and improvement of multiparty controlled quantum secure direct communication protocol, ACTA PHYSICA SINICA, 2008, 57: 7452-7456.

[26] Qin S. J., Wen Q. Y., Meng L. M., Zhu F. C., High Efficiency of Two Efficient QSDC with Authentication Is at the Cost of Their Security, Chin. Phys. Lett., 2009, 26: 020312.

[27] Li B. K., Yang Y. G., Wen Q. Y., Threshold Quantum Secret Sharing of Secure Direct Communication, Chin. Phys. Lett., 2009, 26: 010302.

[28] Wang X., Liu Y. M., Han L. F., Zhang Z. J., Multiparty quantum secret sharing of secure direct communication with high-dimensional quantum superdense coding, Int. J. Quant. Info., 2008, 6: 1155-1163.

[29] Chamoli A., Bhandari C. M., Secure direct communication based on ping-pong protocol, Quant. Info. Proc., 2009, 8: 347-356.

[30] Qin S. J., Wen Q. Y., Meng L. M., Zhu F. C., Quantum secure direct communication over the collective amplitude damping channel Sci. China Ser. G, 2009, 52: 1208- 1212

[31] Li X. H., Deng F. G., Li C. Y., Liang Y. J., Zhou P., and Zhou H. Y., Deterministic secure quantum communication without maximaly entangled states, J. Korean Phys. Soc., 2006, 49: 1354-1359

[32] Yan F. L. and Zhang X., A scheme for secure direct communication using EPR pairs and teleportation Euro. Phys. J. B, 2004, 41: 75-78.

[33] Gao. T., Yan. F. L., and Wang. Z. X., Deterministic secure direct communication using GHZ states and swapping quantum entanglement, J. Phys. A: Math. Gen, 2005, 38: 5761C5770.

[34] Gao T., Yan F. L. and Wang Z. X., Deterministic secure direct communication using GHZ states and swapping quantum entanglement J. Phys. A, 2005, 38: 5761-5770.

[35] Man Z. X., Zhang Z. J., Li Y., Deterministic secure direct communication by using swapping quantum entanglement and local unitary operations, Chin. Phys. Lett. 2005, 22: 18-21.

[36] Wang J., Zhang Q., and Tang C. J., Quantum secure direct communication without a pre-established secure quantum channel, Int. J. Quantum information, 2006, 4: 925-934

[37] Wang H. F., Zhang S., Yeon K. H., and Um C. I., Quantum secure direct communication by using a GHZ state, J. Korean Phys. Soc., 2006, 49: 459-463

[38] Dong L., Xiu X. M., Gao Y. J. and Chi F., Multiparty controlled deterministic secure quantum communication through entanglement swapping, Int. J. Mod. Phys. C, 2008, 19:1673-1681

[39] Han L. F., Chen Y. M., Yuan H., Deterministic Quantum Secure Direct Communication with Dense Coding and Continuous Variable Operations, Comm. Theo. Phys., 2009, 51: 648-652

[40] Deng F. G., Long G. L., Wang Y., and Xiao L., Increasing the efficiencies of random-choice-based quantum communication protocols with delayed measurement, Chin. Phys. Lett., 2004, 21: 2097-2100

[41] Li C.Y., Zhou H.Y., Wang Y., and Deng F.G., Secure quantum key distribution network with Bell states and local unitary operaitons, Chin. Phys. Lett., 2005, 22: 1049-1052

[42] Deng F. G., Li X. H., Li C. Y., Zhou P., Liang Y. J., and Zhou H. Y., Multiparty quantum secret report, Chin. Phys. Lett., 2006, 23: 1676-1679

[43] Bennett C. H., Brassard G., Crépeau C., Jozsa R, Peres A., and Wootters W, K,, Teleporting an unknown quantum state via dual classical and Einstein-Podolsky-Rosen channels, Phys. Rev. Lett., 1993, 70: 1895-1899

[44] Li X. H., Deng F. G. and Zhou H. Y., Improving the security of secure direct communication based on the secret transmitting order of particles, Phys. Rev. A, 2006, 74: 054302

[45] Lin S., Wen Q. Y., Gao F. and Zhu F. C., Quantum secure direct communication with χ-type entangled states, Phys. Rev. A, 2008, 78: 064304

[46] Dong L., Xiu X. M., Gao Y. J. and Chi F., Quantum Secure Direct Communication Using W State, Comm. Theo. Phys. 2008, 49: 1495-1498

[47] Xiu X. M., Dong H. K., Dong L. Gao Y. J. and Chi F., Deterministic secure quantum communication using four-particle genuine entangled state and entanglement swapping, Opt. Comm. 2009, 282: 2457-2459

[48] Zhang Z. J., Multiparty quantum secret sharing of secure direct communication, Phys. Lett. A 2005, 342: 60-66

[49] Bennett C. H., Brassard G., Popescu S., Schumacher B., Smolin J. A., and Wootters W. K., Purification of noisy entanglement and faithful teleportation via noisy channels, Phys. Rev. Lett., 1996, 76: 722-725

[50] Deutsch D., Ekert A., Jozsa R., Macchiavello C. Popescu S., and Sanpera A., quantum privacy amplification and the security of quantum cryptography over noisy channels, Phys. Rev. Lett., 1996, 77: 2818-2821

[51] Wen K. and Long G. L., One-party Quantum Error Correcting Codes for Unbalanced Errors: Principles and Application to Quantum Dense Coding and Quantum Secure Direct Communications, e-print quant-ph/0609207

[52] Deng F. G. and Long G. L., Quantum privacy amplification for a sequence of single qubits, Commun. Theor. Phys., 2006, 46: 443-446

III Applications in Imaging and Light Processing

23

Beating Difraction Limit using Dark States

Hebin Li[1] and Yuri Rostovtsev[2]
[1]*Department of Physics, Texas A&M University,College Station, TX 77843-4242*
[2]*Department of Physics, University of North Texas,1155 Union Circle #311427, Denton, TX 76203-5017*
USA

1. Introduction

For most three centuries after Hooke introduced optical microscopy, refinement of the instrumentation made microscopes more convinient than anything else. As in modern microscopy, a light source, an objective lens, and an eyepiece were used to project an image magnified from a 100-fold to 1,000-fold into a human eye. This range of magnifications and resolution (of the order of 200 nm) has brought cellular morphology and tissue structure into view and made optical microscopy the perfect partner for biological investigations.

A resolution constraint in optical imaging is imposed by the diffraction limit discovered by Lord Rayleigh in the 19th century. Fundamentally, this restricts the ability to resolve two distant point sources, such as stars or planets, with an angular resolution α smaller than the wavelength of light λ divided by the numerical aperture of the imaging system D, $\alpha > \lambda / D$.

The ability to create small images is important for material processing technology and for improving the resolution of microscopy for bio-medical applications [1]. In recent years, there has been a continuous effort to achieve feature sizes smaller than the diffraction limit. One method uses nonlinearity in photoresist processing to generate narrower features (involving a combination of near field and high exposure techniques). Several other methods have been presented that are able to overcome the diffraction limit of the imaging system. Quantum microscopy is based on using a nonclassical optical field approach [2, 3]. Microscopy with classical fields can be enhanced by the nonlinear optical response of the medium [4]. Classical field amplitude and phase arrangements can be used to locate the position of an atom with subwavelength precision in an atomic beam [5, 6, 7], in a cavity [8], and then to apply localization technique to lithography [9, 10, 11], and to achieve subwavelength diffraction and imaging with classical light using the Doppleron-type resonances [12, 13]. Beating diffraction limit was experimentally demonstarted [14] using the dark states.

Another method uses entangled two-photon states to write features of minimum size $\lambda / 4$ in an N-photon absorbing substrate [4,5]. There are technical difficulties associated with the implementation of all these methods, in particular the latter method requires multi-photon absorbing photoresists that are insensitive to single photon effects which is difficult to achieve.

An early discovery in the modern field of quantum optics is that by going to higher-order correlations in the radiation field, one can improve resolution in stellar interferometers, vis a

vis the Hanbury Brown-Twiss effect. Photon correlation interferometry is a powerful tool with applications ranging from the macro- to the micro-cosmos. For example, to measure stellar diameters and resolve binary stars the Hanbury Brown-Twiss effect can be used. As is also known that it is possible to improve the resolution of optical microscopy [2,3,4] by using photon correlation interferometry and entangled quantum states. This has been demonstrated in the elegant experiment [25] via two photon down conversion.
Below we show how to improve microscopy by using higher order quantum correlation function, namely, $G^{(2)}$. Very recently, this concept has been brought down to the microscopic realm and carried the science an important step further by introducing a novel two-photon source, composed of Raman emission pairs, which allow the two-photon correlation to extract source details with a resolution that is far better than that allowed by the Rayleigh criterion [1].
Not long time ago, it has been shown that applying the geometry and physics of the original Raman quantum eraser scheme [2] allows one to resolve molecular markers separated by distances smaller than the wavelength of the probing radiation. Note that in many biophysical studies, the shape or conformation of a protein or a DNA strand is monitored as a function of various parameters. One way of making such measurements involves attaching markers, such as dye molecules or quantum dots, to two known points on the protein and observing their fluorescence as they move apart.
The key features that enable the improved resolution in this scheme is the entanglement in the quantum field and the carefully chosen geometry of the photodetection apparatus, as described below. The space-time correlations between the emitted photons have novel interference properties that show anti-bunching and bunching effects spaced apart by the Rabi period. Experiments analyzing the photon statistics of this scheme [2,3] hint at new methods for the coherent storage and coupling of quantum information between atomic and photonic systems.
In this Chapter, we review physics background of the methods and techniques that allow researchers to beat the so-called diffraction limit using classical as well as quantum fields. In particularly, we review the possibility of creating spatial patterns having subwavelength size by using the so-called dark states formed by the interaction between atoms and optical fields. These optical fields have a specified spatial distribution. Our experiments in Rb vapor display spatial patterns that are smaller than the length determined by the diffraction limit of the optical system used in the experiment. This approach may have applications to interference lithography and might be used in coherent Raman spectroscopy to create patterns with subwavelength spatial resolution.

2. Beating diffraction limit with classical fields

2.1 Two-level atoms

Let us show here how the idea works for two-level atoms. The Hamiltonian of a two-level atom interacting with an optical resonant field (see Fig. 1) is given by

$$H = \hbar\Omega_d |a\rangle\langle b| + adj., \tag{1}$$

where $\Omega_d = \wp_d E_d/\hbar$ is the Rabi frequency of the drive E_d field; $\wp_d$ is the dipole moment of the atomic transitions. Then, the atomic response is given by the set of density matrix equations [17]

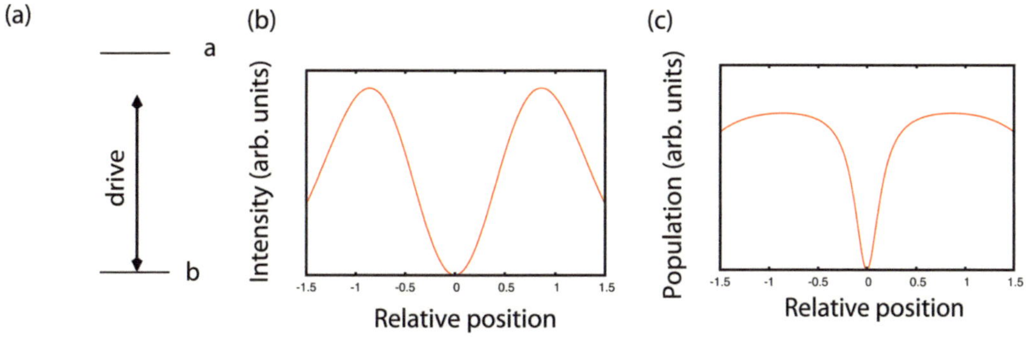

Fig. 1. (a) Energy diagram of a two-level system interacting with a strong drive field. (b) Distribution of the drive field intensity vs. a transverse spatial coordinate. (c) Dependence of the population excited in the atomic medium vs spatial position.

$$\dot{\rho} = -\frac{i}{\hbar}[H,\rho] - \frac{\Gamma\rho + \rho\Gamma}{2} \tag{2}$$

where Γ describes the relaxation processes. In particular for a two-level system the set of equation has the following form

$$\dot{\rho}_{ab} = -\Gamma_{ab}\rho_{ab} + i\Omega_d(n_a - n_b), \tag{3}$$

$$\dot{n}_a = -\gamma n_a + i\Omega_d(\rho_{ab} - \rho_{ba}), \tag{4}$$

where $n_a = \rho_{aa}$, $n_b = \rho_{bb}$; $\Gamma_{ab} = \gamma_{ab} + i(\omega_{ab} - \nu)$, $\gamma_{ab} = 1/T_2$, $\gamma = 1/T_1$ (T_1 and T_2 are corresponding longitudinal and transverse relaxation times). Solving Eqs.(3,4) in a steady-state regime, we obtain

$$\rho_{ab} = i\frac{n_a - n_b}{\Gamma_{ab}}\Omega_d, \tag{5}$$

$$n_a = \frac{2Re\left(\frac{|\Omega_d|^2}{\Gamma_{ab}}\right)}{\gamma + 2Re\left(\frac{|\Omega_d|^2}{\Gamma_{ab}}\right)} n_b \tag{6}$$

Then the population in the upper atomic level is given by

$$n_a = \frac{2Re\left(\frac{|\Omega_d|^2}{\Gamma_{ab}}\right)}{\gamma + 4Re\left(\frac{|\Omega_d|^2}{\Gamma_{ab}}\right)}, \tag{7}$$

for the case of resonance, $\nu = \omega_{ab}$, it is reduced to

$$n_a = \frac{2|\Omega|^2 T_1 T_2}{1 + 4|\Omega|^2 T_1 T_2}. \tag{8}$$

We now assume that the drive field has a spatial distribution of intensity.

$$|\Omega|^2 = |\Omega_0|^2 f(x), \tag{9}$$

where $f(x)$ is the spatial distribution of the intensity of optical drive field. For example, in the case of interference of two waves with wavevectors k_1 and k_2, the optical field is

$$E_d = E_1 e^{ik_1 r} + E_2 e^{ik_2 r}, \tag{10}$$

and intensity distribution is given by

$$|E_d|^2 = |E_1|^2 + |E_2|^2 + 2E_1E_2\cos(k_1 - k_2)r = (|E_1| - |E_2|)^2 + 4|E_1||E_2|\sin^2 2(k_1 - k_2)r, \tag{11}$$

intensity at different spatial position changes between $(|E_1|-|E_2|)^2$ and $(|E_1|+|E_2|)^2$. Introducing $G = 2|\Omega_0|^2 T_1T_2$, we can write

$$\rho_{aa} = \frac{Gf(x)}{1+2Gf(x)}. \tag{12}$$

Then, for the drive field at the position being near to its zero the Rabi frequency is given by

$$|\Omega_d(z,x)|^2 = |\Omega_0|^2 \left(\frac{x}{L}\right)^2, x \ll L, \tag{13}$$

where $\Omega_0 = \Omega_d(z,x_0)$, L is the separation distance between the peaks of the drive field distribution (for interference patern, $L = \lambda/2\sin(\theta/2) > \lambda/2$). A typical excitation profile vs. x shown in Fig. 1(c) demonstrates that the spatial width of excitation can be smaller than the intensity distribution of the optical field, and even smaller than the spot size determined by diffraction limited size. Indeed, the width of spatial distribution of excited atoms is given by

$$\Delta x \simeq \frac{L}{2\sqrt{G}} = \frac{L}{2\Omega_0\sqrt{T_1T_2}}. \tag{14}$$

The most important feature of Eq.(14)is that the width depends on the relaxation parameters and the field strength, but not the diffraction of optical field.

2.2 Using Stark shifts

Three-level atoms provide more flexibility for the localization of the excited atoms or molecules because of different physical mechanisms can be involved. For example, it is shown in Fig. 2 how to use Stark shifts for atomic localization [5]. Level structure of a three-level atom is shown in Fig. 2a (for example ^{152}Sm). Geometry of atomic beam and optical beams can be seen in Fig. 2b. Probe 1 beam is used to optically pump all population in level c. Then atoms reach the region where they have inhomogenious drive beam which is detuned from the atomic resonance and simultaneously this region has a probe beam 2 with frequency ν_2. Due to Stark shift atoms at different spatial location have energy of the excited state a as

$$e_\pm = \frac{\Delta}{2} \pm \sqrt{\left(\frac{\Delta}{2}\right)^2 + |\Omega|^2}, \tag{15}$$

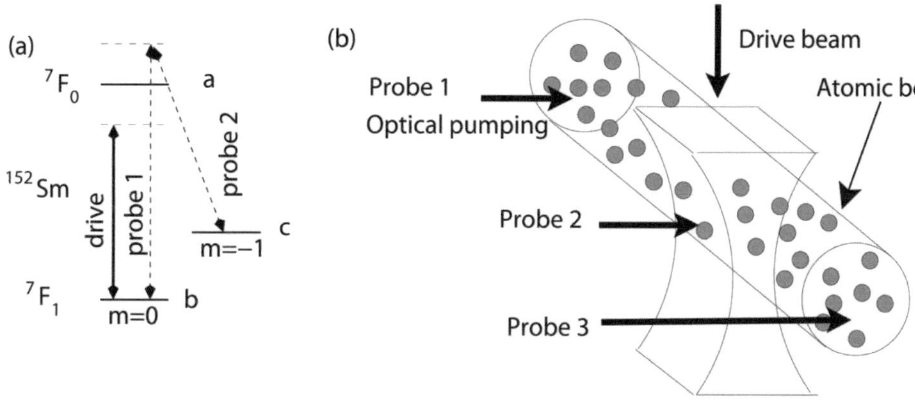

Fig. 2. Qualitative description of the idea of using Stark shifts for atomic localization. (a) Level structure of ^{152}Sm, as an example, and the applied fields. (b) Geometry of atomic beam of ^{152}Sm and optical beams. Probe 1 beam is used to pump all population in level *c*. Drive beam detuned from the atomic resonance and it has spatial distribution such that, at each location it has different Stark shift. Probe 2 beam resonantely interacts with atoms at the particular spatial location where it is resonant to the optical transition. The effect of probe 2 beam is pumping resonant atoms to level *b*. Probe 3 is the field to excite fluorescence from the atoms in the ground state *b*.

where Δ is the detuning of the drive beam, and Ω is the Rabi frequency of the drive beam. The atoms have different detuning from the resonance at different positions, and some of them are at the resonance when the detuning is less than the spontaneous emission rate γ,

$$e_+ - \nu_2 + \frac{1}{\Delta}\frac{\partial|\Omega_0|^2}{\partial x}\Delta x \leq \gamma. \tag{16}$$

The resonant interaction of these atoms with probe 2 beam results in population of the ground state *b*. Then, the probe 3 beam resonantely interacts with atoms at the particular spatial location where it is resonant to the optical transition to cause fluorescence which is detected. The localization of the atoms can be found from Eq.(16)

$$\Delta x = \frac{\gamma\Delta}{\dfrac{\partial|\Omega_0|^2}{\partial x}}, \tag{17}$$

which is also determined by the relaxation rate γ, detuning Δ, and the spatial derivative of drive field intensity, and, the most important is not directly related to the diffraction, and consequently can be smaller than the wavelength of optical radiation as was demonstrated in [5].

2.3 Beating diffraction limit by using Dark states

The Hamiltonian of a three-level atom interacting with optical fields (see the inset in Fig. 4) is given by

$$H = \hbar\Omega_d|a\rangle\langle b| + \hbar\Omega_p|a\rangle\langle c| + adj., \tag{18}$$

where $\Omega_{d,p} = \wp_{d,p}E_{d,p}/\hbar$ are the Rabi frequencies of the drive E_d and the probe E_p fields, respectively; $\wp_{d,p}$ are the dipole moments of the corresponding optical transitions. Then, the atomic response is given by the set of density matrix equations [17]

$$\dot{\rho} = -\frac{i}{\hbar}[H,\rho] - \frac{\Gamma\rho + \rho\Gamma}{2} \tag{19}$$

where Γ describes the relaxation processes.
Here, we present a new approach that is based on coherent population trapping [15, 16, 17, 18, 19]. Optical fields applied to a three-level quantum system excite the so-called *dark state*, which is decoupled from the fields. Similar approaches using coherent population trapping have also been developed by several groups (for example, see [20, 22, 23, 24]).

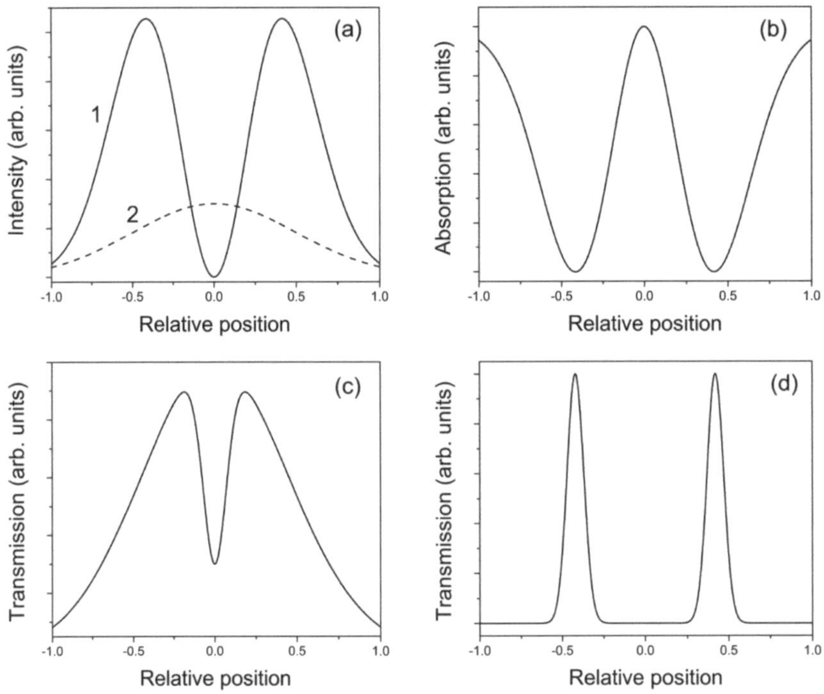

Fig. 3. Qualitative description of the idea. (a) Distribution of the drive (1) and the probe (2) fields vs. a transverse spatial coordinate at the entrance to the cell. (b) Dependence of the absorption coefficient given by Eq.(21) vs position. Plots (c) and (d) show the distribution of the probe beam after propagating through the cell. Case (c) is for a strong drive field and relatively low optical density. Case (d) is for a relatively weak drive field and large optical density.

As a qualitative introduction, assume that the drive field Rabi frequency Ω_d has the particular spatial distribution sketched in Fig. 3(a) by the solid line (1). The weak probe field Rabi frequency Ω_p ($\Omega_p \ll \Omega_d$) has a diffraction limited distribution (shown by the dashed line (2) in Fig. 3(a)). The probe and drive fields are applied to the atom (see the inset in Fig. 4, for the case of ^{87}Rb atoms, where $|a\rangle = |5^2P_{1/2}, F = 1, m = 0\rangle$, $|b\rangle = |5^2S_{1/2}, F = 1, m = -1\rangle$, $|c\rangle = |5^2S_{1/2}, F = 1, m = +1\rangle$). At all positions of nonzero drive field, the dark state, which is given

[15, 16, 17, 18, 19] by $|D\rangle = (\Omega_p|c\rangle - \Omega_d|b\rangle)/\sqrt{\Omega_p^2+\Omega_d^2}$, is practically $|b\rangle$. When the drive field is zero, the dark state is $|c\rangle$, and the atoms at these positions are coupled to the fields and some atoms are in the upper state $|a\rangle$. The size of a spot where the atoms are excited depends on the relaxation rate γ_{cb} between levels $|b\rangle$ and $|c\rangle$. For $\gamma_{cb} = 0$, the size of spot is zero, smaller than the optical wavelength.

The propagation of the probe field Ω_p through the cell is governed by Maxwell's Equations and, for propagation in the z-direction, can be written in terms of the probe field Rabi frequency as

$$\frac{\partial \Omega_p}{\partial z} = -i\eta\rho_{ab} - i\frac{1}{2k}\frac{\partial^2}{\partial x^2}\Omega_p. \tag{20}$$

The first term accounts for the dispersion and absorption of the resonant three-level medium, and the second term describes the focusing and/or diffraction of the probe beam. The density matrix element ρ_{ab} is related to the probe field absorption which in turn depends on the detuning and the drive field. This is characterized by an absorption coefficient:

$$\kappa = \eta\frac{\Gamma_{cb}}{\Gamma_{ab}\Gamma_{cb} + |\Omega_d(z,x)|^2}, \tag{21}$$

where $\Gamma_{cb} = \gamma_{cb} + i\omega$ and $\Gamma_{ab} = \gamma + i\omega$; $\omega = \omega_{ab} - \nu$ is the detuning from the atomic frequency ω_{ab}; γ is the relaxation rate at the optical transition; and $\eta = 3\lambda^2 N\gamma_r/8\pi$; N is the atomic density; γ_r is the spontaneous emission rate. We now assume that the drive field has a distribution of intensity near its extrema given by

$$|\Omega_d(z,x)|^2 = |\Omega_0|^2 \begin{cases} \left[1 - \left(\frac{x-x_0}{L}\right)^2\right], & x \simeq x_0, \\ \left(\frac{x}{L}\right)^2, & x \ll L, \end{cases} \tag{22}$$

where $\Omega_0 = \Omega_d(z,x_0)$, L is the separation distance between the peaks of the drive field distribution, and a typical absorption profile vs. x is shown in Fig. 3(b). Neglecting the diffraction term in Eq.(20), we can write an approximate solution for Eq. (20) as

$$\Omega_p(z,x) = \Omega_p(z=0,x)\exp(-\kappa z). \tag{23}$$

For relatively low optical density ($\kappa z \simeq 1$), nearly all of the probe field propagates through the cell except for a small part where the drive field is zero (see Fig. 3(a)). Absorption occurs there because the probe beam excites the atomic medium. The width of the region of the excited medium, in the vicinity of zero drive field, is characterized by

$$\Delta x = L\sqrt{\frac{\Gamma_{ab}\Gamma_{cb}}{|\Omega|^2}}, \tag{24}$$

where $\Omega = \Omega_d(z=0, x=0)$. This region is small, but its contrast is limited because of the finite absorption of the medium at the center of optical line (Fig. 3(c)).

For higher optical density, this narrow feature becomes broadened (compare Fig. 3(c) and (d)), but two narrow peaks are formed during the propagation of the probe beam (see Fig. 3(d)). For zero detuning, their width is given by

$$\Delta x = L\sqrt{\frac{|\Omega|^2}{\eta \gamma_{cb} z}}. \tag{25}$$

The drive field provides flexibility for creating patterns with sizes smaller than the wavelength of the laser. The distribution of fields is governed by electrodynamics and has a diffraction limit, while the distribution of molecules in their excited states is NOT related to the diffraction limit, but rather determined by the relaxation rates Γ_{ab} and Γ_{cb}, and thus can have spatial sizes smaller than the wavelength.

3. Experimental demonstration

In this section, we report a proof-of-principle experiment in Rb vapor to demonstrate our approach. We have observed that the distribution of the transmitted probe beam intensity has a double-peak pattern, which is similar to that of the drive beam, but the width of the peaks of the probe beam is narrower than that of the drive beam.

The experimental schematic is shown in Fig. 4. We obtain a good quality spatial profile by sending the radiation of an external cavity diode laser through a polarization-preserving single-mode optical fiber. The laser beam is vertically polarized and split into two beams (drive and probe). The probe beam carries a small portion of the laser intensity, and its polarization is rotated to be horizontal.

To create a double-peak spatial distribution for the drive field, the drive beam is split into two beams that cross at a small angle, using a Mach-Zehnder interferometer (shown in the dashed square of Fig. 4). A typical two-peak interference pattern of crossing beams is shown as Fig. 4A.

The probe and drive beams combine on a polarizing beam splitter, arranged so that the probe field and the interference pattern of the drive field are overlapped in a Rb cell. The Rb cell has a length of 4 cm, and is filled with ^{87}Rb. A magnetic shield is used to isolate the cell from any environmental magnetic fields, while a solenoid provides an adjustable, longitude magnetic field. The cell is installed in an oven that heats the cell to reach an atomic density of 10^{12} cm^{-3}. The laser is tuned to the D_1 line of ^{87}Rb at the transition $5^2S_{1/2}(F = 2) \rightarrow 5^2P_{1/2}$ $(F = 1)$.

As stated above, the probe and drive beams have the orthogonal linear polarizations. A quarter-wave plate converts them into left and right circularly polarized beams, which couple two Zeeman sublevels of the lower level and one sublevel of the excited level of the Rb atoms (see the inset of Fig. 4).

After passing through the cell, the probe and drive beams are converted back to linear polarizations by another quarter-wave plate and the separated by a polarizing beam splitter (PBS). The power of transmitted probe field is monitored by a photodiode (PD). The spatial intensity distribution of probe field is recorded by an imaging system, consisting of the lens L3 and a CCD camera.

The intensity of the probe beam is low enough that its transmission through the cell is almost zero without the presence of drive laser. Applying the drive laser makes the atomic medium transparent for the probe laser wherever the EIT condition is satisfied. If the drive laser has a certain transverse spatial distribution, then that pattern can be projected to the transmission profile of the probe laser.

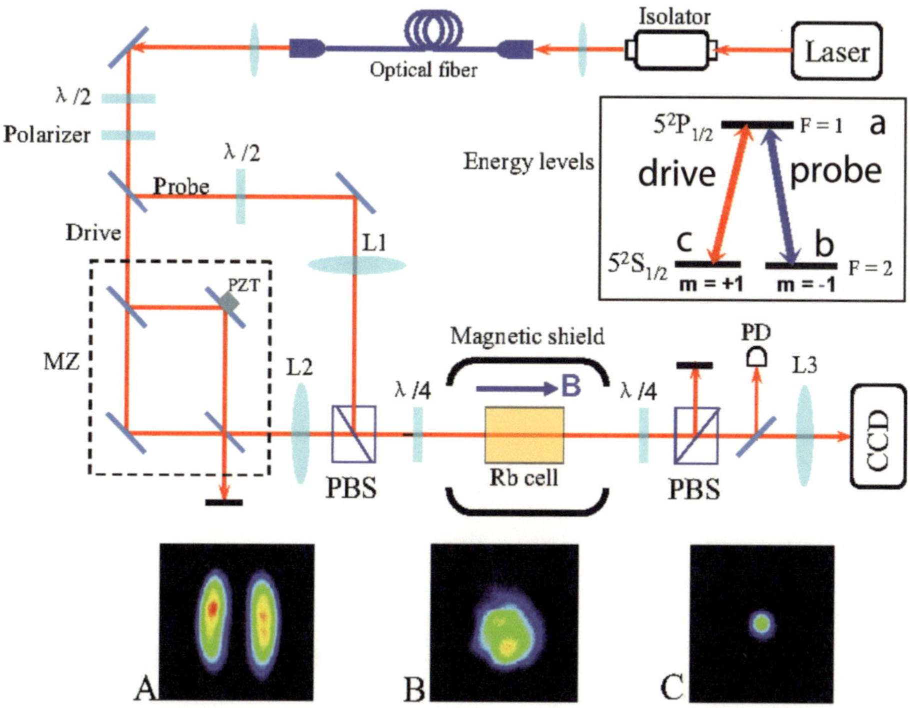

Fig. 4. Experimental schematic. λ /2: half-wave plate; λ /4: quarter-wave plate; L1, L2, L3: lenses; MZ: Mach-Zehnder interferometer; PZT: piezoelectric transducer; PBS: polarizing beam splitter, PD: photo diode; CCD: CCD camera. Picture A is the spatial intensity distribution of the drive field. Picture B is the beam profile of the parallel probe beam without the lens L1. Picture C is the beam profile of the diffraction limited probe beam with the lens L1. All three of pictures have been made with with the camera at the location of the cell, which has temporally been removed. The inset is the energy diagram of the Rb atom, showing representative sublevels.

Two different experiments have been performed. In the first experiment, the lenses L1 and L2 are not used, and the probe beam is a parallel beam with a diameter of 1.4 mm. The image of the drive intensity distribution in the cell is shown in Fig. 5(a). The probe intensity has a Gaussian distribution before entering the cell, and its distribution is similar to the drive intensity distribution after the cell. As shown in Fig. 5(b), however, the transmitted probe intensity has a distribution that has sharper peaks compared with the pattern of the drive intensity. The horizontal cross-sections of the drive and the transmitted probe distributions are shown in Figs. 5(c) and (d) respectively. In the drive intensity profile, the width (FWHM) of the peaks is 0.4 mm. The width (FWHM) of the peaks in the transmitted probe intensity profile is 0.1 mm. The spacing between two peaks is the same for both the drive and transmitted probe fields. We define the *finesse* as the ratio of the spacing between peaks to the width of peaks. The finesse of the transmitted probe intensity distribution is a factor of 4 smaller than that of the drive intensity distribution.

In the second experiment, the lenses L1 and L2 are used. A parallel probe beam (Fig. 4B) with a diameter of 1.4 mm is focused by the lens L1, which has a focal length of 750 mm. The beam size at the waist is 0.5 mm, which is diffraction limited. To assure experimentally that

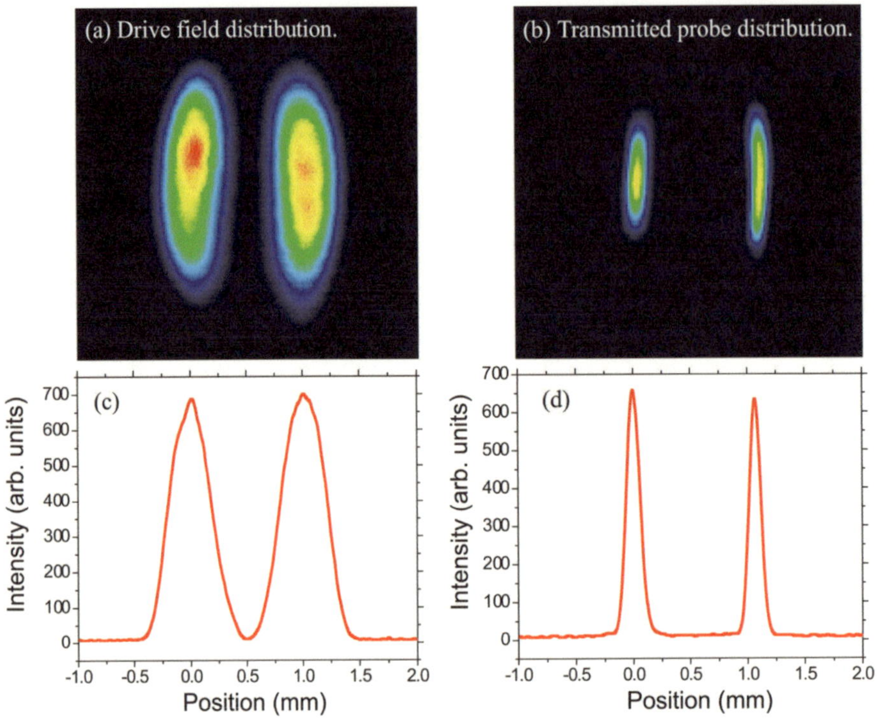

Fig. 5. The results of the experiment with a parallel probe beam. Picture (a) shows the image of the intensity distribution of the drive field in the Rb cell. Picture (b) shows the intensity distribution of the transmitted probe field. Curves (c) and (d) are the corresponding intensity profiles. The widths of the peaks in curves (c) and (d) are 0.4 mm and 0.1 mm, respectively.

the beam is diffraction limited, we increased the beam diameter of the parallel beam by the factor of 2, and the beam size at the waist became two times smaller. The lens L2 is used to make the drive beam smaller in the Rb cell, where the pattern of drive field is spatially overlapped with the waist of the probe beam. Classically, there should be no structures at the waist of the probe beam because it is diffraction limited. Structures can be created in a region smaller than the diffraction limit in our experiment, however. The experimental result is shown in Fig. 6. The drive field still has a double peak intensity distribution (Fig. 6(a)). The transmission of the diffraction limited probe beam also has a double-peak intensity distribution as shown in Fig. 6(b). Curves (c) and (d) are the beam profiles of the drive and transmitted probe beams respectively. The width of the peaks in the drive beam is 165 μm, and the width of the peaks in the transmitted probe beam is 93 μm. The finesse of the transmitted probe beam is 1.8 times greater than that of the drive beam. For the probe beam, the structure created within the diffraction limit has a size characterized by the width of peaks (93 μm). This characteristic size is 5 times smaller than the size of the diffraction limited probe beam (500 μm, see the spot of Fig. 4(C)).

At the end, we would like to stress here that the concept based on dark states successfully works in Rb vapor. One can see that the width of the probe image (C) is at least three times smaller than the width of the drive image (A). Although the diffraction limit is "beaten," the experiment does not violate any laws of optics. The probe beam is diffraction limited, but

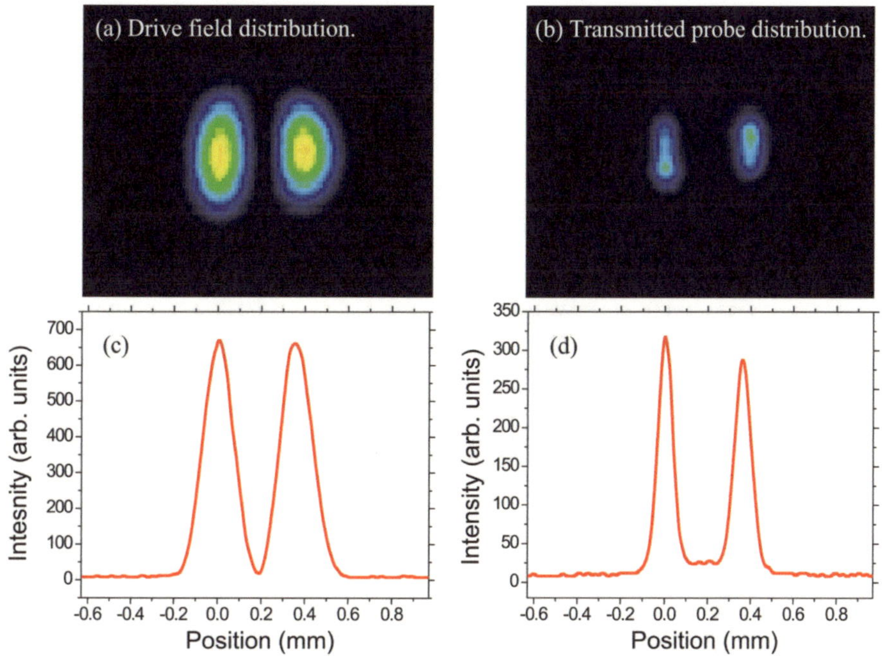

Fig. 6. The results of the experiment with the diffraction limited probe beam. Picture (a) shows the image of the intensity distribution of the drive field in the Rb cell. Picture (b) shows the image of the intensity distribution of the transmitted probe field. Curves (c) and (d) are the corresponding profiles. The widths of the peaks in curves (c) and (d) are 165 μm and 93 μm, respectively.

the atoms are much smaller than the size of diffraction-limited beam. Moreover, due to the strong nonlinearity of the EIT, the characteristic size of the pattern in the transmitted probe beam is much smaller than that of the drive beam and the diffraction limit of the probe beam.

We have also measured the narrowing effect vs. the detuning of the probe field and have performed simulations using the density matrix approach. The results are shown in Fig. 7. The calculations reproduce the data satisfactorily. The dependence on detuning has not been considered in [20, 23, 24, 22]. It is unique for our approach and can be understood in the following way. Absorption by the atomic medium given by Eq.(21) with a drive intensity distribution given by Eq.(22) can be written as

$$\kappa = \eta \left(\frac{\gamma_{cb}}{|\Omega|^2} + \frac{\gamma\omega^2}{|\Omega|^4} + \left(\frac{\gamma_{cb}}{|\Omega|^2} + 2\frac{\gamma\omega^2}{|\Omega|^4} \right) \left(\frac{x}{L} \right)^2 \right). \tag{26}$$

Then, ratio of the width of the probe intensity distribution to the width of the drive intensity distribution is given by

$$R = \frac{L}{\Delta x} = \sqrt{\eta z \left(\frac{\gamma_{cb}}{|\Omega|^2} + 2\frac{\gamma\omega^2}{|\Omega|^4} \right)}. \tag{27}$$

From this we see that the finesse increases with the detuning.

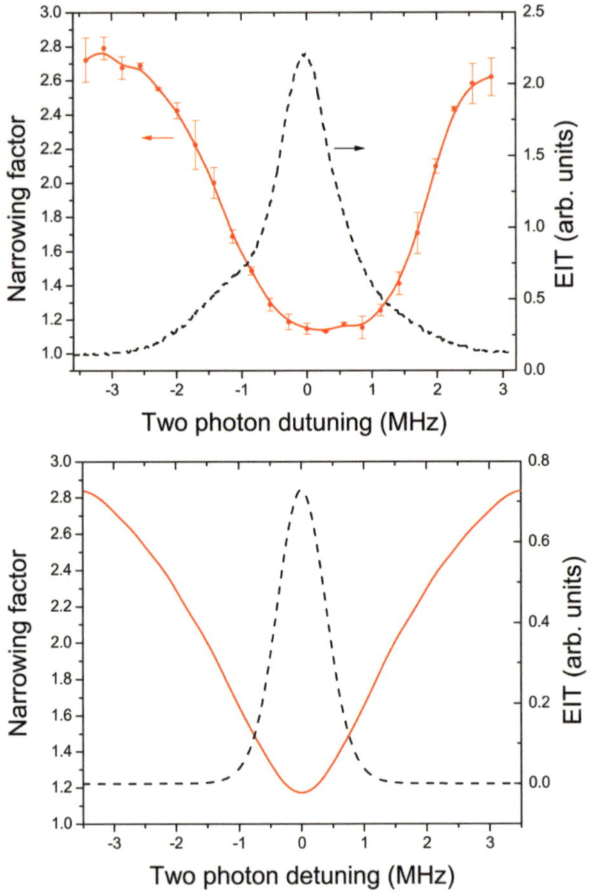

Fig. 7. Narrowing of the transmitted probe intensity distribution as function of the probe detuning: (a) experimental results and (b) theoretical simulation. The transmition of the probe is shown as well.

It is worth to mention here that a proof-of-principle experiment has been already reported in [14] that the concept works in Rb vapor and have experimentally demonstrated the possibility of creating structures having widths smaller than those determined by the diffraction limits of the optical systems. The results obtained here can be viewed as an experimental verification of our approach, as well as evidence supporting the theoretical predictions and results obtained by others [20, 23, 24, 22]. The challenges associated with pushing our method to the subwavelength regime are formidable. In vapor or gaseous medium, transit-time broadening is the dominant dephasing mechanism that limits the smallness of the region in which a dark state can be formed. Solid-state systems may be more appropriate, although, the most difficult aspect of this approach is devising a way to observe subwavelength structures. This technique might be used in microscopy by studying the distribution of molecules with subwavelength resolution or in lithography by manipulating molecules in the excited state. Also, note that it may be possible to apply this approach to coherent Raman scattering (for example, CARS). This may improve the spatial resolution of CARS microscopy.

4 Microscopy with quantum fields

4.1 Simplified model

Let us consider two identical marker molecules that are separated by distance d from each other (see Fig. 8). The level structure of molecules is shown in Fig. 9. We also assume that there is no dipole-dipole interaction between molecules, i.e. the level structure does not depend on the distance between molecules.

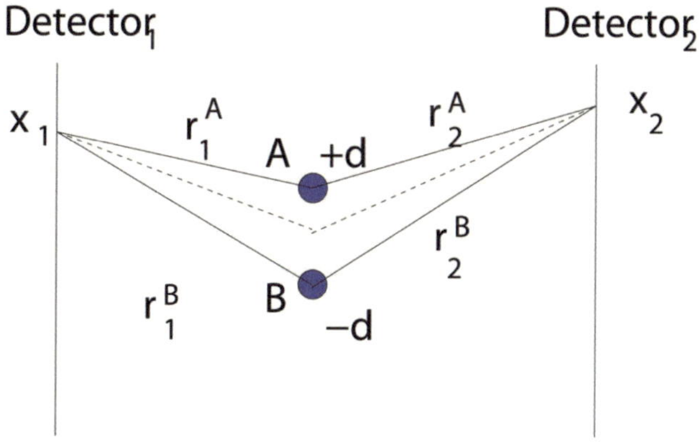

Fig. 8. Two marker molecules at some distance that can be resolved by quantum microscopy analyzing shifts or magnitudes of $G^{(2)}$ fringes.

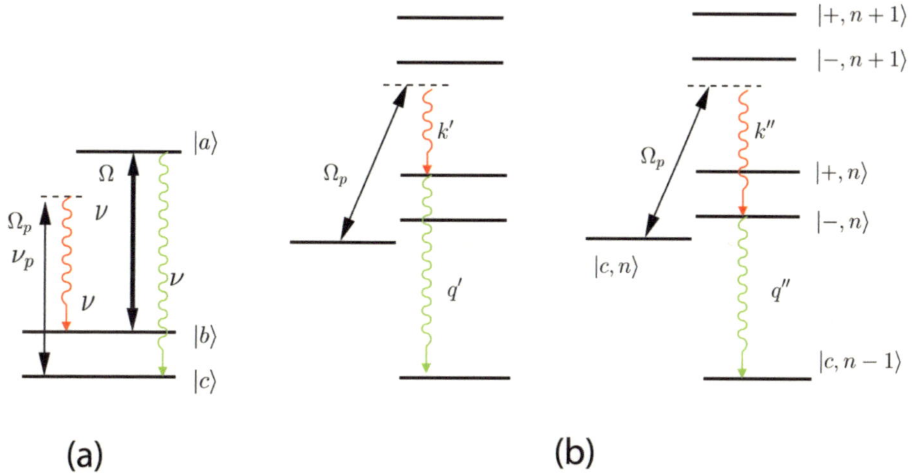

Fig. 9. (a) Raman scheme to generate correlated pair of photons is shown. (b) In dressed state basis, one can see that the system is ladder scheme with splitted intermediate state similarly to eraser scheme studied in [2].

We shine the laser radiation on the system. It consists of two fields that we treat as classical fields: the first field Ω_p is weak and it has large one-photon detuning from the resonance; the second field Ω is much stronger than the first one ($\Omega_p \ll \Omega$) and it is resonant to the electron transition between the ground state and the excited state of two marker molecules.
The first field having frequency ν_p excited the molecule from ground state level b to generate a Raman Stockes photon, and the molecule ends up in the level c. Then the second laser field

having frequency ν_d excites the molecule to level a to generate anti-Stokes photon and ends down to the starting ground level b.

Now, if the two emitted photons are well separated in time, then the two photon state $|\Psi\rangle$ can be factorized as a sum of products of one photon Stokes ($|\nu\rangle$) and anti-Stokes ($|\omega\rangle$) states from molecules at A and B, that is

$$|\Psi\rangle = \frac{1}{\sqrt{2}}[|\nu,\omega\rangle_A + |\nu,\omega\rangle_B]. \tag{28}$$

Then the spontaneous Stokes and anti-Stokes radiation of Fig. 1a will be independent, and the Glauber photon-photon correlation function factorizes. To see this we recall

$$G^{(2)}_{\nu\omega} = \left\langle \Psi \middle| \hat{E}^{(-)}(1)\hat{E}^{(-)}(2)\hat{E}^{(+)}(2)\hat{E}^{(+)}(1) \middle| \Psi \right\rangle \tag{29}$$

in which $G^{(2)}_{\nu\omega} = G^{(2)}_{\nu\omega}(1,1';2,2')$, $\hat{E}^{(+)}(1)$ is the positive frequency (annihilation operator) part of the electric field and $E^{(-)}(1)$ is the corresponding negative frequency (creation operator); 1 stands for $\tilde{\mathbf{r}}_1$, t_1 where $\tilde{\mathbf{r}}_1$ is the vector to detector 1, etc (Fig. 1a). The times t_i are controlled by e.g., shutters.

The two-photon probability $G^{(2)}(1,2) = |\Psi(1,2)|^2$ for a single molecule has been calculated in [2]. The photon-photon correlation function for two molecules has been calculated from the two photon amplitude [2], and it is given by

$$\begin{aligned} G^{(2)}/\mathscr{C} = {} & \mathrm{e}^{-\Gamma\tau_1}\mathrm{e}^{-\gamma\tau_{21}}\sin^2\tilde{\Omega}\tau_{21} + \mathrm{e}^{-\Gamma\tau_1'}\mathrm{e}^{-\gamma\tau_{21}'}\sin^2\tilde{\Omega}\tau_{21}' + \\ & \mathrm{e}^{-(\Gamma/2)(\tau_1+\tau_1')}\mathrm{e}^{-(\gamma/2)(\tau_{21}+\tau_{21}')}\sin\tilde{\Omega}\tau_{21}\sin\tilde{\Omega}\tau_{21}' \times \\ & 2\cos[\nu(\tau_1-\tau_1')+\omega(\tau_2-\tau_2')] \end{aligned} \tag{30}$$

with $\tau_1' > 0, \tau_1 > 0, \tau_2 > \tau_1$ and $\tau_2' > \tau_1'$. Physically, this describes the finite time, governed by $\tilde{\Omega}$, to promote the molecule from b to a (similarly to the driven two-level system, see [17]). Mathematically, the vanishing of $G^{(2)}$ when $\tau_1 = \tau_2$, is a result of quantum interference between the two paths of Fig. 2.

4.2 Determine minimal distance between marker molecules

We are interested to determine how small distance between molecules d can be. To answer this question we present $G^{(2)}$ as series on d. The geometry is shown in Fig. 8. The distance from molecule A to detector 1 is given by

$$r^{A,B}_{1,2} = \sqrt{d^2_{1,2} + (x_{1,2} \mp d/2)^2} \simeq R_{1,2} \mp \frac{dx_{1,2}}{2R_{1,2}}, \tag{31}$$

We can rewrite $\tilde{\Omega}\ \tau_{21}$ as $\frac{\tilde{\Omega}}{c}r_{12} = \alpha - d\beta$, and $\tilde{\Omega}\tau_{21}'$ as $\frac{\tilde{\Omega}}{c}r_{12} = \alpha + d\beta$, where $\alpha = k_\Omega(R_1 - R_2) = k_\Omega R_{12}$ and $\beta = (\frac{x_1}{2R_1} - \frac{x_2}{2R_2})$. Interference term is given by $\cos[\nu(\tau_1 - \tau_1') + \omega(\tau_2 - \tau_2')] = \cos\xi\, d$, where $\xi = k[\frac{x_1}{R_1} + \frac{x_2}{R_2}]$.

The $G^{(2)}(x_1,x_2)$ is given then by

$$G^{(2)}(x_1,x_2) = \sin^2(\alpha - d\beta) + \sin^2(\alpha + d\beta) + 2\sin(\alpha - d\beta)\sin(\alpha + d\beta)\cos\xi d. \tag{32}$$

First, we consider moving two detectors together, $x_1 = x_2$, so

$$G^{(2)}(x_1) = \sin^2[\tilde{k}_\Omega(R_1 - D)]\cos^2[\tilde{k}d\frac{x_1}{R_1}] \tag{33}$$

where we introduce $\tilde{k}_\Omega = \frac{\Omega}{c}(n_1 - n_2)$, $\tilde{k} = k(n_1+n_2)$. The location of the fringes is given by $\frac{\partial G^{(2)}(x_1)}{\partial x_1} = 0$, or

$$\frac{D^2}{R_1^2}\frac{\tilde{k}d}{x_1}\tan[\tilde{k}d\frac{x_1}{R_1}]\tan[\tilde{k}_\Omega(R_1 - D)] = \tilde{k}_\Omega. \tag{34}$$

Using $kd \ll 1$, the position of the first fringe can be given by

$$x_p = \sqrt{(\frac{\pi}{2\tilde{k}_\Omega} + D)^2 - D^2 - \frac{\tilde{k}^2 d^2 D^2}{\tilde{k}_\Omega^2(\frac{\pi}{2\tilde{k}_\Omega} + D)^2\sqrt{(\frac{\pi}{2\tilde{k}_\Omega} + D)^2 - D^2}}}. \tag{35}$$

Minimal distance between molecules can be found at the given accuracy of measurement of fringe position as

$$d_{min} = \frac{\tilde{k}_\Omega(\frac{\pi}{2\tilde{k}_\Omega} + D)\sqrt{(\frac{\pi}{2\tilde{k}_\Omega} + D)^2 - D^2}}{\tilde{k}D}\sqrt{\varepsilon}, \tag{36}$$

where $\varepsilon = \delta x_p/x_p$. While the change of distance between molecules can be measured much more accurately, namely

$$\delta d_{min} = \frac{\lambda}{8\pi^2(n_1+n_2)^2}\left(\frac{\lambda}{d}\right)F(\tilde{k}_\Omega, D)\varepsilon, \tag{37}$$

where we introduce a function

$$F(\tilde{k}_\Omega, D) = \frac{\pi\tilde{k}_\Omega D}{2}\left(\frac{\pi}{2\tilde{k}_\Omega D} + 2\right)\left(\frac{\pi}{2\tilde{k}_\Omega D} + 1\right)^2 \tag{38}$$

optimizing the Rabi frequency of the drive laser field we obtain

$$\delta d_{min} = \frac{\lambda}{8\pi^2(n_1+n_2)^2}\left(\frac{\lambda}{d}\right)\frac{3\sqrt{5}+7}{\sqrt{5}-1}\varepsilon \tag{39}$$

for $\tilde{k}_\Omega D = \pi/(\sqrt{5} - 1)$. Thus optimize driving field we obtain

$$\delta d = \frac{\lambda}{3(n_1+n_2)^2}\left(\frac{\lambda}{d}\right)\varepsilon \tag{40}$$

Second, In the limit of small $d \ll r_{1,2}, R_{1,2}, d_{1,2}, |x_{1,2}|$, we can rewrite Eq.(32) as

$$G^{(2)}(x_1,x_2) \simeq (4-\xi^2 d^2)\sin^2\alpha \tag{41}$$

Then the ratio is given by

$$\frac{G^{(2)}(x_1)}{G^{(2)}(x_1')} = \frac{\sin^2\alpha(x_1)}{\sin^2\alpha(x_1')}\left(1-\frac{\xi^2(x_1)-\xi^2(x_1')}{4}d^2\right) \tag{42}$$

Assume that $x_1+x_2=0$ we obtain

$$\frac{G^{(2)}(x_1)}{G^{(2)}(x_1')} = \frac{\sin^2\alpha(x_1)}{\sin^2\alpha(x_1')}\left(1+\frac{\xi^2(x_1')}{4}d^2\right) \tag{43}$$

In addition, assume that we can measure the ratio with $\varepsilon = 1\%$ accuracy

$$\frac{\xi^2(x_1')}{4}d^2 \simeq 10^{-2} \rightarrow d_{min} = \frac{\sqrt{2}\,\lambda}{\pi\;2}\sqrt{\frac{R_1}{x_1}}\sqrt{\varepsilon} \simeq \frac{\lambda}{20\pi}. \tag{44}$$

Interesting feature of the microscope is that it has high sensitivity to small changing of the distance between molecules. Indeed,

$$\delta d = \frac{\lambda}{2\pi^2(n_1+n_2)^2}\left(\frac{R_1}{x_1}\right)^2\left(\frac{\lambda}{d}\right)\varepsilon. \tag{45}$$

5. Final remarks

We have reviewed the basic physics underlying to localize atoms and molecules better than the the so-called diffraction limit. Precise atomic position measurement has attracted a great deal of interest for many years because of many applications such as atom lithography, microscopy, and atom localization. On the other hand, the resonance fluorescence emitted collectively by many interacting two-level atoms in an external driving laser field has been studied extensively for different parameter ranges [26, 27, 28, 29, 30, 31, 32, 33]. Following the idea of reaching subwavelength resolution for nonidentical, individually addressable objects [15], a relation between collective fluorescent light and the geometry of the setup was also shown experimentally [34].

In [35], motivated by the localization of an atom inside an optical field [36], it has been showed that distance and position information can be obtained by measuring the fluorescence spectrum of a two-atom system inside a standing-wave field, relying entirely on far-field measurement techniques. Typically, this scheme will be limited by the difficulties in fixing the positions of the two atoms rather than by constraints of the measurement scheme itself, which in principle allows one to achieve resolution far below the classical Rayleigh limit of optical microscopy technology.

In addition to the fluorescence spectrum, also the intensity-intensity correlation function of the light emitted by a collection of two- and three-level atom systems subject to driving fields has been investigated [31, 32, 37, 38, 39]. Most of these works, however, focused on nonclassical properties of the emitted field.

The role of the dipole-dipole interaction between atoms has been recently taken into account in [21]. In particular, it was demonstrated how the distance information can be obtained by measuring the intensity-intensity correlation function of the emitted fluorescence field. It turns out that the power spectrum of the intensity-intensity correlation function is well suited to gaining distance information over a wide range of parameters with high accuracy. The obtained results can be applied to physical systems which may be approximated as two-level systems, where the two energy states are connected by an electric-dipole-allowed transition. Possible examples include atoms, molecules, and artificial quantum systems such as quantum dots.

6. Acknowledgement

We gratefully acknowledge the support of the Office of Naval Research, the Robert A. Welch Foundation (Grants # A1261 and A1547), and the CRDF (Award No. AZP-3123-BA-08 of Azerbaijan National Science Foundation).

7. References

[1] C.A. Mack, Fundamental Principles of Optical Lithography : the Science of Microfabrication, (Chichester,West Sussex, England:Wiley, 2007).
[2] M.O. Scully, K. Druhl, Phys. Rev. A25, 2208 (1982);U. Rathe,M.O. Scully, Lett.Math. Phys. 34, 297 (1995);M.O. Scully, U. Rathe, C. Su, G.S. Agarwal, Opt. Commun. 136, 297 (1997).
[3] A. N. Boto, P. Kok, D. S. Abrams, S. L. Braunstein, C. P. Williams, and J. P. Dowling, Phys. Rev. Lett. 85, 2733-2736 (2000).
[4] V. Westphal and S.W. Hell, Phys. Rev. Lett. 94, 143903 (2005). S.W. Hell and J. Wichmann, Opt. Lett. 19, 780 (1994)M. Dyba and S.W. Hell, PRL 88, 163901 (2002)
[5] J. R. Gardner,M. L. Marable, G. R. Welch, and J. E. Thomas, Phys. Rev. Lett. 70, 3404 (1993).
[6] K. D. Stokes, C. Schnurr, J. R. Gardner, M. Marable, G. R. Welch, and J. E. Thomas, Phys. Rev. Lett. 67, 1997 (1991).
[7] J. E. Thomas, Opt. Lett. 14, 1186 (1989). J. E. Thomas, Phys. Rev. A 42, 5652 (1990); K. D. Stokes, C. Schnurr, J. R. Gardner, M. Marable, G. R. Welch and J. E. Thomas, Phys. Rev. Lett. 67, 1997 (1991).
[8] J.-T. Chang, J. Evers, M.O. Scully and M.S. Zubairy, Phys. Rev. A 73, 031803(R) (2006); J.-T. Chang, J. Evers, and M.S. Zubairy, Phys. Rev. A 74, 043820 (2006).
[9] A.P. Chu, K.K. Berggren, K.S. Johnson, M.G. Prentiss, Quant. Semclas. Opt. 8, 521 (1996).
[10] K.S. Johnson, J.H. Thywissen, N.H. Dekker, K.K. Berggren, A.P. Chu, R. Younkin, M.G. Prentiss, Science 280, 1583 (1998).
[11] J.H. Thywissen, M.G. Prentiss, New J. Phys. 7, 47 (2005).
[12] S. Qamar, S.Y. Zhu,M.S. Zubairy, Phys. Rev. A61, 063806 (2000);M. Sahrai, H. Tajalli, K.T. Kapale, M.S. Zubairy, Phys. Rev. A72, 013820 (2005).
[13] P. R. Hemmer, A. Muthukrishnan,M. O. Scully, and M. S. Zubairy, Phys. Rev. Lett. 96, 163603 (2006); Q. Sun, P.R. Hemmer, M.S. Zubairy, Phys. Rev. A75, 065803 (2007).
[14] H. Li, V.A. Sautenkov,M.M. Kash, et al., Phys. Rev. A78, 013803 (2008).
[15] S. E. Harris, Phys. Today 50, 36 (1997).

[16] E. Arimondo, in *Progress in Optics* edited by E. Wolf, Vol. XXXV, p.257 (Elsevier Science, Amsterdam, 1996).
[17] M. O. Scully and M. S. Zubairy, Quantum Optics (Cambridge University Press, Cambridge, UK, 1997).
[18] J. P. Marangos, J. Mod. Opt. 45, 471 (1998).
[19] M. Fleischhauer, A. Imamoglu, J.P. Marangos, Rev. Mod. Phys. 77, 633 (2005).
[20] M. Kiffner, J. Evers, and M. S. Zubairy, Phys. Rev. Lett. 100, 073602 (2008).
[21] J.T. Chang, J. Evers, M. S. Zubairy, Phys. Rev. A74, 043820 (2006).
[22] G.S. Agarwal, K.T. Kapale, J. Phys. B 39, 3437 (2006).
[23] A.V. Gorshkov, L. Jiang, M. Greiner, P. Zoller, M.D. Lukin, Phys. Rev. Lett. 100, 093005 (2008).
[24] D. D. Yavuz, N. A. Proite, Phys. Rev. A 76, 041802 (2007).
[25] M. DAngelo,M.V. Chekhova, and Y. Shih, PRL 87, 013602 (2001).
[26] G. S. Agarwal, A. C. Brown, L. M. Narducci, and G. Vetri, Phys. Rev. A 15, 1613 (1977).
[27] A. S. Jahangir Amin and J. G. Cordes, Phys. Rev. A 18, 1298 (1978).
[28] G. Lenz and P. Meystre, Phys. Rev. A 48, 3365 (1993).
[29] T. Richter, Opt. Acta 30, 1769 (1983).
[30] H. S. Freedhoff, Phys. Rev. A 19, 1132 (1979); R. D. Griffin and S. M. Harris, ibid. 25, 1528 (1982).
[31] Z. Ficek, R. Tanas, and S. Kielich, Opt. Acta 30, 713 (1983).
[32] Z. Ficek and R. Tanas, Phys. Rep. 372, 369 (2002).
[33] M. Macovei and C. H. Keitel, Phys. Rev. Lett. 91, 123601 (2003).
[34] C. Hettich, C. Schmitt, J. Zitzmann, S. Kuhn, I. Gerhardt, and V. Sandoghdar, Science 298, 385 (2002).
[35] J.-T. Chang, J. Evers, M. O. Scully, and M. S. Zubairy, Phys. Rev. A 73, 031803(R) (2006).
[36] T. Azim, M. Ikram, and M. S. Zubairy, J. Opt. B: Quantum Semiclassical Opt. 6, 248 (2004); F. Ghafoor, S. Qamar, and M. S. Zubairy, Phys. Rev. A 65, 043819 (2002); S. Qamar, S.-Y. Zhu, and M. S. Zubairy, ibid. 61, 063806 (2000); F. Le Kien, G. Rempe, W. P. Schleich, and M. S. Zubairy, ibid. 56, 2972 (1997); M. Sahrai, H. Tajalli, K. T. Kapale, and M. S. Zubairy, ibid. 72, 013820 (2005).
[37] G. C. Hegerfeldt and D. Seidel, J. Opt. B: Quantum Semiclassical Opt. 4, 245 (2002).
[38] C. Skornia, J. von Zanthier, G. S. Agarwal, E.Werner, and H.Walther, Phys. Rev. A 64, 063801 (2001).
[39] M. O. Scully and C. H. Raymond Ooi, J. Opt. B: Quantum Semiclassical Opt. 6, S575 (2004).

24

The Physics of Ghost Imaging

Yanhua Shih
Department of Physics
University of Maryland, Baltimore County,
Baltimore, MD 21250,
U.S.A.

1. Introduction

One of the most surprising consequences of quantum mechanics is the nonlocal correlation of a multi-particle system observable in joint-detection of distant particle-detectors. Ghost imaging is one of such phenomena. Taking a photograph of an object, traditionally, we need to face a camera to the object. But with ghost imaging, we can image the object by pointing a CCD camera towards the light source, rather than towards the object. Ghost imaging is reproduced at quantum level by a non-factorizable point-to-point image-forming correlation between two photons. Two types of ghost imaging have been experimentally demonstrated since 1995. Type-one ghost imaging uses entangled photon pairs as the light source. The non-factorizable image-forming correlation is the result of a nonlocal constructive-destructive interference among a large number of biphoton amplitudes, a nonclassical entity corresponding to different yet indistinguishable alternative ways for the photon pair to produce a joint-detection event between distant photodetectors. Type-two ghost imaging uses chaotic light. The type-two non-factorizable image-forming correlation is caused by the superposition between paired two-photon amplitudes, or the symmetrized effective two-photon wavefunction, corresponding to two different yet indistinguishable alternative ways of triggering a join-detection event by two independent photons. The multi-photon interference nature of ghost imaging determines its peculiar features: (1) it is nonlocal; (2) its imaging resolution differs from that of classical; and (3) the type-two ghost image is turbulence-free.[1] Ghost imaging has attracted a great deal of attention, perhaps due to these features for certain applications. Achieving these features, the realization of nonlocal multi-photon interference is a necessary condition. Classical simulations, such as the man-made factorizable speckle-speckle correlation, can never have such features.

Before introducing the concept of ghost imaging, we briefly review the physics of classical optical imaging. Assuming an object that is either self-luminous or externally illuminated, imagining each point on the object surface as a point radiation sub-source, each point sub-source will emit spherical waves to all possible directions. How much chance do we expect to have a spherical wave collapsing into a point or a "speckle" by free propagation? Obviously, the chance is zero unless an imaging system is applied. The concept of optical

---

[1] For instance, any fluctuation of the refraction index or phase disturbance in the optical path has no influence to the type-two ghost image.

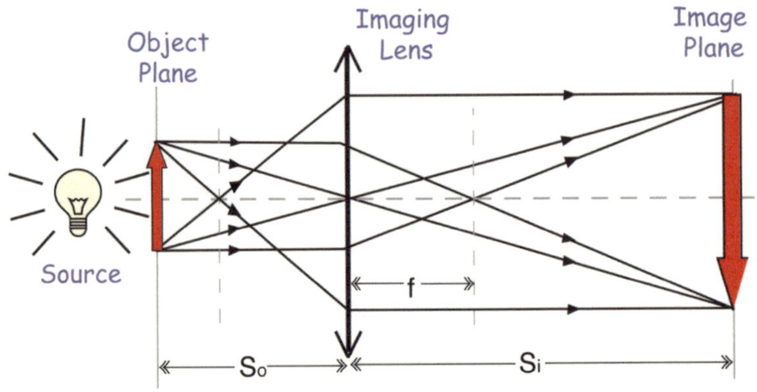

Fig. 1. Optical imaging: a lens produces an *image* of an object in the plane defined by the Gaussian thin-lens equation $1/s_i + 1/s_o = 1/f$. Image formation is based on a point-to-point relationship between the object plane and the image plane. All radiations emitted from a point on the object plane will "collapse" to a unique point on the image plane.

imaging was well developed in classical optics for this purpose. Figure 1 schematically illustrates a standard imaging setup. In this setup an object is illuminated by a radiation source, an imaging lens is used to focus the scattered and reflected light from the object onto an image plane which is defined by the "Gaussian thin lens equation"

$$\frac{1}{s_i} + \frac{1}{s_o} = \frac{1}{f}, \tag{1}$$

where s_o is the distance between the object and the imaging lens, s_i the distance between the imaging lens and the image plane, and f the focal length of the imaging lens. Basically this equation defines a point-to-point relationship between the object plane and the image plane: any radiation starting from a point on the object plane will "collapse" to a unique point on the image plane. It is not difficult to see from Fig. 1 that the point-to-point relationship is the result of *constructive-destructive interference*. The radiation fields coming from a point on the object plane will experience equal distance propagation to superpose constructively at one unique point on the image plane, and experience unequal distance propagations to superpose destructively at all other points on the image plane. The use of the imaging lens makes this constructive-destructive interference possible.

A perfect point-to-point image-forming relationship between the object and image planes produces a perfect image. The observed image is a reproduction, either magnified or demagnified, of the illuminated object, mathematically corresponding to a convolution between the object distribution function $|A(\vec{\rho}_o)|^2$ (aperture function) and a δ-function which characterizes the perfect point-to-point relationship between the object and image planes:

$$I(\vec{\rho}_i) = \int_{obj} d\vec{\rho}_o \left|A(\vec{\rho}_o)\right|^2 \delta(\vec{\rho}_o + \frac{\vec{\rho}_i}{m}) \tag{2}$$

where $I(\vec{\rho}_i)$ is the intensity in the image plane, $\vec{\rho}_o$ and $\vec{\rho}_i$ are 2-D vectors of the transverse coordinates in the object and image planes, respectively, and $m = s_i/s_o$ is the image magnification factor.

In reality, limited by the finite size of the imaging system, we may never obtain a perfect point-to-point correspondence. The incomplete constructive-destructive interference turns the point-to-point correspondence into a point-to-"spot" relationship. The δ-function in the convolution of Eq. (2) will be replaced by a point-spread function:

$$I(\vec{\rho}_i) = \int_{obj} d\vec{\rho}_o \, |A(\vec{\rho}_o)|^2 \, somb^2 [\frac{R}{s_o} \frac{\omega}{c} |\vec{\rho}_o + \frac{\vec{\rho}_i}{m}|], \tag{3}$$

where the sombrero-like function, or the Airy disk, is defined as

$$somb(x) = \frac{2J_1(x)}{x},$$

and $J_1(x)$ is the first-order Bessel function, and R the radius of the imaging lens, and R/s_o is known as the numerical aperture of the imaging system. The sombrero-like point-spread function, or the Airy disk, defines the spot size on the image plane that is produced by the radiation coming from point $\vec{\rho}_o$. It is clear from Eq. (3) that a larger imaging lens and shorter wavelength will result in a narrower point-spread function, and thus a higher spatial resolution of the image. The finite size of the spot determines the spatial resolution of the imaging system.

Type-one and type-two ghost imaging, in certain aspects, exhibit a similar point-to-point imaging-forming function as that of classical except the ghost image is reproducible only in the joint-detection between two independent photodetectors, and the point-to-point imaging-forming function is in the form of second-order correlation,

$$R_{12}(\vec{\rho}_i) = \int_{obj} d\vec{\rho}_o \, |A(\vec{\rho}_o)|^2 \, G^{(2)}(\vec{\rho}_o, \vec{\rho}_i), \tag{4}$$

where $R_{12}(\vec{\rho}_i)$ is the joint-detection counting rate between photodetectors D_1 and D_2. Mathematically, the convolution is taken between the aperture function of the object $|A(\vec{\rho}_o)|^2$ and a nontrivial poin-to-point second-order correlation function $G^{(2)}(\vec{\rho}_o, \vec{\rho}_i)$, corresponding to the probability of observing a joint photo-detection event at coordinates $\vec{\rho}_o$ and $\vec{\rho}_i$. It is the special physics behind $G^{(2)}(\vec{\rho}_o, \vec{\rho}_i)$ made ghost imaging so special.

The first type-one ghost imaging experiment was demonstrated by Pittman *et al.* in 1995 [1] enlightened by the theoretical work of Klyshko [2]. The schematic setup of the experiment is shown in Fig. 2. A continuous wave (CW) laser is used to pump a nonlinear crystal to produce an entangled pair of orthogonally polarized signal (e-ray of the crystal) and idler (o-ray of the crystal) photons in the nonlinear optical process of spontaneous parametric down-conversion (SPDC). The pair emerges from the crystal collinearly with $\omega_s \cong \omega_i \cong \omega_p/2$ (degenerate SPDC). The pump is then separated from the signal-idler pair by a dispersion prism, and the signal and idler are sent in different directions by a polarization beam splitting Thompson prism. The signal photon passes through a convex lens of 400mm focal length and illuminates a chosen aperture (mask). As an example, one of the demonstrations used the letters "UMBC" for the object mask. Behind the aperture is the "bucket" detector package D_1, which is made by an avalanche photodiode placed at the focus of a short focal length collection lens. During the experiment D_1 is kept in a fixed position. The idler photon is captured by detector package D_2, which is basically an optical fiber coupled avalanche photodiode. The input tip of the fiber is scannable in the transverse plane by two step motors (along orthogonal directions). The output pulses of D_1 and D_2, both operate in the photon counting mode, are independently counted as the counting rate of D_1 and D_2, respectively, and counted by a coincidence circuit for the joint-detection events of the pair. The single detector counting rates of D_1 and D_2 are both monitored to be constants during the measurement. Surprisingly, a ghost image of the chosen aperture is observed in coincidences during the scanning of the fiber tip, when the following two experimental conditions are satisfactory: (1) D_1 and D_2 always measure a pair; (2) the distances s_o, which is the optical distance between the aperture to the lens, s_i, which is the optical distance from the imaging lens going backward along the signal photon path to the two-photon source of SPDC then going forward along the idler photon path to the fiber tip, and the focal length of the imaging lens f satisfy the Gaussian thin lens equation of Eq. (1).

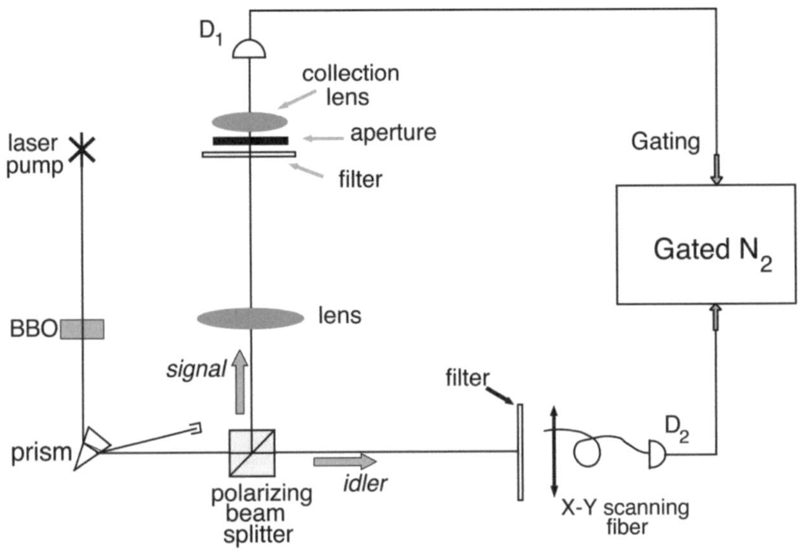

Fig. 2. Schematic set-up of the first "ghost" image experiment. The experimental demonstrations of ghost imaging and ghost interference [4] in 1995 together stimulated the foundation of quantum imaging in terms of geometrical and physical optics.

Figure 3 shows a typical measured ghost image. It is interesting to note that while the size of the "UMBC" aperture inserted in the signal path is only about 3.5mm×7mm, the observed image measures 7mm×14mm. The image is therefore magnified by a factor of 2 which equals the expected magnification $m = s_i/s_o$. In this measurement s_o = 600mm and s_i = 1200mm. When D_2 was scanned on transverse planes other than the ghost image plane the images blurred out.

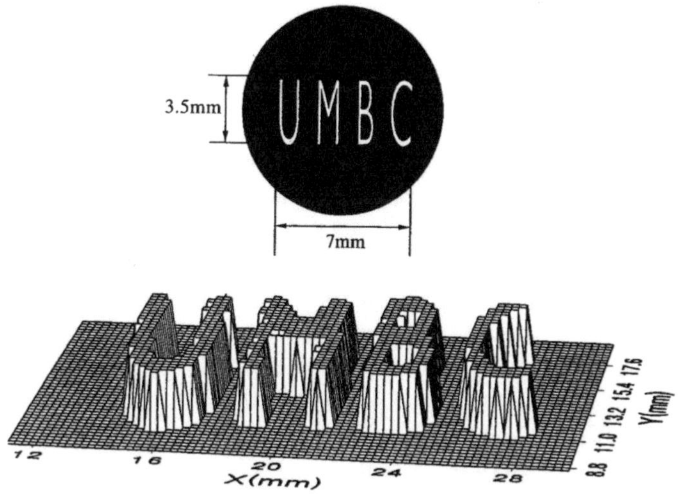

Fig. 3. Upper: A reproduction of the actual aperture "UMBC" placed in the signal beam. Lower: The image of "UMBC": coincidence counts as a function of the fiber tip's transverse coordinates in the image plane. The step size is 0.25mm. The image shown is a "slice" at the half maximum value.

The experiment was immediately given the name "ghost imaging" by the physics community due to its nonlocal feature. In the language of Einstein-Podolsky-Rosen (EPR) [3], the nonfactorizable[2] point-to-point image-forming correlation

$$G^{(2)}(\vec{\rho}_o, \vec{\rho}_i) \sim \delta(\vec{\rho}_o + \vec{\rho}_i/m) \tag{5}$$

observed in this experiment represents a nonlocal behavior of a measured pair of photons: neither the signal photon nor idler photon "knows" precisely where to go when the pair is created at the source. However, if one of them is observed at a point on the object plane, the other one must arrive at a unique corresponding point on the image plane.[3] Although questions regarding fundamental issues of quantum theory still exist, the experimental demonstration of ghost imaging [1] and ghost interference [4] in 1995 together stimulated the foundation of quantum imaging in terms of geometrical and physical optics.

Type-two ghost imaging uses chaotic radiation sources. Different from type-one, the nonfactorizable point-to-point image-forming correlation between the object and image planes is only partial with at least 50% constant background,

$$G^{(2)}(\vec{\rho}_o, \vec{\rho}_i) \sim 1 + \delta(\vec{\rho}_o - \vec{\rho}_i). \tag{6}$$

---

[2] Statistically, a factorizable correlation function $G^{(2)}(\mathbf{r}_1, t_1; \mathbf{r}_2, t_2) = G^{(1)}(\mathbf{r}_1, t_1) G^{(1)}(\mathbf{r}_2, t_2)$ characters independent radiations at space-time $(\mathbf{r}_1, t_1)$ and $(\mathbf{r}_2, t_2)$. In ghost imaging, the light on the object plane and the light at the CCD array is described by a non-factorizeable point-to-point image-forming function, indicating nontrivial statistical correlation between the two measured intensities.

[3] The ghost imaging experiment is thus considered a demonstration of the historical Einstein-Podolsky-Rosen (EPR) experiment.

The first near-field lensless ghost imaging experiment was demonstrated by Scarcelli *et al* in 2005 and 2006 [5][6] after their experimental demonstration of two-photon interference of chaotic light in 2004 [7]. Figure 4 illustrates an improved setup of the type-two ghost imaging experiment by Meyers *et al.* [8]. The thermal radiation of a chaotic source, which has a fairly large size in the transverse dimension, is split into two by a 50%−50% beamsplitter. One of the beams illuminates a toy soldier as shown in Fig. 4. The scattered and reflected photons from the solider (object) are collected and counted by a "bucket" detector D_2. In the other beam a high resolution CCD array, operated at the photon counting regime, is placed toward the radiation source for joint-detection with the "bucket" detector D_2. The counting rate of D_2 and the un-gated output of the CCD are both monitored to be constants during the measurement. Surprisingly, a 1:1 ghost image of the toy soldier is captured in the joint-detection between D_2 and the CCD, when taking $z_1 = z_2$. The 1:1 ghost image of the toy soldier is shown in Fig. 5. The images "blurred out" when the CCD is moved away from $z_1 = z_2$, either to the side of $z_1 > z_2$ or $z_1 < z_2$.

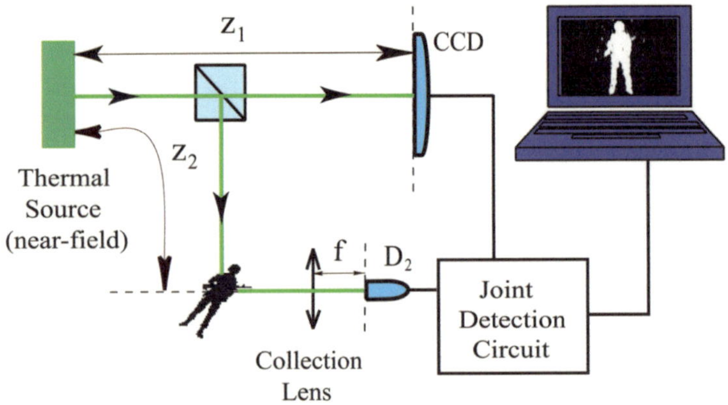

Fig. 4. Near-field lensless ghost imaging of chaotic light demonstrated by Meyers *et al.*. D_2 is a "bucket" photon counting detector that is used to collect and count all random scattered and reflected photons from the object. The joint-detection between D_2 and the CCD array is realized by a photon-counting-coincidence circuit. D_2 is fixed in space. The counting rate of D_2 and the un-gated output of the CCD are both monitored to be constants during the measurement. Surprisingly, a 1:1 ghost image of the object is captured in joint-detection between D_2 and the CCD, when taking $z_1 = z_2$. The images "blurred out" when the CCD is moved away from $z_1 = z_2$, either in the direction of $z_1 > z_2$ or $z_1 < z_2$.

There is no doubt that chaotic radiations propagate to any transverse plane in a random and chaotic manner. A brief discussion for Fresnel free-propagation is given in the appendix. In the lensless ghost imaging experiment, a large transverse sized chaotic light source, as shown in Fig. 4, is usually used for achieving better spatial resolution. The source consists a large number of independent point sub-sources randomly distributed on the source plane. Each point sub-source may randomly radiate independent spherical waves to the object and image planes. Due to the chaotic nature of the source there is no interference between these sub-fields. These independent sub-intensities simply add together, yielding a constant total intensity in space and in time on any transverse plane. In the lensless ghost imaging setup,

Fig. 5. Ghost image of a toy soldier model.

there is no lens applied to force these spherical waves collapsing to a point or a "speckle", and there is no chance to have two identical copies of any "speckle" of the source onto the object and image planes. What is the physical cause of the point-to-point image-forming correlation? Although the non-factorizable point-to-point correlation between the object and image planes is only partial, the type-two ghost imaging looks more surprising than type-one because of the nature of the light source. Unlike the signal-idler photon pair, the jointly measured photons in type-two ghost imaging are just two independent photons that fall into the coincidence time window by chance only. Nevertheless, analogous to EPR, the non-factorizable partial point-to-point correlation represents a nonlocal behavior of a measured pair of independent photons: neither photon-one nor photon-two "knows" precisely where to go when they are created at each independent sub-sources; however, if one of them is observed at a point on the object plane, the other one has twice greater probability of arriving at a unique corresponding point on the image plane.[4]

We have concluded and will show that the partial point-to-point correlation between the object and image planes in type-two ghost imaging is the result of *two-photon interference*. Similar to that of type-one, it involves the nonlocal superposition of two-photon amplitudes, a nonclassical entity corresponding to different yet indistinguishable alternative ways of triggering a joint-detection event [9]. Different from that of type-one, the joint-detection events observed in type-two ghost imaging are triggered by two randomly distributed independent photons. It is interesting to see that the quantum mechanical concept of *two-photon interference* is applicable to "classical" thermal light.[5] In fact, this is not the first time in the history of physics we apply quantum mechanical concepts to thermal light. We should not forget Planck's theory of blackbody radiation originated the quantum physics. The

[4] Similar to the HBT correlation, the contrast of the near-field partial point-to-point image-forming function is 50%, i.e., two to one ratio between the maximum value and the constant background, see Eq. (33).

[5] There exist a number of definitions for classical light and for quantum light. One of the commonly accepted definitions considers thermal light classical because its positive *P*-function.

radiation Planck dealt with was thermal radiation. Although the concept of "two-photon interference" comes from the study of entangled biphoton states [9], the concept should not be restricted to entangled systems. The concept is generally true and applicable to any radiation, including "classical" thermal light. The partial point-to-point correlation of thermal radiation is not a new discovery either. The first set of temporal and spatial far-field intensity-intensity correlations of thermal light was demonstrated by Hanbury Brown and Twiss (HBT) in 1956 [10][11]. The HBT experiment created quite a surprise in the physics community and lead to a debate about the classical or quantum nature of the phenomenon [11][12]. Although the discovery of HBT initiated a number of key concepts of modern quantum optics, the HBT phenomenon itself was finally interpreted as statistical correlation of intensity fluctuations and considered as a classical effect. It is then reasonable to ask: Is the near-field type-two ghost imaging with thermal light a simple classical effect similar to that of HBT? Is it possible that the ghost imaging phenomenon itself, including the type-one ghost imaging of 1995, is merely a simple classical effect of intensity fluctuation correlation?[13][14][15][16] This article will address these important questions and explore the multi-photon interference nature of ghost imaging.
To explore the two-photon interference nature, we will analyze the physics of type-one and type-two ghost imaging in five steps. (1) Review the physics of coherent and incoherent light propagation; (2) Review classical imaging as the result of constructive-destructive interference among electromagnetic waves; (3) analyze type-one ghost imaging in terms of constructive-destructive interference between the biphoton amplitudes of an entangled photon-pair; (4) analyze type-two ghost imaging in terms of two-photon interference between chaotic sub-fields; and (5) discuss the physics of the phenomenon: whether it is a quantum interference or a classical intensity fluctuation correlation.

2. Classical imaging

To understand the multi-photon interference nature of ghost imaging, it might be helpful to see the constructive-destructive interference nature of classical imaging first. We start from a typical classical imaging setup of Fig. 6 and ask a simple question: how does the radiation field propagate from the object plane to the image plane? In classical optics such propagation is usually described by an optical transfer function $h(\mathbf{r}-\mathbf{r}_0, t-t_0)$. We prefer to work with the single-mode propagator, namely the Green's function, $g(\mathbf{k}, \mathbf{r}-\mathbf{r}_0, t-t_0)$ [17][18], which propagates each mode of the radiation from space-time point $(\mathbf{r}_0, t_0)$ to space-time point $(\mathbf{r}, t)$. We treat the field $E(\mathbf{r}, t)$ as a superposition of these modes. A detailed discussion about $g(\mathbf{k}, \mathbf{r}-\mathbf{r}_0, t-t_0)$ is given in the Appendix. It is convenient to write the field $E(\mathbf{r}, t)$ as a superposition of its longitudinal and transverse modes under the Fresnel paraxial approximation,

$$E(\vec{\rho}, z, t) = \int d\vec{\kappa}\, d\omega\, \tilde{E}(\vec{\kappa}, \omega)\, g(\vec{\kappa}, \omega; \vec{\rho}, z)\, e^{-i\omega t}, \tag{7}$$

where $\tilde{E}(\vec{\kappa}, \omega)$ is the complex amplitude for the mode of frequency ω and transverse wave-vector $\vec{\kappa}$. In Eq. (7) we have taken $z_0 = 0$ and $t_0 = 0$ at the object plane as usual. To simplify the notation, we have assumed one polarization.

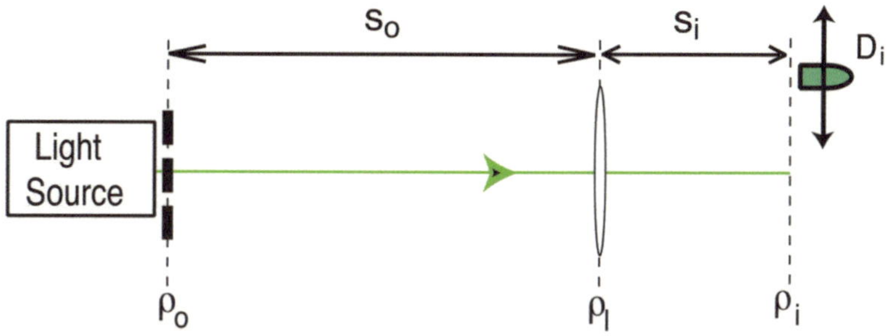

Fig. 6. Typical imaging setup. A lens of finite size is used to produce a magnified or demagnified image of an object with limited spatial resolution.

Based on the experimental setup of Fig. 6 and following the Appendix, $g\,(\vec{\kappa},\,\omega;\,\vec{\rho},\,z)$ is found to be

$$\begin{aligned} & g(\vec{\kappa},\omega;\vec{\rho}_i,s_o+s_i) \\ &= \int_{obj} d\vec{\rho}_o \int_{lens} d\vec{\rho}_l \left\{A(\vec{\rho}_o)\, e^{i\vec{\kappa}\cdot\vec{\rho}_o}\right\} \left\{\frac{-i\omega}{2\pi c}\,\frac{e^{i\frac{\omega}{c}s_o}}{s_o}\, e^{i\frac{\omega}{2cs_o}|\vec{\rho}_l-\vec{\rho}_o|^2}\right\} \left\{e^{-i\frac{\omega}{2cf}\,|\vec{\rho}_l|^2}\right\} \\ &\times \left\{\frac{-i\omega}{2\pi c}\,\frac{e^{i\frac{\omega}{c}s_i}}{s_i}\, e^{i\frac{\omega}{2cs_i}|\vec{\rho}_i-\vec{\rho}_l|^2}\right\}, \end{aligned} \tag{8}$$

where $\vec{\rho}_o$, $\vec{\rho}_l$, and $\vec{\rho}_i$ are two-dimensional vectors defined, respectively, on the object, lens, and image planes. The first curly bracket includes the aperture function $A(\vec{\rho}_o)$ of the object and the phase factor $e^{i\vec{\kappa}\cdot\vec{\rho}_o}$ contributed at the object plane by each transverse mode $\vec{\kappa}$. The terms in the second and fourth curly brackets describe free-space Fresnel propagation-diffraction from the source/object plane to the imaging lens, and from the imaging lens to the detection plane, respectively. The Fresnel propagator includes a spherical wave function $e^{i\frac{\omega}{c}(z_j-z_k)}/(z_j-z_k)$ and a Fresnel phase factor $e^{i\omega|\vec{\rho}_j-\vec{\rho}_k|^2/2c(z_j-z_k)}$. The third curly bracket adds the phase factor introduced by the imaging lens.
We now rewrite Eq. (8) into the following form

$$\begin{aligned} g(\vec{\kappa},\omega;\vec{\rho}_i,z=s_o+s_i) &= \frac{-\omega^2}{(2\pi c)^2 s_o s_i}\, e^{i\frac{\omega}{c}(s_o+s_i)}\, e^{i\frac{\omega}{2cs_i}|\vec{\rho}_i|^2} \int_{obj} d\vec{\rho}_o\, A(\vec{\rho}_o)\, e^{i\frac{\omega}{2cs_o}|\vec{\rho}_o|^2}\, e^{i\vec{\kappa}\cdot\vec{\rho}_o} \\ &\times \int_{lens} d\vec{\rho}_l\, e^{i\frac{\omega}{2c}[\frac{1}{s_o}+\frac{1}{s_i}-\frac{1}{f}]|\vec{\rho}_l|^2}\, e^{-i\frac{\omega}{c}(\frac{\vec{\rho}_o}{s_o}+\frac{\vec{\rho}_i}{s_i})\cdot\vec{\rho}_l}. \end{aligned} \tag{9}$$

The image plane is defined by the Gaussian thin-lens equation of Eq. (1). Hence, the second integral in Eq. (9) reduces to, for a finite sized lens of radius R, the so-called point-spread function, or the Airy disk, of the imaging system:

$$\int_{lens} d\vec{\rho}_l\, e^{-i\frac{\omega}{c}(\frac{\vec{\rho}_o}{s_o}+\frac{\vec{\rho}_i}{s_i})\cdot\vec{\rho}_l} = \frac{2J_1(x)}{x} = somb(x), \tag{10}$$

where the sombrero-like function $somb(x) = 2J_1(x)/x$ with argument $x = [\frac{R}{s_o}\,\frac{\omega}{c}|\vec{\rho}_o + \rho_i/m|]$ has been defined in Eq. (3). Eq. (10) indicates a constructive interference.

Substituting Eqs. (9) and (10) into Eq. (7) enables one to obtain the classical self-correlation function of the field, or, equivalently, the intensity on the image plane

$$I(\vec{\rho}_i, z_i, t_i) = \langle\, E^*(\vec{\rho}_i, z_i, t_i)\, E(\vec{\rho}_i, z_i, t_i)\,\rangle, \tag{11}$$

where ⟨...⟩ denotes an ensemble average. To simplify the mathematics, monochromatic light is assumed as usual.

Case (I): *Incoherent imaging.* The ensemble average yields zeros except when $\vec{\kappa} = \vec{\kappa}'$. The image is thus

$$I(\vec{\rho}_i) \;\; \propto \;\; \int d\vec{\rho}_o \left|A(\vec{\rho}_o)\right|^2 somb^2[\frac{R}{s_o}\frac{\omega}{c}|\vec{\rho}_o + \frac{\vec{\rho}_i}{m}|]. \tag{12}$$

An incoherent image, magnified by a factor of m, is thus given by the convolution between the modulus square of the object aperture function and the point-spread function. The spatial resolution of the image is determined by the finite width of the $|\, somb\,|^2$-function.

Case (II): Coherent imaging. The coherent superposition of the $\vec{\kappa}$ modes in both E $(\vec{\rho}_i, \tau)$ and E$(\vec{\rho}_i, \tau)$ results in a wavepacket. The image, or the intensity distribution on the image plane, is

$$I(\vec{\rho}_i) \propto \Big| \int_{obj} d\vec{\rho}_o\, A(\vec{\rho}_o)\, e^{i\frac{\omega}{2cs_o}|\vec{\rho}_o|^2} somb[\frac{R}{s_o}\frac{\omega}{c}|\vec{\rho}_o + \frac{\vec{\rho}_i}{m}|]\Big|^2. \tag{13}$$

A coherent image, magnified by a factor of m, is thus given by the modulus square of the convolution between the object aperture function (multiplied by a Fresnel phase factor) and the point-spread function.

For $s_i < s_o$ and $s_o > f$, both Eqs. (12) and (13) describe a real demagnified inverted image. In both cases, a narrower *somb*-function yields a higher spatial resolution. Therefore the use of a larger imaging lens and shorter wavelengths will improve the spatial resolution of an imaging system.

3. Biphoton and type-one ghost imaging

In this section we analyze type-one ghost imaging. Type-one ghost imaging uses entangled photon pairs such as the signal-idler biphoton pairs of SPDC [19][9]. The nearly collinear signalidler system generated by SPDC can be described, in the ideal case, by the following entangled biphoton state [9]:

$$|\,\Psi\,\rangle = \Psi_0 \int d\vec{\kappa}_s\, d\vec{\kappa}_i\, \delta(\vec{\kappa}_s + \vec{\kappa}_i) \int d\omega_s\, d\omega_i\, \delta(\omega_s + \omega_i - \omega_p)\, a^\dagger(\vec{\kappa}_s, \omega_s)\, a^\dagger(\vec{\kappa}_i, \omega_i)\,|\,0\,\rangle, \tag{14}$$

where ω_j , $\vec{\kappa}_j$ (j = s, i, p), are the frequency and transverse wavevector of the signal, idler, and pump, respectively. For simplicity a CW single mode pump with $\vec{\kappa}_p = 0$ is assumed. Eq. (14) indicates that the biphoton state of the signal-idler pair is an entangled state. The single-photon state of the signal and the idler can be evaluated by taking a partial trace of its twin,

$$\hat{\rho}_s = tr_i \,|\, \Psi \,\rangle\langle\, \Psi \,| = \int d\vec{\kappa}_s \, d\omega_s \, a^\dagger(\vec{\kappa}_s, \omega_s) |\, 0 \,\rangle\langle\, 0 \,| a(\vec{\kappa}_s, \omega_s),$$
$$\hat{\rho}_i = tr_s \,|\, \Psi \,\rangle\langle\, \Psi \,| = \int d\vec{\kappa}_i \, d\omega_i \, a^\dagger(\vec{\kappa}_i, \omega_i) |\, 0 \,\rangle\langle\, 0 \,| a(\vec{\kappa}_i, \omega_i). \tag{15}$$

Although the signal-idler system is in a pure state, the state of the signal photon and the idler photon, respectively, are both mixed states.

Let us imagine a measurement in which two point-like photon counting detectors (D_1 and D_2) are placed at the output plane of an SPDC source for the detection of the signal photon and the idler photon, respectively, and for the joint-detection of the signal-idler pair. The probability of observing a photo-detection event in the SPDC output plane $\vec{\rho}_j$ at time t_j, $j = s, i$, is calculated from the first-order photo-detection theory of Glauber [20]

$$G^{(1)}(\vec{\rho}_j, t_j) = tr\, \hat{\rho}\, E^{(-)}(\vec{\rho}_j, t_j) E^{(+)}(\vec{\rho}_j, t_j), \tag{16}$$

where we have chosen $z_j = 0$ for the SPDC output plane as usual. It is easy to find that

$$G^{(1)}(\vec{\rho}_s, t_s) \cong \text{constant}, \qquad G^{(1)}(\vec{\rho}_i, t_i) \cong \text{constant}, \tag{17}$$

which means that the signal photon and the idler photon both have equal probability to be observed at any position in the output plane of the SPDC at any time. The probability of observing a jointdetection event between D_1 and D_2 located at $\vec{\rho}_s$ and $\vec{\rho}_i$ in the SPDC output plane of $z_s = z_i = 0$ is calculated from the second-order photo-detection theory of Glauber [20]:

$$\begin{aligned} & G^{(2)}(\vec{\rho}_s, t_s; \vec{\rho}_i, t_i) \\ &= \langle\, \Psi \,|\, E_s^{(-)}(\vec{\rho}_s, t_s) E_i^{(-)}(\vec{\rho}_i, t_i) E_i^{(+)}(\vec{\rho}_i, t_i) E_s^{(+)}(\vec{\rho}_s, t_s) \,|\, \Psi \,\rangle \\ &= |\langle\, 0 \,|\, E_i^{(+)}(\vec{\rho}_i, t_i) E_s^{(+)}(\vec{\rho}_s, t_s) \,|\, \Psi \,\rangle|^2 \\ &\equiv |\, \Psi(\vec{\rho}_s, t_s; \vec{\rho}_i, t_i) \,|^2, \end{aligned} \tag{18}$$

where $\Psi(\vec{\rho}_s, t_s; \vec{\rho}_i, t_i)$ is defined as the effective biphoton wavefunction. The transverse spatial part of the effective biphoton wavefunction is easily calculated to be:

$$\Psi(\vec{\rho}_s, \vec{\rho}_i) \simeq \delta(\vec{\rho}_s - \vec{\rho}_i), \tag{19}$$

under the condition $t_s \simeq t_i$. Equations (14), (17), and (19) suggest that the entangled signal-idler photon pair is characterized by the EPR correlation [3] in transverse momentum and transverse position; hence, similar to the original EPR state, we have [21]

$$\begin{aligned} & \Delta(\vec{\kappa}_s + \vec{\kappa}_i) = 0 \;\;\&\;\; \Delta(\vec{\rho}_s - \vec{\rho}_i) = 0 \\ \text{with} \;\; & \Delta\vec{\kappa}_s \sim \infty, \;\; \Delta\vec{\kappa}_i \sim \infty, \;\; \Delta\vec{\rho}_s \sim \infty, \;\; \Delta\vec{\rho}_i \sim \infty. \end{aligned} \tag{20}$$

In EPR's language, the signal photon and the idler photon may come from any point in the output plane of the SPDC. However, if the signal (idler) is found in a certain position, the idler (signal) must be observed in the same position, with certainty (100%). Simultaneously, the signal photon and the idler photon may have any transverse momentum. However, if a certain value and direction of the transverse momentum of the signal (idler) is observed, the transverse momentum of the idler (signal) will be uniquely determined with equal value and opposite direction.

The EPR δ-functions, $\delta(\vec{\rho}_s - \vec{\rho}_i)$ and $\delta(\vec{\kappa}_s + \vec{\kappa}_i)$ in transverse position and momentum, are the key to understanding the ghost imaging experiment of Pittman *et al.* of 1995. $\delta(\vec{\rho}_s - \vec{\rho}_i)$ indicates that the signal-idler pair is always emitted from the same point on the output plane of the biphoton source. Simultaneously, $\delta(\vec{\kappa}_s + \vec{\kappa}_i)$ defines the angular correlation of the pair: the signal-idler pair always exists at roughly equal but opposite angles relative to the pump for degenerate SPDC. This then allows for a simple explanation of the experiment in terms of "usual" geometrical optics in the following manner: we envision the nonlinear crystal as a "hinge point" and "unfold" the schematic of Fig. 2 into the Klyshko picture [2] of Fig. 7. The signal-idler biphoton amplitudes can then be represented by straight lines (but keep in mind the different propagation directions) and therefore the image is reproduced in coincidences when the aperture, lens, and fiber tip are located according to the Gaussian thin lens equation of Eq. (1). The image is exactly the same as that one would observe on a screen placed at the fiber tip if detector D_1 were replaced by a point-like light source and the nonlinear crystal by a reflecting mirror.

Comparing the "unfolded" schematic of the ghost imaging experiment with that of the classical imaging setup of Fig. 1, it is not difficult to find that any "light point" on the object plane has a unique corresponding "light point" on the image plane. This point-to-point correspondence is the result of the constructive-destructive interference among these biphoton amplitudes that are illustrated as the geometrical rays in Fig. 7. Similar to the situation in classical imaging, these biphoton amplitudes which experience equal optical path propagation will superpose constructively at each pair of one-to-one points of the object plane and the image plane for a joint-detection event, while these that experience unequal distance propagation will superpose destructively at all other points on the object and image planes. The use of the imaging lens makes this constructivedestructive interference possible. It is this unique point-to-point EPR correlation that makes the "ghost" image of the object-aperture function possible. Despite the completely different physics from classical optics, the remarkable feature is that the relationship between the focal length f of the lens, the aperture's optical distance s_o, and the image's optical distance s_i, satisfies the Gaussian thin lens equation of Eq. (1). It is worth emphasizing again that the geometric rays in Fig. 7 represent the biphoton amplitudes of a signal-idler photon pair, and the point-topoint correspondence is the result of the constructive-destructive interference of these biphoton amplitudes.

We now calculate $G^{(2)}(\vec{\rho}_1, \vec{\rho}_2)$ for the "ghost" imaging experiment in detail, where $\vec{\rho}_1$ and $\vec{\rho}_2$ are the transverse coordinates of the point-like photodetector D_1 and D_2, on the object and image planes, respectively. We will show that there exists a δ-function-like point-to-point correlation between the object and image planes, $\delta(\vec{\rho}_1 - \vec{\rho}_2/m)$. We will then show how the object function of $A(\vec{\rho}_o)$ is transferred to the image plane as a magnified image $A(\vec{\rho}_2/m)$.

We first calculate the effective biphoton wavefunction $\Psi(\vec{\rho}_1, z_1, t_1; \vec{\rho}_2, z_2, t_2)$, as defined in Eq. (18). By inserting the field operators into $\Psi(\vec{\rho}_1, z_1, t_1; \vec{\rho}_2, z_2, t_2)$, and considering the commutation relations of the field operators, the effective biphoton wavefunction is calculated to be

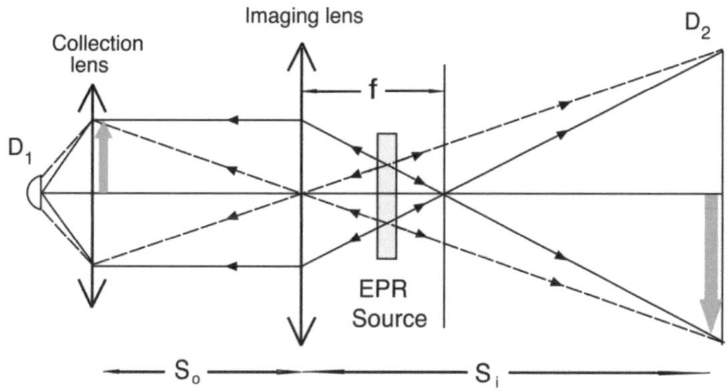

Fig. 7. An unfolded schematic of the 1995 ghost imaging experiment, which is helpful for understanding the physics. Since the biphoton "light" propagates along "straight lines", it is obvious that any point on the object plane corresponds to a unique point on the image plane. Although the placement of the lens, the object, and detector D_2 obeys the Gaussian thin lens equation, it is important to notice that the geometric rays in the figure actually represent the biphoton amplitudes of an entangled photon pair. The point-to-point correspondence is the result of a constructive-destructive interference of these biphoton amplitudes.

$$\Psi(\vec{\rho}_1, z_1, t_1; \vec{\rho}_2, z_2, t_2) = \Psi_0 \int d\vec{\kappa}_s \, d\vec{\kappa}_i \, \delta(\vec{\kappa}_s + \vec{\kappa}_i) \int d\omega_s \, d\omega_i \, \delta(\omega_s + \omega_i - \omega_p) \\ \times \; g(\vec{\kappa}_s, \omega_s; \vec{\rho}_1, z_1) \, e^{-i\omega_s t_1} \, g(\vec{\kappa}_i, \omega_i; \vec{\rho}_2, z_2) \, e^{-i\omega_i t_2}. \tag{21}$$

Equation (21) indicates a coherent superposition of all the biphoton amplitudes shown in Fig. 7.

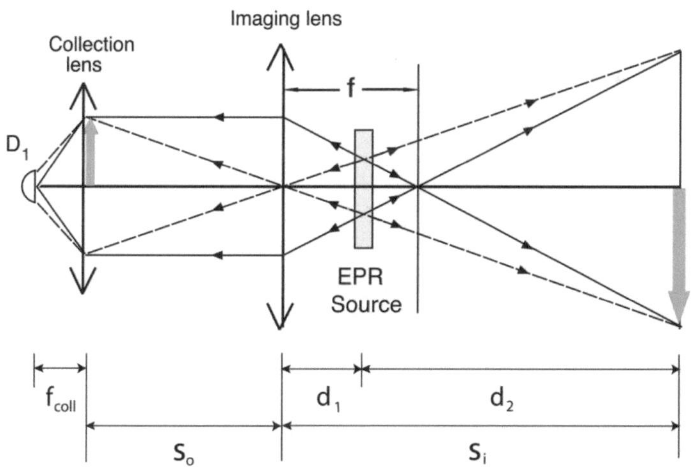

Fig. 8. In arm-1 the signal propagates freely over a distance d_1 from the output plane of the source to the imaging lens, passes an object aperture at distance s_o, and then is focused onto photon-counting detector D_1 by a collection lens. In arm-2 the idler propagates freely over a distance d_2 from the output plane of the source to a point-like photon counting detector D_2.

Next, we follow the unfolded experimental setup of Fig. 8 to establish the Green's functions $g(\vec{\kappa}_s, \omega_s, \vec{\rho}_1, z_1)$ and $g(\vec{\kappa}_i, \omega_i, \vec{\rho}_2, z_2)$. In arm-1 the signal propagates freely over a distance d_1 from the output plane of the source to the imaging lens, passes an object aperture at distance s_o, and then is focused onto photon-counting detector D_1 by a collection lens. We will evaluate $g(\vec{\kappa}_s, \omega_s, \vec{\rho}_1, z_1)$ by propagating the field from the output plane of the biphoton source to the object plane. In arm-2 the idler propagates freely over a distance d_2 from the output plane of the biphoton source to a point-like detector D_2. $g(\vec{\kappa}_i, \omega_i, \vec{\rho}_2, z_2)$ is thus a free propagator.

(I) Arm-1 (source to object):

The optical transfer function or Green's function in arm-1, which propagates the field from the source plane to the object plane, is given by:

$$g(\vec{\kappa}_s, \omega_s; \vec{\rho}_1, z_1 = d_1 + s_o) = e^{i\frac{\omega_s}{c}z_1} \int_{lens} d\vec{\rho}_l \int_{source} d\vec{\rho}_s \left\{ \frac{-i\omega_s}{2\pi c d_1} e^{i\vec{\kappa}_s \cdot \vec{\rho}_s} e^{i\frac{\omega_s}{2cd_1}|\vec{\rho}_s - \vec{\rho}_l|^2} \right\} \\ \times e^{-i\frac{\omega}{2cf}|\vec{\rho}_l|^2} \left\{ \frac{-i\omega_s}{2\pi c s_o} e^{i\frac{\omega_s}{2cs_o}|\vec{\rho}_l - \vec{\rho}_1|^2} \right\}, \tag{22}$$

where $\vec{\rho}_s$ and $\vec{\rho}_l$ are the transverse vectors defined, respectively, on the output plane of the source and on the plane of the imaging lens. The terms in the first and second curly brackets in Eq. (22) describe free space propagation from the output plane of the source to the imaging lens and from the imaging lens to the object plane, respectively. Again, $e^{i\frac{\omega_s}{2cd_1}|\vec{\rho}_s - \vec{\rho}_l|^2}$ and $e^{i\frac{\omega_s}{2cs_o}|\vec{\rho}_l - \vec{\rho}_1|^2}$ are the Fresnel phases as defined in the Appendix. Here the imaging lens is treated as a thin-lens, and the transformation function of the imaging lens is approximated as a Gaussian, $l(|\vec{\rho}_l|, f) \cong e^{-i\frac{\omega}{2cf}|\vec{\rho}_l|^2}$.

(II) Arm-2 (from source to image):

In arm-2, the idler propagates freely from the source to the plane of D_2, which is also the plane of the image. The Green's function is

$$g(\vec{\kappa}_i, \omega_i; \vec{\rho}_2, z_2 = d_2) = \frac{-i\omega_i}{2\pi c d_2} e^{i\frac{\omega_i}{c}d_2} \int_{source} d\vec{\rho}'_s \, e^{i\frac{\omega_i}{2cd_2}|\vec{\rho}'_s - \vec{\rho}_2|^2} e^{i\vec{\kappa}_i \cdot \vec{\rho}'_s} \tag{23}$$

where $\vec{\rho}'_s$ and $\vec{\rho}_2$ are the transverse vectors defined, respectively, on the output plane of the source and the plane of photodetector D_2.

(III) $\Psi(\vec{\rho}_1, \vec{\rho}_2)$ and $G^{(2)}(\vec{\rho}_1, \vec{\rho}_2)$ (object plane - image plane):

For simplicity, in the following calculation we consider degenerate ($\omega_s = \omega_i = \omega$) and collinear SPDC. The effective transverse biphoton wavefunction $\Psi(\vec{\rho}_1, \vec{\rho}_2)$ is then evaluated by substituting the Green's functions $g(\vec{\kappa}_s, \omega; \vec{\rho}_1, z_1)$ and $g(\vec{\kappa}_i, \omega; \vec{\rho}_2, z_2)$ into Eq. (21),

$$\Psi(\vec{\rho}_1, \vec{\rho}_2) \propto \int d\vec{\kappa}_s \, d\vec{\kappa}_i \, \delta(\vec{\kappa}_s + \vec{\kappa}_i) \, g(\vec{\kappa}_s, \omega; \vec{\rho}_1, z_1) \, g(\vec{\kappa}_i, \omega; \vec{\rho}_2, z_2) \\ \propto e^{i\frac{\omega}{c}(s_o + s_i)} \int d\vec{\kappa}_s \, d\vec{\kappa}_i \, \delta(\vec{\kappa}_s + \vec{\kappa}_i) \int_{lens} d\vec{\rho}_l \int_{source} d\vec{\rho}_s \, e^{i\vec{\kappa}_s \cdot \vec{\rho}_s} e^{i\frac{\omega}{2cd_1}|\vec{\rho}_s - \vec{\rho}_l|^2} \\ \times e^{-i\frac{\omega}{2cf}|\vec{\rho}_l|^2} e^{i\frac{\omega_s}{2cs_o}|\vec{\rho}_l - \vec{\rho}_1|^2} \int_{source} d\vec{\rho}'_s \, e^{i\vec{\kappa}_i \cdot \vec{\rho}'_s} e^{i\frac{\omega_i}{2cd_2}|\vec{\rho}'_s - \vec{\rho}_2|^2} \tag{24}$$

where all the proportionality constants have been ignored. After completing the double integral of $d\vec{\kappa}_s$ and $d\vec{\kappa}_i$

$$\int d\vec{\kappa}_s\, d\vec{\kappa}_i\, \delta(\vec{\kappa}_s + \vec{\kappa}_i)\, e^{i\vec{\kappa}_s\cdot\vec{\rho}_s}\, e^{i\vec{\kappa}_i\cdot\vec{\rho}_s'} \sim \delta(\vec{\rho}_s - \vec{\rho}_s'),$$

Eq. (24) becomes

$$\Psi(\vec{\rho}_1,\vec{\rho}_2) \propto e^{i\frac{\omega}{c}(s_0+s_i)} \int_{lens} d\vec{\rho}_l \int_{source} d\vec{\rho}_s\, e^{i\frac{\omega}{2cd_2}|\vec{\rho}_2-\vec{\rho}_s|^2}\, e^{i\frac{\omega}{2cd_1}|\vec{\rho}_s-\vec{\rho}_l|^2}\, e^{-i\frac{\omega}{2cf}|\vec{\rho}_l|^2}\, e^{i\frac{\omega}{2cs_o}|\vec{\rho}_l-\vec{\rho}_o|^2}.$$

Next, we complete the integral for $d\vec{\rho}_s$

$$\Psi(\vec{\rho}_1,\vec{\rho}_2) \propto e^{i\frac{\omega}{c}(s_0+s_i)} \int_{lens} d\vec{\rho}_l\, e^{i\frac{\omega}{2cs_i}|\vec{\rho}_2-\vec{\rho}_l|^2}\, e^{-i\frac{\omega}{2cf}|\vec{\rho}_l|^2}\, e^{i\frac{\omega}{2cs_o}|\vec{\rho}_l-\vec{\rho}_1|^2}, \tag{25}$$

where we have replaced d_1+d_2 with s_i (as depicted in Fig. 8). Although the signal and idler propagate in different directions along two optical arms, interestingly, the Green function in Eq. (25) is equivalent to that of a classical imaging setup, as if the field is originated from a point $\vec{\rho}_1$ on the object plane and propagated the lens and then arrived at point $\vec{\rho}_2$ on the imaging plane. The mathematics is consistent with our previous qualitative analysis of the experiment.

The finite integral on $d\vec{\rho}_l$ yields a point-to-"spot" relationship between the object plane and the image plane that is defined by the Gaussian thin-lens equation

$$\Psi(\vec{\rho}_1,\vec{\rho}_2) \propto \int_{lens} d\vec{\rho}_l\, e^{i\frac{\omega}{2c}[\frac{1}{s_o}+\frac{1}{s_i}-\frac{1}{f}]|\vec{\rho}_l|^2}\, e^{-i\frac{\omega}{c}(\frac{\vec{\rho}_1}{s_o}+\frac{\vec{\rho}_2}{s_i})\cdot\vec{\rho}_l} = somb\Big(\frac{R}{s_o}\frac{\omega}{c}|\vec{\rho}_1+\frac{\vec{\rho}_2}{m}|\Big). \tag{26}$$

If the integral is taken to infinity, by imposing the condition of the Gaussian thin-lens equation the effective transverse biphoton wavefunction can be approximated as a δ function

$$\Psi(\vec{\rho}_1,\vec{\rho}_2) \sim \delta(\vec{\rho}_1 + \vec{\rho}_2/m) \sim \delta(\vec{\rho}_o + \vec{\rho}_I/m), \tag{27}$$

where we have replaced $\vec{\rho}_1$ and $\vec{\rho}_2$ with $\vec{\rho}_o$ and $\vec{\rho}_I$, respectively, to emphasize the point-to-point EPR correlation between the object and image planes. To avoid confusion with the "idler" we have used $\vec{\rho}_I$ to label the image plane.

We now include an object-aperture function, a collection lens and a photon counting detector D_1 into the optical transfer function of arm-1 as shown in Fig. 2. The collection-lens−D_1 package can be simply treated as a "bucket" detector. The "bucket" detector integrates the biphoton amplitudes $\Psi(\vec{\rho}_o,\vec{\rho}_2)$, which are modulated by the object aperture function $A(\vec{\rho}_o)$ into a joint photodetection event. This process is equivalent to the following convolution

$$R_{1,2} \propto \int_{object} d\vec{\rho}_o\, \big|A(\vec{\rho}_o)\big|^2\, \big|\Psi(\vec{\rho}_o,\vec{\rho}_2)\big|^2 \simeq \big|A(\vec{\rho}_2/m)\big|^2 = \big|A(\vec{\rho}_I/m)\big|^2. \tag{28}$$

Again, D_2 is scanned in the image plane ($\vec{\rho}_2 = \vec{\rho}_I$). A ghost image of the object is thus reproduced on the image plane by means of the joint-detection between the point-like-detector D_2 and the bucket detector D_1.

The physical process corresponding to the above convolution is rather simple. Suppose the point detector D_2 is triggered by an idler photon at a transverse position of $\vec{\rho}_I$ in a joint-detection event with the bucket detector D_1 which is triggered by the signal twin that is either transmitted or reflected from a unique point $\vec{\rho}_o$ on the object plane. This unique point-to-point determination comes from the non-factorizable correlation function $\delta(\vec{\rho}_o + \vec{\rho}_I/m)$. Now, we move D_2 to another transverse position $\vec{\rho'}_I$ and register a joint-detection event. The signal photon that triggers D_1 must be either transmitted or reflected from another unique point $\vec{\rho'}_o$ on the object plane which is determined by $\delta(\vec{\rho'}_o+\vec{\rho'}_I/m)$. The chances of receiving a joint detection event at $\vec{\rho}_I$ and at $\vec{\rho'}_I$ would be modulated by the values of the aperture function $A(\vec{\rho}_o)$ and $A(\vec{\rho'}_o)$, respectively. Accumulating a large number of joint-detection events at each transverse coordinates on the image plane, the aperture function $A(\vec{\rho}_o)$ is thus reproduced in the joint-detection as a function of $\vec{\rho}_I$.

The observation of type-one ghost imaging has demonstrated a non-factorizable point-to-point EPR correlation between the object and image planes. This point-to-point correlation is the result of a constructive-destructive interference between biphoton amplitudes,

$$G^{(2)}(\vec{\rho}_o, \vec{\rho}_I) = \Big| \int d\vec{\kappa}_s \, d\vec{\kappa}_i \, \delta(\vec{\kappa}_s + \vec{\kappa}_i) \, g(\vec{\kappa}_s, \vec{\rho}_o) \, g(\vec{\kappa}_i, \vec{\rho}_I) \Big|^2 = somb^2\Big(\frac{R}{s_o}\frac{\omega}{c}|\vec{\rho}_O + \frac{\vec{\rho}_I}{m}|\Big). \quad (29)$$

In this view we consider the ghost imaging experiment of Pittman *et al.* a realization of the 1935 EPR *gedankenexperiment* [21] [22].

Classical theory has difficulties when facing type-one ghost imaging phenomenon. In the classical theory of light, a joint measurement between two photodetectors D_1 and D_2 measures the statistical correlation of intensity fluctuations,

$$G^{(2)}(\vec{\rho}_1, \vec{\rho}_2) = \langle \, I(\vec{\rho}_1) \, I(\vec{\rho}_2) \, \rangle = \bar{I}(\vec{\rho}_1)\bar{I}(\vec{\rho}_2) + \langle \, \Delta I(\vec{\rho}_1)\Delta I(\vec{\rho}_2)\rangle. \quad (30)$$

Therefore, the point-to-point image-forming correlation is considered as a result of the statistical correlation of intensity fluctuations between the object and the image planes. Comparing Eq. (30), which has a constant background $\bar{I}(\vec{\rho}_1)\ \bar{I}(\vec{\rho}_2)$, with Eq. (29), which has a zero background, the mean intensities $\bar{I}(\vec{\rho}_1)$ and $\bar{I}(\vec{\rho}_2)$ must be zero, otherwise the result would lead to non-physical conclusions. The measurements, however, never yield zero mean values of $\bar{I}(\vec{\rho}_1)$ and $\bar{I}(\vec{\rho}_2)$ under any circumstances. In fact, the individual-detector counting rates of D_1 and D_2 were monitored in the experiment of Pittman *et al.* with much greater value than that of the coincidences. It is clear that the classical theory of statistical correlation of intensity fluctuations does not reflect the correct physics behind type-one ghost imaging.

4. Type-two ghost imaging with chaotic radiation

In this section we discuss the physics of type-two ghost imaging. The near-field lensless ghost imaging with chaotic radiation was first demonstrated by Scarcelli *et al.* in the years

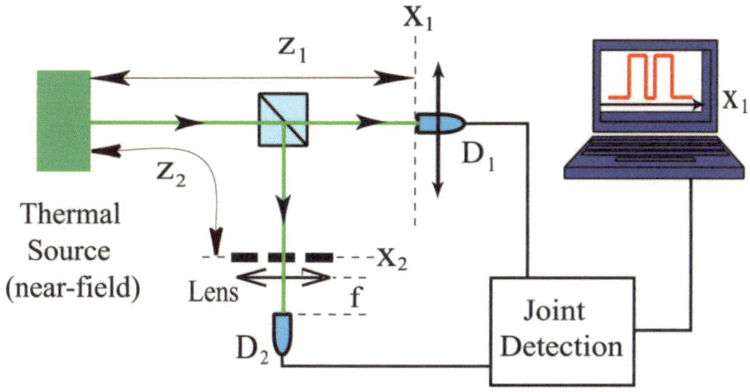

Fig. 9. Near-field lensless ghost imaging with chaotic light demonstrated in 2006 by Scarcelli *et al.* D_1 is a point-like photodetector that is scannable along the x_1-axis. The joint-detection between D_1 and the bucket detector D_2 is realized either by a photon-counting coincidence counter or by a standard HBT linear multiplier (RF mixer). In this measurement D_2 is fixed in the focal point of a convex lens, playing the role of a bucket detector. The counting rate or the photocurrent of D_1 and D_2, respectively, are measured to be constants. Surprisingly, an image of the 1-D object is observed in the joint-detection between D_1 and D_2 by scanning D_1 in the plane of $z_1 = z_2$ along the x_1-axis. The image, is blurred out when $z_1 \neq z_2$. There is no doubt that thermal radiations propagate to any transverse plane in a random and chaotic manner. There is no lens applied to force the thermal radiation "collapsing" to a point or speckle either. What is the physical cause of the point-to-point image-forming correlation in coincidences?

from 2005 to 2006 [5][6] following their experimental demonstration of two-photon interference of chaotic light [7]. The schematic experimental setup of their 2006 demonstration is shown in Fig. 9. Radiation with a narrow spectral bandwidth $\Delta\omega$ of a few millimeters diameter from a chaotic pseudothermal source [23] was equally divided into two by a 50% − 50% non-polarizing beam-splitter. In the reflected arm, a double-slit with slit separation $b = 1.5$ mm and slit width $a = 0.2$ mm, was placed at a distance $z_2 = 139$ mm from the source and a bucket detector D_2 was placed just behind the object. In the transmitted arm a point detector D_1 was scanned in the transverse plane of $z_1 = z_2$. Scarcelli *et al* tested two different joint detection schemes, namely the photon counting coincidence circuit and the standard HBT correlator. In the photon counting regime two Geiger mode avalanche photodiodes were employed for single-photon detection. In the bright light condition, two silicon PIN diodes were used with a standard analog HBT linear multiplier. The bucket detector D_2 was simulated by using a short focal length lens ($f = 25mm$) to focus the light coming from the object onto the active area of the detector while the point detector D_2 was simulated by a pinhole like aperture. After a large number of reaped measurements for different experimental schemes and conditions, Scarcelli *et al* reported the following observations.

Observation (1): A typical measured ghost image of the double-slit is shown in Fig. 10. The measured curve reports the joint-detection counting rate between D_1 and D_2, or the output current of a HBT linear multiplier, as a function of the transverse position of the point detector D_1 along x_1 axis. Notice, in Fig. 10 the constant background has been removed from the correlation.

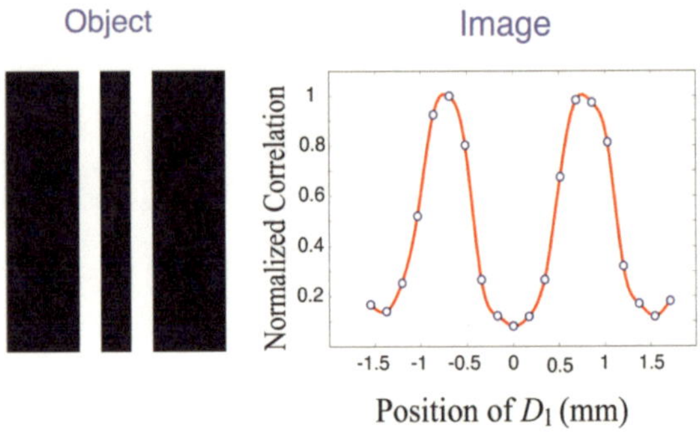

Fig. 10. The double-slit and its ghost image. Notice, the constant background has been removed from the correlation.

Observation (2): The measured contrasts vary significantly under different experimental schemes and conditions. It was found that the image contrast can achieve ~50% in photon counting measurement if no more than one joint-detection event occurring within the time window of the coincidence circuit. 50% is the maximum image contrast we expect for thermal light ghost imaging.

Observation (3): To achieve less than one joint-detection event per coincidence time window, weak light source is not a necessary condition. It can be easily achieved under bright light condition by using adjustable ND-filters with D_1 and D_2.

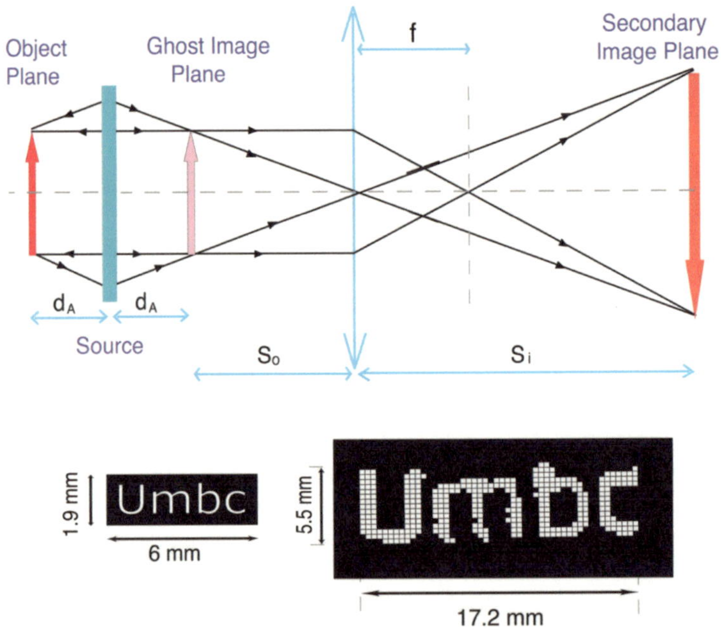

Fig. 11. Unfolded schematic experimental setup of a secondary image measurement of the ghost image and the measured secondary images. By using a convex lens of focal length f, the ghost image is imaged onto a secondary image plane, which is defined by the Gaussian thin-lens equation, $1/s_o + 1/s_i = 1/f$, with magnification $m = -s_i/s_o$. This setup is useful for distant large scale ghost imaging applications.

To confirm the observations are imaged images, and not "projection shadows", Scarcelli *et al.* made two additional measurements. In the first measurement, photodetector D_1 was moved away from the ghost image plane of $z_1 = z_2$. Whether moved in the direction of $z_1 > z_2$ or $z_1 < z_2$, the ghost image became "blurred". The measurement also showed that the depth of the image is a function of the angular size of the thermal source: a larger angular sized source (opening angle $\Delta\theta$ relative to the photodetectors) produces sharper image with shorter image depth. In the second measurement, Scarcelli *et al.* constructed a secondary imaging system, illustrated schematically in Fig. 11. By using a convex lens of focus length f the ghost image is imaged onto a secondary image plane, which is defined by the Gaussian thin-lens equation, $1/s_o + 1/s_i = 1/f$, with magnification $m = -s_i/s_o$. In this measurement the scanning photodetector D_1 is placed on the secondary imaging plane. The secondary image of the ghost image is observed in the joint-detection between D_1 and D_2 by means of either a photon-counting coincidence counter or a HBT linear multiplier.

4.1 What is the physical cause of chaotic light ghost imaging?

It is the partial point-to-point correlation between the object plane and the image plane that makes ghost imaging with thermal light possible. Similar but different from classical imaging and type-one ghost imaging, mathematically, type-two ghost imaging is the result of a convolution between the aperture function $|A(\vec{\rho}_2)|^2$ and a δ-function like partial point-to-"spot" correlation function

$$R_{12} \propto \int_{object} d\vec{\rho}_2 \, \big|A(\vec{\rho}_2)\big|^2 \, \big[1 + somb^2\big(\frac{\pi\Delta\theta\,|\vec{\rho}_1 - \vec{\rho}_2|}{\lambda}\big)\big] \tag{31}$$

in 2-D, where $\Delta\theta$ is the angular diameter of the radiation source viewed from the photodetector, $\vec{\rho}_1$ and $\vec{\rho}_2$ are the transverse coordinates on the object plane and the image plane, respectively, or

$$R_{12} \propto \int_{object} dx_2 \, \big|A(x_2)\big|^2 \, \big[1 + sinc^2\big(\frac{\pi\Delta\theta\,(x_1 - x_2)}{\lambda}\big)\big]. \tag{32}$$

in 1-D. For a chosen wavelength, the spatial resolution of the ghost image is determined by the angular diameter of the light source: the larger the size of the source in transverse dimensions, the higher the spatial resolution of the lensless ghost image. The point-to-"spot" image-forming functions in Eqs. (31) and (32) have been verified experimentally by Scarcelli *et al.*

The physical process corresponding to the convolution of Eq. (31) and (32) is similar to that of the type-one ghost imaging. Suppose the point detector D_1 or a CCD element is triggered by a photon at a transverse position of $\vec{\rho}_1$ in a joint-detection event with the bucket detector D_2 which is triggered by another photon that is either transmitted or reflected from the object. According to Eq. (31) and (32), under condition of $z_1 = z_2$, the photon from the object would have twice greater chance to be found at $\vec{\rho}_{obj} = \vec{\rho}_1$. Now, we move D_1 to another transverse position $\vec{\rho'}_1$, or locate another CCD element at $\vec{\rho'}_1$ for joint-detection. The photon that triggers D_2 would have twice greater chance of been located at $\vec{\rho}'_{obj} = \vec{\rho}'_1$. The probabilities of receiving a joint detection event at $\vec{\rho}_1 = \vec{\rho}_{obj}$ and at $\vec{\rho}' = \vec{\rho}'_{obj}$ would be modulated by the values of the aperture function $A(\vec{\rho}_{obj})$ and $A(\vec{\rho}'_{obj})$, respectively. Accumulating a large number of joint-detection events for each transverse coordinates $\vec{\rho}_1$, or for each CCD element in the image plane, a 50% contrast aperture function $A(\vec{\rho}_1) = A(\vec{\rho}_{obj})$ is thus reproduced in the joint-detection as a function of $\vec{\rho}_1$.[6]

To achieve thermal light ghost image with 50% contrast, we need (1) randomly distributed radiations on the object plane and on the image plane, respectively; and (2) for any photoelectron event at $\vec{\rho}_1$ there exists a unique corresponding point $\vec{\rho}_{obj} = \vec{\rho}_1$ on the object plane which has twice chance of observing another photoelectron event jointly and simultaneously. There is no doubt that random and chaotic radiation would propagate to any transverse plane in a random and chaotic manner. Therefore, condition (1) is satisfied automatically for chaotic thermal radiation. However, it is not easy to understand condition (2). We have been asking ourself a question since the first observation of lensless thermal light ghost image: what is the physical cause of the non-factorizable partial point-to-point image-forming function of $1+\delta(\vec{\rho}_1 - \vec{\rho}_{obj})$? There seems no reason to have such a statistical correlation for thermal light. Figure 12 schematically illustrates this situation. To simplify the picture we assume the source in 1-D with a large number of independent point sub-sources randomly distributed from $-b/2$ to $b/2$. Each point sub-source, such as the jth and the kth sub-source, randomly radiates independent spherical waves to the object and image planes, respectively. Due to the chaotic nature of the source, these independent and incoherent subintensities simply add together yielding a constant total intensity spatially and temporally on any transverse plane. The more chaotic sub-fields that contribute to the intensity sum, the less value of $\Delta I/I$ is expected. For any two transverse planes, such as the object and the image planes in Fig. 9, each with independent and randomly distributed intensities, statistically, there is no reason to expect any spatial or temporal correlations. What is the physical cause that forces a twice large probability for the thermal radiation to jointly appear at $\vec{\rho}_1 = \vec{\rho}_{obj}$?

In fact, we have been facing this question since 1956, after the discovery of Hanbury Brown and Twiss (HBT). The lensless ghost imaging setup looks similar to that of the historical HBT spatial interferometer which was used for measuring the angular size of distant stars. A significant difference is that the lensless ghost imaging measurement is in near-field[7] for imaging purposes [5].

The HBT experiment created quite a surprise in the physics community with an enduring debate about the classical or quantum nature of the phenomenon [11][12]. Figure 13 is a schematic of the historical HBT experiment which measures the transverse spatial

---

[6] To observe thermal light ghost image with maximum 50% contrast requires achieving a necessary experimental condition: no more than one joint detection event within the coincidence time window.

[7] The concept of "near-field" was defined by Fresnel to be distinct from the Fraunhofer far-field. The Fresnel near-field is different from the "near-surface-field" which considers a distance of a few wavelengths from a surface.

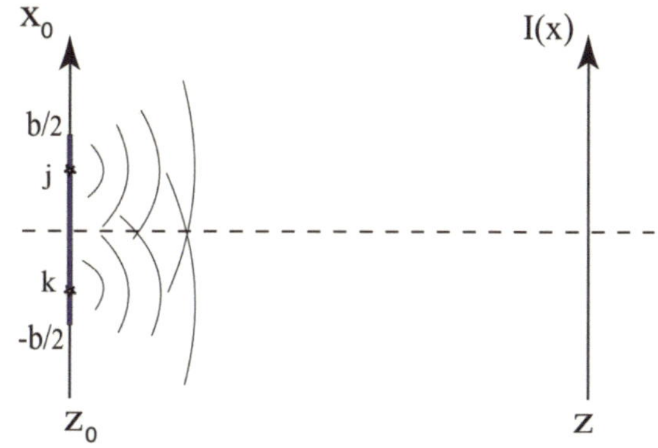

Fig. 12. A large number of independent point sub-sources, such as the jth and kth, are randomly distributed on the plane of a thermal source. These point sub-sources randomly radiate independent spherical waves to the object and image planes, respectively. Due to the chaotic nature of the source, these independent sub-intensities simply added together yielding a constant total intensity in space and in time on any transverse planes.

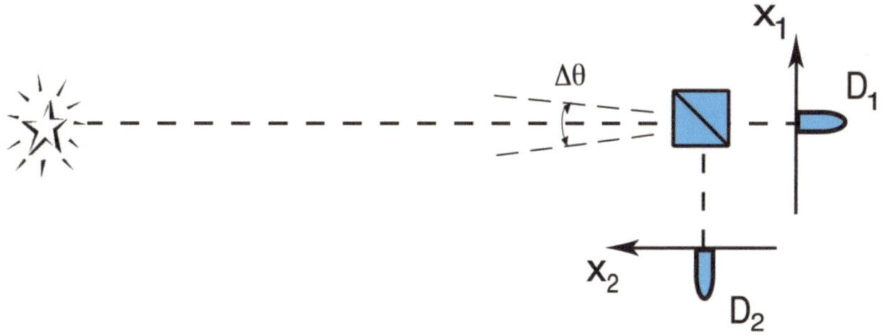

Fig. 13. Schematic of the historical Hanbury Brown and Twiss experiment which measures the transverse spatial correlation of far-field thermal radiation.

correlation of far-field thermal radiation. Performing the measurement in 1-D by scanning photodetectors D_1 and/or D_2 along the axes x_1 and x_2, the second-order transverse spatial correlation function $G^{(2)}(x_1, x_2)$ was found to be

$$G^{(2)}(x_1, x_2) \sim I_0^2 \left\{1 + sinc^2\left[\frac{\pi\Delta\theta(x_1 - x_2)}{\lambda}\right]\right\}, \tag{33}$$

where $\Delta\theta$ is the angular size of the star, λ the wavelength of the radiation. The far-field HBT correlation of Eq. (33) has been interpreted as the result of classical statistical correlation of the intensity fluctuations

$$\langle I_1 I_2 \rangle = \langle (\bar{I}_1 + \Delta I_1)(\bar{I}_2 + \Delta I_2) \rangle = \bar{I}_1 \bar{I}_2 + \langle \Delta I_1 \Delta I_2 \rangle,$$

where $\bar{I}_1$ and $\bar{I}_2$ are the mean intensities of the radiation measured by photodetectors D_1 and D_2, respectively. The second term in Eq. (33), $I_0^2\, sinc^2[\pi\Delta\theta(x_1 - x_2)/\lambda]$, is phenomenologically interpreted as the intensity fluctuation correlation $\langle \Delta I_1 \Delta I_2 \rangle$ in classical theory. For visible wavelengths and large values of $\Delta\theta$ this function quickly drops from its maximum to minimum when $x_1 - x_2$ moves from zero to a value such that $\Delta\theta(x_1 - x_2)/\lambda = 1$. In this situation we effectively have a "point-to-point" relationship between the x_1 and x_2 axes: for each point on the x_1 there exists only one point on the x_2 that may have a nonzero intensity fluctuation correlation.

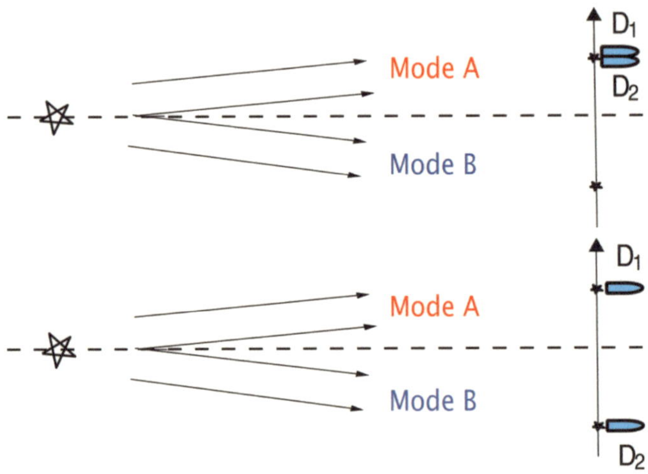

Fig. 14. A phenomenological interpretation of the historical HBT experiment. Upper: the two photodetectors receive identical modes of the far-field radiation and thus experience identical intensity fluctuations. The joint measurement of D_1 and D_2 gives a maximum value of $\langle \Delta I_1 \Delta I_2 \rangle$. Lower: the two photodetectors receive different modes of the far-field radiation. In this case the joint measurement gives $\langle \Delta I_1 \Delta I_2 \rangle = 0$. Unfortunately, this hand-waving interpretation does not reflect the correct physics in the case of $\Delta\theta \neq 0$. For a finite angular sized source, there is no chance, at least realistically, for D_1 and D_2 to receive radiation from a single radiation mode only. Nevertheless, the above theory has convinced us to believe that the observation of the intensity fluctuation correlation only takes place in the far-field zone of the thermal source.

The well-accepted interpretation of the HBT phenomenon is the following: in HBT the measurement is taken in the far-field zone of the radiation source, which is equivalent to the Fourier transform plane. When D_1 (D_2) is scanned in the neighborhood of $x_1 = x_2$, the two detectors measure the same mode of the radiation field. The measured intensities have the same fluctuations and yield a maximum value of $\langle \Delta I_1 \Delta I_2 \rangle$. The two upper curves of $I(t)$ in Fig. 15 schematically illustrate this situation. When the two photodetectors move apart from $x_1 = x_2$, D_1 and D_2 measure different modes of the radiation field. In this case, the measured two modes may have completely different fluctuations. The measurement yields $\langle \Delta I_1 \Delta I_2 \rangle = 0$ and gives $\langle I_1 I_2 \rangle = \bar{I}_1 \bar{I}_2$. This situation is illustrated in the two lower curves of $I(t)$ in Fig. 15. Unfortunately, this handwaving interpretation does not reflect the correct physics in the case of $\Delta\theta \neq 0$. For a finite angular sized source, there is no chance, at least realistically, for D_1 and D_2 to receive radiation from a single radiation mode only. Nevertheless, the above theory has convinced us to believe that the observation of the intensity fluctuation correlation only takes place in the far-field zone of the thermal source.

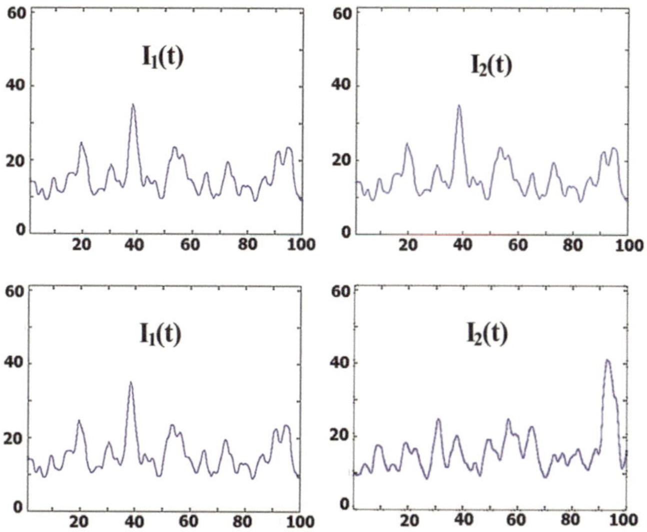

Fig. 15. Schematic illustration of the light intensities $I_1(t)$ at D_1 and $I_2(t)$ at D_2. The two upper (lower) curves of $I(t)$ corresponds to the upper (lower) configuration in Fig 14.

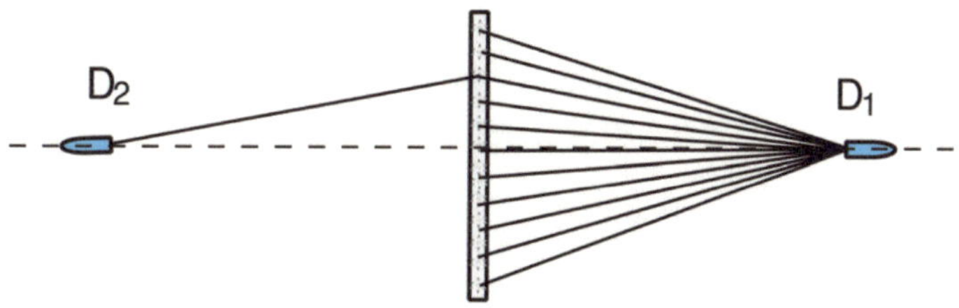

Fig. 16. Modified near-field HBT measurement - an unfolded Klyshko picture of the setup. Assuming a large sized disk-like near-field chaotic source, each point on the disk can be considered as an independent sub-source. It is easy to see that (1) D_1 and D_2 are capable of receiving radiation from a large number of sub-sources; and (2) D_1 and D_2 have more chances to be triggered jointly by radiations from different sub-sources; (3) The ratio between the joint-detections triggered by radiation from a single sub-source and from different sub-sources is roughly $N/N_2 = 1/N$ in any transverse position of D_1 and D_2.

What will happen if we move the photodetectors D_1 and D_2 to the "near-field" as shown in the unfolded schematic of Fig. 16? Does this hand-waving argument still predict the point-to-point correlation in this situation? We consider a disk-like thermal source with a large number of independent and randomly radiating point sub-sources and assume the radiations coming from the same sub-source have the same intensity fluctuation, and the radiations coming from different sub-sources have different intensity fluctuations. It is easy to see that in the near-field, (1) each photodetector, D_1 and D_2, is capable of receiving radiations from a large number of sub-sources; and (2) D_1 and D_2, have more chances to be triggered jointly by radiation from different subsources; (3) The ratio between the joint-detections triggered by radiation from a single sub-source and from different sub-sources is roughly $N/N_2 = 1/N$ in any transverse position of D_1 and D_2. For a large value of N the contribution of joint-detections triggered by radiation from a single subsource in any transverse position of D_1 and D_2 has the same negligible value $\langle \Delta I_1 \Delta I_2 \rangle / \bar{I}_1 \bar{I}_2 \sim 0$. Following the above philosophy, the near-field $G_{(2)}(\vec{\rho}_1, \vec{\rho}_2)$ should be a constant for any chosen transverse coordinates $\vec{\rho}_1$ and $\vec{\rho}_2$. The experimental observations, however, have shown a different story.

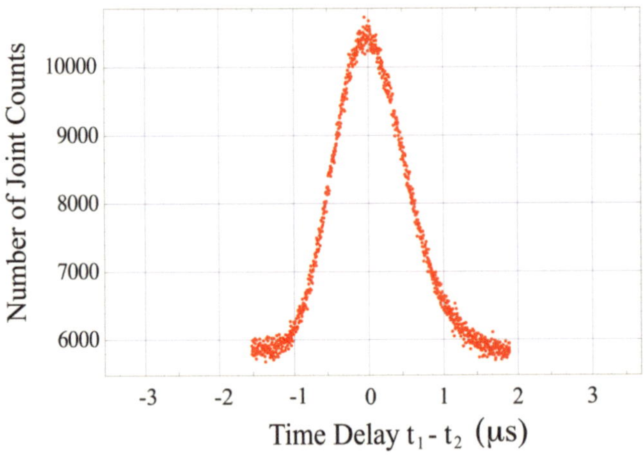

Fig. 17. $G^{(2)}(t_1 - t_2)$ of a chaotic source. The temporal correlation width is measured ~ $0.5\mu s$, which means that unless $t_1 - t_2 > 0.5\mu s$, the value of $G^{(2)}(t_1 - t_2)$will stay at the neighborhood of its maximum.

The nontrivial near-field point-to-point correlation was experimentally observed in a modified HBT experiment by Scarcelli *et al.* in 2005 before the near-field lensless ghost imaging demonstration. The modified HBT has a similar experimental setup as that of the historical HBT of Fig. 13, except replacing the distant star with a near-field disk-like chaotic source. This light source has a considerably large angular diameter from the view of the photodetectors D_1 and D_2. The point photodetectors D_1 and D_2 are scannable along the axes of x_1 and x_2, respectively. The frequency bandwidth $\Delta\omega$ of this thermal source is chosen to be narrow enough to achieve $\sim\mu s$ correlation width of $G^{(2)}(t_1 - t_2)$ which is shown in Fig. 17. This means to change $G^{(2)}$ from its maximum (minimum) value to minimum (maximum) value requires a few hundred meters optical delay in the arm of either D_1 or D_2. The transverse intensity distributions were examined before the measurement of transverse correlation. The counting rate (weak light condition) or the output photocurrent (bright light condition) of each individual photodetector was found to be constant, i.e., $I(\vec{\rho}_1)\sim$ constant and $I(\vec{\rho}_2)\sim$ constant by scanning D_1 and D_2 in the transverse planes of $z_1 = z_0$ and $z_2 = z_0$, respectively. There is no surprise to have constant $I(\vec{\rho}_1)$ and $I(\vec{\rho}_2)$. The physics has been clearly illustrated in Fig. 12. By using this kind of chaotic source, Scarcelli *et al.* measured the 1-D near-field normalized transverse spatial correlation of $g^{(2)}(x_1 - x_2)$ by scanning D_1 in the neighborhood of $x_1 = x_2$. The measurements confirmed the point-to-"spot" correlation of

$$g^{(2)}(x_1 - x_2) \sim 1 + sinc^2\Big[\frac{\pi\Delta\theta(x_1 - x_2)}{\lambda}\Big], \tag{34}$$

where, again, $\Delta\theta$ is the angular diameter of the near-field disk-like chaotic source. It is worth emphasizing that $g^{(2)}(x_1-x_2)$ dependents on x_1-x_2 only. Taking x_1-x_2 = constant, $g^{(2)}(x_1-x_2)$ is invariant under the displacements of transverse coordinates.

A simplified summary of the experimental observation is shown in Fig. 18: (1) In the upper figure, D_1 and D_2 are placed at equal distances from the source and aligned symmetrically on the optical axis. The normalized joint-detection, or the value of $g^{(2)}$ achieved its maximum of ~ 2. (2) In the middle figure, D_1 is moved up a few millimeters to a non-symmetrical position, the normalized joint-detection, or the value of $g^{(2)}$ is measured to be ~ 1. (3) In the lower figure, D_2 is moved a few millimeters up to a symmetrical position with respect to D_1. The normalized joint-detection, or the value of $g^{(2)}$ turned back to its maximum of ~ 2 again.

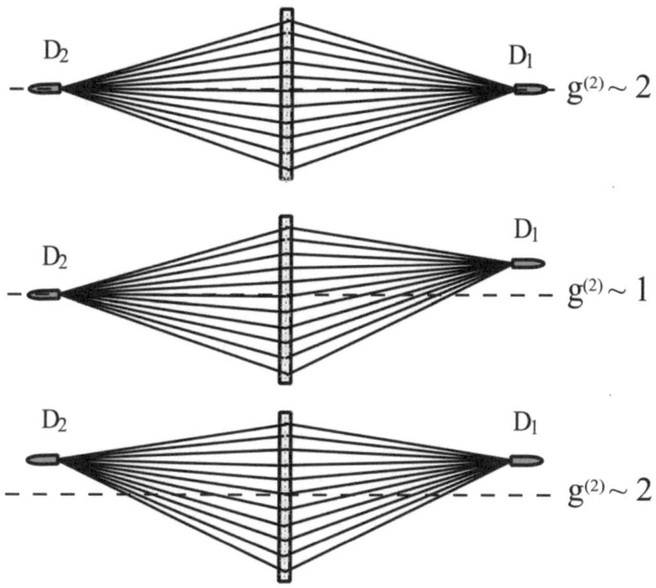

Fig. 18. Schematic of the near-field spatial correlation measurement of Scarcelli *et al.* Upper: $g^{(2)} \sim 2$, where D_1 and D_2 are placed at equal distances from the source and aligned symmetrically on the optical axis. In the spirit of the traditional interpretation of HBT the intensities measured by D_1 and D_2 must have same fluctuations as shown in the figure. Middle: $g^{(2)} \sim 1$, where D_1 is moved up a few millimeters to an asymmetrical position. In the spirit of the traditional interpretation of HBT the intensities measured by D_1 and D_2 must have different fluctuations. Lower: $g^{(2)} \sim 2$, where D_2 is moved up to a symmetrical position with respect to D_1, again. In the spirit of the traditional interpretation of HBT the intensities measured by D_1 and D_2 must have same fluctuations again. What is the physical cause of the changes of the intensity fluctuations then? Remember the $G^{(2)}(t_1 - t_2)$ function has a width of $\sim 0.5\mu$, see Fig. 17.

It is easy to see that the classical theory of statistical correlation of intensity fluctuations is facing difficulties in explaining the experimental results. In near-field D_1 and D_2 receive the same large number of modes at any $\vec{\rho}_1$ and $\vec{\rho}_2$. In the spirit of the traditional interpretation of HBT, there seems no reason to have a different intensity fluctuation correlation between

$\vec{\rho}_1 = \vec{\rho}_2$ and $\vec{\rho}_1 \neq \vec{\rho}_2$ for the $G^{(2)}$ function shown in Fig. 17. In the upper measurement, we have obtained the maximum value of $g^{(2)} \sim 2$ at $z_1 = z_2$ and $x_1 = x_2$, which indicates the achievement of a maximum intensity fluctuation correlation as shown in Fig. 17 with $|t_1 - t_2| \sim 0$. In the middle measurement, $g^{(2)} \sim 1$ indicates a minimum intensity fluctuation correlation by moving D_1 a few millimeters up, which means the intensities measured by D_1 and D_2 must have different fluctuations. In the lower measurement D_2 is moved up a few millimeters to a new symmetrical position with respect to D_1, the measurements obtain $g^{(2)} \sim 2$ again. The intensities measured by D_1 and D_2 must have same fluctuations again. What is the physical cause of the changes of the intensity fluctuations then? Remember the $G^{(2)}(t_1 - t_2)$ function has a width of $\sim 0.5\mu$.

For half a century since 1956, it has been believed that the HBT correlation is observable in the far-field only. It was quite a surprise that in 2005 Scarcelli *et al.* successfully demonstrated a near-field point-to-point transverse correlation of chaotic light, indicating that the nontrivial HBT spatial correlation is observable in the near-field and is useful for reproducing ghost images in a nonlocal manner.[8] The experiment of Scarcelli *et al.* raised a question: "Can two-photon correlation of chaotic light be considered as correlation of intensity fluctuations?" [5] At least, this experiment suggested we reexamine the relationship between the quantum mechanical concept of joint-detection probability with the classical concept of intensity fluctuation correlation. It seems that jointly observing a pair of photons at space-time point $(\mathbf{r}_1, t_1)$ and $(\mathbf{r}_2, t_2)$ is perhaps only phenomenologically connected but not physically caused by the classical statistical correlation of intensity fluctuations. The point-to-point image-forming correlation is more likely the result of an interference. In the view of two-photon interference, far-field is not a necessary condition for observing the partial point-to-point correlation of thermal light. Furthermore, it is quite common in two-photon interference type experiments to observe constant counting rates or intensities in individual photodetectors D_1 and D_2, respectively, and simultaneously observe nontrivial space-time correlation in the joint-detection between D_1 and D_2. These observations are consistent with the quantum theory of two-photon interferometry [9].

4.2 Quantum theory of thermal light ghost imaging

According to the quantum theory of light, the observed partial point-to-point image-forming correlation is the result of multi-photon interference. In Glauber's theory of photo-detection [20], an idealized point photodetector measures the probability of observing a photo-detection event at space-time point $(\mathbf{r}, t)$

$$G^{(1)}(\mathbf{r}, t) = tr\{\hat{\rho}\, E^{(-)}(\mathbf{r}, t) E^{(+)}(\mathbf{r}, t)\}, \tag{35}$$

where $\hat{\rho}$ is the density operator which characterizes the state of the quantized electromagnetic field, $E^{(-)}(\mathbf{r}, t)$ and $E^{(+)}(\mathbf{r}, t)$ the negative and positive field operators at space-time coordinate $(\mathbf{r}, t)$. The counting rate of a point photon counting detector, or the

[8] We cannot help but stop to ask: What has been preventing this simple move from far-field to near-field for half a century?

output current of a point analog photodetector, is proportional to $G^{(1)}(\mathbf{r}, t)$. A joint-detection of two independent point photodetectors measures the probability of observing a joint-detection event of two photons at space-time points $(\mathbf{r}_1, t_1)$ and $(\mathbf{r}_2, t_2)$

$$G^{(2)}(\mathbf{r}_1, t_1; \mathbf{r}_2, t_2) = tr\{\hat{\rho}\, E^{(-)}(\mathbf{r}_1, t_1) E^{(-)}(\mathbf{r}_2, t_2) E^{(+)}(\mathbf{r}_2, t_2) E^{(+)}(\mathbf{r}_1, t_1)\}, \tag{36}$$

where $(\mathbf{r}_j, t_j)$, $j = 1, 2$, is the space-time coordinate of the jth photo-detection event. The coincidence counting rate of two photon counting detectors, or the output reading of a linear multiplier (RF mixer) between two photodetectors, is proportional to $G^{(2)}(\mathbf{r}_1, t_1; \mathbf{r}_2, t_2)$. To calculate the partial point-to-point correlation between the object plane and the image plane, we need (1) to estimate the state, or the density matrix, of the thermal radiation; and (2) to propagate the field operators from the radiation source to the object and the image planes. We will first calculate the state of thermal radiation at the single-photon level for photon counting measurements to explore the physics behind ghost imaging as two-photon interference and then generalize the result to any intensity of thermal radiation.

We assume a large transverse sized chaotic source consisting of a large number of independent and randomly radiating point sub-sources. Each point sub-source may also consist of a large number of independent atoms that are ready for two-level atomic transitions in a random manner. Most of the time, the atoms are in their ground state. There is, however, a small chance for each atom to be excited to a higher energy level E_2 ($\Delta E_2 \neq 0$) and later return back to its ground state E_1. It is reasonable to assume that each atomic transition generates a field in the following single-photon state

$$|\,\Psi\,\rangle \simeq |\,0\,\rangle + \epsilon \sum_{\mathbf{k},s} f(\mathbf{k}, s)\, \hat{a}^{\dagger}_{\mathbf{k},s}\, |\,0\,\rangle, \tag{37}$$

where $|\epsilon| \ll 1$ is the probability amplitude for the atomic transition, $f(\mathbf{k}, s) = \langle\, \Psi_{\mathbf{k},s}\,|\,\Psi\,\rangle$ is the probability amplitude for the radiation field to be in the single-photon state of wave number $\mathbf{k}$ and polarization s: $|\,\Psi_{\mathbf{k},s}\,\rangle = |\,1_{\mathbf{k},s}\,\rangle = \hat{a}^{\dagger}_{\mathbf{k},s}\,|\,0\,\rangle$. For this simplified two-level system, the density matrix that characterizes the state of the radiation field excited by a large number of possible atomic transitions is thus

$$\begin{aligned}
\hat{\rho} &= \prod_{t_{0j}} \Big\{|\,0\,\rangle + \epsilon \sum_{\mathbf{k},s} f(\mathbf{k}, s)\, e^{-i\omega t_{0j}}\, \hat{a}^{\dagger}_{\mathbf{k},s}\,|\,0\,\rangle\Big\} \\
&\times \prod_{t_{0k}} \Big\{\langle\,0\,| + \epsilon^{*} \sum_{\mathbf{k}',s'} f(\mathbf{k}', s')\, e^{i\omega' t_{0k}}\, \langle\,0\,|\hat{a}_{\mathbf{k}',s'}\Big\} \\
&\simeq \Big\{|\,0\,\rangle + \epsilon[\sum_{t_{0j}} \sum_{\mathbf{k},s} f(\mathbf{k}, s)\, e^{-i\omega t_{0j}}\, \hat{a}^{\dagger}_{\mathbf{k},s}\,|\,0\,\rangle] + \epsilon^2[...]\Big\} \\
&\times \Big\{\langle\,0\,| + \epsilon^{*}[\sum_{t_{0k}} \sum_{\mathbf{k}',s'} f(\mathbf{k}', s')\, e^{i\omega' t_{0k}}\, \langle\,0\,|\hat{a}_{\mathbf{k}',s'}] + \epsilon^{*2}[...]\Big\},
\end{aligned} \tag{38}$$

where $e^{-i\omega t_{0j}}$ is a random phase factor associated with the jth atomic transition. Since $|\epsilon| \ll 1$, it is a good approximation to keep the necessary lower-order terms of ϵ in Eq. (38). After summing over t_{0j} (t_{0k}) by taking into account all its possible values we obtain

$$\hat{\rho} \simeq |\,0\,\rangle\langle\,0\,| + |\epsilon|^2 \sum_{\mathbf{k},\mathbf{s}} |f(\mathbf{k},\mathbf{s})|^2 \,|\,1_{\mathbf{k},\mathbf{s}}\,\rangle\langle\,1_{\mathbf{k},\mathbf{s}}\,| \\ + |\epsilon|^4 \sum_{\mathbf{k},s}\sum_{\mathbf{k}',s'} |f(\mathbf{k},\mathbf{s})|^2\, |f(\mathbf{k}',s')|^2 \,|\,1_{\mathbf{k},s}1_{\mathbf{k}',s'}\,\rangle\langle\,1_{\mathbf{k},s}1_{\mathbf{k}',s'}\,| + \ldots \tag{39}$$

Similar to our earlier discussion we will focus our calculation on the transverse correlation by assuming a narrow enough frequency bandwidth in Eq. (39). In the experiments of Scarcelli *et al.* the coherence time of the radiation was chosen ~μs, the maximum achievable optical path differences ~ps by the scanning of D_1 and D_2, and the response time of the photodetectors is much less than the coherence time. The transverse spatial correlation measurement is under the condition of achieving a maximum temporal coherence of $G^{(2)}(t_1 - t_2) \sim 2$ during the scanning of D_1 and D_2 at any $\vec{\rho}_1$ and $\vec{\rho}_2$. In the *photon counting* regime, under the above condition, it is reasonable to model the thermal light in the following mixed state

$$\hat{\rho} \simeq |\,0\,\rangle\langle\,0\,| + |\epsilon|^2 \sum_{\vec{\kappa}} \hat{a}^\dagger(\vec{\kappa})|\,0\,\rangle\langle\,0\,|\hat{a}(\vec{\kappa}) + |\epsilon|^4 \sum_{\vec{\kappa}}\sum_{\vec{\kappa}'} \hat{a}^\dagger(\vec{\kappa})\hat{a}^\dagger(\vec{\kappa}')\,|\,0\,\rangle\langle\,0\,|\,\hat{a}(\vec{\kappa}')\hat{a}(\vec{\kappa}). \tag{40}$$

Basically we are modeling the light source as an incoherent statistical mixture of single-photon states and two-photon states with equal probability of having any transverse momentum. The spatial part of the second-order coherence function is thus calculated as

$$\begin{aligned} G^{(2)}(\vec{\rho}_1, z_1; \vec{\rho}_2, z_2) &= tr[\,\hat{\rho}\, E^{(-)}(\vec{\rho}_1, z_1) E^{(-)}(\vec{\rho}_2, z_2) E^{(+)}(\vec{\rho}_2, z_2) E^{(+)}(\vec{\rho}_1, z_1)\,] \\ &= \sum_{\vec{\kappa},\vec{\kappa}'} \langle 1_{\vec{\kappa}} 1_{\vec{\kappa}'}| E^{(-)}(\vec{\rho}_1, z_1) E^{(-)}(\vec{\rho}_2, z_2) E^{(+)}(\vec{\rho}_2, z_2) E^{(+)}(\vec{\rho}_1, z_1) |1_{\vec{\kappa}} 1_{\vec{\kappa}'}\rangle \\ &\equiv \sum_{\vec{\kappa},\vec{\kappa}'} \left|\Psi_{\vec{\kappa},\vec{\kappa}'}(\vec{\rho}_1, z_1; \vec{\rho}_2, z_2)\right|^2, \end{aligned} \tag{41}$$

where we have defined an effective two-photon wavefunction in transverse spatial coordinates

$$\Psi_{\vec{\kappa},\vec{\kappa}'}(\vec{\rho}_1, z_1; \vec{\rho}_2, z_2) = \langle 0| E^{(+)}(\vec{\rho}_2, z_2) E^{(+)}(\vec{\rho}_1, z_1) |1_{\vec{\kappa}} 1_{\vec{\kappa}'}\rangle. \tag{42}$$

The transverse part of the electric field operator can be written as

$$E^{(+)}(\vec{\rho}_j, z_j) \propto \sum_{\vec{\kappa}} g_j(\vec{\rho}_j, z_j; \vec{\kappa})\, \hat{a}(\vec{\kappa}), \tag{43}$$

again, $g_j(\vec{\rho}_j,\, z_j\,;\, \vec{\kappa})$ is the Green's function. Substituting the field operators into Eq. (42) we have

$$\Psi_{\vec{\kappa},\vec{\kappa}'}(\vec{\rho}_1, z_1; \vec{\rho}_2, z_2) = \frac{1}{\sqrt{2}}\left[g_2(\vec{\rho}_2, z_2; \vec{\kappa}) g_1(\vec{\rho}_1, z_1; \vec{\kappa}') + g_2(\vec{\rho}_2, z_2; \vec{\kappa}') g_1(\vec{\rho}_1, z_1; \vec{\kappa})\right] \tag{44}$$

and

$$G^{(2)}(\vec{\rho}_1, z_1; \vec{\rho}_2, z_2) = \sum_{\vec{\kappa},\vec{\kappa}'} \left| \frac{1}{\sqrt{2}} \left[g_2(\vec{\rho}_2, z_2; \vec{\kappa}) g_1(\vec{\rho}_1, z_1; \vec{\kappa}') + g_2(\vec{\rho}_2, z_2; \vec{\kappa}') g_1(\vec{\rho}_1, z_1; \vec{\kappa}) \right] \right|^2, \quad (45)$$

representing the key result for our understanding of the phenomenon. Eqs. (44) and (45) indicates an interference between two alternatives, different yet indistinguishable, which leads to a joint photo-detection event. This interference phenomenon is not, as in classical optics, due to the superposition of electromagnetic fields at a local point of space-time. This interference is the result of the superposition between $g_2(\vec{\rho}_2, z_2; \vec{\kappa}) g_1(\vec{\rho}_1, z_1; \vec{\kappa}')$ and $g_2(\vec{\rho}_2, z_2; \vec{\kappa}') g_1(\vec{\rho}_1, z_1; \vec{\kappa})$, the socalled two-photon amplitudes, non-classical entities that involve both arms of the optical setup as well as two distant photo-detection events at $(\vec{\rho}_1, z_1)$ and $(\vec{\rho}_2, z_2)$, respectively. Examining the effective wavefunction of Eq. (44), we find this symmitrized effective wavefunction plays the same role as that of the symmitrized wavefunction of identical particles in quantum mechanics. This peculiar nonlocal superposition has no classical correspondence, and makes the type-two ghost image turbulence-free, i.e., any phase disturbance in the optical path has no influence on the ghost image [24]. Fig. 19 schematically illustrates the two alternatives for a pair of mode $\vec{\kappa}$ and $\vec{\kappa}'$ to produce a joint photo-detection event: $\vec{\kappa}1 \times \vec{\kappa}'2$ and $\vec{\kappa}2 \times \vec{\kappa}'1$. The superposition of each pair of these amplitudes produces an individual sub-interference-pattern in the joint-detection space of $(\vec{\rho}_1, z_1, t_1; \vec{\rho}_2, z_2, t_2)$. A large number of these sub-interference-patterns simply add together resulting in a nontrivial $G^{(2)}(\vec{\rho}_1, z_1; \vec{\rho}_2, z_2)$ function. It is easy to see that each pair of the twophoton amplitudes, illustrated in Fig. 19, will superpose constructively whenever D_1 and D_2 are placed in the positions satisfying $\vec{\rho}_1 \simeq \vec{\rho}_2$ and $z_1 \simeq z_2$; and consequently, $G^{(2)}(\vec{\rho}_1, z_1; \vec{\rho}_2, z_2)$ achieves its maximum value as the result of the sum of these individual constructive interferences. In other coordinates, however, the superposition of each individual pair of the two-photon amplitudes may yield different values between constructive maximum and destructive minimum due to unequal optical path propagation, resulting in an averaged sum.

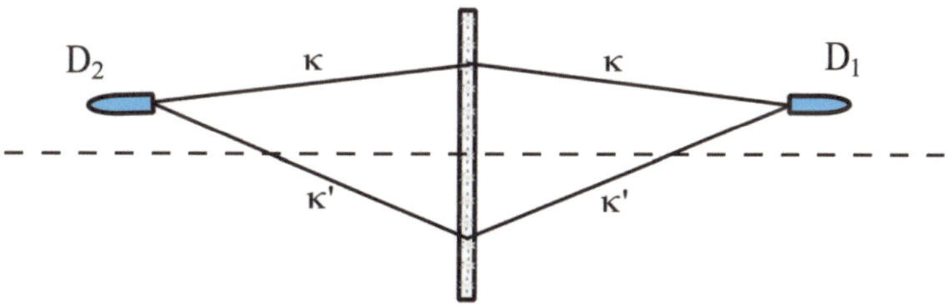

Fig. 19. Schematic illustration of two-photon interference: a superposition between two-photon amplitudes $g_2(\vec{\rho}_2, z_2\vec{\kappa}) g_1(\vec{\rho}_1, z_1; \vec{\kappa}')$ and $g_2(\vec{\rho}_2, z_2; \vec{\kappa}') g_1(\vec{\rho}_1, z_1; \vec{\kappa})$. It is clear that the amplitudes $g_2(\vec{\rho}_2, z_2; \vec{\kappa}) g_1(\vec{\rho}_1, z_1; \vec{\kappa}')$ and $g_2(\vec{\rho}_2, z_2; \vec{\kappa}') g_1(\vec{\rho}_1, z_1; \vec{\kappa})$ will experience equal optical path propagation and superpose constructively when D_1 and D_2 are located at $\vec{\rho}_1 \simeq \vec{\rho}_2$ and $z_1 \simeq z_2$. This nonlocal superposition has no classical correspondence.

Before calculating $G^{(2)}(\vec{\rho}_1, z_1; \vec{\rho}_2, z_2)$ we examine the single counting rate of the point photodetectors D_1 and D_2 which are placed at $(\vec{\rho}_1, z_1)$ and $(\vec{\rho}_2, z_2)$, respectively. With reference to the experimental setup of Fig. 9, the Green's function of free-propagation is derived in the Appendix

$$g_1(\vec{\rho}_1, z_1; \vec{\kappa}) = \int d\vec{\rho}_s \left\{ \frac{-i\omega}{2\pi c} \frac{e^{i\frac{\omega}{c} z_1}}{z_1} e^{i\frac{\omega}{2cz_1}|\vec{\rho}_1 - \vec{\rho}_s|^2} \right\} e^{-i\vec{\kappa}\cdot\vec{\rho}_s},$$
$$g_2(\vec{\rho}_2, z_2; \vec{\kappa}) = \int d\vec{\rho}'_s \left\{ \frac{-i\omega}{2\pi c} \frac{e^{i\frac{\omega}{c} z_2}}{z_2} e^{i\frac{\omega}{2cz_2}|\vec{\rho}_2 - \vec{\rho}'_s|^2} \right\} e^{-i\vec{\kappa}\cdot\vec{\rho}'_s},$$

where $\vec{\rho}_s$ is the transverse vector in the source plane, and the field has propagated from the source to the $\vec{\rho}_1$ plane and $\vec{\rho}_2$ plane in arms 1 and 2, respectively. The single detector counting rate or the output photocurrent is proportional to $G^{(1)}(\mathbf{r}, t)$ as shown in Eq. (35),

$$\begin{aligned} G^{(1)}(\vec{\rho}_j, z_j) &= tr\{\hat{\rho}\, E^{(-)}(\vec{\rho}_j, z_j) E^{(+)}(\vec{\rho}_j, z_j)\} \\ &\propto \sum_{\vec{\kappa}} \langle 0 \,|\, \hat{a}(\vec{\kappa})\, E^{(-)}(\vec{\rho}_j, z_j) E^{(+)}(\vec{\rho}_j, z_j)\, \hat{a}^\dagger(\vec{\kappa}) | \,0\, \rangle \\ &\sim \text{constant}, \end{aligned} \tag{46}$$

where j = 1, 2 indicating the jth photodetector.

Although $G^{(1)}(\vec{\rho}_1, z_1)$ and $G^{(1)}(\vec{\rho}_2, z_2)$ are both constants, $G^{(2)}(\vec{\rho}_1, z_1; \vec{\rho}_2, z_2)$ turns to be a nontrivial function of $(\vec{\rho}_1, z_1)$ and $(\vec{\rho}_2, z_2)$,

$$\begin{aligned} G^{(2)}(\vec{\rho}_1, z_1; \vec{\rho}_2, z_2) &= \sum_{\vec{\kappa}, \vec{\kappa}'} \Big| \frac{1}{\sqrt{2}} \big[g_2(\vec{\rho}_2, z_2; \vec{\kappa}) g_1(\vec{\rho}_1, z_1; \vec{\kappa}') + g_2(\vec{\rho}_2, z_2; \vec{\kappa}') g_1(\vec{\rho}_1, z_1; \vec{\kappa}) \big] \Big|^2 \\ &\equiv G^{(1)}(\vec{\rho}_1, z_1) G^{(1)}(\vec{\rho}_2, z_2) + \big| G^{(1)}(\vec{\rho}_1, z_1; \vec{\rho}_2, z_2) \big|^2, \end{aligned} \tag{47}$$

where

$$\begin{aligned} \big| G^{(1)}(\vec{\rho}_1, z_1; \vec{\rho}_2, z_2) \big|^2 &= \Big| \int d\vec{\kappa}\; g_1^*(\vec{\rho}_1, z_1; \vec{\kappa})\, g_2(\vec{\rho}_2, z_2; \vec{\kappa}) \Big|^2 \\ &\propto \Big| \int d\vec{\rho}_s\, e^{-i\frac{\omega}{c} z_1}\, e^{-i\frac{\omega}{2cz_1}|\vec{\rho}_1 - \vec{\rho}_s|^2}\, e^{i\frac{\omega}{c} z_2}\, e^{i\frac{\omega}{2cz_2}|\vec{\rho}_2 - \vec{\rho}_s|^2} \Big|^2. \end{aligned}$$

If we choose the distances from the source to the two detectors to be equal ($z_1 = z_2 = d$), the above integral of $d\vec{\rho}_s$ yields a point-to-point correlation between the transverse planes $z_1 = d$ and $z_2 = d$,

$$\big| G_{12}^{(1)}(\vec{\rho}_1; \vec{\rho}_2) \big|^2 \propto \Big| \int d\vec{\rho}_s\, e^{i\frac{\omega}{cd}(\vec{\rho}_1 - \vec{\rho}_2)\cdot\vec{\rho}_s} \Big|^2 = somb^2\Big[\frac{R}{d}\frac{\omega}{c}|\vec{\rho}_1 - \vec{\rho}_2|\Big] \sim \delta(\vec{\rho}_1 - \vec{\rho}_2), \tag{48}$$

where the δ-function is an approximation by assuming a large enough thermal source of angular size $\Delta\theta \sim R/d$ and high enough frequency ω, such as a visible light source. The nontrivial $G^{(2)}$ function is therefore,

$$G^{(2)}(\vec{\rho}_1; \vec{\rho}_2) \sim 1 + \delta(\vec{\rho}_1 - \vec{\rho}_2). \tag{49}$$

In the ghost imaging experiment, the joint-detection counting rate is thus

$$R_{12} \propto \int d\vec{\rho}_2 \, |A(\vec{\rho}_2)|^2 \, G^{(2)}(\vec{\rho}_1; \vec{\rho}_2) \sim R_0 + |A(\vec{\rho}_1)|^2, \tag{50}$$

where R_0 is a constant and $A(\vec{\rho}_2)$ is the aperture function of the object.

So far, we have successfully derived an analytical solution for ghost imaging with thermal radiation at the single-photon level. We have shown that the partial point-to-point correlation of thermal radiation is the result of a constructive-destructive interference caused by the superposition of two two-photon amplitudes, corresponding to two alternative ways for a pair of jointly measured photons to produce a joint-detection event. In fact the above analysis is not restricted to single-photon states. The partial point-to-point correlation of $G^{(2)}(\vec{\rho}_1; \vec{\rho}_2)$ is generally true for any order of quantized thermal radiation [25]. Now we generalize the calculation to an arbitrary quantized thermal field with occupation number from $n_{\mathbf{k},\mathbf{s}} = 0$ to $n_{\mathbf{k},\mathbf{s}} \gg 1$ by keeping all higher order terms in Eq. (38). After summing over t_{0j} and t_{0k} the density matrix can be written as

$$\hat{\rho} = \sum_{\{n\}} p_{\{n\}} \, |\{n\}\rangle\langle\{n\}|, \tag{51}$$

where $p_{\{n\}}$ is the probability for the thermal field in the state

$$|\{n\}\rangle \equiv \prod_{\mathbf{k},s} |n_{\mathbf{k},s}\rangle = |n_{\mathbf{k},s}\rangle |n_{\mathbf{k}',s'}\rangle ... |n_{\mathbf{k}''...',s''...'}\rangle.$$

The summation of Eq. (51) includes all possible modes $\mathbf{k}$, polarizations s, occupation numbers $n_{\mathbf{k},s}$ for the mode ($\mathbf{k}$, s) and all possible combinations of occupation numbers for different modes in a set of $\{n\}$. Substituting the field operators and the density operator of Eq. (51) into Eq. (35) we obtain the constant $G^{(1)}(\vec{\rho}_j, z_j, t_j)$, $j = 1, 2$, which corresponds to the intensities $I(\vec{\rho}_1, z_1, t_1)$ and $I(\vec{\rho}_2, z_2, t_2)$,

$$\begin{aligned} & G^{(1)}(\vec{\rho}_j, z_j, t_j) \\ & = \sum_{\{n\}} \int d\vec{\kappa} \int d\vec{\kappa}' g_j^*(\vec{\rho}_j, z_1, t_j; \vec{\kappa}) g_j(\vec{\rho}_j, z_1, t_j; \vec{\kappa}') \, p_{\{n\}} \, \langle\{n\}| \, a(\vec{\kappa}) a^\dagger(\vec{\kappa}') \, |\{n\}\rangle \\ & \propto \sum_{\{n\}} n_{\vec{\kappa}} n_{\vec{\kappa}'} \int d\vec{\kappa} \, \left| \, g_j^*(\vec{\rho}_j, z_j, t_j; \vec{\kappa}) \, \right|^2 \\ & \simeq \text{constant}. \end{aligned} \tag{52}$$

Although $G^{(1)}(\vec{\rho}_1, z_1, t_1)$ and $G^{(1)}(\vec{\rho}_2, z_2, t_2)$ are both constants, substituting the field operators and the density operator of Eq. (51) into Eq. (36), we obtain a nontrivial point-to-point correlation function of $G^{(2)}(\vec{\rho}_1; \vec{\rho}_2)$ at the two transverse planes $z_1 = d$ and $z_2 = d$,

$$
\begin{aligned}
&G^{(2)}(\vec{\rho}_1;\vec{\rho}_2)\\
&=\sum_{\{n\}}\int d\vec{\kappa}\int d\vec{\kappa}'\int d\vec{\kappa}''\int d\vec{\kappa}'''g_1^*(\vec{\rho}_1,z_1;\vec{\kappa})g_2^*(\vec{\rho}_2,z_2;\vec{\kappa}')g_2(\vec{\rho}_2,z_2;\vec{\kappa}'')g_1(\vec{\rho}_1,z_1;\vec{\kappa}''')\\
&\quad\times\, p_{\{n\}}\,\langle\{n\}|\,a(\vec{\kappa})a(\vec{\kappa}')a^\dagger(\vec{\kappa}'')a^\dagger(\vec{\kappa}''')\,|\{n\}\rangle\\
&\propto\sum_{\{n\}}n_{\vec{\kappa}}\,n_{\vec{\kappa}'}\int d\vec{\kappa}\int d\vec{\kappa}'\int d\vec{\kappa}''\int d\vec{\kappa}'''g_1^*(\vec{\rho}_1,z_1;\vec{\kappa})g_2^*(\vec{\rho}_2,z_2;\vec{\kappa}')g_2(\vec{\rho}_2,z_2;\vec{\kappa}'')g_1(\vec{\rho}_1,z_1;\vec{\kappa}''')\\
&\quad\times\, p_{\{n\}}(\delta_{\vec{\kappa}\vec{\kappa}'''}\delta_{\vec{\kappa}'\vec{\kappa}''}+\delta_{\vec{\kappa}\vec{\kappa}''}\delta_{\vec{\kappa}'\vec{\kappa}'''})\\
&=\sum_{n_{\vec{\kappa}}n_{\vec{\kappa}'}}p_{\{\ldots n_{\vec{\kappa}}\ldots n_{\vec{\kappa}'}\ldots\}}\,n_{\vec{\kappa}}\,n_{\vec{\kappa}'}\\
&\quad\times\Big\{\int d\vec{\kappa}\int d\vec{\kappa}'\Big|\frac{1}{\sqrt{2}}\big[g_1(\vec{\rho}_1,z_1;\vec{\kappa})g_2(\vec{\rho}_2,z_2;\vec{\kappa}')+g_2(\vec{\rho}_2,z_2;\vec{\kappa})g_1(\vec{\rho}_1,z_1;\vec{\kappa}')\big]\Big|^2\Big\}\\
&\propto\Big\{1+somb^2\big[\frac{R}{d}\frac{\omega}{c}(\vec{\rho}_1-\vec{\rho}_2)\big]\Big\}.
\end{aligned}
\tag{53}
$$

It is clear that in Eq. (53), the partial point-to-point correlation of thermal light is the result of a constructive-destructive interference between two quantum-mechanical amplitudes. We also note from Eq. (53) that the partial point-to-point correlation is independent of the occupation numbers, $\{n\}$, and the probability distribution, $p_{\{n\}}$, of the quantized thermal radiation.

It is interesting but not surprising to see that the effective two-photon wavefunction in bright light condition

$$
\Psi_{\vec{\kappa},\vec{\kappa}'}(\vec{\rho}_1,z_1;\vec{\rho}_2,z_2)=\frac{1}{\sqrt{2}}\big[g_2(\vec{\rho}_2,z_2;\vec{\kappa})g_1(\vec{\rho}_1,z_1;\vec{\kappa}')+g_2(\vec{\rho}_2,z_2;\vec{\kappa}')g_1(\vec{\rho}_1,z_1;\vec{\kappa})\big]
$$

is the same as that of weak light at single-photon level. In fact, the above effective wavefunction does play the same role in specifying two different yet indistinguishable alternatives for the two annihilated photons contributing to a joint-detection event of D_1 and D_2, which implies that the partial point-to-point correlation is the result of two-photon interference in bright light condition. This nonlocal partial correlation indicates that a 50% contrast ghost image is observable at bright light condition provided registering no more than one coincidence event within the joint-detection time window. This requirement can be easily achieved by using adjustable ND-filters with D_1 and D_2.

Quantum theory predicts and calculates the probability of observing a certain physical event. The output photocurrent of an idealized point photodetector is proportional to the probability of observing a photo-detection event at space-time point (**r**, t). The joint-detection between two idealized point photodetectors is proportional to the probability of observing a joint photo-detection event at space-time points ($\mathbf{r}_1$, t_1) and ($\mathbf{r}_2$, t_2). In most of the experimental situations, there exists more than one possible alternative ways to produce a photo-detection event, or a joint photo-detection event. These probability amplitudes, which are defined as the single-photon amplitudes and the two-photon amplitudes, respectively, are superposed to contribute to the final measured probability, and consequently determine the probability of observing a photo-detection event or a joint photo-detection event. In the view of quantum theory, whenever the state of the quantum system and the alternative ways to produce a photo-detection event or a joint photo-detection event are determined, the result of a measurement is determined. We may consider this as a basic criterion of quantum measurement theory.

4.3 A semiclassical model of nonlocal interference

The multi–photon interference nature of type-two ghost imaging can be seen intuitively from the superposition of paired-sub-fields of chaotic radiation. Let us consider a similar experimental setup as that of the modified HBT experiment of Scarcelli *et al.*. We assume a large angular sized disk-like chaotic source that contains a large number of randomly radiating independent point "sub-sources", such as trillions of independent atomic transitions randomly distributed spatially and temporally. It should be emphasized that a large number of independent or incoherent subsources is the only requirement for type-two ghost imaging. What we need is an ensemble of point-sub-sources with random relative phases so that the sub-fields coming from these sub-sources are able to take all possible values of relative phases in their superposition. It is *unnecessary* to require the radiation source to have either nature or artificial intensity fluctuations at all. In this model, each point sub-source contributes to the measurement an independent spherical wave as a sub-field of complex amplitude $E_j = a_j e^{i\varphi_j}$, where a_j is the real and positive amplitude of the jth sub-field and φ_j is a *random* phase associated with the jth sub-field. We have the following picture for the source: (1) a large number of independent point-sources distribute randomly on the transverse plane of the source (counted spatially); (2) each point-source contains a large number of independently and randomly radiating atoms (counted temporally); (3) a large number of subsources, either counted spatially or temporally, may contribute to each of the independent radiation mode $(\vec{\kappa}, \omega)$ at D_1 and D_2 (counted by mode). The instantaneous intensity at space-time $(\mathbf{r}_j, t_j)$, measured by the jth idealized point photodetector D_j, $j = 1, 2$, is calculated as

$$\begin{aligned} I(\mathbf{r}_j, t_j) &= E^*(\mathbf{r}_j, t_j) E(\mathbf{r}_j, t_j) = \sum_l E_l^*(\mathbf{r}_j, t_j) \sum_m E_m(\mathbf{r}_j, t_j) \\ &= \sum_{l=m} E_l^*(\mathbf{r}_j, t_j) E_l(\mathbf{r}_j, t_j) + \sum_{l \neq m} E_l^*(\mathbf{r}_j, t_j) E_m(\mathbf{r}_j, t_j), \end{aligned} \tag{54}$$

where the sub-fields are identified by the index l and m originated from the l and m sub-sources. The first term is a constant representing the sum of the sub-intensities, where the lth sub-intensity is originated from the lth sub-source. The second term adds the "cross" terms corresponding to different sub-sources. When *taking into account all possible realizations of the fields,* it is easy to find that the only surviving terms in the sum are these terms in which the field and its conjugate come from the same sub-source, i.e., the first term in Eq. (54). The second term in Eq. (54) vanishes if $\varphi_l - \varphi_m$ *takes all possible values*. We may write Eq. (54) into the following form

$$I(\mathbf{r}_j, t_j) = \langle I(\mathbf{r}, t) \rangle + \Delta I(\mathbf{r}, t), \tag{55}$$

where

$$\langle I(\mathbf{r},t)\rangle \equiv \langle \sum_l E_l^*(\mathbf{r}_j,t_j) \sum_m E_m(\mathbf{r}_j,t_j) \rangle = \sum_l E_l^*(\mathbf{r}_j,t_j) E_l(\mathbf{r}_j,t_j). \tag{56}$$

The notation ⟨...⟩ denotes the mathematical expectation, when *taking into account all possible realizations of the fields,* i.e., taking into account all possible complex amplitudes for the large number of sub-fields in the superposition. In the probability theory, the expectation value of a measurement equals the mean value of an ensemble. In a real measurement, the superposition may not take all possible realizations of the fields and consequently the measured instantaneous intensity $I(\mathbf{r}, t)$ may differ from its expectation value $\langle I(\mathbf{r}, t)\rangle$ from time to time. The variation $\delta I(\mathbf{r}, t)$ turns to be a random function of time. The measured $I(\mathbf{r}, t)$ fluctuate randomly in the neighborhood of $\langle I(\mathbf{r}, t)\rangle$ non-deterministically.
In the classical limit, a large number of independent and randomly radiated sub-sources contribute to the instantaneous intensity $I(\mathbf{r}_j, t_j)$. These large number of independent randomly distributed sub-fields may have *taken all possible realizations of their complex amplitudes* in the superposition. In this case the sum of the cross terms vanishes,

$$\Delta I(\mathbf{r},t) = \sum_{l\neq m} E_l^*(\mathbf{r}_j,t_j) E_m(\mathbf{r}_j,t_j) \simeq 0, \tag{57}$$

therefore,

$$I(\mathbf{r}_j,t_j) \simeq \sum_l E_l^*(\mathbf{r}_j,t_j) E_l(\mathbf{r}_j,t_j) = \langle I(\mathbf{r},t)\rangle.$$

Now we calculate the second-order correlation function $G^{(2)}(\mathbf{r}_1, t_1; \mathbf{r}_2, t_2)$, which is defined as

$$G^{(2)}(\mathbf{r}_1,t_1;\mathbf{r}_2,t_2) \equiv \langle \sum_{j,k,l,m} E_j^*(\mathbf{r}_1,t_1)\, E_k(\mathbf{r}_1,t_1) E_l^*(\mathbf{r}_2,t_2)\, E_m(\mathbf{r}_2,t_2) \rangle, \tag{58}$$

where the notation ⟨ ... ⟩, again, denotes an expectation operation by *taking into account all possible realizations of the fields,* i.e., averaging all possible complex amplitudes for the sub-fields in the superposition. In the following calculation we only take into account the random phases of the subfields without considering the amplitude variations. Due to the chaotic nature of the independent sub-sources, after taking into account all possible realizations of the phases associated with the sub-fields, the only surviving terms in the summation are those with: (1) $j = k$, $l = m$, (2) $j = m$, $k = l$. Therefore, $G^{(2)}(\mathbf{r}_1, t_1; \mathbf{r}_2, t_2)$ reduces to the sum of the following two groups:

$$\begin{aligned} G^{(2)}(\mathbf{r}_1,t_1;\mathbf{r}_2,t_2) &= \langle \sum_j E_j^*(\mathbf{r}_1,t_1)\, E_j(\mathbf{r}_1,t_1) \sum_l E_l^*(\mathbf{r}_2,t_2)\, E_l(\mathbf{r}_2,t_2) \\ &\quad + \sum_j E_j^*(\mathbf{r}_1,t_1)\, E_j(\mathbf{r}_2,t_2) \sum_l E_l^*(\mathbf{r}_2,t_2)\, E_l(\mathbf{r}_1,t_1) \rangle \\ &= \langle \sum_j \sum_l \left| \frac{1}{\sqrt{2}} \big[E_j(\mathbf{r}_1,t_1) E_l(\mathbf{r}_2,t_2) + E_l(\mathbf{r}_1,t_1) E_j(\mathbf{r}_2,t_2) \big] \right|^2 \rangle. \end{aligned} \tag{55}$$

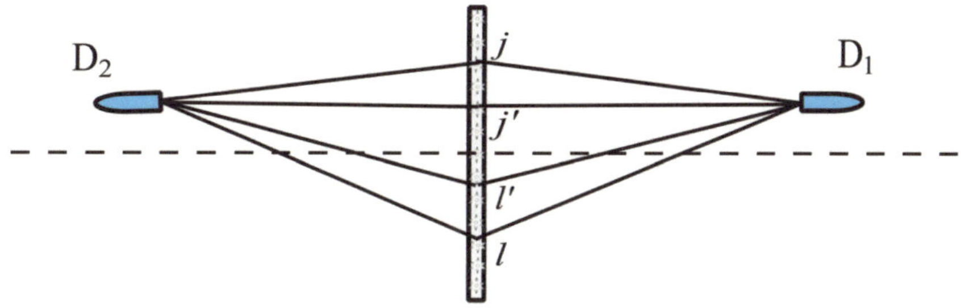

Fig. 20. Schematic illustration of $\sum_{j,l}|E_{j1}E_{l2}+E_{l1}E_{j2}|^2$. It is clear that the amplitude pairs $j1 \times l2$ with $l1 \times j2$, where j and l represent all point sub-sources, pair by pair, will experience equal optical path propagation and superpose constructively when D_1 and D_2 are located at $\vec{\rho}_1 \simeq \vec{\rho}_2$, $z_1 \simeq z_2$. This interference is similar to symmetrizing the wavefunction of identical particles in quantum mechanics.

It is not difficult to see the nonlocal nature of the superposition shown in Eq. (59). In Eq. (59), $G^{(2)}(\mathbf{r}_1, t_1; \mathbf{r}_2, t_2)$ is written as a superposition between the paired sub-fields $E_j(\mathbf{r}_1, t_1) E_l(\mathbf{r}_2, t_2)$ and $E_l(\mathbf{r}_1, t_1)E_j(\mathbf{r}_2, t_2)$. The first term in the superposition corresponds to the situation in which the field at D_1 was generated by the *jth* sub-source, and the field at D_2 was generated by the *lth* sub-source. The second term in the superposition corresponds to a different yet indistinguishable situation in which the field at D_1 was generated by the *lth* sub-source, and the field at D_2 was generated by the *jth* sub-source. Therefore, an interference is concealed in the joint measurement of D_1 and D_2, which physically occurs at two space-time points $(\mathbf{r}_1, t_1)$ and $(\mathbf{r}_2, t_2)$. The interference corresponds to $|E_{j1}E_{l2} + E_{l1}E_{j2}|^2$. It is easy to see from Fig. 20, the amplitude pairs $j1 \times l2$ with $l1 \times j2$, $j'1 \times l'2$ with $l'1 \times j'2$, $j1 \times l'2$ with $l'1 \times j2$, and $j'1 \times l2$ with $l1 \times j'2$, etc., pair by pair, experience equal total optical path propagation, which involves two arms of D_1 and D_2, and thus superpose constructively when D_1 and D_2 are placed in the neighborhood of $\vec{\rho}_1 = \vec{\rho}_2$, $z_1 = z_2$. Consequently, the summation of these individual constructive interference terms will yield a maximum value. When $\vec{\rho}_1 \neq \vec{\rho}_2$, $z_1 = z_2$, however, each pair of the amplitudes may achieve different relative phase and contribute a different value to the summation, resulting in an averaged constant value.

It does not seem to make sense to claim a nonlocal interference between [(E_j goes to D_1) × (E_l goes to D_2)] and [(E_l goes to D_1) × (E_j goes to D_2)] in the framework of Maxwell's electromagnetic wave theory of light. This statement is more likely adapted from particle physics, similar to symmetrizing the wavefunction of identical particles, and is more suitable to describe the interference between quantum amplitudes: [(particle-j goes to D_1) × (particle-l goes to D_2)] and [(particle-l goes to D_1) × (particle-j goes to D_2)], rather than waves. Classical waves do not behave in such a manner. In fact, in this model each sub-source corresponds to an independent spontaneous atomic transition in nature, and consequently corresponds to the creation of a photon. Therefore, the above superposition corresponds to the superposition between two indistinguishable two-photon amplitudes, and is thus called *two-photon interference* [9]. In Dirac's theory, this interference is the result of a measured pair of photons interfering with itself.

In the following we attempt a near-field calculation to derive the point-to-point correlation of $G^{(2)}(\vec{\rho}_1, z_1; \vec{\rho}_2, z_2)$. We start from Eq. (59) and concentrate to the transverse spatial correlation

$$G^{(2)}(\vec{\rho}_1, z_1; \vec{\rho}_2, z_2) = \langle \sum_j \sum_l \Big| \frac{1}{\sqrt{2}} \big[E_j(\vec{\rho}_1, z_1) E_l(\vec{\rho}_2, z_2) + E_l(\vec{\rho}_1, z_1) E_j(\vec{\rho}_2, z_2)\big] \Big|^2 \rangle. \quad (60)$$

In the near-field we apply the Fresnel approximation as usual to propagate the field from each subsource to the photodetectors. $G^{(2)}(\vec{\rho}_1, z_1; \vec{\rho}_2, z_2)$ can be formally written in terms of the Green's function,

$$\begin{aligned} & G^{(2)}(\vec{\rho}_1, z_1; \vec{\rho}_2, z_2) \\ &= \langle \Big| \int d\vec{\kappa}\, d\vec{\kappa}' \Big| \frac{1}{\sqrt{2}} \big[g(\vec{\rho}_1, z_1, \vec{\kappa}) g(\vec{\rho}_2, z_2, \vec{\kappa}') + g(\vec{\rho}_2, z_2, \vec{\kappa}) g(\vec{\rho}_1, z_1, \vec{\kappa}')\big] \Big|^2 \rangle \\ &= \langle \int d\vec{\kappa}\, \big|g(\vec{\rho}_1, z_1, \vec{\kappa})\big|^2 \int d\vec{\kappa}'\, \big|g(\vec{\rho}_2, z_2, \vec{\kappa}')\big|^2 + \Big| \int d\vec{\kappa}\; g^*(\vec{\rho}_1, z_1, \vec{\kappa})\, g(\vec{\rho}_2, z_2, \vec{\kappa}) \Big|^2 \rangle \\ &\equiv G^{(1)}(\vec{\rho}_1, z_1) G^{(1)}(\vec{\rho}_2, z_2) + \big|G^{(1)}(\vec{\rho}_1, z_1; \vec{\rho}_2, z_2)\big|^2. \end{aligned} \quad (61)$$

In Eq. (61) we have formally written $G^{(2)}$ in terms of the first-order correlation functions $G^{(1)}$, but keep in mind that the first-order correlation function $G^{(1)}$ and the second-order correlation function $G^{(2)}$ represent different physics based on different measurements. Substituting the Green's function derived in the Appendix for free propagation

$$g(\vec{\rho}_j, z_j, \vec{\kappa}) = \frac{-i\omega}{2\pi c} \frac{e^{i\frac{\omega}{c} z_j}}{z_j} \int d\vec{\rho}_0\, a(\vec{\rho}_0)\, e^{i\varphi(\vec{\rho}_0)}\, e^{i\frac{\omega}{2cz_j}|\vec{\rho}_j - \vec{\rho}_0|^2}$$

into Eq. (61), we obtain $G^{(1)}(\vec{\rho}_1, z_1) G^{(1)}(\vec{\rho}_2, z_2)$ ~constant and

$$\big|G^{(1)}(\vec{\rho}_1, z_1; \vec{\rho}_2, z_2)\big|^2 \propto \Big|\langle \frac{1}{z_1 z_2} \int d\vec{\rho}_0\, a^2(\vec{\rho}_0)\, e^{-i\frac{\omega}{c} z_1}\, e^{-i\frac{\omega}{2cz_1}|\vec{\rho}_1 - \vec{\rho}_0|^2}\, e^{i\frac{\omega}{c} z_2}\, e^{i\frac{\omega}{2cz_2}|\vec{\rho}_2 - \vec{\rho}_0|^2} \rangle\Big|^2.$$

Assuming $a_2(\vec{\rho}_0)$ ~constant, and taking $z_1 = z_2 = d$, we obtain

$$\begin{aligned} \big|G^{(1)}_{12}(\vec{\rho}_1; \vec{\rho}_2)\big|^2 &\propto \Big| \int d\vec{\rho}_0\, a^2(\vec{\rho}_0)\, e^{-i\frac{\omega}{2cd}|\vec{\rho}_1 - \vec{\rho}_0|^2}\, e^{i\frac{\omega}{2cd}|\vec{\rho}_2 - \vec{\rho}_0|^2} \Big|^2 \\ &\propto \Big| e^{-i\frac{\omega}{2cd}(|\vec{\rho}_1|^2 - |\vec{\rho}_2|^2)} \int d\vec{\rho}_0\, a^2(\vec{\rho}_0)\, e^{i\frac{\omega}{cd}(\vec{\rho}_1 - \vec{\rho}_2)\cdot\vec{\rho}_0} \Big|^2 \\ &\propto somb^2\Big[\frac{R}{d}\frac{\omega}{c}|\vec{\rho}_1 - \vec{\rho}_2|\Big], \end{aligned} \quad (62)$$

where we have assumed a disk-like light source with a finite radius of R. The transverse spatial correlation function $G^{(2)}(\vec{\rho}_1; \vec{\rho}_2)$ is thus

$$G^{(2)}(|\vec{\rho}_1 - \vec{\rho}_2|) = I_0^2\, \Big[1 + somb^2\Big(\frac{R}{d}\frac{\omega}{c}|\vec{\rho}_1 - \vec{\rho}_2|\Big)\Big]. \quad (63)$$

Consequently, the degree of the second-order spatial coherence is

$$g^{(2)}(|\vec{\rho}_1 - \vec{\rho}_2|) = 1 + somb^2\Big(\frac{R}{d}\frac{\omega}{c}|\vec{\rho}_1 - \vec{\rho}_2|\Big). \quad (64)$$

For a large value of $2R/d \sim \Delta\theta$, where $\Delta\theta$ is the angular size of the radiation source viewed at the photodetectors, the point-spread *somb*-function can be approximated as a δ-function of $|\vec{\rho}_1 - \vec{\rho}_2|$. We effectively have a "point-to-point" correlation between the transverse planes of $z_1 = d$ and $z_2 = d$. In 1-D Eqs. (63) and (64) become

$$G^{(2)}(x_1 - x_2) = I_0^2 \left[1 + sinc^2\left(\frac{\pi\Delta\theta(x_1 - x_2)}{\lambda}\right)\right] \tag{65}$$

and

$$g^{(2)}(x_1 - x_2) = 1 + sinc^2\left(\frac{\pi\Delta\theta(x_1 - x_2)}{\lambda}\right), \tag{66}$$

which has been experimentally demonstrated and reported in Fig. 18.
We have thus derived the same second-order correlation and coherence functions as that of the quantum theory. The non-factorizable point-to-point correlation is expected at any intensity. The only requirement is a large number of point sub-sources with random relative phases participating to the measurement, such as trillions of independent atomic transitions. There is no surprise to derive the same result as that of the quantum theory from this simple model. Although the fields are not quantized and no quantum formula was used in the above calculation, this model has implied the same nonlocal two-photon interference mechanism as that of the quantum theory. Different from the phenomenological theory of intensity fluctuations, this semiclassical model explores the physical cause of the phenomenon.

5. Classical simulation

There have been quite a few classical approaches to simulate type-one and type-two ghost imaging. Different from the natural non-factorizable type-one and type-two point-to-point imaging-forming correlations, classically simulated correlation functions are all factorizable. We briefly discuss two of these man-made factoriable classical correlations in the following.
(I) Correlated laser beams.
In 2002, Bennink *et al.* simulated ghost imaging by two correlated laser beams [26]. In this experiment, the authors intended to show that two correlated rotating laser beams can simulate the same physical effects as entangled states. Figure 21 is a schematic picture of the experiment of Bennink *et al.*. Different from type-one and type-two ghost imaging, here the point-to-point correspondence between the object plane and the "image plane" is made artificially by two co-rotating laser beams "shot by shot". The laser beams propagated in opposite directions and focused on the object and image planes, respectively. If laser beam-1 is blocked by the object mask there would be no joint-detection between D_1 and D_2 for that "shot", while if laser beam-1 is unblocked, a coincidence count will be recorded against that angular position of the co-rotating laser beams. A shadow of the object mask is then reconstructed in coincidences by the blocking–unblocking of laser beam-1.
A man-made factorizable correlation of laser beam is not only different from the non-factorizable correlations in type-one and type-two ghost imaging, but also different from the standard statistical correlation of intensity fluctuations. Although the experiment of Bennink *et al.* obtained a ghost shadow, which may be useful for certain purposes, it is clear that the

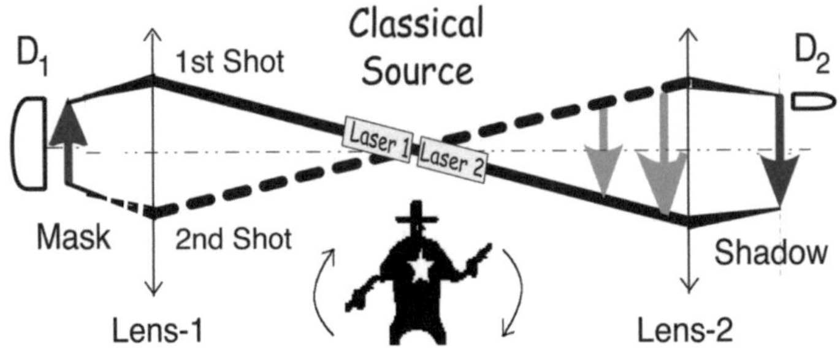

Fig. 21. A ghost shadow can be made in coincidences by "blocking-unblocking" of the correlated laser beams, or simply by "blocking-unblocking" two correlated gun shots. The man-made trivial "correlation" of either laser beams or gun shots are deterministic, i.e., the laser beams or the bullets know where to go in each shot, which are fundamentally different from the quantum mechanical nontrivial nondeterministic multi-particle correlation.

physics shown in their experiment is fundamentally different from that of ghost imaging. In fact, this experiment can be considered as a good example to distinguish a man-made trivial deterministic classical intensity-intensity correlation from quantum entanglement and from a natural nonlocal nondeterministic multi-particle correlation.

(II) Correlated speckles.

Following a similar philosophy, Gatti *et al.* proposed a factorizable "speckle-speckle" classical correlation between two distant planes, $\vec{\rho}_1$ and $\vec{\rho}_2$, by imaging the speckles of the common light source onto the distant planes of $\vec{\rho}_1$ and $\vec{\rho}_2$, [13]

$$G^{(2)}(\vec{\rho}_1, \vec{\rho}_2) \propto \delta(\vec{\rho}_o - \vec{\rho}_1/m)\delta(\vec{\rho}_o - \vec{\rho}_2/m), \tag{67}$$

where $\vec{\rho}_0$ is the transverse coordinate in the plane of the light source.[9]

The schematic setup of the classical simulation of Gatti *et al.* is depicted in Fig. 22 [13]. Their experiment used either entangled photon pairs of spontaneous parametric down-conversion (SPDC) or chaotic light for obtaining ghost shadows in coincidences. To distinguish from

[9] The original publications of Gatti *et al.* choose 2f-2f classical imaging systems with $1/2f + 1/2f = 1/f$ to image the speckles of the source onto the object plane and the ghost image plane. The man-mde speckle-speckle image-forming correlation of Gatti *et al.* shown in Eq. (67) is factorizeable, which is fundamentally different from the natural non-factorizable image-formimg correlations in type-one and type-two ghost imaging. In fact, it is very easy to distinguish a classical simulation from ghost imaging by examining its experimental setup and operation. The man-made speckle-speckle correlation needs to have two sets of identical speckles observable (by the detectors or CCDs) on the object and the image planes. In thermal light ghost imaging, when using pseudo-thermal light source, the classical simulation requires a slow rotating ground grass in order to image the speckles of the source onto the object and image planes (typically, sub-Hertz to a few Hertz). However, to achieve a natural HBT nonfactorizable correlation of chaotic light for type-two ghost imaging, we need to rotate the ground grass as fast as possible (typically, a few thousand Hertz, the higher the batter).

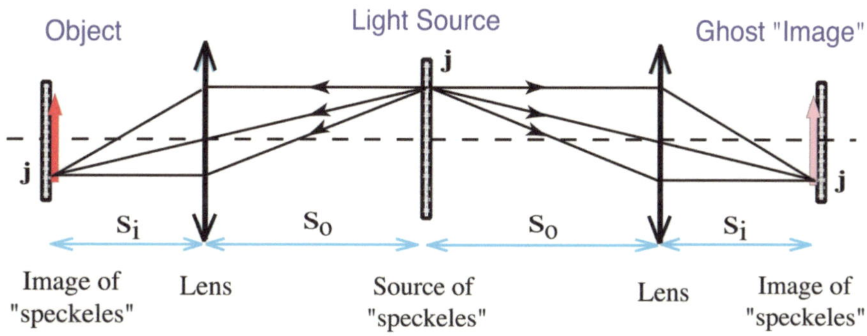

Fig. 22. A ghost "imager" is made by blocking-unblocking the correlated speckles. The two identical sets of speckles on the object plane and the image plane, respectively, are the *classical images* of the speckles of the source plane. The lens, which may be part of a CCD *camera* used for the joint measurement, reconstructs classical images of the speckles of the source onto the object plane and the image plane, respectively. s_o and s_i satisfy the Gaussian thin lens equation $1/s_o + 1/s_i = 1/f$.

ghost imaging, Gatti *et al.* named their work "ghost imager". The "ghost imager" comes from a man-made classical speckle-speckle correlation. The speckles observed on the object and image planes are the classical images of the speckles of the radiation source, reconstructed by the imaging lenses shown in the figure (the imaging lens may be part of a CCD *camera* used for the joint measurement). Each speckle on the source, such as the *j*th speckle near the top of the source, has two identical images on the object plane and on the image plane. Different from the non-factorizeable nonlocal image-forming correlation in type-one and type-two ghost imaging, mathematically, the speckle-speckle correlation is factorizeable into a product of two classical images of speckles. If two point photodetectors D_1 and D_2 are scanned on the object plane and the image plane, respectively, D_1 and D_2 will have more "coincidences" when they are in the position within the two identical speckles, such as the two *j*th speckles near the bottom of the object plane and the image plane. The blocking-unblocking of the speckles on the object plane by a mask will project a ghost *shadow* of the mask in the coincidences of D_1 and D_2. It is easy to see that the size of the identical speckles determines the spatial resolution of the ghost shadow. This observation has been confirmed by quite a few experimental demonstrations. There is no surprise that Gatti *et al.* consider ghost imaging classical [27]. Their speckle-speckle correlation is a man-made classical correlation and their ghost imager is indeed classical. The classical simulation of Gatti *et al.* might be useful for certain applications, however, to claim the nature of ghost imaging in general as classical, perhaps, is too far [27]. The man-made factorizable speckle-speckle correlation of Gatti *et al.* is a classical simulation of the natural nonlocal point-to-point image-forming correlation of ghost imaging, despite the use of either entangled photon source or classical light.

6. Local? Nonlocal?

We have discussed the physics of both type-one and type-two ghost imaging. Although different radiation sources are used for different cases, these two types of experiments demonstrated a similar non-factorizable point-to-point image-forming correlation:

Type-one:

$$\delta(\vec{\rho}_1 - \vec{\rho}_2) \sim \Big| \int d\vec{\kappa}_s \, d\vec{\kappa}_i \, \delta(\vec{\kappa}_s + \vec{\kappa}_i) \, g_1(\vec{\kappa}_s, \vec{\rho}_1) \, g_2(\vec{\kappa}_i, \vec{\rho}_2) \Big|^2, \tag{68}$$

Type-two:

$$\begin{aligned} 1 + \delta(\vec{\rho}_1 - \vec{\rho}_2) &\propto \langle \sum_j \sum_l \big| \frac{1}{\sqrt{2}} [E_j(\vec{\rho}_1) E_l(\vec{\rho}_2) + E_l(\vec{\rho}_1) E_j(\vec{\rho}_2)] \big|^2 \rangle \\ &= \langle \int d\vec{\kappa} \, d\vec{\kappa}' \big| \frac{1}{\sqrt{2}} [g_1(\vec{\kappa}, \vec{\rho}_1) g_2(\vec{\kappa}', \vec{\rho}_2) + g_2(\vec{\kappa}, \vec{\rho}_2) g_1(\vec{\kappa}', \vec{\rho}_1)] \big|^2 \rangle. \end{aligned} \tag{69}$$

Equations (68) and (69) indicate that the point-to-point correlation of ghost imaging, either typeone or type-two, is the results of two-photon interference. Unfortunately, neither of them is in the form of $|\sum_j E_j|^2$ or $|E_1 + E_2|^2$, and neither is measured at a local space-time point. The interference shown in Eqs. (68) and (69) occurs at different space-time points through the measurements of two spatially separated independent photodetectors.

In type-one ghost imaging, the δ-function in Eq. (68) means a typical EPR position-position correlation of an entangled photon pair. In EPR's language: when the pair is generated at the source the momentum and position of neither photon is determined, and neither photon-one nor photon-two "knows" where to go. However, if one of them is observed at a point at the object plane the other one must be found at a unique point in the image plane. In type-two ghost imaging, although the position-position determination in Eq. (69) is only partial, it generates more surprises because of the chaotic nature of the radiation source. Photon-one and photon-two, emitted from a thermal source, are completely random and independent, i.e., both propagate freely to any direction and may arrive at any position in the object and image planes. Analogous to EPR's language: when the measured two photons were emitted from the thermal source, neither the momentum nor the position of any photon is determined. However, if one of them is observed at a point on the object plane the other one must have twice large probability to be found at a unique point in the image plane. Where does this partial correlation come from? If one insists on the view point of intensity fluctuation correlation, then, it is reasonable to ask why the intensities of the two light beams exhibit fluctuation correlations at $\vec{\rho}_1 = \vec{\rho}_2$ only? Recall that in the experiment of Sarcelli *et al.* the ghost image is measured in the near-field. Regardless of position, D_1 and D_2 receive light from all (a large number) point sub-sources of the thermal source, and all sub-sources fluctuate randomly and independently. If $\Delta I_1 \Delta I_2 = 0$ for $\vec{\rho}_1 \neq \vec{\rho}_2$, what is the physics to cause $\Delta I_1 \Delta I_2 \neq 0$ at $\vec{\rho}_1 = \vec{\rho}_2$?

The classical superposition is considered "local". The Maxwell electromagnetic field theory requires the superposition of the electromagnetic fields, either $|\sum_j E_j|^2$ or $|E_1 + E_2|^2$, takes place at a local space-time point ($\mathbf{r}$, t). However, the superposition shown in Eqs. (68) and (69) happens at two different space-time points ($\mathbf{r}_1$, t_1) and ($\mathbf{r}_2$, t_2) and is measured by two independent photodetectors. Experimentally, it is not difficult to make the two photo-detection events space-like separated events. Following the definition given by EPR-Bell, we consider the superposition appearing in Eqs. (68) and (69) *nonlocal*. Although the two-photon interference of thermal light can be written and calculated in terms of a semiclassical model, the nonlocal superposition appearing in Eq. (69) has no counterpart in the classical measurement theory of light, unless one forces a nonlocal classical theory by allowing the superposition to occur at a distance through the measurement of independent photodetectors, as we have done in Eq. (59). Perhaps, it would be more difficult to accept a nonlocal classical measurement theory of thermal light rather than to apply a quantum mechanical concept to "classical" thermal radiation.

7. Conclusion

In summary, we may conclude that ghost imaging is the result of quantum interference. Either type-one or type-two, ghost imaging is characterized by a non-factorizable point-to-point image-forming correlation which is caused by constructive-destructive interferences involving the nonlocal superposition of two-photon amplitudes, a nonclassical entity corresponding to different yet indistinguishable alternative ways of producing a joint photo-detection event. The interference happens within a pair of photons and at two spatially separated coordinates. The multi-photon interference nature of ghost imaging determines its peculiar features: (1) it is nonlocal; (2) its imaging resolution differs from that of classical; and (3) the type-two ghost image is turbulence-free. Taking advantage of its quantum interference nature, a ghost imaging system may turn a local "bucket" sensor into a nonlocal imaging camera with classically unachievable imaging resolution. For instance, using the Sun as light source for type-two ghost imaging, we may achieve an imaging spatial resolution equivalent to that of a classical imaging system with a lens of 92-meter diameter when taking pictures at 10 kilometers.[10] Furthermore, any phase disturbance in the optical path has no influence on the ghost image. To achieve these features the realization of multi-photon interference is necessary.

8. Acknowledgment

The author thanks M. D'Angelo, G. Scarcelli, J.M. Wen, T.B. Pittman, M.H. Rubin, and L.A. Wu for helpful discussions. This work is partially supported by AFOSR and ARO-MURI program.

Appendix: Fresnel free-propagation

We are interested in knowing how a known field $E(\mathbf{r}_0, t_0)$ on the plane $z_0 = 0$ propagates or diffracts into $E(\mathbf{r}, t)$ on another plane z = constant. We assume the field $E(\mathbf{r}_0, t_0)$ is excited by an arbitrary source, either point-like or spatially extended. The observation plane of z = constant is located at an arbitrary distance from plane $z_0 = 0$, either far-field or near-field. Our goal is to find out a general solution $E(\mathbf{r}, t)$, or $I(\mathbf{r}, t)$, on the observation plane, based on our knowledge of $E(\mathbf{r}_0, t_0)$ and the laws of the Maxwell electromagnetic wave theory. It is not easy to find such a general solution. However, the use of the Green's function or the

[10] The angular size of Sun is about 0.53°. To achieve a compatible image spatial resolution, a traditional camera must have a lens of 92-meter diameter when taking pictures at 10 kilometers.

field transfer function, which describes the propagation of each mode from the plane of $z_0 = 0$ to the observation plane of z = constant, makes this goal formally achievable.
Unless $E(\mathbf{r}_0, t_0)$ is a non-analytic function in the space-time region of interest, there must exist a Fourier integral representation for $E(\mathbf{r}_0, t_0)$

$$E(\mathbf{r}_0, t_0) = \int d\mathbf{k}\, E(\mathbf{k})\, w_{\mathbf{k}}(\mathbf{r}_0, t_0)\, e^{-i\omega t_0}, \tag{A-1}$$

where $w_{\mathbf{k}}(\mathbf{r}_0, t_0)$ is a solution of the Helmholtz wave equation under appropriate boundary conditions. The solution of the Maxwell wave equation $w_{\mathbf{k}}(\mathbf{r}_0, t_0)\, e^{-i\omega t_0}$, namely the Fourier mode, can be a set of plane-waves or spherical-waves depending on the chosen boundary condition. In Eq. $(A{-}1)$, $E(\mathbf{k}) = a(\mathbf{k})e^{i\varphi(\mathbf{k})}$ is the complex amplitude of the Fourier mode $\mathbf{k}$. In principle we should be able to find an appropriate Green's function which propagates each mode under the Fourier integral point by point from the plane of $z_0 = 0$ to the plane of observation,

$$\begin{aligned} E(\mathbf{r}, t) &= \int d\mathbf{k}\, E(\mathbf{k})\, g(\mathbf{k}, \mathbf{r} - \mathbf{r}_0, t - t_0)\, w_{\mathbf{k}}(\mathbf{r}_0, t_0)\, e^{-i\omega t_0} \\ &= \int d\mathbf{k}\, g(\mathbf{k}, \mathbf{r} - \mathbf{r}_0, t - t_0)\, E(\mathbf{k}, \mathbf{r}_0, t_0), \end{aligned} \tag{A-2}$$

where $E(\mathbf{k}, \mathbf{r}_0, t_0) = E(\mathbf{k})\, w_{\mathbf{k}}(\mathbf{r}_0, t_0)\, e^{-i\omega t_0}$. The secondary wavelets that originated from each point on the plane of $z_0 = 0$ are then superposed *coherently* on each point on the observation plane with their after-propagation amplitudes and phases. It is convenient to write Eq. $(A{-}2)$ in the following form

$$E(\mathbf{k}, \mathbf{r}_0, t_0) = E(\mathbf{k})\, w_{\mathbf{k}}(\mathbf{r}_0, t_0)\, e^{-i\omega t_0} \tag{A-3}$$

where we have used the transverse-longitudinal coordinates in space-time ($\vec{\rho}$ and z) and in momentum ($\vec{\kappa}, \omega$).
Fig. $A{-}1$ is a simple example in which the field propagates freely from an aperture A of finite size on the plane σ_0 to the observation plane σ. Based on Fig. $A{-}1$ we evaluate $g(\vec{\kappa},\ \omega;\ \vec{\rho},\ z)$, namely the Green's function for free-space Fresnel propagation-diffraction. According to the Huygens-Fresnel principle the field at a given space-time point $(\vec{\rho},\ z,\ t)$ is the result of a superposition of the spherical secondary wavelets that originated from each point on the σ_0 plane (see Fig. $A{-}1$),

$$E(\vec{\rho}, z, t) = \int d\omega\, d\vec{\kappa}\ E(\vec{\kappa}, \omega; 0, 0) \int_{\sigma_0} d\vec{\rho}_0\, \frac{\tilde{A}(\vec{\rho}_0)}{r}\, e^{-i(\omega t - kr)}, \tag{A-4}$$

where we have set $z_0 = 0$ and $t_0 = 0$ at plane σ_0, and defined $r = \sqrt{z^2 + |\vec{\rho} - \vec{\rho}_0|^2}$. In Eq. $(A{-}4)$, $\tilde{A}(\vec{\rho}_0)$ is the complex amplitude or relative distribution of the field on the plane of σ_0, which may be written as a simple aperture function in terms of the transverse coordinate $\vec{\rho}_0$, as we have done in the earlier discussions.

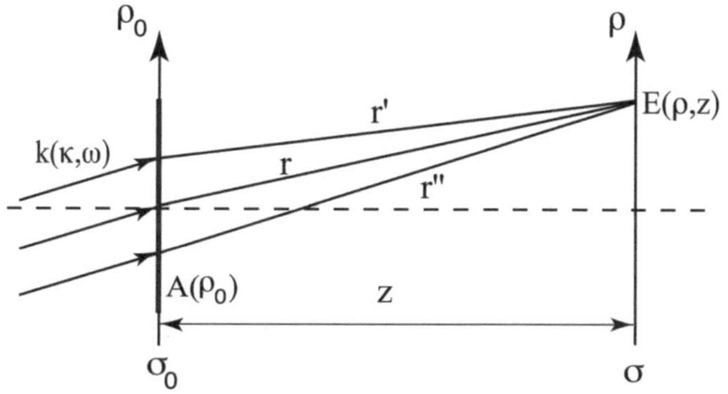

Fig. A–1. Schematic of free-space Fresnel propagation. The complex amplitude $\tilde{A}(\vec{\rho}_0)$ is composed of a real function $A(\vec{\rho}_0)$ and a phase $e^{-i\vec{\kappa}\cdot\vec{\rho}_0}$ associated with each of the transverse wavevectors $\vec{\kappa}$ in the plane of σ_0. Notice: only one mode of wavevector $\mathbf{k}(\vec{\kappa}, \omega)$ is shown in the figure.

In the near-field Fresnel paraxial approximation, when $|\vec{\rho}-\vec{\rho}_0|^2 \ll z^2$ we take the first-order expansion of r in terms of z and $\vec{\rho}$,

$$r = \sqrt{z^2 + |\vec{\rho}-\vec{\rho}_0|^2} \simeq z(1+\frac{|\vec{\rho}-\vec{\rho}_0|^2}{2z^2}), \tag{A-5}$$

so that $E(\vec{\rho}, z, t)$ can be approximated as

$$E(\vec{\rho}, z, t) \simeq \int d\omega\, d\vec{\kappa}\; E(\vec{\kappa}, \omega, 0, 0) \int d\vec{\rho}_0\, \frac{\tilde{A}(\vec{\rho}_0)}{z}\, e^{i\frac{\omega}{c}z}\, e^{i\frac{\omega}{2cz}|\vec{\rho}-\vec{\rho}_0|^2} e^{-i\omega t},$$

where $e^{i\frac{\omega}{2cz}|\vec{\rho}-\vec{\rho}_0|^2}$ is named the Fresnel phase factor.
Assuming that the complex amplitude $\tilde{A}(\vec{\rho}_0)$ is composed of a real function $A(\vec{\rho}_0)$ and a phase $e^{-i\vec{\kappa}\cdot\vec{\rho}_0}$, associated with the transverse wavevector and the transverse coordinate on the plane of σ_0, as is reasonable for the setup of Fig. A–1, we can then write $E(\vec{\rho}, z, t)$ in the form

$$E(\vec{\rho}, z, t) = \int d\omega\, d\vec{\kappa}\; E(\vec{\kappa}, \omega; 0, 0)\, e^{-i\omega t}\, \frac{e^{i\frac{\omega}{c}z}}{z} \int d\vec{\rho}_0\, A(\vec{\rho}_0)\, e^{i\vec{\kappa}\cdot\vec{\rho}_0}\, e^{i\frac{\omega}{2cz}|\vec{\rho}-\vec{\rho}_0|^2}.$$

The Green's function $g(\vec{\kappa}, \omega; \vec{\rho}, z)$ for free-space Fresnel propagation is thus

$$g(\vec{\kappa}, \omega; \vec{\rho}, z) = \frac{e^{i\frac{\omega}{c}z}}{z} \int_{\sigma_0} d\vec{\rho}_0\, A(\vec{\rho}_0)\, e^{i\vec{\kappa}\cdot\vec{\rho}_0}\, G(|\vec{\rho}-\vec{\rho}_0|, \frac{\omega}{cz}). \tag{A-6}$$

In Eq. (A–6) we have defined a Gaussian function $G(|\vec{\alpha}|, \beta) = e^{i(\beta/2)|\alpha|^2}$, namely the Fresnel phase factor. It is straightforward to find that the Gaussian function $G(|\vec{\alpha}|, \beta)$ has the following properties:

$$
\begin{aligned}
G^*(|\vec{\alpha}|,\beta) &= G(|\vec{\alpha}|,-\beta), \\
G(|\vec{\alpha}|,\beta_1+\beta_2) &= G(|\vec{\alpha}|,\beta_1)\,G(|\vec{\alpha}|,\beta_2), \\
G(|\vec{\alpha}_1+\vec{\alpha}_2|,\beta) &= G(|\vec{\alpha}_1|,\beta)\,G(|\vec{\alpha}_2|,\beta)\,e^{i\beta\vec{\alpha}_1\cdot\vec{\alpha}_2}, \\
\int d\vec{\alpha}\;G(|\vec{\alpha}|,\beta)\,e^{i\vec{\gamma}\cdot\vec{\alpha}} &= i\frac{2\pi}{\beta}\,G(|\vec{\gamma}|,-\frac{1}{\beta}).
\end{aligned} \tag{A-7}
$$

Notice that the last equation in Eq. (A−7) is the Fourier transform of the $G(|\vec{\alpha}|,\beta)$ function. As we shall see in the following, these properties are very useful in simplifying the calculations of the Green's functions $g(\vec{\kappa},\,\omega;\,\vec{\rho},\,z)$.

Next, we consider inserting an imaginary plane σ' between σ_0 and σ. This is equivalent to having two consecutive Fresnel propagations with a diffraction-free σ' plane of infinity. Thus, the calculation of these consecutive Fresnel propagations should yield the same Green's function as that of the above direct Fresnel propagation shown in Eq. (A−6):

$$
\begin{aligned}
& g(\omega,\vec{\kappa};\vec{\rho},z) \\
&= C^2\frac{e^{i\frac{\omega}{c}(d_1+d_2)}}{d_1d_2}\int_{\sigma'} d\vec{\rho'}\int_{\sigma_0} d\vec{\rho}_0\;\tilde{A}(\vec{\rho}_0)G(|\vec{\rho'}-\vec{\rho}_0|,\frac{\omega}{cd_1})G(|\vec{\rho}-\vec{\rho'}|,\frac{\omega}{cd_2}) \\
&= C\,\frac{e^{i\frac{\omega}{c}z}}{z}\int_{\sigma_0} d\vec{\rho}_0\;\tilde{A}(\vec{\rho}_0)\,G(|\vec{\rho}-\vec{\rho}_0|,\frac{\omega}{cz})
\end{aligned} \tag{A-8}
$$

where C is a necessary normalization constant for a valid Eq. (A−8), and $z = d_1 + d_2$. The double integral of $d\vec{\rho}_0$ and $d\vec{\rho'}$ in Eq. (A−8) can be evaluated as

$$
\begin{aligned}
& \int_{\sigma'} d\vec{\rho'}\int_{\sigma_0} d\vec{\rho}_0\;\tilde{A}(\vec{\rho}_0)\,G(|\vec{\rho'}-\vec{\rho}_0|,\frac{\omega}{cd_1})\,G(|\vec{\rho}-\vec{\rho'}|,\frac{\omega}{cd_2}) \\
&= \int_{\sigma_0} d\vec{\rho}_0\;\tilde{A}(\vec{\rho}_0)\,G(\vec{\rho}_0,\frac{\omega}{cd_1})\,G(\vec{\rho},\frac{\omega}{cd_2}) \\
&\quad\times\int_{\sigma'} d\vec{\rho'}\;G(\vec{\rho'},\frac{\omega}{c}(\frac{1}{d_1}+\frac{1}{d_2}))\,e^{-i\frac{\omega}{c}(\frac{\vec{\rho}_0}{d_1}+\frac{\vec{\rho}}{d_2})\cdot\vec{\rho'}} \\
&= \frac{i2\pi c}{\omega}\frac{d_1d_2}{d_1+d_2}\int_{\sigma_0} d\vec{\rho}_0\;\tilde{A}(\vec{\rho}_0)\,G(\vec{\rho}_0,\frac{\omega}{cd_1})\,G(\vec{\rho},\frac{\omega}{cd_2}) \\
&\qquad\times G(|\frac{\vec{\rho}_0}{d_1}+\frac{\vec{\rho}}{d_2}|,\frac{\omega}{c}(\frac{d_1d_2}{d_1+d_2})) \\
&= \frac{i2\pi c}{\omega}\frac{d_1d_2}{d_1+d_2}\int_{\sigma_0} d\vec{\rho}_0\;\tilde{A}(\vec{\rho}_0)\,G(|\vec{\rho}-\vec{\rho}_0|,\frac{\omega}{c(d_1+d_2)})
\end{aligned}
$$

where we have applied Eq. (A−7), and the integral of $d\,\vec{\rho'}$ has been taken to infinity. Substituting this result into Eq. (A−8) we obtain

$$
\begin{aligned}
& g(\vec{\kappa},\omega;\vec{\rho},z) \\
&= C^2\,\frac{i2\pi c}{\omega}\frac{e^{i\frac{\omega}{c}(d_1+d_2)}}{d_1+d_2}\int_{\sigma_0} d\vec{\rho}_0\;\tilde{A}(\vec{\rho}_0)\,G(|\vec{\rho}-\vec{\rho}_0|,\frac{\omega}{c(d_1+d_2)}) \\
&= C\,\frac{e^{i\frac{\omega}{c}z}}{z}\int_{\sigma_0} d\vec{\rho}_0\;\tilde{A}(\vec{\rho}_0)\,G(|\vec{\rho}-\vec{\rho}_0|,\frac{\omega}{cz}).
\end{aligned}
$$

Therefore, the normalization constant C must take the value of $C = -i\omega/2\pi c$. The normalized Green's function for free-space Fresnel propagation is thus

$$g(\vec{\kappa},\omega;\vec{\rho},z) = \frac{-i\omega}{2\pi c}\,\frac{e^{i\frac{\omega}{c}z}}{z}\int_{\sigma_0} d\vec{\rho}_0\,\tilde{A}(\vec{\rho}_0)\,G(|\vec{\rho}-\vec{\rho}_0|,\frac{\omega}{cz}). \tag{A-9}$$

9. References

[1] T.B. Pittman, Y.H. Shih, D.V. Strekalov, and A.V. Sergienko, Phys. Rev. A 52, R3429 (1995).

[2] D.N. Klyshko, Usp. Fiz. Nauk, 154, 133 (1988); Sov. Phys. Usp, 31, 74 (1988); Phys. Lett. A 132, 299 (1988).

[3] A. Einstein, B. Podolsky, and N. Rosen, Phys. Rev. 35, 777 (1935).

[4] D.V. Strekalov, A.V. Sergienko, D.N. Klyshko and Y.H. Shih, Phys. Rev. Lett. 74, 3600 (1995). Due to its nonlocal behavior, this experiment was named "ghost" interference by the physics community.

[5] G. Scarcelli, V. Berardi, and Y.H. Shih, Phys. Rev. Lett. 96, 063602 (2006).

[6] A. Valencia, G. Scarcelli, M. D'Angelo, and Y.H. Shih, Phys. Rev. Lett. 94, 063601 (2005).

[7] G. Scarcelli, A. Valencia, and Y.H. Shih, Europhys. Lett. 68, 618 (2004).

[8] R. Meyers, K.S. Deacon, and Y.H. Shih, Phys. Rev. A 77, 041801(2008).

[9] Y.H. Shih, IEEE J. of Selected Topics in Quantum Electronics, 9, 1455 (2003).

[10] R. Hanbury-Brown, and R.Q. Twiss, Nature, 177, 27 (1956); 178, 1046, (1956); 178, 1447 (1956).

[11] R. Hanbury-Brown, *Intensity Interferometer*, Taylor and Francis Ltd, London, 1974.

[12] M.O. Scully and M.S. Zubairy, *Quantum Optics*, Cambridge University Press, Cambridge, 1997.

[13] A. Gatti, E. Brambilla, M. Bache and L.A. Lugiato, Phys. Rev. A 70, 013802, (2004), and Phys. Rev. Lett. 93, 093602 (2004).

[14] K. Wang, D. Cao, quant-ph/0404078; D. Cao, J. Xiong, and K. Wang, quant ph/ 0407065.

[15] Y.J. Cai, and S.Y. Zhu, quant-ph/0407240, Phys. Rev. E, 71, 056607 (2005).

[16] B.I. Erkmen and J.H. Shapiro, Phys. Rev. A 77, 043809 (2008).

[17] M. H. Rubin, Phys. Rev. A 54, 5349 (1996).

[18] J. W. Goodman, *Introduction to Fourier Optics*, McGraw-Hill Publishing Company, New York, NY, 1968.

[19] D.N. Klyshko, *Photon and Nonlinear Optics*, Gordon and Breach Science, New York, 1988.

[20] R.J. Glauber, Phys. Rev. 130, 2529 (1963); Phys. Rev. 131, 2766 (1963).

[21] M. D'Angelo, A. Valencia, M.H. Rubin, and Y.H. Shih, Phys. Rev. A 72, 013810 (2005).

[22] J.C. Howell *et al.*, Phys. Rev. Lett., 92, 210403 (2004).

[23] W. Martienssen and E. Spiller, Am. J. Phys. 32, 919 (1964).

[24] R. Meyers, K.S. Deacon, and Y.H. Shih, to be published.

[25] J.B. Liu, and Y.H. Shih, Phys. Rev. A, 79, 023818 (2009).

[26] R.S. Bennink, S.J. Bentley, and R.W. Boyd, Phys. Rev. Lett. 89, 113601 (2002); R.S. Bennink, *et al.*, Phys. Rev. Lett. 92, 033601 (2004).
[27] A. Gatti *et al.*, Phys. Rev. Lett. 98, 039301 (2007) (comment); G. Scarcelli, V. Berardi, and Y.H. Shih, Phys. Rev. Lett. 98, 039302 (2007) (reply).

25

High Performance Holographic Polymer Dispersed Liquid Crystal Systems Formed with the Siloxane-containing Derivatives and Their Applications on Electro-optics

Yeonghee Cho and Yusuke Kawakami
Japan Advanced Institute of Science and Technology
Japan

1. Introduction

Holography is a very powerful technology for high density and fast data storage, which have been applied to the systems known as holographic polymer dispersed liquid crystal (HPDLC), in which gratings are formed by anisotropic distribution of polymer and LC-rich layers through photopolymerization of monomers or oligomers and following phase separation of LC in the form of interference patterns of incident two laser beams [1-5]. Much attentions have been attracted to HPDLC systems due to their unique switching property in electric field to make them applicable to information displays, optical shutters, and information storage media [6-15].

Many research groups have made efforts to realize useful recording materials for high performance holographic gratings [16-18]. Photo-polymerizable materials, typically multi-functional acrylates, epoxy, and thiol-ene derivatives have been mostly studied because of their advantages of optical transparency, large refractive index modulation, low cost, and easy fabrication and modification[19-25]. T.J. Bunning group has reported investigation that the correlation between polymerization kinetics, LC phase separation, and polymer gel point in examining thiol-ene HPDLC formulations to enable more complete understanding of the formation of thiol-ene HPDLCs [26]. Kim group has developed that the doping of conductive fullerene particles to the formulations based on polyurethane acrylate oligomers in order to reduce the droplet coalescence of LC and operating voltage [27].

Further extensive research has been devoted to the organic-inorganic hybrid materials having the sensitivity to visible laser beam to resolve the drawbacks of photopolymerizable materials such as volume shrinkage, low reliability, and poor long term stability even high reactivity of them as well waveguide materials, optical coatings, nonlinear optical materials, and photochromic materials [28-29]. Blaya et al. theoretically and experimentally analyzed the angular selectivity curves of nonuniform gratings recorded in a photopolymerizable silica glass due to its rigidity suppressing the volume shrinkage [30]. Ramos et al. found that a chemical modification of the matrix with tetramethylorthosilicate noticeably attenuates the shrinkage, providing a material with improved stability for permanent data storage applications [31].

However, those materials still have significant drawbacks such as volume shrinkage, low reliability, and poor long term stability.

Recently, we have focused on the siloxane-containing derivatives by taking advantage of their chemical and physical properties with high thermal stability, high optical clarity, flexibility, and incompatibility[32].

In this research, first, siloxane-containing epoxides were used to induce the efficient separation of LC from polymerizable monomer and to realize high diffraction efficiency and low volume shrinkage during the formation of gratings since the ring-opening polymerization (ROP) systems with increased excluded free-volume during the polymerization suppress the volume shrinkage [33]. Although various epoxide derivatives were used, cyclohexane oxide group should be more suitable to control the volume shrinkage in the polymerization due to their ring structure with more bulky group. Actually, we improved the volume shrinkage causing a serious problem during the photopolymerization, by using the ROP system with novel siloxane-containing spiroorthoester and bicyclic epoxides.

Generally, the performance of holographic gratings in HPDLC systems strongly depends on the final morphologies, sizes, distribution, and shapes of phase-separated LC domains controlled by adjusting the kinetics of polymerization and phase separation of LC during the polymerization. Control of the rate and density of cross-linking in polymer matrix is very important in order to obtain clear phase separation of LC from polymer matrix to homogeneous droplets. Too rapid initial cross-linking by multi-functional acrylate makes it difficult to control the diffusion and phase separation of LC. At the same time, high ultimate conversion of polymerizable double bond leading to high cross-linking is important for long-term stability. These are not easy to achieve at the same time.

Till now optimization of cross-linking process has been mainly pursued by controlling the average functionality of multi-functional acrylate by mixing dipentaerythritol pentaacrylate (DPEPA), trimethylolpropane triacrylate (TMPTA) and tri(propyleneglycol) diacrylate, or by diluting the system with mono-functional vinyl compound like 1-vinyl-2-pyrollidone (NVP) [34-37]. In case of TMPTA, considerably high concentration was used. Mono-functional NVP adjusts the initial polymerization rate and final conversion of acrylate functional groups by lowering the concentration of cross-linkable double bonds [38]. However, the effects were so far limited, and these systems still caused serious volume shrinkage and low final conversion of polymerizable groups. Thus, the gratings are not long-term stable, either. Moreover, the phase separation of LC component during the matrix formation was governed only by its intrinsic property difference against polymer matrix, accordingly not well-controlled. These systems could be called as "passive grating formation" systems.

Thus, if we consider the structure and reactivity of siloxane compounds in relation with the property, it will be possible to propose new systems to improve the performance of HPDLC gratings.

Second, the objective of this research is to show the effectiveness of the simultaneous siloxane network in formation of polymer matrix by radically polymerizable multi-functional acrylate by using trialkoxysilyl (meth)acrylates, and to characterize the application of dense wavelength division multiplexing (DWDM) systems. By loading high concentration of trialkoxysilyl-containing derivatives, volume shrinkage during the formation of polymer matrix should be restrained. The principal role of multi-functional acrylate in grating formation is to make the LC phase-separate by the formation of cross-linked polymer matrix.

Our idea is to improve the property of gratings through importing the siloxane network in polymer matrix, by not only lowering the contribution of initial rapid radical cross-linking of TMPTA and realizing complete conversion of double bonds, but also maintaining the desirable total cross-linking density assisted by hydrolysis-condensation cross-linking of trialkoxysilyl group in the (meth)acrylate component to control the phase separation of LC from polymer matrix [39]. Such cross-linking can be promoted by the proton species produced from the initiating system together with radical species by photo-reaction [40-42]. In our system, phase separation of LC is not so fast compared with simple multi-functional acrylate system, and secondary cross-linking by the formation of siloxane network enforce the LC to completely phase-separate to homogeneous droplets, and high diffraction efficiency could be expected. We named this process as "proton assisted grating formation".

These systems should provide many advantages over traditional systems induced only by radical polymerization by improving: 1) the volume shrinkage by reducing the contribution of radical initial cross-linking by importing the siloxane network in whole polymer networks, 2) the contrast of siloxane network formed by the hydrolysis of ω-methacryloxyalkyltrialkoxysilane against polymer matrix, and 3) the stability of final gratings via combination of the characteristics of siloxane gel and rather loosely cross-linked radically polymerized system.

Finally, poly (propylene glycol) (PPG) derivatives functionalized with triethoxysilyl, hydroxyl, and methacrylate groups were synthesized to control the reaction rate and extent of phase separation of LC, and their effects were investigated on the performance of holographic gratings. The well-constructed morphology of the gratings was evidenced by atomic force microscopy (SEM).

2. Experimental

2.1 Holographic recording materials

Multi-functional acrylates, trimethylolpropane triacrylate (TMPTA) and dipentaerythritol penta-/hexa- acrylate (DPHA), purchased from Aldrich Chemical Co., were used as radically cross-linkable monomers to tune the reaction rate and cross-linking density.

Structures of ring-opening cross-linkable monomers used in this study are shown in **Figure 1**. Bisphenol-A diglycidyl ether (A), neopentyl glycol diglycidyl ether (B), bis[(1,2-epoxycyclohex-4-yl)methyl] adipate (F) from Aldrich Chemical Co. and 1,3-bis(3-glycidoxypropyl)-1,1,3,3-tetramethyldisiloxane (C), 1,5-bis(glycidoxypropyl)-3-phenyl-1,1,3,5,5-pentamethyltrisiloxane (E) from Shin-Etsu Co. were used without further purification. 1,5-Bis(glycidoxypropyl)-1,1,3,3,5,5-hexamethyltrisiloxane (D), 1,3-bis[2-(1,2-epoxycyclohex-4- yl)ethyl]-1,1,3,3-tetramethyldisiloxane (G), and 1,5-bis[2-(1,2-epoxycyclohex-4-yl)ethyl]- 1,1,3,3,5,5-hexamethyltrisiloxane (H) were synthesized by hydrosilylation of allyl glycidyl ether, or 4-vinyl-1-cyclohexene-1,2-epoxide (Aldrich Chemical Co.) with 1,1,3,3,5,5-hexamethyltrisiloxane, or 1,1,3,3-tetramethyldisiloxane (Silar Laboratories) in toluene at 60~70°C for 24h in the presence of chlorotris(triphenylphosphine)rhodium(I) [RhCl(PPh3)3] (KANTO chemical co. Inc.).

Methacryloxymethyltrimethylsilane (M_M-TMS), methacryloxymethyltrimethoxysilane (M_M-TMOS), 3-methacryloxypropyltrimethoxysilane (M_P-TMOS), 3-methacryloxypropyltriethoxysilane (M_P-TEOS), 3-N-(2-

methacryloxyethoxycarbonyl)aminopropyltriethoxysilane (M_U-TEOS), and 3-N-(3-methacryloxy-2-hydroxypropyl)aminopropyltriethoxysilane (M_H-TEOS), purchased from Gelest, Inc., were used as reactive diluents (**Figure 2**). Methacrylate with trialkoxysilane are capable of not only radical polymerization but also hydrolysis-condensation.

To investigate the effects of functional groups of photo-reactive PPG derivatives on performance of holographic gratings, three types of PPG derivatives were functionalized with triethoxysilyl, hydroxyl, and methacrylate groups as shown in **Figure 3**. PPG derivative with difunctional triethoxysilyl groups (PPG-DTEOS) and PPG derivative together with hydroxyl and triethoxysilyl groups (PPG-HTEOS) were synthesized by using 1 mol of PPG (Polyol.co. Ltd.) with 2 mol and 1 mol of 3-(triethoxysilyl)propyl isocyanate (Aldrich), respectively. PPG derivative together with methacrylate and triethoxysilyl groups PPG-MTEOS was synthesized by using 1 mol of PPG-HTEOS with 1 mol of 2-isocyanatoethyl methacrylate (Gelest, Inc.).

1-Vinyl-2-pyrrolidone (NVP) was used as another radically polymerizable reactive diluent. Commercial nematic LC, TL203 (T_{NI}=74.6 °C, n_e=1.7299, n_o=1.5286, Δn=0.2013) and E7 (T_{NI}=61 °C, n_e=1.7462, n_o=1.5216, Δn=0.2246) , purchased from Merck & Co. Inc., were used without any purification.

2.2 Composition of photo-initiator system and recording solution

Photo-sensitizer (PS) and photo-initiator (PI) having sensitivity to visible wavelength of Nd-YAG laser (λ= 532 nm) selected for this study are 3, 3'-carbonylbis(7-diethylaminocoumarin) (KC, Kodak) and diphenyliodonium hexafluorophosphate (DPI, AVOCADO research chemicals Ltd.), respectively, which produce both cationic and radical species [43-45]. The concentrations of the PS and PI were changed in the range of 0.2-0.4 and 2.0-3.0 wt % to matrix components, respectively.

Recording solution was prepared by mixing the matrix components (65 wt%) and LC (35 wt%), and injected into a glass cell with a gap of 14 μm and 20 μm controlled by bead spacer.

2.3 Measurement of photo-DSC and FTIR

The rate of polymerization was estimated from the heat flux monitored by photo-differential scanning calorimeter (photo-DSC) equipped with a dual beam laser light of 532nm wavelength. Matrix compounds were placed in uncovered aluminum DSC pans and cured with laser light by keeping the isothermal state of 30 °C at various light intensities.

Infrared absorption spectra in the range 4000-400 cm^{-1} were recorded on polymer matrix compounds by Fourier Transform Infrared Spectroscopy (FTIR) (Perkin-Elmer, Spectrum One).

2.4 Optical setup for transmission holographic gratings

Nd:YAG solid-state continuous wave laser with 532 nm wavelength (Coherent Inc., Verdi-V2) was used as the irradiation source as shown in Figure 4.

The beam was expanded and filtered by spatial filters, and collimated by collimator lens. s-Polarized beams were generated and split by controlling the two $\lambda/2$ plates and polarizing beam splitter. Thus separated two s-polarized beams with equal intensities were reflected by two mirrors and irradiated to recording solution at a pre-determined external beam angle (2θ) which was controlled by rotating the motor-driven two mirrors and moving the rotation stage along the linear stage. In this research, the external incident beam angle was fixed at 16° (2θ) against the line perpendicular to the plane of the recording cell.

Real-time diffraction efficiency was measured by monitoring the intensity of diffracted beam when the shutter was closed at a constant time interval during the hologram recording. After the hologram was recorded, diffraction efficiency was measured by rotating the hologram precisely by constant angle by using motor-driven controller, with the shutter closed to cut-off the reference light, to determine the angular selectivity. Holographic gratings were fabricated at 20mW/cm^2 intensity for one beam, and the optimum condition was established to obtain the high diffraction efficiency, high resolution, and excellent long-term stability after recording. Diffraction efficiency is defined as the ratio of diffraction intensity after recording to transmitting beam intensity before recording.

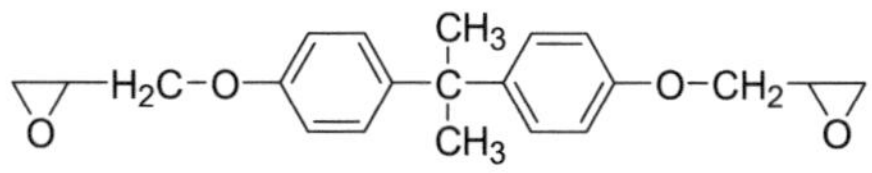

Bisphenol A diglycidy lether (**A**)

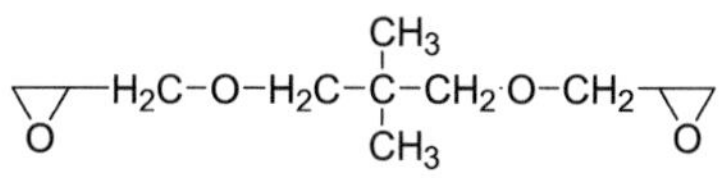

Neopentylglycol diglycidyl ether (**B**)

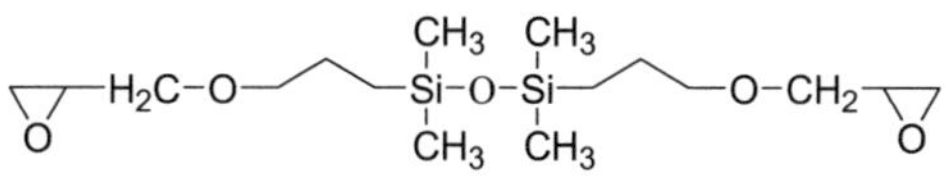

1,3-Bis(3-glycidoxypropyl)-1,1,3,3-tetramethyldisiloxane (**C**)

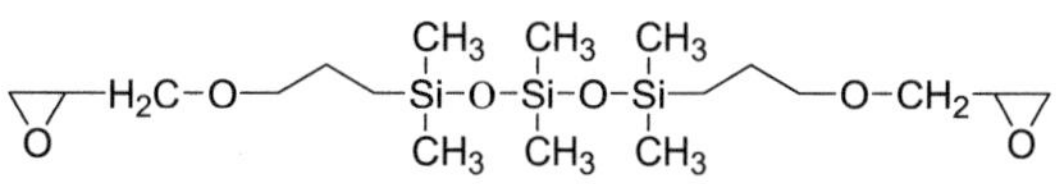

1,3-Bis(3-glycidoxypropyl)-1,1,3,3,5,5-hexamethyltrisiloxane (**D**)

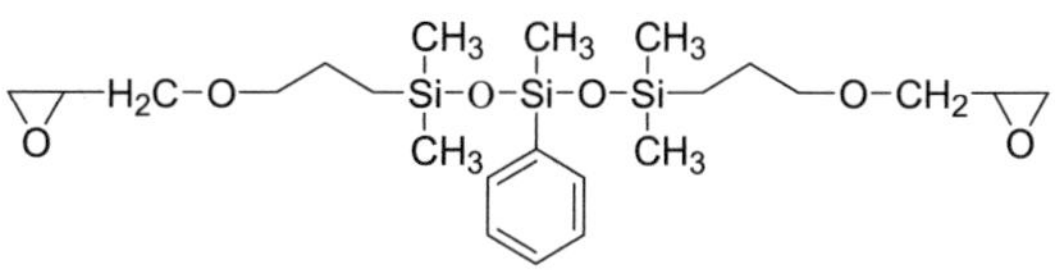

1,5-Bis(3-glycidoxypropyl)-3-phenyl-1,1,3,5,5-pentamethyltrisiloxane (**E**)

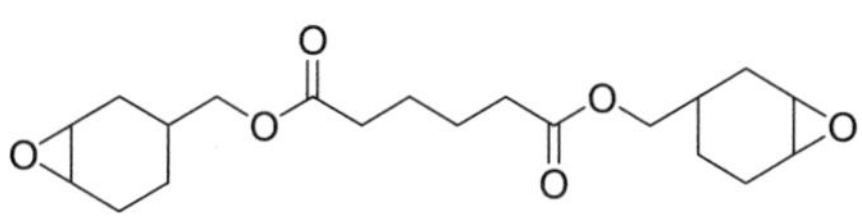

Bis[(1,2-epoxycyclohex-4-yl)methyl] adipate (**F**)

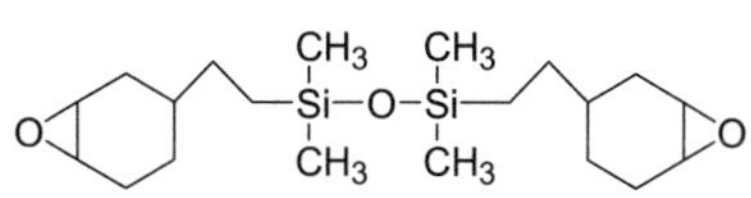

1,3-Bis[2-(1,2-epoxycyclohex-4-yl)ethyl]-1,1,3,3-tetramethyldisiloxane (**G**)

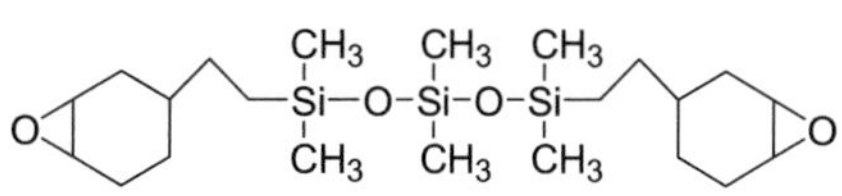

1,5-Bis[2-(1,2-epoxycyclohex-4-yl)ethyl]-1,1,3,3,5,5-hexamethyltrisiloxane (**H**)

Fig. 1. Chemical structures of ring-opening cross-linkable monomers.

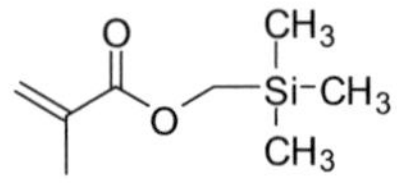

Methacryloxymethyltrimethylsilane (**MM-TMS**)

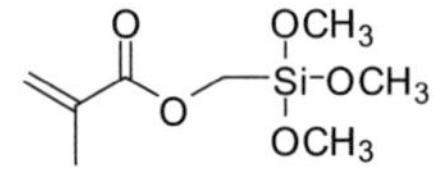

Methacryloxymethyltrimethoxysilane (**MM-TMOS**)

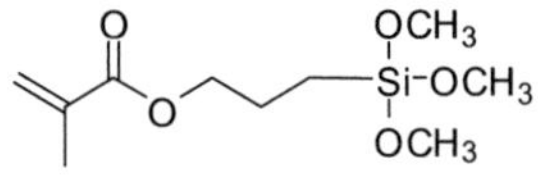

3-Methacryloxypropyltrimethoxysilane (**MP-TMOS**)

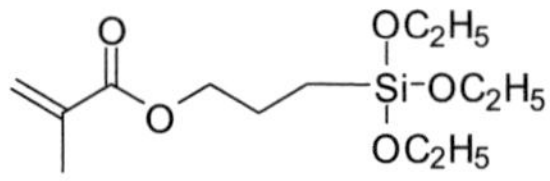

3-Methacryloxypropyltriethoxysilane (**MP-TEOS**)

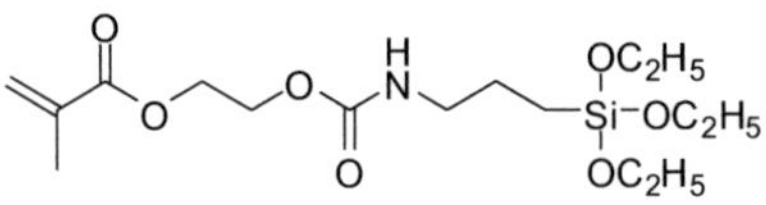

3-*N*-(2-methacryloxyethoxycarbonyl)aminopropyltriethoxysilane (**MU-TEOS**)

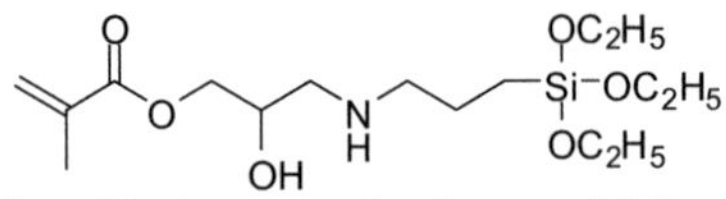

3-*N*-(3-methacryloxy-2-hydroxypropyl)aminopropyltriethoxysilane (**MH-TEOS**)

Fig. 2. Structures of ω-methacryloxyalkyltri-alkyl or -alkoxysilanes.

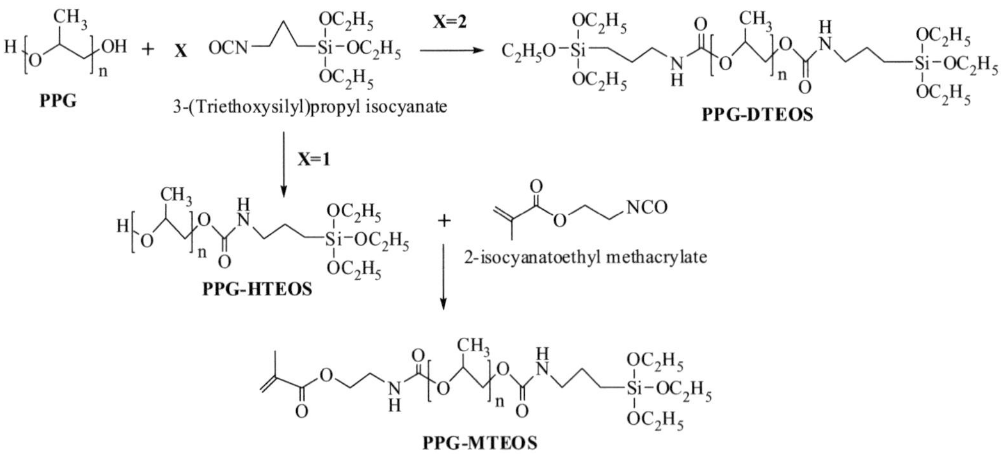

Fig. 3. Chemical structures of PPG derivatives functionalized with triethoxysilyl, hydroxyl, and methacrylate groups as polymer matrix components.

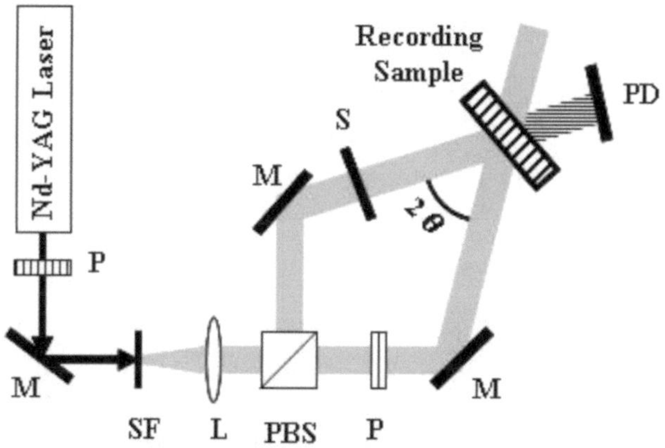

Fig. 4. Experimental setup for the holographic recording and real-time reading; P: 1/2λ plate, M: mirror, SF: spatial filter, L: collimating lens, PBS: polarizing beam splitter, S: shutter, 2θ: external inter-beam angle, PD: power detector.

2.5 Morphology of holographic gratings

Surface morphology of gratings was examined with scanning electron microscope (SEM, HITACHI, S-4100). The samples for measurement were prepared by freeze-fracturing in liquid nitrogen, and washed with methanol for 24h to extract the LC, in case necessary. Exposed surface of the samples for SEM was coated with a very thin layer of Pt-Pd to minimize artifacts associated with sample charging (HITACHI, E-1030 ion sputter). Surface topology of transmission holographic grating was examined with atomic force microscopy (AFM, KIYENCE, VN8000). The samples for measurement were prepared by freeze-fracturing in liquid nitrogen, and washed with methanol for 24h to extract the LC. AFM having a contact mode cantilever (KIYENCE, OP-75042) was used in tapping mode for image acquisition.

3. Results and discussion

3.1 Effects of siloxane-containing bis(glycidyl ether)s and bis(cyclohexene oxide)s on the real-time diffraction efficiency

Real-time diffraction efficiency, saturation time, and stability of holographic gratings according to exposure time were evaluated. Figure 5 shows the effects of chemical structures of bis(glycidyl ether)s (A - E) on real-time diffraction efficiency at constant concentration of E7 (10 wt %) in recording solution [DPHA : NVP : (A - E) = 50: 10: 40 relative wt %].

In general, high diffraction efficiency can be obtained by the formulation of recording solution with large difference in refractive indexes between polymer matrix and LC, and by inducing the good phase separation between polymer rich layer and LC rich layer. As expected, gratings formed with C having siloxane component had remarkably higher diffraction efficiency than gratings formed with A and B without siloxane component, which seemed to have resulted from effects of siloxane component to induce good phase separation of E7 from polymer matrix toward low intensity fringes by its incompatible property against E7. Longer induction period for grating formation of C was attributed to lower viscosity of recording solution, and the diffraction efficiency gradually increased and reached to higher value, which resulted from the further phase separation of E7 due to the flexible siloxane chain that helped migration of E7 toward low intensity fringes.

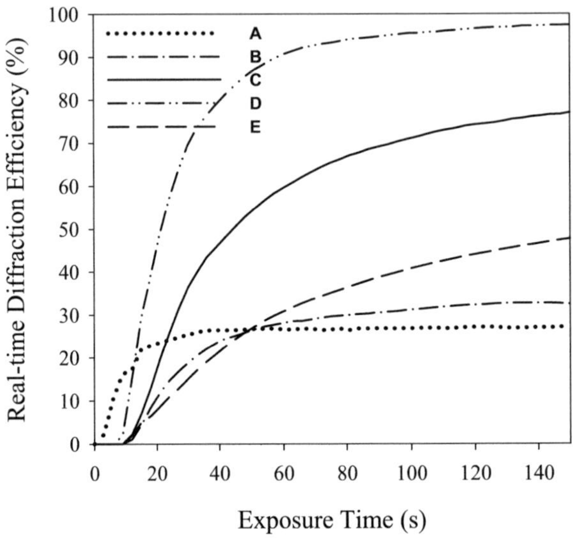

Fig. 5. Real-time diffraction efficiency of the gratings formed with (A - E) with 10 wt % E7 [DPHA: NVP: (A - E) = 50: 10: 40 relative wt %].

All the gratings formed with (C – E) having siloxane component showed high diffraction efficiencies. The highest diffraction efficiency 97% was observed for D with trisiloxane chain, probably due to its incompatible property with E7. However, gratings formed with E, having phenyl group in the trisiloxane chain, showed the lowest diffraction efficiency. Bulky phenyl group attached in the siloxane chain reduced the flexibility of the chain to result in the suppression of phase separation. It might have contributed to the increase of the interaction between polymer matrix with E7 having bi-/terphenyl group.

Figure 6 shows the real-time diffraction efficiency of the gratings formed with bis(cyclohexene oxide) derivatives (F - H) at constant concentration of E7 (10 wt %) [DPHA: NVP: (F - H) = 50: 10: 40 relative wt %].

Gratings formed with G and H having siloxane component had higher diffraction efficiency than F without it, which seemed to indicate that, as mentioned above, siloxane chain in G and H made the solution less viscous, and incompatible with E7, which helped the easy diffusion and good phase separation between polymer matrix and E7 to result in high refractive index modulation, n. Especially, H showed higher diffraction efficiency than E, probably due to flexibility and incompatibility brought about by its longer siloxane chain. However, compared with C and D, G and H did not give higher diffraction efficiency, even with longer siloxane chain. This may be understood because of the difference in the chemical structure of ring-opening cross-linkable group. G and H have bulkier cyclohexene oxide as functional group and have higher viscosity, accordingly its diffusion toward high intensity fringes seems difficult compared with that of C or D.

3.2 Volume shrinkage of the gratings depending on the structure of bis(epoxide)

Photo-polymerizable system as holographic recording material usually causes significant volume shrinkage during the formation of gratings, which can distort the recorded fringe pattern and cause angular deviations in the Bragg profile. Therefore, it is very important to solve the problem of volume shrinkage in photopolymerization systems.

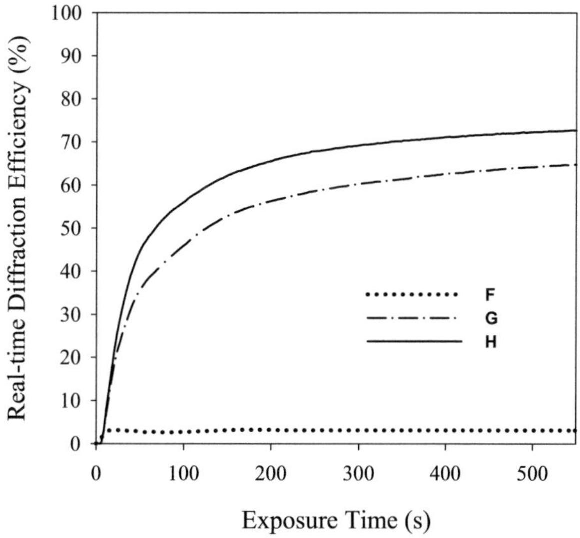

Fig. 6. Real-time diffraction efficiency of the gratings formed with (F – H) and 10 wt % E7 [DPHA: NVP: (F - H) = 50: 10: 40 relative wt %].

For the measurement of volume shrinkage, slanted holographic gratings were fabricated by simply changing the angles of reference (R) and signal (S) beams, as shown in Figure 7 [46].

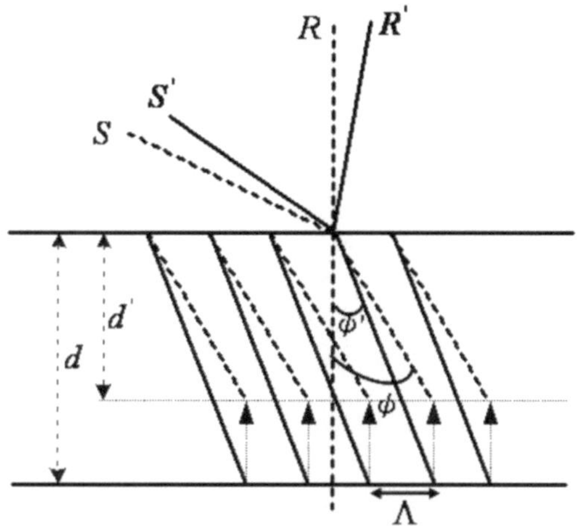

Fig. 7. Fringe-plane rotation model for slanted transmission holographic recording to measure the volume shrinkage.

R and S are recording reference (0°) and signal (32°) beams. φ (16° in this study) is the slanted angle against the line perpendicular to the plane of the recording cell of gratings formed with S and R. Solid line in the grating indicates the expected grating. d is the sample thickness. Actual grating formed by S and R was deviated from the expected grating shown by dashed line by volume shrinkage of the grating. Presumed signal beam (S'), which should have given actual grating was detected by rotating the recorded sample with reference light R off. This rotation of angle was taken as deviation of slanted angle. R′ and S′ are presumed compensation recording reference and signal beams. φ′ is the slanted angle in presumed recording with S′ and R′, and d′ is the decreased sample thickness caused by volume shrinkage. Degree of volume shrinkage can be calculated by following equation;

$$\text{Degree of volume shrinkage} = 1 - \frac{d'}{d} = 1 - \frac{\tan\varphi'}{\tan\varphi} \quad (\tan\varphi' = \frac{\Lambda}{d}, \tan\varphi = \frac{\Lambda}{d'}) \tag{1}$$

Figure 8 shows the angular deviations from the Bragg profile of the gratings formed with C and G having bis(glycidyl ether) and bis(cyclohexene oxide), respectively, at constant concentration of E7 (10 wt %) [DPHA : NVP: (C or G) = 50: 10: 40 relative wt%]. The angular shifts from the Bragg matching condition (0 degree) at both positions of diffracted R and S beams indicates the extent of volume shrinkage of the gratings. Grating prepared from the recording solution containing only radically polymerizable compounds [DPHA : NVP = 50: 50 in relative wt%] was used as the reference.

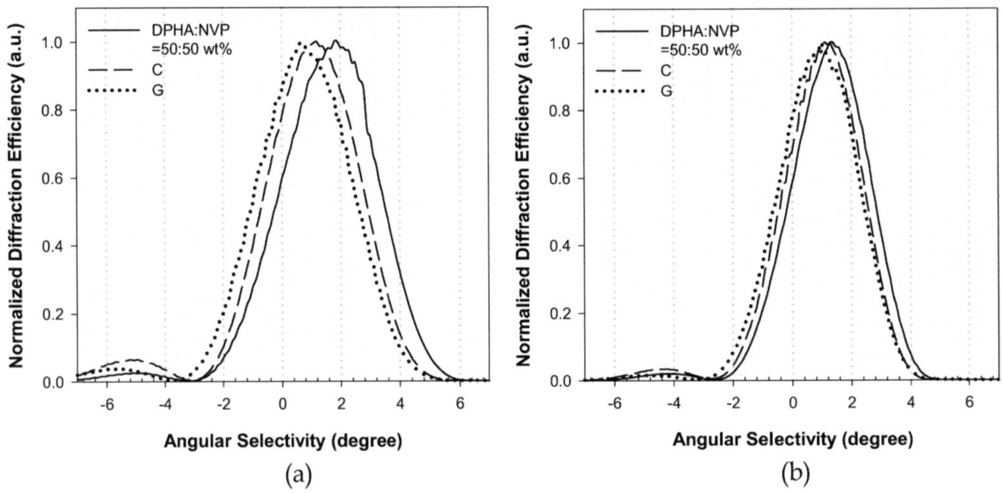

Fig. 8. Angular deviation from the Bragg profile for the gratings formed with C and G [DPHA: NVP : (C or G) = 50: 10: 40 relative wt %] detected by (a) diffracted S beam, and (b) diffracted R beam.

As shown in Figure 8, gratings formed with G having bis(cyclohexene oxide) showed smaller deviation from Bragg matching condition than gratings formed with C having bis(glycidyl ether) for both diffracted R and S beams. The diffraction efficiency after overnight was only slightly changed, which indicated the volume shrinkage after overnight was negligible.

Diffraction efficiency, angular deviation, and volume shrinkage of each system were summarized in Table 1.

Gratings formed with only radically polymerizable multifunctional acrylate (DPHA: NVP = 50:50 relative wt %) showed the largest angle deviation, and the largest volume shrinkage of 10.3% as is well known. Such volume shrinkage could be reduced by combining the ring-opening cross-linkable monomers. Especially, bis(cyclohexene oxide)s were effective to reduce the volume shrinkage (5.6 %), probably due to its cyclic structure, although their diffraction efficiency was lower than those formed with bis(glycidyl ether)s.

Recording solution	Diffraction efficiency (%)a	Angular deviation of diffracted		φ′	Degree of volume shrinkage (%)
		S beam (degree)	R beam (degree)		
DPHA: NVP =50:50 wt %	2	1.8	1.35	14.42	10.3
DPHA: NVP : C = 50: 10: 40 wt %	47	1.2	1.1	14.85	7.5
DPHA: NVP: G = 50:10:40 wt %	29	0.7	1.0	15.15	5.6
DPHA: NVP : D = 50: 10: 40 wt %	54	0.83	0.76	15.21	5.2
DPHA: NVP: H = 50:10:40 wt %	31	0.66	0.70	15.32	4.5

Table 1. Deviations from Bragg angle of diffracted S and R beams (degree) and degree of volume shrinkage and diffraction efficiency determined by S beama.

The shrinkage effect could be caused by mechanical reduction of the grating pitch and a real time change in refractive index of the irradiated mixture. Which factor is playing a major role is not clear at present. Distinction of these factors will be a future problem.

One of the possible reasons for small volume shrinkage is the effective formation of IPN structure in the grating in the recording system DPHA : NVP : G = 50: 10: 40 relative wt %. The balance between the formation of initial cross-linking of DPHA and following cross-linking by G might be proper to produce effective IPN structure.

Good evidence for these was shown in Figure 9 of SEM morphologies.

Figure 9 (a) and (c) show clearly phase-separated polymer layers after the treatment with methanol, which means almost perfect phase separation between polymer rich layers and E7 rich layers. Cross-sectional and surface views of the sample could be observed. When 20 wt % E7 was used, a little incompletely phase separated E7 layers were shown in Figure 9 (d), although much higher E7 was phase separated than the case of 5 wt % E7 [Figure 9 (b)]. Grating spacing was close to the calculated value from the composition of recording solution for the grating prepared with 5 wt % E7.

3.3 Angular selectivity

When the multiplex hologram recording is required, it is necessary to know the angular selectivity. The smaller the value, the more multiplex data or gratings can be recorded [47-49]. Angular selectivity ($\Delta\theta_{ang}$) is defined by Kogelnik's coupled wave theory as follows [50]:

$$\Delta\theta_{ang} = \frac{1}{2n\sin\theta}\sqrt{\left[\left(\frac{\lambda}{T}\right)^2 - \left(\frac{\Delta n}{\cos\theta}\right)^2\right]} \tag{2}$$

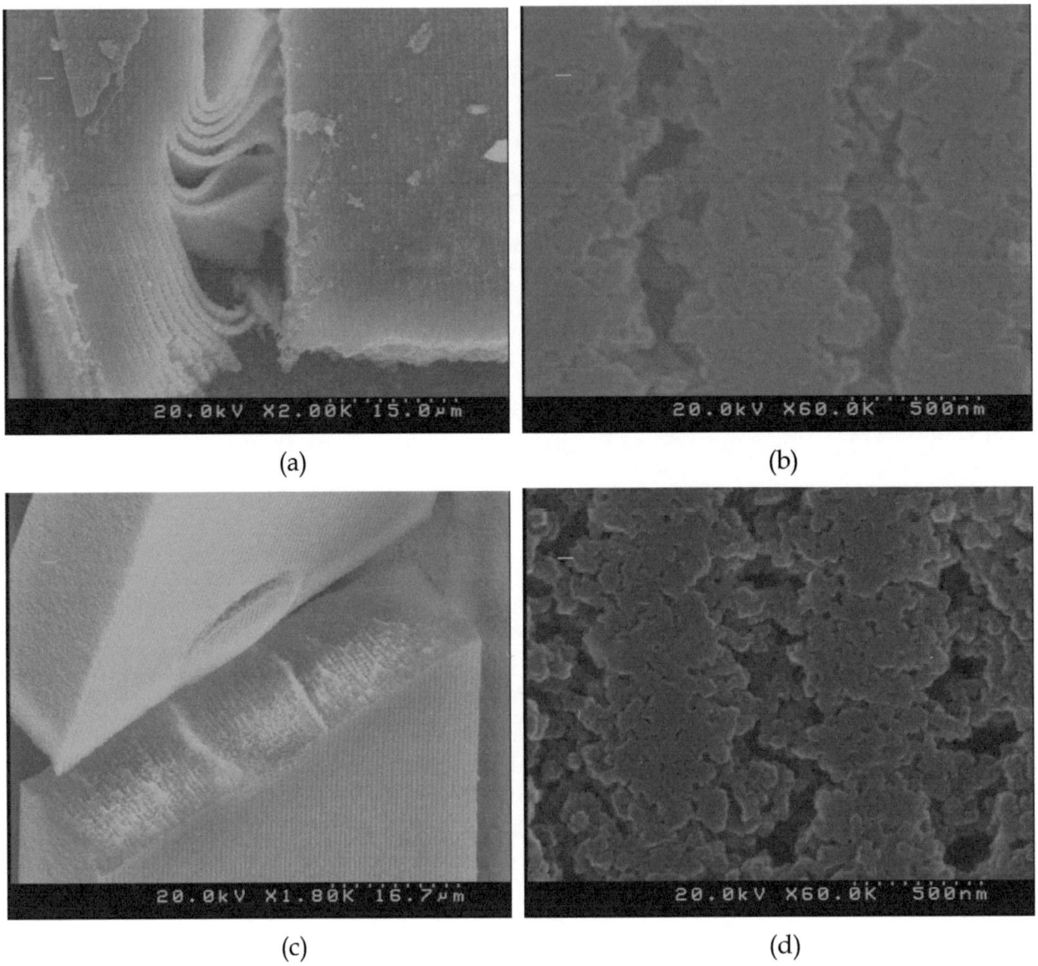

Fig. 9. SEM morphologies of gratings formed with H, TMPTA and various concentration of E7 [TMPTA : NVP: H = 50: 10: 40 relative wt %] (a) 5 wt %, (b) 5 wt %, ×60K, (c) 20 wt %, and (d) 20 wt %, ×60K.

where n is the average refractive index of recording solution, θ is the internal incident beam angle, T is the thickness of the hologram, λ is the recording wavelength, and n is the modulation of refractive index of the recording solution after recording.

Angular selectivity of our samples were similar, irrespective of the structures of epoxides (about 4°) as typically shown in Figure 10. Solid line represents the simulated theory values according to the Kogelnik's coupled wave theory.

G. Montemezzani group reported that the use of Kogelnik's expression assuming fully symmetric beam geometries in highly birefringent materials such as LC leads to a large error [51]. Our experimental data showed only a little deviation from the theoretical values by the Kogelnik's coupled wave theory. This maybe attributed to the slight thickness reduction by small volume shrinkage still existing. The role of both factors should be clarified in the future.

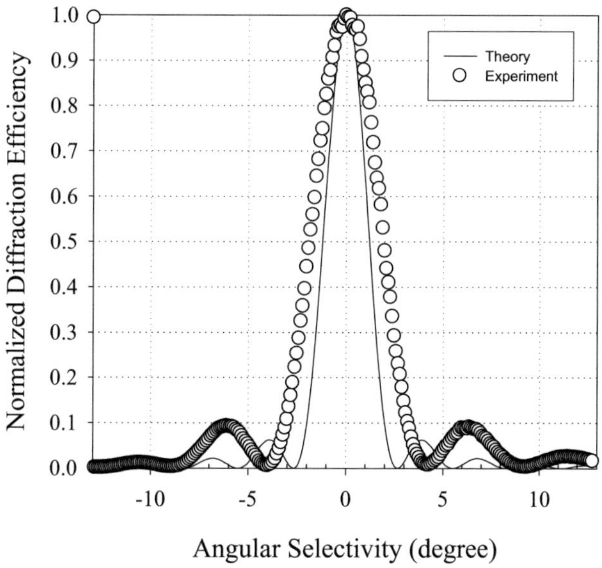

Fig. 10. Angular selectivity of gratings formed with H , TMPTA, and 5 wt % E7 [TMPTA : NVP : H = 50: 10: 40 relative wt %].

3.4 Effectiveness of M_M-TMOS on formation of holographic gratings

As a preliminary experiment, M_M-TMS and M_M-TMOS were compared as a diluent for the polymer matrix component (totally 65 wt%, TMPTA : M_M-TMS, or M_M-TMOS : NVP = 10 : 80 : 10 in wt%, average double bond functionality = 1.1 on mole base), together with 35wt% LC of TL203. As shown in Figure 11 gratings could not be formed with M_M-TMS even with 30 min irradiation of light, because of the low average functionality of the polymerization system. G. P. Crawford reported that HPDLC gratings made with monomer mixtures with average double bond functionality less than 1.3 were mechanically very weak[52]. In general, it is difficult to form holographic gratings with low concentration of multi-functional acrylate (average double bond functionality < 1.2) by dilution with mono-functional component in radical polymerization.

Dramatic enhancing in the diffraction efficiency to about 86% (induction period of 144 sec) was observed in case of M_M-TMOS, even with only 10 wt% TMPTA by using 0.2 wt% KC and 2wt% DPI. Only trimethoxysilyl and trimethylsilyl parts are different in these two formulations. Hydrolysis of trimethoxysilyl group by moisture and following condensation seems responsible for the increased diffraction efficiency.

Effects of Alkyl and Spacer Groups in ω-Methacryloxyalkyltrialkoxysilanes on the Formation and Performance of Gratings

In order to systematically study the influence of alkyl group and spacer group of ω-methacryloxyalkyltrialkoxysilanes on the formation and performance of the formed gratings, their chemical structures were modified as shown in Figure1. The relative concentration was set as TMPTA : ω-methacryloxyalkyltrialkoxysilane : NVP = 10 : 80 : 10 wt% to clearly extract the effects of hydrolysis-condensation of trialkoxysilyl group on the formation of the gratings and the performance of the formed gratings.

Figure 11 shows the real-time diffraction efficiency of holographic gratings formed with various ω-methacryloxyalkyltrialkoxysilanes capable of radical photo-polymerization and hydrolysis- condensation.

When spacer was changed from methylene to propylene (M_P-TMOS), the diffraction efficiency was dropped to 72% with longer induction period (576 sec). This seems to be because of the higher hydrophobicity of the spacer group compared with M_M-TMOS. The rate of the hydrolysis-condensation of trialkoxysilyl functions seems very important.

By changing the trialkoxysilyl functional group from trimethoxy to triethoxy (M_P-TMOS to M_P-TEOS) with the same propylene spacer, not only the diffraction efficiency was decreased to 13%, but the induction period was also elongated to 693 sec, which again strongly suggested that the hydrolysis-condensation process of trialkoxysilane function is playing an essential role in grating formation.

In grating formation, induction period basically depends on the time of the formation of cross-linked polymer matrix. In classical grating formation by radical polymerization of multi-functional acrylates, induction period is observed because polymerization does not start until the complete consumption of oxygen present in the system as an inhibitor. In the present system, the induction period depends on the actual gelation time of recording solution assisted by hydrolysis-condensation of trialkoxysilyl functions. The induction period varies by the physical property of ω-methacryloxyalkyltrialkoxysilane derivatives. The rate of the hydrolysis-condensation of trialkoxysilyl functions by moisture strongly depends on the hydrophobicity of the methacrylate monomer. Polymerization of recording solution leads to changes in the chemical potential of the system, and increases the miscibility gap between LC and polymerized matrix

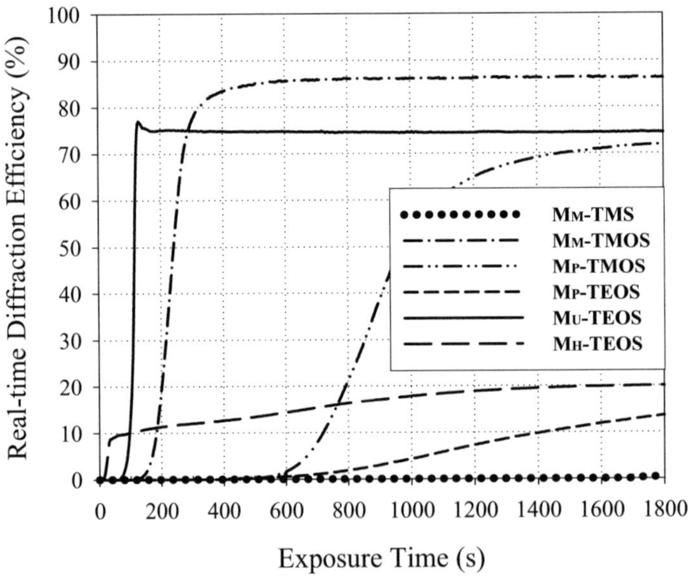

Fig. 11. Real-time diffraction efficiency of the gratings formed with various ω-methacryloxyalkyltrialkoxysilanes in the recording solution with 65 wt% matrix compounds of TMPTA : ω-methacryloxyalkyltrialkoxysilane : NVP = 10 : 80 : 10 wt% and 35 wt% TL203, and KC-DPI (0.2 wt% - 2 wt% to matrix compounds) with one beam intensity of 20 mW/cm^2.

To investigate the effects of chemical structures of spacer between trialkoxysilylalkyl group and methacrylate group, hydrophilic urethane and hydroxylpropylene groups were introduced in the spacer of the monomer structure. The highest diffraction efficiency of 75% and remarkably shorter induction period of 75 sec were obtained for grating formed with M_U-TEOS having urethane linkage in spacer group. In addition, gratings formed with M_H-TEOS having hydroxylpropylene group in the spacer showed the shortest induction period of 18 sec, although the diffraction efficiency was considerably low (20%).

To summarize the results, we may consider that the radically mono-functionally polymerizable 3-methacryloxypropyltrialkoxysilane became apparently multi-functional cross-linkable monomer by hydrolysis and condensation of trialkoxysilyl group as shown in Scheme 1, which induced the high concentration of cross-linking with moderate rate by the hydrolysis.

Scheme 1 proposed matrix formation processes: 1) radical cross-linking by TMPTA, 2) simultaneous radical cross-linking of TMPTA and small amounts of multi-functional methacrylate formed via hydrolysis-condensation of trialkoxysilyl group, followed by further cross-linking by hydrolysis, 3) competing rapid cross-linking of (meth)acrylate functions and sol-gel process of trialkoxysilane function, followed by further cross-linking by radical polymerization and sol-gel process.

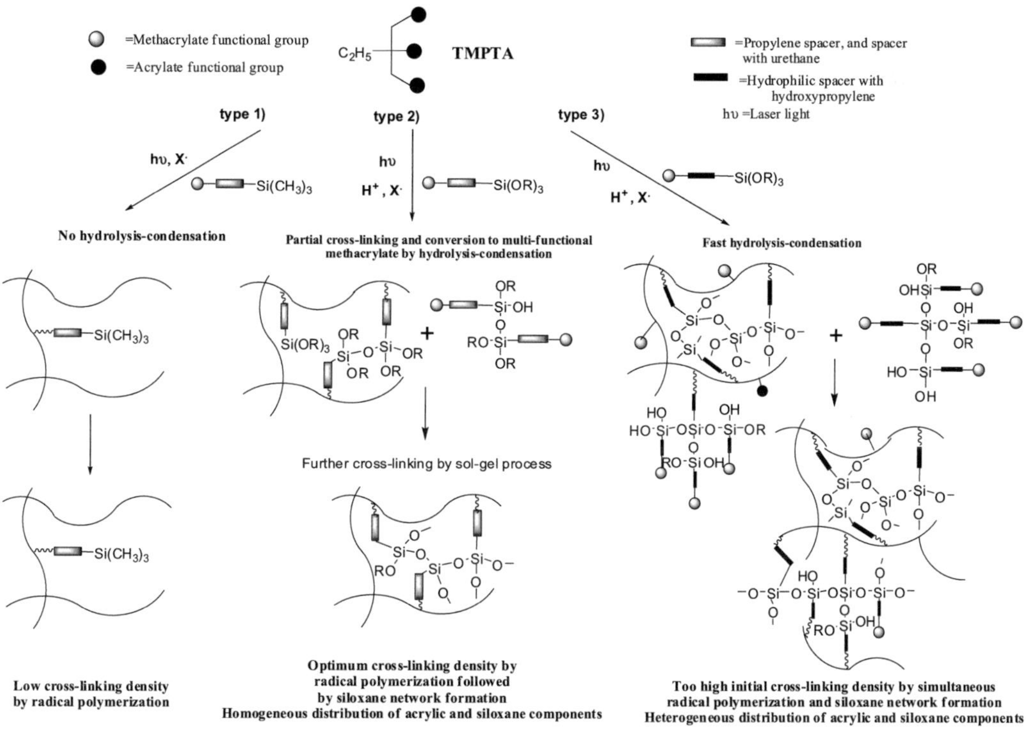

Scheme 1. Proposed matrix formation processes: 1) radical cross-linking by TMPTA, 2) simultaneous radical cross-linking of TMPTA and small amounts of multi-functional methacrylate formed via hydrolysis-condensation of trialkoxysilyl group, followed by further cross-linking by hydrolysis, 3) competing rapid cross-linking of (meth)acrylate functions and sol-gel process of trialkoxysilane function, followed by further cross-linking by radical polymerization and sol-gel process.

In case of methacryloxymethyltrimethylsilane, cross-linking density is not high enough to form grating. This process corresponds to type 1) in Scheme 1. In TMOS or TEOS system, the hydrolysis of trialkoxysilyl group is relatively slow compared with the fast radical polymerization of TMPTA. Thus, grating formation is not rapid, but following cross-linking by hydrolysis assisted the formation of polymer matrix and further diffusion of LC to form gratings with high diffraction efficiency. By the introduction of urethane function in the spacer, the hydrophilic nature of the spacer increases the hydrolysis of triethoxysilyl group by moisture, and converts mono-functional methacrylate to apparently multi-fumctional methacrylate, and assisted the formation of polymer matrix by radical polymerization together with cross-linking by hydrolysis condensation. This process corresponds to type 2) in Scheme 1. In case of the introduction of hydroxypropylene spacer, too much hydrophilic nature of the spacer strongly enhanced the hydrolysis of the trialkoxysilyl group, and created the situation where apparently high concentration of multi-functional (meth)acrylates in the initial polymerization solution, and resulted in rapid formation of grating by radical cross-linking, but low diffraction efficiency. This process corresponds to type 3) in Scheme 1.

The spectral responses of the demultiplexer are measured by use of a broadband erbium-doped fiber amplifier (EDFA) source and are monitored by means of an optical spectrum analyzer for each channel. Figure 12 shows the spectra of the two output channels with the uniform gratings formed with M_U-TEOS at the ratio of TMPTA: M_U-TEOS: PI solution = 40: 50: 10 wt% (35 wt% of TL203 to monomer solution). All the channels had a 3-dB bandwidth of 0.13 nm and a channel spacing of 0.4 nm, and the interchannel cross-talk level, defined as the difference between the maximum power of a channel at the band edge and its power at the adjacent signal's band edge, is reduced by ~18dB.

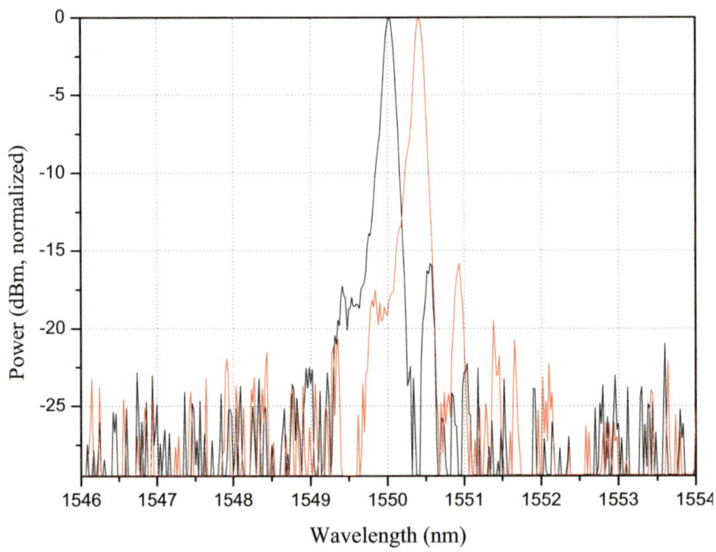

Fig. 12. Spectra of two output channels with 0.4 channel spacing for uniform gratings formed with M_U-TEOS at the ratio of TMPTA: M_U-TEOS: PI solution = 40: 50: 10 wt% (35 wt% of TL203 to monomer solution).

3.5 Effects of triethoxysilyl, hydroxyl, and methacrylate functional groups based on PPG derivatives on performance of holographic gratings

Figure 13 shows the AFM surface topology of the gratings formed with the formulation with 65 wt% polymer matrix compound of the ratio 20: 10: 50: 20 in TMPTA: NVP: Mu-TEOS: PPG-DTEOS and 35wt% of E7. Very regular and well-defined gratings were fabricated as shown in Figure 13(a) scanned in 10 μm length. The grating spacing was approximately 839.8 nm as shown in Figure 13(b) scanned in 3 μm length, which was in good agreement with the calculated spacing value of 965 nm by Bragg's equation (grating spacing $\Lambda = \lambda / 2\sin\theta$, λ is 532 nm wavelength of laser light and θ is 16 degree of incident external half angle in this experiment). Polymer matrix layers are shown as the sinusoidal pattern in profile of AFM topology since the LC layers were washed out by methanol from the positions of the valley parts in sinusoidal pattern, which maybe concluded that polymer matrix with PPG-DTEOS was exactly formed by photo-reaction in high intensity regions and E7 was phase-separated in low intensity regions of interference pattern of two laser beams.

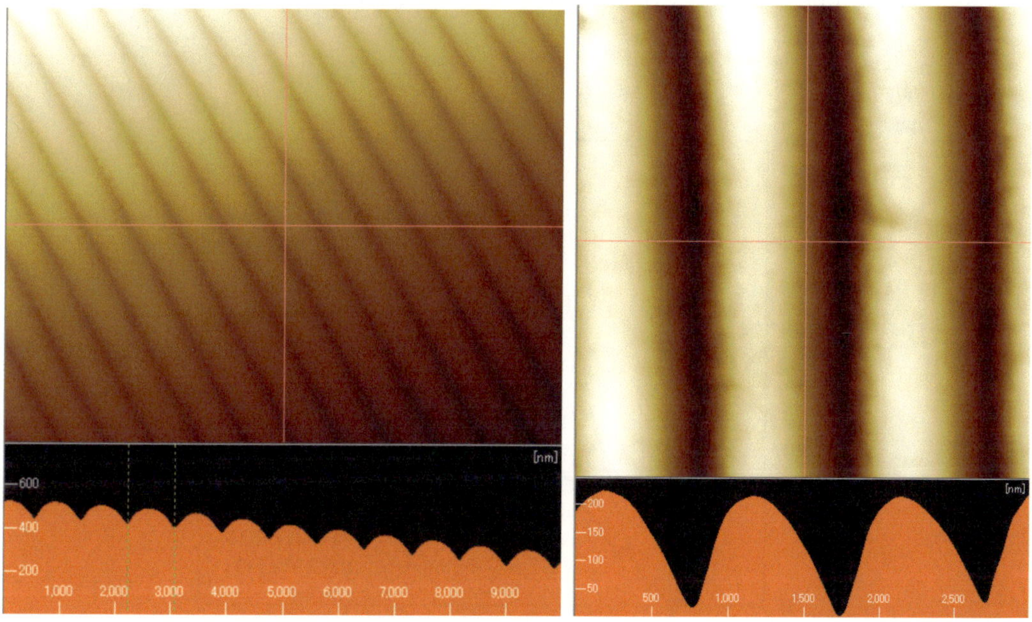

(a) (b)

Fig. 13. AFM surface topologies of the gratings formed with the formulation with 65 wt% polymer matrix compound of the ratio 20: 10: 50: 20 in TMPTA: NVP: Mu-TEOS: PPG-DTEOS and 35wt% of E7 in (a) 10 μm and (b) 3 μm scanning lengths.

The effectiveness of functional groups of PPG derivatives were observed on real-time diffraction efficiency as shown in Figure 14. Photo-reactive solutions (matrix components) were prepared from the ratios of 20: 10: 50: 20 wt% in TMPTA: NVP: Mu-TEOS: PPG derivatives. Holographic recording solutions with E7 were ready to make holographic gratings in the ratio of 65 wt% and 35wt% as photo-reactive solutions and E7, respectively. By changing the functional groups of PPG derivatives as triethoxysilyl, hydroxyl, and methacrylate groups, remarkable differences were observed on diffraction efficiency even though the induction periods for grating formation were similar with 33 second. The highest final diffraction efficiency of 93% was observed in holographic gratings formed from PPG-DTEOS with difunctional triethoxysilyl groups at 240 second of irradiation of light. In the case of PPG-HTEOS together with hydroxyl and triethoxysilyl groups, maximum diffraction efficiency was approximately 78%. When the PPG-MTEOS together with methacrylate and triethoxysilyl groups was used, maximum diffraction efficiency reached at 96% and gradually decreased to 60%. These phenomena may be considered that the functional groups of PPG derivatives affected strongly on the diffraction efficiency attributed to the difference of reaction kinetics and extend of LC phase separation. In the case of PPG-DTEOS and PPG-HTEOS, phase separation of LC should not be so fast compared with the case of PPG-MTEOS, and further cross-linking by the formation of siloxane network helped the LC to gradually phase-separate into low intensity regions of interference patterns, and maximum diffraction efficiency was reached at slower exposure time than PPG-MTEOS.

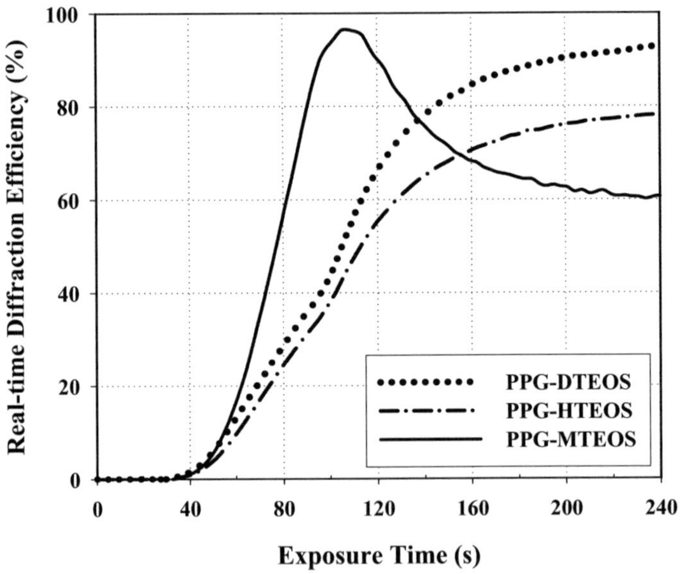

Fig. 14. Real-time diffraction efficiency of the gratings formed with PPG-DTEOS, PPG-HTEOS, and PPG-MTEOS in the recording solution with 65 wt% polymer matrix compounds of the ratios 20 : 10 : 50 : 20 wt% in TMPTA : NVP: Mu-TEOS: PPG derivatives, and 35 wt% E7.

4. Conclusion

We have demonstrated the effectiveness of the introduction of siloxane component to ring opening crosslinkable monomer as polymer matrix component to enhance the performance of HPDLC grating. High diffraction efficiency over 97% was obtained by using 1,3-bis(3-glycidoxypropyl)-1,1,3,3,5,5-hexamethyltrisiloxane (D), which induces a fast and good phase separation due to flexible and incompatible properties of siloxane component even at low concentration of E7 (10wt %). Gratings with much higher performance such as low volume shrinkage with 5.57% could be obtained in the case of 1,3-bis[2-(1,2-epoxycyclohex-4-yl)ethyl]-1,1,3,3-tetramethyldisiloxane (G).

Stable transmission holographic polymer dispersed liquid crystal gratings were efficiently prepared via network formation by radical polymerization of tri-functional acrylate assisted by hydrolysis-condensation reaction or trialkoxysilane functional group of ω-methacryloxyalkyltrialkoxysilane, induced by radical and proton species produced in the photo-decomposition of initiating system composed of 3, 3′-carbonylbis[7′-diethylaminocoumarine] as a photo-sensitizer and diphenyliodonium hexafluorophosphate as a photo-initiator.

The longest grating spacing of 0.9 μm indicated the least volume shrinkage.

At higher concentration of methacrylate, gratings formed with trimethoxysilylmethyl methacrylate capable of siloxane network formation showed remarkably higher diffraction efficiency than that formed with trimethylsilylmethyl methacrylate, which does not have functional groups to be cross-linked by hydrolysis

High diffraction efficiency of 72% was obtained in gratings formed with trimethoxysilylpropyl acrylate and **E7** (35wt%) with 0.2 wt% 3, 3′-carbonylbis(7-diethylaminocoumarin), and 1 wt% diphenyliodonium hexafluorophosphate. In SEM morphology, very regular and well-defined gratings were observed for the gratings formed with trimethoxysilylpropyl acrylate. Although gratings formed with high concentration of trimethoxysilylpropyl acrylate had some cracks in polymer matrix, the largest grating spacing was observed indicating the lowest volume shrinkage.

We developed a very useful holographic recording materials based on polypropylene glycol (PPG) derivatives functionalized with triethoxysilyl, hydroxyl, and methacrylate groups by taking into account the reaction rates and extent of phase separation of E7 of nematic LC in transmission holographic polymer dispersed liquid crystal (HPDLC) systems.

Holographic gratings were clearly formed from the radical polymerization and hydrolysis-condensation reaction of recording solution with PPG derivatives, which was demonstrated by AFM topology with very regular and well-defined morphology having the grating spacing of approximately 839.8 nm.

The highest final diffraction efficiency of 93% was observed in holographic gratings formed from PPG-DTEOS with difunctional triethoxysilyl groups at 240 second of irradiation of light. In the case of PPG-HTEOS together with hydroxyl and triethoxysilyl groups, maximum diffraction efficiency was approximately 78%. When the PPG-MTEOS together with methacrylate and triethoxysilyl groups was used, maximum diffraction efficiency reached at 96% and gradually decreased to 60%. These phenomena may be considered that the functional groups of PPG derivatives affected strongly on the diffraction efficiency attributed to the difference of reaction kinetics and extend of LC phase separation.

Consequently, if I consider the structure and reactivity of siloxane compounds in relation with the property, it will be possible to propose new systems to improve the performance of HPDLC gratings. I believe that these novel recording materials controlled in nanometer scale can be significantly contributed to the development and progress in the optics, electronics, photo-induced patterning, microsystem, and nanotechnololgy. Especially, electrically switchable holographic gratings are very promising for actual application such as 3-D image storage, full color LC display, dense wavelength division multiplexing (DWDM), and polarization-selective element. Moreover, this research will contribute to the establishment of new-generation display or device technology and will encourage the activity of industries.

5. References

[1] Coufal, H. (1998). *Nature,* 393, 628.

[2] Heanue, J.F.; Bashaw, M.C. & Hesselink, L. (1994). *Science,* 265, 749.

[3] Shen, X.A.; Nguyen, A.; Perry, J.W.; Huestis, D.L. & Kachru, R. (1997). *Science,* 278, 96.

[4] Chilling, M.L.; Colvin, V.L.; Dhar, L.; Harris, A.L.; Schilling, F.C.; Katz, H.E.; Wysocki, T.; Hale, A.; Blyler, L.L. & Boyd, C. (1999). *Chem. Mater.* 11, 247.

[5] Park, M.S.; Kim, B.K. & Kim, J.C. (2003). *Polymer* 44(5), 1595.

[6] Meng, S.; Nanjundiah, K.; Kyu, T.; Natarajan, L.V.; Tondiglia, V.P. & Bunning, T.J. (2004). *Macromolecules,* 37, 3792.

[7] Meng, S.; Kyu, T.; Natarajan, L.V.; Tondiglia, V.P.; Sutherland, R. L. & Bunning, T.J. (2005). *Macromolecules,* 38, 4844.

[8] Kyu, T. & Nwabunwa, D. (2001). *Macromolecules,* 34, 9168.

[9] Pikas, D. J.; Kirkpatrick, S.M.; Tomlin, D.W.; Natarajan, L.; Tondiglia, V.P. & Bunning, T.J. (2002). *Appl. Phys. A,* 74, 767.

[10] Sutherland, R. L. (2002). *J. Opt. Soc. Am. B,* 19, 2995.

[11] Kato, K.; Hisaki, T. & Date, M. (1999). *Jpn. J. Appl. Phys.,* 38, 1466.

[12] White, T. J.; Natarajan, L. V.; Tondiglia, V. P.; Bunning, T. J. & Guymon, C. A. (2007). *Macromolecules,* 40, 1112.

[13] White, T. J.; Natarajan, L. V.; Tondiglia, V. P.; Lloyd, P. F.; Bunning, T. J. & Guymon, C. A. (2007). *Macromolecules,* 40, 1121.

[14] Kato, K.; Hisaki, T. & Date, M. (1999). *Jpn. J. Appl. Phys.,* 38, 805.

[15] Natarajan, L. V.; Brown, D. P.; Wofford, J. M.; Tondiglia, V. P.; Sutherland, R. L.; Lloyd, P. F. & Bunning, T. J. (2006). *Polymer,* 47, 4411.

[16] Jazbinsek, M.; Olenik, I.D.; Zgonik, M.; Fontecchio, A.K. & Crawford, G.P. (2001). *J. Appl. Phys.* 90(8), 3831.

[17] Zhang, J.; Carlen, C.R.; Palmer, S. & Sponsler, M.B. (1994). *J. Am. Chem. Soc.,* 116, 7055.

[18] Lucchetta, D.Ee.; Karapinar, R.; Manni, A. & Simoni, F. (2002). *J. Appl. Phys.,* 91, 6060.

[19] Choi, D. H.; Cho, M. J.; Yoon, H.; Kim, J. H. & Paek, S. H. (2004). *Opt. Mater.,* 27, 85.

[20] Jazbinsek, M.; Olenik, I.D.; Zgonik, M.; Fontecchio, A.K. & Crawford, G.P. (2001). *J. Appl. Phys.*, 90 (8), 3831.
[21] Zhang, J.; Carlen, C. R.; Palmer, S. & Sponsler, M. B. (1994). *J. Am. Chem. Soc.*, 116, 7055.
[22] Lucchetta, D. E.; Karapinar, R.; Manni, A. & Simoni, F. (2002). *J. Appl. Phys.*, 91, 6060.
[23] Rosa, M. E. D.; Tondiglia, V. P. & Natarajan, L. V. (1998). *J. Appl. Phys. Sci.*, 68, 523.
[24] Carretero, L.; Blaya, S.; Mallavia, R.; Madrigal, R. F. & Fimia, A. (1998). *J. Modern Opt.*, 45, 2345.
[25] Zhang, J. & Sponsler, M. B. (1992). *J. Am. Chem. Soc.*, 114, 1506.
[26] White, T.J.; Natarajan, L.V.; Tondiglia, V.P.; Lloyd, P.F.; Bunning, T.J. & Guymon, C.A. (2007). *Macromolecules*, 40, 1121-1127.
[27] Woo, J.Y.; Kim, E.H. & Kim, B.K. (2007). *J. of Polym. Sci.* A, 45, 5590-5596.
[28] Lee, L.H.; & Chen, W.C. (2001). *Chem. Mater.* 13, 1137.
[29] Yoshida, M. & Prasad, P.N. (1996). *Chem. Mater.* 8, 235.
[30] Murciano, A.; Blaya, S.; Carretero, L.; Acebal, P.; Perez-Molina, M.; Madrigal, R.F. & Fimia, A. (2008). *J. of Appl.Phys*, 104, 063109.
[31] Ramos, G.; Álvarez-Herrero, A.; Belenguer, T.; del Monte, F. & Levy, D. (2004). *Appl. Opt.*, 43, 4018-4024.
[32] Cho, Y. H.; Shin, C. W.; Kim, N.; Kim, B. K. & Kawakami, Y. (2005). *Chemistry of Materials*, 17,6263.
[33] Ohe, Y.; Kume, M.; Demachi, Y.; Taguchi, T. & Ichimura, K. (1999). *Polym. Adv. Technol.* 10, 544-553.
[34] Liu, Y. J.; Zhang, B.; Jia, Y. & Xu, K. S. (2003). *Opt. Commun.*, 218, 27.
[35] Sarkar, M. D.; Qi, J. & Crawford, G. P. (2002). *Polymer*, 43, 7335.
[36] Escuti, M. J.; Kossyrev, P. & Crawford, G. P. (2000). *Appl. Phys. Lett.*, 77, 4262.
[37] Cairns, D. R.; Bowley, C.C.; Danworaphong, S.; Fontecchio, A. K.; Crawford, G. P.; Li, L. & Faris, S. M. (2000). *Appl. Phys. Lett.*, 77, 2677.
[38] Bunning, T. J.; Liechty, W. B.; Natarajan, L. V.; Tondiglia, V. P.; Bunning, T. J. & Guymon, C. A. (2006). *Polymer*, 47, 2289.
[39] Cho, Y. H. & Kawakami, Y. (2005). *Silicon Chem.*, 3, 219.
[40] Gomurashvili, Z. & Crivello, J. V. (2002). *Macromolecules*, 35, 2962.
[41] Gomurashvili, Z. & Crivello, J. V. (2001). *J. Polym. Sci.: Part A: Polym. Chem.*, 39, 1187.
[42] Crivello, J. V. & Jiang, F. (2002). *Chem. Mater.* 14, 4858.
[43] Crivello, J. V. & Lam, J. H. W. (1997). *Macromolecules*, 10, 1307.
[44] Crivello, J. V. & Lee, J. L. (1989). *J. Polym. Sci.: Part A: Polym. Chem.*, 27, 3951.
[45] Castellanos, F.; Fouassier, J. P.; Priou, C. & Cavezzan, J. (1996). *J. Appl. Polym. Sci.*, 60, 705.
[46] Waldman, D.A.; Ingwall, R.T.; Dhal P.K.; Horner, M.G.; Kolb, E.S.; Li, H.-Y. S.; Minns, R.A. & Schild, H.G. (1995). *SPIE*, 2689, 127
[47] Rhee, U.S.; Caulfield, H.J.; Shamir, J.; Vikram, C.S. & Mirsalehi, M.M. (1993). *Opt. Engin.* 32(8), 1839.
[48] Mok, F.H. (1993). *Opt. Lett.* 18 (11), 915.
[49] Curtis, K.; Pu, A. & Psaltis, D. (1994). *Opt. Lett.*, 19(13).

[50] Kogelnik, H. (1969). *Bell Syst. Tech. J.*, 48, 2909.
[51] Montemezzani, G. & Zgonik, M. (1997). *Physical Review E*, 55(1), 1035.
[52] Sarkar, M.D.; Gill, N.L.; Whitehead; J.B. & Crawford; G.P. (2003). *Macromolecules*, 36, 630.

26

Multicolor Stationary Light

Yi Chen[1], Serguei Andreevich Moiseev[2] and Byoung Seung Ham[1]
[1]*Center for Photon Information Processing, and the Graduate School of IT, Inha University*
[2]*Kazan Physical-Technical Institute of Russian Academy of Sciences*
[1]*South Korea*
[2]*Russia*

1. Introduction

Quantum coherence and interference (Scully & Zubairy, 1997) are leading edge topics in quantum optics and laser physics, and have led to many important novel effects, such as coherent population trapping (CPT) (Arimondo & Orriols, 1976; Alzetta et al., 1976; Gray et al., 1978), lasing without inversion (LWI) (Harris, 1989; Scully et al., 1989; Padmabandu et al., 1996), electromagnetically induced transparency (EIT) (Boller et al., 1991; Harris, 1997; Ham et al., 1997; Phillips et al., 2003; Fleischhauer et al., 2005; Marangos, 1998), high refractive index without absorption (Scully, 1991; Scully & Zhu 1992; Harris et al., 1990), giant Kerr nonlinear effect (Schmidt & Imamoglu, 1996), and so on. In particular, EIT, which can dramatically modify the absorption and dispersion characteristics of an optical medium, plays an important role in quantum optics. In the last two decades, EIT has attracted great attention and has been successfully applied to ultraslow and stopping light (Kasapi et al., 1995; Hau et al., 1999; Kocharovskaya et al., 2001; Liu et al., 2001; Turukhin et al., 2002), quantum switching (Ham & Hemmer, 2000), quantum memory (Ham et al., 1997; ibid., 1998), quantum entanglement generation (Lukin & Imamoglu, 2000), and quantum computing (Lukin & Imamoglu, 2001).
It is well established that light is the fastest information carrier in nature. However, controlling light for localized application is very difficult. Thus, manipulation of light velocity becomes a crucial task in optical and quantum information processing (Nielsen & Chuang, 2000). Recently light localization using EIT has been demonstrated for stationary light (Bajcsy et al., 2003). Stationary light gives novel effects to nonlinear quantum optics in the context of lengthening light-matter interaction time. Compared with ultraslow light, where the medium's length is a limiting factor, stationary light is free from spatial constraint. For example, the interaction time using ultraslow light in a semiconductor quantum dot, whose spatial dimension is less than a few tens of nanometers, is much less than nanosecond. By using a stationary light technique, however, we can enormously increase the interaction time of the light with such a nano optical medium. In this chapter, we discuss stationary light based on the EIT-induced ultraslow light phenomenon. We theoretically investigate how to dynamically manipulate multicolor (MC) stationary light in the multi double lambda-type system by simply changing the parameters of control fields, and demonstrate ultralong trapping of light, which is different from the conventional quantum mapping phenomenon. Quantum coherent control of the stationary light has potential applications to various quantum optical processing such as quantum nondemolition measurement and quantum wavelength conversion.

This chapter is organized as follows. In section 2, quantum coherent control of two-color stationary light is described. In section 3, quantum manipulation of MC stationary light is presented. In section 4, we give the results of MC stationary light, with discussions. Finally, section 5 offers conclusions.

2. Quantum coherent control of light

2.1 Model and theory

In this section, we present dynamic control of two-color stationary light in a double-lambda type four-level system using EIT. Fig. 1 shows the energy level diagram. We assume that initially one weak probe quantum field E_+ which couples the transition |1> – |3> enters the medium in a forward propagation direction k_+, where all the atoms are in the ground state |1>. Under the influence of classical control field Ω_+, which is resonant with the transition |2> – |3> and propagates in the same direction κ_+, the probe field propagates at an ultraslow group velocity. After the probe pulse enters the medium completely, we adiabatically switch on the second control field Ω_-, which is resonant with the transition |2> – |4> and propagates in the backward direction κ_-. The probe pulse will be almost stationary if the stationary light condition is satisfied. The propagation direction k_- of the field E_- is determined by phase matching with the Bragg condition: $k_- = k_+ - \kappa_+ + \kappa_-$. Here Ω_+ and Ω_- are the Rabi frequencies of two control fields. Below we will study the evolution of the stationary light when (1) Ω_- is switched off after a certain time while keeping the control filed Ω_+ on, and (2) Ω_+ is switched off after a certain time while keeping the control filed Ω_- on (see Fig. 2).

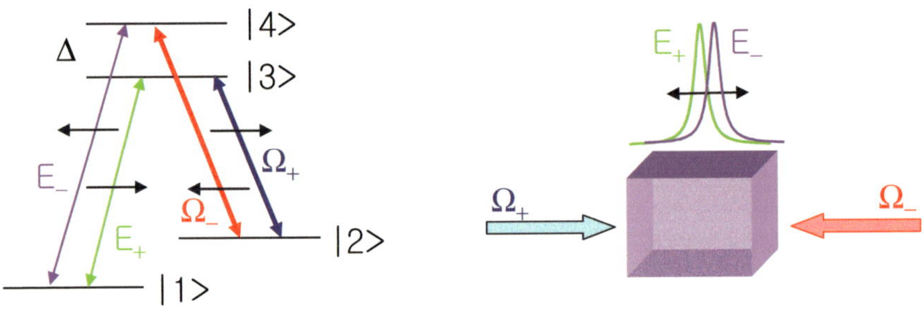

Fig. 1. Energy level diagram for two-color stationary light.

For theoretical analytical purposes, we introduce a quantum field $E_\sigma = \sqrt{\hbar\omega_\sigma / (2\varepsilon_0 V)} A_\sigma e^{-i\omega_\sigma (t-\sigma z/c)} + H.C.$, where $\sigma = + / -$ stands for forward/backward field, respectively, which is traveling in a ±z direction. A_σ is the slowly varying field operator, $\hbar$ is the Planck's constant, and ε_0 is the electric permittivity. We assume that the quantization volume V is 1. Under the rotating-wave approximation, we obtain the following Hamiltonian for the quantum fields and atoms in the interaction picture:

$$H = \hbar g_+ \sum_{j=1} A_+(t,z_j) P_{31}^j e^{i\omega_+ z_j / c} + \hbar g_- \sum_{j=1} A_-(t,z_j) P_{41}^j \, e^{-i\omega_- z_j / c}$$
$$-\hbar \sum_{j=1} \{\Omega_+ P_{32}^j \exp[i(K_+ z_j + \varphi_+)] + \Omega_- P_{42}^j \exp[-i(K_- z_j - \varphi_-)]\} + H.C., \quad (1)$$

where $P^j_{nm} = (P^j_{mn})^+$ are the atomic operators, $g_\sigma = \wp_\sigma \sqrt{\omega_\sigma / (2\varepsilon_o \hbar V)}$ is the coupling constant of photons with atoms, $\wp$ is the dipole moment for each transition, and $\varphi_{\sigma=\pm}$ are the phases of the control fields, $\omega_+ = \omega_{31}$ and $\omega_- = \omega_{41}$. Using Hamiltonian (1) and adding the decay constants γ_3, γ_4, and γ_2 for the atomic transitions |1> – |3>, |1> – |4> and |1> – |2>, we derive equations for the field A_σ and atomic operators P^j_{mn}.
Under the typical adiabatic condition for the slow light propagations (Fleischhauer & Lukin, 2000; Zibrov et al., 2002), we introduce new field operators $\Psi = e^{-i(\phi_\sigma + \sigma\omega_{21} z/c)} \sqrt{N} g_\sigma A_\sigma / \Omega_\sigma$ (N is atomic density). Due to a weak field $E_{\sigma=\pm}$ under the slow-light propagation $v_g \cong c\Omega_+^2 / N g_+^2 << c$, we ignore the atomic population on the excited levels |3> and |4> and assume slowly varying amplitudes of the laser fields. Then we can obtain the following coupled wave equations for the new field operators:

$$\begin{aligned}(\tfrac{\partial}{\partial z} - i\tfrac{\omega_{21}}{c})\Psi_+(\tau,z) = &- \xi_+\alpha_-[\Psi_+(\tau,z) - \Psi_-(\tau,z)] \\ &- \tfrac{\partial}{c\partial\tau}\{\alpha_+\Psi_+(\tau,z) + \alpha_-\Psi_-(\tau,z)\} - (\gamma_2' / c)\Psi_+(\tau,z)\end{aligned} \qquad (2)$$

$$\begin{aligned}(\tfrac{\partial}{\partial z} + i\tfrac{\omega_{21}}{c})\Psi_-(\tau,z) = &- \xi_-\alpha_+[\Psi_+(\tau,z) - \Psi_-(\tau,z)] \\ &+ (g_- / g_+)^2 \tfrac{\partial}{c\partial\tau}\{\alpha_+\Psi_+(\tau,z) + \alpha_-\Psi_-(\tau,z)\} + (g_- / g_+)^2(\gamma_2' / c)\Psi_-(\tau,z)\end{aligned} \qquad (3)$$

where $\xi_\sigma = N g_\sigma^2 / c\gamma$ are the absorption coefficients, $\alpha_\sigma = \Omega_\sigma^2 / [\gamma\gamma_2 + \Omega_\Sigma^2]$, $\Omega_\Sigma^2 = \Omega_+^2 + \Omega_-^2$, $\tau = \int_{-\infty}^{t} dt' [\gamma\gamma_2 + \Omega_\Sigma^2] / (g_+^2 N)$ is a new time scale, and $\gamma_2' = N g_+^2 [\gamma\gamma_2 + \Omega_\Sigma^2]^{-1} \gamma_2$.

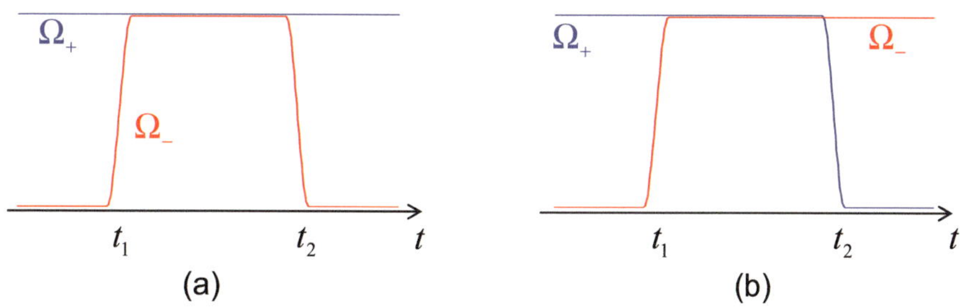

Fig. 2. Temporal scheme of the switching sequence of the two control fields. (a) Ω_- is turned off at $t=t_2$, (b) Ω_+ is turned off at $t=t_2$.

Using spatial Fourier transformation $\Psi_\sigma(\tau,z) = \int_{-\infty}^{\infty} dk e^{ikz} \tilde{\Psi}_\sigma(\tau,k)$, we can find the following analytical solutions of Eqs. (2) and (3):

$$\Psi_\sigma(\tau,z) = \int_{-\infty}^{\infty} dk \exp\{+ikz + i\int_{\tau_o}^{\tau} d\tau' \omega(\tau',k)\} \tilde{\Psi}_\sigma(\tau_o,k) \qquad (4)$$

$$\omega(\tau',k) = i\eta\gamma_2' - \frac{ck\{\eta(\xi_-\alpha_+ - \xi_+\alpha_-) - ik\} - i(\partial/\partial\tau')(\eta^{-1}\xi_- - ik\tilde{\alpha})}{\{\eta^{-1}\xi_- - ik\tilde{\alpha}\}} \qquad (5)$$

where $\tilde{\Psi}_-(\tau_o,k)=\tilde{f}_{-+}(k)\tilde{\Psi}_+(\tau_o,k)$, $\tilde{f}_{-+}(k)=(1+ik/\xi_+)/(1-ik/\xi_-)$, $\eta=(\Omega_\Sigma^2+\gamma\gamma_2)\Omega_\Sigma^{-2}$ and $\tilde{\alpha}=\alpha_+-(g_-/g_+)^2\alpha_-$. The functions $\tilde{\Psi}_{+,-}(\tau_o,z)$ are determined by the initial conditions for the field $A_{+,-}$ which enters the medium ($\Omega_-(t<t_1)=0$). From Eq. (4) we obtain the coupled fields $\Psi_+(t,z')$ and $\Psi_-(t,z)$ expressed by nonlocal spatial relations:

$$\begin{aligned}\Psi_+(t,z)&=\int_{-\infty}^{z+\varepsilon}dz'f_+(z-z')\Psi_-(t,z')\\ \Psi_-(t,z)&=\int_{z-\varepsilon}^{\infty}dz'f_-(z-z')\Psi_+(t,z')\end{aligned}\tag{6}$$

where $f_\sigma(z-z')=\xi_\sigma(1+\frac{\xi_\sigma}{\xi_{\sigma'}})\eta(z'-z)e^{-\xi_\sigma(z'-z)}-\frac{\xi_\sigma}{\xi_{\sigma'}}\delta(z-z')$ ($\sigma=\pm1,\sigma'=\mp1$, $\eta(x\geq0)=1$, $\eta(x<0)=0$ and ε is a small value $\varepsilon\to0$). Eq. (6) points out that spatial quantum correlations between the fields $\Psi_+(t,z)$ and $\Psi_-(t,z')$ spread out within the spatial size ξ_σ^{-1}. This means that the fields $\Psi_+(t,z')$ and $\Psi_-(t,z)$ can be quantum copies of each other if the spatial size l of $\Psi_\pm(t,z')$ is larger than ξ_σ^{-1}. Therefore the spatial correlations of quantum fields E_+ and E_- can be determined by Eq. (6).

Now we study temporal dynamics of the two-color coupled fields. Initially the control field Ω_+ ($\Omega_-=0$) is turn on and a probe pulse A_+ with a Gaussian shape ($A_+(t,z=0)=A_{+,o}\exp\{-t^2/(2T^2)\}$, $A_{+,o}$ and T are the amplitude and temporal duration of the field) enters the medium. The solution of Eq. (4) describes the propagation of the field Ψ_+, that is, $A_+=(g_+^2N)^{-1/2}\Omega_+\Psi_+e^{i(\phi_++\omega_{21}z/c)}$ with an ultraslow group velocity $v_g\equiv c\Omega_+^2(0)/Ng_+^2<<c$, and the initial spatial size $l_o=v_gT$ with an amplitude decay in accordance with $\exp\{-\eta\gamma_2t\}$. Thus, due to the EIT effect, the field A_+ is transparent to the optically dense medium ($\xi_{+,-}l_o>>1$). At time $t=t_1$, we turn on the backward control field Ω_-. Using the dispersion relation $\omega(k)$ in Eq. (5), we obtain the group velocity for the interacting fields A_+ and A_-:

$$v=-[(\Omega_\Sigma^2+\gamma\gamma_2)/Ng_+^2]\partial\omega/\partial k|_{k=0}=c\eta^2[g_-^2\Omega_+^2-g_+^2\Omega_-^2]/[N(g_+g_-)^2]\tag{7}$$

From Eq. (7), the stationary light condition ($v=0$) can be derived:

$$\Omega_+/g_+=\Omega_-/g_-\tag{8}$$

Thus the group velocity of the coupled light can be easily controlled by manipulating the control field Rabi frequencies. If the stationary light condition is satisfied, the two coupled lights can be completely stopped. Ignoring small terms proportional to ck^3/ξ_-^2 in Eq. (5), we find the approximate solution of the integral (Eq. (4)) for arbitrarily varying amplitudes of the control laser fields:

$$\begin{aligned}A_\sigma(\tau,z)&=\frac{l_o}{cB(\tau)}\frac{\eta(\tau)(\Omega_\sigma(\tau)g_+)}{\eta(0)(\Omega_+(0)g_\sigma)}A_{+,o}\\ &\times\exp\{-\int_0^\tau(\eta\gamma_2')d\tau'-(\beta_\sigma(\tau)-z/c)^2/(2B^2(\tau))\}\exp\{i(\phi_\sigma+\sigma\omega_{21}z/c\}\end{aligned}\tag{9}$$

where

$$\beta_+(\tau)=\int_0^{\tau} d\tau' M_1(\tau'),\quad \beta_-=\beta_+-z_o/c,\quad M_1=\xi_-^{-1}\{\eta^2(\xi_-\alpha_+-\xi_+\alpha_-)-\tfrac{\partial}{cd\tau'}(\eta\tilde{\alpha})\} \tag{10}$$

$$B(\tau)=\sqrt{(l_o/c)^2+(2/c)\int_0^{\tau} d\tau' M_2(\tau')},\quad M_2=\xi_-^{-2}\eta\{\xi_- -\eta\tilde{\alpha}(\xi_-\alpha_+-\xi_+\alpha_-)+\eta\tfrac{\partial}{2cd\tau'}\tilde{\alpha}^2)\} \tag{11}$$

and $z_o=(\xi_-\xi_+)^{-1}(\xi_+ +\xi_-)$ is a spatial shift of A_- with respect to the envelope A_+ ($z_o << l_o$).

2.2 Numerical simulations and discussions

Here we numerically demonstrate quantum manipulation of a traveling light pulse for the two-color stationary light by solving Eqs. (2) and (3). For simplicity we ignore the weak decay rate γ_2 between the two ground states |1> and |2>, and assume small level splitting ω_{21} to ignore the phase mismatch between the fields Ψ_+ and Ψ_-. When the backward control field is turned off ($\Omega_-=0$), Eqs. (2) and (3) satisfy slow-light wave equations (Hau et al., 1999; Turukhin et al., 2002). We note that Eqs. (2) and (3) coincide with the standing single-frequency light based on a standing wave grating in a three-level system if $g_+=g_-$ with Doppler broadening (Bajcsy et al., 2003). It should be noted that Eqs. (2) and (3) show general coupled equations of standing lights whether the optical dense medium is Doppler broadened or not. At $t=t_1$, both control fields Ω_+ and Ω_- are turned on. The coupled quantum fields A_σ are therefore generated and they propagate and evolve together in the medium. If the stationary light condition $g_+/g_-=\Omega_+/\Omega_-=1$ is satisfied, the coupled fields A_σ should be standing still with nearly the same amplitude until one control field is turned off. This condition corresponds to the appearance of the so-called dark state in the double-Λ system (Korsunsky & Kosachiov, 1999), which can also dramatically change the interaction of the copropagating light fields.

Compared with the standing-wave grating case (Bajcsy et al., 2003; Andre et al., 2005), the group velocity $v=cM_1(\Omega_\Sigma^2/Ng_+^2)$ in Eq. (11) can be changed by adjusting the two control fields. However, the lowest velocity is realized only if $\left|\frac{\partial}{cd\tau}\eta\tilde{\alpha}\right|\to 0$ and $(\xi_-\alpha_+-\xi_+\alpha_-)\to 0$ occur with complete light stoppage ($v=0$). The fields A+ and A- are strongly coupled and move or stay together with the following amplitude ratio: $(g_-/\Omega_-)A_-(t,z)\cong(g_+/\Omega_+)A_+(t,z)$. This means the field A- can be generated by the frequency shift $\omega_-=\omega_{41}+\delta\omega_+$ if the frequency of the field A+ is tuned to $\omega_+=\omega_{31}+\delta\omega_+$. The standing field $\hat{E}(t,z)\sim g_+^{-1}\Omega_+\Psi_+e^{-i\omega_+(t-z/c)}+g_-^{-1}\Omega_-\Psi_-e^{-i\omega_-(t+z/c)}$ is bound to the coherence moving grating (Moiseev & Ham, 2006; Ham, 2006; Moiseev et al., 2006; Moiseev & Ham, 2007; Ham & Hahn, 2009) which is completely different from the standing-wave grating (Bajcsy et al., 2003; Andre et al., 2005).

At time $t=t_2$, we turn off one control laser Ω_+ or Ω_-. Fig. 2 gives the switching sequence of two control fields for these two cases: (a) the control field Ω_+ is always on, while the Ω_- is on only for $t_1<t<t_2$; (b) the control field Ω_+ is on for $0<t<t_2$, while the control field Ω_- is on for $t_1<t$.

Fig. 3 shows numerical simulation of the two-color stationary light for case (a) mentioned above (also see Fig. 2(1)). For simplification, we set $\gamma_2=0$ and assume the same Rabi

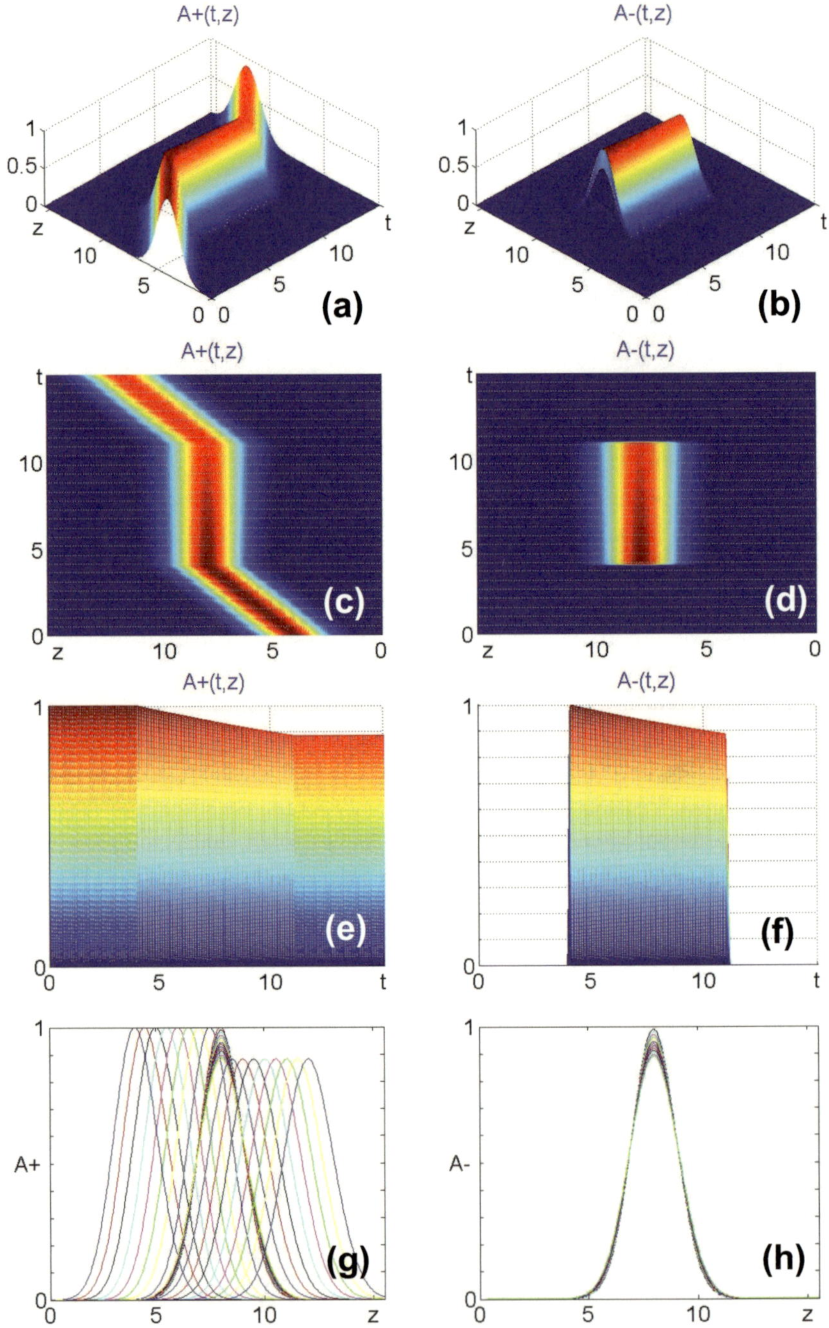

Fig. 3. Numerical simulation of two-color stationary light. The control field Ω_+ is always on, while the Ω_- is on only for $t_1(4) < t < t_2(11)$.

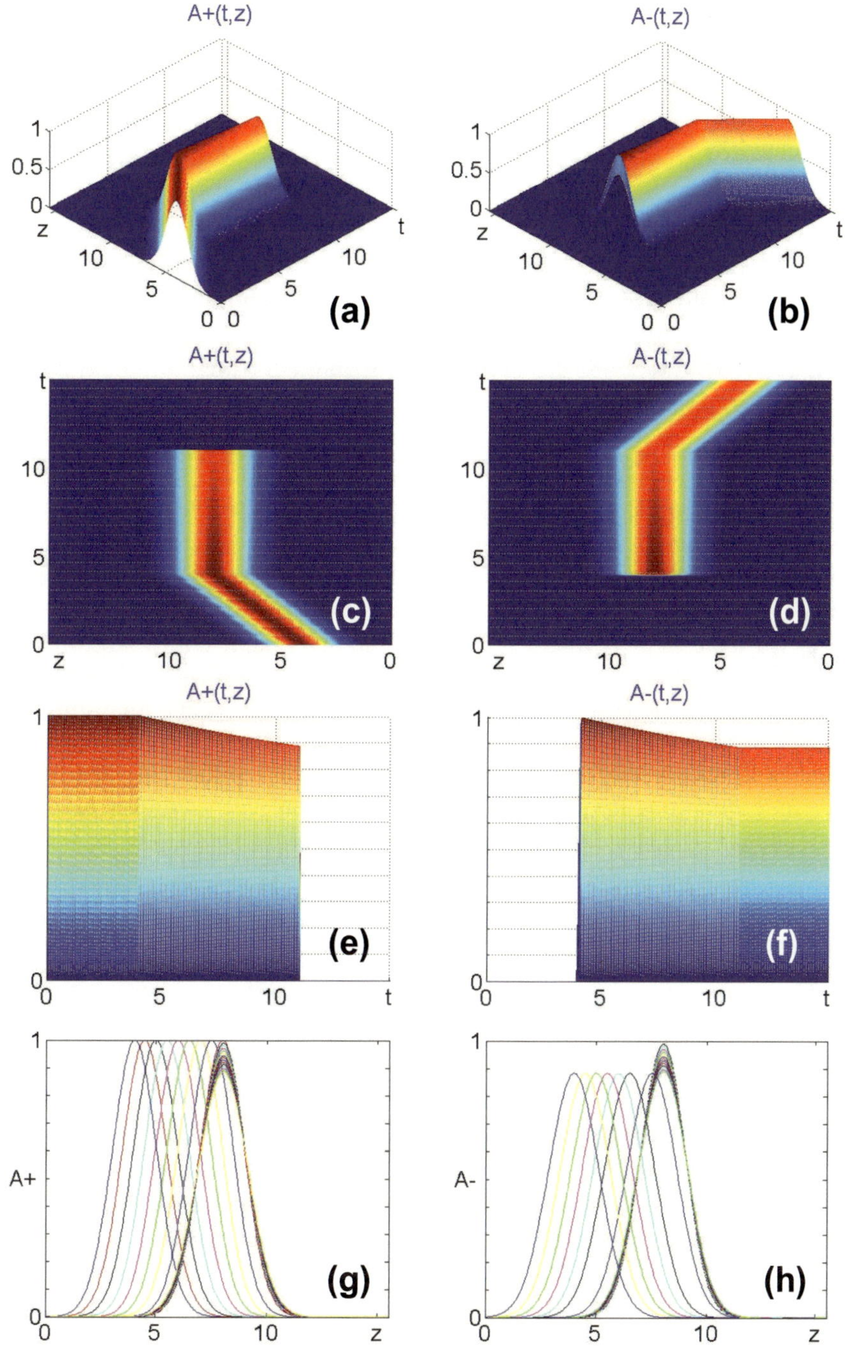

Fig. 4. Numerical simulation of two-color stationary light for wavelength conversion. The control field Ω_+ is on for $0 < t < t_2(11)$, while the Ω_- is on for $t_1(4) < t < 15$.

frequencies for the control fields. Fig. 3(c) and Fig. 3(d) show the top view of the fields A_+ and A_- propagation, respectively. Fig. 3(e) and Fig. 3(f) show the temporal evolutions of the field amplitudes A_+ and A_-, respectively. Fig. 3(g) and Fig. 3(h) show space-time evolution of the fields A_+ and A_-, respectively. These figures show that when the control pulse Ω_- is turned off at $t = t_2$ ($t_2 = 11$), the standing field A_- disappears completely, and the original field A_+ keeps propagation moving in the same direction at the same group velocity as it had for $0 < t < t_1$ ($t_1 = 4$).
Fig. 4 shows numerical simulation of the two-color stationary light for case (b) (see Fig. 2(2)). Similar to Fig. 3, Fig. 4(c) and Fig. 4(d) show the top view of the coupled fields. Fig. 4(e) and Fig. 4(f) show the temporal evolutions of the coupled field amplitudes. Fig. 4(g) and Fig. 4(h) show space-time evolution of the coupled fields. As seen in Figs. 4(c) and 4(d), when Ω_+ is turned off at $t = t_2$ ($t_2 = 11$) while the Ω_- is kept turned on, the quantum field A_+ disappears at $t = t_2$, but the A_- propagates in the backward direction with new carrier frequency ω_-. Figs. 3 and 4 show that by controlling the coupling fields, the coupled fields E_+ and E_- can be manipulated for stationary light or frequency conversion, which has potential application for quantum nonlinear optics which needs a longer interaction time. Here we note that the maximum trapping time of stationary light is determined by the coherence decay rate γ_2 between the two ground states |1> and |2>. Thus for practical application, we may choose an optically dense medium with a long spin decay time.
In summary, we have demonstrated two-color stationary light and quantum wavelength conversion using a quantum-mechanically reversible process between photons and atomic coherence in a double-Λ four-level system. In this scheme, a traveling quantum field can be manipulated by simply adjusting the control fields' parameters for (1) determination of two-color stationary light, (2) selection of propagation direction, either forward or backward, and (3) dynamic quantum frequency conversion. The present quantum manipulation of two-color stationary light can be used for enhanced nonlinear interactions between single photon fields.

3. Quantum manipulation of MC stationary light

If we choose a multilevel system, MC stationary light should be possible. In this section, we show quantum coherent control of multiple travelling light pulses in an optically dense medium by generalizing the approach we used in the previous section. Fig. 5 shows an energy level diagram of the present MC stationary light. The control fields with Rabi frequencies Ω_m copropagate forward along with the weak quantum fields $\hat{E}_m (m \in [1,...,M_+])$, whereas the other quantum fields $\hat{E}_n (n \in [1,...,M_-])$ propagate backward along with their control fields with Rabi frequencies Ω_n. The propagation directions of fields for the effective generation and strong interaction between the fields should satisfy the phase matching condition. As shown in Fig. 5, the quantum fields $\hat{E}_m$ and $\hat{E}_n$ are nearly resonant with optical transitions $|1\rangle - |m\rangle$ and $|1\rangle - |n\rangle$, respectively. The control fields Ω_m and Ω_n have frequencies close to the frequencies of transitions $|2\rangle - |m\rangle$ and $|2\rangle - |n\rangle$, respectively (where $\Omega_m(t) = \Omega_{m,o}(t)e^{i\varphi_m}$, $\Omega_n(t) = \Omega_{n,o}(t)e^{i\varphi_n}$, and $\phi_{m(n)}$ are the constant phases of the fields).
We assume that initially all the atoms stay in the ground state $|1\rangle$ and only one weak probe quantum field $\hat{E}_l$ enters into the medium with a forward propagation direction k_l and the

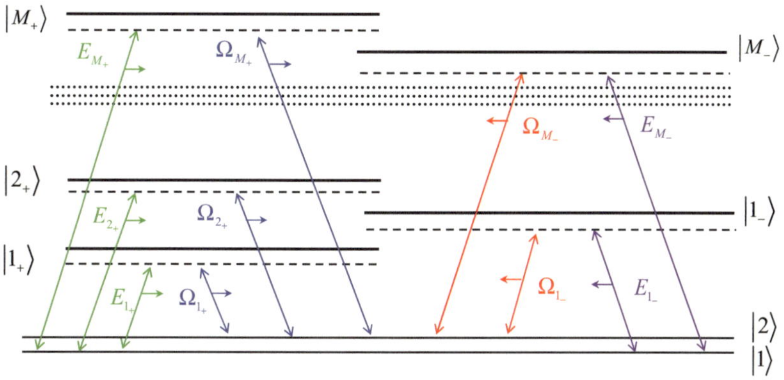

Fig. 5. Energy level diagram for MC stationary light.

atoms are driven by one classical control field $\Omega_l \neq 0$ with same propagation direction K_l. The initial state of light and atoms in the interaction picture is given by $|\varphi(t<<0)\rangle_{in} = |\varphi_l\rangle|1\rangle_{atoms}$ ($|1\rangle_{atoms} = \prod_j |1_j\rangle$ and $|\varphi_l\rangle$ are the ground state of the atoms and initial state of the probe field, respectively). The quantum field $\hat{E}_l$ propagates at a slow group velocity due to the control field Ω_l, and atomic coherent wave $\langle \hat{P}_{12} \rangle$ is created between two ground states $|1\rangle$ and $|2\rangle$. After the probe pulse completely enters the medium, we switch on the additional control fields ($\Omega_m(t>t_1) \neq 0$ and $\Omega_n(t>t_1) \neq 0$). Due to the ultraslow propagation of the probe field, such switching processes can be easily realized. The new atomic polarization waves with wave vectors $k_{m,n} = k_l - K_l + K_{m,n}$ (phase matching condition) which generate the new light field $\hat{E}_{m,n}$ are created. Therefore all the quantum fields become coupled with each other via atomic coherence $\langle \hat{P}_{12} \rangle$ which has the quantum information of the initial probe field $\hat{E}_l$. Below we will study the evolution of MC stationary light in the coherent resonance medium driven by the M_++M_- control fields.

For a theoretical analysis of MC stationary light based on the multi double lambda-type scheme in Fig. 5, we introduce quantum fields $\hat{E}_{m,n}(t,z) = \sqrt{\hbar\omega_{m,n}/(2\varepsilon_0 S)}\hat{A}_{m,n}(t,z)e^{-i(\omega_{m,n}t \mp k_{m,n}z)}$, where down indexes m and n are related to the signs $\mp$ or $\pm$ in the equations. $\hat{A}_{m(n)}$ are slowly varying field operators, where $[\hat{A}_p(t,z), \hat{A}_q(t,z')] = \delta_{p,q}\delta(z-z')$, $\hbar$ is the Planck's constant, ε_0 is the electric permittivity, and S is a cross section of light beams (Loudon 2000). The interaction of the quantum fields with atoms driven by the control laser fields $\Omega_m(t)$ and $\Omega_n(t)$ is given by the following Hamiltonian in the interaction picture:

$$\hat{H}(t) = \hat{V}_f(t) + \hat{V}_b(t) + H.C.,$$
$$\hat{V}_f(t) = -\hbar\sum_j \{\sum_{m=1}^{M_+} g_m \hat{A}_m(t,z_j)e^{-i(\Delta_m t - k_m z_j)}\hat{P}_{m1}^j + \sum_{m=1}^{M_+} \Omega_m(t)e^{-i(\Delta_m t-(k_m-k_0)z_j)}\hat{P}_{m2}^j\} \quad (12)$$
$$\hat{V}_b(t) = -\hbar\sum_j \{\sum_{n=1}^{M_-} g_n \hat{A}_n(t,z_j)e^{-i(\Delta_n t - k_n z_j)}\hat{P}_{n1}^j + \sum_{n=1}^{M_-} \Omega_n(t)e^{-i(\Delta_n t-(k_n-k_0)z_j)}\hat{P}_{n2}^j\}$$

where $\widehat{P}^j_{ll'} = |l\rangle_{jj}\langle l'|$ is an operator of the j-th atom, $g_l = \mu_{l,1}\sqrt{\omega_{l,1}/(2\varepsilon_0\hbar S)}$ is a coupling constant of photons with atoms, $\mu_{l',l}$ is a dipole moment for the transition between the states $|l'\rangle \to |l\rangle$, $v_m = \omega_{m1} + \Delta_m$, $v_n = \omega_{n1} + \Delta_n$, and $k_0 = \omega_{21}/c$.

Using Eq. (12) and adding the relaxation constants and Langevin forces associated with the atomic relaxation processes, we derive the Heisenberg equations for the atomic operators $\widehat{S}^j_{1m} = \widehat{P}^j_{1m}\exp\{i(\Delta_m t - k_m z_j)\}$, $\widehat{S}^j_{1n} = \widehat{P}^j_{1n}\exp\{i(\Delta_n t + k_n z_j)\}$ and $\widehat{P}^j_{12}$:

$$\frac{\partial}{\partial t}\widehat{S}^j_{1m} = -\tilde{\gamma}_{1m}\widehat{S}^j_{1m} + ig_m\widehat{A}_m(t,z_j)(\widehat{P}^j_{11} - \widehat{P}^j_{mm}) + i\Omega_m e^{-ik_0 z_j}\widehat{P}^j_{12} + \widehat{F}^j_{1m;(1)}(t) \tag{13}$$

$$\frac{\partial}{\partial t}\widehat{S}^j_{1n} = -\tilde{\gamma}_{1n}\widehat{S}^j_{1n} + ig_n\widehat{A}_n(t,z_j)(\widehat{P}^j_{11} - \widehat{P}^j_{nn}) + i\Omega_n e^{ik_0 z_j}\widehat{P}^j_{12} + \widehat{F}^j_{1n;(1)}(t) \tag{14}$$

$$\begin{aligned}\frac{\partial}{\partial t}\widehat{P}^j_{12} = &-\gamma_{12}\widehat{P}^j_{12} + i\sum_{m=1}^{M_+}\Omega^*_m(t)e^{ik_0 z_j}\widehat{S}^j_{1m} + i\sum_{n=1}^{M_-}\Omega^*_n(t)e^{-ik_0 z_j}\widehat{S}^j_{1n} \\ &- ig_m\widehat{A}_m(t,z_j)\widehat{S}^j_{m2} - ig_n\widehat{A}_n(t,z_j)\widehat{S}^j_{n2} + \widehat{F}^j_{12;(0)}(t)\end{aligned} \tag{15}$$

where $\tilde{\gamma}_{m,n} = \gamma_{m,n} - i\Delta_{m,n}$, $\widehat{F}^j_{1m;(1)}(t) = \exp\{i(\Delta_m t - k_m z_j)\}\widehat{F}^j_{1m;(0)}(t)$, $\widehat{F}^j_{1n;(1)}(t) = \exp\{i(\Delta_n t + k_n z_j)\}\widehat{F}^j_{1n;(0)}(t)$ and the fluctuations satisfy the usual properties

$$\left\langle \widehat{F}^j_{pq;(0)}(t)\right\rangle = 0 \tag{16}$$

$$\left\langle \widehat{F}^j_{1m(n);(0)}(t)\widehat{F}^j_{m(n)2;(0)}(t')\right\rangle = (\gamma_{1,m(n)} + \gamma_{m(n),2} - \gamma_{12})\left\langle \widehat{P}^j_{12}(t)\right\rangle\delta(t-t') \tag{17}$$

$$\left\langle \widehat{F}^j_{1m(n);(0)}(t)\widehat{F}^j_{m(n)1;(0)}(t')\right\rangle = 2\gamma_{1m(n)}\left\langle \widehat{P}^j_{11}(t)\right\rangle\delta(t-t') \cong 2\gamma_{1m(n)}\delta(t-t') \tag{18}$$

In Eqs. (13)-(15), we ignore the influences of the weak populations of excited states $|m\rangle$ and $|n\rangle$. Thus we have the following equations for the weak field operators $\widehat{A}_{m,n}$:

$$\left(\frac{\partial}{c\partial t} + \frac{\partial}{\partial z}\right)\widehat{A}_m(t,z) = i(\sqrt{n_0 S}g^*_m/c)\widehat{S}_{1m}(t,z) \tag{19}$$

$$\left(\frac{\partial}{c\partial t} - \frac{\partial}{\partial z}\right)\widehat{A}_n(t,z) = i(\sqrt{n_0 S}g^*_n/c)\widehat{S}_{1n}(t,z) \tag{20}$$

where $\widehat{S}_{1m(n)}(t,z) = (n_0 S)^{-1/2}\sum_j \widehat{S}^j_{1m(n)}(t)\delta(z - z_j)$.

We assume a typical adiabatic condition for slow light propagation $\tilde{\gamma}_{m,n}\delta t_0 >> 1$ (δt_0 is large enough temporal duration of the probe pulse) and ignore the fluctuation forces $\widehat{F}^j_{m2;(2)}$ and $\widehat{F}^j_{n2;(2)}$ due to the weak quantum fields. Using Eqs. (13)-(20) and introducing new field operators $\widehat{\Psi}_{m,n}(t,z) = e^{\pm ik_0 z}\sqrt{n_0 S}g_{m,n}\widehat{A}_{m,n}(t,z)/\Omega_{m,n}(t)$, we finally obtain the following equations for $\widehat{\Psi}_{m,n}$ under the slow light propagation condition $v_{m,n}(t) = c\Omega^2_{m,n}(t)/Ng^2_{m,n} << c$:

$$\left(\frac{\partial}{\partial z} - ik_0\right)\hat{\Psi}_m = -\xi_m\{\hat{\Psi}_m - (1 - \frac{\partial}{\mu\partial t})[G(\hat{\Psi})/\mu]\} - (\xi_m/\mu)\hat{F}_{1m}^{(3)} \quad (21)$$

$$\left(\frac{\partial}{\partial z} + ik_0\right)\hat{\Psi}_n = +\xi_n\{\hat{\Psi}_n - (1 - \frac{\partial}{\mu\partial t})[G(\hat{\Psi})/\mu]\} + (\xi_n/\mu)\hat{F}_{1n}^{(3)} \quad (22)$$

where the fluctuation forces are:

$$\hat{F}_{1n,1m}^{(3)}(t,z) = (n_0 S)^{-1/2}\sum_j [\hat{F}_{12,(\Sigma 2)}^j(t) - i\exp(\mp ik_0 z_j)\mu(t)\Omega_{n,m}^{-1}(t)\hat{F}_{1n,1m;(2)}^j(t)]\delta(z - z_j) \quad (23)$$

Eqs. (21) and (22) can be solved by using spatial Fourier transformation. Assuming that initial field $\hat{\Psi}_l(t,z)$ is determined by the probe pulse field $A_l(t,z)$, after some algebraic calculations we finally obtain the following solution:

$$\hat{\Psi}_{m,(n)}(t,z) = \hat{\Psi}_{\det;m(n)}(t,z) + \delta\hat{\Psi}_{st;m(n)}(t,z) \quad (24)$$

where

$$\hat{\Psi}_{\det;m(n)}(t,z) = \int_{-\infty}^{\infty} dk \exp(ikz) T(t - t_0, k, k_0)\hat{\psi}_l(t_0,k) = \int_0^L dz' f_{m,(n);l}(z - z')\hat{\Psi}_{\det;l}(t,z')$$

$$\delta\hat{\Psi}_{st;m,(n)}(t,z) = \int_0^L dz' f_{m,(n);l}(z - z')\delta\hat{\Psi}_{st;l}(t,z') + \delta\hat{\Psi}_{st;m(n);l}(t,z)$$

$$f_{m(n),l}(z - z') = (1/2\pi)\int dk \exp[ik(z - z')][1 + i\eta_{m(n),l}(k;k_0)]$$

$$\eta_{n(m);l}(k;k_0) = [\xi_l^{-1}(k - k_0) + \xi_{n,m}^{-1}(k_0 \pm k)]/[1 - i\xi_{n,m}^{-1}(k_0 \pm k)]$$

4. Results and discussion

In this section, we analyze quantum evolution of the MC field. Under the slow light condition and taking into account weak relaxation processes between the two ground levels $|1\rangle$ and $|2\rangle$, we can get the key information of quantum dynamics control from the dispersion relation:

$$\omega(t,k,k_0) \cong v(t)k - i\delta\omega_{kk}^{''}(t)k^2/2 + ..., \quad (25)$$

$$v(t) = \omega_k^{'} = c[\sum_m^{M_+}\Omega_{m;0}^2/Ng_m^2 - \sum_n^{M_-}\Omega_{n;0}^2/Ng_n^2] = \sum_m^{M_+} v_m - \sum_n^{M_-} v_n \quad (26)$$

$$\delta\omega_{kk}^{''} = i\omega_{kk}^{''} = 2[\sum_{m,m'}^{M_+}\Gamma_m\Gamma_{m'}(\xi_m^{-1} - \xi_{m'}^{-1})^2 + \sum_{n,n'}^{M_-}\Gamma_n\Gamma_{n'}(\xi_n^{-1} - \xi_{n'}^{-1})^2 + \sum_{m,n}^{M_+,M_-}\Gamma_m\Gamma_n(\xi_m^{-1} + \xi_n^{-1})^2]/\mu \quad (27)$$

where $v(t)$ is the united group velocity of all the laser fields, and $-i\delta\omega_{kk}^{''}$ is the second order dispersion term which determines the spatial broadening of the MC light pulse.

4.1 Traveling MC field

If $M_- = 0$ and $M_+ > 1$ we have the traveling MC light field with the group velocity:

$$v(t) = v_{travel}(t) = \sum_m^{M_+} v_m(t) \tag{28}$$

where the spatial dispersion will be

$$\delta\omega''_{kk;travel} = 2\{\sum_{m,m'}^{M_+} \Gamma_m \Gamma_{m'} (\xi_m^{-1} - \xi_{m'}^{-1})^2\} / \mu \tag{29}$$

If $\Delta_{m(m')} / g^2_{m(m')} = const$, the spatial broadening of the traveling MC field can be minimized to

$$\delta\omega''_{kk;travel} = 2\{\sum_{m,m'}^{M_+} \Gamma_m \Gamma_{m'} (1/\xi_m^0 - 1/\xi_{m'}^0)^2\} / \mu \tag{30}$$

If $\gamma_m / g_m^2 = const$, this value can be equal to zero.

4.2 Stationary MC field

Eg. (26) shows that all the laser fields propagate and evolve together with one group velocity $v(t)$ in the medium. When $v(t) = 0$, that is, when all the laser fields are completely stopped, we can obtain MC stationary light:

$$\sum_m^{M_+} \Omega^2_{m;0}(t) / g_m^2 - \sum_n^{M_-} \Omega^2_{n;0}(t) / g_n^2 = 0 \tag{31}$$

This is the MC stationary light condition. Obviously, this relation generalizes the results obtained for two-color stationary light. The amplitudes of these coupled fields can be manipulated by varying the Rabi frequencies of the corresponding control fields: $\hat{A}_m(t,z) / \hat{A}_n(t,z) \cong \Omega_m(t) g_n / (\Omega_n(t) g_m)$. Therefore optical control of group velocity, electric field amplitudes and direction of the MC field can be easily realized by changing the control fields' amplitudes.

For a Gaussian shape of the initial input probe pulse, we have the following amplitudes $A_{m,n}$ of the MC stationary light:

$$\left\langle A_{m(n)}(t,z) \right\rangle_\alpha = \frac{\Omega_{m(n)}(t) g_l}{\Omega_l(t_0) g_{m(n)}} \frac{l_0}{l_{m(n)}(t)} A_{l,0} \exp\{-\gamma_2(t-t_0) - \frac{[z + z_0 - v_l t_0 + \delta z_{m(n)}]^2}{2 l^2_{m(n)}(t)}\} e^{i(\vartheta_l + \varphi_{m(n)} - \varphi_l)} \tag{32}$$

Eq. (32) shows that the amplitude $A_{m,n}$ can be larger than the initial probe pulse amplitude $A_{l,0}$, if $\Omega_{m(n)}(t) / \Omega_l(t_0) > 1$.

The maximum MC field stationary time is determined by relaxation constant γ_{12} and spatial spreading of the stationary pulse shape which is determined by the dispersion term $\delta\omega''_{kk}$. Appling Eq. (31) to Eq. (29), we get the second order dispersion:

$$\delta\omega''_{kk}\Big|_{v=0} = \delta\omega''_{kk;l}(t) = 2\{\sum_m^{M_+} v_m(t) \xi_m^{-1} + \sum_n^{M_-} v_n(t) \xi_n^{-1}\} \tag{33}$$

Using the conditions $\mathrm{Im}(\xi_m^{-1} - \xi_l^{-1}) = 0$ and $\mathrm{Im}(\xi_n^{-1} + \xi_l^{-1}) = 0$, the dispersion can be minimized to zero:

$$\sum_m^{M_+} v_m(t)(\Delta_m / g_m^2) + \sum_n^{M_-} v_n(t)(\Delta_n / g_n^2) = 0 \tag{34}$$

by satisfying the particular relationship to the initial probe field E_l

$$\Delta_m / g_m^2 - \Delta_{l'} / g_l^2 = 0, \quad \Delta_n / g_n^2 + \Delta_{l'} / g_l^2 = 0 \tag{35}$$

From Eqs. (34) and (35), we can know that the condition for minimum spreading is independent of the total number of the control fields M+ and M-. Putting Eq. (31) into Eq. (27), we find the important relation for the minimum spatial spreading of the MC stationary light pulses:

$$\delta\omega_{kk}''\Big|_{v=0} \equiv \delta\omega_{kk;0}''(t) = 2\{\sum_m^{M_+} v_m(t)\xi_m^0 + \sum_n^{M_-} v_n(t)\xi_n^0\} \tag{36}$$

4.3 MC-wavelength conversion

After optical trapping of the initial probe pulse by using MC stationary light, it is possible to generate an arbitrary forward traveling light field A_m from MC stationary fields by controlling the amplitudes of the forward control fields ($m \in [1,....,M_+]$, $\Omega_m(t > t_2) \neq 0$, and $\Omega_{l\neq m}(t > t_2) = 0$), or to generate an arbitrary backward traveling light field A_n by controlling the backward control fields ($n \in [1,....,M_-]$, $\Omega_n(t > t_2) \neq 0$, and $\Omega_{l\neq n}(t > t_2) = 0$). Generation of some number of copropagating quantum fields A_m or A_n is also possible.
The electric field amplitude of the m-th component in the MC light at time t_{out} on the medium output (z=L) is given by

$$\langle A_{m;o}(t)\rangle_\alpha = \sqrt{\frac{v_m(t_{out})}{v_l(t_0)}\frac{l_0}{l_m(t_{out})}} A_{l;0} \exp\{-\gamma_2(t-t_0) + i(\vartheta_l + \varphi_m - \varphi_l)\} \exp\{-\frac{1}{2}(t-t_{out})^2 / (\delta t_{m,out})\} \tag{37}$$

with the temporal duration $\delta t_{m,travel} = l_m(t_{out}) / v_{travel}(t_{out})$. Eq. (37) shows temporal properties of the output MC light field. We note that this MC wavelength conversion may have potential applications in optical communications networks.

5. Conclusion

We first demonstrated the two-color stationary light and quantum wavelength conversion using quantum coherence resulting from strongly coupled slow light through EIT in a double-lambda system. Then we generalized the approach to the MC light fields in the multi double Λ coherent atomic medium driven by the M++M- control intensive laser fields and showed how to manipulate the MC light field within the adiabatic limit. The results show that the MC light fields can be controlled by simply adjusting the control fields' parameters for (1) MC stationary light, (2) selection of propagation direction (forward or backward), and (3) MC wavelength conversion. The maximum stationary time and minimum spatial spreading of the MC field have also been discussed. On-demand quantum manipulation of the MC light field can greatly increase the interaction time of the light and medium, and holds promise for applications in optical buffer, controllable switching, and quantum optical information processing.

We acknowledge that this work was supported by the CRI program (Center for Photon Information Processing) of the Korean Ministry of Education, Science and Technology via National Research Foundation.

6. References

Alzetta, G.; Gozzini, A.; Moi, L. & Oriols, G. (1976). An experimental method for the observation of r.f. transitions and laser beat resonances in oriented sodium vapor. *Nuovo Cimento B*, Vol. 36, (1976) 5-20

Andre, A; Bajcsy, M; Zibrov, A. X. & Lukin, M. D. (2005). Nonlinear optics with stationary pulses of light. *Phys. Rev. Lett.*, Vol. 94, No. 6, (2005) 063902, ISSN 0031-9007

Arimondo, E & Orriols, G. (1976). Nonabsorbing atomic coherences by coherent two-photon transitions in a three-level optical pumping. *Nuovo Cimento Letters*, Vol. 17, No. 6, (1976) 333-338

Bajcsy, M.; Zibrov, A. S. & Lukin, M. D. (2003). Stationary pulses of light in an atomic medium. *Nature* (London), Vol. 426, (2003) 638-641, ISSN 0028-0836

Boller, K. -J; Imamolu, A. & Harris, S. E. (1991). Observation of electromagnetically induced transparency. *Phys. Rev. Lett.*, Vol. 66, No. 20, (1991) 2593-2596, ISSN 0031-9007

Fleischhauer, M. & Lukin, M. D. (2000). Dark-state polaritons in electromagnetically induced transparency. *Phys. Rev. Lett.*, Vol. 84, No. 22, (2000) 5094-5097, ISSN 0031-9007

Fleischhauer, M.; Imamoglu, A. & Marangos, J. P. (2005). Electromagnetically induced transparency: optics in coherent media. *Rev. Mod. Phys.*, Vol. 77, No. 2, (2005) 633-673, ISSN 0034-6861

Gray, H. R.; Whitley, R. M. & Stroud, Carlos R., Jr. (1978). Coherent trapping of atomic populations. *Opt. Lett.*, Vol. 3, (1978) 218-220, ISSN 0146-9592

Ham, B. S.; Hemmer, P. R. & Shahriar, M. S. (1997). Efficient electromagnetically induced transparency in a rare-earth doped crystal. *Opt. Commun.*, Vol. 144, (1997) 227-230, ISSN 0030-4018

Ham, B. S.; Shahriar, M. S.; Kim, M. K. & Hemmer, P. R. (1997). Frequency-selective time-domain optical data storage by electromagnetically induced transparency in a rare-earth-doped solid. *Opt. Lett.*, Vol. 22, No. 24, (1997) 1849-1851, ISSN 0146-9592

Ham, B. S.; Shahriar, M. S.; Kim, M. K. & Hemmer, P. R. (1998). Spin coherence excitation and rephasing with optically shelved atoms. *Phys. Rev. B*, Vol. 58, No. 18, (1998) R11825-R11828, ISSN 1098-0121

Ham, B. S. & Hemmer, P. R. (2000). Coherence switching in a four-level system: quantum switching. *Phys. Rev. Lett.*, Vol. 84, No. 18, (2000) 4080-4083, ISSN 0031-9007

Ham, B. S. (2006). Spatiotemporal quantum manipulation of travelling light: Quantum transport. *Applied Physics Letters*, Vol. 88, No. 12, (2006) 121117, ISSN 0003-6951

Ham, B. S. & Hahn, J. (2009). Ultralong trapping of light using double spin coherence gratings. arXiv: 0901.1082v2, (2009)

Harris, S. E. (1989). Lasers without inversion: Interference of lifetime-broadened resonances. *Phys. Rev. Lett.*, Vol. 62, No. 9, (1989) 1033-1036, ISSN 0031-9007

Harris, S. E.; Fieldm, J. E. & Imamoglu, A. (1990). Nonlinear optical processes using electromagnetically induced transparency. *Phys. Rev. Lett.*, Vol. 64, No. 10, (1990) 1107-1110, ISSN 0031-9007

Harris, S. E. (1997). Electromagnetically induced transparency. *Phys. Today*, Vol. 50, No. 7, (1997) 36-42; ISSN 0031-9228

Hau, L. V.; Harris, S. E.; Dutton, Z. & Behroozi, C. H. (1999). Light speed reduction to 17 metres per second in an ultracold atomic gas. *Nature*, Vol. 397, (1999) 594-598, ISSN 0028-0836

Kasapi, A.; Jain, M.; Yin, G. Y. & Harris, S. E. (1995). Electromagnetically induced transparency: propagation dynamics. *Phys. Rev. Lett.*, Vol. 74, No. 13, (1995) 2447-2450, ISSN 0031-9007

Kocharovskaya, O., Rostovtsev, Y. & Scully, M. O. (2001). Stopping light via hot atoms. *Phys. Rev. Lett.*, Vol. 86, No. 4, (2001) 628-631, ISSN 0031-9007

Korsunsky, E. A. & Kosachiov, D. V. (1999). Phase-dependent nonlinear optics with double-lambda atoms. *Phys. Rev. A*, Vol. 60, No. 6, (1999) 4996-5009, ISSN 1050-2947

Liu, C.; Dutton, Z.; Behroozi, C. H.; Hau, L. V. & Harris, S. E. (2001). Observation of coherent optical information storage in an atomic medium using halted light pulses. *Nature*, Vol. 409, (2001) 490-493, ISSN 0028-0836

Loudon, R. (2000). *The Quantum Theory of Light.* Oxford University Press, Oxford, ISBN 0198501765

Lukin, M. D. & Imamoglu, A. (2000). Nonlinear optics and quantum entanglement of ultraslow single photons. *Phys. Rev. Lett.*, Vol. 84, No. 7, (2000) 1419-1422, ISSN 0031-9007

Lukin, M. D. & Imamoglu, A. (2001). Controlling photons using electromagnetically induced transparency. *Nature*, Vol. 413, (2001) 273-276, ISSN 0028-0836

Marangos, J. P. (1998). Topical review electromagnetically induced transparency. *J. Modern Opt.*, Vol. 45, No. 3, (1998) 471-503, ISSN 0950-0340

Moiseev, S. A. & Ham, B. S. (2006). Quantum manipulation of two-color stationary light: Quantum wavelength conversion. *Phys. Rev. A*, Vol. 73, No. 3, (2006) 033812, ISSN 1050-2947

Moiseev, S. A. & Ham, B. S. (2006). Quantum control and manipulations of three-color stationary light. *J. Korean Phys. Soc.*, Vol. 48, No. 4, (2006) 540-545, ISSN 0374-4884

Moiseev, S. A.; Chen, Y. & Ham, B. S. (2006). Numerical analysis of stationary light for potential applications of a quantum interface. *J. Korean Phys. Soc.*, Vol. 49, No. 6, (2006) 2293-2302, ISSN 0374-4884

Moiseev, S. A. & Ham, B. S. (2007). Quantum control and manipulation of multi-color light fields. *Optics and Spectroscopy*, Vol. 103, No. 2, (2007) 210-218, ISSN 0030-400X

Nielsen, M. A. & Chuang, I. L. (2000). *Quantum Computation and Quantum Information.* Cambridge University Press, Cambridge, ISBN 0521635039

Padmabandu, G. G.; Welch, G. R.; Shubin, I. N.; Fry, E. S.; Nikonov, D. E.; Lukin, M. D. & Scully, M. O. (1996). Laser oscillation without population inversion in a sodium atomic beam. *Phys. Rev. Lett.*, Vol. 76, No. 12, (1996) 2053-2056, ISSN 0031-9007

Phillips, M. C.; Wang, H. L.; Rumyantsev, I.; Kwong, N. H.; Takayama, R. & Binder, R. (2003). Electromagnetically induced transparency in semiconductors via biexcition coherence. *Phys. Rev. Lett.*, Vol. 91, No. 18, (2003) 183602, ISSN 0031-9007

Schmidt, H. & Imamoglu, A. (1996). Giant Kerr nonlinearities obtained by electromagnetically induced transparency. *Opt. Lett.*, Vol. 21, (1996) 1936-1938, ISSN 0146-9592

Scully, M. O.; Zhu, S. Y. & Gravrielides, A. (1989). Degenerate quantum-beat laser: Lasing without inversion and inversion without lasing. *Phys. Rev. Lett.*, Vol. 62, No.24, (1989) 2813-2816, ISSN 0031-9007

Scully, M. O. (1991). Enhancement of the index of refraction via quantum coherence. *Phys. Rev. Lett.*, Vol. 67, No. 14, (1991) 1855-1858, ISSN 0031-9007

Scully, M. O. & Zhu, S. Y. (1992). Ultra-large index of refraction via quantum interference. *Opt. Commun.*, Vol. 87, (1992) 134-138, ISSN 0030-4018

Scully, M. O. & Zubairy, M. S. (1997). *Quantum Optics*. Cambridge University Press, Cambridge, England, ISBN 0521435951

Turukhin, A. V.; Sudarshanam, V. S; Shahriar, M. S.; Musser, J. A.; Ham, B. S. & Hemmer, P. R. (2002). Observation of ultraslow and stored light pulses in a solid. *Phys. Rev. Lett.*, Vol. 88, No. 2, (2002) 023602, ISSN 0031-9007

Zibrov, A. S.; Matsko, A. B.; Kocharovskaya, O; Rostovtsev, Y. V; Welch, G. R. & Scully, M. O. (2002). Transporting and time reversing light via atomic coherence. *Phys. Rev. Lett.*, Vol. 88, No. 10, (2002) 103601, ISSN 0031-9007

27

Fundamentals and Applications of Quantum Limited Optical Imaging

Warwick P. Bowen[1], Magnus T. L. Hsu[1] and Jian Wei Tay[2]
[1]*School of Mathematics and Physics, University of Queensland, QLD 4072*
[2]*Physics Department, University of Otago, Dunedin*
[1]*Australia*
[2]*New Zealand*

1. Introduction

The field of optical imaging is focussed on techniques to extract useful information about a physical system from the spatial structure of light. There are two main themes to research in this field, the resolving of previously unknown structures ranging in size from microns in microscopy to distant galaxies in astronomical telescopy; or if the structure is selected from an *a priori* known set, then its unique discrimination, such as in data read-out from a CD or DVD. In general, both types of imaging involve the collection and focusing of light after interaction with the object. However, the process of information extraction can be quite different. In resolving an unknown structure, a full two dimensional image is usually desired. Here, the metric of success is generally the resolution of the final image. In most cases diffraction is the key concern, presenting the diffraction limit to the resolution of the final image as approximated by Abbe (Born & Wolf, 1999). There are ways to overcome this limit, such as by utilising non-linearities (Hell *et al.*, 2009), or using metamaterials (Pendry, 2000) to form so called superlenses, and this is a vibrant and growing area of research.
The focus of this Chapter, however, is on the second theme of imaging, discrimination between a set of known structures. As we will see, this form of imaging is important, not only for read-out of information from data storage devices, but also in other areas such as microscopy (Fabre *et al.*, 2000; Tay *et al.*, 2009) and satellite navigation (Arnon, 1998; Nikulin *et al.*, 2001). In structure discrimination, the goal is not to achieve a two dimensional image, but rather to generate a signal which unambiguously distinguishes each element of the set. Hence, the diffraction limit and other constraints on imaging resolution are no longer the primary concern, but rather the signal-to-noise ratio with which the discrimination may be performed. To maximise the signal the optical measurement must be matched carefully to the set of structures to be discriminated; whereas the noise typically comes from electronic, environmental, and optical sources. Much engineering effort has been applied to minimising the noise sources for important imaging systems; however, fundamentally the quantisation of light imposes the quantum noise limit (QNL) which is outside of engineering control. In this Chapter we consider a general imaging system, and show how the optical mode carrying full signal information may be determined. We introduce spatial homodyne detection (Beck, 2000; Hsu *et al.*, 2004) as a method to optimally extract this signal, showing

how the QNL to measurement sensitivity may be determined and even surpassed using non-classical states of light. We illustrate the implications of these techniques for two key imaging systems, atomic force microscopy (Binning *et al.*, 1985; Fabre *et al.*, 2000) and particle tracking in optical tweezers (Block, 1992; Tay *et al.*, 2009); comparing optimal spatial homodyne based signal extraction to the standard extraction methods used in such systems today.

2. Quantum formalism for optical measurements

The field of optical measurements has progressed significantly, with photo-detection techniques advancing from the use of the photographic plate in the 19th century to the semiconductor-based photodetectors commonly encountered today. One is now able to measure with high accuracy and speed, the range of parameters that describe an optical field. For example, the amplitude and phase quadratures, the Stokes polarisation parameters, and the transverse spatial profile that are commonly used to parameterise the optical field (Walls & Milburn, 1995). These parameters can be measured and quantified using a range of detection techniques such as interferometry, polarimetry and beam profiling (Saleh & Teich, 1991). However, experimentally measured values for these parameters are estimates due to the presence of classical and quantum noise, and detection inefficiencies.

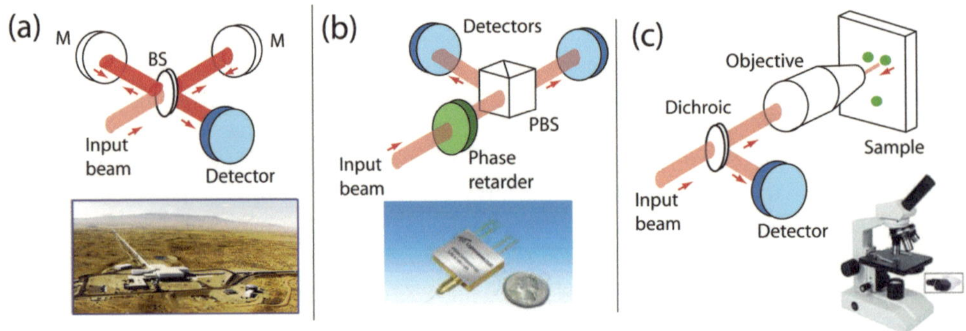

Fig. 1. Schematics of (a) a Michelson interferometer with an inset photo of the Laser Interferometer Gravity-wave Observatory (LIGO), (b) a polarimeter with inset photo of an on-chip polarimeter, and (c) an optical microscope with an inset photo of an optical microscope. M: mirror, BS: beam-splitter and PBS: polarising beam-splitter.

Fig. 1 shows examples of techniques used for the measurement of (a) amplitude and phase quadratures (Slusher *et al.*, 1985), (b) polarization (Korolkova & Chirkin, 1996) and (c) spatial variables (Pawley, 1995). Fig. 1 (a) shows a Michelson interferometer whereby an input field is split using a beam-splitter, followed by propagation of the two output fields through different paths with an effective path difference. These two fields are then interfered to produce an output interference signal. Depending on the effective path difference, destructive or constructive interference is obtained at the output of the interferometer. Variations of this technique include the Mach-Zehnder (Mach, 1892; Zehnder, 1891) and Sagnac (Sagnac, 1913) interferometers. A polarimeter is shown in Fig. 1 (b), where an input field is phase retarded and the different polarisation components of the input field are separated using a polarization beam-splitter. A measurement of the intensity difference between the different polarization components provides information on the Stokes variables that characterise the polarization phase space (Bowen *et al.*, 2002). Interferometry and polarimetry are essentially single spatial mode techniques, since the spatial discrimination of the field structure cannot be characterized with these techniques. In order to reach their measurement sensitivity limits, classical noise sources have to be reduced (or eliminated) sufficiently such that quantum noise becomes the dominant noise source. Consequently, optimal measurements of the amplitude and phase quadratures as well as the polarisation variables are obtained, with measurement sensitivity bounded at the QNL.

Measurements of the spatial properties of light are more complex, since multiple spatial modes are naturally involved. Therefore noise sources are no longer the sole consideration, with the modal selection and filtration processes also becoming critical. Fig. 1 (c) shows a schematic of an optical microscope, where a focused light field is used to illuminate and image a microscopic sample. Existing techniques to resolve the finer spatial details of an optical image include for example the filtration of different spatial frequency components via confocal microscopy (Pawley, 1995); or the collection of non-propagating evanescent modes that decay exponentially over wavelength-scales via near-field microscopy (Synge, 1928).

Here we are interested in the procedure of optimal parameter measurement, as shown in Fig. 2, whereby the detection system is tailored to optimally extract a specific spatial signal. An input field is spatially perturbed (i.e. a spatial signal is applied to the optical field, be it known or unknown), and the resultant field is detected. To be able to optimally measure the perturbation applied to the field, the relevant signal field components have to be identified and resolved.

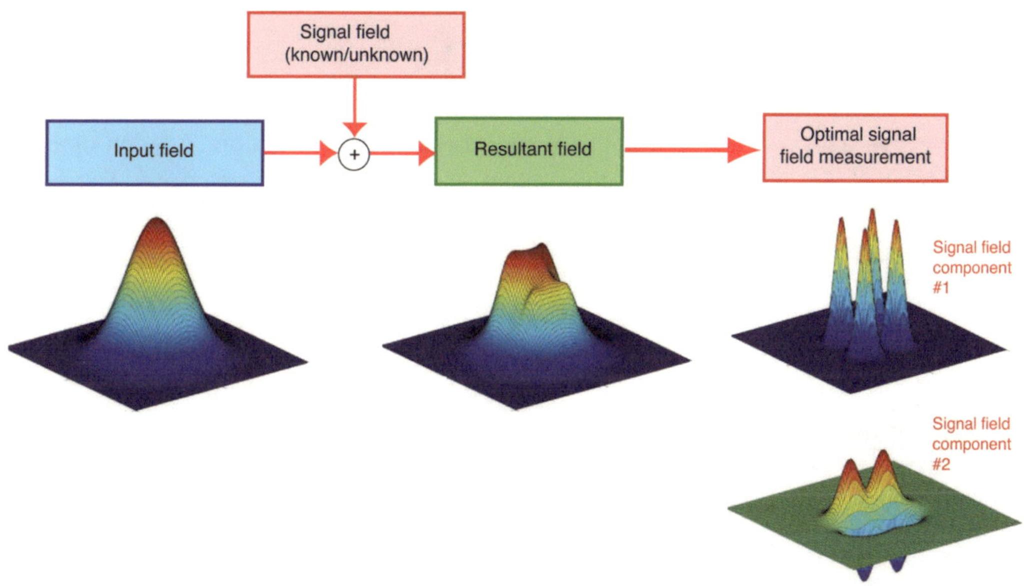

Fig. 2. The optimal parameter measurement procedure. An input field is perturbed by some known or unknown spatial signal and the resultant field is detected. Optimal measurements of the perturbation can be performed by identifying and resolving the relevant signal field components.

We now present a formalism for defining the quantum limits to measurements of spatial perturbations of an optical field. The spatial perturbation, quantified by parameter p is entirely arbitrary, and could for instance be the displacement or rotation of a spatial mode in the transverse plane (Hsu *et al.*, 2004; 2009), or the perturbation introduced by an environmental factor such as scattering from a particle within the field or atmospheric fluctuations.

In general, the optical field requires a full three dimensional description using Maxwell's equations (Van de Hulst, 1981). In systems where all dimensions are significantly larger than the optical wavelength, however, the paraxial approximation can usually be invoked and the field can be described using two dimensional transverse spatial modes in a convenient basis. The spatial quantum states of an optical field exist within an infinite dimensional Hilbert space, and may be conveniently expanded in the basis of the rectangularly-symmetric TEM_{mn} or circularly-symmetric LG_{nl} modes, with the choice of modal basis dependent on the spatial symmetry of the imaged optical field.

A field of frequency ω can be represented by the positive frequency part of the electric field operator $\tilde{\mathbf{E}}^{+}(\rho)e^{i\omega t}$. Following Tay *et al.* (2009), the transverse information of the field is described fully by the slowly varying field envelope operator $\tilde{\mathbf{E}}^{+}(\rho)$, given by

$$\tilde{\mathbf{E}}^{+}(\rho) = i\sqrt{\frac{\hbar\omega}{2\epsilon_0 V}} \sum_{j,m,n} \tilde{a}^{j}_{mn}\mathbf{u}^{j}_{mn}(\rho), \tag{1}$$

where $\rho = (x,y)$ is a co-ordinate in the transverse plane of the field, V is the volume of the optical mode, and the summation over the parameters j, m and n is given by

$$\sum_{j,m,n} \equiv \sum_{j\in\{x,y\}} \sum_{m=0}^{\infty} \sum_{n=0}^{\infty}. \tag{2}$$

The respective transverse beam amplitude function and the photon annihilation operator are given by $\mathbf{u}^{j}_{mn}(\rho)$ and $\tilde{a}^{j}_{mn}$, with polarisation denoted by the superscript j. The $\mathbf{u}_{mn}(\rho)$ mode functions are normalized such that their self-overlap integrals are unity, with the inner product given by

$$\begin{aligned}\left\langle \mathbf{u}^{j}_{mn}(\rho), \mathbf{u}^{j'}_{m'n'}(\rho)\right\rangle &= \int_{-\infty}^{\infty} \left[\mathbf{u}^{j}_{mn}(\rho)\right]^{*} \cdot \mathbf{u}^{j'}_{m'n'}(\rho)d\rho \\ &= \delta_{mm'}\delta_{nn'}\delta_{jj'}.\end{aligned} \tag{3}$$

An arbitrary spatial perturbation, described by parameter p, is now applied to the field. Eq. (1) can therefore be expressed as a sum of coherent amplitude components and quantum noise operators, given by

$$\begin{aligned}\tilde{\mathbf{E}}^{+}(\rho,p) &= i\sqrt{\frac{\hbar\omega}{2\epsilon_0 V}} \sum_{j,m,n} \tilde{a}^{j}_{mn}\mathbf{u}^{j}_{mn}(\rho,p) \\ &= i\sqrt{\frac{\hbar\omega}{2\epsilon_0 V}} \left[\alpha(p)\mathbf{v}(\rho,p) + \sum_{j,m,n} \delta\tilde{a}^{j}_{mn}\mathbf{u}^{j}_{mn}(\rho,0)\right],\end{aligned} \tag{4}$$

where $\alpha(p)\mathbf{v}(\rho,p) = \sum_{j,m,n}\langle\tilde{a}^j_{mn}\rangle\mathbf{u}^j_{mn}(\rho,0) = \sum_{j,m,n}\langle\tilde{a}^j_{mn}\rangle u_{mn}(\rho,0)\hat{\mathbf{j}}$ being the coherent amplitude of mode $\mathbf{v}(\rho, p)$, and $\hat{\mathbf{j}}$ is the unit polarisation vector. From Eq. (4), one can then relate $\alpha(p)$ and $\mathbf{v}(\rho, p)$ to $\tilde{\mathbf{E}}^{+}(\rho, p)$ by

$$\alpha(p) = \sqrt{\frac{2\epsilon_0 V}{\hbar\omega}\left\langle\overline{\mathbf{E}}^{+}(\rho,p),\overline{\mathbf{E}}^{+}(\rho,p)\right\rangle} \tag{5}$$

$$\mathbf{v}(\rho,p) = -iN_v\overline{\mathbf{E}}^{+}(\rho,p) \tag{6}$$

where $\overline{\mathbf{E}}^{+}(\rho,p) = \langle\tilde{\mathbf{E}}^{+}(\rho,p)\rangle$, and the normalization constant N_v is given by

$$\begin{aligned} N_v &= \left\langle\overline{\mathbf{E}}^{+}(\rho,p),\overline{\mathbf{E}}^{+}(\rho,p)\right\rangle^{-1/2} \\ &= \left[\iint_{-\infty}^{\infty}\left[\overline{\mathbf{E}}^{+}(\rho,p)\right]^{*}\cdot\overline{\mathbf{E}}^{+}(\rho,p)d\rho\right]^{-1/2}. \end{aligned} \tag{7}$$

The mean number of photons passing through the transverse plane of the field per second is given by $|\alpha(p)|^2$. We also assume, without loss of generality, that $\alpha(p)$ is real. The quantum noise operator carrying all of the noise on the field in mode $\mathbf{u}_{mn}(\rho) = \mathbf{u}^j_{mn}(\rho,0)$ is given by $\delta\tilde{a}^j_{mn} = \tilde{a}^j_{mn} = \langle\tilde{a}^j_{mn}\rangle$.
In the limit of small estimate parameter p, we can take the first order Taylor expansion of the first bracketed term in Eq. (4), given by

$$\alpha(p)\mathbf{v}(\rho,p) \approx \alpha(0)\mathbf{v}(\rho,0) + p\cdot\left.\frac{\partial[\alpha(p)\mathbf{v}(\rho,p)]}{\partial p}\right|_{p=0}, \tag{8}$$

where the first term on the right-hand side of Eq. (8) indicates that the majority of the power of the field is in the $\mathbf{v}(\rho,0)$ mode. The second term defines the spatial mode $\mathbf{w}(\rho)$ corresponding to small changes in the parameter p, given by

$$\mathbf{w}(\rho) = \frac{1}{N_{\mathbf{w}}}\left.\frac{\partial[\alpha(p)\mathbf{v}(\rho,p)]}{\partial p}\right|_{p=0}. \tag{9}$$

where $N_{\mathbf{w}}$ is the normalisation given by

$$N_{\mathbf{w}} = \left\langle\left.\frac{\partial[\alpha(p)\mathbf{v}(\rho,p)]}{\partial p}\right|_{p=0},\left.\frac{\partial[\alpha(p)\mathbf{v}(\rho,p)]}{\partial p}\right|_{p=0}\right\rangle^{-1/2}. \tag{10}$$

Notice that the first term in Eq. (8) is independent of p; while the second term, and therefore the amplitude of mode $\mathbf{w}(\rho)$, is directly proportional to p. Therefore, by measuring the amplitude of mode $\mathbf{w}(\rho)$ it is possible to extract all available information about p. As a consequence, we henceforth term $\mathbf{w}(\rho)$ the *signal mode.*

3. Detection systems

Several techniques have been developed to experimentally quantify the amplitude of the signal mode. Here we discuss the three most common of such: array detection, split detection, and spatial homodyne detection, as shown in Fig. 3.

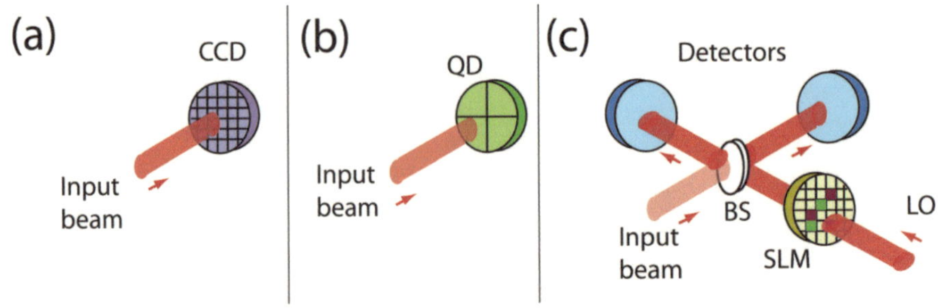

Fig. 3. Detection systems for the measurement of the spatial properties of the field. (a) Array, (b) split and (c) spatial homodyne detection systems. BS: beam-splitter, LO: local oscillator field, CCD: charge-coupled detector, QD: quadrant detector (four component split detector), SLM: spatial light modulator.

3.1 Array detection

As shown in Fig. 3 (a) array detectors in general consist of an $m \times n$ array of pixels each of which generates a photocurrent proportional to its incident optical field intensity. One subclass of array detectors is the ubiquitous charge-coupled device (CCD), which is the most common form of detector used for characterisation of the spatial properties of light beams. To the authors knowledge, the first quantum treatment of optical field detection using array detectors was given in Beck (2000). In this work Beck (2000) proposed the use of two array detectors with a local oscillator in a homodyne configuration to perform spatial homodyne detection. Such techniques will be discussed in detail in section 3.3. Quantum measurements with a simple single array were first considered later in papers by Treps, Delaubert and others (Treps *et al.*, 2005; Delaubert *et al.*, 2008). An ideal array detector consists of a two dimensional array of infinitesimally small pixels, each with unity quantum efficiency, and each registering the amplitude of its incident field with high bandwidth. However, realistic array detectors stray far from this ideal; with efficiencies generally around 70 % due both to the intrinsic inefficiency of the pixels and due to dead zones between pixels, complications in shift register readout, and bandwidth limitations[1]. To date, all quantum imaging experiments utilising array detectors have been performed in the context of spatial homodyne detection. We therefore defer further discussion of these techniques to Section 3.3.

3.2 Split detection

One of the most important spatial parameters of an optical beam is the fluctuation of its mean position, commonly termed *optical beam displacement*, which provides extremely sensitive information about environmental perturbations such as forces exerted on microscopic systems (see Sections 4 and 5), mechanical vibrations, and air turbulence; as well as control information in techniques such as satellite navigation (Arnon, 1998; Nikulin *et al.*, 2001)) and locking of optical resonators (Shaddock *et al.*, 1999), to name but a few. The most convenient means to measure optical beam displacement is through measurement on a split detector (Putman *et al.*, 1992; Treps *et al.*, 2002; 2003), as shown in Figure 3 (b). Such detectors are composed of two or more PIN photodetectors arranged side-by-side. So long

[1] For example, to achieve a typical quantum imaging detection bandwidth of 1 MHz, a 10-bit 10 megapixel CCD camera would require a total bit transfer rate of 100 T-bits/s.

as the optical field is aligned to impinge equally on the two photodetectors, and the optical beam shape is well behaved, the difference between the output photocurrents provides a signal proportional to the beam displacement. Furthermore, since only a pair of PIN photodiodes is used, both the efficiency and bandwidth issues related to array detection are easily resolved. The limitation of split detectors, however, is that they are restricted to measurement of a certain subset of signal modes, and therefore, in general will not be optimal for a given application (Hsu *et al.*, 2004). Here we derive the split detection signal mode following the treatments of Hsu *et al.* (2004) and Tay *et al.* (2009). The sensitivity achievable in the measurement of a general signal mode will be treated later in Section 3.4.

The difference photocurrent output from a split detector can in general be written as

$$\Delta i_{\mathrm{SD}} = i_{x<0} - i_{x>0} \tag{11}$$

$$= \frac{2\epsilon_0 V}{\hbar\omega}\left[\int_{-\infty}^{\infty} dy \int_{-\infty}^{0} dx \tilde{\mathbf{E}}^{+\dagger}\cdot\tilde{\mathbf{E}}^{+} - \int_{-\infty}^{\infty} dy \int_{0}^{\infty} dx \tilde{\mathbf{E}}^{+\dagger}\cdot\tilde{\mathbf{E}}^{+}\right]. \tag{12}$$

This can be shown (Fabre *et al.*, 2000) to be equal to

$$\Delta i_{\mathrm{SD}} = \alpha(p)\tilde{X}_f^{+} \tag{13}$$

where $\tilde{X}_f^{+} = \tilde{a}_f^{\dagger} + \tilde{a}_f$ is the amplitude quadrature operator of a *flipped mode* with mode intensity equal to that of the incident field but a π phase flip about the split between photodiodes. The transverse mode amplitude function of the flipped mode is given by

$$\mathbf{v}_f(\rho) = \begin{cases} \mathbf{v}(\rho,0), & x \geq 0 \\ -\mathbf{v}(\rho,0), & x < 0. \end{cases} \tag{14}$$

It is useful to separate the flipped mode amplitude quadrature operator into a coherent amplitude component

$$\alpha_f(p) = \alpha(p)\left\langle \mathbf{v}_f(\rho), \mathbf{v}(\rho,p)\right\rangle \tag{15}$$

which contains the signal due to the parameter p; and a quantum noise operator $\delta\tilde{X}_f^{+} = \tilde{X}_f^{+} - \langle\tilde{X}_f^{+}\rangle$ which places a quantum limit on the measurement sensitivity, so that

$$\Delta i_{\mathrm{SD}} = \alpha(p)\left[2\alpha_f(p) + \delta\tilde{X}_f^{+}\right]. \tag{16}$$

Hence, we see that split detection measures the signal and noise in a flipped version of the incident mode.

3.3 Spatial homodyne detection

Spatial homodyne detection was first proposed by Beck (2000) using array detectors, and was extended to the case of pairs of PIN photodiodes with a spatially tailored local oscillator field by Hsu *et al.* (2004). Spatial homodyne detection has the significant advantage over split detection in that the detection mode can be optimised to perfectly match the signal mode. The proposal of Beck (2000) has the advantage of allowing simultaneous extraction of multiple signals (Dawes *et al.*, 2001); whilst that of Hsu *et al.* (2004) allows high bandwidth extraction of a single arbitrary spatial mode and is polarization sensitive allowing optimal measurements where the signal is contained within spatial variations of the polarisation of the field. Here, we explicitly treat local oscillator tailored spatial homodyne allowing the inclusion of polarisation effects. However, we emphasise that the two schemes are formally equivalent for single-signal-mode single-polarisation fields.

In a local oscillator tailored spatial homodyne, the input field is interfered with a much brighter local oscillator field on a 50/50 beam splitter; with the two output fields individually detected on a pair of balanced single element photodiodes, as shown in fig. 3 (c). The difference photocurrent between the two resulting photocurrents is the output signal. By shaping the local oscillator field, for example by using a set of spatial light modulators (SLM), an arbitrary spatial parameter of the input field can be interrogated. Spatial homodyne schemes of this kind have been shown to perform at the Cramer-Rao bound (Delaubert *et al.*, 2008), and therefore enable optimal measurement of any spatial parameter p.

The performance of a spatial homodyne detector can be assessed in much the same way as split detection in the previous section. Here we follow the approach of Tay *et al.* (2009), choosing a LO with a positive frequency electric field operator

$$\tilde{\mathbf{E}}^{+}_{\mathrm{LO}}(\rho) \;=\; i\sqrt{\frac{\hbar\omega}{2\epsilon_0 V}}\left[\alpha_{\mathrm{LO}}\mathbf{w}(\rho)+\sum_{j,m,n}\delta\tilde{a}^{j}_{mn,\mathrm{LO}}\mathbf{u}^{j}_{mn}(\rho)\right]e^{i\phi}, \tag{17}$$

with the relative phase between the local oscillator and the input beam given by ϕ and local oscillator mode chosen to match the signal mode. The input beam described in Eq. (4) is interfered with the LO on a 50/50 beam splitter to give the output fields

$$\tilde{\mathbf{E}}^{+}_{\pm}=\frac{1}{\sqrt{2}}\left(\tilde{\mathbf{E}}^{+}\pm\tilde{\mathbf{E}}^{+}_{\mathrm{LO}}\right) \tag{18}$$

where the subscripts + and – distinguish the two output fields. The photocurrents produced when each field impinges on an infinitely wide photodetector are given by

$$i_{\pm} \;=\; \frac{2\epsilon_0 V}{\hbar\omega}\int_{-\infty}^{\infty}\tilde{\mathbf{E}}^{+\dagger}_{\pm}\cdot\tilde{\mathbf{E}}^{+}_{\pm}d\rho \tag{19}$$

$$=\; \frac{2\epsilon_0 V}{\hbar\omega}\int_{-\infty}^{\infty}\frac{1}{2}(\tilde{\mathbf{E}}^{+}\pm\tilde{\mathbf{E}}^{+}_{\mathrm{LO}})^{\dagger}\cdot(\tilde{\mathbf{E}}^{+}\pm\tilde{\mathbf{E}}^{+}_{\mathrm{LO}})\,d\rho, \tag{20}$$

which together with Eqs. (1), (3), and (17) yield the photocurrent difference

$$\begin{aligned}
\Delta i_{\mathrm{SH}} &= i_{+}-i_{-}\\
&= \alpha_{\mathrm{LO}}\int_{-\infty}^{\infty}\left[e^{-i\phi}[\mathbf{w}(\rho)]^{*}\cdot\sum_{j,m,n}\tilde{a}^{j}_{mn}\mathbf{u}^{j}_{mn}(\rho,p)+e^{i\phi}\mathbf{w}(\rho)\cdot\left(\sum_{j,m,n}\tilde{a}^{j}_{mn}\mathbf{u}^{j}_{mn}(\rho,p)\right)^{\dagger}\right]d\rho\\
&= \alpha_{\mathrm{LO}}\left[e^{-i\phi}\sum_{j,m,n}\tilde{a}^{j}_{mn}\left\langle\mathbf{w}(\rho),\mathbf{u}^{j}_{mn}(\rho,p)\right\rangle+e^{i\phi}\left(\sum_{j,m,n}\tilde{a}^{j}_{mn}\left\langle\mathbf{w}(\rho),\mathbf{u}^{j}_{mn}(\rho,p)\right\rangle\right)^{\dagger}\right]\\
&= \alpha_{\mathrm{LO}}\left[e^{-i\phi}\tilde{a}_{w}+e^{i\phi}\tilde{a}^{\dagger}_{w}\right]\\
&= \alpha_{\mathrm{LO}}\tilde{X}^{\phi}_{w}
\end{aligned} \tag{21}$$

where the annihilation operator $\tilde{a}_w$ describes the input field component in mode $\mathbf{w}(\rho)$, and $\tilde{X}_w^{\phi} = e^{-i\phi}\tilde{a}_w + e^{i\phi}\tilde{a}_w^{\dagger}$ is the ϕ-angled quadrature operator of that component. The derivation above is valid in the limit that the local oscillator power is much greater than the signal power ($\alpha_{\rm LO} \gg \langle \tilde{a}_w \rangle$), which enables terms that do not involve $\alpha_{\rm LO}$ to be neglected.

An optimal estimate of the parameter p is obtained since the local oscillator mode is chosen to match the signal mode $\mathbf{w}(\rho)$, as shown in Eq. (21). The spatial homodyne detection scheme then extracts the quadrature of the signal mode with quadrature phase angle given by ϕ.

3.4 Quantifying the efficacy of parameter estimation

Eqs. (16) and (21) provide the output signal from both homodyne and split detection schemes. However we have yet to determine the efficacy of both schemes. To obtain a quantitative measure of the efficacy, we now introduce the signal-to-noise ratio (SNR) and sensitivity measures.

From Eq. (21) we see that the mean signal output from the spatial homodyne detector is given by

$$\langle \Delta i_{\rm SH} \rangle = \alpha_{\rm LO}\alpha_w(p)\left(e^{i\Delta\phi} + e^{-i\Delta\phi}\right), \tag{22}$$

where $\alpha_w(p) = \alpha(p)\ \langle \mathbf{w}(\rho),\mathbf{v}(\rho,\ p)\rangle$. The maximum signal strength occurs when the local oscillator and signal phases are matched, such that $\phi = 0$, and is given by

$$\langle \Delta i_{\rm SH} \rangle = 2\alpha_{\rm LO}\alpha_w(p). \tag{23}$$

The noise can be calculated straight-forwardly, and is given by

$$\sqrt{\langle \Delta i_{\rm SH}^2 \rangle - \langle \Delta i_{\rm SH} \rangle^2} = \alpha_{\rm LO}\Delta\tilde{X}_w^{\phi} = \alpha_{\rm LO}, \tag{24}$$

where $\Delta^2\tilde{X}_w^{\phi} = \langle(\delta\tilde{X}_w^{\phi})^2\rangle$ is the signal mode ϕ-quadrature variance, and we have used the fact that $\Delta^2\tilde{X}_w^{\phi} = \langle(\delta\tilde{X}_w^{\phi})^2\rangle = 1$ for a low noise coherent laser. Clearly, a non-classical squeezed light field can be used to reduce the noise such that $\Delta\tilde{X}_w^{\phi_{\rm LO}} < 1$, however in most cases the resources expended to achieve this outweigh the benefit. Without non classical resources, the signal-to-noise ratio of spatial homodyne detection is therefore limited to

$$\mathrm{SNR}_{\rm SH,coh} = 2\alpha_w(p). \tag{25}$$

Normally, the physically relevant parameter is the sensitivity $\mathcal{S}$ of the detection apparatus, that is the minimum observable change in the parameter p. This is defined as the change in p required to generate a unity signal-to-noise ratio,

$$\mathcal{S}_{\rm SH,coh} = \left[\left.\frac{\partial \mathrm{SNR}}{\partial p}\right|_{p=0}\right]^{-1} = \frac{1}{2}\left[\left.\frac{\partial \alpha_w(p)}{\partial p}\right|_{p=0}\right]^{-1}. \tag{26}$$

Equivalently, one finds a SNR for the split detection scheme in the coherent state limit of

$$\mathrm{SNR}_{\rm SD,coh} = 2\alpha_f(p), \tag{27}$$

with a corresponding sensitivity given by

$$\mathcal{S}_{\text{SD,coh}} = \frac{1}{2}\left[\left.\frac{\partial \alpha_f(p)}{\partial p}\right|_{p=0}\right]^{-1}. \tag{28}$$

The efficacy of both these detection schemes shall be discussed in the following sections, based on the context in which they are employed. However as we shall demonstrate, the spatial homodyne scheme offers significant improvement over the split detection scheme, and is optimal for all measurements of spatial parameter p.

4. Practical applications 1: Laser beam position measurement

Laser beam position measurement has wide-ranging applications from the macroscopic scale involving the alignment of large-scale interferometers (Fritschel *et al.*, 1998; 2001) and satellites to the microscopic scale involving the imaging of surface structures as encountered in atomic force microscopy (AFM) (Binning *et al.*, 1985). In an AFM, a cantilever with a nanoscopic-sized tip is scanned across a sample surface, as shown in Fig. 4 (a). The force between the sample surface and tip (e.g. van Der Waals, electrostatic, etc.) results in the tip being modulated spatially as it is scanned across the undulating sample surface. A laser beam is incident on the back of the cantilever with the spatial movement of the cantilever displacing the incident laser beam. The resultant reflected laser beam is detected on a split detector, providing information on the laser position with respect to the centre of the detector, with this information directly related with the AFM tip position. The use of the split detector is ubiquitous in AFM systems.

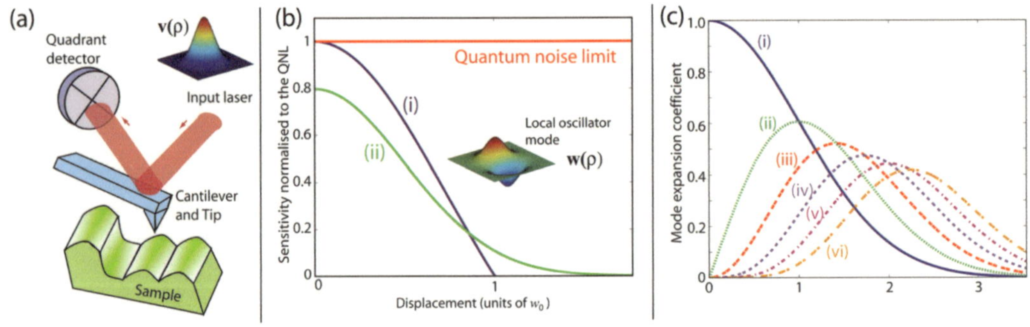

Fig. 4. (a) Schematic diagram illustrating an input field reflected from the back of a cantilever onto a split detector for position sensing of the tip location with respect to a sample. The input laser field has a TEM_{00} spatial profile, given by $\mathbf{v}(\rho)$. (b) Sensitivity of (i) spatial homodyne and (ii) split detection for the measurement of the displacement of a TEM_{00} input field. The local oscillator field had a TEM_{10} mode-shape, given by $\mathbf{w}(\rho)$. (c) The coefficients of the Taylor expansion of $\mathbf{v}(\rho, p)$. The coefficients correspond to the undisplaced (i) TEM_{00}, (ii) TEM_{10}, (iii) TEM_{20}, (iv) TEM_{30}, (v) TEM_{40}, (vi) TEM_{50} modes. Figures (b) and (c) were reproduced from Hsu *et al.* (2004), with permission.

4.1 Split detection

We now formalise the effects from the application of split detection in determining the AFM tip position. We assume that the laser field incident on the AFM cantilever has a TEM_{00}

modeshape. The spatial information of the displaced field, reflected from the AFM cantilever, is described fully by the field operator given in Eq. (1), now expanded to

$$\tilde{\mathbf{E}}^{+}(\rho,p) = i\sqrt{\frac{\hbar\omega}{2\epsilon_0 V}}\left(\alpha(p)\mathbf{u}_{00}(\rho,p) + \sum_{m,n=0}^{\infty}\delta\hat{a}_{mn}\mathbf{u}_{mn}(\rho,p)\right), \tag{29}$$

where $\mathbf{u}_{mn}(\rho, p)$ are the transverse beam amplitude functions for the displaced TEM_{mn} modes and $\delta\hat{a}_{mn}$ are the corresponding quantum noise operators. $\alpha(p)$ is the coherent amplitude of the displaced TEM_{00} field. Using the formalism developed in the previous sections, $\mathbf{v}(\rho, p) = \mathbf{u}_{00}(\rho, p)$ and substituting this into Eqs. (15) and (16), gives the normalised difference photocurrent

$$\langle \Delta i_{SD}\rangle_{d\ll w_0} = \sqrt{\frac{2}{\pi}}\frac{2|\alpha(p)|^2\tau p}{w_0}, \tag{30}$$

where τ is the measurement time. The difference photo-current is linearly proportional to the displacement p. In the regime where the displacement is assumed to be small (whereby $p \ll w_0$ and w_0 is the waist of the incident laser field), the normalised difference photo-current begins to roll off and asymptotes to a constant for larger p. This can be easily understood, since for $p \gg w_0$ the beam is incident almost entirely on one side of the detector. In this regime, large beam displacements only cause small variations in $\langle \Delta i_{SD}\rangle$, making it difficult to determine the beam displacement precisely.

For small displacements, the sensitivity of the displacement measurement is found to be given by (Hsu *et al.* (2004))

$$S_{SD,d\ll w_0} \approx \sqrt{\frac{2}{\pi}}\frac{2|\alpha(p)|\sqrt{\tau}}{w_0}. \tag{31}$$

4.2 Spatial homodyne detection

As discussed earlier, we can use the optimal spatial parameter estimation scheme based on spatial homodyne detection, to detect the beam position in AFM systems. First, the relevant signal mode $\mathbf{w}(\rho)$ of the displaced TEM_{00} input field is identified. A Taylor expansion of the displaced $\mathbf{v}(\rho, p) = \mathbf{u}_{00}(\rho, p)$ input mode provides the relevant displacement signal mode $\mathbf{w}(\rho)$. The coefficients of the Taylor expansion as a function of beam displacement are illustrated in Fig. 4 (c). For small displacements, only the TEM_{00} and TEM_{10} modes have significant non-zero coefficients (Anderson, 1984). This means that for small displacements the TEM_{10} mode contributes most to the displacement signal. For larger displacement, other higher order modes become significant as their coefficients increase. Therefore a spatial homodyne measurement of the displaced TEM_{00} mode using a LO with centred TEM_{10} mode-shape is optimal in the small displacement regime. From Fig. 4 (c), we see that when the input beam is centred, no power is contained in the TEM_{10} mode. Since the Hermite-Gauss modes are orthonormal, the TEM_{10} local oscillator beam only detects the TEM_{10} vacuum noise of the input beam. However as the TEM_{00} beam is displaced, power is coupled into the TEM_{10} mode. This coupled power interferes with the TEM_{10} local oscillator, causing a linear change in the photo-current observed by the homodyne detector.

Using Eq. (1), the electric field operator describing the TEM_{10} local oscillator beam is given by

$$\tilde{\mathbf{E}}^{+}_{\rm LO}(\rho) = \sqrt{\frac{\hbar\omega}{2\epsilon_0 V}}\left(\alpha_{\rm LO}\mathbf{u}_{10}(\rho) + \sum_{m,n=0}^{\infty}\delta\hat{a}^{\rm LO}_{mn}\mathbf{u}_{mn}(\rho)\right) \quad (32)$$

where the first bracketed term is the coherent amplitude which resides in the TEM_{10} mode, the second bracketed term denotes the quantum fluctuations of the beam, and $\alpha_{\rm LO}$ is the coherent amplitude of the LO. The difference photo-current between the two detectors used in the spatial homodyne detection can then be shown from Eqs. (21) to be (Hsu *et al.* (2004))

$$\begin{aligned} \Delta i_{\rm SH} &= \alpha_{\rm LO}\left(2p\cdot\frac{\partial[\alpha(p)\mathbf{u}_{00}(\rho,p)]}{\partial p} + \delta\hat{X}^{+}_{10}\right) \\ &= \alpha_{\rm LO}\left(\frac{2|\alpha(p)|^2\sqrt{\tau}}{w_0}p + \delta\hat{X}^{+}_{10}\right) \end{aligned} \quad (33)$$

where $\delta\hat{X}^{+}_{1} = \delta\hat{a}_1 + \delta\hat{a}^{\dagger}_{1}$ is the amplitude quadrature noise operator of the TEM_{10} component of the displaced beam, and we have assumed that $\alpha_{\rm LO} \gg \alpha(p)$.
The spatial homodyne detection sensitivity, obtained in the same manner as that for split detection, is shown in Figure 4 (b). In the small displacement regime, we obtain

$$S_{{\rm SH},d\ll w_0} = \frac{2|\alpha(p)|\sqrt{\tau}}{w_0} \quad (34)$$

The spatial homodyne detector was shown to be optimal in Section 3.3. Therefore Eq. (34) sets the optimal sensitivity achievable for small displacement measurements. A comparison of the efficiency of split detection for small displacement measurement with respect to the spatial homodyne detector is given by the ratio

$$\epsilon_{\rm SD} = \frac{S_{{\rm SD},d\ll w_0}}{S_{{\rm SH},d\ll w_0}} = \sqrt{\frac{2}{\pi}} \sim 80\%. \quad (35)$$

This $\sqrt{2/\pi}$ factor arises from the coefficient of the mode overlap integral, between $\mathbf{v}(\rho, p) = \mathbf{u}_{00}(\rho, p)$ and $\mathbf{v}_f(\rho, p) = \mathbf{u}_{f00}(\rho, p)$, as shown in Eq. (15), where $\mathbf{u}_{f00}(\rho, p)$ is the flipped TEM_{00} mode. Fig. 4 (b) shows that the maximum sensitivity of split detection is limited at ~80 %. The sensitivity decreases and asymptotes to zero for large displacement, and is below optimal for all displacement values.

4.3 Using spatial squeezing to enhance measurements for split detection systems

The detection mode arising from the geometry of a split detector is the flipped mode $\mathbf{v}_f(\rho)$. Therefore, in order to perform sub-QNL measurements using a split detector, squeezing of the flipped mode is required. In the case of a quadrant detector, since both horizontal and vertical displacements can be monitored, there exist two detection modes. Therefore, two spatial squeezed modes are required to achieve sub-QNL measurements along two different axes in the transverse plane. An experimental demonstration of simultaneous squeezing for quadrant detection along two different axes in the transverse plane was shown by Treps *et al.* (2003). In their experiment, three beams were required - a bright coherent field with a horizontally phase flipped mode-shape, denoted by TEM_{f00}, and two squeezed fields with TEM_{00} and TEM_{f0f0} (i.e. both phase flipped in the horizontal and vertical axes) mode-shapes, as shown in Fig. 5 (a). The mode-shape was obtained via the implementation of phase flips on the quadrants in the transverse field. These modes were then overlapped using low finesse, impedance matched optical cavities, with the resulting field imaged onto on a quadrant detector. Measurements of the beam position fluctuations in the horizontal axis were performed by taking the difference between the photocurrents originating from the left and right halves of the quadrant detector. Correspondingly, the beam position fluctuation in the vertical axis were obtained from the difference of the photocurrents from the top and bottom halves of the detector. Treps *et al.* (Treps *et al.*, 2003) demonstrated that simultaneous sub-QNL fluctuations in both horizontal and vertical beam position are obtainable.

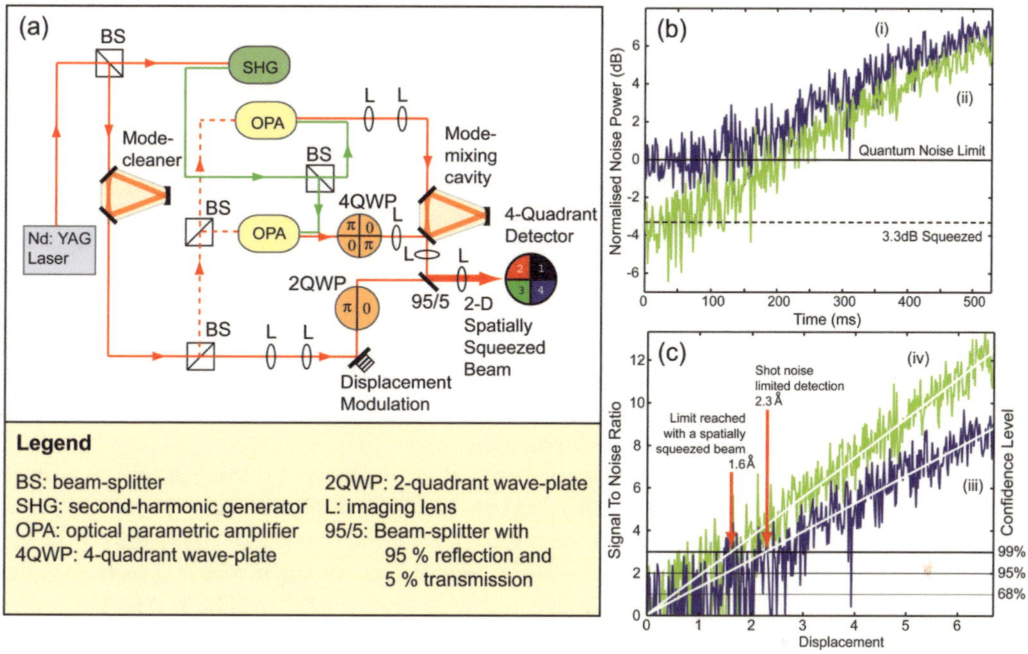

Fig. 5. (a) Schematic of experimental setup for the production of a 2-dimensional spatial squeezed beam for quadrant detection. (b) Measurements of a displacement signal with increasing amplitude in time performed using a beam in the (i) coherent state and (ii) spatial squeezed state. (c) Signal-to-noise ratio (left vertical axis) versus measured displacement. Traces (iii) and (iv) are results obtained from data traces (i) and (ii), respectively. RBW = VBW = 1 kHz, averaged over 20 traces each with detection time Δt = 1 ms per data point. Figures were reproduced from Treps *et al.* (2002), with permission.

Treps *et al.* (2003) also showed that simultaneous sub-QNL measurements of a displacement signal along two different axes can be produced. A displacement modulation at frequency Ω was applied to the spatial squeezed beam via the use of a mirror mounted on a PZT. The amplitude of the displacement modulation was determined by demodulating the photocurrent at frequency Ω and then measuring the power spectral density, using a spectrum analyser. The measured signal consists of the sum of the squares of the quantum noise with and without applied displacement modulation, given by $p_{mod}(\Omega)$ and $p_{noise}(\Omega)$, respectively. For a displacement amplitude modulation that increased with time, the results

of the measurement are shown in Fig. 5 (b). Curve (i) is the result of a displacement measurement performed with a coherent state input beam and sets the classical limit to displacement measurements using quadrant detectors. Curve (ii) is the result of a displacement measurement performed using the spatially squeezed beam.
In order to determine the smallest detectable displacement signal, the results obtained were normalised to the respective noise levels for the coherent and the spatially squeezed beams, shown in Fig. 5 (c). The vertical axis is the signal-to-noise ratio for the displacement measurement. With a 99%confidence level, the smallest detectable displacement is 2.3 Å for a coherent state beam. With the use of the spatially squeezed beam, the smallest detectable displacement was 1.6 Å. Therefore, the spatial squeezed beam provided a factor of 1.5 improvement in the smallest detectable displacement signal, over the coherent state beam.

4.4 Using spatial squeezing to enhance measurements for spatial homodyne detection

Although squeezing of the flipped mode $\mathbf{v}_f(\rho)$ enhances beam displacement measurements on a split detector with sensitivity below the QNL, this scheme remains non-optimal for beam displacement measurements. As shown in previous sections, the QNL for beam displacement measurements is reached in a spatial homodyne detector, assuming the imaging field is in the coherent state. Therefore in order to surpass this QNL, squeezing of the signal mode $\mathbf{w}(\rho)$ responsible for the beam displacement is required. Following the theoretical treatment by Hsu *et al.* (2004), Delaubert *et al.* (2006) performed the first experimental demonstration of squeezing the TEM_{10} displacement signal mode, for an incident TEM_{00} beam, followed by displacement signal detection using a TEM_{10} local oscillator beam in the spatial homodyne detector.
The squeezed TEM_{10} mode of the incident beam was produced by imaging the squeezed TEM_{00} output beam from an optical parametric oscillator (OPO) onto a phase mask, as shown in Fig. 6 (a). The phase mask converts the TEM_{00} squeezed beam into a TEM_{10} squeezed beam with an efficiency of ~80 %. By using an asymmetric Mach-Zehnder interferometer for combining odd and even-ordered modes, the bright TEM_{00} beam (i.e. $\mathbf{v}(\rho)$) was combined with the squeezed TEM_{10} beam (i.e. $\mathbf{w}(\rho)$). The resulting beam was spatially squeezed for optimal detection of small beam displacements. Using a mirror actuated via a PZT, displacement of the beam at RF frequencies was imposed. However, this actuation scheme also introduced a tilt to the beam, therefore the beam was effectively displaced and tilted in the transverse plane, given by

$$\begin{aligned} \mathbf{u}_{00}(\rho, p=d) &= \mathbf{u}_{00}(\rho,0) + d\cdot\frac{\partial \mathbf{u}_{00}(\rho,0)}{\partial d} = \mathbf{u}_{00}(\rho,0) + \frac{d}{w_0}\cdot \mathbf{u}_{10}(\rho,0) \\ \mathbf{u}_{00}(\rho, p=\theta) &= \mathbf{u}_{00}(\rho,0) + \theta\cdot\frac{\partial \mathbf{u}_{00}(\rho,0)}{\partial \theta} = \mathbf{u}_{00}(\rho,0) + i\frac{\pi\theta w_0}{\lambda}\cdot \mathbf{u}_{10}(\rho,0) \end{aligned} \tag{36}$$

where d, θ and w_0 are the displacement, tilt and waist of the beam, respectively. The small beam displacement signal is contained in the amplitude quadrature of the TEM_{10} mode, whilst the small beam tilt signal is contained in the phase quadrature of the TEM_{10} mode. The displacement and tilt of a beam have been shown to be conjugate observables (Hsu *et al.*, 2005).
The resulting modulated beam was subsequently analysed by interference with a TEM_{10} local oscillator beam, produced via an optical cavity resonant on the TEM_{10} mode. Note that

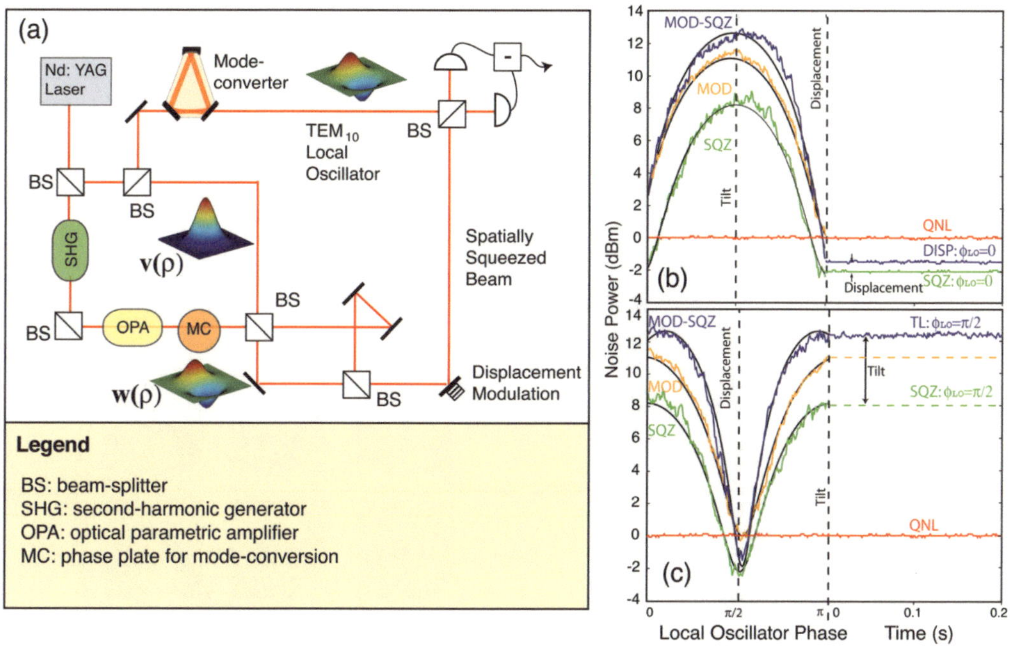

Fig. 6. (a) Schematic diagram of the experimental demonstration of sub-QNL beam displacement measurement using a spatial homodyne detector. Measurements of the (b) displacement and (c) tilt modulation signals using a spatial homodyne detector. The tilt signal was significantly greater than the displacement signal (9:1). Initially the LO and input beam phases were scanned from 0 to π, then was subsequently locked. SQZ: the quadrature noise for the TEM_{10} squeezed mode without modulation signal, resulting in 2 dB of squeezing and 8 dB of anti-squeezing. MOD: the applied modulation signal detected with coherent light only. MOD-SQZ: measured modulation signal using TEM_{10} squeezed light. Since the squeezed TEM_{10} mode was in-phase with the TEM_{00} bright mode component, the displacement measurement was improved, whilst the tilt measurement was degraded. The TEM_{00} waist size was $w_0 = 106\ \mu$m in the PZT plane, beam power 170 μW, RBW=100 kHz, and VBW=100 Hz, corresponding to a minimum resolvable displacement QNL of $d_{QNL} = 0.6$ nm. Figures (b) and (c) were reproduced from Delaubert *et al.* (2006), with permission.

the strength of the spatial homodyne detector is that it can also measure beam tilt, which is not accessible in the plane of a split detector, simply by adjusting the relative phase between the LO and the input beams. The resulting interfered beams were then detected on two photodetectors and their photocurrents analysed on a spectrum analyser. The results are shown in Fig. 6 (b) and (c), for different relative phase values between the TEM_{10} local oscillator and spatial squeezed beams.

The displacement and tilt of the input beam were accessed by varying the phase of the local oscillator. When the TEM_{10} mode was in phase with the bright TEM_{00} mode component, displacement measurements were enhanced below the QNL, as shown in Fig. 6 (c). Since beam displacement and tilt are conjugate observables, an improvement in the beam displacement measurement degraded the tilt measurement, shown in Fig. 6 (b). The minimum resolvable displacement was d_{exp} = 0.15 nm, significantly better than was achievable without the use of spatially squeezed light.

5. Practical applications 2: Particle sensing in optical tweezers

Optical tweezer systems (Ashkin, 1970) have been used extensively for obtaining quantitative biophysical measurements. In particular, particle sensing using optical tweezers provides information on the position, velocity and force of the specimen particles.
A conventional optical tweezers setup is shown in Fig. 7 (a), where a TEM_{00} trapping field is focused onto a scattering particle. The effective restoring/trapping force acting on the particle is due to two force components: (i) the *gradient force* $\mathbf{F}_{grad}$ resulting from the intensity gradient of the trapping beam, which traps the particle transversely toward the high intensity region; and (ii) the *scattering force* $\mathbf{F}_{scat}$ resulting from the forward-direction

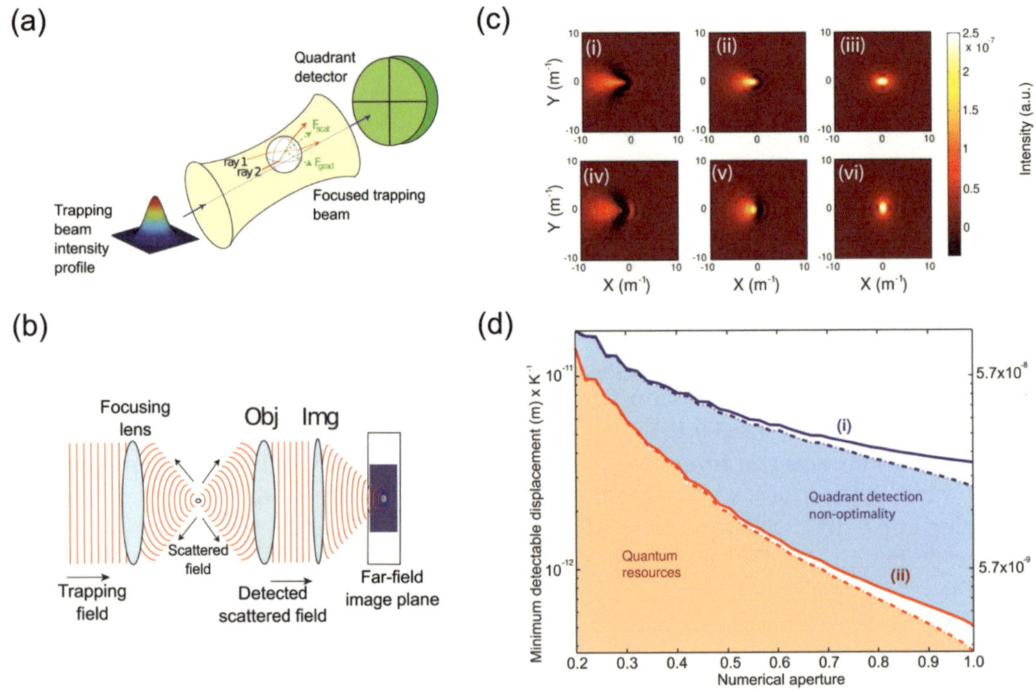

Fig. 7. (a) Illustration showing a TEM_{00} trapping field focussed onto a spherical scattering particle. The gradient and scattering forces are given by F_{grad} and F_{scat}, respectively. (b) Schematic representation of the trapping and scattered fields in an optical tweezers. The trapping field is incident from the left of the diagram. Obj: objective lens, and Img: imaging lens. (c) Interference pattern of the trapping and forward scattered fields in the far-field plane. Figures (i)-(iii) and (iv)-(vi) assume that the trapping field is x-and y-linearly polarised, respectively. The particle displacements are given by (i), (iv): 1 μm; (ii), (v): 0.5 μm; and (iii), (vi): 0 μm. (d) Minimum detectable displacement normalised by K, versus collection lens NA for (i) split and (ii) spatial homodyne detection. The solid and dashed lines are for x- and y-linearly polarised trapping fields, respectively. The axis on the right shows the minimum detectable displacement assuming 200 mW trapping field power, λ = 1064 nm, particle radius a = 0.1 μm, permittivity of the medium ϵ_1 = 1, permittivity of the particle ϵ_1 = 3.8 and trapping field focus of 4 μm. Absorptive losses in the sample were assumed to be negligible. Figures (b), (c) and (d) were reproduced from Tay *et al.* (2009), with permission.

radiation pressure of the trapping beam incident on the particle. In the focal region of the trapping field, the gradient force dominates over the scattering force, resulting in particle trapping.

To obtain a physical understanding of the trapping forces involved, consider the case with a spherical particle, which has a diameter larger than the trapping field wavelength. Rays 1 and 2 are refracted in the particle, and consequently undergo a momentum change resulting in an equal and opposite momentum change being imparted on the particle. Due to the in tensity profile of the beam, the outer ray is less intense than the inner ray which results in the generation of the gradient force (Ashkin, 1992).

If the particle has radius smaller than the wavelength of the trapping laser however, the trapping force is instead generated by an induced dipole moment. In this size regime, the actual shape of the particle is no longer important so long as the particle has no structural deviations greater than the wavelength of the trapping beam. Hence the particle can be treated as a normal dipole with an induced dipole moment along the direction of trapping beam polarisation. The gradient force acting on the particle is then generated due to the interaction of its induced dipole moment with the transverse electromagnetic fields of the trapping field. This force is proportional to the intensity of the beam and has the same net result as before; it acts to return the particle to the centre of the trapping beam focus.

A particle in the beam path will also scatter light. The transverse scattered field profile is dependent on the position of the particle with respect to the centre of the trapping field. By imaging the scattered field onto a position sensitive detector, the position and force of the trapped particle is able to be measured. For these measurements, split detection is most commonly used (Lang & Block, 2003; Gittes & Schmidt, 1998; Pralle *et al.*, 1999), although some direct measurement techniques utilise CCD arrays. To demonstrate the potential enhancement of measurements, we compare split detection and spatial homodyne scheme.

5.1 Modelling

For a single spherical, homogeneous particle with diameter much smaller than the wavelength, the scattered field can be modelled as dipole radiation (Van de Hulst, 1981)[2]. The total field after the objective lens consists of both the scattered and residual trapping fields, given by (Tay *et al.*, 2009)

$$\overline{\mathbf{E}}^{\mathrm{I}+}_{\mathrm{total}}(\rho, p) = \overline{\mathbf{E}}^{\mathrm{I}+}_{\mathrm{scat}}(\rho, p) + \overline{\mathbf{E}}^{\mathrm{I}+}_{\mathrm{trap}}(\rho). \tag{37}$$

where $\overline{\mathbf{E}}^{\mathrm{I}+}_{\mathrm{scat}}(\rho, p)$ and $\overline{\mathbf{E}}^{\mathrm{I}+}_{\mathrm{trap}}(\rho)$ are the respective complex scattered and trapping fields at the image plane. To demonstrate how the changing particle position affected the field at the image plane, Tay *et al.* (2009) calculated the interference between the forward scattered and residual trapping fields for a range of particle displacements in the plane of the trap waist, shown in Fig. 7 (c) for trapping field (i) x- and (ii) y-linearly polarised. Note that the distribution of the field was compressed in the direction of the trapping field polarisation due to dipole scattering along the polarisation axis.

---

[2] In the case where there are multiple inhomogeneous particles scattering the input trapping field, several numerical methods exist to calculate the scattered field - e.g. the finite difference frequency domain and T-matrix hybrid method (Loke *et al.*, 2007); and the discrete-dipole approximation and point matching method (Loke *et al.*, 2009).

As before, the critical parameters for assessing sensitivity of particle monitoring are $\alpha(p)$, $\mathbf{v}(\Gamma, p)$ and $\mathbf{w}(\Gamma)$. Using Eq. (6) we obtain

$$\begin{aligned}\mathbf{v}(\Gamma,p) &= -iN_v\overline{\mathbf{E}}^{\mathrm{I}+}_{\mathrm{total}}(\Gamma,p)\\ &= -i\sqrt{\frac{2\epsilon_0 V}{\hbar\omega}}\frac{1}{\alpha_{\mathrm{trap}}}\left(\overline{\mathbf{E}}^{\mathrm{I}+}_{\mathrm{scat}}(\Gamma,p)+\overline{\mathbf{E}}^{\mathrm{I}+}_{\mathrm{trap}}(\Gamma)\right),\end{aligned} \tag{38}$$

where α_{trap} is the coherent amplitude of the trapping field. Now using Eq. (9) the functional form for the mode that contains information about the particle position is given by

$$\mathbf{w}(\Gamma) = -iN_w\frac{\partial\overline{\mathbf{E}}^{\mathrm{I}+}_{\mathrm{scat}}(\Gamma,p)}{\partial p}\bigg|_{p=0}. \tag{39}$$

Note that this mode is only dependent on the scattered field.
It is then possible to calculate the SNR of the spatial homodyne and split detection schemes for particle sensing in an optical tweezers arrangement. By substituting the expressions obtained in Eq. (39) into Eq. (25), the SNR for the spatial homodyne detection scheme is given by

$$\begin{aligned}\mathrm{SNR}_{\mathrm{SH,coh}} &= -2i\sqrt{\frac{2\epsilon_0 V}{\hbar\omega}}\int_{-\infty}^{\infty}\mathbf{w}(\Gamma)^*\cdot\overline{\mathbf{E}}^{\mathrm{I}+}_{\mathrm{scat}}(\Gamma,p)d\Gamma\\ &= -2K\sqrt{2V}\int_{-\infty}^{\infty}\mathbf{w}(\Gamma)^*\cdot\mathbf{A}(\Gamma)d\Gamma,\end{aligned} \tag{40}$$

where the image plane co-ordinates are given by Γ and

$$K = \alpha_{\mathrm{trap}}k^2a^3\left(\frac{\epsilon_1-\epsilon_2}{\epsilon_1+2\epsilon_2}\right), \tag{41}$$

where ϵ_1 and ϵ_2 are the respective permittivity of the medium and particle; and a is the radius of the particle. In a similar manner, using Eq. (27), the SNR for the split detection scheme is given by

$$\mathrm{SNR}_{\mathrm{SD,coh}} = -2K\sqrt{2V}\int_{-\infty}^{\infty}\mathbf{v}_f(\mathbf{\Gamma})^*\cdot\mathbf{A}(\mathbf{\Gamma})d\mathbf{\Gamma}. \tag{42}$$

Correspondingly, the sensitivities for the spatial homodyne and split detection schemes can be calculated using Eqs. (26) and (28), respectively. The explicit forms for these expressions can be found in Tay *et al.* (2009).

5.2 Results

The performance of both split and spatial homodyne detection schemes were compared by considering the sensing of particle displacement from the centre of the optical tweezers trap (Tay *et al.*, 2009). The SNR for (a) split; and spatial homodyne detection in the (b) small and (c) large displacement regimes are shown in Fig. 8. It was found that the SNR for spatial homodyne detection was maximised at different particle displacement regimes depending on the LO mode used.

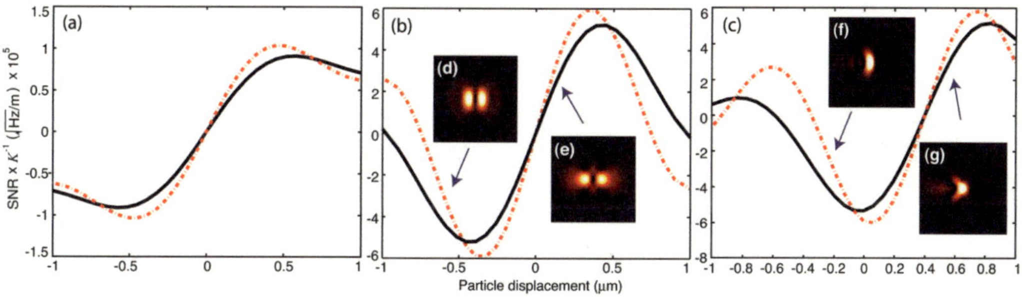

Fig. 8. SNR normalised to K versus particle displacement with respect to the centre of the trapping field, for (a) split detection; (b) spatial homodyne detection with LO spatial mode optimised for small displacement measurements; and (c) spatial homodyne detection with LO spatial mode optimised for larger displacement measurements. The black solid and red dashed lines are for x- and y-linearly polarised trapping fields, respectively. The corresponding LO spatial modes are the inset figures with (d) y- and (e) x-linearly polarised trapping fields for the small displacement regime. For the large displacement regime, the LO spatial modes are correspondingly: (f) y- and (g) x-linearly polarised trapping fields. Model parameters are 200 mW trapping field power, $\lambda = 1064$ nm, particle radius $a = 0.1$ μm, permittivity of the medium $\epsilon_1 = 1$, permittivity of the particle $\epsilon_1 = 3.8$ and trapping field focus of 4 μm. Absorptive losses in the sample were assumed to be negligible. Figures were reproduced from Tay *et al.* (2009), with permission.

Assuming small displacements, the LO field was determined from the first order term in the Taylor expansion of Eq. (9) for the scattered field, with the SNR given in Fig. 8 (b). For particle displacements less than the trapping beam waist, linearity of the SNR was obtained. Optimum sensitivity (i.e. the maximum SNR slope) occurred at zero displacement and surpassed the maximum of split detection by almost an order of magnitude. However, for particle displacements ~ |0.4j| μm, the SNR peaked, indicating that small displacements of a particle around ~ |0.4| μm are not resolvable using the current LO mode. As the particle was displaced further, a drop in the total scattered power was observed due to the particle moving out of the trapping field, resulting in an exponential decay of the SNR. To re-optimise the LO mode for particle displacement around any arbitrary position, a Taylor expansion in p of the scattered field can be taken at that position while only retaining the first order term. For example, for particles fluctuating around ~ |0.4| μm, the re-optimised LO mode resulted in the SNR given in Fig. 8 (c) where the maximum SNR slope was now located around ~ ±0.4 μm. Therefore, it is possible to dynamically adjust the LO mode to optimise the measurement sensitivity whilst the particle moves, resulting in optimum particle sensing for all displacement values.

The corresponding sensitivities for (i) split and (ii) spatial homodyne detection as a function of increasing objective lens NA are shown in Fig. 7 (d). It was observed that the minimum detectable displacement for both split and spatial homodyne detection decreased with increasing NA due to more scattered field being collected, thereby providing more information about the particle position. However, spatial homodyne detection outperforms split detection for all NA values, since spatial homodyne optimally extracts the displacement information from the detected field, whereas the split detection scheme only measures partial displacement information, as shown in Eq. (15). Due to the optimal signal and noise measurement using the spatial homodyne scheme, curve (ii) defines the minimum detectable displacement in optical tweezers systems.

To provide quantitative values for the minimum detectable displacement, the sensitivities for both detection schemes using realistic experimental values are shown in the right-hand side axis of Fig. 7 (d). The *split detection non-optimality* shaded area shows the loss in particle sensing sensitivity due to incomplete mode detection from a split detector. The *quantum resources* shaded area shows that quantum resources such as spatial squeezed light (Treps *et al.*, 2002; 2003) can be used to further enhance the particle sensing sensitivity beyond the QNL.

The ability to tailor the local oscillator mode provides tremendous optimisation ability for particle sensing. Not only is optimal information extraction possible, but it is now possible to perform sensing of multiple inhomogeneous particles, with information extraction of any spatial parameter p, via the modification of the LO mode.

6. Conclusion

We have presented a quantum formalism for the measurement of the spatial properties of an optical field. It was shown that the spatial homodyne technique is optimal and outperforms split detection for the detection of spatial parameter p. Applications of this measurement scheme in enhancing the sensitivities of atomic force microscopes and optical tweezers measurements have been discussed.

7. References

Anderson, D. Z. (1984). Alignment of resonant optical cavities. *Appl. Opt.*, Vol. 23, No. 17, 2944-2949.

Arnon, S. (1998). Use of satellite natural vibrations to improve performance of free-space satellite laser communication. *Appl. Opt.*, Vol. 37, 5031.

Ashkin, A. (1992). Forces of a single-beam gradient laser trap on a dielectric sphere in the ray optics regime. *Biophys. J.*, Vol. 61, No. 2, 569–582.

Ashkin, A. (1970). Acceleration and Trapping of Particles by Radiation Pressure. *Phys. Rev. Lett.*. Vol. 24, No. 4, 156-159.

Beck M. (2000). Quantum state tomography with array detectors *Phys. Rev. Lett.*, Vol. 84, No. 25, 5748.

Binnig, G.; Quate, C. F.; and Gerber, Ch. (1985). Atomic Force Microscope. *Phys. Rev. Lett.*, Vol. 56, No. 9, 930-933.

Block S. M. (1992). Making light work with optical tweezers *Nature*, Vol. 360, 493-495.

Born M. and Wolf E. *Principles of Optics* (Cambridge University Press, Cambridge, UK, 1999)

Bowen W. P., Schnabel R., Bachor H.-A. , and Lam P. K. (2002). Polarization squeezing of continuous variable stokes parameters. *Physical Review Letters* Vol. 88, 093601.

Dawes, A.M.; Beck, M.; and Banaszek, K. (2003). Mode optimization for quantum-state tomography with array detectors. *Phys. Rev. A.*, Vol. 67, 032102.

Dawes, A.M.; and Beck, M. (2001). Simultaneous quantum-state measurements using array detection. *Phys. Rev. A.*, Vol. 63, 040101(R).

Delaubert, V., Treps, N., Harb, C. C., Lam, P. K., and Bachor, H.-A. (2006). Quantum measurements of spatial conjugate variables: displacement and tilt of a Gaussian beam. *Opt. Letts.*, Vol. 31, No. 10, 1537-1539.

Delaubert, V.; Treps, N.; Fabre, C.; Bachor, H.-A.; and R´efr´egier, P. (2008). Quantum limits in image processing. *Europhys. Lett.*, Vol. 81, 44001.

Fabre, C.; Fouet, C. J.; and Maˆıtre, A. (2000). Quantum limits in the measurement of very small displacements in optical images. *Opt. Lett.*, Vol. 25, pp. 76.

Fritschel, P.; Mavalvala, N.; Shoemaker, D.; Sigg, D.; Zucker, M.; and González, G. (1998). Alignment of an interferometric gravitational wave detector. *Appl. Opt.*, Vol. 37, No. 28, 6734–6747.

Fritschel, P.; Bork, R.; Gonz´alez, G.; Mavalvala, N.; Ouimette, D.; Rong, H.; Sigg, D.; and Zucker, M. (2001). Readout and Control of a Power-Recycled Interferometric Gravitational-Wave Antenna. *Appl. Opt.*, Vol. 40, No. 28, 4988–4998.

Gittes, F.; and Schmidt, C. F. (1998). Interference model for back-focal-plane displacement detection in optical tweezers. *Opt. Lett.* Vol. 23, No. 1, 7–9.

Hell, S.W.; Schmidt, R.; and Egner, A. (2009). Diffraction-unlimited three-dimensional optical nanoscopy with opposing lenses *Nature Photonics* 3, 381-387.

Hsu, M. T. L.; Delaubert, V.; Lam, P. K.; and Bowen,W. P. (2004). Optimal optical measurement of small displacements. *Journal of Optics B:Quantum and Semiclassical Optics*, Vol. 6, No. 12, 495-501.

Hsu, M. T. L.; Bowen, W. P.; Treps, N.; and Lam, P. K. (2005). Continuous-variable spatial entanglement for bright optical beams. *Phys. Rev. A*, Vol. 72, No. 1, 013802.

Hsu, M. T. L.; Bowen,W. P. and Lam, P. K. (2009). Spatial-state Stokes-operator squeezing and entanglement for optical beams. *Physical Review A*, Vol. 79, No. 4, 043825.

Korolkova, N. V.; and Chirkin, A. S. (1996). Formation and conversion of the polarization-squeezed light. *Journal of Modern Optics*, Vol. 43, issue 5, 869-878.

Lang, M. J.; and Block, S. M. (2003). LBOT-1: Laser-based optical tweezers. *Am. J. Phys.*, Vol. 71 No. 3, 201–215; and references therein.

Loke, V. L. Y.; Nieminen, T. A.; Parkin, S. W.; Heckenberg, N. R.; and Rubinsztein-Dunlop, H. (2007). FDFD/T-matrix hybrid method. *J. Quant. Spec. Rad. Trans.*, Vol. 106, No. 1, 274–284.

Loke, V. L. Y.; Nieminen, T. A.; Heckenberg, N. R.; and Rubinsztein-Dunlop, H., J. (2009). T-matrix calculation via discrete dipole approximation, point matching and exploiting symmetry. *Quant. Spec. Rad. Trans.*, Vol. 110, No. 14, 1460–1471.

Mach, L. (1892). Z. Instrumentenkunde 12, 89.

Nikulin, V. V.; Bouzoubaa, M.; Skormin, V. A.; and Busch, T. E. (2001). Modeling of an acousto-optic laser beam steering system intended for satellite communication. *Opt. Eng.*, Vol. 40, 2208.

Pawley J. (1995). Handbook of Biological Confocal Microscopy, Springer, Germany.

Pendry, J. B. (2000). Negative refraction makes a perfect lens. *Phys. Rev. Lett.* , Vol. 85, No. 14, 3966-3969.

Pralle, A.; Prummer, M.; Florin, E.-L.; Stelzer, E. H. K.; and Hörber, J. K. H. (1999). Threedimensional high-resolution particle tracking for optical tweezers by forward scattered light. *Microsc. Res, Tech.*, Vol. 44, No. 5, 378-386.

Putman, C. A.; De Grooth, B. G.; Van Hulst, N. F.; and Greve, J. (1992). A detailed analysis of the optical beam deflection technique for use in atomic force microscopy. *J. Appl. Phys.*, Vol. 72, No. 1, 6.

Sagnac, G. (1912). L'éther lumineux démontré par l'effet du vent relatif d'éther dans un interféromètre en rotation uniforme. Comptes Rendus 157 (1913), S. 708-710

Saleh, B. E. A.; and Teich, M. C. (1991). Fundamentals of Photonics, JohnWiley and Sons, New York.

Shaddock, D. A.; Gray, M. B.; and McClelland, D. E., (1999). Frequency locking a laser to an optical cavity by use of spatial mode interference. *Opt. Lett.*, Vol. 24, No. 21, pp. 1499-1501.

Slusher, R. E.; Hollberg, L. W.; Yurke, B.; Mertz, J. C.; and Valley, J. F. (1985). Observation of Squeezed States Generated by Four-Wave Mixing in an Optical Cavity, *Phys. Rev. Lett.*, Vol. 55, No. 22, 2409-2412.

Synge, E. H. (1928). A suggested method for extending the microscopic resolution into the ultramicroscopic region. *Phil. Mag.* 6, 356.

Tay, J. W.; Hsu, M. T. L.; and Bowen, W. P. (2009). Quantum limited particle sensing in optical tweezers. e-print arXiv:0907.4198

Treps, N.; Andersen, U.; Buchler, B.; Lam, P. K.; Maˆıtre, A.; Bachor, H.-A.; and Fabre, C. (2002). Surpassing the Standard Quantum Limit for Optical Imaging Using Nonclassical Multimode Light. *Phys. Rev. Lett.*, Vol. 88, No. 20, 203601 (2002).

Treps, N., Grosse, N.; Bowen,W. P.; Fabre, C.; Bachor, H.-A.; and Lam, P. K. (2003). A Quantum Laser Pointer. *Science.*, Vol. 301, No. 5635, 940–943. See also: Treps, N.; Grosse, N.; Bowen, W. P.; Hsu, M. T. L.; Maître, A.; Fabre, C.; Bachor, H.-A.; and Lam, P. K. (2004). Nano-displacement measurements using spatially multimode squeezed light. *J. Opt. B: Quantum Semiclass.*, Vol. 6, No. 8, S664–S674.

Treps, N.; Delaubert, V.; Maître A.; Courty, J.M.; Fabre, C. (2005). Quantum noise in multipixel image processing. *Phys. Rev. A*, Vol. 71, No. 1, 013820.

Van De Hulst, H. C. (1981). *Light Scattering by Small Particles*. Dover Publications, ISBN:0486642283, New York. See also: Berne, B. J., and R. Pecora. (2000). *Dynamic Light Scattering*. Dover Publications, ISBN:0486411559, New York.

Walls, D. F.; and Milburn, G. J. (1995). Quantum Optics, Springer, Germany. Zehnder, L. (1891). Z. Instrumentenkunde 11, 275.

28

Broadband Light Generation in Raman-active Crystals Driven by Femtosecond Laser Fields

Miaochan Zhi, Xi Wang and Alexei V. Sokolov
Texas A&M University
U. S. A

1. Introduction

Short pulse generation requires a wide phase-locked spectrum. Earlier the short pulses were obtained by expanding the spectrum of a mode locked laser from self phase modulation (SPM) in an optical fiber and then compensating for group velocity dispersion (GVD) by diffraction grating and prism pairs. Pulses as short as 4.4 fs have been generated (Steinmeyer et al., 1999). For ultrafast measurements on the time scale of electronic motion, generation of subfemtosecond pulses is needed. Generation of subfemtosecond pulses with a spectrum centered around the visible region is even more desirable, due to the fact that the pulse duration will be shorter than the optical period and will allow sub-cycle field shaping. As a result, a direct and precise control of electron trajectories in photoionization and high-order harmonic generation will become possible. But to break the few-fs barrier new approaches are needed.

In recent past, broadband collinear Raman generation in molecular gases has been used to produce mutually coherent equidistant frequency sidebands spanning several octaves of optical bandwidths (Sokolov & Harris, 2003). It has been argued that these sidebands can be used to synthesize optical pulses as short as a fraction of a fs (Sokolov et al., 2005). The Raman technique relied on adiabatic preparation of near-maximal molecular coherence by driving the molecular transition slightly off resonance so that a single molecular superposition state is excited. Molecular motion, in turn, modulates the driving laser frequencies and a very broad spectrum is generated, hence the term for this process "molecular modulation". By phase locking, a pulse train with a time interval of the inverse of the Raman shift frequency is generated. While at present isolated attosecond X-ray pulses are obtained by high harmonic generation (HHG) (Kienberger et al., 2004), the pulses are difficult to control because of intrinsic problems of X-ray optics. Besides, the conversion efficiency into these pulses is very low (typically 10^{-5}). On the other hand, the Raman technique shows promise for highly efficient production of such ultrashort pulses in the near-visible spectral region, where such pulses inevitably express single-cycle nature and may allow non-sinusoidal field synthesis (Sokolov et al., 2005).

In the Raman technique ns pulses are applied for preparing maximal coherence when gas is used as a Raman medium. When the pulse duration is shorter than the dephasing time T_2

($T_2 = \frac{1}{\pi c \Delta \nu_R}$, $\Delta \nu_R$ is the Raman linewidth), the response of the medium is a highly transient process, i.e. the Raman polarization of the medium doesn't reach a steady state within the duration of the pump pulse. In this transient stimulated Raman scattering (SRS) regime, a large coherent molecular response is excited. The advantage of using a short pulse is that the number of pulses in the train will be reduced compared with the ns Raman technique. But when a single fs pump is used, the strong SPM suppresses the Raman generation (Kawano et al., 1998). When the pulse duration is reduced to less than a single period of molecular vibration or rotation, an impulsive SRS regime is reached (Korn et al., 1998). In this regime, an intense fs pulse with a duration shorter than the molecular vibrational period prepares the vibrationally excited state and a second, relatively weak, delayed pulse propagates in the excited medium in the linear regime and experiences scattering due to the modulation of its refractive index by molecular vibrations, which results in the generation of the Stokes and anti-Stokes sidebands (Nazarkin et al., 1999). This technique has the advantage of eliminating the parasitic nonlinear processes since they are confined only within the pump pulse duration.

A closely related approach, which is called four-wave Raman mixing (FWRM) for generating ultrashort pulses using two-color stimulated Raman effect, is proposed by Imasaka (Yoshikawa & Imasaka, 1993). It is based on an experimental result his student has stumbled on. Shuichi Kawasaki was trying to develop a tunable source for thermal lens spectroscopy and he noticed bright, multicolored spots out of the Raman cell pressurized with hydrogen, which they called "Rainbow Star" (Katzman, 2001). The applied beam was supposed to be monochromatic but it actually had two colors in it. To confirm the FWRM hypothesis, a nonlinear optical phenomenon in which three photons interact to produce a fourth photon, they used two-color laser beams with frequencies separated by one of the rotational level splitting for hydrogen (590 cm^{-1}). Indeed, they observed the generation of more than 40 colors through the FWRM process, which provided a coherent beam consisting of equidistant frequencies covering more than thousandths cm^{-1} in frequency domain (Imasaka et al., 1989). This FWRM process resulted in the generation of higher-order rotational sidebands at reduced pump intensity compared to the stimulated Raman scattering. The generation of the FWRM fields required phase matching and were coherently phased, and therefore had the potential to be used to generate sub-fs pulses (Kawano et al., 1999).

Later ps pulses (Kawano et al., 1996), ps and fs pump pulses (Kawano et al., 1997), and a single fs pulse (Kawano et al., 1998) were used to find the optimal experimental conditions for efficient generation of high-order rotational lines. Generally speaking, when the additional Stokes field is supplied rather than grown from quantum noise, advantages include: highorder anti-Stokes generation, higher conversion efficiency, and improved reproducibility of the pulses generated, as shown in earlier experiments with gas in ns regime (Gathen et al., 1990). Recently, efficient generation of high-order anti-Stokes Raman sidebands in a highly transient regime is also observed using a pair of 100-fs laser pulses tuned to Raman resonance with vibrational transitions in methane or hydrogen (Sali et al., 2004; 2005). They found that in this transient regime, the two-color set-up permits much higher conversion efficiency, broader generated bandwidth, and suppression of the competing SPM. The high conversion efficiency observed proves the preparation of substantial coherence in the system although the prepared coherence in the medium cannot be near maximal as in the case of the adiabatic Raman technique.

Almost all these works were carried out using a simple-molecule gas medium such as H_2, D_2, SF_6 or methane since the gas has negligible dispersion and long coherence lifetime. Molecular gas also has a few other advantages as a Raman medium. They are easily obtainable with a high degree of optical homogeneity and have high frequency vibrational modes with small spectral broadening, which leads to large Raman frequency shifts and large Raman scattering cross sections. However, a Raman gas cell with long interaction length is needed due to the lower particle concentration (Basiev & Powell, 1999).

What about a solid Raman medium such as a Raman crystal? As we know, the high density of solids results in the high Raman gain. The higher peak Raman cross sections in crystals result in lower SRS thresholds, higher Raman gain, and greater Raman conversion efficiency (Basiev & Powell, 1999). In addition, there is no need for cumbersome vacuum systems when working with room temperature crystals, and therefore a compact system can be designed.

The difficulty in using crystals is the phase matching between the sidebands because the dispersion in solids is significant. Sideband generation using strongly driven Raman coherence in solid hydrogen is reported but the generation process is very close to that of H_2 gas and solid hydrogen is a very exotic material (Liang et al., 2000). Observation of generation with few sidebands (Stokes or anti-Stokes) in other solid material is nothing new. About two decades ago, Dyson *et. al.* has observed one Stokes (S) and one anti-Stokes (AS) generated on quartz during an experiment designed for another purpose (Dyson et al., 1989). Later, there were numerous works of using Raman crystals for building Raman lasers which extended the spectral coverage of solid-state lasers by using SRS (Pask, 2003). A detailed review of crystalline and fiber Raman lasers is given by Basiev (Basiev et al., 2003). The focus of our work is efficient generation over a broad spectral range. Compared to the crystals that are used for building Raman lasers, the sample we use is much thinner (about 1 mm or less). The phenomenon that we use in our work is essentially different from SRS: in our regime the generation process is fully coherent, does not depend on seeding by spontaneous scattering, and occurs on a time scale much shorter than the inverse Raman linewidth. We use two-color pumping (with the frequency difference matching the Raman frequency), so our sideband generation is more similar to multiple-order coherent Anti-Stokes Raman Scattering (CARS) than to SRS.

Therefore this chapter is focused on the development of efficient broadband generation using Raman crystals. Since coherence lifetime in a solid is typically shorter than in a gas, the use of fs (or possibly ps) pulses is inevitable when working with room-temperature solids. We studied broadband sideband generation in a Raman-active crystal lead tungstate ($PbWO_4$) either with two 50 fs pulses or a pair of time-delayed chirped pulses (Zhi & Sokolov, 2007; 2008). Similar broadband generation is also observed in diamond (Zhi et al., 2008). Coherent high-order anti-Stokes scattering has also been observed in many other types of crystals such as $YFeO_3$, $KTaO_3$, $KNbO_3$ and TiO_2 when two-color femtosecond (fs) pulses are used (Takahashi, 2004; Matsubabra et al., 2006; Matsuki et al., 2007). Great progress has been made recently toward synthesis of ultrashort, even few-cycle pulses using Raman crystal. For example, last year Matsubara *et al.* have demonstrated promising Fourier synthesizer using multiple CARS signals obtained in a $LiNbO_3$ crystal at room temperature, and generated isolated pulses with 25 fs duration at 1 kHz repetition rate (Matsubara et al., 2008). Very recently, they just realized the generation of pulses in the 10 fs regime using multicolor Raman sidebands in $KTaO_3$ (Matsubara et al., 2009).

One related technique is the short pulse generation through cascade four-wave mixing (CFWM) of femtosecond pulses in bulk isotropic nonresonant media. Crespo *et al.* were the first ones to observe the cascade of highly nondegenerate FWM in transparent isotropic condensed medium (150 μm thick BK7 glass slide) and find that the process also corresponds to a coherent scattering effect with geometrically minimized phase mismatch (Crespo et al., 2000). Cascaded FWM and multicolored array generation in a sapphire plate was recently reported by Liu *et al.*, who used two crossing beams of a femtosecond laser (Liu & Kobayashi, 2008). Cascaded nondegenerate FWM technique for high-power single-cycle pulse synthesis in the visible and ultraviolet ranges was realized very recently by Weigand *et al.* (Weigand et al., 2009). They generated coherent spectra spanning over two octaves in bandwidths using a thin fused silica slide, which supports a near-single-cycle 2.2 fs visible-UV pulses. The advantage of this method is that tunable multicolor femtosecond pulses can be generated by changing the crossing angle between the two input beams (Liu et al., 2009). Recently, sub-20- fs multicolor laser pulses using cascaded four-wave mixing with chirped incident pulses has been generated in a 1 mm thick fused-silica glass (Liu & Kobayashi, 2009). The generation of sub-fs pulse using a solid medium seems very promising.
In this chapter, we will first describe our experimental work on sideband generation in lead tungstate and diamond driven by two color ultrashort fs pulses. We will explain, for example, how the generation is affected by the angle between the pump beams, the detuning between the two pump frequencies, the polarization, etc. We will describe how we achieve efficient sideband generation by using a pair of time delayed linearly chirped pulses. After that, we will discuss the two dimensional colored array generation in lead tungstate and diamond when three femtosecond pulses are focused in the crystals. We will show some interesting unsolved "mysterious" experimental results we encountered in our experiment, which may inspire theoretical simulations. At last, we will propose future work that can be extended from this research.

2. Experiment setup

The schematics of the setup is shown in Fig. 1 (a). The seed laser (Mira, Coherent) is a modelocked (Kerr Lens Mode-locking) ultrafast laser that produces ultrashort, wide bandwidth (>50 nm), fs pulses (<20 fs) using Ti:Sapphire as the gain medium. The output pulse has a repetition rate of 76 MHz with a wavelength at 800 nm. The pulses are used to seed an amplifier (Legend, Coherent). The pulse is first stretched by a grating to about 200 ps and then is amplified by a Ti:Sapphire regenerative amplifier system pumped by a pulsed, frequency doubled Nd:YLF laser at 523.5 nm. It is compressed afterwards by the grating compressor to a pulse with energy >1.0 mJ at a 1 kHz repetition rate and pulse duration of <35 fs FWHM. The energy stability is <1% RMS. This amplified pulse is used to pump two computer-controlled optical parametric amplifiers (OPerA, Coherent). It is first down-converted to a signal pulse (the short wavelength, 1150-1600 nm) and an idler pulse (the long wavelength, 1600-2630 nm) by a parametric process. These pulses obtained from the two OPAs can be frequency doubled or mixed with the fundamental pulses to produce up to 30 μJ per 50 fs Gaussian pulse at tunable visible wavelengths. For example, the SHG of the signal beam will produce pulses with wavelength ranging from 575 to 800 nm. The sum frequency generation (SFG) will produce pulses with wavelength range from 533 to 613 nm. These pulses are commonly used in our experiment. With the fundamental pulse, we can choose three pulses to use for excitation. For the convenience of description, here we follow

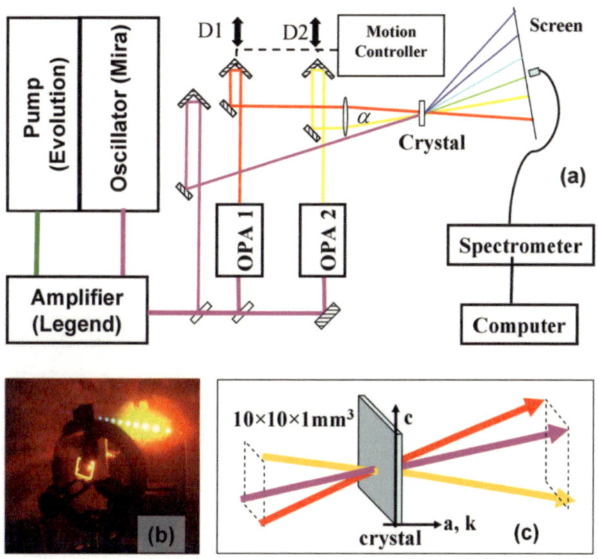

Fig. 1. (a) Experimental setup. D1, D2 are retro-reflectors. They are mounted on a motorcontrolled translation stage. OPA: Optical Parametric Amplifier. The spectra are measured with an Ocean Optics fiber-coupled spectrometer. (b) A picture of the mounted crystal ($PbWO_4$) and the generated sidebands which are projected to a screen behind the crystal. (c) A schematics of the beam geometry with respect to the crystal axises.

the CARS terminology. For the first two pulses, we call the short wavelength one as pump and the longer one as Stokes.We call the third pulses as probe. Normally a small fraction of the fundamental pump pulse (fixed wavelength) is used as the Stokes pulse while the output from the two OPAs (tunable) are either served as a pump or probe pulse. We typically use 1 to 2 μJ per pulse focused to about a 100 μm size spot at the sample. This laser intensity is right below the onset of (strong) SPM. The laser beams are typically sent perpendicular to the large surface and crossed at the crystal after the focal lens with a small angle a, which varies from 2 to 7 degrees. The retro-reflectors in one or two beams (if a probe beam is used) are mounted on a motor controlled translation stage so that the delay between the pulses can be varied with a precision of 1 μm, which is about 6.7 fs. The spectra are measured with a fiber-coupled spectrometer (Ocean Optics, USB 2000, slit size 25 μm, measures wavelength ranging from 180 to 870 nm), which has an average optical resolution of 1.4 nm, with a slightly higher resolution in the visible region. We take the pictures of the sidebands projected on a piece of white paper. In Fig. 1 (b) we show a picture of the crystal together with the sidebands generated and projected on a piece of paper. In Fig. 1 (c) we show a typical geometry of the input beams.

3. Broadband light generation in Raman crystals with two-color laser fields

3.1 Broadband light generation in $PbWO_4$ driven by two fs pulses (Zhi & Sokolov, 2007)

The first sample we choose is $PbWO_4$ (lead tungstate), which exhibits good optical transparency (from 0.33 to 5.5 μm), high damage threshold, and is non-hygroscopic. It is also a popular crystal used for building Raman lasers using ns or ps pulsed pumping (Kaminskii et al., 2000). The $PbWO_4$ sample has a size of 10 × 10 ×1 mm³, with the large surface perpendicular to the a-axis of the crystal. $PbWO_4$ has a strong narrow Raman line at 901 cm^{-1} with linewidth $\Delta\nu_R$=4.3 cm^{-1}, which corresponds to a phonon relaxation time $T_2 \approx$ 2.5 ps. In the past, when much longer single-color 100 ps pulses were applied to $PbWO_4$ (steady state regime, pulse duration $\tau_p >> T_2$), several high-order Stokes and anti-Stokes sidebands were generated (Kaminskii et al., 2000). Another relatively strong Raman line at 325 cm^{-1} has a linewidth of 7.5 cm^{-1}, which is almost comparable to the width of 901 cm^{-1} line for a certain crystal orientation.

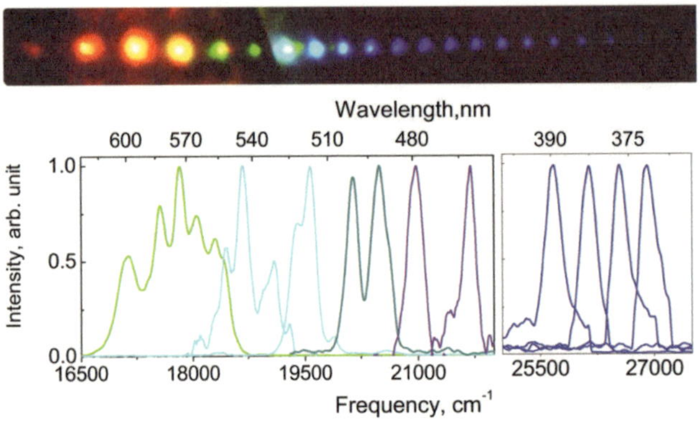

Fig. 2. Broadband generation in $PbWO_4$ crystal with two pulses (at 588 nm and 620 nm) applied at an angle of 4 degrees to each other. Top: generated colors projected on a white paper screen. The two input pulses (bright yellow and red spots), two S and two AS are attenuated by an neutral-density filter. Note that the line connecting the color spots has a slight cusp at intermediate AS orders. Bottom: spectra of the generated sidebands (left: AS 1 to AS6; right: AS 12 to AS 16). The frequency spacing between the sidebands at higher orders decreases gradually.

Using two-color ultrashort pulses (transient regime, $\tau_p >> T_2$), we observe efficient generation of many sidebands in $PbWO_4$ when two pulses (λ_1= 620 nm, λ_2= 588 nm, $\delta\omega$= 930 cm^{-1}) with parallel polarizations are crossed at the crystal with an angle of 4 degrees (Fig. 2). The angles are measured external of the crystal except stated otherwise. The sidebands emerge spatially well-separated and have the same polarization as the two input beams. Up to 20 AS and 2 S sidebands are observed on a white paper screen which is put about 25 cm away from the crystal (Fig. 2 top).

The spectrum of the first 6 AS and the higher order sidebands (AS 12 to AS 16) is shown in Fig. 2 (bottom). Note here we call the frequency upshifted sideband as anti-Stokes (AS) and the frequency downshifted ones as Stokes (S). The spectra of the lower-order sidebands show a rich structure, due to simultaneous excitation of several Raman lines by the large spectral width of the fs laser pulses. The frequency spacing between the sidebands decreases gradually and reaches about 450 cm^{-1} at the highest orders measured, to our surprise.

To prove the Raman-resonant nature of sideband generation, and to separate the effect of instantaneous FWM, we tune the difference between the two applied laser frequencies ($\delta\nu$), and measure the generated AS frequencies. Fig. 3 shows these generated frequencies as a function of the angle at which sidebands emerge from the $PbWO_4$ crystal. We perform this measurement at a relatively large input beam crossing angle of 6 degrees. At this angle, and at sufficiently large $\delta\nu$ (1804 or 2002 cm^{-1}), the generated AS 1 beam splits into two slightly separated distinctively colored beams: one corresponding to (non-resonant) FWM, and the other (which is much brighter) corresponding to Raman-resonant AS generation. By moving the fiber tip of the spectrometer to the location shown in Fig. 3 (insert) by an arrow, we measure the FWM frequency (as opposed to the Raman-shifted frequency measured at the center of the main AS 1 beam). We observe that as we vary $\delta\nu$ from 844 to 2002 cm^{-1}, the Raman sidebands are generated at approximately the same angle and with roughly the same frequency shift from the previous order, while the FWM frequency varies as $\nu_{FWM} = 2\nu_2 - \nu_1$. The observed decrease in the frequency spacing between higher-order sidebands is likely due to phase matching.

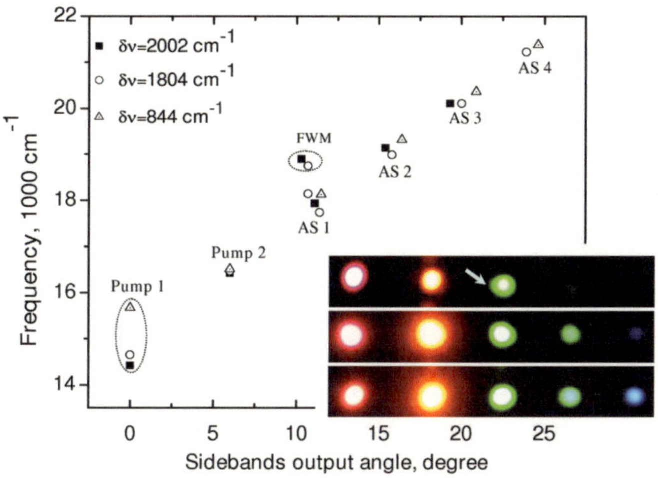

Fig. 3. Peak frequency of the generated sidebands, plotted as a function of the output angle. One input frequency (pump 2) is fixed while the $\delta\nu = \nu_2 - \nu_1$ is tuned to 844 cm^{-1} (triangles), 1804 cm^{-1} (circles), and 2002 cm^{-1} (squares) respectively. The FWM frequency (measured at the point shown on the insert by the arrow) varies as $\nu_{FWM} = 2\nu_2 - \nu_1$ while the Raman sideband frequencies stay approximately fixed. The insert shows the output beams projected onto a screen, for these same values of $\delta\nu$ (varying from 2002 to 844 cm^{-1} top to bottom).

When we vary the angle between the two applied laser beams (while keeping the two wavelengths fixed), we observe substantial changes in both the AS frequency shifts, and in the conversion efficiency. AS conversion is negligible for collinear input beams. The optimum conversion in $PbWO_4$ (for $\delta\nu$= 930 cm^{-1}) is achieved when the angle between the applied beams is 4°. When the angle is further increased, AS conversion decreases, while the frequency separation of the AS sidebands goes up. Apparently, phase matching plays a critical role in the generation of multiple spectral sidebands in Raman-active crystals (as compared to the collinear Raman generation in gasses (Sokolov & Harris, 2003; Sokolov et al., 2005)). Even though at larger beam crossing angles the conversion efficiency is expected to decrease because of the reduced beam overlap, for angles below 7° it is the phase matching, along with the spectrum of exited Raman transitions, that determines the conversion efficiency and the frequencies generated in thin crystals.

The generation is very sensitive to the polarization of the beams. First of all, it has been shown that the Raman gain is strongly peaked when the crystal is excited by a beam whose polarization is parallel to the c-axis of the crystal in the SRS experiment (Kaminskii et al., 2000). Secondly, we find that the generation is best when the pump and Stokes beams have the parallel polarization, which is either parallel or normal to the c-axis. Thirdly, the generation has different frequency spacing when the polarization of the pump beams are parallel with different axis of the crystal as shown in Fig. 4. When we change the polarization of the input beams from S (the polarization is parallel to the plane containing the optical axis c and the wavevector k, see Fig. 4 (a)) to P (perpendicular to S), we observe a significant variation in the generated frequency spacing. We speculate that this is the result of dispersion since when P polarization is used, the crystal sees refractive index of n_o (larger than n_e).

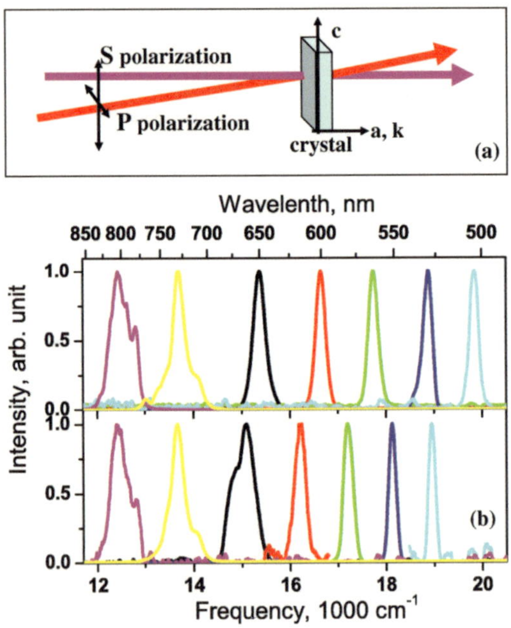

Fig. 4. (a) Crystal orientation and the two possible polarization of the laser beams. (b) The spectra of the two pump beams and the first 5 AS generated in the $PbWO_4$ crystal when two pulses (at 729 nm and 805 nm) are applied at two different polarizations [P (top) and S (bottom)]. The sideband frequencies have different frequency spacings.

3.2 Broadband light generation in diamond driven by two fs pulses (Zhi et al., 2008)

The motivation to use diamond for broadband generation is two-fold. Firstly, diamond has a single strong narrow Raman line at a very large frequency shift (1332 cm^{-1}), compared to the other crystals (Basiev et al., 1999). Secondly, diamond is isotropic and the refractive index is well known, which makes it easier (compared to $PbWO_4$ we studied) to do some theoretical calculation and thus help us understand the complicated experimental results.

Beside the above-mentioned two reasons, diamond also has several remarkable properties which are desirable for broadband generation (Pierson, 1993). First of all, diamond is capable of transmitting over an unusually broad spectral range (from X-ray region to the microwave and mm wavelengths) and has the widest electromagnetic bandpass of any material. This broadband transmission is essential for the sideband generation. Second, it has an extremely high thermal conductivity (five times that of copper) and is extremely chemically inert. Therefore, it is not easy to get damaged by a laser and it does not require much protection from moisture. At last, diamond has the highest atom density of any material. A high density means a high Raman gain.

Back in 1963, diamond was chosen as a representative covalent crystal for the SRS experiment, one AS and two S were observed when a high intensity (20 MW/cm^2) Ruby laser was used for excitation (Eckhardt et al., 1963). Natural diamonds are costly and scarce. Recently, synthesis of large area diamonds at lower pressure using chemical vapor deposition (CVD) technique was developed. Kaminskii *et al.* used CVD diamond for Raman laser converter based on SRS and reported observation of SRS (up to 1 S and 3 AS sidebands) in diamond (using both ns and ps pulses) (Kaminskii et al., 2005; 2006).

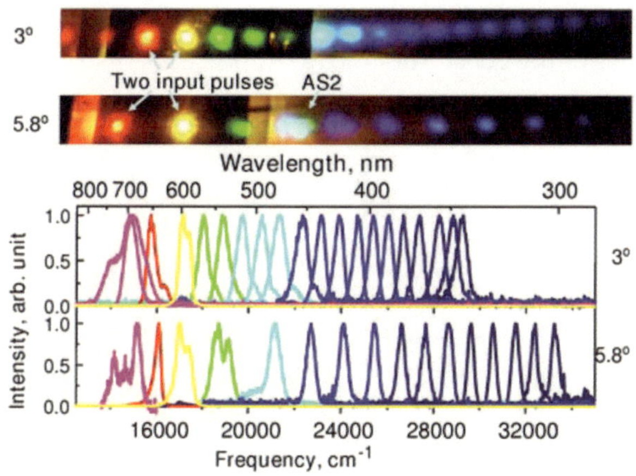

Fig. 5. Broadband generation in diamond with two input pulses (λ_1= 630 nm and λ_2= 584 nm, $\delta\nu$= 1250 cm^{-1}) crossed at angles of 3° and 5.8°. Top: Generated beams projected onto a white screen. The two pump beams, S 1 and the first few AS beams are attenuated (after the sample) by a neutral-density filter. The AS 2 spot clearly shows two different colors, with blue corresponding to the Raman generation and green to the FWM signal. Bottom: Normalized spectra of the generated sidebands.

The experimental setup is about the same as the one we used for $PbWO_4$ crystal. We start with an angle 3.6° (the phase matching angle between the peak wavelengthes) between the pair of beams. By applying two 50 fs pulses (at λ_1= 630 nm, λ_2= 584 nm, and $\delta\nu$= 1250 cm^{-1}), we obtain generation of up to 16 AS and 2 S sidebands. The highest frequency generated is in the UV region at a wavelength of 301 nm. Similar to the sideband generation in $PbWO_4$, the sidebands are not equally spaced in frequency. The energy conversion from pump to AS 1 is 3%, 2% to the S1 and 0.5% to AS 2. The total energy conversion from pump to sidebands is 14%. The sidebands generated using 3° crossing angle overlap substantially in spectrum with each other.

From Fig. 5, we can clearly see that the instantaneous four-wave-mixing (FWM) signal coexists with Raman generation in the lower orders of the sidebands. The AS 2 beam spot shows two colors: blue and green. By measuring the frequency shift from the preceding sideband, we deduce that the blue is due to Raman generation, while the green beam corresponds to the FWM signal.

We find that sidebands generated at 5.8° have a larger (about twice) frequency spacing compared to the 3° case when we plot the sideband frequency versus the sideband order as shown in Fig. 5 top picture. Also the sidebands come out at a much larger angle spacing as shown in the Fig. 5 bottom picture (we define the Stokes output angle as 0 degree). This is more obvious for low-order sidebands which are more affected by the strong FWM signal. This shows that the initial phase matching condition between the two pump beams decides the generated sideband output angles and frequencies.

We keep the pump frequency the same while tuning the Stokes frequency from 820 cm^{-1} to 2608 cm^{-1} and record the sideband frequency. The result is shown in Fig. 6. When $\Delta\nu$ varies from 820 cm^{-1} to 2608 cm^{-1}, the output angle doesn't vary much. Also, the frequency vary much less compared to the change of $\Delta\nu$. These facts are clear indications that the signals are generated mainly through the Raman process. This also shows that the tuning to exact Raman resonance is not necessary as we can obtain generation even when $\Delta\nu$ is twice the Raman frequency. When the phase matching condition is satisfied, the low-order sidebands are very strong because of the boost from the FWM process.

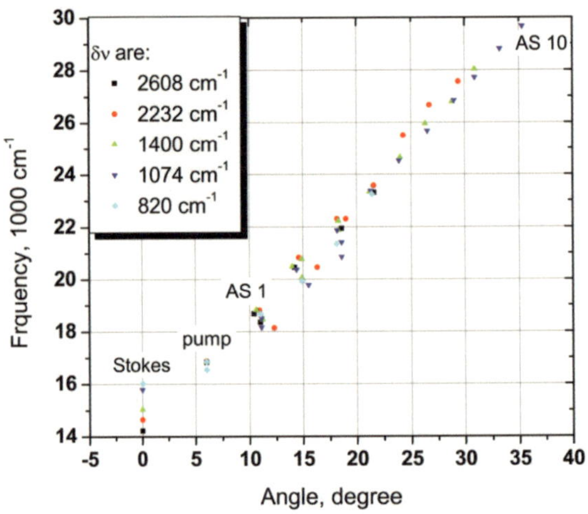

Fig. 6. The Sideband frequency and output angle at the different detuning ($\Delta\nu$ vary from 820 cm^{-1} to 2608 cm^{-1}) between the pump (fixed at 594 nm) and Stokes pulses. The output angles do not vary much, nor does the frequencies of the sidebands, although $\Delta\nu$ varies a lot.

Due to the material dispersion, phase matching is optimized when different frequency components propagate at different angles. We perform calculations that support our qualitative understanding of the role of phase matching using a pair of Gaussian pulses, which are very close to those we measure in experiment using a spectrometer. We take the spectral intensity of an n's AS sideband as given by

$$I_n(\nu) \sim I_{n-1}(\nu-\nu_R)\cdot \mathrm{sinc}^2\{[\vec{k}(\nu)-\vec{k}(\nu-\nu_R)-\vec{k}_R]L/2\}. \tag{1}$$

Here $\vec{k}$ is the wavevector ($k = n\nu/c$), ν_R is the Raman shift and c is the speed of light. The refractive index of diamond n is calculated by using the generalized Cauchy dispersion formula (Smith et al., 2001):

$$n(\lambda) = 2.37 - 1.0\times 10^{-5}/(1.24/\lambda)^2 + 8.0\times 10^{-3}\times(1.24/\lambda)^2 + 1.0\times 10^{-4}\times(1.24/\lambda)^4, \tag{2}$$

with λ in units of μm. This formula fits well with the available experimental data (Piersonf, 1993; Edwards & Ochoa, 1981). The k-vector of the Raman excitation $\vec{k}_R$ is determined by the directions of the two input beams. We assume that the angle between the low-order sidebands is approximately the same.

We find that when the angle is smaller than the phase matching angle, the peak frequency shift of the generated AS 1 from the pump pulse is 1163 cm^{-1}, which is smaller than the Raman shift. The opposite happens when an angle larger than the phase matching angle is applied, as shown in Fig. 7. The peak frequency shift of the generated AS 1 from the pump pulse is 1301 cm^{-1} for input angle of 3.6° and 1466 cm^{-1} at 4.6°. This agrees with our experimental observations in diamond. The peak frequencies of the two input pulses phase match at 3.6°. Therefore, when the two input pulses cross at 5.8°, the generated spectrum has a larger frequency spacing than that of the 3°. This calculation may also explain the gradually decreasing frequency spacing between the sidebands, since due to the normal medium dispersion, the optimum beam-crossing angle increases with increasing sideband frequency.

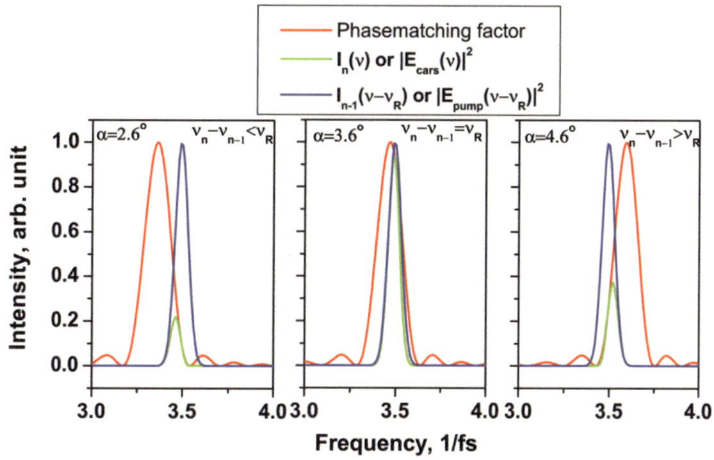

Fig. 7. Theoretical calculation of the generated AS 1 when the two pumps cross at angle 2.6, 3.6 and 4.6 degree respectively, with the 3.6 degree corresponding to the phase matching angle between two input pulses. The input pulses have the central wavelengths of is 630 nm and 581.23 nm, respectively. The frequency difference is exactly the Raman shift 1332 cm^{-1}. The sample thickness used for calculation is 500 μm. The phase matching factor is $\mathrm{sinc}^2(\delta k * L/2)$, where δk is the phase mismatch.

3.3 Broadband light generation in $PbWO_4$ by excitation of the Raman mode at 325 cm^{-1}

3.3.1 Excitation with two fs pulses

With fs pulse pumping, the SRS gain increment explicitly depended on the integral cross section instead of the peak cross section of spontaneous Raman scattering (Basiev et al., 2004) therefore the excitation of 325 cm^{-1} line can dominate when phase matching is satisfied. When we reduce the angle between the pump and Stokes beams to 2.9 degree and tune the frequency difference below 600 cm^{-1}, we observe generation due to excitation of the Raman mode at 325 cm^{-1} as shown in Fig. 9 (a). Due to the small phase matching angle between the two input pulses, the output beams are very close to each other and become inseparable for high-order sidebands. We see the interplay between the phase matching and Raman resonance when we fix the Stokes pulse at 804 nm wavelength while tunning the pump pulse from 760 to 780 nm, with a detuning varying from 408 to 615 cm^{-1}. Although it is far from Raman resonance at 615 cm^{-1} detuning, the phase matching condition between the pump and Stokes is satisfied, which results in generation of 22 AS and 2 S. The phase matching between the pump and probe fields is also good, as can be seen from the FWM signal (blue) shown in the picture. As many as 10 CARS signals are measured, which indicates a significant coherence being built up in the crystal.

Because of the lower frequency spacing between the two pumps and the wide spectral width of the fs pulses, the smaller rotational modes are all excited so that the lower order sidebands have multiple peaks. This is different from the multiple peaks generated when 901cm^{-1} transition is excited. Those multiple peaks are due to the coexistence of the FWM and Raman signals. The high-order sidebands are about equally spaced at a 320 cm^{-1} spacing. The angle between the two applied beams is 2.5 degree, which corresponds to 1.2 degrees inside the crystal. Thus it might be possible to generate sidebands with a collinear configuration.

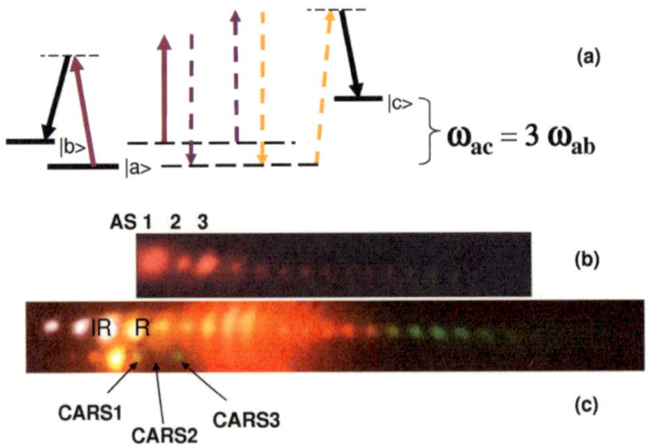

Fig. 8. (a) Energy level schematics; (b) The AS 3 generated by pump and Stokes beams which are tuned to excite the Raman mode at 325 cm^{-1} is stronger than AS 2 due to the excitation of Raman mode at 903 cm^{-1}; (c) CARS 1, 2 and 3 signals generated by a delayed probe pulse.

One interesting feature we noticed in our experiment is that the high order AS 3 is stronger than AS 2 when the 325 cm^{-1} is excited with pump pulse at 770 nm and Stokes pulse at 805 nm, as shown in Fig. 8 (b). What happens can be explained by a schematic energy level diagram as shown in Fig. 8 (a). The Raman mode width in the crystal is larger than the one in a gas. Consequently, although ω_{R1}=903 cm^{-1} is not exactly three times the ω_{R2}=325 cm^{-1}, it is still possible that the AS 3 gets enhanced by the Raman mode at 903 cm^{-1}. A clear indication is shown in Fig. 8 (c) where a third probe beam is used. The picture is taken with a delayed probe, therefore only CARS signal can survive. The CARS 3 signal, which is generated due to excitation of 901 cm^{-1}, is very strong, compared to the CARS 2 signal.

3.3.2 Excitation with a pair of time-delayed linear chirped pulses

In the above experiments we used two nearly transform-limited 50 fs laser pulses tuned such that their frequency difference was approximately equal to the Raman frequency. One complication in those experiments was simultaneous excitation of several Raman lines by the large spectral width of the fs laser pulses. Also, the high-peak intensity of the pulses leads to generation of the strong instantaneous FWM signals. These resulted in a rich structure of the lower-order sidebands' spectra. In addition, the total laser energy fluence in those experiments was limited by "parasitic" nonlinear processes, such as self-focusing and SPM, which can strongly distort the pulses. Only 2 to 3 mW of average beam power can be used.

Femtosecond pulse sequencing and shaping allows selective (narrow-band) Raman excitation by broad-band pulses, as discussed in the past (Zheltikov, 2002; Gershgoren et al., 2003; Dudovich, 2002). At the same time, the shaped pulses are necessarily longer than transform-limited, and therefore have lower peak intensity, helping to avoid the onset of parasitic nonlinear processes at larger energy fluences.

We obtain chirped pulses by a misalignment of the compressor in the amplifier. The pulse is then split into two by an ultrafast beamsplitter and the delay between the two pulses can be varied. The two beams are recombined and focused in the crystal by a 2 inch diameter lens (focal lens f=50 cm). Both beams are attenuated by variable neutral density filters so that the power used is below the threshold for the parasitic nonlinear processes. The power used ranges from 5 mW to 20 mW, depending on the pulse chirp and the focusing conditions.

A linearly chirped pulse can be written as:

$$E(t) = \exp(-at^2) * \exp[I * (bt^2 + \omega_l t)]. \tag{3}$$

Here $1/\sqrt{a}$ is roughly the pulse duration, b is the chirp rate of the pulse, and ω_l is the laser center frequency.

The intensity for two time-delayed linearly chirped pulses is:

$$\begin{aligned} I(t) &= |E(t) + E(t - t_d)|^2 \\ &= \exp(-2at^2) + \exp[-2a(t - t_d)^2] \\ &+ 2\exp[-a(t^2 + (t - t_d)^2]\cos[b(2t - t_d)t_d + \omega_l t_d]. \end{aligned} \tag{4}$$

When $\Delta\omega = \omega_R = bt_d$, the last term in $I(t)$ is proportional to $\cos(2\omega_R t + C)$ [$C = (\omega_l \omega_R - \omega_R^2)/b$, is a constant]. We see that there is a periodic beat in the pulse intensity in the time domain. The periodicity of this pulse train can be matched to the period of this Raman mode at ω_R, permitting selective mode excitation (Gershgoren et al., 2003). One can adjust the excitation frequency simply by adjusting the time delay between the two pulses. A related method of achieving the spectral selectivity is Fourier domain pulse shaping, where a pulse train is created by applying a periodic spectral phase to a single fs pulse using a pulse shaper (Dudovich, 2002).

For a fixed $\Delta\omega = \omega_R$, t_d is proportional to $1/b$, so the slope of t_d vs. the inverse of the chirp rate gives the Raman frequency ω_R (which is to be excited). We calculate the chirp rate b from $b = \omega_R/t_d$ by measuring the t_d using pulses with different chirp rate when the Raman mode at 325 cm^{-1} is excited.

The pulse that is used here is negatively chirped, with a pulse duration full width at half maximum (FWHM) ranging from 80 fs to 2 ps. The chirp is approximately linear. The available spectral width (FWHM) of the pulse is around 460 cm^{-1}. The angle between the pump beams is about 2.4°. We observe as many as 40 AS at two different delays. The output angle of the highest-order sideband with respect to the pump beam is about 80°. For convenience of description, we call the pump beam with variable pulse delay as the variable-delay pump and the other one as the fixed-delay pump beam. We observe AS generation on the variable-delay pump side at a positive pulse delay (here the variable-delay pump acts as the pump beam and the fixed-delay pump acts as the Stokes beam, as conventionally $\omega_{pump} > \omega_{Stokes}$.). When we move the pulse forward from 0 delay to -t_d, we observe AS generation on the side of the fixed-delay pump beam (as expected), since now the fixed-delay beam functions as a pump beam.

When we change the pulse chirp to 620 cm^{-1}/ps, we observe, unexpectedly, that some of the high-order sidebands are stronger than the low-order ones, as shown in Fig. 9 (d). This can be seen even more clearly when a larger angle between the two pump beams is used and when the Raman mode at 903 cm^{-1} is excited (as described below). Kuon Inoue *et al.* have observed a similar non-monotonous variation of spectral intensity of the generated sidebands when using two-color subpicosecond excitation in TiO_2 crystal (Inoue et al., 2007). We believe this maybe due to the large phase mismatch of certain sidebands. Another possible explanation is based on our experimental observation. Since the FWM coexists with the Raman process, certain sideband gets enhanced when the two processes overlap at that order.

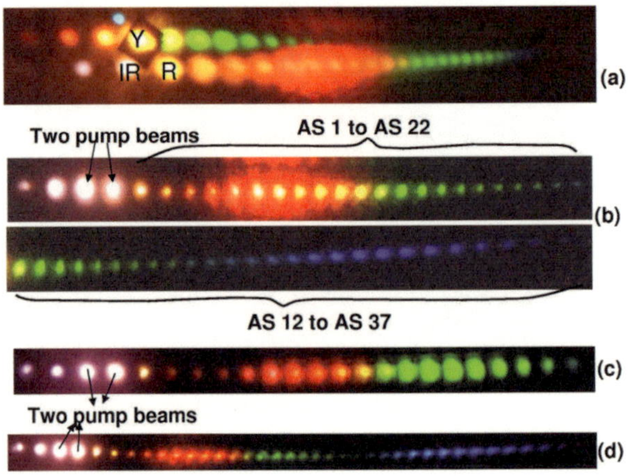

Fig. 9. (a) Broadband generation in $PbWO_4$ pumped by nearly-transform-limited fs pulses (IR, λ_{Stokes}=804 nm, R, λ_{pump}=766 nm). The angle between the pump and Stokes beams is 2.9°. A third probe pulse (Y) leads to generation of many orders of S and AS sidebands. (b) and (c) Broadband generation in $PbWO_4$ using two time-delayed linearly chirped pulses applied at angle of 2.4° to each other. The pulse chirp is about 1280 cm^{-1}/ps for part (b), and 440 cm^{-1}/ps for part (c). (d) The intensity modulation of the generated sidebands is clearly seen when the pulse chirp is about 620 cm^{-1}/ps.

Broadband generation at a chirp rate of 1280 cm^{-1}/ps and a pulse delay of 0.7 ps are shown in Fig. 9 (b). We observe 40 AS and 3 S sidebands. The frequency separation between the sidebands is around 320 cm^{-1} on average. It decreases to 240 cm^{-1} for the high-order sidebands. We measure high nonlinear conversion efficiency. As much as 41% of the pump pulse energy and 21% of the Stokes pulse energy are converted into the generated sidebands. When the chirp rate is increased to 1060 cm^{-1}/ps, the sideband generation seems to become less effective. The number of observed AS sidebands decreases to 34. The conversion efficiency from the pump and Stokes beams decreases to 33% and 19%, respectively. When we introduce more chirp until a chirp rate of 440 cm^{-1}/ps is reached, the efficiency reduces to about 14% for the pump and 11% for the Stokes beam. About 22 AS sidebands are observed. However, the substantial pulse chirping, as well as the relatively loose beam focusing, allow us to use a rather high combined average power of over 15 mW (measured after the crystal) without introducing the parasitic effects. Another feature of the generation using pulses with a high chirp rate is that the FWM signal gets weaker. Consequently, the generated sidebands have good beam profiles, as shown in Fig. 9 (c), and their spectra are mostly single-peaked.

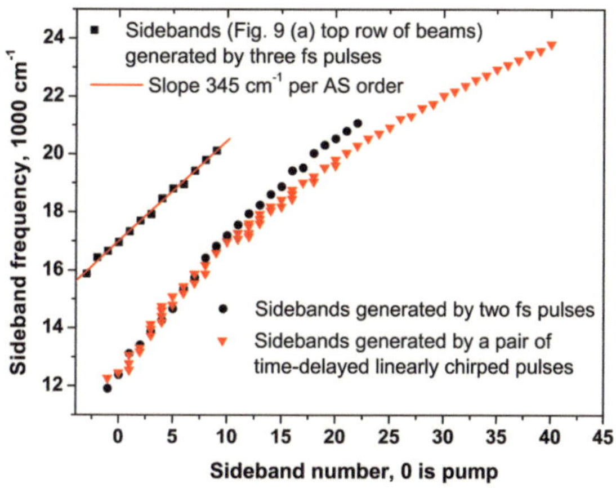

Fig. 10. Comparison of the sideband generation in $PbWO_4$ using two nearly-transform-limited fs pulses (λ_{Stokes}= 804 nm and λ_{pump}= 766 nm) and a pair of time-delayed chirped pulses. Many more sidebands are generated in the latter case.

In Fig. 10 we plot the sideband frequency as a function of the sideband order. The squares show the peak frequencies of the sidebands generated by the pump (at 760 nm) and Stokes pulses (at 804 nm). The frequency spacing decreases slightly as the order goes higher. We also plot the many orders of CARS signals generated by all three pulses. We can see that the CARS signal has a regular frequency spacing of 345 cm^{-1}, which is close to the spontaneous Raman frequency measured at 325 cm^{-1}. The probe beam (Yellow) is labeled as 0 order. Compared to the sidebands generated by two (nearly-transform-limited) fs pulses (Fig. 10, filled circles), there are many more sidebands generated when a pair of time delayed linearly chirped pulses are used (Fig. 10, triangles). The AS sidebands span a range of 12,000 cm^{-1}.

4. Coherence between the generated sidebands (Zhi & Sokolov, 2007)

We investigate the mutual coherence among the generated sidebands. We first generate multiple AS sidebands by focusing Red (λ_R= 718 nm) and IR (λ_{IR}= 812 nm) beams into the $PbWO_4$ crystal. Then a third (Yellow) beam is sent along the direction of the generated AS 3 sideband with a matching wavelength (λ_Y= 574 nm). Once the overlap in frequency, space, and time is achieved, the sidebands (AS 2 to AS 7) start to visibly flicker, due to interference between signals generated through different channels. We measure the pulse energy of AS 5 on a shot-by-shot basis by using a fast photodiode.

The statistics of the AS 5 pulse energy is shown in Fig. 11. Solid black bars give the histogram (number of pulses vs. AS 5 energy), with only Red and IR pulses applied at the input. This histogram shows a typical normal distribution, with about 10% average variations. However, with the addition of the Yellow beam at the input, the histogram of the AS 5 pulse energy (913 pulses total) transforms into a very different distribution (Fig. 11, white bars). We perform a simple calculation, which supports our qualitative understanding of this result. We consider interference of two fields (of the same frequency), whose intensities (I_1 and I_2) fluctuate within 10% of their mean values. We further assume that the relative phase of these two fields varies randomly between 0 and 2π (every value of $\Delta\varphi$ being equally probable). The resultant intensity $I = I_1 + I_2 + 2\sqrt{I_1 I_2}[\cos(\Delta\varphi)]$ is expected to produce a histogram that is inversely proportional to the derivative of I with respect to $\Delta\varphi$ and therefore has two peaks (at $I_1 + I_2 + 2\sqrt{I_1 I_2}$ and $I_1 + I_2 - 2\sqrt{I_1 I_2}$, where $\Delta\varphi$ equals 0 and π respectively). This is exactly what our simulation shows (Fig. 11, dotted curve). In this simulation, we take the average value of $< I_1 >=0.41$ (arbitrary units) from the measurement, and find (from the best fit in Fig. 11) $< I_2 >=0.09$. We repeat the calculation 91300 times (using random number generators), and divide the calculated number of counts (per intensity) by a factor of 100 when we compare the simulation with the experiment. The two peaks in the simulated histogram appear to be broadened by the fluctuations of I_1 and I_2, which are taken to be 10% each (matching the experimental observations). The peak on the right (at higher pulse energy) is calculated to be lower and broader than the peak on the left, in excellent qualitative agreement with the experimental data. This measurement, and its comparison with theory, confirms our expectation that the (highly-coherent) Raman process results in generation of mutually-coherent sidebands.

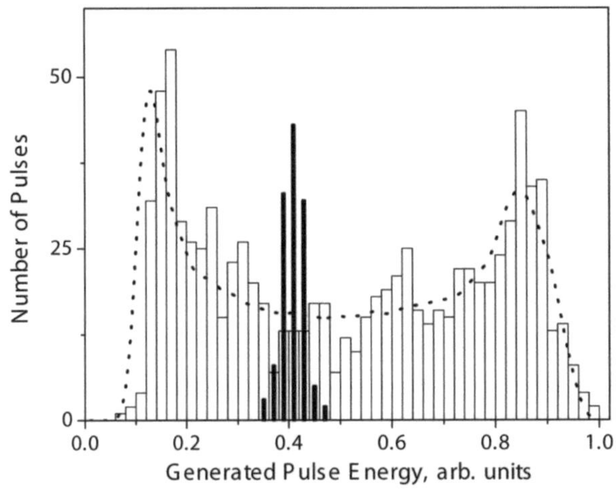

Fig. 11. Histograms of AS 5 pulse energy. Solid black bars: the number of pulses (out of 150) vs. AS 5 pulse energy generated with Red and IR input beams only. White bars: the histogram of AS 5 pulse energy (913 pulses total) with the addition of the third input beam. The dotted curve is a theoretical prediction obtained assuming perfect single-shot coherence of the two interfering fields, and random shot-to-shot variation of their relative phase.

5. Discussion of broadband generation in Raman crystals with two-color laser fields

In general, due to the large dispersion in solids, a non-collinear beam geometry is needed for effective excitation in a Raman crystal. In the CFWM method, the sample is so thin that the optical path is smaller than the Rayleigh range, the nonlinear length and the dispersion length. Therefore the linear dispersion and competing nonlinear optical effects such as SPM are minimized. In SRS and in Raman lasers high conversion efficiency is often achieved by using a much thicker sample or an external cavity. For example, when $PbWO_4$ is used in a Raman laser, a typical thickness is up to 70 mm (Kaminskii et al., 2000). As a comparison, the sample we use is 1 mm thick, which is slightly thicker than the solid medium used in the CFWM method but is much thinner than the one used in the Raman laser.

The requirement of the frequency spacing between the two pump pulses to be tuned to Raman frequency is greatly relaxed since we are using fs pulses which have about 450 cm^{-1} bandwidth. Moreover, the wider Raman spectral linewidth (compared to gas medium) also eases the resonance requirement. We have observed Raman generation in diamond even when we tuned to double the Raman frequency. The best proof is that we observed the simultaneous excitation of 325 cm^{-1} and 901 cm^{-1}, which leads to quantum beating in the CARS signal, as is shown in Fig. 16 later in the chapter.

A large number of adjustable experimental parameters offers increased flexibility, but at the same time requires careful alignment procedures in order to achieve day-to-day reproducibility. We normally start with the beam energy right below the SPM threshold. After we find the overlap and observe generation, we can increase the two pump beams power further without causing SPM. In a way, we can say that the Raman generation process suppresses the SPM process. A similar situation has been observed in the gas experiment using fs pulses (Sali et al., 2004). Both processes—the (Raman non-resonant) instantaneous FWM process and the Raman process often coexist. Each generated sideband is a result of the interplay between the two processes. The sidebands come out at almost equally separated angles but different peak frequency spacings. The intensity may not be monotonously decreasing as the order goes higher due to the large phase mismatch of certain sidebands. The spectral shapes of these lower-order sidebands sometimes show complex spectral shapes with multipeaks, either due to the simultaneous excitation of several Raman modes or the co-existence of the FWM process. The generated sidebands have similar pulse duration, close to those of the pump pulses (Liu & Kobayashi, 2008).

In these experiments, we observe a larger number of AS sidebands than Stokes ones. This asymmetry is typically present since the Raman frequency is only one or two orders of magnitude smaller than the laser frequency, and hence the total generated bandwidth is comparable to the pump laser frequency. At substantially lower frequencies generation is intrinsically less efficient. In addition, infrared sidebands are harder to detect. Finally, in the present experiment phase matching occurs at a considerably larger output angle (and leads to a correspondingly worse spatial overlap with the pump beams) for a Stokes sideband compared to the same order AS sideband.

The sideband generation due to excitation of large- and small-frequency Raman modes shows few distinct differences. In the case of large-frequency Raman mode, a large angle is needed to fulfill the phasematching condition. As a result, the deviation of the frequency spacing from the Raman frequency is getting large as the sideband order goes higher. For broadband generation with 325 cm^{-1} or even lower frequency 191 cm^{-1}, an almost constant frequency spacing can be maintained up to a very high order, therefore, more than 40 sidebands can be generated. Also, the spacing is close to the Raman frequency excited. The ideal case would be the case of using a gas when phase matching is fulfilled and many sidebands are generated (Chen et al., 2008).

Compared to the fourier-transform-limited pulse excitation, temporal stretching of the excitation pulses allows us to use large pulse energies per unit area (while still avoiding parasitic nonlinear effects), and correspondingly increase the Raman coherence. Although the FWM process still creates some multiple peaks in the spectra of the low-order sidebands, the complications due to simultaneous excitation of multiple Raman lines are eliminated. Different Raman modes can be selectively excited by varying the time delay and the angle between the two pump beams.

At last, let's compare the generation between different types of crystals. Diamond has a higher damage threshold so more power can be used. With one single Raman mode, the sideband spectrum is cleaner. Due to the large Raman shift, the higher order sidebands lose phase matching quickly. Therefore, only around 20 sidebands are generated.

6. Multicolor generation in Raman crystals with three-pulse excitation

The generation in solids is complicated yet interesting. For example, broadband two dimensional (2-D) multicolored arrays with more than ten periodic columns and more than ten rows were generated in a sapphire plate using two crossed femtosecond laser beams overlapping in time and space has just been observed (Liu et al., 2009). Here we show two schemes with the additional third probe pulse irradiation on the sample. One scheme also leads to multicolor 2-D array generation. The three beams are arranged either in a planar (linear) configuration or Box CARS configuration (Demtröder, 2003) (see also Fig. 1 (c)).

6.1 Multicolor generation in diamond using three-color laser fields

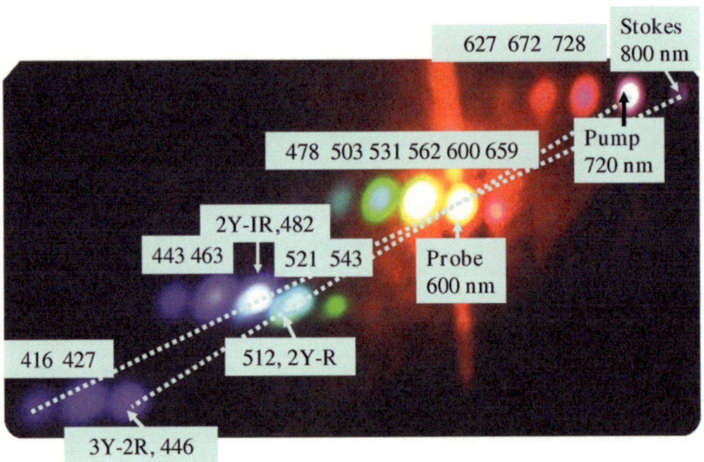

Fig. 12. The 2-D array generation in diamond with three input pulses (λ_{pump}= 720 nm, λ_{Stokes}= 800 nm, and λ_{probe}= 600 nm). The wavelengths of the sidebands are labeled in nm. The degenerated FWM signal (2Y-IR, a shorthand of $2\omega_Y - \omega_{IR}$) from the probe and Stokes pulses and the one (2Y-R) from the probe and pump pulses are much stronger than the Raman generation spots. They either superimpose or shift slightly. The six-wave-mixing signal (3Y-2R) from the pump and probe pulses is also visible.

When we apply a Stokes pulse (IR, 800 nm), a pump pulse (red, 728 nm) and a visible probe pulse (yellow, 600 nm) at the usual boxed-CARS geometry, in addition to the strong CARS and coherent Stokes Raman scattering (CSRS) signals, we obtain a 2-D color array as shown in Fig. 12. The setup used here is similar to the one we used for visible and UV coherent Raman spectroscopy of dipicolinic acid and other biologically-relevant chemicals (Pestov et. al., 2005). The angle between the pump and Stokes beams is 5°. The probe beam is 11° from the pump beam and 16° from the Stokes beam. We propose the following explanation for the 2-D color array generation. The frequency spacing between the pump and Stokes pulses is around 1000 cm^{-1}. As a result, the average sideband spacing is very regular, about 1040 cm^{-1}. The sidebands in the first row are generated by the pump and Stokes pulses. The second row is generated by the CARS/CSRS and the high-order CARS processes. The FWM signal from the probe and Stokes/pump pulses leads to the generation of the third row. The degenerate FWM signal from the probe and Stokes pulses ($2\omega_Y - \omega_{IR}$) and from the probe and pump ($2\omega_Y - \omega_R$) pulses are much stronger than the generation spots due to Raman effect. They either superimpose or shift slightly. The six-wave-mixing signal ($3\omega_Y - 2\omega_R$) from the pump and probe pulses results in the fourth row sideband generation.

6.2 Multicolor generation in $PbWO_4$ using three-color laser fields

We observe two-dimensional (2-D) multicolored Raman generation (Fig. 13 (a)) when we apply pump beam at 730 nm, Stokes beam at 804 nm, and probe beam at 604 nm in a Box CARS configuration. The angle between the pump, Stokes and probe beams are the same as those used in the diamond experiment, as described above. When we turn the crystal 90 degree so that the relative polarization to the crystal axis changes from P to S, the 2-D multicolored Raman generation pattern changes dramatically and the sideband wavelength vary accordingly. The FWM signals produced by the three pulses sometimes overlap with the Raman sidebands, which leads to a generation of 2-D color array of similar nice round beams (Fig. 13 (a)). Although the FWM signal may overlap with one of the sidebands in space, the spectrum normally has two distinct peaks. When we increase the angle between the pump and Stokes input beams to 6 degrees while keeping the probe beam fixed, we observe up to 50 new generated colors (Fig. 13 (b)). We believe this is mainly due to the good phase matching between the pump beams and the resultant sidebands. Now the FWM signals are not overlapped with the Raman sidebands and appear as smaller beams (the smaller blue and green beams in between the regular 2-D color array).

We plot the measured peak frequencies of the Raman sidebands generated in the top 3 rows, the pump-Stokes row, the probe beam row and the bottom 4 rows as shown in Fig. 14 (a). The pump-Stokes row is marked by a dotted line in Fig. 13 (b). We find that the frequency spacing is close to the Raman frequency 901 cm^{-1}. Therefore, it is clear that the sidebands are generated by the Raman processes. The frequency spacing between different rows is around 600 cm^{-1}, which we believe is the result of phase matching of the probe pulse with the sideband beams generated by the pump and Stokes pulses.

We notice that the sidebands generated by pump and Stokes alone are strongly affected by the probe pulse. In Fig. 14 (b) we show the intensities of the first four AS with the probe pulse present and absent. We see that the addition of the probe pulse broadens the spectrum. We observe that the sidebands intensity is increased as well. When we change the probe delay so that the pulses don't overlap in time, the AS fields are not affected by the probe pulse. We plot the pump intensity before and after crystal in Fig. 14 (c). We conclude that more pump energy is converted to the sidebands when both Stokes and probe pulses are applied.

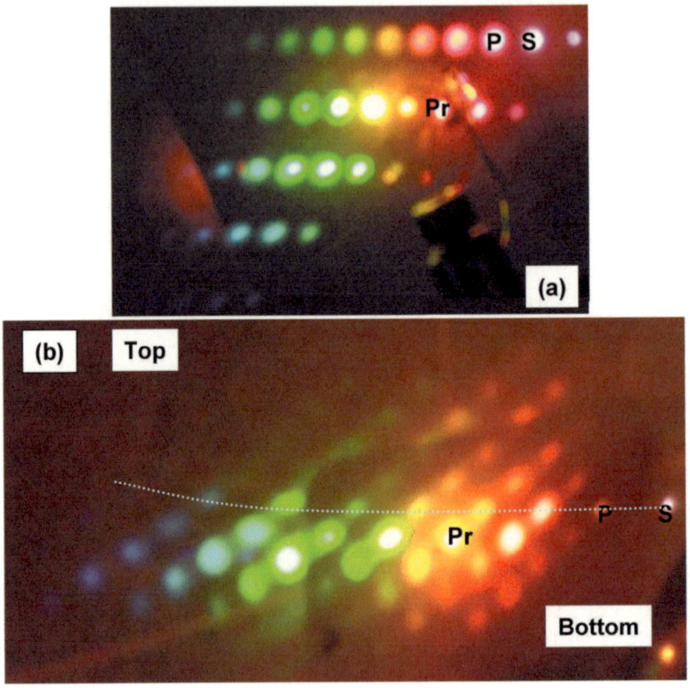

Fig. 13. (a) The Box CARS configuration. Three beams are sent in at the three corners of the box and the CARS signal is generated in the fourth corner. (b) and (c) Different 2-D patterns generated in $PbWO_4$ under different input angles setting different phase matching conditions. In (b), a piece of small filter is used to block the strong probe beam for better picture. In (c), a large square neutral density filter is used to lower the intensity of the low order sidebands and three input beams. P: pump beam; S: Stokes beam; Pr: probe beam.

Next we use a planar configuration for the three input beams. We first generate sidebands using pump pulses at 755 nm and Stokes pulses at 803 nm as shown in the top of Fig. 15 (a). Then we send the third beam (which we also call probe beam) at 590 nm at the same direction and wavelength as AS 6. The configuration is shown in Fig. 15 (b). The probe beam and the pump beam satisfy the phase matching condition so that a strong FWM (green) signal is efficiently generated, which we believe acts as a seed to generate high-order sidebands (Fig. 15 (c)). We measure 18 sidebands with three pulses while with two pulses we generate 11 sidebands. The measured wavelength of the highest-order sideband is extended from 509 nm down to 467 nm. Also the intensity of the high-order sidebands generated by two pulses is enhanced when a third probe beam is applied. We measure a 25% conversion efficiency from the probe beam to the sidebands. With two pulses, the high order sideband frequency spacing is 500 cm^{-1} per order (blue line in Fig. 15 (c)). It decreases to 350 cm^{-1} per order (cyan line in Fig. 15 (c)) when three pulses are used when more high order sidebands are generated. We would like to emphasize that this is a "proof of principle" study. A careful alignment and choice of the pump wavelengths should result in substantial additional improvement in sidebands generation. Because the probe beam doesn't have the exactly same wavelength as the AS 6, we have not observed the beating as described in our earlier paper (Zhi & Sokolov, 2007).

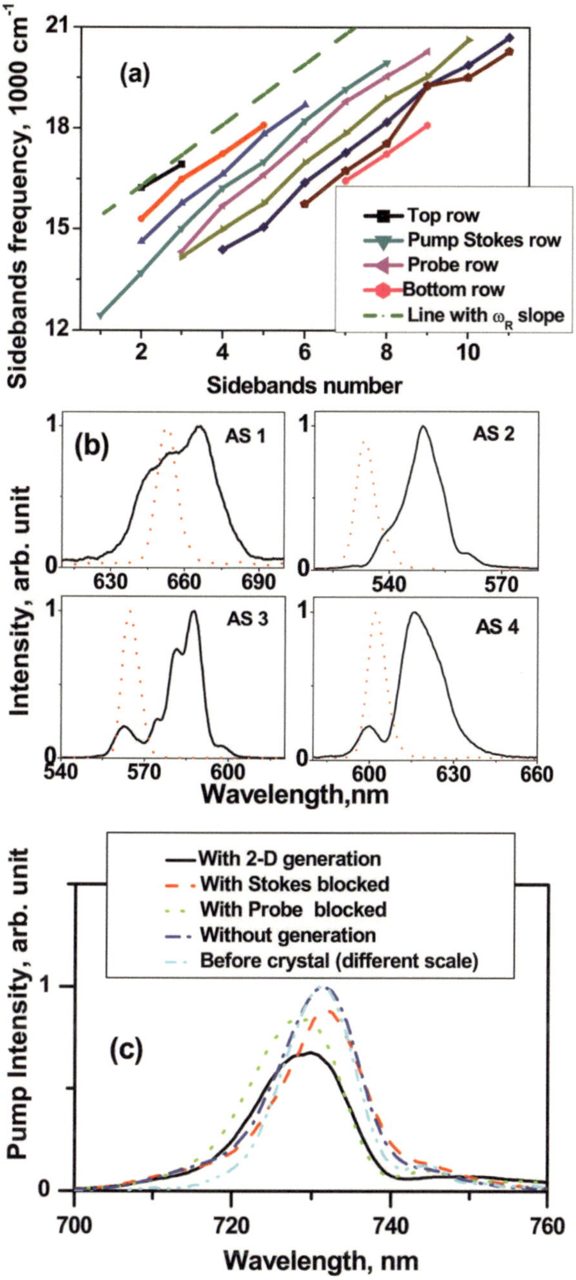

Fig. 14. Sidebands generation in $PbWO_4$ with three pulses excitation in a Box CARS configuration. (a) Peak frequencies of the 2-D color array generation. The frequency is measured along or parallel to the pump-Stokes row (dotted line in Fig. 13 (c)). The dashed line has a slope of Raman frequency 901 cm^{-1}. (b) solid line: AS sidebands generated by the three pulses; dashed line: AS sidebands generated by the pump and Stokes pulses only. (c) Pump intensity with and without sideband generation.

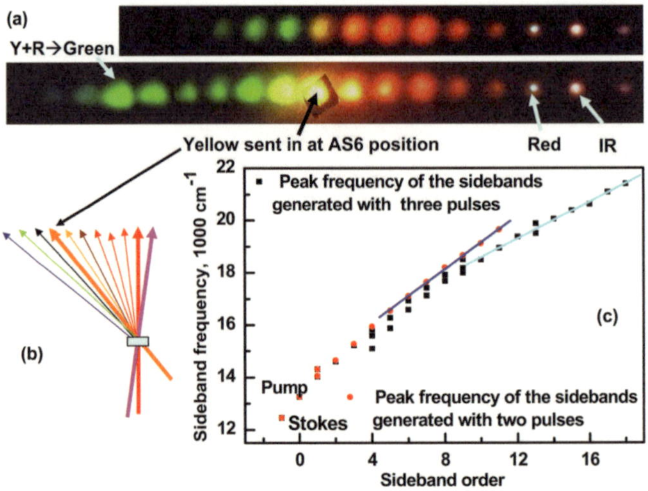

Fig. 15. Sidebands generation in $PbWO_4$ with three pulses excitation in a planar configuration. The probe pulse is sent in at the same direction and wavelength as AS 6. (a) Generation pictures with pump and Stokes beams present (top picture) and all three pulses (pump, Stokes and probe) present (bottom). (b) Input beam geometry. (c) The sideband frequency as a function of the sideband order generated by two/three pulses.

7. Coherence decay measurements in Raman crystals

When a fs pulse are used as a probe, we observe the quantum beating between the two modes, which have a frequency difference of 565 cm^{-1}. The corresponding beating period in the time domain is 59 fs, which is what we measure as shown in Fig. 16 (a). The CARS signal can be described by the following formula:

$$\begin{aligned} I_{CARS} &= |Ae^{-t/T_2} + Be^{i\triangle\omega t - t/T_2' + i\varphi}|^2 \\ &= A^2e^{-2t/T_2} + B^2e^{-2t/T_2'} \\ &+ 2ABe^{-t(1/T_2+1/T_2')}\cos(\triangle\omega t + \varphi). \end{aligned}$$

By fitting the experimental data (Fig. 16 (b)) with the above equation, we get T_2=1.3 ps, T_2' =3.8 ps, and ω=106.5 rad/ps, which is exactly the frequency difference between the two modes ($2\pi/T = 2\pi * (3/100 * 565cm^{-1}) = 106.5$).

When we use a narrow-band shaped probe beam (spectral width of about 1 nm), we are able to measure the coherence decay of the two strong modes in the crystal as shown in Fig. 16 (c). Here the measured decay time for Raman mode at 328 cm^{-1} is 1.5 ps and 2.5 ps for 901 cm^{-1} mode. The discrepancy between our two measurements could be due to the finite pulse width.

We also measure the CARS decay time of the High Pressure High Temperature (HPHT) diamond sample that we used for our experiment. Using a pair of visible pulses as pump (584 nm) and Stokes (635 nm) and using a UV pulse (also about 50 fs width) at 318 nm as a probe, we get the CARS signal at 305.8 nm ($\delta\nu$= 1250 cm^{-1}) and CSRS signal at 332 nm ($\delta\nu$= 1326 cm^{-1}) as shown in Fig. 17. Undoubtedly, this is the strong Raman line in diamond.

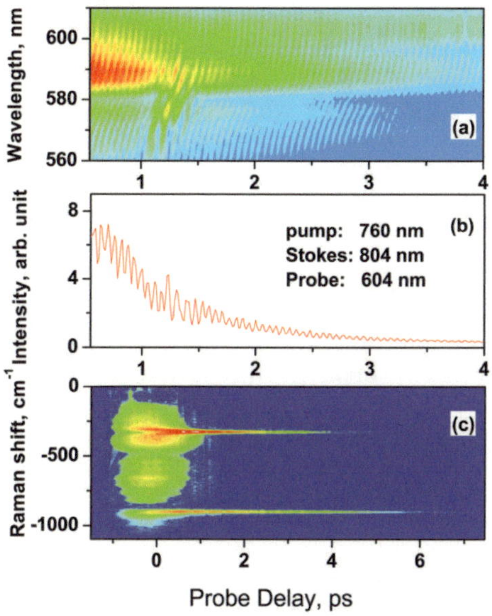

Fig. 16. (a) Quantum beating between the two strong Raman lines in $PbWO_4$ measured by using a fs probe pulse with $\Delta\nu$ between pump and Stokes pulses equal to 720 cm^{-1}. (b) The cross section from (a), which can be used to get the fitting parameters. The beating has a frequency of 106.5 rad/ps, which corresponds to the Raman shift difference of 565 cm^{-1} between the two modes. (c)The coherence decay of the simultaneously excited strong Raman lines measured when fs pump, Stokes pulses and a narrow-band ps probe (about 1 nm spectral width) are applied to the crystal.

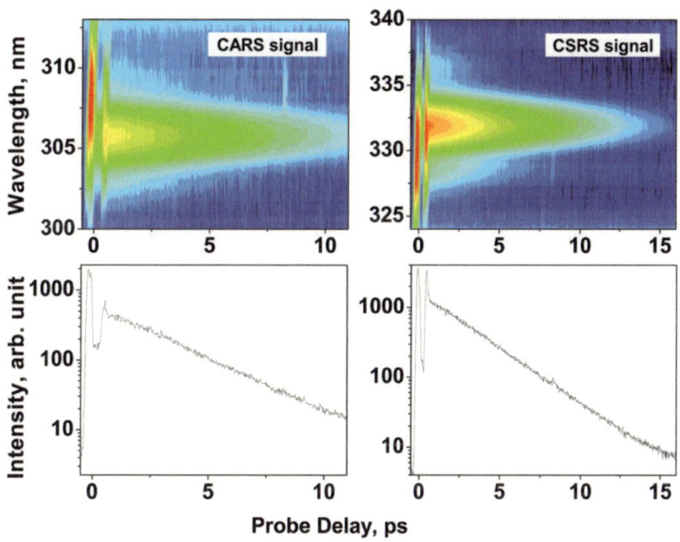

Fig. 17. The CARS and CSRS signals observed in diamond using a UV probe pulse. Top, the spectra of the CARS (left, center wavelength is 305 nm) and CSRS (right, center wavelength is 332 nm) as a function of the probe delay. Bottom, the exponential decay of the CARS (left) and CSRS (right) signals intensity average over their center wavelengths. Log scale is used.

The width of the line in the SRS spectrum related to SRS-promoting vibration transition is $\Delta\nu_R \approx 2.5$ cm^{-1}. Therefore the phonon relaxation time $T_2 = (\pi c \Delta\nu_R)^{-1} \approx 4.2$ ps (Kaminskii et al., 2005). We measure the exponential decay of the CARS/CSRS signal as shown in Fig. 17 (bottom). We get the intensity of the CARS/CSRS signal by averaging over the center wavelengths. To remove the effect of the strong instantaneous FWM signal over the CARS/CSRS signal, we only fit the data about 2 ps after the 0 delay with an exponential decay and obtain decay time T_2 as 2.6 and 2.7 ps for CARS and CSRS, respectively. There is a gap between the FWM and the CARS signal, which remains a mystery to us. Similar gap has also been observed in $PbWO_4$ crystal.

We observe strong absorption of the UV beam when it is used as a probe while a pair of visible pulses are used for excitation. The absorption only happens when the pulses overlap in time. The exact origin of this absorption is unknown, but most likely it is related to a two-photon electronic excitation process. We identify that the width of the absorption dip is related to the group dispersion delay of the different pulses.

8. Conclusion and future work

We realize efficient generation of light frequencies spanning the infrared, visible, and ultraviolet spectral regions in Raman crystals, which provides a possible new way to generate record-short (subfemtosecond) pulses. Our experimental work lends insight to the complicated sideband generation process in crystals. From working with the popular Raman crystal $PbWO_4$, we find that first of all, it is very important to satisfy the phase matching condition, i.e. non-collinear geometry has to be used for efficient generation in the crystal. By varying the detuning, angles, and polarization of the pump and Stokes beams, we find that FWM and Raman effects coexist and the generated sideband frequencies are very sensitive to the angles and the polarization of the pump beams. We design an interference experiment which proves the mutual coherence among the Raman sidebands generated. When a third probe beam is applied in the same plane with the pump and Stokes pulses, more sidebands are generated and the energy of the high-order sidebands is enhanced. When the three laser fields (pump, Stokes and probe) are applied to the crystal in a Box CARS configuration, we observe that up to 50 colors are generated which form a 2-D color array.

We demonstrate similar generation in another crystal, diamond, which shows that the Raman generation method is universal and can be applied to other popular Raman crystals. Since diamond is an isotropic material with a simpler dispersion relationship compared to that of the uniaxial crystal $PbWO_4$, the theoretical phase-matching calculation can easily be performed and can help understand the sideband generation process. The experiment done in diamond also sheds some light on how the 2-D color array is generated when three pulses are applied to the diamond crystal.

Compared to the Raman generation using fs fourier transform limited pulses, we obtain more efficient sidebands generation in $PbWO_4$ crystal with more sidebands generated by using a pair of time-delayed linearly chirped pulses. Up to 40 AS and 5 S are observed. As high as 41 % of pump and 21 % of Stokes conversion efficiency is measured. This shows that this is a good way to build maximal coherence in the crystal. The pulses have ps duration, which is on the order of the coherence decay time. This means ps pulses may be more effective for sideband generation than fourier transform limited fs pulses.

Many other future projects can be extended from this research. Since this is a very new and challenging experiment, there are many "mysteries" remaining to be solved. For example, the splitting in the time domain of the FWM and CARS signal, the exact mechanism of 2- D color array generation when three pulses are used, and the UV absorption when pulses are overlapped. The unequal frequency spacing between the sidebands needs more vigorous theoretical examination. It would be helpful if we could calculate and predict how many sidebands and at what angles are generated. It would not only help us understand how the phase matching condition, Raman generation, FWM interplay with each other, it could also help us find the best condition for efficient sidebands generation. Experimentally, one will learn more of the sideband generation in crystals when more quantitative results are obtained. For example, it will be important to characterize the sidebands, i.e. to measure the absolute sideband pulse energy and the sideband pulse width. The characterization also can help us find the optimal experimental condition for sidebands generation. For example, one can study how the generation efficiency varies as a function of the relative delay between the two pulses. If the sidebands generated in the crystal are chirped when linearly chirped input pulses are used, we might be able to compress the pulses later to get shorter pulses, as has been demonstrated (Liu & Kobayashi, 2009).

In the near future, we envision driving one of these Raman transitions by a pair of time-delayed chirped picosecond pulses, and then using another (compressed or shaped) fs pulse from the same source laser, to probe the Raman coherence prepared by the two excitation pulses. Our experiment has already shown efficient spectral broadening of a (third) probe pulse; future experiments will possibly lead to substantial pulse compression in the same medium, as predicted in our earlier work (Sokolov & Harris, 2003; Sokolov et al., 2005).

Whether the collinear generation in $PbWO_4$ crystal by the excitation of the small-frequency Raman mode at 325 cm^{-1} can be realized or not deserves to be studied, since collinear generation will result in the convenience of combining the sidebands for sub-fs pulse synthesis. Using two separately prepared linearly chirped pulses for the excitation of the Raman mode at 903 cm^{-1} in $PbWO_4$ crystal is also a promising experiment given the efficient sidebands generation from the excitation of the Raman mode with a lower frequency. An alternative way of excitation would be to use a pulse shaper to obtain a pulse train with a period equals to the period of Raman vibrations. If ps pulses are available, a comparison of sideband generation using ps pulses, fs pulses, and linearly chirped pulses would be an interesting subject as well. Optimizing the sideband generation in diamond is possible if a good sample with the right thickness is available. We have observed Stokes generated in the IR region, where diamond has a very small dispersion. Therefore it is a promising candidate material for far IR generation given its wide transmission spectral range. Broadband light generation in other types of crystals such as KGW ($KGd(WO_4)_2$, a very popular Raman crystal due to its high efficiency) and CVD diamond (both polycrystalline and single crystal form) will be studied next. CVD diamond is much cheaper than the natural one, and high quality samples are available now. The high damage threshold and strong single Raman line are very desirable properties for our purpose.

The important future project will be to realize the goal of the generation of sub-fs pulses using the broadband light generated in crystals. First of all we can try to realize the collinear generation in the crystal by exciting the lower Raman frequency in $PbWO_4$. Then we can use a pulse shaper, such as Dazzler (FASTLITE) to combine the sidebands generated by the chirped pulse excitation. For the angled output, one possible scheme is combine the generated sidebands using a 4-f grating setup, as has been demonstrated recently (Matsubara et al., 2008; Weigand et al., 2009).

9. Acknowledgement

We acknowledge support from National Science Foundation (Grants no. PHY354897 and 722800), Texas Advanced Research Program (Grant no. 010366-0001-2007), the Army

Research Office (Grant no. W911NF-07-1-0475), and the Robert A. Welch Foundation (Grant no. A- 1547).

10. References

Basiev, T. T., Sobol, A. A., Zverev, P. G., Osiko, V. V. & Powell, R. C. (1999). Comparative Spontaneous Raman Spectroscopy of Crystals for Raman Lasers, *Applied Optics* Vol. (38): 594-598.

Basiev, T. T., & Powell, R. C. (1999). Introduction, *Optical Materials* Vol. (11): 301-306.

Basiev, T. T., Sobol, A. A., Voronko, Y. K. & Zverev, P. G. (2000). Spontaneous Raman spectroscopy of tungstate and molybdate crystals for Raman lasers, *Optical Materials* Vol. (15): 205-216.

Basiev, T. T., Osiko, V. V., Prokhorov, A. M. & Dianov, E. M. (2003). Crystalline and Fiber Raman Lasers, in *Solid-State Mid-Infrared Laser Sources*, Springer, Berlin / Heidelberg, pp. 359-408.

Basiev, T. T., Zverev, P. G., Karasik, A. Y., Osiko, V. V., Sobol, A. A. & Chunaev, D. S. (2004). Picosecond stimulated Raman scattering in crystals, *Journal of Experimental and Theoretical Physics* Vol. (99): 934-941.

Chen, W., Hsieh, Z., Huang, S. W., Su, H., Lai, C., Tang, T., Lin, C.,Lee, C., Pan, R., Pan, C. & Kung, A. H. (2008). Sub-Single-Cycle Optical Pulse Train with Constant Carrier Envelope Phase, *Phys. Rev. Lett.* Vol. (100): 163906.

Crespo, H., Mendonça, J. T. & Dos Santos, A. (2000). Cascaded highly nondegenerate four-wave-mixing phenomenon in transparent isotropic condensed media, *Opt. Lett.* Vol. (25): 829-831.

Demtröder, W. (2003). Laser spectroscopy: Basic concepts and Instrumentation, 3rd Edition, Springer, New York.

Dudovich, N. Oron, D. & Silberberg, Y. (2002). Single-pulse coherently controlled nonlinear Raman spectroscopy and microscopy, *Nature* Vol. (418): 512-514.

Dyson, A., Dangor, A. E., & Dymoke-Bradshaw, A. K. L. (1989). Sideband generation by coherent anti-Raman scattering in quartz of a two-frequency high-power laser beam, *Journal of Physics B: Atomic, Molecular and Optical Physics* Vol. (22): L231-L233.

Eckhardt, G., Bortfeld D. P. & Geller, M. (1963). Stimulated Emission of Stokes and Anti-Stokes Raman Lines from Diamond, Calcite, and Alpha-sulfur Single Crystals, *Appl. Phys. Lett.* Vol. (3): 137-138.

Edwards, D. F. & Ochoa, E. (1981). Infrared refractive index of diamond, J. Opt. Soc. Am. Vol. (71): 607-608.

Schulz-Von der Gathen, V., Bornemann, T., Kornas, V. & Dobele, H. F. (1990). VUV generation by high-order CARS, *Quantum Electronics, IEEE Journal of* Vol. (26): 739-743.

Gershgoren, E., Bartels, R. A., Fourkas, J. T., Tobey, R., Murnane, M. M. & Kapteyn, H. C. (2003). Simplified setup for high-resolution spectroscopy that uses ultrashort pulses, *Opt. Lett.* Vol. (28): 361-363.

Imasaka, T., Kawasaki, S., & Ishibashi, N. (1989). Generation of more than 40 laser emission lines from the ultraviolet to the visible regions by two-color stimulated raman effect, *Applied Physics B: Lasers and Optics* Vol. (49): 389-392.

Inoue, K., Kato, J., Hanamura, E., Matsuki H., & Matsubara, E. (2007). Broadband coherent radiation based on peculiar multiple Raman scattering by laser-induced phonon gratings in T_iO_2, *Phys. Rev. B* Vol. (76): 041101(R).

Kaminskii, A. A., McCray, C. L., Lee, H. R., Lee, S. W., Temple, D. A., Chyba, T. H., Marsh, W. D., Barnes, J. C., Annanenkov, A. N., Legun, V. D., Eichler, H. J., Gad, G. M. A. & Ueda K. (2000). High efficiency nanosecond Raman lasers based on tetragonal $PbWO_4$ crystals, *Optics Communications* Vol. (183): 277-287.

Kaminskii, A. A., Ralchenko V. G., Konov, V. I. & Eichler, H. J. (2005). High-order Stokes and anti-Stokes Raman generation in CVD diamond, *Physica Status Solidi (B)* Vol. (242): R4-R6.

Kaminskii, A. A., Ralchenko V. G. & Konov, V. I. (2006). CVD-diamond - a novel χ^3-nonlinear active crystalline material for SRS generation in very wide spectral range, *Laser Physics Letters* Vol. (3): 171-177.

Katzman, S. (2001). Rainbow Stars: A spectrum of possibilities, *American Chemical Society magazine News in Brief* Vol. (73): 351.

Kawano, H., Lin, C. H. & Imasaka, T. (1996). Generation of high-order rotational lines by four-wave Raman mixing using a high-power picosecond Ti:Sapphire laser, *Applied Physics B: Lasers and Optics* Vol. (63):121-124.

Kawano, H., Hirakawa, Y., & Totaro, I. (1997). Generation of more than 40 rotational Raman lines by picosecond and femtosecond Ti:sapphire laser for Fourier synthesis, *Applied Physics B: Lasers and Optics* Vol. (65): 1-4.

Kawano, H., Hirakawa, Y. & Imasaka, T. (1998). Generation of high-order rotational lines in hydrogen by four-wave Raman mixing in the femtosecond regime, *Quantum Electronics, IEEE Journal of* Vol. (34): 260-268.

Kawano, H., Mori, T., Hirakawa, Y., & Imasaka, T. (1999). Use of phase-locking in four-wave Raman mixing for generating ultrashort optical pulses, *Physical Review A* Vol. (59): 4703-4715.

Kienberger, R., Goulielmakis, E., Uiberacker, M., Baltuska, A., Yakovlev, V., Bammer, F., Scrinzi, A., Westerwalbesloh, Th., Kleineberg, U., Heinzmann, U., Drescher M., & Krausz, F. (2004) Atomic transient recorder, *Nature* Vol. (427): 817-821.

Korn, G., Dühr, O., & Nazarkin, A. (1998). Observation of Raman Self-Conversion of fs-Pulse Frequency due to Impulsive Excitation of Molecular Vibrations, *Physical Review Letters* Vol. (81): 1215-1218.

Liang, J. Q., Katsuragawa, M., Kien, F. L. & Hakuta, K. (2000). Sideband Generation Using Strongly Driven Raman Coherence in Solid Hydrogen, *Phys. Rev. Lett.* Vol. (85): 2474- 2477.

Liu, J. & Kobayashi, T. (2008). Cascaded four-wave mixing and multicolored arrays generation in a sapphire plate by using two crossing beams of femtosecond laser, *Opt. Express* Vol. (16): 22119-22125.

Liu, J. & Kobayashi, T. (2009). Generation of sub-20-fs multicolor laser pulses using cascaded four-wave mixing with chirped incident pulses, *Optics Letters* Vol. (34): 2402 2404.

Liu, J., Kobayashi, T. & Wang, Z. (2009). Generation of broadband two-dimensional multicolored arrays in a sapphire plate, *Opt. Express* Vol. (17): 9226-9234.

Matsubara, E., Inoue K. & Hanamura, E. (2006). Dynamical symmetry breaking induced by ultrashort laser pulses in $KTaO_3$, *J. Phys. Soc. Jpn.* Vol. (75): 024712 .

Matsubara, E., Sekikawa, T. & Yamashita, M. (2008). Generation of ultrashort optical pulses using multiple coherent anti-Stokes Raman scattering in a crystal at room temperature, *Appl. Phys. Lett.* Vol. (92): 071104.

Matsubara, E., Kawamoto, Y., Sekikawa, T. & Yamashita, M. (2009). Generation of ultrashort optical pulses in the10 fs regime using multicolor Raman sidebands in $KTaO_3$, *Opt. Lett.* Vol. (34): 1837-1839.

Matsuki, H., Inoue, K. & Hanamura, E. (2007). Multiple coherent anti-Stokes Raman scattering due to phonon grating in $KNbO_3$ induced by crossed beams of two-color femtosecond pulses *Phys. Rev. B* Vol. (75): 024102.

Nazarkin, A., Korn, G., Wittmann, M., & Elsaesser, T. (1999). Generation of Multiple Phase-Locked Stokes and Anti-Stokes Components in an Impulsively Excited Raman Medium, *Phys. Rev. Lett.* Vol. (83): 2560-2563.

Pask, H. M. (2003). The design and operation of solid-state Raman lasers, *Progress in Quantum Electronics* Vol. (27): 3-56.

Pestov, D., Zhi, M., Sariyanni, Z. E., Kalugin, N. G., Kolomenskii, A. A., Murawski, R. K., Paulus, G. G., Sautenkov, V. A., Schuessler, H., Sokolov, A. V., Welch, G. R., Rostovtsev, Y. V., Siebert, T., Akimov, D. A., Graefe, S., Kiefer, W. & Scully, M. O. (2005). Visible and UV coherent Raman spectroscopy of dipicolinic acid, *Proc. Nat. Acad. Sci. U. S. A.* Vol. (102): 14976-14981.

Pierson, H. O. (1993). Handbook of Carbon Graphite, Diamond and Fullerenes: Properties, Processing, and Applications, Noyes Publications.

Sali, E., Mendham, K. J., Tisch, J. W. G., Halfmann, T. & Marangos, J. P. (2004). High-order stimulated Raman scattering in a highly transient regime driven by a pair of ultrashort pulses, *Opt. Lett.* Vol. (29): 495-497.

Sali, E., Kinsler, P., New, G. H. C., Mendham, K. J., Halfmann, T., Tisch, J. W. G. & Marangos, J. P. (2005). Behavior of high-order stimulated Raman scattering in a highly transient regime, *Physical Review A* Vol. (72): 013813.

Sokolov, A. V. & Harris, S. E. (2003). Ultrashort pulse generation by molecular modulation, *Journal of Optics B: Quantum and Semiclassical Optics* Vol. (5): R1-R26.

Sokolov, A. V., Shverdin, M. Y., Walker, D. R., Yavuz, D. D., Burzo, A. M. Yin, G. Y. & Harris, S. E. (2005). Generation and control of femtosecond pulses by molecular modulation, *Journal of Modern Optics* Vol. (52): 285-304.

Smith, D. Y., Inokuti M. & Karstens, W. (2001). A generalized Cauchy dispersion formula and the refractivity of elemental semiconductors, *J. Phys.: Condens. Matter* Vol. (13): 3883- 3893.

Steinmeyer, G., Sutter, D. H., Gallmann, L., Matuschek N. & Keller U., (1999). Frontiers in Ultrashort Pulse Generation: Pushing the Limits in Linear and Nonlinear Optics, *Science*, Vol. (286): 1507-1512.

Takahashi, J. (2004), Generatoin of a broadband spectral comb with multiwave mixing by exchange of an impulsively stimulated phonon, *Opt. Express* Vol. (12): 1185-1191.

Weigand, R., Mendonca, J. T. & Crespo, H. M. (2009). Cascaded nondegenerate four-wavemixing technique for high-power single-cycle pulse synthesis in the visible and ultraviolet ranges, *Phys. Rev. A* Vol. (79): 063838.

Yoshikawa, S. & Imasaka, T.(1993). A new approach for the generation of ultrashort optical pulses, *Optics Communications* Vol. (96): 94-98.

Zheltikov, A. M. (2002). Spectroscopic and quantum-control aspects of ultrashort-pulse synthesis through impulsive high-order stimulated Raman scattering, *Journal of Raman Spectroscopy* Vol.(33): 112-120.

Zhi, M. & Sokolov, A. V. (2007). Broadband coherent light generation in a Raman-active crystal driven by two-color femtosecond laser pulses, *Opt. Lett.* Vol.(32): 2251-2253.

Zhi, M. & Sokolov, A. V. (2008). Broadband generation in a Raman crystal driven by a pair of time-delayed linearly chirped pulses, *New J. Phys.* Vol.(10): 025032.

Zhi, M., Wang, X. & Sokolov, A. V. (2008). Broadband coherent light generation in diamond driven by femtosecond pulses, *Opt. Express* Vol.(16): 12139-12147.

29

Holographic 3-D Displays - Electro-holography within the Grasp of Commercialization

Stephan Reichelt, Ralf Häussler, Norbert Leister, Gerald Fütterer, Hagen Stolle, and Armin Schwerdtner
SeeReal Technologies
Germany

1. Introduction

Holography is a diffraction-based coherent imaging technique in which a complex three-dimensional object can be reproduced from a flat, two-dimensional screen with a complex transparency representing amplitude and phase values. It is commonly agreed that real-time holography is the ne plus ultra art and science of visualizing fast temporally changing 3-D scenes. The integration of the real-time or electro-holographic principle into display technology is one of the most promising but also challenging developments for the future consumer display and TV market. Only holography allows the reconstruction of natural-looking 3-D scenes, and therefore provides observers with a completely comfortable viewing experience. But to date several challenges have prevented the technology from becoming commercialized. But those obstacles are now starting to be overcome. Recently, we have developed a novel approach to real-time display holography by combining an overlapping sub-hologram technique with a tracked viewing-window technology (Schwerdtner, Leister & Häussler, 2007; Schwerdtner, Häussler & Leister, 2007). For the first time, this enables solutions for large screen interactive holographic displays (Stolle & Häussler, 2008; Reichelt et al., 2008).
This chapter presents these novel solutions for large real-time holographic 3-D displays in the context of previous and current approaches to electro-holography. The holographic display developed by us combines a tailored holographic recording scheme with active tracking of the observer. This unique approach dramatically reduces the demand for the space-bandwidth product of the hologram and thus allows the use of state-of-the-art spatial light modulators and enables real-time calculation. The fundamentals and challenges of the holographic display technology are described, its implementation in prototypes is demonstrated, and the bright prospects for the 3-D display market are discussed.

2. Real-time holographic display technology

When talking about holographic display technology, a note of caution about commonly used terminology is needed. For marketing or other reasons, the term 'holographic display' is often misused to name systems, which are not truly holographic in the sense of video-holography. Systems, which make use of holographic screens or holographic optical elements to project images are just examples. But even volumetric displays that create light

spots somewhere within their volume are called in many cases 'holographic'. On the other hand, truly holographic recordings are being called displays, whereas there are in fact static holograms or dynamic holograms, which are not yet real-time capable with rewriting-times in the minute range and with large-scale setups (Tay et al., 2008).
What we mean with real-time holographic displays are systems that are based on diffraction to reconstruct the wave field of a 3-D scene in space with coherent light. Such displays must operate at or near video rate to merit the name of video holography. Furthermore, real-time holography must not only display the hologram at video rate but also compute the hologram frames in real time to enable user interaction.

2.1 Why is holography the ultimate 3-D technology?

In human vision, three-dimensional perception is triggered by a large number of cues. Among them monocular cues such as shading, occlusion, relative size, fogging, perspective distortion, and texture gradient as well as binocular ones such as vergence (angular disparity) and stereopsis (horizontal disparity). In natural viewing situations, depth information is an ever-present cue in the visual perception. Generally, and in addition to parallax, the physiological depth cues of accommodation and vergence are considered to be the most important ones for depth perception. *Accommodation* is the mechanism by which the human eye alters its optical power to hold objects at different distances into sharp focus on the retina. The power change is induced by the ciliary muscles, which steepen the crystalline lens' curvature for objects at closer distances. *Vergence*, by contrast, is the simultaneous movement of both eyes toward the point of interest. The optical axes of both eyes converge on this point to image the object onto the respective fovea regions. When the eyes are not properly aligned with each other, strabismus occurs that may adversely impair any 3-D perception. But most importantly, vergence movements and accommodation are closely linked with each other - automatically and subconsciously. That's why the image of an object is sharp and the two perspectives are fused. Together with other monocular and binocular cues, the focus depth cues - accommodation and blur in the retinal image - contributes to our visual ability to perceive the environment in three dimensions.
Over the last decade, various technologies for visualizing three-dimensional (3-D) scenes on displays have been technologically demonstrated and refined, among them such of stereoscopic, multi-view, integral-imaging, volumetric, or holographic type. It is generally believed that the next big thing in the display industry is imminent, namely the transition from 2-D to 3-D visualization. It is seen as nothing less than the third epoch-making change in film industry, after the change from silent to sound movie in the 1920's and from black-and-white to color in the 1950's.
Most of the current approaches utilize the conventional stereoscopic principle, first described by Wheatstone (Wheatstone, 1838). But except of super-multiview displays, they all lack of their inherent conflict between vergence and accommodation since scene depth cannot be physically realized but only feigned by displaying two views of different perspective on a flat screen and delivering them to the corresponding left and right eye. This mismatch requires the viewer to override the physiologically coupled oculomotor processes of vergence and eye focus, which may cause visual discomfort and fatigue.
The difference between normal viewing and stereoscopic viewing with conventional 3-D displays is illustrated in Fig. 1. Natural viewing provides real stimuli; the viewer is both fixated and focused on the object, i.e. accommodation distance and vergence distance are

exactly matched. But the situation changes for stereoscopic 3-D displays. Though the viewer is still fixated to the object with the same vergence as in natural viewing, his eye focus is now at the display and not where the object seems to be. That is because the viewer's eyes always focus on the brightest point or highest contrast. With stereoscopic 3-D displays, depth is only an optical illusion. Hence with stereoscopic displays, the normal physiological correlation between vergence and accommodation is disrupted (Hoffman et al., 2008). When looking at a stereoscopic display for a while, this so-called depth cue mismatch between convergence and focus leads to eye strain and fatigue. This fundamental problem to stereoscopic 3-D is a physiologically one that cannot be solved by technological means. The only workaround for stereoscopic displays is either to limit scene depth to very short sequences (short time viewing) or to artificially reduce the depth of a scene (squeezed, non-proportional depth). The so-called comfort depth range of stereoscopic displays, which can be used for depth illusion without getting eye strain and fatigue is limited to a region close to the display that corresponds approximately 20...30% of the distance between viewer and display. Only within this region the human eye can tolerate a certain amount of mismatch. According to optometrists, this tolerance is in the range of 1/4 diopter. Although stereo 3-D can work well for some applications, for example cinema with far observing distance or cell phones with short viewing time, it causes significant human factor risks for mainstream products as PC monitors and TV. Looking at current stereoscopic 3-D displays and prototypes it can also be observed that even the 1/4 diopter is scarcely utilized, limiting usable depth range even further.

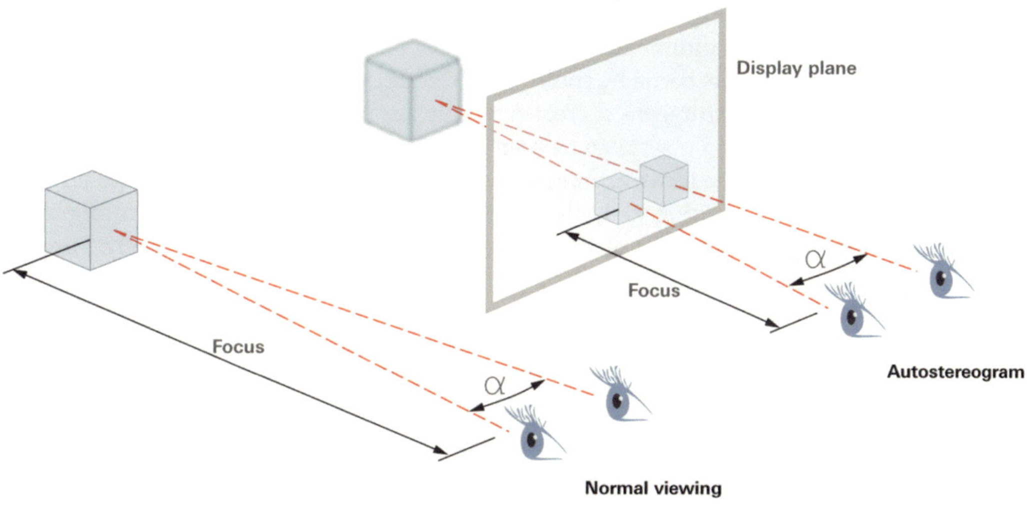

Fig. 1. Comparison between natural viewing or holographic 3-D display (left) and stereoscopic viewing with a 3-D stereo display (right).

Therefore, the inherent limitations of all 3-D stereo display technologies can be summarized as follows:

- Depth cue mismatch of convergence and accommodation leads to eye strain and fatigue,
- Reduced comfort depth range requires a non-proportional depth squeezing or allows only short-time viewing of scenes with large depth,
- Potential for inappropriate use, and therefore a consumer application risk.

It is important to note that the comfort depth range in a display and the content generated for instance within a movie or a game are independent. This leads to a significant risk that even in a well-made stereoscopic 3-D display improper content compromises user comfort or health and may (unfairly but) possibly held against the display manufacturer.
Contrary to stereo 3-D, which inherently causes fatigue and eye strain for natural depth 3-D scenes (i.e. properly scaled depth), holographic 3-D provides all viewing information of a natural scene – including eye focus – and therefore unlimited depth. Whoever can see 3-D in real life can see 3-D on a holographic display without fatigue or other consumer risks. Holographic displays are based on coherent object reconstruction. They deliver full focus cues that are needed to provide the observer with a completely comfortable 3-D viewing experience (Benton & Bove, 2008). We have developed and successfully demonstrated a novel approach to real-time display holography based on a sub-hologram encoding technique and a tracked viewing-window technology. Our solution is capable to fulfill the observer's expectations on real depth perception.

2.2 Classic holography and historic obstacles

Holography was invented by Dennis Gabor in 1947 (Gabor, 1948), but high-quality holograms could only be made on photographic film, which for technical reasons preclude animation. By this classic holography, the 3-D scene information is encoded in the entire hologram, i.e. every tiny region or pixel of the hologram contributes to each object point. The specialty of such holograms is well-known: If the hologram is broken into pieces, each piece will reconstruct the original scene, even though with less resolution and in smaller size. When the hologram is illuminated by the reference wave, the combination of all of its cells reproduces the complete scene by multiple interferences. A classic film hologram has a large diffraction angle, which means it creates a large angular spectrum. The viewing zone from which the reconstructed object can be seen is large, both eyes of the viewer fit into this zone and the viewer can even move around and see different perspectives of the scene.
The difficulties arise when trying to apply the classic approach of holography to digital or electro-holography. The challenges of this approach are twofold: (a) the spatially sampled representation of the hologram by a light modulator (spatial resolution issue) and (b) the fast computation of the hologram (processing issue).
Hence, one of the most serious restrictions of video holography has been the dynamic representation of the hologram by an electrically addressed spatial light modulator (SLM) having a pixelized structure with limited spatial resolution. The complex amplitude distribution that reconstructs the desired object or scene is calculated and represented at regular discrete locations, i.e. at the pixel positions of the spatial light modulator. Since the hologram is sampled, aliasing has to be prevented. Otherwise, improper reconstruction with image artifacts would occur. The amount of information that can be recorded in the hologram is directly related to the spatial resolution and the size of the SLM. This fact is represented by the dimensionless *space-bandwidth product*

$$\mathrm{SBP} = \nu_x b_x \cdot \nu_y b_y = \frac{b_x}{2\delta_x} \cdot \frac{b_y}{2\delta_y} \qquad (1)$$

with ν being the maximum spatial frequency according to the sampling theorem, b the width and δ the pixel pitch of the modulator in x and y direction (Lohmann, 1967; Lohmann et al., 1996). In general, the space-bandwidth product capability of an optical system is directly related to its quality and performance. For example, a present state-of-the-art LCOS microdisplay with 1920 × 1080 pixel resolution, a pixel pitch of 8 µm and a total size of 0.7" gives a space-bandwidth product of 518,400. The Nyquist limit for the maximum spatial frequency is thus 62.5 Lp/mm, which translates into a maximum diffraction angle of 2.27°.

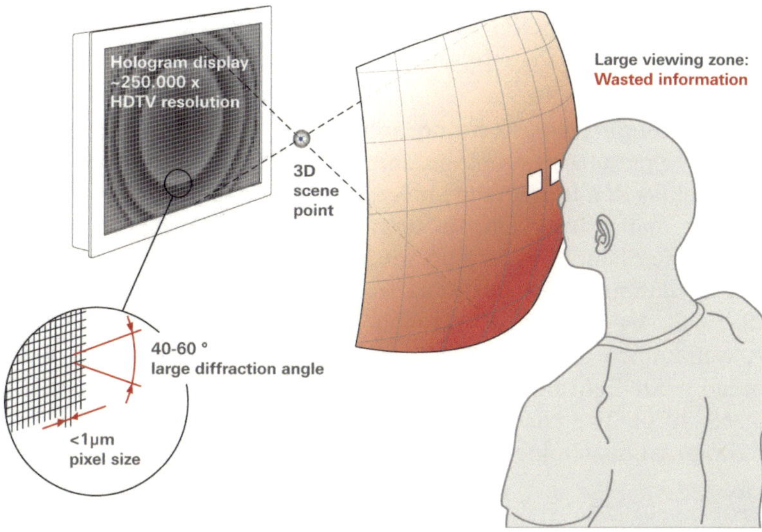

Fig. 2. Principle of classic holography. In conventional holography, every hologram pixel contributes to each object point of the 3-D scene, that is, holographic information existing at a large viewing zone.

Let us recall that in conventional holography the diffraction angle must be large to create a viewing zone that covers at least both eye and that different areas of the hologram encode the wave field originating from another perspective of the object (see Fig. 2). In order words, the primary objective of conventional holography is to reconstruct the 3-D object in space, which can be seen by any viewer binocularly from different view points at different perspectives. To achieve a sufficient viewing zone, pixel sizes in the range of the one micron or less would required. Moreover, to create large objects and fully exploit the 3-D impression the display should be large. However, this corresponds to a huge amount of information that - even if large SLM with tiny pixels would be available - must still be handled in data processing and computing. To give an example, extreme-resolution displays with a pixel size of roughly 0.5 microns would be required, which translates to the huge demand for calculation of billions to trillions of complex values for each of the 2 million scene points (1920×1080) to determine an HDTV scene in 3-D. When considering these requirements, the insurmountable obstacles to realize conventional holography by using today's technology become immediately obvious. The reasons why all past attempts of transferring conventional holography to display and TV applications have heretofore failed can be summarized as follows:

Insufficient display resolution: In order to achieve a viewing angle of ±30°, which is necessary to serve several users, a pixel pitch of about one wavelength or less is required. This means that for a 47-inch holographic display, for example, a resolution of 250,000 times that of HDTV is necessary.

Inadequate data volume and processing requirements: The computation of each display frame requires significantly more steps for a holographic display compared to a 2-D display. Typical hologram computation involves calculations of Fourier transformations. This factor, coupled with the greatly increased number of pixels required, places a demand for enormous amounts of computational power. Thus, real-time videoquality holograms would typically require processing power up to several hundred Peta-FLOPS, i.e. approximately 10^{17} floating-point operations per second. This is far more than the current computation power of super computers.

2.3 Full parallax vs. single parallax holography

With full-parallax holograms, the holographic information is delivered in both x and y direction. When looking at a full parallax hologram, the perspective of the scene varies with the viewpoint - no matter in which direction the observer is moving. In single-parallax holograms, on the other hand, the parallax information is sacrificed in one dimension. That way both the computational effort and data transfer can be substantially reduced. Because of the eyes are side by side, it is common practice to make horizontal-parallax-only holograms. A well-known example of a horizontal-parallax-only (HPO) hologram is the optically recorded white-light rainbow hologram invented by Benton (Benton, 1969; Benton & H. S. Mingace, 1970). The concept of single parallax holograms was later successfully transferred to computational holography (St-Hilaire et al., 1992).

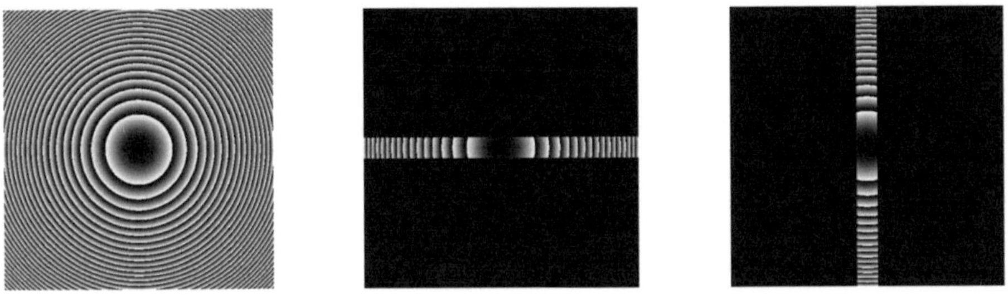

(a) Full parallax (b) HPO (c)VPO

Fig. 3. Examples of full and parallax-limited holograms. The spherical phase of a simple single-point hologram is shown (kinoform of a point or Fresnel zone lens). HPO – horizontal-parallax-only; VPO – vertical-parallax-only.

Both benefits and limitations of full and parallax-limited holograms become obvious from Fig. 3, which shows a very simple hologram and its parallax-limited versions. A full parallax hologram reconstructs the object point from a large area with spatial frequencies in all directions, which comes along with a large information content that must be all calculated, transferred by computer and spatially resolved by the light modulator. In comparison, a parallax-limited hologram that is a sliced version of the full type, diffract the light basically in one dimension. Beside the reduced computational effort, such a configuration is beneficial for other reasons as well. For example, the remaining pixel (or 'saved' bandwidth) could be used for realizing hologram interlacing for different colors or just for simplifying hologram representation with a given display architecture. However, there are also tradeoffs with single-parallax holograms. As the diffraction occurs mainly in one direction, the diffracted wave is slightly elliptical and the spatial resolution of the reconstruction in the non-diffracted direction might be marginally reduced. However, when taking into account the resolution capabilities of the human eye and generate the hologram and the display system accordingly, the benefits of the parallax-limited holograms outweigh its constraints. It should be noted that SeeReal's sub-hologram approach is inherently applicable to both encoding principles with similar gains in efficiency.

2.4 Brief review of previous and current approaches to electro-holography

There have been many practical approaches to electro-holography in the past decades. Several of them are briefly presented in this chapter as examples.

A pioneering holographic display was set up at the MIT Media Lab in S. A. Benton's group and continuously improved (St-Hilaire et al., 1992; Lucente et al., 1993; St-Hilaire et al., 1993). These systems use an acousto-optic modulator (AOM), scanners and an optical imaging system. High-frequency acoustic waves locally modulate the refractive index of the AOM crystal and thus the phase of transmitted light. The AOM generates a horizontal line of the hologram that is vertically continued by a vertical scanner. Recent progress was made with an improved AOM that allows higher bandwidth and a simplified optical setup (Smalley et al., 2007). The system is specified with a cube-like object volume with approximately 80 mm edge length and 24° viewing angle at a frame rate of 30 Hz.

Another approach was made by QinetiQ using a so-called Active-Tiling technique (Stanley et al., 2003; Slinger et al., 2004). A SLM with 1 million pixels is replicated sequentially 25-fold on an optically addressable SLM (OASLM) using 5 x 5 replication optics. Four of these units are stacked horizontally to yield a SLM with 100 million pixels in total at a pixel pitch of 6.6 μm. The modular system design allows stacking of more units to achieve higher numbers of pixels. A replay system with an Active-Tiling SLM with 100 million pixels achieved an object with 140 mm width and a viewing zone width of 85 mm at 930 mm distance.

Direct tiling of SLMs is used for another holographic display (Maeno et al., 1996). Five SLMs with 3 million pixels each are tiled to yield 15 million pixels in total. The object may be as large as 50 mm, 150 mm high and 50 mm deep and can be viewed with both eyes at a distance of 1 m.

Effort was also made to optimize the calculation of holograms. A computing system with dedicated hardware performs hologram calculation much faster than a PC. As an example, the HORN-6 cluster uses a cluster of boards equipped with FPGA chips (Ichihashi et al., 2009). The system needs 1 second to calculate a hologram with 1920 x 1080 pixels if the object is composed of 1 million points and 0.1 second if the object is composed of 100,000 points.

All these approaches have in common that a large number of pixels is needed to reconstruct an object with small or medium size. These requirements for the SLM and the computing system hinder scaling to larger sizes, e.g. 20" object size with unlimited depth for desktop applications or TV.

3. SeeReal's novel solution to real-time holography

3.1 Fundamental idea and overview

The fundamental idea of our concept is fairly simple when considering holography - even literally - from an information point-of-view. All visual acuity is limited by the capabilities of the human eye, i.e. its angular and depth resolution, color and contrast sensitivity, numerical aperture, magnification, etc., where the characteristics of the eye may vary widely from individual from individual. It may additionally be confined by monochromatic and chromatic aberrations.

The majority of optical instruments, such as visual microscopes or telescopes utilize the eye as the final element of the optical system. The eye's specific capabilities are thus taken into account in the optical system design. We view holography in the same way. When considering the human vision system regarding to where the image of a natural environment is received by a viewer, it becomes obvious that only a limited angular spectrum of any object reaches the retina. In fact, it is limited by the pupil's aperture of some millimeters. If the positions of both eyes are known, it therefore would be wasteful to reconstruct a holographic scene or object that has an extended angular spectrum as it is common practice in classical holography. As mentioned above, in every part of a classic hologram the entire object information is encoded, cf. Fig. 2. This means that a large viewing zone with parallax information within this zone exists; by moving within this zone the viewer can "look around" the reconstructed object and thus sees different perspectives of the scene. This approach is historically explained by the interference-based exposure technique onto high-resolution holographic films and is useful for static holograms as known from artistic holographic recordings. The key idea of our solution to electro-holography is to reconstruct a limited angular spectrum of the wave field of the 3-D object, which is adapted laterally in size to about the human's eye entrance pupil, cf. Fig. 4. That is, the highest priority is to reconstruct the wave field at the observer's eyes and not the three-dimensional object itself. The designated area in the viewing plane, i.e. the virtual 'viewing window' from which an observer can perceive the proper holographic reconstruction is located at the Fourier plane of the holographic display. It corresponds to the zero-order extension of the underlying SLM cross grating. The holographic code (i.e. the complex amplitude transmittance) of each scene point is encoded on a designated area on the hologram that is limited in size. This area in the hologram plane is called a sub-hologram. The position and size of the sub-hologram is defined by the position of object point and viewing window geometry. There is one sub-hologram per scene point, but owing to the diffractive nature of holography, sub-holograms of different object points may be overlapping. The complex amplitude transmittances of different sub-holograms can be added without any loss of information.

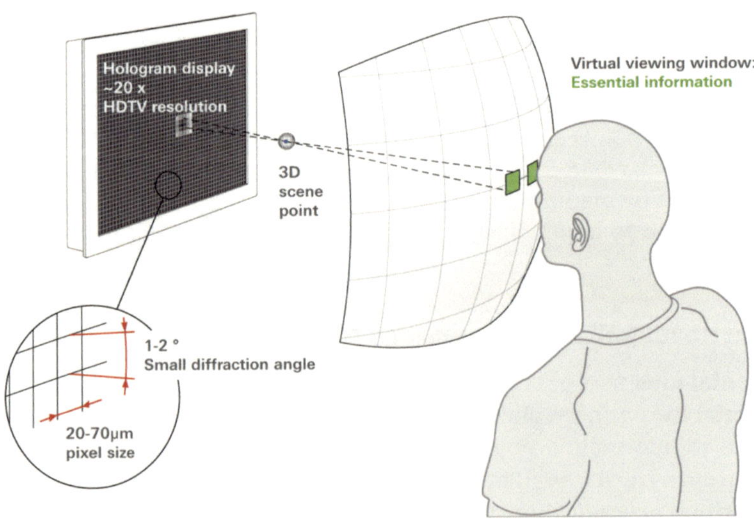

Fig. 4. Principle of viewing-window holography. With viewing-window holography the essential and proper holographic information exists at the eye positions only.

So far, for the sake of simplicity, we have discussed the matter for a single viewing window, which carries the information for one eye only. But how is then parallax information generated? Binocular view can be created by delivering different holographic reconstructions with the proper difference in perspective to left and right eye, respectively. For this, the techniques of spatial or temporal multiplexing can be utilized.

For such a binocular-view multiplexed hologram, the reconstructed 3-D object can be seen from a single pair of viewing windows only. Advantageously, dynamic or real-time video holography offers an additional degree of freedom in system design with respect to temporal-multiplex operation. Given that the computational power is sufficient and the spatial light modulator is fast enough, the hologram can be updated quickly. By incorporating a tracking system, which detects the eye positions of one or more viewers very fast and precisely and repositions the viewing window accordingly, a dynamic 3-D holographic display can be realized that circumvents all problems involved with the classic approach to holography. The steering of the viewing window can be done in different ways, either by shifting the light source and thus shifting the image of the light source, or by placing an additional steering element close to the SLM that realizes a variable prism function. Selected implementations of steering principles will be explained in more detail in section 3.5.

To summarize, the pillars of our holographic display technology are:

Viewing-window holography: By limiting the information of the holographic reconstruction to the viewing windows, the required display resolution is decreased dramatically. Pixel sizes in the range of today's commercially available displays are sufficient.

Real-time computation of sub-holograms: By limiting the encoding to sub-holograms, the computing requirements are greatly reduced. Sub-hologram encoding brings computation into graphics card or ASIC range. The principle also enables temporal color multiplexing, speckle reduction, and suppression of higher orders within the viewing window.

Tracking of viewing windows: An active and real-time tracking of the viewing window allows a free movement of the observer.

3.2 The viewing-window and sub-hologram concept

The optical principle of our holographic approach is schematically depicted in Fig. 5. Coherent light coming from a point light source is imaged by a positive lens (L+) into the observer plane and creates the spherical reference wave for hologram illumination. Very close to the imaging lens, the spatial light modulator (SLM) is positioned.

3.2.1 What is a viewing window?

The inherent regular SLM structure generates a diffraction pattern in the far field whose zero-order extension is the viewing window (VW) where the eye of the observer is located. Given small angles, the size of the viewing window is obtained from the grating equation and trigonometry to

$$w_{x,y} = \frac{\lambda d}{p_{x,y}} \tag{2}$$

with d being the observer distance, λ the wavelength and p the pixel pitch of the SLM in x or y direction, respectively. Only within the viewing window the information of the wave field

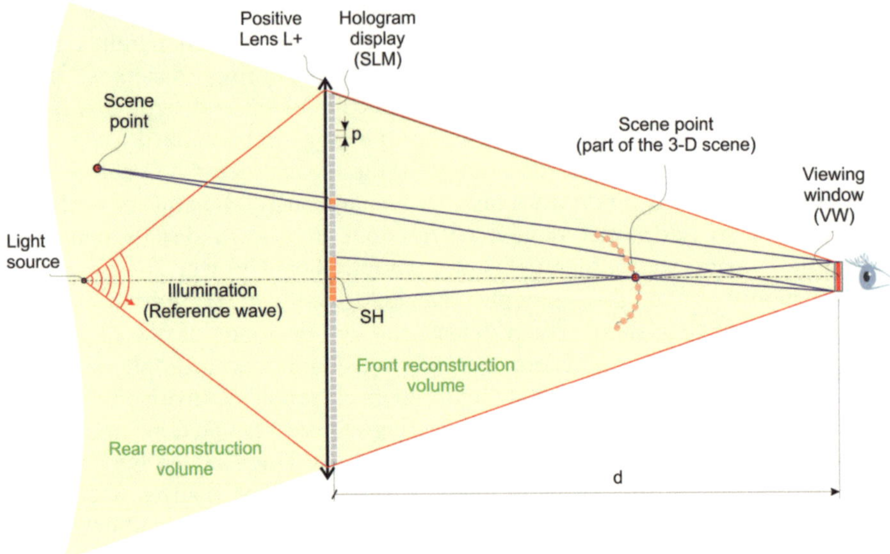

Fig. 5. Schematic principle of the sub-hologram concept (side view). L+, positive lens; SLM, spatial light modulator; SH, sub-hologram; VW, viewing window; other abbreviation defined in the text.

of the 3-D object has to be generated. In Table 1 the wavelength-dependent viewing-window size $w_{x,y}$ for different display types having typical viewing distances d and pixel pitches $p_{x,y}$ are listed. For a proper visual perception of a colored scene, the size of smallest viewing window is the determining factor. For blue light, the viewing window must be therefore at least the same size as the pupil diameter. Depending on the scene luminance, the entrance pupil diameter of the human eye varies from 2...6 mm. Therefore, the required pixel pitches of the holographic display mainly result from the viewing situation and distance (television, desktop, or mobile display), where additionally the wavelength-dependent diffraction has to be considered.

	Desktop-Display with $p_{x,y}$ = 30 µm			TV-Display with $p_{x,y}$ = 70 µm		
	λ = 475 nm (B)	λ = 532 nm (G)	λ = 635 nm (R)	λ = 475 nm (B)	λ = 532 nm (G)	λ = 635 nm (R)
$w_{x,y}$ at d = 750 mm	11.9 mm	13.3 mm	15.9 mm			
$w_{x,y}$ at d = 2500 mm				17.0 mm	19.0 mm	22.7 mm

Table 1. Size of the viewing window $w_{x,y}$ for exemplary holographic display types with typical viewing distances d at RGB-wavelengths.

For the common examples, the viewing window for blue light is about 3/4 of the size for red light. Viewing windows much larger than the pupil diameter provide more tolerances for the tracking system, i.e. the accuracy in pupil detection and viewing-window shifting could be then less stringent. With larger viewing windows, on the other hand, the intensity is distributed over a larger area, which means that only part of it will pass through the pupil. The best compromise between technological issues, tracking accuracy, and reconstructed scene brightness has therefore to be chosen.

3.2.2 What is a sub-hologram?

A direct consequence of the viewing-window holographic scheme is the following. For each reconstructed object point, there is only a limited region in the hologram where data from this object point is coded. Each point of the scene (which can be treated as a point-like emitter for interference based hologram modeling) is associated with a locally limited area of the hologram. This limited region is called a sub-hologram. Size and position of this sub-hologram (SH) is defined via simple projection from the edges of the viewing window through the scene point that has to be encoded. By superimposing all sub-holograms, i.e. strictly speaking the angular limited complex amplitudes $U_n(x,y) = A_n(x,y)\exp[-i\phi_n(x,y)]$ emanating from each of the n scene points, the entire hologram is composed. Since the complex amplitudes have to be calculated only within each sub-hologram area, the computational effort is dramatically reduced, which enables a real-time hologram calculation. A beneficial side effect of that special holographic recording scheme is the reduced demand for the temporal coherence of the light source, which must have at least a coherence length of λN_F, where N_F is the Fresnel number of the maximum sub-hologram. Monocular motion parallax information is delivered within the viewing window, which may be either in full-parallax or single-parallax, depending on the recording scheme of the hologram and the overall optical setup.

In contrast to a common hologram, if an entire viewing window-type hologram is broken into several pieces, each piece will reconstruct only part of the original scene, but with full resolution (apart from object points close to the border of the reconstructed scene fragment). The refractive analogue of one non-overlapping sub-hologram is a small lens having an amplitude and phase distribution that focuses light from the hologram to the object point.

3.3 Hologram synthesis

Hologram synthesis means the *calculation* of the complex wavefield $\mathcal{H}$ at the hologram plane (x,y). Or, which way we know the proper amplitude and phase distribution within the hologram plane that reconstructs the desired object? The complex wave field at the hologram plane $\mathcal{H}(x,y)$ is the superposition of the object wave $\mathcal{O}(x,y)$ with the reference wave $\mathcal{R}(x,y)$. The object wave $\mathcal{O}$ in turn, represents the superposition of all Huygens' elementary waves virtually emerged from the object being reconstructed. In classical holography this is the reflected, refracted or diffracted light from any existing object, whereas with synthetic or computer-generated holography the object exists mathematically as a set of data only (position in space, amplitude, and color) and the superposition can be easily made by computing means. When the complex wave field at the hologram plane $\mathcal{H}(x,y)$ is again illuminated with the reference wave $\mathcal{R}$, the object is holographically reconstructed.

In the following two methods for hologram synthesis are described in more detail, a direct analytic method as well as a Fourier-based propagation method. Both methods assume that the object is being composed of a set of n points, which are defined in position, amplitude, and color.

3.3.1 Direct analytic modeling

A three-dimensional scene is represented by a sufficient number of points defined at discretized locations. The propagation from the point source to the hologram plane is modeled by using exact analytic functions where each object point can be regarded as a coherent point source that emits a spherical wave at a wavelength λ. At the hologram plane this wave is defined as $U_n(x,y) = A_n(x,y)\exp[-i\varphi_n(x,y)]$. If the location of the spherical emitter is given by $P_n(x_0,y_0, z_0)$, the phase at the hologram plane can be written as

$$\varphi_n(x,y) = \frac{2\pi}{\lambda}\left(\sqrt{z_0^2 + (x - x_0)^2 + (y - y_0)^2} - z_0\right) + \varphi_0, \tag{3}$$

which corresponds exactly to the so-called phase function of an ideal lens with a focal length of z_0. The quantity φ_0 ($0 \leq \varphi_0 \leq 2\pi$) is an initial phase assigned to each scene point, which can be used as a further degree of freedom in design of the hologram. It is common practice to randomize this phase offset. To finally obtain the hologram phase function $\phi_n(x,y)$, the phase of the reference wave must be subtracted from $\varphi_n(x,y)$. The phase of the reference wave ϕ_R, which is in fact the illumination wave formed by the lens L+, is again the phase of an ideal spherical lens with focal length of d, see Fig. 5. Thus, the hologram phase for one scene point is given by $\phi_n(x,y) = \varphi_n(x,y) - \phi_R(x,y)$. It might be noted that the hologram phase function $\phi_n(x,y)$ associated to one emitter is always a rotationally symmetric function. The origin is located at the point where a virtual line connecting the center of the viewing window with the scene point intersects the hologram plane.

But in contrast to the classic approach, viewing-window holography computes and encodes the complex amplitudes $U_n(x,y)$ of the each scene points only in a designated area of the hologram plane, the sub-hologram area. Aliasing is prevented by cutting-off the spatial frequency contributions of the object points that exceed the spatial resolution of the light modulator. The spatial frequency of the hologram phase for one point-like emitter can be derived from

$$\vec{\nu}(x,y) = \frac{\nabla\phi_n(x,y)}{2\pi} \tag{4}$$

where the maximum allowable spatial frequency is given by the resolution of the spatial light modulator and must satisfy the relation $\nu(x) \leq \frac{1}{2\delta_x}$ and $\nu(y) \leq \frac{1}{2\delta_y}$, respectively.

3.3.2 Fourier-based modeling

Figure 6 illustrates an alternative method for hologram synthesis that is based on fast Fourier transforms (FFT) of object planes. It shows a side view of a three-dimensional object or three-dimensional scene, the spatial light modulator (hologram), and the viewing window. The viewing window is positioned at or close to an observer eye. The 3-D-scene is located within a frustum that is defined by the edges of the viewing window and the SLM, respectively (drawn as red dashed lines in Fig. 6). This frustum may be approximated by a pyramid if the viewing window is much smaller than the SLM.

For calculation, the 3-D-scene is sliced in layers ($L_1 \ldots L_m$) that are parallel with both the SLM and viewing-window plane. The continuously distributed object points are assigned to the closest layer. The extension of each layer is limited by the frustum and depends on the distance from the viewing window. With the approximation of the viewing window being much smaller than the SLM, the extension of a layer is proportional to its distance from the viewing window.

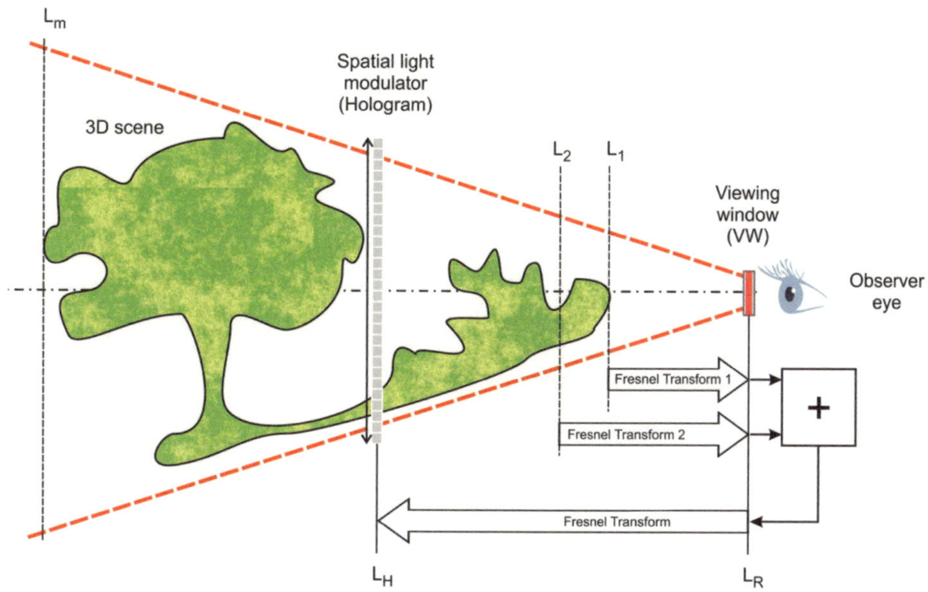

Fig. 6. FFT-based hologram synthesis.

In each layer, the object points are assigned to the nearest sampling point of the layer. The calculation method uses Fresnel transforms between the object layers $L_1 \ldots L_m$, a reference layer L_R in the observer plane and a hologram layer L_H in the plane of the SLM. The calculation of a Fresnel transform can be mathematically performed as a Fourier transform and multiplication with quadratic phase factors (Goodman, 1996). The discrete Fourier transform can be efficiently executed by using the fast Fourier transform (FFT) algorithms. Hologram synthesis comprises three steps:

1. Firstly, the layers L_1 to L_m are transformed subsequently to the reference layer L_R by m Fresnel-Transforms.
2. Secondly, the wave fields calculated in the first step are summed up to a superimposed complex-valued wave field in the viewing window. This superimposed wave field represents the frequency-limited wave field that would be generated by a real existing 3-D scene.
3. Thirdly, the superimposed wave field in the viewing window is back-transformed to the hologram layer L_H by an inverse Fresnel transform. This yields finally the hologram function $\mathcal{H}(x,y)$.

The information in each layer is not continuous but sampled. It is essential that all object layers, the viewing window and the hologram layer L_H contain the same number of sampling points N. This number corresponds to the pixel number of the spatial light modulator.

As mentioned above, the extension of an object layer L_m is proportional to its distance d_m to the reference layer L_R. Hence also the sampling interval p_m in a layer is proportional to its distance d_m to L_R, i.e. $p_m \propto d_m$. As a consequence, the periodicity interval w_m in the reference layer L_R is the same for all object layers L_m, that is $w_m = \lambda d_m / p_m$ = constant. This ensures that the periodicity interval in the L_R has the same extension for all object layers. Therefore, a common viewing window that is located within one periodicity interval can be defined where the wave field within this viewing window is unique.

As explained, in the third step the wave field in the viewing window is transformed to the hologram layer. The hologram layer and the reference layer are related by a direct or an inverse Fresnel transform. The number of sampling points N in each object layer is the same as in the layer L_H on the SLM. As the viewing window is within one periodicity interval of an object layer, it is also within one periodicity interval of the L_H or SLM. Hence, as the hologram reconstructs the wave field in the viewing window, this wave field will be unique therein. Periodic repetitions of this wave field that are inherent for sampled holograms are outside the viewing window. The hologram reconstructs the wave field that would be generated by a real existing 3-D scene in the viewing window. Disregarding reconstruction imperfections, an observer whose eyes are in one or two viewing windows will have the same perception as if the wave field emanates from a real existing object. Periodic repetitions of the reconstructed object are thus not visible.

3.4 Hologram encoding methods

Hologram encoding refers to the *representation* of the complex wavefield $\mathcal{H}$ at the hologram plane (x,y), i.e. to the process of converting the complex wavefield into a format, which can be displayed at the SLM by addressing its pixel. Hologram encoding is therefore directly related to the hardware implementation of the SLM. In synthetic or digital holography, a fully complex representation would be most qualified. But the major challenge is in finding a method and device to record a complex-valued hologram transmission function.

Generally speaking, there are various possibilities for a spatially sampled representation of complex wavefields by spatial light modulators:

Complex representation: A spatial light modulator that provides a full complex-valued modulation would be the ideal, whereas independent, non-coupled amplitude and phase addressing is mandatory. Although one can think of such SLM, which may implement the detour-phase principle for example, but thus far such devices are non-existent. Another possibility would be a sandwich of two active modulation layers, which are independently controlled for amplitude and phase modulation (Gregory et al., 1992). The challenge is then to put them together as close as possible to avoid cross talk. Both concepts seem difficult to realize even with today's enabling technologies.

Decomposition methods: Since the very beginning of computational holography decomposition methods for complex-valued wavefields have been developed[1]. Famous examples of holograms utilizing the detour-phase concept are those from Brown and Lohmann (Brown & Lohmann, 1966), Lee (Lee, 1970), Burckhardt (Burckhardt, 1970) and the double-phase holograms from Hsueh and Sawchuck (Hsueh & Sawchuk, 1978). All methods have in common that the hologram is divided into discrete resolution cells having apertures or stops of different size and position or having a certain number of sub-cells. That way both amplitude and phase quantities can be approximated. Originally developed for static holograms or holographic filters, those methods are also suited for implementation with spatial light modulators. However, to modulate both amplitude and phase two or more sub-pixels have to be combined to one macro-pixel. That means part of the light modulators original resolution has to be sacrificed for the sake of full holographic modulation.

---

[1] For a comprehensive overview, cf. for example the book edited by Schreier (Schreier, 1984)

In the following subsections, two decomposition methods capable for SLM implementation are described in more detail.

3.4.1 Burckhardt amplitude encoding

One method to decompose a complex-valued function is the method suggested by Burckhardt (Burckhardt, 1970), which is a simplified version of Lee's original approach (Lee, 1970). One hologram cell is laterally divided into three amplitude-modulating sub-cells. The lateral shift between the sub-cells represents phase angles of 0°, 120° and 240° and acts as a phase offset, similar to the detour-phase principle. In holograms of this type, a phasor is decomposed into three vectors that run parallel to $\exp(i0) = 1$, $\exp(i2\pi/3) = -0.5 + i\sqrt{3}/2$ and $\exp(i4\pi/3) = -0.5 - i\sqrt{3}/2$. Since the phase values are already represented by the lateral displacement of the sub-cells, any complex amplitude transmittance $\mathcal{H} = A\exp(i\Phi)$ can be encoded in one macro-pixel. The laterally displaced sub-pixels have positive amplitude transparencies of A_1, A_2, A_3, respectively. Hence,

$$\mathcal{H}(x,y) = A_1(x,y)\exp\left[i0\right] + A_2(x,y)\exp\left[i\frac{2\pi}{3}\right] + A_3(x,y)\exp\left[i\frac{4\pi}{3}\right]. \quad (5)$$

The magnitude A_i of one term is always zero. Depending on the phase angle Φ, only the two adjacent vectors are sufficient for representing $\mathcal{H}$. Figure 7 shows the decomposition of $\mathcal{H}$ for the case $0 \leq \Phi \leq 2\pi/3$

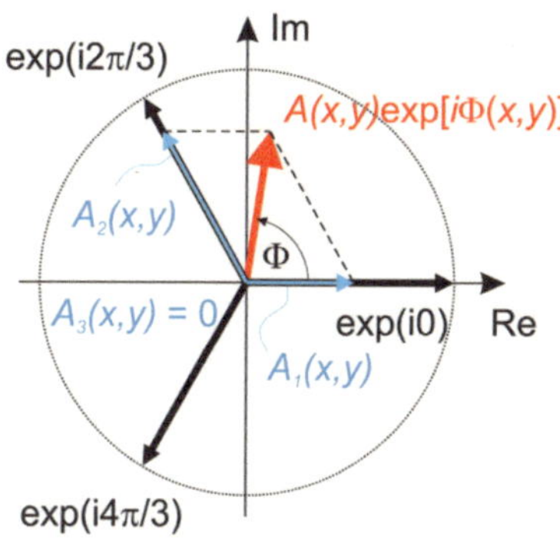

Fig. 7. Geometric representation of Burckhardt's decomposition method into three real and positive components.

Thus, an amplitude modulating light modulator with independently-driven sub-pixels can be employed. The amplitude quantities A_1, A_2, A_3 are are written as grey values into three sub-pixels that form one macro-pixel.

As large, high-resolution amplitude LC-displays are common in medical displays, such panels are commercially available. Another advantage is that only a single active layer is necessary to represent the complex amplitude transmittance $\mathcal{H}(x,y)$ of the hologram. Furthermore, the decomposition can be done in an analytic way, which simplifies the calculation enormously. On the other hand, the diffraction efficiency of Burckhardt holograms is with approx. 1% quite low. Notwithstanding, a fully complex modulation is achieved in a simple and practicable way, i.e. by an amplitude-modulating SLM that comprises of an array of macro-pixels, albeit at the cost of reduced sampling and diffraction efficiency.

3.4.2 Two-phase encoding

The two-phase[2] encoding method is based on the principle that any complex amplitude transmittance $\mathcal{H} = A\exp(i\Phi)$ can be decomposed into the sum of two vectors with constant magnitude of 0.5 and different phase quantities ϕ_1, ϕ_2 (Chu & Goodman, 1972; Hsueh & Sawchuk, 1978), i.e.

$$\begin{aligned} \frac{1}{2}\exp\left[i\phi_1(x,y)\right] + \frac{1}{2}\exp\left[i\phi_2(x,y)\right] &= \cos\left(\frac{\phi_1 - \phi_2}{2}\right)\exp\left[i\left(\frac{\phi_1+\phi_2}{2}\right)\right] \\ &= A(x,y)\exp\left[i\Phi(x,y)\right]. \end{aligned} \tag{6}$$

Figure 8 shows the decomposition of $\mathcal{H}$ in the complex plane. For example, two identical phase values with $\phi_1 = \phi_2$ give a resulting vector with maximum amplitude and phase of Φ (constructive interference), whereas two phases with a phase difference of π result in a vector with zero amplitude with non-defined phase (destructive interference). Arbitrary complex values are generated by combining other phases than those of the special cases $\Delta\phi = 0$ or π.

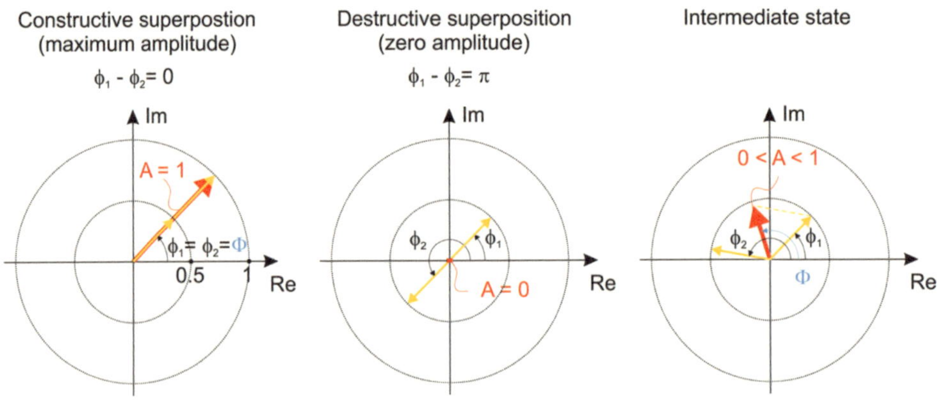

Fig. 8. Geometric representation of the dual-phase decomposition.

The other way round, the decomposed phase values for a given complex amplitude transmittance $\mathcal{H}(x,y)$ can be written as

$$\phi_1(x,y) = \Phi(x,y) + \cos\left[A(x,y)\right] \tag{7}$$

$$\phi_2(x,y) = \Phi(x,y) - \cos\left[A(x,y)\right]. \tag{8}$$

As a result, a fast phase-only LC panel can be used as light modulator. A pair of two pixels of a phase-only modulating SLM is then combined into a complex-valued macro pixel (Birch et al., 2000). Both pixel act as the intended complex-valued macro pixel only if light modulated by both pixel is superimposed. A physical combination of light modulated by two phase subpixels may be achieved by beam-combining micro-elements. The hologram is encoded by first normalizing the amplitudes of the complex amplitude transmittance in a range from 0 to 1 and then calculating the phase quantities ϕ_1, ϕ_2 from the equations above.

[2] Also called as double-phase or dual-phase.

Because of its phase coding, the diffraction efficiency of dual-phase holograms is compared to Burckhardt-type holograms greatly increased to approx. 10%. Again, only a single active layer is required for representing the entire hologram information. Since only two sub-pixels have to be combined to one macro-pixel a better sampling at the hologram plane is being present.

3.5 Tracking methods

For a non-tracked viewing-window-type hologram, the reconstructed 3-D object could be seen from a single or pair of viewing windows only. However, SeeReal's approach to dynamic holography is directly related to eye tracking. In case of a movement of the observer's eyes, the observer window is tracked to the new eye position. Hence it is possible to reduce the size of a viewing window to the size of approximately an eye pupil. Two viewing windows, i.e. one for the left eye and one for the right eye, are always located at the positions of the observer eyes. But then, how to move the viewing window in the observer plane? Advantageously, dynamic holography offers the additional freedom of temporal-multiplex operation. By incorporating a tracking system that detects the eye positions of one or more viewers very fast and precisely and repositions the viewing window accordingly, a dynamic 3-D holographic display can be realized. Thus, the viewing angle of the reconstructed object is beneficially enlarged while maintaining the moderate resolution of the spatial light modulator.

Tracked viewing-window holography must therefore fulfill the following key functions:

- Detection of the current eye position in x,y, z and
- Means for shifting the observer window to this position.

To realize the former task, the holographic displays developed by SeeReal are equipped with an eye position detection system composed of a stereo camera and an imaging processing means. Images from the observer having a different perspective are captured by two cameras as exemplary shown in Fig. 9. Multi-threaded software that comprises of image processing, pattern recognition and artificial intelligence is working in a two step process. In a first step, the face of the observer is recognized within the captured image and afterwards the eyes are detected within the region of the face. Once the face is identified, only the eye detection algorithms have to be executed, which makes the entire recognition process much faster. The results obtained from left and right image of the stereo camera are then combined to a 3-D model that defines the position of the eyes in space.

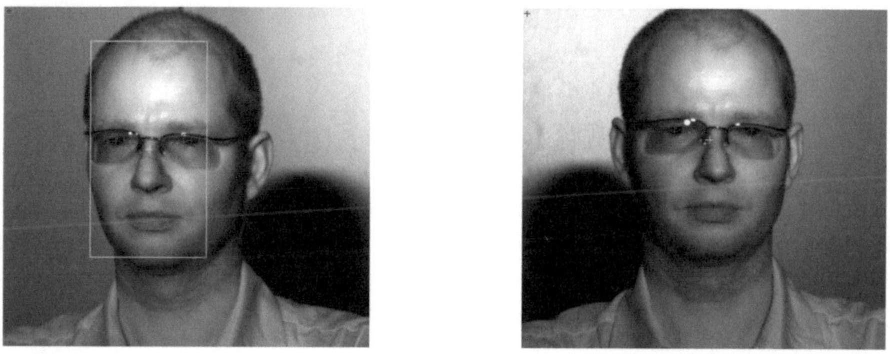

Fig. 9. Images captured by the tracking cameras (left and right view of a stereo camera). The current system is capable to track simultaneously up to 4 viewers in real-time.

We have developed different alternatives for the tracking means, two of them are explained here in more detail. The steering of the viewing window can be done for example by shifting the light source and thus shifting the image of the light source accordingly, or by placing an additional element close to the SLM that realizes a variable deflection. In the following, we discuss implementations of these alternatives.

3.5.1 Light-source shifting

The first principle that was developed and implemented in prototypes is based on light source shifting. The optical principle is schematically sketched in Fig. 10. By imaging through a lens, a shift of any light source in object space results in a shift of its image. Since the viewing window is located within the zero order of the spatial light modulator, the holographic reconstruction as can be viewed from the viewing window will be always correct. From a holographic way of thinking, this corresponds to an illumination of the hologram with a tilted reference wave. When a single lens is used, this method allows in principle for a tracking in *x*,*y* and *z* direction. However, there is a practical limit for *x* and *y* shifting because of the paraxial limit of the lens. A skew ray path introduces aberrations that, if too large, may deteriorate the holographic reconstruction quality. Although aberrations can be compensated by encoding means, a practical limit has been identified at approximately ±10°.

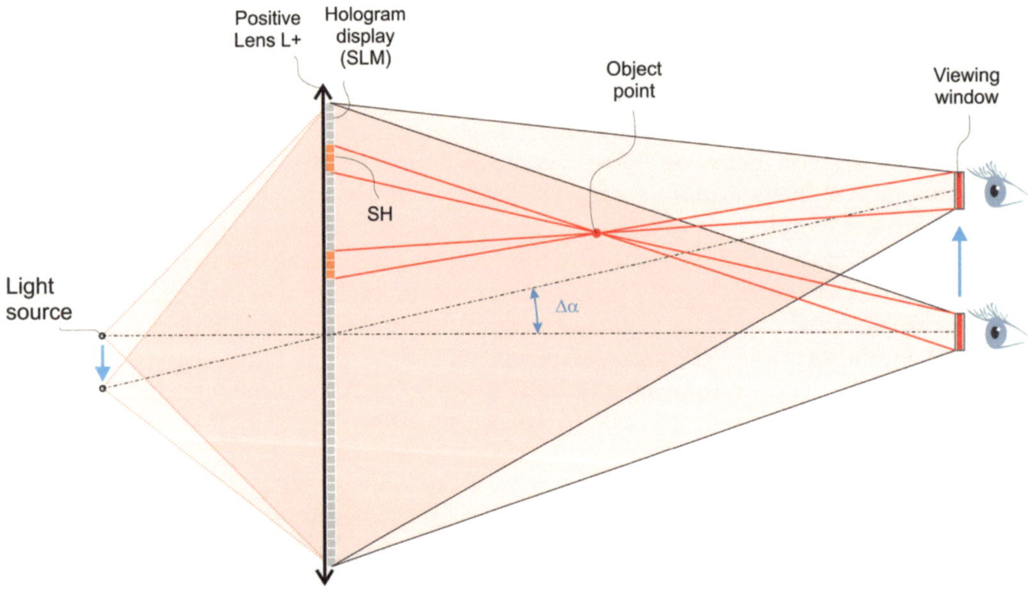

Fig. 10. Schematic principle of the light source tracking method.

The position of the light source does not have to be shifted mechanically. One possibility would be an active array with a large number of light sources only one of them switched on at the same time. Tracking would then be performed by switching between several light sources. Each light source position then corresponds to a distinct tracking position in the viewing plane.

Another possibility is the use of a secondary light source. A secondary light source might be an activated pixel in an additional liquid crystal display (LCD) that is illuminated by a homogenous backlight. By activating a pixel at the desired position on the LCD, the light source can be shifted electronically without mechanical movement.

Tracking by light-source shifting has been successfully implemented in the prototypes by using a homogenous backlight and an LCD-shutter panel, cf. section 4. For the prototypes however, not a single lens and a single (secondary) light source is used, but a matrix of simultaneously emitting light sources and lenses is utilized instead. For a large display a single lens wouldn't be a feasible solution because of its thickness, weight, costs and display compactness. The shutter pixels act as secondary light sources. By switching on different shutter pixel, the light source position can be changed. An LED array (primary light sources) is used for illuminating the shutter. Pitch of the lens array needs to be large compared to pixel pitch of the SLM such as to still have a certain number of pixels with coherent illumination.

Light source tracking has been proven to be a reliable solution. On the other hand, it has also certain disadvantages. For example, the use of secondary light sources is not optimal in terms of light efficiency of the system. Also there may be illumination crosstalk by light from secondary sources passing the wrong lens of the lens array. This does not cause any problem in a single-user system but may be disadvantageous for a multi-user display. The most important drawback of light source tracking is the limitation of the tracking angle by aberrations. Large tracking angles put the need for an oblique optical path from the light source through the lens array. Aberrations may not necessarily degrade the reconstruction of single points, but might somewhat corrupt the observer window leading to vignetting effects in the reconstruction.

While light source tracking may be well suited for a single user display with a tracking range of about ±10°, it is less practicable for multi-user displays and large tracking ranges, as needed for example for TV applications.

3.5.2 Steering of the reconstruction

Since the capabilities of the previous tracking method are limited in terms of tracking range, alternative solutions that enable larger ranges have been developed. The conceptual design of a holographic display that steers the holographic reconstruction is shown in Fig. 11. With a beam steering element placed at the front of the hologram display, the optical path from the light source to the SLM can be kept constant. As an advantage, the hologram is always illuminated by the same planar wavefront, which is ideal in terms of light efficiency and aberrations. The beam-steering element deflects the light after passing the SLM and directs the light toward the observer eyes. In addition to the prism function, it could realize a focusing function.

There exist various promising approaches to nonmechanical beam steering, which are currently at different stages of development (McManamon et al., 2009). The challenge of such beam steering devices is that often both a large deflection angle and a large aperture of the deflector are required. Refractive solutions are thus not suitable because of the thickness a prism would have. But when the optical system operates at coherent or narrow-band light, diffractive approaches can be utilized. For a transmission grating with a local period of Λ, the angle of the diffracted light is given by the grating equation

$$\sin\alpha_m = \frac{m\lambda}{\Lambda} + \sin\alpha_{\text{in}} \tag{9}$$

where m is the diffraction order, λ the wavelength of light and α_{in} the angle of the incident light. Such variable diffractive gratings can be divided into two categories. Either a sawtooth-like grating is adjustable in its period, or in its blaze angle. The variable period grating most often operates at the first diffraction order (m = 1) and the maximum steering angle is defined by the grating's smallest permissible period. The minimum period arises from diffraction efficiency requirements at a given angle as well as the addressing resolution of the grating. The steering up to the maximum angle is beneficially continuous. Variable blaze gratings on the other hand have a fixed period and diffract the light into the designated order by matching the blaze angle to the diffraction order m. Since the variable blaze grating type steers light only at discrete angles, an extra variable period grating stage is required for continuous steering between those angles.

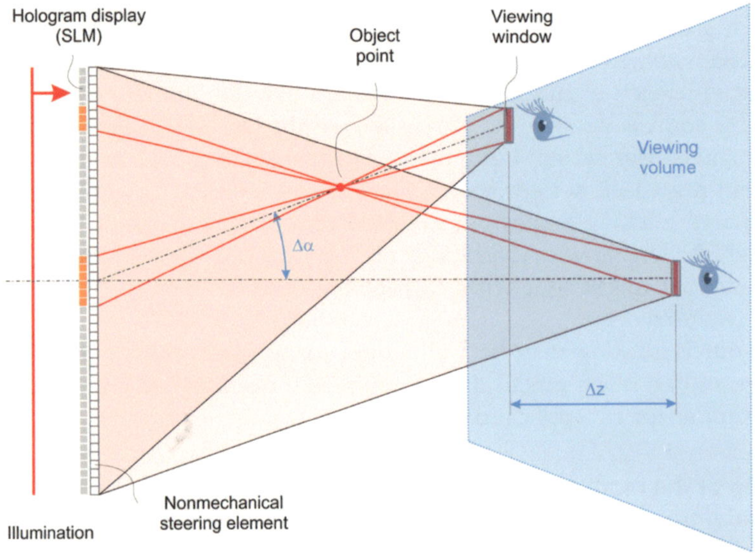

Fig. 11. Schematic side-view of the steering-of-reconstruction principle. $\Delta\alpha$ is the tracking angle and Δz is the focal length variation of the nonmechanical steering element.

Light could for example be steered and focused by writing a phase function including a prism and focus term into a liquid crystal layer. The effective refractive index and hence the deflection angle is controlled by a voltage applied to electrodes at the cells. Embodiments as variable period grating as well as variable blaze gratings can be realized.

Another steering or tracking concept for the holographic reconstruction is based on electrowetting. Electrowetting or exactly electrowetting on dielectrics (EWOD) (Beni et al., 1982; Berge & Peseux, 2000) can be regarded as an electrostatic manipulation of liquids that enables to vary the wettability of a conducting liquid (Mugele & Baret, 2005). The conductive liquid and an electrode are separated by a thin dielectric hydrophobic layer thus forming a parallel plate capacitor. By applying a voltage between the electrode and the conductive liquid, the droplet wets the hydrophobic dielectric. Without a voltage, the droplet returns to the dewetted state, i.e. to its initial contact angle θ_0. Since the thin dielectric layer prevents the liquid from electrolysis, the process is highly reliable. Below a critical saturation threshold, the behavior of electrowetting-on-dielectrics can be well-predicted by the so-called electrowetting equation

$$\cos\theta_V = \cos\theta_0 + \frac{\varepsilon_r\varepsilon_0}{2t\gamma_{la}}V^2, \tag{10}$$

which can be derived from Lippmann's electrocapillary equation and Young's equation for a three-phase contact line. In this equation, θ_0 is the initial contact angle at zero voltage, ε and t are the dielectric constant and the thickness of the dielectric layer, respectively, V is the applied voltage and γ_{la} is the interfacial surface tension of the liquid-ambient (typically electrolyte-oil) interface. In recent years, electrowetting has been successfully applied to various optical applications such as varifocal lenses, amplitude-modulating displays and fiber coupler and switches.

How electrowetting can be applied for realizing a liquid prism is illustrated in Fig. 12 (Kuiper et al., 2005; Smith et al., 2006). At initial state with no voltage, the liquids form a curved meniscus, depending on the interfacial surface tensions between the liquids and the solid. Since here the sidewall is hydrophobically coated, the water-based electrolyte features a large initial contact angle of $\theta_0 > 150°$ (Fig. 12 left, $\theta_L = \theta_R = \theta_0$). If a voltage difference is applied between the insulated sidewall electrode and the electrolyte, the contact angle of a conducting droplet can be decreased. At a certain voltage pair of equal voltages, the contact angle at both left and right electrode reaches 90° resulting in a flat meniscus (Fig. 12 middle, $\theta_L = \theta_R = 90°$). The light passing through such a cell will not be altered in its propagation direction. Prism functionality can be realized if the sum of left and right contact angle equals to 180° (Fig. 12 right, $\theta_L + \theta_R = 180°$, where $\theta_L \neq \theta_R$).

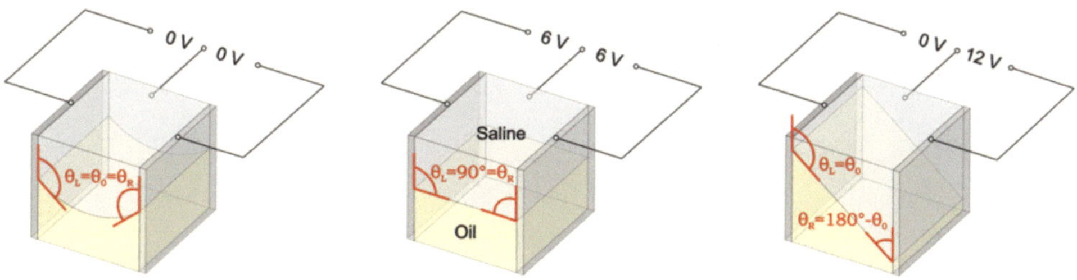

Fig. 12. Operation principle of an electrowetting prism. For simplicity and visualization, here only a prism with 1D-deflection capability (2 sidewall electrodes) has been drawn. A prism capable for 2D-deflection comprises of 4 sidewall electrodes.

It is advantageous to minimize the size of the prisms to get faster response, because dynamic response scales with the volume density product of the used liquids. Therefore, the intended prism size is adapted to the pixel pitch of the SLM. Since the response time of electrowetting cells of that size are in the range of <1 ms, time sequential tracking of several users becomes feasible.

3.6 Color holography

As holography is based on diffraction and as diffraction is wavelength-dependent, the 3-D scene has to be separated in its color components. Usually, these are red, green and blue. Three holograms are computed (one for each color component) and the 3-D scene is reconstructed using three light sources with the corresponding wavelengths. There are several methods to combine the three holograms and the three light sources, for example:

Spatial multiplexing. The red, green and blue holograms are spatially separated. For instance, they may be displayed on three separate SLMs that are illuminated by red, green and blue light sources. An arrangement of dichroic beamsplitters combines the output of the SLMs. The optical setup is bulky, above all for large displays.

Temporal multiplexing. The red, green and blue holograms are displayed sequentially on the same SLM. The red, green and blue light sources are switched in synchronization with the SLM. Fast SLMs are required to avoid color flickering.

4. Implementations and prototypes

Our holographic approach has been successfully demonstrated by prototypes having 20.1-inch diagonal[3]. The prototypes are intended to demonstrate the key principles of our solution for large-sized real-time holography, i.e. viewing-window holography with sub-hologram encoding technique, cheap and interactive real-time computing, and the feasibility with common pixel sizes. However, it should be emphasized that the prototypes do not represent commercial solutions with a flat design and are not at all optimized in intensity and tracking performance. Although commercial solutions that fulfill the latter features have already been developed, they are not described in this section.

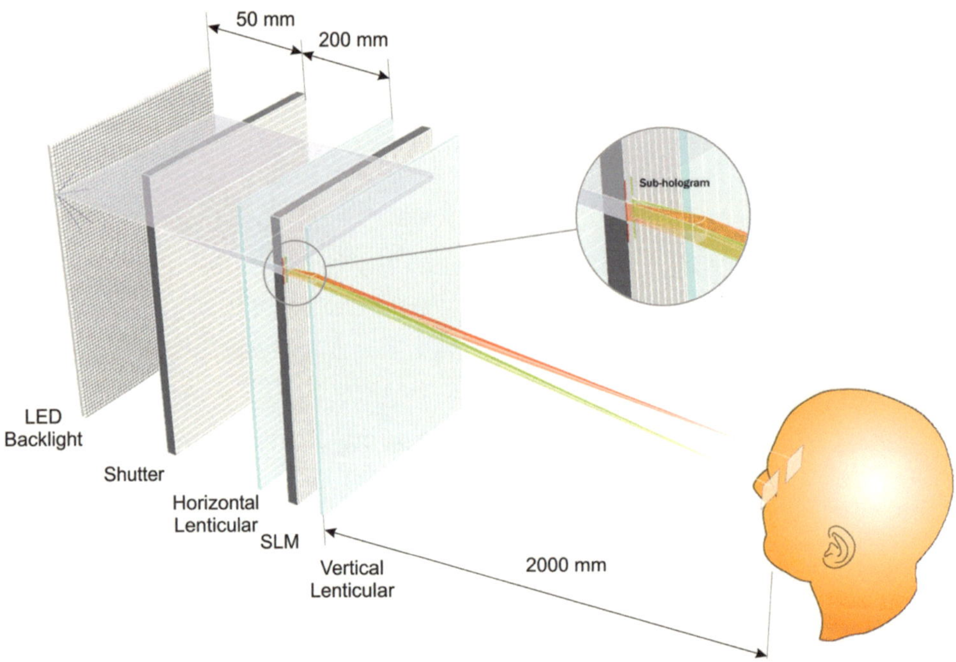

Fig. 13. Optical principle of the 20.1-inch holographic display prototype. Sizes and distances are not to scale.

4.1 General description of components

The second generation of the direct view holographic display prototype ("VISIO 20") comprises a grayscale amplitude-modulating liquid crystal panel (NEC NL256204AM15-01) with a 3 × 5 megapixel resolution at a pixel pitch of $p_x = 156$ µm and $p_y = 52$ µm, an operation

[3] Presented at SID 2007 (Long Beach, USA) and Display 2008 (Tokyo, Japan)

frequency of 60 Hz with a relatively slow response time of 30 ms (Fig. 14). The used 1D hologram encoding (here vertical-parallax only) is a common practice to further reduce bandwidth requirements and is well-suited for the given pixel arrangement and geometry.

(a)

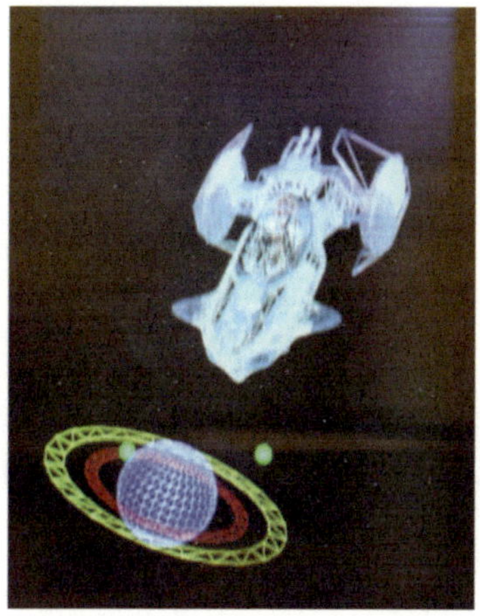

(b)

Fig. 14. (a) 20.1-inch direct view prototype ("VISIO 20") and (b) photograph of a holographic reconstruction.

The optical scheme of the prototype is depicted in Fig. 13. An LED backlight consists of red, green and blue high-brightness LEDs emitting at wavelengths of 627 nm, 530 nm, and 470 nm, respectively. The spectral linewidth (FWHM) of $\Delta\lambda \approx$ 30nm provides sufficient temporal coherence. Light coming from the RGB-LED backlight is mostly blocked by a first LC display that acts as shutter or variable secondary light source array. Only those pixels that are switched on transmit the light, and thus a variable (secondary) line light source is realized having a spatial coherence corresponding to the pixel opening. A lenticular comprising approximately 60 horizontal cylindrical lenses is used for hologram illumination and for imaging the light sources into the viewing window. Each cylindrical lens is illuminated by a horizontal line light source. Furthermore, secondary line light sources and arrayed cylindrical lenses are aligned such that all light source images coincide onto the viewing window.

In the SLM the sum of all complex amplitudes $\Sigma U_n(x,y)$ is encoded by combining three amplitude-modulating pixels for each complex value according to the Burckhardt-encoding scheme described above (cf. section 3.4). Two viewing windows delivering slightly different holographic perspectives of the scene are generated by a vertically aligned lenticular beam-splitter and an interlaced (horizontally multiplexed) hologram. High-precision user tracking is realized by a stereo camera incorporated in the holographic display and advanced eye recognition algorithms combined with active light source shifting by the shutter panel.

4.2 Color implementation

In the prototypes, holographic reconstruction is performed either in monochrome (optionally R, G, or B) or in full-color. Two types of full-color holographic displays have been realized that are based on either a temporal or a spatial multiplexing of colors (Häussler et al., 2009).

At the system for which *temporal color multiplexing* is implemented, the colors are displayed sequentially. The SLM displays the holograms of the red, green and blue 3-D scene components one after the other where the backlight is switched between red, green and blue LEDs. Both processes are synchronized. However, two obstacles have to be taken into account to achieve good reconstruction quality:

- The pixels of the SLM have a finite response time. For a LCD, this is the time the liquid crystals need to align to the electric field applied to the pixel cell. The LCD panel that we use as SLM stems from a medical LCD and has a long response time t_{on} + t_{off} of typically 30 ms.
- The pixels do not switch simultaneously across the SLM as the pixels are addressed in columns and rows. The rows of the SLM are addressed sequentially, with one frame period needed from the first to the last row. As a consequence, there is a time lag of up to one frame period across the SLM.

Both effects have to be taken account as each part of the hologram has to be illuminated with the corresponding wavelength. For instance, if SLM and backlight were switched from red to blue simultaneously, the last rows of the SLM would still display the red hologram when the backlight is already switched to blue. Therefore, we used a scanning backlight and a time lag between switching the SLM and the backlight. Figure 15 illustrates this process.

Fig. 15. State of SLM rows (top) and backlight rows (bottom) versus time. The SLM graph illustrates the effect of finite response time and row-by-row addressing of the SLM after switching from one hologram to the next hologram. The backlight graph shows delayed and row-by-row switching of the backlight to compensate these effects.

The top graph shows the states of the SLM rows versus time and the bottom graph the states of the backlight rows versus time. The gradual color transition along the time axis of the SLM graph illustrates the finite response time after switching from a hologram of one color component to the hologram of the next color component. There is no sharp transition from one hologram to the next hologram but an intermediate interval in which the pixels of the SLM transit to the next state. The color transition along the row axis of the SLM graph illustrates that the SLM is addressed row-by-row. At a point in time at which the last row has just received the data of the current frame, the first row will already receive the data of the next frame. As an example, at the second dotted vertical line, the last row has just settled to the red hologram, whereas the first row already starts to transit to the green hologram. The intermediate states are indicated by the slanted gradual color transition. At these points in time, the state of the respective SLM pixel is undefined, and illumination by the backlight has to be avoided. Therefore, we built a scanning backlight in which the rows of LEDs are grouped in 16 groups. Switching of these groups is illustrated in the backlight graph of Fig. 15. These groups are switched on and off sequentially such that the corresponding parts of the hologram are only illuminated if its pixels are in a settled state of the associated color. A complete cycle comprises three frames with colors red, green and blue and three intermediate transition frames. As the frame rate of the SLM is 60 Hz, the full-color frame rate is 10 Hz. The human vision perceives a full-color holographic reconstruction, albeit with color flickering. Color flickering will disappear and a steady reconstruction will be visible with availability of faster SLMs.

In contrast, the display with *spatial color multiplexing* shows the three backlight colors and the three holograms for red, green and blue color components simultaneously. The three holograms are interlaced on the same SLM. A color filter is used to achieve that each hologram is illuminated with its associated wavelength only. Six holograms are interlaced on the SLM: three red, green and blue holograms that generate the viewing window for the left eye VW_L and also three holograms to generate that for the right eye VW_R. A beam-splitting lenticular is used to separate the light for left and right viewing window, VW_L and VW_R. Color filters are used to separate the wavelengths. Figure 16 illustrates top views of two possible arrangements of color filters. The left arrangement uses color filters that are integrated in the SLM pixels. One lens of the beam-splitting lenticular is assigned to two pixels of the SLM. The light of all left pixels at the lenses coincides in the observer plane and generates VW_R. Vice versa the right pixels generate VW_L. The color filters are arranged in columns such that each column of the filter extends over two columns of the SLM, as illustrated in the left graph of Figure 16. Such an arrangement of color filters integrated in the SLM pixels and two neighboring pixels having the same color is not commercially available. Standard LCD panels have color filters with color changing from pixel to pixel. An external color filter laminated on the cover glass of the panel would have a disturbing separation between pixel and color filter. Therefore, in our prototype we used the arrangement illustrated in the right graph of Figure 16. The color filters are attached directly to the structured surface of the beam-splitting lenticular. This arrangement avoids a disturbing separation between lenticular and color filter and facilitates tracked viewing windows in the same way as with a monochrome display. The functional principle is analog to that of the arrangement in the left graph of Figure 16.

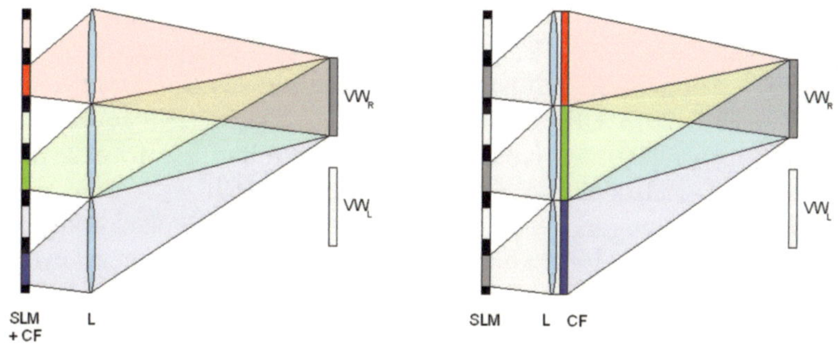

Fig. 16. Top view of an arrangement of color filters in a holographic display with spatial color multiplexing. The left graph shows color filters integrated in the SLM pixels (SLM + CF) and the right graph separate color filters (CF). The graphs show three lenses of the positive lenticular (L+) that splits the light to generate left viewing window (VW_L) and right viewing window (VW_R). For simplification, only the light illuminating VW_R is shown and the light illuminating VW_L is omitted. The light sources and the Fourier-transforming lenses are not shown.

5. Conclusions

In conclusion, a novel approach for real-time holography that has a strong market potential for desktop, TV, as well as mobile displays has been presented. To date, it is the only practical solution known to the authors that is capable of holographic reconstruction of large 3-D scenes of natural size and depth made with commercially available component technologies. The essential idea of the proprietary and patented approach is that for a holographic display the highest priority is to reconstruct the wavefront at the eye position that would be generated by a real existing object and not to reconstruct the object itself. The tracked viewing-window holographic technology limits pixel size to levels already known for commercially available displays. Sub-hologram encoding brings computation into graphics card or ASIC range. The new concept is applicable to desktop, TV, and mobile imaging.

While there have been impressive developments in 3-D display technology in the past decade, the remaining visual conflicts between natural viewing and 3-D stereo visualization have prevented 3-D displays from becoming a universal consumer product. In principle, the only 3-D display capable of completely matching natural viewing is an electro-holographic display.

SeeReal's new approach to electro-holography not only proves that it is possible, but that it is also closer to adoption than many experts imagined. The principles and concepts are already in place. The checks and verifications are completed. Color prototypes are in full use. The technology already exists - it is just a question of time for all the pieces of the puzzle to come together and for the first commercial real-time 3-D holographic displays to hit the market.

6. References

Beni, G., Hackwood, S. & Jackel, J. L. (1982). Continuous electrowetting effect, *Applied Physics Letters* 40(10): 912–914. URL: *http://link.aip.org/link/?APL/40/912/1*

Benton, S. A. (1969). Hologram reconstructions with extended incoherent sources, *J. Opt. Soc. Am.* 59: 1545.

Benton, S. A. & Bove, V. M. (2008). *Holographic Imaging*, John Wiley and Sons.

Benton, S. A. & H. S. Mingace, J. (1970). Silhouette holograms without vertical parallax, *Appl. Opt.* 9(12): 2812–2813. URL: *http://ao.osa.org/abstract.cfm?URI=ao-9-12-2812*

Berge, B. & Peseux, J. (2000). Variable focal lens controlled by an external voltage: an application of electrowetting, *Eur. Phys. J. E* 3: 159–163.

Birch, P. M., Young, R., Budgett, D. & Chatwin, C. (2000). Two-pixel computer-generated hologram with a zero-twist nematic liquid-crystal spatial light modulator, *Opt. Lett.* 25(14): 1013–1015. URL: *http://ol.osa.org/abstract.cfm?URI=ol-25-14-1013*

Brown, B. R. & Lohmann, A. W. (1966). Complex spatial filtering with binary masks, *Appl. Opt.* 5(6): 967–969. URL: *http://ao.osa.org/abstract.cfm?URI=ao-5-6-967*

Burckhardt, C. B. (1970). A Simplification of Lee's Method of Generating Holograms by Computer, *Appl. Opt.* 9(8): 1949–1949. URL: *http://ao.osa.org/abstract.cfm?URI=ao-9-8-1949*

Chu, D. C. & Goodman, J. W. (1972). Spectrum shaping with parity sequences, *Appl. Opt.* 11(8): 1716–1724. URL: *http://ao.osa.org/abstract.cfm?URI=ao-11-8-1716*

Gabor, D. (1948). A new microscopic principle, *Nature* 161: 777–778.

Goodman, J.W. (1996). *Introduction to Fourier Optics*, 2nd edn, McGraw-Hill, New York.

Gregory, D. A., Kirsch, J. C. & Tam, E. C. (1992). Full complex modulation using liquid-crystal televisions, *Appl. Opt.* 31(2): 163–165. URL: *http://ao.osa.org/abstract.cfm?URI=ao-31-2-163*

Häussler, R., Reichelt, S., Leister, N., Zschau, E., Missbach, R. & Schwerdtner, A. (2009). Large real-time holographic displays: from prototypes to a consumer product, Vol. 7237, SPIE, p. 72370S. URL: *http://link.aip.org/link/?PSI/7237/72370S/1*

Hoffman, D. M., Girshick, A. R., Akeley, K. & Banks, M. S. (2008). Vergence–accommodation conflicts hinder visual performance and cause visual fatigue, *J. Vis.* 8(3): 1–30. URL: *http://journalofvision.org/8/3/33/*

Hsueh, C. K. & Sawchuk, A. A. (1978). Computer-generated double-phase holograms, *Appl. Opt.* 17(24): 3874–3883. URL: *http://ao.osa.org/abstract.cfm?URI=ao-17-24-3874*

Ichihashi, Y., Nakayama, H., Ito, T., Masuda, N., Shimobaba, T., Shiraki, A. & Sugie, T. (2009). Horn-6 special-purpose clustered computing system for electroholography, *Opt. Express* 17(16): 13895–13903. URL: *http://www.opticsexpress.org/abstract.cfm?URI=oe-17-16-13895*

Kuiper, S., Hendriks, B. H. W., Hayes, R. A., Feenstra, B. J. & Baken, J. M. E. (2005). Electrowetting-based optics, Vol. 5908, SPIE, p. 59080R. URL: *http://link.aip.org/link/?PSI/5908/59080R/1*

Lee, W. H. (1970). Sampled fourier transform hologram generated by computer, *Appl. Opt.* 9(3): 639–643. URL: *http://ao.osa.org/abstract.cfm?URI=ao-9-3-639*

Lohmann, A.W. (1967). The space-bandwidth product, applied to spatial filtering and holography, *Research Paper RJ-438 (IBM San Jose Research Laboratory, San Jose, Calif.)* pp. 1–23.

Lohmann, A. W., Dorsch, R. G., Mendlovic, D., Zalevsky, Z. & Ferreira, C. (1996). Spacebandwidth product of optical signals and systems, *J. Opt. Soc. Am. A* 13(3): 470–473. URL: *http://josaa.osa.org/abstract.cfm?URI=josaa-13-3-470*

Lucente, M. E., St-Hilaire, P., Benton, S. A., Arias, D. & Watlington, J. A. (1993). New approaches to holographic video, Vol. 1732, SPIE, pp. 377–386. URL: *http://link.aip.org/link/?PSI/1732/377/1*

Maeno, K., Fukaya, N., Nishikawa, O., Sato, K. & Honda, T. (1996). Electro-holographic display using 15mega pixels lcd, Vol. 2652, SPIE, pp. 15–23. URL: *http://link.aip.org/link/?PSI/2652/15/1*

McManamon, P. F., Bos, P. J., Escuti, M. J., Heikenfeld, J., Serati, S., Xie, H. & Watson, E. A. (2009). A review of phased array steering for narrow-band electrooptical systems, *Proceedings of the IEEE* 97(6): 1078–1096.

Mugele, F. & Baret, J.-C. (2005). Electrowetting: from basics to applications, *Journal of Physics: Condensed Matter* 17(28): R705–R774. URL: *http://stacks.iop.org/0953-8984/17/R705*

Reichelt, S., Häussler, R., Leister, N., Fütterer, G. & Schwerdtner, A. (2008). Large Holographic 3D Displays for Tomorrow's TV and Monitors - Solutions, Challenges, and Prospects, *Proceedings of the IEEE LEOS Annual Conference.* (invited).

Schreier, D. (1984). *Synthetische Holografie*, Physik-Verlag,Weinheim.

Schwerdtner, A., Häussler, R. & Leister, N. (2007). A new approach to electro-holographic displays for large object reconstructions, *Adaptive Optics: Analysis and Methods/ Computational Optical Sensing and Imaging/Information Photonics/Signal Recovery and*

Synthesis Topical Meetings on CD-ROM, Optical Society of America, p. PMA5. URL: *http://www.opticsinfobase.org/abstract.cfm?URI=URI=DH-2007-PMA5*

Schwerdtner, A., Leister, N. & Häussler, R. (2007). A new approach to electro-holography for TV and projection displays, *SID-Proc.*, p. 32.3.

Slinger, C. W., Cameron, C. D., Coomber, S. D., Miller, R. J., Payne, D. A., Smith, A. P., Smith, M. G., Stanley, M. &Watson, P. J. (2004). Recent developments in computer-generated holography: toward a practical electroholography system for interactive 3d visualization, Vol. 5290, SPIE, pp. 27–41. URL: *http://link.aip.org/link/?PSI/5290/27/1*

Smalley, D. E., Smithwick, Q. Y. J. & V. Michael Bove, J. (2007). Holographic video display based on guided-wave acousto-optic devices, Vol. 6488, SPIE, p. 64880L. URL: *http://link.aip.org/link/?PSI/6488/64880L/1*

Smith, N. R., Abeysinghe, D. C., Haus, J. W. & Heikenfeld, J. (2006). Agile wide-angle beam steering with electrowetting microprisms, *Opt. Express* 14(14): 6557–6563. URL: *http://www.opticsexpress.org/abstract.cfm?URI=oe-14-14-6557*

St-Hilaire, P., Benton, S. A. & Lucente, M. (1992). Synthetic aperture holography: a novel approach to three-dimensional displays, *J. Opt. Soc. Am. A* 9(11): 1969–1977. URL: *http://josaa.osa.org/abstract.cfm?URI=josaa-9-11-1969*

St-Hilaire, P., Benton, S. A., Lucente, M. E., Sutter, J. D. & Plesniak, W. J. (1993). Advances in holographic video, Vol. 1914, SPIE, pp. 188–196. URL: *http://link.aip.org/link/?PSI/1914/188/1*

Stanley, M., Bannister, R. W., Cameron, C. D., Coomber, S. D., Cresswell, I. G., Hughes, J. R., Hui, V., Jackson, P. O., Milham, K. A., Miller, R. J., Payne, D. A., Quarrel, J., Scattergood, D. C., Smith, A. P., Smith, M. A. G., Tipton, D. L., Watson, P. J., Webber, P. J. & Slinger, C. W. (2003). 100-megapixel computer-generated holographic images from active tiling: a dynamic and scalable electro-optic modulator system, Vol. 5005, SPIE, pp. 247–258. URL: *http://link.aip.org/link/?PSI/5005/247/1*

Stolle, H. & Häussler, R. (2008). A new approach to electro-holography: Can this move holography into the mainstream?, *Information Display* pp. 2–4.

Tay, S., Blanche, P.-A., Voorakaranam, R., Tunc, A. V., Lin, W., Rokutanda, S., Gu, T., Flores, D., Wang, P., Li, G., St-Hilaire, P., Thomas, J., Norwood, R. A., Yamamoto, M. &

Peyghambarian, N. (2008). An updatable holographic three-dimensional display., *Nature* 451(7179): 694–698. URL: *http://dx.doi.org/10.1038/nature06596*

Wheatstone, C. (1838). On some remarkable, and hitherto unobserved, phenomena of binocular vision, *Philosophical Transactions of the Royal Society of London* 128: 371–394.

IV Applications in Biology and Medicine

30

Combining Optical Coherence Tomography with Fluorescence Imaging

Shuai Yuan and Yu Chen
Fischell Department of Bioengineering, University of Maryland
College Park, Maryland 20742,
U. S. A.

1. Introduction

1.1 Overview of optical coherence tomography

Optical coherence tomography (OCT) is a promising imaging technology which can provide subsurface imaging of biological tissues (Huang et al., 1991). OCT images are similar to the ultrasound images, but with significantly higher resolution (5-10 times) than the clinical ultrasound. OCT enables imaging of tissue microstructure with micron-level resolution, approaching that of standard excisional biopsy and histopathology, except that imaging can be performed in real time without the removal of a tissue specimen.

The optical configuration of OCT is the same as that of a low coherence interferometer (LCI). Low-coherence interferometry is a type of optical interferometry based on low-coherent light sources (Mandel & Wolf, 1995). This technique is capable of measuring depth-resolved (axial, z) tissue structure, birefringence, flow (Doppler shift), and spectra at a micrometer-level resolution by precisely measuring the amplitude and the relative phase of reflected or backscattered light. Optical interferometer has been invented for 120 years. But most of the time, it was just laboratory equipments for physics research until 1970s. In the early 1980s the emergence of high brightness semiconductor broadband light sources and single mode optical fibers stimulated the development of LCI. The availability of superluminescent diodes (SLD) made low-coherence reflectometry practical in a fiber-optic system (Wang et al., 1982). SLD could be used in LCI to achieve a distance resolution of about 10 μm. Another innovation was the development of fiber optic devices such as couplers which are important in the development of fiber optic based designs. These technological developments are the prerequisites of current OCT system used for noninvasive optical imaging of living tissues *in vivo*.

In 1980s, the LCI technique was mainly developed by several groups for reflection measurements in telecommunications devices with micron resolution (Danielson & Whittenberg, 1987; Youngquist et al., 1987). The potential of LCI for three-dimensional imaging in biological tissue was first realized in 1991 (Huang et al., 1991). Since the original work, a large number of papers have been published regarding various aspects of OCT. OCT has been successfully translated to various clinical applications including ophthalmology (Schuman et al., 2004), cardiology (Jang et al., 2005), gastroenterology (Bouma et al., 2000; Li et al., 2000; Sivak et al., 2000; Chen et al., 2007a), dermatology (Welzel et al., 2004), dentistry (Otis et al., 2000), urology (D'Amico et al., 2000), among others (Huang et al., 1991; Fujimoto et al., 1995; Brezinski et al., 1996; Tearney et al., 1997; Fujimoto et al., 2000; Fujimoto, 2003).

Figure 1A shows a schematic of the time-domain OCT (TD-OCT) system. Measurements are performed using a Michelson interferometer with a low coherence length light source. One arm of the interferometer directs the light onto the sample and collects the backscattered signal. The other interferometer arm generates a reference signal. Optical interference between the light from the sample and reference mirror occurs only when the optical distances traveled by the light in both the sample and reference paths match to within the coherence length of the light. Low coherence interferometry permits the optical path length and magnitude of the light reflected from tissue microstructures to be measured with extremely high accuracy and sensitivity. Cross-sectional images are constructed by laterally scanning the optical beam and performing sequential axial measurements of backscattered light at different transverse positions (see Figure 1B). Three-dimensional (3D) volumetric images can be obtained similarly by steering the beam in two dimensions.

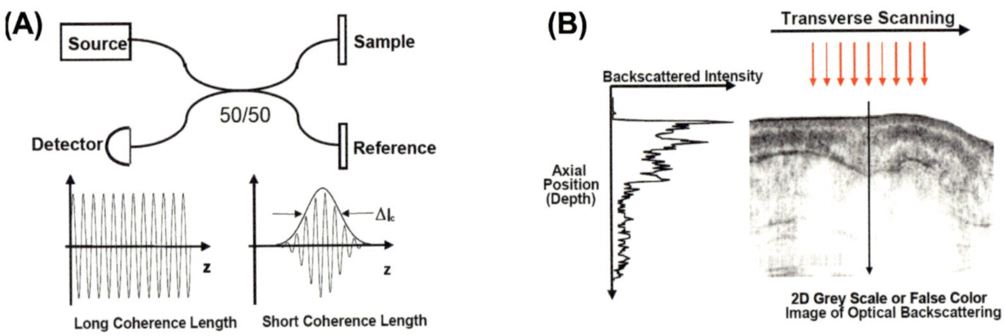

Fig. 1. (A) Optical Coherence Tomography (OCT) generates images by measuring the echo time delay of light. The system uses a fiber optical Michelson interferometer with a low coherence light source. (B) Cross-sectional images are generated by scanning a beam across the tissue and measuring the backscattering intensity as a function of depth. A gray scale or false color image can then be displayed.

Although time domain OCT is successful in many biomedical applications, it is hampered in achieving real-time imaging mainly by the relatively complicated mechanical designs for dual scanning, a mechanical delay line scanning in the reference arm for coherence gating and a position scanning in the sample arm. An alternate approach to coherence gating that does not employ a scanning delay line is to acquire a broad band spectrum of the backscattered interferometric signal generated by mixing sample light with reference light at a fixed group delay and then obtain depth-scan information by an inverse FT of the spectrum. This approach is called Fourier domain OCT (FD-OCT). Two distinct methods have been developed that employ this FD-OCT approach. The first, spectral domain OCT (SD-OCT) (Fercher et al., 1995; Wojtkowski et al., 2002) uses a broadband light source and achieves spectral discrimination with a dispersive spectrometer in the detector arm. The second method, swept source OCT (SS-OCT) (Fercher et al., 1995; Chinn et al., 1997; Haberland et al., 1998) achieves spectral discrimination by rapidly tuning a narrowband source through a broad band in the source arm. Schematics of a SD-OCT system and a SS-OCT system are shown in Figure 2. FD-OCT system can not only achieve a fast frame acquisition rate as it is designed for, but it can also achieve superior sensitivity (defined as the signal-to-noise ratio (SNR) for a perfect sample reflector) over TD-OCT (Huber et al., 2006). Recently, FD-OCT becomes more and more popular in biomedical research. Commercialized FD-OCT systems have been in the market for several years and were used widely in many recent studies. In this chapter, all of our newly developed multi-modality systems are based on SS-OCT.

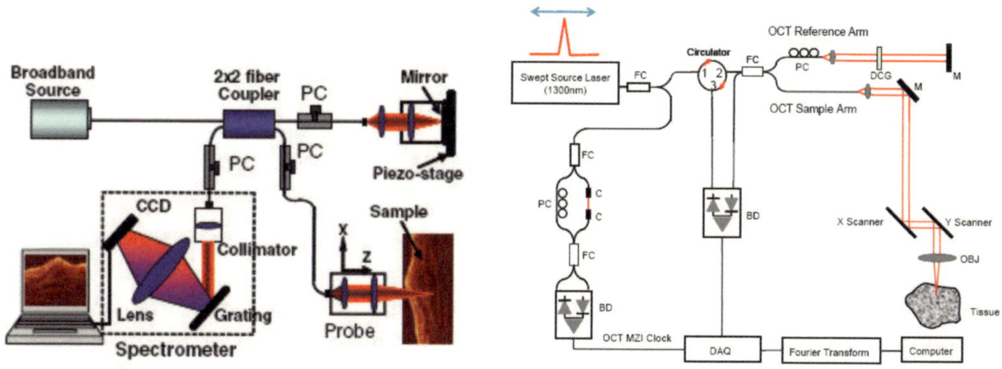

Fig. 2. Typical SD-OCT (left) from (Wang, 2007) with permission and SS-OCT system (right).

1.2 Principles of Swept Source Optical Coherence Tomography (SS-OCT)

Detailed aspects of traditional time-domain OCT theory and applications, such as resolution, sensitivity, noise, signal-to-noise ratio, etc, could be found in many literatures (Schmitt, 1999; Fercher et al., 2003), and will not be covered in this chapter. For FD-OCT, the optical paths of light beam are the same as those in TD-OCT techniques. The light beam from the source is split by a fiber coupler or beam splitter and comes through and back in the reference arm and the sample arm. The beam is then recombined at the coupler or the beam splitter and forms an interferometric signal. In SD-OCT a spectrogram is obtained by using a dispersive spectrometer with a CCD camera in the detector arm, and in SS-OCT a spectrogram is obtained by using a frequency swept laser or tunable laser with just a single detector and without dispersion components.

In SS-OCT, the spectrum of the light source is described as a discrete spectral intensity distribution with a line width σ_k and tuning mode-hop δ_k'. The interferometric signal at the detector can be described as (Liu & Brezinski, 2007):

$$I_D(k) = G(k)[e^{-k^2/2\sigma_k^2} \otimes comb(k/\delta_k')][p_R^2 + \int_0^z p_S^2(z')dz' + 2\mathrm{Re}\{\int_0^z p_S(z')p_R e^{j4\pi k[n_s(k_0)z'-z_R]}dz'\}] \tag{1}$$

where k is the wavenumber, $I_D(k)$ is the detected light intensity at wavenumber k, $G(k)$ is the spectral coefficient combing the intensity spectrum of the light beam and the conversion/coupling factors from the source to the detectors, p_R represents the fixed backscattering coefficient in the reference arm, $p_S(z)$ represents the backscattering coefficient at the position z in the sample, and $n_S(k)$ is the refractive index of the sample at wavenumber k. Both p_R and $p_S(z)$ include coupling factors from the fiber coupler or beam

splitter. The depth variable Z' is measured from O_S, the origin of coordinates in the sample arm. Consequently the origin of coordinates in the reference arm O_R is the point from where the optical group delay to the coupler matches that between O_S and the coupler in the sample arm. The mirror position is Z_R measured off the O_R. In Eq. (1), the first term within the second square brackets is the signal from the reference arm; the second term comes from the sample arm and encodes the depth information of the object; the third term described the mutual interference of all scattering waves.
The Fourier transfer of Eq. (1) can be expressed as (Liu & Brezinski, 2007):

$$\begin{aligned} F^{-1}\{I_D(k)\} = F^{-1}[G(k)] \otimes [e^{-2\sigma_k^2 z''^2} comb(\delta_k' z'')] \otimes \\ [F^{-1}\{p_R^2 + \int_0^z p_S^2(z')dz'\} + p(\frac{z''+2Z_R}{2n_s(k_0)}) + p(\frac{-z''+2Z_R}{2n_s(k_0)})] \end{aligned} \tag{2}$$

From Eq. (2), it is seen that the mode-hop δ_k' will limit the *A*-scan range and the laser line width σ_k will degrade the *A*-scan profile.
In SS-OCT, the sensitivity improvement is obtained through higher spectral intensity of the laser source. In a swept or tunable laser source, the intensity over a small range of the spectrum can be easily kept much higher than that of a wideband light source. Thus possible *SNR* improvement can be gained in SS-OCT (Huber et al., 2006). Furthermore, a bandpass filter technique can be used, by shifting the region of interest (ROI) of the sample fairly far off the plane O_S, to keep all the high frequency spectral oscillation introduced by the interested scatters and filter out much of the low frequency noise (Liu & Brezinski, 2007). Another theoretical advantage of SS-OCT is that the tuning speed of the swept source could be high enough to shift the spectrum of the temporal spectral interferogram higher and out of the low frequency noise range, which is from zero to tens of kilohertz.

2. Combining OCT with fluorescence

Early detection of neoplastic changes before metastasis occurs remains a critical objective in clinical cancer diagnosis and treatment. Excisional biopsy and histopathology is currently the gold standard for cancer diagnosis. However, it can suffer from false negative rates due to sampling errors. Optical imaging technologies can provide real-time imaging of human tissues in vivo with resolutions approaching that of histopathology and have the potential to reveal the biochemical and/or molecular information; therefore they could significantly improve the identification of malignancies at curable stages. The ability to assess tissue architectural morphology (such as the alterations in glandular or stromal morphology) and molecular information (such as up-regulation of receptors and over-expression of enzymes), in vivo and in real time, without the need for tissue excision, would be a major advance in cancer diagnostics and therapy. There is a critical need to develop such an imaging technology with high sensitivity and specificity and with strong translational potential to clinical medicine.
Since OCT provides high-resolution, cross-sectional imaging of tissue microstructure in situ and in real-time, and fluorescence imaging provides the biochemical and metabolism information, there are great interests to combine these two modalities to provide the structural and functional information of the tissues in order to enhance the disease detection capability. Pan *et al.* suggested that ALA fluorescence-guided endoscopic OCT could enhance the efficiency and sensitivity of early bladder cancer diagnosis (Pan et al., 2003). In

an animal model studies, they demonstrated that the specificity of fluorescence detection of transitional cell carcinoma was significantly enhanced by fluorescence-guided OCT (53% vs. 93%), and the sensitivity of fluorescence detection also improved by combination with OCT (79% vs. 100%) (Wang et al., 2005).

Tumlinson *et al.* have developed a combined OCT and laser-induced fluorescence (LIF) spectroscopy imaging catheter for in vivo mouse colon imaging (Tumlinson et al., 2004; McNally et al., 2006), as shown in Figure 3. The source beam from SLD is passed to a 50:50 fiber beam splitter and separated to the reference beam and the sample beam. The reference beam travels, through a neutral-density (ND) filter to a collimator, and then returns to the beam splitter after reflection from a retroreflector. The sample beam is coupled into a dual-modality endoscope. The light from the reference and sample paths is recombined at the beam splitter and travels to a detector. The LIF He–Cd laser source is coupled into a multimode fiber and into the dual-modality endoscope, too. Two adjacent collection fibers in the endoscope collect the emission light and carry it to the CCD spectrometer for analysis.

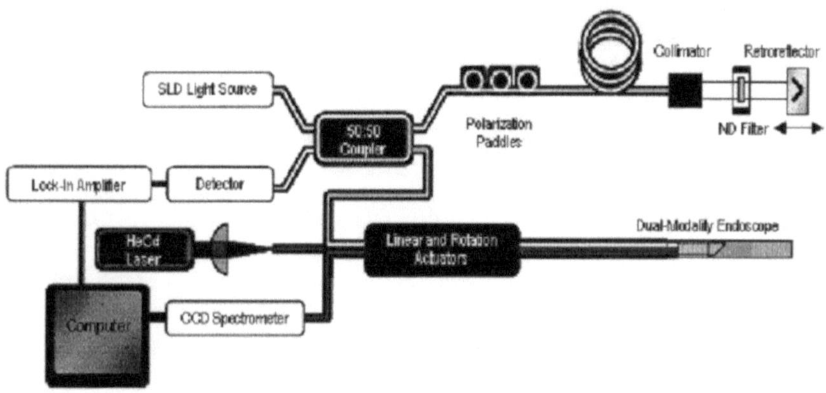

Fig. 3. Schematic of the OCT–LIF system (McNally et al., 2006) with permission.

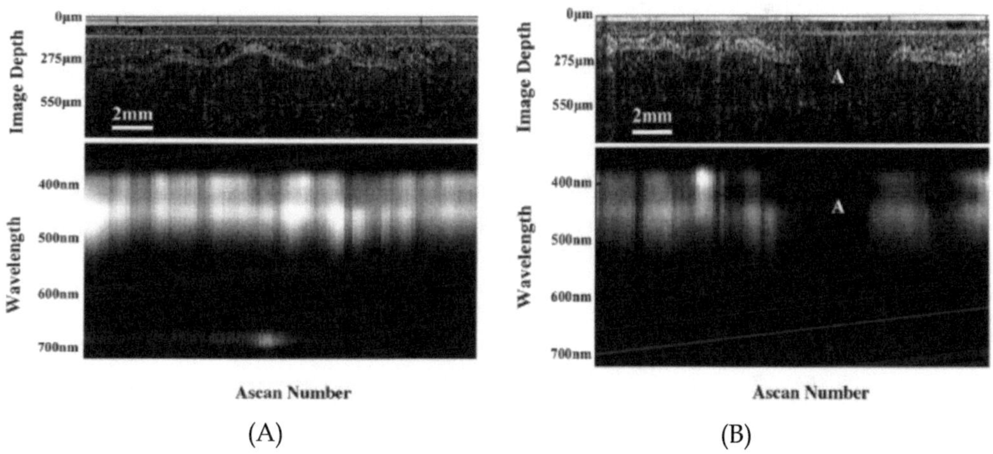

(A) (B)

Fig. 4. OCT images (top) and corresponding LIF spectra (bottom) of healthy colon taken from a high-chlorophyll diet mouse (A) and adenoma (B) (McNally et al., 2006) with permission.

This miniaturized 2 mm diameter catheter-based dual-modality system has been used to monitor the disease progression in mouse colon longitudinally, and is able to identify colorectal adenomas in murine models (Hariri et al., 2006; McNally et al., 2006; Hariri et al., 2007). In one study, McNally *et al.* used the OCT-LIF system to study healthy and adenomatous mouse colon. Their study found close agreement among each of the modalities and with histology. Both modalities were capable of identifying diseased tissue accurately. Here, the OCT images provided structural information about the tissue, while the autofluorescence signal measured by LIF provided biochemical information. Using this multi-modal system, they found that the magnitude of the 680 nm signal correlates with chlorophyll content in the mouse diet, suggesting that the autofluorescent compound is a dietary metabolite. Figure 4 shows their OCT image (top) and corresponding LIF spectra (bottom) of healthy colon taken from a high-chlorophyll diet mouse (left) and adenoma mouse (right). The LIF emission peaks at approximately 390 and 450 nm, with an absorption dip at 420 nm (typical of healthy tissue). The presence of a distinct peak at approximately 680 nm is visible throughout the colon as well. The adenoma can be visualized in the OCT image (Region A in Figure 4B). The corresponding LIF signal shows a significant reduction in the 390 and 450 nm signals and a peak in the 680 nm signal over the adenomatous region.

3. Combining OCT with Fluorescence Molecular Imaging (FMI)

3.1 Introduction

Fluorescence molecular imaging (FMI) could obtain molecular information by measuring fluorescence intensity of fluorescent bio-markers which target specific molecules. Previous researches have shown that cancerous tumors can be identified with fluorescent markers (Tung et al., 2000; Achilefu, 2004). In a study done in mice, it was shown that adenomatous polyps in the colon express 36% more of the proteolytic enzyme cathespin B than normal tissue (Marten et al., 2002). Also, it was shown that mice colons that expressed more cathespin B, when injected with a fluorescent dye, showed significantly higher fluorescent intensities than mice injected with dye that did not have adenomatous polyps.

Although LIF technique and 2-D FMI technique have the same scientific bases and the silimiar abilities in monitoring molecular processes, FMI technique is more convenient in obtaining 2-D images showing flurophore's spatial distribution and molecular process spatial hetergenity using specifically-targeted dye. In our recent study, we combined the OCT with FMI system to investigate the correlation between OCT structural features and FMI molecular information. The system demonstrated that it could co-register en face OCT and FMI images. Relationships of FMI intensity and dye concentration as well as FMI intensity and target fluorescence tube depth were studied. The capability of imaging biological tissue was demonstrated by imaging the mouse colon cancer model ex vivo.

3.2 Methods

3.2.1 Optical Coherence Tomography (OCT)

Figure 5 shows the schematic of the co-registered OCT and fluorescence imaging system. The fiber-based high-speed, high-resolution OCT system utilizes the wavelength-swept laser as the light source. It generates a broadband spectrum of 100 nm at 1310 nm, which provides an axial resolution of 10 μm in the tissue. The laser operates at a sweep rate of 16 kHz (equivalent to an imaging speed of 30 frames per second for a 512 axial-line image) with an average output power of 12 mW. The system sensitivity is 95 dB. A Michelson interferometer composed of one circulator and a fiberoptic 50/50 splitter is used to generate the Fourier-domain OCT signal. The reference arm consists a stationary mirror (M) and a polarization controller (PC). The light from the sample arm passes through a collimator (C) and is steered by a pair of galvanometer mirrors (X and Y) then focused by an objective lens to a spot. The power on the sample is 4 mW with a spot size of 15 μm. The OCT interference signal returned from both the sample and reference arms is detected by a balanced photodetector (BD). A Mach-Zehnder interferometer (MZI) with a fixed path difference is used to generate an optical frequency clock. Data acquisition is triggered by the zero-crossings of the MZI fringes, which are evenly spaced in optical frequency. Discrete Fourier transform (DFT) is performed on the data to generate an axial depth profile of the sample (A-line) with 3 mm imaging depth and 512 pixels (Andrews et al., 2008).

3.2.2 Fluorescence Molecular Imaging (FMI)

The fluorescence molecular imaging (FMI) system uses continuous-wave (CW) laser diodes at 675 nm as the excitation source. The excitation light is combined with the OCT sample arm using a dichroic mirror. The typical illumination power on the sample is approximately 1 mW. The reflectance and fluorescence light is detected by the same fiber and then connected to a fiber splitter to divide the collected light into reflectance and fluorescence signals. The simultaneous measurement of reflectance and fluorescence signals from the same source and detector geometry is important for minimizing the influence of optical coupling variation for both excitation and collection paths. The reflectance signals are detected by an avalanche photodiodes (APD), while the fluorescence signals first pass through an emission filter set (700 ± 10 nm), and then are detected by a photomultiplier tube (PMT). The illumination and filter wavelength are chosen based on the excitation and emission properties of the near-infrared dye Cy5.5. Such a system can be readily adapted to image other fluorescence dyes with different excitation/emission wavelengths.

3.2.3 Molecular imaging contrast agent

Colorectal cancer is the third most common form of cancer, and the second leading cause of cancer death in the United States (Jemal et al., 2008). Recently, Roney *et al.* demonstrated that the legume lectin *Ulex* europaeus agglutinin I (UEA-1) binds the surfaces of adenomatous polyps in specimens of colorectal cancer from the APC^{Min} mouse model. The carbohydrate α-L-fucose, which is over-expressed on the surfaces of polyps, facilitates the bond with UEA-1. Thus, α-L-fucose may be a possible biomarker to target and detect colon adenomas by molecular imaging methods (Roney et al., 2008).
The details of the synthesis of the molecular contrast agent UEA-1 conjugated polymerized liposomes have been described elsewhere (Roney et al., 2008). Basically, the contrast agent contains the fluorescence dye Lissamine Rhodamine PE, therefore, in order to imaging this specific contrast agent, we use a laser source with 532 nm excitation and a fluorescence filter with 600 ± 10 nm.

3.2.4 Animal model of colon cancer

Excised small and large bowels (N=4) of male C57BL/6J $APC^{Min/+}$ mice were commercially purchased from the Jackson Laboratory (stock #002020, 8 wks), and stored at -80 °C until further use. The bowels were thawed to room temperature (RT) for 15-20 mins, cut along the longitudinal axis, and preserved in formaldehyde (RT, 15 mins) before the commencement of staining. The specimens were incubated in UEA-1 conjugated and non-conjugated liposomes (containing 0.4 μM Lissamine Rhodamine PE) at RT for 45 mins. The specimens were then washed in PBS, before being mounted and imaged by OCT/FMI system.

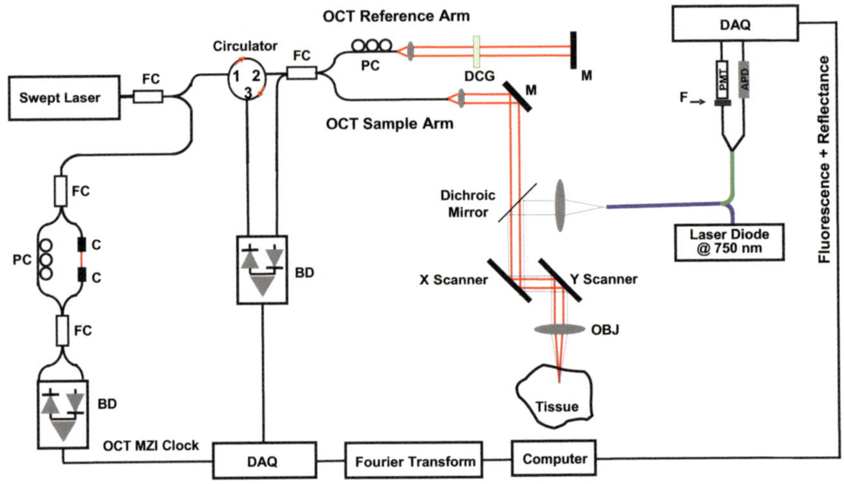

Fig. 5. Schematic of integrated OCT/FMI system. The left side is the OCT imaging subsystem and the right side is the FMI imaging subsystem. They are combined by a dichroic mirror in the sample arm. FC: fiber coupler; PC: polarization control; C: collimator, MZI: Mach-Zehnder interferometer (frequency clocks), M: mirror, BD: balanced detector, DAQ: data acquisition board, DCG: dispersion compensating glasses, OBJ: objective; PMT: photomultiplier tube; APD: avalanche photodiode; F: fluorescence filter (Yuan et al., 2010b) with permission.

3.3 Characterization of FMI and OCT signal

3.3.1 Characterization of tissue scattering coefficients

In a typical OCT A-scan, OCT signal I(z) decays with the depth z. Since 1990s, several theoretical models of OCT have been developed to fit OCT signals as a function of depth. Those models can be used to find the scattering coefficient μ_s of the scattering media (Schmitt et al., 1993; Faber et al., 2004; Levitz et al., 2004; Turchin et al., 2005). Most recent models (Levitz et al., 2004; Turchin et al., 2005a) consider factors from the multi-scattering and the Gaussian beam propagation. They can produce a very good fit on A-scan profile as well as a close match of the scattering coefficient value to the MIE theory. In our study, we used the model from Letivz *et al.* (2004) to fit our experimental data and then find the scattering coefficient:

$$I_t(z) = I_0\{e^{-2\mu_s z} + \frac{4e^{-\mu_s z}[1-e^{-\mu_s z}]}{1+\frac{w_S^2}{w_H^2}} + [1-e^{-\mu_s z}]^2 \frac{w_H^2}{w_S^2}\} \tag{3}$$

The first term in Eq. (3) represents the single scattering contribution, the third term represents the multiple scattering contribution, while the second term is the cross-term. w_H and w_S reflects contribution from the Gaussian beam propagation. w_H, w_S are the 1/e Gaussian beam irradiance radius at the probing depth in the absence of scattering and in the presence of scattering, respectively. They can be defined as:

$$w_H{}^2 = w_0{}^2 (A - \frac{B}{f})^2 + (\frac{B\lambda}{2\pi w_0})^2 \tag{4}$$

$$w_S{}^2 = w_0{}^2 (A - \frac{B}{f})^2 + (\frac{B\lambda}{2\pi w_0})^2 + [\frac{B\lambda}{\pi\rho(z)}]^2 \tag{5}$$

$$\rho(z) = \sqrt{\frac{3}{\mu_S z} \frac{\lambda}{\pi\theta_{rms}} (\frac{nB}{z})} \tag{6}$$

Here ρ(z) is the lateral coherence length. A and B are elements from the ABCD ray-matrix for light propagation from the lens plane to the probing depth in the sample. If the focal plane of the sample beam is fixed on the surface of the sample, A = 1 and B = d + z / n. d is the distance between the lens and the surface. w_0 is the 1/e irradiance radius of the input sample beam at the lens plane, λ is the center wavelength of the source, f is the focal length of the lens, n is the mean index of refraction of the sample and θ_{rms} is the root-mean-square scattering angle. We could find the value of scattering coefficient μ_s by minimizing the mean-square deviation of the logarithms of the experimental curve $I_e(z)$ and fitting theoretical curve $I_t(z)$:

$$[\mu_s, \theta_{rms}] = \arg min_{(\mu_s, \theta_{rms})} \sum_n \log^2 [I_t(z_n) / I_e(z_n)] \tag{7}$$

We have used this method to find scattering coefficients of polystyrene microspheres and intralipid phantom. The obtained values are close to the values estimated by MIE theory or values reported in literatures (Vanstaveren et al., 1991; Flock et al., 1992; Zaccanti et al., 2003), as seen in Table 1 and Figure 6. For comparison, fitting result from Faber's model is also presented in Table 1 (Faber et al., 2004). This model, as shown in Eq. (8), includes single scattering term in Eq (3) and a factor for axial point spread function. The model works better for weakly scattered tissues.

$$I_t(z) = I_0 \frac{1}{\left(\frac{z - z_{cf}}{z_R}\right)^2 + 1} e^{-2\mu_s z} \tag{8}$$

Here Z_{cf} is the position of the confocal gate and Z_R is the apparent Rayleigh length.

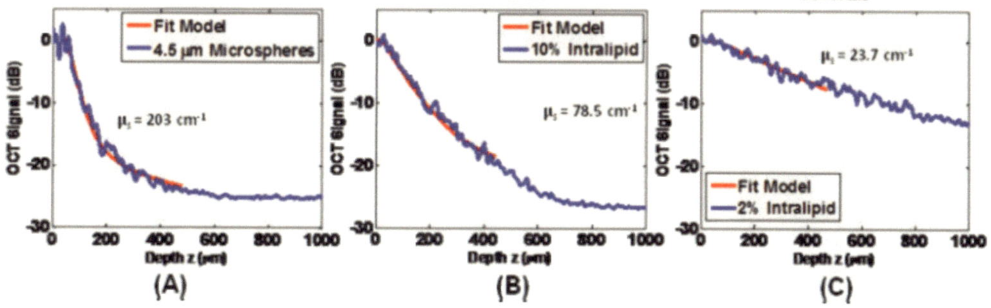

Fig. 6. OCT A-scan profiles and their fit result to the model Eq. (3). 6A is on microsphere phantom (4.5 μm, Polysciences); 6B is on 10% intralipid solution; 6C is on 2% intralipid.

Phantom	μ_s (cm^{-1}) from fitting Eq. (3)	μ_s (cm^{-1}) from fitting Eq. (8)	μ_s (cm^{-1}) from MIE or reported
4.5 μm microsphere	203	160.1	211 [a]
10% Intralipid	78.5	69.7	63 - 82 [b]
2% Intralipid	23.7	22.9	16 - 21 [c]

a, From MIE theory (http://omlc.ogi.edu/software/mie/); b, from Vanstaveren et al. (1991) and Flock et al. (1992); c, values are estimated from values for 10% intralipid and empirical function from Zaccanti et al. (2003)

Table 1. Scattering coefficient values of several phantoms from the model Eq. (3) comparing with theoretical values or reported values (at 1300 nm) and values from Faber model.

3.3.2 Characterization of FMI and OCT signal

A 400 μm inner-diameter and 900 μm outer-diameter capillary tube filled with different concentration of Cy 5.5 was used as a phantom. It was placed in the scattering phantom containing 2% intralipid ($\mu_s \sim 83$ cm^{-1} at FMI wavelengths, the scattering coefficient is close to skin scattering coefficient) (Troy & Thennadil, 2001). The tube filled with fluorescence dye mimics a contrast-agent-labeled "tumor", while the intralipid overlaying at the top of "tumor" mimics the "skin" tissue.

In one set of experiments on the phantom, the center of the tube is set at 500 μm beneath the surface and different concentration of Cy 5.5 was filled into the tube. FMI intensity was then measured for each configuration. Titration of the Cy5.5 concentration reveals a linear relationship between the FMI intensity and the dye concentration as shown in Figure 7(a).

In the second part of the experiment with the phantom, the relationship of the FMI intensity and the tube (dye) depth was studied by setting concentration of Cy 5.5 inside the capillary tube at 1 μM and varying the tube depth (measured from the top surface of the tube) from 0 μm to 200μm. The result, as presented in Figure 7(b), shows that the relationship can be well fitted into exponential function, which is a simplified model of Eq. (3) at small depths. It is consistent with other group's report (Turchin et al., 2005), which expects for small depths:

$$I \propto \exp(-2\mu_t \cdot d). \qquad (9)$$

where $\mu_t = \mu_s + \mu_a$ is the total extinction coefficient. Both the absorption coefficient μ_a of 10% intralipid and water are less than 0.01 cm^{-1} at FMI wavelengths (Flock et al., 1992; Pope & Fry, 1997), while the scattering coefficient μ_s at FMI wavelengths is ~ 83 cm^{-1}, ~ 3-4 orders of magnitude larger than the absorption coefficient. Therefore Eq. (9) can be simplified as:

$$I \propto \exp(-2\mu_s \cdot d). \qquad (9b)$$

The scattering coefficient μ_s can then be further determined from the fitting the measured data to Eq. (9b). From our measurement, μ_s of 2% intralipid at 700 nm is about 88.0 cm^{-1}, close to the reported value 83 cm^{-1} from the literature (Flock et al., 1992; Pope & Fry, 1997).

Although the model Eq. (3) works fine in fitting experimental data and finding the actual μ_s values, it takes much longer time to fit than that using the simplified model Eq. (9b). If the medium is weakly scattering or only A-line profile near the surface are processed, the results from Eq. (3) and Eq. (9b) are close and we can use the simplified model Eq. (9b) to obtain μ_s values. Figure 8 shows the differences between single scattering model Eq. (9b) and the model Eq. (3) for different μ_s values and fitting depths. A simulated A-scan is first generated by Eq. (3) with a known μ_s value, then the linear decay phase is fitted using Eq. (9b) to estimate the μ_s value. Figure 8(A) shows the fitted μ_s values on the 120 μm of a theoretical plot of A-scan profile based on Eq. (3). Figure 8(B) and 8(C) show that of 200 μm and 280 μm depths, correspondingly. In 120 μm case, if the μ_s value is between 15 - 165 cm^{-1},

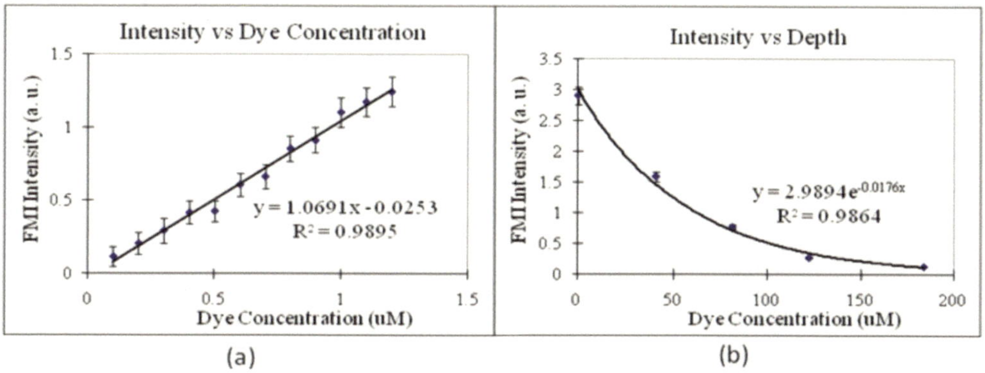

Fig. 7. (a) Relationship of fluorescence intensity versus Cy 5.5 concentration. A 400 μm inner-diameter capillary tube filled with different concentration of Cy 5.5 was placed in 2% intralipid (similar to skin scattering) and with the center at 500 μm beneath the surface. (b) Relationship of fluorescence intensity versus capillary tube depth (depth of the top surface). The concentration of Cy 5.5 inside the tube was fixed at 1 μM. (Yuan et al., 2010b) with permission.

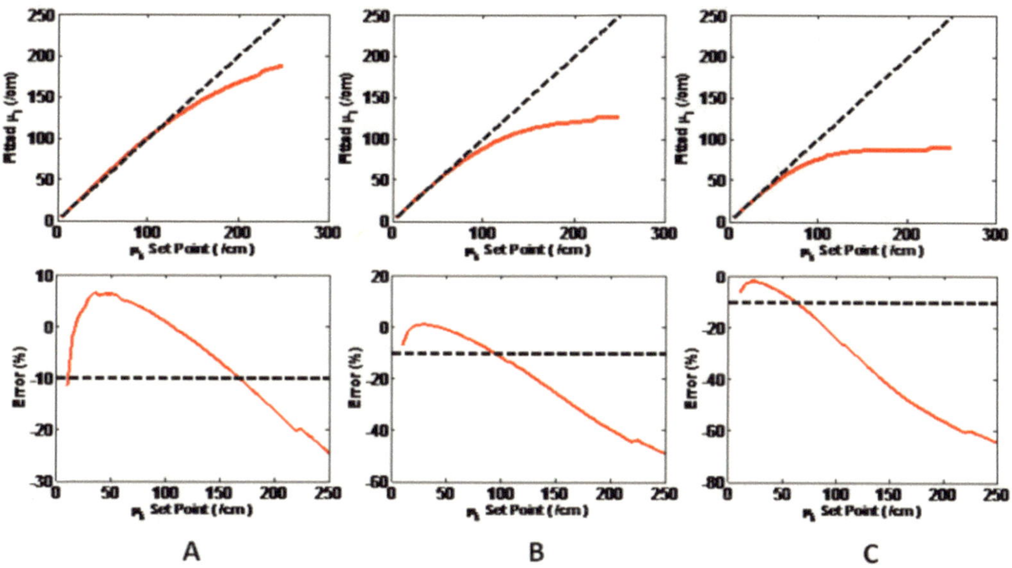

Fig. 8. The difference between the single scattering model from Eq. (9b) and the model from Eq. (3). (A) Fitted μ_s values using single scattering model on theoretical plot of 120 μm A-scan profile from Eq. (3). (B) Fitted μ_s values on theoretical plot of 200 μm A-scan profile. (C) Fitted μ_s values on theoretical plot of 280 μm A-scan profile. The differences in percent between the two models are shown in the lower row.

the difference between the fitting results from two models is less than 10%. In 200 μm case, if the actual μ_s is between 10 - 90 cm^{-1}, the difference is less than 10%; for 280 μm profile, 10 - 60 cm^{-1} μ_s will result in a difference of less than 10%.

3.4 OCT imaging of mouse bowels

Figure 9 shows the representative OCT images of both the normal and polypoid regions of mouse bowel. Figure 9(a) shows the cross-sectional OCT image of the normal bowel, with the characteristic villous and crypt structures. Figure 9(b) shows the en face projection OCT image, which is analogous to the standard endoscope images. Individual villi/crypt can be visualized as well as the gross pattern of their distribution. Figure 9(c) shows the 3D reconstruction of the normal bowel. The OCT image Figure 9(a) agrees well with the histology shown in Figure 9(d).

In contrast, the polyp region shows the irregular mass in the OCT cross-sectional image in Figure 9(e). Projection OCT and 3D OCT in Figure 9(f) and 9(g) respectively show the whole polyp structures. Histology shown in Figure 9(h) confirms the OCT imaging results. These results demonstrated OCT's ability to obtain micron-resolution, cross-sectional and 3D images of mouse bowel microstructures, and can be used to identify abnormalities such as the development of polyps.

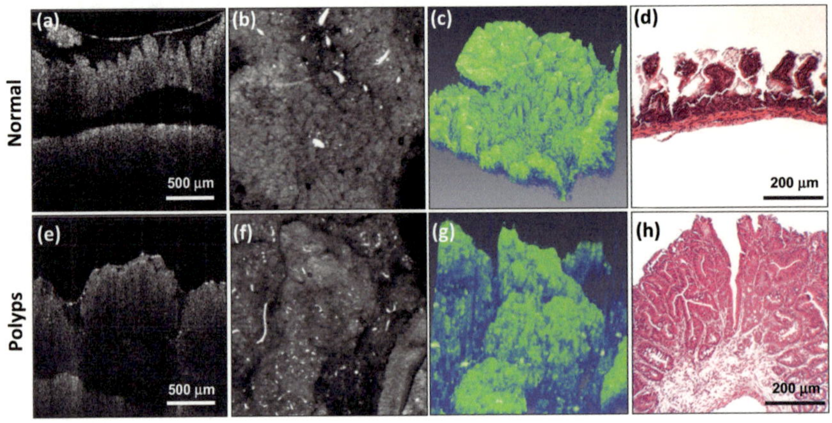

Fig. 9. OCT images of mouse bowel from two regions of interests (ROI), normal region and polypoid region. Normal region: (a) Cross-sectional OCT images. (b) OCT en face projection image. (c) 3-D OCT image (d) Histology. Polyps: (e) Cross-sectional OCT images. (f) OCT en face projection image. (g) 3-D OCT image (h) Histology. (Yuan et al., 2010b) with permission.

3.5 OCT/FMI imaging of mouse bowels

Figure 10 shows the results of the OCT/FMI imaging of mouse bowel polyp treated with UAE-1 conjugated liposome. Figure 10(A) is the height profile image obtained from OCT. The profile shows the height of the tissue surface estimated from OCT. Higher regions show red and lower regions show blue. Polyps regions usually rise up and are shown as red in the profile image. Figure 10(B) and 10(C) shows the corresponding cross-section OCT images of line 1 and 2 in Figure 10(A). Figure 10(D) is the scattering coefficient image. The scattering coefficient is obtained by fitting each OCT A-scan with Eq. (9b). 30 pixels (~ 120 μm in depth) beneath the tissue surface are used for each A-scan profile fitting. Simplified model

Eq. (9b) is used because (1) the algorithm runs much faster than Eq. (3) and (2) μ_s is about 60 -150 cm^{-1} values and the difference of fitted μ_s between two models are less than 10% as shown in Figure 8. Figure 10(E) shows the fluorescence image of the same ROI from FMI and Figure 10(F) is the fused image of Figure 10(D) and 10(E). Comparing Figure 10(A) and 10(F), we could see that the polyp region shows a high scattering coefficient, possibly caused by the alteration of original structures. Fluorescence image further confirmed the polyp region, which shows high fluorescence signal around the edge. The binding mechanism of the UEA-1 contrast agent with polyps is still under investigation. Figure 10(F) shows very good correlation between OCT and FMI.

In conclusion, we have demonstrated a co-registered OCT and FMI imaging system. This system enables simultaneous imaging of tissue morphology and molecular information at high resolution over 2-3 mm field-of-view, and will have potential applications in small animal imaging and clinical imaging. Imaging results from our system on mouse intestinal tissues shows that glycoprotein targeting contrast agent promises to imaging polyps. This imaging technique is translatable to clinics in the forms of endoscopy, laparoscopy, and needle imaging probes, and might enable many biomedical applications.

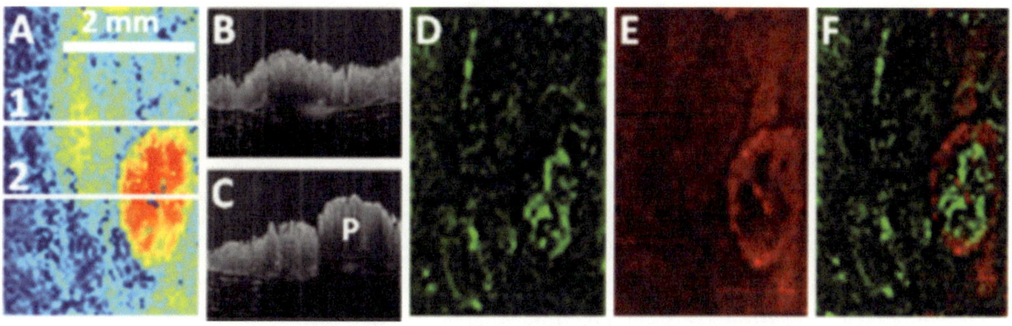

Fig. 10. OCT/FMI imaging of the polyp treated with UAE-1 conjugated liposome. (A) The height profile image obtained from OCT; (B,C) The cross-section OCT images of line 1 ans 2 in Figure 9(A), correspondingly. (D) The scattering coefficient image in the same ROI obtained from OCT. The scale bar is [60 150] cm^{-1}. (E) The fluorescence image of the same ROI. (F) Fused image from D and E, showing very good co-registration. (Yuan et al., 2010a) with permission.

4. Combining OCT with Fluorescence Laminar Optical Tomography (FLOT)

4.1 Introduction

Although FMI technique works great for some applications, the depth-integrated fluorescence imaging has two limitations (Ntziachristos et al., 2002): 1) it cannot provide depth-resolved information, and 2) the detected signals non-uniformly depend on the depth. These present significant challenges in the quantitative investigation of the signal contributions. Diffuse optical tomography (DOT) has been successfully implemented in quantification of fluorescence contrast agents in tissues (Ntziachristos & Weissleder, 2001; Li et al., 2003). But DOT works only for large volume tissues (cm and up). Laminar optical tomography (LOT) has been developed to perform depth-resolved functional imaging with ~100 μm resolution and up to 2.5 mm depth (Dunn & Boas, 2000; Hillman et al., 2004). LOT is an extension of confocal microscopy using the multiple-displaced-confocal-pinhole design to capture the photons traveling through different depths, and the image is obtained through transport-based reconstruction (Dunn & Boas, 2000). Fluorescence LOT (FLOT) has been demonstrated (Hillman et al., 2007). The comparable imaging scale (1–2 mm in depth) of OCT and FLOT as well as their complementary contrast mechanisms motivate the integration of OCT and FLOT as a multimodality imaging platform to provide tissue structural and molecular information in 3D in millimeter imaging scale.

A traditional FLOT system uses a point-focused light illumination and a 1D array detector to detect lights that undergo different depths. 3D images are obtained through 2D raster scanning of the illumination point. Alternatively, a line-field illumination can be utilized, and only 1D scanning is required to acquire 3D information. Line-scanning imaging has been successfully demonstrated in ophthalmoscopy (Iftimia et al., 2006), confocal microscopy (Dwyer et al., 2007), OCT (Yasuno et al., 2006), and optical coherence microscopy (Chen et al., 2007b). Line-scan imaging promises to alleviate the complicated 2D scanner design and therefore has strong potential to be miniaturized into endoscopic imaging devices. In addition, for LOT, line-scan imaging will use a 2D array detector such as a CCD camera as the detector. This may improve the imaging performance by using the high-pixel density CCD and will augment the widely available CCD-based fluorescence microscopes with the FLOT imaging mode. In this session, we will discuss combine OCT with line-scan FLOT.

4.2 Methods

4.2.1 Fluorescence Laminar Optical Tomography (FLOT)

Figure 11 shows the schematic of the co-registered OCT and line-scan FLOT imaging system. The left portion is the OCT subsystem, which is the same as the one used in OCT/FMI system. The line-scan FLOT system is shown in the right portion. The excitation source is a CW laser diode (Power Technology, Inc.) at 670 nm. A cylindrical lens is used to expand the point illumination into a line. The excitation light is combined with the OCT sample arm by a dichroic mirror (DM-1). The reflectance and fluorescence light is separated by another dichroic mirror (DM-2), and the fluorescence signal is detected by an electron-multiplying CCD (EM-CCD) camera (Cooke Corp.) with an emission filter (700 ± 10 nm). These wavelengths are chosen based on the optical properties of the near-IR dye Cy5.5. This system can be readily adapted to image other fluorescence dyes by changing the excitation source and emission filter wavelengths. The line illuminated on the sample measures 2.2 mm in length and 50 μm in width with 8 mW power. The OCT and line-scan FLOT imaging are synchronized by scanning with the same galvanometer (Y). The OCT beam is scanned in two axes (fast axis *X* and slow axis *Y*). The scanning speed for *X* is 25 Hz (with 624 samples in 2.5 mm), and the scanning speed for *Y* is 0.1 Hz (with 256 samples in 2.5 mm). Line-scan FLOT scans in *Y* only at the same speed as OCT, with 256 frames acquired by CCD (acquisition speed of 25 Hz with 30 ms exposure time). The resulting OCT 3D volume size is 624 (*X*) by 256 (*Y*) by 512 (*Z*) pixels (with voxel size of 4 μm × 10 μm × 4.4 μm), and FLOT image raw data is 501 (*X*) by 1024 (*Y*) (2.2 mm × 2.2 mm) by 256 (*T*) frames. The total acquisition time for each 3D volume is approximately 10 s. The rate-limiting factors are the FLOT CCD frame rate, and the high number of OCT images acquired in the *Y* dimension. The 3D imaging time can be significantly reduced by binning the CCD camera.

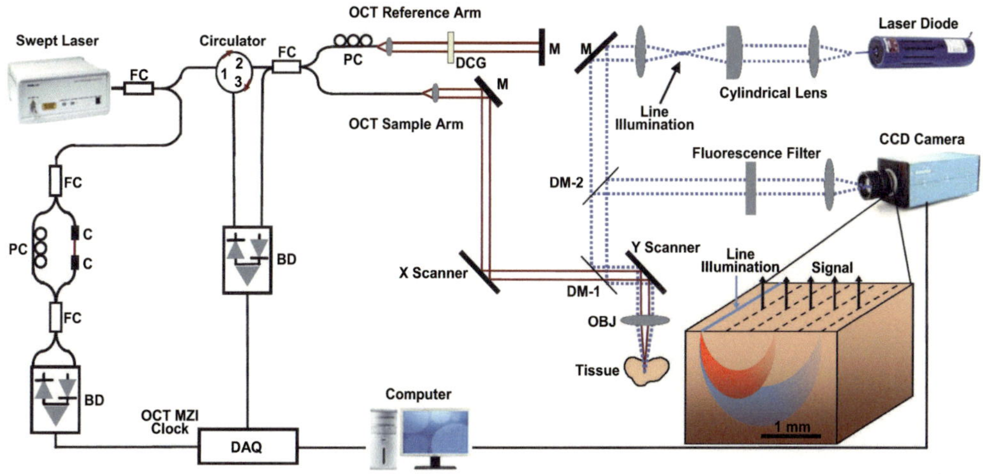

Fig. 11. Schematic of the combined OCT and line-scan FLOT system. (Chen et al., 2010) with permission.

4.2.2 Data processing

The depth-resolved FLOT images were obtained through an image reconstruction similar to those used in CT or DOT (Arridge, 1999). The fluorescence signal can be expressed as (Crilly et al., 1997; Hillman et al., 2007):

$$F_0(\vec{r}_d,\hat{\Omega}_d)=\frac{\sigma_{ex}\cdot\gamma}{4\pi}\int\psi_{ex}(\vec{r}-\vec{r}_s)\cdot C_f(\vec{r})\cdot\psi_{em}(\vec{r}_d-\vec{r})\cdot d^3\vec{r}\ , \tag{10}$$

where σ_{ex} is the absorption cross-section of the fluorophores at the excitation wavelength, γ is the fluorescence quantum yield, $\Psi_{ex}(r-r_s)$ is the excitation fluence distribution calculated from the excitation photon radiance, $C_f(r)$ is the fluorophore concentration at position r, and $\Psi_{em}(r_d-r)$ is the probability that a photon emitted by a source at position r will be detected. Eq. (10) can be discretized into voxels, which yields a matrix equation:

$$F=JC\ , \tag{11}$$

where F is the fluorescence measurements, C is the spatially-distributed fluorophores concentration, and J is the weight function matrix that represent the sensitivity of each measurement to the fluorophore concentration at each voxel, and can be expressed as:

$$J_{s,m}(\vec{r}_s,\vec{r}_d,\vec{r})=D_0\cdot\psi_{ex}(\vec{r}-\vec{r}_s)\cdot\psi_{em}(\vec{r}_d-\vec{r})\ , \tag{12}$$

where D_0 is a constant. Figure 12 shows the sensitivity distribution of FLOT using Monte-Carlo simulation. The sensitivity profile varied as the source-detector separation (Δ) increased. Therefore, the depth-dependent information was encoded in the multiple detector meaurement, and was recovered through image reconstruction. The sensitivity matrix for line illumination was achieved by convolving the point-source sensitivity matrix along the Y dimension.

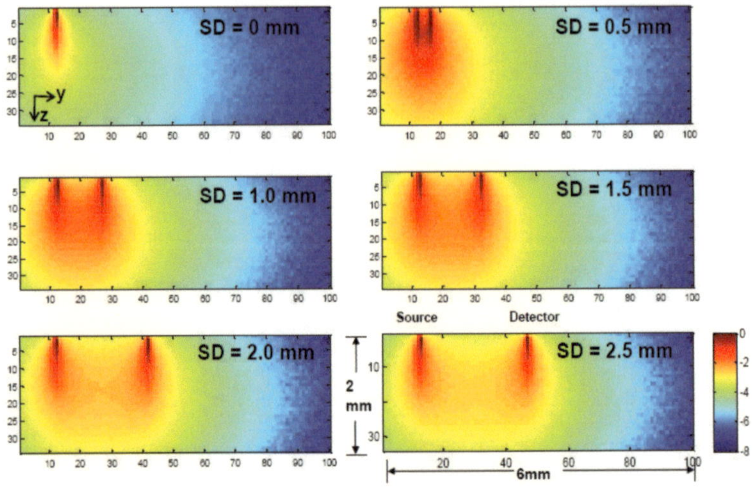

Fig. 12. Monte-Carlo simulated measurement sensitivity distribution of FLOT measurements (log scale). Tissue geometry is 3 mm (lateral) by 2 mm (depth) with scattering coefficient μ_s = 8 mm^{-1} for excitation and 7 mm^{-1} for emission (g = 0.9). (Chen et al., 2010) with permission.

4.2.3 Imaging reconstruction in line-scan FLOT

Image reconstruction is performed within individual planes (*YZ*) perpendicular to the illumination line (*X*). To accelerate the reconstruction, the raw FLOT data are binned to 40 (*X*) by 70 (*Y*) pixels and downsampled to 64 (*T*) frames. The corresponding pixel dimension is 55 μm × 30 μm with a scan step of 40 μm. This is equivalent to FLOT measurements of 70 source-detector pairs with separations from 0 to 2.2 mm. The selection of 70 source-detector pairs gives a sampling density of 30 μm per detector interval, which provides sufficient sampling for the FLOT system. The cross-sectional FLOT image (*YZ*) is 70 (*Y*) × 60 (*Z*) pixels with 30 μm × 30 μm pixel dimensions. The reconstruction is based on the method discussed in session 4.2.2. The background optical properties for reconstruction are μ_a=0.2 mm^{-1} and μ_s =7.2 mm^{-1} at 670 nm excitation, μ_a=0.2 mm^{-1} and μ_s=6.5 mm^{-1} at 700 nm emission, and g = 0.9. The image reconstruction is performed through the simultaneous iterative reconstruction technique approach (Kak et al., 1987), with iteration number N = 1000 to make sure the change between iterations is <0.1%.

4.3 FLOT phantom experiment

We first experimentally checked the sensitivity distribution of FLOT calculated from Monte-Carlo simulation by scanning the phantom: a capillary tube containing the fluorescent dye Cy5.5 using the FLOT system. Figure 13(A) shows a representative sensitivity matrix for line illumination and source-detection separation d = 1.5 mm. As the source detector scans through the object, the expected signal is shown in Figure 13(B), which is the profile at certain depth in the sensitivity matrix. Two peaks are present owing to the high sensitivity beneath either the source or the detector. Objects at different depths show different responses with more widely spread signals from the deeper object, owing to the stronger light scattering. The different profiles for objects with different depths contain the depth information and could be used to reconstruct the depth position of the object. Figures 13(C) shows the FLOT data for the capillary tube at different depths. The composite data is presented as a 2D matrix (*XT*) of line profiles along the illumination axis (*X*) at a specific position (*Y* = *d*, where *d* is the source-detector separation) with respect to the acquired frame time index (*T*, scanning of the line illumination position). FLOT data for different source-detector separations (0, 0.5, 1.0, 1.5 and 2.0 mm) and object depths (~ 0 and ~ 0.8 mm) are presented. The data show two lines with high intensities, which agrees with the simulation pattern in Figure 13(B). The object located at a deeper depth appears more blurred owing to scattering.

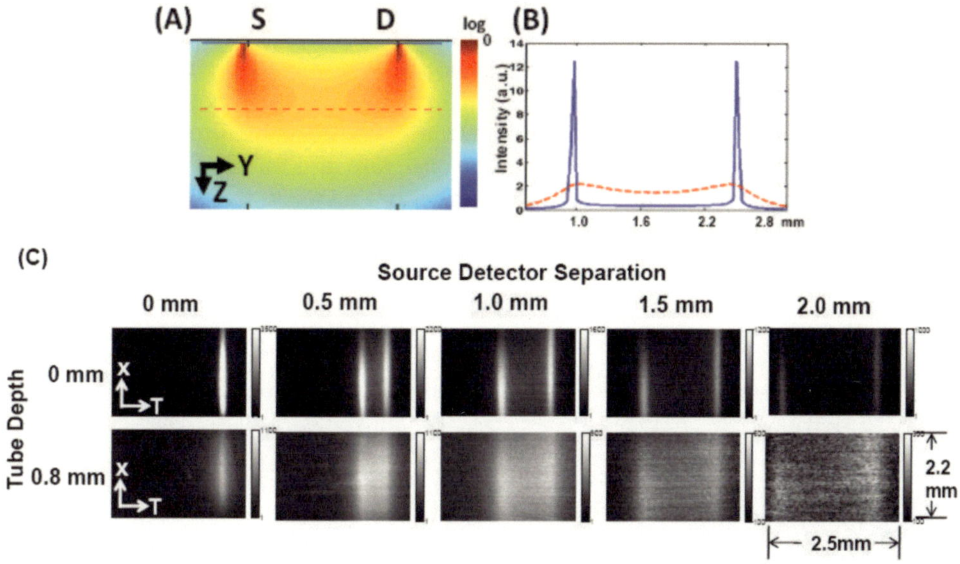

Fig. 13. (A) Monte Carlo simulated sensitivity distributions for line source and detector with separation *d* = 1.5 mm. Log scale from −4 to 0. (B) Intensity profiles for point object scanned through the *Y* axis (solid curve, depth ~ 0 mm; dashed curve, ~ 0.8 mm). (C) FLOT measurement (*XT*) for different source-detector separations from Y = 0 mm to *Y* = 2.0 mm) for object depth ~ 0 mm and ~ 0.8 mm. (Yuan et al., 2009) with permission.

To check the co-registration between OCT and FLOT, we obtained the depth-resolved 3D FLOT images through image reconstruction. Figure 14 shows the comparison of OCT and FLOT images of the capillary tube phantom. OCT readily images the 3D structure of the tube as shown in Figure 14(A). Figure 14(B) shows the comparison of OCT cross-sectional image (YZ) and FLOT image at location "1" denoted in (A). Cross-sectional OCT reveals the capillary tube (indicated by the arrow) with approximately 0.6 mm beneath the surface. FLOT cross-sectional image shows a fluorescence object at the same location, with slightly larger size due to the relatively larger FLOT resolution compared to OCT. Cross-sectional OCT and FLOT images at location "2" also show good agreement with the tube depth approximately 0.9 mm (Figure 14C). Compared to (B), the FLOT image for deeper object shows slightly enlargement in size. In addition, the peak values are slightly reduced when object gets deeper due to the enlargement of the reconstructed object size. Figure 14(D) shows the OCT cross-sectional image along the tube longitudinal dimension (XZ). The tube is placed with an angle with respect to the horizontal axis, and slightly bends flat. The FLOT image (XZ) reveals similar contour of the capillary tube as shown by OCT. Also, this image is averaged over a range of 150 μm in Y dimension along the center of the tube, and the results indicated nearly constant fluorophore concentration. In addition, we noted the decrease in the reconstructed fluorophore concentration (3 μM instead of 10 μM). This could be explained by the enlarged reconstructed object size (~ 300 μm in diameter vs. 100 μm).

Here the OCT image provides the structural information of the phantom and the FLOT reconstructed image provides the fluorescence dye distribution information. In biomedical applications, FLOT would provide fluorescence-dye-targeted molecular information. So in biomedical applications, the hybrid system can be used for concurrent depth-resolved tissue-structural and molecular imaging.

In conclusion, we demonstrates the co-registered OCT and line-scanning FLOT system, which has strong potential to provide depth-resolved tissue structural and molecular information at millimeter imaging scale. This represents a new multimodal imaging regime and could be useful for investigating the tissue structure and function relationship. Future work will focus on the utilization of the OCT structural image as the *a priori* information to improve the FLOT reconstruction, and incorporation of line-scan OCT to co-register with line-scan FLOT for endoscopy imaging.

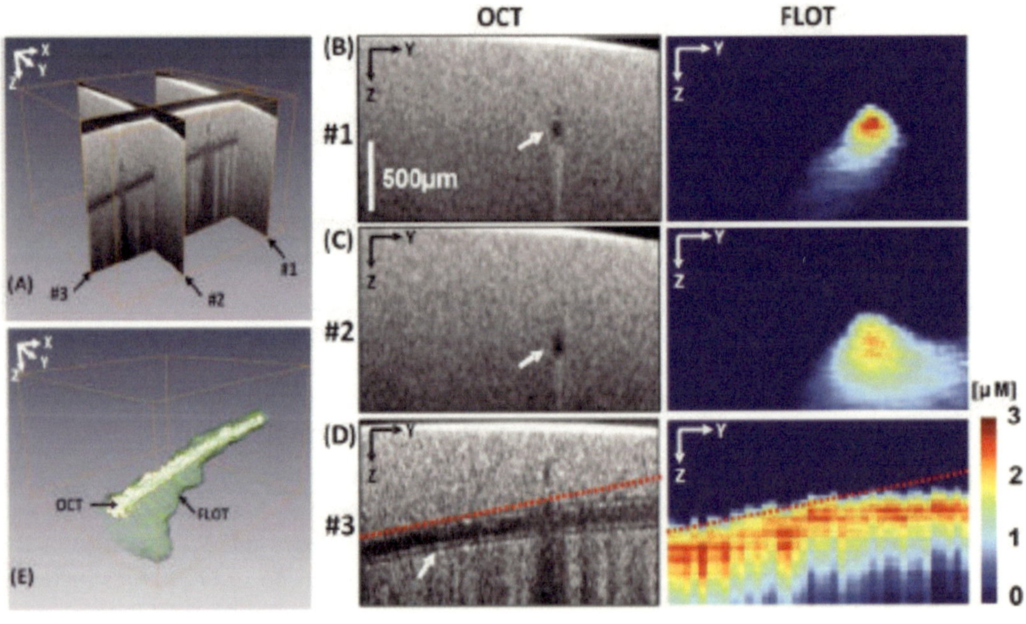

Fig. 14. Co-registration of OCT and line-scan FLOT of a capillary tube filled with fluorescence dye. (A) 3D OCT imaging showing three representative slices. (B) Left, OCT image (slice #1) of the capillary tube (arrow); right, FLOT image reconstruction of the fluorescence object. (C) OCT and FLOT images of different position (slice #2). (D) OCT and FLOT image (slice #3) reveal the curvature of the capillary tube (two identical dotted lines serve as the reference slopes). (E) 3D isosurface of FLOT image shows good co-localization with OCT (Yuan et al., 2009) with permission.

5. Conclusion

In this chapter, we introduce several multimodal optical imaging techniques which combines OCT and fluorescence imaging. OCT provides high-resolution, cross-sectional imaging of tissue microstructure, while fluorescence imaging reveals the biochemical and/or molecular information. Multimodal optical imaging techniques which combine OCT and fluorescence imaging could provide morphological, molecular and functional information simultaneously, and have great potential in disease diagnostics and therapy. Several such multimodal systems as well as their biomedical applications are introduced and discussed in this chapter. An ongoing development of novel multimodal system (OCT/FLOT) is also briefly introduced.

6. Acknowledgement

We thank Dr. David Boas (Massachusetts General Hospital) for providing the Monte-Carlo simulation code, Dr. Heng Lian (NanYang Technical University), Michael Lai (University of Maryland), and Qian Li (University of Maryland) for technical assistance of the imaging system and reconstruction algorithm, Drs. Ronald Summers and Celeste Roney (National Institutes of Health) for developing molecular imaging contrast agents, Dr. Ian White (University of Maryland) for the fabrication of capillary phantom, and Dr. James Jiang (Thorlabs, Inc.) for instrument support of SS-OCT. This research is supported by the UMCP GRB Research Award, Maryland Nano-Biotechnology Award, Minta-Martin Award, UMB-UMCP SEED Grant Program, and the Prevent Cancer Foundation.

7. References

Achilefu, S. (2004) Lighting up tumors with receptor-specific optical molecular probes. Technology in Cancer Research & Treatment, 3, 393-409.

Andrews, P. M., Chen, Y., Onozato, M. L., Huang, S. W., Adler, D. C., Huber, R. A., Jiang, J., Barry, S. E., Cable, A. E. & Fujimoto, J. G. (2008) High-resolution optical coherence tomography imaging of the living kidney. Laboratory Investigation, 88, 441-449.

Arridge, S. R. (1999) Optical tomography in medical imaging. Inverse Problems, 15, R41-R93.

Bouma, B. E., Tearney, G. J., Compton, C. C. & Nishioka, N. S. (2000) High-resolution imaging of the human esophagus and stomach in vivo using optical coherence tomography. Gastrointestinal Endoscopy, 51, 467-474.

Brezinski, M. E., Tearney, G. J., Bouma, B. E., Izatt, J. A., Hee, M. R., Swanson, E. A., Southern, J. F. & Fujimoto, J. G. (1996) Optical coherence tomography for optical biopsy. Properties and demonstration of vascular pathology. Circulation, 93, 1206-1213.

Chen, Y., Aguirre, A. D., Hsiung, P. L., Desai, S., Herz, P. R., Pedrosa, M., Huang, Q., Figueiredo, M., Huang, S. W., Koski, A., Schmitt, J. M., Fujimoto, J. G. & Mashimo, H. (2007a) Ultrahigh resolution optical coherence tomography of Barrett's esophagus: preliminary descriptive clinical study correlating images with histology. Endoscopy, 39, 599-605.

Chen, Y., Huang, S. W., Aguirre, A. D. & Fujimoto, J. G. (2007b) High-resolution line-scanning optical coherence microscopy. Optics Letters, 32, 1971-1973.

Chen, Y., Yuan, S., Li, Q., Naphas, R., Wierwille, J., Blackwell, T. R., Winnard, P. T., Raman, V. & Glunde, K. (2010) Integrated optical coherence tomography (OCT) and fluorescence laminar optical tomography (FLOT). IEEE Journal of Selected Topics in Quantum Electronics (In press).

Chinn, S. R., Swanson, E. A. & Fujimoto, J. G. (1997) Optical coherence tomography using a frequency-tunable optical source. Optics Letters, 22, 340-342.

Crilly, R. J., Cheong, W. F., Wilson, B. & Spears, J. R. (1997) Forward-adjoint fluorescence model: Monte Carlo integration and experimental validation. Applied Optics, 36, 6513-6519.

D'amico, A. V., Weinstein, M., Li, X., Richie, J. P. & Fujimoto, J. (2000) Optical coherence tomography as a method for identifying benign and malignant microscopic structures in the prostate gland. Urology, 55, 783-787.

Danielson, B. L. & Whittenberg, C. D. (1987) Guided-Wave Reflectometry with Micrometer Resolution. Applied Optics, 26, 2836-2842.

Dunn, A. & Boas, D. (2000) Transport-based image reconstruction in turbid media with small source-detector separations. Optics Letters, 25, 1777-1779.

Dwyer, P. J., Dimarzio, C. A. & Rajadhyaksha, M. (2007) Confocal theta line-scanning microscope for imaging human tissues. Applied Optics, 46, 1843-1851.

Faber, D., van der Meer, F., Aalders, M. & van Leeuwen, T. (2004) Quantitative measurement of attenuation coefficients of weakly scattering media using optical coherence tomography. Optics Express, 12, 4353-4365.

Fercher, A. F., Drexler, W., Hitzenberger, C. K. & Lasser, T. (2003) Optical coherence tomography-principles and applications. Reports on Progress in Physics, 66, 239-303.

Fercher, A. F., Hitzenberger, C. K., Kamp, G. & Elzaiat, S. Y. (1995) Measurement of Intraocular Distances by Backscattering Spectral Interferometry. Optics Communications, 117, 43-48.

Flock, S. T., Jacques, S. L., Wilson, B. C., Star, W. M. & Vangemert, M. J. C. (1992) Optical-Properties of Intralipid - a Phantom Medium for Light-Propagation Studies. Lasers in Surgery and Medicine, 12, 510-519.

Fujimoto, J. G. (2003) Optical coherence tomography for ultrahigh resolution in vivo imaging. Nature Biotechnology, 21, 1361-1367.

Fujimoto, J. G., Brezinski, M. E., Tearney, G. J., Boppart, S. A., Bouma, B., Hee, M. R., Southern, J. F. & Swanson, E. A. (1995) Optical biopsy and imaging using optical coherence tomography. Nature Medicine, 1, 970-972.

Fujimoto, J. G., Pitris, C., Boppart, S. A. & Brezinski, M. E. (2000) Optical coherence tomography: an emerging technology for biomedical imaging and optical biopsy. Neoplasia, 2, 9-25.

Haberland, U. H. P., Blazek, V. & Schrnitt, H. J. (1998) Chirp optical coher of layered scattering media. Journal of Biomedical Optics, 3, 259-26

Hariri, L. P., Tomlinson, A. R., Wade, N. H., Besselsen, D. G., Utzinger, U., Barton, J. K. (2007) Ex vivo optical coherence tomography ar fluorescence spectroscopy imaging of murine gastrointestinal tra Medicine, 57, 175-185.

Hariri, L. P., Tumlinson, A. R., Besselsen, D. G., Utzinger, U., Gerner, E. W (2006) Endoscopic optical coherence tomography and laser-indu spectroscopy in a murine colon cancer model. Lasers in Surgery ar 305-313.

Hillman, E. M., Boas, D. A., Dale, A. M. & Dunn, A. K. (2004) Laminar opti demonstration of millimeter-scale depth-resolved imaging in turbi Letters, 29, 1650-1652.

Hillman, E. M. C., Bernus, O., Pease, E., Bouchard, M. B. & Pertsov, A resolved optical imaging of transmural electrical propagation in Optics Express 15, 17827-17841.

Huang, D., Swanson, E. A., Lin, C. P., Schuman, J. S., Stinson, W. G., Chang, Flotte, T., Gregory, K., Puliafito, C. A. & Fujimoto, J. G. (1991) Op tomography. Science, 254, 1178-1181.

Huber, R., Wojtkowski, M. & Fujimoto, J. G. (2006) Fourier Domain Mode Lo A new laser operating regime and applications for optical coherenc Optics Express, 14, 3225-3237.

Iftimia, N. V., Hammer, D. X., Bigelow, C. E., Rosen, D. I., Ustun, T., Ferrante & Ferguson, R. D. (2006) Toward noninvasive measurement of blo using spectral domain low coherence interferometry and retinal tra Express, 14, 3377-3388.

Jang, I. K., Tearney, G. J., Macneill, B., Takano, M., Moselewski, F., Iftima, N., Houser, S., Aretz, H. T., Halpern, E. F. & Bouma, B. (2005) In-vivo ch of coronary atherosclerotic plaque by use of Optical Coherence Circulation, 111, 1551-1555.

Jemal, A., Siegel, R., Ward, E., Hao, Y., Xu, J., Murray, T. & Thun, M. J. (statistics, 2008. CA a Cancer Journal for Clinicians, 58, 71-96.

Kak, A. C., Slaney, M. & Ieee Engineering in Medicine and Biology Society. (19 of computerized tomographic imaging, New York, IEEE Press.

Levitz, D., Thrane, L., Frosz, M. H., Andersen, P. E., Andersen, C. B., Vala Swartling, J., Andersson-Engels, S. & Hansen, P. R. (2004) Determinati scattering properties of highly-scattering media in optical coherence images. Optics Express, 12, 249-259.

Li, A., Miller, E. L., Kilmer, M. E., Brukilacchio, T. J., Chaves, T., Stott, J., Zhang Chorlton, M., Moore, R. H., Kopans, D. B. & Boas, D. A. (2003) Tomogra breast imaging guided by three-dimensional mammography. Appliec 5181-5190.

X. D., Boppart, S. A., Van Dam, J., Mashimo, H., Mutinga, M., Drexler, W., Klein, M., Pitris, C., Krinsky, M. L., Brezinski, M. E. & Fujimoto, J. G. (2000) Optical coherence tomography: advanced technology for the endoscopic imaging of Barrett's esophagus. Endoscopy, 32, 921-930.

ı, B. & Brezinski, M. E. (2007) Theoretical and practical considerations on detection performance of time domain, Fourier domain, and swept source optical coherence tomography. Journal of Biomedical Optics, 12, 044007.

andel, L. & Wolf, E. (1995) Optical Coherence and Quantum Optics, Cambridge University Press, Cambridge, England.

arten, K., Bremer, C., Khazaie, K., Sameni, M., Sloane, B., Tung, C. H. & Weissleder, R. (2002) Detection of dysplastic intestinal adenomas using enzyme-sensing molecular beacons in mice. Gastroenterology, 122, 406-414.

cnally, J. B., Kirkpatrick, N. D., Hariri, L. P., Tumlinson, A. R., Besselsen, D. G., Gerner, E. W., Utzinger, U. & Barton, J. K. (2006) Task-based imaging of colon cancer in the Apc(Min/+) mouse model. Applied Optics, 45, 3049-3062.

tziachristos, V., Tung, C. H., Bremer, C. & Weissleder, R. (2002) Fluorescence molecular tomography resolves protease activity in vivo. Nature Medicine, 8, 757-760.

ıtziachristos, V. & Weissleder, R. (2001) Experimental three-dimensional fluorescence reconstruction of diffuse media by use of a normalized Born approximation. Optics Letters, 26, 893-895.

)tis, L. L., Everett, M. J., Sathyam, U. S. & Colston, B. W., Jr. (2000) Optical coherence tomography: a new imaging technology for dentistry. The Journal of the American Dental Association, 131, 511-4.

'an, Y. T., Xie, T. Q., Du, C. W., Bastacky, S., Meyers, S. & Zeidel, M. L. (2003) Enhancing early bladder cancer detection with fluorescence-guided endoscopic optical coherence tomography. Optics Letters, 28, 2485-2487.

'ope, R. M. & Fry, E. S. (1997) Absorption spectrum (380-700 nm) of pure water. II. Integrating cavity measurements. Applied Optics, 36, 8710-8723.

Roney, C. A., Xie, J., Xu, B., Jabour, P., Griffiths, G. & Summers, R. M. (2008) Glycoprotein expression by adenomatous polyps of the colon. Proceedings of SPIE, 6916, 69161O.

Schmitt, J. M. (1999) Optical coherence tomography (OCT): a review. IEEE Journal of Selected Topics in Quantum Electronics, 5, 1205-1215.

Schmitt, J. M., Knuttel, A. & Bonner, R. F. (1993) Measurement of Optical-Properties of Biological Tissues by Low-Coherence Reflectometry. Applied Optics, 32, 6032-6042.

Schuman, J. S., Puliafito, C. A. & Fujimoto, J. G. (2004) Optical coherence tomography of ocular diseases (2nd Edition), Thorofare, NJ, Slack Inc.

Sivak, M. V., Jr., Kobayashi, K., Izatt, J. A., Rollins, A. M., Ung-Runyawee, R., Chak, A., Wong, R. C., Isenberg, G. A. & Willis, J. (2000) High-resolution endoscopic imaging of the GI tract using optical coherence tomography. Gastrointestinal Endoscopy, 51, 474-479.

Tearney, G. J., Brezinski, M. E., Bouma, B. E., Boppart, S. A., Pitvis, C., Southern, J. F. & Fujimoto, J. G. (1997) In vivo endoscopic optical biopsy with optical coherence tomography. Science, 276, 2037-2039.

Troy, T. L. & Thennadil, S. N. (2001) Optical properties of human skin in the near infrared wavelength range of 1000 to 2200 nm. Journal of Biomedical Optics, 6, 167-176.

Tumlinson, A. R., Hariri, L. P., Utzinger, U. & Barton, J. K. (2004) Miniature endoscope for simultaneous optical coherence tomography and laser-induced fluorescence measurement. Applied Optics, 43, 113-121.

Tung, C. H., Mahmood, U., Bredow, S. & Weissleder, R. (2000) In vivo imaging of proteolytic enzyme activity using a novel molecular reporter. Cancer Research, 60, 4953-4958.

Turchin, I. V., Sergeeva, E. A., Dolin, L. S., Kamensky, V. A., Shakhova, N. M. & Richards-Kortum, R. (2005) Novel algorithm of processing optical coherence tomography images for differentiation of biological tissue pathologies. Journal of Biomedical Optics, 10, 064024.

Vanstaveren, H. J., Moes, C. J. M., Vanmarle, J., Prahl, S. A. & Vangemert, M. J. C. (1991) Light-Scattering in Intralipid-10-Percent in the Wavelength Range of 400-1100 nm. Applied Optics, 30, 4507-4514.

Wang, C. S., Cheng, W. H., Hwang, C. J., Burns, W. K. & Moeller, R. P. (1982) High-Power Low-Divergence Super-Radiance Diode. Applied Physics Letters, 41, 587-589.

Wang, R. K. (2007) Fourier domain optical coherence tomography achieves full range complex imaging in vivo by introducing a carrier frequency during scanning. Physics in Medicine and Biology, 52, 5897-5907.

Wang, Z. G., Durand, D. B., Schoenberg, M. & Pan, Y. T. (2005) Fluorescence guided optical coherence tomography for the diagnosis of early bladder cancer in a rat model. Journal of Urology, 174, 2376-2381.

Welzel, J., Reinhardt, C., Lankenau, E., Winter, C. & Wolff, H. H. (2004) Changes in function and morphology of normal human skin: evaluation using optical coherence tomography. British Journal of Dermatology, 150, 220-225.

Wojtkowski, M., Leitgeb, R., Kowalczyk, A., Bajraszewski, T. & Fercher, A. F. (2002) In vivo human retinal imaging by Fourier domain optical coherence tomography. Journal of Biomedical Optics, 7, 457-463.

Yasuno, Y., Endo, T., Makita, S., Aoki, G., Itoh, M. & Yatagai, T. (2006) Three-dimensional line-field Fourier domain optical coherence tomography for in vivo dermatological investigation. Journal of Biomedical Optics, 11, 014014.

Youngquist, R., Carr, S. & Davies, D. (1987) Optical coherence-domain reflectometry: a new optical evaluation technique. Optics Letters, 12, 158-160.

Yuan, S., Li, Q., Jiang, J., Cable, A., & Chen, Y. (2009) Three-dimensional coregistered optical coherence tomography and line-scanning fluorescence laminar optical tomography. Optics Letters, 34, 1615-1617.

Yuan,S., Roney, C. A., Li, Q., Jiang, J., Cable, A., Summers, R. M. & Chen, Y. (2010a) Correlation of morphological and molecular parameters for colon cancer. Proceedings of SPIE 7555. (Submitted)

Yuan,S., Roney, C. A., Wierwille, J., Chen, C. W., Xu, B., Griffiths, G., Jiang, J., Ma, H., Cable, A., Summers, R. M. & Chen, Y. (2010b) Co-registered optical coherence tomography

and fluorescence molecular imaging for simultaneous morphological and molecular imaging. Physics in Medicine and Biology, 55, 191-206.
Zaccanti, G., Del Bianco, S. & Martelli, F. (2003) Measurements of optical properties of high-density media. Applied Optics, 42, 4023-4030.

31

Polarization-Sensitive Optical Coherence Tomography in Cardiology

Wen-Chuan Kuo
Institute of Electro-optical Science and Technology, National Taiwan Normal University Taiwan

1. Introduction

Atherosclerotic vascular disease is a common cause of morbidity and mortality in developed countries (Arroyo & Lee, 1999). In particular, the rupture of atherosclerotic plaques is the most common event initiating acute ischemic heart disease (Shah, 2003). Therefore, it is crucial to detect vulnerable coronary atheromatous plaques prior to their rupture or erosion to prevent irreversible myocardial damage. Autopsy studies have identified several histological characteristics of these vulnerable plaques, such as a large lipid pool, thin fibrous cap (<65 μm), and activated macrophages near the fibrous cap (Falk et al., 1995). Therefore, modalities capable of visualizing the vessel wall might help in detecting lesions with high risks for acute events (Pasterkamp et al., 2000; Peters et al., 1994). There are several plaque imaging modalities. The oldest and most widely used technology is X-ray angiography, which can detect narrowing of the coronary blood vessels. The first imaging technique to demonstrate the benefits of imaging inside the arterial wall is intravascular ultrasound (IVUS). However, the current resolution is not sufficient to visualize the thin fibrous caps and small disruptions within the intimal and medial dissections. In the 1980s, coronary angioscopy, which allows direct visualization of the surface color and superficial morphology of atherosclerotic plaque, thrombus, neointima, and stent struts, was introduced. However, it cannot help in the assessment of subsurface lesions. Other proposed techniques include electron beam computed tomography (EBCT), magnetic resonance imaging (MRI), or positron emission tomography (PET); these are noninvasive screening tools that do not subject the patient to catheterization. In addition to the aforementioned techniques, which are merely a selection of the imaging modalities currently used in vivo or that are in the validation stage, the use of optical techniques for biomedical imaging is gaining considerable attention. This is largely due to the potential of optical techniques to provide high-resolution imaging without the need for ionizing radiation and associated risks.

Optical coherence tomography (OCT), which is based on a low-coherence interferometer, has emerged as a rapid, non-contact and noninvasive, high-resolution imaging tool (Huang et al., 1991). From the mid-1990s, the ability of intravascular OCT to provide high-resolution (10–20 μm) cross-sectional images of both *in vitro* human aorta and coronary arteries has been demonstrated (Brezinski et al., 1996; Fujimoto et al., 1995). The resolution of OCT images was up to 10 times better than that of conventional ultrasound, MRI, and computed tomography (CT) (Jang et al., 2002; Yabushita et al., 2002). Therefore, using OCT, small structural details (such as the width of intimal caps and the presence of fissures in atherosclerotic plaques (Bresinski et al., 1997) could be resolved and intramural collections of lipid within the intima of a vessel wall could be detected (Brezinski et al., 1996; Fujimoto et al., 1995). Furthermore, the objective OCT image criterion for risk-stratifying plaque characterization has been established on the basis of the intrinsic optical properties of a typical plaque, whose constituents are lipid, calcium, and fibrous tissue (Bresinski et al., 1997; Jang et al., 2002; Stamper et al., 2006; Tearney et al., 2006; Yabushita et al., 2002). On this basis, OCT has a detection sensitivity and specificity of 71%–79% and 97%–98% for fibrous plaques, 95%–96% and 97% for fibrocalcific plaques, and 90%–94% and 90%–92% for lipid-rich plaques, respectively (Tearney et al., 2006; Yabushita et al., 2002). Moreover, OCT has also been shown to quantify plaque macrophage content (Tearney et al., 2003) in lipid-rich plaques and to assess the success of intracoronary stent implantation in patients with coronary artery disease during percutaneous intra-arterial procedures (Bouma et al., 2003). At present, a company, LightLab Imaging, is targeting the cardiovascular market using commercializing intravascular OCT technology by providing dedicated imaging wires and occlusion balloon catheters.

In general, OCT images are obtained from measurements of the echo time delay and the intensity of the backscattered light from a specimen. Further, OCT employs the inherent differences in the index of refraction, rather than enhancement with dyes, to differentiate tissue types. However, since the plaque components are heterogeneous, they may sometimes generate reflected signals that confuse or obscure the identity of these components; multiple scattering by the cap also creates difficulties in identifying the plaque due to the diffuse nature of the plaque border (Stamper et al., 2006). Polarization-sensitive OCT (PS-OCT), a functional mode of OCT, combines the advantages of OCT with additional image contrasts obtained by using the birefringence of the specimen as a contrast agent. Many biological tissues have a microscopic fibrous structure and so exhibit intrinsic birefringence. Moreover, changes in birefringence may indicate changes in functionality, structure, or viability of tissues in the early stages of the disease (de Boer et al., 1997).

From 2004, we have been presenting the application of PS-OCT in human atherosclerosis, and have proposed approaches to characterize a plaque lesion on the basis of its birefringence property (Kuo et al., 2004; 2005; 2007). Moreover, in a recent study, our laboratory has assessed the arterial characteristics in human atherosclerosis by quantitatively determining both scattering and birefringence properties of vessel tissue from PS-OCT images (Kuo et al., 2007; 2008). Based on our findings, a quantitative PS-OCT image criterion for plaque characterization was constructed. In the remainder of this chapter, the results that we obtained using the PS-OCT system for imaging human atherosclerosis *in vitro* are summarized. We hope that our results, along with the results from other investigators, will construe a step forward in the application of PS-OCT imaging technology for clinically diagnosing atherosclerosis in the near future.

2. Principle of polarization-sensitive optical coherence tomography (PS-OCT) system

The optical setup of the PS-OCT system used in this study is shown in Fig. 1. A collimated beam from a superluminescent diode (SLD) centered at a wavelength of 837 nm with a spectral bandwidth of 17.5 nm was used as a low-coherence light source in a Michelson interferometer. The axial resolution, which depends on the temporal coherence properties of the SLD), was 17 μm, while the lateral resolution (determined by the numerical aperture of the objective) was 10 μm. The incident beam was vertically polarized by a polarizer placed in the interferometer. A nonpolarization beam splitter (BS) was used to split the light wave into signal and reference beams. In the Michelson interferometer, a quarter-wave plate (QWP) with an azimuth angle set at 45° to the horizontal was used to focus the circular polarized light onto the examined specimen. On the other hand, the reference beam light was directed to a plane mirror mounted on a linear translator, which repetitively scanned the reference arm optical path length at a constant speed (1 mm/s). Another QWP (set at 22.5° to the horizontal) in the reference beam path rotated the polarization of the incident laser beam by 45°, thereby becoming the reflected reference beam.

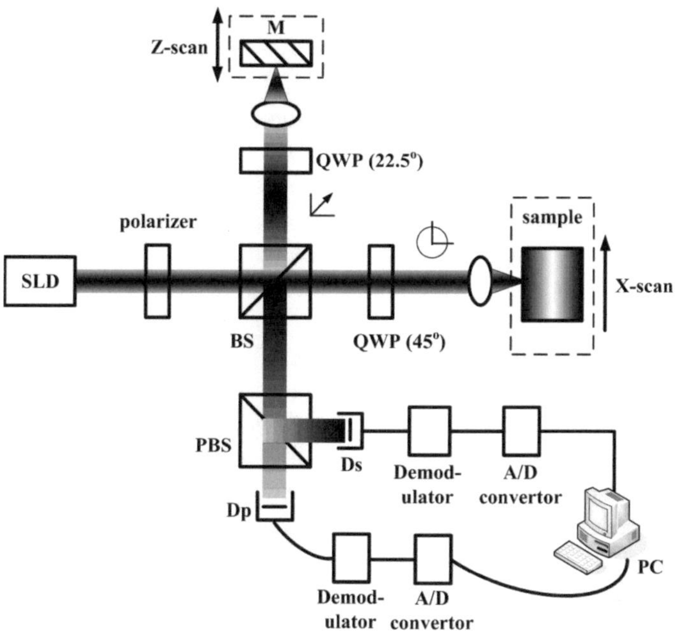

Fig. 1. Schematic of the conventional PS-OCT system: SLD, superluminescent diode; QWP, quarter wave plate; M, reference mirror; BS: beam slitter; PBS, polarized beam splitter; Dp and Ds, photo-detectors; PC, personal computer.

The laser beam was reflected from the specimen and recombined with the reflected reference beam, and then both the horizontal (*P* wave) and vertical components (*S* wave) were independently directed toward two photodetectors *Dp* and *Ds*, respectively, using a polarized BS (PBS). From the ac coupling of the detector signals, the full interferometric signals were recorded. The amplitudes A_i (*z*) and phases ϕ_i (*z*) of the interference signals at different depths (*z*) were determined using the Hilbert transform; *i* = *P* and *S* represent the P and S polarization states, respectively. Three parameters—the backscatter intensity *R*(*z*), phase retardation $\Phi(z)$, and fast-axis angle β(*z*) of a specimen—were calculated using the amplitude and phase of the interference signal (Hitzenberger et al., 2001):

$$R(z) \sim A_P(z)^2 + A_S(z)^2 \quad (1)$$

$$\Phi(z) = \tan^{-1}\left(A_S(z)/A_P(z)\right) \tag{2}$$

$$\beta(z) = 1/2 \times (180° - \Delta\phi) \tag{3}$$

Here, $\Delta\phi = \phi_P - \phi_S$ is the phase difference between the P- and S-polarized heterodyne signals. Finally, 2D images of the above three parameters were obtained simultaneously by using repeated A-scan acquisition and mechanically scanning the specimens laterally through a focused 0.5 mW signal beam. In this experiment, the system sensitivity was obtained as 100 dB using a highly reflective plane mirror as the test object in this setup. The following section demonstrates our preliminary *in vitro* investigations of human aortic specimens using PS-OCT.

In this study, we adapted a free-space PS-OCT system to precisely control the polarization state of the laser beam used in birefringent imaging. Several other groups have developed a high speed fiber-based PS-OCT system for application as a medical instrument *in vivo* (Guo et al., 2004; Park et al., 2001; 2004; Saxer et al., 2000). Moreover, an optically clear hemoglobin-based blood substitute has also been used to displace blood and enable OCT imaging with minimal patient discomfort (Villard et al., 2002). Further, several Fourier domain PS-OCT techniques (Park et al., 2005; Yamanari et al., 2006; Zhang et al., 2004) have been reported recently and have received considerable attention due to the high data acquisition rates (e.g., acquisition at 80 to 110 fps), which can eliminate motion artifacts and reduce ischemia during blood-free optical imaging. This allows for comprehensive scanning of long arterial segments during a short balloon occlusion or even 1 bolus liquid flush without occlusion. The first clinical study using this technology is being initiated in order to investigate vulnerable plaque hypothesis in a prospective multicenter manner. By combining the above features, PS-OCT can be used to measure reflected intensity, phase retardation, and fast-axis angle distributions, and thereby provide a greater contrast than is available with conventional OCT systems.

3. In vitro PS-OCT imaging of human atherosclerosis

Specimens of the aorta with white or yellow plaque were obtained from heart transplant recipients at the National Taiwan University Hospital, Taiwan. The photographs of some specimens are shown in Fig. 2. The protocol was approved by the ethics committees of the National Taiwan University Hospital. The specimens were dipped in saline (4°C), cut into segments smaller than 1 × 1 cm, and examined. Each segment was mounted in a cuvette and moistened with a normal saline bath maintained at 37°C during the imaging. Only the intimal surface was exposed for PS-OCT imaging. The aortic specimen regions imaged with PS-OCT were marked for subsequent histopathological examination. After PS-OCT imaging, all the specimens were fixed in 10% neutral formalin for 24 h and then processed for standard paraffin embedding. Serial sections with 4 μm thickness were cut within the region of the PS-OCT examination, and stained with hematoxylin and eosin (H and E) for routine examination. The distribution of the collagen structure in the plaque lesion was also examined using Masson trichrome and picrosirius red staining procedures as well as a polarization microscope. Finally, the entire specimens were classified into normal vessel (N), lipid (L), fibrocalcific (C), and fibrous lesions (F) by a pathologist (J. J. Shyu).

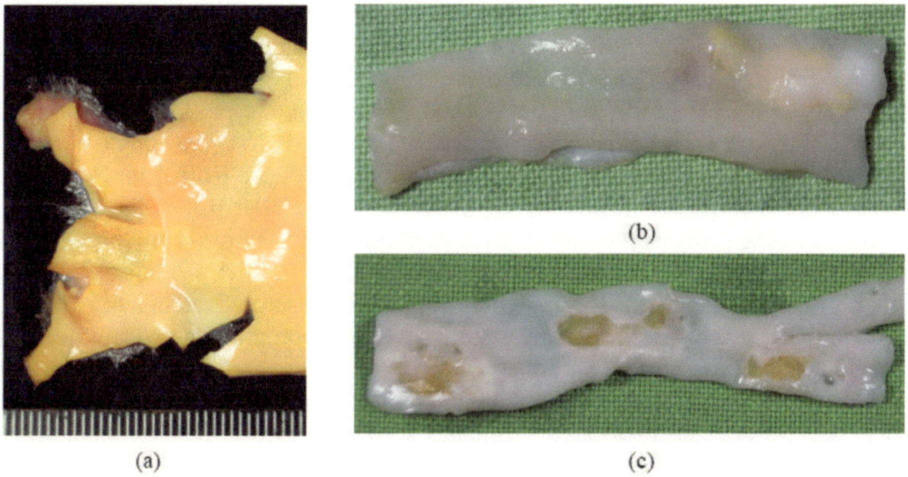

Fig. 2. Photographs of the aorta with white or yellow plaque.

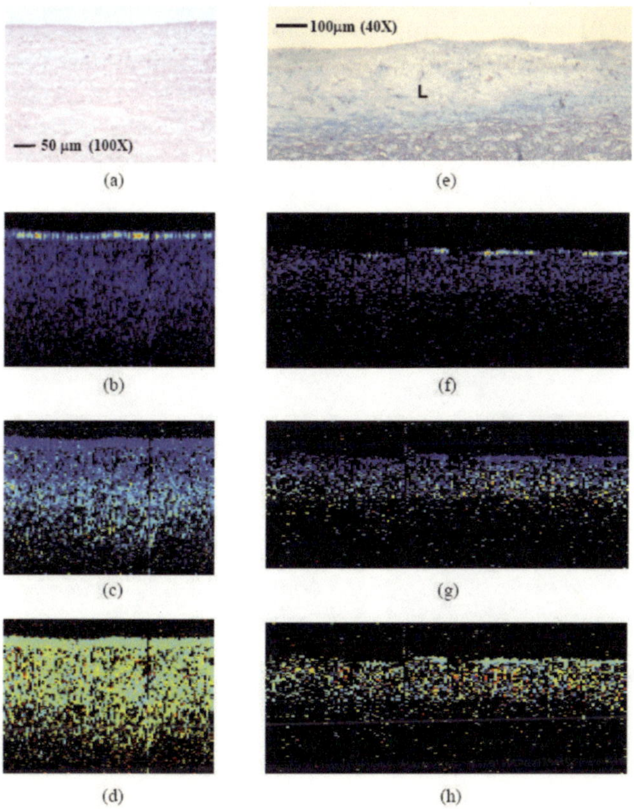

Fig. 3. Histological and PS-OCT images of a normal aortic wall (left column) and a plaque with lipid-loaded lesion (right column): (a) Histology (H and E; magnification ×100); (e) Histology (Masson's trichrome; magnification ×40); (b), (f) Back-scattered intensity image; (c), (g) Phase retardation image (linear color scale degrees); (d), (h) Fast-axis angle image (linear color scale degrees).

The PS-OCT images of representative specimens are shown in Figs. 3–6. The histological image of the normal vessel wall [Fig. 3(a)] showing a medial layer below the intima is compared with the PS-OCT image of the same specimen [Fig. 3(b)]. The signal-rich layer closest to the lumen is the intima. In the normal vessel wall, the phase retardation increases uniformly [Fig. 3(c)], and the pseudocolor distribution of the fast-axis angle signals is also uniform [Fig. 3(d)]. The pale area in Fig. 3(e) is a subintimal lipid-loaded region (L), which is morphologically composed mostly of the necrotic debris of foamy cells. Because of the paraffin embedding process, the solvent treatment removes the lipid from these lipid-loaded structures, which therefore appear as empty spaces in stained sections [Fig. 3(e)]. The corresponding PS-OCT image [Fig. 3(f)] reveals a decreased signal density under a thin homogeneous surface band. Moreover, the phase retardation and fast-axis angle signals are distributed in a slightly more random manner in the atherosclerotic lesion [Figs. 3(g) and 3(h), respectively] than in a normal vessel wall [Figs. 3(c) and 3(d)].

Moreover, the PS-OCT and histological images showed a plaque having small amounts of fibrous connective tissue (blue stain; black arrows) within a lipid-loaded area [Fig. 4(a)]. The signal density (arrows) was stronger, the backscattering signal was more heterogeneous [Fig. 4(b)], and the variation in the phase retardation [Fig. 4(c)] and fast-axis angle distribution [Fig. 4(d)] was more abrupt in the fibrous tissue than in the lipid-loaded region (L). Figure 4(e) shows a typically advanced plaque within the vascular intima; it is characterized by a necrotic lipid core covered by a thicker fibrous cap (CF ~250 μm; stained blue with Masson's trichrome). Plaque development in the vascular wall involves a reorganization of intimal collagen fibers (Rekhter, 1999). Figure 4(f) shows a relatively deep

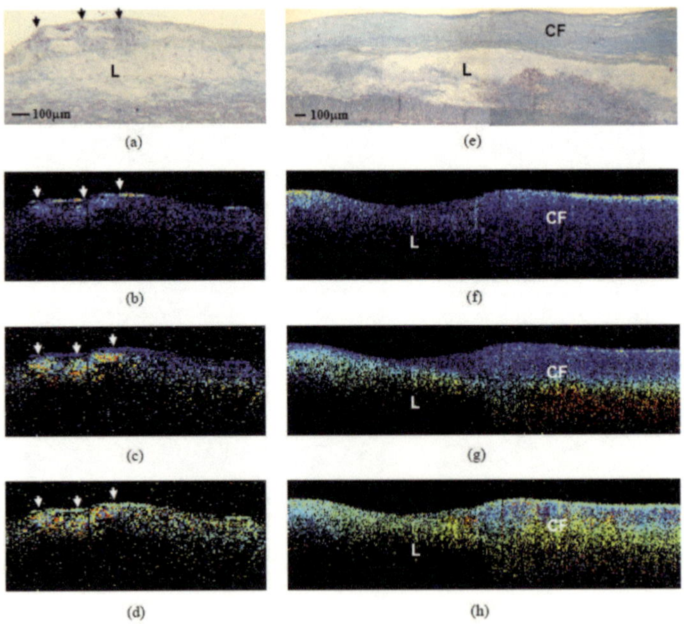

Fig. 4. Histological and PS-OCT images of vessel wall with a small fibrous lesion in the lipid-loaded area (left column) and a lipid-loaded fibroatheroma with a thick fibrous cap (right column): (a), (e) Histology (Masson's Trichrome; ×40); (b), (f) Back-scattered intensity image; (c), (g) Phase retardation image (linear color scale degrees); (d), (h) Fast-axis angle image (linear color scale degrees).

lipid-loaded (*L*) area close to the media. The medial layer had a low backscattering intensity, and hence, the interface between the plaque and the media was not well defined. A comparison of the PS-OCT [Figs. 4(g) and 4(h)] images with the histological images [Figs. 4(e)] showed gradual changes in phase retardation and fast-axis angle signals, which were due to the accumulation of collagen fiber in the plaque. Further, the changes in the pseudocolor in Fig. 4(g) were more uniform within the vessel wall than in those regions indicated by the arrows in Fig. 4(c).

Fig. 5 also shows an atheroma plaque (*) of a coronary artery stained with trichrome (a, 40×) and picrosirius red (b, 100X), which was examined under a polarization microscope (c, 100×). The structure above the mark (*) is the fibrous cap in the tunica intima, and the structure below the mark (*) is the tunica media. Picrosirius polarized microscopy reveals birefringence regions (e.g., organized collagen in a vessel wall). The intense birefringence of the collagen fiber represented in Fig. 5(e), left region, is confirmed by Figure 5(c) wherein the thick collagen fiber can be observed (in orange color). The fine collagen fiber (green color) of Fig. 5(c) is also consistent with small changes in the phase retardation shown in the right region of Fig. 5(e).

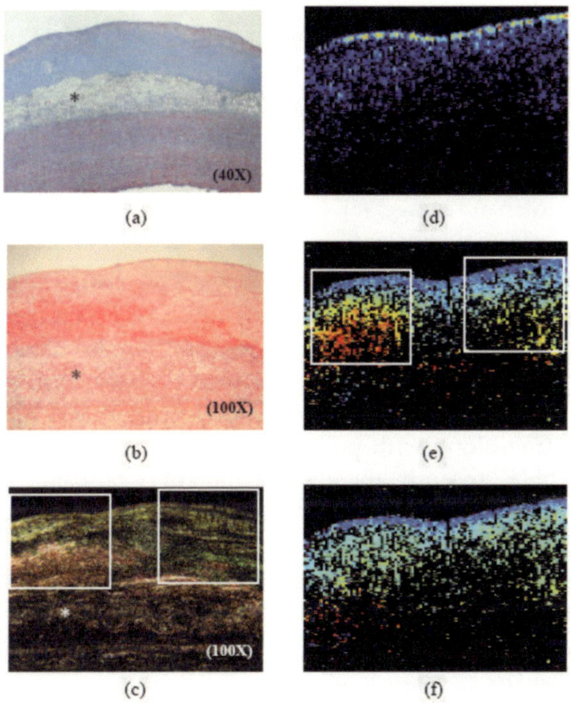

Fig. 5. Lipid-loaded fibroatheroma with a thick fibrous cap. (a) Histology (Masson's trichrome; ×40); (b) histology and (c) examined under polarization microscope (Picrosirius polarization; ×40); (d) back-scattered intensity image; (e) phase retardation image (linear color scale degrees); (f) fast-axis angle (linear color scale degrees).

Finally, two fibrocalcific plaques are shown in Fig. 6. The PS-OCT image showed a large sharply delineated, signal-rich area of heterogeneous backscattering [Fig. 6(b) and 6(f)], as well as strong birefringence [Fig. 6(c) and 6(g)]. Different structural orientations were also indicated by the PS-OCT image [i.e., different orientations of a fast-axis angle signal in three parts of the tomogram; see Fig. 6(h)] but not by the H and E stained specimen [Fig. 6(e)]. Since the calcified lesion was damaged during the sectioning process, only a large empty hole with a few calcified fragments appeared within the calcified plaque.

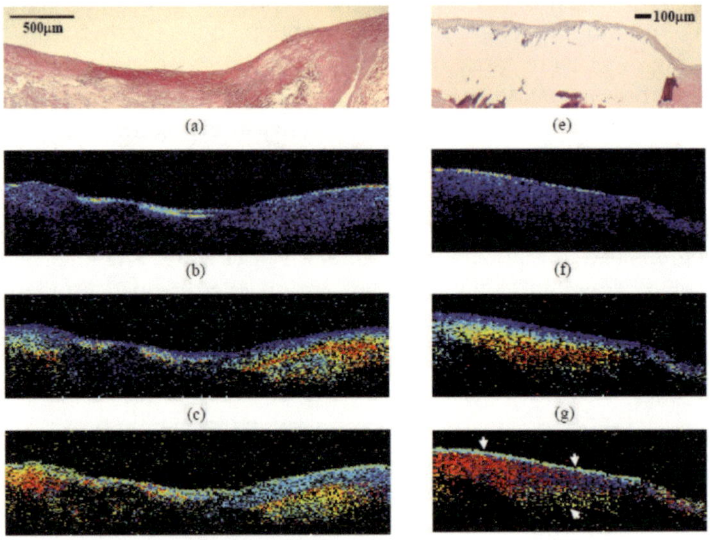

Fig. 6. Histological and PS-OCT images of fibrocalcific plaques: (a), (e) Histology (H and E); (b), (f) Back-scattered intensity image; (c), (g) Phase retardation image (linear color scale degrees); (d), (h) Fast-axis angle image (linear color scale degrees).

Using the above experiments, the capability of PS-OCT for imaging atherosclerotic plaques in human specimens has been evaluated. We have demonstrated that the normal vascular intima has a low intrinsic birefringence property, while changes in birefringence characteristics were apparent in fibrous and calcified plaques; moreover, the birefringence characteristics were different from those in normal vessels and lipid-loaded lesions. By using picrosirius staining along with polarization microscopy, we could also identify the thickness of collagen fiber. Recently, the identification of organized collagen fiber in arteries has also been demonstrated by using a single-detector PS-OCT (Giattina et al., 2006). In addition, another report showed that the PS-OCT measurements of birefringence have a strong positive correlation with thick collagen fiber content ($r = 0.76$, $p < 0.001$) and also a smooth muscle cells density ($r = 0.74$, $p < 0.01$) (Nadkarni et al., 2007).

4. Extracting optical properties from PS-OCT images

It is well known that optical properties can be used to indicate whether a tissue is in a normal or pathological state (Kortum & Muraca, 1996). Further, accurate knowledge of optical properties is essential for the optimum use of light in diagnosis and the treatment of diseases. In this study, we constructed a quantitative PS-OCT image criterion for plaque characterization. Following PS-OCT imaging, an algorithm was used to determine both scattering (i.e., μ_s and g_{eff}) and birefringence properties (i.e. Δn and β) of vessel tissue from the above PS-OCT images. The μ_s can be thought of as the reciprocal of the average distance a photon travels between scattering events. The g_{eff} factor describes how isotropic or anisotropic the scattering is, and is related to the particle size in the specimen. The Δn value characterizes the differential speed of propagation between two orthogonal polarized states of light in the specimen; it may change with derangement and mechanical failure of the collagen network in the vessel. And the β value could be thought of as a parameter of the fiber orientation in fibrous tissues where birefringence is caused by form birefringence.

First, the user selected regions (such as the white rectangle shown in the left column of Fig. 7) corresponding to those evaluated by histopathology. The regions were then automatically divided into several regions of interest (ROIs) (e.g., green dashed inset in the left column of Fig. 7) beginning from the intimal surface and including approximately 25 A-scans. Further, the size of each ROI was kept constant. The R, Φ, and β signals within each ROI were laterally (i.e., along the x-axis) delineated and averaged. Subsequently, μ_s and the root-mean-square scattering angle (θ_{rms}), which can be used to calculate the effective anisotropy factor ($g_{eff} = \cos(\theta_{rms})$), were extracted by fitting the reflectivity signals as a function of depth to an extended Huygens-Fresnel model (Kuo et al., 2008; Levitz et al., 2004; Thrane et al., 2000). This is shown in the right column of Fig. 7.

$$\langle i^2(z)\rangle = \left\{\frac{\alpha^2 P_R P_S \sigma_b}{\pi\omega_H^2}\right\}\cdot\left\{\exp(-2\mu_s z) + \frac{4\exp(-\mu_s z)[1-\exp(-\mu_s z)]}{1+\omega_S^2/\omega_H^2} + [1-\exp(-\mu_s z)]^2\frac{\omega_H^2}{\omega_S^2}\right\} \tag{4}$$

Here $\langle i^2(z)\rangle$ is the mean square of the heterodyne signal current; α, the power to current conversion ratio; P_R and P_S, the power of the reference and input sample beams; σ_b, the effective backscattering cross-section; and ω_H and ω_S, the $1/e$ irradiance radius at the probing depth in the absence and presence of scattering, respectively. The pixels near the interface, which was due to the specular reflection between the scattering and non-scattering media, were excluded from the fit (Levitz et al., 2004). Furthermore, the profiles of the averaged phase retardation signals have three layers (black arrows in the right column of Fig. 7). Δn can be calculated by linear least-squares fitting through the averaged Φ data over the depth of the ROI, and then its slope can be determined from the formula:

$$\Phi = (360/2\pi)\cdot k_0\cdot d\cdot\Delta n \tag{5}$$

Here k_0 is the wave vector and d is the thickness of the fitting range. In addition, the mean fast-axis angle calculated by averaging across the width of the ROI at each depth can be determined from Equation (3).

Statistical analyses were performed using SPSS (version 14.0; SPSS Inc.). A p-value < 0.05 was considered to be statistically significant. The test of significant difference of optical parameters was performed by Kruskal-Wallis statistics and used to evaluate whether the four optical properties contributed to the differentiation between different kinds of vessels. After performing a significant test, multiple comparison procedures were then used to determine which means are different. The following equation was used:

$$\left|R_i - R_j\right| > Z_{\alpha/k(k-1)} \times \sqrt{\frac{n_t(n_t+1)}{12}} \times \left(\frac{1}{n_i} + \frac{1}{n_j}\right) \tag{6}$$

Here R_i is the mean rank of the i th group; R_j, the mean rank of the j th group; k, the number of independent variables; n_t, the total number of samples; n_i and n_j, the sample numbers of the i th and j th group, respectively; $Z_{\alpha/k(k-1)}$, the critical value at the significance level α; and

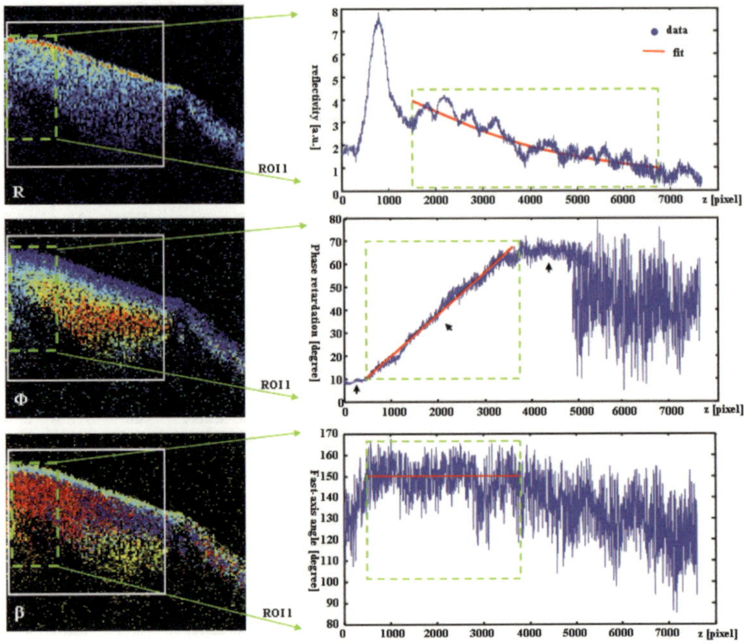

Fig. 7. Procedure of the PS-OCT extraction algorithm.

k(k-1), the number of comparisons. A Spearman's ρ correlation test was also used to evaluate whether these four properties have correlations with each other.

Finally, multinomial logistic regressions were used to generate a predictive model based on a linear combination of weights ($X_{r\rho}$) of optical properties (ρ= scattering coefficient, effective anisotropy factor, birefringence, and fast-axis angle) as shown in this equation:

$$\mathrm{logit}[\frac{OR(\text{diseased vessel type})}{OR(\text{Normal vessel})}] = A_r + \sum B_{r\rho} X_{r\rho} \tag{7}$$

Here OR = Odds Ratio, $r = {}^{C}\!/_{N}, {}^{F}\!/_{N}, {}^{L}\!/_{N}$, A_r is a constant, and $B_{r\rho}$ is an adjustable coefficient for each optical property. This model was used to classify the artery specimens into four diagnostic classes. The accuracy of this model for plaque characterization was evaluated using receiver operating characteristic (ROC) analysis (Metz, 1978).

Figures 3-6, given in previous pages, show illustrative PS-OCT images with the corresponding histopathology of normal, lipid, fibroatheroma, and fibrocalcific plaques. Altogether, 30 aortic specimens and therefore 135 ROIs from each region across totally R, Φ, and β images were collected. The extracted data, μ_s, g_{eff}, Δn, and β, are summarized in Fig. 8, where each box shows the median, 25th and 75th percentiles, and the extreme values within a category. Open circles and stars indicate outlier data.

Kruskal-Wallis statistics shows that μ_s ($p = 0.022$), Δn ($p < 0.001$), and β ($p < 0.001$) have significant differences in normal vessels and three types of atherosclerotic vessels, by measuring how much the ranks of the four groups differ from the mean rank of all groups. The g_{eff} value does not show any significant difference ($p = 0.104$). From the multiple comparison test, we found that F to C shows significant difference in μ_s; Δn between C and

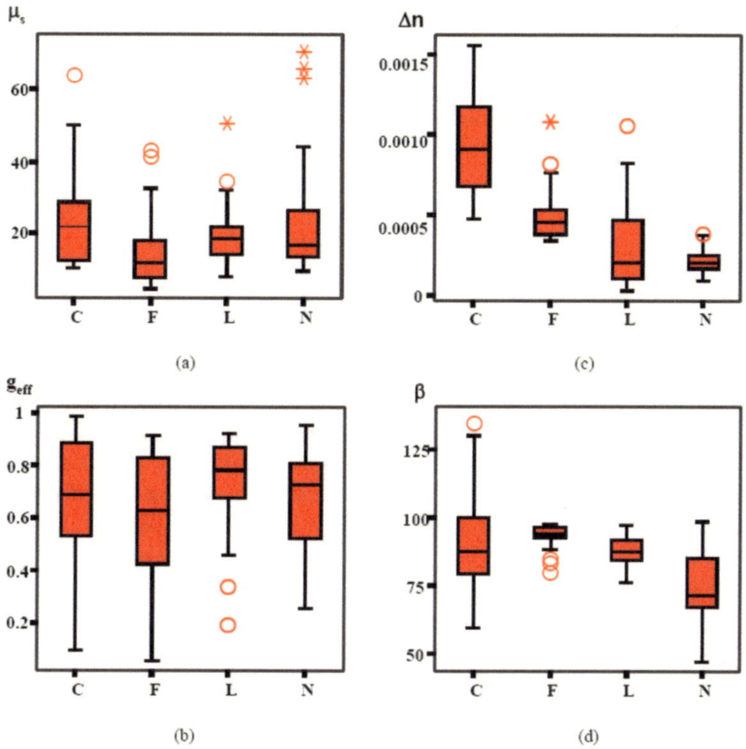

Fig. 8. Distributions of μ_s, g_{eff}, Δn, and β in normal vascular intima (N), lipid laden (L), fibrous (F), and fibrocalcific (C) plaques.

N, F and N, L and C, and L and F has significant differences; and β between C and N, F and N, L and N, and L and F has significant differences.

Spearman's ρ correlation test shows that only g_{eff} correlates with the scattering coefficient (r = -0.584, p = 0.003) in fibrocalcific plaque, while this value correlates with the birefringence value (r = -0.563, p = 0.008) in fibrous lesions. Finally, three regression models, Equations (8)-(10), were used to predict the odds ratio of C to N, F to N, and L to N, respectively.

$$\text{logit}\left[\frac{OR(type=C)}{OR(type=N)}\right] = -19.182 + 0.044X_{\mu_s} + 0.119X_{\Delta n} + 0.141X_{\beta} \tag{8}$$

$$\text{logit}\left[\frac{OR(type=F)}{OR(type=N)}\right] = -19.377 - 0.047X_{\mu_s} + 0.068X_{\Delta n} + 0.201X_{\beta} \tag{9}$$

$$\text{logit}\left[\frac{OR(type=L)}{OR(type=N)}\right] = -9.746 - 0.008X_{\mu_s} + 0.028X_{\Delta n} + 0.113X_{\beta} \tag{10}$$

The prediction results are given in Table 1. This method identified that 17 of 23 lesions are fibrocalcific and that 105 of 112 lesions are not fibrocalcific. In the case of fibrous plaque, 7 of 21 lesions were identified as fibrous and 110 of 114 as not fibrous lesions. Finally, the method identified 33 of 48 lesions as lipid regions and 55 of 87 as not lipid regions. The constructed regression model achieved 90%, 87%, and 65% prediction accuracy for C, F, and L, respectively.

Classification				
Histology observed	Model predicted			
	N	C	F	L
N	27	0	0	16
C	1	17	2	3
F	0	1	7	13
L	7	6	2	33

Table 1. Plaque characterization by quantitative PS-OCT and histology

Our preliminary data indicated that more than 80% normal arterial samples had μs value between 10 and 39 mm^{-1} and have significant differences from other different types of plaques ($p < 0.05$); this is consistent with the results obtained by Levitz (Levitz et al., 2004). From the multiple comparison tests, we also noticed that a significant difference in scattering property exists between fibrous and fibrocalcific plaques. These findings are consistent with the results obtained with qualitative image-based plaque characterization methods where fibrous and fibrocalcific plaques can be distinguished by the signal-rich and signal-poor regions respectively (Stamper et al., 2006; Yabushita et al., 2002). However, the effective anisotropy factor demonstrates no significant difference between normal and other atherosclerotic lesions ($p = 0.104$), perhaps because g_{eff} of the fibrocalcific and fibrous lesions were correlated with μ_s and Δn, respectively. In the case of the birefringence property of the vessel that has not been quantitatively analyzed previously, i.e., β values, they were maximum in the most atherosclerotic lesion at over 70 degrees. Smaller β values were present in the best-fit areas of normal vascular intima. The Δn values were small and more concentrated in normal intima, but they demonstrated larger variations in the entire atherosclerotic lesion. The birefringence coefficient was larger in abundant thicker collagen fibers ($\Delta n = 9.409 \times 10^{-4}$; bright yellow to orange color, constituting >60% of the left region of histology in Fig. 5c) than in thin collagen fibers ($\Delta n = 5.386 \times 10^{-4}$; green color in right region of histology in Fig. 5c). Both β and Δn values have significant differences between the normal arterial vessel and other different types of plaques ($p < 0.05$).

In this study, no attempt has been made to differentiate a necrotic core from a lipid pool. Since the signal from the necrotic cores may be too weak for reliable measurements, future studies based on histological stains that can differentiate the two are needed. It is also noteworthy that the Φ and β signals are distributed in a slightly more random manner in the lipid lesion than in the normal vessel wall and fibrous and fibrocalcific plaques. This may be due to the polarization state of light that is to be randomized by multiple scattering in lipid-rich tissue, which reduces the accuracy of birefringence measurements. Alternatively, further modifications of these PS-OCT criteria, such as the addition of a threshold limit for the signal-poor region and incorporation of the standard deviation of the birefringence signal within one ROI, may be required to differentiate lipid lesions better.

5. Conclusion

Collagen fiber constitutes up to 60% of the total atherosclerotic plaque protein. Uncontrolled collagen accumulation leads to vascular stenosis, whereas excessive collagen breakdown weakens plaques making them prone to rupture (Falk et al., 1995; Rekhter, 1999). Assessing the phase retardation change may be a method to quantify the collagen content in atherosclerotic lesions, and it may provide significant pathophysiological information that can influence clinical decision-making in patients with risk factors. Furthermore, computer-based quantitative analysis can automatically determine the plaque type; this will eliminate the training time for each reader and disparity between different diagnoses. The quantitative information on both arterial scattering and birefringence properties can also be integrated with the qualitative visual information provided by PS-OCT images, and this can support the facilitation of image-based plaque characterization methods. Our preliminary results present an important step in validating this new imaging modality and can provide a basis for the interpretation of PS-OCT images obtained from human specimens. However, an analysis from a considerably larger set of specimens as well as an analysis taking the effect of cluster data (i.e., specimens from the same person) into consideration will be required for developing a more suitable prediction model in the future. Moreover, it is likely that the combination of other functional modalities such as optical coherence elastography (Rogowska et al., 2004; 2006) or spectroscopic OCT (Morgner et al., 2000), which can provide additional indexes (such as cellular and molecular components and mechanical properties of arterial walls), will have a greater predictive value for constructing a risk-stratifying plaque characterization criterion that can be applied in future clinical utilities.

6. Acknowledgments

The authors thank Dr. N. K. Chou of the Department of Surgery of National Taiwan University Hospital for providing aortic tissues and Prof. J. J. Shyu of the Department of Veterinary Medicine of National Taiwan University for histology examinations. This research was supported by the National Science Council of Taiwan.

7. References

Arroyo, L. H. & Lee, R. T. (1999). Mechanisms of plaque rupture: mechanical and biologic interactions. *Cardiovascular Research*, Vol. 41, No. 2, (369-375), ISSN 0008-6363

Bouma, B. E.; Tearney, G. J.; Yabushita, H.; Shishkov, M.; Kauffman, C. R.; DeJoseph Gauthier, D.; MacNeill, B. D.; Houser, S. L.; Aretz, H. T.; Halpern, E. F. & Jang, I. K. (2003). Evaluation of intracoronary stenting by intravascular optical coherence tomography. *Heart*, Vol. 89, (317-321), ISSN 1355-6037

Brezinski, M. E.; Tearney, G. J.; Bouma, B. E.; Izatt, J. A.; Hee, M. R.; Swanson, E. A.; Southern, J. F. & Fujimoto, J. G. (1996). Optical coherence tomography for optical biopsy: properties and demonstration of vascular pathology. *Circulation*, Vol. 93, (1206–1213), ISSN 0009-7322

Brezinski, M. E.; Tearney, G. J.; Weissman, N. J.; Boppart, S. A.; Bouma, B. E.; Hee, M. R.; Weyman, A. E.; Swanson, E. A.; Southern, J. F. & Fujimoto, J. G. (1997). Assessing atherosclerotic plaque morphology: comparison of optical coherence tomography and high frequency intravascular ultrasound. *Heart*, Vol. 77, (397-403), ISSN 1355-6037

de Boer, J. F.; Milner, T. E.; van Gemert, M. J. C. & Nelson, J. S. (1997). Two-dimensional birefringence imaging in biological tissue by polarization-sensitive optical coherence tomography. *Opt. Lett.*, Vol. 22, (934-936), ISSN 0146-9592

Falk, E.; Shah, P. K. & Fuster, V. (1995) Coronary plaque disruption. *Circulation*, Vol. 110, (657–671), ISSN 0009-7322

Fujimoto, J. G.; Bresinski, M. E.; Tearney, G. J.; Boppart, S. A.; Bouma, B. E.; Hee, M. R.; Southern, J. F. & Swanson, E. A. (1995). Optical biopsy and imaging using optical coherence tomography. *Nature Medicine*, Vol. 1, (970-972), ISSN 1078-8956

Giattina, S. D.; Courtney, B. K.; Herz, P. R.; Harman, M.; Shortkroff, S.; Stamper, D. L.; Liu, B.; Fujimoto, J. G.; Brezinski, M. E. (2006). Assessment of coronary plaque collagen with polarization sensitive optical coherence tomography (PS-OCT). *Int J Cardiol*, Vol. 107, (400-409), ISSN 0167-5273

Guo, S.; Zhang, J.; Wang, L.; Nelson, J. S. & Chen, Z. (2004). Depth-resolved birefringence and differential optical axis orientation measurements with fiber-based polarization-sensitive optical coherence tomography. *Opt. Lett.*, Vol. 17, (2025-2027), ISSN 0146-9592

Hitzenberger, C. K.; Gotzinger, E.; Sticker, M.; Pircher, M. & Fercher, A. F. (2001). Measurement and imaging of birefringence and optic axis orientation by phase resolved polarization sensitive optical coherence tomography. *Opt. Express*, Vol. 9, (780-790), ISSN 1094-4087

Huang, D.; Swanson, E. A.; Lin, C. P.; Schuman, J. S.; Stinson, W. G.; Chang, W.; Hee, M. R.; Flotte, T.; Gregory, K.; Pufialito, C. A. & Fujimoto, J. G. (1991). Optical coherence tomography. *Science*, Vol. 254, (1178–1181), ISSN 0036-8075

Jang, I. K.; Bouma, B. E.; Kang, D. H.; Park, S. J.; Park, S. W.; Seung, K. B.; Choi, K. B.; Shishkov, M.; Schlendorf, K.; Pomerantsev, E.; Houser, S. L.; Aretz, H. T. & Tearney, G. J. (2002). Visualization of coronary atherosclerotic plaques in patients using optical coherence tomography: comparison with intravascular ultrasound. *J. Am. Coll. Cardiol.*, Vol. 39, (604–609), ISSN 0735-1097

Kortum R. R. & Muraca, E. S. (1996). Quantitative optical spectroscopy for tissue diagnosis. *Ann. Rev. Phys. Chem.*, Vol. 47, (555-606), ISSN 0066-426X

Kuo, W. C.; Shyu, J. J.; Chou, N. K.; Lai, C. M.; Huang, H. C.; Chou, C.; & Jan, G. J. (2004). Imaging of Human Aortic Atherosclerotic plaques by Polarization-Sensitive Optical Coherence Tomography, *Proceedings of IEEE Conference on Engineering in Medicine and Biology*, pp. 1111-1114, ISBN 0879425598, San Francisco, Aug 2004, EMBS, CA

Kuo, W. C.; Shyu, J. J.; Chou, N. K.; Lai, C. M.; Tien, E. K.; Huang, H. J.; Chou, C. & Jan, G. J. (2005). Correlation of collagen synthesis with polarization-sensitive optical coherence tomography imaging of in vitro human atherosclerosis, *Proceedings of SPIE* 5690, pp. 563-571, ISBN 9780819456649, San Jose, January 2005, SPIE, CA.

Kuo, W. C.; Chou, N. K.; Chou, C.; Lai, C. M.; Huang, H. J. & Shyu, J. J. (2007). Polarization-sensitive optical coherence tomography for imaging human atherosclerosis. *Appl. Opt.*, Vol. 46, (2520-2527), ISSN 0003-6935

Kuo, W. C.; Hsiung, M. W. & Yang, P. N. (2007). Extracting quantitative optical properties of human vessel from PS-OCT images. *Journal of Medical and Biological Engineering*, Vol. 27, (191-197), ISSN 1609-0985

Kuo, W. C.; Hsiung, M. W.; Shyu, J. J.; Chou, N. K. & Yang, P. N. (2008). Quantitative analysis on optical properties of human atherosclerosis by using polarization-sensitive optical coherence tomography, *Proceedings of SPIE* 6842, pp. 684223-1~684223-9, ISBN 9780819470171, San Jose, January 2008, SPIE, CA.

Kuo, W. C.; Hsiung, M. W.; Shyu, J. J.; Chou, N. K. & Yang, P. N. (2008). Assessment of arterial characteristics in human atherosclerosis by extracting optical properties from polarization-sensitive optical coherence tomography. *Opt. Express*, Vol. 16, (8117-8125), ISSN 1094-4087

Levitz, D.; Thrane, L.; Frosz, M. H.; Andersen, P. E.; Andersen, C. B.; Valanciunaite, J.; Swartling, J.; Andersson-Engels S. & Hansen, P. R. (2004). Determination of optical properties of highly-scattering media in optical coherence tomography. *Opt. Express*, Vol. 12, (249-259), ISSN 1094-4087

Metz, C. E. (1978). Basic principles of ROC analysis. *Seminars in Nuclear Medicine*, Vol. 8, (283-298), ISSN 0001-2998

Morgner, U.; Drexler, W.; Kartner, F. X.; Li, X. D.; Pitris, C.; Ippen, E. P. & Fujimoto, J. G. (2000). Spectroscopic optical coherence tomography. *Opt. Lett.*, Vol. 25, No. 2, (111-113), ISSN 0146-9592

Nadkarni, S. K.; Pierce, M. C.; Park, B. H.; de Boer, J. F.; Whittaker, P.; Bouma, B. E.; Bressner, J. E.; Halpern, E.; Houser, S. L.; & Teaeney, G. J. (2007). Measurement of collagen and smooth muscle cell content in atherosclerotic plaques using polarization sensitive optical coherence tomography. *J. Am. Coll. Cardiol.*, Vol. 49, (1474-1481), ISSN 0735-1097

Park, B. H.; Pierce, M. C.; Cense, B.; Yun, S. H.; Mujat, M.; Tearney, G. J.; Bouma, B. E. & de Boer, J. F. (2005). Realtime fiber-based multi-functional spectral-domain optical coherence tomography at 1.3 mm. *Opt. Express*, Vol. 13, (3931–3944), ISSN 1094-4087

Park, B. H.; Pierce, M. C.; Cense, B. & de Boer, J. F. (2004). Jones matrix analysis for a polarization-sensitive optical coherence tomography system using fiber-optic components. *Opt. Lett.*, Vol. 21, (2512-2514), ISSN 0146-9592

Park, B. H.; Saxer, C.; Srinivas, S. M.; Nelson, J. S. & de Boer, J. F. (2001). In vivo burn depth determination by high-speed fiber-based polarization sensitive optical coherence tomography. *J. Biomed. Opt.*, Vol. 6, No. 4, (474-479), ISSN 1083-3668

Pasterkamp, G.; Falk, E.; Woutman, J. & Borst, C. (2000). Techniques characterizing the coronary atherosclerotic plaque: Influence on clinical decision making. *J. Am. Coll. Cardiol.*, Vol. 36, (13–21), ISSN 0735-1097

Peters, R. J. G.; Kok, W. E. M.; Havenith, M. G.; Rijsterborgh, H. & vanderWal, A. C. (1994). Histopathologic validation of intracoronary ultrasound imaging. *J. Am. Soc. Echocardiography*, Vol. 7, (230-241), ISSN 0894-7317

Rekhter, M. D. (1999). Collagen synthesis in atherosclerosis: too much and not enough. *Cardiovascular Research*, Vol. 41, (376-384), ISSN 0008-6363

Rogowska, J.; Patel, N. A.; Fujimoto, J. G. & Brezinski, M. E. (2004). Optical coherence tomographic elastography technique for measuring deformation and strain of atherosclerotic tissues. *Heart*, Vol. 90, No. 5, (556–562), ISSN 1355-6037

Rogowska, J.; Patel, N. A.; Plummer, S. & Brezinski, M. E. (2006). Quantitative optical coherence tomographic elastography: method for assessing arterial mechanical properties. *Br. J. Radiol.*, Vol. 79, No. 945, (707–711), ISSN 1748-880X

Saxer, C. E.; de Boer, J. F.; Park, B. H.; Shao, Y.; Chen, Z. & Nelson, J. S. (2000). High-speed fiber-based polarization-sensitive optical coherence tomography of in vivo human skin. *Opt. Lett.*, Vol. 18, (1355-1357), ISSN 0146-9592

Shah, P. K. (2003). Mechanisms of plaque vulnerability and rupture. *J. Am. Coll. Cardiol.*, Vol. 41, No. 4, (15S–22S), ISSN 0735-1097

Stamper, D.; Weissman, N. J.; Brezinski, M. (2006). Plaque characterization with optical coherence tomography. *J. Am. Coll. Cardiol.*, Vol. 47, (C69–C79), ISSN 0735-1097

Tearney, G. J.; Jang, I-K & Bouma, B. E. (2006). Optical coherence tomography for imaging the vulnerable plaque. *J. Biomed. Opt.*, Vol. 11, No. 2, (021002), ISSN 1083-3668

Tearney, G. J.; Yabushita, H.; Houser, S. L.; Aretz, H. T.; Janf, I. K.; Schlendorf, K. H.; Kauffman, C. R.; Shishkov, M.; Halpern, E. F. & Bouma, B. E. (2003). Quantification of macrophage content in atherosclerotic plaques by optical coherence tomography. *Circulation*, Vol. 107, (113–119), ISSN 0009-7322

Thrane, L.; Yura, H. T. & Andersen, P. E. (2000). Analysis of optical coherence tomography systems based on the extended Huygens-Fresnel principle. *J. Opt. Soc. Am. A*, Vol. 17, (484-490), ISSN 1464-4258

Villard, J. W.; Feldman, M. D.; Kim, J.; Milner, T. E. & Freeman, G. L. (2002). Use of a Blood Substitute to determine instantaneous murine right ventricular thickening with optical coherence tomography. *Circulation*, Vol. 105, (1843-1849), ISSN 0009-7322

Yabushita, H.; Bouma, B. E.; Houser, S. L.; Aretz, H. T.; Jang, I. K.; Schlendorf, K. H.; Kauffman, C. R.; Shishkov, M.; Kang, D. H.; Halpern, E. F. & Tearney, G. J. (2002). Characterization of human atherosclerosis by optical coherence tomography. *Circulation*, Vol. 106, (1640–1645), ISSN 0009-7322

Yamanari, M.; Makita, S.; Madjarova, V. D.; Yatagai, T. & Yasuno, Y. (2006). Fiber-based polarization-sensitive Fourier domain optical coherence tomography using B-scan-oriented polarization modulation method. *Opt. Express*, Vol. 14, (6502–6515), ISSN 1094-4087

Zhang, J.; Jung, W.; Nelson, J. S. & Chen, Z. (2004). Full range polarization-sensitive Fourier domain optical coherence tomography. *Opt. Express*, Vol. 12, (6033–6039), ISSN 1094-4087

32

Two-photon Fluorescence Endomicroscopy

Yicong Wu and Xingde Li
Department of Biomedical Engineering, Johns Hopkins University
Baltimore, Maryland 21205
U.S.A.

1. Introduction

Two-photon fluorescence (TPF) microscopy is a powerful technique for high-resolution imaging of biological tissues, enabling depth-resolved morphological and functional assessment of biological tissues via a non-invasive route (Denk et al., 1990; Helmchen & Denk, 2005; Konig, 2000; Zipfel et al., 2003). In TPF microscopy, a molecule (i.e., the fluorophore) can absorb two photons quasi-simultaneously (10^{-15} - 10^{-18} s) and emit a single photon during relaxation from the excited state to the ground state. The probability for the fluorescent emission is thus quadratically dependent on the excitation light intensity. With a focusing unit, much more two-photon fluorescence is generated from the focal spot than where the beam is diffused. Effectively, excitation is restricted to the very small focal volume (~1 femtoliter), resulting in the inherent optical sectioning ability without the need for a pinhole to reject out-of-focus photons. This optical sectioning capability permits whole-field fluorescence collection and thus enhances the collection efficiency in highly scattering tissues. In TPF microscopy, near-infrared (NIR) femtosecond laser is generally employed for effective excitation, which increases the penetration depth and reduces image deterioration due to the less scattering in turbid tissues. In addition, NIR excitation likely causes less photodamage outside the focal volume.

With the advances in micro-optics and micro-mechanical components, a TPF endomicroscopy system is becoming attractive as a basic research tool with a much smaller form factor and lower cost compared to a conventional TPF microscope. Moreover, the TPF endomicroscopy system has a great potential to transform the powerful TPF technology for *in vivo* studies and clinical applications. Recently, increasing interests have been focusing on the development of TPF endomicroscope with a small size which can go through the accessory port of a standard endoscope for *in vivo* and clinical studies while maintaining the TPF imaging ability similar to a standard TPF microscope. Major challenges for TPF endomicroscopy devices are efficient delivery of single-mode ultrashort pulses, wide-field collection of the TPF signals, fast 2-D/3-D beam scanning with a miniature objective lens of good optical properties, and overall miniaturization of the probe assembly (Bao et al., 2008; Engelbrecht et al., 2008; Flusberg et al., 2005a; Flusberg et al., 2005b; Fu et al., 2006; Gobel et al., 2004a; Helmchen et al., 2001; Hoy et al., 2008; Jung & Schnitzer, 2003; Jung et al., 2008; Konig et al., 2007; Le Harzic et al., 2008; Levene et al., 2004; Myaing et al., 2006; Wu et al., 2009a; Wu et al., 2009b).

This book chapter offers a review of fiber-optic TPF endomicroscopy technologies with emphasis on major technological development challenges. The advantages and limitations associated with various TPF endomicroscopy systems are discussed. Special design and engineering considerations are presented with our recently developed all-fiber-optic rapid scanning TPF imaging endomicroscopy system as an example. Some representative endomicroscopic TPF imaging results are illustrated, demonstrating that the emerging TPF endomicroscopy systems are very promising for basic laboratory research and for early disease detection and image-guided interventions.

2. Challenges in two-photon fluorescence endomicroscopy

2.1 Single-mode femtosecond laser delivery and large-area TPF signal collection

The first major issue in TPF endomicroscopic implementation is how to efficiently deliver single-mode femtosecond excitation light and collect multimode two-photon fluorescence signals. It is well known that a single-mode fiber (SMF) can be used to deliver and focus single-mode femtosecond excitation light to a near diffraction limited spot. However, the TPF collection efficiency severely suffers due to the small core diameter of a SMF. Some embodiments utilize a separate multimode fiber for effective TPF collection (Helmchen et al., 2001), as shown in Fig. 1(a). The multimode fiber with large core diameter (e.g. 1-2 mm) and high NA (e.g. 0.4-0.8) increases the collection area and it also makes the collection efficiency less sensitive to the spherical and chromatic aberration of the objective lens. Such configuration can be further improved by replacing the common SMF with a hollow-core photonic bandgap fiber (HC-PCF) with zero dispersion at the selected excitation wavelength (Engelbrecht et al., 2008; Flusberg et al., 2005b; Gobel et al., 2004b; Hoy et al., 2008; Le Harzic et al., 2008). Owing to the dramatically reduced group-velocity dispersion (GVD) and nonlinear optical effects (such as self-phase modulation, SPM) in the specially designed HC-PCF, femtosecond pulses in HC-PCF experience negligible temporal distortion, and no additional pulse prechirping is required (Agrawal, 2007).

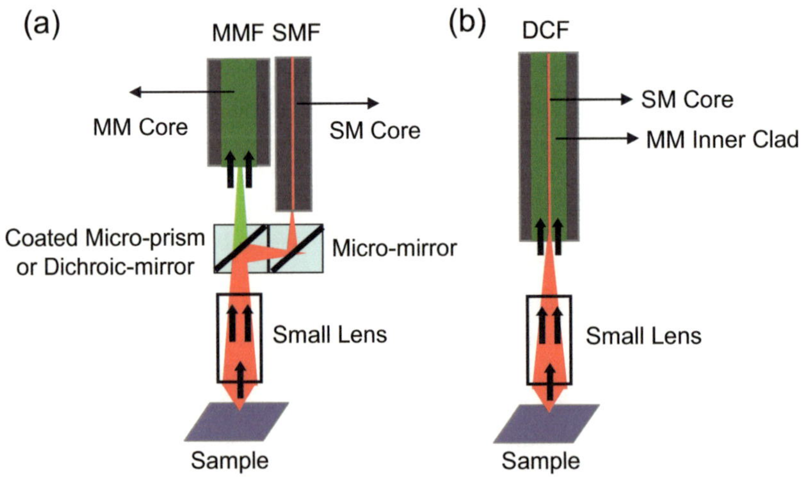

Fig. 1. Schematic of fiber-optic two-photon fluorescence endomicroscope: (a) Two-fiber configuration; (b) Single-fiber configuration. SMF: single-mode fiber; MMF: multi-mode fiber; DCF: double-clad fiber; SM: single-mode; MM: multi-mode.

As shown in Fig. 1(a), the two-fiber configuration involves a dichroic mirror and a prism and it is difficult to minimize the endomicroscope. In order to create a more compact and flexible endomicroscope, single double-clad fibers (DCFs, see Fig. 2) have been employed in TPF endomicroscopes for both single-mode laser excitation delivery with the single-mode core and efficient TPF collection with the multimode inner cladding layer (Bao et al., 2008; Fu et al., 2007; Fu et al., 2006; Jung et al., 2008; Myaing et al., 2006; Wu et al., 2009a; Wu et al., 2009b), as shown in Fig. 1(b). With the advance of fiber fabrication technology, double-clad fiber could be developed with high performance including large inner clad diameter and NA and less nonlinear optical effects.

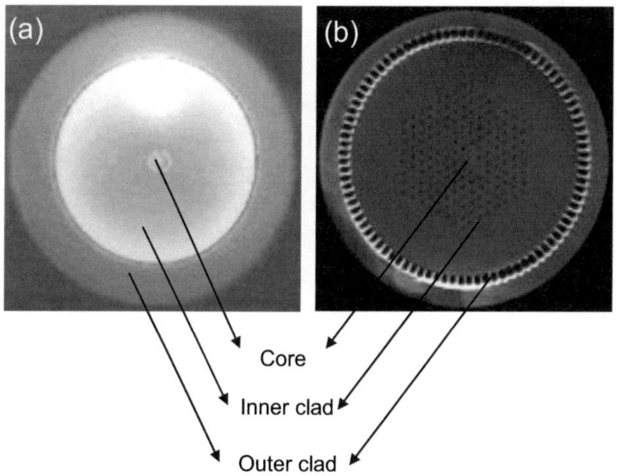

Fig. 2. Schematic of double-clad fibers generally employed in two-photon fluorescence endomicroscopy systems: (a) Conventional double-clad fiber; (b) Photonic crystal double-clad fiber.

A conventional DCF, as shown in Fig. 2(a), is a step-index fiber composed of a single-mode core, a multi-mode inner cladding layer and an outer cladding layer. The materials for the three layers are typically germanium-doped silica, pure silica and fluorine-doped silica, respectively. The DCF, allowing single-mode delivery of fs excitation light through the single-mode core and collection of multimode TPF signals via the inner clad, is commercially available (Fibercore Ltd., SMM900) and has been successfully implemented in a scanning fiber-optic TPF endomicroscope with an excellent imaging ability (Bao et al., 2008; Myaing et al., 2006; Wu et al., 2009a; Wu et al., 2009b). Compared to a single-mode fiber, the DCF (with core/inner clad diameter of 3.5/103 μm and NA of 0.19/0.24) greatly improves the collection efficiency of TPF signals by 2-3 orders. Another type of DCF is photonic crystal double-clad fiber (PC-DCF) as shown in Fig. 2(b). The PC-DCF comprises a single-mode core with pure silica and inner and outer cladding layers with hybrid air-silica structures (Bjarklev et al., 2003; Knight, 2003). PC-DCF is also commercially available (Crystal Fiber, DC-165-16-Passive) and has been used for developing TPF endomicroscopy technologies (Fu et al., 2005; Fu et al., 2007; Fu et al., 2006; Jung et al., 2008). The PC-DCF has a core/inner clad diameter of 16/165 μm and NA of 0.04/0.6. The large core of the PC-DCF reduces the nonlinear optical effects up to a certain excitation power (Bao & Gu, 2009). But the large core diameter and the related low NA make it challenging to focus the excitation beam to a small spot size with a given miniature objective lens. The use of a PC-DCF would also increase the rigid length of an endomicroscope at its distal end due to the requirement of beam expansion and refocusing mechanisms. Generally speaking, in engineering a compact fiber-optic TPF endomicroscope, the core size of the DCF has to be carefully chosen with a tradeoff among the excitation/collection efficiency, the nonlinear effects, the overall diameter and the rigid length of the probe.

Since the SMF, DCF and PC-DCF have normal dispersion, ultrashort pulses transmitting in these fibers will be temporally broadened due to GVD and nonlinear effects such as SPM (Agrawal, 2007), resulting in the reduction of TPF excitation efficiency. Therefore, pre-chirping is required for fiber-optic TPF endomicroscopes with such fibers. A conventional pulse stretcher based on a grating and lens pair can be utilized for negative prechirping before the pulses are launched into the fibers (Bao et al., 2008; Helmchen et al., 2001; Myaing et al., 2006; Treacy, 1969). However, the grating/lens pulse stretcher consists of bulky optics with a double-pass configuration which is generally sensitive to alignment and has suboptimal throughput. Recently, photonic crystal fibers based on photonic bandgap effects to guide light propagation have been developed. These fibers exhibit anomalous dispersion over certain wavelength range and can be used for prechirping (Bjarklev et al., 2003; Reeves et al., 2003). For example, the hollow-core photonic bandgap fiber (PBF) from Crystal Fibre (HC-800-02) offers negative GVD with the wavelength longer than 800 nm. It has been employed for dispersion compensation in the endomicroscopes (Wu et al., 2009a; Wu et al., 2009b). Table 1 summarizes the measured GVD parameter (β_2) and dispersion parameter (D) for excitation pulses at 810±18 nm with an initial pulse width of 60 fs. The reference values of a conventional silica core single-mode fiber (SMF) at 810 nm are listed (Agrawal, 2007). As can be seen, the measured GVD of the DCF is ~43,065 fs^2/m, whereas the PBF offers a negative GVD of ~35,246 fs^2/m. As a result, the positive dispersion of a DCF can be compensated by a PBF when the length ratio of the PBF to DCF is ~1.1 at 810±18 nm. The achievable pulse width is about 130 fs with 20 mW delivered through the DCF core. As the power laser transmitting in the DCF core increases (e.g. up to 50 mW), the pulses suffer self-phase modulation and other nonlinear effects, and the temporal pulse duration broadens to about 200 fs.

	β_2 (fs^2/m)	D (ps/nm/km)
DCF (810 ± 18 nm)	43,065	-123.7
PBF (810 ± 18 nm)	-35,246	101.2
SMF (810 nm) *	34,120	-98.0

Table 1. Measured GVD parameter (β_2) and dispersion parameter (D) of a conventional double-clad fiber (DCF) and hollow-core photonic bandgap fiber (PBF). For reference, the β_2 and D values of a single-mode fiber (SMF) are cited from Ref. (Agrawal, 2007).

2.2 Miniature high-speed scanning head

The second challenge in developing a fiber-optic TPF endomicroscope is the beam scanner at the distal end which has to be in a small footprint. Current endomicroscope embodiments are mainly based on micro-electro-mechanical system (MEMS) scanning mirrors (Bao et al., 2008; Fu et al., 2006; Hoy et al., 2008; Jung et al., 2008; Piyawattanametha et al., 2006) (Fig. 3(a)) and piezoelectric resonant fiber-optic scanners (Engelbrecht et al., 2008; Flusberg et al., 2005b; Helmchen et al., 2001; Myaing et al., 2006; Wu et al., 2009a; Wu et al., 2009b) (Fig.

3(b)). 1-D or 2-D scanning mirrors can be micro-fabricated on a single silicon plate with torsional hinges, supporting substrates and control circuits integrated on the same chip (Hagelin & Solgaard, 1999; Lin & Fang, 2003; Yao & MacDonald, 1997). The use of electrostatic actuation, in particular those with a comb drive structure, permits low power consumption and strong actuation force (Hah et al., 2004). A wide range of frequency response from 100 Hz to 10 kHz can be achieved with MEMS scanners. Typical MEMS mirrors with a 0.5-2 mm diameter can have a mechanical scanning angle up to ~30° with reasonably low driving voltages (~10-120 V) (Lang et al., 1999; Schenk et al., 2000). Using MEMS techniques, a raster scanning pattern can be easily created, as shown in Fig. 3(a). Overall, MEMS scanners have a great potential to be integrated in a compact endomicroscope yet the relatively large substrates with the drive circuits still present significant engineering challenges in their endomicroscopic applications. A TPF endomicroscope based on a 2-D MEMS mirror with a size of ~3.2 mm x 3 mm has been firstly developed with a line acquisition rate of 3.5 kHz (Piyawattanametha et al., 2006). Later, another 2-D MEMS mirror with a size of ~5 mm in diameter, with a speed of 7 lines/s over an area of 80 x 130 μm^2 has been assembled in a TPF endomicroscope (Fu et al., 2007). Recently, higher TPF imaging rate up to 10 frames per second has been demonstrated in an endomicroscope prototype but with a large dimension of 10 × 15 × 40 mm^3 (Hoy et al., 2008).

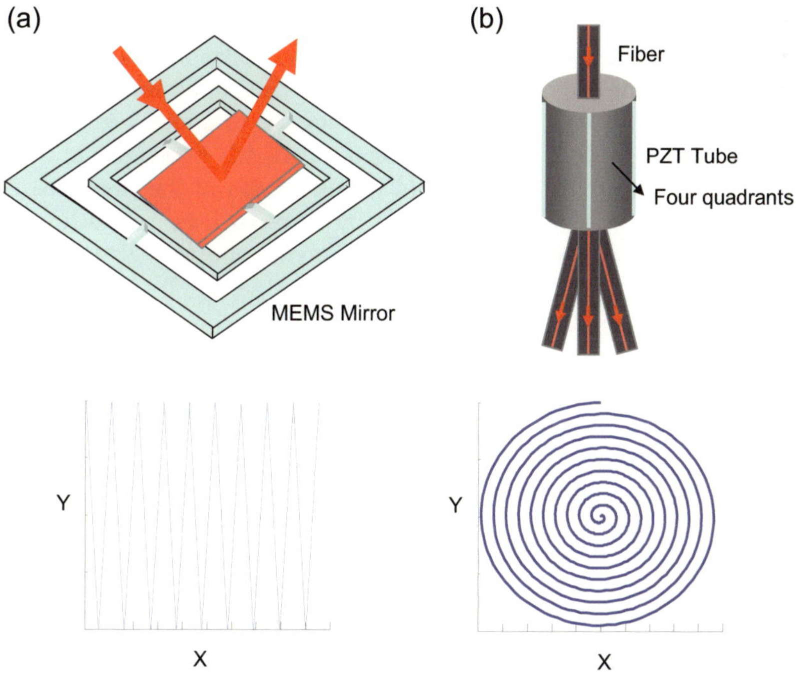

Fig. 3. Schematic of miniature 2D scanning mechanism and pattern: (a) MEMS mirror; (b) PZT based fiber resonant scanner.

Although MEMS scanners can achieve a large lateral beam scan at a high speed, they require complicated fabrication processes and control mechanisms. A simpler method has been developed by scanning the distal tip of an optical fiber cantilever at its mechanical resonant frequency with a piezoelectric actuator (Cobb et al., 2005; Engelbrecht et al., 2008; Flusberg et al., 2005b; Helmchen et al., 2001; Liu et al., 2004; Myaing et al., 2006; Seibel & Smithwick, 2002; Wu et al., 2009a; Wu et al., 2009b). When a piezo actuator is driven at the mechanical resonant frequency of the attached fiber cantilever, small vibration at the cantilever base (on the order of a few microns) can be dramatically amplified at the sweeping fiber tip and a mechanical scanning angle larger than 40 degrees can be conveniently achieved. One type of resonant fiber-optic piezoelectric scanners is based on tubular piezoelectric actuators, as shown in Fig. 3(b) (Engelbrecht et al., 2008; Myaing et al., 2006; Wu et al., 2009a; Wu et al., 2009b). The cylindrical symmetry makes this type of actuators very suitable for endomicroscopic settings. The outer surface of the PZT tube is divided into four quadrants, forming two pairs of drive electrodes. The fiber-optic cantilever is glued to the actuator tip. One-dimensional resonant line scan is achieved when one pair of electrodes are driven with a sinusoidal waveform at the mechanical resonant frequency of the cantilever, which is given by (Kinsler et al., 1982)

$$f = \frac{\beta}{4\pi}\left(\frac{E}{\rho}\right)^{1/2}\left(\frac{R}{L^2}\right) \tag{1}$$

Here f is the resonant frequency of the cantilever; β is a constant of 3.52 at the zeroth-order vibration mode; E and ρ are the Young's modulus and mass density of the cantilever, respectively; R and L are the radius and length of the cantilever, respectively. The equation tells that the cantilever tip scanning frequency is inversely proportional to the square of the cantilever length. For a single-mode fiber cantilever, the rule of thumb is that a 10-mm long cantilever corresponds to a resonant scanning frequency of about 1 kHz. The deflection angle of the fiber cantilever (α) is determined by the PZT drive voltage and frequency. The deflection range of the fiber tip is dependent on the length of the fiber, i.e., 2Lsinα. The sweeping fiber tip is imaged onto the sample by a micro lens, producing lateral beam scanning. Finally, the lateral beam scanning range D on the sample is given by

$$D = 2LM\sin\alpha \tag{2}$$

where M is the magnification of the lens from the fiber tip to the sample. When using a 8.5-mm long fiber cantilever and a lens with a magnification of ~1.8, an approximately 2.5 mm scanning range on the sample can be achieved with a peak-to-peak drive voltage of ~60 V near the resonance frequency of 1.4 kHz (Liu et al., 2004).

To generate a circular scan, one pair of the electrodes will be driven with a sine wave while the other pair driven by a cosine wave with the same amplitude and frequency. By triangularly modulating the drive voltage, an open-close spiral scanning pattern can be formed (see Fig. 3(b)). The method has been successfully adopted in TPF endomicroscopes with a frame rate of 2.6-3.3 Hz composing of 512x512 pixels with a scanning range of 100-220 µm (Myaing et al., 2006). Using the same method, a TPF endomicroscope prototype has been developed with a video rate of 25 Hz imaging speed, but with lower sampling pixels (128x128 pixels for a field of view of 200 µm) (Engelbrecht et al., 2008). Another kind of resonant PZT scanners utilizes a bi-morph piezoelectric plate to vibrate a freely-standing fiber-optic cantilever in Lissajous scanning pattern (Flusberg et al., 2005b; Helmchen et al., 2001). This resonant fiber scanner can also achieve large lateral deflections at the fiber tip (e.g. ~1 mm) and a good frame rate of 2 fps with a reasonably compact form (e.g. 8 mm long x 2 mm wide x 0.5 mm thick) (Helmchen et al., 2001).

2.3 Miniature objective lens

Another key component in a fiber-optic TPF endomicroscope is the high-quality miniature imaging optics to handle the large separation between the NIR excitation and visible TPF wavelengths. In addition to a compact size, the objective lens should have good optical performance over a broad spectral range to achieve a tight excitation focus (i.e., high lateral and axial resolution as well as excitation efficiency) and reduce chromatic aberration (to prevent the loss in coupling the TPF signals back to the fiber-optic endomicroscope). Gradient-index (GRIN) lenses, owning to their small diameter and cylindrical geometry, have been commonly used as the focusing unit in TPF endomicroscopes (Flusberg et al., 2005b; Fu et al., 2007; Fu et al., 2006; Wu et al., 2009a). Generally, a GRIN objective lens with a diameter of 0.35-3 mm and a NA of 0.3~0.6, can produce lateral and axial resolutions (FWHM) at an order of 1-2 μm and 5-20 μm, respectively. However, a GRIN lens suffers severe chromatic aberration, causing a considerable focal shift between the NIR excitation and the visible TPF signals. For a compact, single fiber configuration, a large focal shift will result in dramatic reduction in the TPF collection efficiency (Wu et al., 2009b).

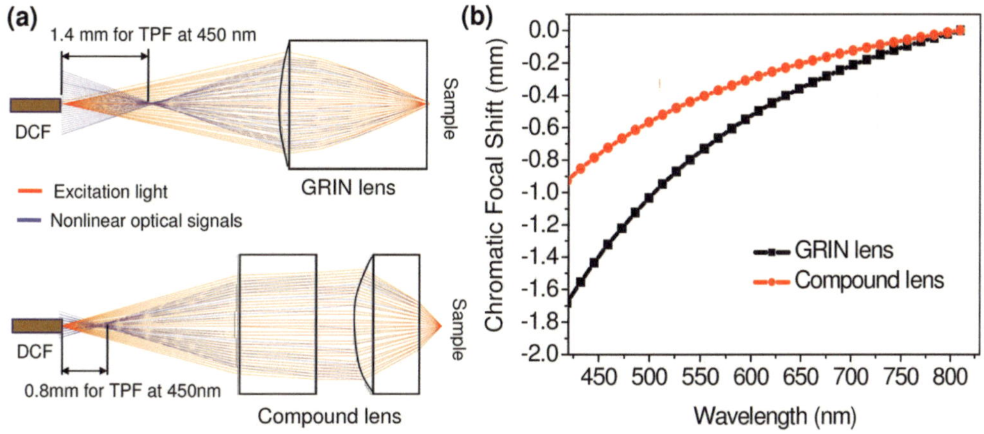

Fig. 4. (a) Representative optical ray plots for excitation light and nonlinear optical signals when using a GRIN lens and an aspherical compound lens; (b) Chromatic longitudinal focal shift from the DCF tip for the TPF signals with 810 nm excitation when using a GRIN lens and an aspherical compound lens. Images are adapted from Ref. (Wu et al., 2009b) with permission.

A miniature compound lens with less chromatic aberration compared to a GRIN lens, has been proposed and employed in a fiber-optic TPF endomicroscope to increase the TPF collection (Wu et al., 2009b). The compound lens was made of a pair of aspherical lenses (2.2-mm effective diameter) with an overall magnification of 0.32 from the fiber tip to the sample. Ray tracing analyses were performed for the compound lens and compared with a GRIN lens (of a 1.8 mm diameter) for quantitative description of the adverse effect of chromatic aberration in a TPF endomicroscope. Both configurations have the same magnification of 0.32 and the same NA of 0.5 on the sample. Fig. 4(a) shows representative ray plots for excitation light at 810 nm (in the forward direction, i.e. from the DCF to the sample) and the TPF signals around 450 nm (in the backward direction). As can be seen, the back focus of the TPF signals from the excitation focal volume within the sample falls in

front of the DCF tip due to the lens chromatic aberration. Fig. 4(b) quantitatively describes the chromatic longitudinal focal shift from the DCF tip at the TPF wavelengths with an excitation wavelength at 810 nm. It is found that the back focal points of TPF signals (420-600 nm) when using a GRIN lens are ~1.7-0.6 mm in front of the DCF tip, which will dramatically reduce the TPF collection efficiency. In comparison, the aspherical compound lens exhibits less chromatic aberration (e.g. with an ~0.9-0.3 mm chromatic longitudinal focal shift over 420-600 nm), resulting in an increased TPF collection efficiency.
Table 2 lists the collection efficiency calculated by ray tracing for TPF signals at some typical wavelengths - 450, 500, 550 and 600 nm when using a GRIN lens and a compound lens, where 100% representing full collection of the back-reflected TPF signals by a perfect achromatic lens. Here, assuming the collection is based on the commercial double-clad fiber (Fibercore Ltd., SMM900) with an inner clad of ~103 um diameter and 0.24 NA. The quantitative analyses indicate that the compound lens increases the collection efficiency of the DCF over the GRIN lens by a factor ~4-6 for the TPF signals. This simulation result has been confirmed by the experiments on TPF imaging of fluorescent beads (Wu et al., 2009b). It is expected that objective lens with a smaller diameter, higher NA and less chromatic aberration can be fabricated to further improve the excitation and collection efficiency of fiber-optic TPF endomicroscopes.

Different Configurations (810 nm excitation)	TPF (600 nm)	TPF (550 nm)	TPF (500 nm)	TPF (450 nm)
GRIN and DCF	21.5%	9.9%	5.3%	2.1%
Compound Lens and DCF	78.3%	45.7%	22.9%	12.7%

Table 2. Calculated collection efficiency of TPF signals based on geometrical ray tracing.

To achieve accurate depth-resolved imaging in fiber-optic TPF endomicroscopy, a built-in mechanism for changing the focal plane has to be implemented (preferably at the distal end). A hydraulic system has been demonstrated to move the end of a fiber relative to the objective lens (Rouse et al., 2004). An alternative approach is to utilize a miniature motor to alter the distance between the fiber and the objective lens (Flusberg et al., 2005b). In addition, deformable MEMS lenses and reflecting membranes have also demonstrated a great potential for varying the beam focus at the distal end of the probe by using electrowetting or electrostatic force (Berge & Peseux, 2000; Kwon et al., 2002). Overall, many challenges remain with distal end beam focus tracking (or depth scanning) in endomicroscopy including system miniaturization, reduction in the drive voltage and improvement on focus tracking accuracy and repeatability over a sufficient range.

3. A new generation two-photon fluorescence endomicroscopy system

By addressing the above major challenges for fiber-optic TPF endomicroscope devices, extremely compact all-fiber-optic scanning endomicroscopy systems have been developed for two-photon fluorescence imaging of biological samples (Wu et al., 2009a; Wu et al., 2009b). A conventional DCF was employed for single-mode fs pulse delivery and multimode collection of TPF signals. The DCF also serve as a resonant cantilever driven by a tubular PZT actuator for fast 2D beam scanning. A single PBF with anomalous dispersion at ~ 810 nm was used for pulse prechirping in replacement of a bulky grating/lens-based pulse stretcher to compensate the normal dispersion of the DCF. The combination of DCF and PBF in the endomicroscopy system made the system all-fiber-optic and basically a plug-and-play unit. For beam focusing, a miniature compound lens was developed to achieve a tight excitation focus, reduce chromatic aberration and improve the excitation and collection efficiency of TPF signals. In addition, a short multimode fiber was introduced at the tip of the DCF to further mitigate the adverse effect of chromatic aberration. *Ex vivo* TPF imaging of cells and various tissues samples including pig cornea and rat oral cavity was conducted along with this extremely compact all-fiber-optic TPF endomicroscopy system. The imaging results strongly suggest this newly developed endomicroscopy technology has the great potential for *in vivo* and real-time superficial tissue imaging.

3.1 Scanning fiber-optic TPF endomicroscope probe

Fig. 5(a) shows the schematic of the endomicroscope probe. The fiber-optic probe basically consisted of a conventional double-clad fiber, a small tubular piezoelectric actuator and a micro lens for beam focusing. A 300-µm long MMF (105/125 µm core/clad and 0.22 NA) was spliced at the distal end of the DCF. The length (300-µm) of the MMF collector was chosen such that it would shorten the distance between its end surface and the back focus of the TPF signals as much as possible. Here, the MMF had almost the same core size and NA as the inner clad of the DCF and its short length did not significantly influence the excitation beam path. As confirmed with ray-tracing simulations and experiments on fluorescent bead imaging, the short MMF collector guided more TPF signals back to the DCF.

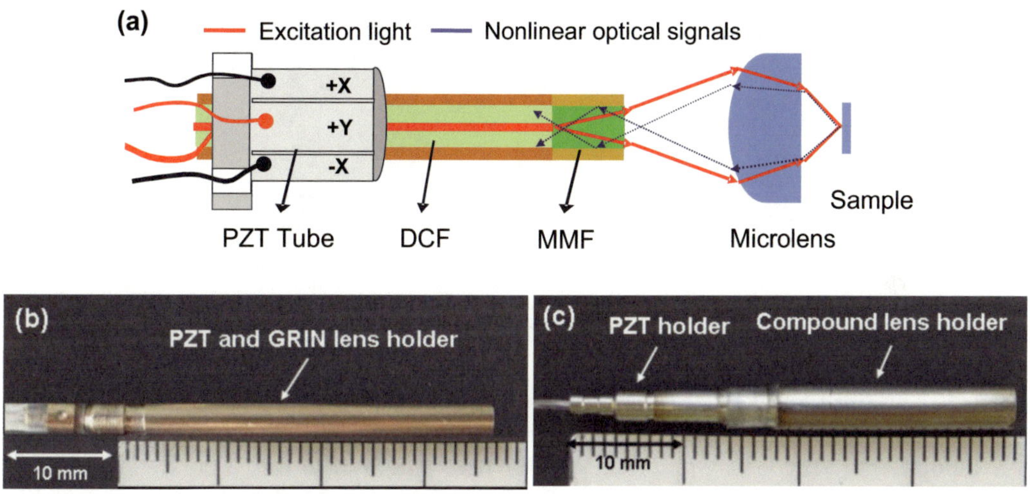

Fig. 5. (a) Schematic of the distal end of the fiber-optic scanning TPF endomicroscope; (b) Photo of a prototype with a GRIN lens; (c) Photo of a prototype with a compound lens. The tubular piezoelectric actuator (PZT), fiber-optic cantilever, and microlens were encased in a hypodermic tube with an overall outer diameter of 2.4 mm and 3.5 mm for the GRIN lens and compound lens, respectively. Image (b) is adapted from Ref. (Wu et al., 2009a) with permission; and Image (c) is adapted from Ref. (Wu et al., 2009b) with permission.

For beam focusing, two kinds of miniature optics were implemented, i.e., a GRIN lens and a compound lens, respectively. The GRIN lens with a 0.22 pitch and a 1.8-mm diameter (NSG America, Inc.) had a magnification of 0.5 from the DCF tip to the sample. The compound lens was made of a pair of 3-mm miniature aspherical lenses (modified from Archer OpTx L150 and L110 lenses). The purpose of the use of the miniature compound lens is to achieve a tighter focus and increase the TPF excitation efficiency, as well as to minimize the chromatic aberration and increase the TPF collection efficiency.

The endomicroscope had a resonant scanning frequency of ~1,690 Hz (i.e. 1,690 scanning circles per second would be generated), resulting in a frame rate of 3.3 frames/second with each frame consisting of 512 circular scans. An ~320-μm scanning diameter traced by the DCF tip could be achieved with a relatively low peak-to-peak drive voltage of ~±30 V. This corresponded to a beam scanning area of ~160 μm and ~100 μm over the sample when using the GRIN lens with a magnification of 0.5 and the compound lens with a magnification of 0.32, respectively.

The distal end of the probe, including the piezoelectric actuator, fiber scanner and the lenses, was housed in a thin-wall hypodermic tube with an overall diameter of 2.4 mm (with the GRIN lens) and 3.5 mm (with the compound lens), as shown in Fig. 5(b) and Fig. 5(c), respectively. Fig. 6(a) and (b) respectively show a photo of the entire fiber-optic scanning TPF endomicroscope along and a photo of the 2.4 mm endomicroscope going through the accessory port (3.8 mm) of a standard endoscope.

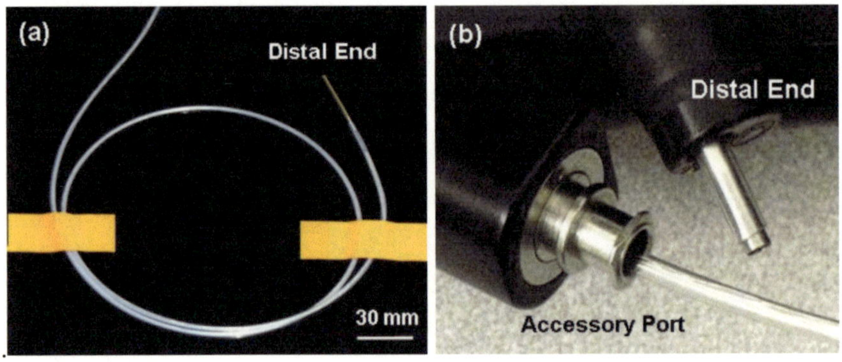

Fig. 6. (a) Photo of the entire fiber-optic scanning TPF endomicroscope probe based on a GRIN lens; (b) Photo of the entire probe with a standard endoscope showing that the fiber-optic endomicroscope can go through the accessory port of a standard endoscope.

3.2 Entire all-fiber-optic TPF endomicroscopy system

Fig. 7(a) shows the schematic of the entire endomicroscopy system with a home-built Ti:Sapphire laser. The laser pulses (of a center wavelength ~810 nm, a bandwidth ~36 nm and a measured temporal pulse width ~60 fs at the laser output) were coupled into PBF (Crystal Fibers Ltd., HC-800-02) by a fiber-launching lens with an effective NA of ~0.2 and then directed into the core of the DCF. Here, the PBF was used to compensate the temporal pulse broadening caused by the positive dispersion of femtosecond pulses in the single-mode core of the DCF. The length ratio of the PBF to DCF was kept at ~1.1 to achieve optimal dispersion compensation for this excitation band (810±18 nm) as previously discussed (Wu et al., 2009a). A pulse width of less than 200 fs was achieved even at a power of 50-70 mW delivered through the DCF core. In addition to its compactness (compared to a grating-lens based pulse stretcher), a single PBF also greatly simplified the system alignment and improved the system stability. Furthermore, the PBF helped to reshape the laser beam from the laser output, thus improving the coupling efficiency from the PBF to the DCF core by almost a factor of 2 compared to the use of a pulse stretcher (i.e. from 20-30% with a pulse stretcher to ~50% with a PBF). PBF works conveniently for dispersion compensation when the excitation wavelength is fixed to avoid the need for changing the length of the PBF, which is ideal when an extremely compact fs fiber laser is used for excitation. This simple dispersion compensation approach made the entire nonlinear endomicroscopy system compact, reliable and essentially plug-and-play. As shown in Fig. 7(b), the entire endoscope system, including all the fiber-launchers, dispersion unit and the photon detector (except the excitation light source), can be put together within a box of a small footprint (53 x 38 x 18 cm), leaving the flexible scanning endomicroscope probe extending outside the box and freely accessible. The compact size and the plug-and-play feature greatly simply the operation of the endomicroscope system and facilitate their applications for both laboratory research and future *in vivo* studies.

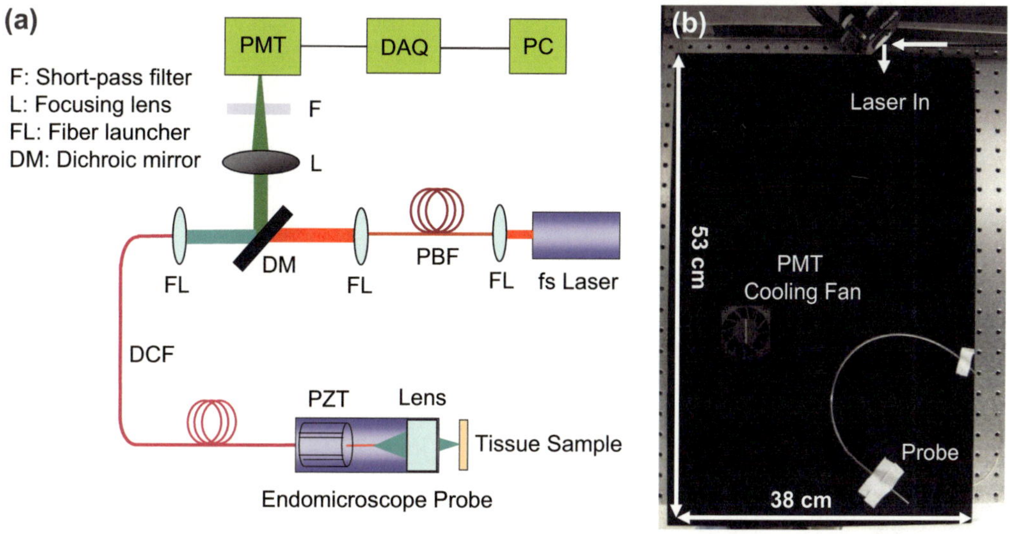

Fig. 7. (a) Schematic of the all-fiber-optic scanning TPF endomicroscope imaging system. The combination of a DCF and PBF made the system all-fiber-optic, in which the DCF was used for both optical nonlinear excitation and TPF (or second harmonic generation) collection, whereas the PBF was employed for dispersion management. PBF: Photonic bandgap fiber; DCF: Double-clad fiber; PMT: Photomultiplier tube; DAQ: Data acquisition. (b) Photo of the all-fiber-optic scanning TPF endomicroscope imaging system inside a box with the flexible endomicroscope probe extended outside the box and placed on the cover. The endomicroscope can be easily plugged into the system and becomes freely accessible.

In imaging experiments, the laser power delivered to the samples was controlled at about 10-15 mW for TPF imaging of stained biological samples. The TPF signals were collected back through the same DCF (core and inner clad), separated from the excitation light at the proximal end of the endomicroscope by a dichroic mirror and then directed to a photomultiplier tube (PMT). The residual excitation light was further blocked by a short-pass filter mounted in front of the PMT. Appropriate band-pass filters were also employed for the detection of the TPF signals. The photo current from the PMT was amplified, converted to voltage and digitized.

Fluorescent beads were used to test the system performance. Using a phantom made of 0.1-μm fluorescent beads, the lateral and axial resolution (FWHM) was measured by scanning through the center of a single bead along the lateral and axial direction, respectively. Gaussian fits to the fluorescence intensities show that the endomicroscopy system with a GRIN lens has ~1.6 μm lateral resolution and ~11.4 μm axial resolution, whereas the endomicroscopy system with the compound lens has ~1.2 μm lateral resolution and ~9.0 μm axial resolution, as shown in Figs. 9(a) and (b). The theoretical lateral resolution (FWHM) and axial resolution (FWHM) can be calculated by (Zipfel et al., 2003):

$$\text{Lateral resolution (FWHM)} = D \times M \times \sqrt{\ln 2} / 2 \,, \tag{3}$$

$$\text{Axial resolution (FWHM)} = D^2 \times M^2 \times \sqrt{2}\pi / 4\lambda \,, \tag{4}$$

where D is the mode-filed diameter of the DCF, M is the magnification of the endomicroscope from the fiber tip to sample, and λ is the excitation wavelength. According to the mode-filed diameter of the DCF (D = 4.1 μm), the theoretical lateral resolution should be 0.85 μm and 0.54 μm for the endomicroscope with the GRIN lens (M = 0.5) and the compound lens (M = 0.32), respectively; the theoretical axial resolution should be 5.75 μm and 2.36 μm for the endomicroscope with GRIN lens and the compound lens, respectively. This discrepancy is probably mainly attributed to the chromatic aberration and geometrical aberration of the imaging optics, which can be improved with customized achromatic objective lenses.

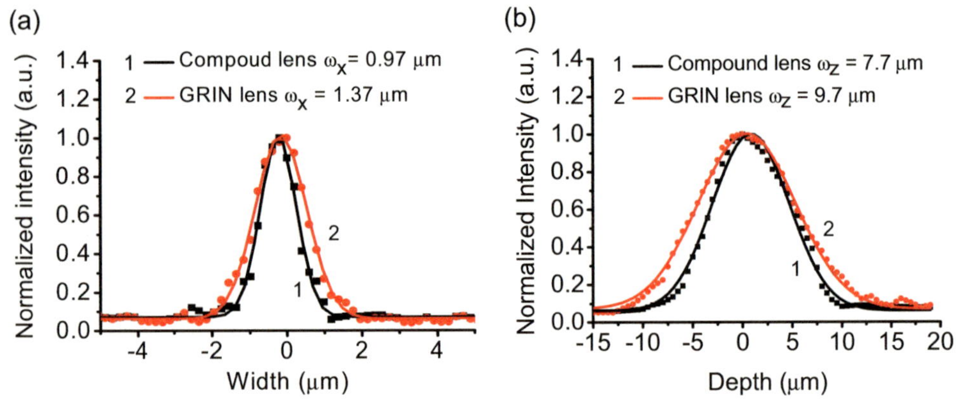

Fig. 8. Fluorescence intensity profiles (dots) across the center of a 0.1-μm fluorescent bead along (a) the lateral and (b) axial dimension. Solid traces are Gaussian-fitted curves.

3.3 Typical fiber-optic endomicroscopic TPF images

The imaging capability of the flexible and compact all-fiber-optic scanning endomicroscope was demonstrated on *in vitro* TPF imaging of stained cancer cells and depth-resolved *ex vivo* TPF imaging of stained biological samples, as shown in Figs. 9-10, respectively. Fig. 9(a) shows a typical TPF image of breast cancer cells SK-BR-3 that over expressed epidermal growth factor receptors (EGFR). The cells were targeted by fluorescein isothiocyanate (FITC) tagged anti Here2/Nu monoclonal antibodies. The single-photon excitation/emission maximum of FITC are 494/521 nm, and the two-photon fluorescence spectrum with maximally excitation at 700-850 nm is similar to the single-photon fluorescence spectrum (Xu et al., 1996). As shown in Fig. 9(a), strong TPF signals of FITC from cell membranes were easily observed. Fig. 9(b) shows a typical TPF image of ovarian cancer cells SK-OV-3. The cells were stained with AO (acridine orange), the dye for enhancing nuclei contrast (Adams, 1974; Fu et al., 2007). The single- and two-photon fluorescence characteristics of AO when bound to DNA are similar to those of FITC (Neu et al., 2002). As can be seen, cellular nuclei can be clearly identified from the TPF image.

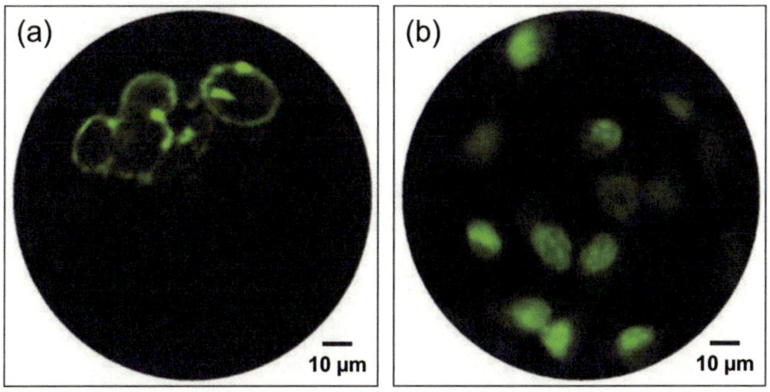

Fig. 9. Typical TPF images obtained with the scanning all-fiber-optic endomicroscope based on the compound lens: (a) Breast cancer cells SK-BR-3 targeted by FITC-tagged monoclonal antibodies (anti Here2/Nu); (b) Ovarian cancer cells SK-OV-3 stained with AO dye; Scale bar: 10 µm.

Depth-resolved TPF imaging of epithelial tissues was performed using the scanning all-fiber-optic endomicroscope based on a GRIN lens, as shown in Fig. 10. All the images were averaged for 10 frames. Figs. 10(a) and (b) show typical images of a formalin fixed pig corneal tissue at ~100 µm and ~300 µm below the surface, respectively. The tissue sample was stained with 1% AO in saline solution after fixation in 10% formalin. The sparsely distributed keratocytes at different depths can be clearly identified from the images. Figs. 10(c) and (d) show representative images of a rat oral tissue sample at depths of 15 and 55 µm below the tissue surface, respectively. The fresh dissected tissue sample was directly stained with 1% AO and the epithelial cell nuclei can be identified by the TPF images. As seen from the depth-resolved images, the nucleus density increases from the superficial layer to the basal layer. Figs. 10(e) and (f) show typical images of a chick cochlear section expressing GPF (green fluorescence protein). The single-photon excitation/emission maximum of GFP are 491/510 nm, whereas its two-photon excitation/emission maxima are about 900/510 nm (Xu et al., 1996). Although the excitation efficiency is relatively weak at 810 nm excitation, the structures (hair cells) revealed by the TPF signals are still observable. Different distribution of hair cells can be observed at different depths (see Fig. 10(e) from the surface and Fig. 10(f) from 40 µm depth below the surface). Overall, the depth-resolved TPF images clearly reveal the tissue layered structures that are consistent with text book histology (Ross et al., 1989).

Overall, the extremely compact, all-fiber-optic and plug-and-play scanning endomicroscopy system has shown excellent ability in two-photon fluorescence imaging of biological samples. The system employed a conventional double-clad fiber for femtosecond pulse delivery, TPF signals collection and fast beam scanning. By replacing a bulky grating/lens-based pulse stretcher with a single photonic bandgap fiber for pulse prechirping, the system

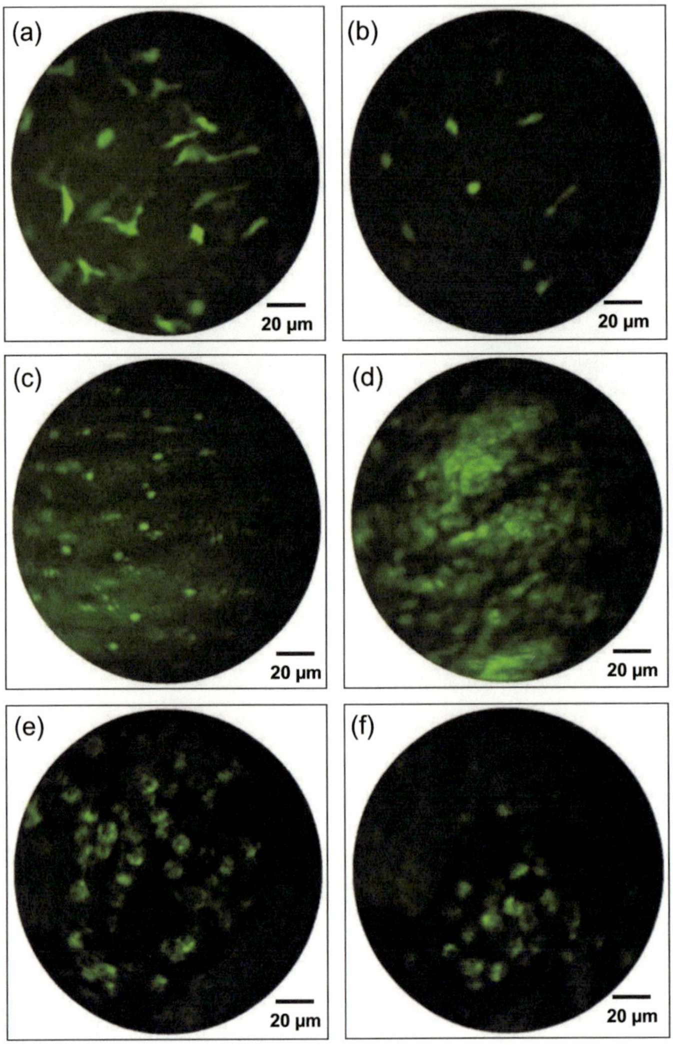

Fig. 10. Typical depth-resolved TPF images obtained with the scanning all-fiber-optic endomicroscope bases on a GRIN lens: (a & b) Pig cornea sample stained with AO dye (a) at the depth of 100 µm and (b) at the depth of 300 µm; (c & d) Rat oral tissue stained with AO dye (c) at the depth of 15 µm and (d) at the depth of 55 µm; (e & f) Chick cochlear expressing GFP (e) from the tissue surface and (b) at the depth of 40 µm. Scale bar: 20 µm.

size is significantly reduced and the power delivery efficiency through the endomicroscope is improved. Real-time *ex vivo* 2D and depth-resolved TPF imaging of biological samples were performed with the scanning endomicroscope at cellular and subcellular resolution. The preliminary results strongly suggest the promising potential of this all-fiber-optic endomicroscopy technology as a basic laboratory research tool and a clinical tool for real-time assessment of epithelium and stroma under various clinically relevant conditions. The performance of the endomicroscope system can be further improved to increase the signal-to-noise ratio and resolve more detailed intracellular structures by introducing better compound lenses with higher NAs.

4. Summary

In this book chapter, the general technological and engineering challenges in developing a scanning two-photon fluorescence fiber-optic endomicroscopy system has been discussed, including single-mode femtosecond laser delivery, large-area TPF signal collection, dispersion management, fast two-dimensional beam scanning and high-quality miniature imaging optics. Detailed design issues and imaging performance has been illustrated using our recently developed scanning all-fiber-optic TPF endomicroscope as an example. The two-photon fluorescence endomicroscope has shown great flexibility and reliability for high-resolution imaging of internal luminal organs.

In summary, endomicroscopy technologies for two-photon fluorescence imaging have been under rapid development, aiming for non-invasive, high-resolution and high-speed imaging of tissue microstructures and assessment of tissue pathology *in vivo*. The development of optical fibers (such as double-clad fibers and photonic bandgap fibers), miniature beam scanning mechanisms (such as MEMS scanners and fiber-optic resonant scanners) and miniature imaging optics (such as compound lens with chromatic aberration and high NA) greatly facilitates probe miniaturization and functional integration for high-resolution and high-speed TPF imaging of biological samples. With further technology innovation, miniature fiber-optic TPF endomicroscope can potentially generate high-quality images approaching to those achieved by standard bench-top two-photon fluorescence microscopy in the near future.

5. Acknowledgement

The authors greatly thank the invaluable contributions of Dr. Michael J. Cobb, Dr. Yuxin Leng, Daniel J. MacDonald, Jeff Magula, Dr. Mon T. Myaing and Jiefeng Xi in the development of the fiber-optic two-photon fluorescence endomicroscopy technologies. This work has been supported in part by the National Institutes of Health and National Science Foundation (Career Award XDL).

6. References

Adams, L.R. (1974). Acridine-Orange Staining of Epithelial-Cells in Strong Salt Solution. *Journal of Histochemistry & Cytochemistry*, Vol. 22, No. 7, pp. 492-494.

Agrawal, G.P. (2007). *Nonlinear Fiber Optics*. Academic Press: San Diego.

Bao, H.C.; Allen, J.; Pattie, R.; Vance, R. & Gu, M. (2008). Fast handheld two-photon fluorescence microendoscope with a 475 mu m X 475 mu m field of view for in vivo imaging. *Optics Letters*, Vol. 33, No. 12, pp. 1333-1335.

Bao, H.C. & Gu, M. (2009). Reduction of self-phase modulation in double-clad photonic crystal fiber for nonlinear optical endoscopy. *Optics Letters*, Vol. 34, No. 2, pp. 148-150.

Berge, B. & Peseux, J. (2000). Variable focal lens controlled by an external voltage: An application of electrowetting. *European Physical Journal E*, Vol. 3, No. 2, pp. 159-163.

Bjarklev, A.; Broeng, J. & Bjarklev, A.S. (2003). *Photonic Crystal Fibers*. Kluwer Academic Publishers: Norwell, MA.

Cobb, M.J.; Liu, X.M. & Li, X.D. (2005). Continuous focus tracking for real-time optical coherence tomography. *Optics Letters*, Vol. 30, No. 13, pp. 1680-1682.

Denk, W.; Strickler, J.H. & Webb, W.W. (1990). 2-Photon Laser Scanning Fluorescence Microscopy. *Science*, Vol. 248, No. 4951, pp. 73-76.

Engelbrecht, C.J.; Johnston, R.S.; Seibel, E.J. & Helmchen, F. (2008). Ultra-compact fiber-optic two-photon microscope for functional fluorescence imaging *in vivo*. *Optics Express*, Vol. 16, No. 8, pp. 5556-5564.

Flusberg, B.A.; Cocker, E.D.; Piyawattanametha, W.; Jung, J.C.; Cheung, E.L.M. & Schnitzer, M.J. (2005a). Fiber-optic fluorescence imaging. *Nature Methods*, Vol. 2, No. 12, pp. 941-950.

Flusberg, B.A.; Lung, J.C.; Cocker, E.D.; Anderson, E.P. & Schnitzer, M.J. (2005b). In vivo brain imaging using a portable 3.9 gram two-photon fluorescence microendoscope. *Optics Letters*, Vol. 30, No. 17, pp. 2272-2274.

Fu, L.; Gan, X.S. & Gu, M. (2005). Nonlinear optical microscopy based on double-clad photonic crystal fibers. *Optics Express*, Vol. 13, No. 14, pp. 5528-5534.

Fu, L.; Jain, A.; Cranfield, C.; Xie, H.K. & Gu, M. (2007). Three-dimensional nonlinear optical endoscopy. *Journal of Biomedical Optics*, Vol. 12, No. 4, pp. 040501.

Fu, L.; Jain, A.; Xie, H.K.; Cranfield, C. & Gu, M. (2006). Nonlinear optical endoscopy based on a double-clad photonic crystal fiber and a MEMS mirror. *Optics Express*, Vol. 14, No. 3, pp. 1027-1032.

Gobel, W.; Kerr, J.N.D.; Nimmerjahn, A. & Helmchen, F. (2004a). Miniaturized two-photon microscope based on a flexible coherent fiber bundle and a gradient-index lens objective. *Optics Letters*, Vol. 29, No. 21, pp. 2521-2523.

Gobel, W.; Nimmerjahn, A. & Helmchen, F. (2004b). Distortion-free delivery of nanojoule femtosecond pulses from a Ti : sapphire laser through a hollow-core photonic crystal fiber. *Optics Letters*, Vol. 29, No. 11, pp. 1285-1287.

Hagelin, P.M. & Solgaard, O. (1999). Optical raster-scanning displays based on surface-micromachined polysilicon mirrors. *Ieee Journal of Selected Topics in Quantum Electronics*, Vol. 5, No. 1, pp. 67-74.

Hah, D.; Huang, S.T.Y.; Tsai, J.C.; Toshiyoshi, H. & Wu, M.C. (2004). Low-voltage, large-scan angle MEMS analog micromirror arrays with hidden vertical comb-drive actuators. *Journal of Microelectromechanical Systems*, Vol. 13, No. 2, pp. 279-289.

Helmchen, F. & Denk, W. (2005). Deep tissue two-photon microscopy. *Nature Methods*, Vol. 2, No. 12, pp. 932-940.

Helmchen, F.; Fee, M.S.; Tank, D.W. & Denk, W. (2001). A miniature head-mounted two-photon microscope: High-resolution brain imaging in freely moving animals. *Neuron*, Vol. 31, No. 6, pp. 903-912.

Hoy, C.L.; Durr, N.J.; Chen, P.Y.; Piyawattanametha, W.; Ra, H.; Solgaard, O. & Ben-Yakar, A. (2008). Miniaturized probe for femtosecond laser microsurgery and two-photon imaging. *Optics Express*, Vol. 16, No. 13, pp. 9996-10005.

Jung, J.C. & Schnitzer, M.J. (2003). Multiphoton endoscopy. *Optics Letters*, Vol. 28, No. 11, pp. 902-904.

Jung, W.Y.; Tang, S.; McCormic, D.T.; Xie, T.Q.; Ahn, Y.C.; Su, J.P.; Tomov, I.V.; Krasieva, T.B.; Tromberg, B.J. & Chen, Z.P. (2008). Miniaturized probe based on a microelectromechanical system mirror for multiphoton microscopy. *Optics Letters*, Vol. 33, No. 12, pp. 1324-1326.

Kinsler, L.E.; Frey, A.R.; Coppens, A.B. & Sanders, J.V. (1982). *Fundamentals of Acoustics*. Wiley: New York.

Knight, J.C. (2003). Photonic crystal fibres. *Nature*, Vol. 424, No. 6950, pp. 847-851.

Konig, K. (2000). Multiphoton microscopy in life sciences. *Journal of Microscopy-Oxford*, Vol. 200, pp. 83-104.

Konig, K.; Ehlers, A.; Riemann, I.; Schenkl, S.; Buckle, R. & Kaatz, M. (2007). Clinical two-photon microendoscopy. *Microscopy Research and Technique*, Vol. 70, No. 5, pp. 398-402.

Kwon, S.; Milanovic, V. & Lee, L.P. (2002). Large-displacement vertical microlens scanner with low driving voltage. *Ieee Photonics Technology Letters*, Vol. 14, No. 11, pp. 1572-1574.

Lang, W.; Pavlicek, H.; Marx, T.; Scheithauer, H. & Schmidt, B. (1999). Electrostatically actuated micromirror devices in silicon technology. *Sensors and Actuators a-Physical*, Vol. 74, No. 1-3, pp. 216-218.

Le Harzic, R.; Weinigel, M.; Riemann, I.; Konig, K. & Messerschmidt, B. (2008). Nonlinear optical endoscope based on a compact two axes piezo scanner and a miniature objective lens. *Optics Express*, Vol. 16, No. 25, pp. 20588-20596.

Levene, M.J.; Dombeck, D.A.; Kasischke, K.A.; Molloy, R.P. & Webb, W.W. (2004). In vivo multiphoton microscopy of deep brain tissue. *Journal of Neurophysiology*, Vol. 91, No. 4, pp. 1908-1912.

Lin, H.Y. & Fang, W.L. (2003). A rib-reinforced micro torsional mirror driven by electrostatic torque generators. *Sensors and Actuators a-Physical*, Vol. 105, No. 1, pp. 1-9.

Liu, X.M.; Cobb, M.J.; Chen, Y.C.; Kimmey, M.B. & Li, X.D. (2004). Rapid-scanning forward-imaging miniature endoscope for real-time optical coherence tomography. *Optics Letters*, Vol. 29, No. 15, pp. 1763-1765.

Myaing, M.T.; MacDonald, D.J. & Li, X.D. (2006). Fiber-optic scanning two-photon fluorescence endoscope. *Optics Letters*, Vol. 31, No. 8, pp. 1076-1078.

Neu, T.R.; Kuhlicke, U. & Lawrence, J.R. (2002). Assessment of fluorochromes for two-photon laser scanning microscopy of biofilms. *Applied and Environmental Microbiology*, Vol. 68, No. 2, pp. 901-909.

Piyawattanametha, W.; Barretto, R.P.J.; Ko, T.H.; Flusberg, B.A.; Cocker, E.D.; Ra, H.J.; Lee, D.S.; Solgaard, O. & Schnitzer, M.J. (2006). Fast-scanning two-photon fluorescence imaging based on a microelectromechanical systems two-dimensional scanning mirror. *Optics Letters*, Vol. 31, No. 13, pp. 2018-2020.

Reeves, W.H.; Skryabin, D.V.; Biancalana, F.; Knight, J.C.; Russell, P.S.; Omenetto, F.G.; Efimov, A. & Taylor, A.J. (2003). Transformation and control of ultra-short pulses in dispersion-engineered photonic crystal fibres. *Nature*, Vol. 424, No. 6948, pp. 511-515.

Ross, M.H.; Reith, E.J. & Romrell, L.J. (1989). *Histology: A text and atlas*. Williams & Wilkins.

Rouse, A.R.; Kano, A.; Udovich, J.A.; Kroto, S.M. & Gmitro, A.F. (2004). Design and demonstration of a miniature catheter for a confocal microendoscope. *Applied Optics*, Vol. 43, No. 31, pp. 5763-5771.

Schenk, H.; Durr, P.; Haase, T.; Kunze, D.; Sobe, U.; Lakner, H. & Kuck, H. (2000). Large deflection micromechanical scanning mirrors for linear scans and pattern generation. *Ieee Journal of Selected Topics in Quantum Electronics*, Vol. 6, No. 5, pp. 715-722.

Seibel, E.J. & Smithwick, Q.Y.J. (2002). Unique features of optical scanning, single fiber endoscopy. *Lasers in Surgery and Medicine*, Vol. 30, No. 3, pp. 177-183.

Treacy, E.B. (1969). Optical Pulse Compression with Diffraction Gratings. *Ieee Journal of Quantum Electronics*, Vol. Qe 5, No. 9, pp. 454-&.

Wu, Y.C.; Leng, Y.X.; Xi, J.F. & Li, X.D. (2009a). Scanning all-fiber-optic endomicroscopy system for 3D nonlinear optical imaging of biological tissues. *Optics Express*, Vol. 17, No. 10, pp. 7907-7915.

Wu, Y.C.; Xi, J.F.; Cobb, M.J. & Li, X.D. (2009b). Scanning fiber-optic nonlinear endomicroscopy with miniature aspherical compound lens and multimode fiber collector. *Optics Letters*, Vol. 34, No. 7, pp. 953-955.

Xu, C.; Zipfel, W.; Shear, J.B.; Williams, R.M. & Webb, W.W. (1996). Multiphoton fluorescence excitation: New spectral windows for biological nonlinear microscopy. *Proceedings of the National Academy of Sciences of the United States of America*, Vol. 93, No. 20, pp. 10763-10768.

Yao, Z.J. & MacDonald, N.C. (1997). Single crystal silicon supported thin film micromirrors for optical applications. *Optical Engineering*, Vol. 36, No. 5, pp. 1408-1413.

Zipfel, W.R.; Williams, R.M. & Webb, W.W. (2003). Nonlinear magic: multiphoton microscopy in the biosciences. *Nature Biotechnology*, Vol. 21, No. 11, pp. 1368-1376.

33

Quantitative Phase Imaging using Multi-wavelength Optical Phase Unwrapping

Nilanthi Warnasooriya and Myung K. Kim
University of South Florida
U.S.A.

1. Introduction

Quantitative phase imaging is a vital technique in many areas of science. Studying properties and characteristics of biological and other microscopic specimens has been facilitated with new quantitative phase imaging microscopy methods. In quantitative phase imaging, phase images are obtained by interfering two light beams - one reflected from, or traversed through, the specimen and the other reflected from a reference mirror. This can be achieved by two methods; holography or phase-shifting interferometry. In holography, one interferogram is used to produce the phase image, while phase-shifting interferometry uses three or more interferograms.

Each fringe in an interferogram represents an area of data ranging from 0 to 2π . Therefore, the final phase map obtained from a series of interferograms also contains 2π ambiguities. Such phase maps are called 'wrapped' phase maps, and are needed to be 'unwrapped' by removing 2π ambiguities. Once these 2π ambiguities are removed, a continuous surface profile of the test object can be obtained. Such a surface profile provides height information of surface features. Generally, phase unwrapping is done by using numerical algorithms. Most of these numerical algorithms are computationally intensive and can fail when there are irregularities in the test object.

In the basic phase unwrapping method, the phase image is divided to horizontal lines and these lines are unwrapped separately by scanning pixels and adding an offset to each pixel. At each discontinuity a 2π offset is added or subtracted. After all horizontal lines are unwrapped, they are connected vertically and the unwrapping process is done along vertical lines. There are many phase unwrapping methods to remove 2π ambiguities and most can be categorized into two types; path-dependent methods and path-independent methods. Path-dependent methods detect positions of edges and phase ambiguities in images and use this information to calculate phase offset values. In path-independent methods, areas that can cause errors in unwrapping are identified and eliminated before the unwrapping process starts.

In 1994 Ghiglia and Romero used a least squares integration method with phase unwrapping. In this method, which is known as least squares integration of phase gradient method, the phase gradient is obtained as wrapped phase differences along two perpendicular directions and the gradient field is least squares integrated to obtain continuous phase. However, this method is not effective for phase maps with high noise. P.

G. Charette and I. W. Hunter proposed a robust phase unwrapping method for phase images with high noise content. The basic concept behind this method is to identify contiguous areas that are not on or close to a fringe boundary by locally fitting planes to the phase data. Then these areas are phase shifted with respect to one another by multiples of 2π to unwrap the phase.

Software algorithms that exist for detecting and removing 2π ambiguities are mostly computational-intensive and prone to errors when the phase profile is noisy or when the object has irregularities. Multi-wavelength optical phase unwrapping is an easy method that can be used to eliminate 2π ambiguities in phase maps without such problems.

In this chapter we will present quantitative phase images of cells and other microscopic samples, using multi-wavelength optical phase unwrapping. Three types of light sources are used in a standard four-frame phase shifting interferometer to obtain phase profiles with larger beat wavelengths, thus removing 2π ambiguities without increasing phase noise. The effectiveness of multi-wavelength optical phase unwrapping with both incoherent and coherent light sources will be demonstrated.

2. Multi-wavelength optical phase unwrapping

When an object is imaged by a wavelength smaller than the object's height, phase image of the object contains 2π ambiguities as shown in Fig. 1. It is clear that there are many distance values for a given phase value. In order to obtain an unambiguous optical thickness profile, there should be only one z distance for a given phase.

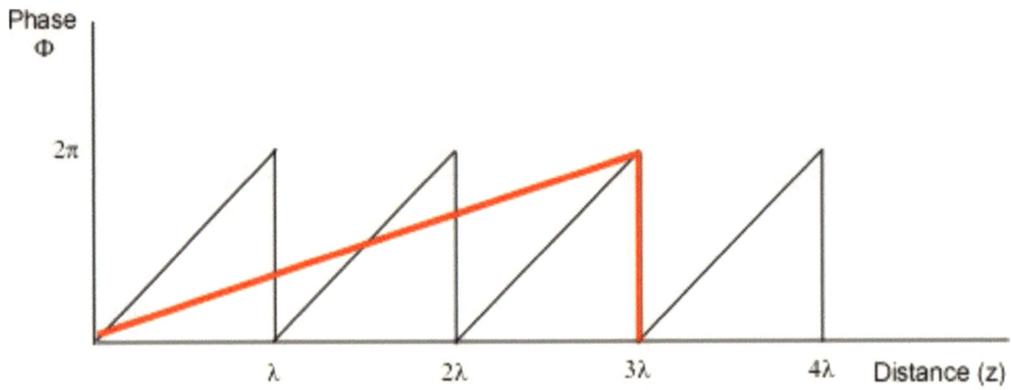

Fig. 1. Phase Vs distance. 2π ambiguities occur when the distance is a multiple of the wavelength. In the axial range Z= 3λ, for a wavelength λ, there are 3 discontinuities (black line), while for a wavelength 3λ, there are no discontinuities (red line).

For years it has been known that a longer wavelength light source produces fewer fringes over a given object than will a short wavelength light source, thus reducing the number of 2π ambiguities in the phase image. However, the drawback is the need of infrared light sources instead of visible light sources. J. C. Wyant has shown that two wavelengths of visible region can be used in the context of holography to produce a longer beat wavelength. Using various pairs of wavelengths from an Argon and HeNe lasers an aspheric optic element was tested. First, a hologram of the test target was obtained by using a wavelength λ_1. Then the hologram was processed and placed at the original position of the interferometer and illuminated by a second wavelength λ_2. The resultant interferogram was identical to the interferogram that would have been illuminated by a light source of Λ_{12}; the beat wavelength of wavelengths λ_1 and λ_2, where $\Lambda_{12} = \lambda_1\lambda_2 / |\lambda_1 - \lambda_2|$.

The term 'two-wavelength interferometry' was first used by C. Polhemus in a paper where he introduced a two-wavelength technique for interferometric testing. In the method of static interferometry, a fringe pattern obtained with a light source of wavelength λ_1 is recorded and replaced into the system as a moiré reference mask. Then the light source is replaced by one with wavelength λ_2. The resultant moiré fringe pattern is identical to the fringe pattern that would have been obtained with a light source of Λ_{12}. Polhemus modified the method to apply in real time systems. In the method of dynamic interferometry, light sources of λ_1 and λ_2 are operated simultaneously in the interferometer setup, giving a resultant fringe pattern of Λ_{12}.

In 1984, two-wavelength phase shifting interferometry was introduced as an optical phase unwrapping method. In this method, phase shifting interferometry and two-wavelength interferometry were combined to extend the phase measurement range of single-wavelength phase shifting interferometry. Two methods were introduced to solve 2π ambiguities by using two-wavelength phase shifting interferometer. In the first method, two sets of wrapped phase data are obtained with wavelengths λ_1 and λ_2. The data is then used to calculate the phase difference between pixels for beat wavelength Λ_{12}. Then all phase difference values are integrated to calculate the relative surface height of the test object. In the second method, two phase maps of different wavelengths are used to produce a phase map of beat wavelength. The beat wavelength phase map is then used as a reference to correct 2π ambiguities in the single wavelength phase map. Both methods were used to measure 1-D surface heights. Two-wavelength phase shifting interferometry has also been used to obtain three dimensional contour maps of aspheric surfaces with an accuracy of $\Lambda_{12}/100$. However, a disadvantage of this optical phase unwrapping is that the phase noise in each wavelength is magnified by a factor equal to the magnification of the wavelengths. This problem has been addressed by J. Gass, A. Dakoff and M. K. Kim. In the context of digital holography, two phase maps of wavelengths λ_1 and λ_2 are used to produce a phase map called "coarse map" with beat wavelength Λ_{12}. Then one of the single wavelength phase maps is used to reduce the amplified phase noise of the coarse map. The resultant 'fine map' has noise similar to the noise of the single wavelength phase map, with a larger axial range free of 2π ambiguities.

The two-wavelength phase unwrapping method has been extended to multiple wavelengths; enabling measurements of steep surfaces without software phase unwrapping. A hierarchical phase unwrapping algorithm that chooses a minimum number of wavelengths to increase the accuracy of optical unwrapping has been introduced by C. Wagner, W. Osten and S. Seebacher. The basic principle of this method is to start with a larger beat wavelength. Then a systematic reduction of beat wavelengths is used to improve the accuracy of the measurement while the information of the preceding measurements is used to eliminate 2π ambiguities. A similar version of hierarchical phase unwrapping has been presented by U. Schnars and W. Jueptner, however it has not been used experimentally. Three-wavelength phase unwrapping algorithms have been introduced in both interferometry and digital holography enabling measurements of steep surfaces without software phase unwrapping. While the principle of multi-wavelength phase unwrapping has been known in interferometry, until recently known applications had been confined to optical profilers with raster-scanned point-wise interferometry. Other than recent digital holography experiments, the first known application of multi-wave phase unwrapping to full-frame phase images in interferometry has been presented by N. Warnasooriya and M. K. Kim.

2.1 Principle of two-wavelength optical phase unwrapping

The basis of multi-wavelength optical phase unwrapping method is the idea of beat wavelength. For two wavelengths λ_1 and λ_2, the beat wavelength Λ_{12} is defined by

$$\Lambda_{12} = \frac{\lambda_1 \lambda_2}{|\lambda_1 - \lambda_2|} \tag{1}$$

For the mth wavelength λ_m, the surface profile Z_m of an object is related to the phase difference φ_m as follows;

$$Z_m = \frac{\lambda_m \varphi_m}{2\pi} \tag{2}$$

It is apparent that unambiguous range of Z can be increased by using a longer λ.
Consider two single wavelength phase maps φ_1 and φ_2 with wavelengths λ_1 = 530 nm and λ_2 = 470 nm respectively. The beat wavelength Λ_{12} for λ_1 and λ_2 is Λ_{12} = 4.151 µm. The Λ_{12} can be increased by choosing closer values of λ_1 and λ_2. The phase map for Λ_{12} is obtained by subtracting one single wavelength phase map from the other and then adding 2π whenever the resultant value is less than zero. This phase map is called "coarse map" φ_{12}. The surface profile for coarse map φ_{12} is given by $Z_m = \Lambda_{12}\varphi_{12} / 2\pi$. However, the phase noise in each single wavelength phase map is magnified by the same factor as the magnification of wavelengths. In the two-wavelength optical phase unwrapping method introduced by J. Gass *et.al.*, the phase noise is reduced by using the following steps.
First, the surface profile Z_{12} is divided into integer multiples of a single wavelength, say λ_1. Then, the result is added to the single wavelength surface profile Z_1. This significantly reduces the phase noise in the coarse map. However, at the boundaries of wavelength intervals λ_1 the noise of the single wavelength phase map appears as spikes. These spikes can be removed by comparing the result with the coarse map surface profile Z_{12}. If the difference is more than half of λ_1, addition or subtraction of one λ_1 depending on the sign of the difference removes the spikes. The final result 'fine map' has a noise level equal to that of single wavelength surface profile. If a single wavelength phase map φ_m contains a phase noise of $2\pi\varepsilon_m$ the two-wavelength phase unwrapping method works properly for $\varepsilon_m < \lambda_m / 4\Lambda_{12}$. Using a lager beat wavelength reduces the maximum noise limit.

2.2 Principle of three-wavelength optical phase unwrapping

The advantage of three wavelength phase unwrapping method is that the beat wavelength can be increased without reducing the maximum noise limit. Suppose the three chosen wavelengths are λ_1 = 625 nm, λ_2 = 590 nm, and λ_3 = 530 nm. The first two wavelengths give beat wavelength Λ_{12} = 10.53 µm. Instead of using the surface profile of Z_{12}, which has a high noise, an identical surface profile can be produced by using surface profiles Z_{13} and Z_{23} with beat wavelengths Λ_{13} = 3.49 μm and Λ_{23} = 5.21 μm. The resultant "coarse map of coarse maps" φ_{13-23} with surface profile Z_{13-23} also has the same beat wavelength $\Lambda_{13-23} = \Lambda_{13}\Lambda_{23} / |\Lambda_{13} - \Lambda_{23}| = 10.53$ μm.

The noise reduction is done as follows. In the first step, the quantity of integer multiples of Λ_{13} present in the range Z_{13-23} is calculated. The result $Z(a)$ is given by

$$Z(a) = \text{int}\left[\frac{Z_{13-23}}{\Lambda_{13}}\right]\Lambda_{13} \tag{3}$$

In the next step, the result is added to the surface profile Z_{13}.

$$Z(b) = Z(a) + Z_{13} \tag{4}$$

The resultant map is then compared with Z_{13-23}. If the difference $Z(c)$ is more than half of Λ_{13}, one Λ_{13} is added or subtracted depending on the sign difference.

$$Z(d) = \begin{cases} Z(c) + \Lambda_{13} & \text{if } Z(c) > \Lambda_{13}/2 \\ Z(c) & \text{if } -\Lambda_{13}/2 \le Z(c) \le \Lambda_{13}/2 \\ Z(c) - \Lambda_{13} & \text{if } Z(c) < -\Lambda_{13}/2 \end{cases} \tag{5}$$

The resultant surface profile $Z(d)$ is called "intermediate fine map" and has significantly reduced noise. Any remaining noise is due to the noise in the phase map φ_{13}. The remaining noise in $Z(d)$ is reduced by using a single wavelength, say λ_1. First, the intermediate fine map $Z(d)$ is divided into integer multiples of λ_1.

$$Z(e) = \text{int}\left[\frac{Z(d)}{\lambda_1}\right]\lambda_1 \tag{6}$$

Then the result is added to the single wavelength surface profile Z_1.

$$Z(f) = Z(e) + Z_1 \tag{7}$$

The resultant map $Z(f)$ is then compared with Z_1. If the difference is more than half of λ_1, one λ_1 is added or subtracted depending on the sign difference. The noise in the final map is equal to the noise in the single wavelength surface profile Z_1. The maximum noise level ε_m in the single wavelength phase map for the three wavelength phase unwrapping to work is given by the smaller value of $\Lambda_{13}/4\Lambda_{12} \sim 8.3\%$ or $\lambda_1/4\Lambda_{13} \sim 4.5\%$. Therefore, the three wavelength phase unwrapping method increases the beat wavelength without magnifying the noise in the final phase map.

The method can be applied to phase images obtained with any type of light source, regardless of the coherence length of the source.

3. Multi-wavelength optical phase unwrapping experiments

In this experiment, four step phase shifting algorithm is applied to the interference microscope. Though the minimum number of intensity values needed for phase calculation is three, even a small error in measurements can cause a large phase error. Taking four intensity measurements can reduce this effect. In order to acquire phase images, the Michelson interferometer is used as the experimental setup as shown in Figure (2).

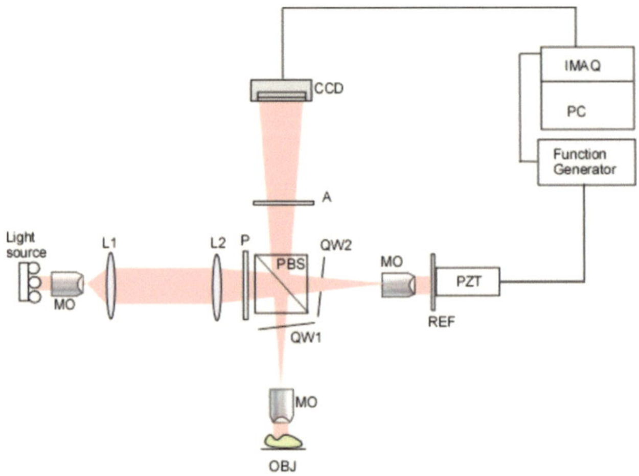

Fig. 2. A schematic diagram of the experimental setup. See text for details.

The light is expanded and collimated by the microsocpe objective MO and the lens L1, respectively. The light is then linealry polarized by the polarizer P. The polarized beam splitter (PBS) splits the incoming beam into an S-polarized (polarization plane is perpendicular to polarization axis) ray and a P-polarized (polarization plane is parallel to polarization axis) ray. The S-polarized beam is reflected at the PBS to illuminate the sample object OBJ and the P-polarized beam is transmitted throught the PBS to illuminate the reference mirror REF. When the S-polarized light passes through the quarter wave plate (QW1), the phase changes by 90° and it becomes circulary polarized. After reflcting at the mirror and going through another 90° phase shift at QW1, the light becomes P-polarized. This change from S-polarization to P-polarization avoids light traveling back to the light source and directs all reflected light to the charge-coupled device (CCD). Similarly, P-polarized light illuminating the REF changes to S-polarized light and travels to the CCD. At the analyzer A, the two S-polarized and P-polarized light beams are changed into a common polarization state so that the interference can occur on the CCD plane.

The polarizer-analyzer pair also controls the variation of the realtive intensity between the two arms. The reference mirror is mounted on a piezo-electric transducer (PZT). A function generator supplies a ramp signal to the PZT to dither the reference mirror by a distance of one wavelength. Images are recorded at quarter wavelength intervals.

Images acquired by the CCD are sent to an image acquisition board (National Instruments IMAQ PCI-1407) installed in the computer. The Intensity $I(x,y)$ of the light captured by CCD can be written as;

$$I(x,y) = I_O(x,y) + I_B(x,y) + I_R(x,y) + 2\sqrt{I_O(x,y)I_R(x,y)}\cos\left[\phi_i + \phi(x,y)\right] \tag{8}$$

Here $I_O(x,y)$ is the part of the beam reflected by the object that is coherent with respect to $I_R(x,y)$, the intensity of the beam reflected by the reference mirror. $I_B(x,y)$ is part of reflection from the object that is incoherent with respect to the reference – i.e. outside the coherence length. $\phi(x,y)$ the relative phase between the object and the reference mirror and ϕ_i is the phase shift introduced by moving the reference mirror by quarter wavelength intervals. Intensity distributions corresponding to the four images, acquired at $\phi_i = 0, \pi/2, \pi$ and $3\pi/2$, can be given as follows;

$$\begin{aligned} I_0 &= I_O + I_B + I_R + 2\sqrt{I_O I_R}\cos\phi \\ I_{\pi/2} &= I_O + I_B + I_R - 2\sqrt{I_O I_R}\sin\phi \\ I_\pi &= I_O + I_B + I_R - 2\sqrt{I_O I_R}\cos\phi \\ I_{3\pi/2} &= I_O + I_B + I_R + 2\sqrt{I_O I_R}\sin\phi \end{aligned} \tag{9}$$

The phase map of the object is given by;

$$\phi = Tan^{-1}\left(\frac{I_{3\pi/2} - I_{\pi/2}}{I_0 - I_\pi}\right) \tag{10}$$

Once a phase profile of the specimen is obtained, it can be used to determine the height profile of the specimen.
The optical path difference (OPD) between the object wave and reference wave is given by

$$OPD = \frac{\lambda\phi}{2\pi} \tag{11}$$

Here λ is the wavelength of the light beam and ϕ is the relative phase between the object and reference mirror. In the given Michelson type interferometer, the height profile of the object is half the OPD because the light travels towards the object, reflects and travels back. Therefore, the height profile h is related to the phase ϕ s follows.

$$h = \frac{1}{2}\left(\frac{\lambda}{2\pi}\right)\phi \tag{12}$$

4. Multi-wavelength optical phase unwrapping using light emitting diodes

Experimental results of multi-wavelength optical phase unwrapping using light emitting diodes (LED) are presented below. In interferometry, LEDs have been used as light sources in order to reduce the speckle noise inherent to lasers. Since LEDs have coherence lengths in micron range, speckle noise is greatly reduced. All the LEDs used in the experiment are Luxeon™ Emitter diodes from Lumileds Lighting LLC and have a Lambertian (high dome) radiation pattern. The peak wavelength, luminous flux, calculated and measured coherence lengths for red, red-orange, amber and green LEDs used in this experiment are shown in the Table 1. The calculated coherence length of a light source is given by $l_C = (2\ln 2/\pi)(\overline{\lambda}^2/\Delta\lambda)$, where $\overline{\lambda}$ is the mean wavelength and $\Delta\lambda$ is the full width half maximum (FWHM) of Gaussian spectrum. The coherence length was directly measured here by counting the number of fringes in the interference pattern of the tilted mirror object.

Colour	Luminous Flux Φ (lm) *	Peak Wavelength λ (nm)	Spectral Width (nm)	Calculated Coherence Length (μm)	Measured Coherence Length (μm)
Red	44	653.83±0.07	27.24±0.15	6.91±0.04	9.15±2.45
Red-Orange	55	643.42±0.07	23.21±0.14	7.85±0.05	10.29±2.57
Amber	36	603.48±0.03	17.53±0.05	9.14±0.03	10.86±2.56
Green	25	550.18±0.09	38.39±0.19	3.42±0.02	3.85±1.46

Table 1. Characteristics of LEDs. Luminous flux values are at 350 mA, Junction Temperature T_J= 25°C. * LuxeonTM Emitter and Star sample information AB11, 2 (Feb 2002).

4.1 Results for two-wavelength optical phase unwrapping

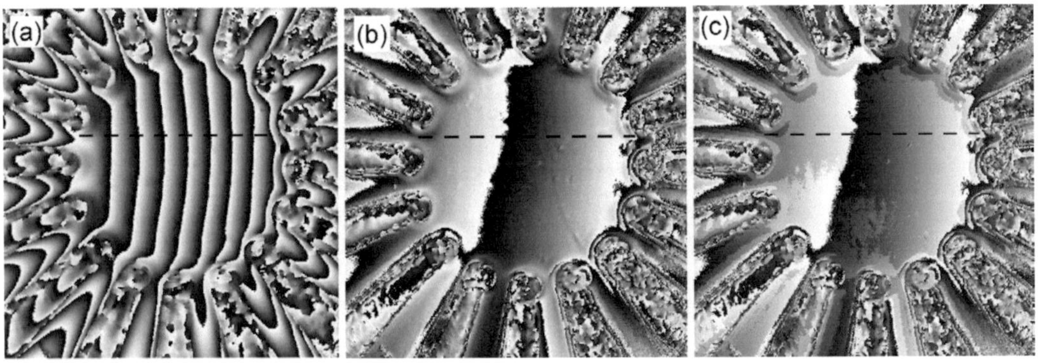

Fig. 3. Results of two-wavelength optical phase unwrapping. (a) a single wavelength phase map; (b) two-wavelength coarse map; (c) two-wavelength fine map with reduced noise.

The object here in the Fig. 3 is a micro-electrode array biosensor. It consists of 16 gold electrodes on a Pyrex glass substrate. The center is a 125 μm diameter circle with an approximate thickness of 2 μm. The center of the device was imaged and the experimental results for two wavelength optical phase unwrapping are shown in Fig. 3. Red (λ_1 = 653.83 nm) and green (λ_2 = 550.18 nm) LEDs are used as the two wavelengths. The beat wavelength Λ_{12} = 3.47 μm. Images are of a 184 μm × 184 μm area. Fig. 3(a) shows a single wavelength phase map φ_1 with λ_1 = 653.83 nm. The coarse map φ_{12} with Λ_{12} = 3.47 μm is shown in Fig. 3(b) and the final phase map with the reduced noise is shown in Fig. 3(c). Figure 4 shows cross section profiles of phase maps along the lines shown in Fig. 3 and the phase noise in the chosen regions. Figure 4(a) is a cross section of the single wavelength phase map with λ_1 = 653.83 nm and Fig. 4(b) is a cross section of coarse map with Λ_{12} = 3.47 μm. A cross section of the fine map with reduced phase noise is shown in Fig. 4(c). For maps (a), (b) and (c), the vertical axis is 4 μm. The root mean square (RMS) noise of the coarse map is 43.27 nm. This is shown in Fig. 4(d). Figure 4(e) shows the reduced noise in the fine phase map. Since the center of the device has a curvature, a paraboloid is fitted to the data. The red dotted line is the best-fit parabolic curve. After subtracting the curvature from the data, the Fig. 4(f) shows the corrected phase noise of 10.29 nm.

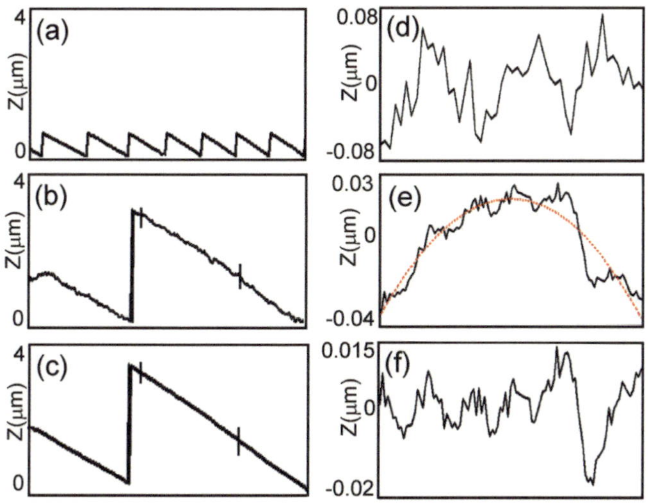

Fig. 4. Surface profiles for two-wavelength optical phase unwrapping. (a) single wavelength surface profile; (b) surface profile of coarse map; (c) surface profile of final unwrapped phase map with reduced noise; (d) noise of the coarse map in the region between the two markers in plot (b). RMS noise is 43.27 nm; (e) noise of final unwrapped phase map in the area shown in (c). Red dotted line is the best fit parabolic curvature and black solid line is data; (f) corrected phase noise of the unwrapped phase map, after subtracting the curvature of the object. RMS noise is 10.29 nm.

4.2 Results for three-wavelength optical phase unwrapping

The experimental results for three-wavelength optical phase unwrapping are presented in Fig. 5. Red (λ_1 = 653.83 nm), amber (λ_2 = 603.48 nm) and green (λ_3 = 550.18 nm) are used as the three wavelengths. The beat wavelength Λ_{13-23} = 7.84 μm. The object is the same micro-electrode array biosensor used in the previous section. Images are of a 184 μm × 184 μm area. Figure 5(a) is the single wavelength phase map φ_1 with λ_1 = 653.83 nm. The three wavelength coarse map is shown in Fig. 5(b) with a beat wavelength Λ_{13-23} = 7.84 μm. The final phase map with the reduced noise is shown in Fig. 5(c).

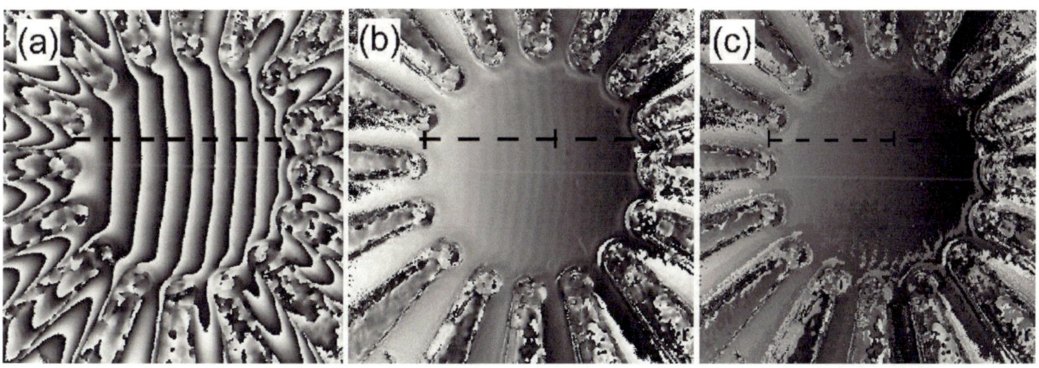

Fig. 5. Results of three-wavelength optical phase unwrapping. (a) single wavelength phase map; (b) three-wavelength coarse map; (c) three-wavelength fine map with reduced noise.

Cross section profile of each phase map is taken along the lines shown in Fig. 5. These cross section profiles and phase noise of coarse and fine maps are shown in Fig. 6. Figures 6 (a)-6(c) show surface profiles of single wavelength phase map, coarse map and fine map respectively. The vertical axis is 11 μm. Figure 6(d) shows 105.79 nm RMS noise of the coarse map. Because of the curvature of the object surface, a paraboloid is fitted with the final fine map data as shown in Fig. 6(e). The black line is data and the red dotted line shows the best-fit parabolic curve. Corrected phase noise in the final fine map is 4.78 nm, which is shown in Fig. 6(f).

The comparison of the two-wavelength optical phase unwrapping method to the three-wavelength optical phase unwrapping method shows that the three-wavelength phase unwrapping increases the axial range of the object, without increasing phase noise. These results show that the two-wavelength phase unwrapping method produced a 3.47 μm unambiguous range with 10.29 nm phase noise, while the three-wavelength phase unwrapping method produced much larger 7.48 μm unambiguous range with smaller 4.78 nm phase noise.

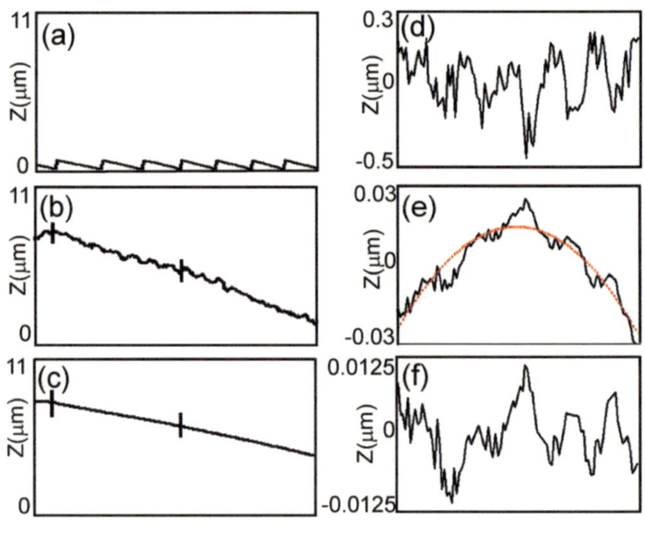

Fig. 6. Surface profiles for three-wavelength optical phase unwrapping. (a) single wavelength surface profile; (b) surface profile of coarse map; (c) surface profile of final unwrapped phase map with reduced noise; (d) noise of the coarse map in the region between the two markers in plot (b). RMS noise is 105.79 nm; (e) noise of final unwrapped phase ma in the area shown in (c). Red dotted line is the best fit parabolic curvature and black solid like is data; (f) corrected phase noise of the unwrapped phase map, after subtracting the curvature of the object. RMS noise is 4.78 nm.

Multi-wavelength optical phase unwrapping methods can also be used for biological cells as shown in Fig. 7. Here, Fig. 7 shows a single wavelength phase map, a coarse map and a fine map of onion cells using red (653.83 nm), amber (603.48 nm) and green (550.18 nm) wavelengths. The beat wavelength is 7.48 μm. Image size is 184 μm × 184 μm. The final fine map clearly shows the cell walls by eliminating the 2π ambiguities that would exist in a single wavelength phase image.

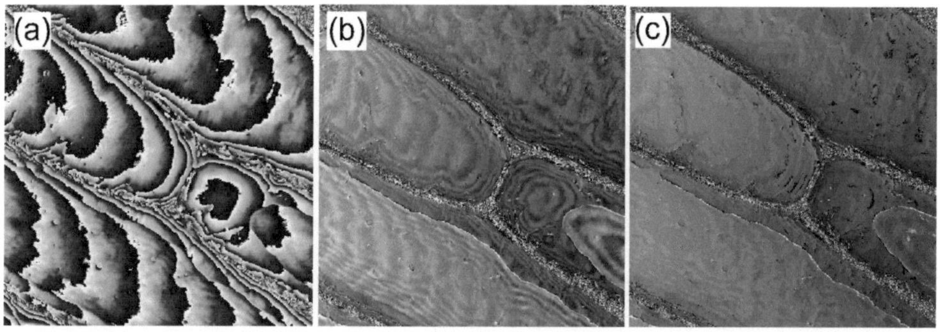

Fig. 7. Results of three-wavelength optical phase unwrapping. (a) a single wavelength phase map; (b) three-wavelength coarse map; (c) three-wavelength fine map with reduced noise.

5. Three-wavelength optical phase unwrapping using laser diodes & lasers

In the previous section, incoherent light sources (light emitting diodes) were used to reduce the speckle noise inherent in lasers. However, light emitting diodes are available in only several different wavelengths. Therefore, wavelength combinations that produce large beat wavelengths are limited. Because of small coherence lengths of light emitting diodes, imaging phase profiles of samples with features larger than the coherence range is not possible. In this section, the effectiveness of the three-wavelength optical phase unwrapping method is tested by using laser diodes and a ring dye laser. Laser diodes have been frequently used as a light source in interferometry due to their frequency tunability, smaller size and cost, compared to those of lasers. They also have shorter coherence lengths, typically few centimeters, compared to coherence length of lasers. However, laser diodes also have a limited availability of wavelength choices. Using a ring dye laser, the beat wavelength can be extended to more than a hundred micrometers. In this section, the effectiveness of the optical phase unwrapping method with any type of light source is presented. The results of three-wavelength optical phase unwrapping using laser diodes are shown in Fig. 8 and the results obtained with a ring dye laser as the light source are shown in Fig. 9, Fig. 10 and Fig. 11.

Figure 8 shows experimental results of three-wavelength optical phase unwrapping method using laser diodes. The object here is a micro-electrode array biosensor with 16 gold electrodes on a Pyrex glass substrate. The three wavelengths are λ_1 = 677.81 nm, λ_2 = 639.37 nm and λ_3 = 636.89.81 nm with a beat wavelength of Λ_{13-23} = 11.27 μm. Figure 8(a) is the single wavelength phase map with λ_1 = 677.81 nm. The three-wavelength coarse map is shown in Fig. 8 (b). The final fine map with reduced noise is show in Fig. 8 (c), and the 3-D rendering in Fig. 8 (d). The unwrapped phase map shows the grainy surface of electrodes.

In Fig. 9, results show phase images of a sample of cheek cells (basal mucosa) illuminated with a ring dye laser. The sample is illuminated by using wavelengths λ_1 = 579 nm, λ_2 = 577 nm and λ_3 = 574 nm. The beat wavelength is 167 μm. The image size is 102 μm per side. Here, Fig. 9(a) is the direct image of the cheek cell. Figure 9(b) shows the single wavelength phase map obtained using λ_1 = 579 nm. The coarse map produced by the three wavelengths is shown in Fig. 9(c). The final fine map with reduced noise is shown in Fig. 9(d) and the 3-D rendering of the final fine map is shown in Fig. 9(e).

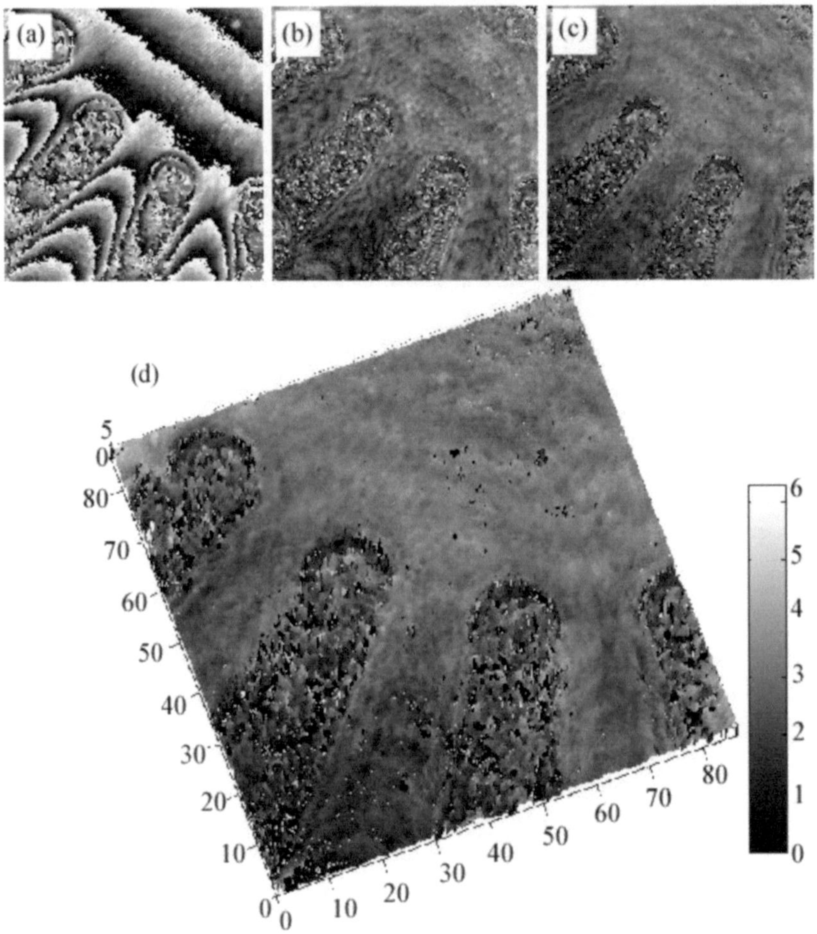

Fig. 8. Three-wavelength optical phase unwrapping of the biosensor using laser diodes. Image size is 86.6 µm per side. (a) a single wavelength phase map with λ_1 = 677.81 nm; (b) three-wavelength coarse map with Λ_{13-23} = 11.27 µm; (c) three-wavelength fine map with reduced noise; (d) 3-D rendering of (c).

In Fig. 10 and Fig. 11, the sample is a piece of 33 1/3 rpm long playing (LP) record. For 33 1/3 rpm records the typical width at the top of the groove ranges from 25.4 µm to 76.2 µm and the groove depth is ~15 µm. The sample is coated with a layer of 200 nm aluminum for better reflectivity. The three wavelengths used for the optical phase unwrapping process is λ_1 = 577 nm, λ_2 = 575 nm and λ_3 = 570 nm, with a beat wavelength of 166 µm. Figure 10(a) is the single wavelength phase map with λ_1 = 577 nm. The three-wavelength coarse map is shown in Fig. 10(b) with beat wavelength Λ_{13-23} = 166 µm. The bottom of the grooves

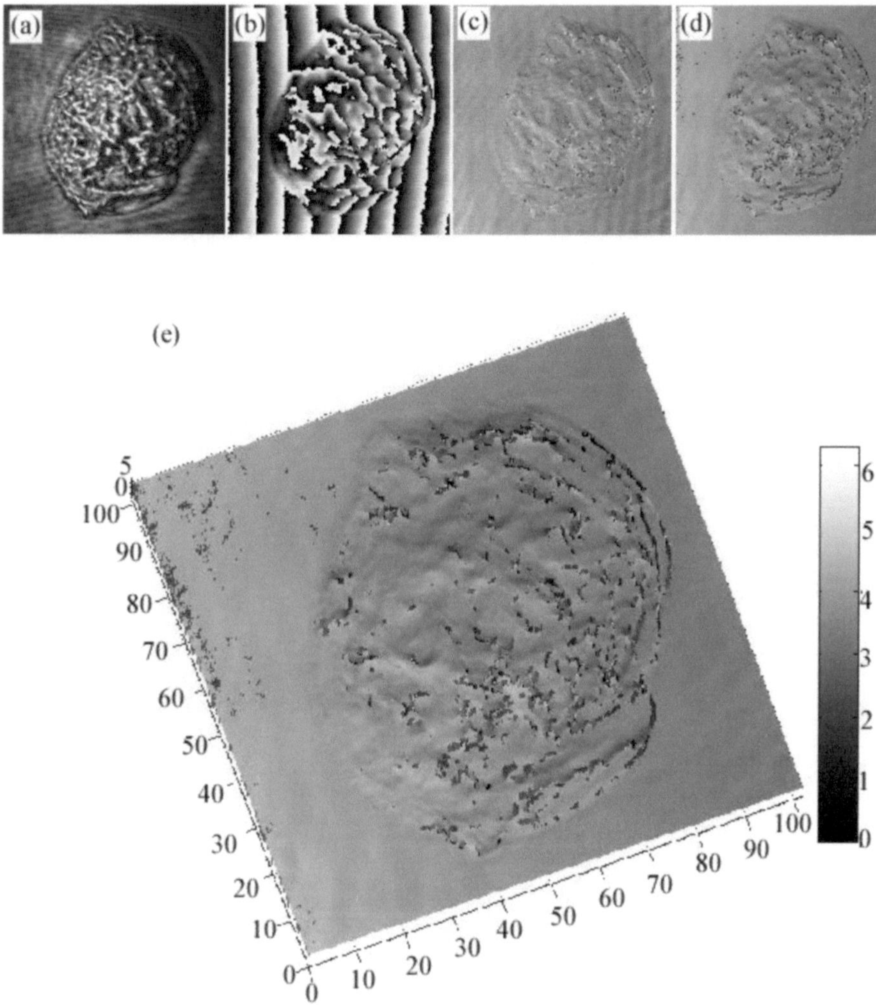

Fig. 9. Three-wavelength optical phase unwrapping of cheek cells using ring dye laser. Image size is 103 μm per side. (a) direct image of the cheek cell; (b) a single wavelength phase map; (c) three-wavelength coarse map; (d) three-wavelength fine map with reduced noise; (e) 3-D rendering of (d).

appears in darker color. The final fine map with reduced noise is shown in Fig. 10(c). Figure 10(d) is the 3-D rendering of the final fine map. In the final unwrapped phase map, the width of the top of the groove is measured along the line shown in Fig. 4(d). The measured width is 44 μm.

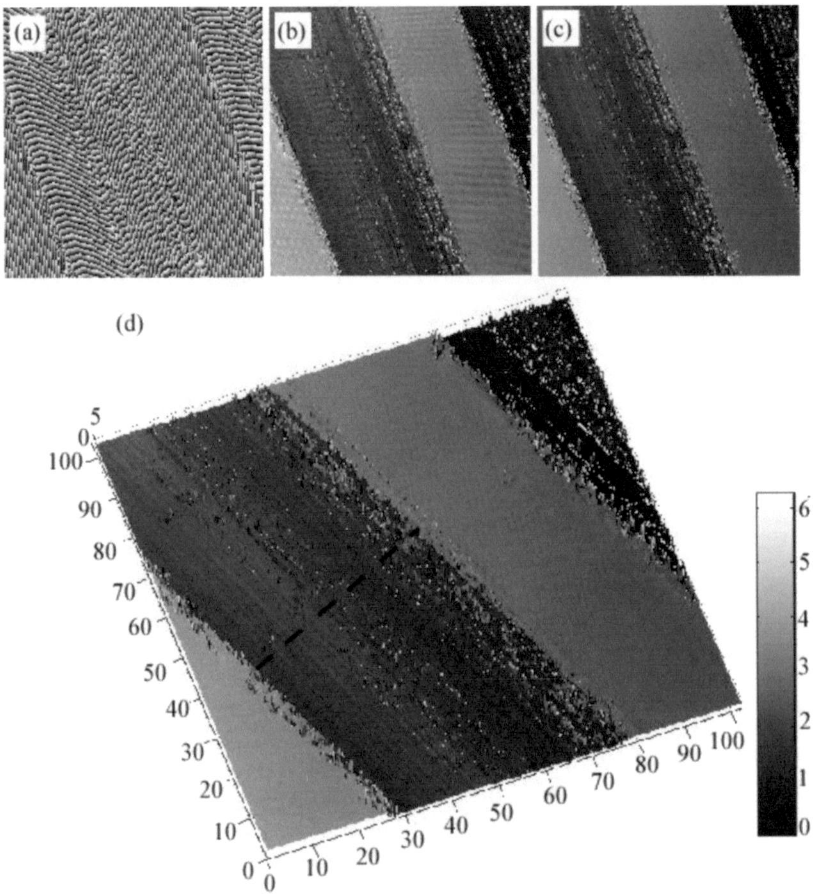

Fig. 10. Three-wavelength optical phase unwrapping of LP record grooves. Image size is 102 µm per side; (a) a single wavelength phase map; (b) three-wavelength coarse map; (c) three-wavelength fine map with reduced noise; (d) 3-D rendering of (c). The grove width is 44 µm.

Cross-sections and phase noise of the coarse and fine maps are shown in Fig.11. Figure 11(a) is the unwrapped coarse map and Fig. 11(b) is the final fine map with reduced noise. Figure 11(c) is the surface profile of the coarse map along the line shown in Fig. 11(a). The RMS noise in the coarse map in the area shown is 2.12 µm and this is shown in Fig. 11(d). Figure 11(e) shows the surface profile of fine map along the line shown in Fig. 11 (b). The groove depth h = 18 µm.

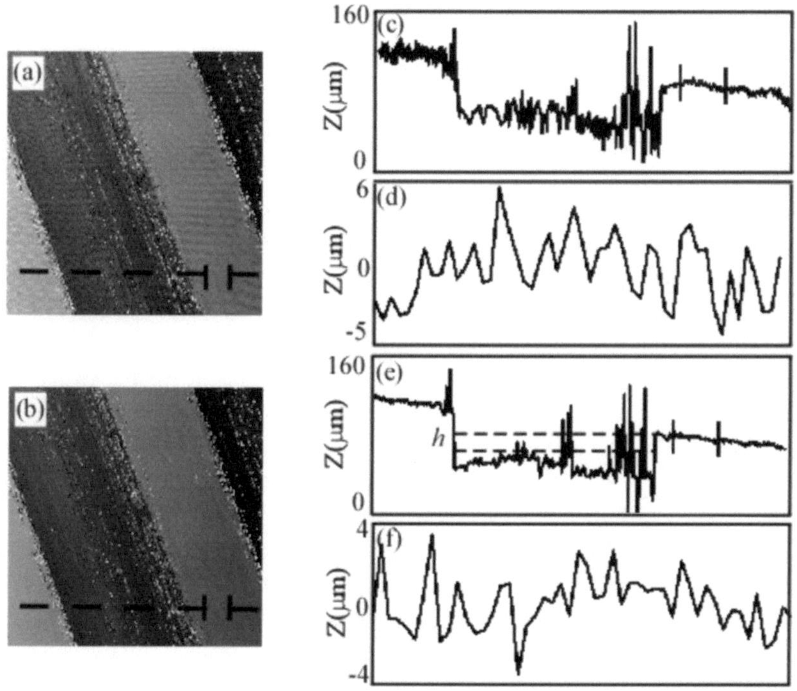

Fig. 11. Surface profiles of LP record grooves. (a) three-wavelength coarse map, (b) three-wavelength fine map with reduced noise, (c) surface profile of coarse map along the line shown in (a); (d) noise in coarse map in the area shown in (a). RMS noise is 2.12 μm ; (e) surface profile of fine map along the line shown in (b). The groove depth h = 18 μm ; (f) noise in the fine map in the area shown in (b). RMS noise is 1.36 μm.

5. Summary

In summary, this chpater demonstrates the effectiveness of the multi-wavelength optical unwrapping method. To our knowledge this is the first time that three wavelengths have been used in interferometry for phase unwrapping without increasing phase noise. Unlike conventional software phase unwrapping methods that fail when there is high phase noise and when there are irregularities in the object, the multi-wavelength optical phase unwrapping method can be used with any type of object. Software phase unwrapping algorithms can take more than ten minutes to unwrap phase images. This is a disadvantage when one needs to study live samples in real time or near - real time. The multi-wavelength optical unwrapping method is significantly faster than software algorithms and can be effectively used to study live samples in real time. Another advantage is that the optical phase unwrapping method is free of complex algorithms and needs less user intervention.

The method is a useful tool for determining optical thickness profiles of various microscopic samples, biological specimens and optical components. The optical phase unwrapping method can be further improved by adding more wavelengths, thus obtaining beat wavelengths tailored for specific samples.

6. References

Charette, P. G.; Hunter, I. W. (1996). Robust phase-unwrapping method for phase images with high noise content. *Applied Optics*, Vol. 35, Issue 19, (July 1996), pp. 3506-3513, ISSN 0003-6935.

Cheng, Y; Wyant, J. C. (1984). Two-wavelength phase shifting interferometry. *Applied Optics*, Vol. 23, Issue 24, (December 1984), pp. 4539-4543, ISSN 0003-6935.

Cheng, Y.; Wyant, J. C. (1985). Multiple-wavelength phase-shifting interferometry. *Applied Optics*, Vol. 24, Issue 6, (March 1985), pp. 804-807, ISSN 0003-6935.

Creath, K.; Cheng, Y.; Wyant, J. C. (1985). Contouring aspheric surfaces using two-wavelength phase-shifting interferometry. *Journal of Modern Optics*, Vol. 32, No. 12, (December 1985), pp. 1455-1464, ISSN 0950-0340.

Dilhaire,S.; Grauby, S.; Jorez, S.; Lopez, L. D. P.; Rampnoux, J.; Claeys, W. (2002). Surface displacement imaging by interferometry with a light emitting diode. *Applied Optics*, Vol. 41, Issue 24, pp.4996-5001, (August 2002),ISSN 0003-6935.

De Groot, P.; Kishner, S. (1991). Synthetic wavelength stabilization for two color laser-diode interferometry. *Applied Optics*, Vol. 30, Issue 28, pp 4026-4033 (October 1991), ISSN 0003-6935.

Fedeyev, V.; Haber, C. C. (2003). Reconstrcution of mechanically recorded sound by image processing. *LBNL Report 51983*, 2003.

Fercher, A.; Drexler, W.; Hitzenberger, C. K.; Lasser, T. (2003). Optical coherence tomography - principles and applications. *Reports on Progress in Physics*, Vol. 66, pp. 239-303 (January 2003).

Gass, J.; Dakoff, A.; Kim, M. K. (2003). Phase imaging without 2π ambiguity by multiwavelength digital holography. *Optics Letters*, Vol. 28, Issue 13, (July 2003), pp. 1141-1143, ISSN 0146-9592.

Ghiglia,D. C.; Romero, L. A. (1994). Robust two-dimensional weighted and unweighted phase unwrapping that uses fast transforms and iterative methods. *Journal of the Optical Society of America A*, Vol. 11, No. 1, (January 1994), pp. 107-117, ISSN 1084-7529.

Ishii, Y.; Onodera, R. (1995). Phase-extraction algorithm in laser-diode phase shifting interferometry. *Optics Letters*, Vol. 20, Issue 18, pp. 1883-1885 (September 1995), ISSN 0146-9592.

Liu, J.; Yamaguchi, I. (2000). Surface profilometry with laser-diode optical feedback interferometer outside optical benches. *Applied Optics*, Vol. 39, Issue 1, pp. 104-107 (January 2000), ISSN 0003-6935.

Lukashkin,A. N.; Bashtanov, M. E.; Russell, I. J. (2005). A self-mixing laser diode interferometer for measuring basilar membrane vibrations without opening the cochlea. *Journal of Neuroscience Methods,* Vol. 148, Issue 2, pp. 122-129 (October 2005), ISSN 0735-7044.

LuxeonTM Emitter and Star sample information AB11, 2 (Feb 2002).

Meiners-Hagen, K.; Burgarth, V.; Abou-Zeid, A. (2004). Profilometry with a multi-wavelength diode laser interferometer. *Measurement Science & Technology,* Vol. 15, No. 4, (April 2004), pp. 741-746, ISSN 0957-0233.

Montfort, F.; Colomb, C.; Charriere, F.; Kuhn, J.; Marquet, P.; Cuche, E.; Herminjard, S.; Depeursinge, C. (2006). Submicrometer optical tomography by multi-wavelength digital holographic microscopy. *Applied Optics,* Vol. 45, Issue 32, (November 2006), pp. 8209-8217, ISSN 0003-6935.

Onodera, R.; Ishii, Y. (1996). Phase-extraction analysis of laser-diode phase shifting interferometry that is insensitive to changes in laser power. *Journal of the Optical Society of America A,* Vol. 13, Issue 1, pp. 139-146 (January 1996), ISSN 1084-7529.

Parshall, D; Kim, M. K. (2006). Digital holographic microscopy with dual wavelength phase unwrapping. *Applied Optics,* Vol. 45, Issue 3, (January 2006), pp. 451-459, ISSN 0003-6935.

Polhemus,C. (1973). Two-wavelength interferometry. *Applied Optics,* Vol. 12, Issue 9, (September 1973), pp. 2071-2074, ISSN 0003-6935.

Repetto, L.; Piano, E.; Pontiggia, C. (2004). Lensless digital holographic microscope with light-emitting diode illumination. *Optics Letters,* Vol. 29, Issue 10, pp. 1132-1134 (May 2004), ISSN 0146-9592.

Schnars, U; Jueptner, W. (2005). *Digital Holography - Digital Hologram Recording, Numerical Reconstruction, and Related Techniques,* Springer, ISBN 354021934X, Berlin Heidelberg.

Servin, M.; Marroquin, J. L.; Malacara, D; Cuevas, F. J. (1998). Phase unwrapping with a regularized phase-tracking system. *Applied Optics,* Vol. 37, No. 10, (April 1998), pp. 1917-1923, ISSN 0003-6935.

Tziraki, M.; Jones, R.; French, P. M. W.; Melloch, M. R.; Nolte, D. D. (2000). Photorefractive holography for imaging through turbid media using low coherent light. *Applied Physics B,* Vol. 70, No. 1, (January 2000), pp. 151-154, ISSN 0946-2171.

Wagner, C.; Osten, W.; Seebacher, S. (2000). Direct shape measurements by digital wavefront reconstruction and multiwavelength countoutring. *Optical Engineering,* Vol. 39, Issue 1, (January 2000), pp. 79-85, ISSN 0091-3286.

Warnasooriya, N.; Kim, M. K. (2006). Multi-wavelength Phase Imaging Interference Microscopy. *Proceedings of SPIE - Volume 6090 Three-Dimensional and Multidimensional Microscopy: Image Acquisition and Processing XIII,* pp. 60900U-1 - 60900U-8, SPIE, January 2006, San Jose, California, USA.

Warnasooriya, N.; Kim, M. K. (2007). LED-based multi-wavelength phase imaging interference microscopy. *Optics Express,* Vol. 15, Issue 15, (July 2007), pp. 9239-9247, ISSN 1094-4087.

Warnasooriya, N.; Kim, M. K. (2009). Quantitative phase imaging using three-wavelength optical phase unwrapping, *Journal of Modern Optics,* Vol. 56, No. 1, (January 2009), pp; 85-92, ISSN 0950-0340.

Wyant,J. C. (1971). Testing aspherics using two-wavelength holography. *Applied Optics*, Vol. 10, Issue 9, (September 1971), pp. 2113-2118, ISSN 0003-6935.

34

Synchrotron-Based Time-Resolved X-ray Solution Scattering (Liquidography)

Shin-ichi Adachi[1], Jeongho Kim[2] and Hyotcherl Ihee[2]
[1]*Photon Factory, High Energy Accelerator Research Organization, 1-1 O-ho, Tsukuba, Ibaraki 305-0801,*
[2]*Center for Time-Resolved Diffraction, Department of Chemistry and Graduate School of Nanoscience & Technology (WCU), KAIST, 305-701,*
[1]*Japan*
[2]*Republic of Korea*

1. Introduction

Visualizing molecular structures in the course of a reaction process is one of the major grand challenges in chemistry, biology and physics. In particular, most chemical and biologically relevant reactions occur in solution, and solution-phase reactions exhibit rich chemistry due to the solute-solvent interplay. Studying photo-induced reactions in the solution phase offers opportunities for understanding fundamental molecular reaction dynamics and interplay between the solute and the solvent, but at the same time the interactions between solutes and solvents make this task challenging. Ultrafast emission, absorption and vibration spectroscopy in ultraviolet, visible and infrared regions have made possible the investigation of fast time-evolving processes. However, such time-resolved optical spectroscopic tools generally do not provide direct and detailed structural information such as bond lengths and angles of reaction intermediates because the spectroscopic signals utilizing light in the ultraviolet to infrared range cannot be directly translated into a molecular structure at the atomic level. In contrast, with the advance of X-ray synchrotron sources that can generate high-flux, ultrashort X-ray pulses, time-resolved X-ray diffraction (scattering) and absorption techniques have become general and powerful tools to explore structural dynamics of matters. Accordingly, the techniques have been successfully applied to studying various dynamics of chemical and biological systems (Coppens, 2003; Coppens *et al.*, 2004; Ihee, 2009; Ihee *et al.*, 2005b; Kim *et al.*, 2002; Schotte *et al.*, 2003; Srajer *et al.*, 1996; Techert *et al.*, 2001; Tomita *et al.*, 2009) and of condensed matters (Cavalieri *et al.*, 2005; Cavalleri *et al.*, 2006; Collet *et al.*, 2003; Fritz *et al.*, 2007; Gaffney *et al.*, 2005; Lee *et al.*, 2005; Lindenberg *et al.*, 2005). On one hand, time-resolved X-ray diffraction enables us to access to the mechanism of structural transformations at the atomic level in crystalline state (Collet *et al.*, 2003; Schotte *et al.*, 2003; Srajer *et al.*, 1996; Techert *et al.*, 2001). On the other hand, time-resolved X-ray absorption fine structure (XAFS) (Chen *et al.*, 2001; Saes *et al.*, 2003; Sato *et al.*, 2009) and time-resolved solution scattering (Davidsson *et al.*, 2005; Ihee, 2009; Ihee *et al.*, 2005a; Plech *et al.*, 2004) can probe structural dynamics in non-crystalline states of materials, complementing the X-ray diffraction technique.

In particular, time-resolved X-ray liquidography (TRXL), which is also known as time-resolved X-ray solution scattering (TRXSS), provides rather direct information of transient molecular structures because scattering signals are sensitive to all chemical species present in the sample and can be compared with the theoretical scattering signal calculated from three-dimensional atomic coordinates of involved chemical species. Accordingly, time-resolved X-ray liquidography using 100-picosecond X-ray pulses from a synchrotron source has been effective in elucidating molecular geometries involved in photoinduced reaction pathways, elegantly complementing ultrafast optical spectroscopy (Cammarata *et al.*, 2008; Cammarata *et al.*, 2006; Christensen *et al.*, 2009; Davidsson *et al.*, 2005; Georgiou *et al.*, 2006; Haldrup *et al.*, 2009; Ichiyanagi *et al.*, 2009; Ihee, 2009; Ihee *et al.*, 2005a; Kim *et al.*, 2006; Kong *et al.*, 2008; Kong *et al.*, 2007; Lee *et al.*, 2008a; Lee *et al.*, 2006; Lee *et al.*, 2008b; Plech *et al.*, 2004; Vincent *et al.*, 2009; Wulff *et al.*, 2006).

Time-resolved X-ray liquidography has been developed by combining the pulsed nature of synchrotron radiation and of lasers. In a typical experiment, a reaction is initiated by an ultrashort optical laser pulse (pump), and the time evolution of the induced structural changes is probed by the diffraction of a time-delayed, short X-ray pulse as a function of the time delay between the laser and X-ray pulses. In other words, the X-ray pulse replaces the optical probe pulse used in time-resolved optical pump-probe spectroscopy. X-ray pulses with a temporal duration of 50 ~ 150 ps are generated by placing an undulator in the path of electron bunches in a synchrotron storage ring.

In this chapter, we aim to review the experimental details and recent applications of time-resolved X-ray liquidography. Especially, we describe the details of the TRXL setup in NW14A beamline at KEK, where polychromatic X-ray pulses with an energy bandwidth of $\Delta E/E \sim 1 - 5\%$ are generated by reflecting white X-ray pulses ($\Delta E/E = 15\%$) through multilayer optics made of W/B_4C or depth-graded Ru/C on silicon substrates. Unlike in conventional X-ray scattering/diffraction experiments, where monochromatic X-rays are used to achieve high structural resolution, polychromatic X-ray pulses containing more photons than monochromatic X-ray pulses are used at the expense of the structural resolution because a higher signal-to-noise ratio is desirable in the TRXL experiment. In addition, we describe in detail the principle of synchronization between the laser and synchrotron X-ray pulses, which is one of the key technical components needed for the success of time-resolved X-ray experiments, and has been vigorously implemented in well-established experimental techniques using synchrotron radiation, such as diffraction, scattering, absorption and imaging. Finally, some examples of applications to various reaction systems ranging from small molecules to proteins are described as well.

2. Experimental

2.1 Optical-pump and X-ray-scatter scheme

In a typical TRXL experiment, an ultrashort optical laser pulse initiates photochemistry of a molecule of interest in the solution phase, and an ultrashort x-ray pulse from a synchrotron facility, instead of an ultrashort optical pulse used in the optical pump-probe experiment, is sent to the reacting volume to probe the structural dynamics inscribed on the time-resolved x-ray diffraction signals as a function of reaction time. TRXL data have been collected using an optical-pump and x-ray-probe diffractometer in the beamline ID09B at ESRF (Bourgeois *et al.*, 1996; Wulff *et al.*, 1997) and the beamline NW14A of PF-AR at KEK (Nozawa *et al.*, 2007). The beamline 14IDB at APS also has the capability of collecting TRXL data. The experimental setup is schematically illustrated in Fig. 1. It comprises a closed capillary jet or open-liquid jet to supply the solution that are pumped by laser pulses and scatter X-rays, a pulsed laser system to excite the sample, a pulsed synchrotron source to produce ultrashort X-ray pulses to scatter from the sample, a synchronized high-speed chopper that selects single X-ray pulses, and an integrating charge-coupled device (CCD) area detector.

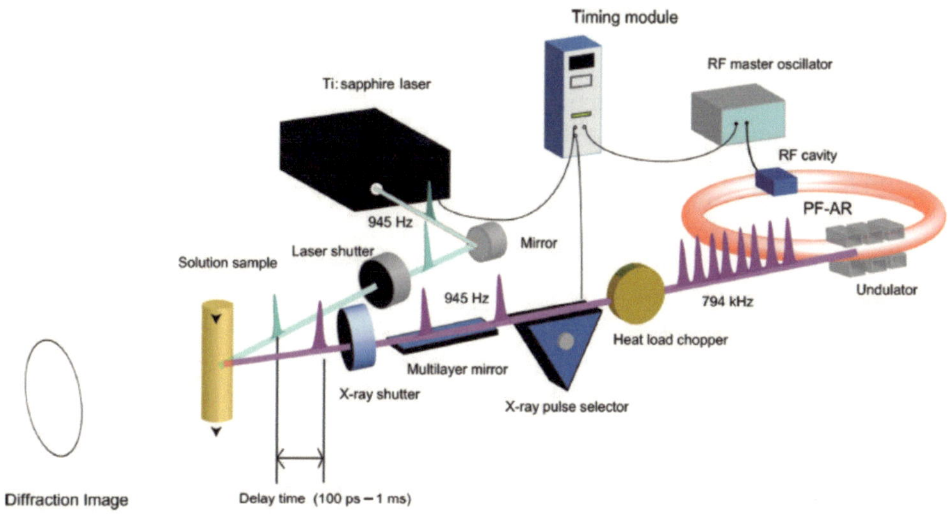

Fig. 1. Schematic drawing of the experimental setup for time-resolved X-ray liquidography. The liquid jet is irradiated by an optical laser pulse. After a well-defined time delay (t), the X-ray pulses generated by a synchrotron and selected by a high-speed chopper are sent to the sample and scatter. The reference diffraction data collected at -3 ns is subtracted from the diffraction data collected at positive time delays to extract the structural changes only.

2.2 Pulsed nature of synchrotron radiation

Synchrotron radiation is described as the radiation from charged particles accelerated at relativistic velocities by classical relativistic electrodynamics. It provides excellent characteristics as an X-ray source such as small divergence, short wavelength, linear or circular polarization, etc. Synchrotron radiation has another useful feature for time-resolved X-ray technique, short-pulsed nature, due to the periodic acceleration of charged particles in storage ring. Electrons circulating in storage ring irradiate synchrotron radiation and lose their energy. In order to compensate for the energy loss, a radio frequency (RF) oscillator accelerates electrons periodically at a harmonic frequency of the revolution frequency $f=c/L$, where c is the speed of light and L is the circumference of the storage ring. In order to keep electrons circulated stably in the storage ring, electrons need to pass through the RF oscillator at the appropriate timing, which is called the stable phase. Electrons stay and oscillate around the stable phase as a group, which is called electron bunch. Due to this equilibration process of the electron bunch, the length of the electron bunch is typically 15 - 45 mm (rms) that corresponds to X-ray pulse duration of 50 - 150 ps. Thus, the timing of the synchrotron X-ray pulse is synchronized with the timing of the RF oscillator. If the laser is externally triggered by the same RF master clock that accelerates electrons, both laser and X-ray pulses can be stably synchronized. This is the basis of time-resolved X-ray experiments using synchrotron radiation.

2.3 X-ray source characteristics and isolation of a single X-ray pulse

Synchrotron radiation is operated at MHz to GHz repetition rate depending on the bunch-filling modes of the storage ring. In particular, time-resolved experiments at synchrotron radiation facilities primarily require sparse bunch-filling mode of the storage ring operation such as single-bunch or hybrid modes. In general, X-ray detectors have a relatively slow response time and, furthermore, two-dimensional X-ray area detectors (e.g. CCD) have no fast gating capabilities. Due to such limitation of X-ray detectors, isolation of a single X-ray pulse from a pulse train is crucial for the success of time-resolved X-ray experiments. Since a single pulse can be readily isolated by using a fast chopper in sparse bunch-filling mode, the operation in the single-bunch or hybrid mode is highly desirable for time-resolved X-ray experiments.

The 6.5 GeV PF-AR is fully operated in a single-bunch mode for about 5000 hours/year. Electrons with a ring current of 60 mA (75.5 nC per bunch) are stored in a single electron bunch with a life time of around 20 hours. The RF frequency and harmonic number of the PF-AR are 508.58 MHz and 640, respectively. Therefore, the X-ray pulses are delivered at a frequency of 794 kHz (= 508.58 MHz / 640) with a pulse duration of about 60 ps (rms). A schematic drawing of the beam line NW14A is shown in Fig. 2.

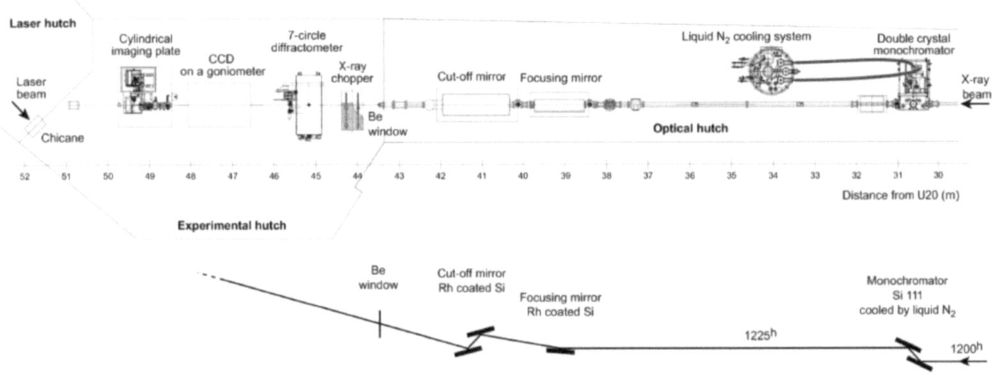

Fig. 2. Schematic drawing of the beamline NW14A of PF-AR at KEK. The X-ray beam is monochromized by a double-crystal monochromator and then focused using a bent-cylindrical mirror. Higher-order harmonics are cut off by a pair of flat mirrors.

The beam line has two undulators with a period length of 20 mm (U20) and 36 mm (U36). The U20 gives the 1st harmonic in the energy range of 13–18 keV. The energy bandwidth of the 1st harmonic is $\Delta E/E = 15\%$, which is utilized as a narrow-bandwidth white beam for TRXL experiments. The U36 covers an energy range of 5–20 keV with 1st, 3rd, and 5th harmonics, and useful for X-ray spectroscopy experiments. The measured photon flux from U36 and U20 at several gaps is shown in Fig. 3.

In order to isolate a single X-ray pulse from the sources, double X-ray choppers are equipped at the NW14A. The first chopper, called as heat-load chopper, has an opening time of 15 μs and is used to isolate 10-pulse train at 945 Hz (Gembicky *et al.*, 2007). The second X-ray chopper, made by Forschungszentrum Jülich (Lindenau *et al.*, 2004), consists of a rotor furnished with a narrow channel for the beam passage and isolates a single X-ray pulse from the 10-pulse train. The Jülich chopper realizes continuous phase locking with timing jitter less than 2 ns. The opening time of the channel at the center of the tapered aperture is

1.64 μs. If the repetition frequency of the pump-probe experiment is lower than 945 Hz, as is the case of using 10-Hz YAG laser system, a millisecond X-ray shutter (UNIBLITZ, XRS1S2P0) is set up between the X-ray chopper and the sample.

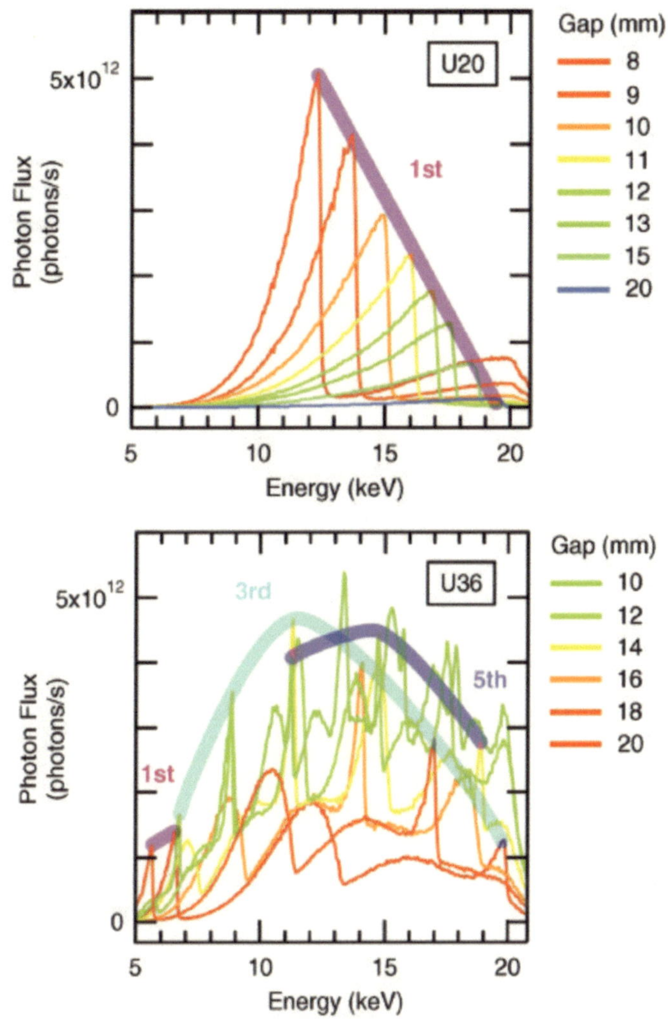

Fig. 3. Measured photon flux from U20 and U36 at several gaps. The intensity was normalized by 60 mA of the ring current and 0.318 mrad (H) x 0.053 mrad (V) beam divergence.

2.4 Energy bandwidth of the incident X-ray beam

In order to gain maximum X-ray photon flux at 1 kHz repetition rate, energy bandwidth of the incident X-ray is the key issue. The X-ray pulse with 3% energy bandwidth of the first harmonics of the undulator has been used for TRXL experiments in the beamline ID09B at ESRF (Cammarata *et al.*, 2008; Cammarata *et al.*, 2006; Christensen *et al.*, 2009; Davidsson *et al.*, 2005; Georgiou *et al.*, 2006; Ichiyanagi *et al.*, 2009; Ihee, 2009; Ihee *et al.*, 2005a; Kim *et al.*, 2006; Kong *et al.*, 2008; Kong *et al.*, 2007; Lee *et al.*, 2008a; Lee *et al.*, 2006; Lee *et al.*, 2008b;

Plech *et al.*, 2004; Vincent *et al.*, 2009; Wulff *et al.*, 2006). For example, the structural dynamics of $C_2H_4I_2$ in methanol were studied at the ID09B beamline (Ihee *et al.*, 2005a), and the reaction pathways and associated transient molecular structures in solution were resolved by the combination of theoretical calculations and global fitting analysis.

On the other hand, high-flux white X-ray at NW14A has $\Delta E/E$ = 15% energy bandwidth when the undulator U20 is used due to relatively large electron beam emittance of PF-AR. In order to examine the feasibility of time-resolved liquidography with such a large bandwidth and to search for the optimal bandwidth, we simulated the Debye scattering curves for the reaction $C_2H_4I_2 \rightarrow C_2H_4I + I$ using (i) a 15% bandwidth with the default X-ray energy distribution, such as the undulator spectrum at the NW14A beamline, (ii) a Gaussian spectrum with a 5% bandwidth, (iii) a Gaussian spectrum with a 1% bandwidth, and (iv) a Gaussian spectrum with a 0.01% energy bandwidth, as shown in Fig. 4.

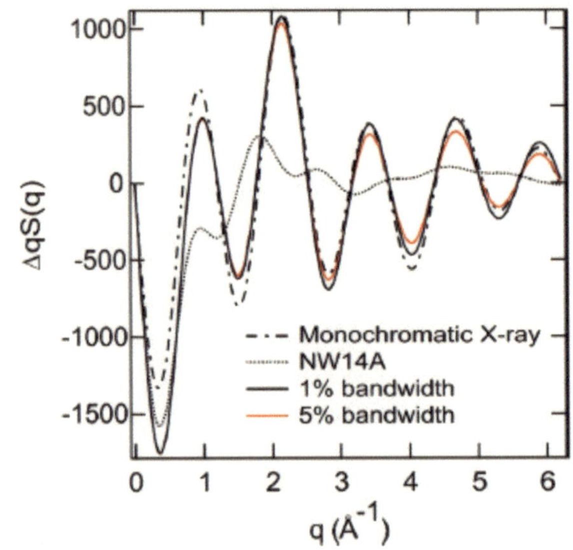

Fig. 4. Debye scattering curves calculated for the model reaction $C_2H_4I_2 \rightarrow C_2H_4I + I$ using a 0.01% (monochromatic) Gaussian X-ray energy profile (dot-dashed line), 5% Gaussian X-ray energy profile (red solid line), 1% Gaussian X-ray energy profile (solid line), and 15% default X-ray energy profile with a long tail (dotted line).

Although the photon flux of X-ray pulse increases with the energy bandwidth of the X-ray, the simulation shows that the default X-ray spectrum that has a 15% energy bandwidth as well as a long tail is not suitable for the time-resolved liquidography experiment owing to deteriorated structural resolution. Especially, the long tail of the default X-ray spectrum further blurs the scattering pattern at high scattering angles than when a symmetric Gaussian spectrum of the same bandwidth is assumed. As a result of the asymmetric lineshape, the X-ray spectrum with a long tail at ID09B of ESRF with a 3% bandwidth is effectively comparable to a symmetric Gaussian spectrum with a 10% bandwidth. In contrast, the scattering curve calculated from the Gaussian spectrum with a 5% energy bandwidth is similar in its structural resolution to that obtained from a 0.01% energy bandwidth (monochromatic) Gaussian spectrum. Furthermore, the total flux of the 5% energy bandwidth X-ray beam is higher than that of the monochromatic X-ray (a 0.01%

energy bandwidth) generated from a Si single crystal by a factor of 500. These estimations clearly suggest that the X-ray pulses with ΔE/E of 5% is appropriate for time-resolved X-ray liquidography experiment since it can provide a strong scattering signal without much sacrificing the structural resolution. Thus, we reduced the bandwidth of the X-ray pulses from the default 15% to less than the 5 % energy bandwidth.

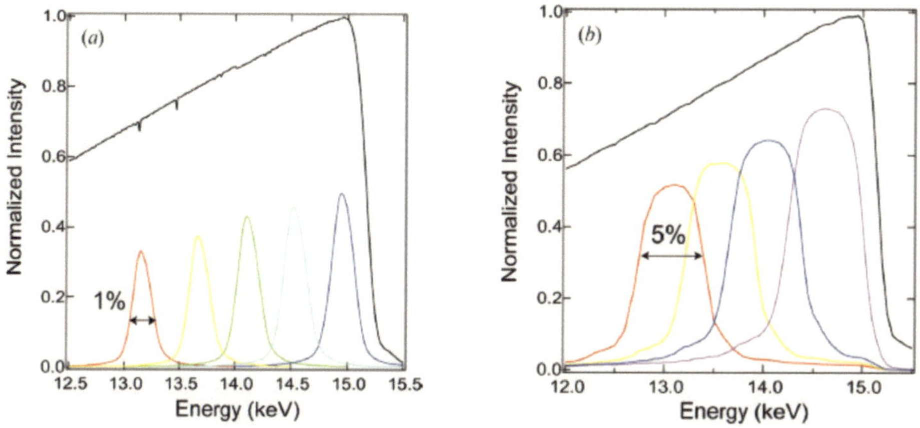

Fig. 5. Broadband X-ray pulses were produced by multilayer optics from the undulator spectrum. The peak energy position is controlled by changing the incident angle. The black curve is the X-ray spectrum from the undulator U20, with a gap of 11 mm. (a) X-ray spectra using the W/B_4C multilayer optics. The X-ray bandwidth is about 1%. (b) X-ray spectra using the depth-graded Ru/C multilayer optics. The X-ray bandwidth is 5%.

The multilayer optics produces X-ray pulses with a 1% to 5% energy bandwidth and allows us to measure TRXL with the undulator at the NW14A beamline. We used two types of multilayer optics. The first optics, made of W/B_4C (d =27.7 Å, X-ray Company, Russia) on a Si single crystal with a size of 50×50×5 mm^3, provides an X-ray spectrum with a 1% energy bandwidth, as shown in Fig. 5(a). The peak energy of the X-ray spectrum can be changed by tilting the angle of the multilayer optics. The second multilayer optics, which is made of depth-graded Ru/C layer (average d = 40 Å, NTT Advanced Technology, Japan), produces a 5% energy bandwidth X-ray spectrum, as shown in Fig. 5(b). A white X-ray with a photon flux of 1×10^9 photons/pulse is produced at a 1 kHz repetition rate. When multilayer optics with 1% and 5% energy bandwidths are used at the downstream of the Jülich chopper, the photon flux of 6×10^7 and 3×10^8 photons/pulse is obtained, respectively.

2.5 Synchronization of laser and X-ray pulses

NW14A is equipped with a 150-fs Ti:sapphire regenerative amplifier laser system (Spectra Physics, Millenia, Tsunami, Spitfire, Empower). The Ti:sapphire laser system produces optical pulses at 800 nm at a 945-Hz repetition rate, with the pulse energy reaching up to 800 μJ/pulse . The laser is installed in a laser booth next to the experimental hutch. An optical parametric amplifier (Light Conversion, TOPAS-C) is also installed in the laser booth for conversion of 800 nm light to broad spectral range from visible to mid-infrared region. The laser beam is brought to the sample in the experimental hutch through a beam duct for the laser. The synchronization of X-ray and laser pulses is based on the RF master clock, by which an electron bunch is driven in the storage ring. When the X-ray experiment is conducted with a 945 Hz Ti:sapphire-laser and a detector that has no gating capabilities (e.g. CCD), an X-ray chopper is required to synchronize the X-ray and laser pulses at a 1:1 ratio. The timing chart of the synchronization is shown in Fig. 6.

The X-ray pulse is emitted every 1.26 μs (794 kHz = 508 MHz / 640) from the PF-AR. After the RF amplifier, the RF master clock signal of PF-AR is split into two major timing components: one for the laser system and the other for the X-ray chopper system. In the X-ray chopper system, the 508 MHz RF and the 794 kHz revolution signals are used as the clock and the reference signals, respectively. A 945 Hz (794 kHz / 840) repetition frequency of the X-ray pulses is then selected to trigger the Ti:sapphire 150-fs laser system running at the same repetition frequency. In the laser system, the mode-locked Ti:sapphire oscillator operating at 85 MHz (508 MHz / 6) synchronized with the X-ray pulses provides seed pulses to the regenerative amplifier. The seed pulses trigger the regenerative amplifier pumped by the Q-switched Nd:YLF laser at 945 Hz (85 MHz / 89600). Then, 945 Hz laser pulses are directed to the sample position by a series of mirrors. The pulse trains of pumping laser and probing X-ray pulses at the sample are shown together in Fig. 6. The timing of the delay between the two pulse trains is controlled by changing the ejection timing of the laser pulses from the regenerative amplifier using a phase shifter (Candox). The timing of the X-ray and the laser is measured with an InGaAs metal-semiconductor-metal (MSM) photodetector (Hamamatsu, G7096) coupled to a high-frequency preamplifier and a 2.5 GHz digital oscilloscope (Tektronix, DPO7254). The rise time of the MSM photodetector is typically 40 ps, which is faster than the X-ray pulse duration, and the photodetector is set at the sample position.

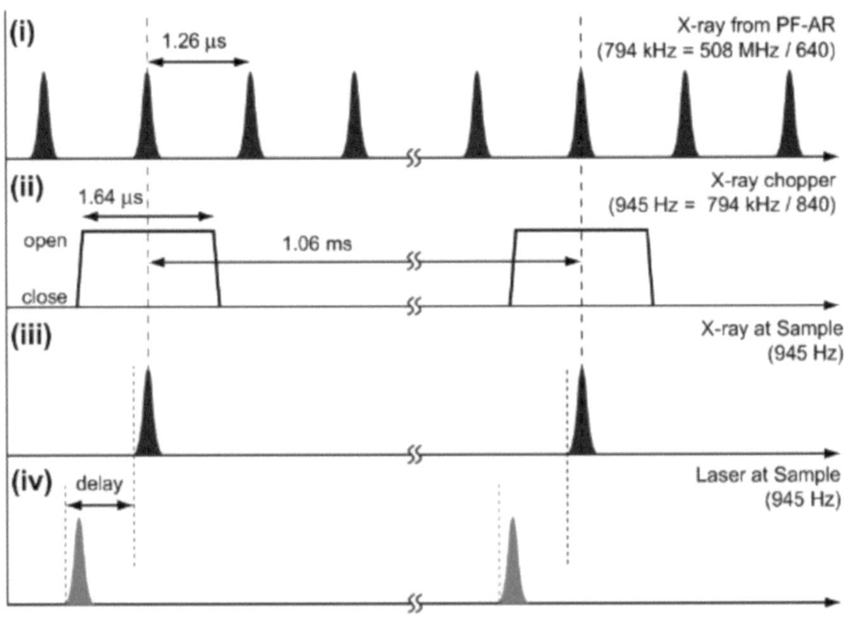

Fig. 6. The timing chart of the synchronization system at NW14A when using the X-ray chopper to synchronize 794 kHz X-ray and 945 Hz laser pulses at a 1:1 ratio. Timing settings of the X-rays from (i) PF-AR, (ii) the X-ray chopper, (iii) the X-rays at the sample, and (iv) the laser at the sample are shown.

2.6 Spatial and temporal overlaps

In order to increase the signal-to-noise ratio of the TRXL data and define accurate time delay between laser and X-ray pulses, the laser and X-ray pulses have to be overlapped at the sample both spatially and temporally. To check the temporal overlap, we place a fast InGaAs detector at the sample position and record the time traces of the laser and X-ray pulses along a single time axis monitored by a 2.5 GHz digital oscilloscope. By adjusting the laser firing time, it is possible to adjust the relative timing between the two pulses within a few picoseconds. During an experiment, the time traces of the laser and X-ray pulses are monitored by fast photodiodes simultaneously and non-intrusively.

The spatial overlap between X-ray and laser pulses is achieved by using a 50 μm diameter pinhole placed at the sample position. The pinhole is located at the center of X-ray beam, and then the laser beam is moved across the pinhole by scanning the position of the focusing lens until it passes through the center of the pinhole. To ensure precise spatial overlap, we monitor the intensity of scattering induced by thermal expansion in a liquid solvent, which typically occurs in 1 μs with our beam sizes. Specifically, the ratio of scattered intensities in the inner and outer disks of the solvent signal is monitored. Once the sample expands, the solvent signal shifts to lower scattering angles, leading to the increase of low-angle scattering and the decrease of high-angle scattering. Therefore, the ratio between the inner and outer part of the solvent signal changes in proportion with the laser excitation. The X-ray beam is typically vertically 200 μm and horizontally 250 μm. The laser spot is of circular shape with a diameter of 300 - 400 μm.

2.7 Sample environment and data acquisition

Two different types of sample cell systems have been used: a diluted solution of 0.5~100 mM concentration or pure solvent is prepared and circulated through either a capillary or through an open-jet sapphire nozzle. Such flow systems provide a stable liquid flow of ~0.3 mm thickness at a speed ensuring the refreshment of probe volume for every laser pulse (typically ~3 m/s). In the capillary-based system, the solution is flowed through a quartz capillary of 0.3 mm diameter. In the open-jet system, the capillary is removed and the solution is passed between two flat sapphire crystals with a spacing of 0.3 mm (Kyburz), which produces a stable naked liquid sheet directly exposed to the laser/x-ray beams. The open-jet system producing a bare liquid jet has the advantage over the closed capillary system in terms that the scattering background arising from the glass material of the capillary is eliminated and thus the signal-to-noise-ratio substantially improves. The lower background also helps to enhance the accuracy of the normalization process. In addition, the capillary jet often encounters a problem that the excitation laser drills a hole in the capillary.

The molecules in the jet are excited by laser pulses from the femtosecond laser system described above. To maximize the population of transients and photoproducts, the laser pulse energy (typically 25 - 100 μJ depending on the excitation wavelength) is set to be relatively higher than that used in typical time-resolved optical spectroscopy, and thus multi-photon excitation often occurs. In general, one wants to follow photochemistry induced by only one-photon absorption that the laser pulse duration of ~100 fs is stretched to ~2 ps by introducing chirp from a pair of fused-silica prisms inserted before the sample. To probe slow photoinduced dynamics, a nanosecond laser system is used instead of the femtosecond laser system.

The laser beam is generally directed to the sample with a 10 degree tilt angle relative to the X-ray beam. The scattered X-ray diffraction signal is recorded by an area detector

(MarCCD165, Rayonics, 2048 × 2048, ~80 μm effective pixel size) with a sample-to-detector distance of ~45 mm. A typical exposure time is ~5 s, and, given the ~1 kHz repetition rate of the laser/X-ray pulses, the detector receives 5×10^3 X-ray pulses and $\sim 5 \times 10^{12}$ X-ray photons per image. Diffraction data are collected for typically 10 or more time-delays (t) from -100 ps up to 1 μs (for example, –100 ps, 0 ps, 30 ps, 100 ps, 300 ps, 1 ns, 3 ns, 10 ns, 30 ns, 100 ns, 300 ns, and 1 μs). Each time-delay is interleaved by a measurement of the unperturbed sample (typically at –3 ns).

3. Data processing and analysis

3.1 Conversion of 2D images into 1D curves

The two-dimensional diffraction images are radially integrated into one-dimensional intensity curve, $S(q, t)^{exp}$, as a function of the momentum transfer q ($q = (4\pi/\lambda)\sin(2\theta/2)$ where λ is the wavelength of the X-rays, 2θ is the scattering angle, and t is the time delay). The curves are averaged and normalized by scaling them to absolute scale of the total (elastic and inelastic) scattering from one solution unit-cell molecule in the isosbestic point at high q values, where the scattering is insensitive to structural changes. After normalization, the diffraction data for the unperturbed sample measured at a negative time delay (typically at –3 ns) is subtracted from the diffraction data collected at positive time delays to extract the diffraction change only. The difference diffraction intensities $\Delta S(q,t)$ contain direct information on the structural changes of the solute and solvent in the probed solution. The relative laser induced diffraction signal change $\Delta S/S$ is quite small. It depends on both time and scattering angle, and is typically less than 0.1 %. Standard deviations as a function of q are calculated in the process of conversion from a 2D image to a 1D curve by taking into account the distribution of the intensities at the same q value. The error of the averaged $\Delta S(q,t)$ can be obtained from the error propagation of standard deviations or by taking another standard deviation from the mean value of individual difference curves. The signal-to-noise ratio of a typical $\Delta S(q,t)$ depends on q and t and oscillates resembling the shape of $\Delta S(q,t)$ except that the negative values of $\Delta S(q,t)$ become positive in the plot of signal-to-noise ratio. A typical averaged $\Delta S(q,t)$ from about 50 - 100 repetitions has a signal-to-noise ratio up to 15. The signal-to-noise ratio is zero when ΔS is zero and reaches a maxima in the peaks and valleys of $\Delta S(q,t)$. To magnify the oscillatory feature at high q, $\Delta S(q,t)$ is often multiplied by q to yield $q\Delta S(q,t)$. Although $q\Delta S(q,t)$ contains direct information on the structural changes, often the result in reciprocal space is not intuitive. For this reason $q\Delta S(q,t)$ is transformed to real space where the changes are more readily interpretable: positive and negative peaks means formation and depletion, respectively, of the corresponding interatomic distance. Obtained through sine-Fourier transforms of $q\Delta S(q,t)$, the difference radial distribution function ($r\Delta R(r,t)$) represents the experimental atom-atom pair distribution function during the course of the reaction.

$$r\Delta R(r,t) = \frac{1}{2\pi^2}\int_0^\infty q\Delta S(q,t)\sin(qr)\exp(-q^2\alpha)dq \qquad (1)$$

where the constant α is a damping constant to account for the finite experimental q range. In principle, the errors in the r-space can be also obtained from the same procedure as the one described for the q-space data: The sine-Fourier transform of every single $q\Delta S(q,t)$ is taken and then averaged over all $r\Delta R(r,t)$ curves, which defines a meaningful standard deviation.

3.2 Data analysis

We fit the experimental difference intensities ($\Delta S(q,t)^{exp}$) for all time delays against theoretical difference intensities ($\Delta S(q,t)^{theory}$), including the changes from three principal components that are mutually constrained by energy conservation in the X-ray illuminated volume: (i) the solute-only term, (ii) the solute-solvent cross term (also called as the cage term), and (iii) the solvent-only term from hydrodynamics as in the following expression.

$$\begin{aligned}\Delta S(q,t)^{theory} &= \Delta S(q,t)_{solute-only} + \Delta S(q,t)_{solute-solvent} + \Delta S(q,t)_{solvent-only} \\ &= \Delta S(q,t)_{solute-related} + \Delta S(q,t)_{solvent-only} \\ &= \frac{1}{R}\left[\sum_k c_k(t)S_k(q) - S_g(q)\sum_k c_k(0)\right] + (\partial S/\partial T)_\rho \Delta T(t) + (\partial S/\partial \rho)_T \Delta\rho(t)\end{aligned} \quad (2)$$

where R is the ratio of the number of solvent molecules to that of solute molecules, k is the index of the solute (reactants, intermediates and products), $c_k(t)$ the fraction of molecules in k as a function of time t, $S_k(q)$ is the solute-related (the solute-only plus the cage components) scattering intensity of species k, $S_g(q)$ is the scattering intensity of the reactants (k = reactants), $(\partial S(q)/\partial T)_\rho$ is the solvent scattering change in response to a temperature rise at constant volume, $(\partial S(q)/\partial \rho)_T$ is the response to a density change at constant temperature, $\Delta T(t)$ and $\Delta\rho(t)$ are the solvent temperature and density changes as a function of time. The equation indicates that, to calculate $\Delta S(q,t)^{theory}$, two types of basis components are needed: time-independent functions such as $S_k(q)$, $(\partial S(q)/\partial T)_\rho$ and $(\partial S(q)/\partial \rho)_T$ and time-dependent functions such as $c_k(t)$, $\Delta T(t)$ and $\Delta\rho(t)$. In the following, the steps involved in the calculation of time-independent and time-independent basis functions are described with the photochemistry of CHI_3 in CH_3OH as an example. Fig. 7 presents an overall scheme for data analysis.

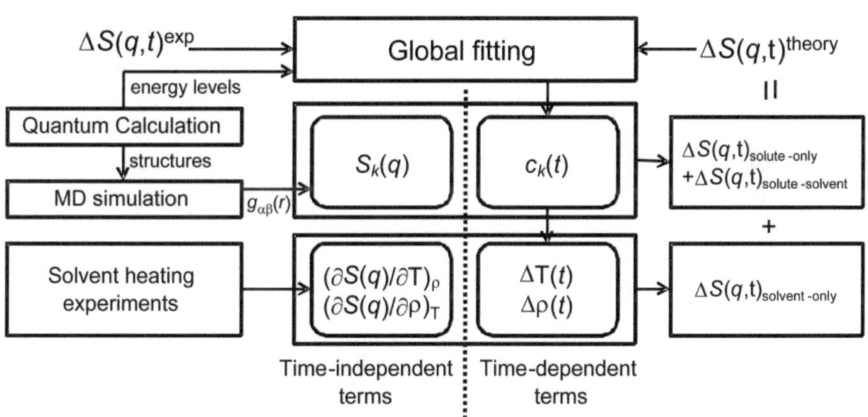

Fig. 7. Schematic of the data analysis. A theoretical difference scattering curve is represented as a sum of the three terms contributions: the solute-only term, the solute-solvent cross term, and the solvent-only term. The discrepancy between the theory and experiment is minimized in global fitting analysis by considering data at all positive time delays simultaneously. See the text for details.

$S_k(q)$ are calculated from MD simulations combined with quantum calculations. The possible structures of the parent molecule, the transient intermediates and the products in solution are provided by fully optimizing the molecular geometry with the ab initio and/or density functional theory (DFT) methods with solvent effects included. In case of the photochemistry of CHI_3 in CH_3OH, the molecular structures of all putative species such as CHI_3, CHI_2, CHI_2-I isomer, I_2, I, and CH_3OH are calculated. The charge on each atom of all related species is also calculated via the natural bond orbital analysis. These structures and the charges of all species are used as starting points for the MD simulations, where one solute molecule is placed in a box containing 512 or more rigid solvent molecules. After MD simulations, pair correlation functions for atom pairs ($g_{\alpha\beta}(r)$ for the atom pair α and β) are calculated. The $S_k(q)$ curves are then computed by

$$S(q)=\sum_{\alpha\beta} f_\alpha(q)\, f_\beta(q)\left(N_\alpha\, \delta_{\alpha\beta} + \frac{N_\alpha N_\beta}{V}\int_0^\infty (g_{\alpha\beta}(r)-1)\,\frac{\sin(qr)}{qr}\,4\pi r^2\, dr \right) \tag{3}$$

where $f_\alpha(q)$ is the atomic formfactor of the α atom, N_α is the number of α atoms in the MD simulation, $\delta_{\alpha\beta}$ is Kronecker delta, and V is the volume of the MD box. Including $g_{\alpha\beta}(r)$ for only the pairs within the solute molecule (for example, CHI_2-I isomer has one type of C···H, three types of C···I, three types of H···I and three types of I···I) results in the solute-only term, which can be also described by Debye scattering of isolated solute molecules as in the gas phase. The cage term is calculated when $g_{\alpha\beta}(r)$ for the solvent-solute cross pairs (for example, the CHI_2 radical in CH_3OH has C_{solute}···$C_{solvent}$, C_{solute}···O, I··· $C_{solvent}$, and I···O, and many other pairs including H) are used in the integration. In practice, $g_{\alpha\beta}(r)$ for both solute-only and solute-solvent cross pairs are used to yield the solute-only plus cage terms, that is, the solute-related terms, $S_k(q)$. The solvent differential functions, $(\partial S(q)/\partial T)_\rho$ and $(\partial S(q)/\partial \rho)_T$, can be obtained either from MD simulations or determined in a separate experiment where the pure solvent is vibrationally excited by near-infrared light (Cammarata *et al.*, 2006). The latter gives superior agreement than the former. In general, the $g_{\alpha\beta}(r)$ from MD simulation for a particular atom pairs α and β can be used to calculate the contribution from that particular atom-atom pair to the overall signal, thereby aiding the peak assignment (for example, the atom pair of I and O gives the I···O interatomic contribution, which is one of the major solute-solvent cross terms).

The basic strategy of the least square fits to the experimental data is to minimize the total χ^2 iteratively in a global fitting procedure, simultaneously minimizing the differences between the experimental and theoretical curves at all positive time delays. The definition of chi-square (χ^2) used is as follows.

$$\chi^2=\sum_t \chi_t^2=\sum_t\sum_q\left(\frac{\left(\Delta S^{theory}(q,t)-\Delta S^{exp}(q,t)\right)}{\sigma_{q,t}}\right)^2 \tag{4}$$

The polychromacity of the X-ray beam has to be taken into account when a $\Delta S(q,t)^{theory}$ curve is compared with the $\Delta S(q,t)^{exp}$ curve by weighting the X-ray spectrum into the $\Delta S(q,t)^{theory}$ curve. A result of global fitting analysis for CHI_3 is shown in Fig. 8. The time-dependent basis functions ($c_k(t)$, $\Delta T(t)$ and $\Delta\rho(t)$) depend on the fitting parameters of the global fitting analysis. A set of rate equations for a reaction kinetic model including all reasonable candidate reaction pathways is set up to extract a reaction mechanism. As a candidate reaction model for CHI_3, the rate constants for dissociation ($CHI_3 \rightarrow CHI_2 + I$), geminate and non-geminate recombination ($CHI_2 + I \rightarrow CHI_3$), and the non-geminate formation of molecular iodine ($I + I \rightarrow I_2$) can be considered. Integrating the rate equations provides $c_k(t)$ to be used to construct the theoretical scattering signal. The $\Delta T(t)$ and $\Delta\rho(t)$ are mathematically linked to $c_k(t)$ and to each other by energy and mass conservation and hydrodynamics. From $c_k(t)$, the time-dependent heat released from solutes to the solvent, $Q(t)$, is calculated and used to compute $\Delta T(t)$ and $\Delta\rho(t)$ via thermodynamic and hydrodynamics relations (Bratos *et al.*, 2004).

The fitting parameters include the rate coefficients, the fraction of the excited molecules, the fraction of the molecules undergoing structural changes, and the laser beam size. Structural parameters such as bond lengths and angles and energy levels of chemical species can be included as fitting parameters.

3.3 Example: Photochemistry of CHI_3

Fig. 8A shows a comparison of $q\Delta S(q, t)^{exp}$ and $q\Delta S(q, t)^{theory}$ from global fitting analysis of TRXL data of CHI_3 in CH_3OH, and Fig. 8B shows the corresponding $r\Delta R(r,t)^{exp}$ and $r\Delta R(r,t)^{theory}$. Fig. 8E summarizes the final fit values. Upon irradiation of 20 mM iodoform in methanol, 24(±1)% of the solute molecules are excited by the laser pulse at 267 nm. Among the excited iodoform, 28(±1)% dissociate into CHI_2 + I within the time resolution of 100 ps, and the remaining 72(±1)% decay into the ground state via vibrational cooling and release their energy to the solvent. The iodine atoms recombine to form I_2 with the bimolecular rate constant of $1.55(\pm0.25) \times 10^{10}$ $M^{-1}s^{-1}$. Based on these values from global fitting analysis, chemical population changes (as shown in Fig. 8C) and the temperature and density change of the solvent (as shown in Fig. 8D) as a function of time can be drawn. Initially, the temperature and the pressure of the solvent increase at a constant volume due to the energy transfer from the solute to solvent. Then, the thermal expansion occurs with a time constant of ~50 ns, returning the sample to ambient pressure. Due to the thermal expansion, the density of the solvent decreases by 1.2 kg/m^3 (0.15%) at 1 μs, leading to a temperature increase of 1.02 K. After the analysis, the whole signal can be decomposed into each component. For example, the solute-only term, the cage term, and the solvent-only term in real space are shown in Figs. 9D, 9E, and 9F along with the assignment of the peaks in the real space. The prominent negative peak around 3.6 Å of the solute-only term (Fig. 9D) is due to the depletion of the I···I distance in CHI_3 and the shoulder at 2.7 Å in late time delays is due to the formation of a new I-I bond in I_2. Most positive and negative peaks located at distances larger than the size of the solute molecule in Fig. 9E and 9F are related to the solvent rearrangement due to temperature and density changes.

4. Applications

4.1 On the issue of isomer formation from CHI_3 in methanol

TRXL has been used to capture the molecular structures of intermediates and their reaction kinetics for various photochemical processes. In the following, we present some application examples ranging from small molecules to proteins, which illustrate the wide applicability of TRXL.

The first example is the photochemistry of iodoform (CHI_3). According to previous time-resolved spectroscopic studies (Wall *et al.*, 2003; Zheng *et al.*, 2000), the CHI_2 radical and I atom generated upon excitation at 267 nm geminately recombine to form iso-iodoform within the solvent cage as the main species (quantum yield of at least 0.5) with a rise time of 7 ps and this iso-iodoform survives for up to microseconds. To investigate the possibility of the isomer formation, we performed the global fitting analysis on the TRXL data with two candidate reaction pathways ([$CHI_3 \rightarrow CHI_2 + I$; simple dissociation channel] and [$CHI_3 \rightarrow CHI_2\text{-}I$; isomer formation channel]). As shown in Fig. 8, the isomer channel reaction model is not compatible with the TRXL data, but a simple dissociation channel gives good agreement

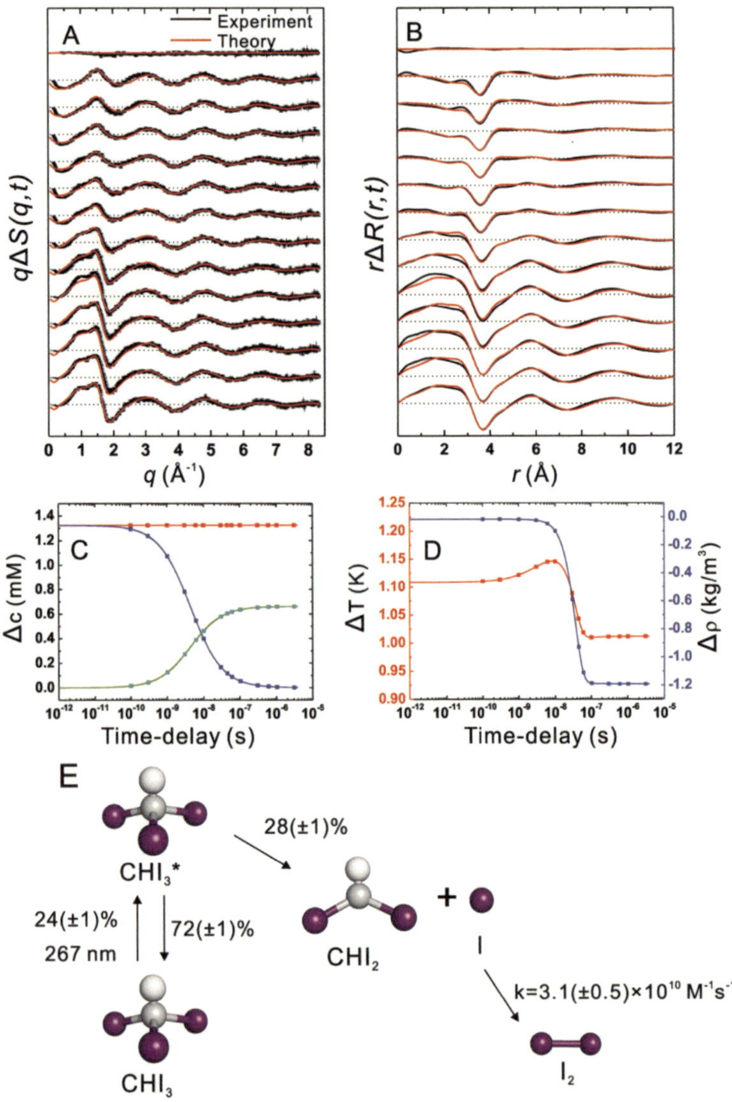

Fig. 8. Structural dynamics of the photochemistry of CHI_3 in methanol upon photolysis at 267 nm determined by TRXL. (A) Experimental difference-diffraction intensities, $q\Delta S(q, t)$ (black) and theoretical curves (red) as a result of global fitting analysis. (B) Difference radial distribution curves, $r\Delta R(r, t)$, corresponding to (A). (C) The population changes of the various chemical species as a function of time delay determined from global fitting analysis. (D) The change in the solvent density (red) and temperature (blue) determined from global fitting analysis. (E) A reaction mechanism determined by TRXL.

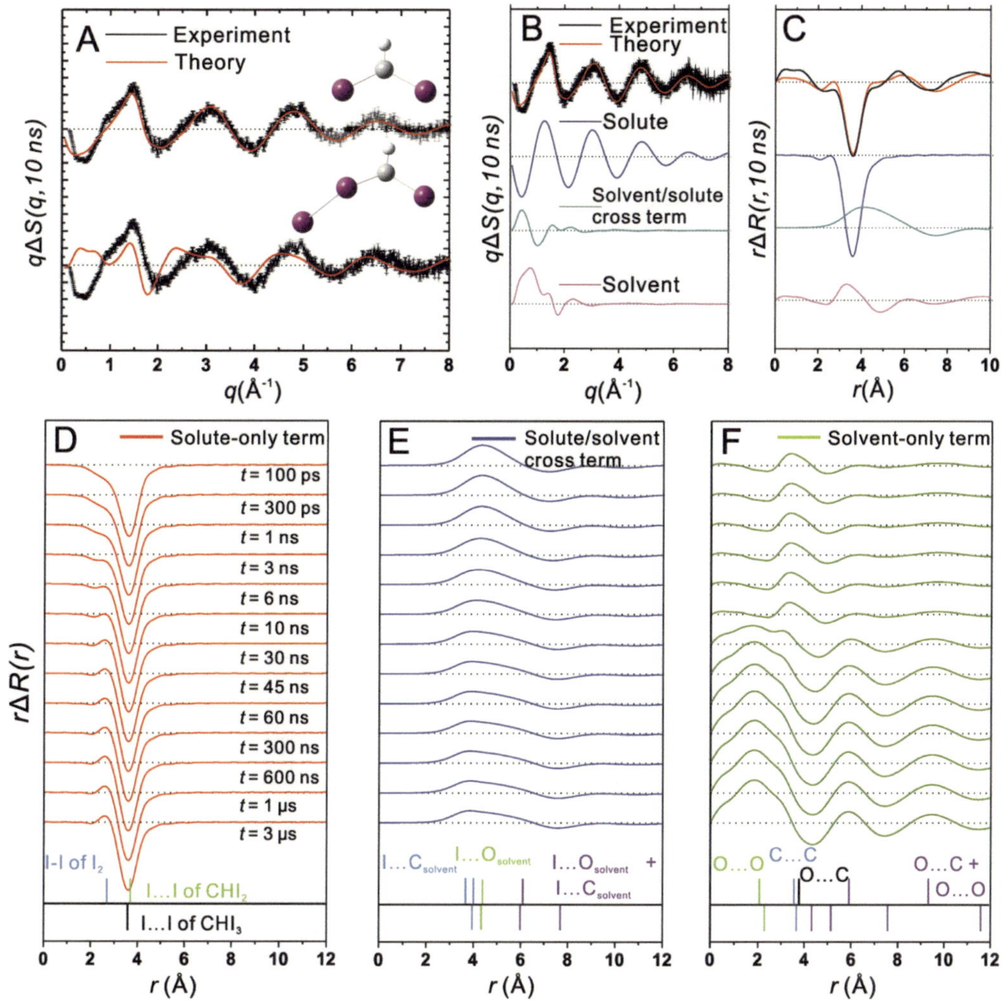

Fig. 9. Determining the major reaction channel for CHI_3 in methanol excited at 267 nm and decomposition into three components for peak assignment. (A) $q\Delta S(q,t)$ for two candidate reaction pathways, CHI_2 formation versus CHI_2-I isomer formation, are compared. Experimental curves with experimental errors are shown in black and theoretical curves are in red. The CHI_2 formation channel gives superior agreement between experiment and theory, confirming that simple dissociation is the major reaction pathway and the isomer formation is negligible. (B) The $q\Delta S(q,10\ ns)$ curve is decomposed into the solute-only, cage, and solvent-only contributions. (C) The same decomposition in the real space for $r\Delta R(r,10\ ns)$ corresponding to (B). (D) The solute-only component of $r\Delta R(r,t)$. (E) The cage component of $r\Delta R(r,t)$. (F) The solvent-only component of $r\Delta R(r,t)$.

(Lee *et al.*, 2008a). Furthermore, when both reaction models are included in the fit, the fraction of the isomer-formation process converges to zero, confirming that the iso-iodoform should be a minor species if it forms at all. Since the X-ray pulse width used in this study is

~100 ps (fwhm), the formation of iso-iodoform as a major species on time scales shorter than our experimental time resolution cannot be ruled out. The subsequent kinetics obtained from TRXL was detailed in the previous section (Data Analysis). It should be noted that the data show that the formation of I_2 is dominant over other possible recombination products such as CHI_3 (from CH_3 and I) and C_2H_6 (from two CH_3).

4.2 Protein folding of cytochrome c

Protein structural changes in solution have been mainly characterized by time-resolved optical spectroscopic methods that, despite their high time resolution (<100 fs), are only indirectly related to three-dimensional structures in space. For protein crystals, a combination of high time resolution and structural sensitivity has become readily available with the advent of sub-nanosecond Laue crystallography (Ihee *et al.*, 2005b; Moffat, 2001; Schotte *et al.*, 2003; Srajer *et al.*, 1996), but its applicability has been limited to a few model systems due to the stringent prerequisites such as highly-ordered and radiation-resistant single crystals. More importantly, crystal packing constraints might hinder biologically relevant motions. Owing to such limitations, the time-resolved X-ray crystallography has been applied to only reversible reactions in single crystals, and it cannot be simply used to study irreversible reactions such as protein folding. To obtain information about protein motions in a more natural environment, X-ray scattering and nuclear magnetic resonance (NMR) methods have been mainly used as direct structural probes of protein structures in solution (Grishaev *et al.*, 2005; Schwieters *et al.*, 2003). Due to the inverse relationship between the interatomic distance and the scattering angle, the scattering from macromolecules is radiated at smaller scattering angles and is typically called as small-angle X-ray scattering (SAXS) or wide-angle X-ray scattering (WAXS) for scattering angles larger than conventional SAXS angles. The SAXS is sensitive to overall structure, for example, overall size and shape, of the protein, while wide-angle X-ray scattering (WAXS) gives more detailed information on the tertiary and quaternary structure such as the fold of helices and sheets. However, thus far, the time resolution had been limited to 160 μs at best (Akiyama *et al.*, 2002). As well, NMR is a powerful technique for structure determination in solution, but it works best for small proteins and needs properly labeled samples (Kainosho *et al.*, 2006). More importantly, due to the nature of microwave pulses, the time resolution of protein NMR is inherently limited to milliseconds.

In case of protein solutions, the relatively low concentration (only a few mM or less) make TRXL measurements non-trivial, and the large molecular size of proteins (more than thousand times larger than small molecules) complicates the structural analysis. However, recent TRXL data from model proteins in solution have demonstrated that the medium to large-scale dynamics of proteins is rich in information on time scales from nanoseconds to milliseconds (Cammarata *et al.*, 2008). TRXL methodology has been applied to human haemoglobin (Hb), a tetrameric protein made of two identical αβ dimers, that is known to have at least two different quaternary structures (a ligated stable "relaxed" (R) state and an unligated stable "tense" (T) structures) in solution. The tertiary and quaternary conformational changes of human hemoglobin triggered by laser induced ligand dissociation have been identified using the TRXL method. A preliminary analysis by the allosteric kinetic model gives a time scale for the R-T transition of ~1–3 μs, which is shorter than the time scale derived with time-resolved optical spectroscopy. The optically induced tertiary relaxation of myoglobin (Mb) and refolding of cytochrome c (Cyt-c) have been also studied with TRXL. As previously mentioned, the advantage of TRXL over time-resolved X-ray protein crystallography is that it can probe irreversible reactions as illustrated with the folding of cytochrome c as well as reversible reactions such as ligand reactions in heme proteins.

The basic idea of protein folding is that the three-dimensional structure of proteins is mainly determined by their amino acid sequences. Unfolded polypeptide chains use this information to accurately and quickly fold into their native structures (Fig. 10a). The optically triggered folding of horse heart Cyt-c has been extensively studied with spectroscopic techniques (Chen *et al.*, 1998; Jones *et al.*, 1993) and also by using fast-mixing SAXS (Akiyama *et al.*, 2002). Cyt-c is a single domain protein similar to Mb. Unlike Hb and Mb, Cyt-c does not usually bind external ligands such as CO since the iron atom of the heme group is covalently coordinated to the Met-80 residue of the protein. However, if Cyt-c is partially unfolded with a denaturing agent, it is possible to replace the Met-80 residue with CO and the CO ligand can be optically dissociated, thereby initiating the re-folding process. The time-dependent evolution of the TR-WAXS signal of Cyt-c after photolysis is evident, especially in the small-angle region (Fig. 10b). As a preliminary analysis, we fitted the observed signal as a linear combination of one pattern at the earliest time delay, 32 μs, and

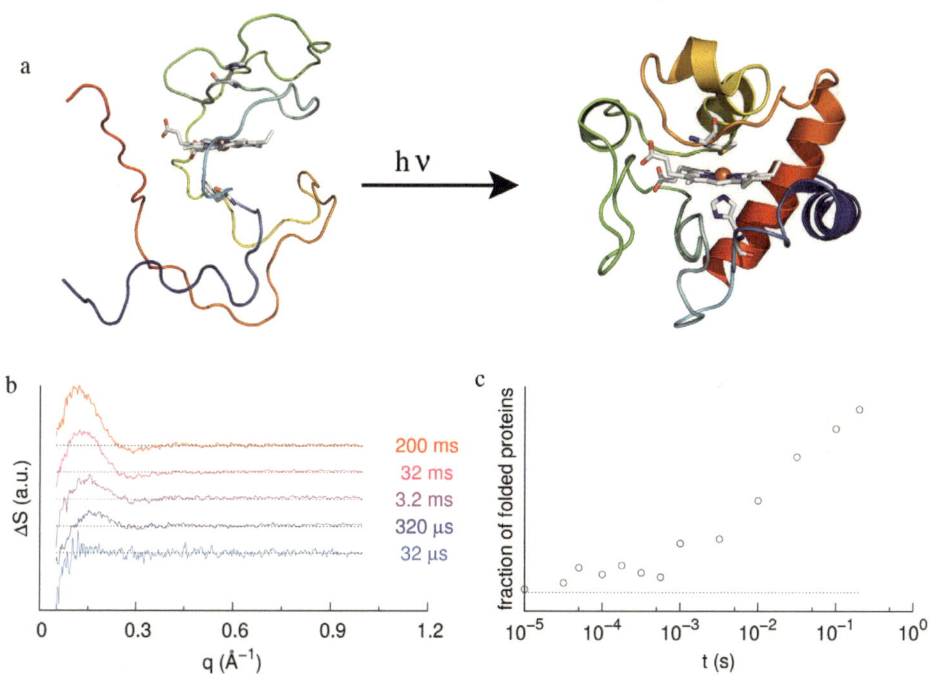

Fig. 10. Application of TRXL to track the folding of cytochrome c. (a) Schematic representation of light-induced folding of cytochrome c. (b) Time-resolved WAXS data relative to CO-photolysis-induced folding of cytochrome c. A 200 ns laser pulse at 532 nm was used to initiate photodissociation of the CO ligand, which in turn initiates the folding process. Experimental data at representative time delays are shown. (c) Population of the folded state as a function of time estimated from a linear combination of the experimental signal at 32 μs and 0.2 s (open symbols). A simple exponential analysis yields a time constant of about 25 ms.

the other at the latest time delay, 0.2 s. This simple approach reproduces the experimental data at all times very well. The plot of the weighting factor of the late time component against time is shown in Fig. 10c and a simple exponential analysis yields a time scale of about 25 ms for the CO-photolysis-triggered folding.

5. Summary and future perspectives

In this chapter, we have described the principle and experimental details of TRXL technique with recent examples of its applications. With 100-ps X-ray pulses readily available from synchrotron radiation, TRXL has been established as a powerful tool for characterizing fast structural transition dynamics of chemical reactions and biological processes, ranging from small molecules to proteins in solution. In particular, the technique provides rather direct information on transient molecular structures since scattering signals are sensitive to all chemical species present in the sample unlike in the optical spectroscopy. Although there still remain challenges to overcome, for example, limited structural and time resolution, TRXL is expected to play an important role in revealing transient structural dynamics in many other systems in solution and liquid phases, especially with the aid of next-generation X-ray sources. At the frontier of the technical advances supporting such bright prospects of TRXL is the advent of linac-based X-ray light sources, which can generate X-ray pulses of femtosecond duration. They include self-amplified spontaneous emission X-ray free electron lasers (SASE-XFEL) and energy recovery linacs (ERL) that are currently under development will be available in the near future.

Among these novel X-ray sources, the high-gain XFEL using SASE promises to generate highly coherent, femtosecond X-ray pulses on the order of 100 fs with a high photon flux up to 10^{13} photons per pulse. The superb time resolution of XFEL will enable us to access reaction dynamics in femtosecond time regime, elucidating much more details of ultrafast structural dynamics. Also, the high flux of XFEL provides the potential for single-shot collection of the XFEL signal. On the other hand, ERL can be operated at a high repetition rate on the order of MHz to GHz. Such high repetition rate capability of ERL will be able to significantly improve the signal-to-noise ratio of TRXL signal since TRXL is basically a perturbative, pump-probe type experiment. With such a high-repetition rate X-ray source, TRXL can be implemented combined with a high-repetition rate oscillator instead of femtosecond amplified lasers, which is commonly operated at only a kHz rate. Furthermore, the nanometer-scale size of the X-ray beam from the ERL (typically 100-nm diameter) will allow tight focusing of the laser beam down to the order of micrometers, enabling the collection of signal from a small volume of sample. Since the scattering signal from the small area will be relatively weak, low-noise and fast-gatable two-dimensional detectors are desirable for future ERL-applied TRXL experiments. The development of pixel detectors using silicon-on-insulator technology will pave the way for such high-performance two-dimensional detectors.

The excellent beam characteristics of the ERL will be further extended to develop the coherent X-ray source, for example, oscillator-type XFEL (XFEL oscillator or XFEL-O) (Kim *et al.*, 2008). The X-ray source generating fully coherent X-ray pulses will serve as the ultimate X-ray light source with superb spatial and temporal coherence. Then, what kind of potential applications can we expect once fully coherent X-ray pulses become available? For example, by making an analogy to the ultrafast optical spectroscopy that fully takes advantage of the temporal coherence of ultrashort optical laser pulses, one could imagine phase-coherent spectroscopy in the X-ray regime with controlled timing, phase, and intensity among multiple, coherent X-ray pulses (Mukamel *et al.*, 2009). As X-ray radiation has the sub-nm wavelength, which corresponds to the sub-attosecond period in the time domain, X-ray pulses offer much higher spatial and temporal resolution than achievable in the optical regime. Thus, the development of X-ray sources that can generate coherent X-ray pulses will revolutionize the whole X-ray science.

6. References

Akiyama, S.; Takahashi, S.; Kimura, T.; Ishimori, K.; Morishima, I.; Nishikawa, Y. & Fujisawa, T. (2002). Conformational landscape of cytochrome c folding studied by microsecond-resolved small-angle x-ray scattering. *Proc. Nat. Acad. Sci.*, 99, 1329-1334.

Bourgeois, D.; Ursby, T.; Wulff, M.; Pradervand, C.; LeGrand, A.; Schildkamp, W.; Laboure, S.; Srajer, V.; Teng, T.-Y.; Roth, M. & Moffat, K. (1996). Feasibility and realization of single-pulse Laue diffraction on macromolecular crystals at ESRF. *J. Synchrotron Rad.*, 3, 65-74.

Bratos, S.; Mirloup, F.; Vuilleumier, R.; Wulff, M. & Plech, A. (2004). X-ray "filming" of atomic motions in chemical reactions. *Chem. Phys.*, 304, 245-251.

Cammarata, M.; Levantino, M.; Schotte, F.; Anfinrud, P. A.; Ewald, F.; Choi, J.; Cupane, A.; Wulff, M. & Ihee, H. (2008). Tracking the structural dynamics of proteins in solution using time-resolved wide-angle X-ray scattering. *Nat. Methods*, 5, 881-886.

Cammarata, M.; Lorenc, M.; Kim, T. K.; Lee, J. H.; Kong, Q. Y.; Pontecorvo, E.; Lo Russo, M.; Schiro, G.; Cupane, A.; Wulff, M. & Ihee, H. (2006). Impulsive solvent heating probed by picosecond x-ray diffraction. *J. Chem. Phys.*, 124, 124504.

Cavalieri, A. L.; Fritz, D. M.; Lee, S. H.; Bucksbaum, P. H.; Reis, D. A.; Rudati, J.; Mills, D. M.; Fuoss, P. H.; Stephenson, G. B.; Kao, C. C.; Siddons, D. P.; Lowney, D. P.; MacPhee, A. G.; Weinstein, D.; Falcone, R. W.; Pahl, R.; Als-Nielsen, J.; Blome, C.; Dusterer, S.; Ischebeck, R.; Schlarb, H.; Schulte-Schrepping, H.; Tschentscher, T.; Schneider, J.; Hignette, O.; Sette, F.; Sokolowski-Tinten, K.; Chapman, H. N.; Lee, R. W.; Hansen, T. N.; Synnergren, O.; Larsson, J.; Techert, S.; Sheppard, J.; Wark, J. S.; Bergh, M.; Caleman, C.; Huldt, G.; van der Spoel, D.; Timneanu, N.; Hajdu, J.; Akre, R. A.; Bong, E.; Emma, P.; Krejcik, P.; Arthur, J.; Brennan, S.; Gaffney, K. J.; Lindenberg, A. M.; Luening, K. & Hastings, J. B. (2005). Clocking femtosecond x rays. *Phys. Rev. Lett.*, 94.

Cavalleri, A.; Wall, S.; Simpson, C.; Statz, E.; Ward, D. W.; Nelson, K. A.; Rini, M. & Schoenlein, R. W. (2006). Tracking the motion of charges in a terahertz light field by femtosecond X-ray diffraction. *Nature*, 442, 664-666.

Chen, E.; Wood, M. J.; Fink, A. L. & Kliger, D. S. (1998). Time-Resolved Circular Dichroism Studies of Protein Folding Intermediates of Cytochrome c. *Biochemistry*, 37, 5589-5598.

Chen, L. X.; Jager, W. J. H.; Jennings, G.; Gosztola, D. J.; Munkholm, A. & Hessler, J. P. (2001). Capturing a Photoexcited Molecular Structure Through Time-Domain X-ray Absorption Fine Structure. *Science*, 292, 262-264.

Christensen, M.; Haldrup, K.; Bechgaard, K.; Feidenhans'l, R.; Kong, Q. Y.; Cammarata, M.; Lo Russo, M.; Wulff, M.; Harrit, N. & Nielsen, M. M. (2009). Time-Resolved X-ray Scattering of an Electronically Excited State in Solution. Structure of the (3)A(2u) State of Tetrakis-mu-pyrophosphitodiplatinate(II). *J. Am. Chem. Soc.*, 131, 502-508.

Collet, E.; Lemee-Cailleau, M. H.; Buron-Le Cointe, M.; Cailleau, H.; Wulff, M.; Luty, T.; Koshihara, S. Y.; Meyer, M.; Toupet, L.; Rabiller, P. & Techert, S. (2003). Laser-induced ferroelectric structural order in an organic charge-transfer crystal. *Science*, 300, 612-615.

Coppens, P. (2003). What can time-resolved diffraction tell us about transient species? excited-state structure determination at atomic resolution. *Chem. Comm.*, 1317-1320.

Coppens, P.; Vorontsov, II; Graber, T.; Kovalevsky, A. Y.; Chen, Y. S.; Wu, G.; Gembicky, M. & Novozhilova, I. V. (2004). Geometry changes of a Cu(I) phenanthroline complex on photoexcitation in a confining medium by time-resolved x-ray diffraction. *J. Am. Chem. Soc.*, 126, 5980-5981.

Davidsson, J.; Poulsen, J.; Cammarata, M.; Georgiou, P.; Wouts, R.; Katona, G.; Jacobson, F.; Plech, A.; Wulff, M.; Nyman, G. & Neutze, R. (2005). Structural determination of a transient isomer of CH2I2 by picosecond x-ray diffraction. *Phys. Rev. Lett.*, 94, 245503.

Fritz, D. M.; Reis, D. A.; Adams, B.; Akre, R. A.; Arthur, J.; Blome, C.; Bucksbaum, P. H.; Cavalieri, A. L.; Engemann, S.; Fahy, S.; Falcone, R. W.; Fuoss, P. H.; Gaffney, K. J.; George, M. J.; Hajdu, J.; Hertlein, M. P.; Hillyard, P. B.; Hoegen, M. H. V.; Kammler, M.; Kaspar, J.; Kienberger, R.; Krejcik, P.; Lee, S. H.; Lindenberg, A. M.; McFarland, B.; Meyer, D.; Montagne, T.; Murray, E. D.; Nelson, A. J.; Nicoul, M.; Pahl, R.; Rudati, J.; Schlarb, H.; Siddons, D. P.; Sokolowski-Tinten, K.; Tschentscher, T.; von der Linde, D. & Hastings, J. B. (2007). Ultrafast bond softening in bismuth: Mapping a solid's interatomic potential with X-rays. *Science*, 315, 633-636.

Gaffney, K. J.; Lindenberg, A. M.; Larsson, J.; Sokolowski-Tinten, K.; Blome, C.; Synnergren, O.; Sheppard, J.; Caleman, C.; MacPhee, A. G.; Weinstein, D.; Lowney, D. P.; Allison, T.; Matthews, T.; Falcone, R. W.; Cavalieri, A. L.; Fritz, D. M.; Lee, S. H.; Bucksbaum, P. H.; Reis, D. A.; Rudati, J.; Macrander, A. T.; Fuoss, P. H.; Kao, C. C.; Siddons, D. P.; Pahl, R.; Moffat, K.; Als-Nielsen, J.; Duesterer, S.; Ischebeck, R.; Schlarb, H.; Schulte-Schrepping, H.; Schneider, J.; von der Linde, D.; Hignette, O.; Sette, F.; Chapman, H. N.; Lee, R. W.; Hansen, T. N.; Wark, J. S.; Bergh, M.; Huldt, G.; van der Spoel, D.; Timneanu, N.; Hajdu, J.; Akre, R. A.; Bong, E.; Krejcik, P.; Arthur, J.; Brennan, S.; Luening, K. & Hastings, J. B. (2005). Observation of structural anisotropy and the onset of liquidlike motion during the nonthermal melting of InSb. *Phys. Rev. Lett.*, 95.

Gembicky, M.; Adachi, S. & Coppens, P. (2007). A kHz heat-load shutter for white-beam experiments at synchrotron sources. *J. Synch. Rad.*, 14, 295-296.

Georgiou, P.; Vincent, J.; Andersson, M.; Wohri, A. B.; Gourdon, P.; Poulsen, J.; Davidsson, J. & Neutze, R. (2006). Picosecond calorimetry: Time-resolved x-ray diffraction studies of liquid CH2Cl2. *J. Chem. Phys.*, 124, 234507.

Grishaev, A.; Wu, J.; Trewhella, J. & Bax, A. (2005). Refinement of multidomain protein structures by combination of solution small-angle X-ray scattering and NMR data. *J. Am. Chem. Soc.*, 127, 16621-16628.

Haldrup, T.; Lemke, H. T.; Haldrup, K.; Nielsen, T. N.; Arms, D. A.; Walko, D. A.; Miceli, A.; Landahl, E. C.; Dufresne, E. M. & Nielsen, M. M. (2009). Picosecond time-resolved laser pump/X-ray probe experiments using a gated single-photon-counting area detector. *J. Synch. Rad.*, 16, 387-390.

Ichiyanagi, K.; Sato, T.; Nozawa, S.; Kim, K. H.; Lee, J. H.; Choi, J.; Tomita, A.; Ichikawa, H.; Adachi, S.; Ihee, H. & Koshihara, S. (2009). 100-picosecond time-resolved solution scattering utilizing wide-bandwidth X-ray beam from multilayer optics. *J. Synch. Rad.*, 16, 391-394.

Ihee, H. (2009). Visualizing Solution-Phase Reaction Dynamics with Time-Resolved X-ray Liquidography. *Acc. Chem. Res.*, 42, 356-366.

Ihee, H.; Lorenc, M.; Kim, T. K.; Kong, Q. Y.; Cammarata, M.; Lee, J. H.; Bratos, S. & Wulff, M. (2005a). Ultrafast x-ray diffraction of transient molecular structures in solution. *Science*, 309, 1223-1227.

Ihee, H.; Rajagopal, S.; Srajer, V.; Pahl, R.; Anderson, S.; Schmidt, M.; Schotte, F.; Anfinrud, P. A.; Wulff, M. & Moffat, K. (2005b). Visualizing reaction pathways in photoactive yellow protein from nanoseconds to seconds. *Proc. Natl. Acad. Sci.*, 102, 7145-7150.

Jones, C. M.; Henry, E. R.; Hu, Y.; C.-K.Chan; Luck, S. D.; A, A. B.; Roder, H.; Hofrichter, J. & Eaton, W. A. (1993). Fast events in protein folding initiated by nanosecond laser photolysis. *Proc. Natl. Acad. Sci.*, 90, 11860-11864.

Kainosho, M.; Torizawa, T.; Iwashita, Y.; Terauchi, T.; Ono, A. M. & P, P. G. (2006). Optimal isotope labelling for NMR protein structure determinations. *Nature*, 440, 52-57.

Kim, C. D.; Pillet, S.; Wu, G.; Fullagar, W. K. & Coppens, P. (2002). Excited-state structure by time-resolved X-ray diffraction. *Acta Cryst. A*, 58, 133-137.

Kim, K.-J.; Shvyd'ko, Y. & Reiche, S. (2008). A proposal for an X-ray free-electron laser oscillator with an energy-recovery linac. *Phys. Rev. Lett.*, 100, 244802.

Kim, T. K.; Lorenc, M.; Lee, J. H.; Russo, M.; Kim, J.; Cammarata, M.; Kong, Q. Y.; Noel, S.; Plech, A.; Wulff, M. & Ihee, H. (2006). Spatiotemporal reaction kinetics of an ultrafast photoreaction pathway visualized by time-resolved liquid x-ray diffraction. *Proc. Natl. Acad. Sci. USA*, 103, 9410-9415.

Kong, Q.; Lee, J. H.; Plech, A.; Wulff, M.; Ihee, H. & Koch, M. H. J. (2008). Ultrafast X-ray Solution Scattering Reveals an Unknown Reaction Intermediate in the Photolysis of $Ru_3(CO)_{12}$. *Angew. Chem. Int. Ed.*, 47, 5550-5553.

Kong, Q. Y.; Wulff, M.; Lee, J. H.; Bratos, S. & Ihee, H. (2007). Photochemical reaction pathways of carbon tetrabromide in solution probed by picosecond X-ray diffraction. *J. Am. Chem. Soc.*, 129, 13584-13591.

Lee, J. H.; Kim, J.; Cammarata, M.; Kong, Q.; Kim, K. H.; Choi, J.; Kim, T. K.; Wulff, M. & Ihee, H. (2008a). Transient X-ray diffraction reveals global and major reaction pathways for the photolysis of iodoform in solution. *Angew. Chem. Int. Ed.*, 47, 1047-1050.

Lee, J. H.; Kim, K. H.; Kim, T. K.; Lee, Y. & Ihee, H. (2006). Analyzing solution-phase time-resolved x-ray diffraction data by isolated-solute models. *J. Chem. Phys.*, 125, 17450.

Lee, J. H.; Kim, T. K.; Kim, J.; Kong, Q.; Cammarata, M.; Lorenc, M.; Wulff, M. & Ihee, H. (2008b). Capturing transient structures in the elimination reaction of haloalkane in solution by transient X-ray diffraction. *J. Am. Chem. Soc.*, 130, 5834-5835.

Lee, S. H.; Cavalieri, A. L.; Fritz, D. M.; Swan, M. C.; Hegde, R. S.; Reason, M.; Goldman, R. S. & Reis, D. A. (2005). Generation and propagation of a picosecond acoustic pulse at a buried interface: Time-resolved X-ray diffraction measurements. *Phys. Rev. Lett.*, 95.

Lindenau, B.; Räbiger, J.; Polachowski, S. & Fremerey, J. K. (2004). X-ray pulse selector with 2 ns lock-in phase setting and stability. *AIP Conf. Proc.*, 705, 1019-1022.

Lindenberg, A. M.; Larsson, J.; Sokolowski-Tinten, K.; Gaffney, K. J.; Blome, C.; Synnergren, O.; Sheppard, J.; Caleman, C.; MacPhee, A. G.; Weinstein, D.; Lowney, D. P.; Allison, T. K.; Matthews, T.; Falcone, R. W.; Cavalieri, A. L.; Fritz, D. M.; Lee, S. H.; Bucksbaum, P. H.; Reis, D. A.; Rudati, J.; Fuoss, P. H.; Kao, C. C.; Siddons, D. P.; Pahl, R.; Als-Nielsen, J.; Duesterer, S.; Ischebeck, R.; Schlarb, H.; Schulte-Schrepping, H.; Tschentscher, T.; Schneider, J.; von der Linde, D.; Hignette, O.; Sette, F.; Chapman, H. N.; Lee, R. W.; Hansen, T. N.; Techert, S.; Wark, J. S.; Bergh, M.; Huldt, G.; van der Spoel, D.; Timneanu, N.; Hajdu, J.; Akre, R. A.; Bong, E.; Krejcik, P.; Arthur, J.; Brennan, S.; Luening, K. & Hastings, J. B. (2005). Atomic-scale visualization of inertial dynamics. *Science*, 308, 392-395.

Moffat, K. (2001). Time-resolved biochemical crystallography: a mechanistic perspective. *Chem. Rev.*, 101, 1569-1581.

Mukamel, S.; Abramavicius, D.; Yang, L.; Zhuang, W.; Schweigert, I. V. & Voronine, D. V. (2009). Coherent multidimensional optical probes for electron correlations and exciton dynamics: From NMR to X-rays. *Acc. Chem. Res.*, 42, 553-562.

Nozawa, S.; Adachi, S.; Takahashi, J.; Tazaki, R.; Guerin, L.; Daimon, M.; Tomita, A.; Sato, T.; Chollet, M.; Collet, E.; Cailleau, H.; Yamamoto, S.; Tsuchiya, K.; Shioya, T.; Sasaki, H.; Mori, T.; Ichiyanagi, K.; Sawa, H.; Kawata, H. & Koshihara, S. (2007). Developing 100 ps-resolved X-ray structural analysis capabilities on beamline NW14A at the photon factory advanced ring. *J. Synch. Rad.*, 14, 313-319.

Plech, A.; Wulff, M.; Bratos, S.; Mirloup, F.; Vuilleumier, R.; Schotte, F. & Anfinrud, P. A. (2004). Visualizing chemical reactions in solution by picosecond x-ray diffraction. *Phys. Rev. Lett.*, 92, 125505.

Saes, M.; Bressler, C.; Abela, R.; Grolimund, D.; Johnson, S. L.; Heimann, P. A. & Chergui, M. (2003). Observing Photochemical Transients by Ultrafast X-Ray Absorption Spectroscopy. *Phys. Rev. Lett.*, 90, 047403.

Sato, T.; Nozawa, S.; Ichiyanagi, K.; Tomita, A.; Chollet, M.; Ichikawa, H.; Fujii, H.; Adachi, S. & Koshihara, S. (2009). Capturing molecular structural dynamics by 100 ps time-resolved X-ray absorption spectroscopy. *J. Synch. Rad.*, 16, 110-115.

Schotte, F.; Lim, M.; Jackson, T. A.; Smirnov, A. V.; Soman, J.; Olson, J. S.; Phillips, G. N. J.; Wulff, M. & Anfinrud, P. (2003). Watching a protein as it functions with 150ps time-resolved X-ray crystallography. *Science*, 300, 1944-1947.

Schwieters, C. D.; Kuszewski, J. J.; Tjandra, N. & Clore, G. M. (2003). The Xplor-NIH NMR molecular structure determination package. *J. Mag. Res.*, 160, 65-73.

Srajer, V.; Teng, T.-Y.; Ursby, T.; Pradervand, C.; Ren, Z.; Adachi, S.; Schildkamp, W.; Bourgeois, D.; Wulff, M. & Moffat, K. (1996). Photolysis of the carbon monoxide complex of myoglobin: Nanosecond time-resolved crystallography. *Science*, 274, 1726-1729.

Techert, S.; Schotte, F. & Wulff, M. (2001). Picosecond X-Ray Diffraction Probed Transeint Structural Changes in Organic Solids. *Phys. Rev. Lett.*, 86, 2030-2033.

Tomita, A.; Sato, T.; Ichiyanagi, K.; Nozawa, S.; Ichikawa, H.; Chollet, M.; Kawai, F.; Park, S.-Y.; Tsuduki, T.; Yamato, T.; Koshihara, S. & Adachi, S. (2009). Visualizing breathing motion of internal cavities in concert with ligand migration in myoglobin. *Proc. Natl. Acad. Sci. USA*, 106, 2612-2616.

Vincent, J.; Andersson, M.; Eklund, M.; Wohri, A. B.; Odelius, M.; Malmerberg, E.; Kong, Q. Y.; Wulff, M.; Neutze, R. & Davidsson, J. (2009). Solvent dependent structural perturbations of chemical reaction intermediates visualized by time-resolved x-ray diffraction. *J. Chem. Phys.*, 130, -.

Wall, M.; Tarnovsky, A. N.; Pascher, T.; Sundstrom, V. & Akesson, E. (2003). Photodissociation Dynamics of Iodoform in Solution. *J. Phys. Chem. A*, 107, 211-217.

Wulff, M.; Bratos, S.; Plech, A.; Vuilleumier, R.; Mirloup, F.; Lorenc, M.; Kong, Q. & Ihee, H. (2006). Recombination of photodissociated iodine: A time-resolved x-ray-diffraction study. *J. Chem. Phys.*, 124, 034501.

Wulff, M.; Schotte, F.; Naylor, G.; Bourgeois, D.; Moffat, K. & Mourou, G. (1997). Time-resolved structures of macromolecules at the ESRF: Single-pulse Laue diffraction, stroboscopic data collection and femtosecond flash photolysis. *Nucl. Instrum. Methods*, A398, 69-84.

Zheng, X. M. & Phillips, D. L. (2000). Solvation effects on the iodoform ultraviolet direct photodissociation reaction. opening the photoisomerization channel. *Chem. Phys. Lett.*, 324, 175-182.

Application of Ultrafast Laser Optoperforation for Plant Pollen Walls and Endothelial Cell Membranes

Sae Chae Jeoung, Mehra Singh Sidhu, Ji Sang Yahng,
Hyun Joo Shin and GuYoun Baik
Korea Research Institute of Standards and Science
Korea

1. Introduction

Laser beam, first introduced in the field of ophthalmology in 1964 has been a powerful tool in medicine, plastic or general surgery, dermatology, gynaecology, oncology, *in vitro* - fertilization and numerous other clinical areas (Nelson & Berm, 1989; Senz & Miiller, 1989). Furthermore, it is also a useful tool for studying basic cellular and biochemical processes. Fluence, energy per unit area, is one of the most important and critical aspects of the laser treatment (Sidhu et al., 2009). Depending upon the laser fluence, either transient or permanent changes can be induced in various cellular compartments including cell walls, plasma membranes, and even organelles with high resolution under physiological environment. For instance, Higashiyama et al. performed a selective ablation of a target compartment of the embryo sac with ultraviolet (UV) pulsed laser to investigate the contribution of each gametophytic cell to pollen tube attraction and found that the synergic cells adjacent to the egg cell plays an important role in attracting the pollen tube (Higashiyama et al., 2001).

Laser pulses with a sufficient spatial resolution can introduce a physical gate in either a membrane or a cell wall under physiological environment such that one could deliver a foreign material into the cell and bring about selective manipulation of subcellular components without demolishing the cell's integrity. It has been reported that in Japonica rice (*Oryza sativa*), which was pre-treated in a hypertonic buffer containing an exogenous *gusA/npt II* chimeric gene, however only five out of 1000 treated cells exhibits the *gusA* gene expression in embryogenic calli (Hoffmann, 1996). In case of tobacco leaves, it has been recently reported that the transferred gene into the tobacco cells via nanosecond laser pulse was stably maintained even during the meiosis. However, the previous reports from nanosecond laser pulses treatment in UV region illustrated a rather limited transformation efficiency of less than 0.5 %. The shortage in transformation efficiency can be overcome by utilizing opto-acoustic shock pressure induced by fundamental output from Nd:YAG pulsed laser (Tang et al., 2006). Since pollen grain is a haploid and is known to be regenerated into a plant, this laser application draws a great interest from the application point of view in plant transformation and genetic engineering. Nevertheless, the cell wall of plant cell including pollen is not easily accessible because it forms a potential barrier for the active or passive transport for the molecules. This problem can be resolved via utilizing the cell protoplast as a carrier for the molecules, but this step may add the numerous other problems such as, tissue disintegration and difficulties in plant regeneration (Krautwig & Lörz, 1995). More recently, ultrafast laser-based gene transfer has been considered as a break-through in the current hiatus of non-viral mediated gene therapy (Greulich & Weber, 1992). Compared to relative intensive work on mammalian cells, application of femtosecond-laser optoperforation in plant cells is quite limited. However, the application of the recently found ultrafast laser ablation technique in the plant cells is expected to bring about a significant increase in gene transfection efficiency by eliminating physical barriers such as the hard cell wall or the turgor pressure (Tirlapur & König, 2002).

It is important to understand the developmental process of a pollen including germination of a receptive stigma and consequent tube growth as a basic sexual mechanism in flowering plants for the manipulation of plant production (Taylor & Hepler, 1997). Pollen transformation is also considered as an alternative strategy for the genetic engineering. Pollen acts as a natural vector for direct gene transfer due to its involvement in the normal sexual process in flowering plants and hence, transformed pollen has been used to generate numerous transgenic plants such as Tobacco, Scots pine etc., (Van der Leede-Plegt et al., 1995; Touraev et al., 1997; Aronen et al., 2003). In addition to several other potential advantages of this technology like rapid propagation with reduced somaclonal variation, pollen transformation does not require any *in vitro* regeneration process that is typically essential for genetic engineering. For instance, for the restoration of American chestnut which is difficult to regenerate, *in vitro*, pollen transformation based on particle bombardment method was successfully applied to transiently express green fluorescence protein (GFP) with an efficiency of 4.1 % (Fernado et al., 2006).

In first section of this chapter, we are discussing the application of ultrafast laser optoperforation to make a transient hole on lily (*Lilium hybrid* cv Shiberia) pollen wall to deliver the desired foreign material into the pollen, which furthermore improve its transfection efficiency and also eliminates the tedious *in vitro* regeneration or tissue disintegration problems. The presence of apparent holes on the cell wall, after single-shot laser irradiation, was confirmed by measuring the AFM topography of the processed pollen surface. Laser exposer with a moderate to high laser fluence induces a massive extrusion of the pollen's interior contents through the hole. However, the *in vitro* germinability of perforated pollens was comparable to those of the untreated ones. If one of the two nuclei compartments of the pollen grains, which are tentatively assigned to be a generative cell, was excluded out from the pollen wall through the transient hole, the second one (vegetative cell) moved along the tube tip endorsed to follow the normal pollen tube growth dynamics. This reveals that during the germination, coherent movement of both the nuclei is not a necessary requisite for normal pollen tube growth. Meanwhile, we have tried and successfully able to deliver a foreign materials into the pollen grain, which results in a transient red fluorescence protein (RFP) expression with a high efficiency. The use of such an optoperforation followed by the delivery of foreign material into the pollens holds an imperative approach in the field of pollen germination, growth of the pollen tube and the pollen transformation.

We will devote the latter part of this chapter to explain the ultrafast laser-assisted optoperforation of endothelial cell membrane. In order to induce the transient perturbation on the live-cell plasma membrane for the efficient delivery of foreign materials, including drugs or genetic material, into the live cells (such as embryonic stem cell), numerous chemical or physical methods have been employed (Kobayashi et al., 2005). Also, it's an ever chasing challenge waiting for us to elucidate the mechanism involved in the self-healing process of the perforated live cell plasma membranes. The mechanism for laser-mediated foreign material transfer as well as other perturbation methods was hypothesized with a limited understanding of the structure and functions of membrane lipids (Engelman, 2005). Recently, the fundamental mechanisms of fs-laser interaction with an aqueous media are extensively studied by Vogel et al. (Vogel et al., 2005). The most frequently invoked explanation was based on the assumption that the physicochemical methods used in the specific cases may either perforate the surface of the cell membranes or enhance the permeability of the foreign materials, enabling a transient transfer of foreign molecules into the cytosol (Kobayashi et al., 2005; Zeira et al., 2003). Manipulation and operation using ultrafast laser pulses have been introduced as a promising tool for the cellular nanosurgery and the transfer of foreign molecules into live cells under *in vitro* and *in vivo* conditions (Zeira et al., 2003; Tirlapur & König, 2002; Shen et al., 2005; König et al., 1999). Ultrafast laser-based gene transfer has been considered as a break-through in the current hiatus of non-viral-mediated gene therapy (Zeira et al., 2003). Lastly, in this chapter we are discussing the experimental results on the optoperforation of retina blood vessels with minimizing any probable damage outside laser focusing area.

2. Ultrafast laser optoperforation of lily (*Lilium hybrid* cv Shiberia) pollen

2.1 Materials and method

Mature pollens were obtained from the anthers of *Lilium hybrid* cv Shiberia. After collection, the pollens were air- dried and stored at 4o C. Before in vitro culture, pollens were rehydrated in deionized water under humid atmosphere for about 30 min and spread on the culture medium (pH 5.8) comprising of 145 mM sucrose, 1.6 mM H_3BO_3, 1.8 mM $Ca(NO_3)_2$ and 0.4 % phytagel (Sigma).
Near-infrared laser beam from amplified Ti:sapphire femtosecond laser at a repetition rate of 1 kHz (Quantronix) was focused on the pollen exine through objective lens (N.A. = 0.6) so that we could selectively puncture a wall inside an ornament. Petri dish containing the pollen grains was mounted on a motorized XYZ - translation stage, which is used to manipulate the sample to expose a fresh pollen grain surface at each laser pulse. The images obtained during the fs-laser irradiation on the pollen's surface were fed into a personal computer for further data analysis. Laser fluence was attenuated by a variable neutral density filter. All the laser exposers were conducted under a single shot configuration by using a fast mechanical shutter with a opening time constant of 0.6 ms, which is synchronized with the laser. After optoperforation, the viability of the operated pollen grains was checked by measuring the germination rate of pollen tube by intrusion of propidium iodide (PI). The glass coverslips with grids marked on the surface, were used to cover over the pollens for the localization of the irradiated pollen grains.
For the pollen transformation study, a foreign DNA with a construct, encoding F1TPae-γ mitochondrial targeting sequence fused to RFP; was added into the culture medium before the laser optoperforation (Lee et al., 2001). MitoTracker Green FM (Molecular probes, invitrogen) was further used to confirm the RFP expression. MitoTracker Green FM is a conventional fluorescent marker used for tracking of the mitochondria (Lovy-Wheeler et al., 2007).
Pollen germination as well as its response to optoperforation was characterized with an optical microscope and a fluorescent microscope (Axioskop, Zeiss). In order to confirm the

hole size, the intact pollen surface treated with laser beam was imaged by atomic force microscope (XE-120, PSIA) without any further treatments.

2.2 Results and discussion

It was observed that size of lily pollen grain increased during pollen re-hydration process. Before pollen tube starts to grow, a laser beam was precisely focused into the interior of the ornament of pollen surface. Figure 1 shows sequential optical microscopic images captured from the movie file obtained during fs-laser optoperforation of lily pollen walls. Immediately, after laser pulse irradiation on pollen wall surface, a cytoplasmic content of the pollen oozes out from the pollen grains.

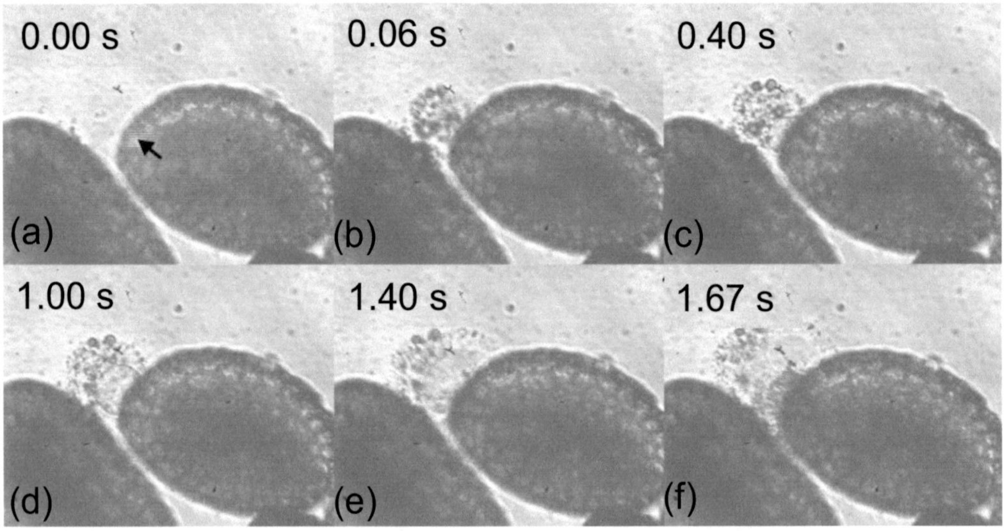

Fig. 1. Sequential images of hydrated lily pollens before (a) and after fs-laser treatment (b-f). The contents of pollens were burst out just after laser irradiation on the selected surface with a time span of 0 to 1.67 seconds. The arrow indicates the point of focusing.

This observation is consistent with the previous reports stating that UV laser pulses treatment causes release of cytoplasmic content either from the pollen itself or its tube (Hoffmann, 1996). This abrupt extrusion of cytoplasm from the treated pollen can be explained in terms of a high turgor pressure inside the hydrated pollen, which ranges between 0.1 and 0.4 MPa, and the mean value being 0.21 ± 0.06 MPa. (Benkert et al., 1997). Unless the buffering osmotic pressure is against the turgor pressure, the pressure inside the cell will no longer sustain due to the presence of hole on the wall. Following the cytoplasmic extrusion, a large vesicle with a diameter of about 10 μm was also flushed out from the pollen grain through the hole. The excluded large structure was apparently distinct from the concomitant flushed contents, as seen under optical microscope. The distinct boundary was rather rigid in nature and it remains intact without any bursting phenomenon.

The extrusion of pollen's cytoplasmic contents and the large organelle was strongly dependent on laser fluence. Figure 2 shows a series of optical microscopic images and a graph that represents the normalized germination percent of the pollen grains as a function of the laser fluence (0.11 - 7.10 J/cm^2). With changing the laser fluence energy from 0.11 J/cm^2 to 7.10 J/cm^2, we perforate the pollen walls with single fs-laser pulse. Figure 2(a)

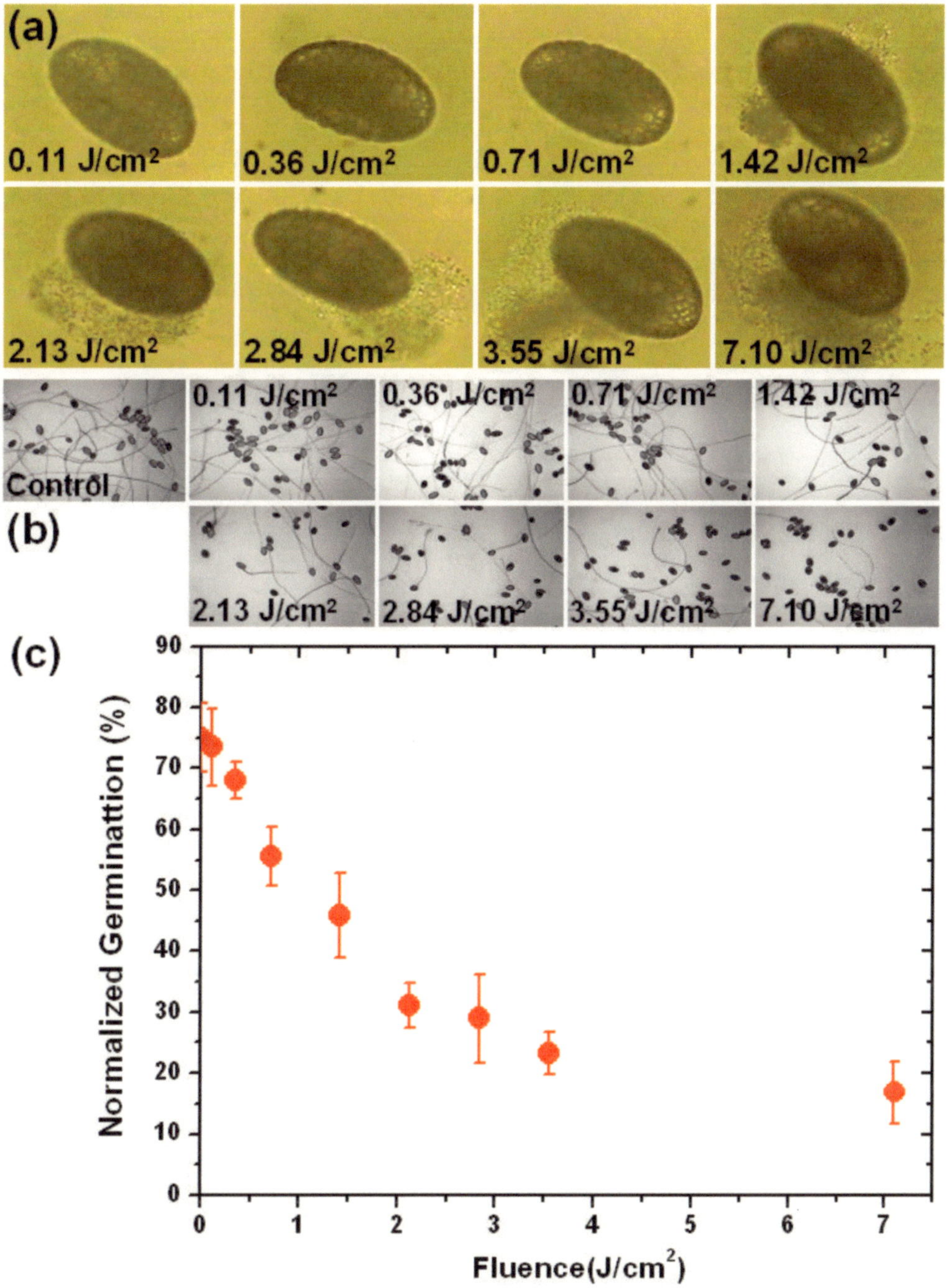

Fig. 2. The percent germination threshold of the hydrated (single pulse laser irradiated) lily pollen grains (a) Optical microscope images (b) Germination of the pollen grains after laser irradiation at varied fluence range (0.11 ~ 7.10 J/cm^2). (c) Graphical representation of Normalized germination percent of the pollen grains as a function of laser fluence. The pollen wall ablation threshold was found to between 0.71 ~ 1.42 J/cm^2.

shows the exclusion of pollen's cytoplasmic contents at progressive laser fluence energy levels. We observe that the hydrated lily pollens mostly begin to exclude cytoplasmic contents at the laser fluence energy level of 1.42 J/cm^2. Sometimes we can observe the extrusion of the cytoplasmic contents even at the laser fluence less than 1.42 J/cm^2. We have also compared the germinabilty rates of treated *vs* non-treated pollen grains at above mentioned laser fluence energy levels. With the optical setup used in the current work, the threshold laser energy for making a physical hole on the wall was found to be about 0.5 J/pulse , where there is no observable extrusions of any cytoplasmic contents were seen out the pollen grains. Figure 3 shows the AFM topography of perforated pollen surface. The diameter and depth of the hole is found to be 1.1 µm and 0.9 µm, respectively.

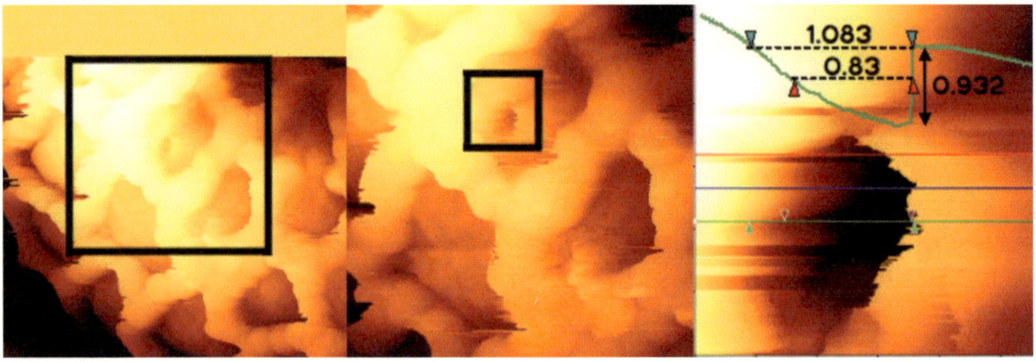

Fig. 3. Inset represents the AFM image of hole after being perforated at 0.5 mW fs-laser pulse. The extrusion of any cytoplasmic contents out of the pollen grains was not observed at the said fluence.

We have cross-checked the viability of the pollen grains after optoperforation by the observation of intrusion of the PI (Propodium Iodide). About 150 out of 300 opto-perforated pollens were not stained with PI despite of the massive exclusion of the cytoplasmic contents, indicating that the fs-laser optoperforation of lily pollen wall hardly harms its viability. More interestingly, the excluded large organelle was not stained with PI, even though the organelle was not likely surrounded by pollen wall. This observation led us to assume that the organelle excluded from the pollen grains by the fs-laser processing still maintained its biological integrity without any observable perturbation**.**

Figure 4 exhibit a series of optical microscopic images during the *in vitro* germination of pollen after optoperforation at a time lapse of 0, 4, and 11 hrs. The pollens operated with the fluence of 8 µJ/pulse are indicated with asterisk. Even though the contents of pollen including the large organelle were massively excluded just after optoperforation of cell wall, still 19 % of the operated pollens (480 pollens in total) were tend to germinate the pollen tubes successfully. This germination efficiency is very close to the value of 24 % as observed under the control experiment (420 pollens in total).

Subsequently, we emphasized our study on nucleus staining of the treated pollens to have a deeper understanding on the nature of excluded large organelles. Figure 5 shows bright-field optical image (a) and fluorescence image (b) of pollens stained with nucleic acid specific dye acridine orange after the fs-laser treatment (1, 2, and 3 labelled pollens). Cellular contents were massively extruded from all of the treated pollens upon illuminating rather at the high laser fluence (~8 µJ/pulse). Amongst them, only pollen "1" has been successfully

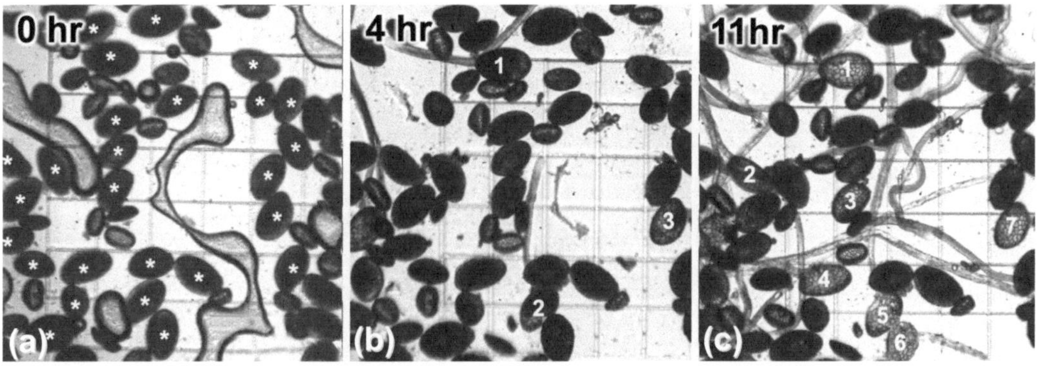

Fig. 4. Sequential images of pollen tube growth before (a) and after fs-laser treatment (b-c). fs-laser beam was focused into the interior of the wall inside pollen ornament. The pollens marked with asterisks were irradiated with laser beam with the fluence of 8 μJ/pulse. Seven pollens out of 26 treated pollens in this run were eventually geminated to grow the pollen tube.

germinated to grow the pollen tube. It should be noted that an apparent fluorescent spot (arrow), indicating the presence of nucleic acid, was observed in the excluded contents in all the treated pollen grains. A faint emission indicated by an arrow head, which supposed to be a vegetative nucleus, was clearly observed in the pollen tube geminating from pollen "1". Meanwhile, the nuclei excluded from the same pollen were obviously seen outside the pollen grain. Thus, it is most likely to state that the excluded large vesicle from the pollen "1" must either be a generative cell with its own nucleus or a sperm cell after the second mitotic division. In other words, the vegetative nucleus of treated pollen "1" is well trapped/tracked inside the tube, while the regenerative nucleus or sperm cell stayed outside the pollen. Vervaeke et al. suggested that the vegetative nucleus (VN) and the generative cell (GC) form a functional assemblage to generate callose plugs (Vervaeke et al., 2005; Strubinska & Sniezko, 2000). However, our findings clearly indicate that the close physical contact between the vegetative nucleus and the generative cell is not a necessary requisite for the pollen germination. Instead, it suggests a kind of chemical connection between the two constituents. Even if it is not known as how the two constituents are connected physically or chemically, the extrusion of generative cell from lily pollen grains by fs-laser optoperforation does not alter the growth of pollen tube or vegetative nucleus movement to a greater extent.

We have further performed the transformation studies for the fs-laser treated *lilium* pollen grains. Initially the pollen grains were soaked in hypotonic solution having a foreign DNA with a construct encoding F1TPae-γ mitochondrial targeting sequence fused to RFP. An apparent expression of RFP from the optoperforated pollens as shown in fluorescence microscopic images (Figure 6) reveals that the current optoperforation method could be successfully applied to transfer the foreign genetic materials into the lily pollens. Figure 7(a), after 72 hours, represents the successful germination of the pollens, treated with an fs-laser beam. From the fluorescence image of germinated pollens (Figure 7(b)) measured at 650 nm (wavelength corresponding to the emission from RFP), we deduced that the DNA delivered into the pollen grain was effectively expressed in pollen tube during the germination due to successful transformation. We have further confirmed transformation of the lily pollen by staining the transformed pollen tube with MitoTrack Green FM, which selectively stains the

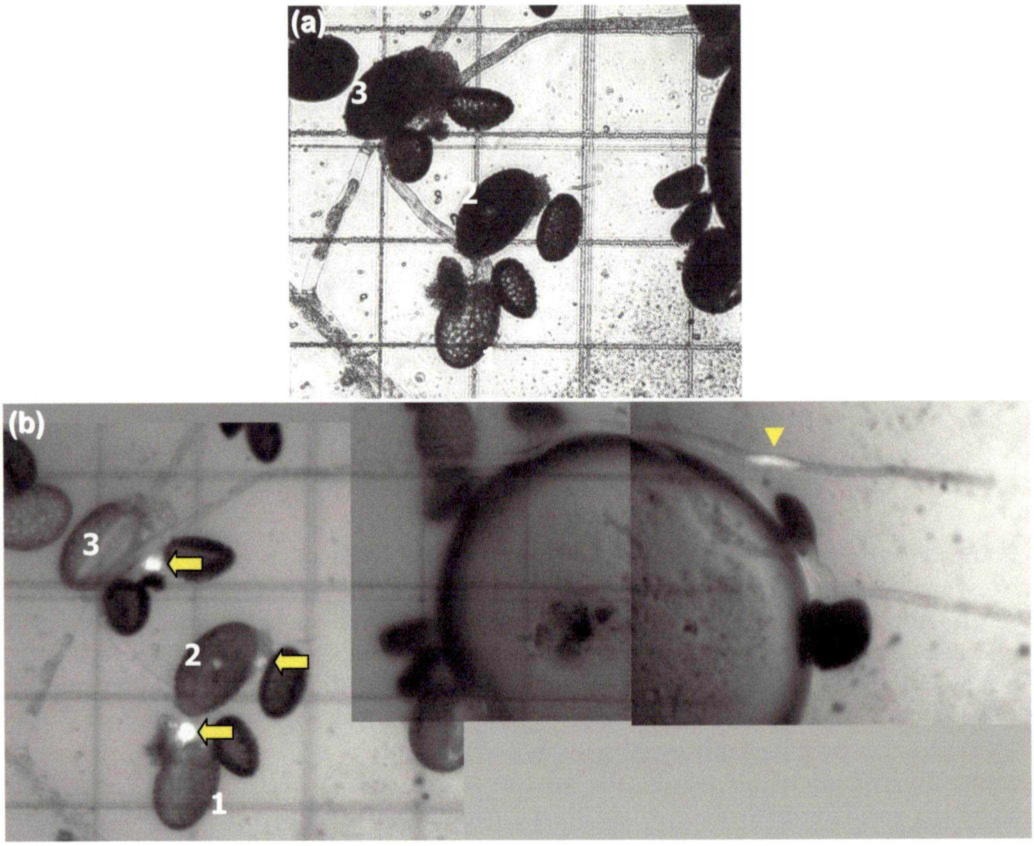

Fig. 5. Bright field (a) and fluorescence label (b) of germinated pollens ("1", "2", and "3") stained with nucleic specific dye, acridine orange after fs-laser treatment. Interior cytoplasmic contents were massively excluded from all the treated pollens due to rather high laser fluence. Amongst them, only pollen "1" successfully germinated to grow the pollen tube. It should be noted that an apparent fluorescent spot (arrow) due to the presence of nucleic acid was observed from the excluded contents from all the treated pollen grains. A dim emission part (arrow head), which should be vegetative nucleus, was observed in the pollen tube geminating from treated pollen "1". The nucleus excluded from the same pollen was very well seen.

mitochondria and results in the expression of green fluorescence. Exact matching of RFP fluorescence image with that of mitochondria in pollen tube led us to conclude that, the foreign DNA with a construct encoding F1TPae-γ mitochondrial targeting sequence fused RFP was delivered into pollen and expressed as designed. Statistically, 42 pollens out of 206 treated pollens were germinated. Amongst the germinated pollens, foreign DNA expression in mitochondria delivered by fs-laser optoperforation was observed in 9 pollen tubes out of 42 germinated pollens. As a result, the transformation efficiency of lily pollen by fs-laser optoperforation was about 4.4 %. However, in case of germinated pollens, the efficiency was about 21.4 %. This transformation efficiency was higher than that of American chestnut pollen transformation based on particle bombardment method to transiently express GFP with an efficiency of 4.1 % (Fernado et al., 2006).

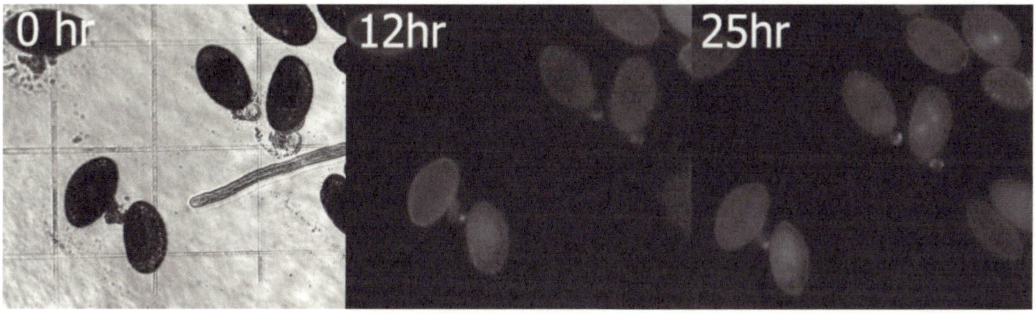

Fig. 6. The red fluorescence images were taken at 0, 12 and 25 hr periodic intervals of time. The fluorescence signal at the membrane site was thought to be due to expression of foreign DNA.

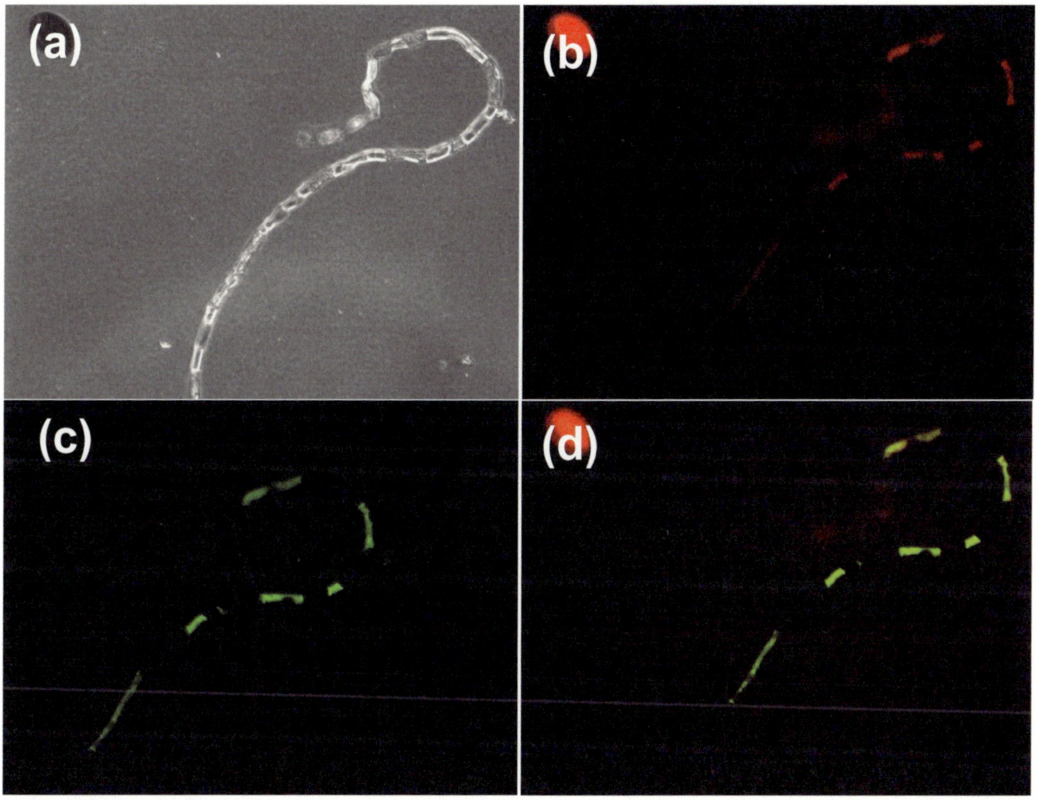

Fig. 7. Direct proof for the selective transformation of a foreign DNA with a construct encoding F1TPae-γ mitochondrial targeting sequence fused to RFP in the mitochondria of a pollen tube. (a) DIC microscopy, (b) RFP fluorescence, (c) MitoTracker green fluorescence and (d) merge b and c

3. Ultrafast laser optoperforation of endothelial cell membranes

3.1 Experimental techniques and materials

The perforation of the plasma membrane of live cells was performed with a femtosecond amplifier system (Integra, Quantronix) with maximum output energy of 1 mJ/pulse at 780 nm, of which the pulse width and the repetition rate is 150 fs and 1 kHz, respectively (Jeoung et al., 2005). The laser beam was tightly focused on the membrane with a high-numerical-aperture objective lens (Nikon, X100, N.A = 1.3, oil immersion lens) mounted to an inverted microscope (Axiovert 200M, Zeiss). All the experiments were done under single-shot configuration by using a fast electrical shutter with an opening time constant less than 0.5 ms. In order to eliminate any desperate loss of laser energy during these experiments, the fast electrical shutter was perfectly synchronized with the laser. The laser energy before the objective lens was kept to be about 5 ~ 7 nJ/pulse by attenuating it with a neutral density filter.

The changes in the topography and the optical images of plasma membrane induced by the ultrafast laser irradiations were simultaneously monitored with both an atomic force microscope (AFM) and a charge coupled device (CCD) detector, respectively. AFM (XE-120, PSIA) was slightly modified and combined to an optical microscope. AFM probe was carefully aligned to be aimed at same position of the focus of fs-laser beam. Contact imaging of the AFM was performed using a silicon nitride bio-cantilever (Bio-lever, Olympus, Japan), of which nominal force constant is 0.03 nN/m. The topographies have been obtained by optimising the AFM running parameters that gives the identical results to that of the two-way scanning configuration (Parpura et al., 1993 & Velegol et al., 2003). The set point for loading force was kept to be very weak between 0.65 nN and 1 nN, to minimize the damage or the attractional movement caused by attraction force of the AFM tip. A successive scanning of the live cells exhibit no observable changes in its phenotypes.

The measurements of the velocity field responsible for the membrane motility after fs-laser perforation have been measured by a micro particle image velocimetry (μ-PIV) (Yahng et al., 2005). The visualization of membrane movements was figured out by attaching a fluorescent polystyrene bead (diameter ~ 200 nm) on the plasma membrane of a target cell. Just after the fs-laser irradiation on the surface of plasma membrane, two successive ns-laser pulses with a time lapse of 10 ms were employed to capture the two fluorescence images. Thereafter, the velocity fields of membrane motility were obtained by a cross correlation calculation with the two images. We have confirmed that there is no damage on the cells with the current volumetric excitation with the ns-laser pulses, where the laser fluence is kept explicitly low.

Neuroblastoma (Neuro2A) cells were used to investigate the dynamics of its plasma membrane and self-healing process. During the experiment, live cells were kept inside a sterile miniaturized cell chamber filled with Dulbecco's Modified Eagle's Media (DMEM, Gibco BRL) containing 10 % Fetal Bovine Serum (FBS, Gibco) and 1 % penicillin/streptomycin (Gibco) under humid atmosphere at 5 % CO_2 partial pressure. Although the lid glass of the cell chamber was removed for AFM imaging or other measurements, no apparent change in its phenotypes was observed within time span of (1.5 hr) the current experiments.

3.2 Results

As a consequence of single-pulse ultrafast laser irradiation on the cells surface (ca. 5 ~ 7 nJ), a circular area appears to be observed on its plasma membrane as shown in Fig. 8(a). The injured area increased relatively fast over several tens of seconds comparable to the percentage of cellular size. It is interesting to note that the initial circular type of injured membrane area changes to an elliptical one with a long axis parallel to the cell. After

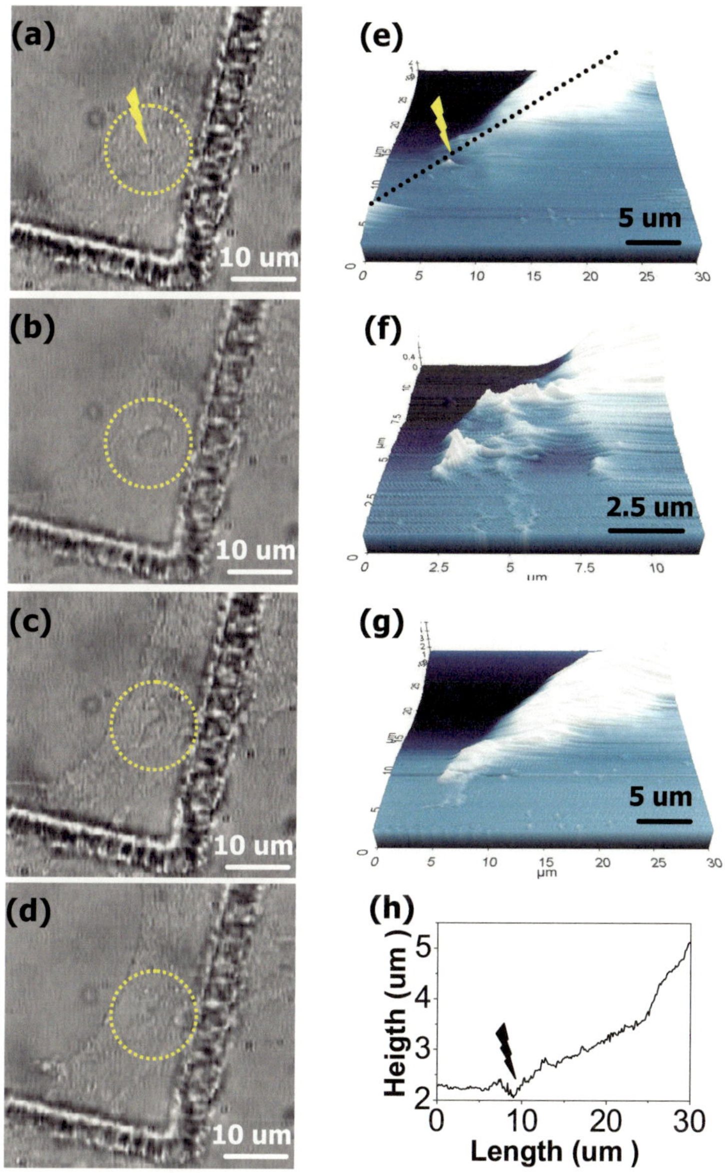

Fig. 8. Transparent photomicrographs (a-d), which were taken 0 sec, 38 sec, 229 sec, and 414 sec, respectively, and topographical images (e-g) taken 1, 30 and 43 minutes, respectively, after ultrafast laser-induced injury in a single Neouro2A cell. The height profile along the broken line in (e) is shown in (h). The image sizes in (a-d) were 50 μm x 50 μm. The hollowed black area is a remnant of fs-laser ablated region, which was processed on the substrate as a guideline before seeding of the cells.

reaching at its maxima of about 6.4 x 4.9 μm^2, within several tens of seconds (Fig. 8(b)), the area starts decreasing spontaneously and then completely faded out in next 7 minutes (Fig. 8(c-d)). We have checked the cell viability using trypan blue-dye extrusion and found no evident change in laser treated cells compared to that of intact cells. AFM topography shown in Fig. 8(e-g) were taken with a time-laps after irradiation. Fig. 8(h) exhibit a typical cross-sectional height profiles along dotted line in Fig. 8(e). It shows the presence of a deep hole on the irradiated plasma membrane, of which diameter is about 1.6 μm at a full width at maximum. Meanwhile, the estimated initial size must be the upper limit due to the limitation of spatial resolution for optical detection as well as the limitation in scanning speed for topographical imaging by the AFM. However, the laser beam irradiations on the fixed cells under the similar experimental conditions leads to a submicron-sized selective ablation (for dissection) and hole drilling on the plasma membrane surface. With a repeated scanning of the injured membrane, the topographical image shows not only the spontaneous movement of the ruptured region on the plasma membrane but also the dynamic changes in its shape as well as its size.

While the presence of injured area was completely faded out in the optical images within a time span of several minutes after the perforation, further changes on the membrane surface were still observed in AFM topographical images as shown in Fig. 8 (f). The observations from the topographical images reveals that the highly localized irregular ruffled features along with small irregular movements after the laser irradiations are in-part an experimental observations for the self-healing process of the plasma membrane (Fig. 8 (f)). After about 1 hour, the plasma membrane was found to be fully recovered in its topographic feature and looks almost similar as that before the perforation. Accompanied with these dynamic changes in plasma membrane, a trace tailed to the operated cellular parts, which is initially absent as shown in Fig. 8 (e), appeared on the substrate surface after several minutes of the perforation. Close examination of the shape and position of the trace for laser perforations indicates a little cellular or cytosolic extrusion along with the progressive increment in hole-size.

Figure 9 exhibits the temporal behavior of the geometrically averaged radius of the hole induced by fs laser ablation. The temporal behavior of the injured area was found to be strongly dependent on the pulse energy and the irradiated position.

For all the cases, where the laser energy is in between 5 nJ and 7 nJ, a very small quantity of the cellular contents are extruded from the injured cells before complete disappearance of the changes at site of injured area. With the progressive rise in the laser fluence, initially there is a successive increase in the injured area verified by the increments in radius of perturbation. However, at the higher fluence levels, the cell contents are massively extruded from the cells results in the complete rupturing of the injured cells without any initiation of self healing process. Upon irradiating the cells towards the site of nucleus, the laser fluence even lower than 5 nJ results in complete rupturing of the plasma membrane. These observations suggest that geometrical factors of the plasma membrane of the live cells plays an important role in the dynamics of ruptured membrane of live cells, assuming that the physicochemical nature of the plasma membrane is isotropic across the entire live cell.

We have an access to the velocity field by attaching the fluorescent beads (diameter ~200 nm) on the surface of the plasma membrane with several time lapses (Fig. 10). After incubation of the cells with the beads for about 12 hours, a strong correlation between fluorescence and topographical image from the beads, led us to know that the beads were

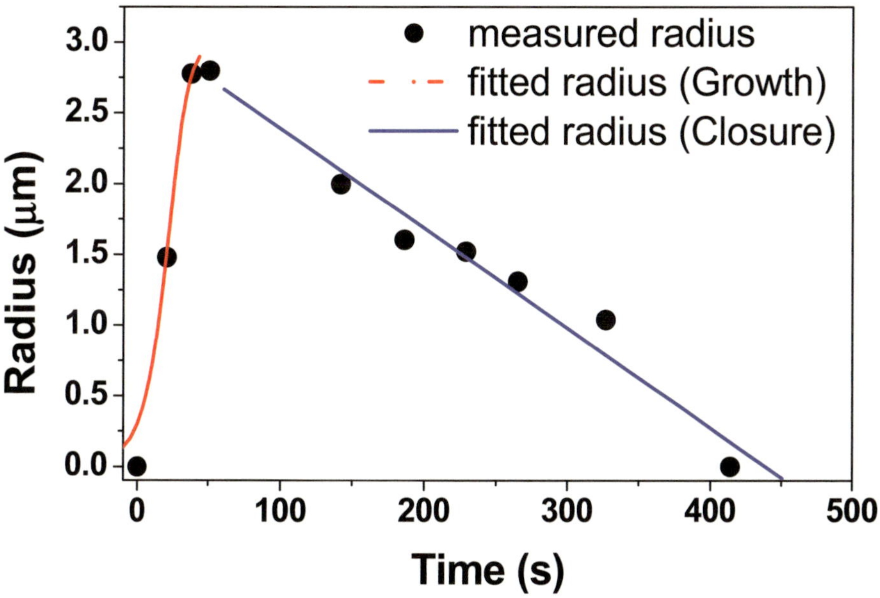

Fig. 9. Temporal behavior of the geometrically averaged radius of hole on the plasma membrane perforated by fs laser. The growth of the holes was fitted with the proposed model using Eq. (3) with the parameters of r_i, r_c, and time constant t of 0.3 μm, 3 μm and 12 sec, respectively. For the closure of the hole, the least-square fitting procedure on the experimental data gives the speed of about 0.007 μm/sec.

not delivered into the cytosol but remain attached on the surface of plasma membranes. With an aid of applied force to the beads by AFM tip, we found that interaction force between the beads and plasma membrane is fairly high enough to visualize the membrane dynamics without any detachment of the beads during the dynamic membrane movements. Just after the laser irradiation, the velocity field is found to be radial in approximately with a direction towards the edge of the cells. The magnitude of the velocity field decreased with increasing the distance from the position of irradiation. This suggests that all the lipids collected from the growing hole are almost homogeneously spread throughout the remaining area of plasma membrane. Furthermore, the absence of any rim collecting the lipids associated with a long-range flow of the membrane indicates a viscoelastic behavior of the membrane opening process. The velocity field of the membrane closer to the initial ruptured area turns to the opposite direction in next 0.7 seconds after the perforation, while the plasma membrane at a long distance from the irradiated area is keeps on expanding. Afterwards, the movement of velocity fields for the membrane looks randomized, as shown in Fig. 10(d) and (e). Followed to these interesting dynamical behaviors of the live cell membrane, overall pattern of the velocity field in both direction and magnitude resumed again, as observed just after the perforations previously (Fig. 10(f)). At a short span of time, the repeated change in direction of velocity field accompanied with an ultimate hole-size increment seems to be due to an elastic propagation. It behaves like a rubber, with fairly a high elastic modulus. After a long overall hole opening time, the velocity fields of membrane movement turned its direction towards the initially injured area, possibly be associated with the self-healing process of the created holes.

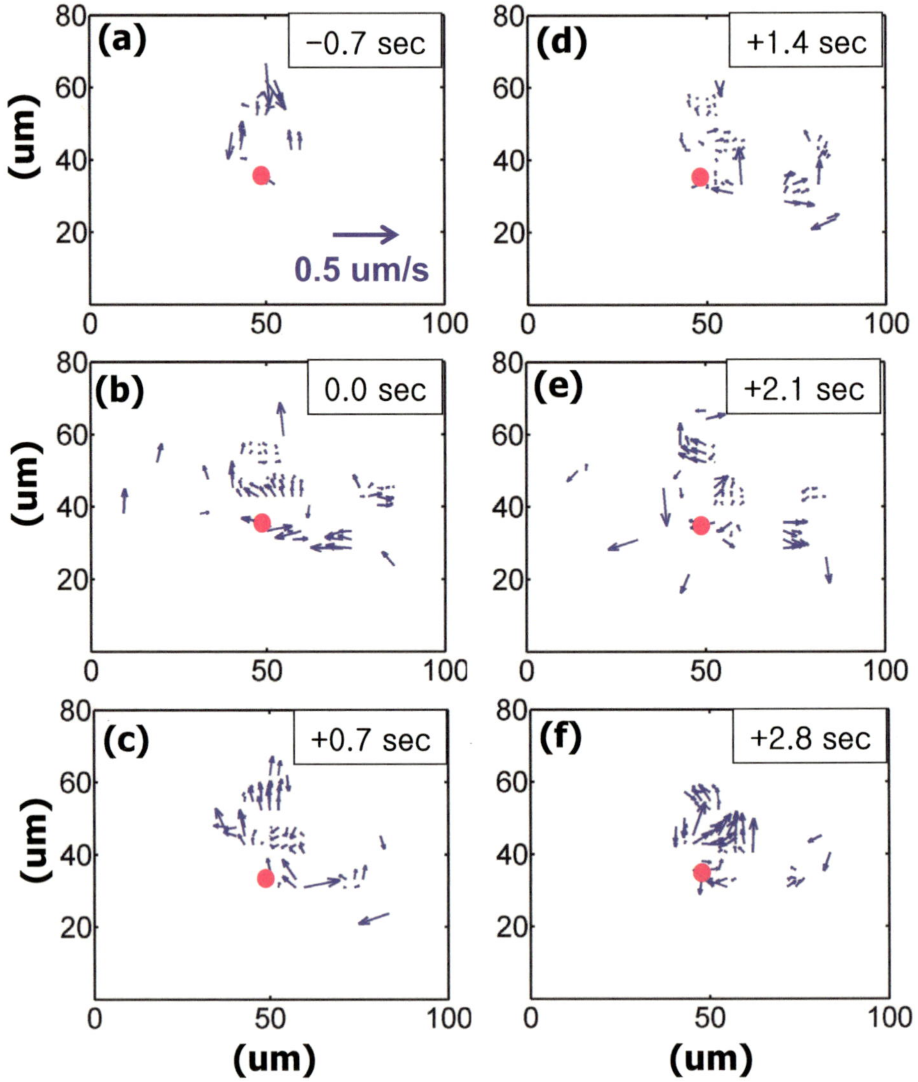

Fig. 10. Time sequence of the measured velocity components and contours of the plasma membrane associated with the hole opening of the plasma membrane of live N2A cells. The red circle indicates the point of laser irradiation.

3.3 Discussion

The simplest explanation for the mechanism of laser-mediated foreign material transfer is based on the assumption that the foreign material added to the transforming medium passively enters into the cell through the laser-induced membrane perforation. Furthermore, it successfully undergone with a rapid self-sealing process follows immediately with no subsequent and irreversible injury to the cell (Zeira et al., 2003). The results demonstrated in the current study made the above assumptions true for the Neuro 2A cells, where the cells were irradiated with an ultrafast laser in near infrared (NIR) region.

The dynamic behavior of the rapidly generated pores, either in organic films or in lipid vesicles has been widely studied (Debregeas et al., 1995 & Sandre et al., 1999). These synthetic materials are quite different in many aspects from the current plasma membranes of the live cell. Especially, they are quite inhomogeneous in distribution of viscosity, having a different elastic modulus and three-dimensional shape, and lastly have a different lipid composition. Meanwhile the previous investigation suggests that the viscosity of the films plays a very important role in opening and closing dynamics of the perforated holes (Sandre et al., 1999). If viscous dissipation is associated with the generation of transient pores dominates the inertia due to the very high viscosity, the leak out or extrusion of the cytosolic components slows down; the pore reaches the size up to 10 μm. The immersion of vesicles in a viscous environment allows visualization of transient pores in a membrane stretched by intense illumination or laser irradiations. Therefore, we have adapted a simple model to describe the dynamics of holes in the plasma membrane of live cells by an analogy with the opening of holes in viscous suspended long-chain polymer films and the dynamics of transient pores in stretched giant unilamellar vesicles (Sandre et al., 1999). The growth law of the hole is derived from a transfer of surface energy into viscous losses:

$$4\pi\eta d\frac{dr}{dt} = 2\pi r f \tag{1}$$

where ηd is the surface viscosity, and f is the surface tension acting on the film. The growth of pores in tense vesicles is more complex, because (*i*) the tension of the membrane relaxes to zero as a pore expands, and (*ii*) the line tension cannot be neglected. It is the driving force to close the pores. Including these two effects, we write the driving force f per unit length of the hole as

$$f = \sigma_0(1 - \frac{r^2}{r_c^2}) - \frac{S}{r} \tag{2}$$

where σ_0 is the membrane tension before the opening of the hole and r_c is the maximum radius of the hole. The first term describes that the surface tension decreases linearly with the area of the pore. The second term is the restoring force that tends to close the pore because of the line energy S per unit length.

Following the model, we are proposing the dynamics of live-cell plasma membrane associated with perforation of a hole with an initial radius of r_i, with further increment, closing and finally complete sealing of hole in the ruptured plasma membrane (Fig. 11). In a period of opening of the hole, the dynamics could be described in terms of a relaxation of surface tension, which is the plasma membrane tension before the perforation, due to the size increment of the hole. The growth of a hole is ruled by Eq. (1) with $f \approx \sigma_0 (1- r^2/r_c^2)$. This leads to:

$$\ln\frac{r}{r_i} - \frac{1}{2}\ln\frac{r_c^2 - r^2}{r_c^2 - r_i^2} = \frac{t}{\tau} \tag{3}$$

where $\tau^{-1} = \sigma_0/\eta d$.

With opening of a hole at the radius r, extending to, in the presence of surface tension leads to an extrusion of the intracellular contents (Fig. 8(c)); finally reducing with the relaxation of the surface tension to zero. The increase in the radius as well as relaxation of the tension is

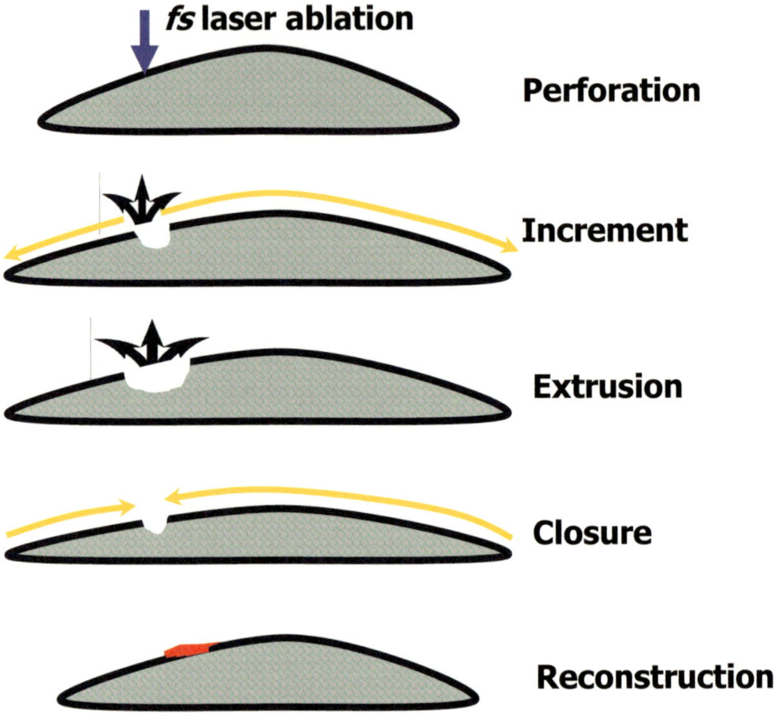

Fig. 11. Schematic diagram for the self-healing process after the fs- laser irradiation on live-cell plasma membrane. Plasma membrane healing process consists of the following four successive processes. **Opening**: when the plasma membrane is ablated by fs-laser, a submicron-sized hole was created on the membrane surface. **Increment**: due to the presence of surface tension of the films, which consist of plasma membrane and cytoskeleton network such as F-actin, the radius is increased with radial flow of the plasma membrane. **Extrusion**: with growth of the hole size, the cellular contents extrude with a relaxation of the initial surface tension. **Closure**: after reaching the maximum radius of the hole, the restoring force for the surface line energy dominates the dynamics to reduce the radius of the hole. **Sealing**: Finally the cellular structures come close to each other and support the closure of ruptured hole where the intracellular supporting networks as well as the lipid molecules have been reconstituted completely.

associated with the domination of restoring force of the hole due to the increment of line energy S per unit length. This finally induces the closure of the transient hole in order to reduce the line energy. Taking Eq. (1) into account, with the restoring force $f = -S/r$, the closing speed shall be approximately described as following:

$$V = -\frac{dr}{dt} = \frac{S}{2\eta d} \tag{4}$$

Based on the above simple model, the dependence of the radius r of the hole on time-lapse t was fitted in both region of opening and closing of the holes using Equations (3) and (4), respectively. It is quite appealing that the observed temporal behavior of the hole radius is fairly well fitted with the parameters of r_i, r_c, and time constant t of 0.3 µm, 3 µm and 12 sec, respectively. In the first order of approximation, the estimated speed of plasma membrane in opening region is found to be 0.2 μm/sec. For the closing period, the fitting of the experimental data results in the speed of about 0.007 μm/sec.

For comparison, the opening time constant and closing speed for a synthetic unilamellar vesicles made of 1,2-dioleoyl-sn-glycero-3-phosphocholine (DOPC) adhering on the substrate were reported to be 0.8 sec with $r_i \sim 1.3$ μm and $r_c \sim 8.0$ μm and 1.8 μm/sec, respectively (Sandre et al., 1999). It is remarkable to compare the closing speed ($S/2\eta d$) of the live cell membrane, which is found to be 250 times slower than the synthetic vesicles. And also, the comparison on the time constant responsible for the opening of hole indicates that the opening process observed from live cell membrane is much slower than that for unilamellar stretched vesicles. These differences might be explained in terms of higher surface viscosity and/or the lower surface tension as well as lower surface line energy in the live cells. Without any exact physical constants for live cells including the above parameters, it is difficult to give any conclusive remarks on the dynamics observed from the current work. Assuming the viscosity of the viscous films on the surface of live cell and its energy loss per unit length to be curved at the edge of the holes is not quite different from those for the previous synthetic vesicles, however, the films in the flow field associated with the hole closing of the live cell must be of two orders of magnitude thicker than that of a synthetic unilamellar vesicles. This interpretation is contrary to the fact that the thickness of cellular membrane based on the fluid mosaic model should be of same order of magnitude as that of the synthetic one (Singer & Nicolson, 1972). It's a well known fact that the cytoskeleton bound to membrane is a deformable intracellular support network made of interconnected filamentous biopolymers and its main constituent is the network of helical actin-filaments (F-actin) (Engelman, 2005; Singer & Nicolson, 1972). So, it is reasonable to suppose that the nature of the films composed with membranes itself and the bound cytoskeleton structure due to its relatively long filament half-life of order of $10^{-1} \sim 1$ min, must be considered as parameters to explain the live cell membrane dynamics induced by the current ultrafast perforations (Oliver et al., 2005). In other words, the presence of the cytoskeleton bound to cell membrane results in the increment of effective thickness in the flow field of live cell membrane for hole growth and closure processes. In fact, typical thickness of actin cortex in motile osteoblasts was reported to be the order of $10^{-1} \sim 1$ μm (Oliver et al., 2005). The rather thick intracellular supporting network compared to a unilamellar membrane of synthetic vesicles should play an important role in the growth and closure dynamics of transient hole in the live cells and thus results in much higher surface viscosity of the films.

Finally, we would like to discuss some explanations for the irregular features observed from the topographical studies during the self-healing process (Fig. 8(f)). The topographical features were maintained for quite a long period even after the disappearance of optical signature of presence of holes on the cell membranes. The long lasting irregular ruffled structures of the membrane are likely to be associated with the final reconstruction of live cell membrane, to that of similar nature as present at the initial stages. Following the previous interpretations on the nature of films in the dynamic hole opening and closing behavior, the intracellular support network as well as the lipid molecules have been not reconstructed completely even though the cellular structures get close enough to do so. The chemistry and physics associated with the reconstruction process is still far from its complete understanding. We can suppose that a non-homogeneous reconstruction process at a relative slow reconstruction speed may result in the generation of irregular topographies at

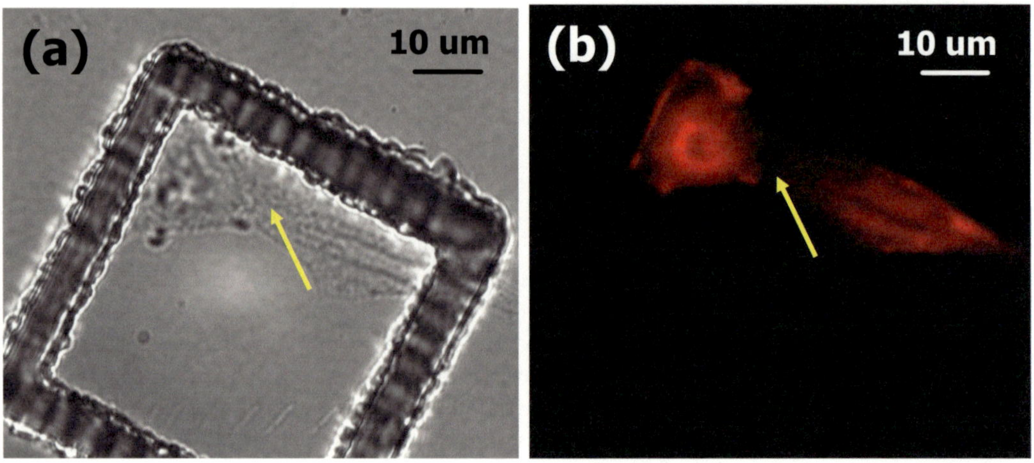

Fig. 12. Bright field (a) and fluorescence image (b) of injured cell stained with Alexa fluor 635 Palloidin (Molecular Probe) to visualize the distribution of F-actin. The cells were fixed and stained after completing the hole closing process under bright field images. The emission intensity near to the ablated region (arrowed) is less than that for other parts. The hollowed black area in the bright-field image (a) is a remnant of fs-laser ablated region, which was processed on the substrate as a guideline before seeding of the cells.

the edge of colliding films on the surface of cells. In order to obtain qualitative information concerning on this supposition, the cells just after completing the closing process under bright field images were stained to know the actin network distribution. As shown in Fig. 12, the fluorescence image shows the lower intensity of Alexa Flour 635 Palloidin expression near to the perforated region rather than other parts of the cells. This observation partly supports our proposition that the reconstruction process of F-actin network might results in several tiny slices of lipid during sealing and interconnecting the dissected cell membrane at the final stage of self-healing process. Another possible interpretation on the observations might be the self-healing process of damaged surface membrane of live cells such as urchin's egg and muscle cell by, so called, surface precipitation reaction mechanism proposed several tens of years ago (Heilbrunn, 1956; Gonzalez-Serrators et al., 1996). If this is the case, the piece of the lipid assembly should be together spontaneously to recover the injured surface membrane of live cells and resulted in the surface irregularity detected with only AFM techniques as shown in Fig. 8 (e)-(f). Even though no any conclusive remarks on the nature of the final reconstruction process of the damaged surface of the live cells will be given with the available information on the process, the irregular features observed should be strongly associated with reconstruction of not only the plasma membrane itself but also the actin network as an cytoskeleton structure of the live cells.

4. Ultrafast laser- assisted optoperforation of primary retinal blood vessels

4.1 Experimental techniques and materials

Furthermore, these extensive studies on single cell and plasma membrane dynamics open up the new scenarios, to apply this technology in a real surgical world, such as retinal blood vessel occlusions.

Experimental porcine eyes obtained from a local abattoir were deposited in ice cold Hanks's Balanced Salt Solution (HBSS) and transported to the laboratory on ice (4 °C) within 1 hr after the slaughter. After a precise incision of the porcine eyes behind the ora serrata, cornea and lens, the retina tissues were carefully peeled off from pigmented epithelium and put it onto the Whatmann 2 paper. Before proceeding further for the fs-laser irradiations, the retina segments with size of about 3 X 5 mm^2 were prepared and to closely mimic the *in vivo* retinal conditions and to prevent the drying off the samples, we covered the tissues with 100 μl of vitreous humour during the experiments.

The fs-laser system used in this study consists of a regenerative amplified Ti: Sapphire (λ = 810 nm) laser, delivering a 150-fs pulse at a repetition rate of 1 kHz (Quantronix, USA). The laser beam is focused on the retina surface with an objective lens of N.A. 0.4 and the estimated laser beam diameter at the retinal surface is about 1.3 μm. Similar to the previous setups, the retina tissues are mounted onto a motorized XYZ translation stage, which is used to manipulate the samples to expose a fresh area of tissue to each laser pulse (Fig. 13). The laser fluence used in the current work is in the range of 1.4 J/cm^2 ~ 99.4 J/cm^2. With 13 different porcine eyeballs, 20 - 25 laser treatments for each eye ball were conducted at intervals of 100 μm on the surface of retinal blood vessels. Blood vessels used in the experiments were primary in nature and were classified on the basis of the location and the

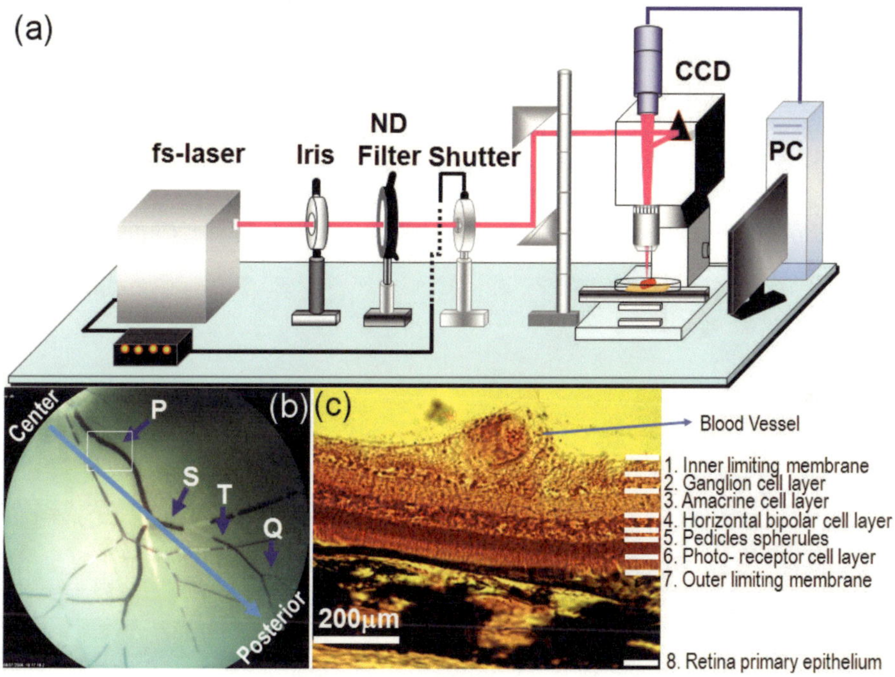

Fig. 13. (a) Schematic diagram of the experimental setup for fs-laser treatment of porcine retina blood vessels (b) The image from a fundus camera (Carl Zeiss Meditec, AG, Germany) shows the organization of blood vessels following the direction of blood flow from the center of the retina, i.e., the fovea region, to the peripheral retina: Key: P = Primary, S = Secondary, T = Tertiary and Q = Quarternary Blood Vessels. (c) The inset shows a typical cross-sectional image of the primary blood vessel, including the underlying multilayered structure of the porcine retina.

diameter of the vessel. All the experiments were conducted under a single-shot configuration by using a mechanical shutter with an opening time constant of 0.6 ms, which is synchronized with the laser, allows an accurate delivery of single pulse- irradiations into the samples. (Sidhu et al., 2009)

After optoperforation of blood vessel epithelium, the location of each laser lesion was carefully mapped. Just after laser treatment, the retina segments were immersed and fixed in 2% *p*-formaldehyde at 4 °C for one hour. In next step, the retina tissues were submersed in 30% sucrose for about one hour to retain the cell shape. The fixed and retained samples were embedded in Jung's Tissue Freezing medium for an hour at room temperature and later on transferred to a refrigerator at -20 °C for 4-6 hrs. Transverse cryo-sections with a thickness of 20 μm (Microtome, LEICA) were made of the lesion area at -15 °C. The slices were stained with Haemotoxylin and Eosin Red (H & E) to assay the laser lesions at the peripheral Inner Limiting Membrane (ILM : upper layer of cells covering the entire retina surface along with the blood vessels) and the blood vessels epithelium of the retina tissue. In order to determine the percent probability of damage for the blood vessels, we examined about ~1002 sampled (6 times) at various laser fluences in the range of 1.4 J/cm^2 - 99.4 J/cm^2.

4.2 Results

With changing laser fluence from 1.4 J/cm^2 to 99.4 J/cm^2, we produce a series of lesions on the blood vessel walls with a single fs-laser pulse. Laser irradiation on the surface of the retinal blood vessels with the fluence higher than about several tens of J/cm^2 results in an apparent disruption of the vessel walls, followed by a continuous outflow of blood within 2 ~ 3 seconds, as shown in Fig. 14. The red color of the vessels fades out after about 5 minutes. After ~ 10 minutes, the bleeding is almost stopped. However, at lower laser fluence, the bleeding is stopped within one minute after extrusion of a little blood. (Sidhu et al., 2009)

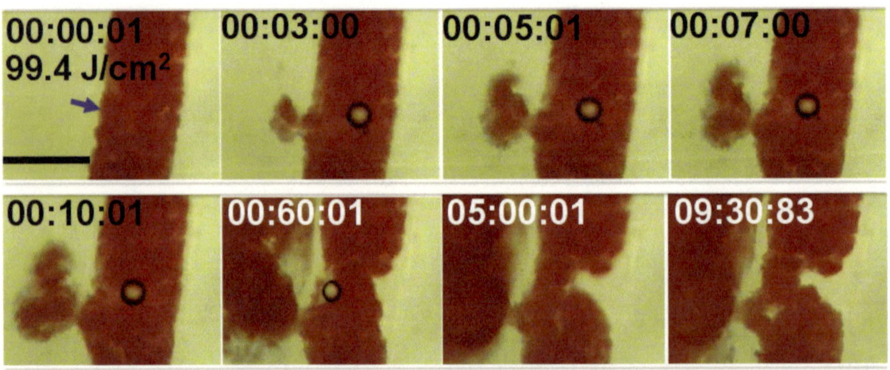

Fig. 14. Captured optical images of blood flow from a retinal blood vessel after exposure to a single fs-laser pulse at 810 nm with time scale from 0 to 10min. The laser beam is focused at the position indicated by the arrow head. Just after laser irradiation, a continuous outflow of blood is observed for 2 ~ 3 seconds. The scale bar is 100 μm.

To examine the blood vessels of a porcine retina after fs-laser irradiation, we sequentially sectioned the retina segment with primary blood vessels with a thickness of 20 μm. Figure 15 represents histological images of the sectioned slices. The laser irradiations at lower fluence cause damage on the surface of the blood vessels with a low probability. At higher laser fluence, however, almost all the retinal segments exhibit apparent disruption of the ILM, which mainly consists of a Glial cell layer. Apparent disruptions of the ILM in the sectioned slices of the retina treated at 14.2 J/cm^2 could be observed in the first and the fifth slices without any visible alterations in other sequential slices. No apparent disruption in the

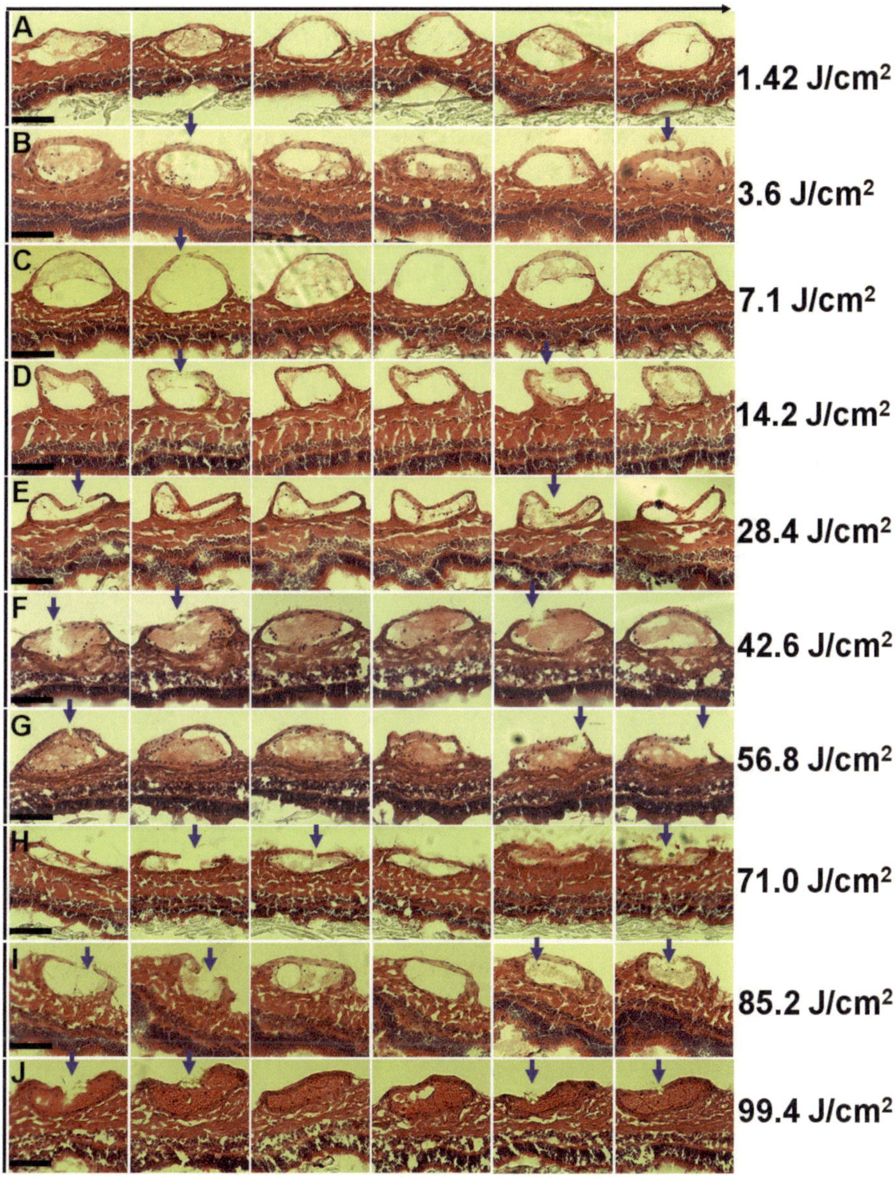

Fig. 15. Sequential sections of a blood vessel lesion caused by a single fs- laser pulse irradiation. The thickness of the sectioned slices is 20 μm. Figures on the right-hand side represent laser fluence. The arrow head indicates a lesion observed on the blood vessels. The scale bar is 100 μm.

surface of the blood vessels endothelial layer and no bleeding after laser irradiation were observed. If the slice thickness of 20 μm in the sectioning procedure is considered, the interval between slices with an apparent disruption is 100 μm, which is quite consistent with the distance between laser-irradiated points on the retina tissue. This observation suggests that from the histological analysis, the lateral damage of the retina upon fs-laser exposure is limited to 20 μm range. In fact, the diameter of the laser lesions on the surface of treated retina, examined via scanning electron microscopy, is found to be less than 4 μm (Fig. 16). With increasing laser fluence to 28.4 J/cm^2, single-pulse laser irradiation induces complete optoperforation on the blood vessel wall. With further increment in the laser fluence to 52.6 J/cm^2, the frequency of sequential slices with observed prominent damage on the blood vessels increases. This reveals that the diameter of the induced lesion is larger than 20 μm under these rather higher laser fluences. The data suggest that up to 85.2 J/cm^2, there is no observable alteration in underlying retina cell layers, except for disruptions in both the blood vessel walls and the ILM. Under a higher fluence of 99.4 J/cm^2, however, a slight distortion is observed in the underlying tissue (Fig. 15(J)). (Sidhu et al., 2009)

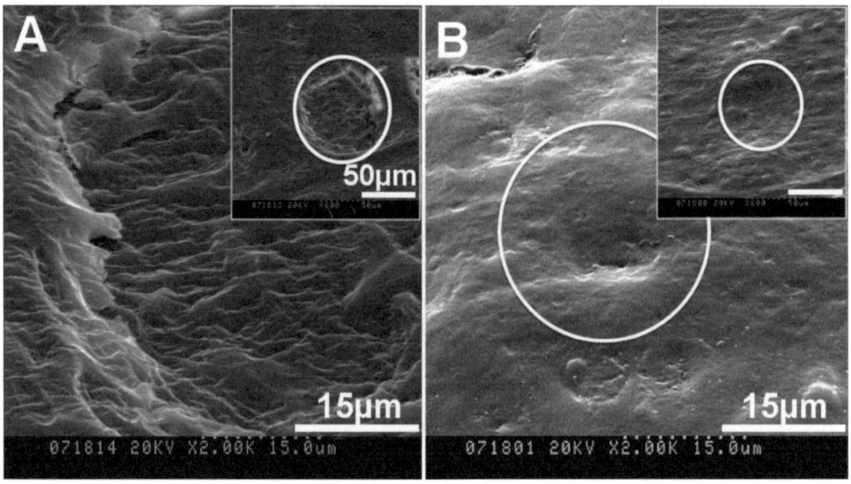

Fig. 16. Scanning electron microscopic images for a laser-irradiated retina surface (white circles) with two different laser fluences 99.4 J/cm^2 (A) and 7.1 J/cm^2 (B). The insets show SEM images with lower magnification.

The cryosections of the retina segments irradiated with a single-shot ultrafast laser were tentatively grouped into three types of lesions: no change, ablation only at the ILM and optoperforation of blood vessel walls (Fig 17). Correlation statistics of these different types of lesions is shown in Fig. 18 as a function of laser fluence. The optoperforation probability of blood vessels increased in the fluence range of 3.6 J/cm^2 ~ 99.4 J/cm^2. We determined the laser fluence to damage the ILM layer, found to be in the range between 1.4 J/cm^2 and 3.6 J/cm^2. However, perforation of the blood vessels is apparently caused by femtosecond-laser irradiation with a higher fluence between 3.6 J/cm^2 and 7.1 J/cm^2. If the percent probability at low fluence is extrapolated, the fluence at zero probability of alteration of the ILM and the perforation of the blood vessels are estimated to be 2.19 ± 1.08 J/cm^2 and 5.85 ± 1.49 J/cm^2, respectively. (Sidhu et al., 2009)

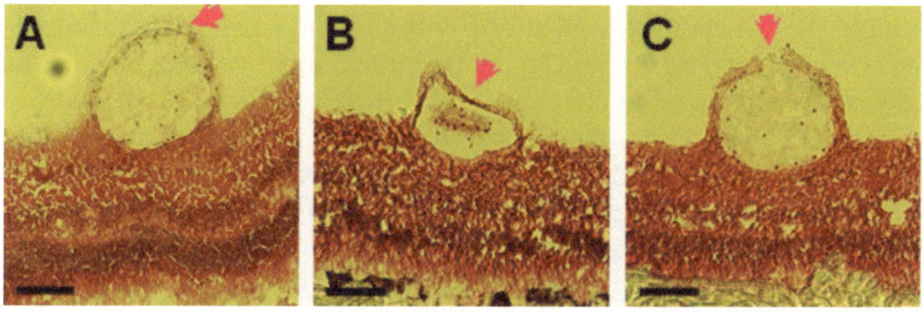

Fig. 17. Representation of retinal segments irradiated with a single-shot ultrafast lasers. They were tentatively grouped into three types of lesions: **A**. No change, **B.** The ablations at the ILM and **C.** The optoperforation of blood vessel walls. The arrow head indicates the point of irradiations on the blood vessels.

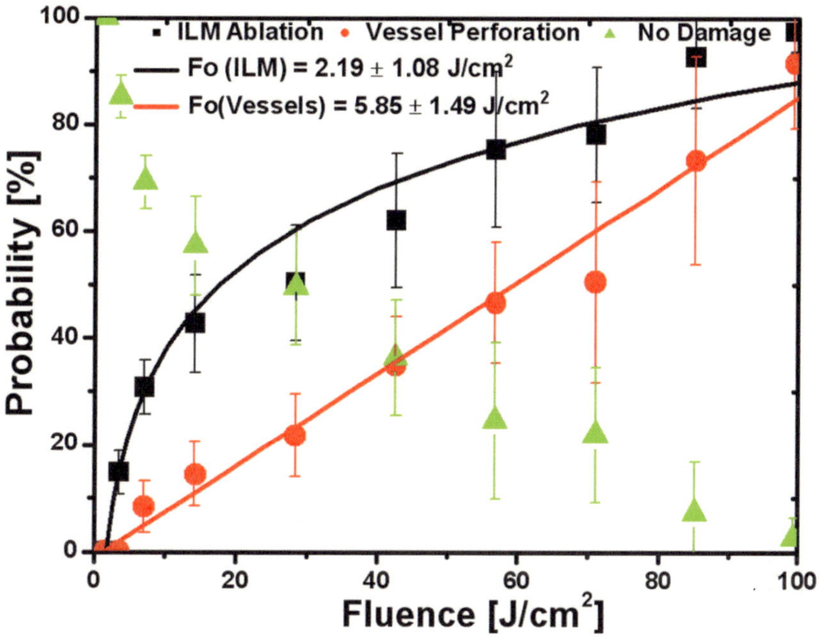

Fig. 18. Linear plot of the percent probability for inner limiting membrane (ILM) damage (solid rectangles) and vessel perforation (solid circles) as a function of the laser fluence. The ablation threshold fluence for ILM and blood vessels was found to be 2.19 ± 1.08 J/cm^2 and 5.85 ± 1.49 J/cm^2, respectively. With increasing fluence, the percent probability of blood vessel perforation monotonically increases. The lines represent an extrapolation to determine the ablation thresholds for perforation of retinal primary blood vessels and for ILM damage of a porcine eye.

4.3 Discussion

Recent development in advanced laser technology transiently facilitates to perform transaction, ablation, and coagulation of tissues via delivery of laser irradiation into a small focal volume are providing an attractive possibilities for new laser surgical technologies. The laser beam is a potential candidate that has already undergone a multi-center clinical trial to evaluate the feasibility for its use in vitreoretinal surgery (Schastak et al., 2007). Limited precision and significant damage by lasers with relative long pulse durations does not allow partial or selective tissue ablation with high precision. If such damage is to be overcome, infrared laser sources, such as CO_2, Er:YAG and Holmium:YAG lasers, have undergone several trials via optical fiber delivery in intraocular surgery. However, apparent collateral damage to surrounding tissue due to significant thermal and shockwave effects have been reported (Paula- Yu et al., 2006).

Laser ablation of tissues could be described using either an optical breakdown model or a thermal confinement models. The optical breakdown model considers plasma formation and subsequent shock wave formation, cavitation, and tissue disruption. The thermal confinement model recognizes the competing thermal effects of the vaporization of water driving an explosive ablation and thermal diffusion leading to collateral damage. This model accounts for the observation that collateral damage is limited if the pulse duration is less than the thermal relaxation time of the ablated tissue volume (Vogel et al., 2003; Apitz et al., 2005).

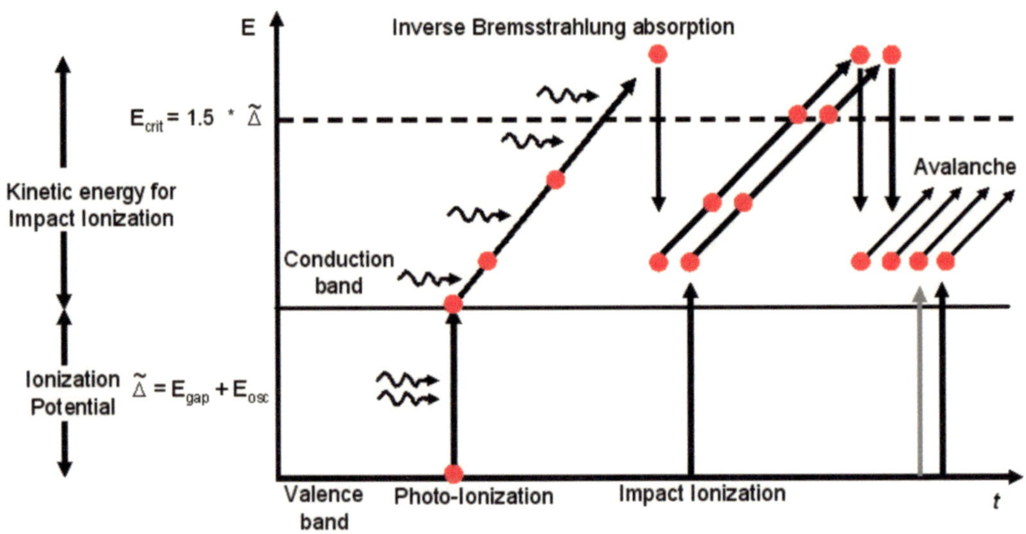

Fig. 19. Interplay of photoionization, inverse Bremsstrahlung absorption, and impact ionization in the process of plasma formation. Recurring sequences of inverse Bremsstrahlung absorption events and impact ionization lead to an avalanche growth in the number of free electrons. (Vogel et al., 2005)

The process of plasma formation through laser induced breakdown in transparent biological media is schematically depicted in Fig. 19. It essentially consists of the formation of quasi-free electrons by interplay of photoionization and avalanche ionization. It's a well known fact that the optical breakdown threshold in water is very similar to that in ocular and other biological media (Docchio et al., 1986). Irradiation by an intense ultrafast laser beam further leads to multiphoton excitation of a target material. The absorbed energy might be transported to the electrons without thermal diffusion to adjacent material because the pulse width is shorter than the vibrational relaxation time constant of several picoseconds. As a result, thermal damage on the surrounding tissues could be minimized, and the biological tissue remains unaffected by the subsequent photoinduced mechanical shock process. This effectively renders the fs-laser surgical process non-thermal. The formation of a high density of free electrons could result in a local plasma formation in the targeted materials. This hot plasma formation results in a permanently damaged region, even inside a cell with a sub-micron size (Vogel et al., 2005). Furthermore, a previous on tissues like the corneal stroma revealed that the ablation threshold fluence decreased with increasing pulse width of the applied laser (Preuss et al., 1995). These uniquely show that ultrafast lasers can be utilized for precise treatment of tissues while minimizing any apparent thermal damage or shock pressure to biological tissues (Kohli et al., 2005). The results illustrated in the current work made the above hypothesis true for the retinal tissues, where retinal blood vessels were selectively perforated with wide range of laser fluence (1.42 ~ 99.4 J/cm^2) with an ultra fast laser in near infra red region.

From the past literature values for the ablation thresholds for various tissues, including the corneal stroma, axons, the eye's anterior chamber, and hard tissue (under a single-shot configuration, as in current work), the ablation threshold of the corneal stroma for an ultrafast laser is in the range of 1 J/cm^2 to 2 J/cm^2. Meanwhile, the ablation threshold for axons of *C. elegans* is reported to be about 4.4 J/cm^2. It is of great interest to note that the value for the femtosecond laser ablation threshold of the ILM of the porcine retina, 2.19 ± 1.08 J/cm^2 as determined in the current work, is in the same range of reported values for the soft tissues. It is also interesting to compare the ablation threshold of the retina upon irradiation by a femtosecond laser to the values for irradiation with an ultraviolet (UV) laser with a nanosecond pulse width, including ArF excimer lasers and higher-harmonic Nd:YAG lasers. The ablation threshold is reported to be in the range of between 0.6 J/cm^2 and 1 J/cm^2 when irradiating single-pulsed UV light into the retina tissue, which is slightly lower than that for femtosecond laser ablation threshold. Considering the remarkable difference in the linear optical absorption coefficients of the retina tissue in the UV and the NIR ranges, it is reasonable to suppose that an ultrafast laser operating in the NIR region would be able to ablate the ILM layer in the retina with much lower deposited energy per unit volume compared to UV nanosecond lasers. The perforation threshold of the underlying primary retinal blood vessels (5.85 ± 1.49 J/cm^2) is significantly higher than the literature values.

The thickness of the ILM, which is essentially a basement membrane consists of retina müller cells, is only 6 μm to 10 μm. The thickness of the ILM is thinnest at the fovea region of the retina. However, the thickness is larger at the posterior pole of retina (Hoerauf et al., 2006). Furthermore, the ILM is also present over the retinal blood vessels. If only the ILM is to be ablated selectively without any alterations in the underlying layers, the energy delivered by the laser irradiation must be confined in thin layers without any apparent diffusion of the deposited energy into other parts of the retina. To evoke this topic, we have examined the dependence of the ablation depth for transparent materials, like retinal tissue, on the laser fluence (Fig. 20). If there is high free electron density due to optical absorption processes, we suppose that the underlying mechanism for the ablation by fs-laser irradiation is not directly governed by the optical and the electronic properties of the materials. Even if the absorption mechanism of the NIR fs-laser is dependent on the optical band gap of each material, two different slopes under fs laser irradiation have already been reported for metals, semiconductors, and dielectrics (Nolte et al., 1997; Furusawa et al., 1999). For a lower laser fluence, F, the ablation depth can be described by the expression $L = \delta ln(F/F_{th}^{(\delta)})$, where δ is the optical penetration depth and $F_{th}^{(\delta)}$ is the threshold laser fluence of ablation [Preuss et al., 1995, Jia et al., 2006]. A fit to the experimental data results in F^{δ}_{th} = 2.2 ± 0.9

J/cm^2 and δ = 8.2 ± 2.2 μm. It should be notified that the optical penetration depth is governed by a nonlinear optical transition, if multi-photon absorption plays an important role in photo-excitation of the materials. Therefore, the optical penetration depth estimated from the current work is difficult to reconcile with the literature value of the optical absorbance of retina tissue at a wavelength of 810 nm. Due to the strong dependence of the multiphoton absorption on the energy density, the value of δ should be relatively small. At any rate, it is of great interest to compare the observed optical penetration depth with the thickness of the ILM in the porcine retina. This comparison led us to propose that the energy delivered by femtosecond laser irradiation under the controlled laser fluence can be confined in the ILM layer, followed by a selective ablation of the layers only if the optical penetration depth of 8.2 ± 2.2 μm is comparable to the thickness of the ILM of the retina.

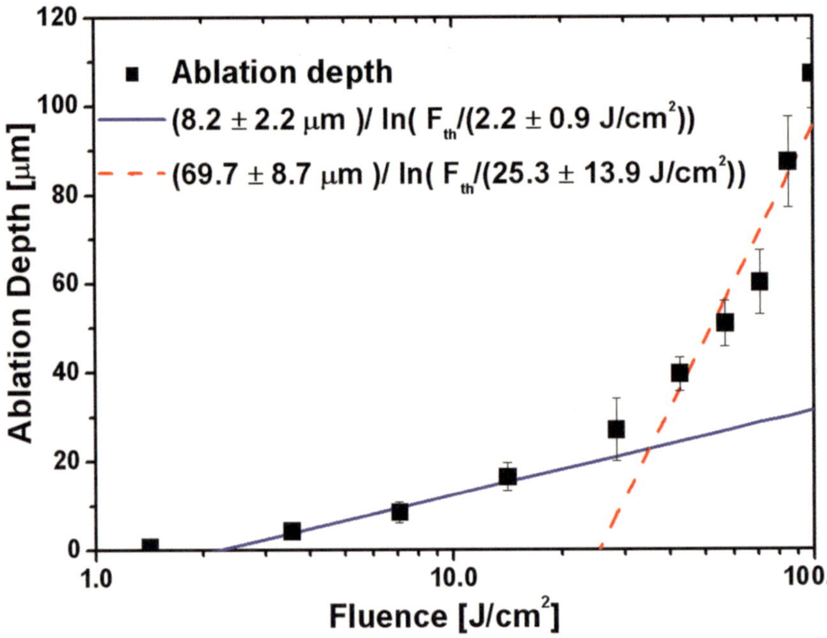

Fig. 20. The lesion depth of a porcine retina caused by fs-laser irradiation as a function of the laser fluence. The blue solid and the red dotted lines represent linear fit. About 300 sectioned slices from more than 10 eyeballs were examined for each laser fluence.

With increasing laser fluence, however, the mechanism underlying the retina ablation can no longer be expressed by the optical penetration depth. As shown in Figure 16, the retina surface treated with a high laser fluence of 99.4 J/cm^2 is very much roughened compared to the surface treated with a low fluence of 7.1 J/cm^2. Based on the changes in the slopes of the semi-logarithmic plot of the ablation depth as function of the fluence, we have supposed that at laser fluence higher than 25.3 J/cm^2, the electronic heat diffusion process plays an important role, even in an ultrafast laser ablation. The ablation depth in this region can be described with the expression of $L = l \ln(F/F_{th}^{(l)})$, where l is the electronic heating depth, and $F_{th}^{(l)}$ is the corresponding threshold fluence. The electronic heating depth and $F_{th}^{(l)}$ are estimated to be 69.7 ± 8.7 μm and 25.3 ± 13.9 J/cm^2, respectively, which means that the thickness of the retina tissue affected by fs-laser irradiation might be abruptly increased for the laser fluence higher than 25.3 J/cm^2. As a result, we have to control the laser fluence very precisely to achieve a selective peeling of the ILM layer without any visible thermal damage being induced by the laser irradiation.

The probability of retina blood vessel damage shows a linear relationship with the laser fluence. With the progressive increase in the laser fluence, selective ablations of concerted retina layers even including primary blood vessels is possible without any apparent damage to the underlying layers of the porcine retina. The threshold fluence to perforate the walls of the primary blood vessels embedded in the porcine retina is estimated to be 5.85 ± 1.49 J/cm^2. If the ablation depth depends on the laser fluence as $\delta ln(F/F_{th}^{(\delta)})$, the thickness of the tissues ablated by a single-shot fs-laser pulse can be estimated to be 8.0 ± 3.0 μm, by using the parameters of δ and $F_{th}^{(\delta)}$ from this work. Meanwhile, the thickness of the tissues covering the primary blood vessels is tentatively determined to be about 25 μm by examining the sectioned slices shown in Fig. 15. If the current interpretation for the ablation depth of the tissues by fs-laser irradiation is correct, the fluence to perforate the primary blood vessels should be about 46 J/cm^2. However, the ablation depth per pulse in the high-laser-fluence region should be described in terms of electronic heating depth with the relation of $L = \delta ln(F/F_{th}^{(l)})$. With the parameters of l and $F_{th}^{(l)}$, we are able to estimate the fluence to fully perforate the primary blood vessels of the retina to be 36.2 J/cm^2. This value for blood vessel perforation is very close to the laser fluence at 1/e^2 percent perforation probability, as shown in Fig. 18.

5. Conclusion

In summary, all the observations from the present work reveals that fs-laser irradiation on pollen walls to make an evident physical hole with an outside diameter of about 1 μm well conserves the physiological state of the cell including its viability and pollen tube germination capability. Furthermore, from the successful delivery of foreign DNA into pollen through the hole reveals that the current method has an evident potential in the field of plant genetic engineering.

Topographical imaging as well as optical imaging of the plasma membranes led us to observe a self-healing process for live cells within several minutes of time after the fs-laser ablation on the live cells. A simple viscoelastic model for both the hole opening and closing process was found to be applicable to interpret its dynamics. The very slow dynamics could be explained in terms of high surface viscosity due to the presence of cytoskeleton network bound to the plasma membrane. The irregular feature in plasma topography observed in the final stage of the healing process might be due to a slice of the assembled lipid, which resulted from the reconstruction of not only the plasma membrane itself but also F-actin network as a cytoskeleton structure of live cells. Although two-dimensional plug flow model adapted in the current work fairly well interpret the experimental observations in macroscopically, the presence of transmembrane proteins, transbilayer interactions, and adhesion sites, etc., in addition to the bound cytoskeleton structure, produces a variety of restrictions on the flow dynamics of the plasma membrane through an alterations in many microscopic physico-chemical properties including thickness and hydrodynamic properties of the fluidic films.

We have developed a new method for elucidating more exact mechanism on the interesting topic of self-healing process based on ultrafast laser perforation of the plasma membrane of the animal cell. A mechanical stimulus to live-cell plasma membrane by the induced surface tension as well as surface line energy can be also applied by the current methods with high spatial resolution and unattainable speed of perforation. So interesting is the spatiotemporal characterization of the plasma membrane movement associated with the healing process that is closely related with the cell migration and transmission of the mechanical stimuli into biochemical signals, which might be mainly governed by cytoskeleton structure (Wang et al., 2005; Yamazaki et al., 2005; Supatto et al., 2005).

We have also successfully applied the current fs- laser technology to selectively perforate the retinal blood vessels without any apparent damage in the concerted retina layers. It provides a major breakthrough for the retinal vein occlusion therapy and removal of abnormal blood vessels (Choroidal Neovascularization (CNV)) grown during numerous retinal diseases.

6. Acknowledgements

This work was financially supported by the Ministry of Knowledge Economy of Korea and KRISS program.

7. References

Apitz, I. & Vogel, A. (2005) Material ejection in nanosecond Er: YAG laser ablation of water, liver, and skin, *Appl. Phys.* A 81, 329–338

Aronen, T. S.; Nikkanen, T. O. & Haggman, H. M. (2003). The production of transgenic Scots pine (Pinus sylvestris L.) via the application of transformed pollen in controlled crossing, *Transgenic Res.* 12, 375-378

Benkert, R.; Obermeyer, G. & Bentrup, F. W. (1997). The turgor pressure of growing lily pollen tubes, *Protoplasm 198*, 1-8

Debregeas, G.; Martin, P. & Brochard-Wyart, F. (1995). Viscous bursting of suspended films, *Phys. Rev. Lett.* 75, 3886-3889

Docchio, F.; Sachhi, C. A. & Marshall, J. (1986) Experimental investigation of optical breakdown thresholds in ocular media under single pulse irradiation with different pulse durations, *Lasers Ophthalmol.* 1, 83-91

Engelman, D. M. (2005). Membrane are more mosaic than fluid, *Nature* 438, 578-570

Fernado, D. D.; Richards, J. L. & Kikkert, J. R. (2006). In vitro germination and transient GFP expression of American chestnut (Castanea dentate) pollen, *Plant Cell Rep.* 25, 450-456

Furusawa, K.; Takahashi, K.; Kumagai, H.; Midorikawa, K. & Obara, M. (1999) "Ablation characteristics of Au, Ag, and Cu metals using a femtosecond: Ti Sapphire laser, *Appl. Phy. A.* 69(7), S359-S366

Gonzalez-Serratos, H.; Rozycka, M.; Cordoba-Rodriguez, R. & Ortega, A. (1996). Membrane healing and restoration of contractility after mechanical injury in isolated skeletal muscle fibers of the frog, *Proc. Natl. Acad. Sci. USA* 93, 5996-6001

Greulich, K. O. & Weber, G. (1992). The light microscope on its way from an analytical to a preparative tool, *J. Microsc.* 167, 127-151

Heilbrunn, L. V. (1956). The surface precipitation reaction, In: *The Dynamics of Living Protoplasm* 62-84, Academic, New York

Higashiyama, T.; Yabe, S.; Sasaki, N.; Nishimura, Y.; Miyagishima, S.; Kuroiwa, H. & Kuroiwa, T. (2001). Pollen Tube Attraction by the Synergid Cell, *Science* 293, 1480-1483

Hoerauf, H.; Brix, A.; Winkler, J.; Droege, G.; Winter, C.; Birngruber, R.; Laqua, H.; and Vogel, A. (2006) A Photoablation of inner limiting membrane and inner retinal layers using the Erbium: YAG-laser: An *in vitro* study, *Lasers Surg. Med.* 38(1), 52-61

Hoffmann, F. (1996). Laser microbeams for the manipulation of plant cells and subcellular structures, *Plant Science* 113, 1-11

Jeoung, S. C.; Kim, H. S.; Park, M. I.; Lee, J.; Kim, C. S. & Park, C. O. (2005). Preparation of room-temperature photoluminescent nanoparticles by ultrafast laser processing of single-crystalline Ge, *Jap. J. Appl. Phys.* 44, 5278-5281

Jia, T. Q.; Chen, H. X.; Huang, M.; Zhao, F. L.; Li, X. X.; Xu, S. Z.; Sun, H. Y.; Feng, D. H.; Li, C. B.; Wang, X. F.; Li, R. X.; Xu, Z. Z.; He, X. K. and Kuroda, H. (2006) Ultraviolet-infrared femtosecond laser-induced damage in fused silica and CaF_2 crystals *Phys. Rev. B* 73 (5) 054105-1 - 054105-9

Kobayashi, N.; Rivas-Carrillo, J. D.; Soto-Gutierrez, A.; Fukazawa, T., Chen, Y.; Navarro-Alvarez, N. & Tanaka, N. (2005). Gene delivery to embryonic stem cells, *Birth Defects Research (Part C)* 75, 10-18

Kohli, V.; Elezzabi, A. Y.; & Acker, J. P. (2005) Cell nanosurgery using ultrashort (femtosecond) laser pulses: applications to membrane surgery and cell isolation, *Lasers Surg. Med.* 37, 227–230

König, K.; Riemann, I.; Fischer, P. & Halbhuber, K. J. (1999). Intracellular nanosurgery with near infrared femtosecond laser pulses, *Cell. Mol. Biol.* 45, 195-201

Krautwig, B & Lörz, H. (1995). Cereal protoplasts, *Plant Science* 111, 1-10

Lee, Y. J.; Kim, D. H.; Kim, Y. & Hwang, I. (2001). Identification of a signal that distinguishes between the chloroplast outer envelope membrane and the endomembrane system in vivo," *Plant Cell* 13, 2175–2190

Lovy-Wheeler, A.; Cardenas, L.; Kunkel, J. G. & Hepler, P. K. (2007). Differential organelle movement on the actin cytoskeleton in lily pollen tubes. *Cell Motil Cytoskeleton.* 64, 217-232

Nelson, J. S. & Berm, M. W. (1989). Laser application in biomedicine. Part II: Clinical applications, *J. Laser Appl.* 1, 9-20

Nolte, S.; Momma, C.; Jacobs, H.; Tünnermann, A.; Chichkov, B. N.; & Wellegehausen, B. (1997) Ablation of metals by ultrashort laser pulses , *J. Opt. Soc. Am.* B 14, 2716-2722

Oliver, J. M.; King, J. R.; Mckinlay, K. J.; Brown, P. D.; Grant, D. M.; Scotchford, C. A. & Wood, J. V. (2005). Thin-film theories for two-phase reactive flow models of active cell motion, *Mathematical Medicine and Biology* 22, 53-98

Parpura, V.; Haydon, P. G. & Henderson, E. (1993). Three-dimensional imaging of living neurons and glia with the atomic force microscope, *J. Cell Sci.* 104, 427-432

Paula-Yu, K.; Miller, J.; Cringle, S. J.; & Yu, D-Y. (2006) Experimental retinal ablation using a fourth-harmonic 266 nm laser coupled with an optical fiber probe, *Invest. Ophthalmol. Vis. Sci.* 47(4), 1587-1593.

Preuss, S.; Demchuk, A.; & Stuke, M. (1995) Sub-picosecond UV laser ablation of metals, *Appl. Phys.* A 61, 33–37

Sandre, O.; Moreaux, L. & Brochard-Wyart, F. (1999). Dynamics of transient pores in stretched vesicles, *Proc. Natl. Acad. Sci. USA* 96, 10591-10596

Schastak, S.; Yafai, Y.; Yasukawa, T.; Wang, Y. S.; Hillrichs, G. & Wiedemann, P. (2007) Flexible UV light guiding system for intraocular laser microsurgery, *Lasers Surg. Med.* 39, 353-357

Senz, R. & Miiller, G. (1989). Laser in Medicine, *Ber. Bunsenges. Phys. Chem.* 93, 269 –277

Shen, N.; Datta, D.; Schaffer, C. B.; LeDuc, P.; Ingber, D. E. & Mazur, E. (2005). Ablation of cytoskeletal filaments and mitochondria in live cells using a femtosecond laser nanoscissor, *Mechanics and Chemistry of Biosystems* 2, 17-26

Sidhu, M. S.; Kim, E. K; Woo, S. Y; Song. M. C.; Jeoung, S. C. & Park, Y. I. (2009) Femtosecond - laser - assisted optoperforation of the primary retinal blood vessel and retina tissue of porcine eyes, *J. Kor. Phys. Soc.* 55(2) (in Press)

Singer, S. J. & Nicolson, G. L. (1972). The fluid mosaic model of the structure of cell membranes, *Science* 175, 720-731

Strubinska, J. & Sniezko, R. (2000). Localization of vegetative nucleus and generative cell nuclei in branching pollen tubes of Oenothera hookeri L. grown in vitro, *Acta Biologica Cracoviensia Series Botanica* 42, 107-112

Supatto, W.; Debarre, D.; Moulia, B.; Brouzes, E.; Martin, J. L.; Farge, E. & Beaurepaire, E. (2005). In vivo modulation of morphogenetic movements in Drosophila embryos with femtosecond laser pulses, *Proc. Natl. Acad. Sci. USA* 102, 1047-1052

Tang, W.; Weidner, D. A.; Hu, B. Y.; Newton, R. J. & Hu, X. H. (2006). Efficient delivery of small interfering RNA to plant cells by a nanosecond pulsed laser-induced stress wave for posttranscriptional gene silencing, *Plant Science* 171, 375-381

Taylor, L. P. & Hepler, P. K. (1997). Pollen germination and tube growth, *Anuu. Rev. Plant Physiol. Plant Mol. Biol.* 48, 461-491

Tirlapur, U. K. & König, K. (2002). Targeted transfection by femtosecond laser, *Nature* 418, 290-291

Touraev, A.; Stoger, E.; Voronin, V. & Heberle-Bors, E. (1997). Plant male germ line transformation, *Plant Journal* 12, 949-956

Van der Leede-Plegt, L. M.; van den Ven, B. C. E.; Schilder, M.; Franker, J. & van Tunen, A. J. (1995). Development of a pollen-mediated transformation method for Nicotiana glutinosa, *Transgenic Res.* 4, 77-86

Velegol, S. B.; Pardi, S.; Li, X.; Velegol, D. & Logan, B. E. (2003). AFM imaging artifacts due to bacterial cell height and AFM tip geometry, *Langmuir* 19, 851-857

Vervaeke, I.; Londers, E.; Piot, G.; Deroose, R. & Deproft, M. P. (2005). The division of the generative nucleus and the formation of callose plugs in pollen tubes of Aechmea fasciata (Bromeliaceae) cultured in vitro. *Sexual plant reproduction* 18, 9-19

Vogel, A. & Venugopalan, V. (2003) Mechanism of pulsed laser ablation of biological tissues, *Chem. Rev.* 103, 577-644

Vogel, A.; Noack, J.; Huttman, G. & Paltauf, G. (2005) Mechanism of femtosecond laser nanosurgery of cells and tissues, *Appl. Phys.* B 81, 1015–1047

Wang, Y.; Botvinick, E. L.; Zhao, Y.; Berns, M. W.; Usami, S.; Tsien, R. Y. & Chien, S. (2005). Visualizing the mechanical activation of Src, *Nature* 434, 1040-1045

Yahng, J. S.; Jeoung, S. C.; Choi, D. S.; Cho, D.; Kim, J. H.; Choi, H. M. & Paik, J. S. (2005). Fabrication of microfluidic devices by using a femtosecond laser micromachining techniques and μ-PIV studies on its fluid dynamics, *J. Korean Phys. Soc.* 47, 977-981

Yamazaki, D.; Kurisu, S. & Takenawa, T. (2005). Regulation of cancer cell motility through actin reorganization, *Cancer. Sci.* 96, 379-386

Zeira, E.; Manevitch, A.; Khatchatouriants, A.; Pappo, O.; Hyam, E.; Darash-Yahana, M.; Tavor, E.; Honigman, A.; Lewis, A. & Galun, E. (2003). Femtosecond infrared laser - An efficient and safe in vivo gene delivery system for prolonged expression, *Molecular Therapy* 8, 342-350